TM 9-2320-289-34
CUCV
Commercial Utility Cargo Vehicle
Direct Support / General Support Manual

May 1992 w/Change 1&2

The CUCV or Commercial Utility Cargo Vehicle is a US Military vehicle based on readily available commercial trucks. Originally intended to augment the purpose-built, but expensive GAMA Goat 6x6 and older Jeeps. The first generation was based on Dodge / Chrysler trucks.

This book is focused on the M1008 series second generation CUCV which was General Motor's first major light-truck military vehicle production since World War II. They began production in 1984 and ended production in 1996 with most units being produced as 1984 model year units. Later production was focused on replacements for existing CUCV's. The majority of units were built from existing heavy duty light truck commercial parts. The M1009 was an upgraded/up-rated Chevy K5 Blazer with a 3/4 ton capacity. The M1008 series trucks were a 1-1/4 ton or 5/4 ton rated truck. In all 70,000 units were produced with three power trains.

This manual is the Direct Supprot and General Support repair instructions for these vehicles. It is published as a convenience to enthusiasts who may wish to have a quality professionally printed copy of the manual.

This publisher has also printed other manuals for this series of vehicles.

Should you have suggestions or feedback on ways to improve this book please send email to Books@OcotilloPress.com

---

Edited 2021 Ocotillo Press
ISBN 978-1-954285-72-9

Cover Photo Credit: Alf Van Beem / Wikipedia, Public Domain Image M1008A1 Chevrolet K30 Pickup with Box

Ocotillo Press
Houston, TX 77017
Books@OcotilloPress.com

TM 9-2320-289-34

ARMY TM 9-2320-289-34
AIR FORCE TO 36A12-1A-2082-2
MARINE CORPS TM 2320-34/4

Supersedes Copy Dated April 1983
See Page i For Details

INTERMEDIATE DIRECT SUPPORT/GENERAL SUPPORT MAINTENANCE MANUAL
FOR

TRUCK, CARGO, TACTICAL, 1 -1/4 TON, 4x4, MI 008
(2320-01-123-6827)

TRUCK, CARGO, TACTICAL, 1 -1/4 TON, 4x4, M1008A1
(2320-01-123-2671)

TRUCK, UTILITY, TACTICAL, 3/4 TON, 4x4, MI 009
(2320-01-123-2665)

TRUCK, AMBULANCE, TACTICAL, 1-1/4 TON, 4x4, MI 010
(2310-01-123-2666)

TRUCK, SHELTER CARRIER, TACTICAL, 1-1/4 TON, 4x4, M1028
(2320-01-127-5077)

TRUCK, SHELTER CARRIER W/PTO, TACTICAL, 1-1/4 TON, 4x4, M1028A1
(2320-01-158-0820)

TRUCK, CHASSIS, TACTICAL, 1 -1/4 TON, 4x4, MI 031
(2320-01-133-5368)

DEPARTMENTS OF THE ARMY, THE AIR FORCE, AND
HEADQUARTERS, MARINE CORPS

12 JANUARY 1989

CHANGE

NO, 2

DEPARTMENT OF THE ARMY, THE AIR FORCE,
AND HEADQUARTERS, MARINE CORPS
Washington, D.C. 1 May 1992

INTERMEDIATE DIRECT SUPPORT/GENERAL SUPPOR T MAINTENANCE MANUAL FOR

TRUCK, CARGO, TACTICAL, 1 -1/4 TON, 4X4, MI 008
(2320-01-123-6827)

TRUCK, CARGO, TACTICAL, 1-1 /4 TON, 4X4, M1008A1
(2320-01-123-2671)

TRUCK, UTILITY, TACTICAL, 3/4 TON, 4X4, MI 009
(2320-01-123-2665)

TRUCK, AMBULANCE, TACTICAL, 1-1/4 TON, 4X4, M1010
(2310-01-123-2666)

TRUCK, SHELTER CARRIER, TACTICAL, 1-1/4 TON, 4X4, M1028
(2320-01-127-5077)

TRUCK, SHELTER CARRIER W/PTO, TACTICAL, 1 -1/4 TON, 4X4, M1028A1
(2320-01-158-0820)

TRUCK, SHELTER CARRIER W/PTO, TACTICAL 1-1/4 TON, 4X4, M1028A2
(2320-01-295-0822)

TRUCK, SHELTER CARRIER, TACTICAL, 1-1/4 TON, 4X4, M1028A3
(2320-01-325-1937)

TRUCK, CHASSIS, TACTICAL, 1 -1/4 TON, 4X4, MI 031
(2320-01-133-5368)

TM 9-2320-289-34, 12 January 1989, is changed as follows:

1. The manual title is changed to read as shown above.
2. Remove old pages and insert new pages.
3. New or changed material is indicated by a vertical bar in the margin.

Remove Pages	Insert Pages
i and ii	i and ii
1-1 and 1-2	1-1 and 1-2
1-5 through 1-7/(1-8 blank)	1-5 through l-7/(1-8 blank)
2-19 through 2-24	2-19 through 2-24
3-29 and 3-30	3-29 and 3-30
2-33 and 2-34	2-33 and 2-34
3-33 through 3-36	3-33 through 3-36
3-41 and 3-42	3-41 and 3-42

Remove Pages (Con't)	Insert Pages (Con't)
3-53 through 3-56	3-53 through 3-56
3-99 through 3-106	3-99/(3-100 blank) and 3-106
3-109 and 3-770	3-109 and 3-110
3-115 and 3-116	3-115 and 3-116
3-119 through 3-122	3-119 through 3-122
3-163 and 3-164	3-163 and 3-164
3-175 through 3-178	3-178 and 3-178 .1/(3-178.2 blank)
4-1 and 4-2	4-1 and 4-2
4-7 and 4-8	4-7 and 4-8
4-37 and 4-38	4-37 and 3-38
4-51 and 4-52	4-51 and 4-52
5-1 and 5-2	5-1 and 5-2
5-5 through 5-16	5-5 through 5-12, and 5-16
5-131 through 5-136	5-131 through 5-136
5-157 through 5-164	5-157 through 5-164
5-187 and 5-188	5-187 and 5-188
5-191 and 5-192	5-191 and 5-192 .
6-7 and 6-8	6-7 and 6-8
6-83 through 6-86	6-83 through 6-86
6-131 through 6-136	6-131 through 6-136
6-139 through 6-14 1/(6-142 blank)	6-139 through 6-14 1/(6-142 blank)
8-1 and 8-2	8-1 and 8-2
8-13 through 8-16	8-13 and 8-14, and 8-16
10-1 and 10-2	10-1 and 10-2
10-7 through 10-10	10-7/(10-8 blank) and 10-10
10-33 through 10-38	10-33 through 10-38
10-41 through 10-42.6	10-41 through 10-42.6
10-73 and 10-74	10-73 and 10-74
10-79 and 10-80	10-79 and 10-80
A-3 and A-4	A-3 and A-4
B-1 and B-2	B-1 and B-2
B-7 and B-8	B-7 and B-8
C-1/(C-2 blank)	C- 1/(C-2 blank)
C-5 and C-6	C-5 and C-6
E-1 and E-2	E-1 and E-2
E-9 and E-10	E-9 and E-10
" Index 1 and Index 2	Index 1 and Index 2
Index 5 through Index 8	Index 5 through Index 8
index 11 through Index 23/(Index 24 blank)	Index 11 through Index 23/(Index 24 blank)

4. File this change sheet in front of the publication for reference purposes.

By Order of the Secretary of the Army:

GORDON R. SULLIVAN
General, United States Army
Chief of Staff

Official:

MILTON H. HAMILTON
Administrative Assistant to the
Secretary of the Army
02692

By Order of the Secretary of the Air Force;

MERRILL A. McPEAK
General, United States Air Force
Chief of Staff

Official;

CHARLES C. McDONALD
General, United States Air Force
Commander, Air Force Logistics Command

By Order of the Marine Corps:

H. E. REESE
Deputy for Support
Marine Corps Research, Development and
Acquisition Command

Distribution:

To be distributed in accordance with DA Form 12–38–E, Block No. 372, Unit, Direct Support and General Support maintenance requirements for TM 9–2320–289–34.

CHANGE

NO, 1

DEPARTMENT OF THE ARMY, THE AIR FORCE,
AND HEADQUARTERS, MARINE CORPS
Washington D. C., 7 *JuLy 1991*

INTERMEDIATE DIRECT SUPPORT/GENERAL SUPPORT MAINTENANCE MANUA L FOR

TRUCK, CARGO, TACTICAL, 1 -1/4 TON, 4X4, M1008
(2320-01 -1 23-6827)

TRUCK, CARGO, TACTICAL, 1 -1/4 TON, 4X4, MI 008A1
(2320-01-123-2671)

TRUCK, UTILITY, TACTICAL, 3/4 TON, 4X4, MI 009
(2320-01-123-2665)

TRUCK, AMBULANCE, TACTICAL, 1 -1/4 TON, 4X4, MI 010
(231 0-01-1 23-2666)

TRUCK, SHELTER CARRIER, TACTICAL, 1 -1/4 TON, 4X4, MI 028
(2320-01-127-5077)

TRUCK, SHELTER CARRIER W/PTO, TACTICAL, 1 -1/4 TON, 4X4, M1028A1
(2320-01 -1 58-0820)

TRUCK, SHELTER CARRIER W/PTO, TACTICAL 1-1/4 TON, 4X4, MI028A2
(2320-01-295-0822)

TRUCK, CHASSIS, TACTICAL, 1-1/4 TON, 4X4, MI 031
(2320-01 -1 33-5368)

TM 9-2320-289-34, 12 January 1989, is changed as follows:

1. The manual title is changed to read as shown above.
2. Remove old pages and insert new pages.
3. New or changed material is indicated by a vertical bar in the margin,

Remove Pages	Insert Pages
i and ii	i and ii
1-1 and 1-2	1-1 and 1-2
1-5/(l-6 blank)	1-5 through 1-7/(1-8 blank)
4-1 and 4-2	4-1 and 4-2
None	4-68.1 and 4-68.2
5-131 and 5-132	5-131 and 5-132
6-7 and 6-8	6-7 and 6-8
6-83 through 6-86	6-83 through 6-86
6-103 and 6-104	6-103 through 6-104
None	6-122.1 through 6-122.9/(6-122,10 blank)
6-131 through 6-138	6-131 through 6-141l(6-142 blank)
10-35 through 1042	10-35 through 1042.6

Remove Pages (Con't)	Insert Pages (Con't)
A-1 through A-4	A-1 through A-4
E-1 and E-2	E-1 and E-2
None	E-16. 1/(E-16.2 blank)
index 1 through Index 6	Index 1 through Index 6
Index 9 through Index 12	Index 9 through Index 12
Index 15 throough Index 23/(index 24 blank)	Index 15 throough Index 23/(index 24 blank)

4. File this change sheet in front of the publication for reference purposes.

By Order of the Secretary of the Army:

GORDON R. SULLIVAN
General, United States Army
Chief of Staff

Official:

PATRICIA P. HICKERSON
Brigadier General, United States Army
The Adjutant General

By Order of the Secretary of the Air Force;

MERRILL A. McPEAK
General, United States Air Force
Chief of Staff

Official;

CHARLES C. McDONALD
General, United States Air Force
Commander, Air Force Logistics Command

Distribution:

By Order of the Marine Corps:

H. E. REESE
Deputy for Support
Marine Corps Research, Development and
Acquisition Command

Distribution:

To be distributed in accordance with DA Form 12-38-E (Block 0372) Direct Support and General Support maintenance requirements for TM9-232-289-34.

WARNING

CARBON MONOXIDE (EXHAUST GASES) CAN KILL

Carbon monoxide is without color or smell, but can kill you. Breathing air with carbon monoxide produces symptoms of headache, dizziness, loss of muscular control, a sleepy feeling, and coma. Brain damage or death can result from heavy exposure. Carbon monoxide occurs in the exhaust fumes of fuel-burning heaters and internal combustion engines. Carbon monoxide can become dangerously concentrated under conditions of no air movement. Precautions must be followed to ensure crew safety when personnel heater or main or auxiliary engines of any truck are operated for any purpose.

1. DO NOT operate personnel heater or engine of truck in enclosed areas.
2. DO NOT idle truck engine without ventilator blower operating and truck windows open.
3. BE ALERT at all times for exhaust odors.
4. BE ALERT for exhaust poisoning symptoms. They are:
    - Headache
    - Dizziness
    - Sleepiness
    - Loss of muscular control
5. If you see another person with exhaust poisoning symptoms:
    - Remove person from area.
    - Expose to fresh air.
    - Keep person warm.
    - Do not permit physical exercise.
    - Administer artificial respiration, if necessary. *
    - Notify a medic.
6. **BE AWARE:** The field protective mask for chemical-biological-radiological (CBR) protection will not protect you from carbon monoxide poisoning.

The Best Defense Against Carbon Monoxide Poisoning Is Good Ventilation

I For First Aid, refer to FM 21-11.

WARNING

AIR CONDITIONING SYSTEM

- Always wear goggles while testing or servicing air conditioning system. DO NOT allow refrigerant to contact your skin. Failure to follow this warning may result in injury to personnel.
- DO NOT test or service air conditioning system while smoking or near open flame. Refrigerant will become a poisonous gas when burned. Failure to follow this warning may result in serious illness or death to personnel.
- DO NOT allow direct flame or electric heaters to contact refrigerant container. Localized heat can raise pressure to a dangerous level and cause refrigerant container to explode, resulting in serious injury or death to personnel.
- Never attempt to connect servicing equipment while air conditioning system is running. Failure to follow this warning may result in injury to personnel.
- Never open high side valve on air conditioner manifold gage set while air conditioning system is operating. Refrigerant may be pumped back into refrigerant container and cause refrigerant container to explode, resulting in serious injury or death to personnel.
- Never attempt to charge system on the high side while air conditioning system is running. Failure to follow this warning may result in injury to personnel.

WARNIN G

AXLES, STEERING, AND SUSPENSION

- Front and rear axle assemblies, and rear spring assembly are heavy. Use extreme caution, provide support, and use assistance during removal and installation. Failure to follow this warning may result in serious injury to personnel or damage to axle assembly.
- Front spring assembly is heavy. Use caution during removal. Failure to follow this warning may result in serious injury to personnel or damage to axle assembly.
- Ensure that "C" lock is correctly seated in differential side gear. Failure to correctly seat " C" lock may result in loss of truck's wheel and serious injury or death to personnel.
- Nuts at steering arm or bolts at upper kingpin bearing cap must be removed alternately and with caution. Compression spring underneath steering arm or bearing cap could fly up during removal, causing serious injury to personnel.
- Tightly hold differential case assembly together to absorb spring pressure during removal and installation of ring gear bolts. Failure to follow this warning may 'result in injury to personnel or damage to differential assembly components.

WARNING

AXLES, STEERING, AND SUSPENSION (Continued)

- Use extreme caution when releasing tension from spring assembly and auxiliary spring. Springs are under considerable pressure and could cause injury to personnel if suddenly released.

- Steering column lower bearing retainer clip is under spring tension, Use caution during removal. Failure to follow this warning may result in injury to personnel.

WARNING

BODY AND ACCESSORIES

- Use caution when handling sheet metal. Sheet metal has numerous sharp edges and splinters. Failure to follow this warning may result in injury to personnel.

- Use only effective chip guarding and personnel protective equipment (goggles/shield, gloves, etc.) when using drill or grinder. Failure to follow this warning may result in injury to personnel.

- Wear heavy gloves and safety goggles when handling window glass or windshield to avoid risk of injury to personnel,

WARNING

BRAKE SYSTEM

- DO NOT use a dry brush or compressed air to clean brake shoes, brake pads, or brake components. There may be asbestos dust on brake shoes, brake pads, or brake components which can be dangerous to you if you breathe it. Brake shoes, brake pads, and brake components must be wet, and a soft brush must be used. Failure to follow this warning may result in serious illness or death to personnel.

- Always wear goggles while performing power booster repair. Power booster spring retainer is under spring pressure. Failure to use correct tools and procedures may result in injury to personnel.

- Accumulator contains compressed gas. DO NOT apply heat to accumulator, DO NOT attempt to repair an inoperative accumulator, always replace an inoperative accumulator with a new one. Drill a 1/16 in. (1.6 mm) hole through end of accumulator can opposite "O" ring to dispose of an inoperative accumulator. Failure to follow this warning may result in serious injury or death to personnel.

WARNING

BRAKE SYSTEM (Continued)

- DO NOT use a drum or rotor that will not meet minimum wear specifications. Failure to follow this warning may result in brake failure and serious injury or death to personnel.
- DO NOT place hand in front of caliper piston when removing with compressed air. Failure to follow this warning may result in injury to personnel.

WARNING

CLEANING AGENTS

- Alcohol used for cleaning is highly flammable and poisonous. Drinking of this alcohol can cause blindness and death. Avoid inhaling its fumes. Keep away from open flame.
- Carburetor cleaning compound is highly flammable. Keep away from open flame. Failure to follow this warning may result in serious injury or death to personnel.
- Dry cleaning solvent P-D-680 is toxic and flammable. Always wear protective goggles and gloves and use only in a well-ventilated area. Avoid contact with skin, eyes, and clothes and DO NOT breathe vapors. DO NOT use near open flame or excessive heat. The solvent's flash point is 100°F-1 38°F (38°C-59°C). If you become dizzy while using cleaning solvent, immediately get fresh air and medical help. If solvent contacts eyes, immediately wash your eyes with water and get medical aid.

WARNING

COMPRESSED AIR

- Use only effective chip guarding and personnel protective equipment (goggles/shield, gloves, etc.) when using compressed air. Compressed air used when checking for restrictions or cleaning purposes should never exceed 30 psi (207 kPa). Failure to follow this warning may result in serious injury to personnel.

WARNING

ELECTRICAL SYSTEM

- Both battery negative cables must be disconnected before removing any electrical system components. (See TM 9-2320-289-20) Failure to follow this warning may result in serious injury or death to personnel.

WARNING

ENGINE ASSEMBLY

- Use extreme caution to ensure that clothing or tools DO NOT get caught in truck's operating drivebelts. Failure to follow this warning may result in serious injury to personnel or equipment damage.

- DO NOT place hands between engine and transmission when installing engine. Failure to follow this warning may result in injury to personnel.

- Avoid skin contact with anaerobic sealing compound. Immediately wash off skin if contact is made. Failure to follow this warning may result in injury to personnel.

- Always wear goggles while using valve spring compressor. Valve springs are tightly compressed. Failure to follow this warning may result in serious injury to personnel.

- Wear goggles while testing fuel injector nozzles. Failure to follow this warning may result in serious eye injury.

- Ensure that governor cover is installed before operation of fuel injector pump. Pressurized fuel may cause injury to personnel and/or damage to equipment.

- DO NOT place hand or arms near nozzle tip of fuel injector nozzle during testing. Nozzle tip must be enclosed in a receptacle, preferably transparent, to contain fuel spray. High-pressure atomized fuel spray from nozzle tip can puncture skin and destroy tissue. Failure to follow this warning may also result in blood poisoning.

- Diesel fuel is flammable, DO NOT perform this procedure near fire, flames, or sparks. A fire extinguisher must be on hand in work area. Failure to follow this warning may result in serious injury or death to personnel.

- Always wear goggles when removing piston pin retainers. Failure to follow this warning may result in serious injury to personnel.

WARNING

TRANSMISSION AND TRANSFER CASE

- Transmission is heavy, Use caution and provide support during installation to, and removal from, holding fixture. Failure to follow this warning may result in serious injury to personnel or damage to transmission.

- Allow transmission oil cooler pipes to cool before attempting to service. Failure to follow this warning may result in serious burns.

WARNING

TRANSMISSION AND TRANSFER CASE (Continued)

- Transmission, torque converter, and transfer case are heavy. Use caution and provide support during removal and installation, Failure to follow this warning may result in serious injury to personnel or equipment damage,

- All valves are under spring tension, Wear goggles and use caution during removal and installation of springs. Maintain pressure against valves while removing and installing retaining pins. Failure to follow this warning may result in injury to personnel.

- Wear goggles and use caution during removal and installation of regulator boost valve spring, Spring is tightly compressed and could fly out of bore. Failure to follow this warning may result in serious eye injury,

- Use only effective chip guarding and personnel protective equipment (goggles/shield, gloves, etc.) when using drill, Failure to follow this warning may result in injury to personnel,

TECHNICAL MANUAL
TM 9-2320-289-34

DEPARTMENTS OF THE ARMY, THE AIR FORCE,
AND HEADQUARTERS, MARINE CORPS
Washington, D.C. 12 January 1989

INTERMEDIATE DIRECT SUPPORT/GENERAL SUPPORT MAINTENANCE MANUAL FOR

TRUCK, CARGO, TACTICAL, 1 -1/4 TON, 4x4, M1008
(2320-01-123-6827)

TRUCK, CARGO, TACTICAL, 1 -1/4 TON, 4x4, M1008A1
(2320-01-123-2671)

TRUCK, UTILITY, TACTICAL, 3/4 TON, 4x4, M1009
(2320-01-123-2665)

TRUCK, AMBULANCE, TACTICAL, 1 -1/4 TON, 4x4, M1010
(2310-01-123-2666)

TRUCK, SHELTER CARRIER, TACTICAL, 1 -1/4 TON, 4x4, M1028
(2320-01-1 27-5077)

TRUCK, SHELTER CARRIER W/PTO, TACTICAL, 1 -1/4 TON, 4x4, M1028A1
(2320-01-158-0820)

TRUCK, SHELTER CARRIER W/PTO, TACTICAL 1 -1/4 TON, 4X4, M1028A2
(2320-01-295-0822)

TRUCK, SHELTER CARRIER, TACTICAL, 1 -1/4 TON, 4X4, M1028A3
(2320-01 -325-1 937)

TRUCK, CHASSIS, TACTICAL, 1 -1/4 TON, 4x4, M1031
(2320-01-1 33-5368)

REPORTING ERRORS AND RECOMMENDING IMPROVEMENTS

(Army) You can help improve this manual, If you find any mistakes or if you know of a way to improve the procedures, please let us know, Mail your letter, DA Form 2028 (Recommended Changes to *Publications and Blank Forms),* or DA Form 2028-2, located in the back of this manual, direct to: Commander, U.S. Army Tank-Automotive Command, ATTN: AMSTA-MB, Warren, MI 48397-5000. *(Marine Corps)* Submit NAVMC 10772 to the Commanding General, Marine Corps Logistic Base (Code 850), Albany, GA 31704. A reply will be furnished to you.

Approved for public release; distribution is unlimited.

TABLE OF CONTENTS

* This publication supersedes TM 9-2320-289-34 dated April 1983.

TABLE OF CONTENTS - Continued

TABLE OF CONTENTS - Continued

HOW TO USE THIS MANUAL

This manual is designed to help operate and maintain the CUCV Series trucks. This manual describes in detail the Intermediate Direct Support and General Support Maintenance prescribed by the Maintenance Allocation Chart (TM 9-2320-289-20) and the Source, Maintenance, and Recoverability (SMR) Codes (TM 9-2320-289-34P).

FEATURES OF THIS MANUAL:

- Bleed-to-edge indicators on the cover and on the edge of applicable manual pages provide quick access to chapters and sections most often used.
- A table of contents is provided for all chapters, sections, and appendices.
- WARNINGS, CAUTIONS, NOTES, subject headings, and other important information are highlighted in BOLD print as a visual aid.
- Statements and words of particular importance are printed in capital letters to create emphasis.
- Instructions are located together with figures that illustrate the specific task you are working on, In many cases, the task steps and illustrations are located side-by-side, making identification and procedure sequence easier to follow.
- Dashed leader lines used in illustrations indicate that called out items are not visible (i. e., they are located within structure). The example illustrates that the torque converter (16) is located within the transmission,

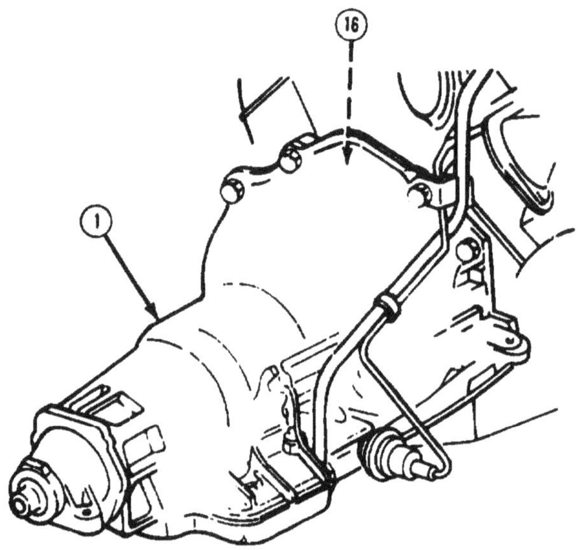

Dashed Leader Line Example

- An alphabetical index is provided at the end of the manual to assist in locating information not readily found in the table of contents.
- Technical instructions include metric in addition to standard units. A metric conversion chart is provided on the inside back cover,

FOLLOW THESE GUIDELINES WHEN YOU USE THIS MANUAL:

- Quickly read through this manual and become familiar with its contents before proceeding to specific maintenance tasks.

TA5014

- A warning summary is provided at the beginning of this manual and should be read before performing any maintenance tasks.
- In the actual maintenance tasks, follow all WARNINGS, CAUTIONS, and NOTES. These are given immediately preceding the procedural steps to which they apply. If these instructions are not followed, or care is not taken, you may injure yourself or cause equipment damage.
- Within a chapter, section, or paragraph, headings are used to help group the material and assist you in quickly finding tasks. Read all preliminary information found at the beginning of each task. After completing a task, ALWAYS perform the follow-on maintenance at the end of the task.

USING YOUR MANUAL: AN EXAMPLE

The operator of a CUCV truck complains that the truck engine cranks normally but will not start. The truck has been assigned to you for repair. To correct the problem you will need to follow these steps:

- Turn to the cover of your manual. On the right margin, find the listing for Troubleshooting and turn to the page listing given. When you reach the troubleshooting section, locate the paragraph entitled Troubleshooting Symptom Index (paragraph 2-3).
- Both mechanical and electrical troubleshooting symptoms are listed in the Troubleshooting Symptom Index and are organized according to equipment category. Read through the index until you find a symptom listing which most closely matches the problem. The symptom "Cranks Normally But Will Not Start" is found under the "ENGINE" category of mechanical troubleshooting. Turn to the page for the troubleshooting procedure that will help solve your problem.
- Follow the instructions for each "Step" listed in the troubleshooting procedure. Start with Step 1 and proceed in order until you reach the step where a problem is actually found. In this case, let's assume that steps 1-4 are performed without difficulty. Step 5 indicates "Test the fuel injector pump. (See paragraph 344)"
- Paragraph 3-44 is located in Chapter 3. Using the Tab/e of Contents, find the listings for Chapter 3. Because the fuel injector pump is part of the fuel system, you would proceed to Section III, Fuel System Maintenance. The index at the beginning of Section III will tell you on which page paragraph 3-44 is located.
- Paragraph 3-44 is the maintenance procedure for testing the fuel injector pump. Carefully read through the procedure, follow all instructions, and perform all steps in the proper sequence. When you have finished the last step, YOU will have tested the fuel injector pump and will know if it should be replaced.

USE OF RTV SEALANT:

When maintenance instructions in this manual instruct you to use RTV sealant (Item 57, Appendix B), follow instructions on the tube or box for proper component assembly time and RTV curing time.

TRUCKS EQUIPPED WITH SWING FIRE HEATERS:

Maintenance tasks in this manual do not allow for trucks which may have swing fire heaters installed. If the truck you are working on has a swing fire heater, you may have to disconnect the heater's components to obtain access to other components if interference exists.

CHAPTER 1
INTRODUCTION

Section 1. GENERAL INFORMATION

1-1. SCOPE.

a. This manual contains instructions for the performance of maintenance of CUCV Series trucks at the intermediate direct support and general support level, Models included are:

(1) MI 008, Truck, Cargo, Tactical, 1 1/4 Ton, 4x4
(2) M1008A1, Truck, Cargo, Tactical, 1 1/4 Ton, 4x4
(3) MI 009, Truck, Utility, Tactical, 3/4 Ton, 4x4
(4) MI 010, Truck, Ambulance, Tactical, 11/4 Ton, 4x4
(5) M1028, Truck, Shelter Carrier, Tactical, 11/4 Ton, 4x4
(6) MI028A1, Truck, Shelter Carrier w/PTO, Tactical, 1 1/4 Ton, 4x4
(7) MI028A2, Truck, Shelter Carrier w/PTO, Tactical, 1 1/4 Ton, 4x4
(8) M1028A3, Truck, Shelter Carrier, Tactical, 11/4 Ton, 4x4
(9) M1031, Truck, Chassis, Tactical, 1 11/4 Ton, 4x4

b. Other manuals which may be referred to should be considered a part of this manual.

1-2. MAINTENANCE FORMS AND RECORDS.

Department of the Army forms and procedures used for equipment maintenance will be those prescribed by DA Pam 738-750, The Army Maintenance Management System (TAMMS).

1-3, DESTRUCTION OF ARMY MATERIEL TO PREVENT ENEMY USE,

Procedures outlined in TM 750-244-6 (Procedures for Destruction of Tank-Automotive Equipment to Prevent Enemy Use) are applicable to these trucks.

1-4. PREPARATION FOR STORAGE OR SHIPMENT.

For information on storage or shipment of these trucks, see TM 9-2320-289-20.

1-5. OFFICIAL NOMENCLATURE, NAMES, AND DESIGNATIONS.

The nomenclature, names, and designations used in this manual are consistent with official usage. All hardware, assemblies, and subassemblies are named according to the nomenclature used in TM 9-2320-289-34P,

1-6, REPORTING EQUIPMENT IMPROVEMENT RECOMMENDATIONS (EIRs).

If your CUCV Series truck needs improvement, let us know. Send us an EIR. You, the user, are the only one who can tell us what you don't like about your equipment. Let us know why you don't like the design or performance. Put it on an SF 368 (Quality *Deficiency Report).* Mail it to us at: Commander, U.S. Army Tank-Automotive Command, ATTN: AMSTA-QRD, Warren, Ml 48397-5000, We'll send you a reply.

1-7. WARRANTY INFORMATION.

To determine if your CUCV Series truck is under warranty, see TB 9-2300-295-15/24.

1-8. METRIC SYSTEM.

The equipment described herein contains metric components and requires the use of metric tools; therefore, metric units, in addition to standard units, will be used throughout this manual, Standard units will be given first with the equivalent metric unit in parentheses, for instance: 7000 lb. (31 78 kg). A metric conversion chart is located on the inside back cover of this manual.

1-9, COMMON TOOLS, SPECIAL TOOLS, AND REPAIR PARTS.

a, For authorized common tools, refer to the Common Table of Allowance (CTA) and the Modified Table of Organization and Equipment (MTOE) applicable to your unit.

b. For special tools, and Test, Measurement, and Diagnostic Equipment (TMDE), refer to the *Repair* Parts *and Special Tools* List *(RPSTL)* manual, TM 9-2320-289-34P, and the *Maintenance Allocafion Chart (MAC)* in TM 9-2320-2879-20.

c. Repair parts are listed and illustrated in the *Repair Parts and Special Too/s List (RPSTL)* manual, TM 9-2320-289-34P.

d. Torque wrenches are normally marked with a measurement term *ft.-lb.* or *in. -lb.* The terms used within this manual are *lb,-ft,* or *lb. -in.,* with the equivalent metric unit in parentheses.

Section II. EQUIPMENT DESCRIPTION AND DATA

1-10. DIFFERENCES BETWEEN MODELS.

	M1008	M1008A1	M1009	M1010	M1028	M1028A1	M1031
a. Model 208 Transfer Case	X	X	X	X	X		
b. Model 205 Transfer Case						X	X
c. 100 amp, 12/24 volt Electrical System	X	X	X		X	X	X
d. 200 amp, 12/24 volt Electrical System				X	X**		
e. Troop Seat Kit*	X	X					
f. Communications Kit		X			X	X	
g. Radio Mounting			X	X			
h. Weapon Brackets	X	X	X		X	X	X
i. Cargo Tie-downs	X	X					
j. Shelter Tie-downs					X	X	
k. Air Conditioner				X			
l. Gas-Particulate Filter Unit (GPFU) System				X			
m. Cargo Cover Kit*	X	X					
n. Rear Passenger Seating			X				
o. Spotlight				X			
p. Floodlights				X			
q. 10.00R-15C Tires			X				
r. LT235/85R-16E Tires	X	X		X	X	X	X
s. Locking Differential			X				
t. No-spin Differential	X	X		X	X	X	X

* These items are not standard equipment, but the indicated trucks may be equipped as shown.

** Selected M1028 trucks only .

1-11. EQUIPMENT DATA.

Vehicle performance data for the CUCV Series trucks is listed in the table below.

Table 1-1. Tabulated Data

DATA	MODEL				
	M1008 M1008A1	M1009	M1010	M1028 M1028A1	M1031
Weights:					
Curb	5900 lb (2679 kg)	5200 lb (2361 kg)	7370 lb (3346 kg)	5800 lb (2633 kg)	5250 lb (2384 kg)
Payload/ Passengers	2900 lb (1317 kg)	1200 lb (545 kg)	2080 lb (944 kg)	3600 lb (1634 kg)	3950 lb (1793 kg)
GVWR	8800 lb (3995 kg)	6400 lb (2906 kg)	9450 lb (4290 kg)	9400 lb (4268 kg)	9200 lb max (4177 kg)
GAWR (front)	4500 lb (2043 kg)	3600 lb (1634 kg)	4500 lb (2043 kg)	4500 lb (2043 kg)	4500 lb (2043 kg)
GAWR (rear)	7500 lb (3405 kg)	3750 lb (1703 kg)	7500 lb (3405 kg)	7500 lb (3405 kg)	7500 lb (3405 kg)
Wheelbase	131.5 in (334 cm)	106.5 in (271 cm)	131.5 in (334 cm)	131.5 in (334 cm)	131.5 in (334 cm)
Track (front)	67.8 in (172 cm)	57.4 in (146 cm)	67.8 in (172 cm)	67.8 in (172 cm)	67.8 in (172 cm)
Track (rear)	65.8 in (167 cm)	54.4 in (138 cm)	65.8 in (167 cm)	65.8 in (167 cm)	65.8 in (167 cm)
Ground Clearance (to T/C skid plate @ GVWR)	10.7 in (27.2 cm)	9.6 in (24.4 cm)	10.5 in (26.7 cm)	10.5 in (26.7 cm)	10.5 in (26.7 cm)
Rear Axle @ GVWR	7.8 in (19.8 cm)	8.6 in (21.8 cm)	7.8 in (19.8 cm)	7.8 in (19.8 cm)	7.8 in (19.8 cm)
Front Axle @ GVWR	8.6 in (21.8 cm)	8.4 in (21.3 cm)	8.6 in (21.8 cm)	8.6 in (21.8 cm)	8.6 in (21.8 cm)
Height (overall) @ Curb	75.4 in (191.5 cm)	75.0 in (190.5 cm)	101.6 in (258 cm)	107.1 in* (272 cm)	76.3 in (193.8 cm)
Length (overall)	220.7 in (560.6 cm)	191.8 in (487.1 cm)	227.7 in (578.4 cm)	220.7 in (560.6 cm)	212.9 in (504.8 cm)
Width (overall)	81.2 in (206.2 cm)	79.6 in (202 cm)	81.2 in (206.2 cm)	81.2 in (206.2 cm)	81.2 in (206.2 cm)
Engine:					
Type	90-degree V8 Diesel	90-degree V8 Diesel	90-degree V8 Diesel	90-degree V8 Diesel	90-degree V8 Diesel
Piston Displacement	379 cu in (6.2 l)	379 cu in (6.2 l)	379 cu in (6.2 l)	379 cu in (6.2 l)	379 cu in (6.2 l)
SAE Net Horsepower	135 hp @ 3600 rpm	135 hp @ 3600 rpm	135 hp @ 3600 rpm	135 hp @ 3600 rpm	135 hp @ 3600 rpm

* Height will vary with different shelters and loading.

Table 1-1. Tabulated Data (Continued)

DATA	MODEL				
	M1008 M1008A1	M1009	M1010	M1028 M1028A1	M1031
Fuel System:					
Low-pressure	Mechanical Lift Pump	Mechanical Lift Pump	Mechanical Lift Pump	Mechanical Lift Pump	Mechanical Lift Pump
High-pressure	Stanadyne Model DB2	Stanadyne Model DB2	Stanadyne Model DB2	Stanadyne Model DB2	Stanadyne Model DB2
Injection Nozzle	Bosch	Bosch	Bosch	Bosch	Bosch
Fuel Tank Capacity	20 gal (75.7 l)	27 gal (102.2 l)	20 gal (75.7 l)	20 gal (75.7 l)	20 gal (75.7 l)
Electrical System	12/24 volt	12/24 volt	12/24 volt	12/24 volt	12/24 volt
Charging System	100 amp/ 28 volt	100 amp/ 28 volt	200 amp/ 28 volt	100 amp/ 28 volt	100 amp/ 28 volt
Transmission:					
Type	Automatic GM THM 400	Automatic GM THM 400	Automatic GM THM 400	Automatic GM THM 400	Automatic GM THM 400
Lubrication Capacity	19 pt (9 l)	19 pt (9 l)	19 pt (9 l)	19 pt (9 l)	19 pt (9 l)
Transfer Case:					
Model	New Process Model 208	New Process Model 208	New Process Model 208	New Process Model 208 or Model 205	New Process Model 205
Speed	2-speed	2-speed	2-speed	2-speed	2-speed
Ratios	2.61 & 1.00	2.61 & 1.00	2.61 & 1.00	2.61 & 1.00 or 1.96 & 1.00	1.96 & 1.00
Lubrication Capacity	5 qt (4.8 l)	5 qt (4.8 l)	5 qt (4.8 l)	5 qt (4.8 l) or 2 qt (2.5 l)	2 qt (2.5 l)
Front Axle:					
Type	Full-floating	Semi-floating	Full-floating	Full-floating	Full-floating
Differential Type	Two-pinion	Two-pinion	Two-pinion	Two-pinion No-spin	Two-pinion No-spin
Lubrication Capacity	4.5 pt (2.13 l)	4.5 pt (2.13 l)	4.5 pt (2.13 l)	4.5 pt (2.13 l)	4.5 pt (2.13 l)
Rear Axle:					
Type	Full-floating	Semi-floating	Full-floating	Full-floating	Full-floating
Differential Type	Four-pinion No-spin	Two-pinion Locking	Four-pinion No-spin	Four-pinion No-spin	Four-pinion No-spin
Lubrication Capacity	5.4 pt (2.1 l)	5.4 pt (2.1 l)	5.4 pt (2.1 l)	5.4 pt (2.1 l)	5.4 pt (2.1 l)
Service Brake:					
Front	Disc	Disc	Disc	Disc	Disc
Rear	Drum	Drum	Drum	Drum	Drum

1-12. DIFFERENCES BETWEEN MODELS.

	M1028A2	M1028A3		M1028A2	M1028A3
a. Model 208 Transfer Case		X	k. Air Conditioner		
b. Model 205 Transfer Case	X		l. Gas-Particulate Filter Unit (GPFU) System		
c. 100 amp, 12/24 volt Electrical System	X	X	m. Cargo Cover Kit		
d. 200 amp, 12/24 volt Electrical System			n. Rear Passenger Seating		
e. Troop Seat Kit*			o. Spotlight		
f. Communications Kit	X	X	p. Floodlights		
g. Radio Mounting			q. 10.00R-15C Tires		
h. Weapons Bracket	X	X	r. LT235/85R-16E Tires	X	X
i. Cargo Tie-downs	X	X	s. Locking Differential		
j. Shelter Tie-downs	X	X	t. No-spin Differential	X	X
			u. Dual Rear Wheels	X	X

1-13. EQUIPMENT DATA.

Table 1-1.1. Tabulated Data

DATA	M1028A2	M1028A3	DATA	M1028A2	M1028A3
Weights:			Rear Axle @ GVWR	7.7 in (19.6 cm)	7.7 in (19.6 cm)
Curb	6120 lb (2778 kg)	6120 lb (2778 kg)	Front Axle @ GVWR	8.6 in (21.8 cm)	8.6 in (21.8 cm)
Payload/ Passengers	3940 lb (1789 kg)	3940 lb (1789 kg)			
GVWR	10,120 lb max (4594 kg)	10,120 lb max (4594 kg)	Height (overall) @ Curb	107.1 in * (272 cm)	107.1 in * (272 cm)
GAWR (front)	3900 lb (1770 kg)	3900 lb (1770 kg)	Length (overall)	220.7 in (560.6 cm)	220.7 in (560.6 cm)
GAWR (rear)	6220 lb (2824 kg)	6220 lb (2824 kg)	Width (overall)	95.8 in (243.3 cm)	95.8 in (243.3 cm)
Wheelbase	131.5 in (334 cm)	131.5 in (334 cm)	Engine:		
			Type	90-degree V8 Diesel	90-degree V8 Diesel
Track (front)	67.8 in (172 cm)	67.8 in (172 cm)	Piston Displacement	379 cu in (6.2 l)	379 cu in (6.2 l)
Track (rear)	75.8 in (193 cm)	75.8 in (193 cm)	SAE Net Horsepower	135 bhp @ 3600 rpm	135 bhp @ 3600 rpm
Ground Clearance (to T/C skid plate @ GVWR)	10.7 in (27.2 cm)	10.7 in (27.2 cm)			

* Height will vary with different shelters and loading.

Table 1-1.1, Tabulated Data (Continued)

DATA			DATA		
	M1028A2	M1028A3		M1028A2	M1028A3
Fuel System:			**Transfer Case (Con't):**		
Low-pressure	Mechanical Lift Pump	Mechanical Lift Pump	**Ratios**	1.96 & 1.00	2.61 & 1.00
High-pressure	Stanadyne Model DB2	Stanadyne Model DB2	**Lubrication Capacity**	2 qt (2.5 l)	5 qt (4.8 l)
Injection Nozzle	Bosch	Bosch	**Front Axle:**		
Fuel Tank Capacity	20 gal (75.7 l)	20 gal (75.7 l)	**Type**	Full Floating	Full Floating
			Differential Type	Two-pinion	Two-pinion
Electrical System	12/24 volt	12/24 volt	**Lubrication Capacity**	4.5 pt (2.1 l)	4.5 pt (2.1 l)
Charging System	100 amp/ 28 volt	100 amp/ 28 volt	**Rear Axle:**		
Transmission:			**Type**	Full Floating	Full Floating
Type	Automatic GM THM 400	Automatic GM THM 400	**Differential Type**	Four-pinion No-spin	Four-pinion No-spin
Lubrication Capacity	19 pt (9 l)	19 pt (9 l)	**Lubrication Capacity**	5.4 pt (2.1 l)	5.4 pt (2.1 l)
Transfer Case:			**Service Brake:**		
Model	New Process Model 205	New Process Model 208	**Front**	Disc	Disc
			Rear	Drum	Drum
Speed	2-speed	2-speed			

CHAPTER 2
TROUBLESHOOTING INSTRUCTIONS

2-1. GENERAL.

a. This section provides information for identifying malfunctions which may develop in the CUCV. Because of the complexity of the CUCV, troubleshooting has been divided into two major areas: mechanical troubleshooting (Table 2-1) and electrical troubleshooting (Table 2-2). The troubleshooting symptom index (paragraph 2-3) provides a list of possible malfunctions, grouped alphabetically by major truck system, with the location of steps that can be taken to correct the malfunction.

b. For a better understanding of how a system operates, see *Principles of Operation* (TM 9-2320-289-20). If you're unsure about the location of an item mentioned in troubleshooting, refer to the maintenance task where the item is replaced. DO NOT perform the maintenance task unless the troubleshooting table tells you to do so.

c. When troubleshooting a malfunction:

(1) Ensure that both batteries are fully charged.

(2) Ensure that components on the truck are in accordance with TM 9-2320-289-34P. Use of improper components may result in any of the malfunctions described in Tables 2-1 or 2-2.

(3) Question the operator to obtain any information that might help determine the cause of the problem. Also ensure that all applicable operator and unit maintenance troubleshooting was performed before beginning troubleshooting procedures in this manual.

(4) Locate the symptom or symptoms in paragraph 2-3 that describe the malfunction. Check both mechanical and electrical symptoms.

(5) Turn to the page in the troubleshooting table where the troubleshooting procedures are listed. Headings at the top of each page show how each troubleshooting task is organized: MALFUNCTION, TEST OR INSPECTION (in step number order), and CORRECTIVE ACTION.

(6) Perform each step in the order listed until the malfunction is corrected, Steps are listed in most probable cause sequence and continue through the least probable cause.

2-2. TROUBLESHOOTING ELECTRICAL MALFUNCTIONS.

WARNING

Both battery negative cables must be disconnected before removing any electrical system components, (See TM 9-2320-289-20) Failure to follow this warning may result in serious injury or death to personnel.

CAUTION

DO NOT leave key in "RUN" position for more than 2 minutes for any electrical system test. Failure to follow this caution may result In damage to glow plugs.

a. Introduction. As a general rule, when troubleshooting malfunctions of electrical systems, check for continuity through the component and any applicable switches and fuses. Use the wiring diagrams in Appendix E and the following instructions to determine the routing of the wiring, Replace any component, switch, or fuse that does not have continuity.

b. Abbreviations, Lead Descriptions, and Symbols.

(1) Abbreviations. The following abbreviations of wire colors are used throughout Appendix E:

BLK . . . . Black
BLK/LT BLU . . . . Black with one light blue stripe
BLK/WHT . . . . Black with one white stripe
BLK/YEL . . . . Black with one yellow stripe
BLUE . . . . Blue
BAN . . . . Brown
BRN/RED . . . . Brown with one red stripe
BRN/WHT . . . . Brown with one white stripe
DK BLU . . . . Dark blue
DK GRN . . . . Dark green
DK GRN/WHT . . . . Dark green with one white stripe
LT BLU . . . . Light blue
LT BLU/RED . . . . Light blue with one red stripe
LT GRN . . . . Light green
GRA . . . . Gray
GIN . . . . Green
ON . . . . Orange
ORN/BLK . . . . Orange with one black stripe
PIN K . . . . Pink
PNK/BLK . . . . Pink with one black stripe
PPL . . . . Purple
PPL/WHT . . . . Purple with one white stripe
RED/WHT . . . . Red with one white stripe
TAN/WHT . . . . Tan with one white striPe
WET . . . . White
WHT/BLK . . . . White with one black stripe
YE . . . . yellow
YEL/BLK . . . . Yellow with one black stripe

(2) Lead Descriptions. Each lead on a wiring diagram is designated according to wire size, wire color, and circuit. For example, lead 3 RED-2J (SXL) is designated as follows:

Wire Color

3 RED-2J (SXL)

Wire Size — Circuit identifier

(a) Wire size is the diameter of the wire in millimeters. Use the following table to convert wire size into wire gage:

Wire Sizse	Wire gage
.5	20
.8	18
1	16
2	14
3	12
5	10
8	8
1 3	6
19	4
32.	2

(b) Wire color is as previously described.

(c) The circuit identifier is only found on the wiring diagram and is used to help distinguish between leads of the same wire size and color. The following abbreviations may appear as part of the circuit identifier (temperature in parentheses is maximum temperature that the wire can withstand):

NOTE

- A fusible link is connected to a lead by a splice and is always constructed of a wire 4 gages higher than the lead it connects to.
- Any wire that is not labeled will be poly-vinyl chloride wire.

HDT	Heavy-walled, high-abrasion, poly-vinyl chloride wire (175°F or 97°C)
HW	Heavy-walled, high-abrasion, Poly-vinyl chloride wire (275°F or 152 'C)
SGT	Poly-vinyl chloride; negative battery cable
SGX	Cross-link polyethylene; positive battery cable
SXL	Heavy-walled, high-abrasion, cross-link polyethylene (275°F or 152°C)
THERMO HW	Fusible link; heavy-walled silicone over cross-link polyethylene.

(3) Symbols. Major items appearing in wiring diagrams and schematics are labeled. Other items that may require explanation are defined below:

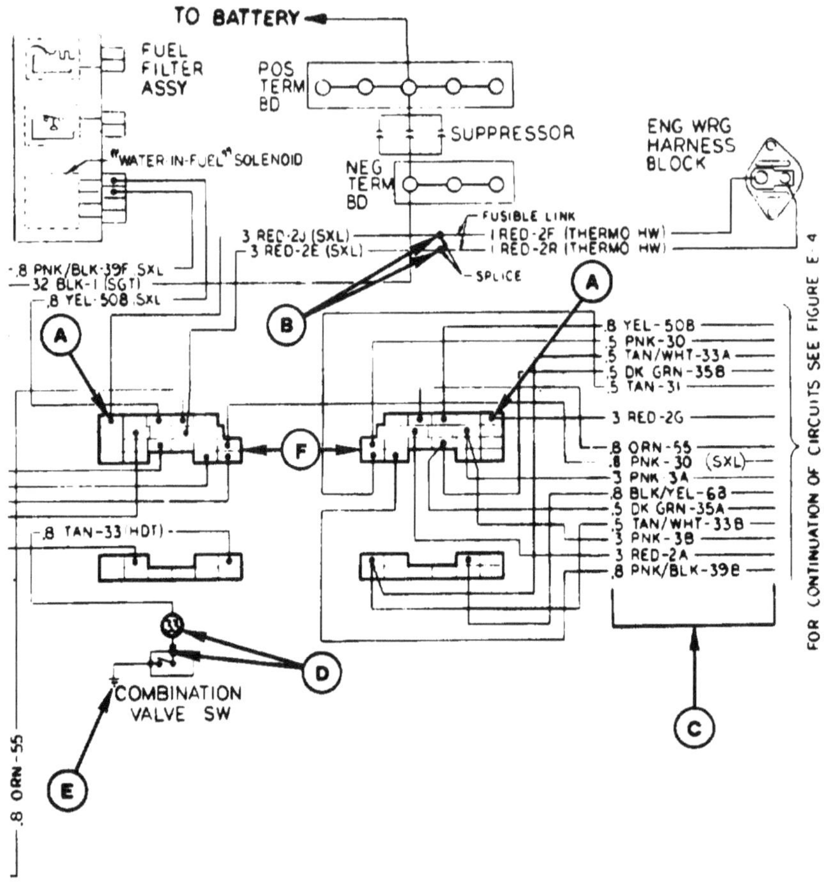

A. Wiring Harness Terminals. Will be found In "mirror Image" locations on the wiring harness male/female connector (F), The circuit through lead 3 RED-2J(SXL) is shown at points marked " (A). n Note that only terminals for the applicable circuit will be shown on the wiring diagram. Terminals shown on other wiring diagrams will be represented by empty blocks on the wiring harness connectors.

B. Spllce. Indicates where a lead is soldered or otherwise connected to 1 or more other leads. Repair any broken soldedered splice using soldering gun.

c. Wiring Harness. Can be Identified on a wiring diagram by a large number of leads going in the same direction. Wring harnesess can be Identified on the truck by a conduit (plastic sleeve) or a loom (string mesh) that wraps around the leads. The leads of each wiring harness are exposed at the base of the connectors.

D. Leads and Connector, Transmit electricity to major components. When disconnecting more than one lead and connector from a major component, tag leads and connectors for Installation.

E. Ground Lead. Connects to frame or body to complete electrical circuit. A disconnected or damaged ground lead can make a major component inoperative, or operative at lower efficiency.

F, Wiring Harness Male/Female Connectors. Are shown on the wiring diagram according to their actual shape. Anywhere that a female connector appears, the "mirror Image" shown opposite It will be Its male connector.

TA50149

C. General Electrical Troubleshooting Instructions.

(1) Any components that have been removed for testing must be installed again if they are not defective,

(2) Fuses are identified by amperage in the troubleshooting procedures, as well by as a number printed on the fuse itself. An easy way to identify fuses is by their color. The following is a list of fuses and their colors:

Amperage Fuse	Color
5 amps	Tan
10 amps	Red
15 amps	Light Blue
20 amps	Yellow
25 amps	White
30 amps	Light Green

(3) Many electrical troubleshooting tasks will require that you check for voltage or continuity through a lead or connector terminal, Leads will be identified by wire color if their location is unclear, Connector terminals will usually be identified by the leads that connect to the back of them.

2-3. TROUBLESHOOTING SYMPTOM INDEX.

MECHANICAL TROUBLESHOOTING

2-3. TROUBLESHOOTING SYMPTOM INDEX (Con't).

MECHANICAL TROUBLESHOOTING

2-3. TROUBLESHOOTING SYMPTOM INDEX (Con't).

MECHANICAL TROUBLESHOOTING

2-3. TROUBLESHOOTING SYMPTOM INDEX (Con't).

ELECTRICAL TROUBLESHOOTING

Troubleshooting Prooedure Page

Table 2-1. Mechanical Troubleshooting

MALFUNCTION
TEST OR INSPECTION
CORRECTIVE ACTION

AIR CONDITIONER

1. COOLING INSUFFICIENT

Step 1. Test air conditioning system, (See paragraph 10-20)

If leaks are found, tighten loose fittings and replace any damaged components.

Step 2. Start engine. Check air conditioner belt for slipping.

If air conditioner belt is still slipping, repair compressor. (See paragraph 10-24)

Step 3. Start engine. Inspect compressor clutch for slipping. If compressor clutch is slipping, shut off engine and check gap between compressor clutch plate and compressor pulley. Gap should be 0.022-0.057 in. (0,560-1 .450 mm).

If gap is not within specifications, disassemble compressor and check for mispositioned shaft or key. (See paragraph 10-24)

Step 4. Check clutch coil housing for improper installation or damage. (See paragraph 10-24)

Reposition or replace clutch coil housing if necessary.

Step 5. Remove air conditioner inlet grille. Inspect condenser coils and air conditioner inlet grille for air blockage due to buildup of foreign material.

Clear air blockage.

CAUTION

DO NOT oporate air conditioner for extended periods with condenser cover assembly removed. Failure to follow this caution may result In damage to air conditioning system due to high head pressure.

Step 6. Remove condenser cover assembly. (See TM 9-2320-289-20) Operate air conditioning system. Feel condenser coil bends. Bends at equal elevation should be about the same temperature.

If bends have significantly different temperatures, cooler bend has a restriction, Replace condenser if there are any restrictions. (See paragraph 10-26)

Table 2-1. Mechanical Troubleshooting (Con't)

MALFUNCTION
TEST OR INSPECTION
CORRECTIVE ACTION

Step 7. Feel evaporator outlet line for warmth. If evaporator outlet line is warm, check for frost spots on line. Frost spots on line indicate restrictions.

If evaporator outlet line is restricted, clear restrictions.

If evaporator outlet line is not restricted, perform step 8.

Step 8. Check expansion valve temperature sensing bulb positioning. It should be wrapped securely to evaporator outlet line with thermal insulation tape. (Item 67, Appendix B) Thermal insulation tape should completely cover temperature sensing bulb.

Adjust positioning if not correct.

If positioning is correct, perform step 9.

WARNING

Compressed air used to test expansion valve should never exceed 30 psi (207 kPa). Use only effective chip guarding and Personnel protective equipment (goggles/shield, gloves. etc.). Failure to follow this warning may result in serious injury to personnel.

Step 9. Remove expansion valve. (See paragraph 10-27) Check expansion valve for clogged inlet screen.

Clear inlet screen if clogged.

If inlet screen is not clogged, attempt to blow through expansion valve using compressed air not to exceed 30 psi (207 kPa). If unable to blow through expansion valve, replace expansion valve.

2. LOW OR NO AIR FLOW

Step 1. Remove condenser cover assembly, (See TM 9-2320-289-20) Check for ice buildup on evaporator coils. If there is ice buildup, test air conditioning system. (See paragraph 10-20)

If low pressure gage indicates that low side is below specifications, perform step 2.

If low pressure gage indicates that low side is within specifications, replace receiver-dehydrator. (See paragraph 10-27)

Table 2-1. Mechanical Troubleshooting (Con't)

MALFUNCTION
TEST OR INSPECTION
CORRECTIVE ACTION

WARNING

Compressed air used to test expansion valve should never exceed 30 psi (207 kPa). Use only effective chip guarding and personnel protective equipment (goggles/shield, gloves, etc.). Failure to follow this warning may result in serious injury to personnel.

Step 2. Remove expansion valve. (See paragraph 10-27) Check expansion valve for clogged inlet screen.

Clear inlet screen if clogged.

if inlet screen is not clogged, attempt to blow through expansion valve using compressed air not to exceed 30 psi (207 kPa). if unable to blow through expansion valve, replace expansion valve.

3. COMPRESSOR ENGAGED BUT NOT OPERATIONAL

Step 1. Start engine. inspect compressor clutch for slipping. if compressor clutch is slipping, shut off engine and check gap between compressor clutch plate and compressor pulley. Gap should be 0.022-0.057 in. (0.560-1 .450 mm).

if gap is not within specifications, disassemble compressor and check for mispositioned shaft or key. (See paragraph 10-24)

Step 2. Test air conditioning system, (See paragraph 10-20)

if system has proper charge, repair compressor. (See paragraph 10-24)

AXLES

4. VIBRATION

Step 1. Check differential lubricating oil level. (See LO 9-2320-289-12)

Step 2. Check companion flanges for excessive rundout (looseness). (See paragraphs 6-7, 6-9, and 6-10)

Titghten loose components and replace any damaged components.

Table 2-1. Mechanical Troubleshooting (Con't)

MALFUNCTION
TEST OR inspection
Corrective ACTION

5. EXCESSIVE NOISE WHILE TURNING

Step 1. Listen for excessive axlie noise while driving straight.

if axle noise is normal, inspect differential side gears and drive pinions for damage. (See paragraphs 6-6, 6-7, 6-8, 6-9, 6-15, 6-16, and 6-17)

If axle noise is excessive, and noise increases while turning: remove, clean, and inspect wheel bearings (see TM 9-2320-289-20) and side bearings (see paragraphs 6-6, 6-8, 6-15, and 6-16).

6. CONTINUOUS LOW-PITCHED " WHIRRING" OR SCRAPING SOUND

Step 1. Remove and inspect differential drive pinion bearings. (See paragraph 6-7, 6-9, or 6-17)

Replace bearings if damaged,

7. "KLUNKING" SOUND DURING INITIAL ACCELERATION

Step 1, Remove and inspect front axle shaft universal joint. (See paragraph 6-4)

Replace any damaged universal joints.

Step 2. Remove pinion shaft from differential assembly. (See paragraph 6-6, 6-8, or 6-16) Inspect pinion shaft for damage.

Replace pinion shaft if damaged.

6. WHINE IN ONE SPEED RANGE

Step 1. Remove appropriate differential cover. (See paragraph 6-5, 6-8, or 6-15) Inspect ring gear and drive pinion for proper gear tooth contact pattern.

Adjust gear tooth contact pattern if it is not correct.

9. WHINE IN ALL SPEED RANGES

Step 1. Remove appropriate differential cover, (See paragraph 6-5, 6-8, or 6-15) Inspect ring gear and drive pinion for damage.

Replace if damaged.

Table 2-1. Mechanical Troubleshooting (Con't)

MALFUNCTION
TEST OR INSPECTION
CORRECTIVE ACTION

BRAKE SYSTEM

10, FRONT BRAKES DRAG

Step 1. Remove front wheels and check operation of caliper piston.

If caliper piston operation is frozen or sluggish, disassemble caliper assembly. (See paragraph 7-4) Check for improper lubrication or dirt in caliper piston or caliper bore. Replace any damaged components.

11, BRAKES PULL WHEN APPLIED

Step 1. Remove front and rear wheels. Inspect rotors and drums for heat spotting and scoring.

Machine to specifications if heat spotted or scored. (See paragraph 7-2)

Step 2. Check rear backing plate shoe guides for insufficient lubrication.

Apply grease (Item 37, Appendix B) to shoe guides if insufficiently lubricated.

Step 3. Check operation of caliper piston.

If caliper piston operation is frozen or sluggish, disassemble caliper assembly. (See paragraph 7-4) Check for improper lubrication or dirt in caliper piston or caliper bore. Replace any damaged components.

Step 4. Check components of suspension system for looseness or damage.

Tighten any loose components. Replace any damaged components.

12. BRAKES SQUEAK

Step 1. Inspect rotors and drums for heat spotting and scoring.

Machine to specifications if heat spotted or scored. (See paragraph 7-2)

13. PEDAL EFFORT EXCESSIVE

Step 1. Remove front wheels and check operation of caliper piston.

If caliper piston operation is frozen or sluggish, disassemble caliper assembly. (See paragraph 7-4) Check for improper lubrication or dirt in caliper piston or caliper bore. Replace any damaged components.

Step 2. Repair power booster. (See paragraph 7-3)

Step 3. Repair power steering pump. (See paragraph 8-5)

Step 4. Replace combination valve. (See TM 9-2320-289-20)

Table 2-1. Mechanical Troubleshooting (Con't)

MALFUNCTION
TEST OR INSPECTION
CORRECTIVE ACTION

14. PEDAL RETURNS SLOWLY

Step 1, Repair power booster. (See paragraph 7-3)

15. PEDAL PULSATES

WARNING

DO NOT use a dry brush or compressed air to clean brake shoes, brake pads, or brake components. There may be asbestos dust on brake shoes, brake pads, or brake components which can be dangerous to you if you breathe it. Brake shoes, brake pads, and brake components must be wet, and a soft brush must be used. Failure to follow this warning may result in serious illness or death to personnel,

Step 1. Remove calipers. (See TM 9-2320-289-20) Install dial indicator so that plunger contacts rotor about 1 in. (25 mm) from outside edge of rotor. Turn rotor one complete rotation, Variation in dial indicator readings (lateral runout) should not be more than 0.004 in, (0, 102 mm). Repeat for other rotor,

If lateral runout is excessive, adjust wheel bearings (See TM 9-2320-289-20) and retest. If lateral runout is still excessive, machine rotor. (See paragraph 7-2)

Step 2. Remove drums. (See TM 9-2320-289-20) Measure inside diameter of drums at open and closed edges and at right angles to each other, Do this at 8 evenly-spaced points.

If there is a difference between measurements (out-of-round), machine drums to specifications. (See paragraph 7-2)

Step 3. Inspect rotors and drums for heat spotting or scoring.

Machine to specifications if heat spotted or scored. (See paragraph 7-2)

Step 4. Adjust wheel bearings. (See TM 9-2320-289-20)

Step 5. Check tires for improper balancing. (See TM 9-261 0-200-24)

Step 6. Check components of suspension system for looseness or damage.

Tighten any loose components. Replace any damaged components.

Table 2-1. Mechanical Troubleshooting (Con't)

MALFUNCTION
TEST OR INSPECTION
CORRECTIVE ACTION

ENGINE

16. WILL NOT CRANK OR CRANKS SLOWLY BUT WILL NOT START

Step 1. Remove torque converter cover and inspect flywheel for chipped or damaged teeth.

Replace flywheel if teeth are chipped or damaged. (See paragraph 3-15)

Step 2. Test starter, (See paragraph 4-5)

17, CRANKS NORMALLY BUT WILL NOT START

Step 1. Check for voltage at pink lead on fuel injector pump fuel shutoff solenoid while assistant cranks engine. There should be at least 9 volts.

If voltage is not correct, trace circuit. (See wiring diagram E-3 or E-5)

If voltage is correct, perform step 2.

Step 2. Disconnect pink lead from fuel injector pump fuel shutoff solenoid. Turn key to "RUN" position, Touch pink lead to fuel injector pump terminal and listen for clicking sound.

If clicking sound is not heard, remove governor cover. (See paragraph 3-38) Connect 12 volt power source to shutoff solenoid terminal and ground. Observe solenoid plunger for freedom of movement. Clean or replace solenoid if plunger sticks.

If clicking sound is heard, perform step 3.

Step 3. Check fuel injector pump timing. (See paragraph 3-44)

Step 4. Remove any glow plug. (See TM 9-2320-289-20) Have assistant depress accelerator pedal halfway and crank engine for 5 seconds.

If no fuel mist is observed in glow plug hole, check for restrictions in fuel return system. Clear any restrictions.

If fuel mist is observed in glow plug hole, remove remaining glow plugs and repeat test, Test fuel injector nozzle for each glow plug hole that does not have fuel vapors. (See paragraph 3-36)

Step 5. Test fuel injector pump. (See paragraph 3-44)

Step 6. Test compression of each cylinder. (See paragraph 3-2)

Table 2-1. Mechanical Troubleshooting (Con't)

MALFUNCTION
TEST OR INSPECTION
CORRECTIVE ACTION

18. WHITE SMOKE (AIR TEMPERATURE WARM)

NOTE

Perform step 1 if white smokes only appears during starting.

Step 1. Check for inoperative glow plug module. (See TM 9-2320-289-20)

Step 2. Check fuel injector pump timing, (See paragraph 3-44)

WARNING

Use extreme caution to ensure that clothing or tools DO NOT get caught in truck's operating drivebelts. Failure to follow this warning may result in serious injury to personnel or equipment damage.

Step 3. Start engine, Depress bottom of rocker lever on right side of fuel injector pump,

If engine sound doesn't change, remove and inspect fuel injector pump servo advance piston assembly. (See paragraph 3-40)

Step 4. Test compression of each cylinder. (See paragraph 3-2)

Step 5. Check timing chain free play. (See paragraph 3-1 1)

19. EXCESSIVE BLACK SMOKE

Step 1. Check fuel injector pump timing. (See paragraph 3-44)

Step 2. Test each fuel injector nozzle. (See paragraph 3-36)

Step 3. Test compression of each cylinder. (See paragraph 3-2)

20. IDLES ROUGH ON COLD-STARTS BUT CLEARS UP AFTER WARM-UP

Step 1. Check fuel injector pump timing. (See paragraph 3-44)

Step 2. Test fuel injector pump, (See paragraph 3-44)

Table 2-1. Mechanical Troubleshooting (Con't)

MALFUNCTION
TEST OR INSPECTION
CORRECTIVE ACTION

WARNING

Use extreme caution to ensure that clothing or tools DO NOT get caught in truck's operating drivebelts. Failure to follow this warning may result in serious injury to personnel or equipment damage.

Step 3. Start engine, Depress bottom of rocker lever on right side of fuel injector pump.

If engine sound doesn't change, remove and inspect fuel injector pump servo advance piston assembly. (See paragraph 3-40)

Step 4. Start engine. Loosen fuel injector line fitting at any fuel injector nozzle until fuel starts to come from fitting.

If there is no engine rpm drop, test fuel injector nozzle. (See paragraph 3-36)

If there is engine rpm drop, tighten fitting to 20 lb. -ft. (27 N.m) with crowfoot attachment J-29698-A and repeat check for each fuel injector nozzle.

21. IDLES ROUGH WITHOUT ABNORMAL NOISE OR SMOKE, WILL NOT CLEAR UP

Step 1. Check fuel injector pump timing. (See paragraph 3-44)

Step 2. Test each fuel injector nozzle. (See paragraph 3-36)

Step 3. Remove governor cover. (See paragraph 3-38) Check for black particles.

If black particles are evident, replace governor weight retaining ring. (See paragraph 3-42)

Step 4. Test fuel injector pump, (See paragraph 3-44)

Step 5. Test compression of each cylinder. (See paragraph 3-2)

22. STARTS BUT WILL NOT CONTINUE TO RUN AT IDLE

Step 1. Check fuel injector pump timing. (See paragraph 3-44)

Step 2. Have assistant start engine. Check fast idle solenoid to see if solenoid holds fuel injector pump lever in fast idle position.

If fast idle solenoid does not operate properly, perform Electrical Troubleshooting malfunction #7.

Step 3. Test each fuel injector nozzle. (See paragraph 3-36)

Step 4. Test fuel injector pump, (See paragraph 3-44)

Step 5. Test compression of each cylinder. (See paragraph 3-2)

Table 2-1, Mechanical Troubleshooting (Con't)

MALFUNCTION
TEST OR INSPECTION
CORRECTIVE ACTION

23. KNOCKS AT IDLE WHEN HOT

Step 1, Inspect torsional damper for looseness.

If torsional damper is loose , remove crankshaft pulley. (See TM 9-2320-289-20) Insert tool through slot in torsional damper, against front lip of engine oil pan, to keep crankshaft from turning. Tighten bolt to 200 Ib.-ft. (271 Nom). Install crankshaft pulley.

Step 2. Perform Mechanical Troubleshooting malfunction #34.

NOTE

If pistons are not defective, retarding fuel injector pump timing will quiet down combustion knock.

Step 3. Loosen fuel injector pump nuts. Rotate fuel injector pump to retard timing as far as it will go. Start engine. If knocking is not reduced, loosen any fuel injector line fitting at fuel fuel injector. Knocking tone will change at fuel line feeding defective piston. Tighten fitting to 20 Ib.-ft, (27 N.m) with crowfoot attachment J-29698-A and repeat check until all defective pistons are located. Reset fuel injector pump timing. (See paragraph 3-44)

Remove and inspect any defective pistons, connecting rods, and connecting rod bearings. (See paragraph 3-1 3)

Step 4. Remove engine oil pan. (See TM 9-2320-289-20) Check clearance between connecting rod bearing caps. (See paragraph 3-1 3)

24, EXCESSIVE OIL LOSS

Step 1. Test compression of each cylinder. (See paragraph 3-2)

Step 2. Remove, clean, and inspect valves. (See paragraph 3-8)

Repair valves if necessary.

25. OIL PRESSURE LIGHT ON AT IDLE

Step 1. Remove, disassemble, and inspect engine oil pump and oil filter valve. (See paragraph 3-33)

Step 2. Check main bearing clearance. (See paragraph 3-1 4)

Step 3. Remove hydraulic valve lifter clamps and guide plates. (See paragraph 3-9) Measure hydraulic valve lifter clearance where it contacts its bore, parallel with roller and 90 degrees to roller. Clearance should not be more than 0,005 in. (O. 130 mm). Repeat for each hydraulic valve lifter.

Replace hydraulic valve lifters that have excessive clearance.

Table 2-1. Mechanical Troubleshooting (Con't)

MALFUNCTION
TEST OR INSPECTION
CORRECTIVE ACTION

step 4. Drain cooling system. (See TM 9-2320-289-20) Remove engine mounts. (See paragraph 3-3) Remove starter. (See TM 9-2320-289-20) Inspect 2 oil galley plugs at front of engine block and 7 oil galley plugs at rear of engine block. Oil galley plugs should be flush with engine block.

If oil galley plugs are damaged or recessed in engine block, cut slot in oil galley plug with chisel, pry out, and discard. Apply sealant (Item 56, Appendix B) to new oil galley plug, and using oil galley plug installer, drive into engine block until flush.

If oil galley plugs protrude from engine block, drive into engine block until flush using oil galley plug installer.

26. OVERHEATS

Step 1. Attempt to spin fan and fan clutch by hand.

If fan and fan clutch revolve more than five times without drag, replace fan clutch. (See TM 9-2320-289-20)

Step 2. Visually check to see if lateral movement of fan, as measured at fan blade, exceeds approximately 0.25 in. (6.35 mm).

If lateral movement exceeds approximately 0.25 in. (6,35 mm), replace fan clutch. (See TM 9-2320-289-20)

Step 3. Check for silicone fluid leakage around fan clutch bearing assembly. Small leakage will not normally affect operation.

If silicone fluid leakage appears excessive, replace fan clutch. (See TM 9-2320-289-20)

Step 4. Start engine, allow to reach operating temperature, then shut off, Carefully feel radiator from right side to left side. Radiator should be warm on right side and hot on left side, with an even temperature rise from right to left. Cold spots indicate restrictions.

Repair or replace radiator if there are any restrictions . (See TM 9-2320-289-20 and TM 750-254)

Step 5. Start engine and check for leakage from water pump and water pump plate.

If leakage is evident, replace water pump. (See paragraph 3-48)

Step 6. Inspect each cylinder head gasket for leakage.

If there is leakage, remove and inspect cylinder head. (See paragraph 3-7) If cylinder head does not require replacement, replace cylinder head gasket and install cylinder head.

Table 2-1, Mechanical Troubleshooting (Con't)

MALFUNCTION
TEST OR INSPECTION
CORRECTIVE ACTION

27. NOTICEABLE LOSS OF POWER

Step 1. Check fuel injector pump timing. (See paragraph 3-44)

Step 2. With engine running, check for compression leaks at all fuel injector nozzles and glow plugs.

Tighten any leaking glow plugs to 10 lb, -ft. (14 N.m). Tighten any leaking fuel injector nozzles to 50 lb, -ft. (68 Nom). Repeat check. If leaks are still found, replace glow plug (see TM 9-2320-289-20) or fuel injector nozzle gasket (see paragraph 3-5).

Step 3. Test each fuel injector nozzle. (See paragraph 3-36)

Step 4. Test compression of each cylinder, (See paragraph 3-2)

28. HEAVY KNOCK WITH ENGINE TORQUE APPLIED

Step 1. Check to see if flywheel is contacting torque converter cover. Check flywheel-to-torque converter bolts for looseness.

Reposition torque converter cover if interfering with flywheel. Tighten bolts to 50 lb.-ft. (68 N.m) if loose.

Step 2. Inspect crankshaft pulley, fan pulley, and torsional damper for looseness.

Tighten crankshaft pulley or fan pulley if loose. If torsional damper is loose, remove crankshaft pulley. Insert tool through slot in torsional damper, against front lip of engine oil pan, to keep crankshaft from turning. Tighten bolt to 200 lb.-ft. (271 N.m), Install crankshaft pulley. Tighten crankshaft pulley bolts to 30 lb, -ft. (41 Nom).

Step 3. Check main bearing clearance. (See paragraph 3-1 4)

Step 4. Check connecting rod bearing clearance. (See paragraph 3-1 3)

29. KNOCKS WHEN COLD, BUT CLEARS UP AFTER WARM-UP, INCREASES WITH ENGINE TORQUE

Step 1. Check to see if flywheel is contacting torque converter cover.

Reposition torque converter cover if interfering with flywheel.

Step 2. Inspect crankshaft pulley, fan pulley, and torsional damper for looseness.

Tighten crankshaft pulley or fan pulley if loose. If torsional damper is loose, remove crankshaft pulley. Insert tool through slot in torsional damper, against front lip of engine oil pan, to keep crankshaft from turning. Tighten bolt to 200 lb. -ft. (271 N.m), Install crankshaft pulley. Tighten crankshaft pulley bolts to 30 lb.-ft. (41 N.m).

Step 3. Check fuel injector pump timing. (See paragraph 3-44)

Table 2-1. Mechanical Troubleshooting (Con't)

MALFUNCTION
TEST OR INSPECTION
CORRECTIVE ACTION

Step 4. Remove engine oil pan, (See TM 9-2320-289-20) Inspect connecting rods to see if any are bent.

Replace any bent connecting rods. (See paragraph 3-1 3)

Step 5. Remove pistons (see paragraph 3-1 3) and check piston-to-cylinder bore clearance. (See paragraph 3-13, CLEANING AND INSPECTION, steps 9 and 10)

30. STALLS UNDER DECELERATION OR HEAVY BRAKING

Step 1. Check idle speed, (See TM 9-2320-289-20)

Step 2. Remove governor Cover . (See paragraph 3-38) Check for black particles,

If black particles are evident, replace governor weight retaining ring. (See paragraph 3-42)

Step 3. Remove governor cover. (See paragraph 3-38) Operate throttle shaft and observe governor assembly and metering valve for binding.

If there is binding, remove governor assembly and metering valve. (See paragraph 3-39) Replace any damaged components. Clean metering valve with calibration fluid. (Item 13, Appendix B) Reinstall governor assembly and metering valve. If binding continues, replace metering valve.

31. EXCESSIVE SURGE AT LIGHT THROTTLE

Step 1. Check fuel injector pump timing. (See paragraph 3-44)

Step 2. Check for restrictions in fuel lines.

Clear or replac e restricted fuel lines, (See paragraph 3-5 or TM 9-2320-289-20) If no restrictions are obvious, replace fuel filter. (See TM 9-2320-289-20)

Step 3. Test each fuel injector nozzle, (See paragraph 3-36)

32. MISFIRES ABOVE IDLE OR RUNS ROUGH WHILE DRIVING (IDLES NORMALLY)

Step 1. Check fuel injector pump timing. (See paragraph 3-44)

Step 2. Check for restrictions in fuel lines.

Clear or replac e restricted fuel lines. (See paragraph 3-5 or TM 9-2320-289-20) If no restrictions are obvious, replace fuel filter. (See TM 9-2320-289-20)

Table 2-1, Mechanical Troubleshooting (Con't)

MALFUNCTION
TEST OR INSPECTION
CORRECTIVE ACTION

---

33. BLUE SMOKE (ENGINE OIL BURNING IN CYLINDERS)

NOTE

Foamy or milky oil or rust on engine oil level indicator are evidence of contamination,

Step 1. Check engine oil level indicator for evidence of contamination.

Replace engine oil and engine oil filter if contaminated . (See TM 9-2320-289-20)

Step 2. Replace crankcase depression regulator valve assembly. (See TM 9-2320-289-20)

Step 3. Check compression of each cylinder. (See paragraph 3-2)

Step 4. Remove cylinder head. (See paragraph 3-7) Inspect valve oil seals and valve guides for damage.

Replace any damaged valve oil seals. (See paragraph 3-8) Repair valve guides if damaged. (See paragraph 3-8)

34. VALVE TRAIN NOISE

Step 1. Start engine and allow to reach operating temperature. Connect STE/ICE tester to STE/ICE DCA connector under dash. (See TM 9-4910-571-12&P) Check engine oil pressure at idle.

If oil pressure is less than 10 psi (69 kPa), perform Mechanical Troubleshooting malfunction #25.

If oil pressure is 10 psi (69 kPa) or more, perform step 2.

Step 2. Check oil pressure at 2000 rpm.

If oil pressure is less than 40 psi (276 kPa), perform Mechanical Troubleshooting malfunction #25.

If oil pressure is 40 psi (276 kPa) or more, perform step 3.

Step 3. Remove rocker arm covers. (See paragraph 3-6) Inspect rocker arm assemblies and mounting bolts for looseness and damage.

Tighten mounting bolts to 40 lb, -ft, (54 N.m) if loose.

If rocker arms appear loose or damaged, see paragraph 3-6.

Step 4. Remove cylinder heads. (See paragraph 3-7) Check to see if any valve springs are broken.

Replace any broken valve springs. (See paragraph 3-8)

Table 2-1. Mechanical Troubleshooting (Con't)

MALFUNCTION
TEST OR INSPECTION
CORRECTIVE ACTION

Step 5. Remove and inspect hydraulic valve lifters. (See paragraph 3-9)

Step 6. Remove valves and inspect valve guides. (See paragraph 3-8)

Step 7. Remove and inspect camshaft. (See paragraph 3-1 2)

35. "RAPPING" SOUND FROM ONE OR MORE CYLINDERS

Step 1. Check fuel injector pump timing. (See paragraph 3-44)

Step 2. Start engine, Loosen fuel injector line fitting at any fuel injector nozzle until fuel starts to come from fitting.

If there is no engine rpm drop, test fuel injector nozzle, (See paragraph 3-36)

If there is engine rpm drop, tighten fitting to 20 lb,-ft. (27 N.m) with crowfoot attachment J-29698-A and repeat check for each fuel injector nozzle.

Step 3. Remove engine oil pan. (See TM 9-2320-289-20) Check clearance between connecting rod bearing caps. (See paragraph 3-1 3)

36. WILL NOT SHUT OFF USING KEY

WARNING

Use extreme caution to ensure that clothing or tools DO NOT get caught in truck's operating drivebelts. Failure to follow this warning may result in serious injury to personnel or equipment damage.

Step 1. With engine at idle, pinch flexible part of fuel return line at fuel injector pump to stop engine,

Step 2. Disconnect pink lead from fuel injector pump fuel shutoff solenoid. Turn key to " RUN" position. Touch pink lead to fuel injector pump terminal and listen for clicking sound.

If clicking sound is not heard, remove governor cover. (See paragraph 3-38) Connect 12 volt power source to shutoff solenoid terminal and ground. Observe shutoff solenoid plunger for freedom of movement. Clean or replace shutoff solenoid if plunger sticks.

If clicking sound is heard, perform step 3.

Step 3. Remove governor cover. (See paragraph 3-38) Operate throttle shaft and observe governor assembly and metering valve for binding.

If there is binding, remove governor assembly and metering valve. (See paragraph 3-39) Replace any damaged components. Clean metering valve with calibration fluid. (Item 13, Appendix B) Reinstall governor assembly and metering valve. If binding continues, replace metering valve.

Table 2-1. Mechanical Troubleshooting (Con't)

MALFUNCTION
TEST OR INSPECTION
CORRECTIVE ACTION

HEATER (ALL EXCEPT ENGINE COOLANT HEATER)

37. WILL NOT START

Step 1. Replace igniter, (See paragraph 11-3)

Step 2. Replace ignition control, (See paragraph 11-3)

Step 3. Remove overheat switch. (See paragraph 11-3) Check for continuity through switch contacts.

If there is no continuity, replace overheat switch.

Step 4. Remove frame detector switch. (See paragraph 11-3) Check condition of ceramic rod.

Replace ceramic rod if damaged.

Step 5. Remove fuel control valve. (See paragraph 11-3) inspect all components for looseness or damage. Check fuel line connectors for clogs.

Tighten any loose components. Replace any damaged components. Clean fuel line connectors if clogged.

Step 6. Remove burner. (See paragraph 11-3) Inspect vaporizer and wick for damage.

If only vaporizer or wick are damaged, replace. If both vaporizer and wick are damaged, repair burner using rear vaporizer kit.

38. SMOKY HEATER EXHAUST AFTER 5 MINUTES CONTINUOUS OPERATION

Step 1. Remove fuel control valve. (See paragraph 11-3) Inspect all components for looseness or damage.

Tighten any loose components. Replace any damaged components.

39. BLOWER MOTOR CONTINUES TO RUN AFTER HEATER IS TURNED OFF

Step 1. Remove flame detector switch. (See paragraph 11-3) Check condition of ceramic rod.

Replace ceramic rod if damaged.

Table 2-1. Mechanical Troubleshooting (Con't)

MALFUNCTION
TEST OR INSPECTION
CORRECTIVE ACTION

ENGINE COO/ANT HEATER

40. WILL NOT START

Step 1. Replace igniter. (See paragraph 11-2)

Step 2. Remove limit switch. (See paragraph 11 -2) Check for continuity through switch contacts.

If there is no continuity, replace limit switch.

Step 3. Replace flame switch. (See paragraph 11-2)

Step 4. Remove thermostat valve. (See paragraph 11-2) Inspect all components for looseness or damage. Check fuel line connectors for clogs.

Tighten any loose components, Replace any damaged components. Clean fuel line connectors if clogged.

41. WILL NOT SHUT OFF

Step 1. Disconnect electrical connector at electric fuel pump to shut off engine coolant heater, Remove thermostat valve. (See paragraph 11-2) Inspect all components for looseness or damage, Check fuel line connectors for clogs.

Tighten any loose components. Replace any damaged components. Clean fuel line connectors if clogged.

42. HEATER OPERATES; ENGINE REMAINS COLD

Step 1. Check engine coolant level.

Fill cooling system if low. (See LO 9-2320-289-12)

Step 2. Remove motor. (See paragraph 11-2) Check to see if coolant pump hoses, coolant pump passages, and blade slots are clogged. Inspect all components for damage.

Disassemble and clean motor to remove clogs. Replace any damaged components.

Table 2-1. Mechanical Troubleshooting (Con't)

MALFUNCTION
TEST OR INSPECTiON
Corrective ACTION

STEERING

43. EXCESSIVE PLAY IN STEERING

Step 1. Inspect steering shaft couplings and flexible coupling for looseness and wear.

Tighten nuts and bolts if loose. Repair steering shaft (see paragraph 8-3) or replace flexible coupling if worn.

Step 2. Inspect upper ball joints for wear (MI 009 only),

Replace ball joints if worn. (See paragraph 6-12)

Step 3. Adjust pitman shaft preload and bearing preload of power steering gear. (See paragraph 8-4)

Step 4. Disconnect power steering pump-to-power steering gear line at power steering gear. Check for worn power steering gear check valve.

Replace power steering gear check valve if worn. (See paragraph 8-4) Bleed power steering system. (See TM 9-2320-289-20)

Step 5. Adjust power steering gear bearing preload. (See paragraph 8-4)

Step 6. Check for damaged or worn power steering gear.

Repair power steering gear if damaged or worn. (See paragraph 8-4)

44. HARD STEERING, LACK OF ASSIST, OR SLOW RETURN OF STEERING

Step 1. Check alinement of power steering gear to steering column. Check to see if flexible coupling is rubbing against power steering gear adjuster plug.

Adjust position of steering column if not in line with steering gear. Loosen pinch bolt of flexible coupling and assemble so that flexible coupling does not rub against power steering gear adjuster plug. Tighten pinch bolt to 30 lb.-ft. (41 N.m).

Step 2. Adjust pitman shaft preload and bearing preload of power steering gear. (See paragraph 8-4)

Step 3. Perform Mechanical Troubleshooting malfunction #45.

Step 4. Disassemble power steering gear (see paragraph 8-4) and check for sticking or plugged spool valve.

Replace spool valve if sticking or plugged.

Step 5. Remove, disassemble, and inspect steering column. (See paragraph 8-2)

Replace any damaged components.

Table 2-1. Mechanical Troubleshooting (Con't)

MALFUNCTION
TEST OR INSPECTION
CORRECTIVE ACTION

45. POWER STEERING SYSTEM DOES NOT OPERATE PROPERLY

NOTE

Use suitable container to catch hydraulic fluid.

Step 1. Disconnect power steering pump-to-power steering gear line at power steering gear. Install power steering system analyzer with valve on power steering gear side. Open valve. Start engine and run at idle. Check for leaks at fittings.

Tighten any leaking fittings.

Step 2. Bleed power steering system. (See TM 9-2320-289-20)

Step 3. Start engine and run at idle. Record flow rate and pressure.

If pressure is over 150 psi (1034 kPa), perform step 4.

If pressure is 150 psi (1034 kPa) or less, perform step 5.

Step 4. Shut off engine. Check for restrictions in power steering lines. Disconnect power steering pump-to-power steering gear line at power steering gear. Inspect check valve for free operation. (See paragraph 8-4)

Clear any restrictions. Reposition or replace check valve if not operating freely.

Step 5. Partially close system analyzer valve to build pressure up to 700 psi (4827 kPa). Subtract current flow rate from flow rate in step 3.

If flow rate has dropped 1 gal. (3.785 1) per minute or more, repair power steering pump. (See paragraph 8-5)

If flow rate has dropped less than 1 gal. (3.785 1) per minute, perform step 8.

CAUTION

DO NOT leave system analyzer valve fully closed for more than 5 seconds. Failure to follow this caution may result in damage to power steering pump.

Step 6. Close and partially open system analyzer valve three times. Note highest pressure each time valve is closed. All three readings must be 1350-1450 psi (9308-9998 kPa).

If all readings are within specifications, perform step 7.

If any reading is not within specifications, replace power steering pump control valve. (See paragraph 8-5)

Table 2-1. Mechanical Troubleshooting (con't)

MALFUNCTION
TEST OR inspection
Corrective ACTION

Step 7. Connect STE/ICE tester to STE/ICE DCA connector under dash. (See TM 9-4910-571-12&P) Raise engine speed to 1500 rpm. Subtract current flow rate from flow rate in step 3.

if flow rate has dropped 1 gal. (3.785 1) per minute or more, remove and clean power steering pump control valve. (See paragraph 8-5) Remove any burrs with crocus cloth (Item 18, Appendix B). If hydraulic fluid is dirty, repair power steering pump. (See paragraph 8-5)

If flow rate has dropped less than 1 gal. (3.785 1) per minute, perform step 8.

Step 8. Record current flow rate. Have assistant turn steering wheel all the way to left and all the way to right. Record flow rate at each stop.

If flow rate drops to less than 1 gal. (3.785 1) per minute at each stop, perform step 9.

If flow rate does not drop to less than 1 gal. (3.785 1) per minute at each stop, repair power steering gear. (See paragraph 8-4)

Step 9. Run engine at idle speed. Have assistant turn steering wheel slightly to left and to right, Observe change in pressure readings. Pressure should drop immediately as steering wheel is turned, and raise immediately as steering wheel is straightened.

If pressure drops slowly; remove, disassemble, and clean power steering gear valve assembly. (See paragraph 8-4) If hydraulic fluid is dirty, repair power steering pump, (See paragraph 8-5)

46. NOISY STEERING

Step 1. Disassemble, clean, and inspect power steering pump. (See paragraph 8-5)

Step 2. Adjust pitman shaft preload of power steering gear. (See paragraph 8-4)

Step 3. Remove power steering gear valve assembly. (See paragraph 8-4) Inspect seals for damage.

Replace seals if damaged.

TRANSFER CASE

47. DIFFICULT TO SHIFT

Step 1. Disassemble transfer case. (See paragraph 5-21 or 5-22) Check for binding, worn, or damaged components inside transfer case.

Free any binding components if possible. Replace any worn or damaged components.

Table 2-1. Mechanical Troubleshooting (Con't)

MALFUNCTION
TEST OR inspection
Corrective ACTION

48. NOISY

Step 1. Check for loose transfer case mounting bolts.

Tighten transfer case-to-adapter bolts if loose. Tighten 2 adapter-to-transmission support bolts to 40 lb. -ft. (54 N.m) if loose. Tighten 4 transfer case-to-front propeller shaft bolts to 15 lb.-ft. (20 N.m) if loose.

Step 2. Disassemble transfer case (see paragraph 5-21 or 5-22) and check for worn or damaged bearings.

Replace any worn or damaged bearings.

49. EXCESSIVE FLUID LOSS

Step 1. Check for loose transfer case housing bolts.

Tighten front output shaft bolts to 35 lb.-ft. (47 N.m) if loose. Tighten rear extension bolts to 30 lb.-ft. (41 N.m) if loose.

Step 2. Check to see if transfer case vent is plugged or restricted.

Clear if plugged or replace if restricted.

Step 3. Disassemble transfer case (see paragraph 5-21 or 5-22) and check for damaged seals and gaskets.

Replace any damaged seals and gaskets.

SO. CONTROL LEVER DISENGAGES FROM POSITION

Step 1. Inspect selector rod at transfer case to see if it is bent or worn.

Replace selector rod if bent or worn. (See TM 9-2320-289-20)

Step 2. Disassemble transfer case (see paragraph 5-21 or 5-22) and check for worn or damaged gears. Check for missing cap plug, shift ball, and poppet spring.

Replace worn or damaged gears. Install cap plug. shift ball, and poppet spring if missing.

Table 2-1, Mechanical Troubleshooting (Con't)

MALFUNCTION
TEST OR INSPECTION
CORRECTIVE ACTION

TRANSMISSION

51. EXCESSIVE OIL LOSS

WARNING

Dry cleaning solvent P-D-680 is toxic and flammable. Always wear protective goggles and gloves and us. only in a well-ventilated I rea. Avoid contact with skin, I yes, and clothes I nd DO NOT breathe vapors. DO NOT us. near open flame or I xcesslvo heat. The solvent's flash point Is 100°F-1 38°F (38°C-S9°C). If you became dizzy while using cleaning solvent, immediately get fresh air and medical help. if solvent contacts eyes, immediately wash your eyes with water and get medical aid.

Step 1. Thoroughly clean exterior of transmission and transfer case with dry cleaning solvent (Item 23, Appendix B), and allow to air dry. Operate truck for about 10 minutes, then shut off. Inspect transmission case and transfer case for leakage.

If hydraulic fluid is leaking from anywhere except front of transmission, inspect components for damage. Replace any damaged components, gaskets, or seals.

If hydraulic fluid is leaking from front of transmission, perform step 2.

Step 2. Remove torque converter cover. Thoroughly clean and dry torque converter and transmission oil pump. Chock rear wheels. Raise front of truck and support on jack stands. Have assistant start engine in " P" (Park). Observe torque converter and transmission oil pump for area of leakage.

Replace any leaking components.

Step 3. Check for loose transmission case-to-transfer case adapter bolts.

Tighten bolts to 30 lb.-ft. (41 N.m) if loose.

NOTE

Thor. is *no* approved way to chock transmission oil filter for blockage or restrictions.

Step 4. Remove transmission oil filter (see TM 9-2320-289-20) and check for water in hydraulic fluid. Inspect intake pipe and "O" ring for improper positioning and damage.

Replace hydraulic fluid and transmission oil filter if any water is present. Replace intake pipe and " O" ring if damaged, Replace transmission oil filter if blockage or restrictions are suspected.

Table 2-1. Mechanical Troubleshooting (Con't)

MALFUNCTION
TEST OR INSPECTION
CORRECTIVE ACTION

Step 5. Remove transmission. (See paragraph 5-1 2) Inspect transmission case for porosity and damage.

Replace transmission case if porous or damaged,

52. NOISY IN PARK, NEUTRAL, AND ALL DRIVING RANGES

Step 1. Remove torque converter cover. (See paragraph 5-1 2) Check flywheel-to-torque converter bolts for looseness.

Tighten bolts to 50 lb.-ft. (68 N.m) if loose.

NOTE

There Is no approved way to check transmission oil filter for blockage or restrictions.

Step 2. Remove transmission oil filter (see TM 9-2320-289-20) and check for water in hydraulic fluid. Inspect intake pipe and "O" ring for improper positioning and damage.

Replace hydraulic fluid and transmission oil filter if any water is present, Replace intake pipe and " O" ring if damaged, Replace transmission oil filter if blockage or restrictions are suspected,

Step 3. Remove torque converter and inspect for damage, (See paragraph 5-1 2)

Step 4. Remove, disassemble, clean, and inspect transmission oil pump. (See paragraph 5-1 5)

53. NOISY DURING ACCELERATION IN ALL DRIVING RANGES

Step 1, Check engine mounts, insulators, and attaching hardware for looseness and damage.

Replace damaged components. (See paragraph 3-3) Tighten loose engine mount-to-insulator locknuts to 55 lb. -ft. (75 N.m), Tighten loose engine mount-to-engine bolts to 35 lb.-ft, (47 N.m).

Step 2. Check to see if transmission oil cooler lines are rubbing underbody.

Position transmission oil cooler lines clear of underbody if rubbing.

54. NOISY WITH TRANSMISSION GEARSHIFT SELECTOR IN " 1," "2," AND/OR REVERSE

Step 1. Remove, disassemble, clean, and inspect transmission intermediate clutch, front internal gear ring, and gear units. (See paragraph 5-1 9)

Table 2-1, Mechanical Troubleshooting (Con't)

MALFUNCTION
TEST OR INSPECTION
CORRECTIVE ACTION

55. SQUEAL AT LOW SPEEDS, ESPECIALLY WHEN HOT

Step 1. Remove speedometer driven gear from transfer case. (See TM 9-2320-289-20) Inspect seal for damage. Inspect speedometer driven gear shaft to see if it is twisted.

Replace speedometer driven gear if shaft is twisted.

Step 2. Remove propeller shafts from transfer case. (See TM 9-2320-289-20) Check for damage or lack of lubrication at output shaft seals.

Lubricate seals with hydraulic fluid (Item 40, Appendix B) or replace if damaged.

56. NO UPSHIFTS, DELAYED UPSHIFTS, OR FULL THROTTLE UPSHIFTS ONLY

Step 1. Disconnect downshift electrical connector from side of transmission. Operate truck and observe upshift performance.

If upshift is normal, replace transmission downshift switch on side of accelerator bracket.

If upshift performance is still poor, perform step 2.

Step 2. Connect downshift electrical connector to side of transmission. Test transmission oil pressure with brakes applied, " D" (Drive), engine at 1000 rpm. (See paragraph 5-11)

If oil pressure is normal; remove, disassemble, clean, and inspect governor assembly. (See paragraph 5-5)

If oil pressure is high, perform step 3.

Step 3. Test transmission oil pressure with brakes applied, "N" (Neutral), engine at 1000 rpm. (See paragraph 5-11)

If oil pressure is normal, perform step 4.

If oil pressure is high, perform step 5.

Step 4. Check for leaking or damaged detent solenoid.

Tighten detent solenoid if loose and replace if damaged. (See paragraph 5-7)

If detent solenoid is not loose or damaged, clean transmission control valve assembly. (See paragraph 5-7)

Step 5. Check for loose or damaged vacuum lines and hoses.

Connect vacuum hoses or vacuum lines if loose and replace if damaged. (See TM 9-2320-289-20)

Table 2-1, Mechanical Troubleshooting (Con't)

MALFUNCTION
TEST OR INSPECTION
CORRECTIVE ACTION

Step 6. Disconnect vacuum hose from vacuum pump. Install hose and vacuum gage on vacuum pump. Start engine. Calculate proper vacuum gage reading as shown in chart. Shut off engine.

If vacuum drops 1 in. Hg or more in less than 1,5 seconds, or if vacuum never reaches proper gage reading, replace vacuum Dump. (See TM 9-2320~289-20)

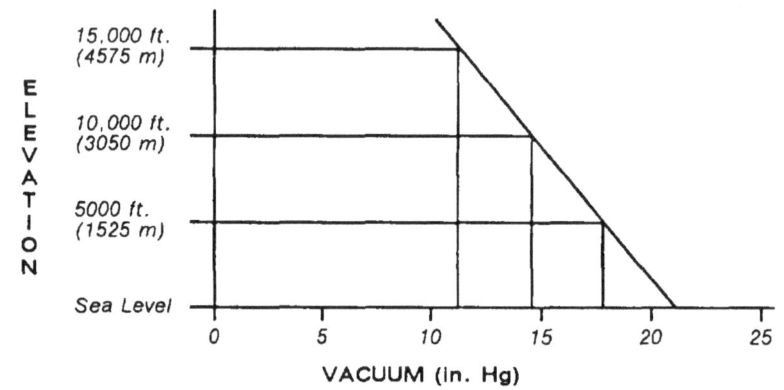

Minimum Acceptable Vacuum vs. Altitude

Step 7. Remove, clean, and inspect vacuum modulator assembly and modulator valve, (See paragraph 5-3)

57. ROUGH 1-2 SHIFT OR 2-3 SHIFf

Step 1. Test transmission oil pressure with brakes applied, " D" (Drive), engine at 1000 rpm. (See paragraph 5-11)

If oil pressure is normal, perform step 2.

If oil pressure is high, perform step 3.

Step 2. Check for loose or damaged detent solenoid,

Tighten detent solenoid if loose and replace if damaged. (See paragraph 5-7)

If detent solenoid is not loose or damaged, clean transmission control valve assembly. (See paragraph 5-7)

Step 3. Check for loose or damaged vacuum lines and hoses.

Connect vacuum hoses or vacuum lines if loose and replace if damaged, (See TM 9-2320-289-20)

Table 2-1, Mechanical Troubleshooting (Con't)

MALFUNCTION
TEST OR INSPECTION
CORRECTIVE ACTION

Step 4. Disconnect vacuum hose from vacuum pump. Install hose and vacuum gage on vacuum pump. Start engine. Calculate proper vacuum gage reading as shown in chart. Shut off engine.

If vacuum drops 1 in. Hg or more in less than 1.5 seconds, or if vacuum never reaches proper gage reading, replace vacuum pump. (See TM 9-2320-289-20)

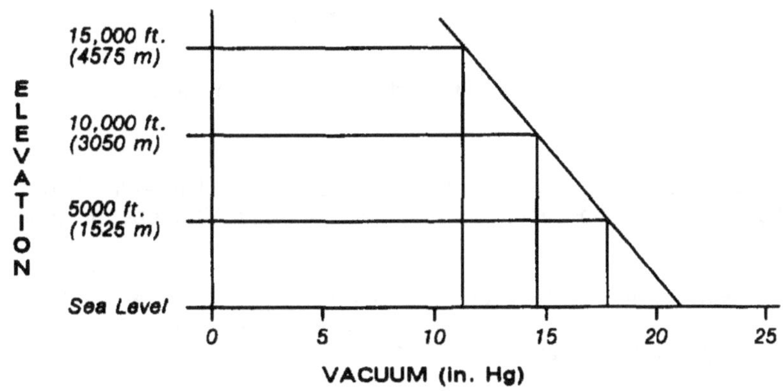

Minimum Acceptable Vacuum vs. Altitude

Step 5. Remove, clean, and inspect vacuum modulator assembly and modulator valve. (See paragraph 5-3)

Step 6. Remove and air check direct clutch for leak to outer area of direct clutch piston. (See paragraph 5-1 7)

Repair direct clutch if it is leaking.

Step 7. Remove, disassemble, clean, and inspect transmission oil pump. (See paragraph 5-1 5)

Step 8. Check for damage to intermediate clutch piston seals, second oil seal ring on center support, and center support assembly.

Replace any damaged components. (See paragraph 5-1 9)

58. LONG 1-2 SHIFT WITH END BUMP, 1-2 SLIPPAGE, OR EARLY 1-2 SHIFT

Step 1. Test transmission oil pressure with brakes applied, " D" (Drive), engine at 1000 rpm. (See paragraph 5-11)

If oil pressure is normal, perform step 2.

If oil pressure is low, perform step 3.

Table 2-1. Mechanical Troubleshooting (Con't)

MALFUNCTION
TEST OR INSPECTION
CORRECTIVE ACTION

Step 2. Check oil pressure test gage response to opening throttle. There should be rapid gage response to opening throttle.

If pressure response is good; remove, disassemble, clean, and inspect transmission control valve assembly. (See paragraph 5-7)

If pressure response is poor; check for restricted vacuum hoses and lines. Clear any restrictions. Remove, clean, and inspect vacuum modulator assembly and modulator valve. (See paragraph 5-3)

NOTE

There is no approved way to check transmission oil filter for blockage or restrictions.

Step 3. Remove transmission oil filter and check for water in hydraulic fluid. (See TM 9-2320-289-20) Inspect intake pipe and " O" ring for improper positioning and damage.

Replace hydraulic fluid and transmission oil filter if any water is present. Replace intake pipe and " O" ring if damaged. Replace transmission oil filter if blockage or restrictions are suspected.

Step 4. Remove, disassemble, clean, and inspect transmission oil pump. (See paragraph 5-15)

Step 5. Check center support bolt for looseness. (See paragraph 5-1 9)

Tighten bolt to 25 lb.-ft. (34 N.m) if loose.

Step 6. Remove, disassemble, clean, and inspect transmission intermediate clutch. (See paragraph 5-1 9)

Step 7. Disassemble, clean, and inspect forward clutch assembly. (See paragraph 5-1 6)

Step 8. Remove and inspect center support seals and center support assembly, (See paragraph 5-1 9)

59. LONG 2-3 SHIFT WITH END BUMP, 2-3 SLIPPAGE, OR EARLY 2-3 SHIFT

Step 1. Perform Mechanical Troubleshooting malfunction #58.

Step 2. Remove, disassemble, clean, and inspect direct clutch assembly. (See paragraph 5-1 7)

Step 3. Remove, disassemble, clean, and inspect front servo assembly, (See paragraph 5-10)

Table 2-1. Mechanical Troubleshooting (Con't)

MALFUNCTION
TEST OR inspection
Corrective ACTION

60. FIRST AND SECOND SPEEDS ONLY; NO 2-3 SHIFT

Step 1. Check tailpipe, muffler, exhaust manifolds, and exhaust pipe for restrictions.

Clear any restrictions.

Step 2. Check for damaged or loose vacuum hoses. Remove, clean, and inspect vacuum modulator assembly and modulator valve. (See paragraph 5-3)

Step 3. Remove, disassemble, clean, and inspect governor assembly. (See paragraph 5-5)

Step 4. Remove, disassemble, clean, and inspect transmission control valve assembly, (See paragraph 5-7)

Step 5. Remove, disassemble, clean, and inspect direct clutch assembly, (See paragraph 5-17)

Step 6. Remove, disassemble, clean, and inspect front servo assembly (see paragraph 5-1 O) and rear servo assembly (see paragraph 5-9).

61. NO DRIVE OR SLIPS IN DRIVE

Step 1. Inspect transmission linkage for damage.

Replace any damaged components. (See paragraph 5-2)

Step 2. Test transmission oil pressure with brakes applied, "D" (Drive)) engine at 1000 rpm, (See paragraph 5-11)

If oil pressure is low, perform step 3.

If oil pressure is normal, perform step 5.

NOTE

There is no approved way to check transmission oil filter for blockage or restrictions.

Step 3. Remove transmission oil filter and check for water in hydraulic fluid. (See TM 9-2320-289-20) Inspect intake pipe and "O" ring for improper positioning and damage.

Replace hydraulic fluid and transmission oil filter if any water Is present. Replace intake pipe and " O" ring if damaged. Replace transmission oil filter if blockage or restrictions are suspected.

Step 4. Remove, clean, and inspect vacuum modulator assembly and modulator valve. (See paragraph 5-3)

Table 2-1, Mechanical Troubleshooting (Con't)

MALFUNCTION
TEST OR INSPECTION
Corrective ACTION

Step 5. Remove, disassemble, clean, and inspect transmission oil pump. (See paragraph 5-15)

Step 6. Remove, disassemble, clean, and inspect forward clutch assembly. (See paragraph 5-1 6)

Step 7. Remove, clean, and inspect roller clutch assembly. (See paragraph 5-1 9)

62. NO REVERSE OR SLIPS IN REVERSE

Step 1. Inspect transmission linkage for damage,

Replace any damaged components. (See paragraph 5-2)

Step 2. Test transmission oil pressure with brakes applied, " R" (Reverse), engine at 1000 rpm. (See paragraph 5-11)

If oil pressure is normal, perform step 3.

If oil pressure is low, perform step 4.

NOTE

There is no approved way to check transmission oil filter for blockage or restrictions.

Step 3. Remove transmission oil filter and check for water in hydraulic fluid. (See TM 9-2320-289-20) Inspect intake pipe and " O" ring for improper positioning and damage.

Replace hydraulic fluid and transmission oil filter if any water is present. Replace intake pipe and "O" ring if damaged. Replace transmission oil filter if blockage or restrictions are suspected.

Step 4. Remove, disassemble, clean, and inspect transmission oil pump, (See paragraph 5-1 5)

Step 5. Remove, disassemble, clean, and inspect transmission control valve assembly. (See paragraph 5-7)

Step 6. Remove, disassemble, clean, and inspect rear servo assembly. (See paragraph 5-9)

Step 7. Remove, disassemble, clean, and inspect direct clutch assembly. (See paragraph 5-17)

Step 8. Remove and inspect rear band, (See paragraph 5-1 9)

Step 9. Remove, clean, and inspect center support assembly and seals. (See paragraph 5-1 9)

Table 2-1. Mechanical Troubleshooting (Con't)

MALFUNCTION
TEST OR INSPECTION
CORRECTIVE ACTION

63. SLIPS IN SECOND OR SHIFTS 1-3 (MISSES SECOND)

Step 1. Remove, disassemble, clean, and inspect intermediate clutch assembly. (See paragraph 5-1 9)

Step 2. Inspect center support assembly, intermediate clutch, and intermediate clutch piston seals. (See paragraph 5-1 9)

64. UPSHITS PROPERLY; WILL NOT DOWNSHIFT

Step 1. Disconnect downshift electrical connector from side of transmission. Connect test light to terminal and ground. Turn key to "RUN" position. Have assistant depress accelerator pedal fully.

If test light lights, check for loose or damaged detent solenoid. Tighten if loose and replace if damaged. (See paragraph 5-7)

If test light does not light, perform step 2.

Step 2. Trace circuit from transmission downshift switch on side of accelerator cable bracket. (See wiring diagram E-4 or E-6)

If circuit is good, replace transmission downshift switch.

Step 3. Remove, disassemble, clean, and inspect transmission control valve assembly. (See paragraph 5-7)

65. NO ENGINE BRAKING WITH TRANSMISSION GEARSHIFT SELECTOR IN "2"

Step 1. Remove, disassemble, clean, and inspect front servo assembly. (See paragraph 5-10)

Step 2. Remove, disassemble, clean, and inspect direct clutch assembly and front band. (See paragraph 5-1 7)

660 NO ENGINE BRAKING WITH TRANSMISSION GEARSHIFT SELECTOR IN "1"

Step 1. Remove transmission control valve assembly. (See paragraph 5-7) Check to see if any check balls are missing or mispositioned, Inspect for damage at check ball seat pockets in transmission case.

Install check balls in correct position if missing or mispositioned. Transmission case must be replaced if check ball seat pockets are damaged.

Step 2. Remove, disassemble, clean, and inspect rear servo assembly. (See paragraph 5-9)

Step 3. Remove and inspect rear band. (See paragraph 5-1 9)

Table 2-1. Mechanical Troubleshooting (Con't)

MALFUNCTION
TEST OR INSPECTION
CORRECTIVE ACTION

67. TRUCK DRIVES WITH TRANSMISSION GEARSHIFT SELECTOR IN NEUTRAL

Step 1. Inspect transmission linkage for improper positioning and damage.

Adjust if out of position, Replace any damaged components, (See paragraph 5-2)

Step 2. Remove parking pawl and bracket. (See paragraph 5-8) Inspect transmission internal linkage for damage.

Step 3. Remove, disassemble, clean, and inspect transmission oil pump. (See paragraph 5-15)

Step 4. Remove, disassemble, clean, and inspect forward clutch assembly. (See paragraph 5-16)

68. WILL NOT HOLD IN PARK OR WiLL NOT RELEASE FROM PARK

Step 1. Inspect transmission linkage for improper positioning and damage. (See paragraph 5-2)

Step 2. Remove parking pawl and bracket. (See paragraph 5-8) Inspect transmission internal linkage for damage.

Table 2-2. Electrical Troubleshooting

MALFUNCTION
TEST OR INSPECTION
CORRECTIVE ACTION

AIR CONDITIONER

1. COOLING INSUFFICIENT

Step 1. Test air conditioning system, (See paragraph 10-20) Record low and high pressure readings,

If low and high pressure readings are within specifications, system is operational.

If high pressure reading is not within specifications, perform step 2.

Step 2. Check for 24 volts at light blue lead terminal at 24 volt air conditioner relay.

If there is voltage, system is operational.

If there is no voltage, perform step 3.

CAUTION

DO NOT operate air conditioner for extended periods with condenser cover assembly removed. Failure to follow this caution may result in damage to air conditioning system due to high head pressure.

Step 3. Remove condenser cover assembly. (See TM 9-2320-289-20) Attempt to operate air conditioning system. Trace light blue lead from 24 volt air conditioner relay to red lead at low pressure switch. (See wiring diagram E-26)

If there is voltage, repair circuit.

If there is no voltage, perform step 4.

Step 4. Trace circuit to black lead on high pressure switch. (See wiring diagram E-26)

If there is voltage, replace low pressure switch. (See paragraph 10-27)

If there is no voltage, perform step 5.

Step 5. Remove air conditioner control panel. (See paragraph 10-22) Trace circuit to "A/C COLD" switch. (See wiring diagram E-26)

If there is voltage, replace high pressure switch. (See paragraph 10-27)

If there is no voltage, perform step 6.

Step 6. Trace blue lead to "VENT-A/C" switch. (See wiring diagram E-26)

If there is voltage, replace "A/C COLD" switch. (See paragraph 10-22)

If there is no voltage, perform step 7.

Table 2-2. Electrical Troubleshooting (Con't)

MALFUNCTION
TEST OR INSPECTION
CORRECTIVE ACTION

Step 7. Trace yellow lead to outside air door switch. (See wiring diagram E-26)

If there is voltage, replace "VENT-A/C" switch, (See paragraph 10-22)

If there is no voltage, perform step 8.

Step 8. Trace yellow lead to 12 volt air conditioner relay, (See wiring diagram E-26)

If there is voltage, replace outside air door switch. (See paragraph 10-22)

If there is no voltage, repair circuit.

2. ONE AIR CONDITIONER CONTROL INOPERATIVE (OTHER CONTROLS OPERATE)

Step 1. Pull out air conditioner control panel. (See paragraph 10-22, REMOVAL, steps 2 and 3) Connect both battery negative cables. (See TM 9-2320-289-20) Check for voltage to inoperative switch.

If there is voltage, disconnect both battery negative cables (see TM 9-2320-289-20) and replace switch, (See paragraph 10-22)

If there is no voltage, trace circuit. (See wiring diagram E-26)

BATTERIES/CHARGING SYSTEM

3. GENERATOR LIGHT REMAINS ON OR TURNS ON DURING OPERATION (ALL EXCEPT MI 010 AND MI 028 WITH 200 amp SYSTEM)

Step 1. Test appropriate alternator. (See paragraph 4-2)

4. VOLTMETER IN RED (OVERCHARGE) ZONE (ALL EXCEPT MI 010 AND MI 028 WITH 200 amp SYSTEM)

Step 1. Test appropriate alternator, (See paragraph 4-2)

5. VOLTMETER IN YELLOW (UNDERCHARGE) ZONE (MI 010 AND MI 028 WITH 200 amp SYSTEM)

Step 1. Remove current/voltage regulator cover. (See paragraph 4-4) With key in "OFF" position, check voltage at current/voltage regulator "1" terminal.

If there is no voltage, repair circuit to front battery positive (+) terminal. (See wiring diagram E-10)

If voltage is less than 12.0, perform step 2.

If voltage is 12.0-12.6, perform step 3.

Table 2-2. Electrical Troubleshooting (Con't)

MALFUNCTION
TEST OR INSPECTION
CORRECTIVE ACTION

Step 2. Check front battery for proper charge and electrolyte level. (See TM 9-2320-289-20) Remove and clean front battery cable and clamp connections.

Service or replace front battery. Replace worn or damaged battery cables and clamps. (See TM 9-2320-289-20)

Step 3. Check voltage at current/voltage regulator "2" terminal.

If there is no voltage, repair circuit to rear battery positive terminal. (See wiring diagram E-10)

If voltage is less than 24.0, perform step 4.

If voltage is 24.0-25.2, perform step 5.

Step 4. Check rear battery for proper charge and electrolyte level. (See TM 9-2320-289-20) Remove and clean rear battery cable and clamp connections.

Service or replace rear battery. Replace worn or damaged battery cables and clamps. (See TM 9-2320-289-20)

Step 5. Turn key to "RUN" position. Check voltage at upper alternator positive (+) terminal.

If voltage is above 9.5, perform step 8.

If there is no voltage, perform step 6.

Step 6. Check voltage at current/voltage regulator " A" terminal.

If voltage is above 9.5, repair circuit to alternator positive (+) terminal. (See wiring diagram E-10)

If there is no voltage, perform step 7.

Step 7. Check voltage at current/voltage regulator relay "IGN" terminal.

If there is no voltage, replace 20 amp engine control fuse (see TM 9-2320-289-20) and trace circuit (see wiring diagram E-10).

If there is voltage, repair circuit to current/voltage regulator "A" terminal. (See wiring diagram E-10)

Step 8. Check voltage at upper alternator "IGN" terminal.

If voltage is above 9.5, perform step 11.

If there is no voltage, perform step 9.

Step 9. Check voltage at current/voltage regulator relay "FLD" terminal.

If voltage is above 10.5, repair circuit to upper alternator "IGN" terminal. (See wiring diagram E-10)

If there is no voltage, perform step 10.

Table 2-2. Electrical Troubleshooting (Con't)

MALFUNCTION
TEST OR INSPECTION
CORRECTIVE ACTION

Step 10. Check voltage at current/voltage regulator relay "A+" terminal.

if voltage is above 11.5, replace current/voltage regulator relay. (See paragraph 4-4)

if there is no voltage, trace circuit to front battery positive (+) terminal. (See wiring diagram E-10)

CAUTION

DO NOT continue rotating adjustment screw after resistance is felt. Adjustment screw adjustment should never exceed 3/4 turn. Failure to follow this caution may damage adjustment screw and require replacement of current/voltage regulator.

Step 11. Start engine, bring to full operating temperature, then shut off. Remove 10 amp/28 v (voltmeter) fuse. (See TM 9-2320-289-20) Adjust current/voitage regulator adjustment screw completely clockwise. Connect tachometer in accordance wit h manufacturer's instructions. Start engine and operate at 1200-1500 rpm. Check voltage at upper alternator positive (+) terminal for at least 30 seconds.

if voltage is 15.5-16.5, replace current/voltage regulator. (See paragraph 4-4) Repeat this step.

if voltage is 28.5-29,1, perform step 12.

if voltage is not 28.5-29.1, perform step 16.

CAUTION

DO NOT continue rotating adjustment screw after resistance is felt. Adjustment screw adjustment should never exceed 3/4 turn. Failure to follow this caution may damage adjustment screw and require replacement of current/voltage regulator.

Step 12. Adjust current/voltage regulator adjustment screw completely counterclockwise and check voltage at upper alternator positive (+) terminal for at least 30 seconds.

if voltage drops to 15,5-16.5, seal adjustment screw (see paragraph 4-4, iNSTALLATION, step 1) and perform step 15.

if voltage does not drop to 15.5-16.5, perform step 13.

Table 2-2. Electrical Troubleshooting (Con't)

MALFUNCTION
TEST OR INSPECTION
CORRECTIVE ACTION

Step 13. Check voltage between rear battery positive (+) and negative (-) terminals.

If voltage is above 14,2, and voltage in step 12 does not drop, replace current/voltage regulator, (See paragraph 4-4) Perform step 11,

If voltage is above 14,2, and voltage in step 12 drops to 16.5-26.5 or below 15.5, replace current/voltage regulator relay. (See paragraph 4-4) Perform step 12.

If voltage is not above 14.2, perform step 14.

Step 14. Check rear battery for proper charge and electrolyte level. (See TM 9-2320-289-20) Remove and clean rear battery cable and clamp connections.

Service or replace rear battery. Replace worn or damaged battery cables and clamps. (See TM 9-2320-289-20) Perform step 12.

Step 15. Check voltage at current/voltage regulator relay "FLD" terminal for several minutes, Voltage should be 6,0-11.0, drop to 0.0 for about 2 seconds each minute, and return to 6.0-11.0.

If current/voltage regulator relay does not operate properly, replace. (See paragraph 4-4) Repeat this step.

If current/voltage regulator relay operates properly, perform step 190

Step 16. Shut off engine. Remove upper alternator voltage regulator (se e TM 9-2320-289-20) and ensure that adjustment screw is in "HI" position.

If adjustment screw is in "HI" position, perform step 18.

If adjustment screw is not in "HI" position, perform step 17.

Step 17. Install adjustment screw in "HI" position and install voltage regulator. (See TM 9-2320-289-20) Start engine and operate at 1200-1500 rpm. Check voltage at upper alternator positive (+) terminal.

If voltage is 28,5-29,1, perform step 12.

If voltage is not 28.5-29.1, perform step 18.

Step 18. Replace upper alternator voltage regulator. (See TM 9-2320-289-20) Start engine and operate at 1200-1500 rpm. Check voltage at upper alternator positive (+) terminal.

If voltage is 28.5-29.1, perform step 12.

If voltage is not 28.5-29.1, remove and test upper alternator. (See paragraph 4-3) Perform step 11,

Step 19. Shut off engine and install 10 amp/28 v (voltmeter) fuse. (See TM 9-2320-289-20) Start engine and check voltmeter.

If voltmeter is not in green zone, replace voltmeter. (See TM 9-2320-289-20)

Table 2-2. Electrical Troubleshooting (Con't)

MALFUNCTION
TEST OR INSPECTION
CORRECTIVE ACTION

6. VOLTMETER IN RED (OVERCHARGE) ZONE (MI 010 AND MI 028 WITH 200 amp SYSTEM)

Step 1. Perform Electrical Troubleshooting malfunction #5, step 11.

STARTING SYSTEM

7. ENGINE STARTS BUT WILL NOT CONTINUE TO RUN AT IDLE

Step 1. Turn key to "RUN" position. Check for voltage at fast idle solenoid.

If there is voltage and fast idle solenoid plunger does not extend, replace fast idle solenoid. (See TM 9-2320-289-20)

If there is no voltage, perform step 2.

Step 2. Disconnect connector from cold advance switch at right rear of engine block. (See wiring diagram E-3 or E-5) Jump light green and pink/black leads at connector, Check for voltage at fast idle solenoid.

If there is voltage and fast idle solenoid plunger does not extend, replace fast idle solenoid. (See TM 9-2320-289-20)

If there is voltage, fast idle solenoid plunger extends, and engine is cold, replace cold advance switch.

If there is voltage, fast idle solenoid plunger extends, and engine is hot, system operates properly.

If there is no voltage, trace circuit. (See wiring diagram E-3 or E-5)

WINDSHIELD WIPER/WASHER SYSTEM

8. WIPER MOTOR INOPERATIVE IN ONE SPEED

Step 1. Turn key to " RUN" position, Turn wiper switch to inoperative speed position, Connect jumper wire from gray lead terminal on wiper motor to ground (if low speed inoperative) or from double white lead terminal on wiper motor to ground (if high speed inoperative).

If wiper motor runs, trace circuit from wiper motor to switch. (See wiring diagram E-20 or E-21)

If wiper motor does not run, remove end cap assembly from wiper motor. (See paragraph 10-8) Inspect brushes for damage. Replace any damaged brushes in end cap assembly.

Table 2-2. Electrical Troubleshooting (Con't)

MALFUNCTION
TEST OR INSPECTION
CORRECTIVE ACTION

9. WIPER SWITCH "OFF," WIPER BLADES DO NOT RETURN TO PARK POSITION

Step 1. Turn key to "RUN" position, Turn wiper switch to "OFF" position. Connect jumper wire between park switch terminals (black/light blue lead terminal and white lead terminal beside it) on wiper motor,

If wiper motor runs, replace park switch, (See paragraph 10-8)

If wiper motor does not run, trace circuit through black/light blue lead from wiper motor to switch, (See wiring diagram E-20 or E-21)

10. WIPER WILL NOT TURN OFF

Step 1. Turn key to "RUN" position. Turn wiper switch to "OFF" position. Disconnect connector from park switch terminals (black/light blue lead terminal and white lead terminal beside it) on wiper motor.

If wiper motor stops, replace park switch. (See paragraph 10-8)

If wiper motor still runs, perform step 2.

Step 2. Disconnect 3-wire connector from windshield wiper motor. Connect 12 volt power source to windshield wiper motor terminal #l.

If wiper motor does not run, trace all 3 circuits from 3-wire connector at wiper motor to switch for short to ground, (See wiring diagram E-20 or E-21)

If wiper motor still runs, remove end cap assembly from wiper motor. (See paragraph 10-8) Check for ground between brushes in end cap assembly. Replace brushes if grounded.

11. WIPER OPERATES IN ONE SPEED ONLY (SAME SPEED IN "LO" AND "HI")

Step 1. Disconnect 3-wire connector from wiper motor, Operate wiper in "LO" and "H1."

If wiper operates correctly in each wiper position, trace circuit from 3-wire connector to switch. (See wiring diagram E-20 or E-21)

If wiper does not operate correctly in each wiper position, remove end cap assembly from wiper motor. (See paragraph 10-8) check for short between brushes in end cap assembly. Replace brushes if shorted.

Table 2-2. Electrical Troubleshooting (Con't)

MALFUNCTION
TEST OR INSPECTION
CORRECTIVE ACTION

12. INTERMITTENT WIPER OPERATION

NOTE

Ensure that windshield is dry when performing test In step 1.

Step 1. Remove 25 amp windshield wiper fuse. (See TM 9-2320-289-20) Turn key to "RUN" position. Check current draw across fuse block terminals where fuse was removed with wiper in "HI" position. Record lowest current draw reading.

If current draw is less than 5 amps, replace windshield WIPER motor end cap assembly. (See paragraph 10-8)

If current draw is more than 5 amps, perform step 2.

Step 2. Remove windshield wiper arms and blades. (See TM 9-2320-289-20) Repeat test in step 1.

If current draw is less than 5 amps, replace windshield wiper blades. (See TM 9-2320-289-20)

If current draw is more than 5 amps, Perform step 3.

Step 3. Disconnect windshield wiper linkage from windshield wiper crank arm (see TM 9-2320-289-20) and repeat test in step 1.

If current draw is less than 5 amps, check windshield wiper linkage for binding, Repair or replace windshield wiper linkage if binding. (See TM 9-2320-289-20)

If current draw is more than 5 amps, disassemble windshield wiper motor (see paragraph 10-8) and replace armature.

13. WINDSHIELD WIPER RUNS BUT BLADES WILL NOT MOVE

Step 1. Check windshield wiper linkage connection to wiper crank arm.

If linkage is connected, replace windshield wiper motor. (See TM 9-2320-289-20)

If linkage is not connected, connect linkage and check operation of windshield wiper.

14. WASHER INOPERATIVE (WIPER OPERATIVE)

Step 1. Check washer solvent container for proper level.

Step 2. Check washer nozzles and hoses for restrictions and damage.

Clear restrictions or replace if damaged. (See TM 9-2320-289-20)

Table 2-2. Electrical Troubleshooting (Con't)

MALFUNCTION
TEST OR INSPECTION
CORRECTIVE ACTION

Step 3. Turn key to "RUN" position, Turn wiper to " LO" position, then push washer button and listen for relay click.

If relay clicks, perform step 4.

If relay does not click, perform step 5.

NOTE

Ensure that 2-wire connector remains connected in step 4.

Step 4. Using test light, probe pink lead and white lead on 2-wire connector to washer solenoid.

If test light does not go on at both leads, trace white lead circuit from wiper motor to engine bulkhead connector. (See wiring diagram E-20 or E-21)

If test light goes on at one terminal only, replace park switch assembly. (See paragraph 10-8)

If test light goes on at both terminals and pink one is dim, ground pink terminal at wiper motor. If washer pump operates, trace circuit from pink terminal from wiper motor to switch. (See wiring diagram E-20 or E-21)

If washer pump does not operate, perform step 5.

Step 5. Disconnect all leads from windshield wiper motor. Remove washer pump cover. Reconnect 3-wire connector. Turn on wiper motor. Check operation of pawl on washer pump.

If pawl moves, perform step 6.

If pawl does not move, check to see if washer pump dog spring is properly connected or pawl is not binding. Connect or replace if damaged. (See paragraph 10-8)

Step 6. Connect 12 volt power source to either washer pump solenoid terminal and ground other terminal for 2 seconds. Pawl should be pulled toward washer solenoid coil, pawl leg should drop down on gear ramp, and pawl should start rotating ratchet gear,

If system operates properly, perform step 7.

If pawl operates properly but ratchet gear does not rotate, check to ensure that pawl engages gear teeth. Inspect pawl and gear for damage. Reposition or replace if damaged. (See paragraph 10-8)

If pawl leg does not clear rim on ratchet gear, check for burr on pawl leg. Remove burr with crocus cloth (Item 18, Appendix B) if present, Reposition or replace washer pump solenoid if improperly positioned or damaged. (See paragraph 10-8)

Table 2-2, Electrical Troubleshooting (Con't)

MALFUNCTION
TEST OR INSPECTION
Corrective ACTION

Step 7. Check to see if piston actuator plate moves back and forth with cam follower pin.

If piston actuator plate moves, replace valve assembly (see paragraph 10-8) and recheck. If still inoperative, replace washer pump piston and housing assembly. (See paragraph 10-8)

If piston actuator plate does not move, replace washer pump piston and housing assembly. (See paragraph 10-8)

15. WASHER WILL NOT SHUT OFF

Step 1. Operate windshield wiper and disconnect pink lead and white lead connector from washer pump.

If washer stops, trace circuit from pink lead and white lead for ground from connector at wiper motor switch to wiper switch. (See wiring diagram E-20 or E-21)

If washer still operates, perform step 2.

Step 2. Disconnect all leads from windshield wiper motor. Remove washer pump cover. Reconnect ail leads. Turn on windshield wiper switch. Check to see if ratchet gear dog spring engages ratchet gear tooth. Check to see if retainer spring on ratchet gear is weak. Check to see if ratchet gear teeth or pawl are damaged.

Replace any damaged components. (See paragraph 10-8)

2-4. STE/ICE TROUBLESHOOTING.

Simplified Test Equipment for Internal Combustion Engines (STE/ICE) for use with the CUCV Series trucks is in the process of being developed. When development is complete, STE/ICE troubleshooting procedures will be added to this manual.

CHAPTER 3
ENGINE MAINTENANCE

Section L ENGINE ASSEMBLY MAINTENANCE

3-1. ENGINE ASSEMBLY MAINTENANCE INDEX.

3-2. COMPRESSION TESTING.

This task covers: a. Testing

INITIAL SETUP:

Equipment Condition

- Air cleaner removed. (See TM 9-2320-289-20)
- Glow plugs removed. (See TM 9-2320-289-20)

Personnel Required

- Mos 63W (2)

Tools/Test Equipment

- Compression adapter, J-26999-30
- Compression gage

a. TESTING

CAUTION

DO NOT add oil to any cylinder for compression testing. Failure to follow this caution may result in extensive damage to engine.

1. Disconnect pink lead from fuel injector pump.
2. Install compression adapter and compression gage in glow plug hole of cylinder to be tested.

CAUTION

Covering should be removed from air Intake opening before cranking engine. Failure to follow this caution may result in damage to engine.

3. Record compression gage reading while assistant cranks engine. Allow six puffs per cylinder.

NOTE

Normal compression should be 380-400 psi (2620-2758 kPa).

4. Repeat steps 2 and 3 for each cylinder. Lowest cylinder reading should not be less than 80% of highest cylinder reading, and no cylinder reading should be less than 380 psi (2620 kPa). Remove and inspect valves and piston rings for any cylinder that fails test. (See paragraphs 3-8 and 3-1 3)
5. Connect pink lead to fuel injector pump.

FOLLOW-ON TASKS:

- Install glow plugs. (See TM 9-2320-289-20)
- Install air cleaner. (See TM 9-2320-289-20)

3-3. ENGINE AND ENGINE MOUNT REMOVAL.

This task covers: a. Removal

INITIAL SETUP:

Equipment Condition

- Hood removed (if removing engine). (See TM 9-2320-289-20)
- Radiator and radiator hoses removed (if removing engine). (See TM 9-2320-289-20)
- Vacuum modulator pipe removed (if removing engine). (See paragraph 5-3)
- Engine coolant heater hoses removed (if removing engine equipped with winterization kits). (See TM 9-2320-289-20)
- Air conditioner compressor removed from brackets and tied aside (if removing Ml010 engine). (See paragraph 10-24)
- Power steering pump dismounted from engine (if removing engine). (See TM 9-2320-289-20)
- Engine oil cooler lines disconnected from engine (if removing engine). (See TM 9-2320-289-20)
- Engine coolant heater front exhaust pipe and heat exchange pipe disconnected from engine oil pan (if removing engine equipped with winterization kits). (See TM 9-2320-289-20)
- Engine wiring harness disconnected from engine (if removing engine). (See paragraph 4-10)

Personnel Required

MOS 63W (2)

NOTE

- If only removing engine mount or engine mount brackets, perform steps 5-8.
- A suitable container should be used to catch fuel from disconnected fuel lines (1 and 3).
- Ensure that all hoses, Pipes, and leads are tagged for installation.

1. Disconnect fuel line (3) from fuel pump (2) and fuel return hose (1).

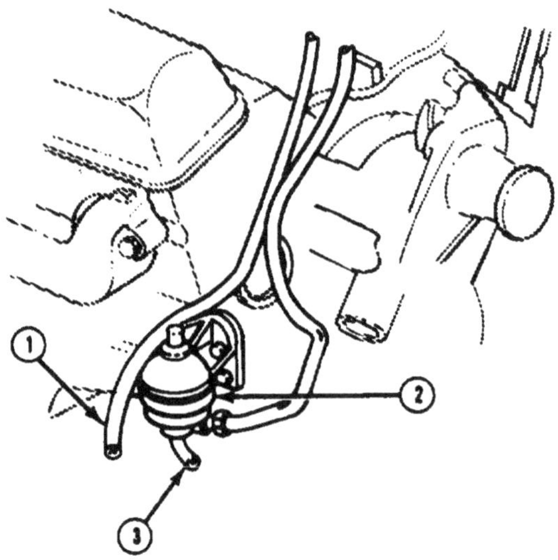

TA60616

3-3. ENGINE AND ENGINE MOUNT REMOVAL (Con't)

NOTE

Mark relationship of flywheel (9) to torque converter (8) for installation.

2. Remove 6 bolts (4) and torque converter cover (6) from transmission assembly (5). Rotate flywheel and remove 6 bolts (7).

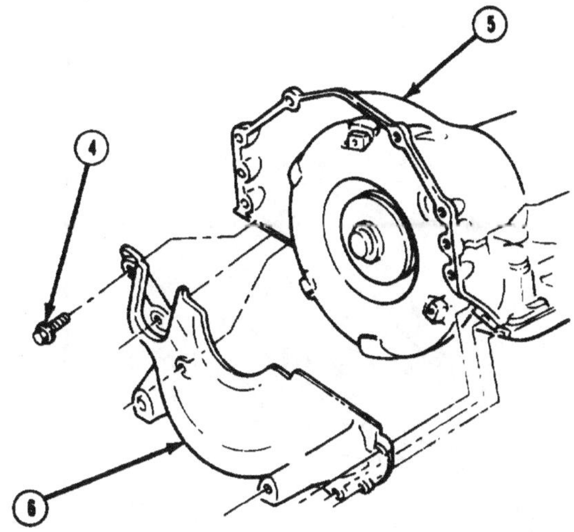

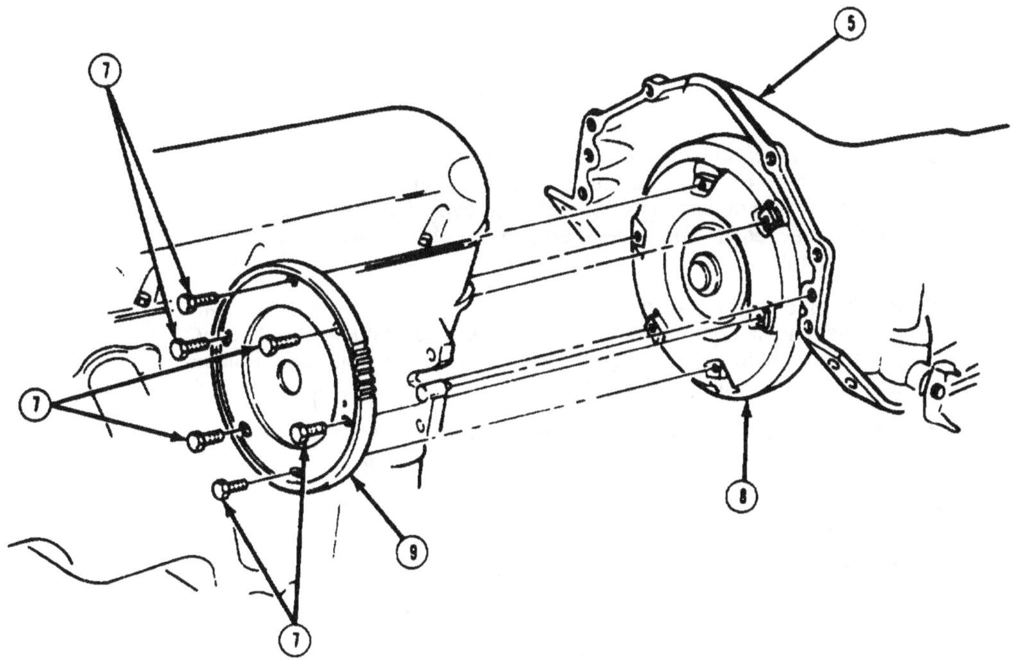

3. Remove 3 nuts (14), washers (13), and springs (12) and disconnect exhaust pipe from exhaust manifold (11). Repeat for other exhaust manifold.

4. Disconnect transmission oil cooler pipes (10) at engine oil pan clip (15).

TA50617

3-3. ENGINE AND ENGINE MOUNT REMOVAL (Con't).

5. Remove locknut (20) and screw (16) from left side engine mount (21) and engine mount bracket (1 7), Repeat for right side. Discard locknuts. Disconnect positive starter cable from clip on oil p-an.

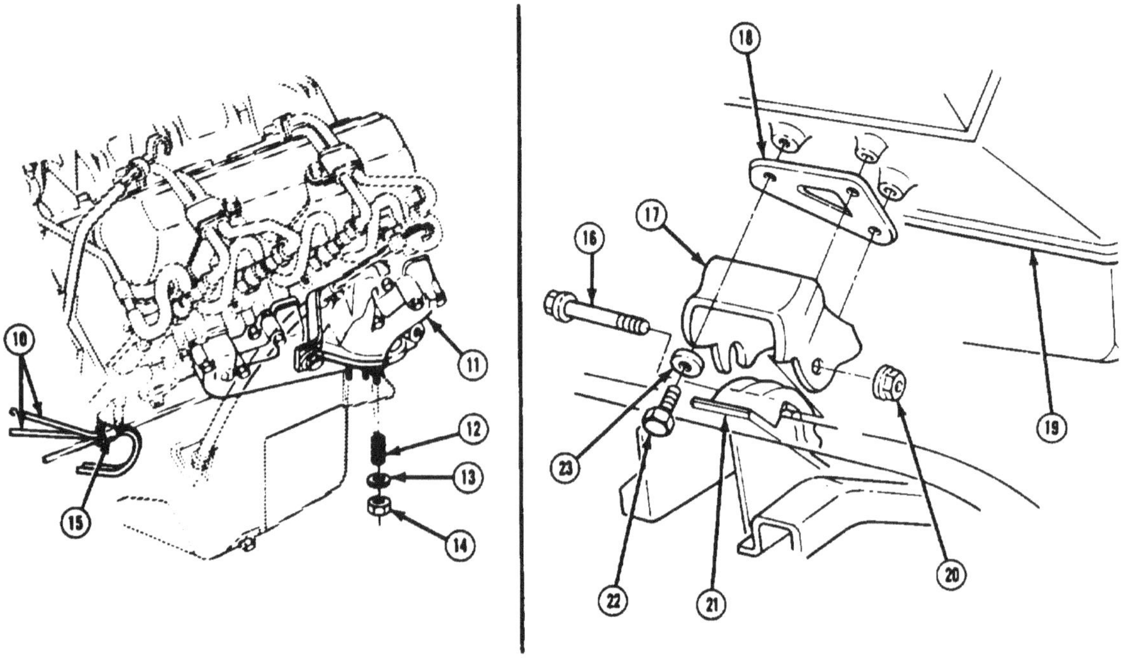

CAUTION

- When raising or supporting engine for any reason, DO NOT use jack under engine oil pan, crankshaft pulley, or any sheet metal. Use suitable lifting device installed at lifting eyes. Failure to follow this caution may result in damage to engine oil pan, crankshaft pulley, or sheet metal.
- If only removing engine mount (21) or engine mount brackets (17), and no other engine components have been removed, engine should only be raised until there is enough clearance to remove bolts (22). Raising engine too far may result in damage to engine components.
- If removing engine mounts (21) or engine mount brackets (17)) lifting device should be installed to right or left front of engine.

6. Install a suitable lifting device on engine lift brackets at left front and right rear of engine. Position wiring harness over transmission oil filler tube. Position vacuum modulator lines to right of lifting device.

7. Raise side of engine (19) where engine mount (21) is to be removed.

8. Remove 3 bolts (22), washers (23), engine mount bracket (17), and spacer (18) if damaged. Remove 3 locknuts, bolts, washers, and engine mount (21). Leave lift in place and repeat step for other side. Discard. locknuts and engine mount if damaged.

TA50618

3-3. ENGINE AND ENGINE MOUNT REMOVAL (Con't).

9. Remove accelerator cable (24) retainer and disconnect accelerator cable from support (25).

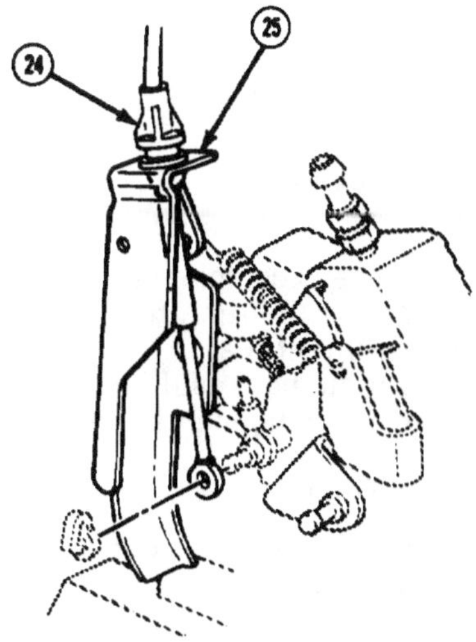

10. Loosen clamp (27) and disconnect heater inlet hose (26) from crossover pipe (28).

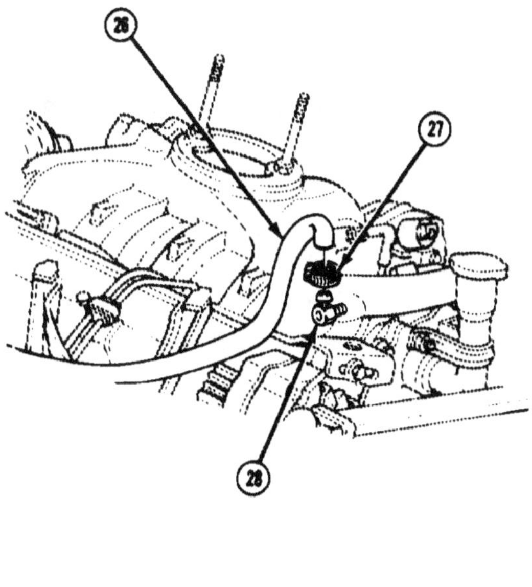

TA50619

3-3. ENGINE AND ENGINE MOUNT REMOVAL (Con't).

11. Lower engine to gain access to upper transmission bolts (29). Support transmission at transmission oil pan using suitable support.

12. Remove 6 transmission bolts (29).

CAUTION

Ensure that engine is completely disconnected prior to lifting from engine compartment. Failure to follow this caution may result in damage to engine components.

NOTE

An assistant will guide engine out of engine compartment as it is being raised.

13. Slide engine straight forward to disengage locating pins from transmission. Remove engine from engine compartment and install on engine stand. Remove lifting device.

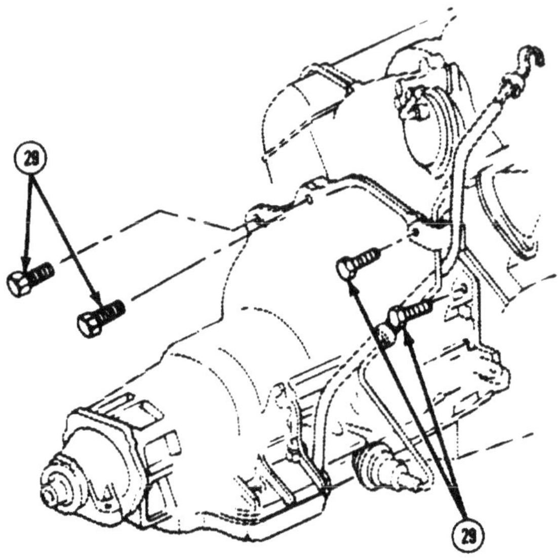

FOLLOW-ON TASKS:

- Install engine or engine mounts. (See paragraph 3-29)

TA60620

3-4. INTAKE MANIFOLD REMOVAL.

This task covers: a, Removal b. Cleaning

INITIAL SETUP:

Equipment Condition

- Both battery negative cables disconnected (if being performed in truck). (See TM 9-2320-289-20)
- Crankcase depression regulator valve assembly and hoses removed (if being performed in truck). (See TM 9-2320-289-20)

Tools/Test Equipment

- Two manifold covers, J-29664-1

Manual References

- TM 9-2320-289-20

a. REMOVAL

NOTE

Step 1 applies to all except M1010.

1. Remove bolt (2), nut (4), and brace (3) from left side alternator bracket (1) and long stud (5).

NOTE

Ensure that position of vacuum pump (7) is noted before rotating.

2. Loosen bolt (6) and rotate vacuum pump (7) to allow access to rear intake manifold bolt.
3. Disconnect 2 fuel filter hoses.

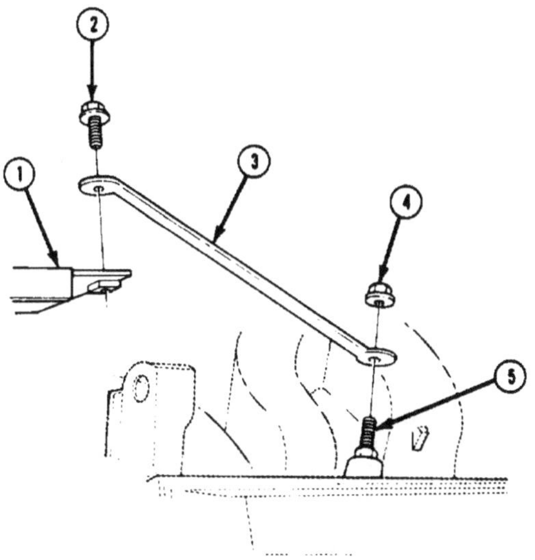

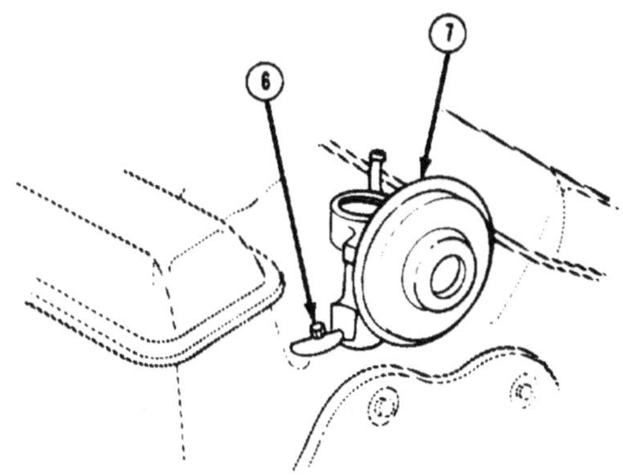

TA60621

3-4. INTAKE MANIFOLD REMOVAL (Con't).

4. Remove long bolt (1 O), 7 short bolts (9), 4 long studs (5), and 4 short studs (11). Move fuel injector pipe clips (12) clear of intake manifold (8). Remove intake manifold and 2 gaskets (13). Discard gaskets.

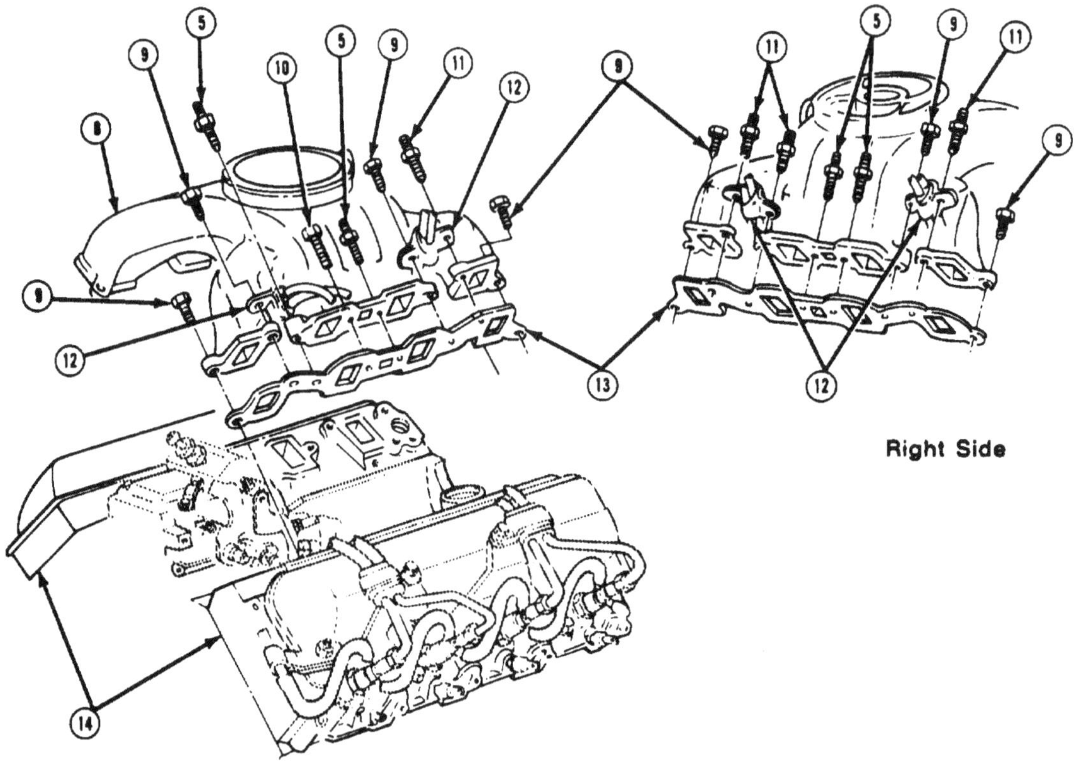

b. CLEANING

CAUTION

Clean rags should be placed in cylinder head (14) openings before cleaning. Ensure that old gasket material does not fall into cylinder head during clearing. Failure to follow this caution may result in damage to engine.

1. Clean all gasket material from mating surfaces of intake manifold (8) and cylinder head (14).
2. Cover cylinder head (14) openings with manifold covers.

TA50622

3-4. INTAKE MANIFOLD REMOVAL (Con't).

FOLLOW-ON TASKS:

- Install intake manifold. (See paragraph 3-28)

3-5. FUEL INJECTOR PIPES AND NOZZLES REMOVAL.

This task covers: a, Removal

INITIAL SETUP:

Equipment Condition

- *Intake* manifold removed. (See paragraph 3-4)

Tools/Test Equipment

- Crowfoot attachment, J-29698-A
- Socket wrench adapter, J-29873

a, REMOVAL

NOTE

- If removing fuel injector pipes (8), perform steps 1-4.
- If removing fuel injector nozzles (6), perform steps 1-3, 5, and 6.

1. Remove 2 screws (1) and clips (2) from fuel injector pipe bracket (4) on one side. Remove clips and inserts (3) from fuel injector pipes (8). Discard inserts if damaged. Repeat step for other side.

CAUTION

Ensure that all openings and connections are Immediately plugged after disconnection to prevent contamination of fue l system.

NOTE

Ensure that fuel injector pipes (8), fuel injector nozzles (6), and Fuel injector pump connectors are tagged for installation.

2. Disconnect 8 fuel injector pipes (8) from 8 fuel injector nozzles (6). Remove 4 nuts (7) and 2 fuel injector pipe brackets (4) from cylinder head studs (5) if damaged.

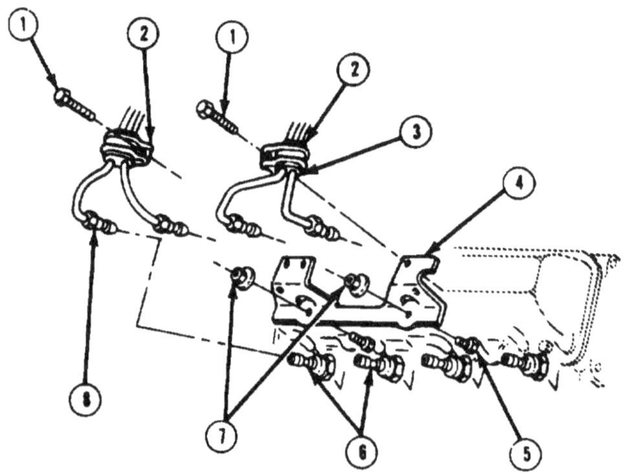

TA50623

3-5. FUEL INJECTOR PIPES AND NOZZLES REMOVAL (Con't).

3. Disconnect 8 fuel injector pipes (8) from fuel injector pump (9) connectors and remove fuel injector pipes.

NOTE

Ensure that locations of inserts are marked for installation.

4. Remove 4 clips (1 O) and inserts from fuel injector pipes (8). Remove 4 screws (13), clamps (1 2), and inserts (11) from fuel injector pipes. Discard inserts if damaged.

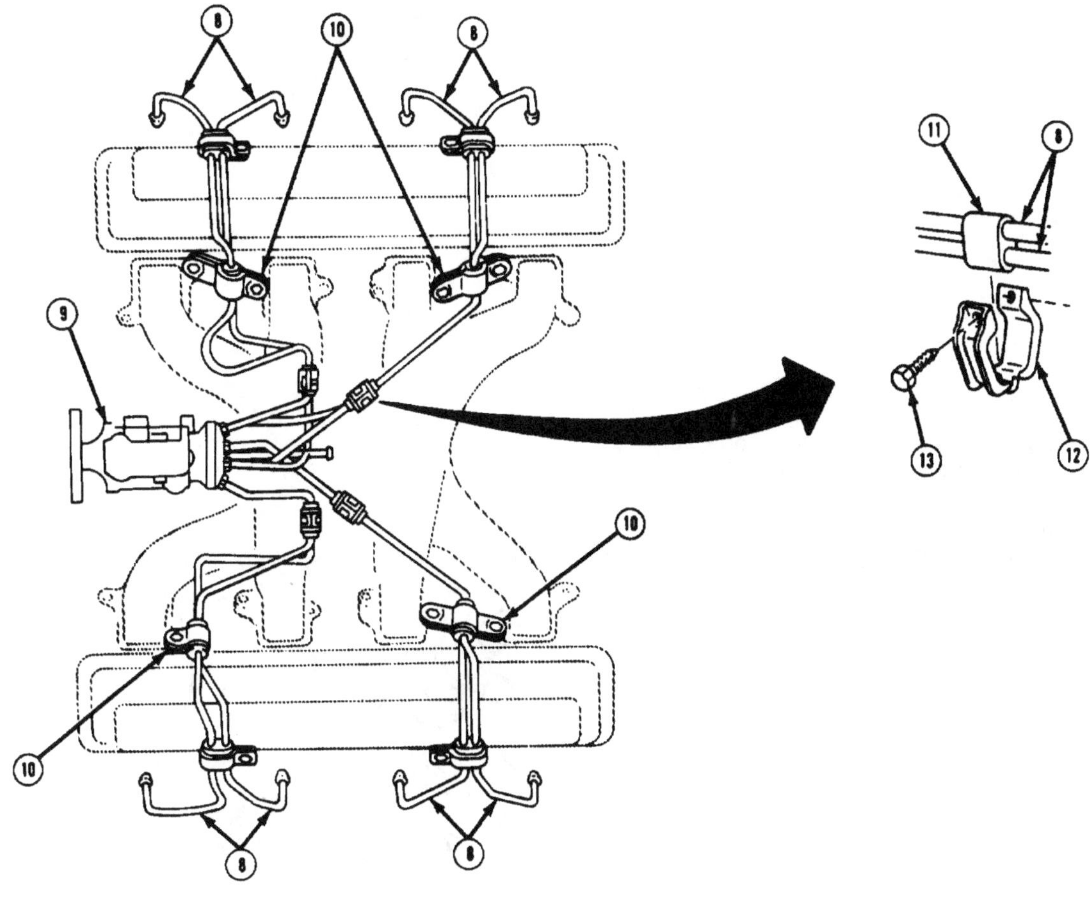

TA50624

3-5. FUEL INJECTOR PIPES AND NOZZLES REMOVAL (Con't).

5. Remove nut (15) and clip (14). Loosen clamps and disconnect 8 fuel drainback hoses (16) from fue l injector nozzles (6). Loosen clamps and remove 2 caps (17).

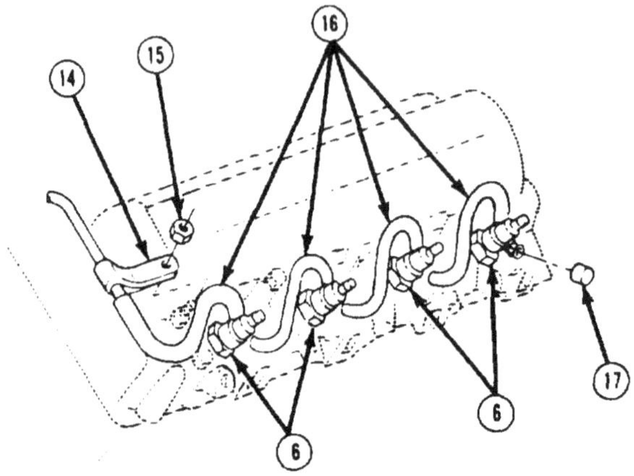

6. Using socket wrench adapter, remove 8 fuel injector nozzles (6) and fuel injector nozzle gaskets (18) from cylinder heads (1 9). Discard fuel injector nozzle gaskets.

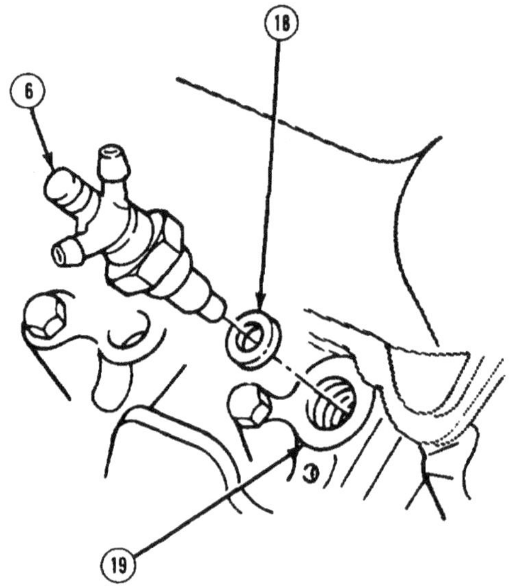

FOLLOW-ON TASKS:

- Install fuel injector pipes or nozzles. (See paragraph 3-27)

TA50625

3-6. ROCKER ARM AND ROCKER ARM COVER REMOVAL AND REPAIR.

This task covers:

a. Removal
b. Disassembly
c. Cleaning and Inspection
d. Assembly

INITIAL SETUP:

Equipment Condition

- Fuel injector pipes removed. (See paragraph 3-5)

MaterialslParts

- Cotter pins (as needed)
- Rocker arm retainers (as needed)
- Dry cleaning solvent (Item 23, Appendix B)

General Safety Instructions

- Dry cleaning solvent is flammable and must not be used near open flame. Use only in a well-ventilated area.

a. REMOVAL

NOTE

- IF removing right rocker arm cover (3), perform steps 1-4.
- iF removing left rocker arm cover (8), perform steps 2-4.

1. Remove 2 nuts (2) and disconnect 2 vacuum line clamps (1) from studs (4) securing right rocker arm cover (3).
2. Disconnect wiring harness from wiring harness clips along rocker arm cover (3 or 8). If removing left rocker arm cover (8), remove wiring harness bracket.
3. Remove 2 nuts (6) and fuel injector pipe bracket (5).

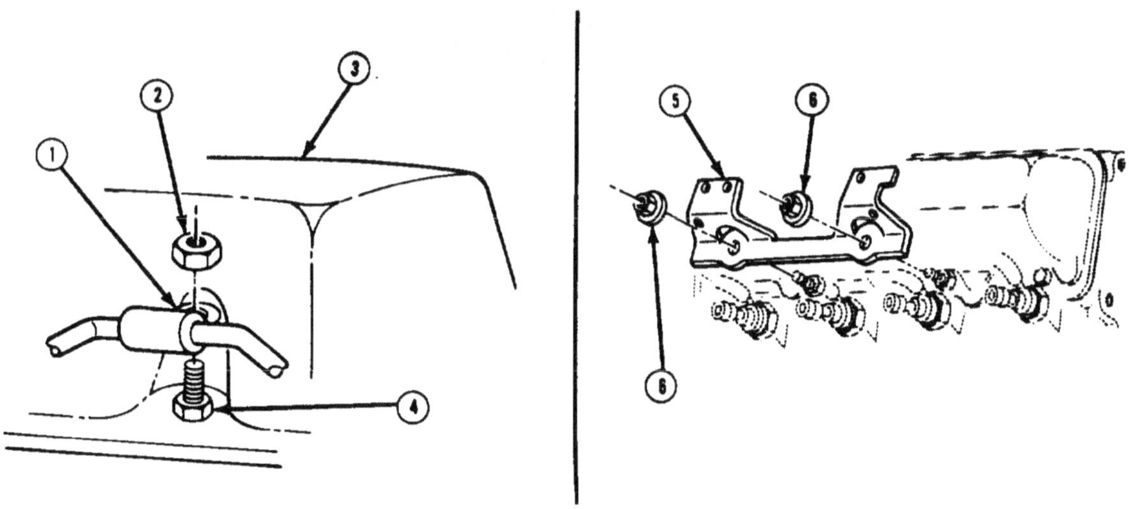

TA60626

3-6. ROCKER ARM AND ROCKER ARM COVER REMOVAL AND REPAIR (Con't).

CAUTION

DO NOT pry on rocker arm covers (3 and 8). Ensure that a soft-faced hammer Is used when removing rocker arm covers. Failure to follow this caution may result in damage to rocker arm covers or cylinder heads (11 and 12).

4. For right rocker arm cover (3), remove 2 bolts (7), 6 studs (4), and rocker arm cover from cylinder head (12). For left rocker arm cover (8), remove 5 bolts (9), 3 studs (10), and rocker arm cover from cylinder head (11).

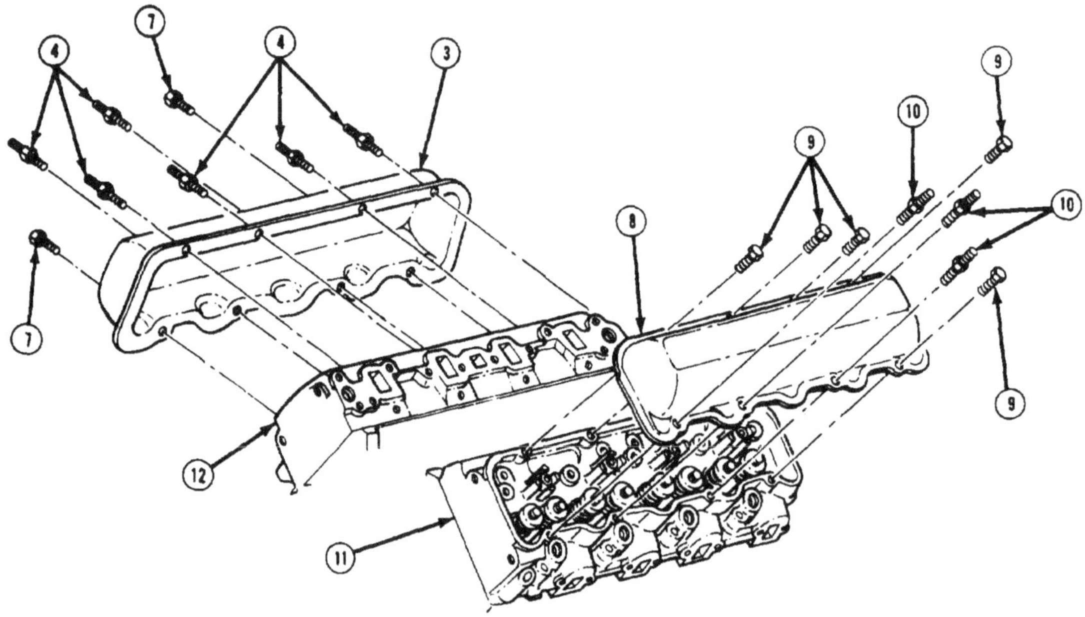

TA50627

3-6. ROCKER ARM AND ROCKER ARM COVER REMOVAL AND REPAIR (Con't).

CAUTION

Ensure that rocker arm assemblies (13) and pushrods are installed in same position and location as when they are removed, A paint stripe identifies upper end of pushrod. If paint stripe is not visible, ensure that upper end of pushrods are marked as they are removed. Failure to follow this caution may result in damage to rocker arm assembly and pushrod.

5. Remove 2 bolts (15) and rocker retainers (14) on each rocker arm assembly (13), and remove rocker arm assemblies. Remove 4 pushrods through holes (16) in cylinder head.

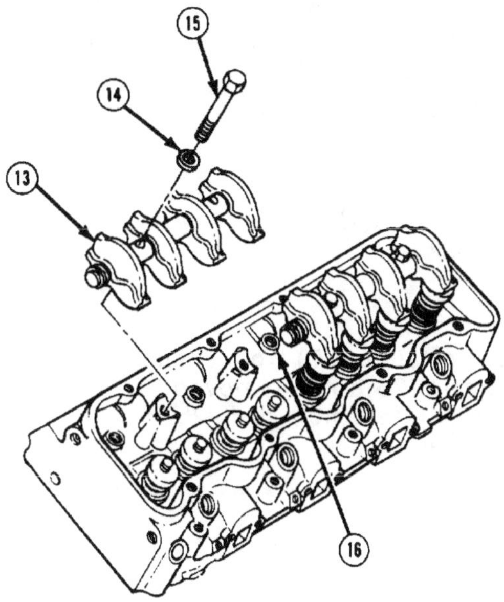

b. DISASSEMBLY

NOTE

- If disassembling 1984 model truck rocker arm assembly (13), perform step 1.
- If disassembling 1985-87 model truck rocker arm assembly (13), perform step 2.

1. Remove cotter pin (1 7), washer (24), and spring washer (23) from one end of shaft (22). Remove outside rocker arm (18). Remove spacer (19) and inside rocker arm (20). Remove spring (21), and repeat this step for other rocker arms if damaged. Discard cotter pins and spring washers.

TA50628

3-6. ROCKER ARM AND ROCKER ARM COVER REMOVAL AND REPAIR (Con't).

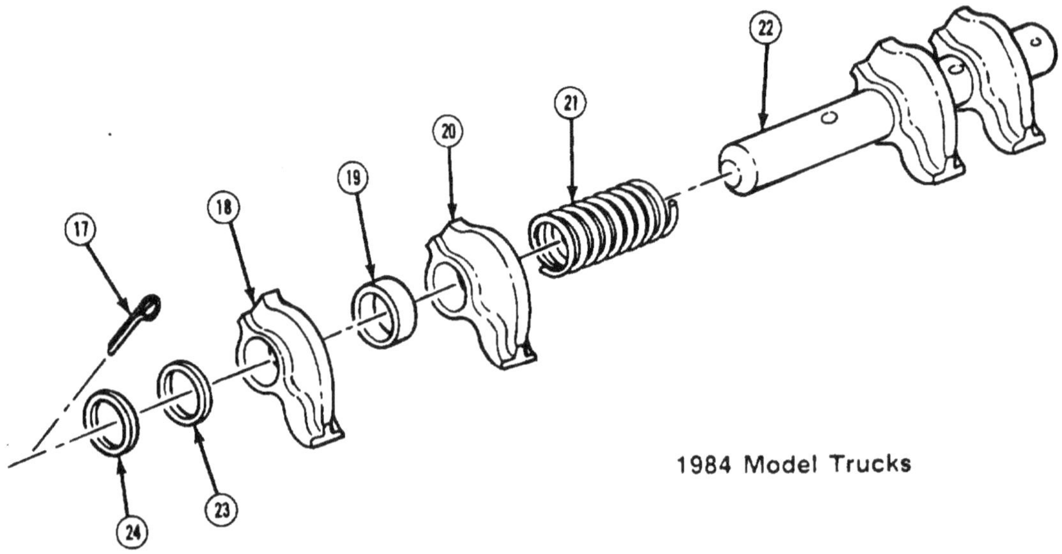

1984 Model Trucks

2. Remove retainer (27) from each rocker arm (26) to be removed. Remove rocker arm from shaft (25). Discard retainers.

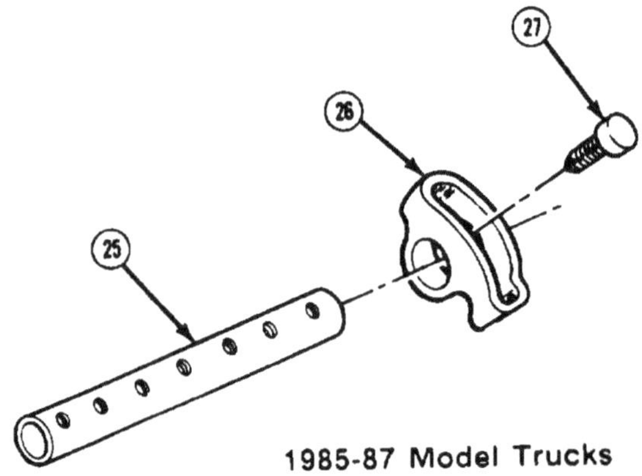

1985-87 Model Trucks

TA50629

3-6. ROCKER ARM AND ROCKER ARM COVER REMOVAL AND REPAIR (Con't).

c. CLEANING AND INSPECTION

WARNING

Dry cleaning solvent P-D-680 is toxic and flammable. Always wear protective goggles and gloves and use only in a well-ventilated area. Avoid contact with skin, eyes, and clothes and DO NOT breathe vapors. DO NOT use near open flame or excessive heat. The solvent's flash point is 100°F-138°F (38°C-59°C). If you become dizzy while using cleaning solvent, immediately get fresh air and medical help. If solvent contacts eyes, Immediately wash your eyes with water and get medical aid.

1. Using dry cleaning solvent, clean RTV sealant, oil, and grease from sealing surfaces of rocker arm cover and cylinder head.
2. Using dry cleaning solvent, clean all carbon and sludge from rocker arms and pushrods.
3. Inspect sealing flanges of rocker arm cover for distortion and damage. Inspect all rocker arm components for damage, Replace any damaged components.

d . ASSEMBLY

NOTE

I If assembling 1964 model truck rocker arm assembly, perform step 1.

I If assembling 1985-67 model truck rocker arm assembly, perform step 2.

1. Install spring (21), inside rocker arm (20), and spacer (19) on shaft (22), Install outside rocker arm (18), new spring washer (23), washer (24), and new cotter pin (17) on shaft, Repeat this step until rocker arm assembly is assembled.

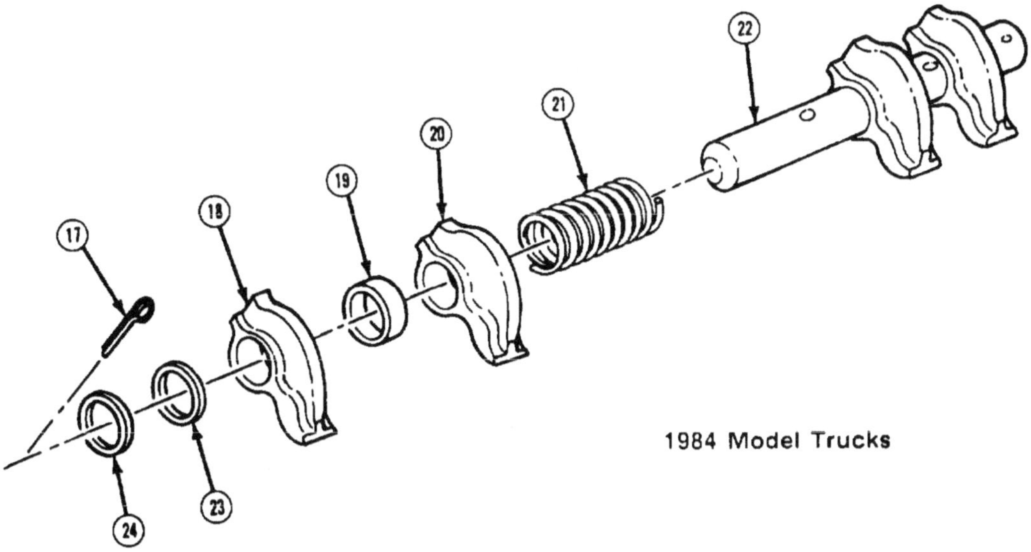

TA50630

3-6. ROCKER ARM AND ROCKER ARM COVER REMOVAL AND REPAIR (Con't).

2. Center rocker arm (26) on hole in shaft (25) and install new retainer (27). Repeat this step until rocker arm assembly is assembled.

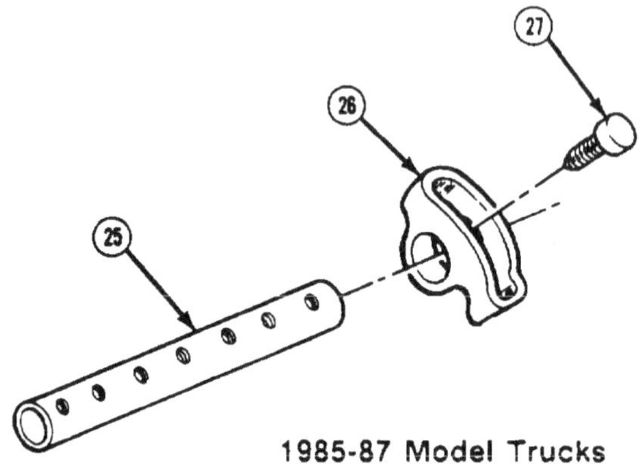

1985-87 Model Trucks

FOLLOW-ON TASKS:

l Install rocker arm and rocker arm cover. (See paragraph 3-26)

TA50631

3-7. CYLINDER HEAD REMOVAL.

This task covers: a. Removal b. Cleaning and Inspection

INITIAL SETUP:

Equipment Condition

- Rocker arm cover, rocker arm assemblies, and pushrods on applicable cylinder head removed, (See paragraph 3-6)
- Oil level indicator tube removed (if removing left cylinder head). (See TM 9-2320-289-20)
- Air conditioner compressor brackets removed (if removing left cylinder head from MI 01 O). (See paragraph 10-24)
- Left side alternator bracket removed (if removing left cylinder head). (See TM 9-2320-289-20)
- Right side alternator or MI 010 alternator bracket removed (if removing right cylinder head), (See TM 9-2320-289-20)
- Exhaust manifold removed. (See TM 9-2320-289-20)
- Radiator, bypass, and heater hoses disconnected (if being performed in truck). (See TM 9-2320-289-20)
- Thermostat and crossover pipe removed. (See TM 9-2320-289-20) " '

Materlals/Parts

- Fuel injector nozzle gaskets (as needed)

Tools/Test Equipment

- Socket wrench adapter, J-29873

Personnel Required

- MOS 63W (2)

Manual References

- TM 9-2320-289-20

a. REMOVAL

NOTE

- If engine is out of truck, perform step 2.
- Ensure that leads are tagged for installation.

1. Tag and disconnect glow plug leads from glow plugs on cylinder head (5) to be removed. If removing left cylinder head, disconnect glow plug wiring harness (2) from temperature sensor (1), Remove nut, and disconnect leads, ground strap, and lockwasher from engine stud. Discard lockwasher.

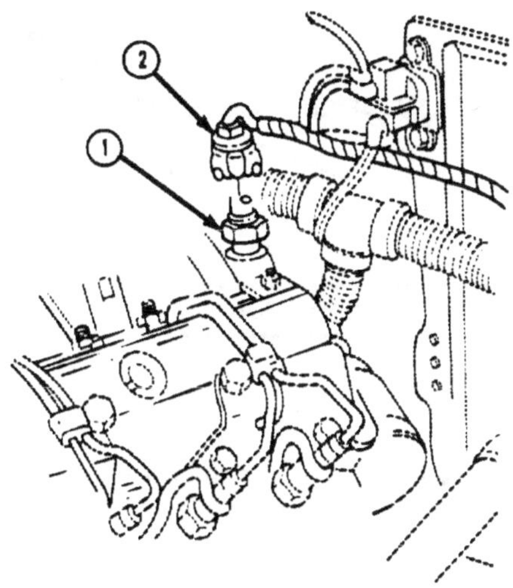

TA50632

3-7. CYLINDER HEAD REMOVAL (Con't)

NOTE

- Rear bolt (6) in left cylinder head (5) may have to remain in cylinder head during removal.
- An assistant will be required for step 2.

2. Remove 17 bolts (6), starting with outside bolts and working toward center. Remove cylinder head (5) and cylinder head gasket (4) from engine block (3). Discard cylinder head gasket,

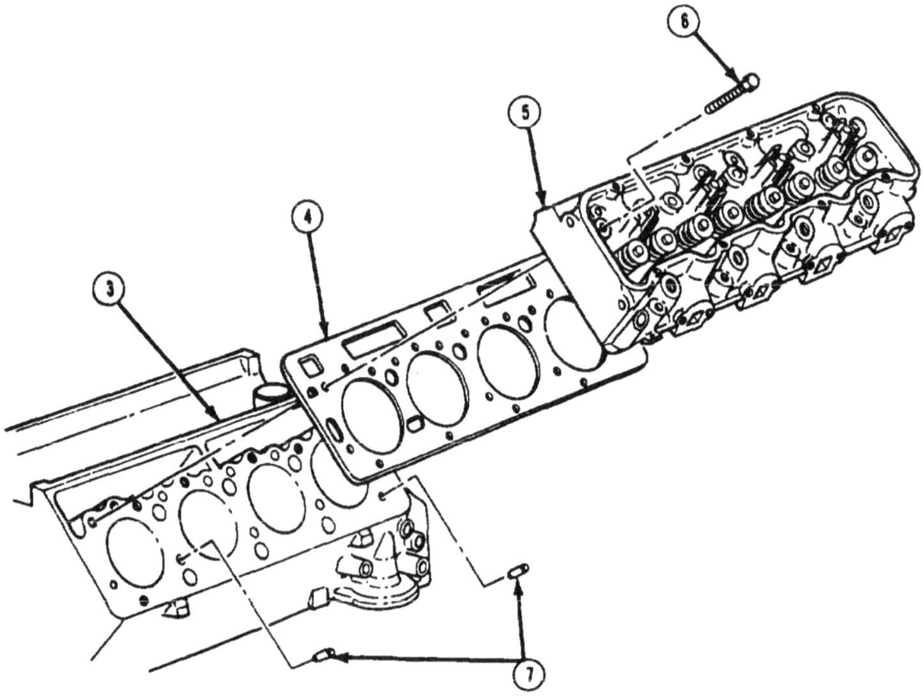

b. CLEANING AND INSPECTION

CAUTION

DO NOT use a motorized wire brush to clean cylinder head (5). Failure to follow this caution may result In damage to cylinder head.

1. Inspect dowel pins (7) for damage. Remove if damaged.
2. Clean carbon deposits, sealing compound, and all gasket material from cylinder head (5) and engine block (3). Remove all sealant and dirt from threads in engine block.
3. Clean sealing compound from bolt (6) threads.

TA50633

3-7. CYLINDER HEAD REMOVAL (Con't).

NOTE

If replacing cylinder head, ensure that ail glow plugs, switches (see TM 9-2320-289-20), fuel injector nozzles (see paragraph 3-5), and valves (see paragraph 3-8) are removed from damaged cylinder head and installed in new cylinder head.

4. Inspect cylinder head (5) for cracks between each intake port (8) and exhaust port (10). Inspect cylinder head where cylinder head gasket will be installed for nicks, heavy scratches, or other damage. Replace cylinder head if any of these conditions exists.

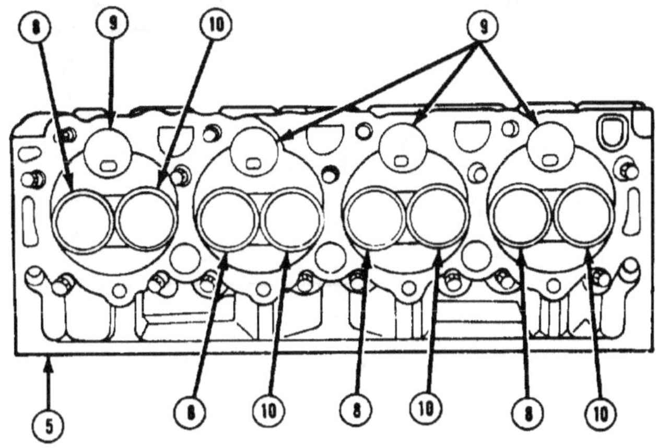

5. Using straightedge (12) and feeler gage (11), measure cylinder head (5) for warpage. If warped more than 0.006 in. (O. 15 mm) lengthwise or 0.003 in. (0.08 mm) widthwise, replace cylinder head.

6. Check each prechamber (9) for proper installation. Prechamber must not be recessed in cylinder head (5). Prechamber should protrude from surface of cylinder head no more than 0.002 in. (0.05 mm) at any point, Using straightedge (12) and feeler gage (11), take this measurement at two or more points on each prechamber. If prechamber protrudes too far, tap into place with socket and light mallet. If prechamber is recessed or damaged, perform step 7.

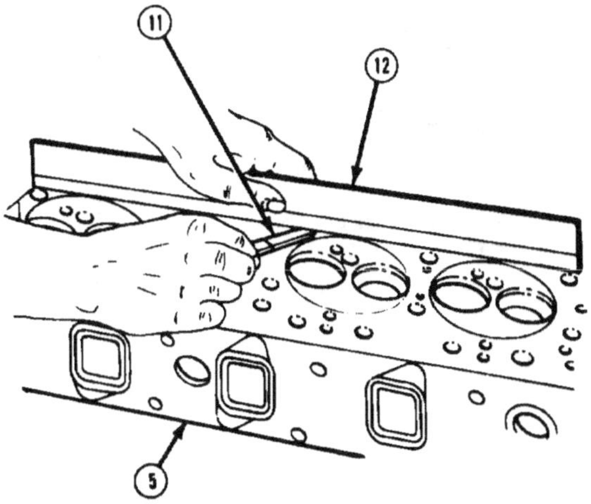

TA50634

3-7. CYLINDER HEAD REMOVAL (Con't).

7. If prechamber (9) is cracked or damaged, remove fuel injector nozzle and gasket on same cylinder as prechamber. Remove old prechamber by placing punch into fuel injector nozzle hole and lightly tapping with hammer. Install new prechamber by aligning locator notches (13) on prechamber and cylinder head (5). Using a 11/4 in. socket and light mallet, tap into place. Repeat measurements in step 6, then install new gasket and fuel injector nozzle.

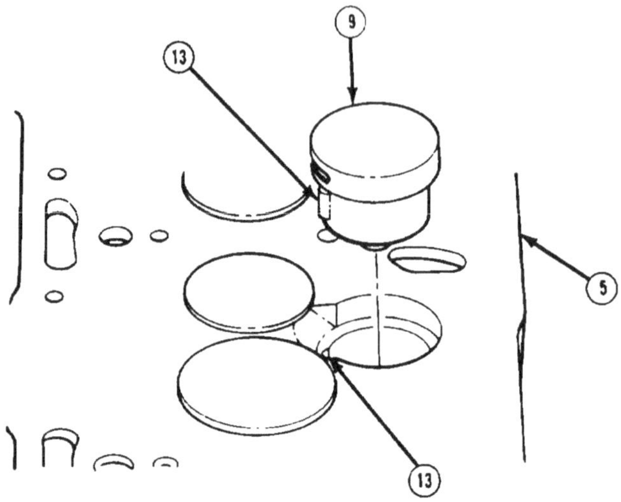

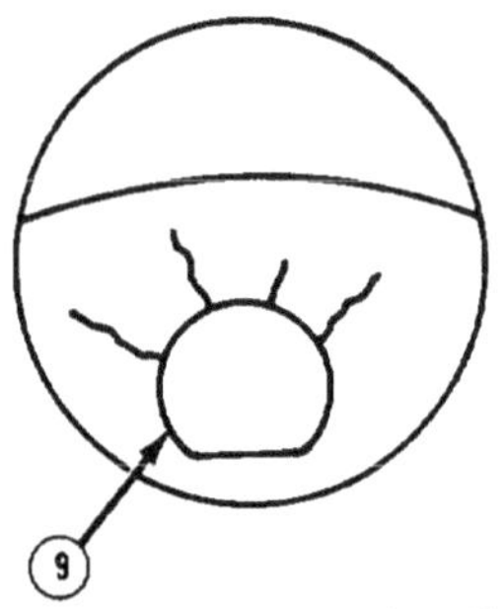

Acceptable Crack Length NOT over 3/16 in. (5 mm)

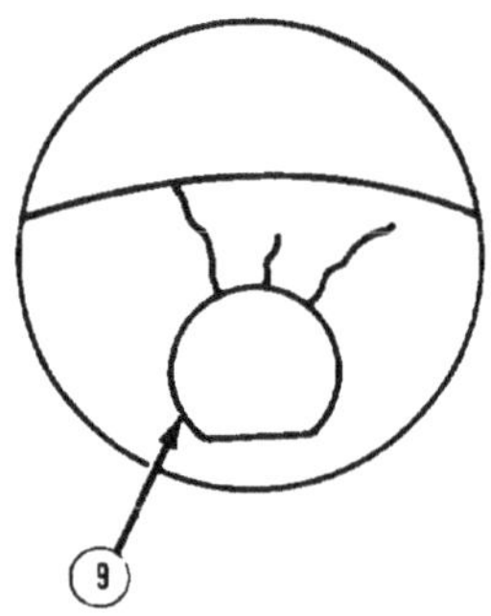

Unacceptable Crack Length Over 3/16 in. (5 mm)

FOLLOW-ON TASKS:

- Install cylinder head. (See paragraph 3-25)

TA50635

3-8. VALVE REMOVAL AND REPAIR.

This task covers: a. Removal b, Cleaning and Inspection c. Repair

INITIAL SETUP:

Equipment Condition

- Cylinder head removed. (See paragraph 3-7)

General Safety Instructions

- Always wear goggles when using valve spring compressor.

Toola/Test Equipment

- Buffing wheel
- Dial indicator
- Torque wrench
- Valve guide cleaning tool
- Valve guide reaming tool
- Valve spring compressor
- Valve spring tester

a. REMOVAL

WARNING

Always wear goggles while using valve spring compressor. Valve springs (6) are tightly compressed. Failure to follow *This warning may result in* serious injury to personnel.

1. Compress exhaust valve (3) spring or intake valve (2) spring and remove 2 keys (10).

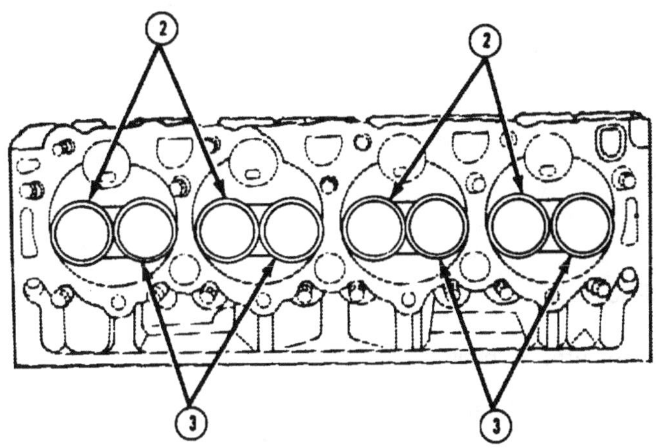

TA50636

3-8. VALVE REMOVAL AND REPAIR (Con't).

NOTE

- Components of exhaust valve (3) assemblies and intake valve (2) assemblies are similar, except that exhaust valve uses rotater (11) and intake valve uses cap (11).
- Ensure that location of each valve (12) and valve components are tagged as they are removed to ensure that they are installed in same port during installation.

2. Remove rotater (11) or cap (11). Remove shield (8) and valve spring (6) with damper (7). Remove oil seal (9) and valve spring shim (5). Remove valve (12). Discard oil seal.

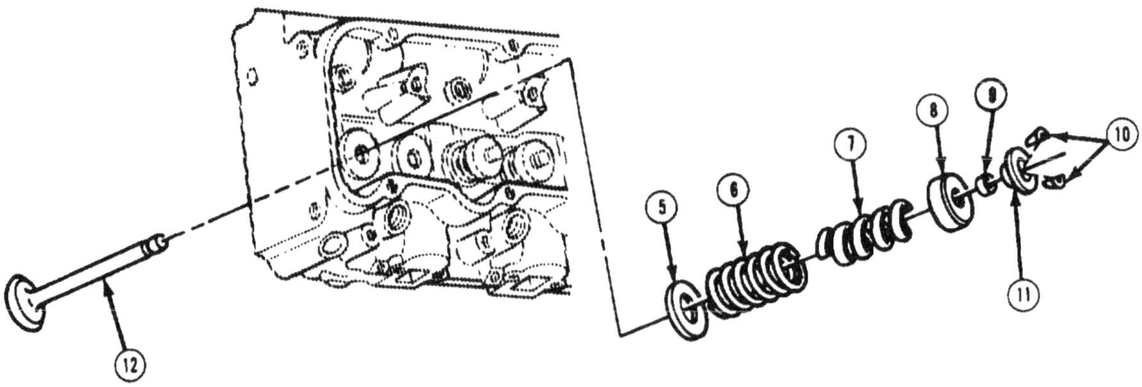

b. CLEANING AND INSPECTION

1. Clean valve (12) using buffing wheel.
2. Inspect valves (12) for burned heads, cracked faces, and damaged stems. Repair if burned or cracked. (See REPAIR, step 1) Replace valves if damaged. if repair will not bring cracked valve within specifications, replace valve.

TA50637

3-8. VALVE REMOVAL AND REPAIR (Con't).

3. Thoroughly clean valve guides (4). Insert valve (12) in valve guide until valve head seats. Install dial indicator (13) on rocker arm cover gasket rail (14) so that it contacts valve stem just above valve guide.

4. Move valve (12) until valve head is 0,06 in. (1,6 mm) off valve seat. Using light pressure, move valve stem from side to side. If clearance exceeds 0,0027 in. (0.06858 mm), repair valve and valve guide. (See REPAIR)

5. Set valve spring tester to 1.81 in. (46 mm). Put pressure on valve spring (6) with torque wrench until height is 1,81 in. (46 mm), If pressure is not 70-80 lb,-ft. (95-1 08 Nom), replace valve spring.

6. Set valve spring tester to 1.39 in. (35.3 mm), Put pressure on valve spring (6) with torque wrench until height is 1.39 in. (35.3 mm). If pressure is not 220-230 lb.-ft. (298-31 2 N.m), replace valve spring.

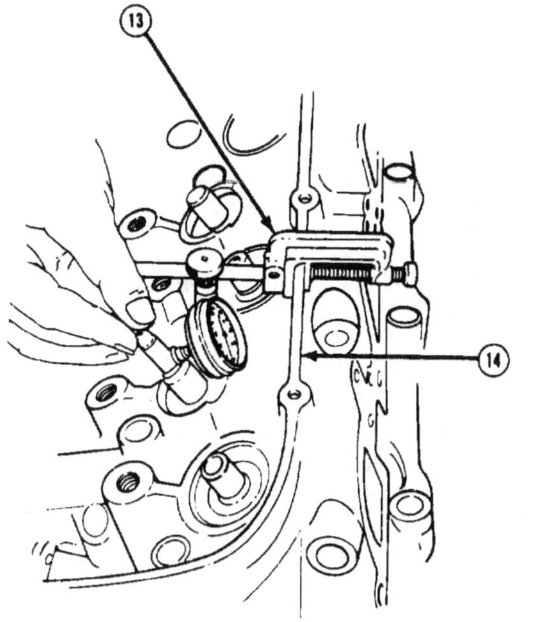

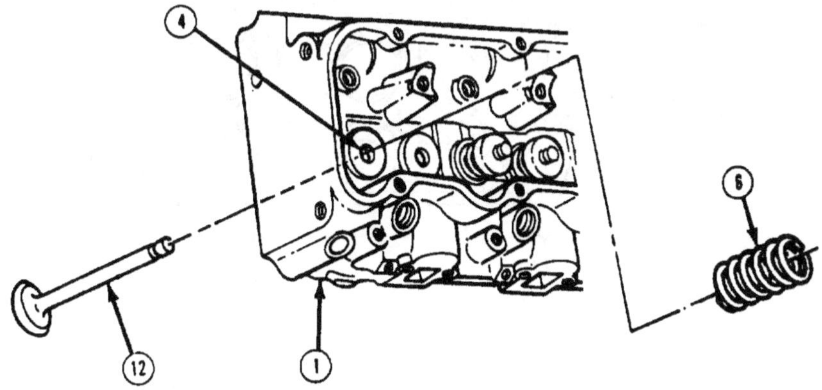

TA50636

3-8. VALVE REMOVAL AND REPAIR (Con't).

NOTE

Valves (12) and seats should be reconditioned if seat width exceeds 0.080 in. (2.03 mm),

1. Grind valve face (15) to obtain a 45 degree angle completely around. Measure all valve heads (1 6). If valve head is less than (0.80 mm) thick at any point, replace valve (12).

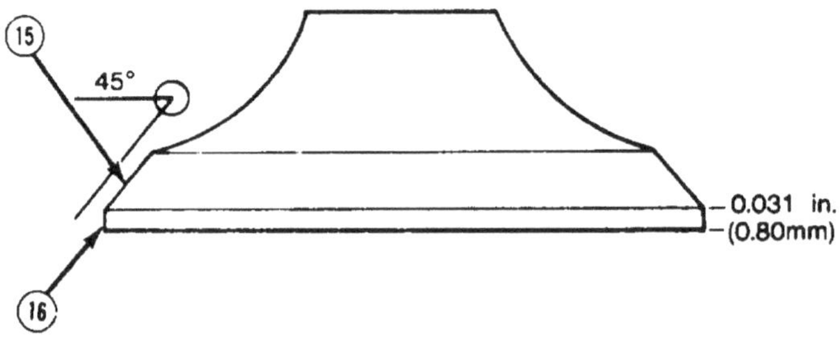

CAUTION

Valve seats are induction hardened. DO NOT grind valve seats in excess of dimensions given in step 2. Failure to follow this caution may result in damage to valve seat and require replacement of cylinder head (1).

2. Grind valve seats to obtain a 46 degree angle. Intake valve (2) seats must be 0.035-0.060 in. (0.889-1 .52 mm) wide. Exhaust valve (3) seats must be 0.060-0.080 in. (1 .52-2 .03 mm) wide. If any valve seat is too wide, replace cylinder head (1).

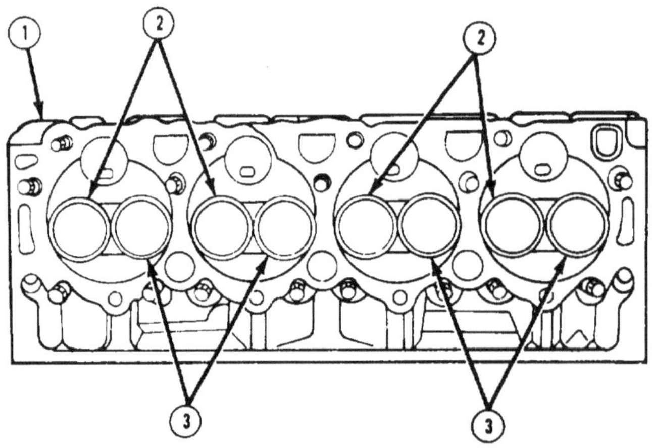

3. Obtain valve (12) with next oversize stem. Using appropriate reaming tool for oversize valve stem, ream valve guide (4).

FOLLOW-ON TASKS:

- Install valve. (See paragraph 3-24)

TA50639

3-9. HYDRAULIC VALVE LIFTER REMOVAL.

This task covers: a. Removal b. Inspection

INITIAL SETUP:

Equipment Condition

- Cylinder head removed. (See paragraph 3-7)

Tools/Test Equipment

- Magnet

a. REMOVAL

1 Loosen bolt (5) and remove clamp (4). Remove guide plate (3).

NOTE

- If more than 1 hydraulic valve lifter (2) is removed, ensure that location of each hydraulic valve lifter is tagged, es it is removed, to ensure that it is installed in same location during installation.
- For 1984 model truck engines, It may be necessary to use mechanical fingers to remove hydraulic valve lifter (2),

Using magnet, remove hydraulic valve lifter (2) from engine block (1),

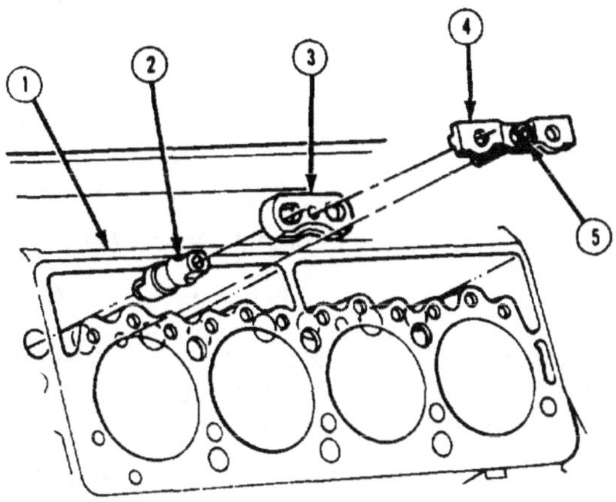

b. INSPECTION

1.. Inspect hydraulic valve lifter (2) body for scuffing or scoring. Replace if scuffed or scored.

2. Inspect hydraulic valve lifter (2) roller for looseness and missing or broken needle bearings. Replace hydraulic valve lifter if these conditions exist.

3. Inspect hydraulic valve lifter (2) roller surface for pits and roughness. If these conditions exist, replace hydraulic valve lifter and inspect mating camshaft lobe for pits and roughness, If camshaft lobe is pitted or rough, replace camshaft. (See paragraph 3-12)

FOLLOW-ON TASKS:

- Install hydraulic valve lifter. (See paragraph 3-23)

TA50640

3-10. TORSIONAL DAMPER AND CRANKCASE COVER REMOVAL.

This task covers: a. Removal b. Cleaning and Inspection

INITIAL SETUP:

Equipment Condition

- Crankshaft pulley removed. (See TM 9-2320-289-20)
- Engine oil drained (if removing crankcase cover). (See TM 9-2320-289-20)
- water pump removed (if removing crankcase cover). (See paragraph 3-48)
- Right side alternator bracket removed (if removing crankcase cover). (See TM 9-2320-289-20)

Tools/Test Equipment

- Torsional damper puller

General Safety Instructions

- Avoid skin contact with anaerobic sealing compound.

a. REMOVAL

NOTE

- If removing torsional damper (7), perform step 2.
- If removing crankcase cover seal (9), perform steps 2 and 3.

1. Aline timing marks (5) on fuel injector pump driven gear (1) and fuel injector pump drive gear (4). Scribe a mark alining fuel injector pump flange (3) and crankcase cover (2). Remove 3 bolts (6) and fuel injector pump driven gear.

2. Remove bolt (11) and washer (1 O) from torsional damper (7), Install bolt in torsional damper. Using torsional damper puller, remove torsional damper from crankshaft (8).

2.1. Remove bolt (11) from crankshaft (8).

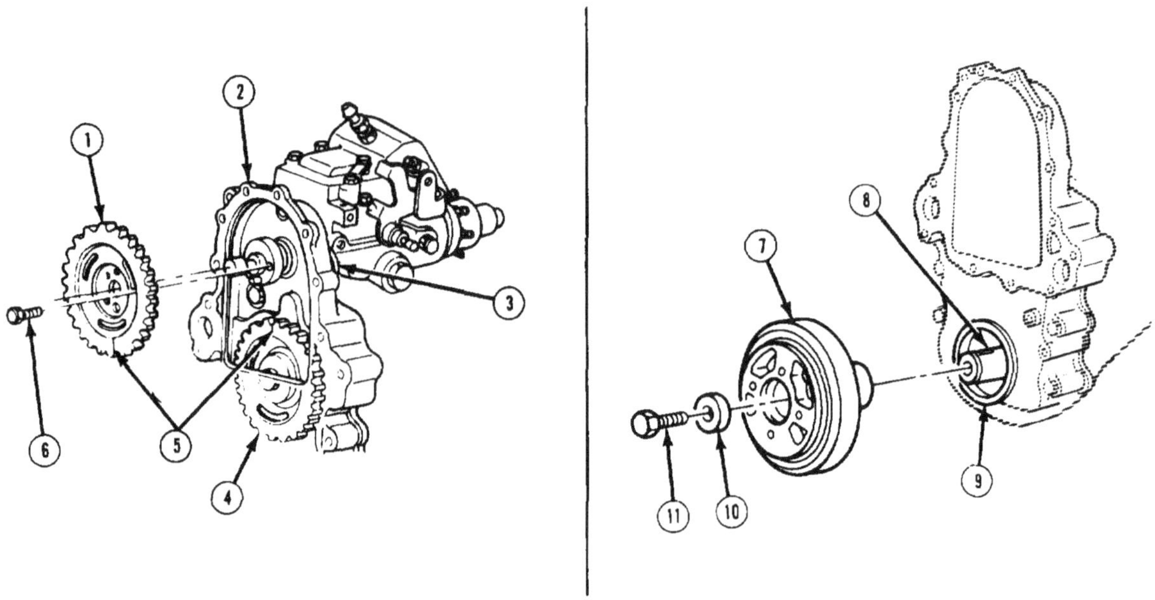

TA50641

3-10. TORSIONAL DAMPER AND CRANKCASE COVER REMOVAL (Con't).

CAUTION

Be careful not to scratch crankshaft (8) or crankcase cover (2) when removing crankcase cover seal (9). Failure to follow this caution may result in oil leakage.

3. Pry out and discard crankcase cover seal (9).

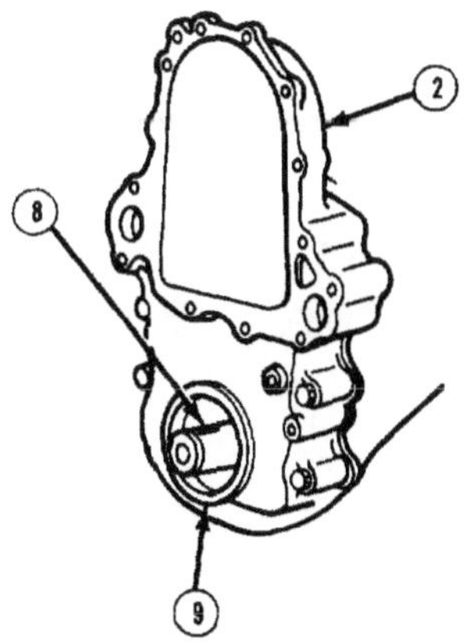

4. Remove 3 nuts (14) from fuel injector pump (13) at rear of crankcase cover (2).

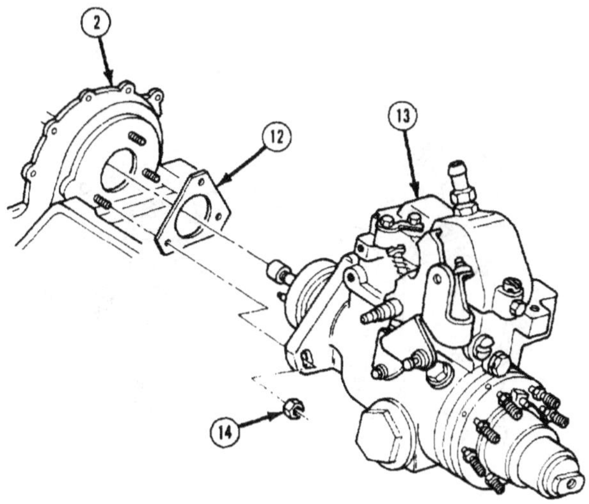

TA50642

3-10 TORSIONAL DAMPER AND CRANKCASE COVER REMOVAL (Con't).

5. Remove bolt (15) and disconnect clip (17) with fuel return line (16) from crankcase cover (2),

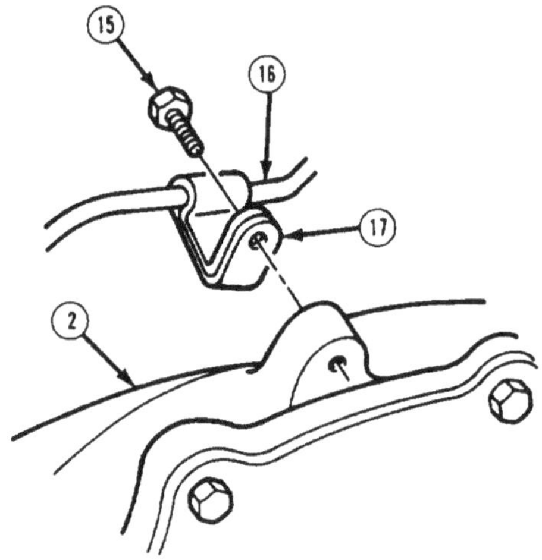

6. Remove nut (19), 2 bolts (18), and baffle (20).

7. Remove 4 bolts securing crankcase cover (2) to engine oil pan. Remove bolt (22). Remove 4 bolts (23) and crankcase cover from engine block (21). Remove and discard fuel injector pump gasket (12).

8. Remove 2 screws (25) and probe holder (24) from crankcase cover (2) if damaged or if replacing crankcase cover.

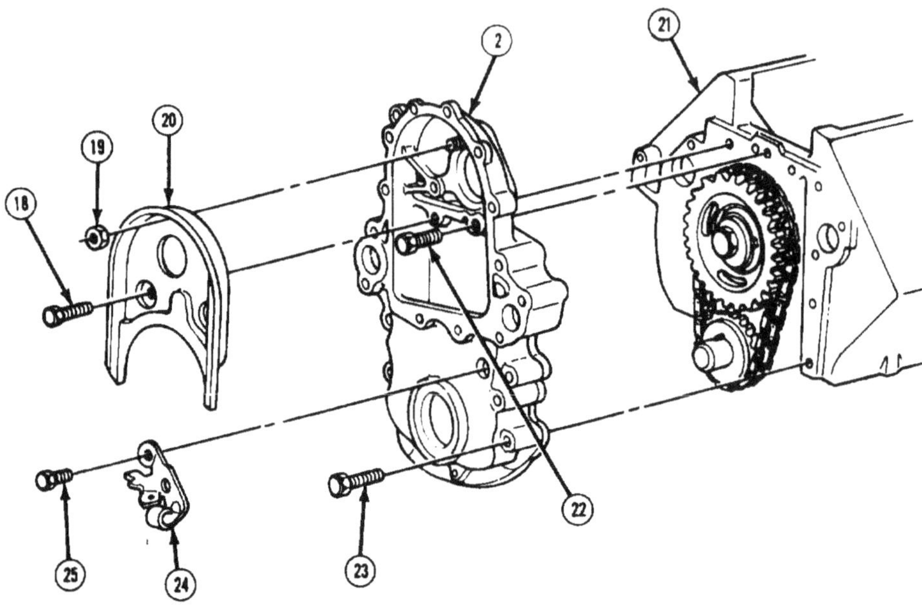

TA50643

3-10. TORSIONAL DAMPER AND CRANKCASE COVER REMOVAL (Con't).

b. CLEANING AND INSPECTION

WARNING

Avoid skin contact with anaerobic sealing compound. Immediately wash off skin if contact is made. Failure to follow this warning may result in injury to personnel.

1. Clean RTV sealant and anaerobic sealing compound from all sealing surfaces, Clean oil from all surfaces.
2. Inspect crankcase cover (2) for cracks, damage to sealing surface, and damage to crankcase cover seal opening. Replace crankcase cover if cracked or damaged.
3. Inspect fuel injector pump driven gear (1) for cracks and damaged teeth. Replace if cracked or teeth are damaged.

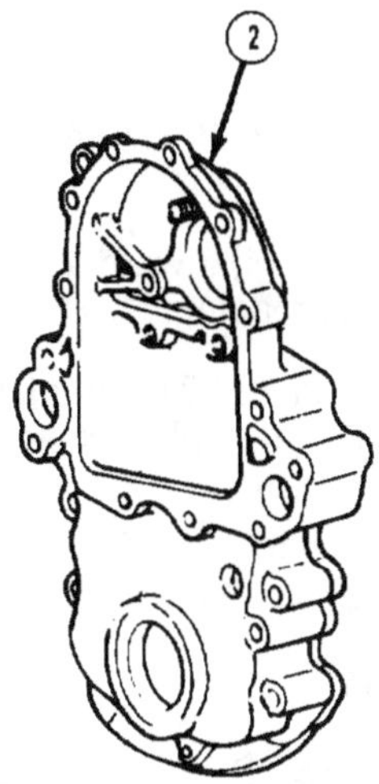

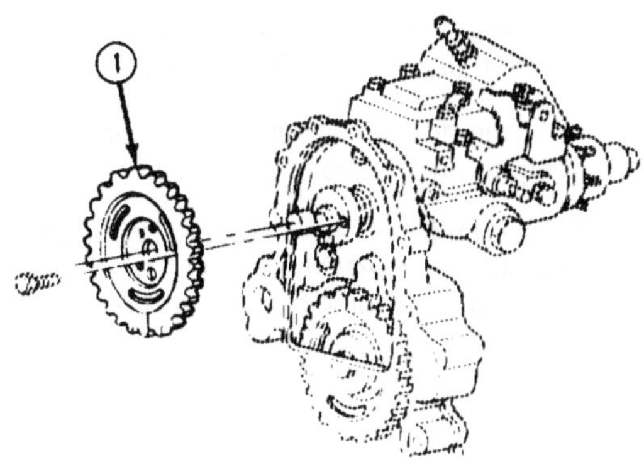

FOLLOW-ON TASKS:

- Install torsional damper and crankcase cover. (See paragraph 3-22)

TA50644

3-11. TIMING CHAIN AND SPROCKETS REMOVAL.

This task covers: a. Checking Timing Chain Free Play b. Removal c. Inspection

INITIAL SETUP:

Equipment Condition

- Crankcase cover removed. (See paragraph 3-10)

Tools/Test Equipment

- Dial indicator

a. CHECKING TIMING CHAIN FREE PLAY

1. Mount dial indicator to front of engine block (5). Position dial indicator plunger so that it contacts timing chain (4) midway between camshaft sprocket (8) and crankshaft sprocket (9).

NOTE

When moving timing chain (4) to check free play, ensure that it is moved parallel to front face of engine block (5).

2. Applying finger pressure on inside of timing chain, move timing chain (4) outward as far as it will go. With timing chain in maximum outward position, set dial indicator to "O."
3. Applying finger pressure on outside of timing chain, move timing chain (4) inward as far as it will go. With timing chain in maximum inward position, note dial indicator reading.
4. Maximum free play with used timing chain (4) must not exceed 0.80 in. (20.3 mm). Maximum free play with new timing chain must not exceed 0.50 in. (12. 7 mm). If free play exceeds this limit, replace timing chain and sprocket.

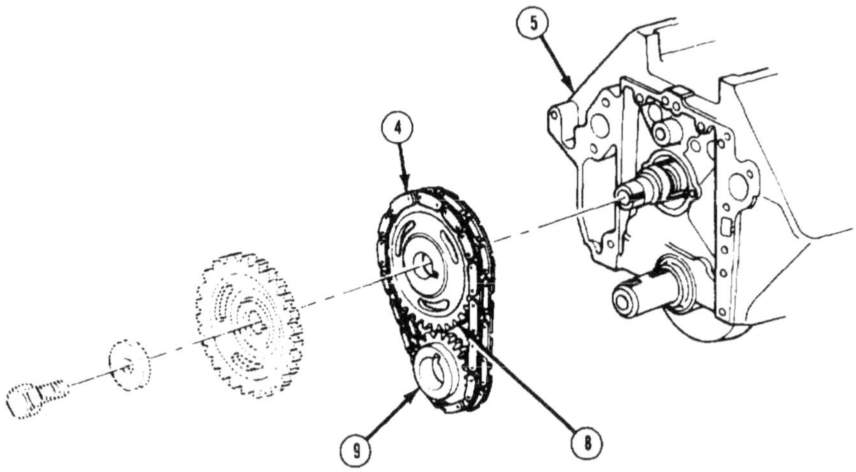

TA50645

3-11. TIMING CHAIN AND SPROCKETS REMOVAL (Con't).

b. REMOVAL

1. Remove bolt (1), thrust washer (2), and fuel injector pump drive gear (3).
2. Aline timing marks on camshaft sprocket (8) and crankshaft sprocket (9). Remove camshaft sprocket from camshaft (6), and crankshaft sprocket from crankshaft (7) with timing chain (4) attached.
3. Disassemble camshaft sprocket (8), crankshaft sprocket (9), and timing chain (4).

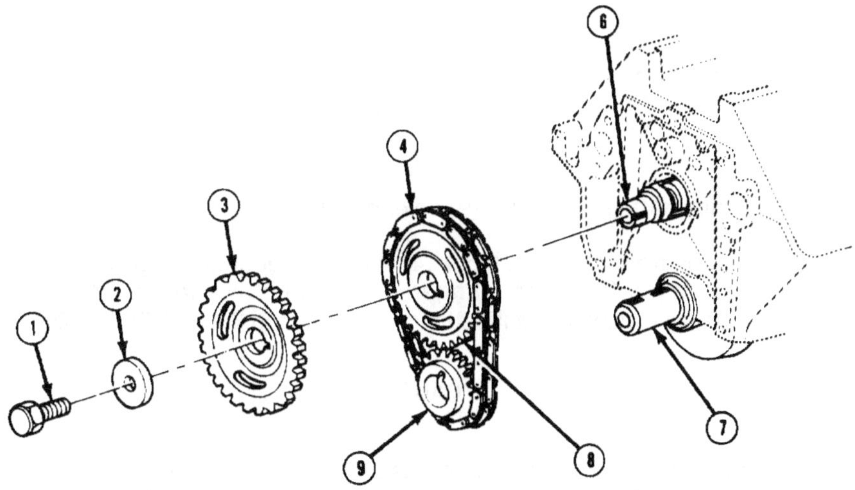

c. INSPECTION

1. Inspect timing chain (4) for cracks and damage. Replace if cracked or damaged.
2. Inspect fuel injector pump drive gear (3), camshaft Sprockclet (8),. and crankshaft sprocket (9) for cracks, damage," and" rounded gear teeth. Replace if cracked, damaged, or gear teeth are rounded,

FOLLOW-ON TASKS:

- Install timing chain and sprockets. (See paragraph 3-21)

TA50646

3-12. CAMSHAFT REMOVAL.

This task covers: a. Removal b. Inspection

INITIAL SETUP:

Equipment Condition

- Engine removed. (See paragraph 3-3)
- Hydraulic valve lifters removed. (See paragraph 3-9)
- Timing chain and sprockets removed. (See paragraph 3-1 1)
- Fuel pump removed. (See TM 9-2320-289-20)
- Vacuum pump removed, (See TM 9-2320-289-20)

Tools/Test Equipment

- Dial indicator
- Two "V" blocks
- Micrometer

a. REMOVAL

1. Mark position of camshaft sprocket key (9) on engine block (4) for installation.

2. Install dial indicator on front of engine block (4) with plunger contacting end of camshaft (3). Attempt to move camshaft toward plunger. End play should not exceed 0.0140 in. (0.356 mm). If end play exceeds specifications, perform step 3 and INSPECTION.

CAUTION

Care should be taken when pulling camshaft (3) from engine block (4). Failure to follow this caution may result in damage to camshaft bearings.

3. Remove 2 bolts (1), plate (2), ring (8), and camshaft sprocket key (9). Remove camshaft (3).

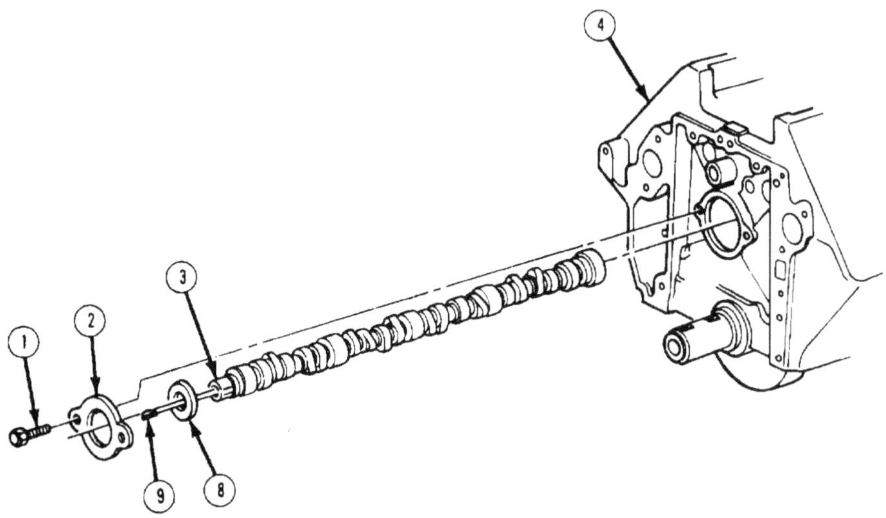

TA50647

3-12. CAMSHAFT REMOVAL (Con't).

b. INSPECTION

1. Measure diameter of camshaft journals. Rear camshaft journal (5) must be at least 2,0053 in. (50.935 mm) diameter. All other camshaft journals (7) must be at least 2.1628 in. (54.935 mm) diameter. If any camshaft journal is not within specifications, replace camshaft and appropriate camshaft bearing. (See paragraph 3-1 6)
2. Support camshaft (3) with front camshaft journal (7) and rear camshaft journal (5) in "V" blocks,
3. Secure dial indicator with plunger contacting low side of camshaft lobe (6). Set dial indicator to "o."

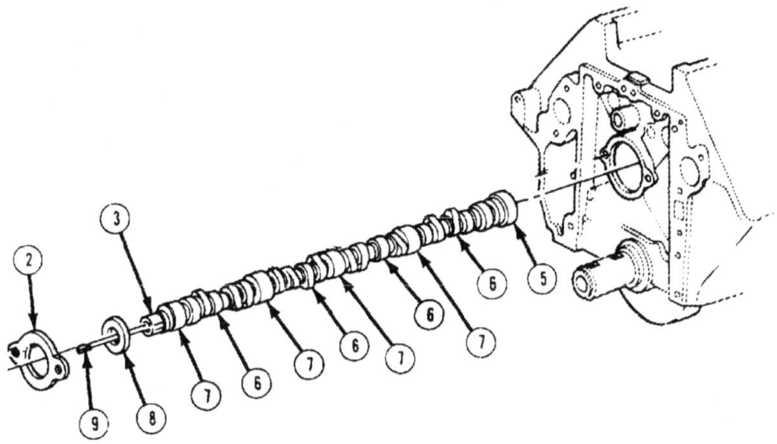

4. Turn camshaft (3) until high side of camshaft lobe (6) contacts dial indicator plunger, Reading must be at least 0.2769 in. (7,0330 mm).
5. Repeat steps 3 and 4 for each camshaft lobe (6), If any reading is not within specifications, replace camshaft.
6. Inspect plate (2), ring (8), and camshaft sprocket key (9) for wear or damage. Replace if worn or damaged,

FOLLOW-ON TASKS:

- Install camshaft, (See paragraph 3-20)

TA50648

3-13. PISTON, CONNECTING ROD, AND CONNECTING ROD BEARING REMOVAL AND REPAIR.

This task covers:

a. Removal
b. Disassembly
c. Cleaning and Inspection
d. Assembly

INITIAL SETUP:

Equipment Condition

- Engine removed. (See paragraph 3-3)
- Engine oil pump removed. (See paragraph 3-33)
- Cylinder head removed (if removing piston and connecting rod). (See paragraph 3-7)

Tools/Test Equipment

- Micrometer
- Ridge reamer
- Snapring pliers

Mater/a/s/Parts

- Dry cleaning solvent (Item 23, Appendix B)
- 3/8 in. I.D. hose (Item 39, Appendix B)
- Lubricating oil (Item 49, Appendix B)

General Safety Instructions

- Dry cleaning solvent is flammable and must not be used near open flame. Use only in a well-ventilated area.
- Always wear goggles when removing piston pin retainers.

a. REMOVAL

NOTE

- Ensure that cylinder bore and components of connecting rod and piston assembLy (5) are marked for installation.
- If removing connecting rod bearing, perform step 3.

1, Turn crankshaft (1) until piston to be removed is at bottom of its stroke.

2. Place a cloth over piston. Remove any deposits from upper end of cylinder bore. Use ridge reamer to remove wear ridge if present.

3. Remove 2 nuts (3) and connecting rod cap (4) from crankshaft (1) journal. Remove lower connecting rod bearing. DO NOT discard nuts at this time.

4. Install short piece of 3/8 in. I.D. hose (2) over threads of connecting rod bolts. Remove piston and connecting rod assembly (5) through top of cylinder bore. Remove upper connecting rod bearing. Remove hoses and retain for installation.

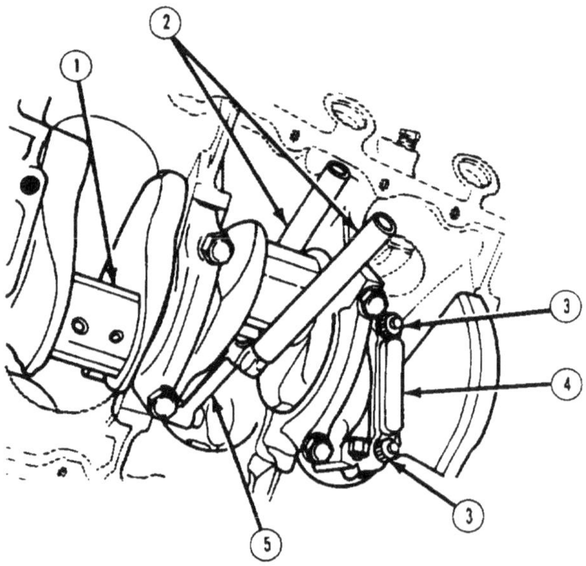

TA50649

3-13. PISTON, CONNECTING ROD, AND CONNECTING ROD BEARING REMOVAL AND REPAIR (Con't).

b. DISASSEMBLY

WARNING

Always wear goggles when removing piston pin retainers (6). Failure to follow this warning may result in serious injury to personnel.

1. Remove 2 piston pin retainers (6). Remove piston pin (7) and separate piston (8) from connecting rod (14), Remove 2 bolts (13) only if damaged.

2. Expand upper compression ring (12) and slide from groove in piston (8). Repeat for lower compression ring (11), oil control ring (1 O), and oil ring expander (9).

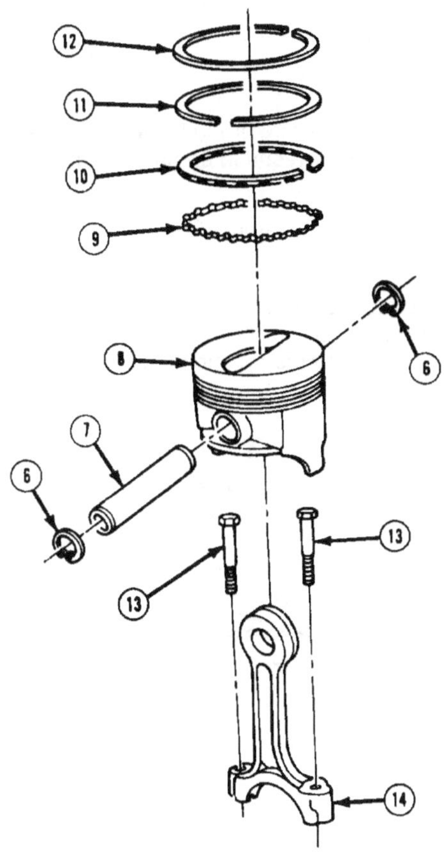

TA50650

c. CLEANING AND INSPECTION

1. Wipe oil from crankshaft journals, and upper and lower connecting rod bearing shells.

2. Inspect connecting rod bearings for scoring or damage. Replace if scored or damaged.

3. Inspect connecting rod (14) for bends or twists. Replace connecting rod and connecting rod cap if connecting rod is bent or twisted,

4. Measure connecting rod journal diameter. Diameter should b e 2.3970-2.3981 in. (60.875-60.913 mm), Replace crankshaft if diameter is not correct. (See paragraph 3-15)

5. Measure diameter of crankshaf t journal at several points and compare readings. Maximum difference in diameter (out-of-round) should be 0.0007 in. (0.01 77 mm). Replace crankshaft if out-of-round. (See paragraph 3-1 5)

3-13. PISTON, CONNECTING ROD, AND CONNECTING ROD BEARING REMOVAL AND REPAIR (Con't).

6. Measure diameter of crankshaft journal at each end and middle, and compare readings. Maximum difference between ends and middle (taper) should be 0.0007 in. (0.0177 mm). Replace crankshaft if there is excessive taper. (See paragraph 3-15)

WARNING

Dry cleaning solvent P-D-68. is toxic and flammable. Always wear protective goggles and gloves and use only in a well-ventilated area. Avoid contact with skin, eyes, and clothes and DO NOT breathe vapors. DO NOT use near open flame or excessive heat, The solvent's flash point is 100°F-138°F (38°C-59°C). If you become dizzy while using cleaning solvent, immediately get fresh air and medical help. If solvent contacts eyes, immediately wash your eyes with water and get medical aid,

CAUTION

DO NOT use wire brush to clean any part of piston. Failure to follow this caution may result in damage to piston.

7. Clean varnish from piston skirts (15) and piston pin (7) with dry cleaning solvent. Clean piston ring grooves with dry cleaning solvent, ensuring that oil ring passage and slot are clean.

8. Inspect piston (8) for cracked, rounded, or wavy piston ring lands (1 7). Inspect piston skirts (15) for cracks, scuffs, or damage. Check for cracks at piston pin hole (16) and erosion at top of piston. Replace piston and piston pin if any of these conditions exist.

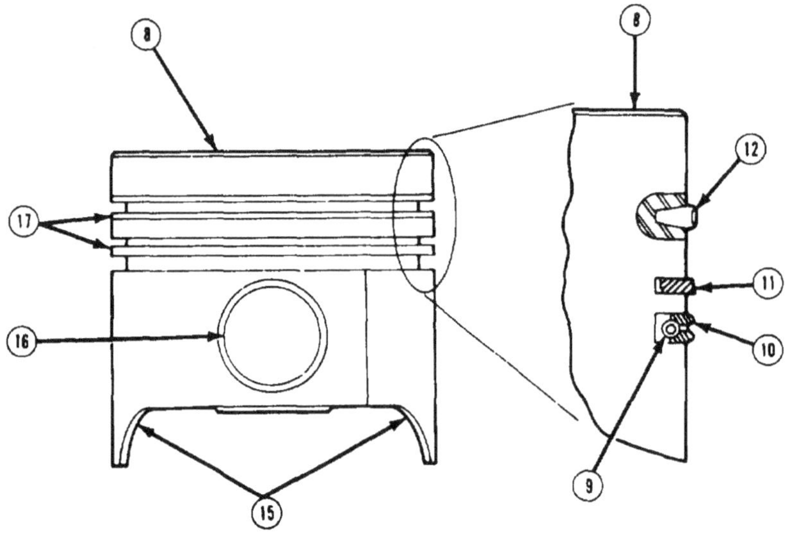

TA50651

3-13. PISTON, CONNECTING ROD, AND CONNECTING ROD BEARING REMOVAL AND REPAIR (Con't).

NOTE

Pistons (8) in 2 rear cyLinder bores (18) run hotter than other 6 pistons. Greater cylinder bore clearance is provided to allow for greater heat expansion.

9. Measure diameter of cylinder bore (18). Measure diameter of piston (8) 90 degrees from piston pin hole (16) at centerline, Subtract piston diameter from cylinder bore diameter to determine clearance. Maximum clearance is 0.005 in. (0. 127 mm) for 6 front pistons and 0.0055 in. (0.140 mm) for 2 rear pistons.

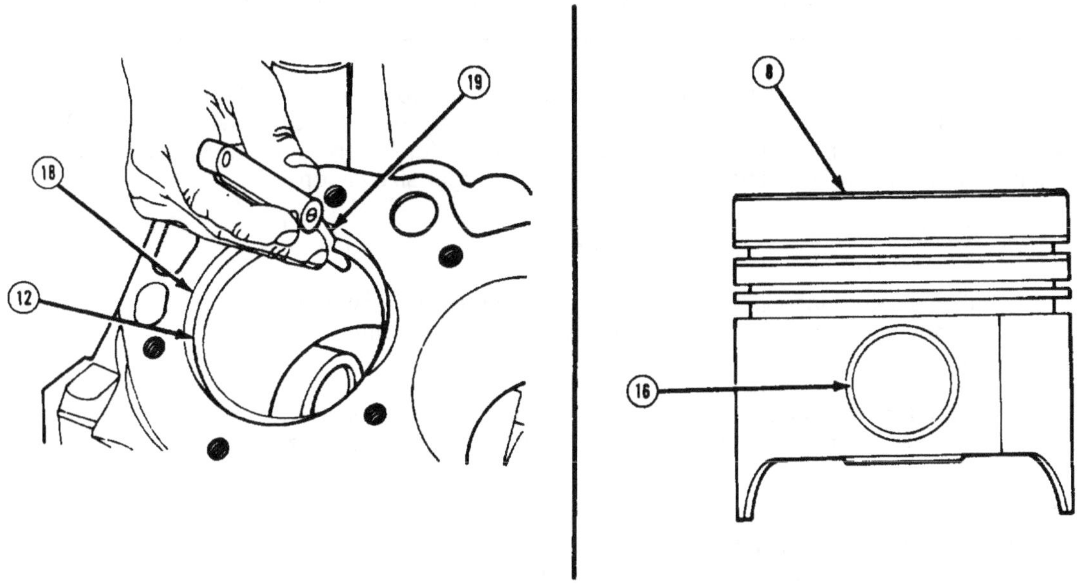

TA50652

3-13. PISTON, CONNECTING ROD, AND CNNECTING ROD BEARING REMOVAL AND REPAIR (Con't).

10. If clearance in step 9 exceeds specifications, refer to Table 3-1:

Table 3-1. Piston Diameters

Piston Type	New Piston Diameter	
Standard	3.9729 -3.9754 in.	(100,911-100.976 mm)
1st Oversize	.0034-4.0039 in.	(101 ,687-101.700 mm)
2nd Oversize	4.0039-4.0044 in.	(101.700-101.713 mm)

If a piston (8) type shown in Table 3-1 will allow cylinder bore (18) clearance of 0.0035-0.0045 in. (0.089-0.1 14 mm) for 6 front pistons and 0.004-0.0050 in . (O. 112-0,127 mm) for 2 rear pistons, use that size piston for installation. It maybe necessary to machine cylinder bore to obtain proper clearance for larger size piston.

NOTE

- If it was determined in steps 9 and 10 that a "standard" piston will be used, perform steps 11-14. If an "oversize" piston will be used, use appropriate oversize piston rings for installation.
- If cylinders are found to exceed specifications, It may be necessary to machine cylinder bore to obtain proper clearance for larger size piston.

11. Place upper compression ring (12) into cylinder bore (18) and press down about 0.25 in, (6.5 mm) into cylinder bore. Ensure that upper compression ring is square to wall of cylinder bore. Using feeler gage (19), measure gap between ends of upper compression ring.

3-13. PISTON, CONNECTING ROD, AND CONNECTING ROD BEARING REMOVAL AND REPAIR (Con't).

12. Repeat step 11 for other piston rings. Gap should be 0.01 18–0.0226 in. (0.300-0.575 mm) for upper compression ring (1 2), 0.0295-0,0403 in, (0.750-1 ,025 mm) for lower compression ring (11), and 0.0098-0.0211 in, (0.249-0,535 mm) for oil control ring (1 O). If any measurement is not correct, replace piston rings as a set.

13. Place upper compression ring (12) and lower compression ring (1 1) in piston (8) ring grooves and roll entirely around groove. If there is binding, remove ring and inspect grooves for nicks or burrs, Smooth out nicks or remove burrs with fine cut file.

14. Install piston rings (9, 10, 11, and 12) in piston (8) ring grooves. Using feeler gage, measure side clearance of each ring. Clearance should be 0.0030-0.0080 in. (0.076-0,203 mm) for upper compression ring (1 2), 0.0015-0.0041 in. (0.039-0.105 mm) for lower compression ring (11), and 0,0016-0.0048 in. (0,040-0,121 mm) for oil control ring (1 O). If any measurement is not correct, replace piston.

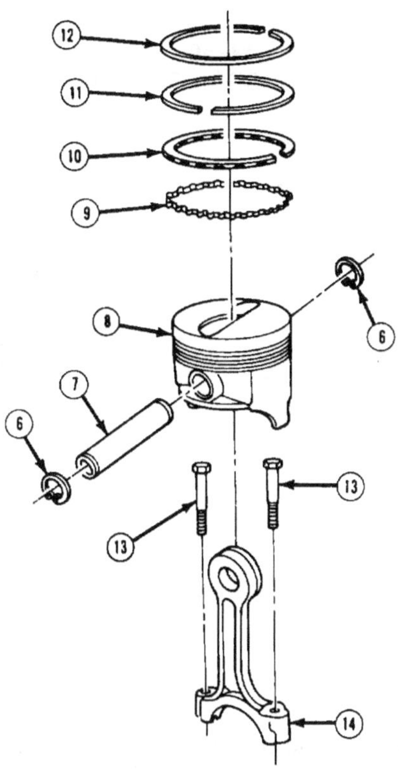

d. ASSEMBL Y

NOTE

Ensure that marked side of upper compression ring (12) and lower compression ring (11) face toward top of piston (8).

1. Expand upper compression ring (12) and lower compression ring (11) and slide into grooves in piston (8),

2. Install oil ring expander (9) and oil control ring (1 O) in piston (8) ring groove,

TA50653

3-13. PISTON, CONNECTING ROD, AND CONNECTING ROD BEARING REMOVAL AND REPAIR (Con't) .

3. Rotate piston rings (9, 10, 11, and 12) until piston ring gaps are positioned as shown.

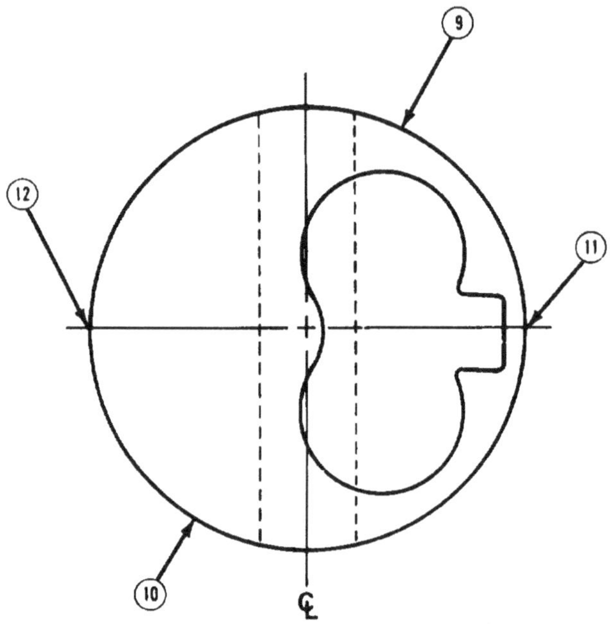

4. Assemble piston (8) and connecting rod (14), and install piston pin (7). Install 2 piston pin retainers (6) in piston. Install 2 bolts (13) in connecting rod if removed.

FOLLOW-ON TASKS:

- Install piston, connecting rod, and connecting rod bearing. (See paragraph 3-19)

TA50654

3-14. CHECKING MAIN BEARING CLEARANCE.

This task covers: a. Checking Main Bearing Clearance

INITIAL SETUP:

Equipment Condition

- Engine removed. (See paragraph 3-3)
- Engine oil pump removed. (See paragraph 3-33)

Tools/Test Equipment

- Torque wrench

Materials/Parts

- Gaging plastic (Item 35, Appendix B)
- Lubricating oil (Item 49, Appendix B)

a. CHECKING MAIN BEARING CLEARANCE

NOTE

- DO NOT remove main bearings (4) from main bearing caps (1) until main bearing clearancence has been checked.
- Ensure that main bearing caps (1) are marked for installation.
- Ensure that crankshaft (5) is facing up for this task.

1. Remove 2 inner bolts (3), 2 outer bolts (2), and main bearing #3 cap (1). Wipe oil from crankshaft (5) journal and main bearing #3 cap.
2. Cut a piece of gaging plastic (6) the width of main bearing (4). Install gaging plastic parallel to crankshaft (5) on crankshaft journal.

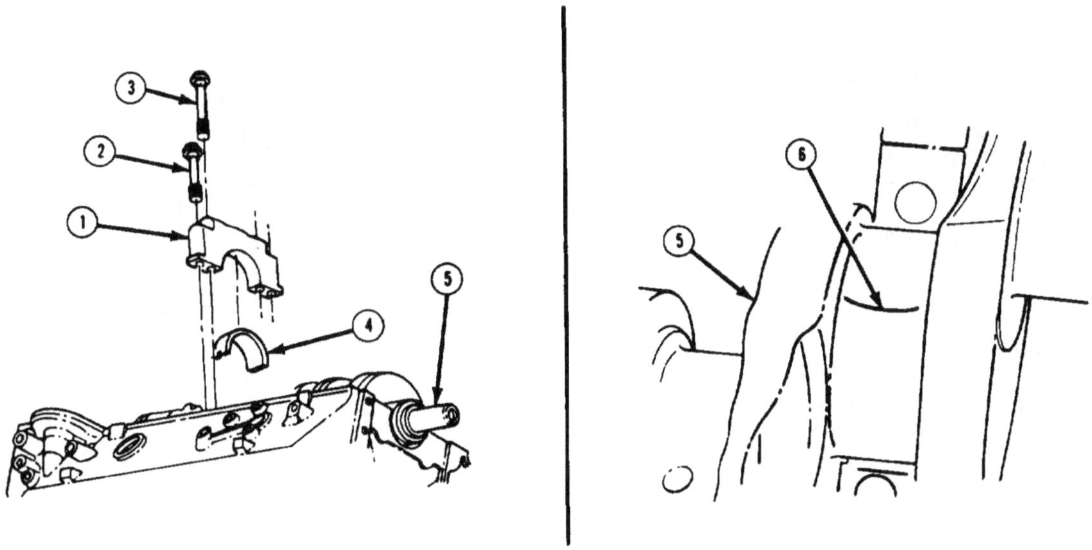

TA60655

3-14. CHECKING MAIN BEARING CLEARANCE (Con't).

CAUTION

Ensure that main bearing caps (1) are fully seated before installing bolts (2 and 3). Failure to follow this caution may result in damage to main bearing caps.

NOTE

- Inner bolts (3) are longer than outer bolts (2).
- Ensure that lubricating oil is applied to bolt (2 and 3) threads before installation,

3. Using soft-faced hammer, tap main bearing #3 cap (1) into place and install 2 inner bolts (3) and 2 outer bolts (2). Tighten all bolts to 40 lb. -ft. (54 N-m). Evenly tighten inner bolts again to 110 lb.-ft. (149 N.m). Evenly tighten outer bolts again to 100 lb.-ft. (136 N.m).

NOTE

- Scale on envelope (7) of gaging plastic (6) should be used to measure clearance.
- DO NOT rotate crankshaft (5) while gaging plastic (6) is between main bearing (4) and crankshaft. This will result in an inaccurate measurement.

4. Remove 2 inner bolts (3), 2 outer bolts (2), and main bearing #3 cap (1). Measure flattened gaging plastic (8) at widest point to determine main bearing clearance. Preferred main bearing clearance is 0.0018 in. (0.0457 mm) for main bearings #1-4 and 0.0033 (0.0838 mm) for main bearing #5. Clearance should be no more than 0.005 in. (0.127 mm).

5. Measure flattened gaging plastic (8) at each end and middle, and compare readings. Maximum difference between ends and middle (taper) should be 0.001 in. (0.025 mm). If there is taper , measure crankshaft (5) journal. Replace crankshaft if there is excessive taper. (See paragraph 3-1 5) Discard gaging plastic.

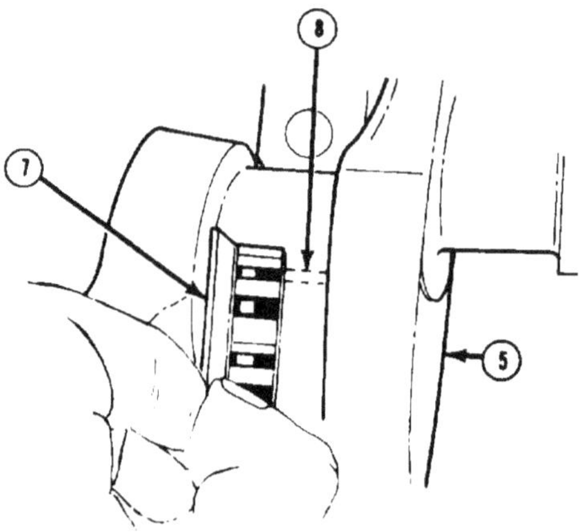

6. If measurements in steps 4 and 5 are not correct, replace main bearing (4) (see paragraph 3-1 5) and repeat steps 2-5. If still not within specifications, use undersized main bearing. If smallest undersized bearing has been used, replace crankshaft (5). (See paragraph 3-1 5)

TA50656

3-14. CHECKING MAIN BEARING CLEARANCE (Con't).

7. Using soft-faced hammer, force crankshaft (5) to extreme rearward position, then to extreme forward position. Using feeler gage (9), measure end play at front end of main bearing #3 (4). Maximum end play should be 0.015 in. (0.381 mm). If end play exceeds specifications, and main bearing #3 was replaced in step 6, replace main bearing #3 and retest. If end play still exceeds specifications, replace crankshaft. (See paragraph 3-1 5)

8. Repeat steps 1-6 for other main bearings (4).

9. Rotate crankshaft (5) to ensure that main bearings (4) are not dragging.

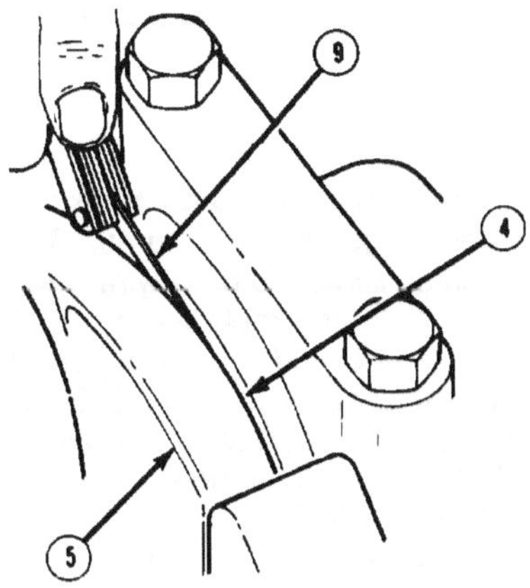

FOLLOW-ON TASKS:

- Install engine oil pump. (See paragraph 3-33)
- Replace rear main seal. (See paragraph 3-34)
- Install engine. (See paragraph 3-29)

TA50657

3-15. CRANKSHAFT, MAIN BEARINGS, AND FLYWHEEL REMOVAL.

This task covers: a. Removal b. Cleaning and Inspection

INITIAL SETUP:

Equipment Condition

- Engine removed (if removing crankshaft or main bearings). (See paragraph 3-3)
- Transmission removed (if only removing flywheel). (See paragraph 5-1 2)
- Glow plugs removed (if removing crankshaft or main bearings). (See TM 9-2320-289-20)
- Engine oil pump removed (if removing crankshaft or main bearings). (See paragraph 3-33)
- Timimg chain and Sprockets removed (if removing crankshaft), (See paragraph' 3-11)

Materials/Parts

- Carburetor cleaning compound (Item 15, Appendix B)
- Dry cleaning solvent (Item 23, Appendix B)
- 3/8 in. I.D. hose (Item 39, Appendix B)

Tools/Test Equipment

- Dial indicator
- Micrometer
- Two " V " blocks

General Safety *Instructions*

- Dry cleaning solvent is flammable and must not be used near open flame. Use only in a well-ventilated area.
- Compressed air used for cleaning purposes should never exceed 30 psi (207 kPa).
- Carburetor cleaning compound is highly flammable. Keep away from open flame.

3-15. CRANKSHAFT, MAIN BEARINGS, AND FLYWHEEL REMOVAL (Con't).

a. REMOVAL

NOTE

- If removing flywheel (1), perform step 1.
- If removing main bearings (8) without removing crankshaft (3), perform steps 2 and 3. If removing rear main bearings, also perform step 6.
- If removing main bearings (8) with crankshaft (3), perform steps 1, 2, and 4-7.
- Ensure that main bearings (8) and main bearing caps (5) are marked for installation.

1. Remove 6 bolts (2) and flywheel (1) from crankshaft (3) flange.

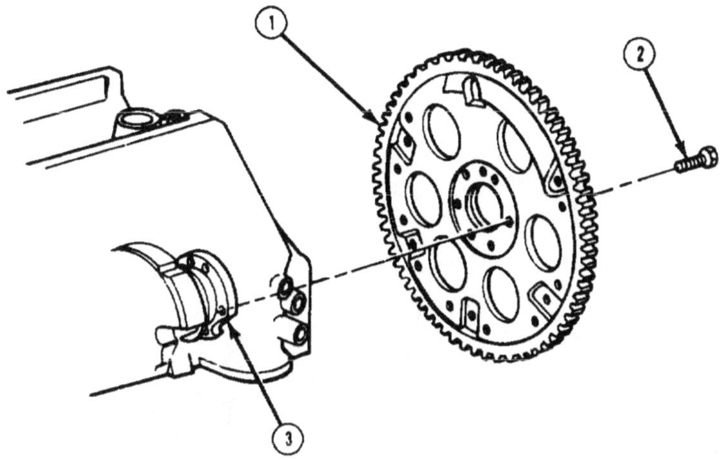

2. Remove 2 inner bolts (7) and 2 outer bolts (6), and main bearing cap (5). Remove lower main bearing (8) from main bearing cap. If removing crankshaft (3), repeat this step for each main bearing cap.

TA50658

3-15. CRANKSHAFT, MAIN BEARINGS, AND FLYWHEEL REMOVAL (Con't).

CAUTION

Care should be taken when using blunt-edged tool to remove upper main bearing. Failure to follow this caution may result in damage to crankshaft (3).

NOTE

No more than one main bearing (8) should be removed at one time. Always replace one main bearing before replacing any other main bearing.

3. Using blunt-edged tool, unseat upper main bearing by tapping into engine block (4). Rotate crankshaft until upper main bearing is rolled out of engine block.

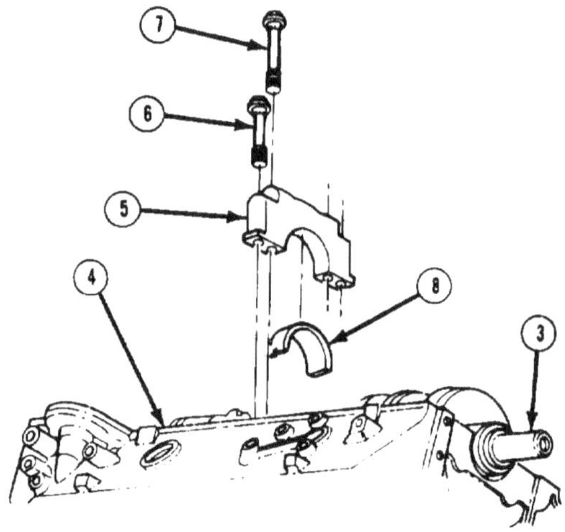

NOTE

Ensure that connecting bearings and connecting rod caps (11) are marked for installation.

4. Remove 16 nuts (10) and 8 connecting rod caps (11) from crankshaft (3) journals. Discard nuts.

5. Install short piece of 3/4 in. I.D. hose (9) over threads of all connecting rod (12) bolts. Push pistons completely into cylinders. Remove crankshaft (3).

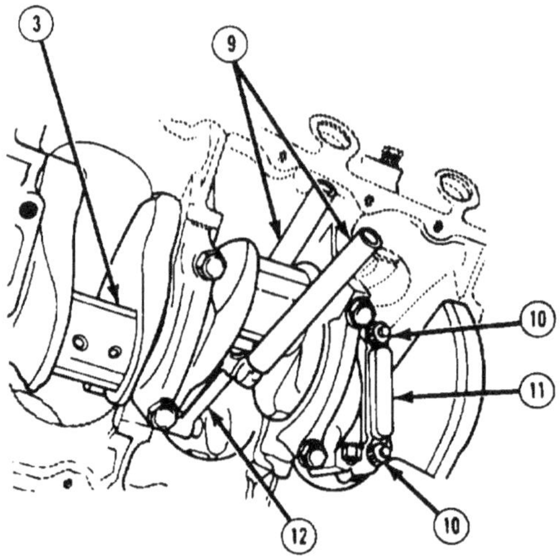

TA50659

3-15. CRANKSHAFT, MAIN BEARINGS, AND FLYWHEEL REMOVAL (Con't)

6. Remove upper rear main seal half (14) from engine block (4). Remove lower rear main seal half (15) from rear main bearing cap (5). Discard rear main seal halves.

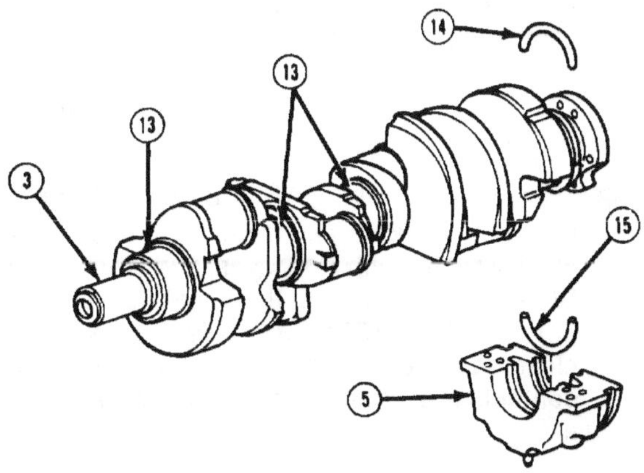

7. Remove upper main bearings from engine block (4).

TA50660

3-15. CRANKSHAFT, MAIN BEARINGS, AND FLYWHEEL REMOVAL (Con't).

b. CLEANING AND INSPECTION

WARNING

- Dry cleaning solvent P-D-680 is toxic and flammable. Always wear protective goggles and gloves and use only in a well-ventilated area. Avoid contact with skin, eyes, and clothes and DO NOT breathe vapors. DO NOT use near open flame or excessive heat. The solvent's flash point is 100°F-138°F (38°C-59°C). If you become dizzy while using cleaning solvent, immediately get fresh air and medical help. If solvent contacts eyes, Immediately wash your eyes with water and get medical aid.
- Compressed air used for cleaning purposes should never exceed 30 PSI (207 kPa). Use only effective chip guarding and personnel protective equipment (goggles/shield, gloves, etc.). Failure to follow this warning may result in serious injury to personnel.

1. Wash crankshaft (3) in dry cleaning solvent and dry with compressed air.
2. Wipe oil from rear main seal grooves in engine block and rear main bearing cap (5). Remove all rear main seal material from mating surfaces.

WARNING

Carburetor cleaning compound is highly flammable, Keep away from open flame. Failure to follow this warning may result in serious injury or death to personnel.

3. Using carburetor cleaning compound, clean bearing caps, rear main seal grooves, and surface of engine block that mates with rear main bearing cap.
4. Measure diameter of main journals (13) at several points and compare readings. Maximum difference in diameter (out-of-round) should be 0.001 in. (0.025 mm). Replace crankshaft if any main journal is beyond specifications.
5. Measure diameter of main journals (13) at each end and middle and compare readings. Maximum difference between ends and middle (taper) should be 0.001 in. (0.025 mm). Replace crankshaft if any main journal is beyond specifications.

NOTE

Production diameter of rear main journal is 2.9493-2.9502 in . (74.91 2-74.936 mm). All other main journals are 2.9495-2.9504 in. (74.917-74.914 mm),

6. Using measurements found in steps 4 and 5, determine if main journals (13) are undersized. Minimum allowable diameter is 2.947 in. (74.874 mm) for rear main journal and 2.948 in. (74.879 mm) for all other main journals. Replace crankshaft if any main journal is less than minimum allowable diameter.

3-15. CRANKSHAFT, MAIN BEARINGS, AND FLYWHEEL REMOVAL (Con't).

7. Check connecting rod journals, (See paragraph 3-13, CLEANING AND INSPECTION, steps 4-6).

8. Support crankshaft (3) with front and rear main journals (13) *in "V"* blocks,

9. Using dial indicator at main journal #2 and main journal #4, rotate crankshaft and check for runout . Replace crankshaft if runout exceeds 0.001 in. (0.025 mm),

10. Inspect connecting rod bearings and main journal bearings for damage. Replace if damaged.

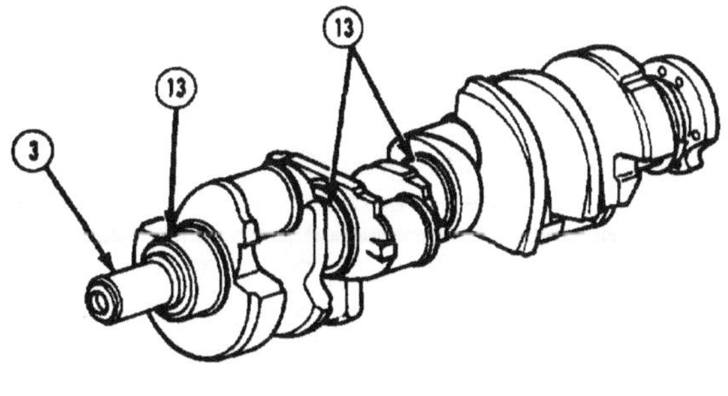

FOLLOW-ON TASKS:

- Install crankshaft, main bearings, and flywheel. (See paragraph 3-18)

TA50661

3-16 CAMSHAFT BEARINGS REMOVAL.

This task covers: a, Removal

INITIAL SETUP:

Equipment Condition

- Camshaft removed. (See paragraph 3-12)
- Crankshaft removed. (See paragraph 3-15)

Tools/Test Equipment

- Camshaft bearing puller, J-35178
- Micrometer

a. REMOVAL

1. Inspect camshaft bearing for scoring and pitting. Measure inside diameter of all camshaft bearings. Replace rear camshaft bearing if inside diameter is greater than 2.0113 in. (51 .087 mm). Replace any other camshaft bearings if inside diameter is greater than 2.1688 in. (55.089 mm). Replace any scored or pitted camshaft bearings. Camshaft bearing-to-journal clearance must not exceed 0.006 in. (0.152 mm).

2. Remove camshaft plug (2) from rear of engine block (1). Discard camshaft plug.

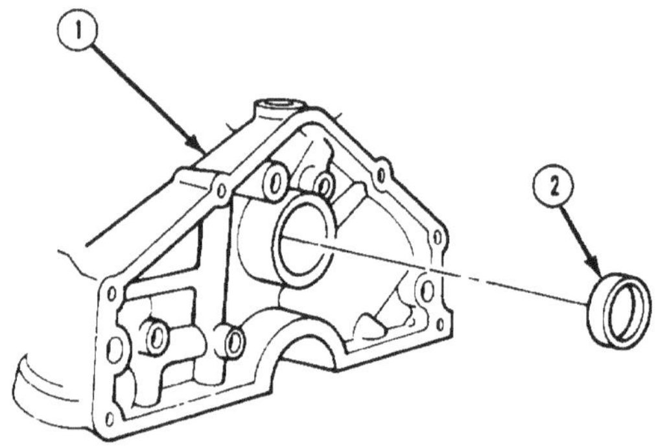

TA50662

3-16. CAMSHAFT BEARINGS REMOVAL (Con't).

3. Install nut (5) and thrust washer (4) on puller screw (6), Index front camshaft bearing pilot (3) to front camshaft bearing, and install puller screw through front camshaft bearing pilot,

NOTE

- Ensure that shoulder of pilot (7) is toward camshaft bearing and that at least 2 threads are exposed at end of puller screw (6).
- Ensure that 3 inner camshaft bearings are removed first. Front and rear camshaft bearings will act as guides for camshaft bearing puller.

4. Install correct pilot (7) to remove 1 inner camshaft bearing. Hold end of puller screw (6) while turning nut (5) until inner camshaft bearing is removed, Repeat this step until other inner damaged camshaft bearings are removed. Discard camshaft bearings.

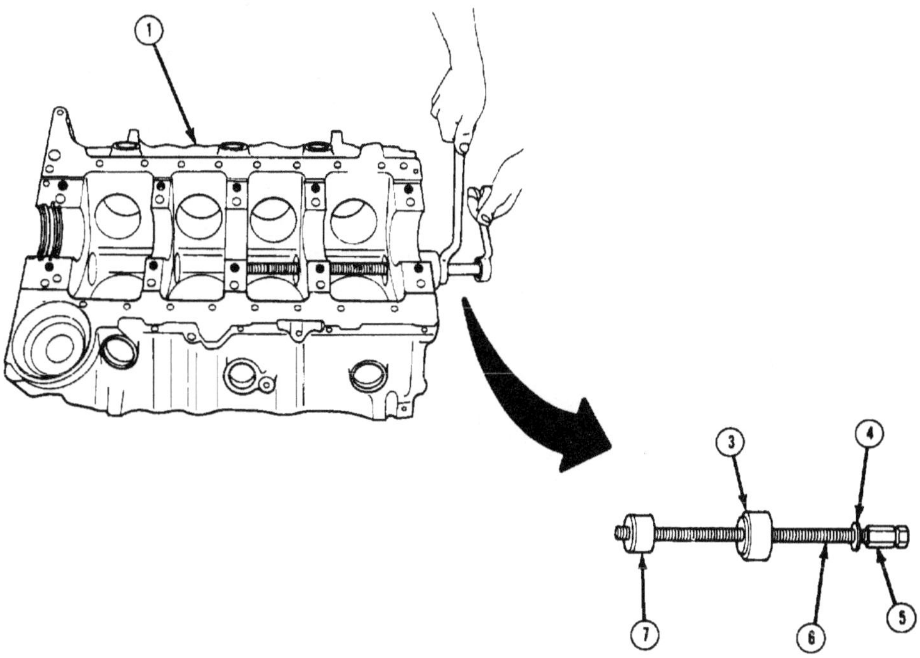

TA701846

3-16. CAMSHAFT BEARINGS REMOVAL (Con't) .

5. Install front camshaft bearing pilot (3) on driver handle (8). Drive front camshaft bearing toward center of engine block (1) to remove. Install correct pilot (7) on driver handle and repeat this step to remove rear camshaft bearing, Discard camshaft bearings.

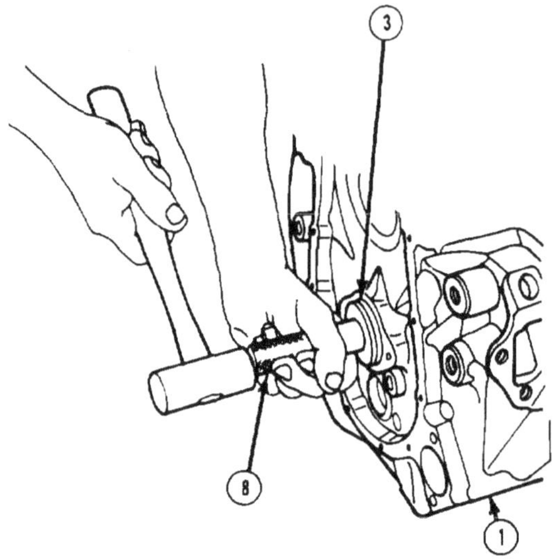

FOLLOW-ON TASKS:

- Install camshaft bearings. (See paragraph 3-17)

TA50664

3-17. CAMSHAFT BEARINGS INSTALLATION.

This task covers: a. Installation

INITIAL SETUP:

Equipment Condition

- Camshaft bearings removed. (See paragraph 3-1 6)

Too/s/Test Equipment

- Camshaft bearing puller, J-35178

Materials/Parts

- One camshaft plug
- Lubricating oil (Item 49, Appendix B)
- Sealant (Item 56, Appendix B)

a. INSTALLATION

NOTE

Seam in each camshaft bearing must be located in upper half of engine face.

All camshaft bearings have an oil passage that must be installed at the 4 o'clock position. Front camshaft bearing has an additional oil passage that must be positioned between 12 o'clock and 1 o'clock positions. Notch in front camshaft bearing must be facing front of engine block (l).

1. Clean all sealant from rear of engine block (1) and camshaft bore.

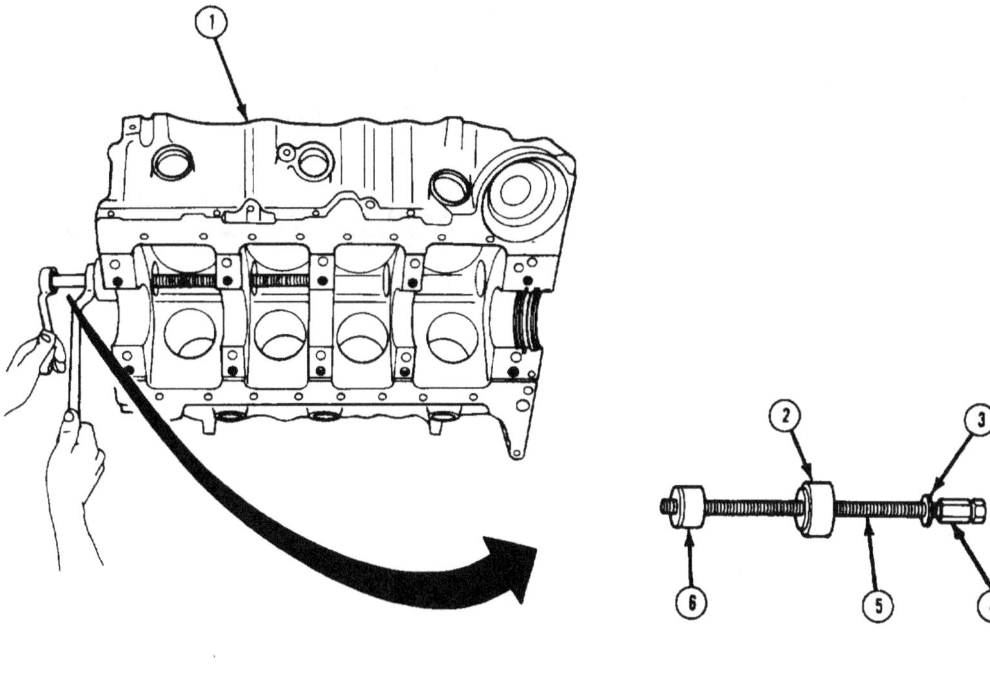

TA701847

3-17. CAMSHAFT BEARINGS INSTALLATION (Con't).

2. Install front camshaft bearing pilot (2) on driver handle (7). Drive front camshaft bearing toward center of engine block (1) to install. Ensure that oil passage is properly alined. Install correct pilot (6) on driver handle and repeat this step to install rear camshaft bearing.

3. Install nut (4) and thrust washer (3) on puller screw (5). Index front camshaft bearing pilot (2) to front camshaft bearing, and install puller screw through front camshaft bearing pilot.

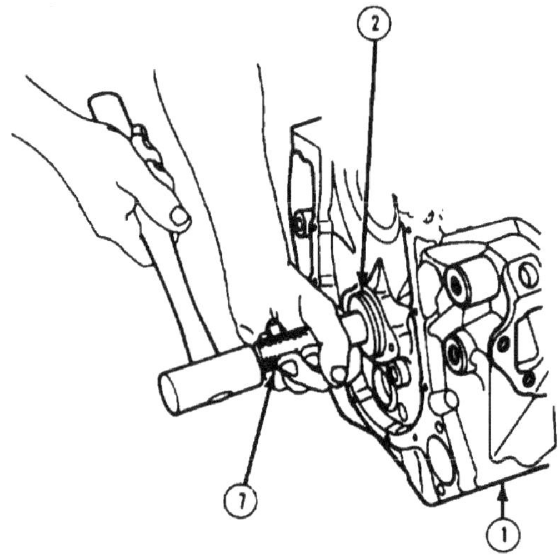

NOTE

Ensure that shoulder of pilot (6) is toward camshaft bearing and" that at least 2 threads are exposed at end of puller screw (5).

4. Install correct pilot (6) to install 1 inner camshaft bearing. Ensure that oil passage is properly alined. Hold end of puller screw (5) while turning nut (4) until inner camshaft bearing is installed. Repeat this step until other inner camshaft bearings are installed.

5. Coat new camshaft plug (6) with sealant and install in rear of engine block (1). Camshaft plug must be flush with engine block to within 0.031 in. (0.80 mm) deep.

6. Apply lubricating oil to all camshaft bearings.

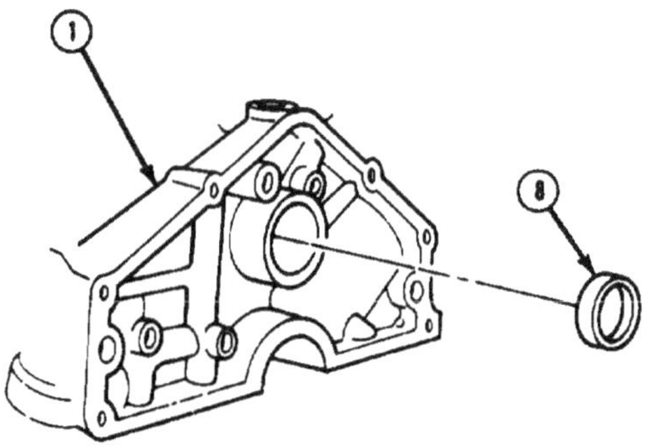

FOLLOW-ON TASKS:

- Install crankshaft, (See paragraph 3-18)
- Install camshaft. (See paragraph 3-20)

TA60666

3-18. CRANKSHAFT, MAIN BEARINGS, AND FLYWHEEL INSTALLATION.

This task covers: a. Installation

INITIAL SETUP:

Equipment Condition

- Crankshaft, main bearings, and flywheel removed, (See paragraph 3-1 5)

Tools/Test Equipment

- Torque wrench

Materials/Parts

- One rear main seal kit
- Sixteen connecting rod nuts
- Lubricating oil (Item 49, Appendix B)

a. INSTALLATION

NOTE

- If installing main bearings (5) and crankshaft (6), perform steps 1, 2, and 4 through 9.
- If installing main bearings (5) and crankshaft (6) was not removed, perform steps 3, 5, and 6. if installing rear main bearings, also perform step 4. If installing #3 main bearings, also perform step 7.
- if installing flywheel, perform step 9.

1. Coat inner surface of upper main bearings with lubricating oil and install in engine block (1) as marked.
2. Position crankshaft (6) in engine block (1) .

CAUTION

Care should be taken when using biunt-edged tool to install upper main bearing, Failure to follow this caution may result in damage to crankshaft (6).

3. Coat surface of upper main bearing with lubricating oil and insert plain (unnotched) end between crankshaft (6) and indented (notched) side of engine block (1). Rotate crankshaft until upper main bearing is rolled into engine block. Using blunt-edged tool, tap upper main bearing into engine block until it seats.

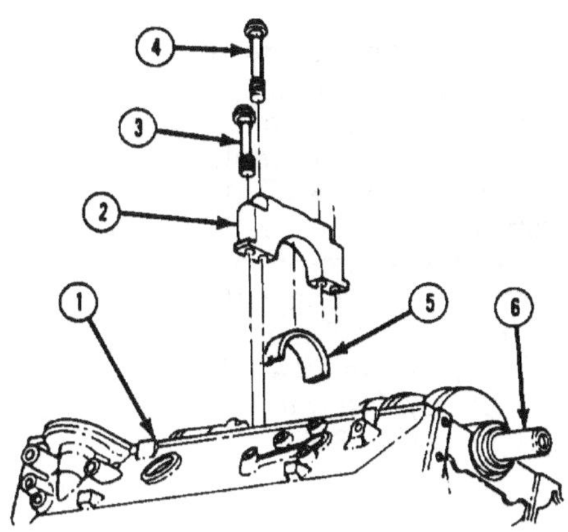

TA50667

3-18. CRANKSHAFT, MAIN BEARINGS, AND FLYWHEEL INSTALLATION (Con't).

4. Install rear main seal. (See paragraph 3-34)

CAUTION

Ensure that main bearing caps (2) are fully seated before installing bolts (3 and 4). Failure to follow this caution may result in damage to main bearing caps.

5. Coat inner surface of lower main bearing (5) with lubricating oil and install in main bearing cap (2) as marked. Using soft-faced hammer, tap main bearing cap into place and install with 2 inner bolts (4) and 2 outer bolts (3). Repeat this step for each main bearing cap that was removed.

NOTE

- Bearing cap #3 bolts should not be tightened to full specifications yet,
- Inner bolts (4) are longer than outer bolts (3),

6. Tighten inner bolts (4) and outer bolts (3) to 40 lb. -ft. (54 N.m). Evenly tighten inner bolts again to 110 lb. -ft. (149 N.m). Evenly tighten outer bolts again to 100 lb.-ft. (136 N.m).

7. Tighten bearing cap #3 inner bolts (4) and outer bolts (3) to 10 lb. -ft, (14 N.m). Using soft-faced hammer, tap end of crankshaft (6) rearward, then forward, to line up #3 main bearing (5) with crankshaft thrust surface. Perform step 6 for bearing cap #3.

8. Coat surface of connecting rod bearings with lubricating oil. Push pistons down to engage connecting rods on connecting rod journals. Remove short pieces of 3/8 in. I.D. hose (7) from threads of connecting rod (1 O) bolts immediately before installing connecting rod caps (9). Install 8 connecting rod caps on crankshaft (6) journals as marked with 16 new nuts (8). Evenly tighten nuts to 45-50 lb.-ft. (61-68 N.m).

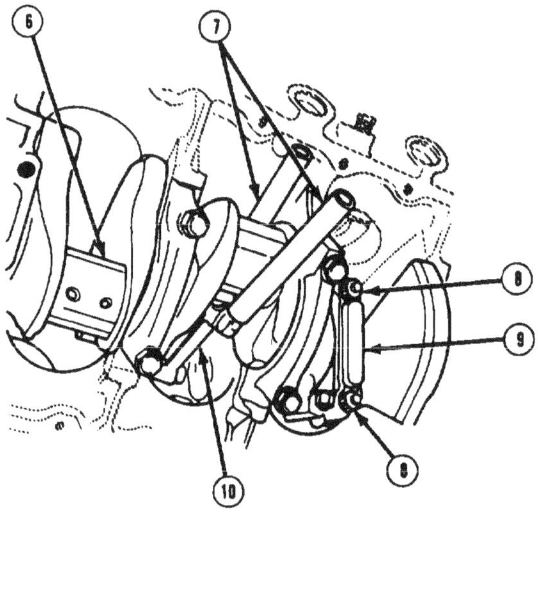

TA50668

3-18. CRANKSHAFT, MAIN BEARINGS, AND FLYWHEEL INSTALLATION (Con't).

NOTE

Ensure that weighted side of flywheel (11) is facing crankshaft (6) when installed,

I9. Install flywheel (11) on crankshaft(6) flage with 6 bolts (12). Thighten bolts to 65 lb.-ft. (88 N.m).

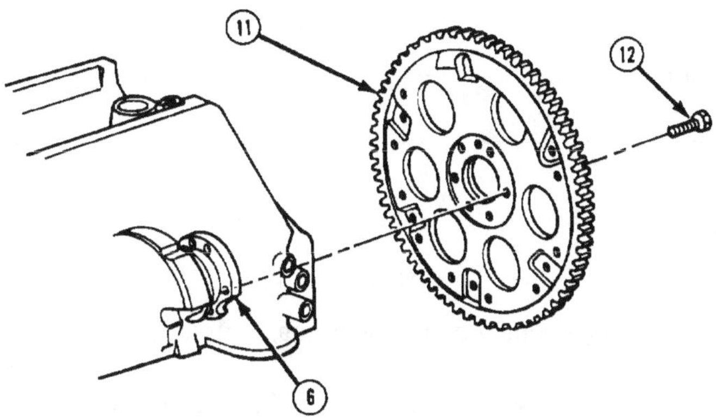

FOLLOW-ON TASKS:

- Install timing chain and sprockets if removed, (See paragraph 3-21)
- Install engine oil pump if removed. (See paragraph 3-33)
- Install glow plugs if removed. (See TM 9-2320-289-20)
- Install transmission if removed. (See paragraph 5-12)
- Install engine. (See paragraph 3-29)

TA50669

3-19. PISTON, CONNECTING ROD, AND CONNECTING ROD BEARING INSTALLATION.

This task covers: a. Installation

INITIAL SETUP:

Equipment Condition

- Piston, connecting rod, and connecting rod bearing removed. (See paragraph 3-1 3)

Tools/Test Equipment

- Piston ring compressor
- Torque wrench

Materials/Parts

- Two connecting rod nuts
- Gaging plastic (Item 35, Appendix B)
- 3/8 in. I.D. hose (Item 39, Appendix B)
- Lubricating oil (Item 49, Appendix B)

a. INSTALLATIO N

NOTE

If only installing connecting rod bearings, perform steps 4-9.

1. Install short piece of 3/8 in. I.D. hose (2) over threads of connecting rod bolts.Lubricate piston. piston rings, and cylinder bore (7) with lubricating oil,
2. See paragraph 3-13, ASSEMBLY, step 3, to ensure that piston ring gaps are properly positioned.

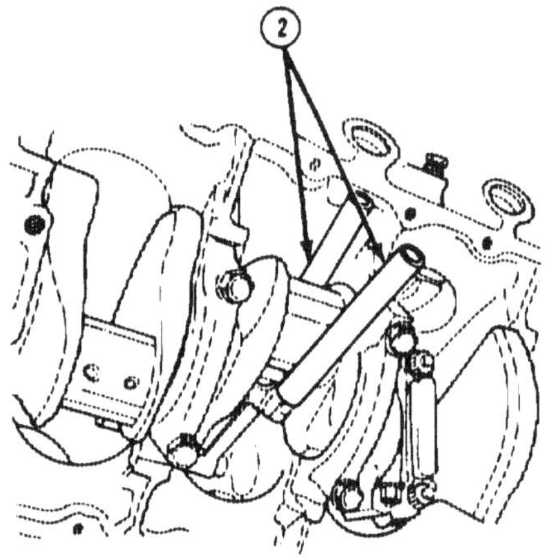

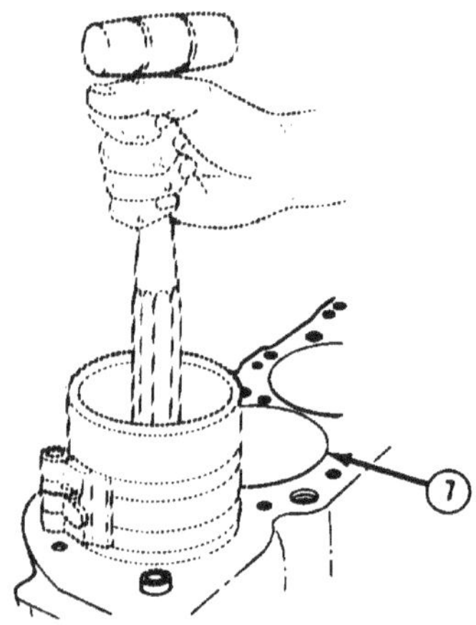

TA50670

3-19 PISTON CONNECTING ROD, AND CONNECTING ROD BEARING INSTALLATION (Con't).

CAUTION

Ensure that piston rings remain compressed while performing step 3. If piston rings catch on cylinder block, remove piston from cylinder bore (7) and repeat steps 2 and 3. Failure to follow this caution may result in damage to piston rings.

NOTE

Ensure that depression on top of piston faces toward outside of engine.

3. Install piston ring compressor (6) about 1/4 in. (6.35 mm) above bottom of piston skirt on piston. Ensure that piston ring compressor is flush with cylinder bore. Tap piston and connecting rod assembly (5) into cylinder bore.

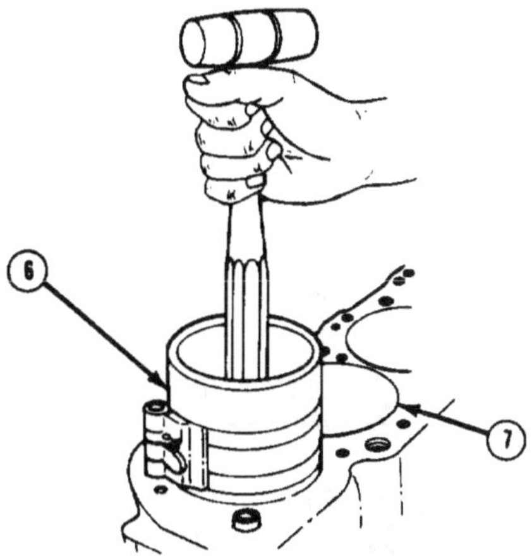

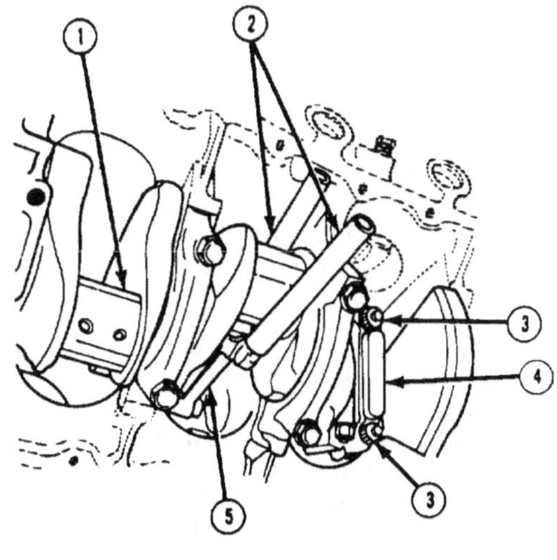

NOTE

Ensure that connecting rod bearings are installed with tang slots toward outside of engine.

4. Install upper connecting rod bearing on piston and connecting rod assembly (5). Guide connecting rod onto crankshaft (1) journal. Remove hoses (2).

TA50671

3-19. PISTON, CONNECTING ROD, AND CONNECTING ROD BEARING INSTALLATION (Con't).

NOTE

Ensure that connecting rod bearings and crankshaft (1) journal are clean of oil before performing step 5.

5. Cut a piece of gaging plastic (8) the length of connecting rod bearing. Install gaging plastic parallel to crankshaft (1) in middle of lower connecting rod bearing.

6. Install connecting rod bearing in connecting rod cap (4), Install connecting rod cap with 2 nuts (3), Evenly tighten nuts to 45-50 lb.-ft, (61-68 N.m).

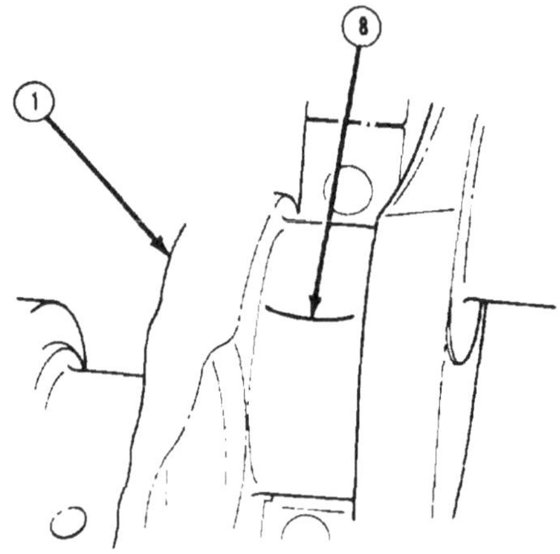

NOTE

Scale on envelope (9) of gaging plastic (8) should be used to measure clearance.

7. Remove 2 nuts (3) and connecting rod bearing cap (4). Measure flattened gaging plastic (10) at widest point to determine connecting rod bearing clearance. Clearance should be 0.0017-0.0039 in. (0.04318-0.09906 mm). Discard gaging plastic,

8. If clearance in step 7 is not correct, use undersized connecting rod bearing and repeat steps 5-7. If clearance is still not correct, replace crankshaft (1). (See paragraph 3-15)

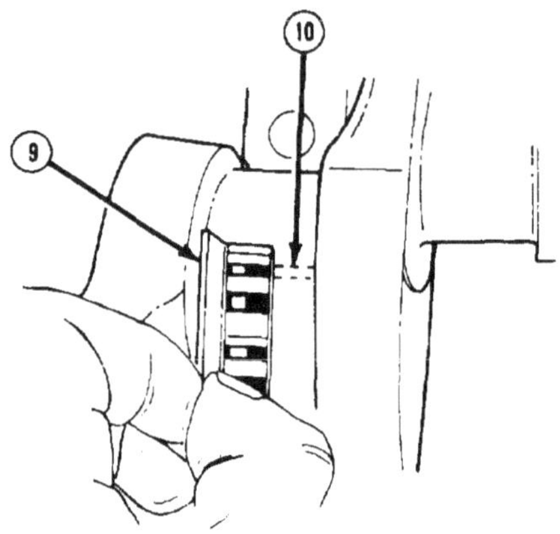

TA50672

3-19. PISTON, CONNECTING ROD, AND CONNECTING ROD BEARING INSTALLATION (Con't).

9. If clearance in step 7 is correct, lubricate connecting rod bearing surfaces with lubricating oil. Discard 2 nuts (3). Install connecting rod cap (4) with 2 new nuts (3). Evenly tighten nuts to 45-50 lb.-ft. (61-68 N.m).

10. Gently tap each connecting rod parallel to crankshaft (1) with soft-faced hammer to ensure that there is clearance.

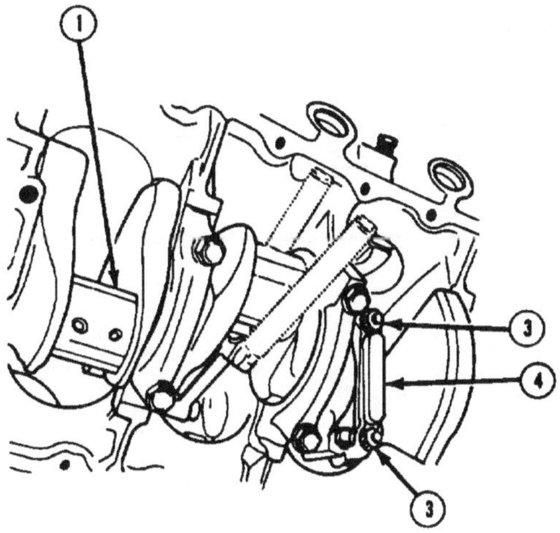

11. Using feeler gage (11), measure clearance between connecting rod caps (4). If clearance is less than 0.0067 in. (0.1 7 mm), or more than 0.025 in. (0.63 mm), replace connecting rod and cap. A clearance of 0.0279 in. (0.71 mm) is allowable with used connecting rods.

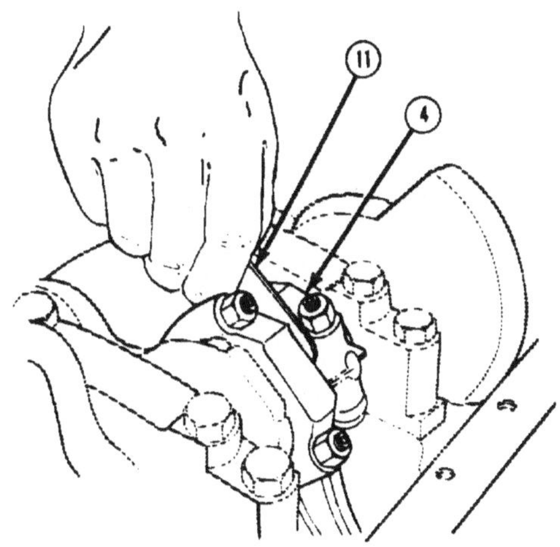

TA50673

3-19. PISTON, CONNECTING ROD, AND CONNECTING ROD BEARING INSTALLATION (Con't).

FOLLOW-ON TASKS:

- Install cylinder head if removed. (See paragraph 3-25)
- Install engine oil pump. (See paragraph 3-33)
- Install engine. (See paragraph 3-29)

3-20. CAMSHAFT INSTALLATION.

This task covers: a. Installation

INITIAL SETUP:

Equipment Condition

- Camshaft removed. (See paragraph 3-12)

Materials Parts

- Molybdenum grease (Item 37, Appendix B)
- Lubricating oil (Item 49, Appendix B)

Tools/Test Equipment

- Torque wrench

a. INSTALLATION

CAUTION

Care should be taken during installation of camshaft (3) into engine block (4). Failure to follow this caution may result in damage to camshaft bearings,

NOTE

If Installing new camshaft (3), ensure that molybdenum grease Is applied to camshaft lobes (6).

1. Apply light coat of engine oil to camshaft journals (5 and 7) and vacuum pump drive gear. Install camshaft (3) in engine block (4). Install ring (8), camshaft sprocket key (9), and plate (2) with 2 bolts (1). Tighten bolts to 20 lb.-ft. (27 N.m).

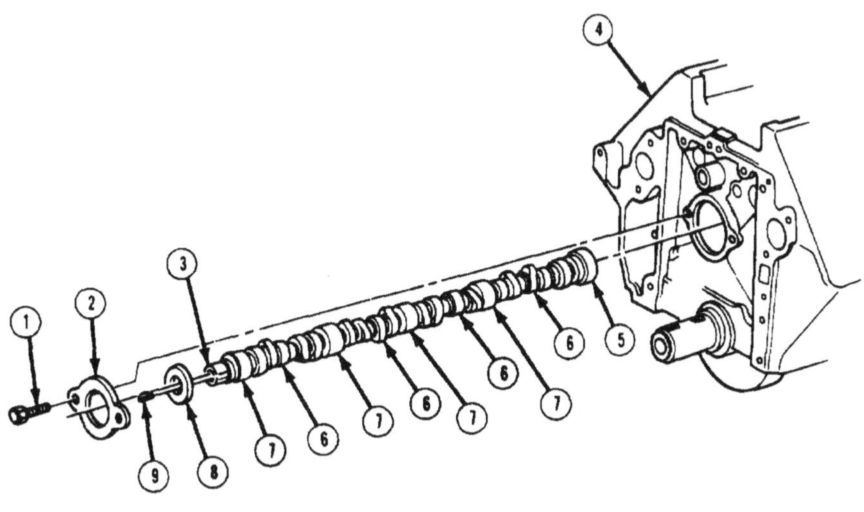

TA50674

3-20. CAMSHAFT INSTALLATION (Con't).

2. Rotate camshaft (3) to aline camshaft sprocket key (9) with mark on engine block (4).

NOTE

If new camshaft (3) was installed, ensure that engine oil filter and all hydraulic valve lifters are replaced.

FOLLOW-ON TASKS:

- Install vacuum pump, (See TM 9-2320-289-20)
- Install fuel pump. (See TM 9-2320-289-20)
- Install hydraulic valve lifters. (See paragraph 3-23)
- Install timing chain and sprockets. (See paragraph 3-21)
- Install new engine oil filter (if new camshaft was installed). (See TM 9-2320-289-20)
- Install engine, (See paragraph 3-29)

3-21. TIMING CHAIN AND SPROCKETS INSTALLATION.

This task covers : a. Installation

INITIAL SETUP:

Equipment Condition

- Timing chain and sprockets removed, (See paragraph 3-1 1)

Tools/Test Equipment

- Torque wrench

Materials/parts

- Lubricating oil (Item 49, Appendix B)

a. INSTALLATIO N

1. Lubricate timing chain (4) with lubricating oil. Assemble crankshaft sprocket (8), camshaft sprocket (7), and timing chain. Ensure that timing marks on crankshaft sprocket and camshaft sprocket are alined.

2. Install crankshaft sprocket (8) on crankshaft (6) and camshaft sprocket (7) on camshaft (5) with timing chain (4) installed.

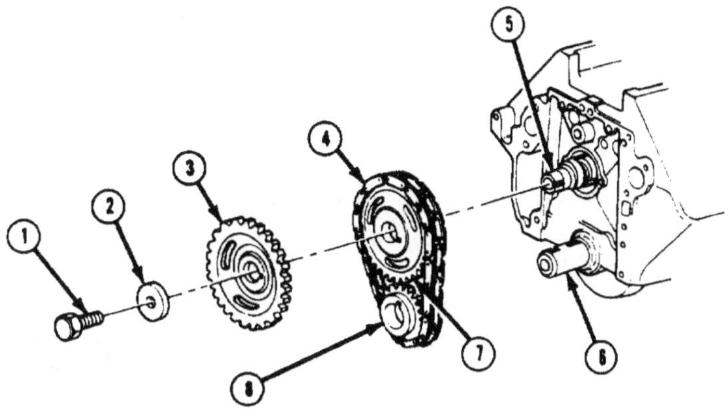

3. Check timing chain free play. (See paragraph 3-11)

4. Install fuel injector pump drive gear (3) with thrust washer (2) and bolt (1). Tighten bolt to 75 lb.-ft. (102 N.m).

FOLLOW-ON TASKS:

- Install crankcase cover, (See paragraph 3-22)

TA50675

3-22. TORSIONAL DAMPER AND CRANKCASE COVER INSTALLATION.

This task covers: a. Installation

INITIAL SETUP:

Equipment Condition

- Torsional damper and crankcase cover removed. (See paragraph 3-10)

Tools/Test Equipment

- Torque wrench

Materials/Parts

- One crankcase cover seal
- One fuel injector pump gasket
- Lubricating oil (Item 49, Appendix B)
- RTV sealant (Item 57, Appendix B)
- Anaerobic sealing compound (Item 59, Appendix B)

General Safety instructions

- Avoid skin contact with anaerobic sealing compound.

a. INSTALLATION

WARNING

Avoid skin contact with anaerobic sealing compound. immediately wash off skin if contact is made. Failure to follow this warning may result in injury to personnel.

NOTE

- if installing crankcase cover seal (7), perform steps 9 and 10.
- if installing torsional damper, perform step 10.

1. Install probe holder (9) on crankcase cover (4) with 2 screws (10) if removed.

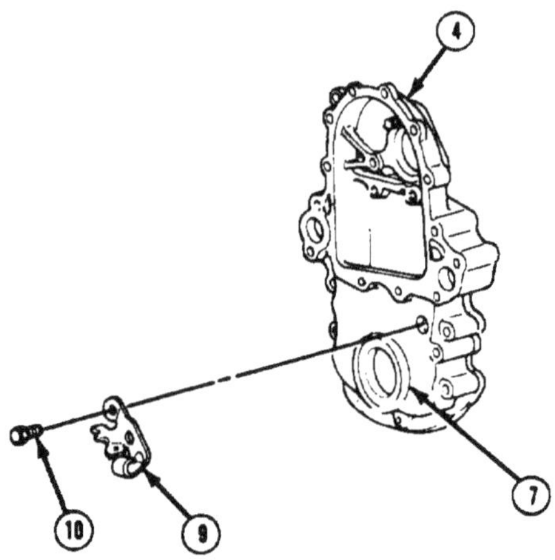

TA50676

3-22. TORSIONAL DAMPER AND CRANKCASE COVER INSTALLATION (Con't).

2. Apply a 0.09 in. (2 mm) bead of anaerobic sealing compound (11) to crankcase cover (4) in pattern shown. Apply a 0.18 in. (5 mm) bead of RTV sealant to lower surface of crankcase cover where it mates with engine oil pan.

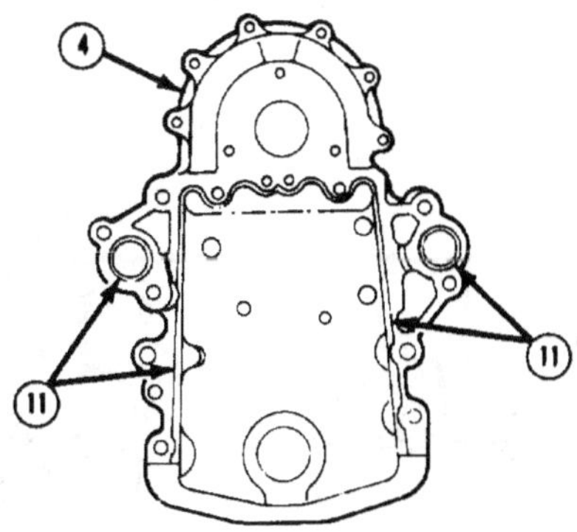

3. Install new fuel injector pump gasket (12) on studs at rear of crankcase cover (4). Install crankcase cover on engine block (5) with 4 bolts (8). Install bolt (6). Tighten bolts to 35 lb.-ft. (47 N.m).

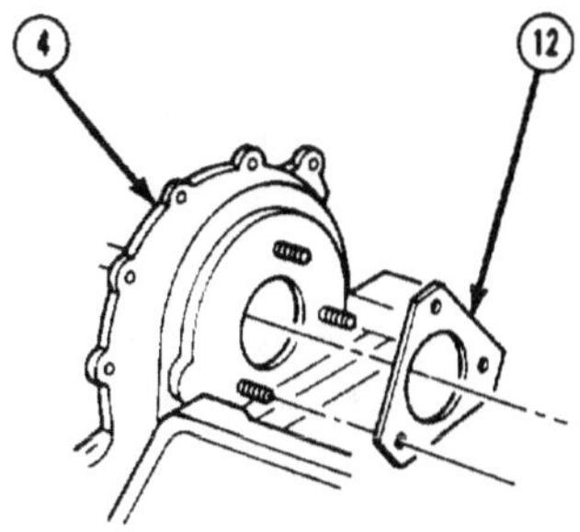

TA50677

3-22. TORSIONAL DAMPER AND CRANKCASE COVER INSTALLATION (Con't).

4 . Secure crankcase cover (4) to engine oil pan with 4 bolts.

5 . Install baffle (3) with 2 bolts (1) and nut (2). Tighten bolts and nut to 35 lb.-ft. (47 Nom).

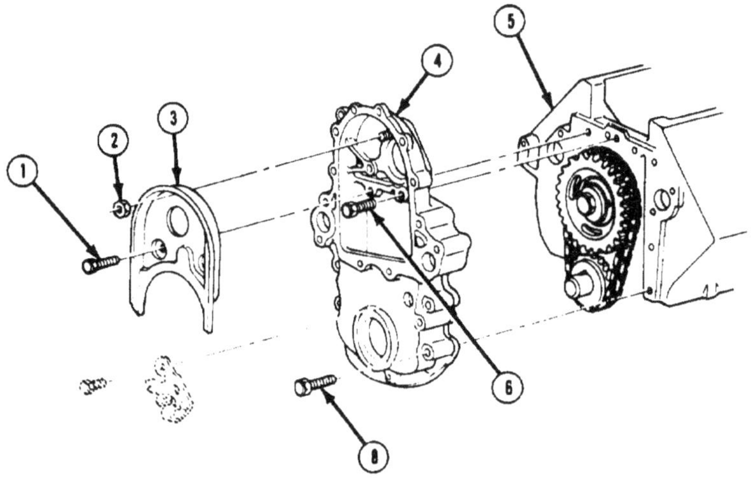

6. Install clip (17) with fuel return line (16) to crankcase cover (4) with bolt (15).

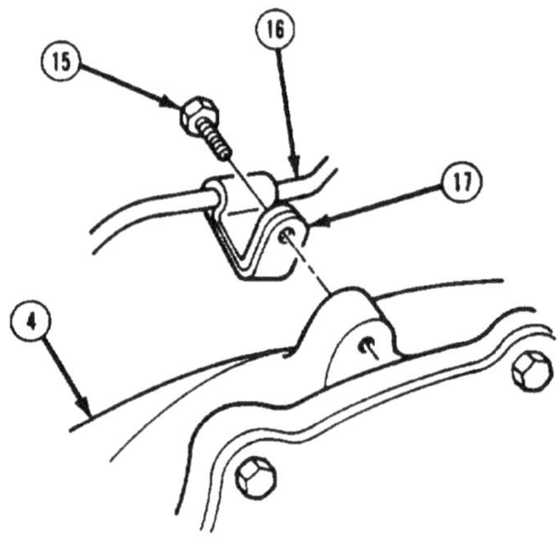

TA50676

3-22. TORSIONAL DAMPER AND CRANKCASE COVER INSTALLATION (Con't).

7. Aline scribe marks on fuel injector pump flange (19) and rear of crankcase cover (4). Install 3 nuts (14) on fuel injector pump (13) at rear of crankcase cover. Tighten nuts to 30 lb.-ft. (41 N.m).

8. Position fuel injector pump driven gear (18) at fuel injector pump drive gear (20) with timing marks (21) alined. Install fuel injector pump driven gear with 3 bolts (22). Tighten bolts to 20 lb.-ft. (27 N.m).

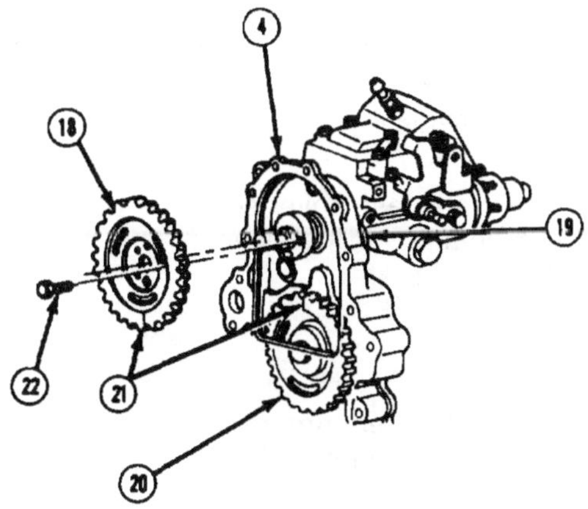

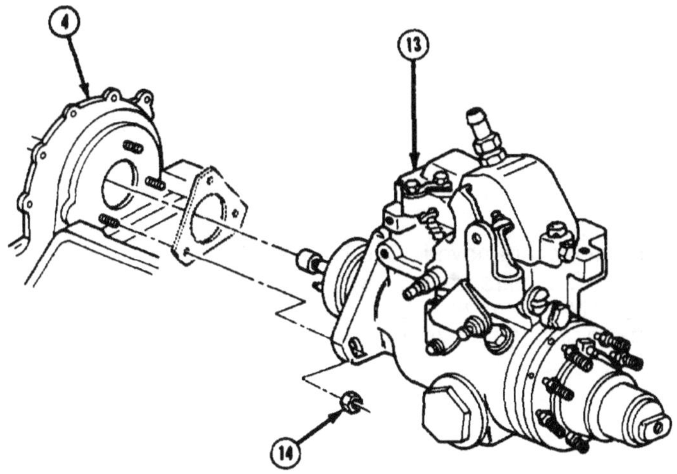

TA50679

3-22 TORSIONAL DAMPER AND CRANKCASE COVER INSTALLATION (Con't).

CAUTION

Be careful not to scratch crankshaft (25) or crankcase cover (4) when installing crankcase cover seal (7). Failure to follow this caution may result In oil leaking.

9. Lubricate new crankcase cover seal (7) with lubricating oil, Install new crankcase cover seal,

NOTE

Ensure that key (24) remains In place while torsional damper (23) is being Installed.

10. Aline torsional damper (23) with key (24) on crankshaft (25), Using block of wood and hammer, drive torsional damper in far enough to start bolt (27).

11. Apply lubricating oil to threads of bolt (27). Install washer (26) and bolt. Tighten bolt until torsional damper (23) completely seats. Tighten bolt to 200 lb.-ft. (271 N.m),

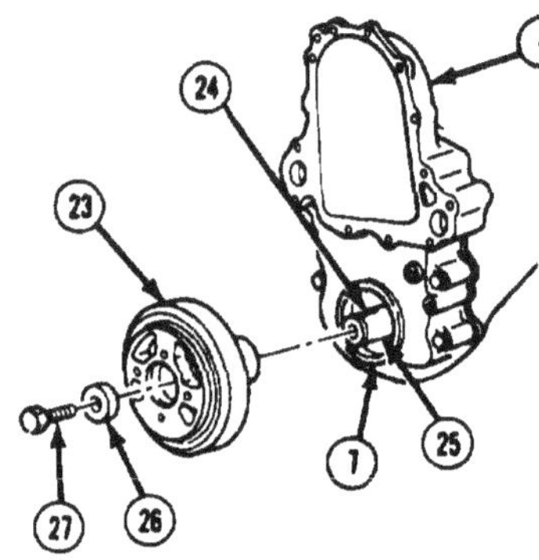

FOLLOW-ON TASKS:

- Install right side alternator bracket if removed. (See TM 9-2320-289-20)
- Install water pump if removed. (See paragraph 3-48)
- Fill engine oil if drained. (See TM 9-2320-289-20)
- Install crankshaft pulley, (See TM 9-2320-289-20)

TA60680

3-23. HYDRAULIC VALVE LIFTER INSTALLATION

This task covers: a. Installation

INITIAL SETUP:

Equipment Condition

- Hydraulic valve lifter removed. (See paragraph 3-9)

Tools/Test Equipment

- Torque wrench

Materials/Parts

- Fuel oil (Item 25, 26, or 27, Appendix B)
- Grease (Item 36, Appendix B)

a. INSTALLATION

CAUTION

New hydraulic valve lifters (2) must be primed with clean diesel fuel oil before installation. Failure to follow this caution may result in damage to hydraulic valve lifters.

NOTE

Perform step 1 only if new hydraulic valve lifter (2) will be installed.

1. Prime hydraulic valve lifter (2) by submerging in clean diesel fuel oil and manually operating hydraulic valve lifter plunger, Lubricate hydraulic valve lifter roller and needle bearings with grease.
2. Install hydraulic valve lifter (2) in proper engine block (1) location as tagged during removal.
3. Install guide plate (3), ensuring that guide plate fits over and does not bind on hydraulic valve lifters (2). Install clamp (4) and tighten bolt (5) to 20 lb.-ft, (27 N.m).

TA50681

3-23. HYDRAULIC VALVE LIFTER INSTALLATION.

4. Rotate torsional damper (6) clockwise 2 complete turns while checking for free movement of hydraulic valve lifters (2). If torsional damper will not turn easily, check all hydraulic valve lifters to ensure proper alinement with guide plates (3). Aline hydraulic valve lifters with guide plates and repeat this step.

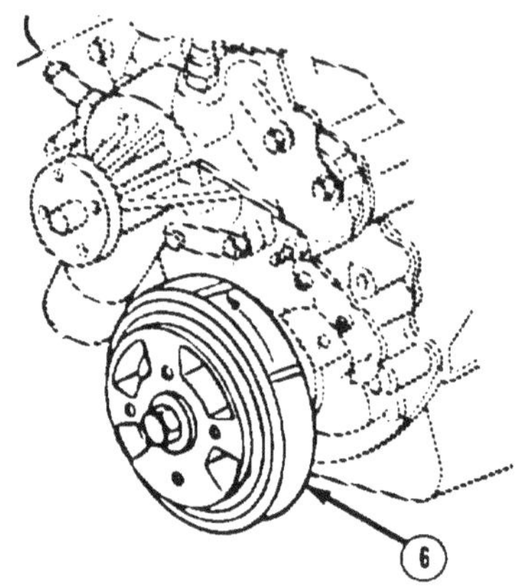

FOLLOW-ON TASKS:

- Install cylinder head, (See paragraph 3-25)

TA50682

3-24. VALVE INSTALLATION.

This task covers: a. Installation

INITIAL SETUP:

Equipment Condition

- Valve removed, (See paragraph 3-8)

Tools/Test Equipment

- Valve spring compressor

Materials/Parts

- One oil seal
- Valve spring shims (as needed)

General Safety Instructions

- Always wear goggles when using valve spring compressor,

a. INSTALLATION

NOTE

Appropriate oversize valves (1 and 2) should be used for any repaired valve guides (3).

1. Install intake valve (1) or exhaust valve (2) in proper valve guide (3) as tagged during removal. Install valve spring shim (4) on valve spring seat. Install valve spring (5) with damper (6) and shield (7).

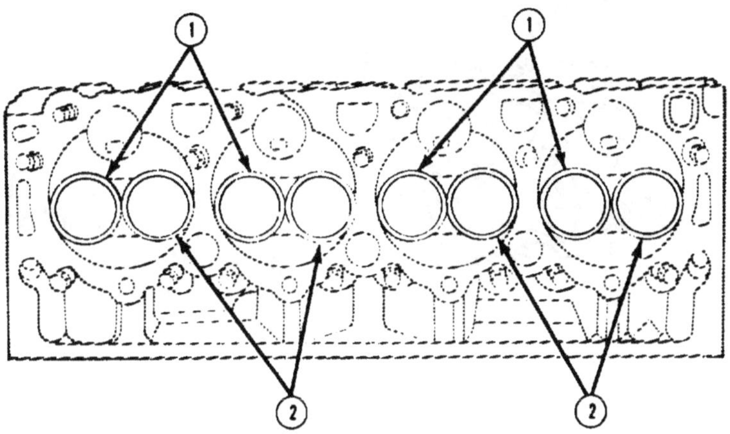

TA50683

3-24. VALVE INSTALLATION (Con't).

WARNING

Always wear goggles while using valve spring compressor. Valve springs (5) are tightly compressed. Failure to follow this warning may result in serious injury to personnel.

NOTE

Components of exhaust valve (2) assemblies and intake valve (1) assemblies are similar, except that exhaust valve uses rotater (1 O) and intake valve uses cap (1 O).

2. Install rotator (1 O) or cap (1 O). Compress valve spring (5). Install new oil seal (8) in lower groove on valve (11) stem. Install 2 keys (9) in grooves in valve stem. Ensure that keys are properly seated, then release valve spring.

3. Ensure that height from top of valve spring shim (4) to top of valve spring (5) is no more than 1.81 in. (46 mm). If height is more than 1.81 in. (46 mm), remove valve assembly (see paragraph 3-8), repeat INSTALLATION, steps 1 and 2 using second valve spring shim, and measure clearance again.

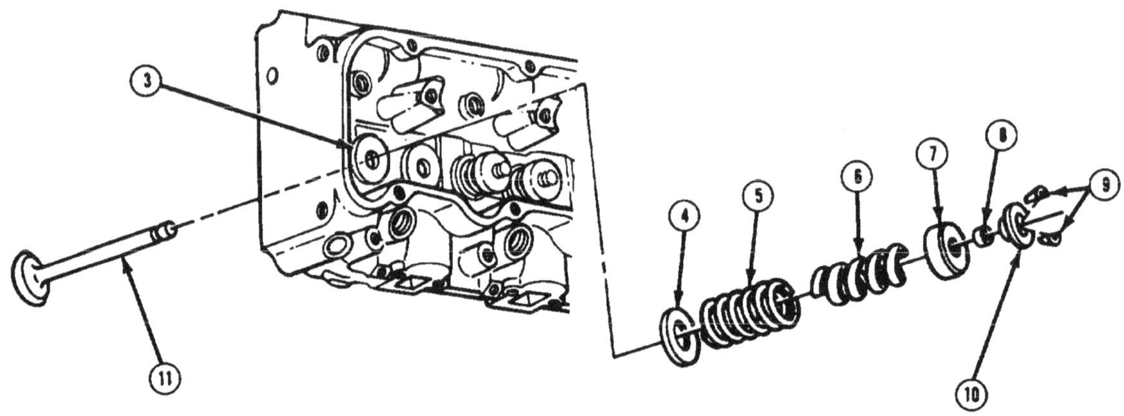

FOLLOW-ON TASKS:

- Install cylinder head. (See paragraph 3-25)

TA50684

3-25. CYLINDER HEAD INSTALLATION.

This task covers: a. Installation

INITIAL SETUP:

Equipment Condition

- Cylinder head removed. (See paragraph 3-7)

Materials/Parts

- One cylinder head gasket
- One lockwasher
- Gasket sealing compound (Item 58, Appendix B)

Tools/Test Equipment

- Torque wrench

a. INSTALLATION

NOTE

If engine is out of truck, perform steps 1-3.

1. Install dowel pins (5) if removed.

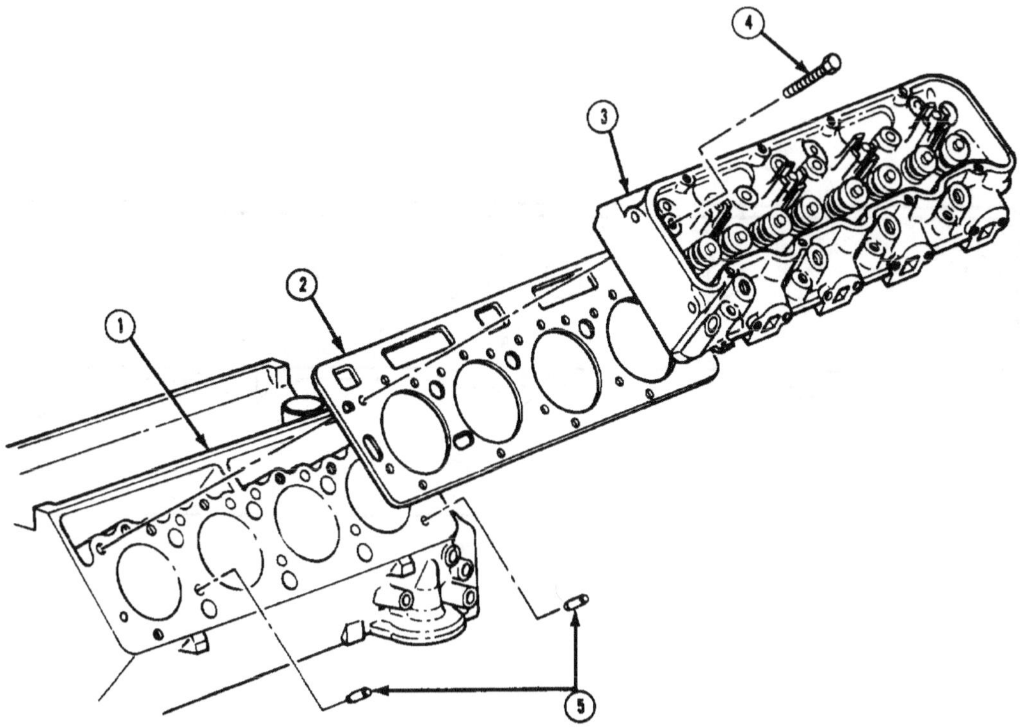

TA50685

3-25. CYLINDER HEAD INSTALLATION (Con't).

CAUTION

DO NOT use any kind of sealant on cylinder head gasket (2). Sealant can cause leaks and result in damage to engine.

NOTE

Ensure that cylinder head (3) is installed before sealing compound dries.

2. Apply sealing compound to threads and underside of bolt (4) heads before installing each bolt.

NOTE

An assistant will be required for step 3.

3. Install new cylinder head gasket (2) to engine block (1) over dowel pins (5). Install rear bolt (4) in cylinder head (3). install cylinder head with remaining 16 bolts. Tighten bolts to 20 lb.-ft. (27 N.m) in alphabetical sequence shown. Tighten bolts again to 50 lb.-ft. (68 N.m) in alphabetical sequence shown. Tighten bolts again an additional 1/4 turn in alphabetical sequence shown.

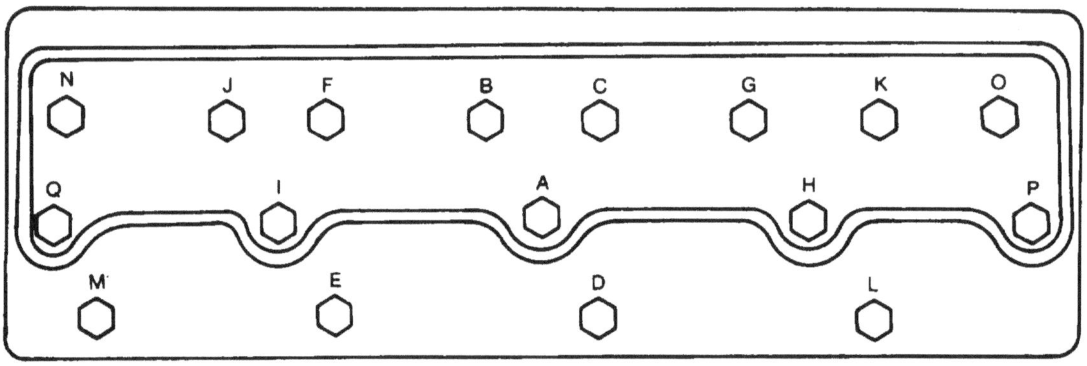

TA50686

3-25. CYLINDER HEAD INSTALLATION (Con't).

4. Install new lockwasher, ground strap, and leads on engine stud with nut. Connect glow plug wiring harness (7) to temperature sensor (6). Connect glow plug leads.

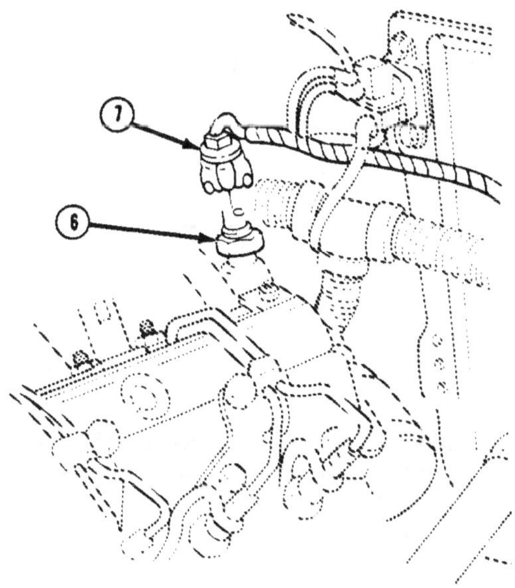

FOLLOW-ON TASKS:

- Install pushrods, rocker arm assemblies, and rocker arm covers. (See paragraph 3-26)
- Install thermostat and crossover pipe. (See TM 9-2320-289-20)
- Connect radiator, bypass, and heater hoses (if being performed in truck). (See TM 9-2320-289-20)
- Install exhaust manifold. (See TM 9-2320-289-20)
- Install right side alternator or MI 010 alternator bracket if removed. (See TM 9-2320-289-20)
- Install left side alternator bracket if removed. (See TM 9-2320-289-20)
- Install air conditioner compressor brackets if removed. (See paragraph 10-24)
- Install oil level indicator tube if removed. (See TM 9-2320-289-20)

TA50687

3-26. ROCKER ARM AND ROCKER ARM COVER INSTALLATION.

This task covers: a. Installation

INITIAL SETUP:

Equipment Condition

- Rocker arm and rocker arm cover removed. (See paragraph 3-6)

Too/s/Test Equipment

- Torque wrench

Materials/Parts

- Lubricating oil (Item 49, Appendix B)
- RTV sealant (Item 57, Appendix B)

a. INSTALLATION

CAUTION

Ensure that rocker arm assemblies (1) and pushrods are installed in same position and location as when they were removed. A paint stripe identifies upper end of pushrod. If paint stripe is not visible, use mark made during removal. Failure to follow this caution may result in damage to rocker arm assembly and pushrod.

NOTE

- If installing right rocker arm cover, perform steps 3-6.
- if installing left rocker arm cover, perform steps 3-5.

1. Install 4 pushrods through holes (4) in cylinder head, Lubricate rocker arm assembly (1) and position, ensuring that pushrods seat on rocker arms.

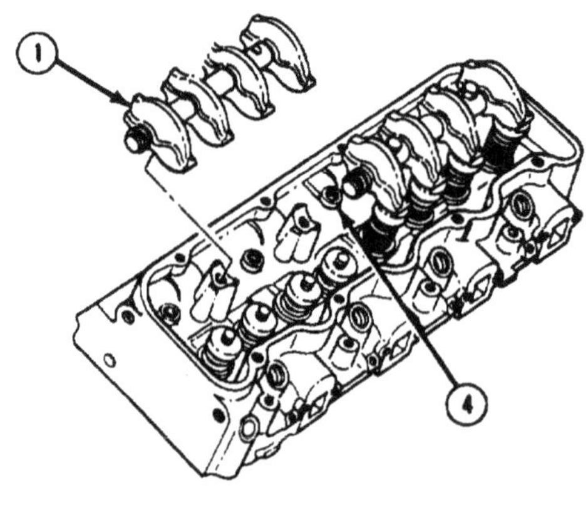

TA50688

3-26. ROCKER ARM AND ROCKER ARM COVER INSTALLATION (Con't).

2. Rotate torsional damper (5) clockwise until mark (8) alines with "O" mark (7) on timing tab. Rotate torsional damper counterclockwise 3.5 in. (88 mm), or until mark is alined with lower water pump bolt (6). Install 2 rocker retainers (2) and bolts (3). Tighten bolts to 40 lb.-ft. (54 N.m).

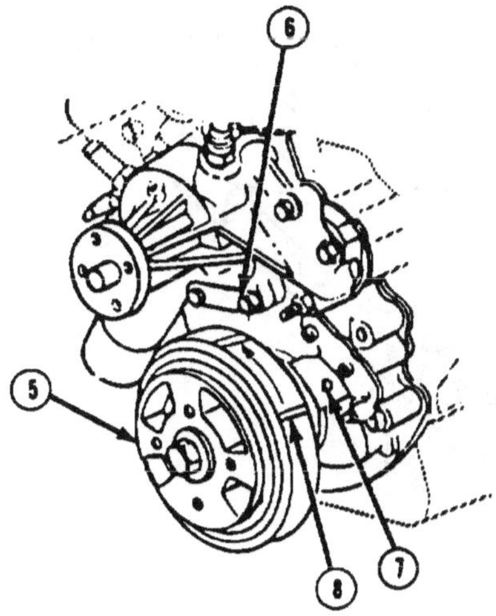

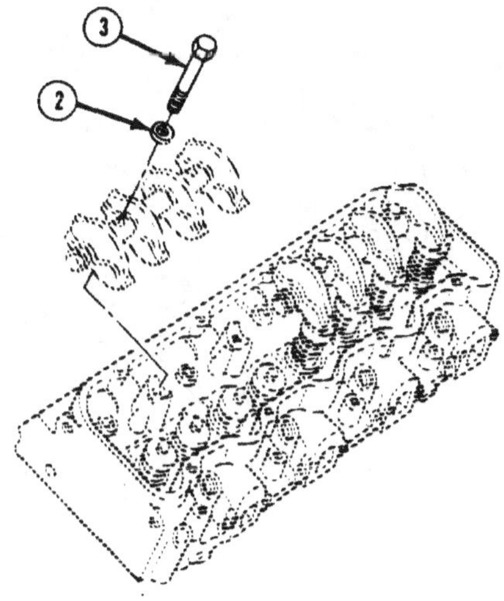

3. Apply a 3/8 in. (5 mm) bead of RTV sealant to rocker arm cover (13 or 14) around inside edge of bolt holes. Install right rocker arm cover (13) with 2 bolts (12) and 6 studs (11). Install left rocker arm cover (14) with 5 bolts (15) and 3 studs (16). Tighten bolts and studs to 15 lb.-ft. (20 N.m).

4. nstall fuel injector pipe bracket (9) with 2 nuts (10).

5. If installing left rocker arm cover (14), install wiring harness bracket. Connect wiring harness to wiring harness clip.

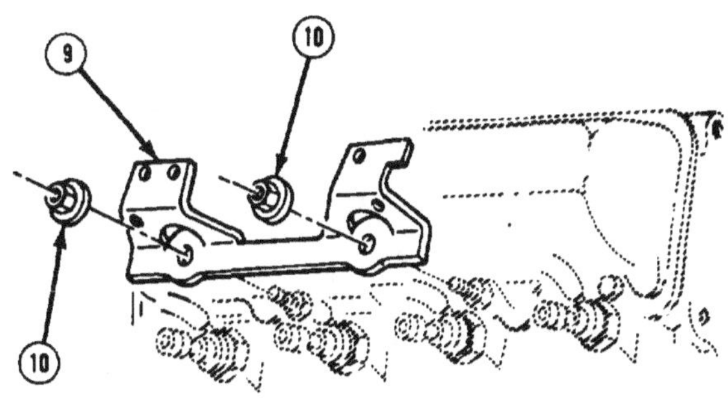

TA60689

3-26. ROCKER ARM AND ROCKER ARM COVER INSTALLATION (Con't),

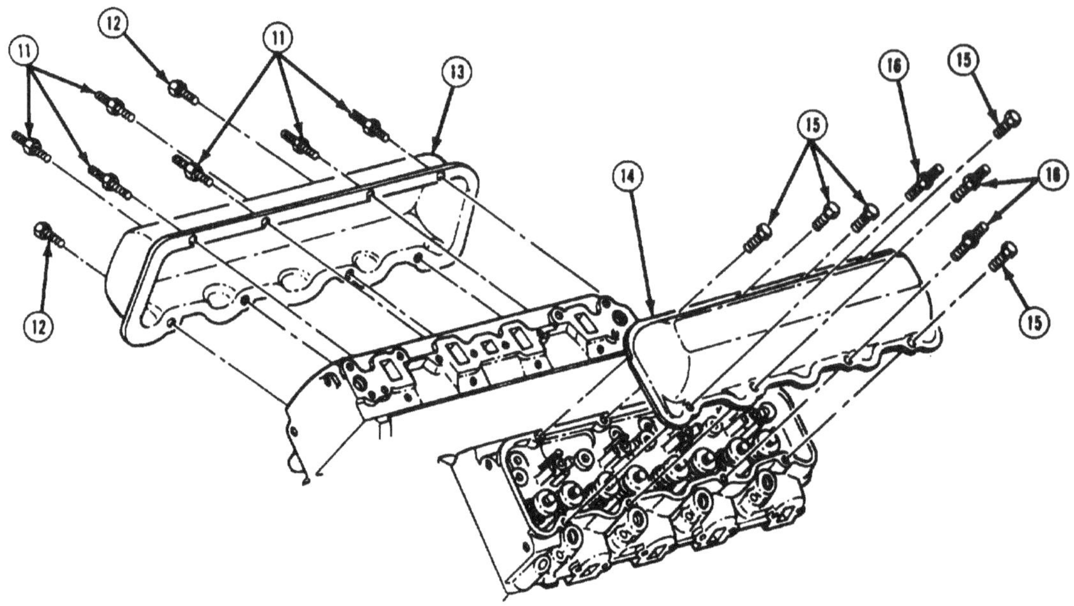

6. Connect 2 vacuum line clamps (17) to studs (11) securing right rocker arm cover (13) and install 2 nuts (18).

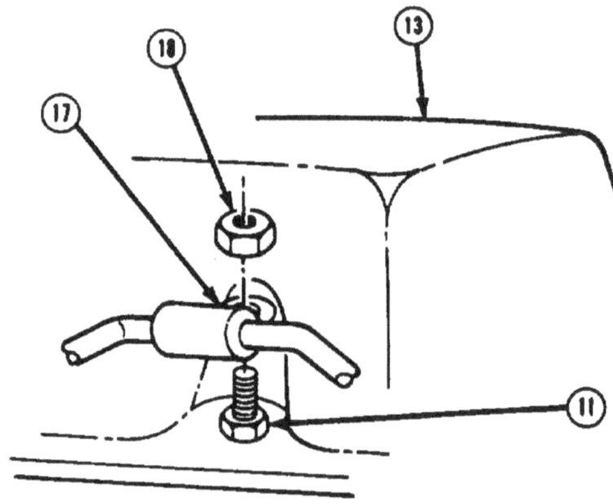

TA50690

3-26. ROCKER ARM AND ROCKER ARM COVER INSTALLATION (Con't).

FOLLOW-ON TASKS:

- Install fuel injector pipes, (See paragraph 3-27)

3-27. FUEL INJECTOR PIPES AND NOZZLES INSTALLATION.

This task covers: a. Installation

INITIAL SETUP:

Equipment Condition

- Fuel injector pipes and nozzles removed, (See paragraph 3-5)

Tools/Test Equipment

- Crowfoot attachment, J-29698-A
- Socket wrench adapter, J-29873
- Torque wrench

Materials/Parts

- Fuel injector nozzle gaskets (as needed)

a. INSTALLATION

NOTE

- Ensure that all plugs are removed before connection,
- If installing fuel Injector nozzles (1), perform steps 1, 2, 4, 5, and 7,
- If installing fuel injector pipes, perform steps 3-5, and 7.

1. Using socket wrench adapter, install new fuel injector nozzle gaskets (2) and fuel injector nozzles (1) in cylinder heads (3). Tighten fuel injector nozzle to 50 lb. -ft. (68 N.m).
2. Install 2 caps (7) and clamps on rear fuel injector nozzles (1). Connect 8 fuel drainback hoses (6) and clamps to fuel injector nozzles. Install clip (4) with nut (5).

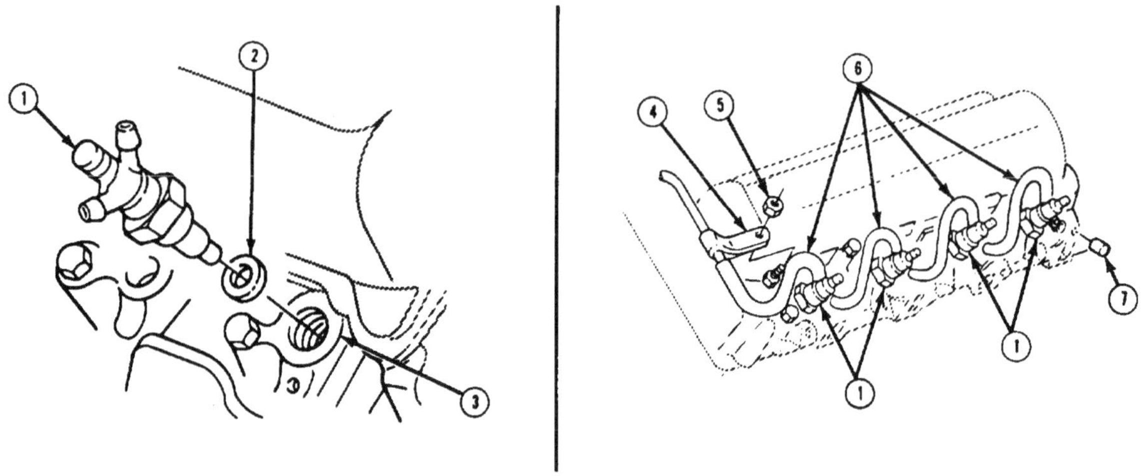

TA50691

3-27. FUEL INJECTOR PIPES AND NOZZLES INSTALLATION (Con't).

3. Install 4 inserts (11), clamps (1 2), and screws (13) on fuel injector pipes (9). Install 4 inserts and lower half clips (1 O) on fuel injector pipes. Bend lower half clips straight up to allow clearance for intake manifold.

4. Connect 8 fuel injector pipes (9) to fuel injector pump (8) connectors. Tighten fittings to 20 lb.-ft, (27 N.m).

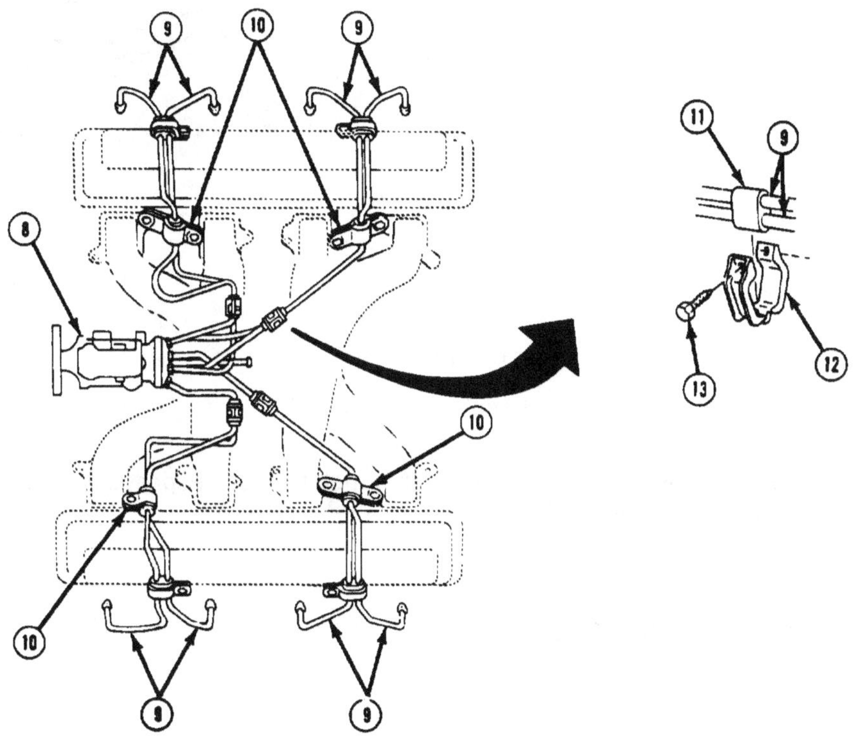

TA50692

3-27. FUEL INJECTOR PIPES AND NOZZLES INSTALLATION (Con't).

5. Install 2 fuel injector pipe brackets (17) on cylinder head studs (18) with 4 nuts (19) if removed. Loosely connect 8 fuel injector pipes (9) to 8 fuel injector nozzles (1).

6. Install 2 inserts (16) and clips (15) on fuel injector pipes (9) on one side. Install clips on fuel injector pipe bracket (17) with 2 screws (14) on one side. Repeat step for other side.

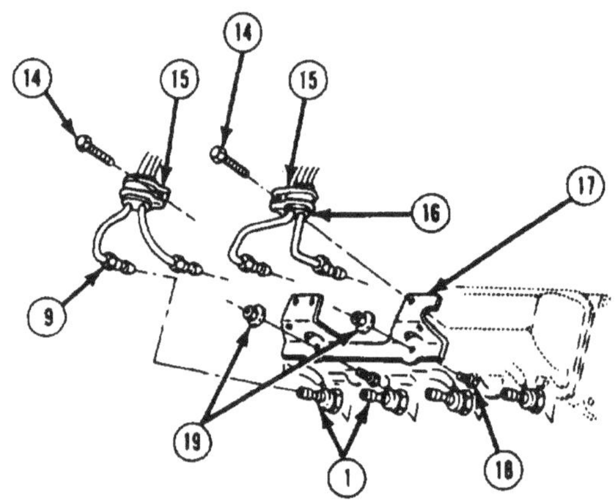

NOTE

Perform step 7 if engine is in truck.

7. Crank engine for 10-15 seconds, then wait 1 minute for starter motor to cool. Repeat until fuel is observed at all fittings. Tighten fittings to 20 lb. -ft. (27 N.m).

FOLLOW-ON TASKS:

- Install intake manifold. (See paragraph 3-28)

TA50693

3-28. INTAKE MANIFOLD INSTALLATION.

This task covers: a. Installation

INITIAL SETUP:

Equipment Condition

- Intake manifold removed. (See paragraph 3-4)

Too/s/Test Equipment

- Torque wrench

Materials/Parts

- Two gaskets

a. INSTALLATION

1. Remove manifold covers from cylinder head (8) openings.
2. Install 2 new gaskets (7) on cylinder head openings, ensuring that holes are alined.

NOTE

- It may be necessary to loosen fuel injector pipes to properly position fuel injector pipe clips (6). Tighten fuel injector pipe fittings to 20 lb.-ft. (27 N.m) if this Is done,
- Intake manifold (1) has cylinder number printed on Intake manifold arms, Ensure that intake manifold Is installed with odd numbers on left side.

3. Install intake manifold (1). Bend lower half fuel injector pipe clips (6) back into position over intake manifold holes. Install upper half fuel injector pipe clips.
4. Install long bolt (4), 7 short bolts (2), 4 long studs (3), and 4 short studs (5). Tighten bolts and studs to 30 lb.-ft. (41 N.m) in alphabetical sequence shown,

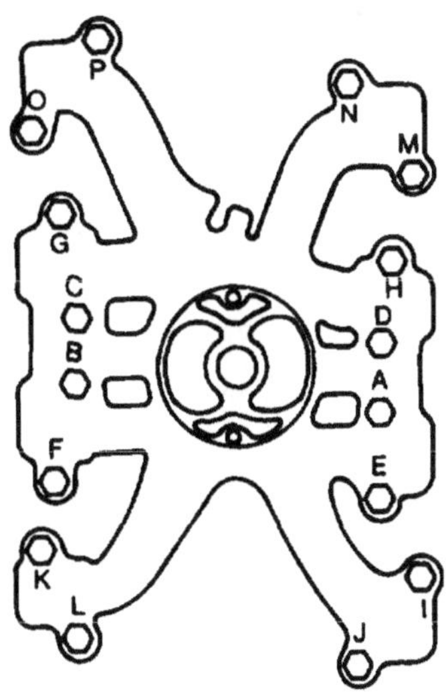

TA60694

3-28. NTAKE MANIFOLD INSTALLATION (Con't).

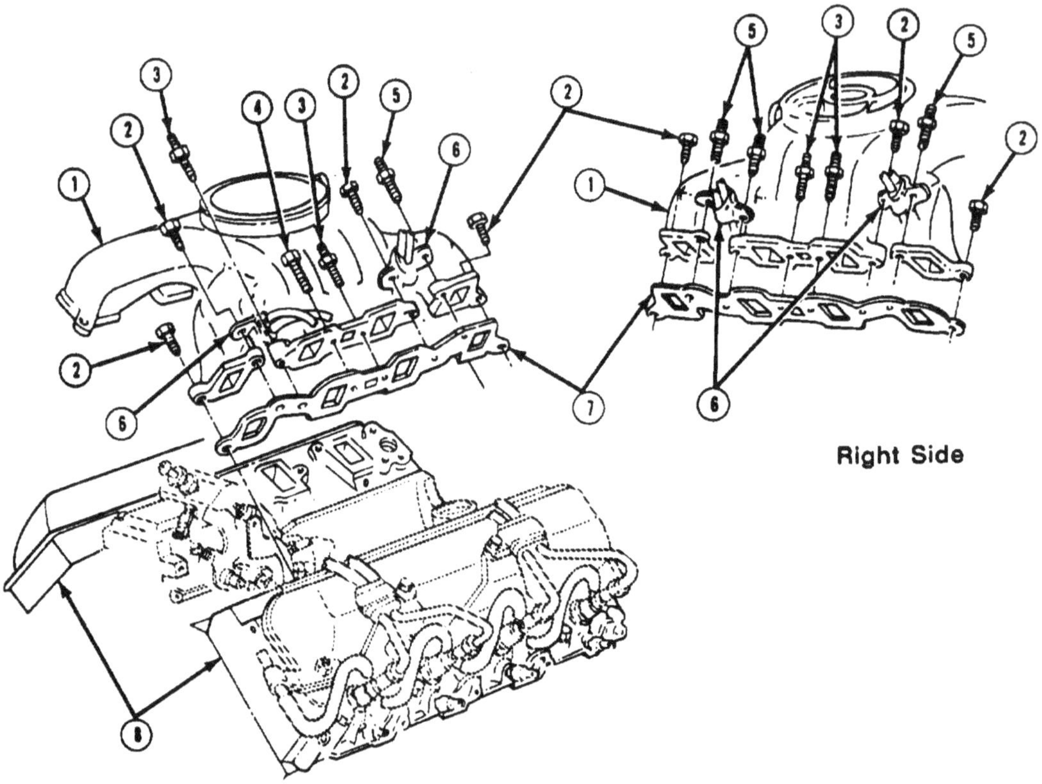

TA50695

3-28. INTAKE MANIFOLD INSTALLATION (Con't).

5. Rotate vacuum pump (10) into position and tighten bolt (9) to 240 lb.-in, (27 N.m).

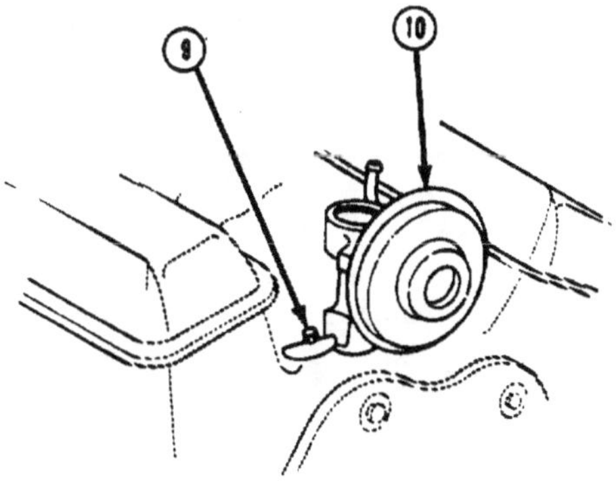

NOTE

Step 6 applies to all except M1010.

6. Instail brace (13) on left side alternator bracket (11), and long stud (3) with bolt (12) and nut (14).

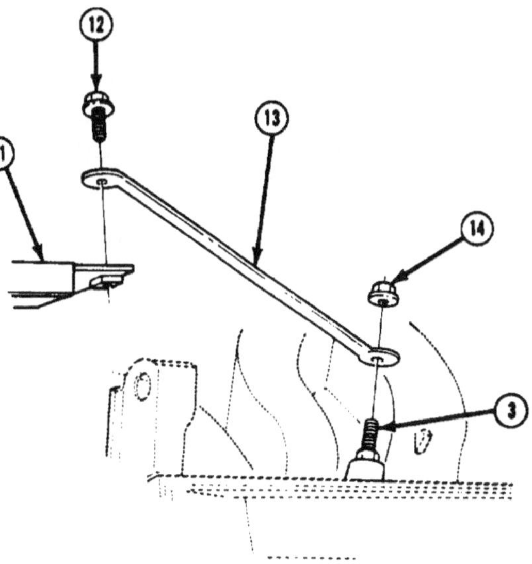

TA50696

3-28. INTAKE MANIFOLD INSTALLATION (Con't).

FOLLOW-ON TASKS:

- Install crankcase depression regulator valve assembly and hoses (if being performed in truck). (See TM 9-2320-289-20)
- Connect both battery negative cables (if being performed in truck), (See TM 9-2320-289-20)

3-29. ENGINE AND ENGINE MOUNT INSTALLATION.

This task covers: a. Installation

INITIAL SETUP:

Equipment Condition

- Engine or engine mounts removed. (See paragraph 3-3)

Tools/Test Equipment

- Torque wrench

Materials/Parts

- Eight locknuts

PersonneL Required

- MOS 63W (2)

General Safety Instructions

- DO NOT place hands between engine and transmission when installing engine.

a. INSTALLATION

WARNING

DO NOT place hands between engine (4) and transmission when installing engine. Failure to follow this caution may result In injury to personnel.

CAUTION

Ensure that wiring, pipes, and hoses are clear of engine (4) during Installation. Failure to follow this caution may result In damage to wiring, pipes, and hoses.

NOTE

- An assistant will guide engine (4) into engine compartment as it is being lowered. Guid e exhaust manifold studs over exhaust pipes.
- If only installing engine mount (6) or engine mount brackets (2), perform steps 4 and 5.

1. Install suitable lifting device on engine lift brackets at left front and right rear of engine (4). Install engine in engine compartment.

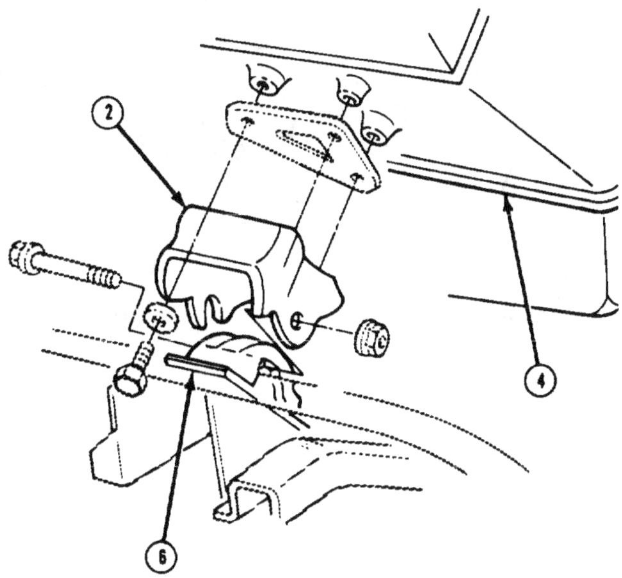

TA50697

3-29. ENGINE AND ENGINE MOUNT INSTALLATION (Con't).

2. Aline engine locating pins (15) to transmission holes (14). Install 6 transmission bolts (9).

3. Remove support at transmission oil pan.

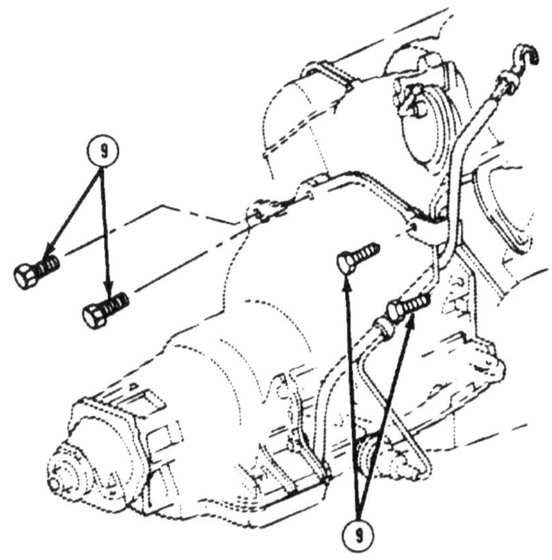

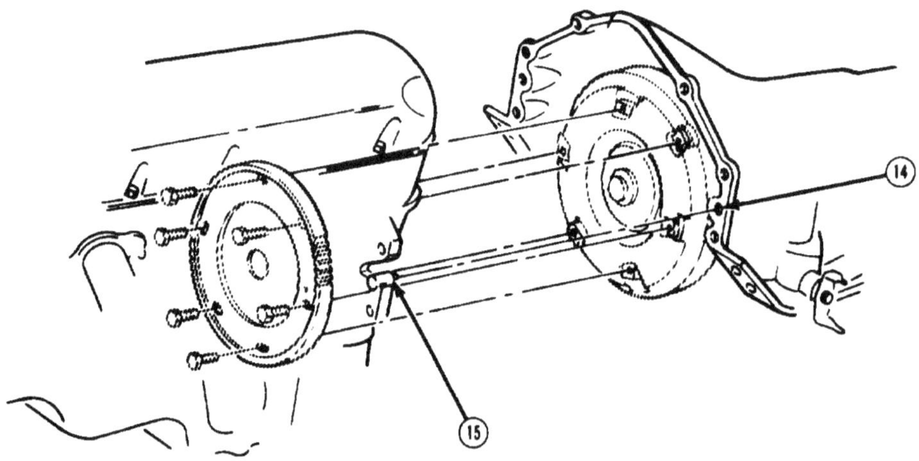

TA50698

3-29. ENGINE AND ENGINE MOUNT INSTALLATION (Con't).

4. Install new engine mount (6) with 3 washers, bolts, and new locknuts if removed. Install spacer (3) and engine mount bracket (2) with 3 washers (8) and bolts (7) if removed, Tighten engine mount locknuts to 30 lb.-ft. (41 N.m). Tighten engine mount bracket bolts to 35 lb.-ft. (47 N.m). Lower side of engine (4) and remove jack. Repeat for other side.

5. Install screw (1) through left side engine mount bracket (2) and engine mount (6). Install new locknut (5). Tighten locknut to 55 lb.-ft. (75 Nom). Repeat this step for right side. Remove engine lifting device.

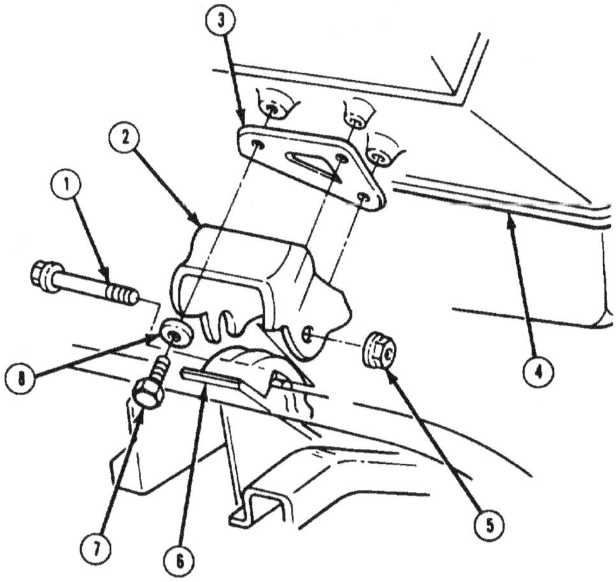

CAUTION

Ensure that torque converter (13) is flush against flywheel (12) and able to rotate freely by hand . Failure to follow this caution will result in damage to engine (4) and transmission.

6. Ailne marks made during removal on flywheel (12) and torque converter (13). Rotate flywheel and install 6 bolts (11) to fiywheel and torque converter finger tight only. Once all bolts are installed, tighten to 50 ib.-ft. (68 N.m).

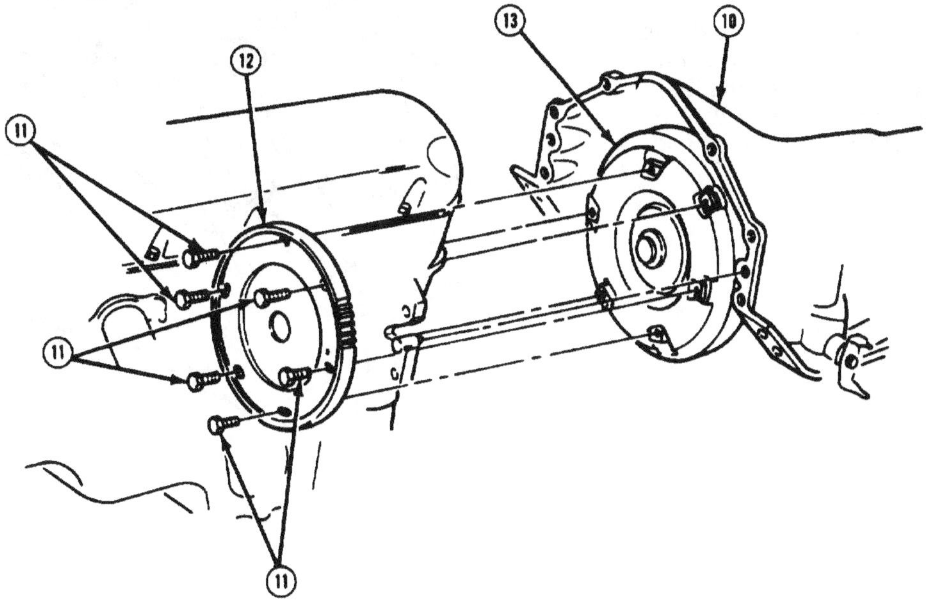

'TA50699

3-29. ENGINE AND ENGINE MOUNT INSTALLATION (Con't).

7. Hook torque converter cover (17) under lip of engine oil pan and install to transmission assembly (1 O) with 6 bolts (16) .

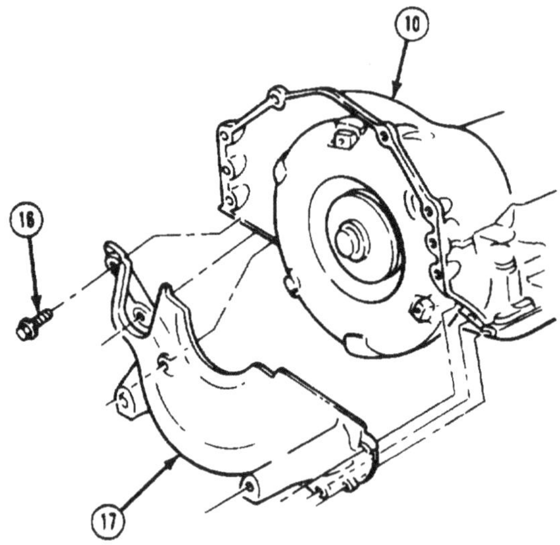

8. Connect transmission oil cooler pipes (18) at engine oil pan clip (23). Connect positive starter cable to other engine oil pan clip.

9. Connect exhaust pipe to exhaust manifold (19) and install 3 springs (20), washers (21), and nuts (22). Tighten nuts to 180 lb.-in. (20 N.m). Repeat for other exhaust manifold.

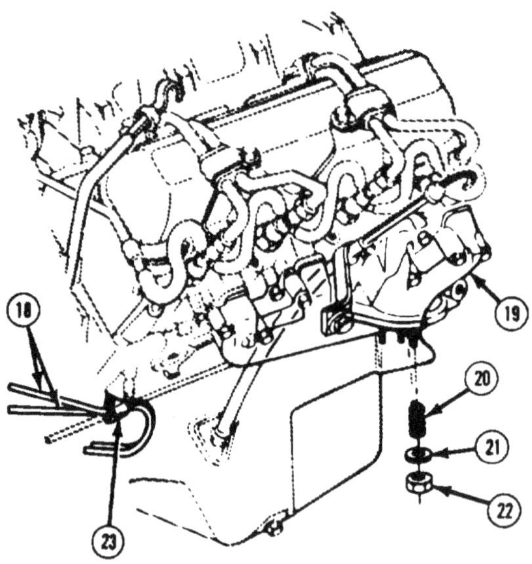

TA50700

3-29. ENGINE AND ENGINE MOUNT INSTALLATION (Con't).

10. Install heater inlet hose (24) on crossover pipe (26) and tighten clamp (25),

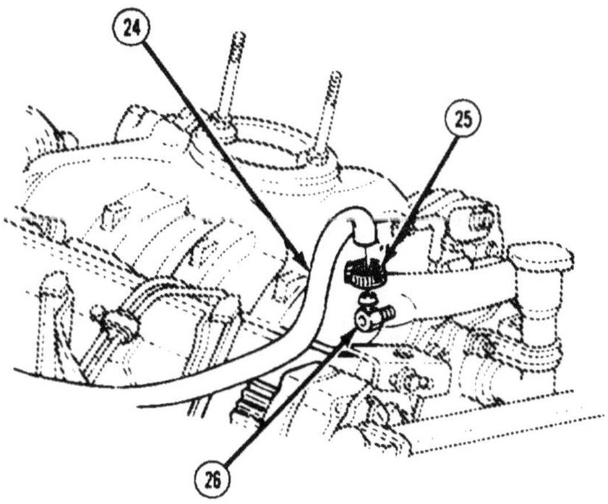

11. Connect accelerator cable (27) to support (28). Install accelerator cable with accelerator cable retainer.

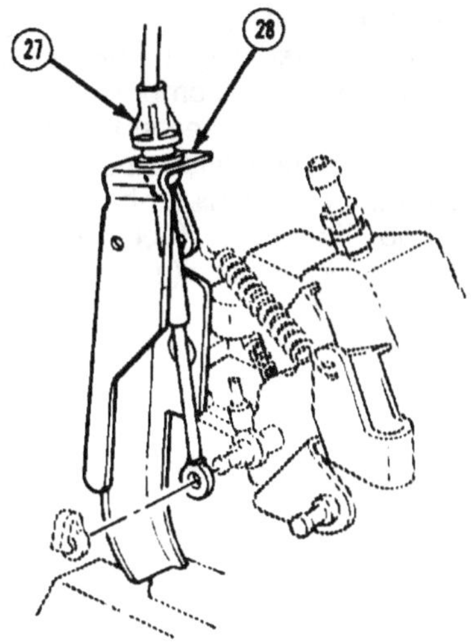

TA50701

3-29. ENGINE AND ENGINE MOUNT INSTALLATION (Con't).

12. Connect fuel line (31) to fuel pump (30) and fuel return hose (29).

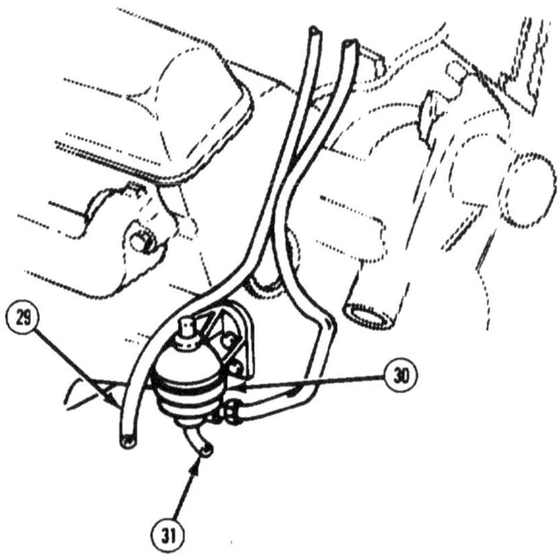

FOLLOW-ON TASKS:

- Connect engine wiring harness if disconnected. (See paragraph 4-1 O)
- Connect engine coolant heater front exhaust pipe and heat exchange pipe to engine oil pan if disconnected. (See TM 9-2320-289-20)
- Connect engine oil cooler lines to engine if disconnected. (See TM 9-2320-289-20)
- Install power steering pump on engine if removed. (See TM 9-2320-289-20)
- Install air conditioner compressor on brackets if removed (Ml 01 O). (See paragraph 10-24)
- Install engine coolant heater hoses if removed. (See TM 9-2320-289-20)
- Install vacuum modulator pipe if removed. (See paragraph 5-3)
- Install radiator and radiator hoses if removed. (See TM 9-2320-289-20)
- Install hood if removed. (See TM 9-2320-289-20)
- Bleed fuel injector pipes (if engine was removed). (See paragraph 3-27, step 7)

TA50702

3-30. ENGINE ASSEMBLY REPLACEMENT.

This task covers: a. Removal b. Installation

INITIAL SETUP:

Equipment Condition

- Engine or engine mounts removed. (See paragraph 3-3)

Manual References

- TM 9-2320-289-20

a. REMOVA L

1. Remove alternators and brackets. (See TM 9-2320-289-20)

NOTE

Step 2 only applies to M1010.

2. Remove air conditionecompressor brackets. (see TM9-2320289-20

NOTE

Step 3 only applies to trucks equipped with winterization kits.

3. Remove, clean, and inspect winterization kit engine oil pan. (See paragraph 3-23)

b. INSTALLATION

NOTE

Step 1 only applies to trucks equipped with winterization kits,

1. Install winterization kit engine oil pan if necessary. (See paragraph 3-32)

NOTE

Step 2 only applies to M1010

2. Install air conditioner compressor brackets. (See paragraph 10-24)
3. Install alternators and brackets. (See TM 9-2320-289-20)

FOLLOW-ON TASKS:

- Install engine and engine mounts. (See paragraph 3-29)

Section II. LUBRICATION SYSTEM MAINTENANCE

3-31. LUBRICATION SYSTEM MAINTENANCE INDEX.

Pages 3-100 through 3-105 have been rescinded.

3-32. ENGINE OIL PAN REPLACEMENT (Con't).

2. Remove 6bolts (6) and torqu e converte cover (8) fro m transmission assembly (7) .

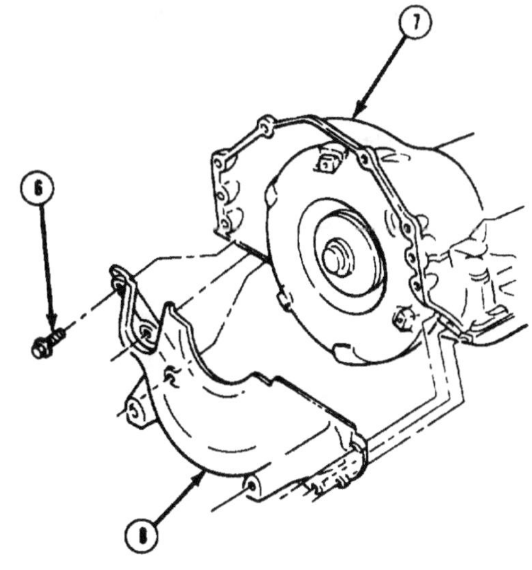

NOTE

Ensure that location of transmission oil cooler line clip is noted for installation.

3. Remove 2 1 bolts (12) and transmission oil cooler line clip. Remove 2 rear bolts (11).

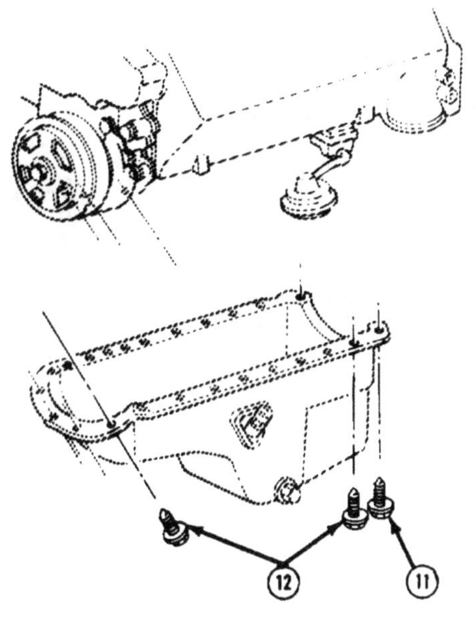

TA50704

3-32. ENGINE OIL PAN REPLACEMENT (Con't).

NOTE

Ensure that location of engine oil pan stud (13) is noted for installation.

4 . Remove nut (15) and starter lead clip (14) from engine oil pan stud (13). Remove engine oil pan stud.

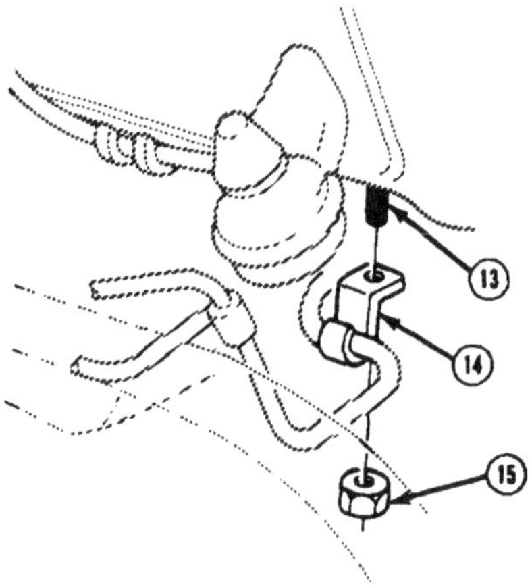

NOTE

Early 1984 model trucks have engine oil pan (5) with a ridged lip which requires an engine oil pan gasket. Late 1984-87 model truck engine oil pans and replacement engine oil pans have a flat lip which requires RTV sealant.

5 . Remove engine oil pan (5) and engine oil pan seal (1 O). Remove and discard engine oil pan gasket if present. Discard engine oil pan seal.

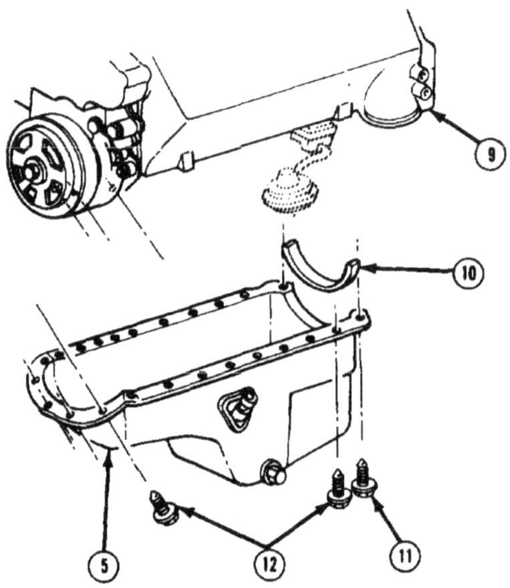

TA50705

3-32. ENGINE OIL PAN REPLACEMENT (Con't).

b. CLEANING AND INSPECTION

WARNING

Dry cleaning solvent P-D-680 is toxic and flammable. Always wear protective goggles and gloves and use only in a well-ventilated area. Avoid contact with skin, eyes, and clothes and DO NOT breathe vapors. DO NOT use near open flame or excessive heat. The solvent's flash point is 100°F-138°F (38°C-59°C). If you become dizzy while using cleaning solvent, immediately get fresh air and medical help. If solvent contacts eyes, immediately wash your eyes with water and get medical aid.

1. Clean oil and all RTV sealant or gasket material from mating surfaces of engine oil pan (5) and engine (9) block with dry cleaning solvent.
2. Inspect engine oil pan (5) for damage. Discard engine oil pan if damaged.

c. INSTALLATION

NOTE

If engine is out of truck, perform steps 1-5.

1. Install new engine oil pan seal (1 O) in groove at rear of engine (9) block. Apply RTV sealant to each end of engine oil pan seal.

NOTE

- If installing new engine oil pan gasket with engine oil pan (5), perform step 2.
- If applying RTV sealant or installing new engine oil pan (5), perform step 3.

2. Install new engine oil pan gasket on lip of engine oil pan (5) and aline with bolt holes.
3. Apply a 0,19 in. (5 mm) bead of RTV sealant (16) along engine oil pan (5) sealing surface as shown.
4. Position engine oil pan (5) under engine (9) block and install engine oil pan stud (13). Install starter lead clip (14) on engine oil pan stud with nut (15).
5. Install 2 rear bolts (11). Install transmission oil cooler line clip and 21 bolts (12).

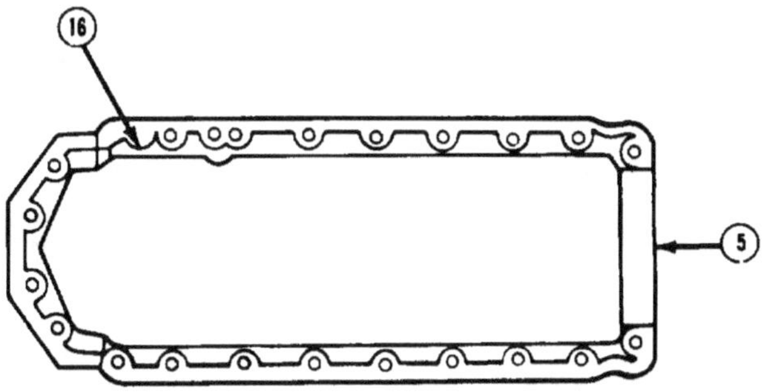

TA50706

3-32. ENGINE OIL PAN REPLACEMENT (Con't).

6. nstall torque converter cover (8) on transmission assembly (7) with 6 bolts (6) ,

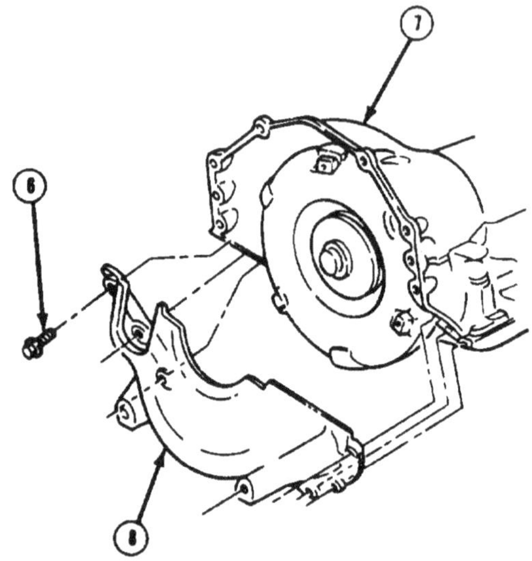

NOTE

Step 7 only applies to trucks equipped with winterization kits.

7. Install heat exchange pipe (2) with 4 screws (1). Connect front exhaust pipe (3) to engine oil pan (5) and tighten clamp (4),

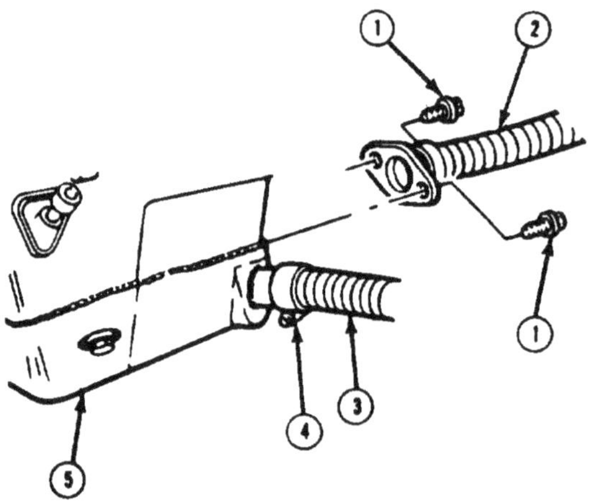

TA50707

3-32. ENGINE OIL PAN REPLACEMENT (Con't).

FOLLOW-ON TASKS:

- Install engine oil filter. (See TM 9-2320-289-20)
- Install engine oil level indicator tube and seal, (See TM 9-2320-289-20)
- Fill engine oil. (See TM 9-2320-289-20)
- Remove jack stands and lower truck, if raised,
- Install starter (if being performed in truck). (See TM 9-2320-289-20)

3-33. ENGINE OIL PUMP MAINTENANCE.

This task covers:

a. Removal
b. Disassembly
c. Cleaning and Inspection
d. Assembly
e. Installation

INITIAL SETUP:

Equipment Condition

- Engine oil pan removed. (See paragraph 3-32)

TooLs/Test Equipment

- Torque wrench

Materials/parts

- Dry cleaning solvent (Item 23, Appendix B)

General Safety Instructions

- Dry cleaning solvent is flammable and must not be used near open flame. Use only in a well-ventilated area.
- Compressed air used for cleaning purposes should never exceed 30 psi (207 kPa).

a. REMOVAL

1. Remove bolt (3) and remove oi pump (4) and shaft (1) from rear main bearing cap under engine block (5).

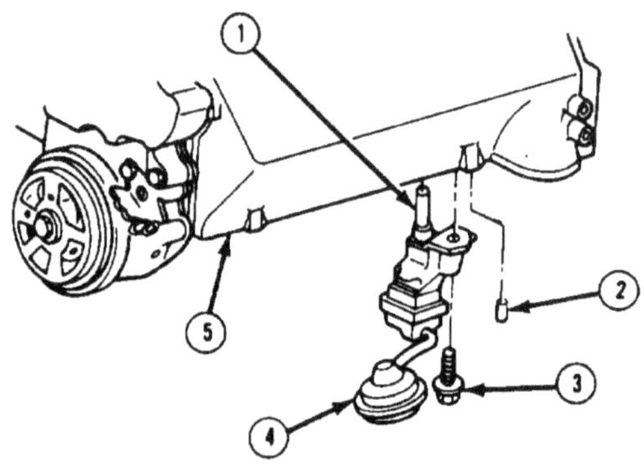

b. DISASSEMBLY

1. Remove 4 cover bolts (9) and separate cover (10) from oil pump body (14).

2. Remove pin (13), spring (1 2), and oil pressure regulator valve (11). Remove oil plug (7) if damaged,

TA50708

3-33. ENGINE OIL PUMP MAINTENANCE.

This task covers:

a. Removal
b. Disassembly
c. Cleaning and Inspection
d. Assembly
e. Installation

INITIAL SETUP:

Equipment Condition

- Engine oil pan removed. (See TM 9-2320-289-20)

Too/s/Test Equipment

- Torque wrench

Materials/Parts

- Dry cleaning solvent (Item 23, Appendix B)

General Safety Instructions

- Dry cleaning solvent is flammable and must not be used near open flame. Use only in a well-ventilated area.
- Compressed air used for cleaning purposes should never exceed 30 psi (207 kPa).

a. REMOVAL

1. Remove bolt (3) and remove oil pump (4) and shaft (1) from rear main bearing cap under engine block (5).

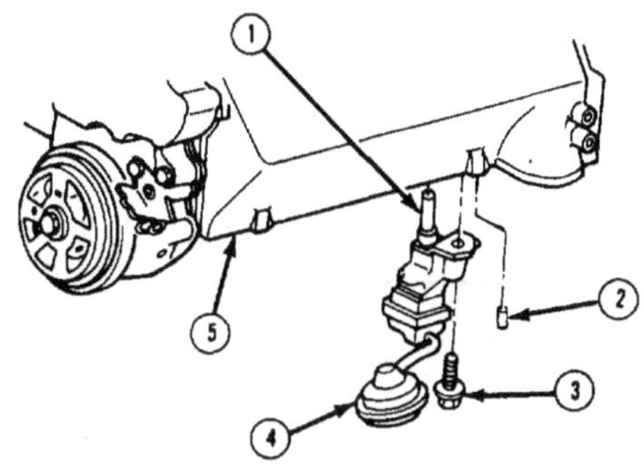

b. DISASSEMBLY

1. Remove 4 cover bolts (9) and separate cover (10) from oil pump body (14).
2. Remove pin (13), spring (1 2), and oil pressure regulator valve (11). Remove oil plug (7) if damaged.

TA50708

3-33. ENGINE OIL PUMP MAINTENANCE (Con't).

3. Remove oil pump screen (8) from cover (10) if damaged.

4. Remove 2 oil pump pins (2) from rear main bearing cap if damaged.

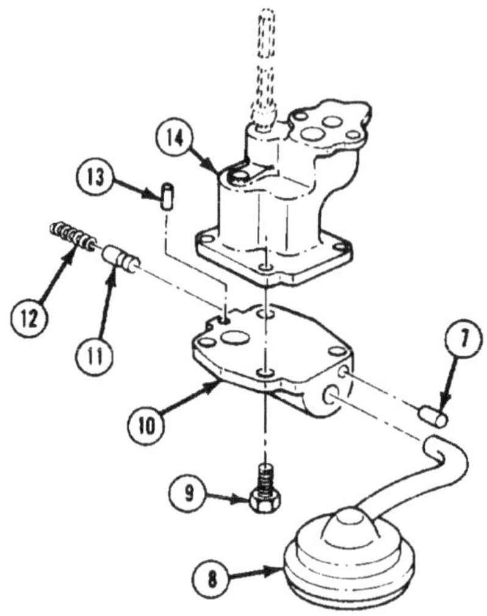

c. CLEANING AND INSPECTION

WARNING

- Dry cleaning solvent P-D-680 is toxic and flammable. Always wear protective goggles and gloves and use only in a well-ventilated area. Avoid contact with skin, eyes, and clothes and DO NOT breathe vapors. DO NOT use near open flame or excessive heat. The solvent's flash point is 100°F-138°F (38°C-59°C). If you become dizzy while using cleaning solvent, immediately get fresh air and medical help. If solvent contacts eyes, immediately wash your eyes with water and get medical aid.
- Compressed air used for cleaning purposes should never exceed 30 psi (207 kPa). Use only effective chip guarding and personnel protective equipment (goggles/shield, gloves, etc.). Failure to follow this warning may result in serious injury to personnel.

1. Clean all components with dry cleaning solvent and dry with compressed air.

2. Inspect oil pump body (14) and cover (10) for cracks, Inspect cover for wear that would permit oil to leak past ends of oil pump gears. If oil pump body is cracked, replace engine oil pump (4) as an assembly. If cover is cracked or worn, replace.

TA50709

3-33. ENGINE OIL PUMP MAINTENANCE (Con't).

3. Check shaft (15) for looseness in oil pump body (14). If loose, replace engine oil pump (4) as an assembly. Check to see if gear teeth on shaft are cracked or worn. If gear teeth are cracked or worn, replace engine oil pump as an assembly,

4. Inspect shaft (1). If worn, remove retainer (6), and replace shaft and retainer.

5. Place oil pressure regulator valve (11) in cover (10) to check for proper fit. If oil pressure regulator valve binds or fits loosely, replace.

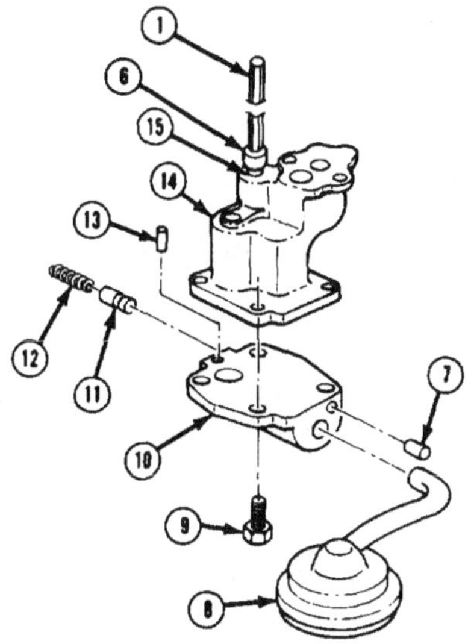

d. ASSEMBLY

1. Install 2 oil pump pins (2) in rear main bearing cap if removed.

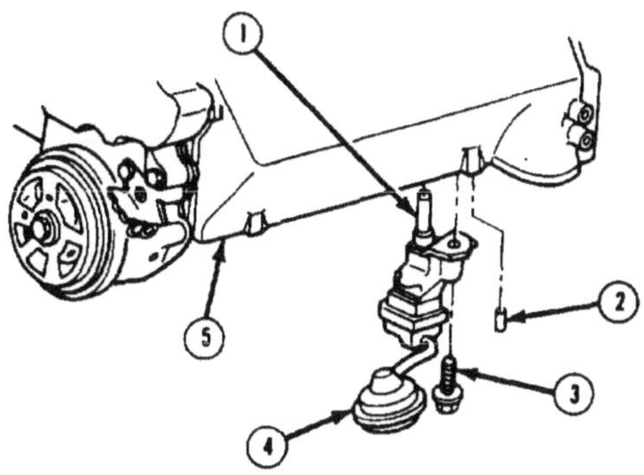

TA50710

3-33. ENGINE OIL PUMP MAINTENANCE (Con't).

2. Install oil plug (7) if removed. Install oil pressure regulator valve (11) and spring (12) in cover (10) and secure with pin (13).

3. Assemble cover (10) to oil pump body (14) with 4 cover bolts (9). Tighten bolts to 20 lb.-ft. (27 N.m),

4. Install oil pump screen (8) in cover (10) if removed,

5. Turn shaft by hand to check for smooth operation.

e. INSTALLATION

1. Aline hex on end of shaft (1) with hex on vacuum pump drive shaft under engine block (5). Install oil pump (4) on rear main bearing cap with bolt (3). Tighten bolt to 75 lb.-ft. (102 N.m).

FOLLOW-ON TASKS:

- Install engine oil pan. (See TM 9-2320-289-20)

3-34. REAR MAIN SEAL REPLACEMENT.

This task covers:

a. Removal
b. Cleaning and Inspection
c. Installation

INITIAL SETUP:

Equipment Condition

- Engine oil pump removed, (See paragraph 3-33)

Tools/Test Equipment

- Torque wrench

Materials/Parts

- One rear main seal
- 0.004 in. (0.102 mm) shim stock
- Carburetor cleaning compound (Item 15, Appendix B)
- Lubricating oil (Item 49, Appendix B)
- Anaerobic sealing compound (Item 59, Appendix B)
- Pipe sealant (Item 61, Appendix B)

General Safety Instructions

- Avoid skin contact with anaerobic sealing compound.
- Carburetor cleaning compound is highly flammable. Keep away from open flame.

a. REMOVAL

1. Remove 2 inner bolts, outer bolts, and rear main bearing cap (1).

2. Remove lower rear main seal half (5) and upper rear main seal half (6). Discard rear main seal halves.

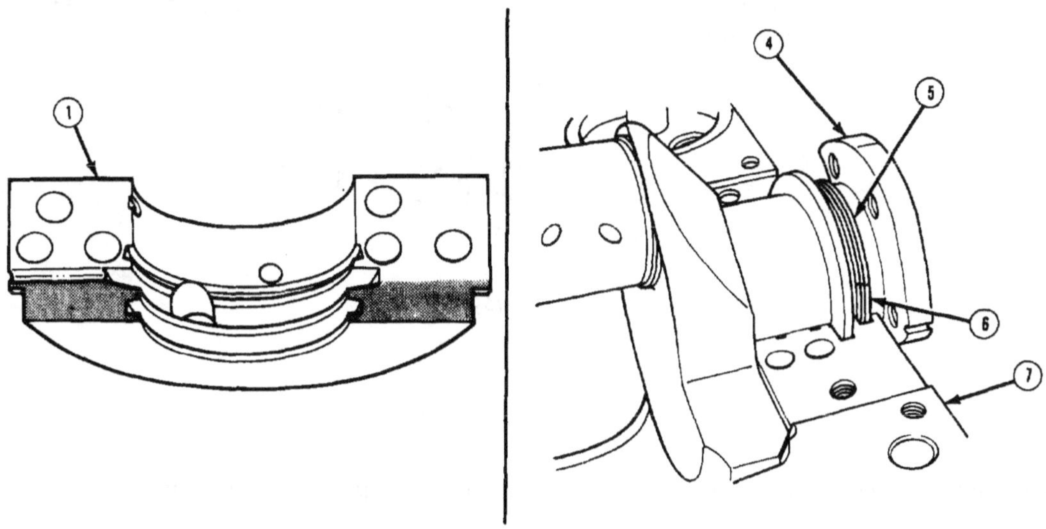

TA50711

3-34. REAR MAIN SEAL REPLACEMENT (Con't).

b. CLEANING AND INSPECTION

1. Wipe oil from rear main seal grooves in engine bloc and rear main bbearing cap (1). Remove all rear main seal material from mating surfaces.

WARNING

Carburetor cleaning compound is highly flammable. Keep away from open flame. Failure to follow this warning may result in serious injury or death to personnel.

2. Using carburetor cleaning compound, clean rear main bearing cap (1), rear main seal grooves, and surface of engine block that mates with rear main bearing cap.

3. Check rear main bearing clearance. (See paragraph 3-1 4)

c. INSTALLATION

1. Loosen inner bolts and outer bolts of all main bearing caps 2 turns and allow crankshaft (4) to drop.

2. Manufacture oil seal installation tool from 0.004 in. (0,102 mm) shim stock as shown.

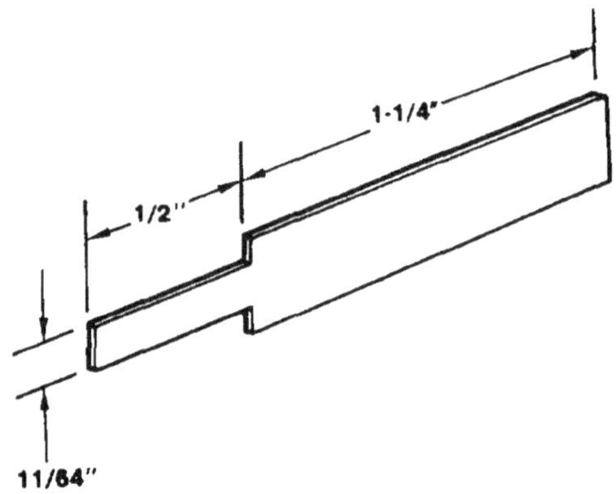

3. Apply lubricating oil to crankshaft (4) journal and rear main seal lips where they contact crankshaft.

NOTE

Ensure that rear main seal halves (5 and 6) are installed with word "OUTSIDE" facing toward rear of engine block (7).

4. Position lip of oil seal installation tool where upper rear main seal half (6) will enter engine block (7).

TA50712

3-34. REAR MAIN SEAL REPLACEMENT (Con't).

5. Place end of upper rear main seal half (6) in engine block (7) groove, using oil seal installation tool as a "shoehorn," and push upper rear main seal half in until 0.5 in. (12.7 mm) extends from engine block groove. Remove oil seal installation tool.

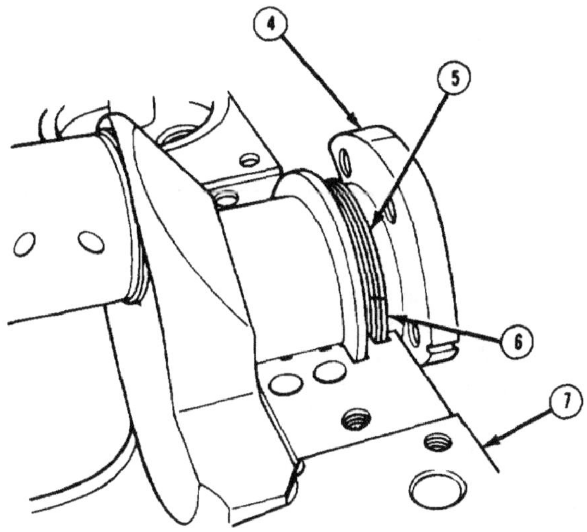

CAUTION

Contact points of rear main seal halves (5 and 6) should be at 4 o'clock and 10 o'clock positions or 8 o'clock and 2 o'clock positions. Failure to follow this caution may result in oil leakage and damage to rear main seal.

6. Insert end of lower rear main seal half (5) in open engine block (7) groove. Place lower rear main seal half along crankshaft (4) so that end touches upper rear main seal half (6).

7. Lightly coat rear main seal groove in rear main bearing cap (1) with pipe sealant.

WARNING

Avoid skin contact with anaerobic sealing compound. Immediately wash off skin if contact is made, Failure to follow this warning may result in injury to personnel.

CAUTION

Ensure that sealing compound (2) does not contact rear main seal groove, rear main bearing, or oil relief slot (3). Failure to follow this caution may result in damage to rear main seal and rear main bearing, or contamination of engine oil.

8. Apply a thin film of anaerobic sealing compound (2) to rear main bearing cap (1) in pattern shown.

TA50713

3-34. REAR MAIN SEAL REPLACEMENT (Con't).

CAUTION

Ensure that main bearing cap (1) is tapped into place before installing bolts. Failure to follow this caution may result in damage to main bearing cap.

NOTE

Inner bolts are longer than outer bolts.

9. Using soft-faced hammer, tap rear main bearing cap (1) into place and install with 2 inner bolts and outer bolts. Tighten all bolts to 40 lb,-ft. (54 Nom). Evenly tighten inner bolts again to 110 lb.-ft. (149 N.m). Evenly tighten outer bolts again to 100 lb.-ft. (136 N.m). Repeat for other main bearing caps.

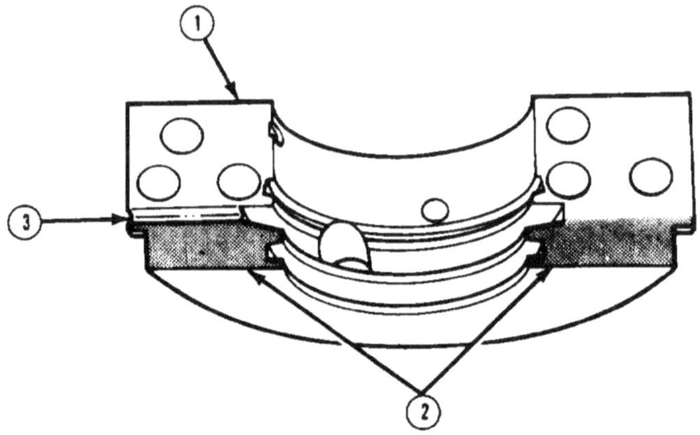

FOLLOW-ON TASKS:

- Install engine oil pump. (See paragraph 3-33)

TA50714

Section III. FUEL SYSTEM MAINTENANCE

3-35. FUEL SYSTEM MAINTENANCE INDEX.

3-36. FUEL INJECTOR NOZZLE TESTING.

This task covers:

a. Preparation
b, Pressure Test
c. Leakage Test
d. Chatter Test

INITIAL SETUP:

Equipment Condition

- Fuel injector nozzle removed. (See paragraph 3-5)

Tools/Test Equipment

- Fuel injector nozzle tester

Materials/Parts

- Calibration fluid (Item 13, Appendix B)

General Safety instructions

- DO NOT place hand or arms near tip of fuel injector nozzles during testing.
- Wear goggles while testing fuel injector nozzles.
- Compressed air used for drying purposes should never exceed 30 psi (207 kPa),

a. PREPARATION

WARNING

- DO NOT place hand or arms near nozzle tip (4) of fuel injector nozzle (3) during testing, Nozzle tip must be enclosed in a receptacle, preferably transparent, to contain fuel spray. High-pressure atomized fuel spray from nozzle tip can puncture skin and destroy tissue. Failure to follow this warning may also result in blood poisoning.
- Wear goggles while testing fuel injector nozzles (3). Failure to follow this warning may result in serious eye injury.

NOTE

Ensure that each of the following tests is performed in order listed until a malfunction is found, A malfunction will only be what is identified in appropriate subparagraph as a malfunction. DO NOT attempt to identify malfunctions that are not covered by test that you are performing, even if there is a symptom that seems to indicate a malfunction.

1, Connect high-pressure inlet (2) to fuel injector nozzle tester, Connect clear plastic tubes to fuel return ports (1) to prevent fuel leakoff from being confused with an actual leak. Close pressure gage shutoff valve.

2. Fill and purge fuel injector nozzle (3) with calibration fluid.

3-36. FUEL INJECTOR NOZZLE TESTING (Con't).

b. PRESSURE TEST

1. Open pressure gagshutoff valve 1/4 turn.

NOTE

Some fuel injector nozzles (3) may drip during this test. This does not indicate leakage.

2. Depress lever of fuel injector nozzle tester slowly, Note at which pressure needle of pressure gage stopped. Minimum pressure should be 1500 psi (1 0,343 kPa). Replace fuel injector nozzle (3) if below minimum pressure,

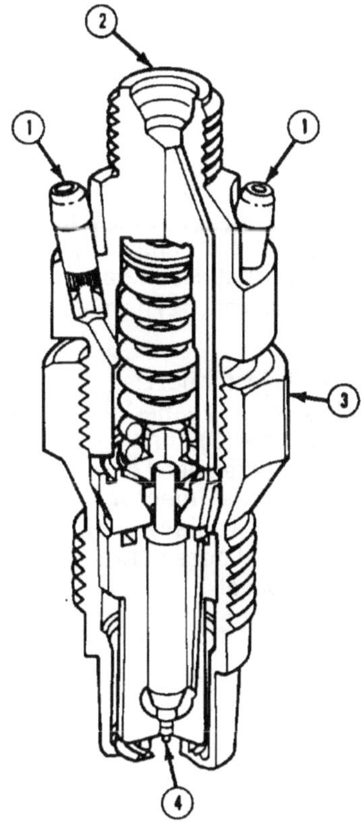

c. LEAKAGE TEST

1. Open pressure gage shutoff valve an additional 1/2 1 1/2 turns.

WARNING

Compressed air used for drying purposes should never exceed 30 psi (207 kPa). Use only effective chip guarding and personnel protective equipment (goggles/shield, gloves, etc.), Failure to follow this warning may result in serious injury to personnel.

2. Dry nozzle tip (4) with compressed air.

3. Depress lever of fuel injector nozzle tester slowly until gage reads 1400 psi (9653 kPa). If a drop forms at nozzle tip (4) and falls off within 10 seconds, replace fuel injector nozzle (3).

TA50716

3-36. FUEL INJECTOR NOZZLE TESTING (Con't).

d. CHATTER TEST

1. Close shutoff lever at pressure gage.

NOTE

"Chatter" from new and used fuel injector nozzles (3) will vary. When speed of lever movement is increased, fuel injector nozzle may "hiss" or "squeal" rather than "chatter"; this Is also acceptable. Any of these sounds indicates that needle valve (6) moves freely and that nozzle seat, guide, and nozzle pintle (5) have no mechanical defects.

2. Depress lever of fuel injector nozzle tester slowly and listen for "chatter" from fuel injector nozzle (3). If no "chatter" is heard, increase speed of lever movement until fuel injector nozzle "chatters." Replace fuel injector nozzle if it does not "chatter," "hiss," or "squeal."

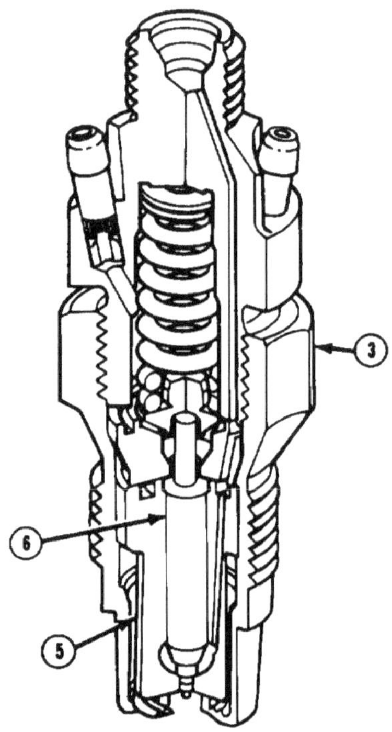

FOLLOW-ON TASKS:

- Install fuel injector nozzle, (See paragraph 3-27)

TA50716

3-37. FUEL INJECTOR PUMP AND TIMING COVER REPLACEMENT.

This task covers: a. Removal b. Installation

INITIAL SETUP:

Equipment *Condition*

- Fuel injector pipes removed. (See paragraph 3-5)
- Transmission vacuum valve removed (if removing fuel injector pump). *(See* TM 9-2320-289-20)

Materials/Parts

- One fuel injector pump gasket
- One timing cover gasket

Tools/Test *Equipment*

- Torque wrench

a. REMOVAL

NOTE

- If removing fuel injector pump (3), perform steps 1-7.
- If removing timing cover and gasket, perform step 8..

1. Remove accelerator cable retainer (5) and accelerator cable assembly (1) from throttle shaft pin (4) and support (6).

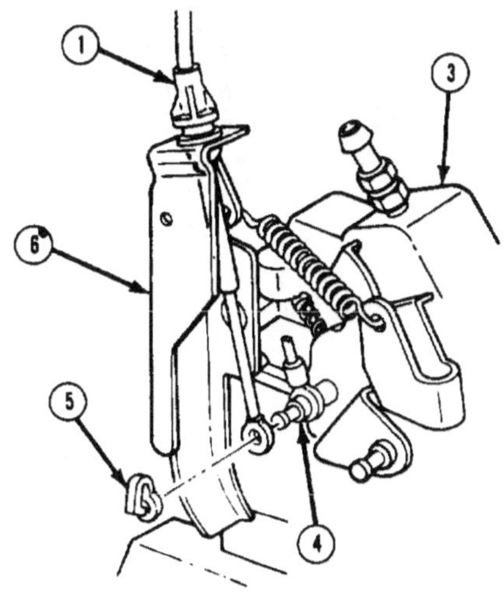

TA50717

3-37. FUEL INJECTOR PUMP AND TIMING COVER REPLACEMENT (Con't).

CAUTION

Ensure that all openings and connections are immediately plugged after disconnection to prevent contamination of fuel system and fuel injector pump (3) internal components.

2, Disconnect fuel return line from connector (2).Discommect fuel supply from end cap assembly (9).

NOTE

Ensure that all electrical leads are tagged for installation .

3. Disconnect electrical leads from fuel injector pump (3),

4. Scribe a mark alining fuel injector pump flange (13) and crankcase cover (7).

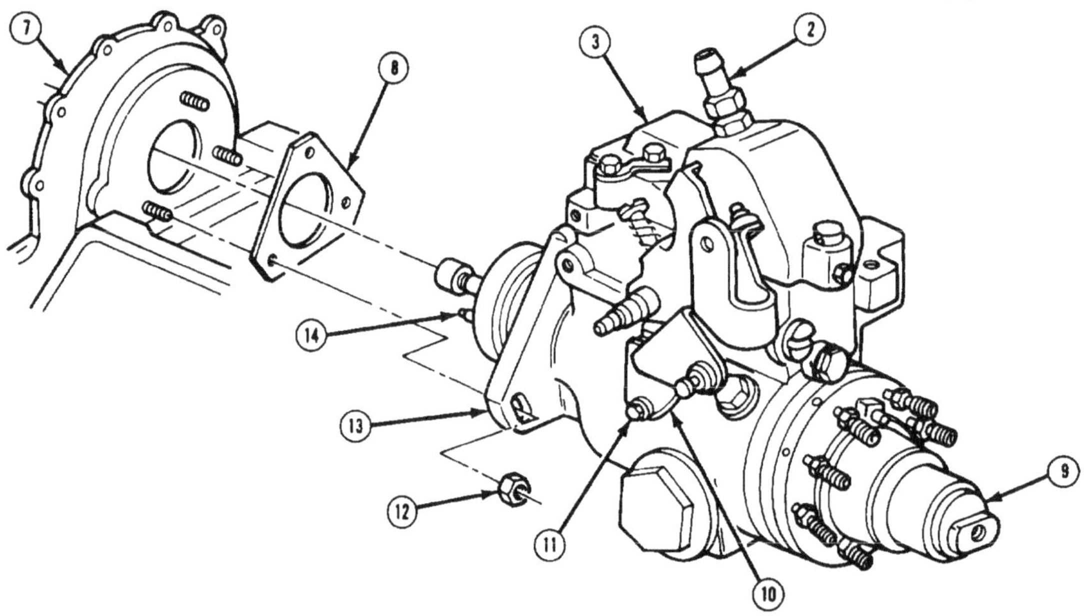

5. Loosen clamp and disconnect crankcase depression regulator valve hose from oil filler tube (15). Remove 2 nuts (18), oil filler tube, and grommet (17) from water pump plate (1 6),

NOTE

Crankshaft pulley will have to be rotated to gain access to bolts (20) on fuel injector pump driven gear (19).

6. Through hole in crankcase cover (7), remove 3 bolts (20) from fuel injector pump driven gear (19).

TA50718

3-37. FUEL INJECTOR PUMP AND TIMING COVER REPLACEMENT (Con't).

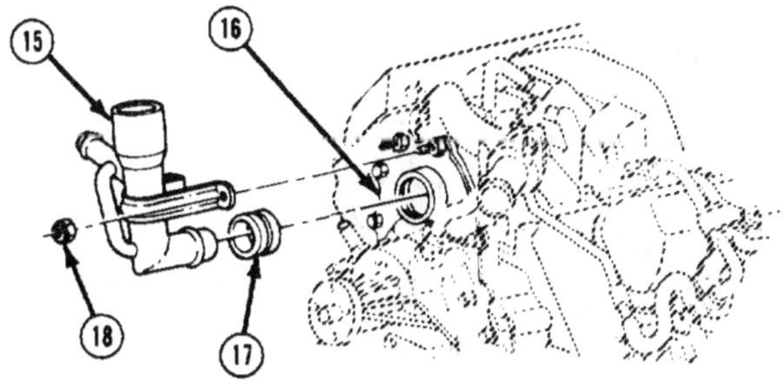

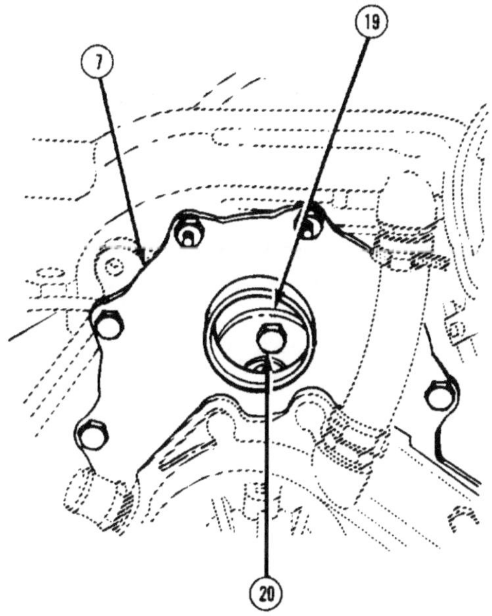

7. Remove 3 nuts (1 2), fuel injector pump (3), and fuel injector pump gasket (8), Discard fuel injector pump gasket.

8. Remove 2 screws (1 1), timing cover (10), and gasket. Discard gasket.

b. INSTALLATION

NOTE

- If installing fuel injector pump (3), perform steps 1-6.
- If installing timing cover (10) and gasket, perform step 7.
- Ensure that all protective plugs are removed before connection.

1. Install new fuel injector pump gasket (8) on crankcase cover (7). Aline locating pin (14) with hole in fuel injector pump driven gear (19) and aline scribe marks on fuel injector pump flange (13) and crankcase cover. Install 3 nuts (1 2). Tighten nuts to 30 lb.-ft. (41 N.m).

NOTE

Crankshaft pulley will have to be rotated to install bolts (20) on fuel injector pump driven gear (19).

2. Through hole in crankcase cover (7), install 3 bolts (20) on fuel injector pump driven gear (19). Tighten bolts to 20 lb.-ft. (27 N.m).

3. Install grommet (17) on water pump plate (1 6). Install oil filler tube (15) on water pump plate with 2 nuts (18), Connect crankcase depression regulator valve hose to oil filler tube and tighten clamp.

TA50719

3-37. FUEL INJECTOR PUMP AND TIMING COVER REPLACEMENT (Con't).

4. Connect electrical leads to fue injector pump (3).
5. Connect fuel supply line to end cap assembly (9). Connect fuel return line to connector (2).

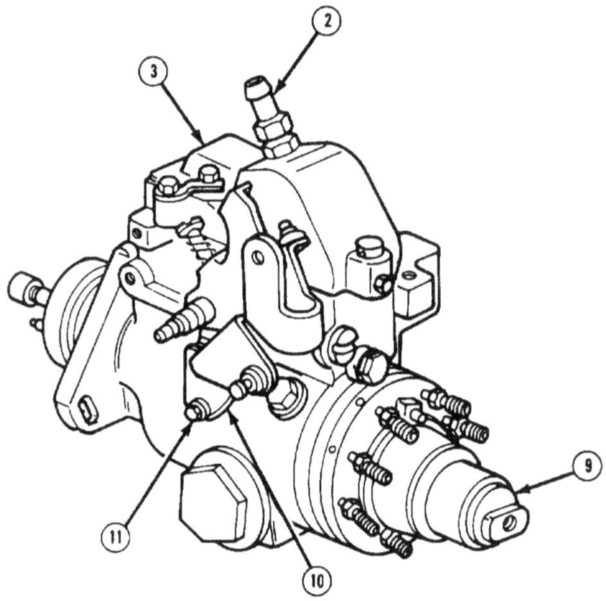

6. Install accelerator cable assembly (1) to support (6) and throttle shaft pin (4) with accelerator cable retainer (5),
7. Install timing cover (10) and new gasket with 2 screws (11).

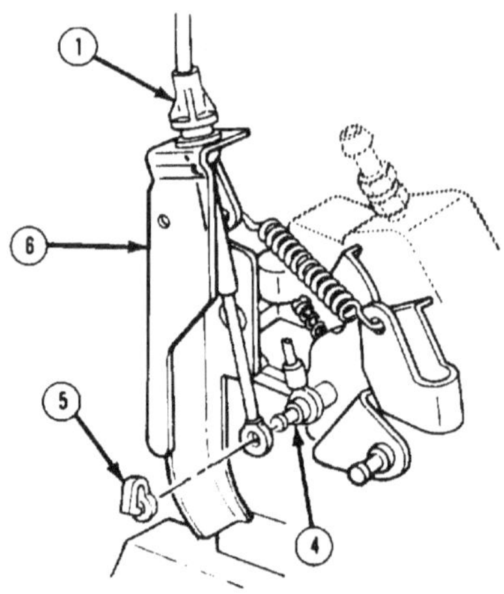

FOLLOW-ON TASKS:

- Install ransmission vacuum valve if removed. (See TM 9-2320-289-20)
- Install fuel injector pipes. (See paragraph 3-27)

TA50720

3-38. FUEL INJECTOR PUMP SHUTOFF SOLENOID AND COLD ADVANCE SOLENOID REPLACEMENT.

This task covers: a. Removal b. Cleaning and Inspection c. Installation

INITIAL SETUP:

Equipment Condition

- Air cleaner removed. (See TM 9-2320-289-20)

General Safety Instructions

- Compressed air used for cleaning purposes should never exceed *30* psi (207 kPa).
- Do NOT perform this procedure near fire, flames, or sparks,

Materials/Parts

- One fuel return line connector seal
- One governor cover gasket
- Three locknuts
- Ten lockwashers
- Calibration fluid (Item 13, Appendix B)
- Fuel oil (Item 25, 26, or 27, Appendix B)

a. REMOVAL

WARNING

Diesel fuel is flammable. DO NOT perform this procedure near fire, frames, or sparks. A fire extinguisher must be on hand in work area. Failure to follow this warning may result in serious injury or death to personnel.

CAUTION

Ensure that work area, tools, and hands are kept clean when repairing fuel injector pump. Failure to follow this caution may result in damage to fuel injector pump components.

3-38. FUEL INJECTOR PUMP SHUTOFF SOLENOID AND COLD ADVANCE SOLENOID REPLACEMENT (Con't).

NOTE

- Ensure that each component is placed in calibration fluid as it is removed.
- If removing governor cover (2), perform steps 1-3.
- Ensure that location of washers and lockwashers is noted for installation,
- Perform step 1 only if fuel injector pump (4) is in truck.
- Ensure that all leads are tagged for installation.

1. Disconnect leads from fuel injector pump (4). Disconnect fuel return line from connector (5).
2. Remove fast idle solenoid (1),

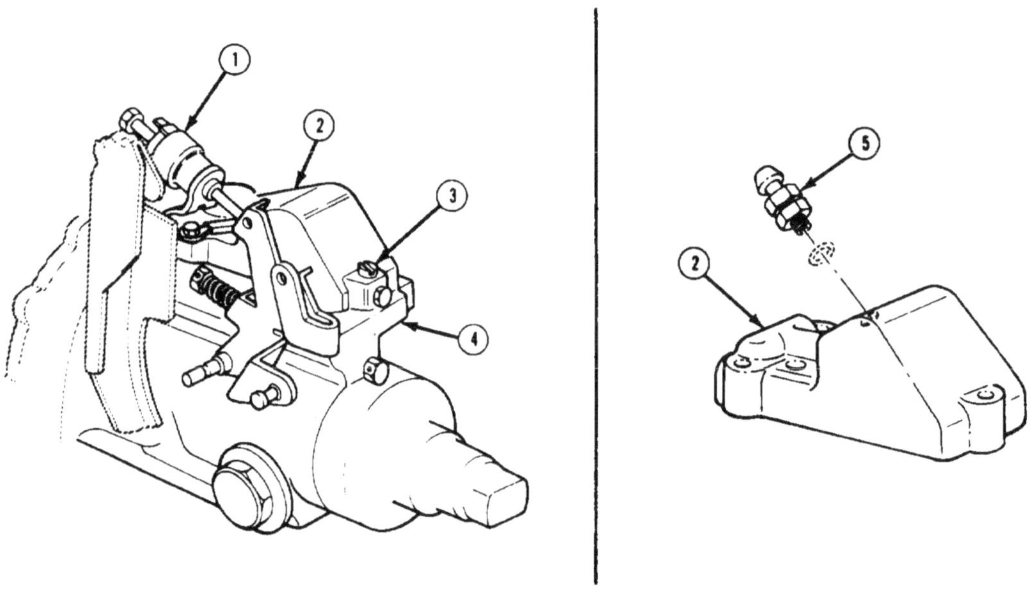

TA50721

3-38. FUEL INJECTOR PUMP SHUTOFF SOLENOID AND COLD ADVANCE SOLENOID REPLACEMENT (Con't).

CAUTION

When governor cover (2) is removed, ensure that rags are placed in fuel injector pump (4) body. Ensure that objects do not fall inside fuel injector pump body. Failure to follow this caution may result in damage to engine or fuel injector pump.

3. Remove 3 screws (3), lockwashers (26), and washers (27). Remove governor cover (2) from fuel injector pump (4). Remove and discard governor cover gasket (22). Discard lockwashers.

4. Remove 2 locknuts (15), terminal insulator (17), terminal ground strap (1 6), lockwasher (25), 4 lockwashers (18), 2 nuts (19), 2 washers (20), 2 insulating washers (21), and shutoff solenoid (24), Discard locknuts and lockwashers.

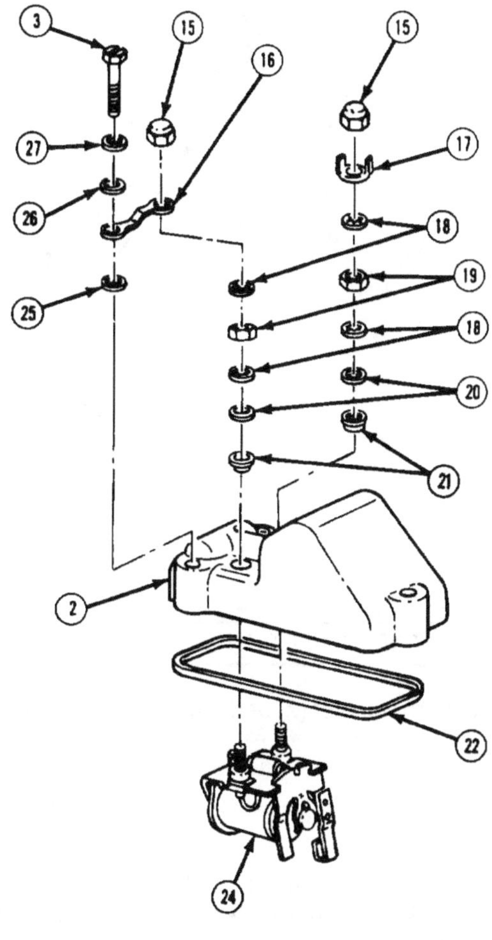

TA50722

3-38. FUEL INJECTOR PUMP SHUTOFF SOLENOID AND COLD ADVANCE SOLENOID REPLACEMENT (Con't).

5. Remove locknut (11), terminal insulator (1 2), 2 lockwashers (9), nut (1 0), washer (8), and insulating washer (7). Remove connector (5), seal (6), and cold advance solenoid (13). Discard seal, locknut, and lockwashers.

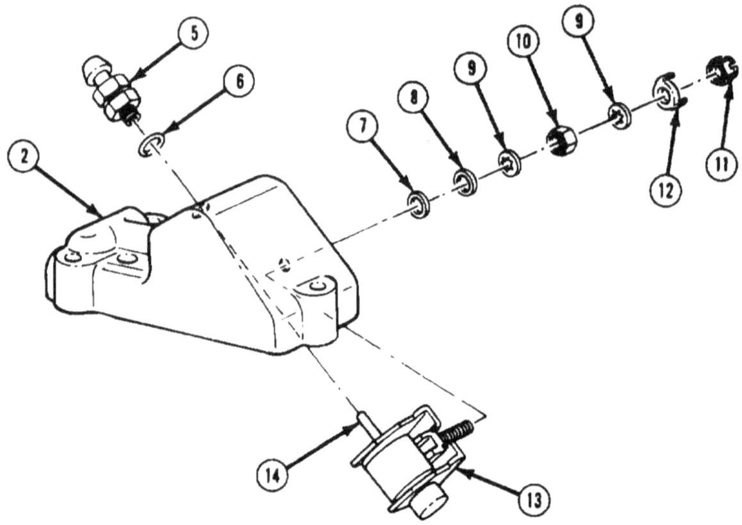

b. CLEANING AND INSPECTION

1. Clean governor cover (2) and mating surfaces of fuel injector pump with fuel oil.

WARNING

Compressed air used for cleaning purposes should never exceed 30 psi (207 kPa). Use only effective chip guarding and personnel protective equipment (goggles/shield, gloves, etc.). Failure to follow this warning may result in serious injury to personnel.

2. Inspect shutoff solenoid (24) and cold advance solenoid (13) for debris and damage. Remove debris with compressed air. Replace if damaged.

3. Depress linkage (23) and plunger (14) to ensure that they operate freely. Replace shutoff solenoid (24) or cold advance solenoid (13) if linkage or plunger do not operate freely.

TA50723

3-38. FUEL INJECTOR PUMP SHUTOFF SOLENOID AND COLD ADVANCE SOLENOID REPLACEMENT (Con't).

c. INSTALLATION

NOTE

If installing governor cover (2), perform steps 3-6.

1. Install cold advance solenoid (13) in governor cover (2). Install new seal (6) and connector (5), ensuring that connector is centered on plunger (14), Install insulating washer (7), washer (8), nut (10), 2 new lockwashers (9), terminal insulator (12), and new locknut (11).

2. Install shutoff solenoid (24) in governor cover (2) with 2 insulating washers (21), 2 washers (20), 2 nuts (19), new lockwasher (25), 4 new lockwashers (18), terminal ground strap (16), terminal insulator (17), and 2 new locknuts (15).

NOTE

Movement from linkage (23) or plunger (14) when cold advance solenoid (13) or shutoff solenoid (24) are energized indicate s proper operation,

3. Connect negative lead of 12 volt power source to governor cover (2) and positive lead to locknut (11) to check operation of cold advance solenoid (13). Repeat for shutoff solenoid (24) by connecting positive lead to either locknut (15). Replace cold advance solenoid or shutoff solenoid if not operating properly.

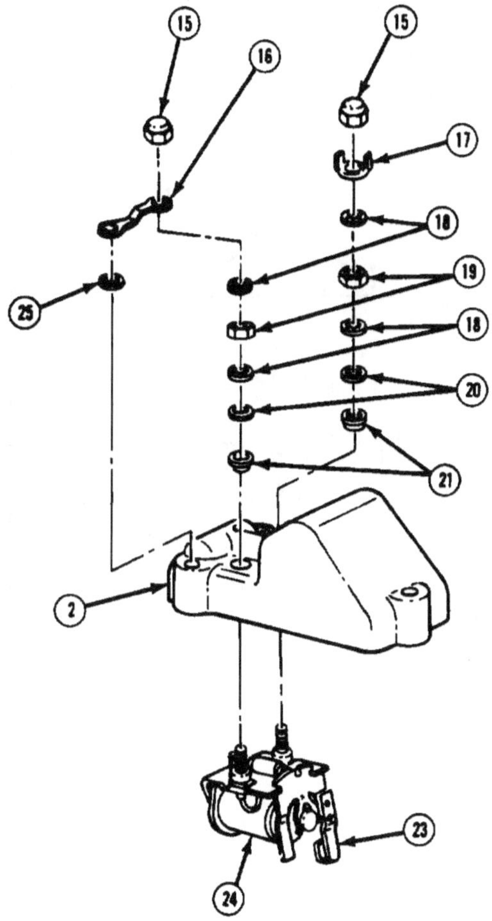

TA50724

3-38. FUEL INJECTOR PUMP SHUTOFF SOLENOID AND COLD ADVANCE SOLENOID REPLACEMENT (Con't).

NOTE

Ensure that rags are removed from fuel injector pump (4) body before installing governor cover (2).

4. Install new governor cover gasket (22) on governor cover (2). Depress throttle lever (28) and hold in idle position. Place governor cover slightly to rear of its correct mounting, and slide into position. Install 3 washers (27), new lockwashers (26), and screws (3).

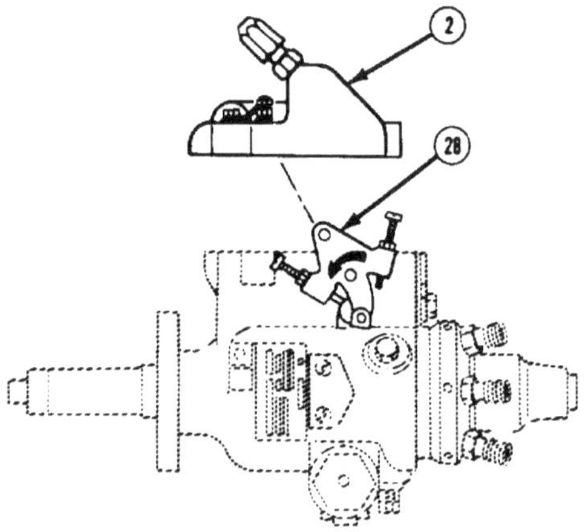

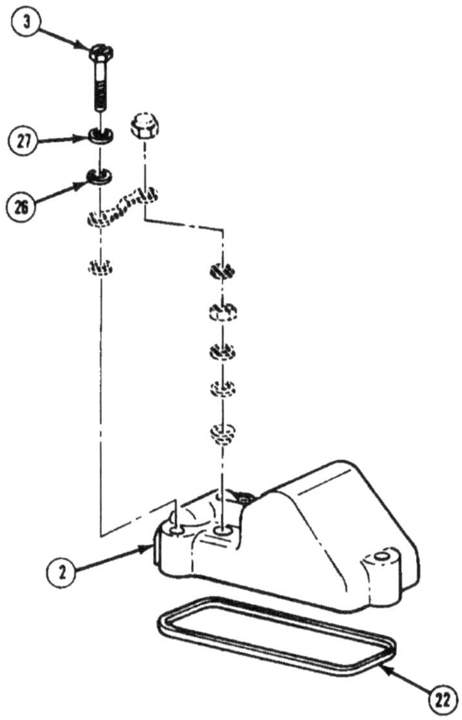

TA50725

3-38. FUEL INJECTOR PUMP SHUTOFF SOLENOID AND COLD ADVANCE SOLENOID REPLACEMENT (Con't).

5. Install fast idle solenoid (1).

NOTE

Perform step 6 only If fuel injector pump (4) Is in truck.

6. Loosely connect fuel return line to connector (5). Connect leads to fuel injector pump (4), Crank engine for 10-15 seconds, then wait 1 minute for starter motor to cool, Repeat until fuel is observed at fuel return line fitting, then tighten fuel return line fitting.

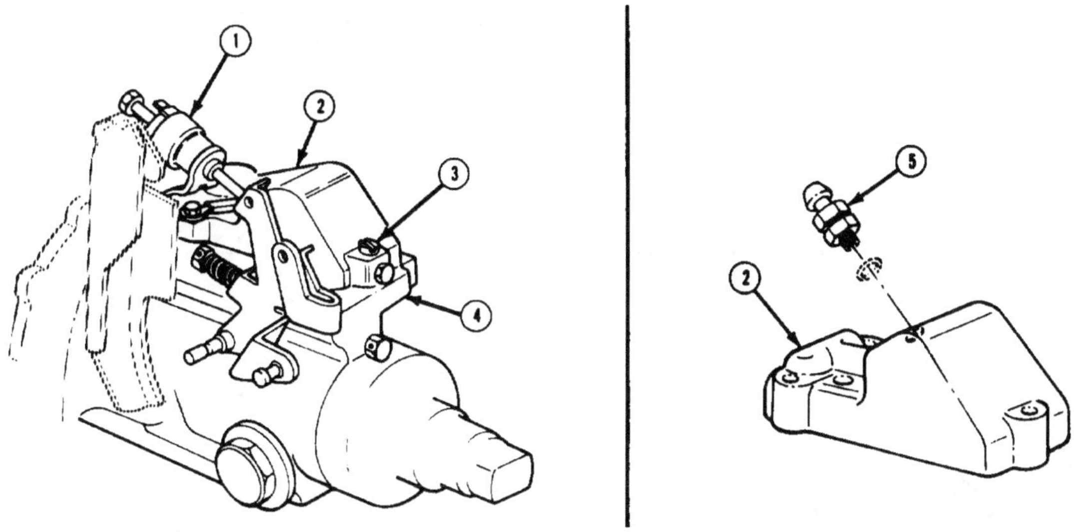

FOLLOW-ON TASKS:

- Install air cleaner, (See TM 9-2320-289-20)

TA50726

3-39. FUEL INJECTOR PUMP THROTTLE SHAFT AND GOVERNOR ASSEMBLY MAINTENANCE.

This task covers:

a. Removal
b. Disassembly
c. Inspection
d. Assembly
e. Installation

INITIAL SETUP:

Equipment Condition

- Fuel injector pump removed. (See paragraph 3-37)
- Governor cover removed. (See paragraph 3-38)

Materials/Parts

- One face cam pin
- One guide stud washer
- Two retaining rings
- Two seals
- Calibration fluid (Item 13, Appendix B)
- Grease (Item 36, Appendix B)
- Pipe sealant (Item 61, Appendix B)

Too/s/Test Equipment

- Screwdriver bit set, J-29843
- Timing gage assembly, J-29601
- Feeler gage
- Torque wrench

a. REMOVAL

CAUTION

Ensure that work area, tools, and hands are kept clean when repairing fuel injector pump . Failure to follow this caution may result in damage to fuel injector pump components.

NOTE

- Ensure that each component is placed in calibration fluid as it is removed.
- Timing gage assembly is used to provide proper face cam (3) alinement for installation.

1. Install timing gage assembly on outside of throttle shaft (2) with slots engaging face cam pin (4). Place spring clip of timing gage assembly over face cam (3) and tighten wingnut. Without loosening wingnut, remove timing gage assembly,

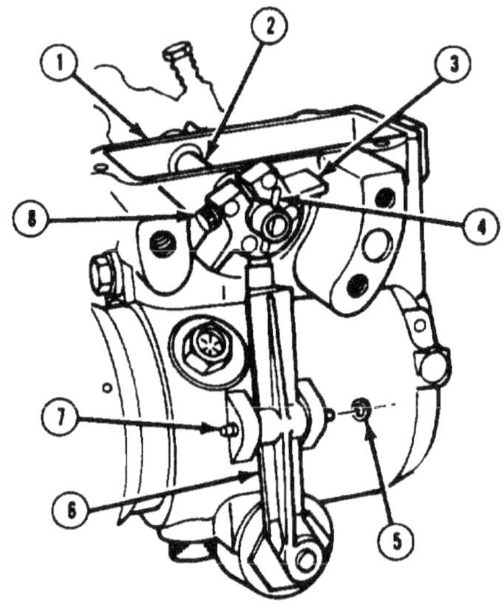

TA50727

3-39. FUEL INJECTOR PUMP THROTTLE SHAFT AND GOVERNOR ASSEMBLY MAINTENANCE (Con't).

2. Remove and discard 2 retaining rings (5). Remove pin (7) and rocker lever (6). Drive out and discard face cam pin (4). Remove screw (8), face cam (3), and face cam washer.

NOTE

Ensure that position of metering spring (12) over top of guide stud (10) is noted. This position must be exactly duplicated during installation.

3. Remove guide stud (10) and washer (11) from fuel injector pump housing (1). Discard washer.

4. Rotate throttle shaft (2) until governor assembly (15) clears fuel injector pump housing (1). Remove governor assembly from throttle shaft.

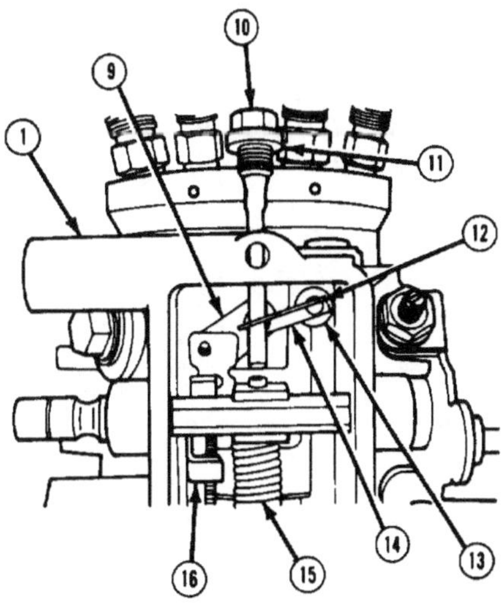

CAUTION

DO NOT cut seals (18) to remove from throttle shaft (2). Nicks in seal seats may result in leakage from fuel Injector pump housing (1).

5. Remove throttle shaft (2) from fuel injector pump housing (1). Remove 2 seals (18) and spacer (17) from throttle shaft. Discard seals.

NOTE

Ensure that position of governor linkage (16) assembly is noted for installation.

6. Disengage linkage hook (9) from metering arm (14). Remove governor linkage (16) assembly, Remove metering arm, metering spring (12), and metering valve (13) as an assembly.

7. Remove metering screw located in metering valve (13) hole.

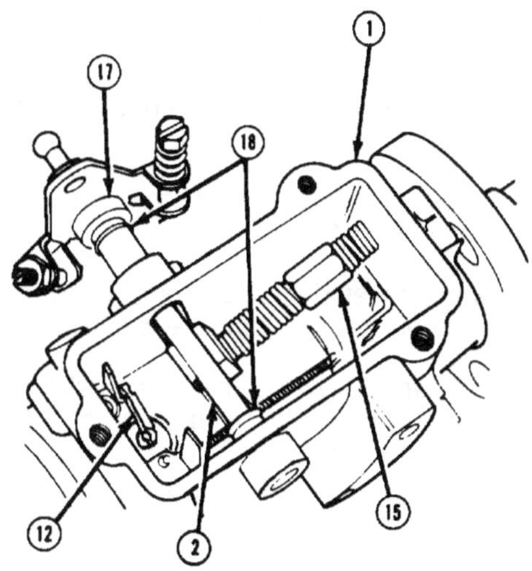

TA5028

3-39. FUEL INJECTOR PUMP THROTTLE SHAFT AND GOVERNOR ASSEMBLY MAINTENANCE (Con't).

b. DISASSEMBLY

1. Remove idle spring (20) and sleeve (21) from rod (22),
2. Place block (19) in soft-jawed vise, Unscrew rod (22) from block . Remove helical spring (24) and washer (23) from rod.

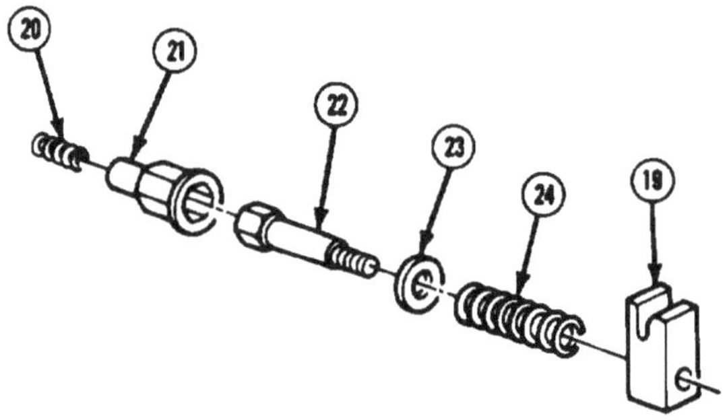

c. INSPECTION]

1. Inspect all springs for distortion, corrosion, or breakage. Replace any spring that is distorted, corroded, or broken.
2. Inspect throttle shaft seal seats for damage or wear that might cause leakage. Replace throttle shaft if these conditions exist.
3. Inspect all bores and grooves for obstructions or damage. Remove any obstructions and replace any damaged components.
4. Check vent wire on metering screw for freedom of movement. Replace metering screw with same size metering screw if vent wire is sticking or damaged.
5. Inspect all components for rust, nicks, scratches, and wear that appears uneven or may result in looseness or leakage. Replace any component that is rusted. "nicked, scratched, 'or has excessive or unusual wear.

d. ASSEMBLY

1. Install washer (23) and helical spring (24) on rod (22). Screw rod onto block (19).
2. Install idle spring (20) on sleeve (21), Install sleeve on rod (22).

e. INSTALLATION

1. Install metering screw in metering valve (13) hole and tighten to 25-30 lb.-in. (3-4 N.m).

TA50729

3-39. FUEL INJECTOR PUMP THROTTLE SHAFT AND GOVERNOR ASSEMBLY MAINTENANCE (Con't).

CAUTION

Never use an abrasive to clean metering valve (13). Failure to follow this caution may result in damage to fuel injector pump housing (l).

2. Place metering arm (14), metering spring (12), and metering valve (13) assembly in its bore and rotate several times to ensure freedom of movement. If metering valve sticks, clean with calibration fluid and try again. Replace metering valve if it continues to stick.

3. Position governor linkage (16) assembly in fuel injector pump housing (1) and allow to hang over side.

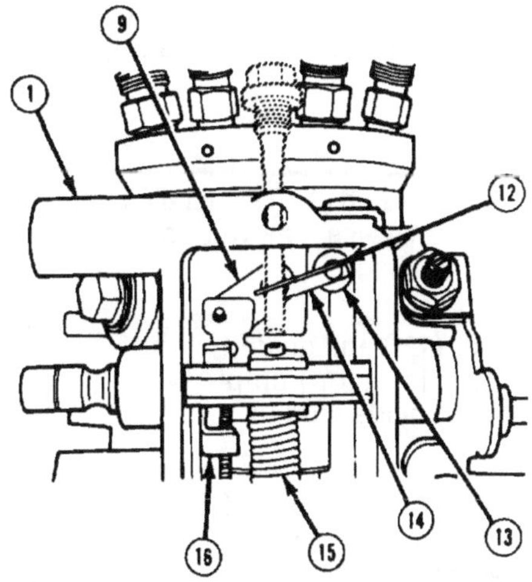

5. Install spacer (17) and 2 new seals (18) on throttle shaft (2). Ensure that seals completely seat on throttle shaft grooves. Install throttle shaft in fuel injector pump housing (1).

6. Rotate throttle shaft (2) to high idle throttle position. With governor assembly (15) between thumb and forefinger, install block (19) on throttle shaft. Rotate throttle shaft back to low idle position until governor assembly is in position,

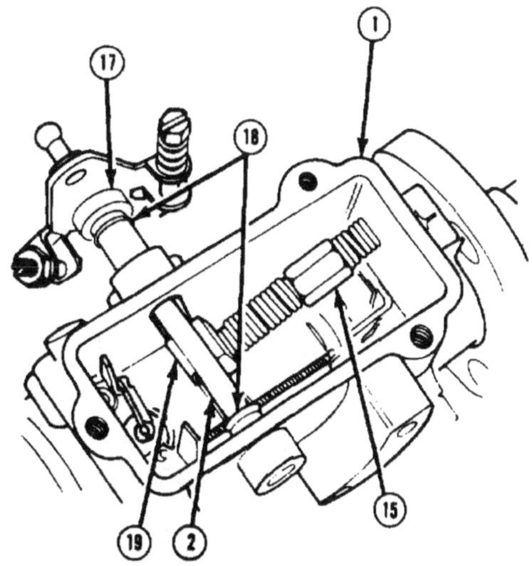

TA50730

3-39. FUEL INJECTOR PUMP THROTTLE SHAFT AND GOVERNOR ASSEMBLY MAINTENANCE (Con't).

CAUTION

DO NOT tighten guide stud (1 O) to more than 85 lb. -in. (9 N.m). Failure to follow this caution may result in damage to fuel injector pump housing (I).

NOTE

Ensure that guide stud (1 O) is installed under metering spring (12) in exact position as when it was removed, and that it engages governor assembly (15) without binding.

7. install new washer (11) and guide stud (10) into fuel injector pump housing (1), Tighten guide stud to 85 lb.-in. (9 N.m).

8 . Rotate throttle shaft (2) to rear of fuel injector pump housing (1). Check to ensure that governor assembly (15) compresses and that there is no binding. Reassemble or reinstall throttle shaft or governor assembly if governor assembly does not compress or there is binding,

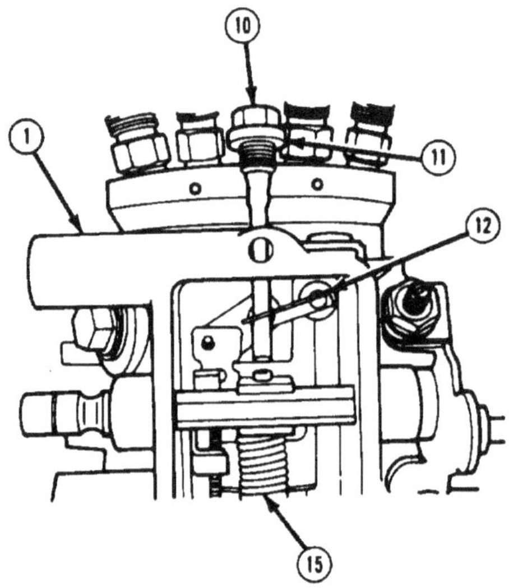

9. Clean pipe sealant from face cam (3) and screw (8), if present. Install face cam washer, face cam, and screw on throttle shaft (2). DO NOT tighten screw. Install new face cam pin (4).

10. install timing gage assembly on throttle shaft (2) with slots engaging face cam pin (4). Insert 0.005 in. feeler gage between spacer (17) and fuel injector pump housing (1). Push throttle shaft into fuel injector pump housing and squeeze face cam (3) to remove clearance. Apply pipe sealant to exposed threads of screw (8) and tighten to 30 lb.-in. (3 N.m). Loosen wingnut and remove timing gage assembly.

11. Install rocker lever (6) with pin (7) and secure with 2 new retaining rings (5).

TA60731

3-39. FUEL INJECTOR PUMP THROTTLE SHAFT AND GOVERNOR ASSEMBLY MAINTENANCE (Con't).

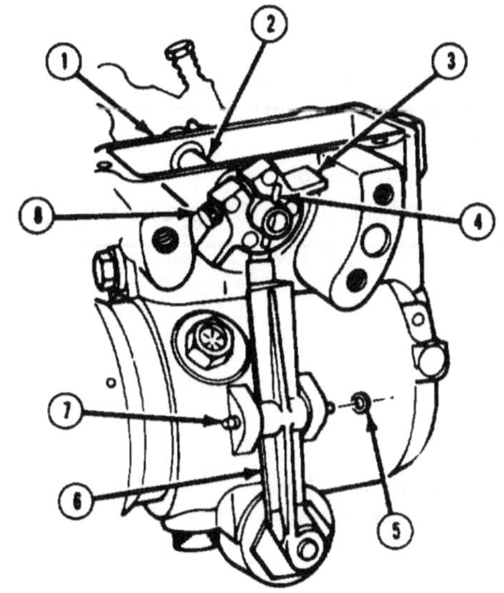

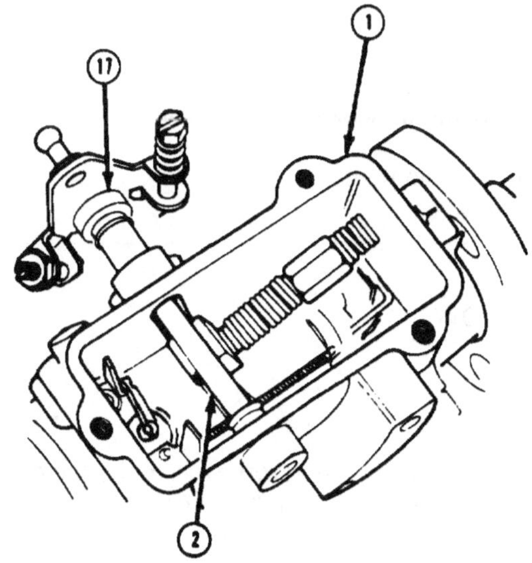

FOLLOW-ON TASKS:

- Install governor cover. (See paragraph 3-38)
- Install fuel injector pump. (See paragraph 3-37)

TA50732

3-40. FUEL INJECTOR PUMP SERVO ADVANCE PISTON REPLACEMENT.

This task covers: a. Removal b. Inspection c. Installation

INITIAL SETUP:

Equipment Condition
- Fuel injector pump removed. (See paragraph 3-37)

Tools /Test Equipment
- Torque wrench

Materials/Parts
- One advance screw hole seal
- One servo plunger seal
- Two plug seals
- Two retaining rings
- Calibration fluid (Item 13, Appendix B)

a. REMOVAL

CAUTION

- Ensure that work area, tools, and hands are kept clean when repairing fuel injector pump. Failure to follow this caution may result in damage to fuel injector pump components.
- DO NOT attempt to adjust or remove servo adjusting screw (4). Failure to follow this caution ‘may result in damage to fuel injector pump.

NOTE

Ensure that each component is placed in calibration fluid as it is removed.

1. Remove and discard 2 retaining rings (3). Remove pin (5) and rocker lever (2) from fuel injector pump housing (l).
2. Carefully invert fuel injector pump.

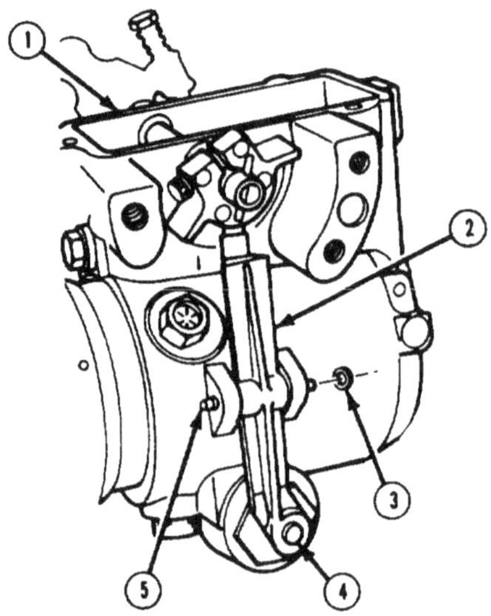

TA50733

3-40. FUEL INJECTOR PUMP SERVO ADVANCE PISTON REPLACEMENT (Con't).

3. Tap advance screw hole plug (6) lightly, and remove advance screw hole plug and seal (7). Discard seal.

4. Remove plunger (17). Remove spring side plug (15) and plug seal (14). Remove seal (16) from inside of spring side plug. Discard seals.

5. Remove cam advance pin (8).

6. Remove piston side plug (13), plug seal [12), and servo advance piston (11). Remove helical sprlng (9) amd servp advamce va;ve (10) from servo advance piston. Discard piston plug seal.

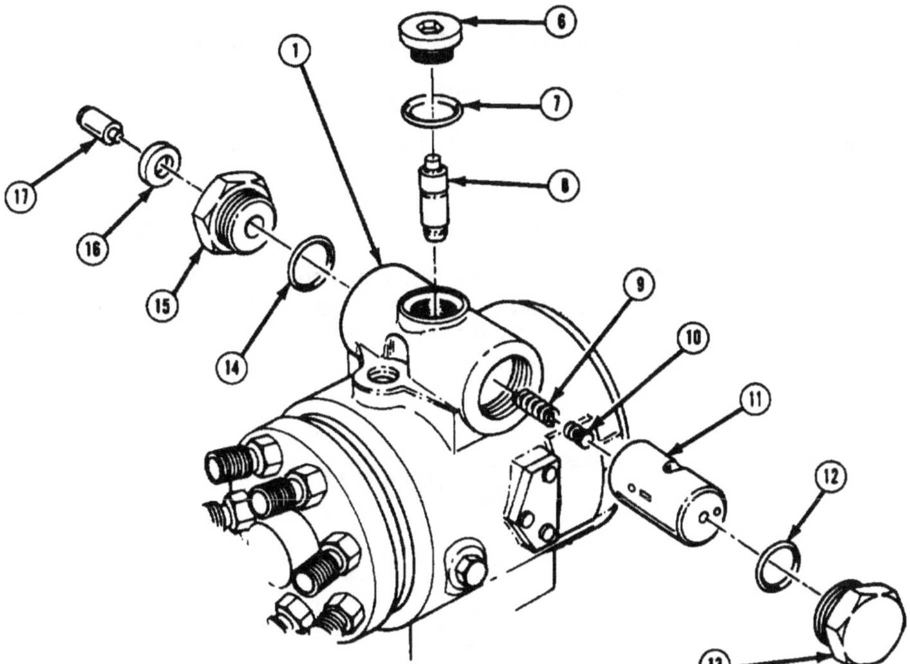

b. INSPECTION

1. Inspect helical spring for distortion, corrosion, or breakage. Replace if distorted, corroded, or broken.

2. Inspect all components for rust, nicks, scratches, and wear that appears uneven or may result in looseness or leakage. Replace any component that is rusted, nicked, scratched, or has excessive or unusual wear.

TA50734

3-40. FUEL INJECTOR PUMP SERVO ADVANCE PISTON REPLACEMENT (Con't).

c. INSTALLATION

CAUTION

DO NOT attempt to adjust or remove servo adjusting screw (4), Failure to follow this caution may result in damage to fuel injector pump.

NOTE

Ensure that hole in servo advance piston (11) faces head locating screw.

1. Instali servo advance piston (11), bore end first, into fuel injector pump housing (1).
2. nsert cam advance pin (8), ensuring that it engages in bore of cam ring.
3. Install new plug seal (12) and piston side plug (13), Tighten piston side plug to 20 lb.-ft. (27 N.m),
4. Instaii new seal (1 6), with lip of seal toward inside of bore, in spring side plug (15).
5. Install servo advance valve (1 O) on helical spring (9). Install servo advance valve and helical spring into servo advance piston (1 1) bore. Install new plug seal (14) and spring side plug (15). Ensure that holes in servo advance valve are in line with holes in servo piston bore. Tighten spring side plug to 20 lb.-ft. (27 N.m).

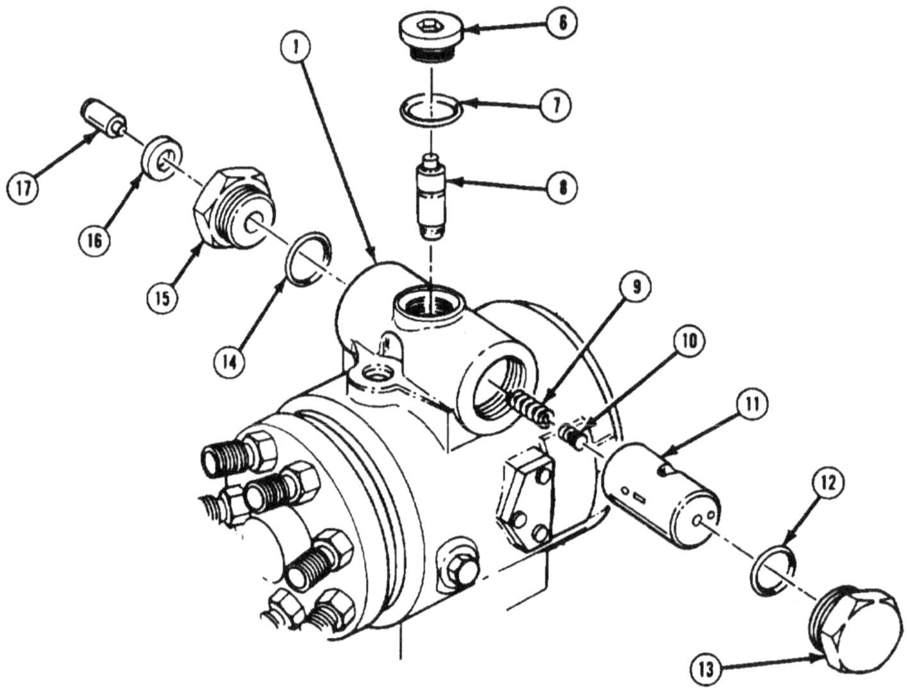

TA50735

3-40. FUEL INJECTOR PUMP SERVO ADVANCE PISTON REPLACEMENT (Con't).

6. Install plunger (17) in spring side plug (15).
7, Install new seal (7) and advance screw hole plug (6). Tighten advance screw hole plug to 75-100 lb.-in. (9-11 N.m).
8. Carefully invert fuel injector pump.
9. Install rocker lever (2) with pin (5) and secure with 2 new retaining rings (3),

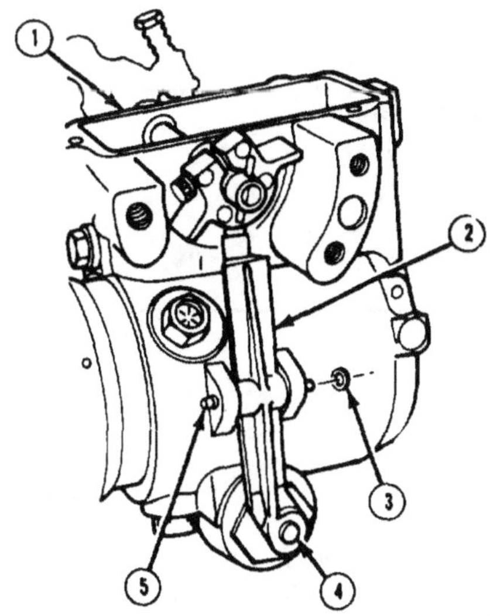

FOLLOW-ON TASKS:

- Install fuel injector pump. (See paragraph 3-37)

TA50736

3-41. FUEL INJECTOR PUMP REGULATOR AND TRANSFER PUMP BLADES MAINTENANCE.

This task covers:

a. Removal
b. Disassembly
c. Inspection
d. Assembly
e. Installation

INITIAL SETUP:

Equipment Condition

- *Fuel* injector pump removed. (See paragraph 3-37)

Tools/Test Equipment

- Crowfoot wrench
- Torque wrench

Materials/Parts

- One end cap seal
- One filter screen seal
- One locking seal
- One regulator seal
- Calibration fluid (Item 13, Appendix B)
- Grease (Item 36, Appendix B)

a. REMOVAL

CAUTION

Ensure that work area, tools, and hands are kept clean when repairing fuel injector pump. Failure to follow this caution may result in damage to fuel injector pump components.

NOTE

- Ensure that each component is placed in calibration fluid as it is removed.
- If removing end cap (3) and regulator (6) assembly, Perform steps 1 and 2.

1. Remove screw (4), plate (1), and locking seal from hydraulic head (2). Discard locking seal.
2. Remove end cap (3) assembly, perfprm steps 1 hydraulic head (2).

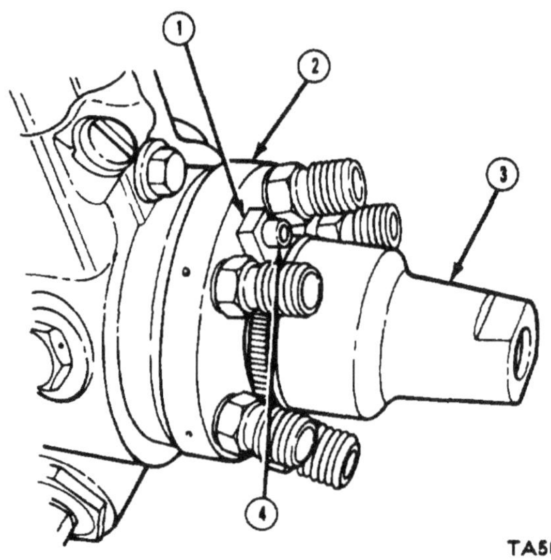

TA50737

3-41. FUEL INJECTOR PUMP REGULATOR AND TRANSFER PUMP BLADES MAINTENANCE (Con't).

3. Remove retaining ring (7), inlet filter screen (8), and plate (9) from regulator (6) assembly. Remove regulator assembly from hydraulic head (2).

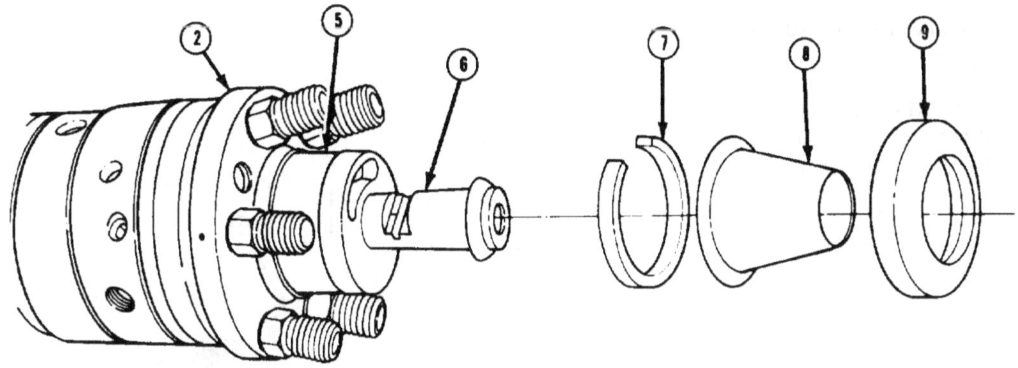

NOTE

Ensure that position of potmark (12), and location of transfer pump blades (11) and springs, are noted for installation.

4. Remove liner (5). Remove 4 transfer pump blades (11) with 2 springs. Remove and discard end cap seal (1 O) from' groove in hydraulic head (2) .

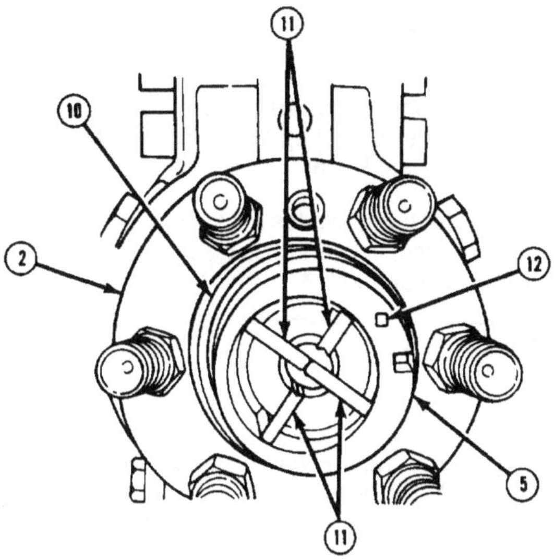

TA50738

3-41. FUEL INJECTOR PUMP REGULATOR AND TRANSFER PUMP BLADES MAINTENANCE (Con't).

b . DISASSEMBLY

NOTE

Ensure that position of roll pin (17) is noted for installation, if removed.

1. Separate roll pin (17) from regulator (6) if damaged.
2. Remove and discard filter screen seal (13). Remove plug (1 6), regulating spring (15), and regulating piston (14),

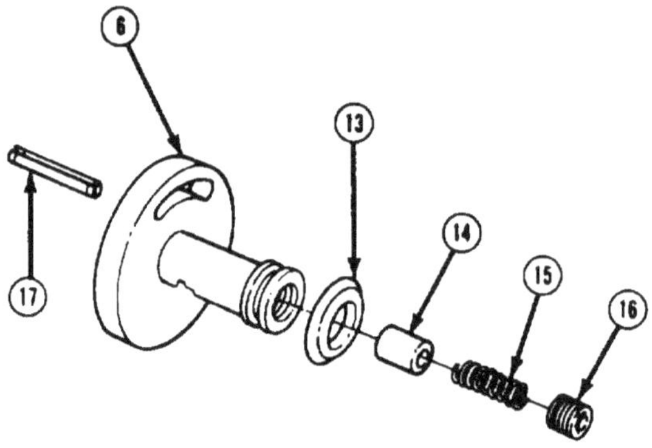

c. INSPECTION

1. Inspect all springs for distortion, corrosion, or breakage. Replace any spring that is distorted, corroded, or broken.
2. Inspect all components for rust, nicks, scratches, and wear that appears uneven or may result in looseness or leakage. Replace any component that is rusted, nicked, scratched, or has excessive or unusual wear.
3. Inspect transfer pump blades for scoring or pitting. Measure length of each blade. If any blade is scored or pitted, or if length is less than 0.538 in. (13.67 mm), replace all transfer pump blades and both springs.

d. ASSEMBLY

NOTE

Ensure that hollow end of regulating piston (14) faces threaded end of regulator (6) bore.

1. Install regulating piston (14) and regulating spring (15) in regulator (6), Install plug (16) until flush with end of regulator.

TA50739

3-41. FUEL INJECTOR PUMP REGULATOR AND TRANSFER PUMP BLADES MAINTENANCE (Con't).

2. Insert a brass rod through unthreaded end of regulator (6) and apply moderate pressure to check for free movement of regulating piston (14). If regulating piston does not move in its bore, replace regulator, regulating piston, and regulating spring (15).

3. Install new filter screen seal (13) on regulator groove.

4. Assemble roll pin (17) and regulator (6) if removed.

e. INSTALLATION

NOTE

If installing end cap and regulator (6) assembly, perform steps 4-7.

1. Install new end cap seal (10) in groove in hydraulic head (2).

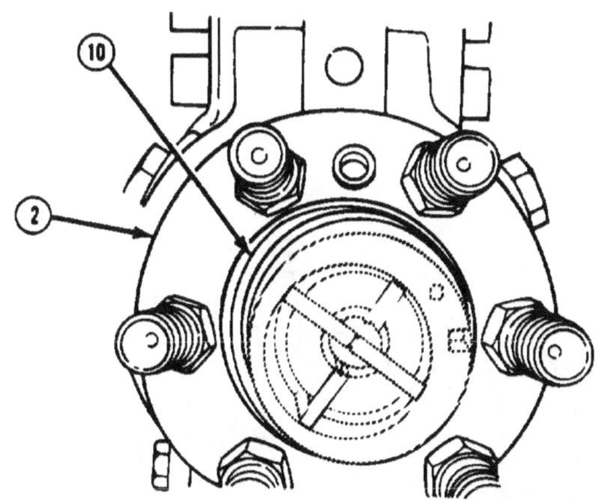

TA50740

3-41. FUEL INJECTOR PUMP REGULATOR AND TRANSFER PUMP BLADES MAINTENANCE (Con't).

2. Install liner (5) in hydraulic head (2), ensuring that slot alines with roll pin (17) hole in hydraulic head.

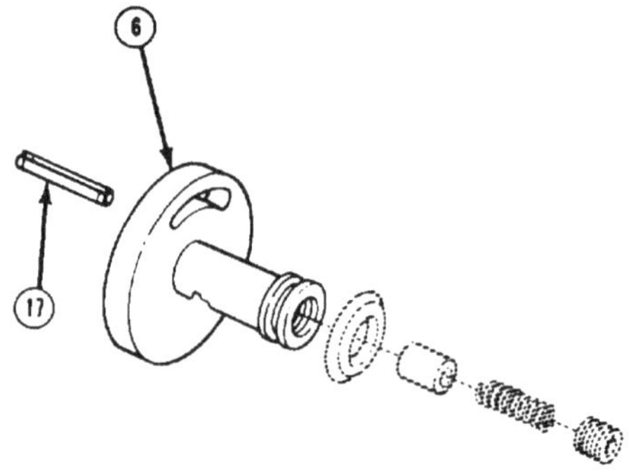

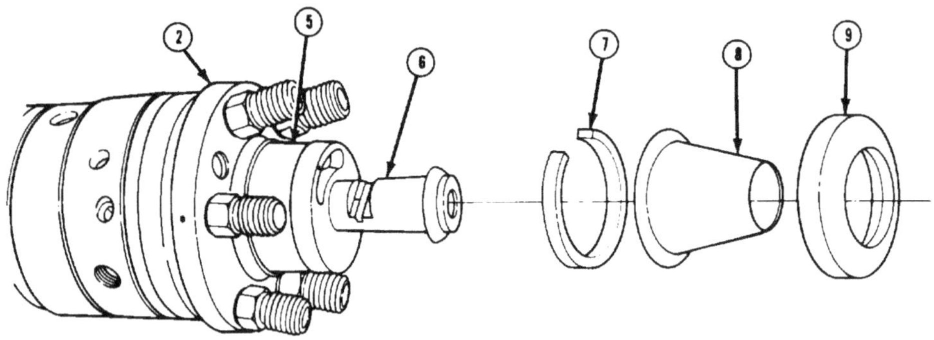

TA50741

3-41. FUEL INJECTOR PUMP REGULATOR AND TRANSFER PUMP BLADES MAINTENANCE (Con't).

CAUTION

Transfer pump blades (11) must be fully compressed during installation. Care must be taken not to score transfer pump blades on sharp edge of liner (5). Failure to foliow this caution may result in damage to transfer pump blades.

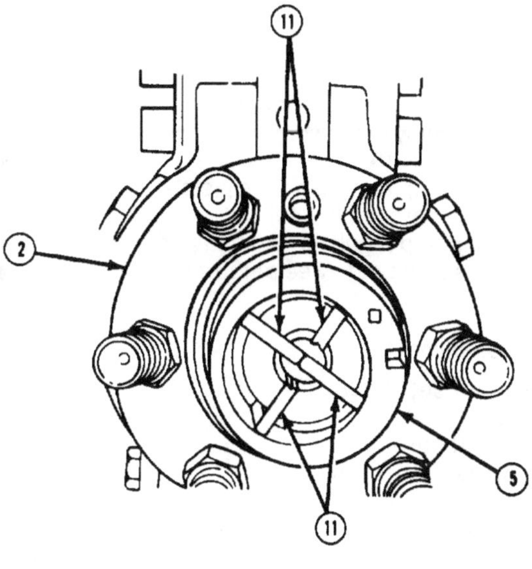

3. Assemble 2 springs and 4 transfer pump blades (11), and install transfer pump blades in hydraulic head (2) rotor slots.

4. Install regulator (6) assembly on hydraulic head (2) with roll pin (17) in liner slot, Install inlet filter screen (8) in plate (9) with retaining ring (7). Install inlet filter screen assembly onto regulator.

5. Appiy grease to mating surfaces of end cap (3) and plate (9). Apply light pressure at end of end cap and rotate counterclockwise until a slight click is heard. Turn end cap clockwise by hand until tight.

6. Tighten end cap (3) to 35 lb.-ft . (47 N.m).

7. Install new locking seal, plate (1), and screw (4), on hydraulic head (2). Tighten screw to 70-80 lb.-in. (8-9 N.m).

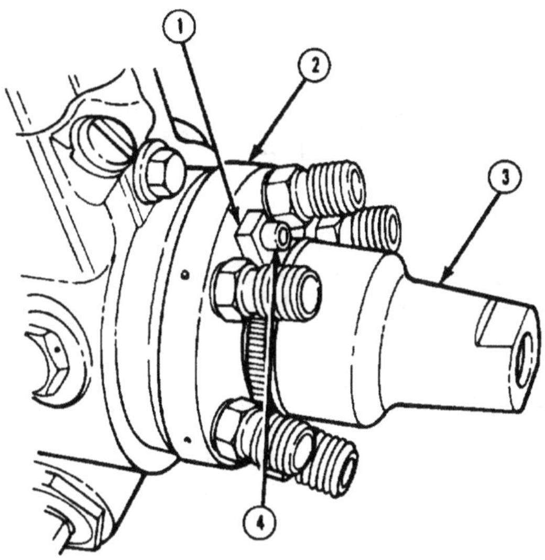

FOLLOW-ON TASKS:

- Install fuel injector pump. (See paragraph 3-37)

TA50742

3-42. FUEL INJECTOR PUMP HYDRAULIC HEAD MAINTENANCE.

This task covers:

a. Removal
b. Disassembly
c. Inspection
d, Assembly
e. Installation

INITIAL SETUP:

Equipment Condition

- Fuel injector pump throttle shaft and governor assembly removed. (See paragraph 3-39)
- Fuel injector pump servo advance piston removed. (See paragraph 3-40)
- Fuel injector pump regulator and transfer pump blades removed. (See paragraph 3-41)

TooLs/Test Equipment

- Screwdriver bit set, J-29843
- Micrometer
- Roller testing fixture
- Snapring pliers
- Torque wrench

Materialist*Parts*

- One delivery valve stop
- One head locating screw seal
- One hydraulic head seal
- Acetone (Item 1, Appendix B)
- Calibration fluid (Item 13, Appendix B)
- Grease (Item 36, Appendix B)
- "O" ring lubricant (Item 42, Appendix B)

a. REMOVAL

CAUTION

Ensure that work area, tools, and hands are kept clean when repairing fuel injector pump (2), Failure to follow this caution may result in damage to fuel injector pump components.

NOTE

Ensure that each component is placed in calibration fluid as it is removed.

1. Remove 2 head locking screws (1),

2. Remove head locating screw (3). Remove seal from head locating screw and discard seal.

3. Install fuel injector pump (2) in soft-jawed vise, and slightly tilt forward.

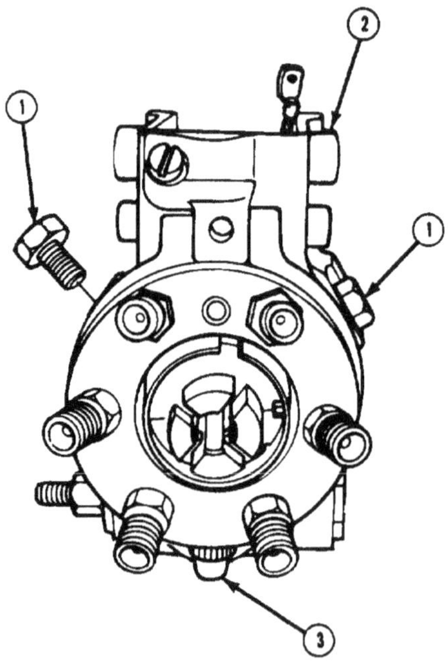

TA50743

3-42. FUEL INJECTOR PUMP HYDRAULIC HEAD MAINTENANCE (Con't).

4. Remove hydraulic head assembly (10) by grasping with both hands and withdrawing with slight rotary motion. Remove and discard hydraulic head seal (9).

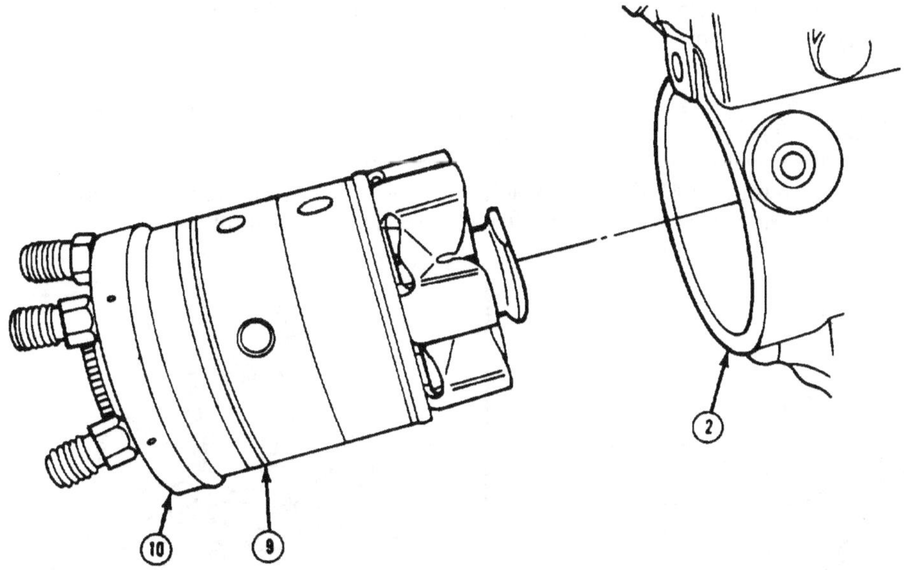

b. DISASSEMBLY

1. Invert hydraulic head assembly (1 O) and allow thrust sleeve (6), thrust sleeve washer (1 1), and 6 weights (5) to fall into your hand.

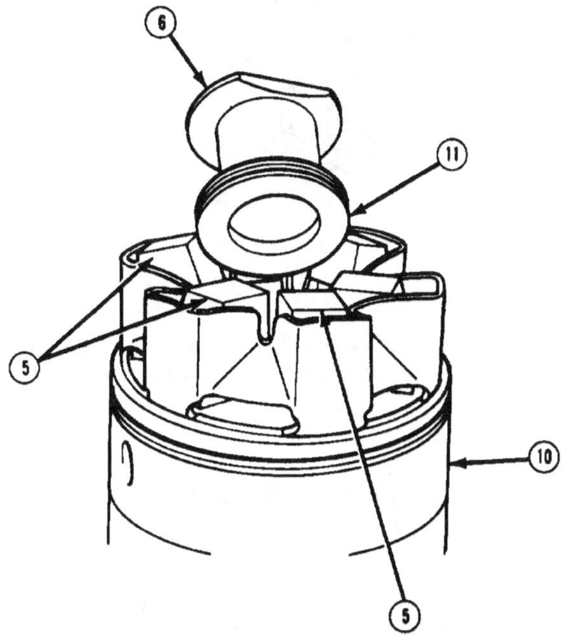

TA50744

3-42. FUEL INJECTOR PUMP HYDRAULIC HEAD MAINTENANCE (Con't).

CAUTION

Never handle precision-ground surface of rotor (1 6), Failure to follow this caution may result In damage to rotor,

2. Remove locating ring (12) and 2 retainers (14) from hydraulic head assembly (10). Invert hydraulic head assembly and place on connectors (13).
3. Remove snapring (17) from rotor (1 6). Remove weight retainer (7) and weight retainer hub (18) from cam ring (8), Remove cam ring from hydraulic head assembly (10).

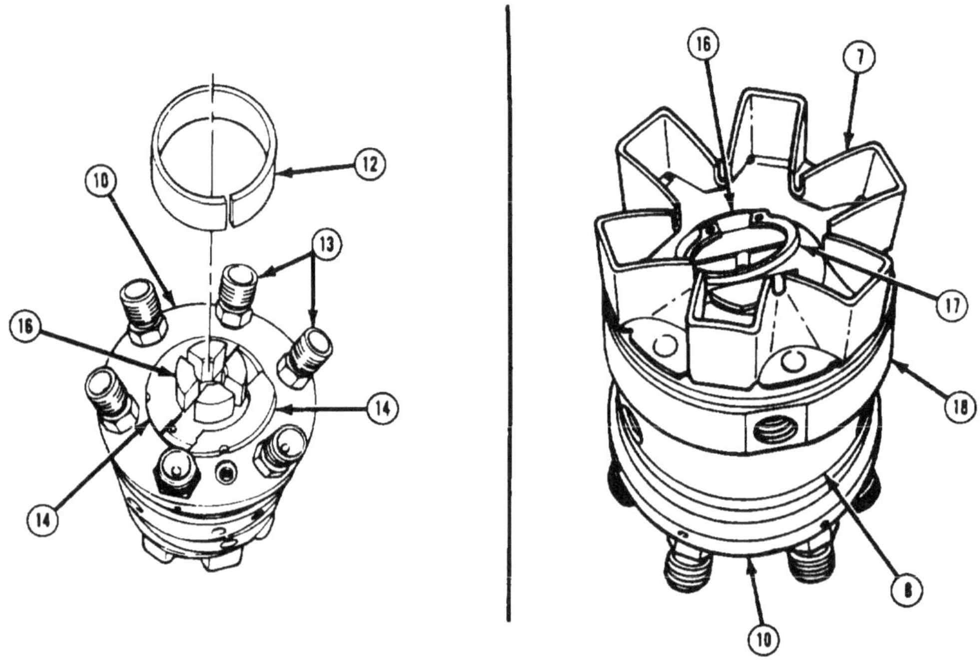

4. Install roller testing fixture (20) in vise, ensuring that air inlet hole is not covered. Connect air hose (22) to air inlet hole, and supply 40-100 psi (276-690 kPa) of compressed air.

NOTE

Ensure that rollers (19) and shoes (23) remain in their slots when installing rotor (16) assembly In roller testing fixture (20).

5. Remove rotor (16) assembly from hydraulic head assembly (10). Install rotor assembly in roller testing fixture (20) on air inlet side,
6. Check dimension between between rollers (19), Dimension should be 1.9765-1.9775 in. (50.203-50.229 mm). Adjust leaf spring screw (21) to see if proper dimension can be obtained. If proper dimension cannot be obtained, replace rollers and shoes (23). Disconnect air hose (22) from air inlet hole.

TA50745

3-42. FUEL INJECTOR PUMP HYDRAULIC HEAD MAINTENANCE (Con't).

7. Remove rotor (16) assembly from roller testing fixture (20). Holding rotor assembly at base, remove screw (28), delivery valve stop (27), spring (26), and delivery valve (24). Discard delivery valve stop.

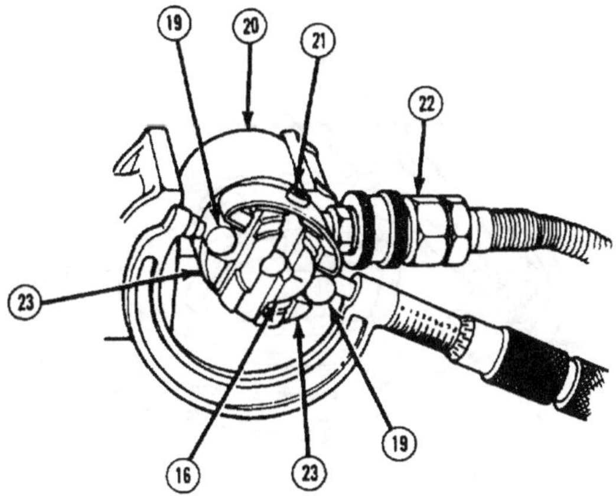

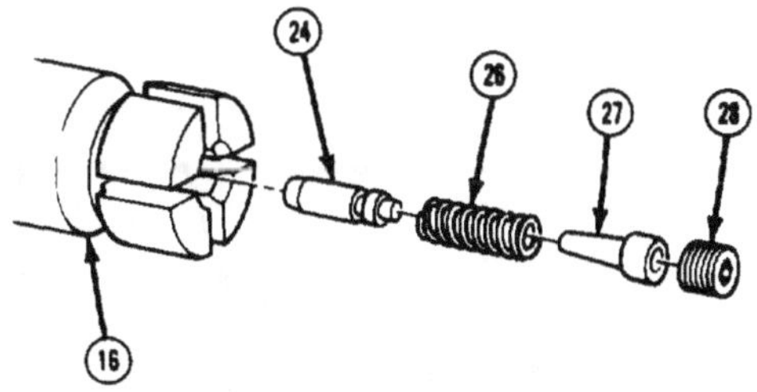

8. Install rotor (16) assembly in roller testing fixture (20). Remove leaf spring screw (21) and leaf spring (30). If present, remove leaf spring shim and save for installation.

NOTE

Hands should be wet with calibration fluid to catch plungers (29) as they are pushed out of rotor (16).

9. Remove 2 roller (19) and shoe (23) assemblies from rotor (16). Slide each roller from its shoe. Using smooth-tipped brass tool, push 2 plungers (29) from rotor bore Remove rotor from roller testin9 fixture (20).

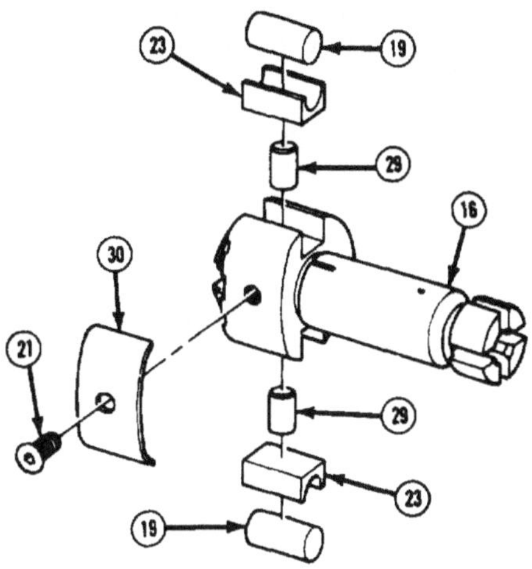

TA60746

3-42. FUEL INJECTOR PUMP HYDRAULIC HEAD MAINTENANCE (Con't).

c. INSPECTION

1. Inspect weights (5) for wear at heel (31) and toe (32). If weight appears worn, place in weight retainer (7). Weight should fit snugly. Replace any weight that does not fit snugly in weight retainer.

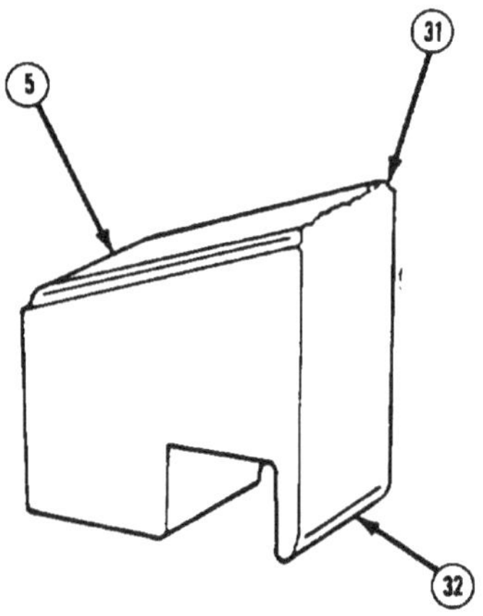

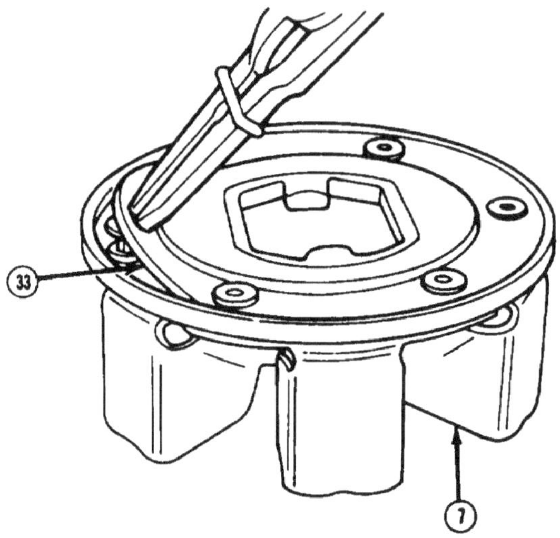

2. Inspect lobes (34) on inside of cam ring (8) for flaking. Inspect edges of cam ring for roundness. Replace cam ring if lobes are flaking or edges are rounded.

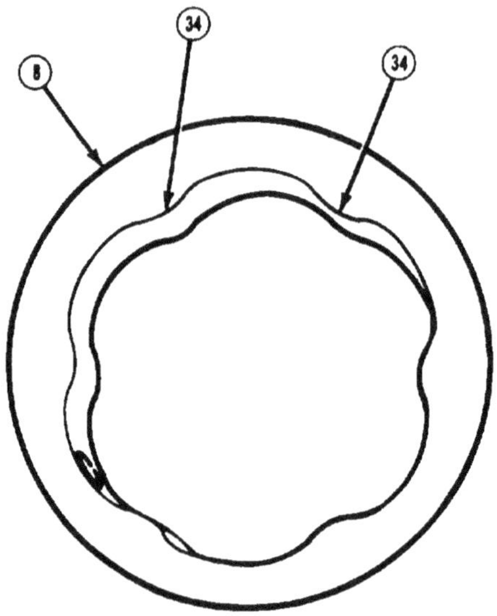

TA50747

3-42. FUEL INJECTOR PUMP HYDRAULIC HEAD MAINTENANCE (Con't).

3. Inspect weight retaining ring (33) on weight retainer (7) for damage. Replace weight retaining ring if damaged,

4. Inspect delivery valve (24) for chipping or erosion at edges of retractor cuff (25), Edges of retractor cuff should be sharp. Replace delivery valve if retractor cuff is chipped or eroded, or if edges are not sharp.

CAUTION

Never handle precision-ground surface of rotor (16). Failure to follow this caution may result in damage to rotor.

5. Submerge rotor (16) in clean calibration fluid and insert plungers (29) into their bores. With thumb and forefinger over shoe (23) slots, tilt rotor from side to side to determine if plungers move freely. If plungers are sticking, but not visibly damaged, clean both plungers and rotor bore with a soft brush and acetone.

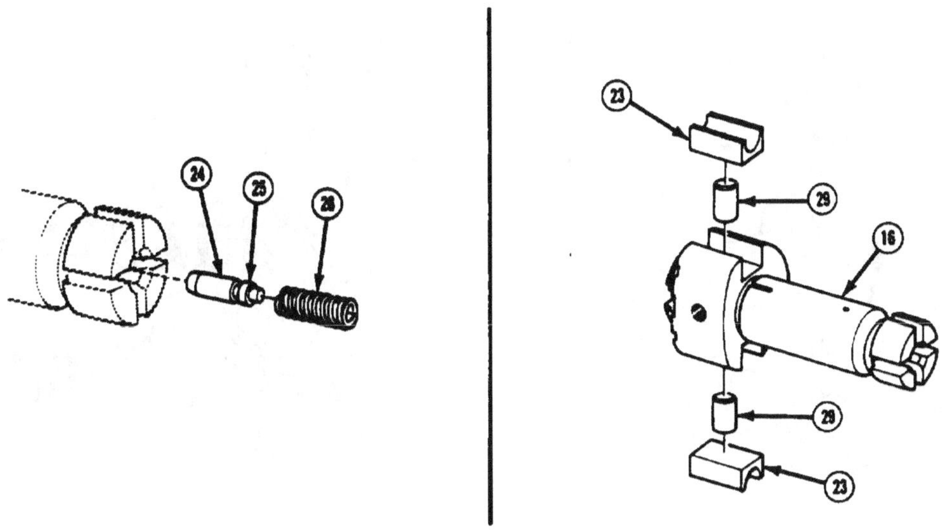

6. Inspect spring (26) for distortion, corrosion, or breakage. Replace if distorted, corroded, or broken.

7. Inspect all components for rust, nicks, scratches, and wear that appears uneven or may result in looseness or leakage. Replace any component that is rusted, nicked, scratched, or has excessive or unusual wear.

TA50748

3-42. FUEL INJECTOR PUMP HYDRAULIC HEAD MAINTENANCE (Con't).

d. ASSEMBL Y

CAUTION

Never handle precision-ground surface of rotor (16). Failure to follow this caution may result In damage to rotor.

1. Install rotor (16) in roller testing fixture (20). Install 2 plungers (29) in rotor bores. Slide 2 rollers (19) into their shoes (23). Install 2 roller and shoe assemblies on rotor.
2. Install leaf spring shim, if removed, and leaf spring (30) on rotor (16) with leaf spring screw (21).

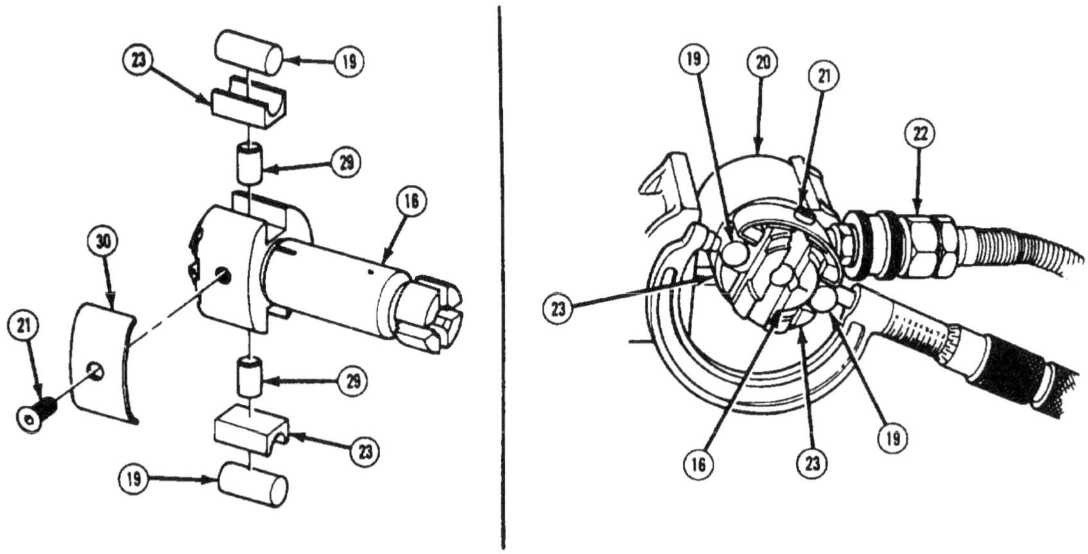

TA50749

3-42. FUEL INJECTOR PUMP HYDRAULIC HEAD MAINTENANCE (Con't).

3. Remove rotor (16) assembly from roller testing fixture (20), Holding rotor assembly at base , install delivery valve (24), spring (26), and new delivery valve stop (27) with screw (28).

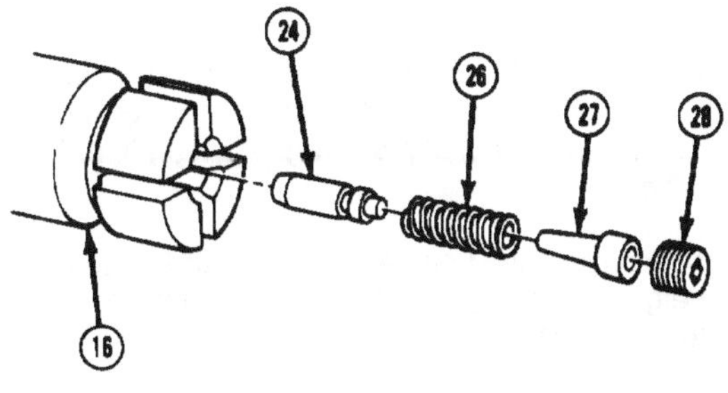

NOTE

Ensure that rollers (19) and shoes (23) remain in their slots when installing rotor (16) assembLy in roller testing fixture (20).

4. Connect air hose (22) to air inlet hole in roller testing fixture (20), and supply 40-100 psi (276-690 kPa) of compressed air. Install rot or assembly in roller testing fixture on air inlet side.

5. Check dimension between between rollers (19). Dimension should be 1.9765-1.9775 in. (50.203-50.229 mm). Adjust leaf spring screw (21) until proper dimension is obtained. Disconnect air hose (22) from air inlet hole.

NOTE

Ensure that rollers (19) and shoes (23) remain in their slots when installing rotor (16) assembly in hydraulic head assembly (10).

6. Install rotor (16) assembly i n hydraulic head assembly (10).

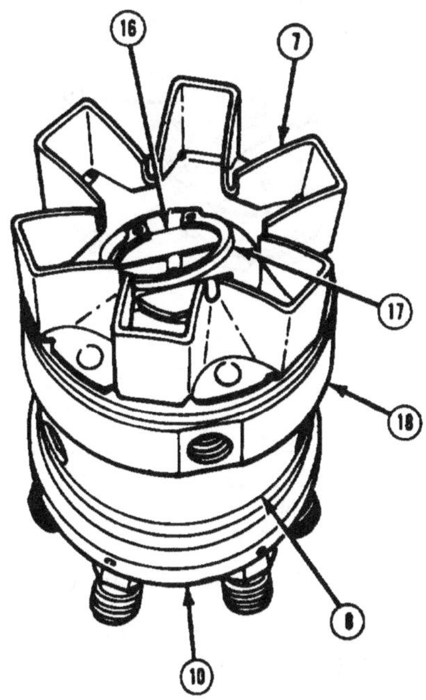

NOTE

When installing cam ring (8), ensure that arrow on cam ring is pointing clockwise and that notch on ca m ring faces front of hydraulic head assembly (10).

7. Install cam ring (8) on hydraulic head assembly (10). Install weight retainer hub (18) and weight retainer (7) on cam ring. Install snapring (17) on rotor (16).

TA50750

3-42. FUEL INJECTOR PUMP HYDRAULIC HEAD MAINTENANCE (Con't).

CAUTION

Carefully hold hydraulic head assembly (10) to keep rotor (16) from falling out. Failure to follow this caution may result in damage to rotor.

8. Invert hydraulic head assembly (10) and place on weight retainer (7). Install 2 retainers (14) in hydraulic head assembly with cutout portions meeting over roll pin locating hole (1 5), Install locating ring (12) at a 90 degree angle to split in retainers, and bottom out against hydraulic head.

9. Invert hydraulic head assembly (10) and place on connectors (13). Install 6 weights (5) in weight retainer (7), with slots facing inward. Install thrust sleeve washer (11) and thrust sleeve (6) into slots of weights.

10. Check to ensure that all weights (5) are level and collapsed against thrust sleeve (6). Reposition any weights that are not properly positioned.

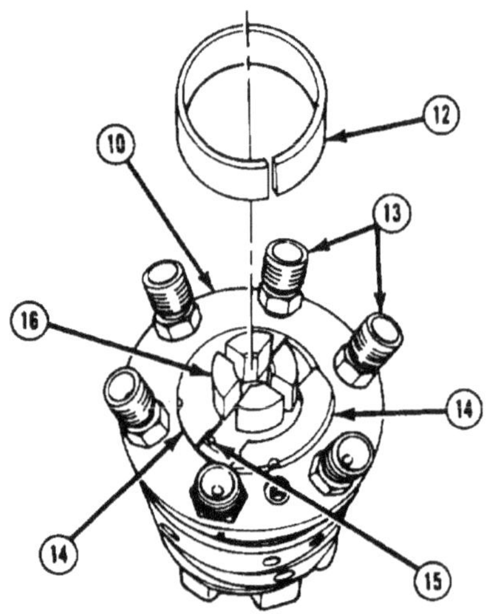

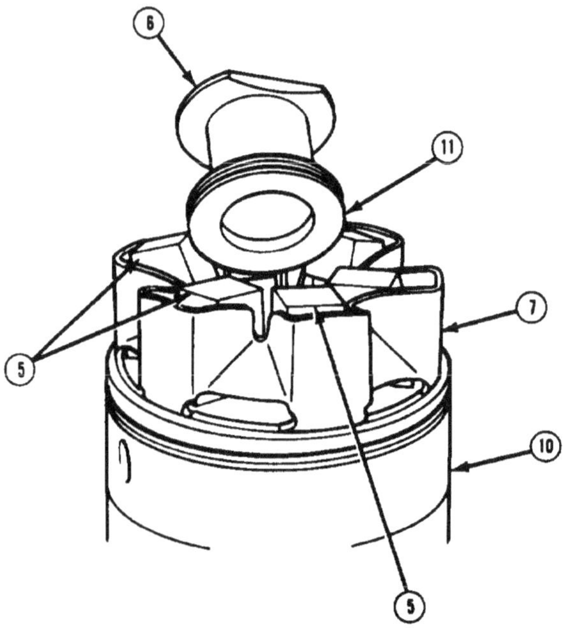

e. INSTALLATION

1. Install new hydraulic head seal (9) on hydraulic head assembly (10).

2. Apply grease to exterior of hydraulic head assembly (10). Apply a light film of "O" ring lubricant to inside edge of fuel injector pump (2) housing.

TA50751

3-42. FUEL INJECTOR PUMP HYDRAULIC HEAD MAINTENANCE (Con't).

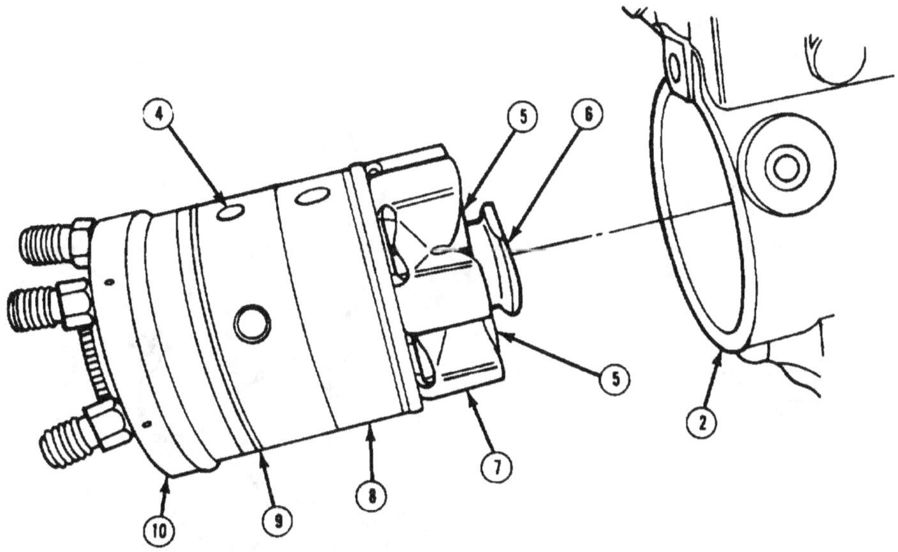

CAUTION

DO NOT force hydraulic head assembly (10) into fuel injector pump (2) housing. Withdraw hydraulic head assembly and try again if it jams. Failure to follow this caution may result in damage to hydraulic head assembly and fuel injector pump housing.

Only install hydraulic head assembly (10) far enough to aline head locking screw (1) holes. Failure to follow this caution may result in damage to hydraulic head seal (9).

NOTE

Ensure that hole in rotor (16) slot is alined with mark on drive shaft tang to ensure proper installation of rotor assembly to drive shaft.

3. Rotate cam ring (8) until smallest hole is 180 degrees opposite metering valve bore (4) and facing downward. Install hydraulic head assembly (1 O) by grasping with both hands and inserting with slight rotary motion. Install 2 head locking screws (1), but DO NOT tighten.

4. Carefully invert fuel injector pump (2).

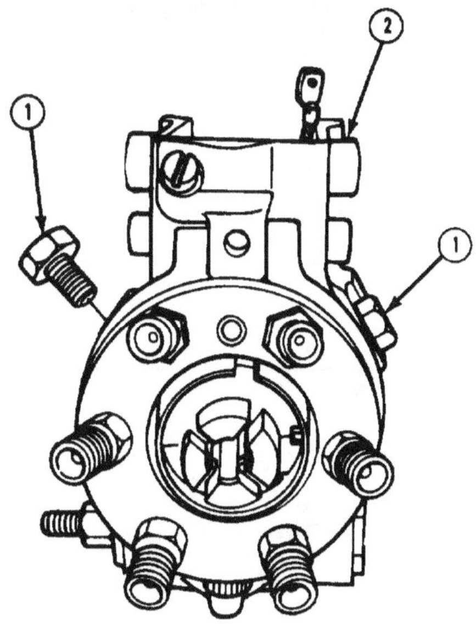

TA50752

3-42. FUEL INJECTOR PUMP HYDRAULIC HEAD MAINTENANCE (Con't).

5. Install new seal on head locating screw (3). Apply a light film of grease to head locating screw. Install new seal and head locating screw and tighten to 15-20 lb.-ft, (20-27 N.m).

6. Check head locating screw (3) to ensure that seal was not sheared during installation. If seal was sheared, remove head locating screw, discard seal, and repeat step 5.

7. Carefully invert fuel injector pump (2). Tighten 2 head locking screws (1) to 15-20 lb.-ft, (20-27 N.m).

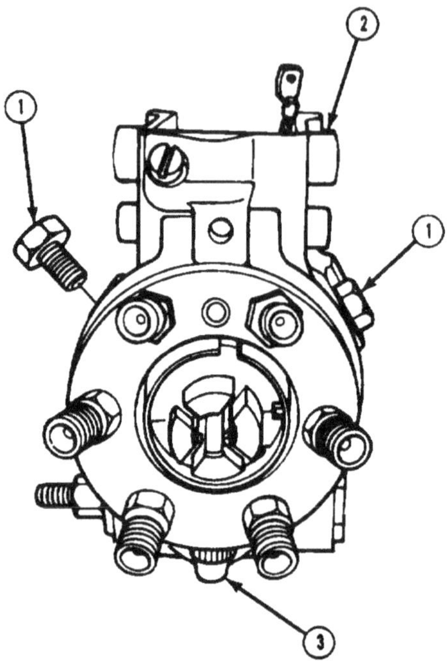

FOLLOW-ON TASKS:

- Install fuel injector pump regulator and transfer pump blades. (See paragraph 3-41)
- Install fuel injector pump servo advance piston. (See paragraph 3-40)
- Install fuel injector pump throttle shaft and governor assembly. (See paragraph 3-39)
- Calibrate fuel injector pump. (See paragraph 3-45)

TA50753

3-43. FUEL INJECTOR PUMP DRIVE SHAFT REPLACEMENT.

This task covers:
a. Removal
b. Inspection
c. Installation

INITIAL SETUP:

Equipment Condition

- Fast idle solenoid and bracket removed. (See TM 9-2320-289-20)
- Fuel injector pump removed. (See paragraph 3-37)
- Throttle shaft and governor assembly removed (if removing drive shaft from 1984 model truck). (See paragraph 3-39)

Materials/Parts

- One retaining ring
- Three seals
- " O" ring lubricant (Item 42, Appendix B)

Tools/Test Equipment

- Drive shaft seal protector

a. REMOVAL

CAUTION

Ensure that work area, tools, and hands are kept clean when repairing fuel injector pump. Failure to follow this caution may result in damage to fuel injector pump components.

NOTE

- If removing drive shaft from 1984 model trucks, perform steps 1, 2, and 4.
- If removing drive shaft from 1985-87 model trucks, perform steps 2-4.

3-43. FUEL INJECTOR PUMP DRIVE SHAFT REPLACEMENT (Con't).

NOTE

- Drive shaft (1) is retained by retaining clip on 1984 model trucks. Drive shaft is retained by spiral-shaped retaining ring (4) on 1985-87 model trucks.
- Drive shaft (1) and retaining clip on 1984 model trucks are not provisioned, If drive shaft is removed from 1984 model truck, 1985-87 model truck drive shaft assembly must be used for installation.

1. Rotate drive shaft (1) until one of raised portions of retaining clip is visible through top of fuel injector pump housing (5). Using long, thin, hook-shaped tool, hook retaining ring and pull slightly to rear. Using needle nose pliers, remove and discard retaining clip.
2. Rotate drive shaft (1) until dowel pin (6) is on top. Pull drive shaft from fuel injector pump housing (5) using rotating motion, Remove and discard dowel pin if damaged. Discard 1984 model truck drive shaft.

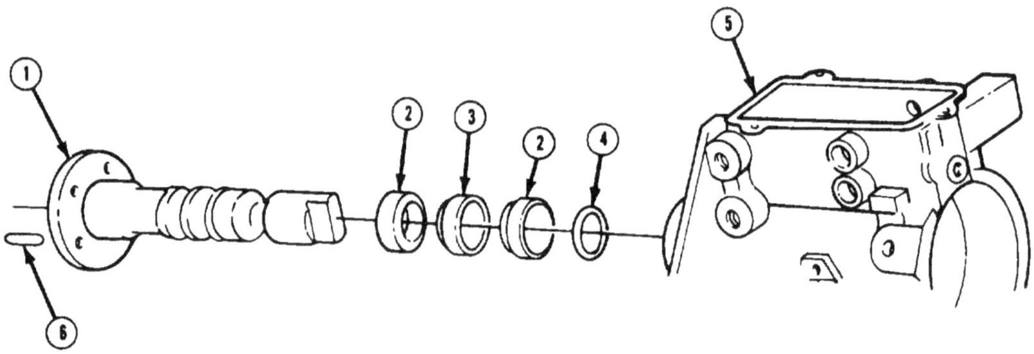

CAUTION

Ensure that any pieces of retaining ring (4) that have broken off inside fuel injector pump housing (5) are removed before installation. Failure to follow this caution may result in damage to fuel injector pump components.

3. Remove and discard retaining ring (4) from drive shaft (1).
4. Remove and discard 2 black seals (2) and red seal (3).

TA50754

3-43. FUEL INJECTOR PUMP DRIVE SHAFT REPLACEMENT (Con't).

b. INSPECTION

1. Measure distance across flat of drive shaft tang (7). Replace drive shaft if measurement is less than 0.305 in, (7,75 mm).

2. Inspect retaining ring groove (8) and seal grooves (9) to ensure that they are free of nicks and scratches. Replace drive shaft if nicked or scratched.

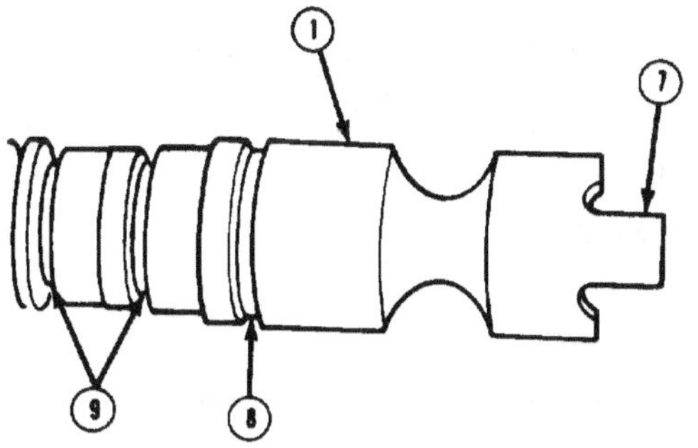

c. INSTALLATION

NOTE

Ensure that drive shaft seal protector is used to install seals (2 and 3). Drive shaft seal protector should be lubricated with "O" ring lubricant before installing each seal.

1. Install 1 black seal (2) with cup end facing drive shaft flange. Install red seal (3) with cup end facing drive shaft tang (7). Install remaining black seal with cup end facing drive shaft tang. Ensure that all seals completely seat in seal grooves (9).

CAUTION

Ensure that retaining ring (4) securely fits on retaining ring groove (8). Failure to follow this caution may cause drive shaft (1) to be loose.

2. Install new retaining ring (4) on retaining ring grooves (8)

NOTE

Ensure that mark on drive shaft tang (7) is alined with hole in rotor slot to ensure proper installation of drive shaft (1) to rotor.

3. Install new dowel pin (6) on drive shaft (1) if removed. Rotate drive shaft until dowel pin is on top. Push drive shaft into fuel injector pump housing (5) using rotating motion.

FOLLOW-ON TASKS:

- Install throttle shaft and governor assembly if removed. (See paragraph 3-39)
- Install fuel injector pump. (See paragraph 3-37)
- Install fast idle solenoid and bracket. (See TM 9-2320-289-20)

TA50755

3-44. FUEL INJECTOR PUMP TESTING AND TIMING.

This task covers: a. Testing b. Timing

INITIAL SETUP:

Equipment Condition

- Fuel injector pump removed (if testing fuel injector pump). (See paragraph 3-37)

Tools/Test Equipment

- Torque wrench

MeterialsIParts

- Calibration fluid (Item 13, Appendix B)

General Safety Instructions

- Use only effective chip guarding and personnel protective equipment (goggles/shield, gloves, etc.) when using compressed air.

a . TESTING

WARNING

Use only effective chip guarding and personnel protective equipment (goggles/shield, gloves, etc.) when using compressed air, failure follow this warning may result in serious injury to personnel.

1. Drain all fuel from fuel injector pump into a suitable container.
2. Remove protective plug from fuel inlet pipe and connect compressedair hose.
3. Submerge fuel injector pump in clean calibration fluid, and apply 20 psi (138 kPa) of clean, dry, compressed air to fuel inlet pipe. Keep fuel injector pump submerged for 10 minutes to allow any trapped air to escape.
4. Check for hubbies, indicating leakage, after 10 minutes. If there is no leakage, reduce air pressure to 2 psi (14 kPa) for 30 seconds. If there still is no leakage, increase air pressure to 20 psi (138 kPa). If there still is no leakage, fuel injector pump is ready for use.
5. If fuel injector pump leaks at any time after 10 minutes, note location of leakage. Remove fuel injector pump from calibration fluid. Inspect leaking area for damage or looseness of mounting hardware. Perform appropriate maintenance task to repair leakage.
6. Disconnect compressed air hose from fuel inlet pipe and install plug.

b . T I M I N G

1. Loosen 3 nuts retaining fuel injector pump to crankcase cover. Aline scribe marks on fuel injector pump flange and crankcase cover. Tighten nuts to 30 Ib.-ft. (41 N.m).

FOLLOW-ON TASKS:

- Install fuel injector pump if removed, (See paragraph 3-37)

3-45. FUEL INJECTOR PUMP CALIBRATION.

This task covers:

a. Testing Equipment Requirements
b. Preparation for Calibration
c. Adjustment
d. Preliminary Testing
e. Calibration
f. After Calibration
g. Checking and Adjusting Timing Mark

INITIAL SETUP:

Equipment condition

- Fuel injector pump removed. (See paragraph 3-37)

Materials/Parts

- One face cam pin
- One timing cover gasket
- Calibration fluid (Item 13, Appendix B)
- Fuel oil (Item 25, 26, or 27, Appendix B)
- Pipe sealant (Item 61, Appendix B)

Tools/Test Equipment

- Adjustable dc voltage source
- Air-timing fixture
- Fuel flowmeter
- Fuel injector pump test stand
- Orifice fitting
- Protractor
- Retention torque adapter
- Throttle advance gage
- Throttle advance gage gasket
- Torque wrench
- Transfer pressure tap adapter

General Safety Instructions

- Ensure that governor cover is installed before operation of fuel injector pump.
- DO NOT perform this procedure near fire, flames, or sparks.
- Compressed air used for cleaning purposes should never exceed 30 psi (207 kPa).

a. TESTING EQUIPMENT REQUIREMENTS

1. Injector lines used for testing must be 0,093 in. (2.36 mm) inside diameter and 25 in. (63.5 cm) long.

2. Test stand injector nozzles must include a 0.020 in. (0.5 mm) orifice plate.

3. Calibration fluid temperature must be checked as close to transfer pump inlet as possible. Operating temperature should be 110°F-115°F (43.3°C-46.1°C).

4. Fuel supply pressure to fuel injector pump must be set at 4.5-5.5 psi (31-38 kPa).

5. The fuel injector pump test stand must include the following:

   (a) Self-alining zero-backlash drive coupling.

   (b) Digital tachometer.

   (c) Adjustable dc voltage source. Used to check operation of cold advance solenoid and shutoff solenoid.

   (d) Transfer pressure gage. Used to measure transfer pump, pressure. Operating range should be 0-160 psi (0-1103 kPa), calibrated in 1 psi (6,895 kPa) increments.

   (e) Housing pressure gage. Used to measure housing pressure. Operating range should be 0-30 psi (0-207 kPa).

3-45. FUEL INJECTOR PUMP CALIBRATION (Con't).

(f) Temperature gage,

(9) Fuel supply valve,

(h) Orifice fitting. Located in fuel supply line between fuel supply pump and fuel supply valve. Orifice fitting is necessary to maintain correct calibration fluid temperature.

(i) Shutoff valve .

(j) Fuel supply pump.

(k) Fuel return line,

(1) Variable speed control.

b. PREPARATION FOR CALIBRATION

1. Flush fuel injector pump with calibration fluid.
2. Check fuel injector pump drive shaft for freedom of rotation.
3. Engage fuel injector pump drive shaft on test stand drive coupling. Install fuel injector pump on test stand mounting bracket with 3 washers and nuts, Tighten nuts to 30 lb.-ft. (41 N.m).
4. Install transfer pump inlet connector (8).
5. Loosely connect 8 test stand injector lines (4) to fuel injector pump connectors.
6. Install transfer pressure tap adapter (3), Install shutoff valve (5) between transfer pressure tap and transfer pressure gage (6).
7. Remove 2 screws, timing cover, and timing cover gasket. Install throttle advance gage gasket and throttle advance gage (13) in timing cover opening. Set throttle advance gage to indicate zero. Discard timing cover gasket.
8. Connect housing pressure gage line (11) to housing pressure connector (12).
9. Connect fuel supply line (9) to transfer pump inlet connector (8).

NOTE

Ensure that fuel return line (2) is connected so that fuel is directed through flowmeter.

10. Connect test stand fuel return line (2) to fuel injector pump return line fitting (1).
11. Connect positive test lead (14) from adjustable dc voltage source (16) to shutoff solenoid terminal. Connect ground lead (15) to mounting plate.
12. Open fuel supply valve (1 O) and turn to test stand "ON" position. Allow fuel to bleed from loosened fittings of test stand injector lines (4) for several seconds, then tighten all fittings.

3-45. FUEL INJECTOR PUMP CALIBRATION (Con't).

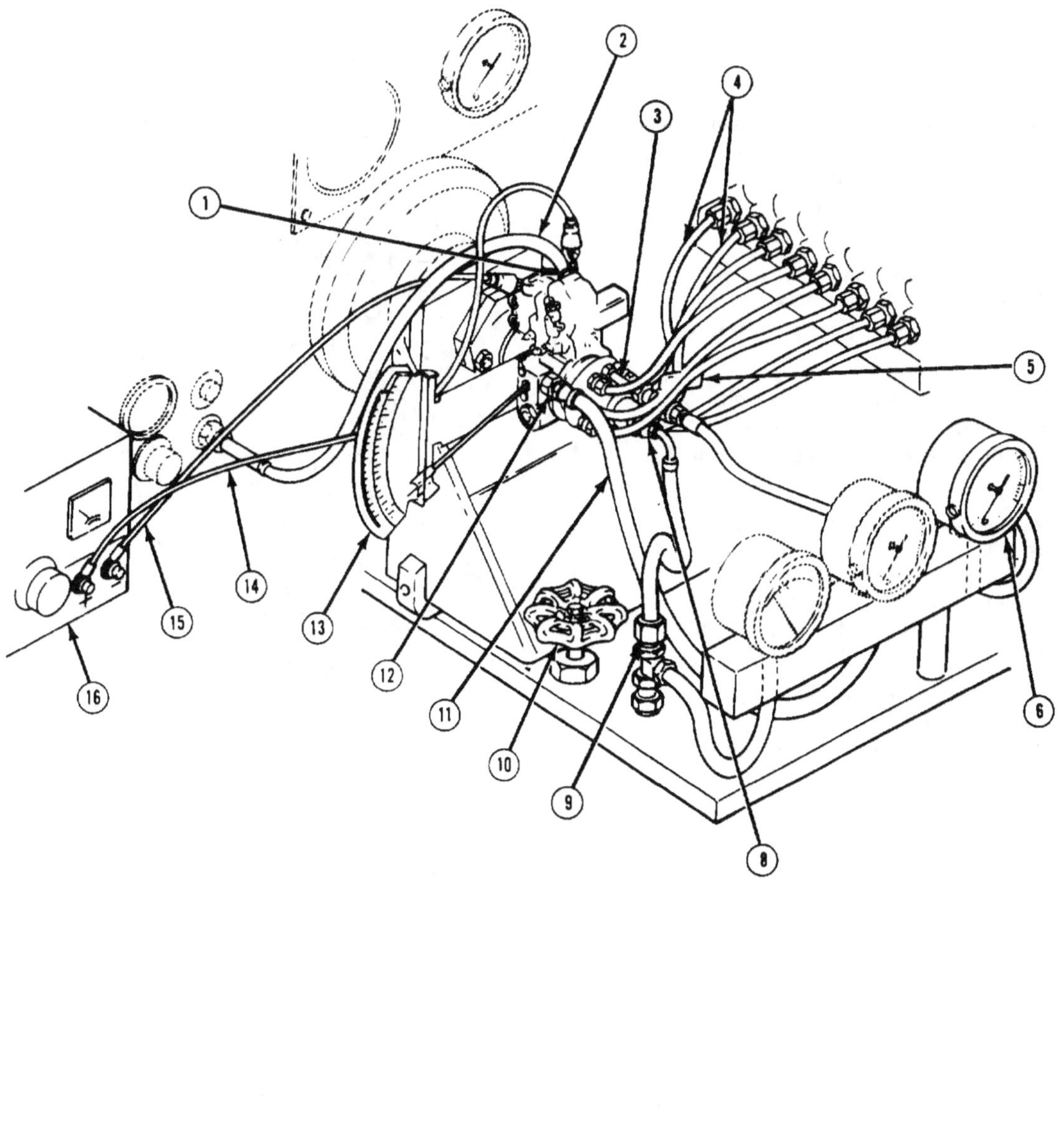

TA50756

3-45. FUEL INJECTOR PUMP CALIBRATION (Con't).

c . ADJUSTMENT

NOTE

Adjustment in step 1 should be made during applicable step, Adjustments in steps 2-5 should be made after all calibration is completed, unless otherwise stated,

1. If any throttle advance gage (13) reading is not correct, adjust position of servo adjusting screw (22) to obtain proper reading. After all testing is complete, apply pipe sealant to servo adjusting screw.
2. If any flowmeter reading is too low, replace metering screw with a smaller metering screw. (See paragraph 3-39)
3. If any flowmeter reading is too high, replace metering screw with a larger metering screw. (See paragraph 3-39)
4. If any housing pressure reading is not correct, test fuel injector pump. (See paragraph 3-44) If no leaks are detected, remove hydraulic head and check for incorrectly assembled components. (See paragraph 3-42)
5. If any transfer pump pressure reading is not correct, remove transfer pump blades and regulator and check for incorrectly assembled components. (See paragraph 3-41)

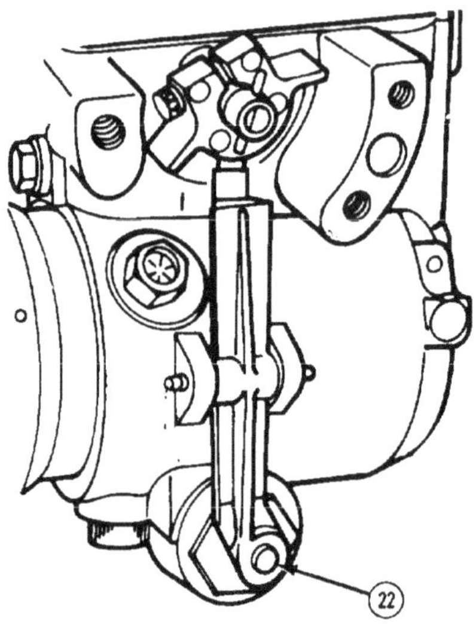

d . PRELIMINARY TESTING

1. Set test stand supply pump boost pressure at 4.5-5.5 psi (31-38 kpa).

NOTE

Test specifications refer to engine rpm. Most test stand tachometers register fuel injector pump rpm, which is half of engine rpm. Ensure that proper adjustments are made during calibration.

2. Set rpm at 2000, with throttle lever (17) in wide open throttle position. Allow fuel injector pump to run for 10 minutes to clear remaining air from system and bring fuel injector pump to operating temperature.
3. Set rpm at 2000, with throttle lever (17) in wide open throttle position. Set transfer pressure gage (6) to indicate 58-64 psi (400-441 kPa). Flowmeter reading must be 225-375 cc, Perform ADJUSTMENT if necessary.

TA50757

3-45. FUEL INJECTOR PUMP CALIBRATION (Con't).

4, Set rpm at 600, with throttle lever (17) in low idle position. Housing pressure gage (7) reading should be 8–12 psi (55–83 kPa).

5 . Set adjustable dc voltage source (16) to 9.0 volts. Connect positive test lead (14) to cold advance solenoid terminal. Housing pressure gage (7) reading should be 0-1 psi (0-6.9 kPa). Disconnect positive test lead.

6 . Set rpm at 600, with throttle lever (17) in low idle position. Flowmeter reading must be 12.5-16.5 cc. If flowmeter reading is not correct, perform step 7.

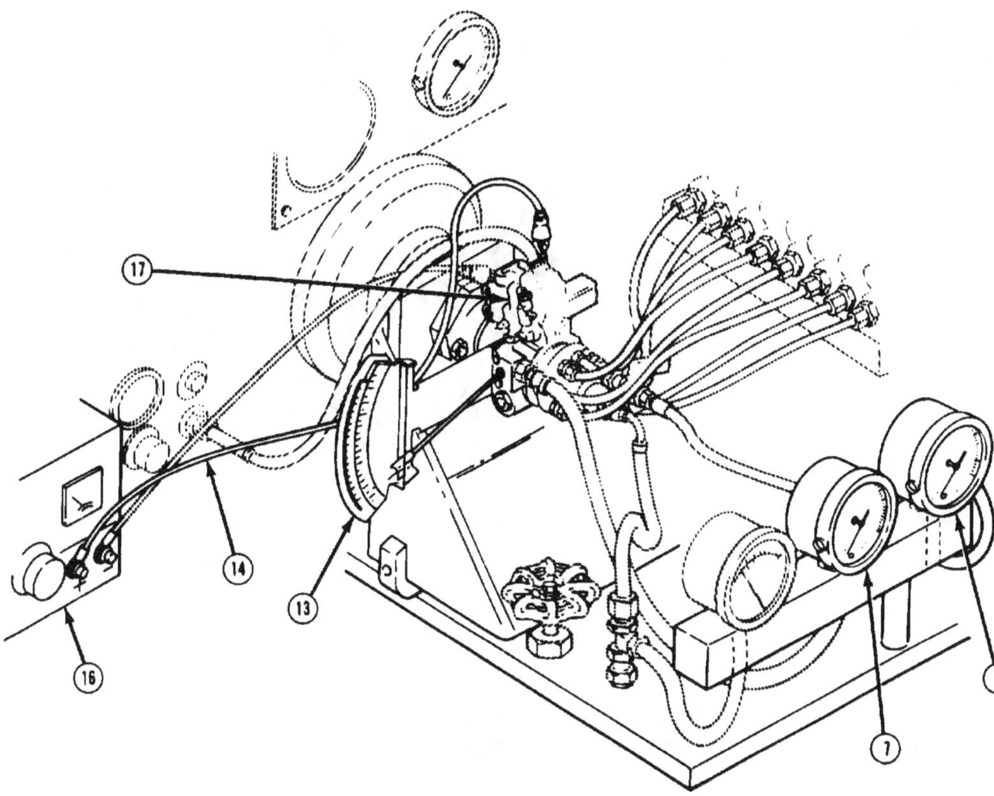

TA50758

3-45. FUEL INJECTOR PUMP CALIBRATION (Con't).

WARNING

Ensure that governor cover is installed before operation of fuel injector pump. Pressurized fuel may cause injury to personnel and/or damage to equipment.

7. Remove governor cover. (See paragraph 3-38) Place fuel injector pump in low idle position, Rotate adjusting nut (24) as needed to obtain specification. Install governor cover and repeat step 6.

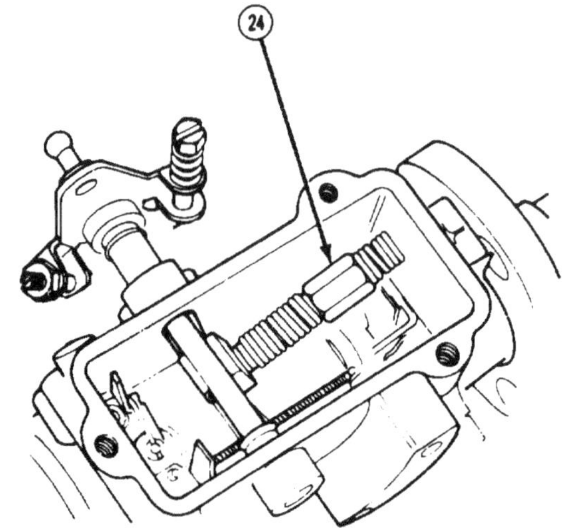

8. Set rpm at 400, with throttle lever (17) in low idle position. Connect positive test lead (14) to shutoff solenoid terminal. Listen for "clicking" sound from shutoff solenoid, indicating proper operation.

9. Shut off fuel supply valve (1 0) and check vacuum gage for minimum lift of 18 in. Hg. Turn on fuel supply valve.

TA50759

3-45. FUEL INJECTOR PUMP CALIBRATION (Con't).

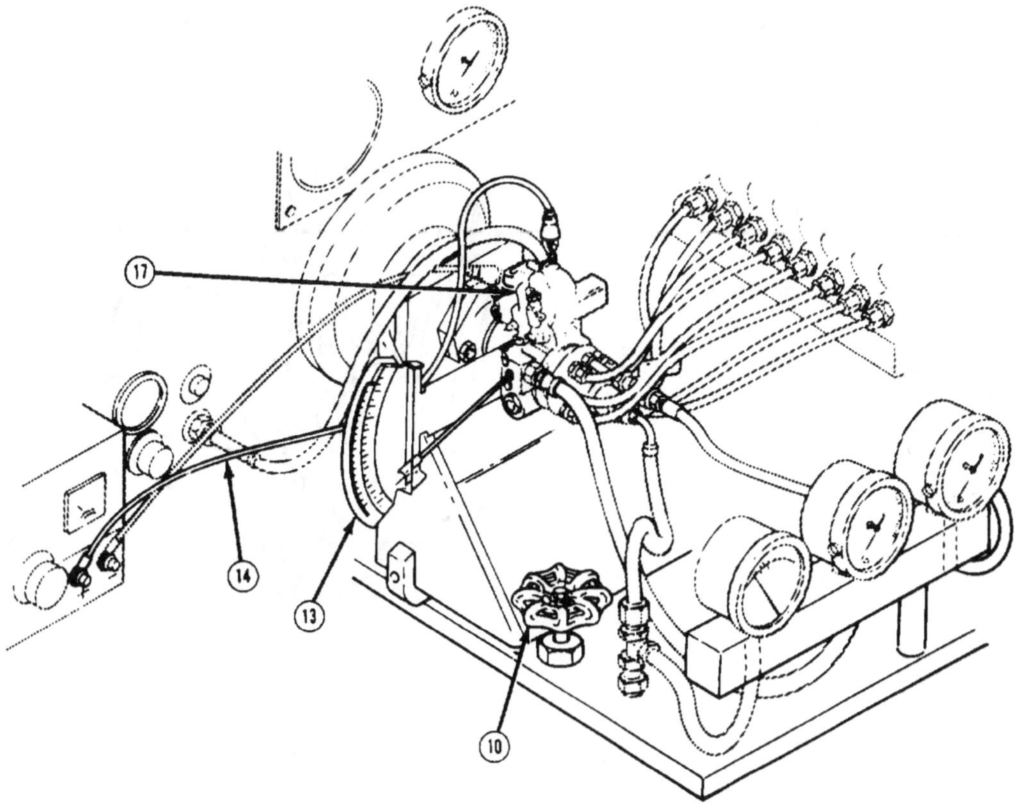

10. Set mechanical light load advance as follows:

    (a) Perform paragraph 3-39, INSTALLATION, step 10.

    (b) Place throttle lever (17) in wide open throttle position. Rotate face cam (19) to maximum lift position in contact with cam roller (23) on rocker lever (21).

    (c) Set rpm at 2000. Adjust position of servo adjusting screw (22) to obtain 0.5-1.5 degrees ca m advance indication on throttle advance gage (13).

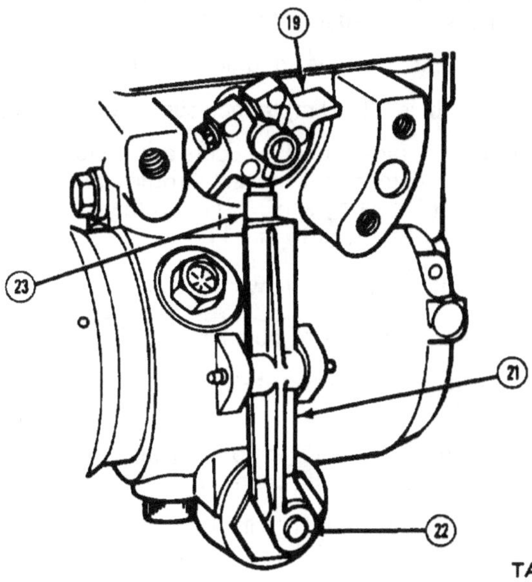

TA50760

3-45. FUEL INJECTOR PUMP CALIBRATION (Con't).

(d) Set rpm at 1500 and rotate throttle lever (17) to obtain a flowmeter reading of 21,5-23.5 cc. Rotate face cam (19) to obtain 2.5-4.5 degrees cam advance indication on throttle advance gage (13).

(e) Repeat paragraph 3-39, INSTALLATION, step 10.

11. Check face cam retention torque as follows:

(a) Turn test stand to "OFF" position. Drive out and discard face cam pin (20).

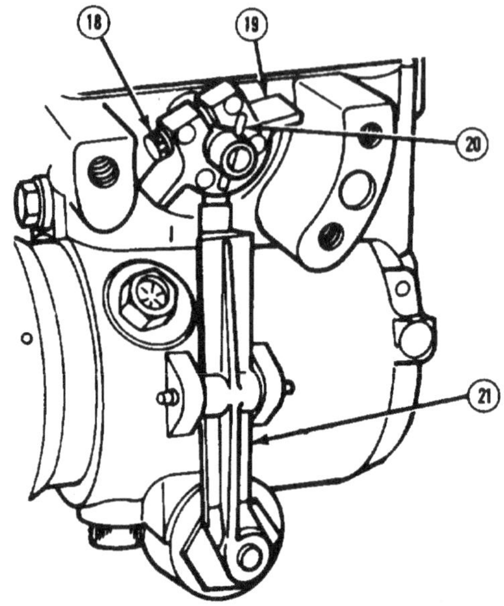

NOTE

Scribe mark will provide a reference point for checking face cam (19) movement.

(b) Position throttle lever (17) against low idle stop and scribe a line on throttle shaft housing in line with rear edge of face cam (19),

(c) Install retention torque adapter over throttle shaft end and engage face cam (19).

(d) Apply 22 lb.-in. (2.5 N.m) to face cam (19). Face cam must not rotate on throttle shaft.

(e) If rotation of face cam (19) occurs at or below 22 lb.-in. (2.5 N.m), loosen screw (18) and remove face cam. Inspect throttle shaft for damage. If throttle shaft is damaged, replace. (See paragraph 3-39) If throttle shaft is not damaged, replace face cam and screw, Repeat steps 10 and 11.

12. Set total throttle travel as follows:

(a) Turn test stand to "ON" position.

(b) Set rpm at 1500 and rotate throttle lever (17) to obtain a flowmeter reading of 19.5-25.5 CC.

(c) Lock throttle lever (17) in position and install protractor (25) onto rocker lever (21). Aline zero-degree mark with center of rib on rocker lever.

(d) Unlock throttle lever (17) and rotate to wide open throttle position, Adjust maximum travel screw to indicate 56.5 degrees on protractor (25).

TA50761

3-45. FUEL INJECTOR PUMP CALIBRATION (Con't).

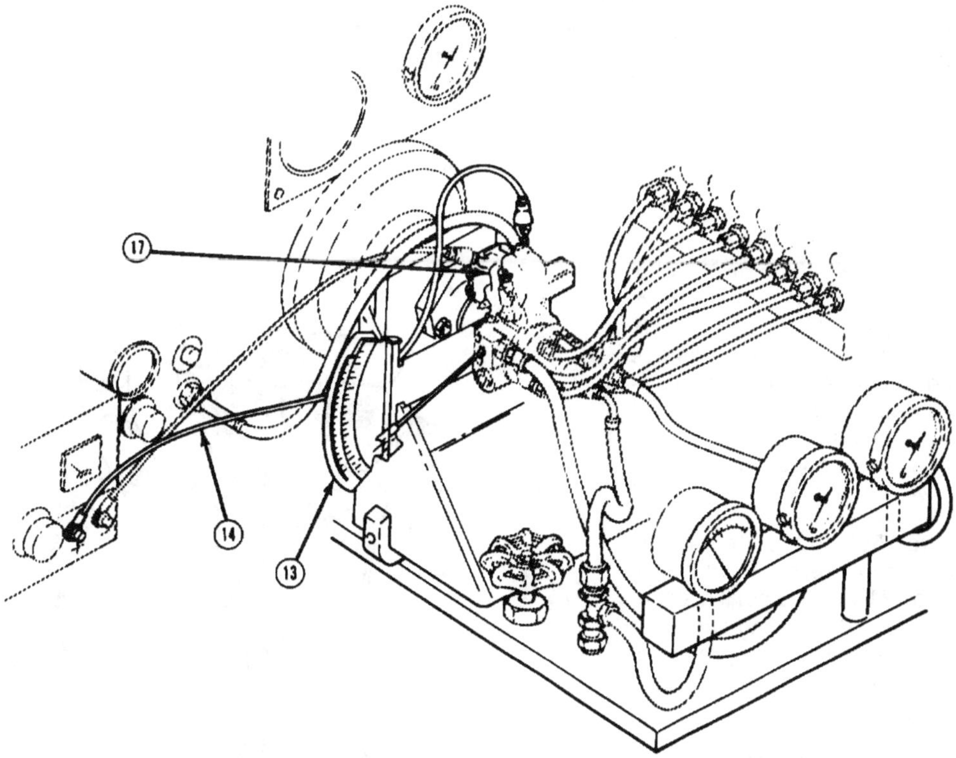

(e) Place throttle to low idle position and aline zero-degree mark on protractor (25) with center of rib on rocker lever (21).

(f) Rotate throttle lever (17) to wide open throttle position. Reading on protractor (25) should be 73-82 degrees.

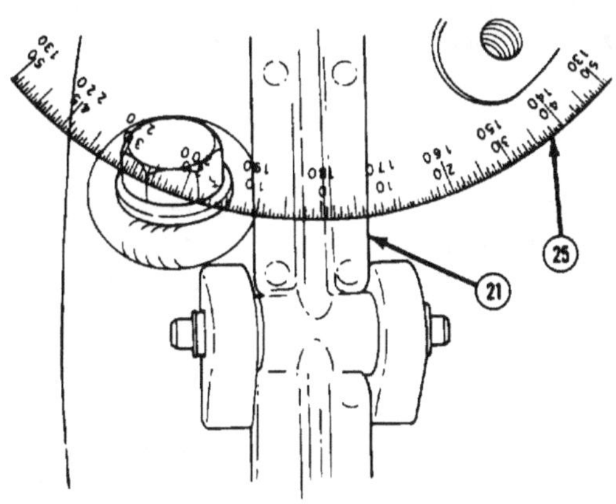

e . CALIBRATION]

1. Set rpm to 1500, with throttle lever (17) in wide open throttle position. Disconnect positive test lead (14) from -- shutoff solenoid terminal. Maximum flowmeter reading should be 4.0 cc. Connect positive test lead to shutoff solenoid terminal.

TA50762

3-45. FUEL INJECTOR PUMP CALIBRATION (Con't).

2 . Set rpm to 150, with throttle lever (17) in wide open throttle position. Minimum flowmeter reading should be 35.0 cc. Open shutoff valve (5). Minimum transfer pressure gage (6) reading should be 12 psi (83 kPa). Close shutoff valve.

3 . Set rpm to 600, with throttle lever (17) in low idle position. Flowmeter reading should be 12.5-16.5 cc. Minimum throttle advance gage (13) reading should be 1.5 degrees.

4 . Adjust position of servo adjusting screw (22) to obtain 2.25 degrees cam advance indication on throttle advance gage (13). Disconnect positive test lead (14) from shutoff solenoid terminal, Connect positive test lead to cold advance solenoid terminal. Housing pressure gage (7) reading should 0-1 psi (0-6.9 kPa). Disconnect positive test lead.

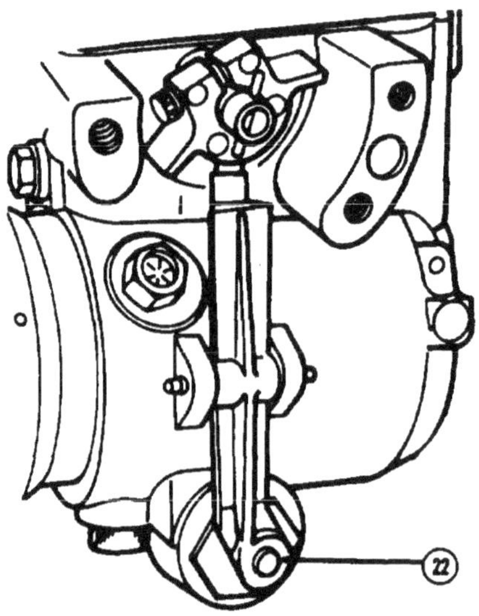

5 . Set rpm to 1500 and rotate throttle lever (17) to obtain a flowmeter reading of 21 .5-23.5 cc. Throttle advance gage (13) reading should be 2.5-4.5 degrees. Housing pressure gage (7) reading should be 8-12 psi (55-83 kPa).

6 . Set rpm to 2000, with throttle lever (17) in wide open throttle position. Flowmeter reading should be 47.5-51.5 cc. Throttle advance gage (13) reading should be 0.5-2.5 degrees. Housing pressure gage (7) reading should be 8-12 psi (55-83 kPa).

7 . Set rpm to 3200, with throttle lever (17) in wide open throttle position. Throttle advance gage (13) reading should be 4.25-6.75 degrees. Housing pressure gage (7) reading should be 8-12 psi (55-83 kPa).

CAUTION

Ensure that throttle lever (17) remains in low idle position no longer than length of time required to record gage readings. Failure to follow this caution may result in damage to fuel injector pump.

8 . Place throttle lever (17) in low idle position, Maximum throttle advance gage (13) reading should be 10.0 degrees. Housing pressure gage (7) reading must be 8-12 psi (55-83 kPa).

9 . Set rpm to 4000, with throttle lever (17) in wide open throttle position. Minimum flowmeter reading should be 40.0 cc.

TA50763

3-45. FUEL INJECTOR PUMP CALIBRATION (Con't).

10. Open shutoff valve (5), Set rpm to 4400, with throttle lever (17) in wide open throttle position. Maximum flowmeter reading must be 4.0 cc. Transfer pressure gage (6) reading must be 125 psi (862 kPa). Close shutoff valve,

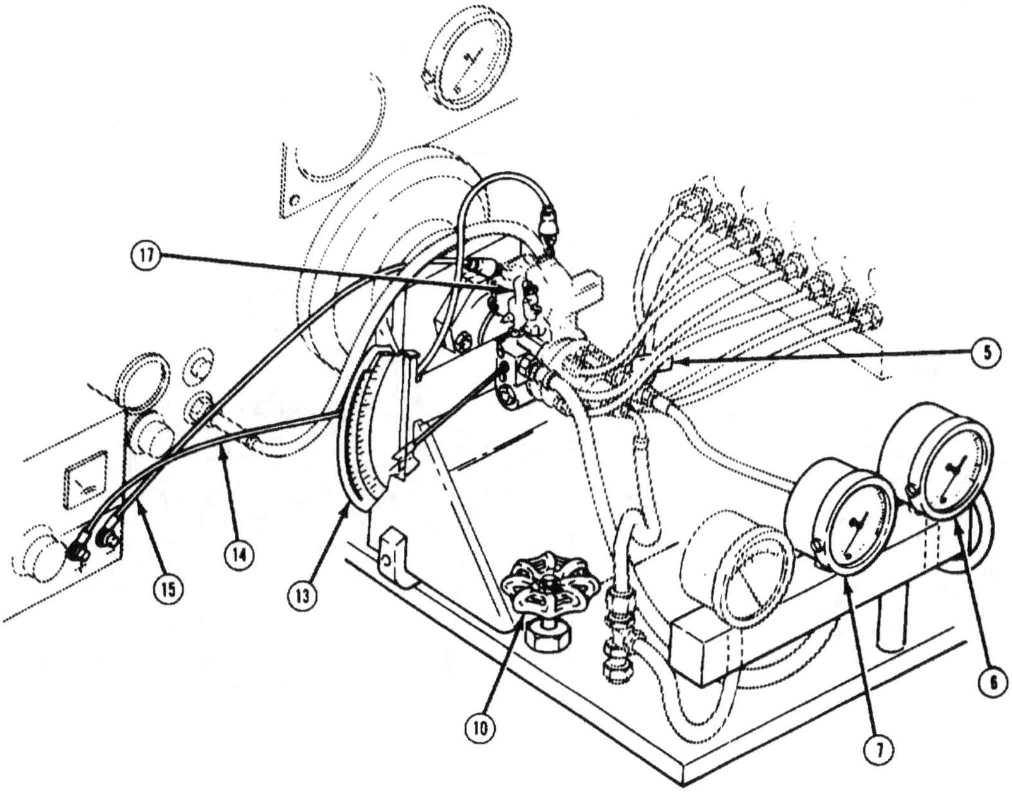

f. AFTER CALIBRATION

1. Return test stand to "OFF" position.
2. Disconnect ground lead (15) and positive test lead (14),
3. Shut off fuel supply valve (10).
4, Disconnect all fuel and gage connections from fuel injector pump.
5. Remove throttle advance gage (13), Discard throttle advance gage gasket. Install new timing cover gasket and timing cover with 2 screws.
6. Remove 3 nuts and washers and fuel injector pump from test stand mounting bracket.

TA50764

3-45. FUEL INJECTOR PUMP CALIBRATION (Con't).

g. CHECKING AND ADJUSTING TIMING MARK

1. Loosen 2 adjusting screws (31) and set air-timing fixture (26) to 0,25-0,75 degrees, Tighten adjusting screws,
2. Rotate fuel injector pump drive shaft until locating pin (34) is at the 7 o'clock position.
3. Install air-timing fixture (26) on fuel injector pump, ensuring that locating pin (34) engages mating hole (30). Secure air-timing fixture by tightening knurled screw (29) finger tight against fuel injector pump flange (33).

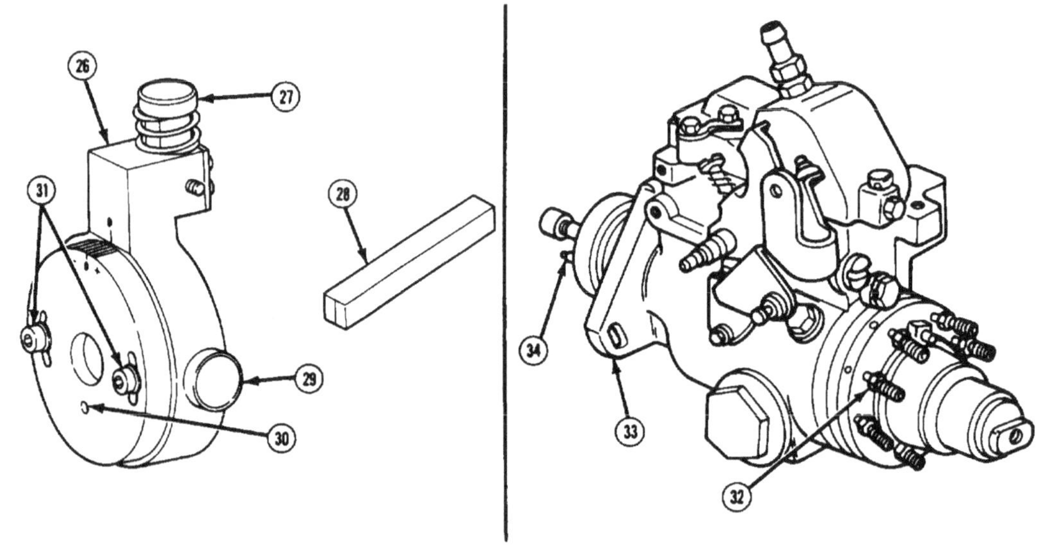

NOTE

#1 discharge fitting (32) can be identified by a number "1" cast on fuel injector pump housing.

4. Supply 60-100 psi (414-689 kPa) of dry, filtered air to #1 discharge fitting (32).

NOTE

The point when cam rollers contact cam ring is point when injection begins at #l discharge fitting (32).

5. Loosen knurled screw (29) and rotate air-timing fixture (26) counterclockwise until cam rollers can be felt making contact with cam ring. Rotate air-timing fixture back to its original position.

TA50765

3-45. FUEL INJECTOR PUMP CALIBRATION (Con't).

6. Repeat step 5 several times to get a definite feel of when rollers first make contact with cam ring.

7. Rotate air-timing fixture (26) counterclockwise until cam rollers can be felt making contact with cam ring, Tighten knurled screw (29).

8. Remove marking scribe adapter (27) and spring from air-timing fixture (26) and install optical sight gage (28).

9. Sight through optical sight gage (28) and check for position of existing timing mark on fuel injector pump flange (33). If existing mark is not in line with timing line on optical sight gage, perform steps 10-15. If existing mark is correct, perform steps 13-15.

10. Remove optical sight gage (28) from air-timing fixture (26). Loosen knurled screw (29) and remove air-timing fixture.

CAUTION

Drive shaft and pilot area of fuel injector pump must be covered with a clean rag during step 11. Failure to follow this caution may result in damage to drive shaft bearings and seals.

11. File off old timing mark.

12. Repeat steps 2, 3, and 5 through 7. Tap marking scribe adapter (27) to place a new timing mark on fuel injector pump flange (33). Remove air-timing fixture (26) and disconnect air supply from #1 discharge fitting (32).

3-45. FUEL INJECTOR PUMP CALIBRATION (Con't).

WARNING

- Diesel fuel is flammable, DO NOT perform this procedure near fire, flames, or sparks.' A fire extinguisher must be on hand in work area. Failure to follow this warning may result in serious injury or death to personnel,
- Compressed air used for cleaning purposes should never exceed 30 psi (207 kPa). Use only effective chip guarding and personnel protective equipment (goggles/shield, gloves, etc.). Failure to follow this warning may result In serious injury to personnel.

13. Clean exterior of fuel injector pump with fuel oil and dry with compressed air.
14. Apply pipe sealant to servo adjusting screw (22).
15. Install new face cam pin (20),

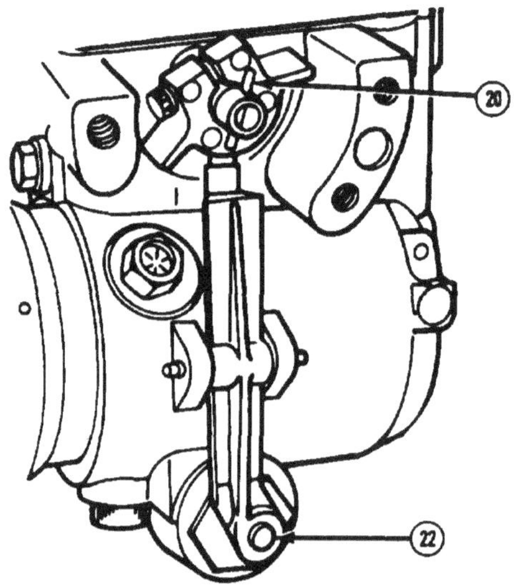

FOLLOW-ON TASKS:

- Install fuel injector pump. (See paragraph 3-37)

TA50766

3-46. TRANSMISSION VACUUM VALVE REPLACEMENT.

This task covers: a, Removal b. Installation

INITIAL SETUP:

Equipment Condition

- Air cleaner removed. (See TM 9-2320-289-20)

Tools/Test Equipment

- Valve gage block, J-33043
- Vacuum gage
- Vacuum pump

a. REMOVAL

NOTE

If only adjusting transmission vacuum valve (5), perform REMOVAL, step 1, and INSTALLATION, steps 2-7.

1. Disconnect vacuum hose connector (2) from transmission vacuum valve (5).
2. Remove 2 screws (6) and transmission vacuum valve from fuel injector pump (1).

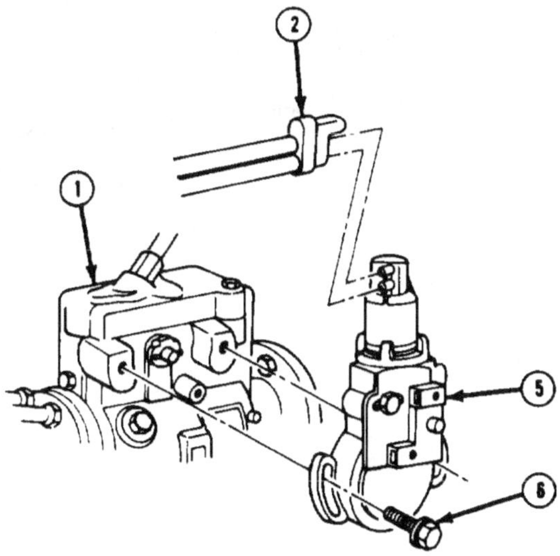

TA50767

3-46. TRANSMISSION VACUUM VALVE REPLACEMENT (Con't).

b. INSTALLATION

NOTE

Transmission vacuum valve (5) must always be adjusted when it is installed.

1. Install transmission vacuum valve (5) on fuel injector pump (1) with 2 screws (6). DO NOT fully tighten screws. Transmission vacuum valve must be free to rotate on fuel injector pump.
2. Attach vacuum source to lower vacuum nipple (4). Attach vacuum gage to upper vacuum nipple (3), Bring vacuum down to 20 in. Hg.

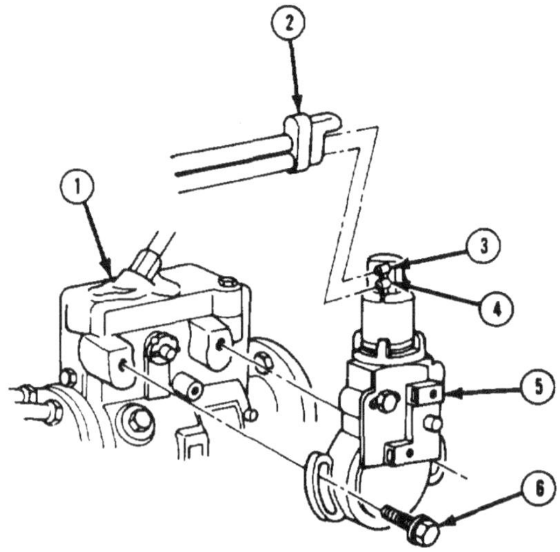

TA50768

3-46. TRANSMISSION VACUUM VALVE REPLACEMENT (Con't).

3. Insert valve gage block (9) between gage boss (1 O) and wide-open stop screw (8) on throttle lever (7). Rotate and hold throttle lever against valve gage block.

4. Loosen 2 screws (6), if not already loose. Slowly rotate transmission vacuum valve (5) clockwise (toward front of truck) until vacuum gage reads 8 in. Hg. Hold transmission vacuum valve in this position, and tighten 2 screws.

5. Check adjustment by releasing throttle lever (7) then rotating throttle lever against valve gage block (9). If vacuum gage does not read 8 in. Hg, loosen 2 screws (6) and repeat steps 2-5.

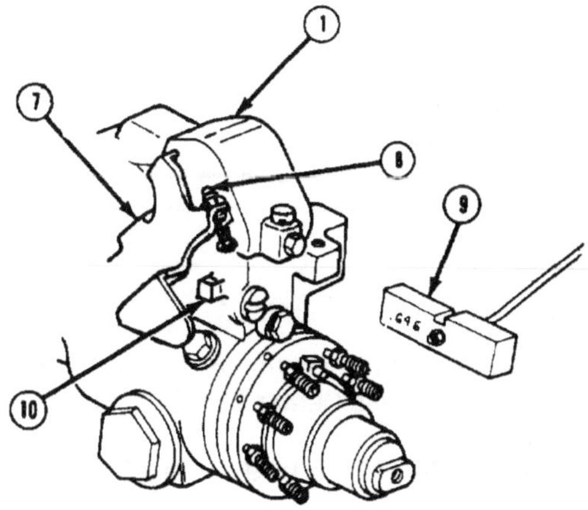

6. Disconnect vacuum source and vacuum gage.

7. Connect vacuum hose connector (2) to transmission vacuum valve (5).

FOLLOW-ON TASKS:

- Install air cleaner. (See TM 9-2320-289-20)

TA50769

3-47. BROKEN GLOW PLUG REMOVAL.

This task covers: a. Removal

INITIAL SETUP:

Equipment Condition

I Fuel injector nozzle at broken glow plug removed. (See paragraph 3-5)

General Safety Instructions

. Compressed air used to remove glow plug should never exceed 30 psi (207 kPa).

a. REMOVAL

1. Place finger over fuel injector nozzle port. Rotate torsional damper to raise piston in affected cylinder to "Top Dead Center." Compression will change when "Top Dead Center" is reached,

2. Unscrew and move glow plug back and forth to break off glow plug tip into cylinder head prechamber.

WARNING

Compressed air used to remove glow plug should never exceed 30 psi (207 kPa). Use only effective chip guarding and personnel protective equipment (goggles/shieid, gloves, etc.). Failure to follow this warning may result in serious injury to personnel.

3. Direct compressed air into glow plug port to remove glow plug tip through fuel injector nozzle port,

4. If preceding procedure does not remove broken glow plug, remove cylinder head. (See paragraph 3-7)

FOLLOW-ON TASKS:

- Install fuel injector nozzle. (See paragraph 3-27)
- Install new glow plug. (See TM 9-2320-289-20)

Pages 3-175 through 3-177 have been rescinded,

3-47. BROKEN GLOW PLUG REMOVAL.

This task covers: a. Removal – Method #1 b. Removal – Method #2

INITIAL SETUP:

Equipment Condition

- All fuel injector nozzles removed. (See paragraph 3-5)

General Safety Instructions

- Compressed air used to remove glow plug should never exceed 30 psi (207 kPa).

WARNING

Failure to remove ail eight fuel injector nozzles before rotating crankshaft may result in injury to personnel.

a . REMOVAL – METHOD #1

1. Place finger over fuel injector nozzle port, Rotate torsional damper to raise piston in affected cylinder to "Top Dead Center. " Compression will change when "Top Dead Center" is reached.
2. Unscrew and move glow plug back and forth to break off glow plug tip into cylinder head prechamber.

WARNING

Compressed air used to remove glow plug should never exceed 30 psi (207 kPa). Use only effective chip guarding and personnel protective equipment (goggles/shield, gloves, etc.). Failure to follow this warning may result in serious injury to personnel.

3. Direct compressed air into glow plug port to remove glow plug tip through fuel injector nozzle port.
4. If preceding procedure does not remove broken glow plug, remove cylinder head. (See paragraph 3-7)

3-47. BROKEN GLOW PLUG REMOVAL (Con't).

b. REMOVAL – METHOD #2

1. Place finger over fuel injector nozzle port. Rotate torsional damper to raise piston in affected cylinder to "Top Dead Center. " Compression will change when "Top Dead Center" is reached.

2. Coil one end of flat 18 in. (45.7 cm) piece of tie wire to match size of injector hole.

3. Place wire coiled end over preheater to combustion chamber hole to prevent glow plug tip from falling into combustion chamber,

4. Using long nose pliers with 90° bend inserted through injector nozzle hole, grasp glow plug tip, break it off, and remove pieces from prechamber.

5. If preceding procedure does not remove broken glow plug, remove cylinder head. (See paragraph 3-7)

FOLLOW-ON TASKS:

- Install fuel injector nozzle. (See paragraph 3-27)
- Install new glow plug. (See TM 9-2320-289-20)

Section IV. COOLING SYSTEM MAINTENANCE

3-48. WATER PUMP REPLACEMENT.

This task covers: a. Removal b. Installation

INITIAL SETUP:

Equipment Condition

- Cooling system drained (if being performed in truck). (See TM 9-2320-289-20)
- Fan shroud removed (if being performed in truck). (See TM 9-2320-289-20)
- Right side alternator bracket removed (all except MI 01 O). (See TM 9-2320-289-20)
- Power steering pump removed (if being performed in truck). (See TM 9-2320-289-20)

Materials/Parts

- One water pump gasket
- Carburetor cleaning compound (Item 15, Appendix B)
- Anaerobic sealing compound (Item 59, Appendix B)
- Pipe sealant (Item 61, Appendix B)

Tools/Test Equipment

- Torque wrench

General Safety Instructions

- Avoid skin contact with anaerobic sealing compound.
- Carburetor cleaning compound is highly flammable. Keep away from open flame.

NOTE

Perform step 1 for M1010 only, if engine is in truck.

1. Remove 2 nuts at upper part of bracket (2). Loosen nut (3), and rotate compressor (1) and bracket off water pump studs.

TA50770

3-48. WATER PUMP REPLACEMENT (Con't).

2. Loosen clamp and disconnect crankcase depression regulator valve hose from oil filler tube (4). Remove 2 nuts (12), oil filler tube, and grommet (11) from water pump plate (5).

3. Loosen clamp (7) and disconnect thermostat bypass hose (6) from water pump (8). Loosen clamp (9) and disconnect radiator outlet hose (1 O) from water pump.

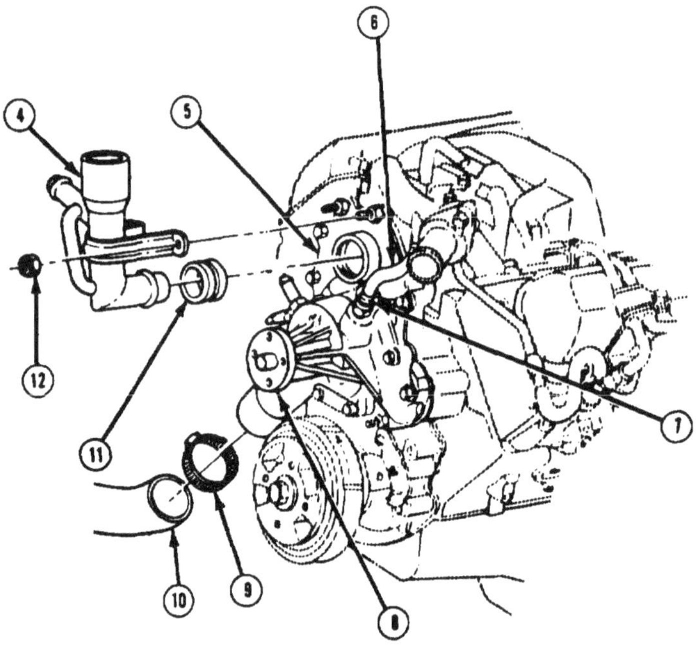

NOTE

- Ensure that a suitable container is used to catch coolant as water pump plate (5) ●nd water pump (8) are removed.
- Ensure that position of all bolts, studs, and screws is noted for Installation.

4. Remove 2 oil filler tube studs (16), 4 bolts (17), stud (23), 2 screws (24), long screw (21), long bolt (22), and 2 studs (13), Remove water pump plate (5) and water pump (8) from crankcase cover (18).

5. Remove 7 bolts (19) and separate water pump (8), water pump gasket (20), and water pump plate (5). Remove nipple (15) and plug (14) if damaged or if replacing water pump. Discard water pump gasket.

TA60771

3-48. WATER PUMP REPLACEMENT (Con't).

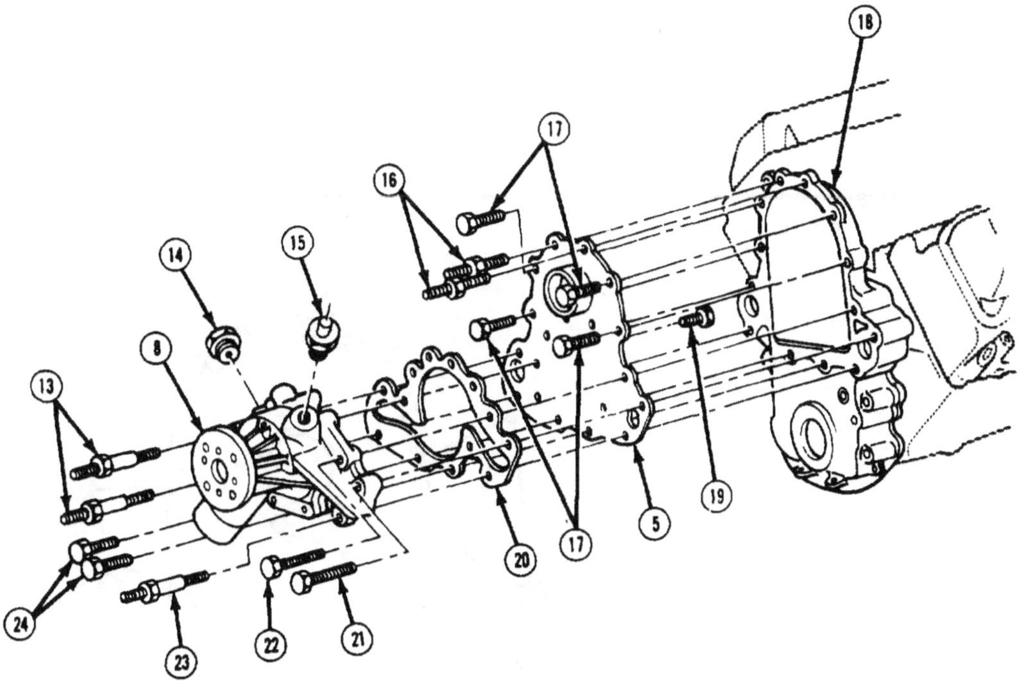

WARNING

Carburetor cleaning compound is highly flammable. Keep away from open flame, Failure to follow this warning may result in serious injury or death to personnel,

6. Clean anaerobic sealing compound from all sealing surfaces. Clean oil from all surfaces with carburetor cleaning compound, Clean pipe sealant from crankcase cover (18) and threads of all bolts, screws, and studs,

b . INSTALLATION

1. Install nipple (15) and plug (14) on water pump (8) with pipe sealant, if removed. Assemble water pump plate (5), new water pump gasket (20), and water pump with 7 bolts (19). Tighten bolts to 20 lb.-ft. (27 N.m).

TA50772

3-48. WATER PUMP REPLACEMENT (Con't).

WARNING

Avoid skin contact with anaerobic sealing compound. Immediately wash off skin if contact is made. Failure to follow this warning may result in injury to personnel.

2. Apply anaerobic sealing compound (25) to crankcase cover (18) in pattern shown.

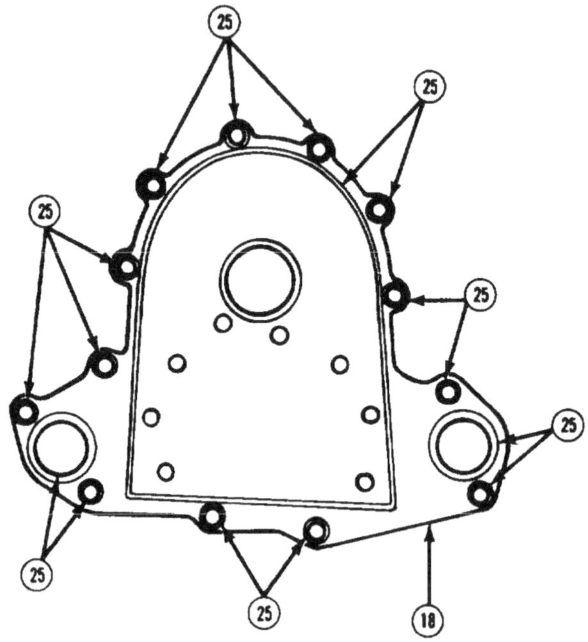

3. Apply pipe sealant to threads of bolts (17). Apply pipe sealant to oil filler tube studs (16) and stud (23) on portion of threads installed on crankcase cover (18).

4. Install water pump (8) and water pump plate (5) on crankcase cover (18) with long bolt (22), long screw (21), 2 studs (13), 2 screws (24), stud (23), 4 bolts (17), and 2 oil filler tube studs (16). Tighten bolts (17) and stud (23) to 35 Ib.-ft. (47 N.m). Tighten all other bolts, studs, and screws to 20 Ib.-ft. (27 Nom).

5. Connect radiator outlet hose (1 O) to water pump (8) and tighten clamp (9). Connect thermostat bypass hose (6) to water pump and tighten clamp (7).

6. Install grommet (11) on water pump plate (5). Install oil filter tube (4) on water pump plate with 2 nuts (1 2). Connect crankcase depression regulator valve hose to oil filler tube and tighten clamp.

TA50773

3-48. WATER PUMP REPLACEMENT (Con't).

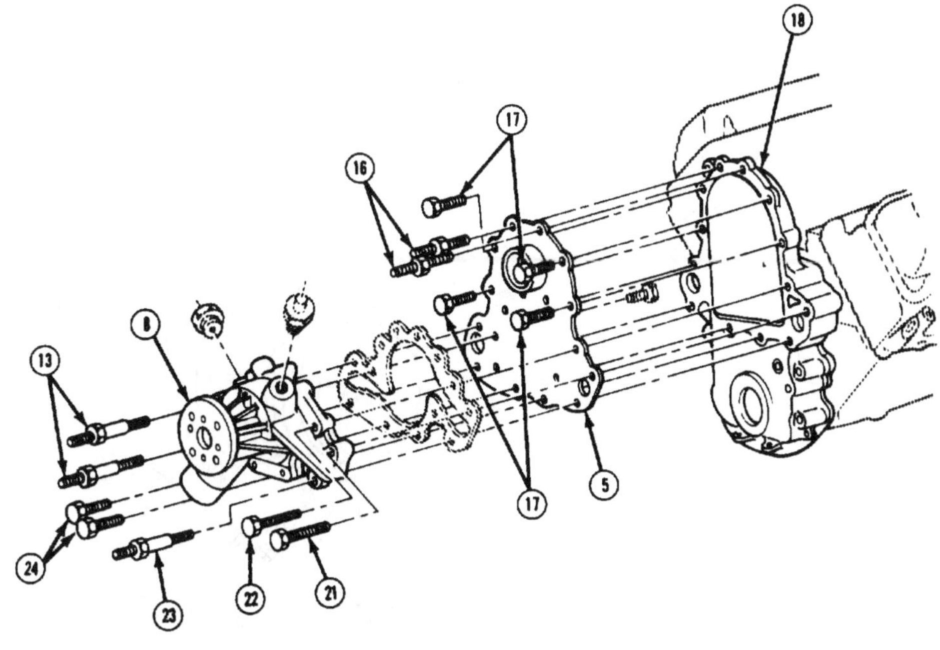

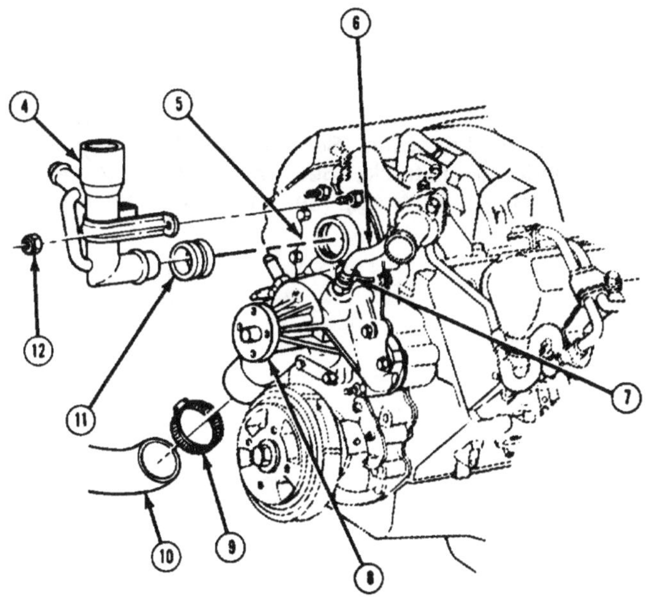

TA50774

3-48. WATER PUMP REPLACEMENT (Con't).

NOTE

Perform step 7 for MI 010 only.

7. Position bracket (2) and compressor (1) on water pump studs, Tighten nut (3). Install 2 nuts at upper part of bracket.

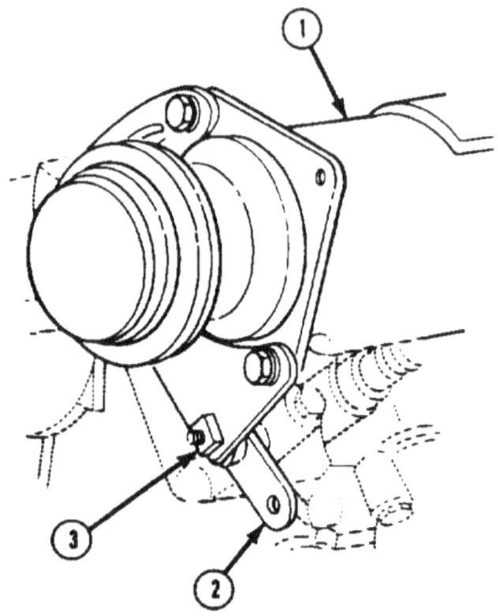

FOLLOW-ON TASKS:

- Install power steering pump (if being performed in truck). (See TM 9-2320-289-20)
- Install right side alternator bracket if removed. (See TM 9-2320-289-20)
- Install fan shroud (if being performed in truck). (See TM 9-2320-289-20)
- Fill cooling system (if being performed in truck). (See TM 9-2320-289-20)

TA50775

CHAPTER 4
ELECTRICAL SYSTEMS MAINTENANC E

4-1. ELECTRICAL SYSTEMS MAINTENANCE INDEX.

4-2. ALTERNATOR TESTING AND REPAIR (ALL EXCEPT M1010).

This task covers:

a. Alternator Output Testing
b. Disassembly
c. Cleaning and Inspection
d. Component Testing
e. Assembly

INITIAL SETUP:

Materials/Parts

- One generator positive (+) terminal package
- One generator negative (-) terminal package
- One generator relay terminal package
- Crocus cloth (Item 18, Appendix B)
- Dry cleaning solvent (Iterm 23, Appendix B)
- Grease (Item *36,* Appendix B)
- Masking tape (Item 68, Appendix B)
- Insulating varnish (Item 71, Appendix B)

Too/s/Test Equipment

- *500* amp test stand
- Belt tensioning gage
- Multimeter
- Test light
- Torque wrench

Manual References

- TM 9-214

General Safety Instructions

- *Compressed* air used for cleaning purposes should never exceed *30* psi (207 kPa).
- Dry cleaning solvent is flammable and must not be used near open flame. Use only in a well-ventilated area.

a. ALTERNATOR OUTPUT TESTING

1. Mount pivot arm of 500 amp test stand to high speed side, and screw mounting flange adapter to pivot arm, Connect pulley drive shaft with shaft of high speed head. Screw pulley drive shaft to mounting flange adapter,

2. Mount alternator to generator/starter mounting bracket on 500 amp test stand. Connect " V" belt from 500 amp test stand pulley to alternator pulley. Adjust belt tension, (See Appendix D)

3. Connect cable from alternator " BAT" terminal to 500 amp test stand "G+" terminal, Connect cable from alternator ground terminal to 500 amp test stand "G-" terminal,

4. Make a jumper wire with an insulated female spade terminal on one end and ring terminal on other end. Jump wire from alternator " NO. 2" terminal to alternator " BAT" terminal.

5. Using test light, connect one probe to alternator " BAT" terminal and other probe to alternator " NO, 1" terminal.

6. Set 500 amp test stand as follows:

   (a) DC ammeter switch to "x10"

   (b) DC voltmeter range switch to "x2"

   (c) DC voltmeter select switch to " BAT"

   (d) Tachometer switch to " DIRECT DRIVE"

   (e) Battery switch and master load switch to "OFF"

4-2. ALTERNATOR TESTING AND REPAIR (ALL EXCEPT M1010) (Con't).

7. Turn 500 amp test stand master power switch and master load switch to "ON" and battery switch to " 12V DC. " Green lamp on 500 amp test stand must illuminate.

NOTE

When tachometer reaches 6500 rpm, test light should go out to indicate proper charging from alternator.

8. Raise speed until tachometer reads 6500 rpm.

9. Set one 100 amp load switch on 500 amp test stand to "ON. "

10. Readings should be 14.0 volts and 90-100 amps. If readings are not correct, repair alternator.

b. DISASSEMBLY

1. Remove 4 bolts (4) and separate end frame (1) and stator frame (3) from drive end housing (2).

2. Place a piece of tape over end frame (1) bearing to prevent entry of dirt and foreign material if bearing will not be replaced.

CAUTION

Stator (5) should be handled with caution to avoid damage to coils. Failure to follow this caution may result in shorted stator and low alternator output.

3. Remove 3 nuts (6) and stator (5) from end frame (1).

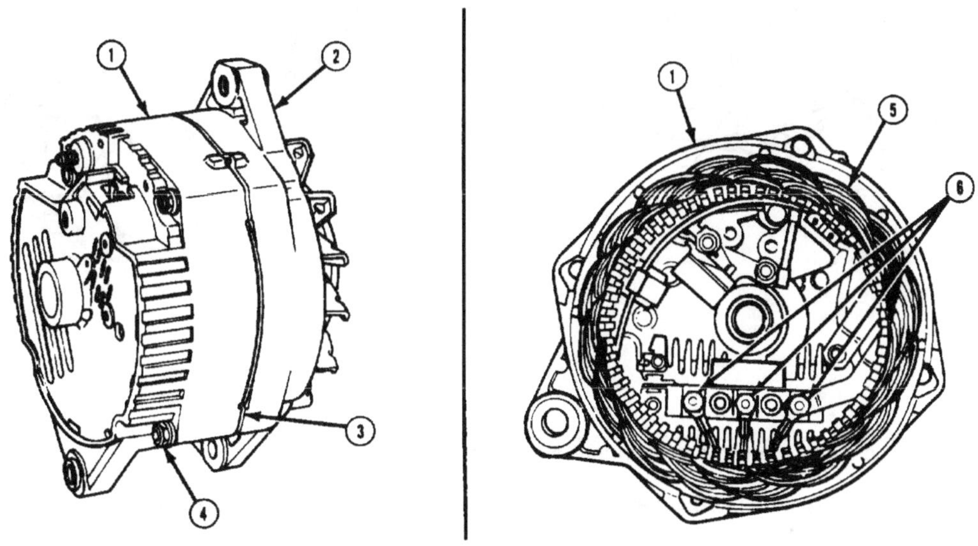

TA50776

4-2. ALTERNATOR TESTING AND REPAIR (ALL EXCEPT M1010) (Con't).

4. Remove nut (16), " R" terminal bar (19), fiber insulator (30), plastic insulator (29) and "R" terminal stud (28). Discard all components.

5. Remove 3 insulated screws (11), resistor (13), diode trio (20), brush holder assembly (9), and voltage regulator (12).

6. Remove screw (21) and disconnect capacitor lead (23) from rectifier bridge (7). Remove screw (24), wire retainer, and capacitor (8).

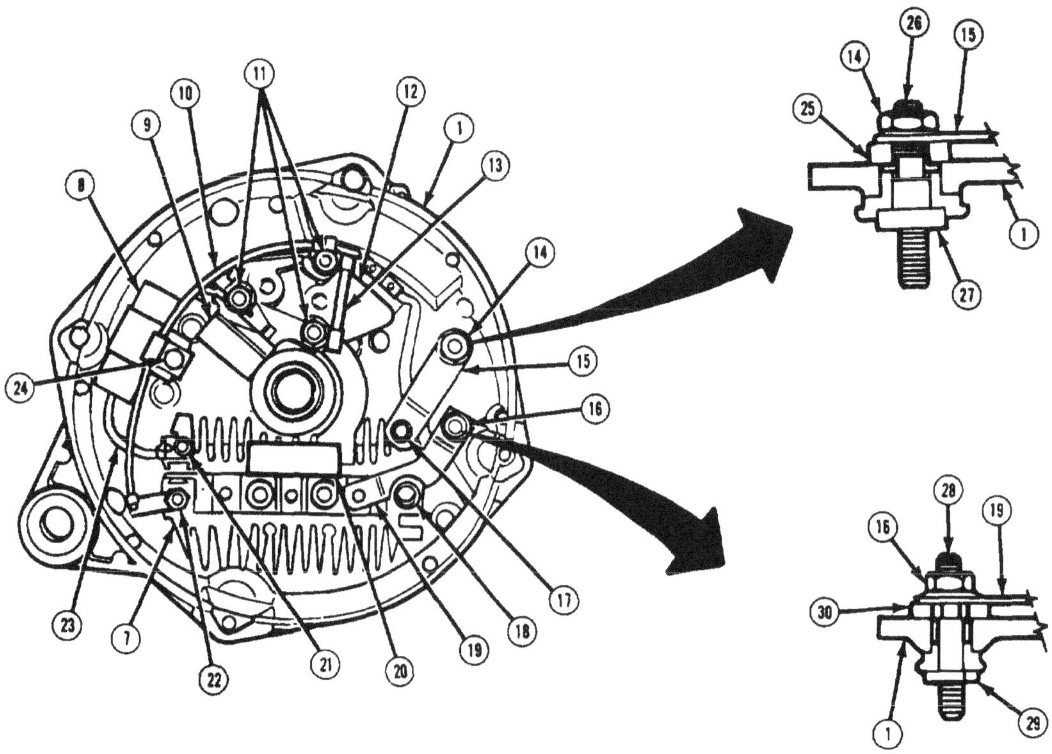

7, Remove screw (22) and rectifier bridge-to-regulator jumper wire (10).

8. Remove nut (17), nut (14), bus bar (15), and fiber insulator (25). Remove battery positive (+) terminal (26) and plastic insulator (27) from outside of end frame (1). Discard all components.

9. On outside of end frame (1), remove nut, lockwasher, flat washer, and fiber insulator from negative (-) terminal stud (18). Discard all components.

TA50777

4-2. ALTERNATOR TESTING AND REPAIR (ALL EXCEPT M1010) (Con't).

10. Remove rectifier bridge (7) and remove negative (-) terminal from rectifier bridge. Remove negative (-) terminal plastic insulator and rectifier bridge fiber insulator.

11. Press out end frame bearing (31) from end frame (1) by pressing toward inside.

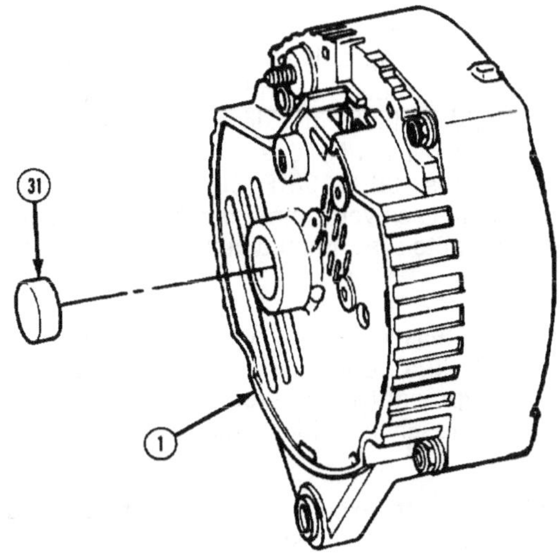

CAUTION

Rotor (32) should not be secured in vise at slip rings (42) on shaft. Failure to follow this caution may result in damage to slip rings or shaft.

12. Place rotor (32) in soft-jawed vise. Remove nut (40), wavewasher (39), pulley (38), fan (37), and collar (36) from drive end housing (2).

13. Remove rotor (32).

NOTE

Dowel pin (35) may be found In end frame (1).

14. Remove 3 screws (41), plate (33), and drive end bearing (34) from drive end housing (2). Remove dowel pin (35) if damaged.

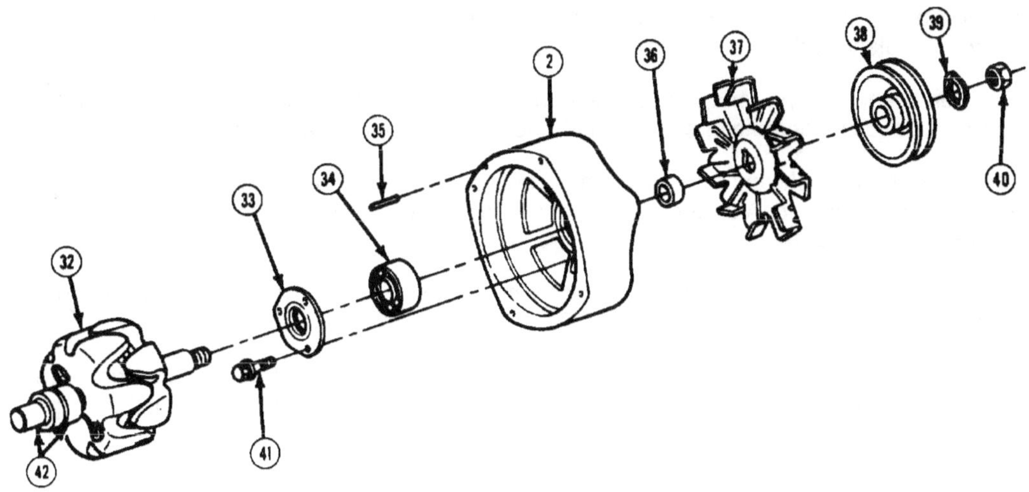

TA50778

4-2. ALTERNATOR TESTING AND REPAIR (ALL EXCEPT M1010) (Con't).

c. CLEANING AND INSPECTION

WARNING

- Dry cleaning solvent P-D-680 is toxic and flammable. Always wear protective goggles and gloves and use only in a well-ventilated area. Avoid contact with skin, eyes, and clothes and DO NOT breathe vapors. DO NOT use near open flame or excessive heat. The solvent's flash point is 100°F-138°F (38°C-59°C). if you become dizzy while using cleaning solvent, immediately get fresh air and medical help. if solvent contacts eyes, immediately wash your eyes with water and get medical aid.
- Compressed air used for cleaning purposes should never exceed 30 psi (207 kPa). Use only effective chip guarding and personnel protective equipment (goggles/shield, gloves, etc.). Failure to follow this warning may result in serious injury to personnel.

1. Clean all metal parts except regulator, rectifier bridge, stator assembly, rotor, and bearing assemblies in dry cleaning solvent. Dry with compressed air.
2. inspect brush holder assembly for damage. Replace brush holder assembly if damaged or if brushes are worn to a length of 0.630 in. (16 mm) or less, If brushes are to be reused, clean with a soft dry cloth.
3. Inspect insulating sleeves on insulated screws for splits or wear. Replace if split or worn.
4. Inspect end frame bearing for damage in accordance with TM 9-214. If bearing is damaged, or if grease supply is used up, replace bearing. Do NOT lubricate bearing.
5. Inspect slip ring end of rotor shaft for overheating or scoring. If there are signs of overheating or scoring, replace end frame bearing and rotor.
6. Inspect drive end bearing for damage in accordance with TM 9-214. Inspect seal in plate for damage. If drive end bearing or seal are damaged, replace both bearing and plate.
7. Inspect stator assembly and rotor winding for burned insulation. Burned insulation appears as a dark or blackened wiring. A strong acrid odor will be apparent. Replace stator assembly or rotor if either looks burned.
8. Inspect stator assembly terminal connector for corrosion or breaks. If corroded or broken, reploace stator.
9. Inspect windings on stator assembly for chipped insulation. Repair stator insulation with insulating varnish if possible. Replace stator if insulator cannot be repaired with insulating varnish.

4-2. ALTERNATOR TESTING AND REPAIR (ALL EXCEPT M1010) (Con't).

10. Inspect rotor slip rings (42) for scoring, wear, or pitting.

    (a) If rings are dirty, clean with 400 grain or finer crocus cloth. Spin rotor and hold crocus cloth against slip rings until clean.

    (b) If scored, worn, or pitted, true rings in lathe to 0.002 in. (0.05 mm). Finish with 400 grain or finer crocus cloth. Blow away all dust.

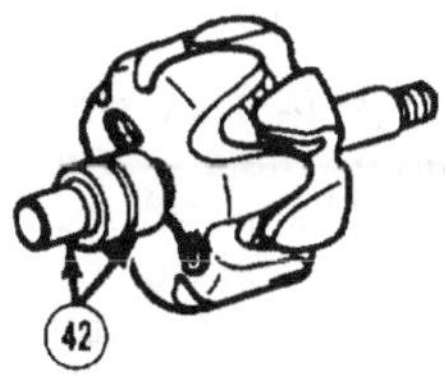

11. Inspect end frame and drive end housing for damage. Replace if damaged.

d. COMPONENT TESTING

1. Using test light or ohmmeter, check continuity between each stator assembly lead terminal with scale of 200 ohms or less. If there is no continuity, replace stator assembly.

2. Using test light or ohmmeter, check continuity between each stator assembly lead and stator frame (3). If continuity is not infinite, replace stator assembly.

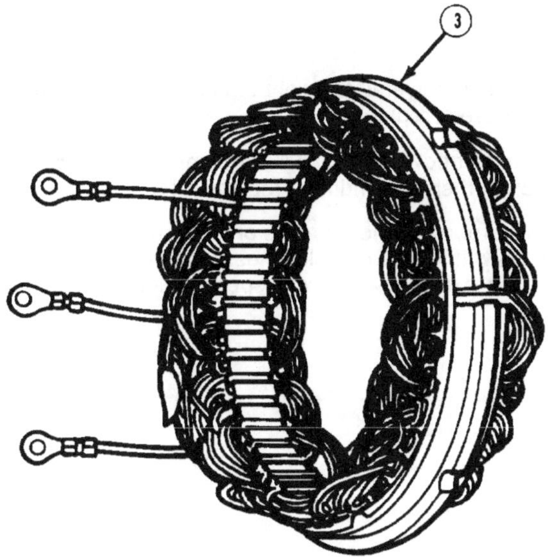

TA50779

4-2. ALTERNATOR TESTING AND REPAIR (ALL EXCEPT M1010) (Con't).

3. Connect voltmeter (44) and fast charger (43) to a 12 volt battery.
4. Connect voltage regulator (12) and test lamp (46) as shown.
5. Observe battery (45) polarity. Test lamp (46) should be on.
6. Turn on fast charger (43) and slowly increase charge rate. Observe voltmeter (44), Test lamp (46) should go out at voltage regulator (12) setting. Voltage regulator setting should be a minimum of 13,5 volts and a maximum of 15,5 volts,
7. Remove test lamp (46), and disconnect voltmeter (44) and fast charger (43).

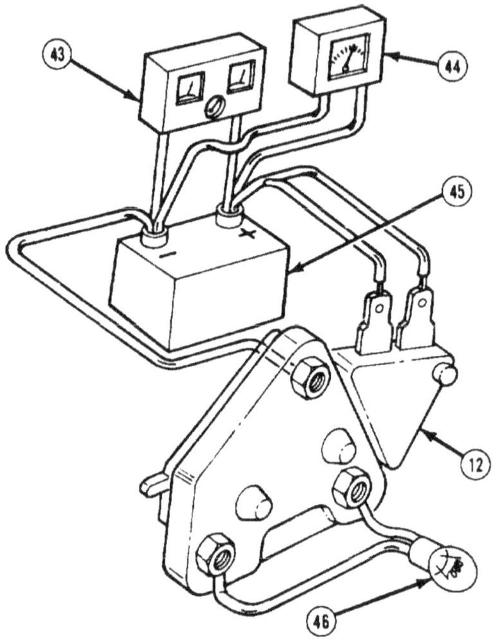

8. Using self-powered test light or ohmmeter, check for continuity between slip rings. If continuity is not present, replace rotor.
9. Using self-powered test light or ohmmeter, check for continuity between each slip ring and rotor shaft. If there is continuity, replace rotor.
10. Check for rotor short circuits or extreme resistance by connecting a battery and ammeter in series with edges of 2 slip rings. Ammeter reading at 70°F (21°C) should be 4.5-5.0 amps, If ammeter reading is not with specification, replace rotor.

TA50780

4-2. ALTERNATOR TESTING AND REPAIR (ALL EXCEPT M1010) (Con't).

11. Connect ohmmeter, with a 1 1/2 volt cell, to single connector (47) and to 1 of 3 connectors (48) on diode trio (20). Set ohmmeter to lowest range scale. Observe reading. Reverse ohmmeter leads to same 2 connectors. In one direction, readings should be infinite, in other direction readings should be finite.

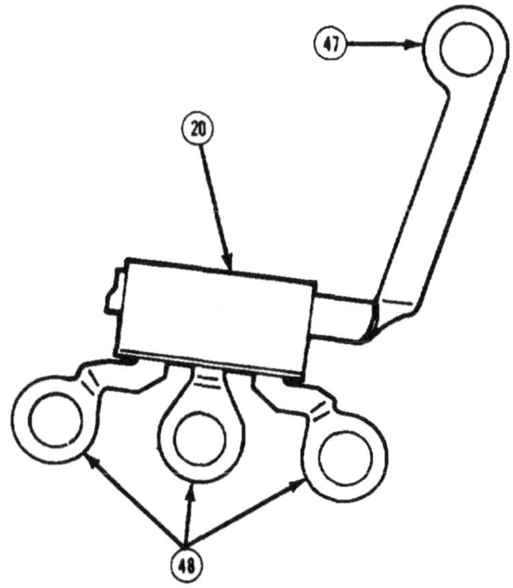

12. Place rectifier bridge (7) on nonconductive work surface, Slightly separate 6 tabs at 3 diode terminals for testing. Connect an ohmmeter to negative heat sink and to each of 3 terminals on negative (-) heat sink side. Record readings. Reverse connectors. One test should give infinite readings and other test should give finite readings.

13. Check 3 terminals on positive (+) heat sink side in same manner. One test should give infinite readings and other test should give finite readings.

14. If any reading does not meet specifications, replace rectifier bridge.

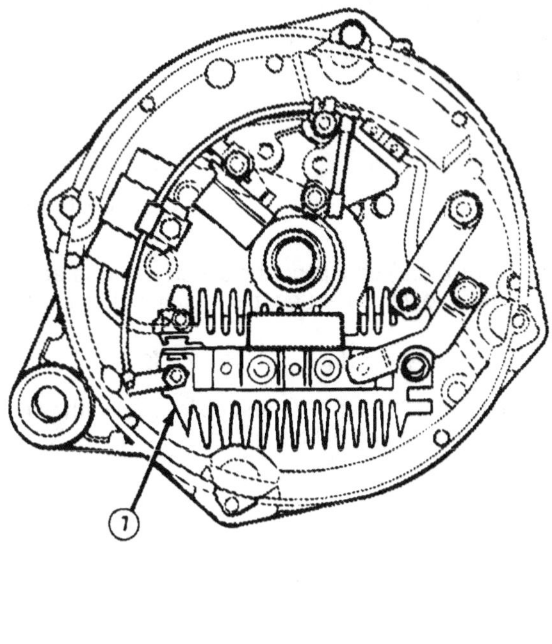

TA50781

4-2. ALTERNATOR TESTING AND REPAIR (ALL EXCEPT M1010) (Con't).

e. ASSEMBLY

NOTE

Ensure that drive end bearing (34) is lubricated with grease as it is assembled.

1. Support drive end housing (2). Position drive end bearing (34) with sealed end facing outward and tap into place.
2. Install plate (33) with 3 screws (41).
3. Install dowel pin (35) if removed.

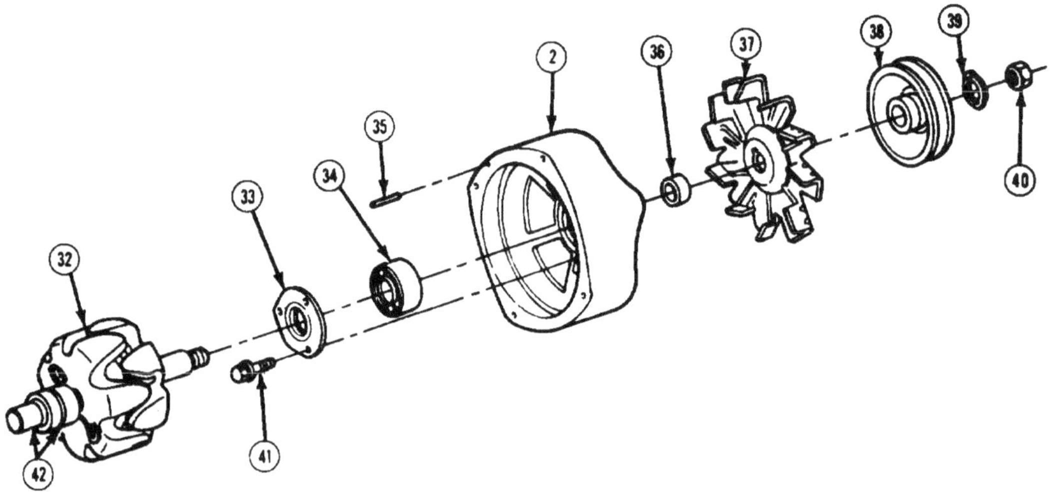

CAUTION

Rotor (32) should not be secured in vise at slip rings (42) on shaft. Failure to follow this caution may result in damage to slip rings or shaft.

4. Install rotor (32) in drive end housing (2).
5. Install collar (36), fan (37), and pulley (38) with wavewasher (39) and nut (40). Place rotor in soft-jawed vise and tighten nut to 50 lb.-ft. (68 N.m).

TA50782

4-2. ALTERNATOR TESTING AND REPAIR (ALL EXCEPT M1010) (Con't).

6. Install end frame bearing (31) to end frame (1) by supporting inner face of end frame with a socket and pressing bearing in from outside to inside until flush with end frame.

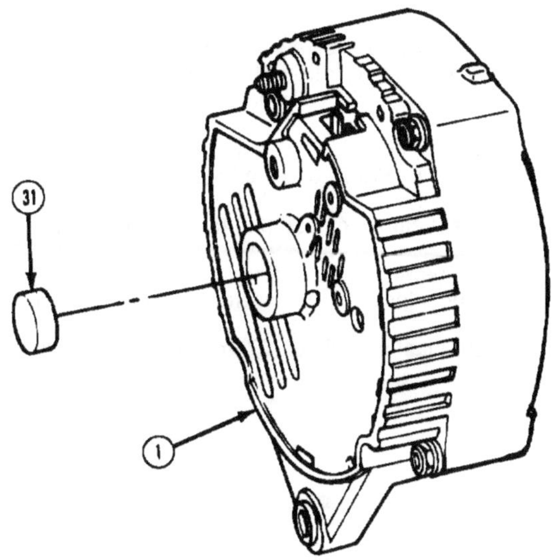

7. Install capacitor (8), rectifier bridge-to-regulator jumper wire (10), wire retainer, and screw (24) .

8. Position rectifier bridge fiber insulator in end frame (1) hole where screw (22) is to be installed. Position negative (-) terminal plastic insulator in end frame hole where negative (-) terminal (18) is to be installed.

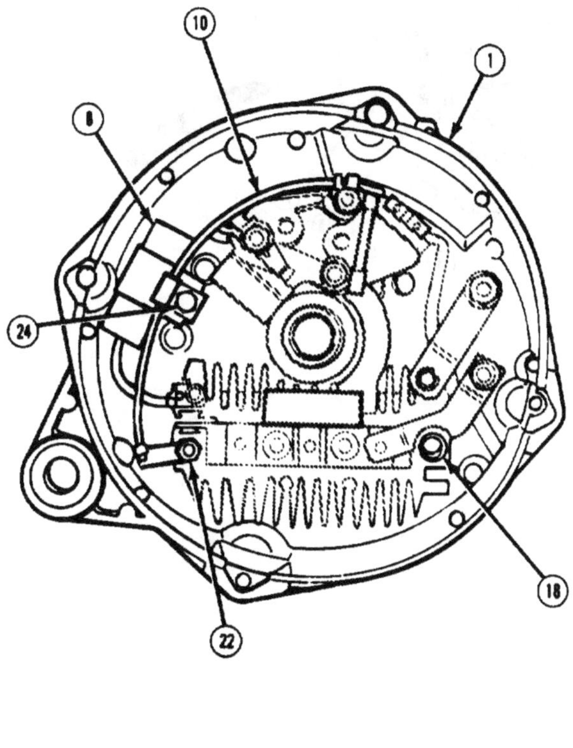

TA50783

4-2. ALTERNATOR TESTING AND REPAIR (ALL EXCEPT M1010) (Con't).

CAUTION

Ensure that screw (22) is installed through rectifier bridge fiber insulator. Failure to follow this caution will result in damage to truck electrical system when alternator is installed.

9. Place rectifier bridge (7) over insulation. install screw (22) through rectifier bridge-to-regulator jumper wire (10), rectifier bridget and rectifier bridge fiber insulator.
10. Connect capacitor lead (23) to rectifier bridge (7) with screw (21).
11. Install new negative (-) terminal stud (18) through rectifier bridge. From outside, secure with new fiber insulator fiat washer, new lockwasher, and new nut.
12. Install plastic insulator (27) to square hole in end frame (1).

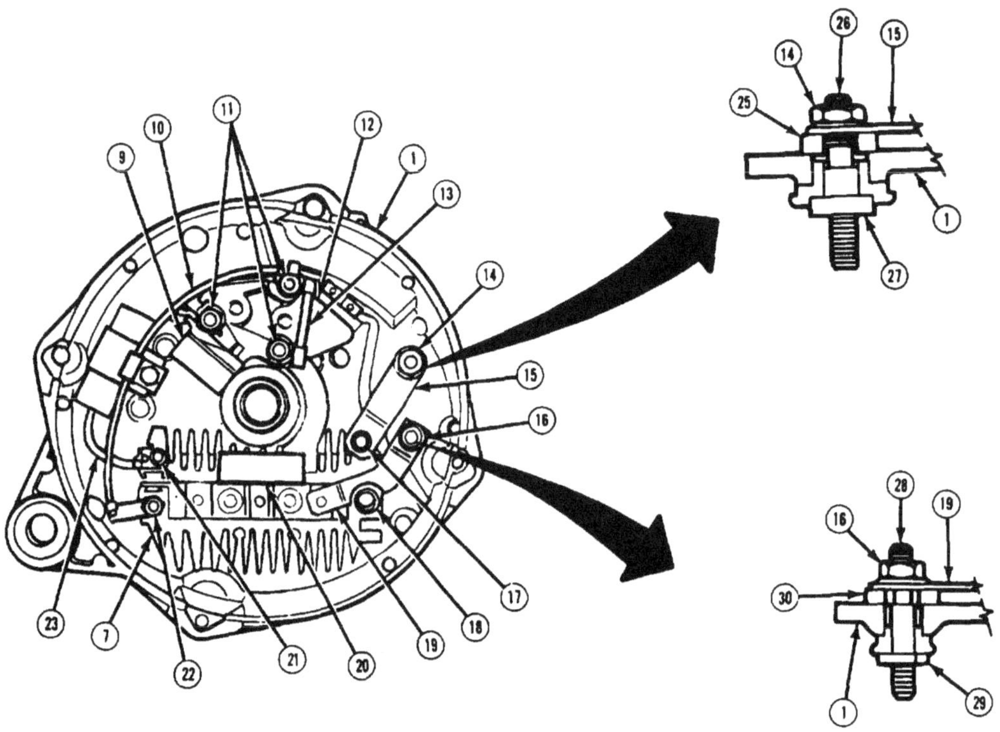

TA50784

4-2. ALTERNATOR TESTING AND REPAIR (ALL EXCEPT M1010) (Con't).

13. Install new battery positive (+) terminal (26) from outside. On inside, install new fiber insulator (25), new bus bar (15), new nut (14), and nut (17).

14, Position voltage regulator (12). Retract brushes into brush holder assembly (9) and retain retracted brushes with a suitable retainer. Ensure that retainer extends through end frame (1) when brush holder assembly is installed.

15, Install brush holder assembly (9) over voltage regulator (1 2).

16, Install diode trio (20) and resistor (13). Position rectifier bridge-to-regulator jumper wire (10) at top screw (11) hole. Install 3 screws (11).

17, Position plastic insulator (29) in hole in outside of end frame (1).

18. Install new "R" terminal stud (28) from outside. On inside, install new fiber insulator (30) on stud .

19. Install new "R" terminal bar (19) and secure with new nut (16).

20. Install stator (5) to end frame (1) alining 3 stator leads with rectifier bridge (7) terminals. Install 3 nuts (6) securely with brushes retracted.

21. Aline dowel pin with dowel pin hole and install drive end frame (2) onto end frame (1), Install 4 bolts (4) ,

22. Turn pulley and fan to ensure that they spin freely and without noise. Remove retainer which was used 'to secure brushes in their retracted position.

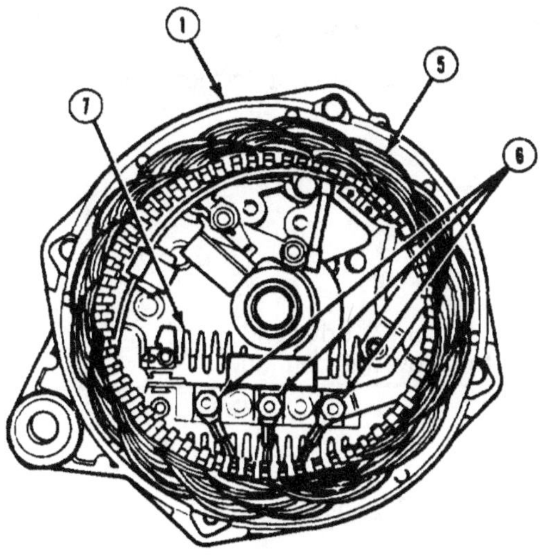

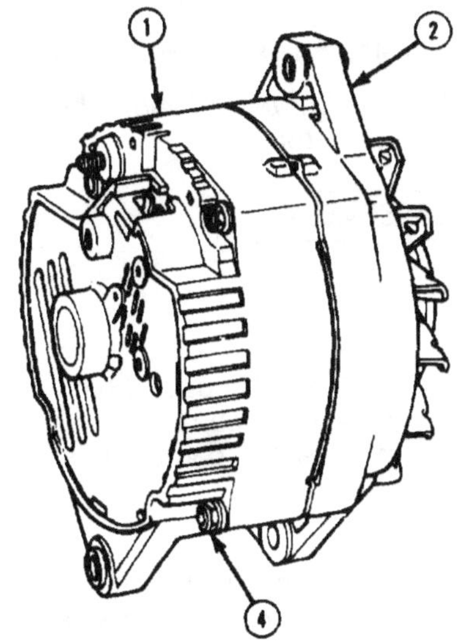

TA50785

4-3. ALTERNATOR TESTING AND REPAIR (M1010).

This task covers:

a. Alternator Output Testing
b. Disassembly
c. Cleaning and Inspection
d. Assembly

INITIAL SETUP:

Materials/Parts

- Two gaskets
- Two "O" rings
- Three locknuts
- Three seals
- Ten lockwashers
- Crocus cloth (Item 18, Appendix B)
- Dry cleaning solvent (Item 23, Appendix B)
- Grease (Item 36, Appendix B)
- Pipe sealant (Item 61, Appendix B)
- Masking tape (Item 68, Appendix B)
- Insulating varnish (Item 71, Appendix B)

Tools/Test Equipment

- 500 amp test stand
- Belt tensioning gage
- Micrometer
- Multimeter
- Test light

Manual References

- TM 9-214

General Safety Instructions

- Compressed air used for cleaning purposes should never exceed 30 psi (207 kPa).
- Dry cleaning solvent is flammable and must not be used near open flame. Use only in a well-ventilated area.

a. ALTERNATOR OUTPUT TESTING

1. Perform ALTERNATOR OUTPUT TESTING in paragraph 4-2, with the following exceptions:

   (a) In step 6(b), set voltmeter range switch to "x4" instead of "x2."

   (b) In step 8, raise speed until tachometer reads 5000 rpm instead of 6500 rpm.

   (c) In step 10, voltage should be 28.0-29.1 instead of 14.0. If readings are not correct, repair alternator.

b. DISASSEMBLY

CAUTION

If drive end housing (3) does not easily separate from stator (2), drive end housing may be tapped with plastic hammer. Failure to follow this caution may result in damage to stator.

NOTE

Ensure that relationship of slipring end housing (1), stator (2), and drive end housing (3) are marked for assembly.

1. Remove 3 locknuts, washers, and screws (6). Separate slipring end housing (1) and stator (2) from drive end housing (3). Discard locknuts.

2. Tag stator (2) leads for assembly, Remove 3 nuts and stator from slipring end housing (1),

4-3. ALTERNATOR TESTING AND REPAIR (M1010) (Con't).

3. Remove 4 screws (14), lockwashers (13), washers (12), brush cover (11), and brush holder gasket (15), Discard lockwashers and gasket.

4. Remove brush holder (9). Remove 2 brushes (10) and "O" ring (16) from brush holder. Discard "O" ring.

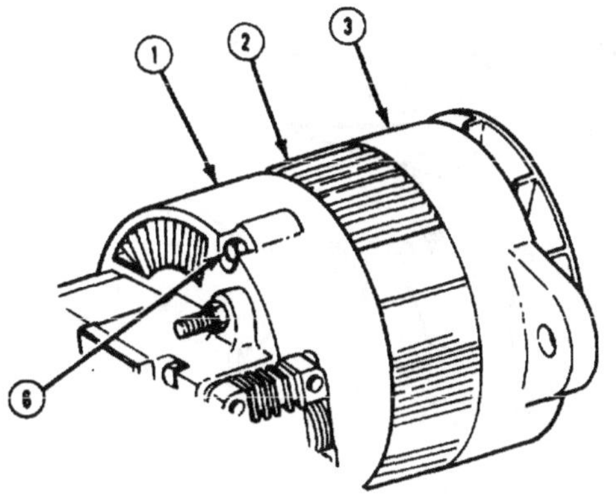

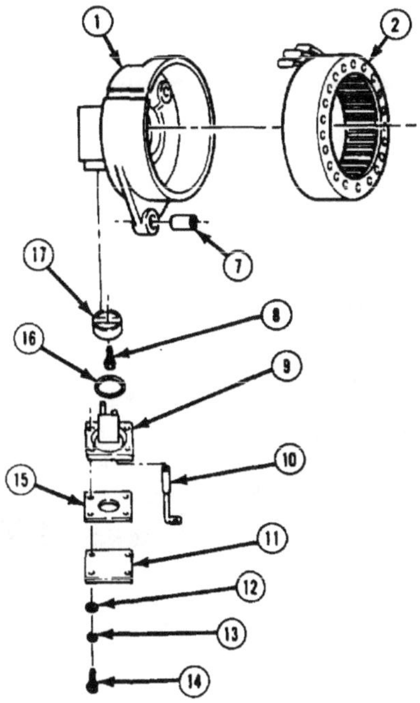

NOTE

Perform step 5 only if brush holder adapter (17) or connector are damaged. Leads connected to brush holder adapter will have to be cut from locking terminals to remove brush holder adapter.

5. Remove 2 screws (8) and pull out brush holder adapter (17). Tag and disconnect 2 leads from brush holder adapter, and remove brush holder adapter.

6. Remove bushing (7) if damaged.

TA50786

4-3. ALTERNATOR TESTING AND REPAIR (M1010) (Con't).

7. Check resistance between output terminal screw (5) on positive rectifier (28) and slipring end housing (1), If resistance is not infinite, replace upper housing bushing (18) and output terminal bushing (19) in step 10. Repeat for negative rectifier (20).

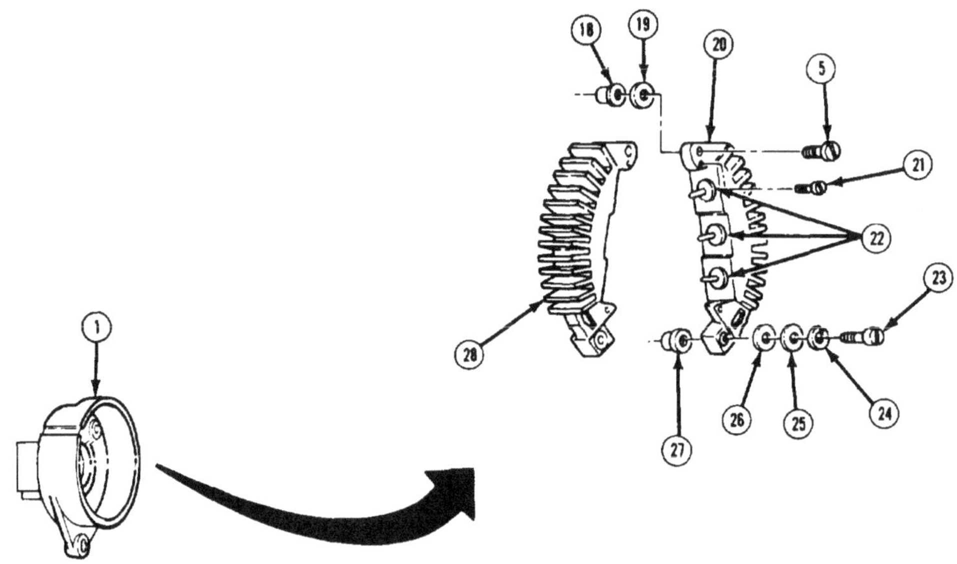

8. Check resistance between output terminal screw (5) on positive rectifier (28) and loop from each lead (22), Reverse multimeter probes. If resistance does not change, replace positive rectifier in step 10. Repeat for negative rectifier (20).

NOTE

Ensure that leads and output terminal screws (5) are tagged for assembly.

9. Remove 3 screws (21) and disconnect 2 leads from positive rectifier (28) and 1 lead from negative rectifier (20). Remove 2 nuts (4) and output terminal screws (5). Remove 2 screws (23), lockwashers (24), guard washers (25), and insulation washers (26). Discard lockwashers. Clean pipe sealant from screw (23) threads.

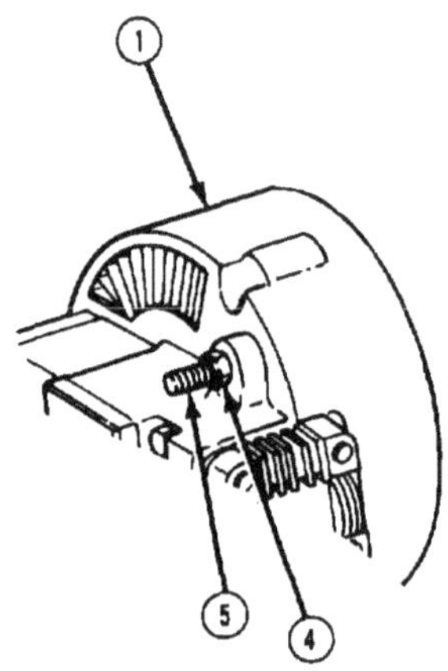

TA50787

4-3. ALTERNATOR TESTING AND REPAIR (M1010) (Con't).

NOTE

If 1 rectifier was determined to be good in step 8, but other rectifier had to be replaced, terminal loops at each lead (22) will have to be cut to separate positive rectifier (28) from negative rectifier (20). Ensure that leads are tagged for assembly if this is done,

10. Remove positive rectifier (28) and negative rectifier (20) from slipring end housing (1). Remove 2 upper housing bushings (18), output terminal bushings (19), and lower housing bushings (27) from slipring end housing,

11, Tag connector (34) leads and disconnect from AC "terminal studs (45) and regulator holder (44).

CAUTION

DO NOT pry on regulator (31) to remove. Failure to follow this caution may result in damage to regulator.

12. Remove 4 screws (29), lockwashers (30), and regulator (31). Remove and discard gasket (35). Discard lockwashers.

CAUTION

Before changing position of adjustment screw (32), ensure that "HI" position hole in regulator (31) is clear of obstructions. Failure to follow this caution may result in damage to regulator.

13. Ensure that adjustment screw (32) is in "HI" position if regulator (31) will not be replaced.

14. Remove 3 nuts (38), washers (37), and AC terminal studs (45).

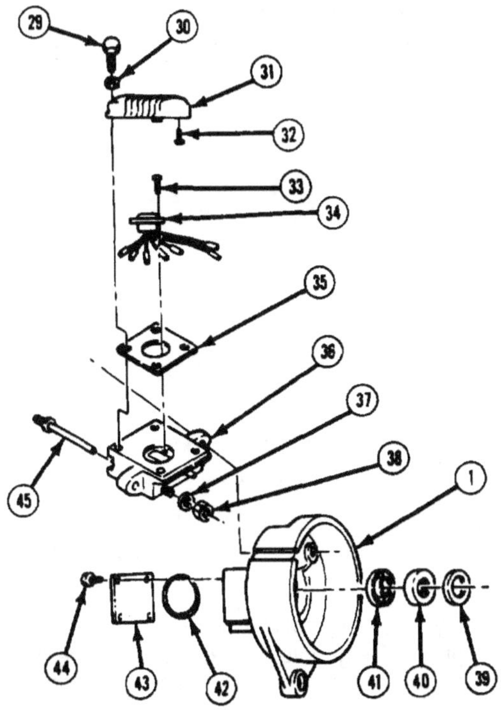

NOTE

Perform step 15 only if brush holder adapter was removed.

15. Pull connector (34) lead through access hole and grommet in slipring end housing (1). Remove regulator holder (36). Remove 2 screws (33) and connector.

16. Remove 4 screws (44), cover plate (43), and " O" ring (42) from slipring end housing (1). Remove inner seal (39). Insert punch in pilot hole on cover plate side of slipring end housing. Drive out slipring bearing (40) and outer seal (41). Discard seals and "O" ring.

TA50788

4-3. ALTERNATOR TESTING AND REPAIR (M1010) (Con't).

CAUTION

Sliprings (55) should not contact any hard surface while pressing rotor (54) from drive end housing (3). Failure to follow this caution may result in damage to rotor,

17. Mark relationship of woodruff key (50) to fan (49) for installation. Remove nut (51), and press rotor (54) and woodruff key out of drive end housing (3). Remove fan and fan spacer (52).

18. Remove 4 screws (53). Press bearing retainer (46), seal (47), and drive end bearing (48) from drive end housing (3). Discard seal,

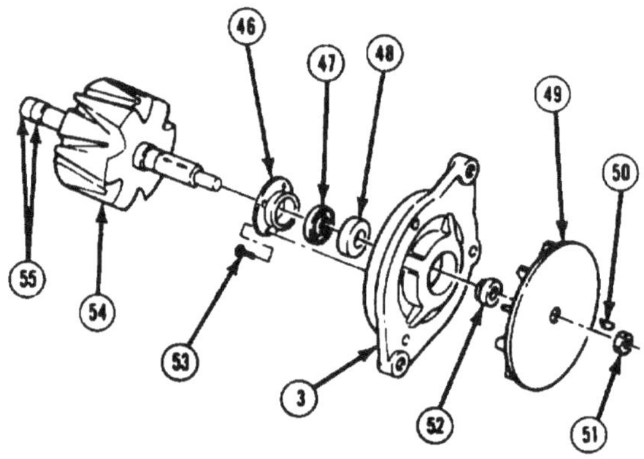

c. CLEANING AND INSPECTION

WARNING

- Dry cleaning solvent P-D-680 is toxic and flammable. Always wear protective goggles and gloves and use only in a well-ventilated area. Avoid contact with skin, eyes, and clothes and DO NOT breathe vapors. DO NOT use near open flame or excessive heat. The solvent's flash point Is 100°F-138°F (38°C-59°C). If you become dizzy while using cleaning solvent, immediately get fresh air and medical help, If solvent contacts eyes, immediately wash your eyes with water and get medical aid.

- Compressed air used for cleaning purposes should never exceed 30 psi (207 kPa). Use only effective chip guarding and personnel protective equipment (goggles/shield, gloves, etc.). Failure to follow this warning may result in serious injury to personnel.

1. Clean all components with dry cleaning solvent and dry with compressed air,

TA50789

4-3. ALTERNATOR TESTING AND REPAIR (M1010) (Con't).

2. Inspect all components for damage or corrosion. Replace cracked components. Repair or replace any component that is nicked or scratched, Remove any corrosion with crocus cloth.

3. Inspect stator for burned, charred, or broken wires. Replace stator if wires are burned, charred, or broken.

4. Check resistance between each stator lead and stator frame. If there is not infinite resistance, replace stator.

5. Inspect connector terminals to see if any are recessed into connector. Replace connector if any terminals are recessed.

6. Inspect bearings in accordance with TM 9-214.

7. Measure slipring (55) diameter, If diameter is less than 0.767 in. (19.48 mm) replace rotor.

8. Measure slipringwina bearing inner race diameter at several points and compare readings. Maximum difference in diameter (out-of-round) should be 0.002 in. (0.051 mm). If diameter is less than 0.871 in. (22. 12 mm), or if bearing race is out-of-round, replace slipring bearing.

d. ASSEMBLY]

into drive end housing. Install 4 screws (53).

CAUTION

Sliprings (55) should not contact any hard surface while pressing rotor (54) into drive end housing (3). Failure to follow this caution may result in damage to rotor.

2. Press rotor (54) into drive end housing (3) and install woodruff key (50). Install fan spacer (52) on rotor shaft. Aline fan (49) to woodruff key and install with nut (51).

4-3. ALTERNATOR TESTING AND REPAIR (M1010) (Con't).

3. Install new outer seal (41) in slipring end housing (1) with flat side facing pilot holes. Press slipring bearing (40) into slipring end housing. Install new inner seal (39) flush with slipring end housing. Install new "O" ring (42) and cover plate (43) with 4 screws (44),

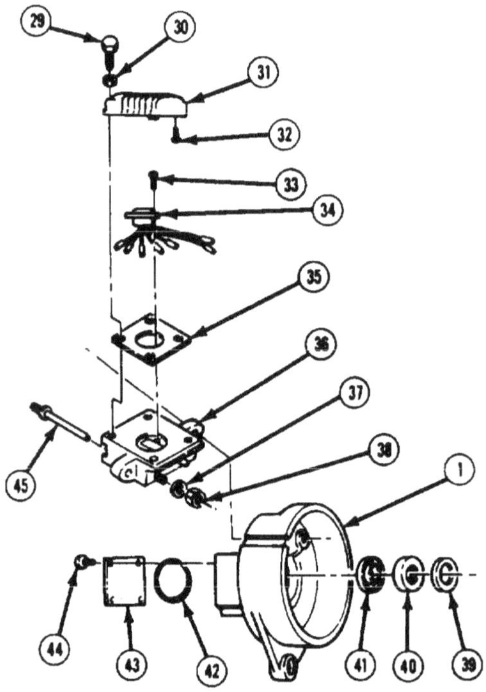

NOTE

Perform step 4 only if connector (34) was removed.

4. Install connector (34) on regulator holder (36) with 2 screws (33). Install regulator with connector leads through slipring end housing (1), Install connector lead through grommet and brush holder adapter hole.

5. Install 3 AC terminal studs (45) with 3 washers (37) and nuts (38).

CAUTION

Before changing position of adjustment screw (32), ensure that "HI*' position hole in regulator (31) is clear of obstructions . Failure to follow this caution may result in damage to regulator.

6. Ensure that adjustment screw (32) is in "HI" position if regulator (31) was replaced.

7. Install new gasket (35) on regulator holder (36). Install regulator (31) with 4 new lockwashers (30) and screws (29).

8. Connect connector (34) leads to AC terminal studs (45) and regulator holder (36).

9. Apply pipe sealant to screw (23) threads. Install 2 insulation washers (26), guard washers (25), new lockwashers (24), screws (23), and output terminal screws (5) on positive rectifier (28) and negative rectifier (20).

TA50790

4-3. ALTERNATOR TESTING AND REPAIR (M1010) (Con't).

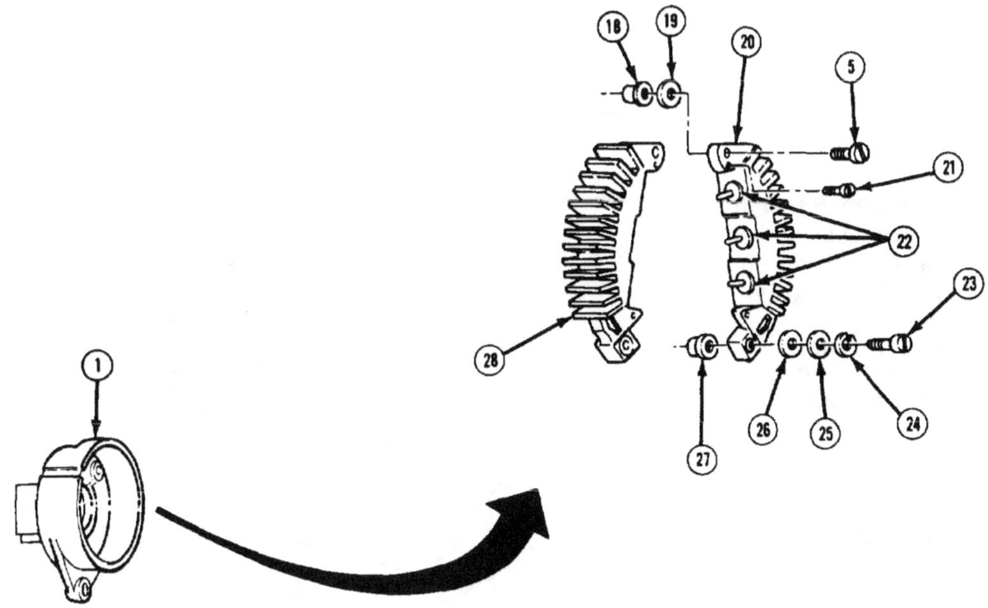

NOTE

If either rectifier was replaced, leads (22) will have to be spliced together with terminal loops before installing positive rectifier (28) and negative rectifier (20).

10. Install 2 lower housing bushings (27) on screws (23). Install output terminal bushings (19) and upper housing bushings (18) on output terminal screws (5). Install rectifier assembly in slipring end housing (1) and install 2 nuts (4) on output terminal screws,

11. Connect 2 leads to positive rectifier (28) and 1 lead to negative rectifier (20), and install 3 screws (21). Connect rectifier leads (22) to AC terminal studs (45).

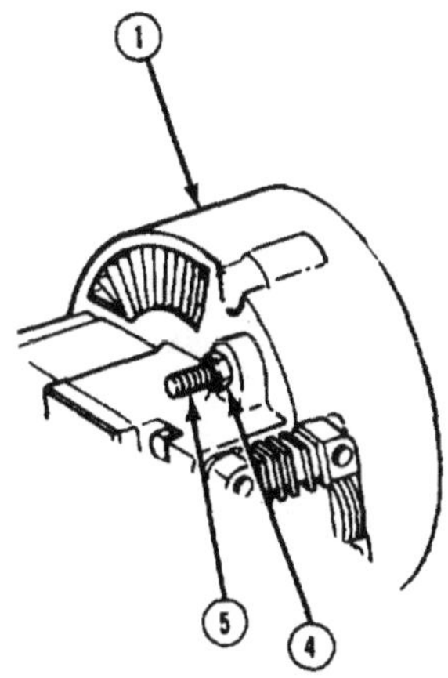

TA50791

4-3. ALTERNATOR TESTING AND REPAIR (M1010) (Con't).

12. Install bushing (7) if removed.

NOTE

Perform step 13 only if brush holder adapter (17) was removed. Ensure that locking terminals are installed on leads before installing brush holder adapter.

13. Connect 2 leads to brush holder adapter (17). Ensure that leads lock on brush holder adapter. Install brush holder adapter with 2 screws (8).

14. Install new "O" ring (16) and 2 brushes (10) in brush holder (9). Press brush holder onto brush holder adapter (17).

15. Install new brush holder gasket (15) and brush cover (11), with 4 washers (12), new lockwashers (13), and screws (14).

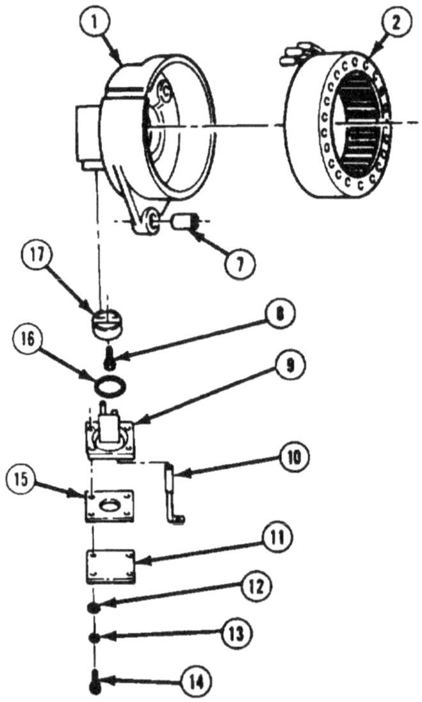

NOTE

Perform step 16 only if stator (2) was not replaced.

16. Mask machined edges of stator (2) with tape, Paint stator with insulating varnish. Wipe excess insulating varnish from stator inside diameter. Remove tape,

17. Install stator (2) on slipring end housing (1) and aline marks made during disassembly. Install stator leads with 3 nuts.

TA50792

4-3. ALTERNATOR TESTING AND REPAIR (M1010) (Con't).

18. Assemble drive end housing (3), stator (2), and slipring end housing (1), Install 3 screws (6), washers, and new locknuts.

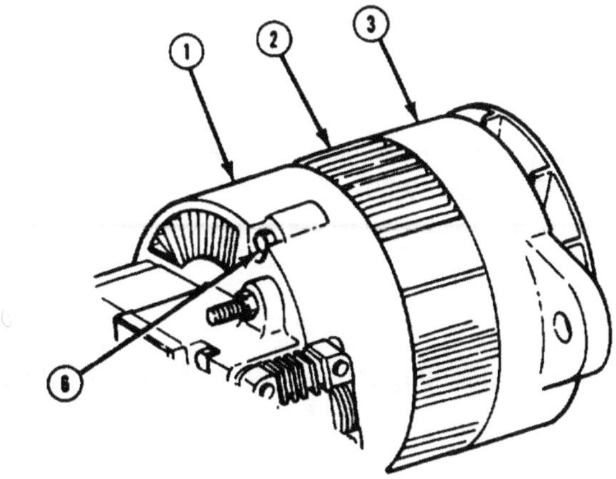

TA50793

4-4. CURRENT/VOLTAGE REGULATOR AND RELAY REPLACEMENT (M1010).

This task covers: a. Removal b. Installation

INITIAL SETUP:

Equipment condition

. Both battery negative cables disconnected, (See TM 9-2320-289-20)

Materials/Parts

Ž Lockwashers (as needed)

● RTV sealant (Item 57, Appendix B)

1. Remove 2 mounting screws (10) and current/voltage regulator cover (11) from inside of left fender.

NOTE

Ensure that leads and current/voltage regulator filter assembly (5) are tagged for installation,

2. Remove 5 nuts (9) and lockwashers, and disconnect leads and current/voltage regulator filter assembly (5). Discard lockwashers.
3. Remove 3 bolts (8) and mounting bracket (7).
4. Remove nuts and lockwashers, and disconnect leads from current/voltage regulator relay (2). Remove 2 retaining screws (3) and current/voltage regulator relay from mounting bracket (7), Discard lockwashers.
5. Loosen 3 lower retaining screws. Remove 3 upper retaining screws (4) and nuts, and current/voltage regulator (6) from mounting bracket (7).

b. INSTALLATION

CAUTION

DO NOT continue rotating adjustment screw (12) after resistance is felt. Adjustment screw adjustment should never exceed 3/4 turn. Failure to follow this caution may damage adjustment screw and require replacement of current/voltage regulator (6),

1. Remove tape covering adjustment screw (12). Gently rotate adjustment screw counterclockwise until slight resistance is felt, If no further testing is required, apply RTV sealant to completely cover adjustment screw,
2. Position current/voltage regulator (6) on 3 lower retaining screws and mounting bracket (7), and hold in place by tightening screws. Install 3 upper retaining screws (4) on current/voltage regulator with nuts.

NOTE

Ensure that 1 bolt (8) is installed through washer solvent container (1),

3. Install mounting bracket (7) with 3 bolts (8),

4-4. CURRENT/VOLTAGE REGULATOR AND RELAY REPLACEMENT (M1010) (Con't).

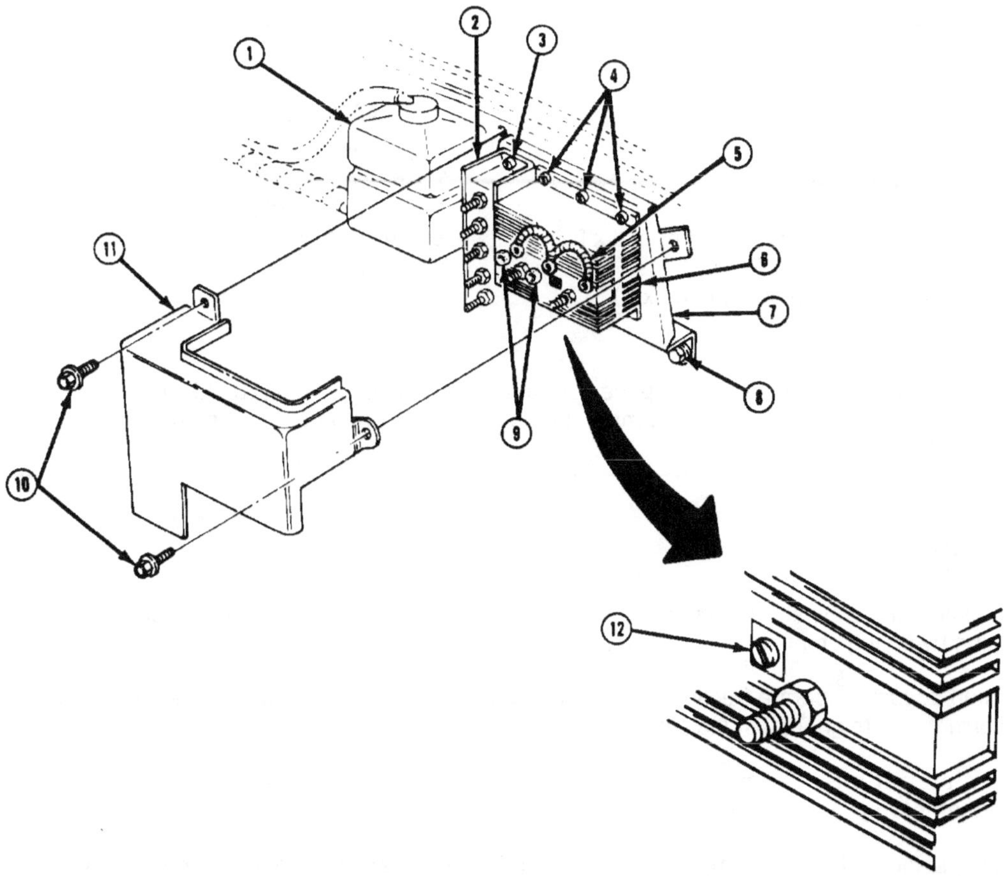

4. Install current/voltage regulator relay (2) on mounting bracket (7) with 2 retaining screws (3). Connect leads to current/voltage regulator relay, and install new lockwashers and nuts.
5. Install current/voltage regulator filter assembly (5) and leads with 5 new lockwashers and nuts (9).
6. Install current/voltage regulator cover (11) with 2 mounting screws (10).

FOLLOW-ON TASKS:

- Connect both battery negative cables. (See TM 9-2320-289-20)

TA50794

4-5. STARTER TESTING AND REPAIR.

This task covers: a. Testing b. Disassembly c. Cleaning and Inspection d. Assembly

INITIAL SETUP:

MaterialsIParts

- One fiber washer
- Two electrical parts kits
- Six lockwashers
- Dry cleaning solvent (Item 23, Appendix B)
- Grease (Item 36, Appendix B)
- Sealing compound (Item 60, Appendix B)

Too/s/Test Equipment

- 500 amp test stand
- Growler
- Hacksaw blade
- Multimeter
- Test light

General Safety Instructions

- Compressed air used for cleaning purposes should never exceed 30 psi (207 kPa).
- Dry cleaning solvent is flammable and must not be used near open flame. Use only in a well-ventilated area.

a. TESTING

1. Turn pinion on screw shaft to check for freedom of operation, Check armature for freedom of rotation by prying on pinion with screwdriver.
2. If pinion does not operate freely, or if armature does not rotate freely, repair starter. If pinion and armature are good, test starter.
3. Mount starter on 500 amp test stand as shown. Supply starter with 24 volts, and record amperage and starter rpm.
4. Readings should be 40-80 amps and 8000-12500 rpm, If readings are not correct, repair starter using guidelines in steps 5-9.
5. If there is low rpm and high amperage, check for:

   (a) Tight, dirty, or worn bearings

   (b) Bent or shorted armature

   (c) Grounded armature or fields
6. If there is high amperage and starter does not operate, check for direct ground in terminal or fields.
7. If there is no amperage and starter does not operate, check for:

   (a) Open field circuit; check connections through field with test light.

   (b) Open armature coils; inspect commutator for badly burned bars.

   (c) Broken brush springs, worn brushes, high insulation between commutator bars, or other conditions that would prevent good contact between brushes and commutator.
8. If there is low rpm and low amperage, check for poor connections, defective leads, or dirty commutator.

4-5. STARTER TESTING AND REPAIR (Con't).

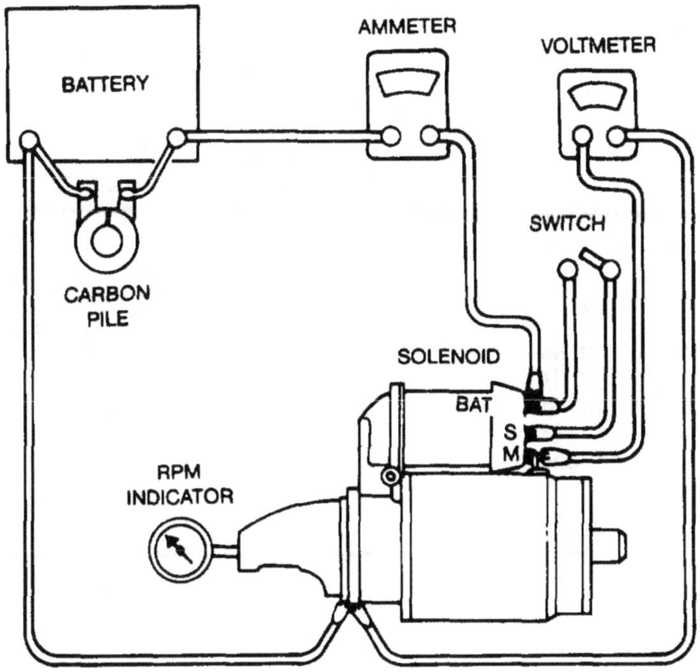

No-load Test

9. If there is high rpm and high amperage, check for:

(a) Shorted fields; replace field coil assembly.

(b) Shorted armature.

TA50796

4-5. STARTER TESTING AND REPAIR (Con't).

b. DISASSEMBLY

1. Remove screw (8) from field coil connector (7). Remove 2 screws (9) and lockwashers. Rotate solenoid (2) 90 degrees, and remove solenoid and plunger spring (10). Discard lockwashers.
2. Remove stud (4), bolt (5), and commutator end frame (6). Remove field housing (3) assembly from drive gear housing (1).
3. Remove snapring (15) and shift lever shaft (14), Remove shift lever (16) and plunger (11) assembly from drive assembly (21), Remove roll pin (13), plunger, and boot (12) from shift lever.

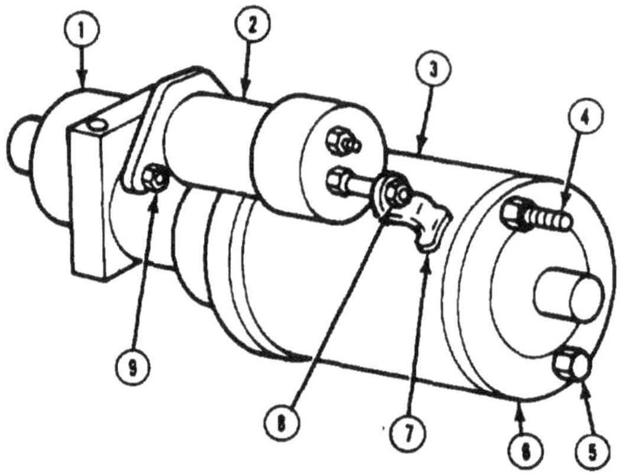

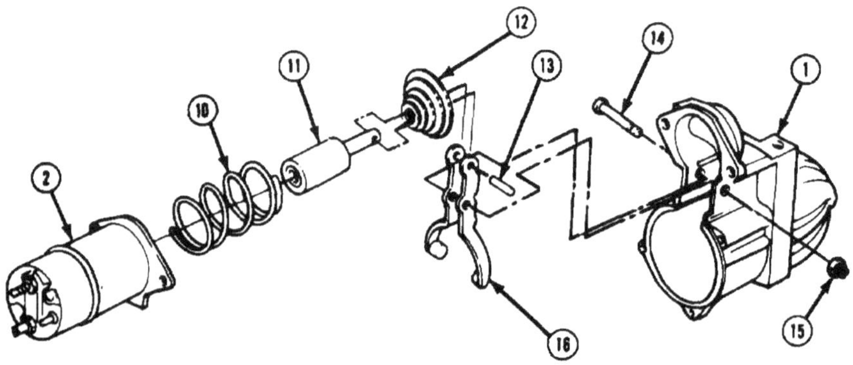

TA50796

4-5. STARTER TESTING AND REPAIR (Con't).

4. Remove 3 screws (26), washers (25), and center bearing (19) and armature (17) assembly from drive gear housing (1). Remove and discard fiber washer (27).

5. Remove thrust washer (23). Slide a socket or piece of pipe over end of armature (17) shaft against collar (22). Drive collar off stopring (24), and remove stopring and collar.

6. Separate armature (17), center bearing (19), washer (20), and drive assembly (21). Remove bushing (18) if damaged.

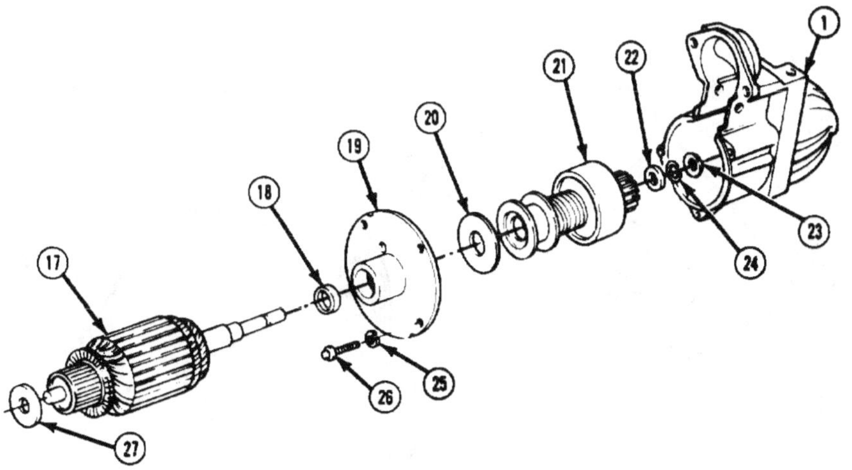

TA50797

4-5. STARTER TESTING AND REPAIR (Con't).

7. Check continuity between field housing (3) and brushes (34) at grounded brush holders (35). Check continuity between field coil connector (7) and brushes at insulated brush holders (31).

8. If there is no continuity in step 7, check connections at brushes (34) and brush holders (31 and 35). Replace any damaged components. If connections are good, and no components appear damaged, replace field coil assembly (38).

9. Check for continuity between brushes (34) at grounded brush holders (35) and brushes at insulated brush holders (31). If there is continuity, replace field coil assembly (38).

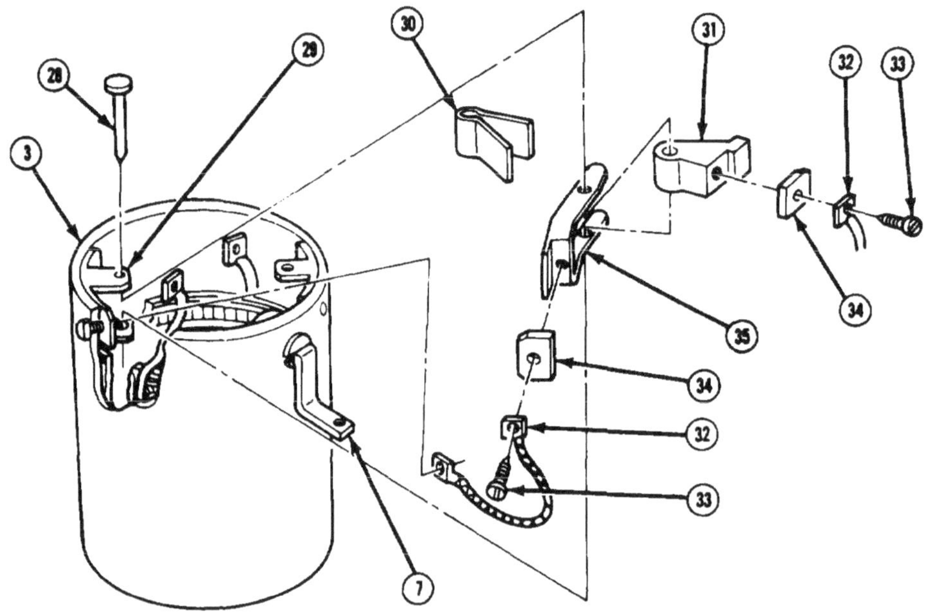

NOTE

Ensure that all leads are tagged for installation,

10. Remove brush support pin (28). Separate brush spring (30), grounded brush holder (35), and insulated brush holder (31) from brush support (29).

11. if removing field coil assembly (38) or brushes (34), remove 2 screws (33) at insulated brush holders (31) and disconnect field coil assembly leads (32), Remove screw and disconnect field coil assembly lead from field coil connector (7). Remove field coil connector and grommet if damaged.

TA50798

4-5. STARTER TESTING AND REPAIR (Con't).

NOTE

Ensure that position of field coil assembly (38) is noted for assembly.

12. Remove 8 pole shoe screws (36) and 4 pole shoes (37). Remove field coil assembly (38) from field housing (3). Remove alinement pin (39) if damaged.

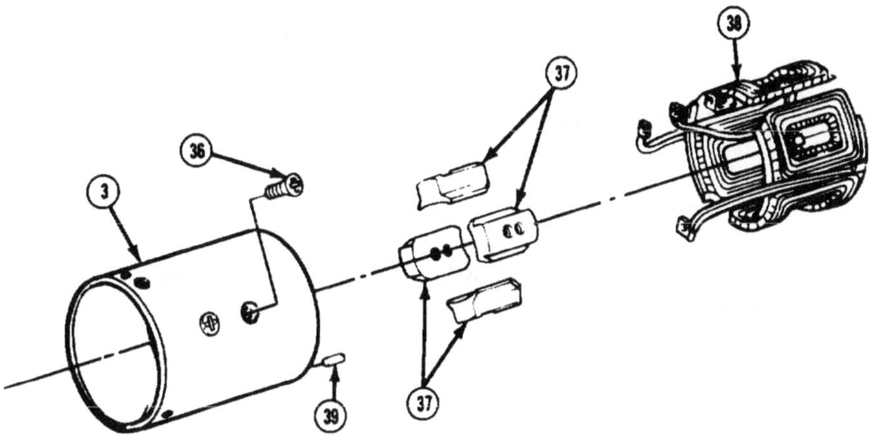

c. CLEANING AND INSPECTION

WARNING

Ž Dry cleaning solvent P-D-680 is toxic and flammable. Always wear protective goggles and gloves and use only in a well-ventilated area. Avoid contact with skin, eyes, and clothes and DO NOT breathe vapors. DO NOT use near open flame or excessive heat. The solvent's flash point is 100°F-138°F (38°C-59°C). If you become dizzy while using cleaning solvent, immediately get fresh air and medical help. If solvent contacts eyes, immediately wash your eyes with water and get medical aid.

- Compressed air used for cleaning purposes should never exceed 30 psi (207 kPa). Use only effective chip guarding and personnel protective equipment (goggles/shield, gloves, etc.). Failure to follow this warning may result in serious injury to personnel.

TA50799

4-5. STARTER TESTING AND REPAIR (Con't).

CAUTION

Dry cleaning solvent should not contact drive assembly, armature, brushes, field coil. or nonmetallic parts. Failure to follow this caution may result in damage to these components.

1. Clean drive assembly, armature, brushes, field coils, and all nonmetallic components with clean, dry brush or cloth. Remove any loose particles with compressed air.
2. Clean all remaining components with cloth dampened with dry cleaning solvent. Dry with compressed air.
3. Inspect all components for discoloration and damage. Replace any discolored or damaged component.

NOTE

Short circuit will pull hacksaw blade (40) against armature (17) and make hacksaw blade vibrate.

4. Install armature (17) on growler (41). Hold hacksaw blade (40) just above armature core while rotating armature. if hacksaw blade vibrates, clean between armature bars and try again. If hacksaw blade still vibrates, replace armature.

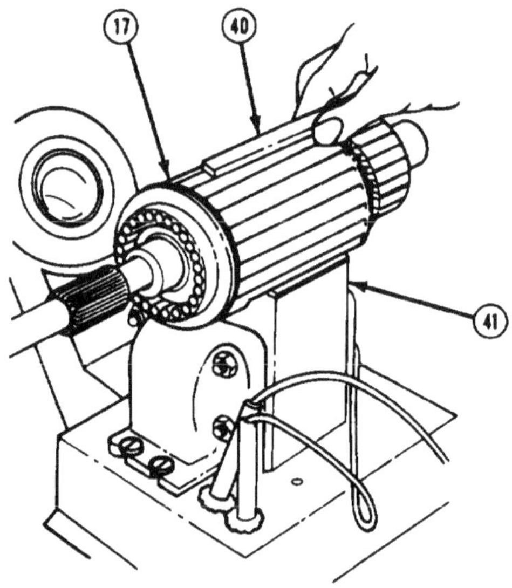

d. ASSEMBLY

1. Install alinement pin (39) if removed. install field coil assembly (38) in field housing (3). Install 4 pole shoes (37) on field coil assembly with 8 pole shoe screws (36).

TA50800

4-5. STARTER TESTING AND REPAIR (Con't).

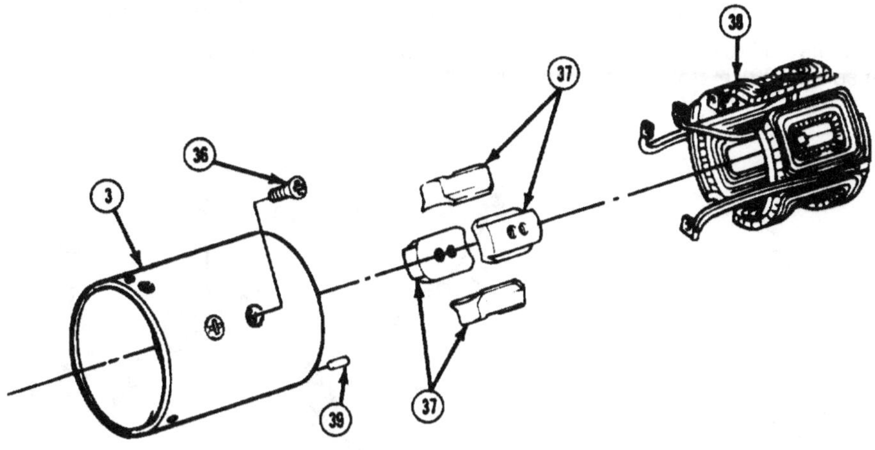

TA50801

4-5. STARTER TESTING AND REPAIR (Con't).

NOTE

If field coil assembly was removed, ensure that 2 field coil assembly leads (32) are installed with brushes (34) at insulated brush holders (31).

2. Install grommet and field coil connector (7) in field housing (3) if removed. Connect field coil assembly lead (32) to field coil connector with screw (33), Connect remaining field coil assembly leads at insulated brush holders (31) and install brushes (34) with remaining screws,

3. Assemble insulated brush holder (31) and grounded brush holder (35) on brush support (29) with brush spring (30), Install brush support pin (28).

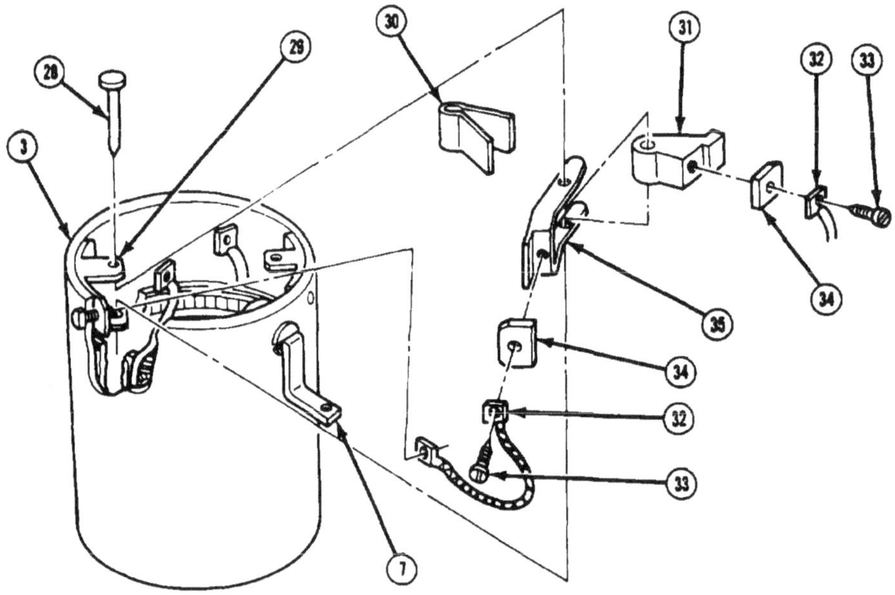

4. install bushing (18) in center bearing (19) if removed. Lubricate shaft end of armature (17) with grease. Assemble drive assembly (21), washer (20), center bearing, and armature.

5. install collar (22) on armature (17) shaft, with cupped end facing end of shaft, Install stopring (24) in armature shaft groove, Install thrust washer (23). Squeeze collar and thrust washer together until collar is forced over retainer,

6. Install new fiber washer (27) on armature (17). install center bearing (19) and armature assembly on drive gear housing (1) with 3 washers (25), and screws (26),

TA50802

4-5. STARTER TESTING AND REPAIR (Con't).

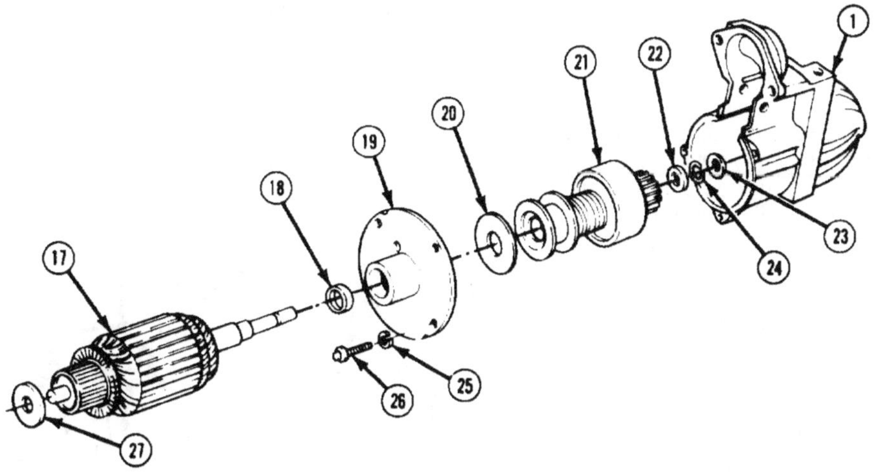

7. Install boot (12) and plunger (11) on shift lever (16) with roll pin (13). Install shift lever and plunger assembly on drive assembly (21). Install shift lever shaft (14) on shift lever with snapring (15).

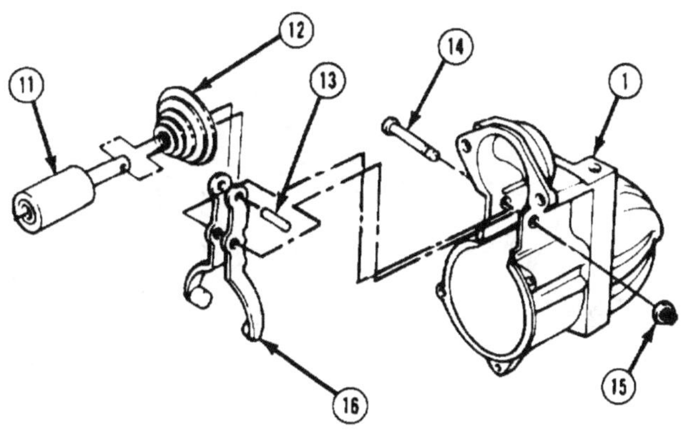

TA50803

4-5. STARTER TESTING AND REPAIR (Con't).

CAUTION

Ensure that field housing (3) assembly is carefully positioned on drive gear housing (1). Failure to follow this caution may result in damage to brushes.

8. Position field housing (3) assembly on drive gear housing (1) ensuring that alinement pin engages drive gear housing hole. install commutator end frame (6) on field housing with bolt (5) and stud (4).

9. Apply sealing compound to solenoid (2) flange where it contacts field housing (3). install plunger spring (10) in solenoid. Rotate solenoid 90 degrees and install on drive gear housing (1) with 2 new lockwashers and screws (9).

10. Carefully insulate field coil connector (7) from solenoid (2) terminal . Connect one 24 volt battery lead from 500 amp test stand to solenoid terminal and other battery lead to starter frame.

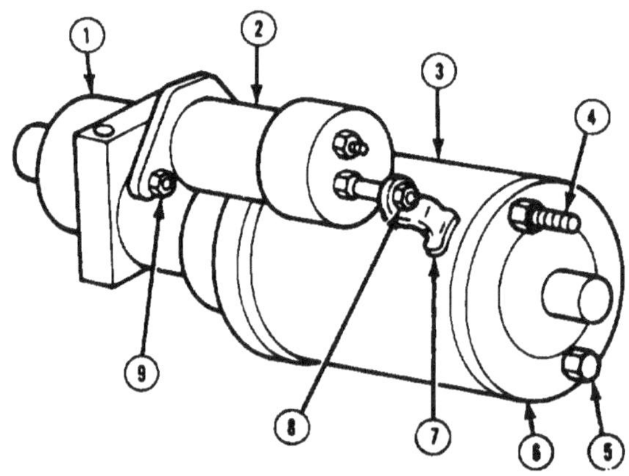

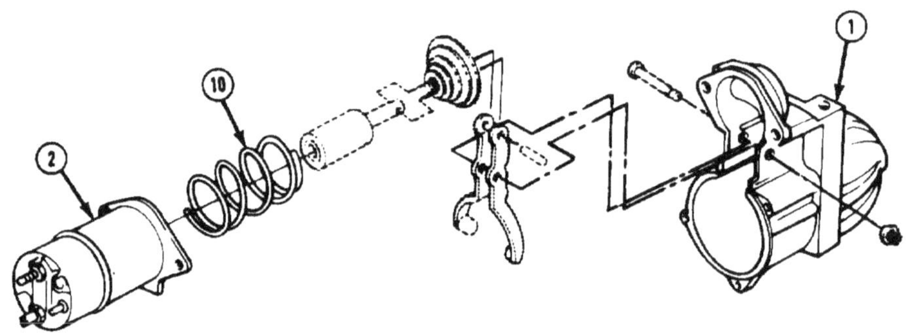

TA50804

4-5. STARTER TESTING AND REPAIR (Con't).

11. Connect jumper lead momentarily from solenoid (2) terminal to starter frame. This will move pinion (42) into cranking position until battery leads are disconnected.

12, Check clearance between pinion (42) and collar (22). If clearance is not 0.010-0.140 in. (0.25-3.56 mm), disconnect test stand battery leads, disassemble starter, and check for incorrect installation. Replace any worn components and retest,

13. Disconnect test stand battery leads. Remove insulation between field coil connector (7) and solenoid (2) terminal, Install screw (8) on field coil connector,

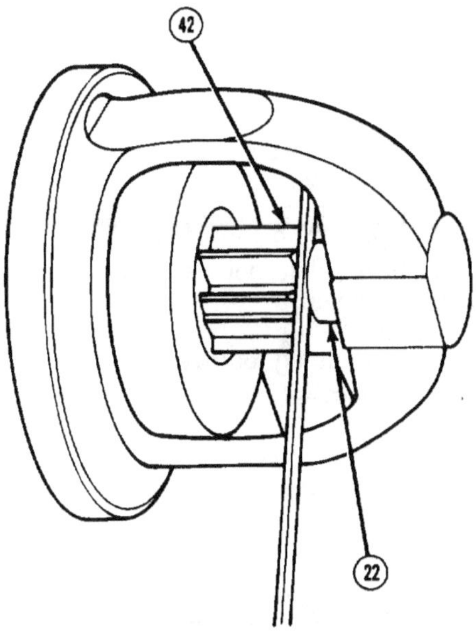

TA50805

4-6. INSTRUMENT PANEL WIRING HARNESS AND EXTENSION REPLACEMENT.

This task covers: a. Removal b. Installation

INITIAL SETUP:

Equipment Condition

- Instrument cluster removed, (See TM 9-2320-289-20)
- Voltmeter relay, generator 2 relay, and starter relay removed, (See TM 9-2320-289-20)
- Transfer case bezel, plate, boot, and control lever housing removed. (See TM 9-2320-289-20)
- Instrument panel padding removed. (See TM 9-2320-289-20)

Materials /Parts

- Grease (Item 36, Appendix B)
- Antiseize tape (Item 65, Appendix B)

Manual References

- TM 9-2320-289-20

a. REMOVAL

NOTE

- If removing all except M1010 instrument panel wiring harness, perform steps 1 through 7, 9 through 12, and 14.
- If removing M1010 instrument panel wiring harness, perform steps 1-14.
- If removing instrument panel extension wiring harness, perform steps 13 and 15.
- Voltmeter lamp connector (8), "WATER-IN-FUEL" light connector (11), "WAIT" light connector (12), and instrument cluster connector (13) were disconnected when instrument cluster was removed.
- Ensure that all leads and connectors are tagged for installation.

1. At underside of steering column (1), disconnect wiper/washer switch connector (4) from wiper/washer switch (3). Unplug wiper/washer switch from bracket (2).

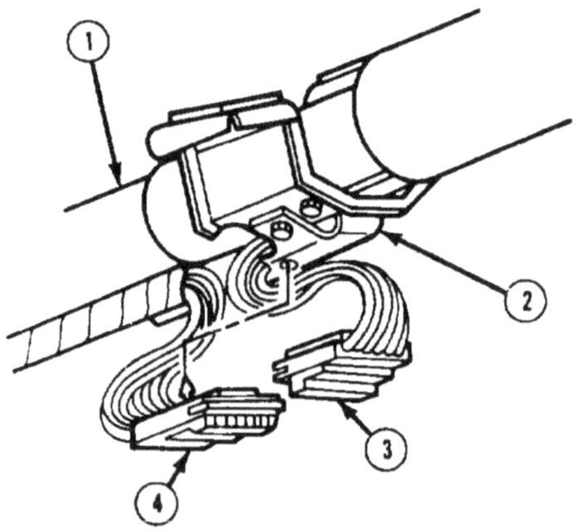

TA50806

4-6. INSTRUMENT PANEL WIRING HARNESS AND EXTENSION REPLACEMENT (Con't).

2. Disconnect headlight switch connector (14), 2 blackout light switch connectors (15), and horn relay (20). Remove horn rectifier (21) if damaged.
3. Remove screw (16), ground lead (17), and wiring harness terminal (18).

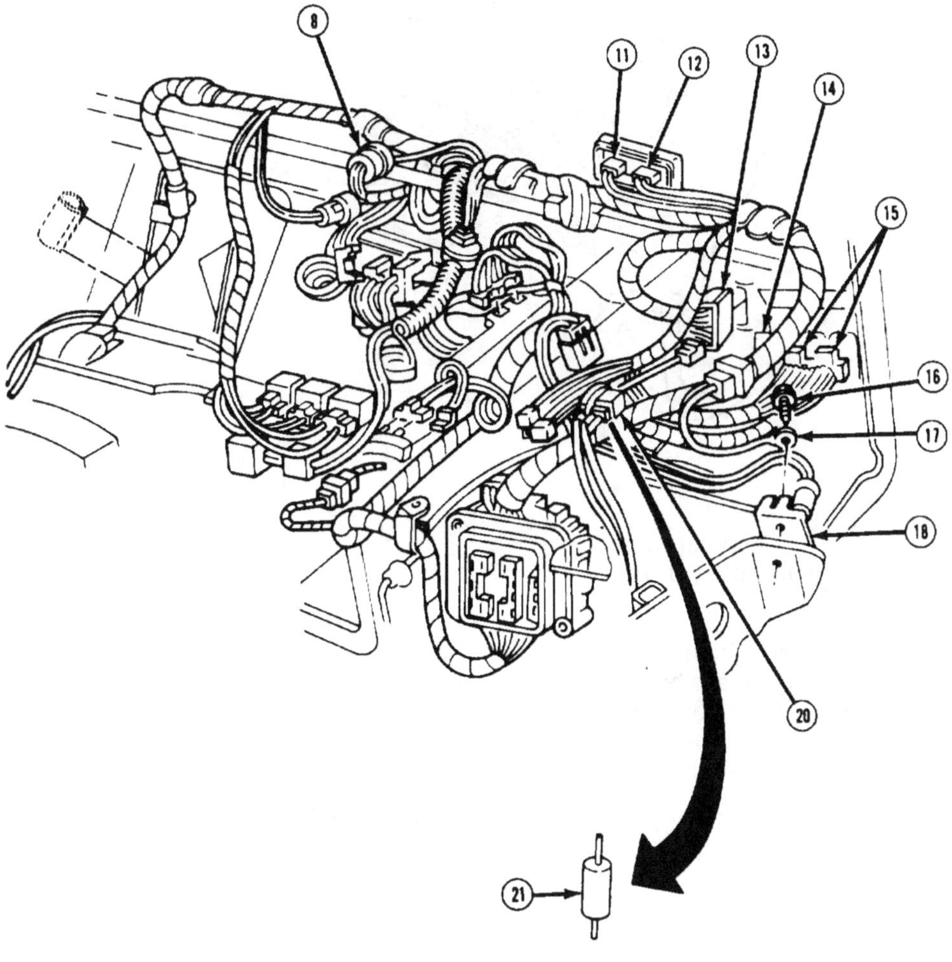

TA50807

4-6. INSTRUMENT PANEL WIRING HARNESS AND EXTENSION REPLACEMENT (Con't).

4. At lower part of steering column (1), disconnect directional signal connector (27), ignition switch connector (28), back-up light switch connector (26), and dimmer switch connector (29),

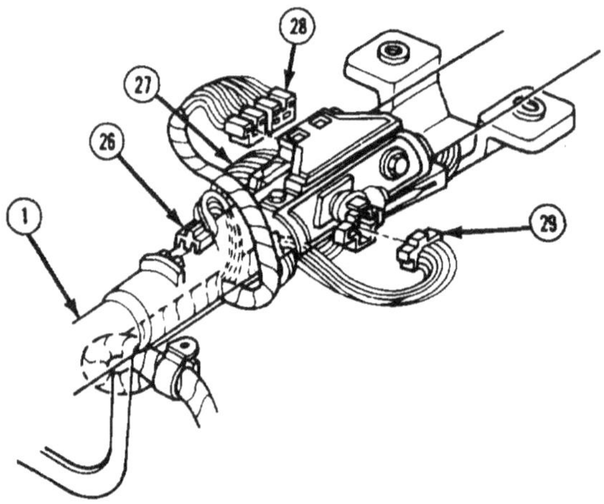

5. Inside engine compartment, disconnect wiring harnesses from rear of fuse box (22).

6. Remove turn signal flasher from fuse box. Remove 2 screws (30) and bolt (31), and pull out fuse box (22). Wipe grease from exterior of fuse box if instrument panel wiring harness will not be replaced.

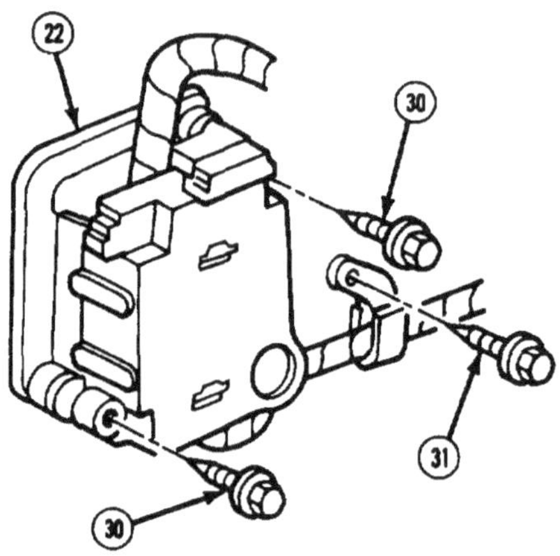

TA50808

4-6. INSTRUMENT PANEL WIRING HARNESS AND EXTENSION REPLACEMENT (Con't).

7. Disconnect heater lamp connector (7), heater connector (10), and glow plug wiring harness connector (23).

8. Disconnect spotlight switch connector (19), 2 floodlight switch connectors (9), and GPFU switch connector (25). Disconnect open door buzzer wiring harness connector from below "WATER-IN-FUEL" connector (11).

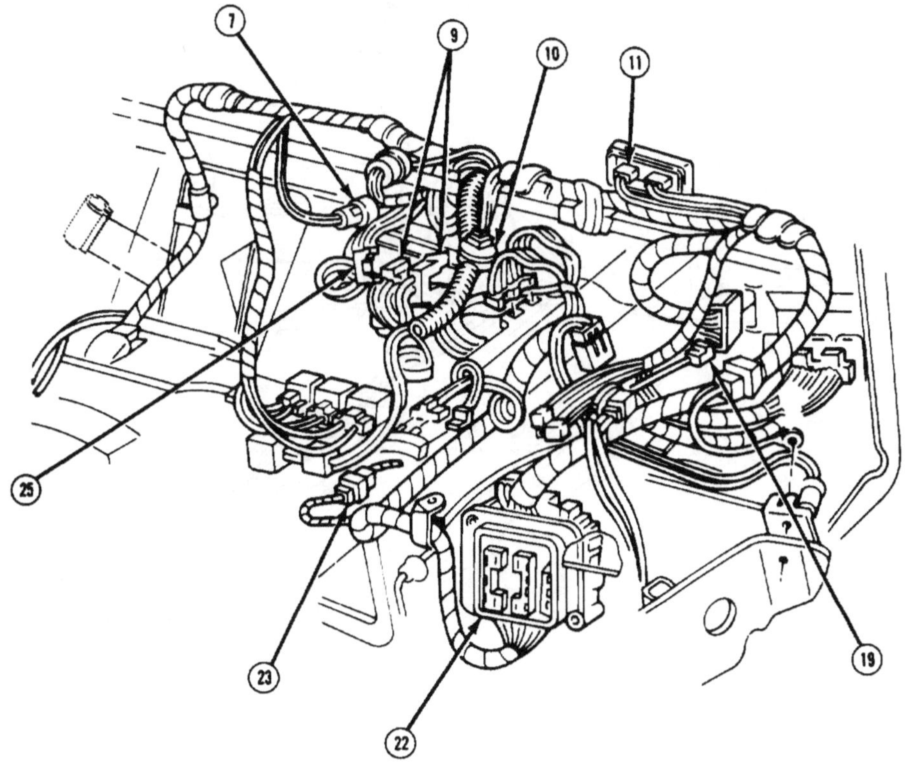

TA50809

4-6. INSTRUMENT PANEL WIRING HARNESS AND EXTENSION REPLACEMENT (Con't).

9 . Disconnect transfer case connector (33) from 4-wheel drive switch (34), Disconnect courtesy light connector (32),

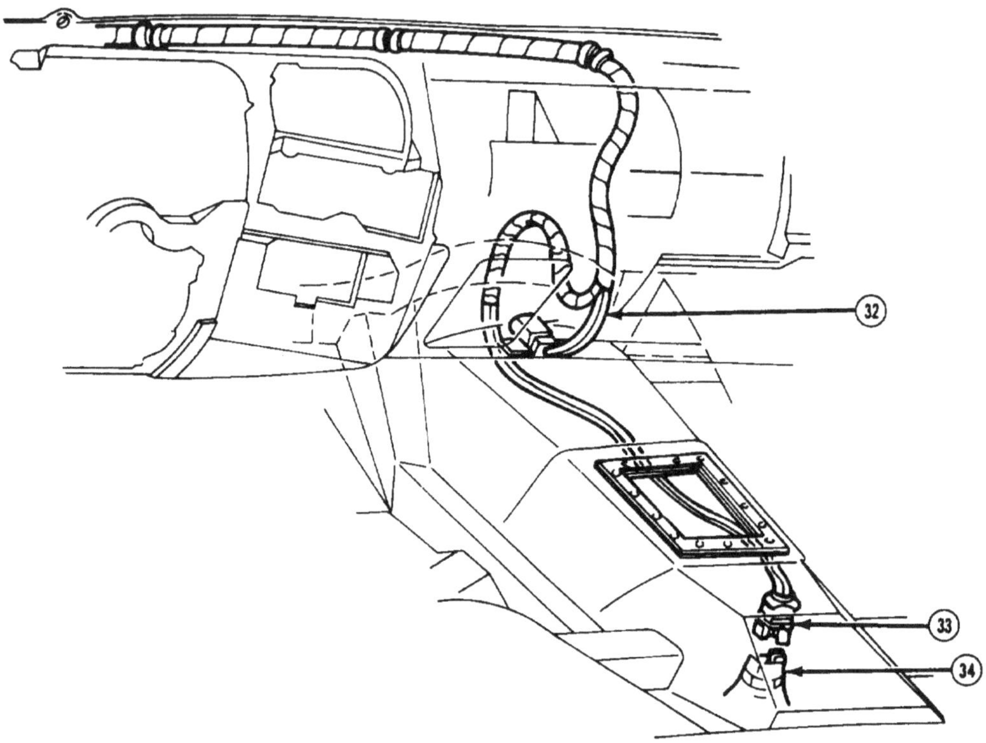

TA50810

4-6. INSTRUMENT PANEL WIRING HARNESS AND EXTENSION REPLACEMENT (Con't).

10. On underside of transfer case control lever reinforcement (36), remove tape (35) and pull transfer case connector (33) lead up through floor.

11. Disconnect leads from accelerator bracket and seatbelt retractor connector.

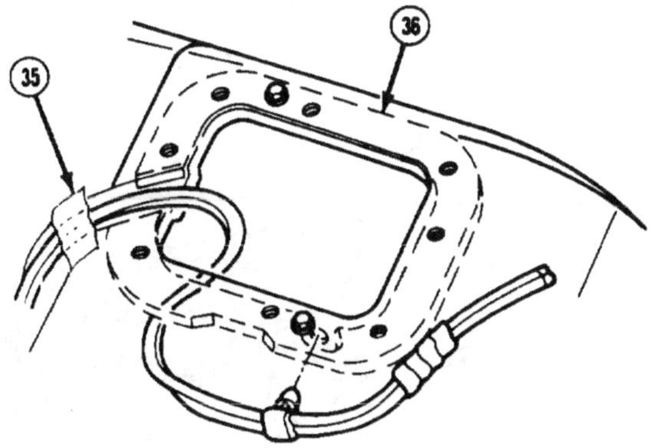

12. Remove 2 nuts (39), washers (38), and support (37). Lower steering column (1).

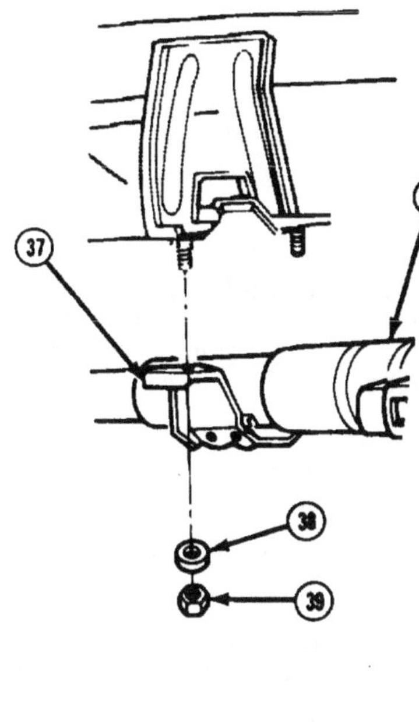

TA50811

4-6. INSTRUMENT PANEL WIRING HARNESS AND EXTENSION REPLACEMENT (Con't).

13. Disconnect extension wiring harness connector (43) from extension wiring harness (40).

14. Remove insulator (5), Unsnap clips along dash (6) and remove instrument panel wiring harness. Remove blackout light relay (24) if damaged. If replacing instrument panel wiring harness, remove hazard flasher and all fuses, and save for installation.

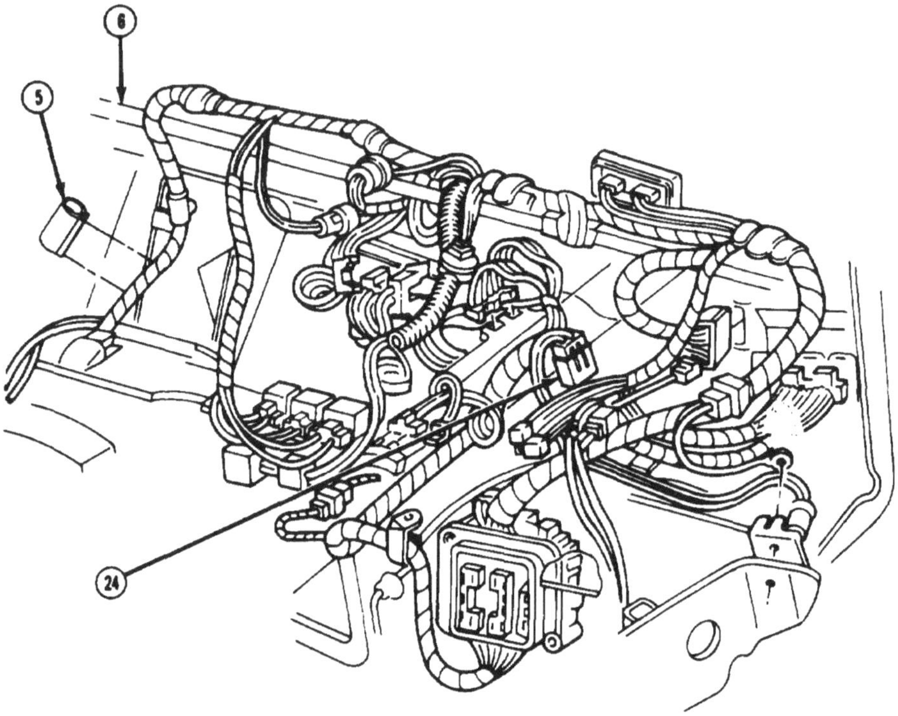

15. Remove 5 screws (45) from clamps (44), Unsnap extension wiring harness (40) clips from frame (41). Disconnect ambulance body accessory connectors (42). Pull extension wiring harness through grommet (46) into engine compartment to remove. Remove grommet if damaged.

b. INSTALLATION

NOTE

- If installing instrument panel extension wiring harness (40), perform steps 1 and 4.
- If Installing all except MI1010 instrument panal wiring harness, perform steps 2, 3, 5 through 8, and 10 through 16.
- If installing M1010 instrument panel wiring harness, perform steps 2-16.

TA50812

4-6. INSTRUMENT PANEL WIRING HARNESS AND EXTENSION REPLACEMENT (Con't).

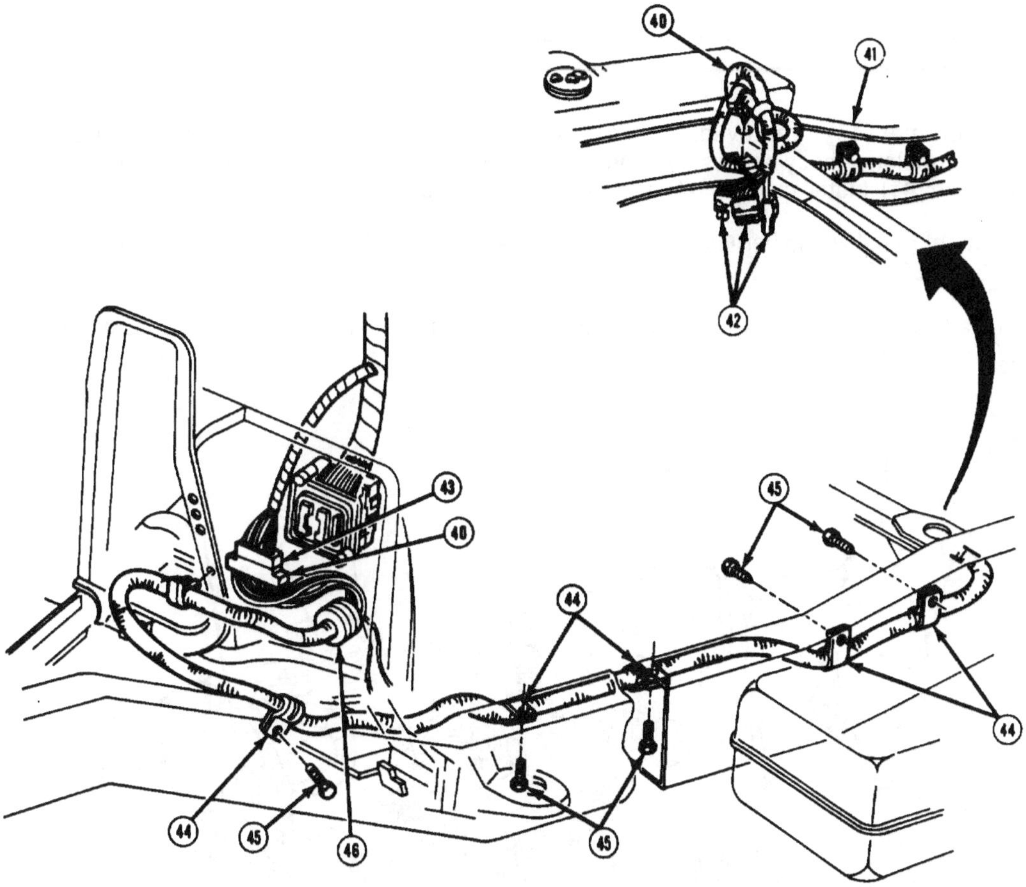

TA50813

4-6. INSTRUMENT PANEL WIRING HARNESS AND EXTENSION REPLACEMENT (Con't).

NOTE

- Voltmeter lamp connector (8), "WATER-IN-FUEL" light connector (11), "WAIT" light connector (12), and instrument cluster connector (13) will be connected when instrument cluster is installed.
- Ensure that all connectors are seated and locked in place as they are installed.

1. Install grommet (46) if removed. Push extension wiring harness (40) through grommet. Connect ambulance body accessory connectors (42). Snap extension wiring harness clips along frame (41), Install clamps (44) along frame with 5 screws (45).
2. If instrument panel wiring harness was replaced, install hazard flasher and all fuses. (See TM 9-2320-289-20)
3. Install blackout light relay (24) if removed. Snap instrument panel wiring harness clips along dash (6). Install insulator (5).

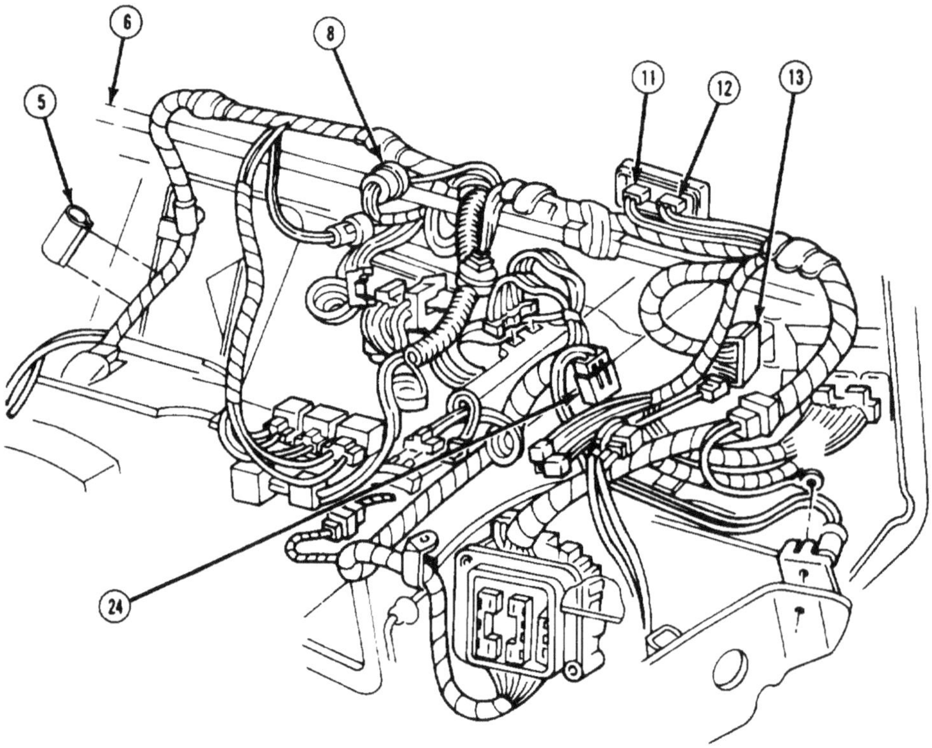

TA50814

4-6. INSTRUMENT PANEL WIRING HARNESS AND EXTENSION REPLACEMENT (Con't).

4 . Connect extension wiring harness connector (43) to extension wiring harness (40).

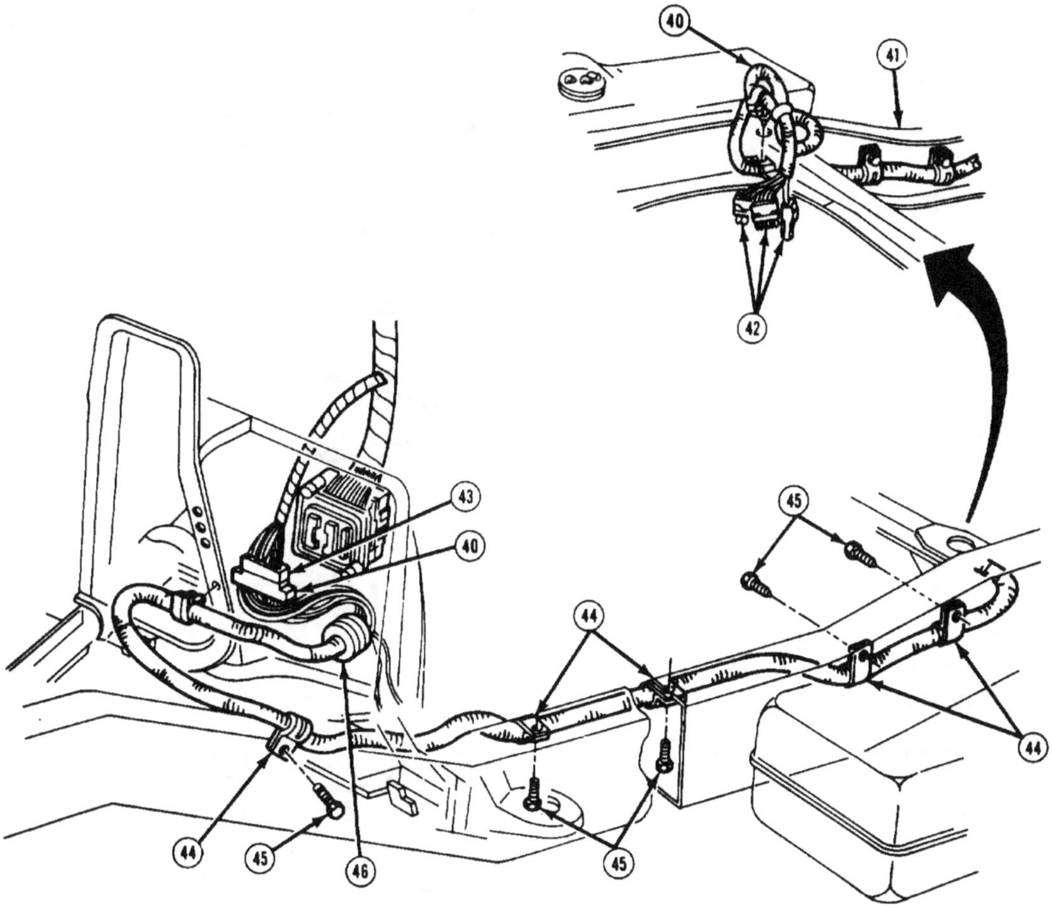

TA50815

4-6. INSTRUMENT PANEL WIRING HARNESS AND EXTENSION REPLACEMENT (Con't).

5. Raise steering column (1), and install support (37) with 2 washers (38) and nuts (39).
6. Connect leads to accelerator bracket and seatbelt retractor connector.

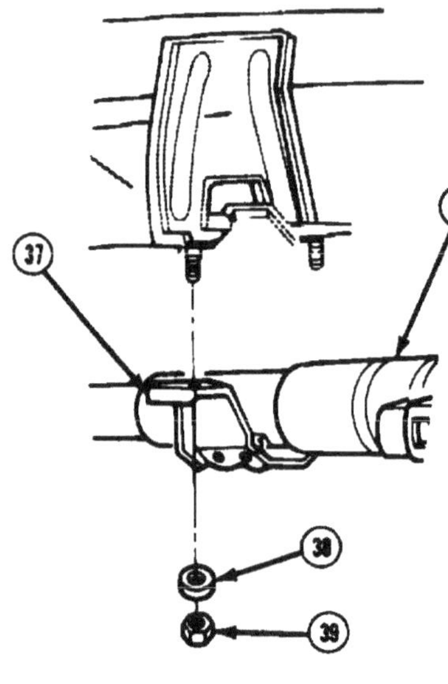

7. Push transfer case connector (33) lead through floor and connect to 4-wheel drive switch (34), Install transfer case connector lead to transfer case control lever reinforcement (36) with tape (35).
8. Connect courtesy light connector (32).

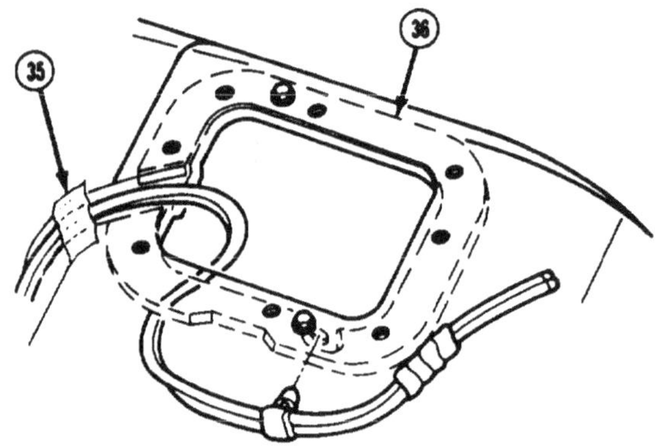

TA50816

4-6. INSTRUMENT PANEL WIRING HARNESS AND EXTENSION REPLACEMENT (Con't).

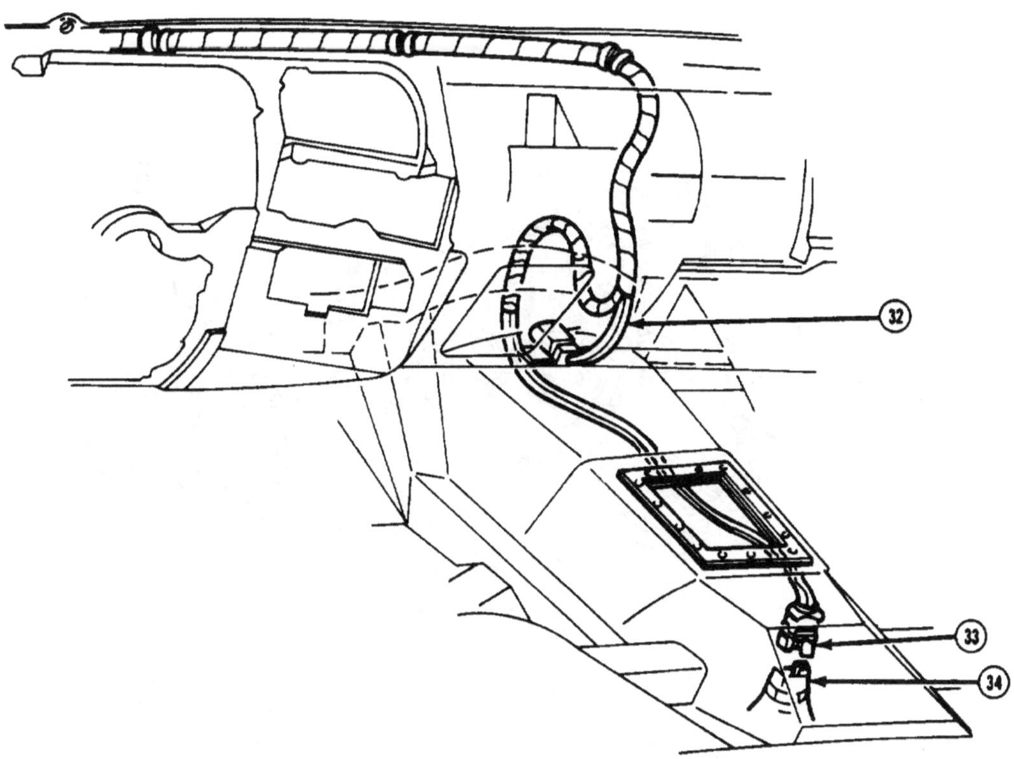

TA50817

4-6. INSTRUMENT PANEL WIRING HARNESS AND EXTENSION REPLACEMENT (Con't).

9. Connect open door buzzer wiring harness connector below "WATER-IN-FUEL" light connector (11). Connect spotlight switch connector (19), 2 floodlight switch connectors (9), and GPFU switch connector (25).

10. Connect heater lamp connector (7), heater connector (10), and glow plug wiring harness connector (23).

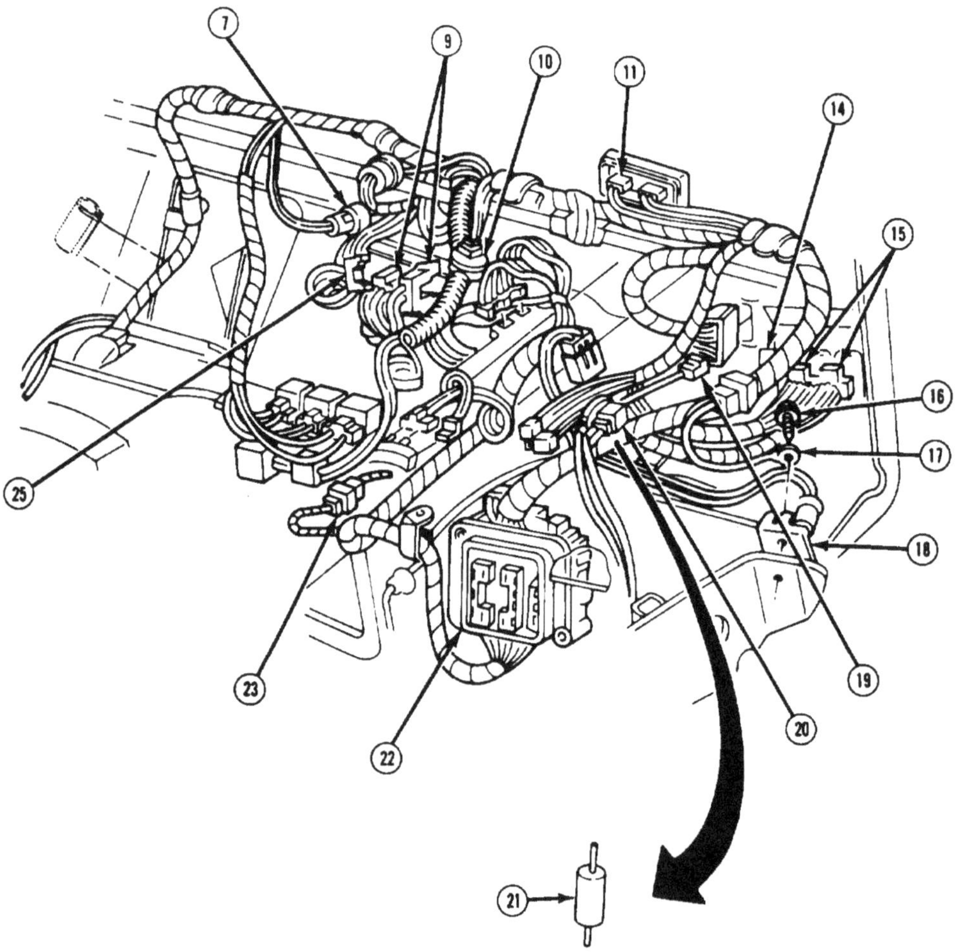

TA50818

4-6. INSTRUMENT PANEL WIRING HARNESS AND EXTENSION REPLACEMENT (con't).

11. Apply grease to sealing surfaces of fuse box (22). Install fuse box with bolt (31) and 2 screws (30). Install turn signal flasher.

12. Inside engine compartment, connect wiring harnesses to rear of fuse box (22).

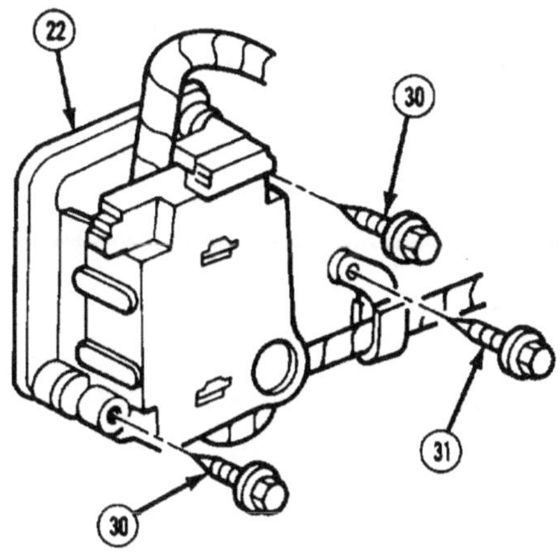

13. At lower part of steering column (1), connect directional signal connector (27), ignition switch connector (28), back-up light switch connector (26), and dimmer switch connector (29).

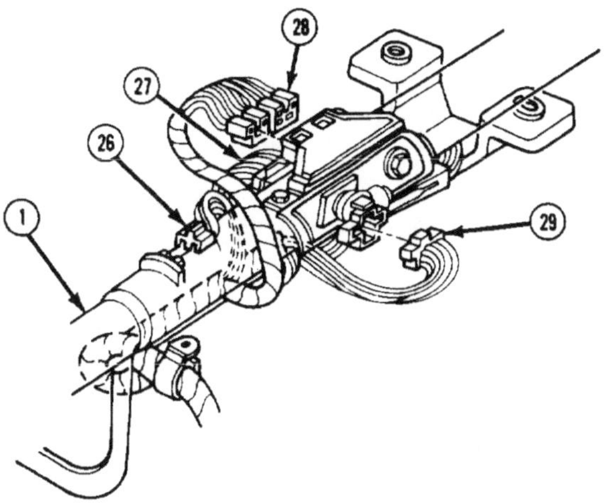

14. Install wiring harness terminal (18) and ground lead (17) with screw (16).

15. Install horn rectifier (21) if removed. Connect headlight switch connector (14), 2 blackout light switch connectors (15), and horn relay (20).

TA50819

4-6. INSTRUMENT PANEL WIRING HARNESS AND EXTENSION REPLACEMENT (Con't).

16. At underside of steering column (1), connect wiper/washer switch connector (4) to wiper/washer switch (3). Plug wiper/washer switch onto bracket (2).

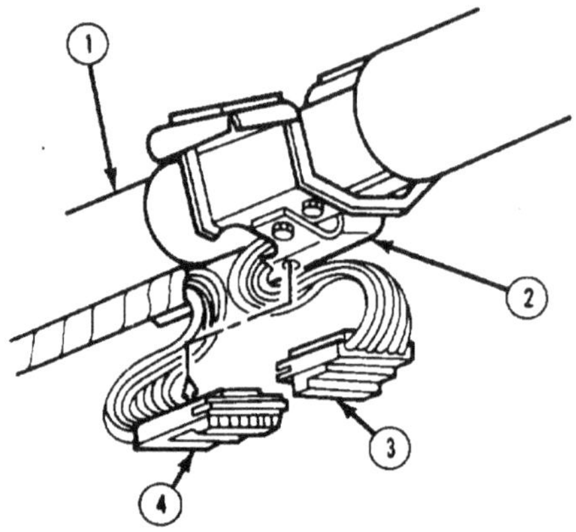

FOLLOW-ON TASKS:

- Install Instrument panel padding, (See TM 9-2320-289-20)
- Install transfer case bezel, plate, boot, and control lever housing. (See TM 9-2320-289-20)
- Install voltmeter relay, generator 2 relay, and starter relay. (See TM 9-2320-289-20)
- Install instrument cluster. (See TM 9-2320-289-20)

TA50820

4-7. FRONT WIRING HARNESS REPLACEMENT.

This task covers: a. Removal b. Installation

INITIAL SETUP:

Equipment Condition

- Front battery~ removed. (See TM 9-2320-289-20)
- Front battery box removed (if truck is equipped with winterization kits). (See TM 9-2320-289-20)
- Radiator grille removed. (See TM 9-2320-289-20)
- Front blackout marker lights removed. (See TM 9-2320-289-20)

Materials/Parts

- Two lockwashers
- Three starwashers
- Tie-down straps (as needed) (Item 63, Appendix B)

a. REMOVAL

1. Remove bolt and disconnect fuse box connector (1) from rear of fuse box. Separate f rent wiring harness connectors from fuse box connector assembly.
2. Open clips retaining front wiring harness (2) to left fenderwell (4).
3. Cut and discard tie-down straps retaining front wiring harness (2) to windshield washer hose. Loosen screws securing washer solvent container (3) and pull front wiring harness free.

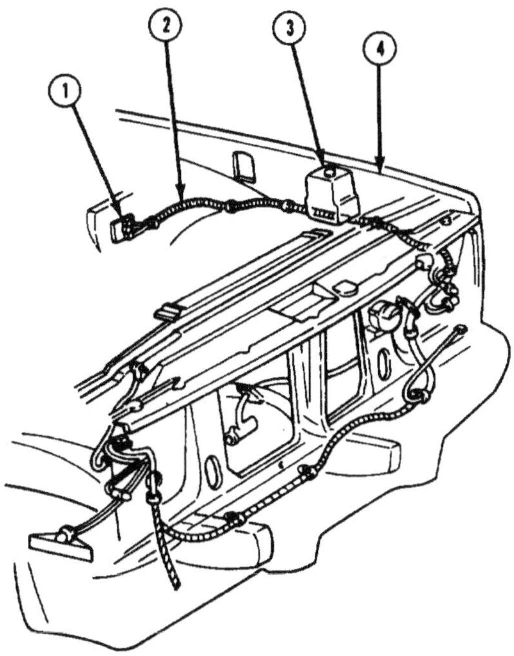

TA50821

4-7. FRONT WIRING HARNESS REPLACEMENT (Con't).

4. Remove bolt (7), lockwasher (10), ground lead (8), and starwasher (9). Disconnect headlight connector (11), parking light connector (12), and marker light connector (13). Discard lockwasher and starwasher,

5. Disconnect blackout drive light connector (5) and horn capacitor connector (6).

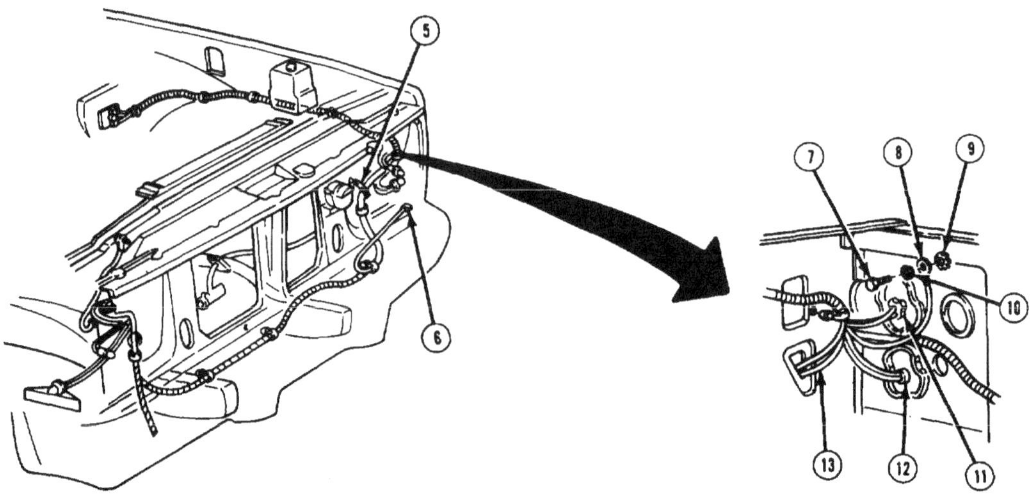

TA50822

4-7. FRONT WIRING HARNESS REPLACEMENT (Con't).

6. Disconnect combination valve connector (16) from brake indicator light bracket (14).

7. Disconnect brake indicator light wire (17) from combination valve (18) and brake indicator light bracket (14) if damaged. Remove screw (15) and brake indicator light bracket if damaged.

8. Remove 2 bolts (23), lockwasher (22), battery lead (21), ground lead (20), and 2 starwashers (19). Discard lockwasher and starwashers.

9. Disconnect headlight connector (28), parking light connector (25), and marker light connector (24). Disconnect low coolant sensor connector (26) from low coolant sensor (27).

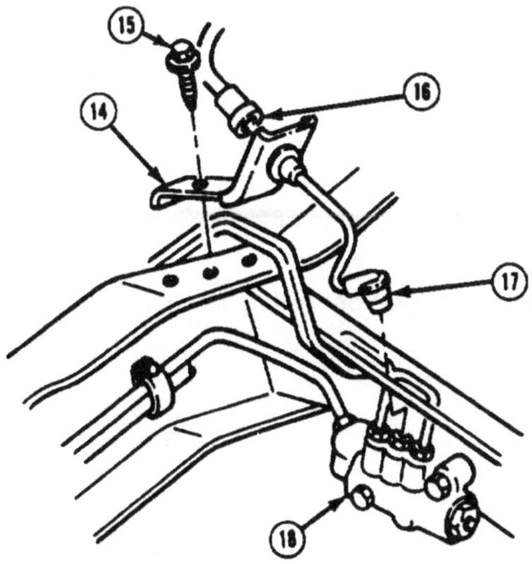

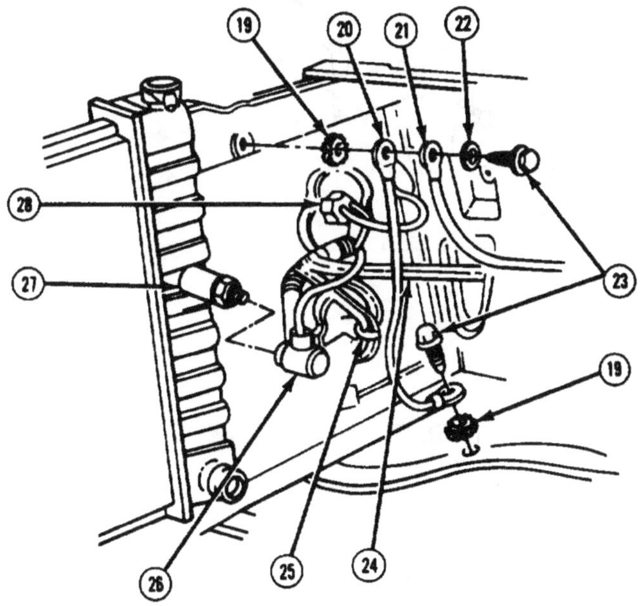

TA60823

4-7. FRONT WIRING HARNESS REPLACEMENT (Con't).

CAUTION

Care should be taken during removal of front wiring harness (2) to prevent snagging it. Failure to follow this caution may result in damage to front wiring harness.

NOTE

Ensure that position of front wiring harness (2) is noted for installation.

10. Open clips retaining front wiring harness (2) to radiator support. Remove front wiring harness.

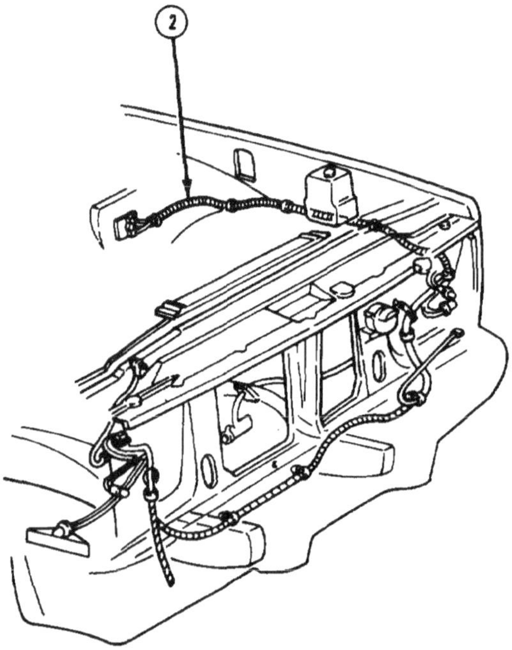

b. INSTALLATION

NOTE

Ensure that all connectors are seated and locked in place as they are installed.

1. Route front wiring harness (2) through radiator support into engine compartment. Install front wiring harness into clips along radiator support.

2. Connect low coolant sensor connector (26) to low coolant sensor (27). Connect headlight connector (28), parking light connector (25), and marker light connector (24).

TA50824

4-7. FRONT WIRING HARNESS REPLACEMENT (Con't).

3. Install ground lead (20) and battery lead (21) with 2 new starwashers (19), new lockwasher (22), and 2 bolts (23).

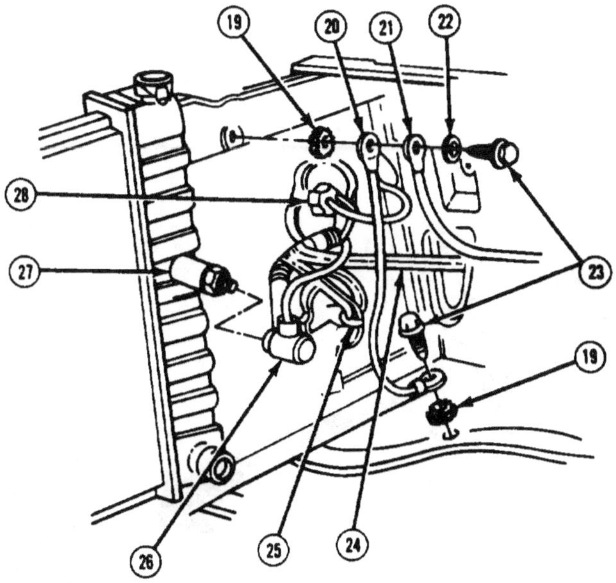

4. Install brake indicator light bracket (14) with screw (15) if removed. Connect brake indicator light wire (17) to combination valve (18) and brake indicator light bracket if removed.

5. Connect combination valve connector (16) to brake indicator light bracket (14).

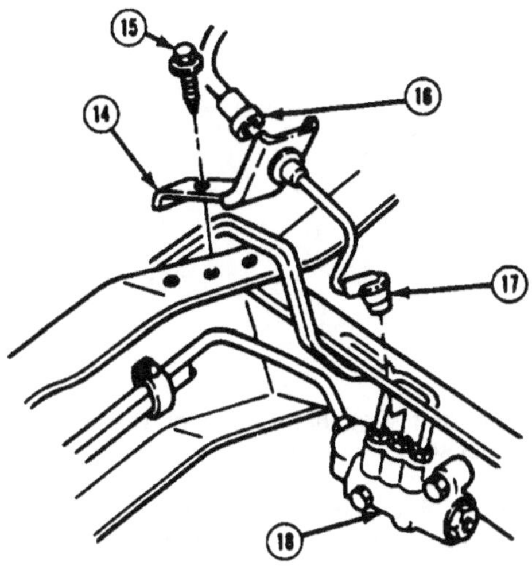

TA60826

4-7. FRONT WIRING HARNESS REPLACEMENT (Con't).

6. Connect blackout drive light connector (5) and horn capacitor connector (6).

7. Connect headlight connector (11), parking light connector (1 2), and marker light connector (13). Install ground lead (8) with new starwasher (9), new lockwasher (10), and bolt (7).

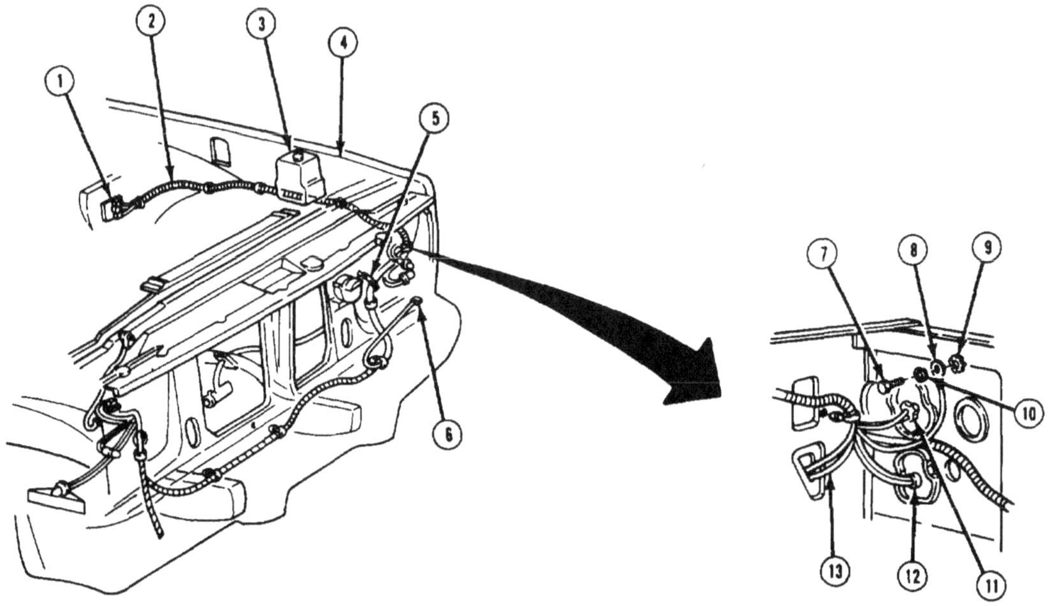

8. Route front wiring harness under washer solvent container (3) and tighten screws securing washer solvent container.

9. Install front wiring harness (2) into clips along left fenderwell (4). Install front wiring harness onto windshield washer hose with new tie-down straps.

10. Assemble front wiring harness (2) to fuse box connector. Connect fuse box connector (1) to rear of fuse box and install bolt.

TA50826

4-7. FRONT WIRING HARNESS REPLACEMENT (Con't).

FOLLOW-ON TASKS:

- Install front blackout marker lights. (See TM 9-2320-289-20)
- Install radiator grille. (See TM 9-2320-289-20)
- Install front battery box if removed. (See TM 9-2320-289-20)
- Install front battery. (See TM 9-2320-289-20)
- Check operation of front lights.

4-8. BODY WIRING HARNESS REPLACEMENT.

This task covers: a. Remova l b. Installation

INITIAL SETUP:

Equipment Condition

- Both battery negative cables disconnected. (See TM 9-2320-289-20)

Materiels/Parts

- One starwasher

a. REMOVA L

NOTE

- If removing body wiring harness (1) from all except M1009 , perform steps 1-3, 5, and 6,
- If removing body wiring harness (1) from Ml 009, perform steps 1, 2, 4, 5, and 7 through 10.
- Ensure that all leads are tagged for Installation.

1 . Remove bolt and disconnect body wiring harness (1) at rear of fuse box (3). Disconnect red lead at fuse box. Depress sides of terminal "30" on fuse box , and disconnect pink sending unit lead by pulling lead toward front of connector.

2 . Open clip (8) retaining body wiring harness (1) to engine bulkhead (2). Unsnap body wiring harness from engine bulkhead,

3 . Remove nut (5), open clip (6), and pull body wiring harness (1) free. Pull speedometer cable free if retained by clip. Remove bolt (4) and retainer plate (7) from inside cab if damaged.

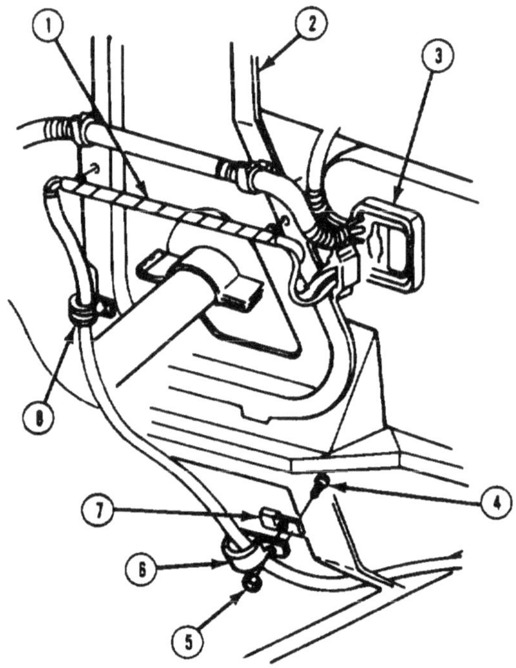

4 . Open adjustable retainer (9). Remove 5 bolts (10), 3 small clips (12), axle vent hose clip (13), and large clip (11) from frame (14). Remove 2 nuts and ground wire, from right and left blackout markers.

TA50827

4-8. BODY WIRING HARNESS REPLACEMENT (Con't).

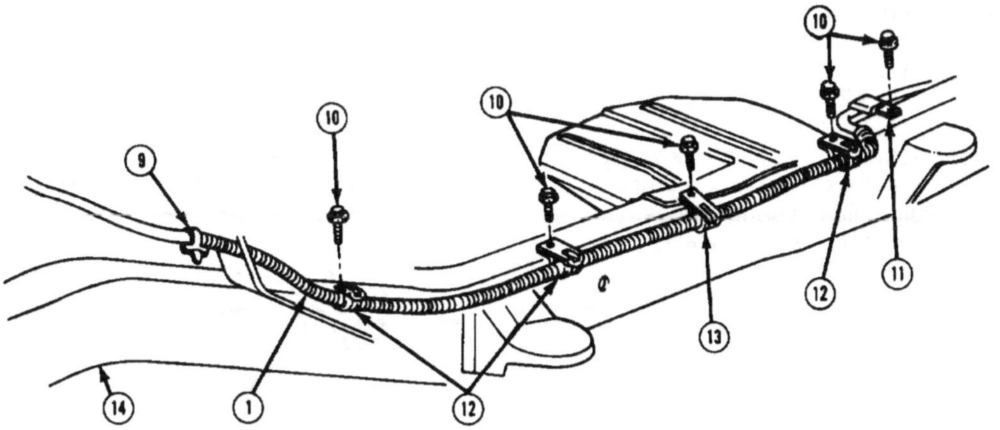

5. Remove screw (19) and starwasher (18), and disconnect ground lead (15) from rear frame (14). Disconnect fuel tank sending unit lead (16). Discard starwasher.

6. Disconnect body wiring harness (1) from taillight wiring harness. Unsnap 9 clips (17) from along frame (14) and remove body wiring harness.

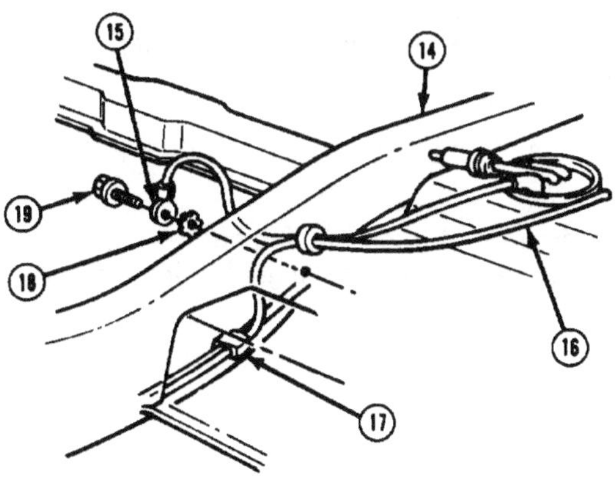

All Except M1009

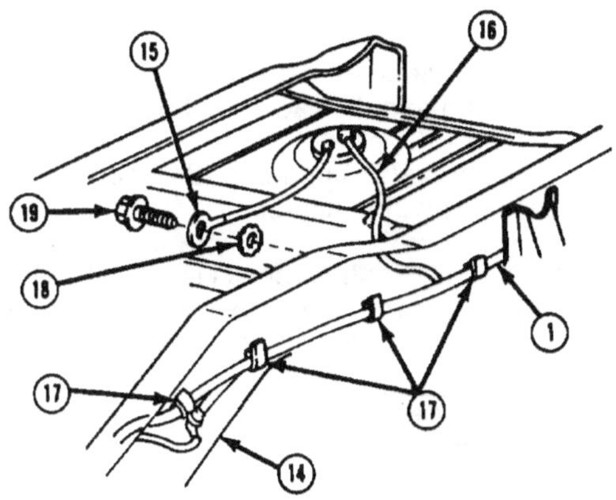

M1009

TA50828

4-8. BODY WIRING HARNESS REPLACEMENT (Con't).

7. Unsnap 1 clip (17) from frame (14), Disconnect body wiring harness (1) from right and left taillight wiring harness, and trailer wiring harness, Remove 2 nuts (21) and clips (22) from frame studs (20). Unsnap clips securing right and left blackou t marker harnesses t o bumper bracket. Pull taillight wiring harness through holes in frame on right and left sides,

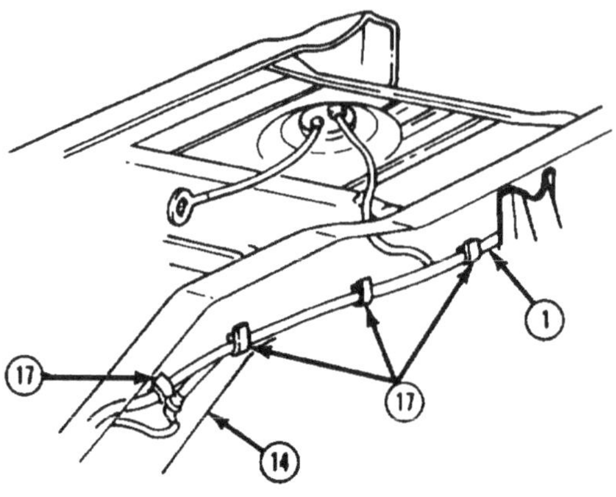

M1009

8. Lower body wiring harness from engine compartment. Attach a guide wire approximately 7 ft. (2. 1 m) long to bulkhead end of body wiring harness (1).

9. Remove screw (25) and clip (24) from frame (14). Unsnap 4 clips (23) from frame and remove body wiring harness (1) from frame.

10. Remove body wiring harness (1) by pulling harness from rear of truck, guiding harness through opening in rear frame member. Remove harness from guide wire and leave guide wire in place in frame to aid in installation.

b. INSTALLATION

NOTE

- If installing body wlrlng harness (1) for all except Ml 009, perform steps 3, 4, and 6 through 8.
- If Installing body wiring harness (1) for Ml 009, perform steps 1, 2, 4, 5, 7, and 8.
- Ensure that all connectors are seated and locked in place as they are installed.

1. Attach bulkhead end of body wiring harness (1) to guide wire at rear of vehicle. Guide wiring harness through frame and disconnect guide wire. Route wiring harness to bulkhead connector in engine compartment,

TA50829

4-8. BODY WIRING HARNESS REPLACEMENT (Con't).

2. Snap 4 clips (23) with body wiring harness (1) into place along frame (14). Install clip (24) on frame with screw (25).

3. Route taillight wiring harness through holes in frame on right and left sides. Install 2 clips (22) on frame studs (20) with 2 nuts (21). Connect body wiring harness (1) to taillight wiring harness and trailer wiring harness.

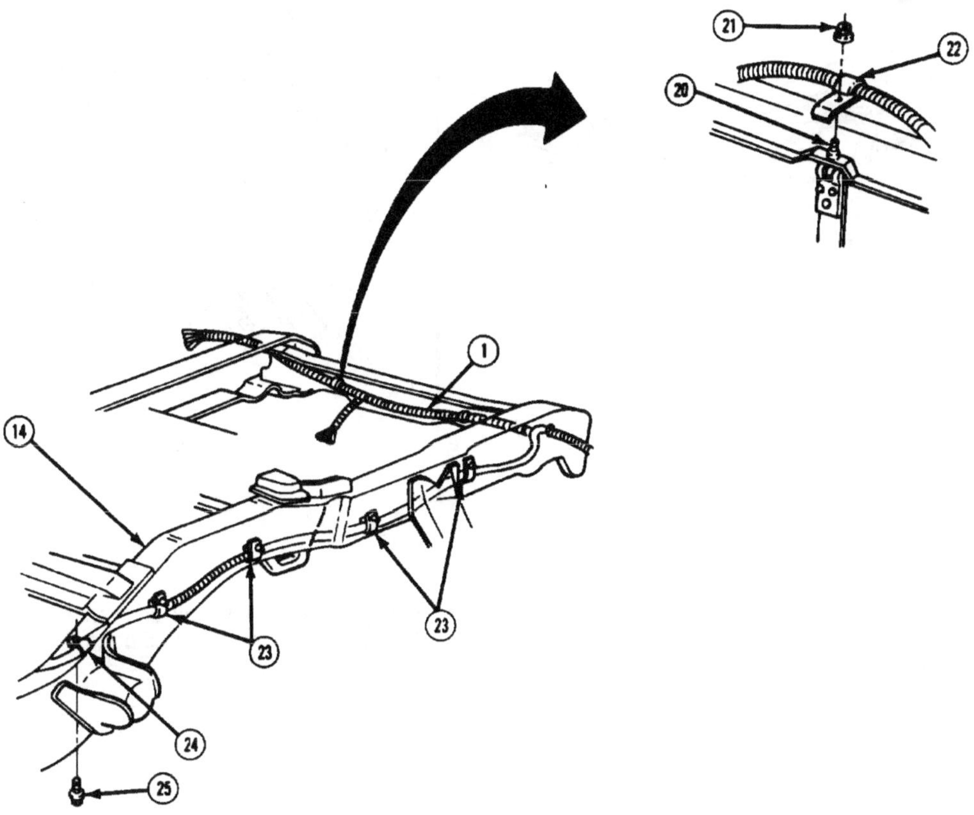

TA502830

4-8. BODY WIRING HARNESS REPLACEMENT (Con't).

4. Snap 9 clips (17) with body wiring harness (1) along frame (14). Connect body wiring harness to taillight wiring harness.

5. Connect fuel tank sending unit lead (16). Connect ground lead (15) and secure with screw (19), and new starwasher (18).

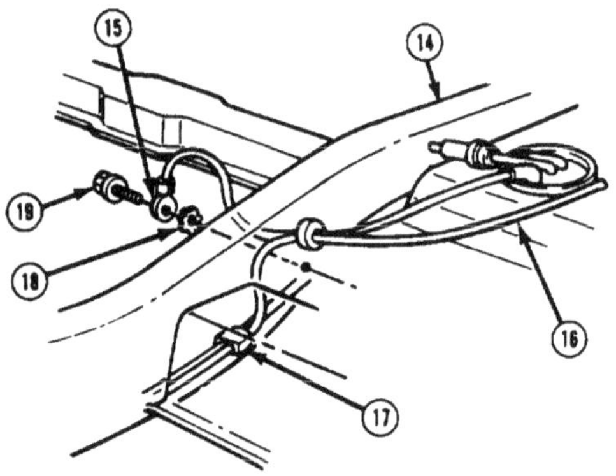

All Except M1009

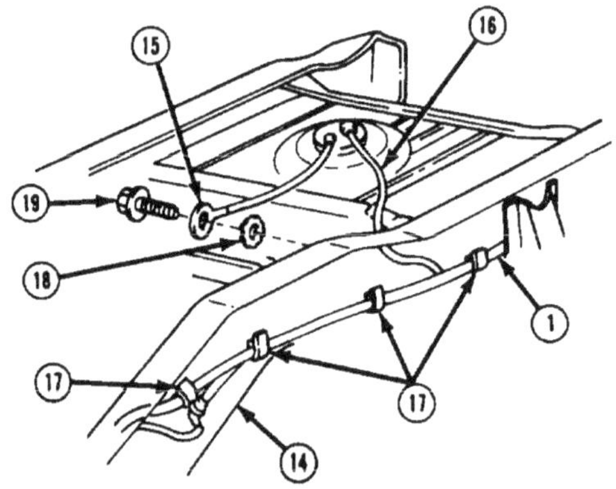

M1009

6. Install large clip (11) and 3 small clips (12) on body wiring harness (1). Install axle vent hose clip (13) on body wiring harness and axle vent hose. Install clips on frame with 5 bolts (10). Install body wiring harness in adjustable retainer (9).

TA50831

4-8. BODY WIRING HARNESS REPLACEMENT (Con't).

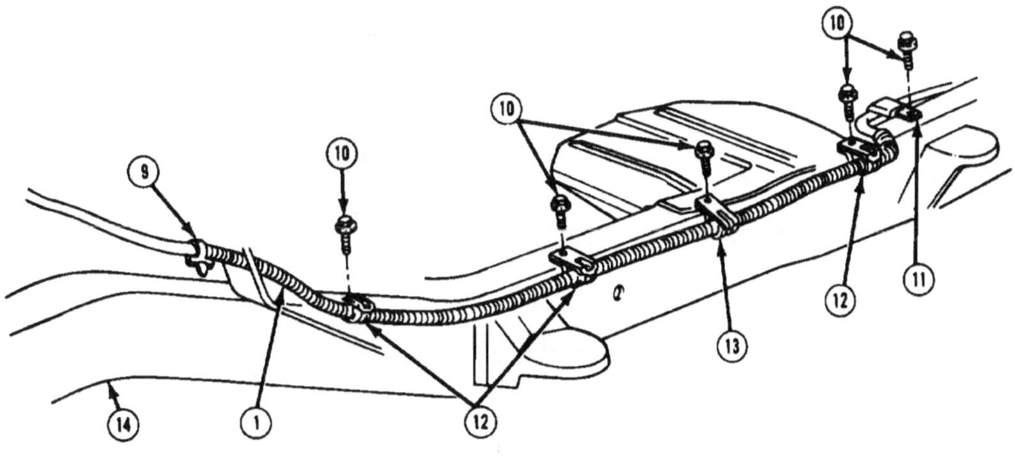

7. Install retainer plate (7) and bolt (4) from inside cab if removed. Install body wiring harness (1) and speedometer cable in clip (6). Install clip on retainer plate with nut (5).

8. Snap body wiring harness (1) to engine bulkhead (2), Install body wiring harness in clip (8).

9. Install pink sending unit lead from body wiring harness into terminal "30" on fuse box, Connect red lead at fuse box. Connect body wiring harness (1) at rear of fuse box (3).

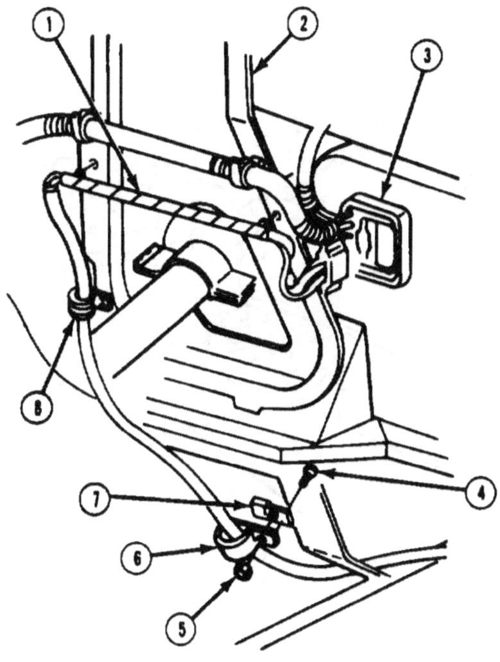

FOLLOW-ON TASKS:

- Connect both battery negative cables. (See TM 9-2320-289-20)

TA50832

4-9. TAILLIGHT WIRING HARNESS REPLACEMENT (M1010 AND M1031).

This task covers : a. Remova l b. Installation

INITIAL SETUP:

Equipment Condition

- Rear blackout marker lights removed. (See TM 9-2320-289-20)

Materials/Parts

- One starwasher

a. REMOVAL

NOTE

- Ensure that all connectors are tagged for installation.
- Perform steps 2 and 3 if removing taillight wiring harness (4) from M1031.

1. Remove connectors (3) from side marker light housings. Remove side marker light bulbs (2),

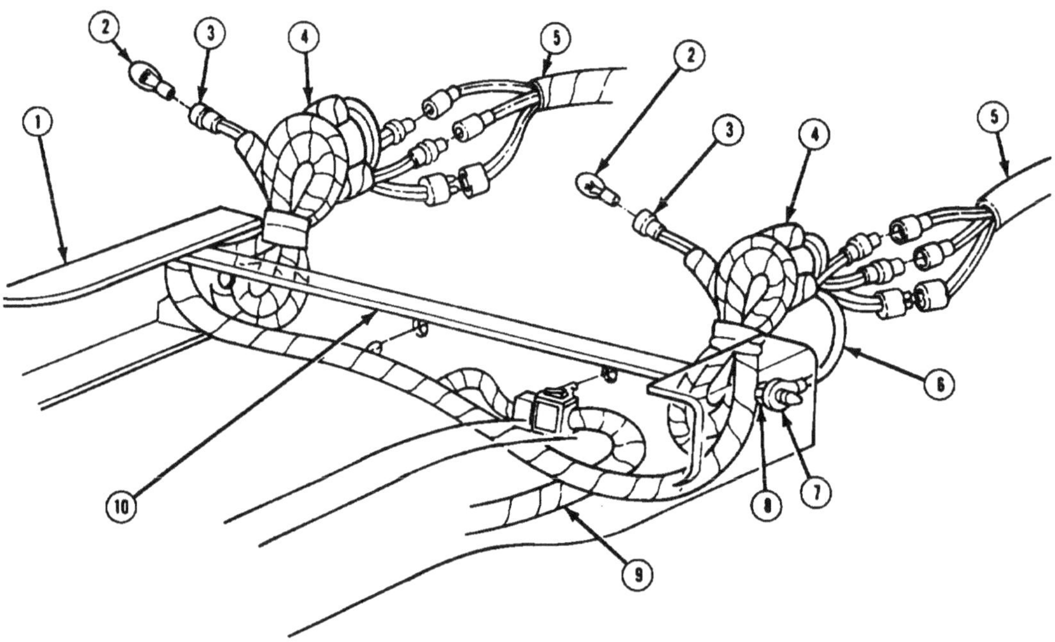

M1010

TA50833

4-9. TAILLIGHT WIRING HARNESS REPLACEMENT (M1010 AND M1031) (Con't).

2. Disconnect taillight wiring harness (4) connectors from taillight connectors (5). Remove screw (8) and starwasher (7), and disconnect ground lead (6) from frame (1). Discard starwasher.

3. Disconnect taillight wiring harness (4) from body wiring harness (9). Unsnap taillight wiring harness from frame rail (10) and remove.

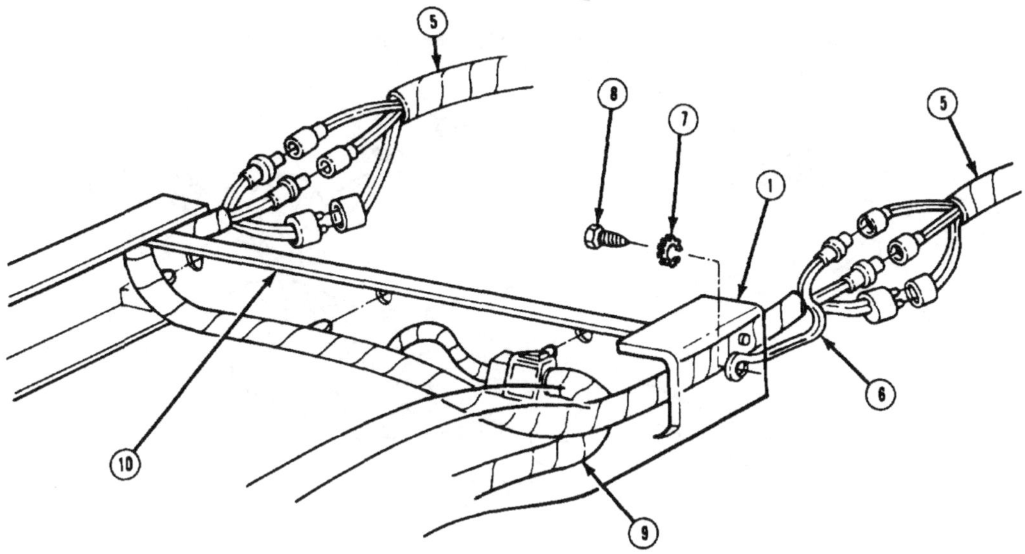

M1031

b. INSTALLATION

NOTE

- Perform steps 1 and 2 if installing taillight wiring harness (4) on M1031.
- Ensure that all connectors are seated and locked in place as they are installed.

1. Snap taillight wiring harness (4) to frame rail (10). Connect taillight wiring harness to body wiring harness (9).

2. Connect ground lead (6) to frame (1) with new starwasher (7) and screw (8). Connect taillight wiring harness (4) connectors to taillight connectors (5).

TA50834

4-9. TAILLIGHT WIRING HARNESS REPLACEMENT (M1010 AND M1031) (Con't).

3 . install bulbs (2). Install connectors (3) to side marker light housings.

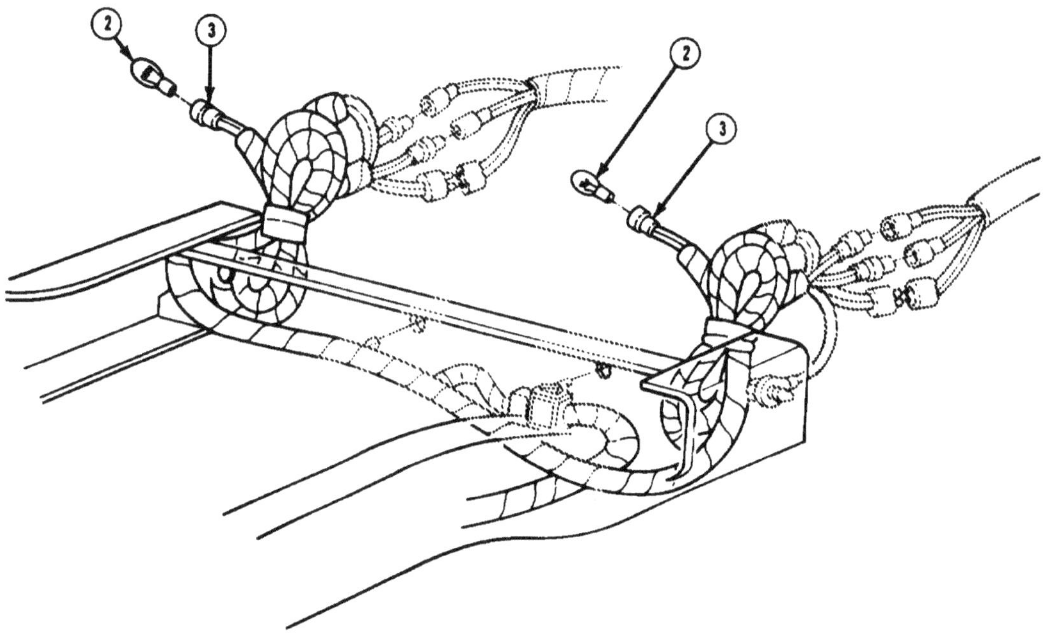

M1010

FOLLOW-ON TASKS:

- Install rear blackout marker lights. (See TM 9-2320-289-20)
- Check operation of taillights.

TA50835

4-9.1. REAR FENDER WIRING HARNESS REPLACEMENT (M1028A2).

This task covers: a. Removal b. Installation

INITIAL SETUP:

Equipment Condition

- Rear fender side marker lights removed. (See TM 9-2330-289-20).
- Taillight and lamps removed. (See TM 9-2320-289-20)

a. REMOVAL

NOTE

Rear fender wiring harness is removed from both sides in same manner.

1. Remove 14 bolts (12), 1 nut (13), and inner filler panel (11).
2. Disconnect rear fender wiring harness (8) from rear lamp wiring harness (7).
3. Remove screw (5) and disconnect ground wire (6).
4, Remove 2 nuts (1), washers (2), bolts (4), and shields (10) from inner fender (9).

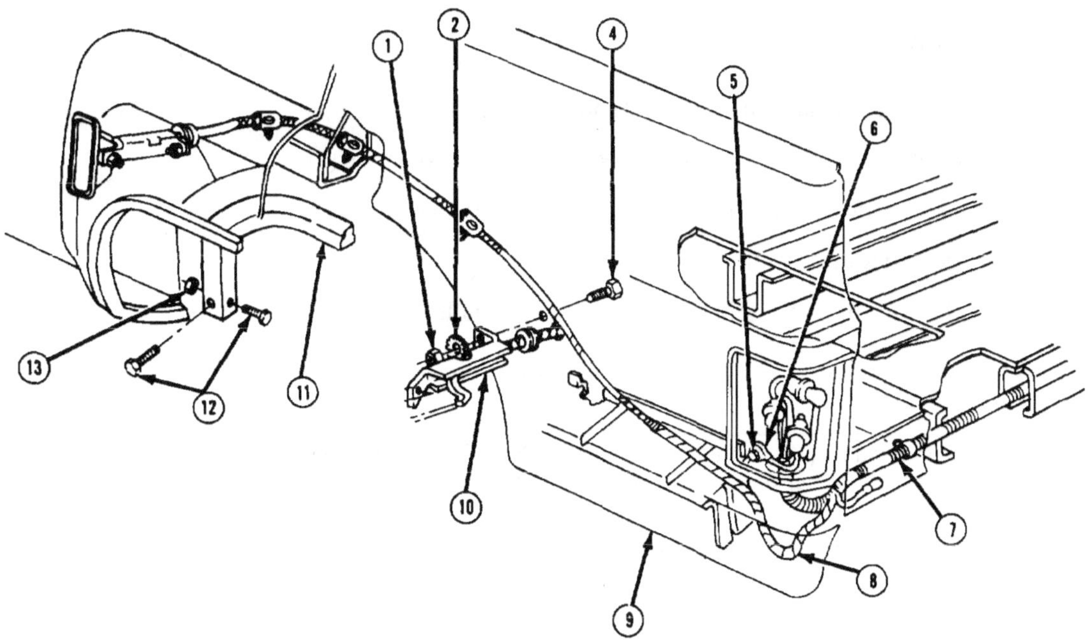

4-9.1. REAR FENDER WIRING HARNESS REPLACEMENT (M1028A2) (Con't).

5. Remove 2 grommets (3), Unsnap rear fender wiring harness (8) from body and remove.

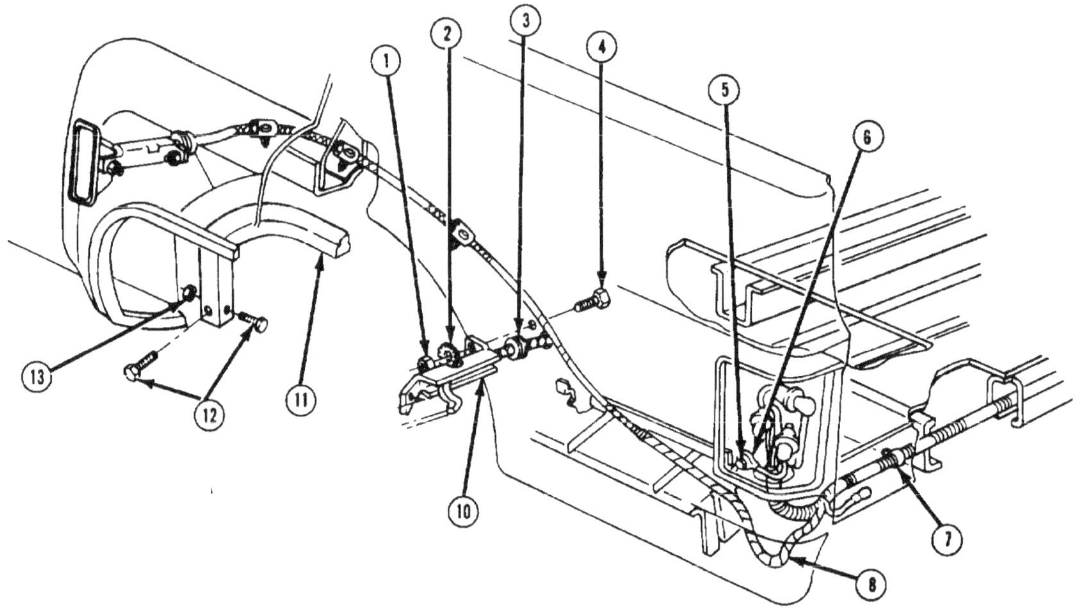

b. INSTALLATION

1. Install rear fender wiring harness (8) and snap to body,
2. Push wiring harness (8) ends through body and install 2 grommets (3).
3. Connect rear fender wiring harness (8) to rear lamp wiring harness (7).
4. Connect ground wire (6) with screw (5).
5. Install 2 shields (10) with 2 bolts (4), washers (2), and nuts (1).

CAUTION

Inner filler panel (11) is made from plastic. Care should be taken when installing bolts (12), or inner filler panel could crack.

6. Install inner filler panel (11) with 14 bolts (12) and 1 nut (13).

FOLLOW-ON TASKS:

- Install taillight and lamps. (See TM 9-2320-289-20)
- Install rear fender side marker lights. (See TM 9-2320-289-20)

4-10. ENGINE WIRING HARNESS REPLACEMENT.

This task covers: a. Removal b. Installation

INITIAL SETUP:

Equipment Condition

- Starter removed. (See TM 9-2320-289-20)
- Air cleaner removed. (See TM 9-2320-289-20)
- Rear battery box removed (if truck is equipped with winterization kits). (See TM 9-2320-289-20)

MaterialslParts

- Eight starwashers
- Fifteen lockwashers
- Six tie-down straps (Item 63, Appendix B)

a. REMOVA L

NOTE

- If disconnecting all except M1010 engine wiring harness from engine assembly, perform steps 1, 2, 6, 7, 11, and 18.
- If disconnecting M1010 engine wiring harness from engine assembly, perform steps 1, 2, 6, 8 through 11, and 18.
- If removing all except M1010 engine wiring harness, perform steps 1-4, 6, 7, 11 through 13, and 16 through 19.
- If removing M1010 engine wiring harness, perform steps 1-6 and steps 8-19.

4-10. ENGINE WIRING HARNESS REPLACEMENT (Con't).

NOTE

Ensure that all leads are tagged for installation.

1. Disconnect 4 right side glow plug connectors (5).
2. Disconnect cold and fast idle switch connector (6), fuel solenoid connector (2), cold advance switch connector (3), and fast idle solenoid connector (4).

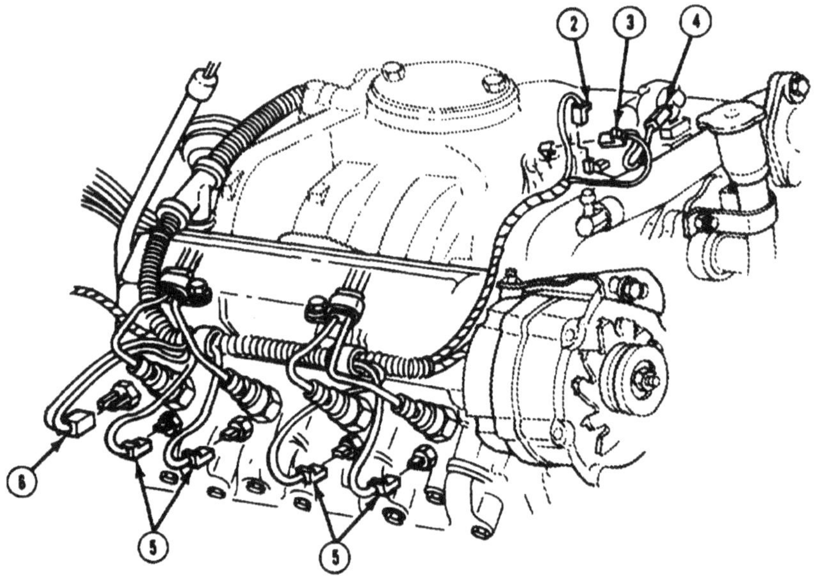

3. Disconnect fuel filter heater connector (10), vacuum switch connector (11), and water sensor connector (13).
4. Remove terminal board (9) covers. Remove 2 nuts and starwashers, and disconnect leads from positive terminal board studs (7). Remove nut and starwasher, and disconnect leads from negative terminal board center stud (8). Discard starwashers.
5. Disconnect diagnostic wiring harness connector (12).

TA50836

4-10. ENGINE WIRING HARNESS REPLACEMENT (Con't).

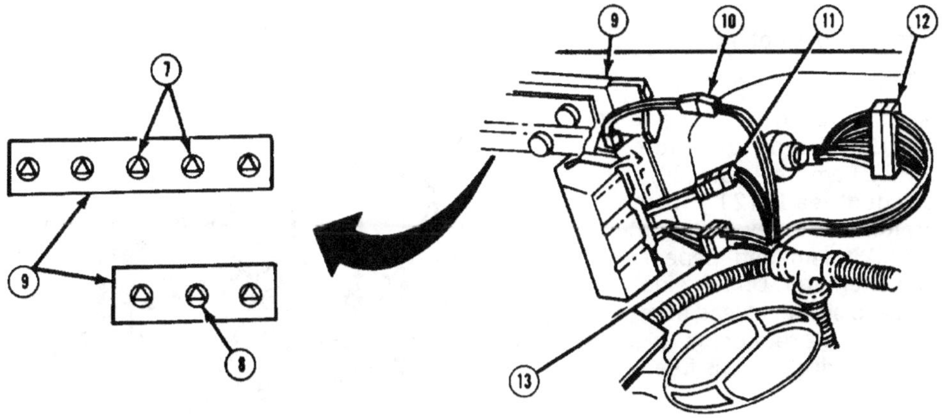

6. Remove nut (16), and disconnect terminal board ground (15), engine wiring harness lead (14), and engine ground strap (17) from intake manifold stud (19). Remove and discard starwasher (18).

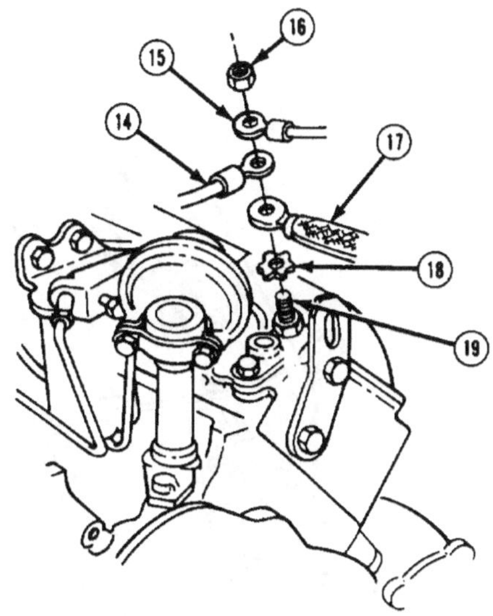

TA50837

4-10 ENGINE WIRING HARNESS REPLACEMENT (Con't).

NOTE

Left side alternator does not have a capacitor (25).

7. Disconnect voltage regulator connector (23) from right side alternator (24). Remove 2 nuts (20) and disconnect leads (21) from alternator. Remove screw (22) and capacitor (25) if damaged. Repeat for left side alternator. On left side alternator, also disconnect lead from "R" terminal. Open 2 retainers (1) and pull engine wiring harness free.

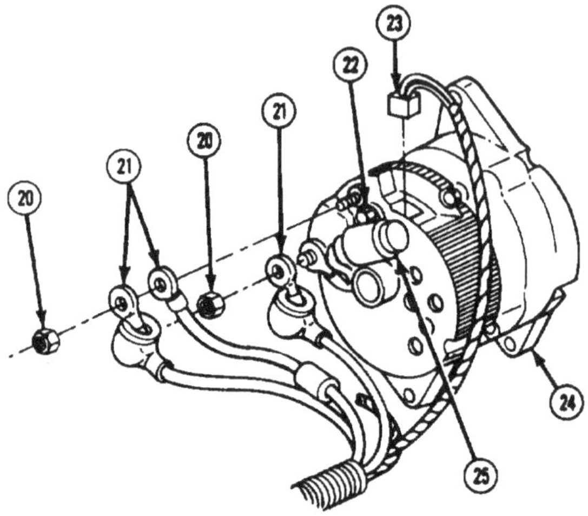

All Except M1010

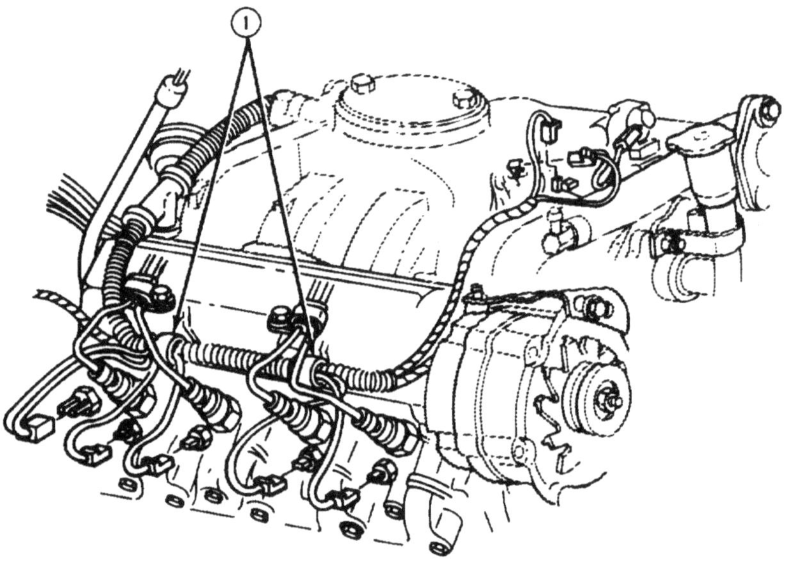

TA50838

4-10. ENGINE WIRING HARNESS REPLACEMENT (Con't).

NOTE

Ensure that location of lockwashers (26) and starwashers (28) is noted for installation.

8 . Remove 7 nuts (27), 4 lockwashers (26), and 2 starwashers (28), and disconnect leads from upper alternator (31) and lower alternator (32). Loosely install nuts (27) on connector studs. Discard lockwashers and starwashers. Open 2 retainers (1) and pull engine wiring harness free.

9 . Remove nut (29), and disconnect engine wiring harness lead and 2 ground wires (30) from intake manifold stud.

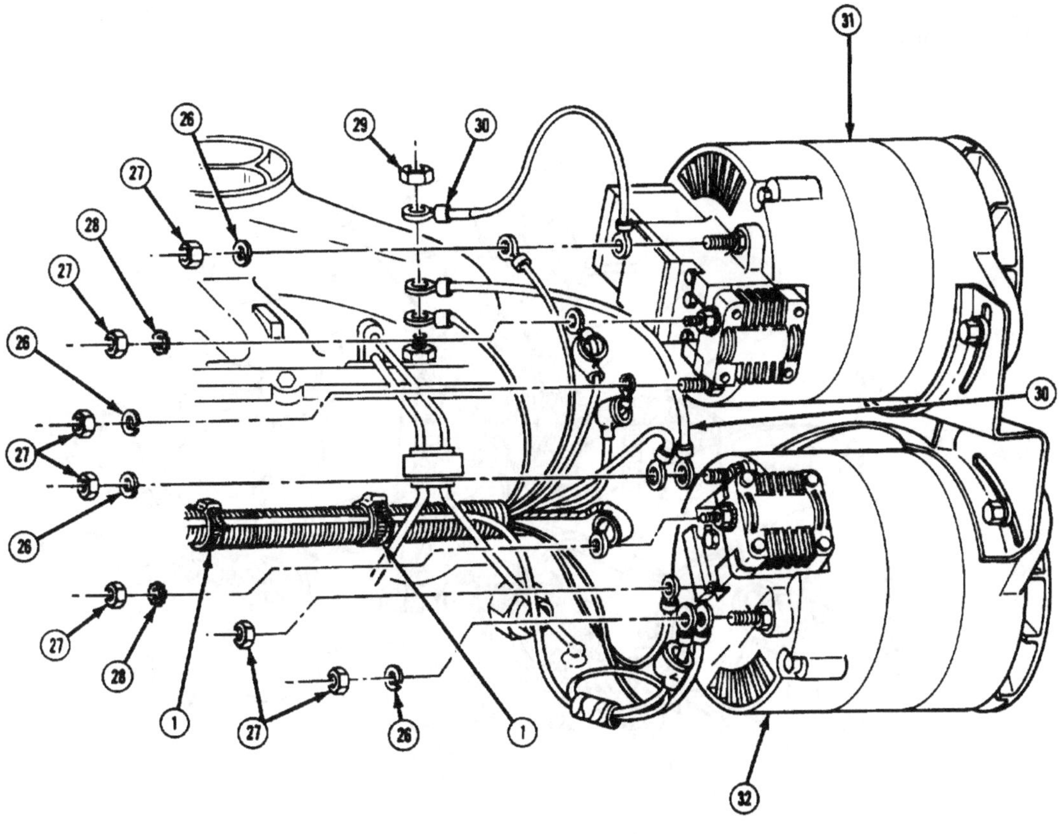

M1010

TA50839

4-10. ENGINE WIRING HARNESS REPLACEMENT (Con't).

10. Remove bolt and lockwasher from clip (34), and disconnect lead from compressor (33). Discard lockwasher.

11. Disconnect water temperature connector (46) and 4 left side glow plug connectors (45). Disconnect glow plug wiring harness (36), Disconnect oil pressure sending unit connector (35) located behind glow plug wiring harness connector.

12. Loosen bolt (44) and disconnect engine wiring harness from rear of fuse box (42). Depress sides of terminal "30" on fuse box, and disconnect pink sending unit lead by pulling lead toward front of connector. Disconnect leads from windshield wiper motor (41). Disconnect capacitor connector (43).

13. Remove small nut (37) and large nut (39), and disconnect leads from engine wiring harness block (38). Remove 4 nuts and 2 starwashers, and disconnect leads from glow plug relay (40). Discard starwashers.

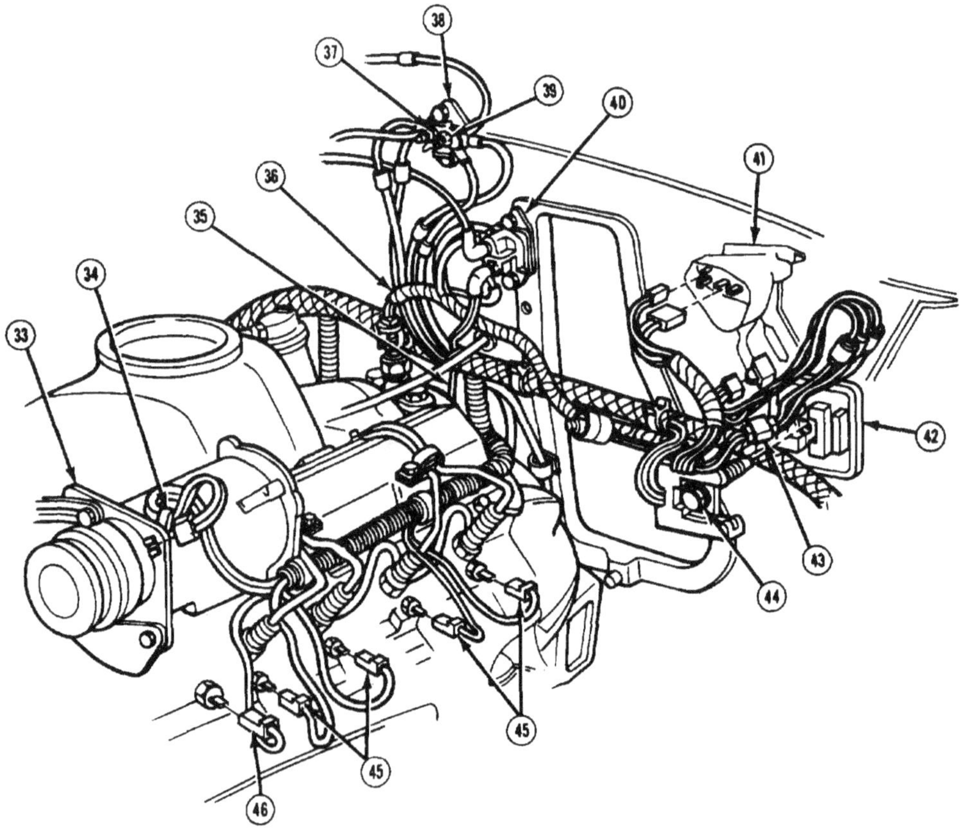

TA60840

4-10. ENGINE WIRING HARNESS REPLACEMENT (Con't).

NOTE

M1010 trucks with retrofit wiring will have additional wiring from engine wiring harness block (38) to current/voltage regulator (49). Ensure that position of retrofit wiring is noted for installation.

14. Remove and discard 6 tie-down straps securing retrofit wiring to engine wiring harness.

15. Remove 2 screws and current/voltage regulator cover. Remove 5 nuts (51) and lockwashers, and disconnect leads from current/voltage regulator relay (48). Remove 5 nuts (50) and lockwashers, and disconnect leads from current/voltage regulator (49). Unsnap engine wiring harness from left fenderwell (47). Discard lockwashers.

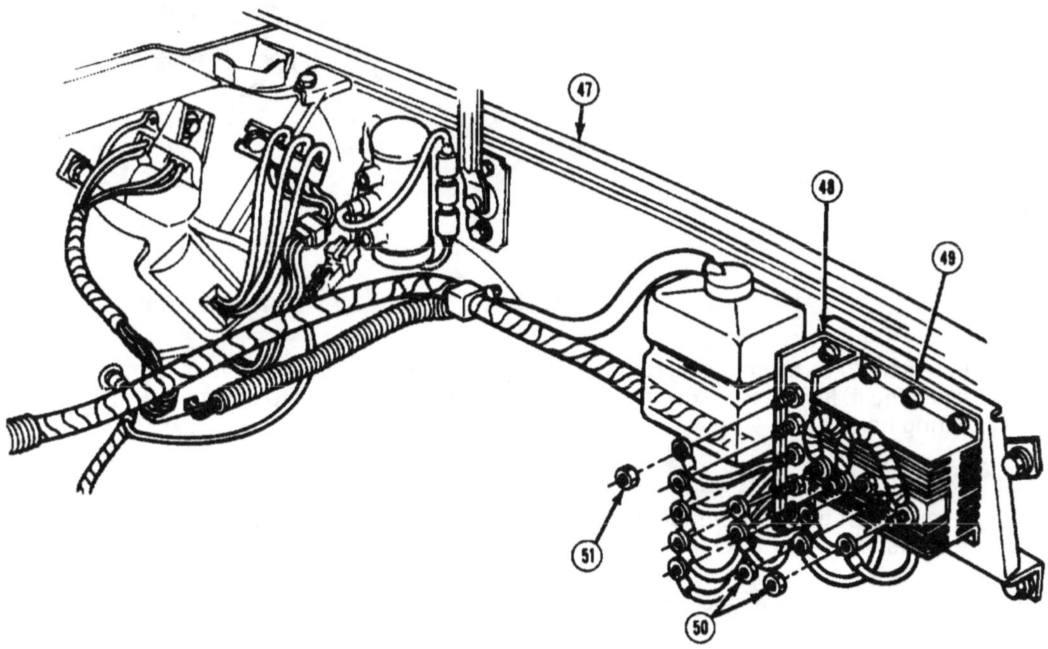

TA50841

4-10. ENGINE WIRING HARNESS REPLACEMENT (Con't).

16. Remove 2 screws (53) from clips at transmission (52). Disconnect transmission downshift switch connector (54).

17. Unsnap all retainers and open all clamps that hold engine wiring harness,

18. Slide right side of engine wiring harness rearward and out from below fuel injector lines . Pull starter connector out from between rear of cylinder head next to transmission fill tube. Slide left side of engine wiring harness from under fuel injector lines.

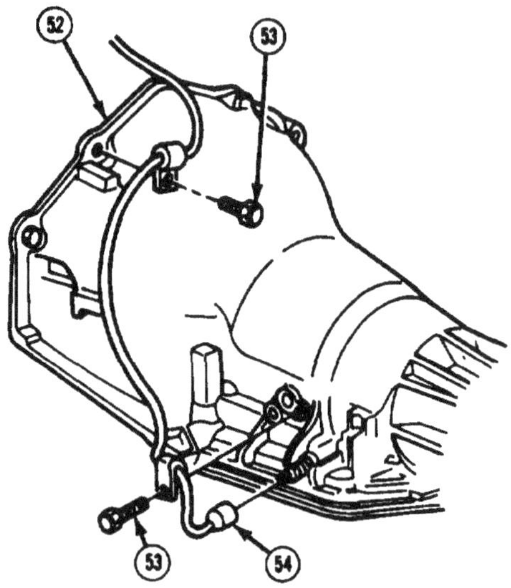

CAUTION

Care should be taken during removal of engine wiring harness to prevent snagging it. Failure to follow this caution may result in damage to engine wiring harness.

NOTE

Ensure that position of main lower engine wiring harness is noted for installation.

19. Remove engine wiring harness.

b . INSTALLATION

NOTE

- if connecting all except M1010 engine wiring harness to engine assembly, perform steps 2, 8, 12, 13, 17, and 18.
- if connecting M1010 engine wiring harness to engine assembly, perform steps 2, 8 through 11, 13, 17, and 18.
- If installing ail except M1010 engine wiring harness, perform steps 1-39 6 through 8, 12, 13, and 15 through 18.
- if installing M1010 engine wiring harness, perform steps 1-11 and steps 13-18.

TA60642

4-10. ENGINE WIRING HARNESS REPLACEMENT (Con't).

NOTE

- Ensure that all connector are seated and locked in place as they are installed.
- Ensure that engine wiring harness is secured in appropriate retainers and clamps as it is installed.

1. Position engine wiring harness in engine compartment.
2. Slide engine wiring harness under fuel injector lines on left and right side of cylinder head, and crankcase depression regulator valve bracket on right side. Route starter connector between rear of cylinder head next to transmission fill tube.
3. Connect transmission downshift switch connector (54). Install clips at transmission (52) with 2 screws (53).
4. Snap engine wiring harness to left fenderwell (47). Connect leads to current/voltage regulator (49), and install 5 new lockwashers and nuts (50). Connect leads to current/voltage regulator relay (48), and install 5 new lockwashers and nuts (51). Install current/voltage regulator cover with 2 screws.

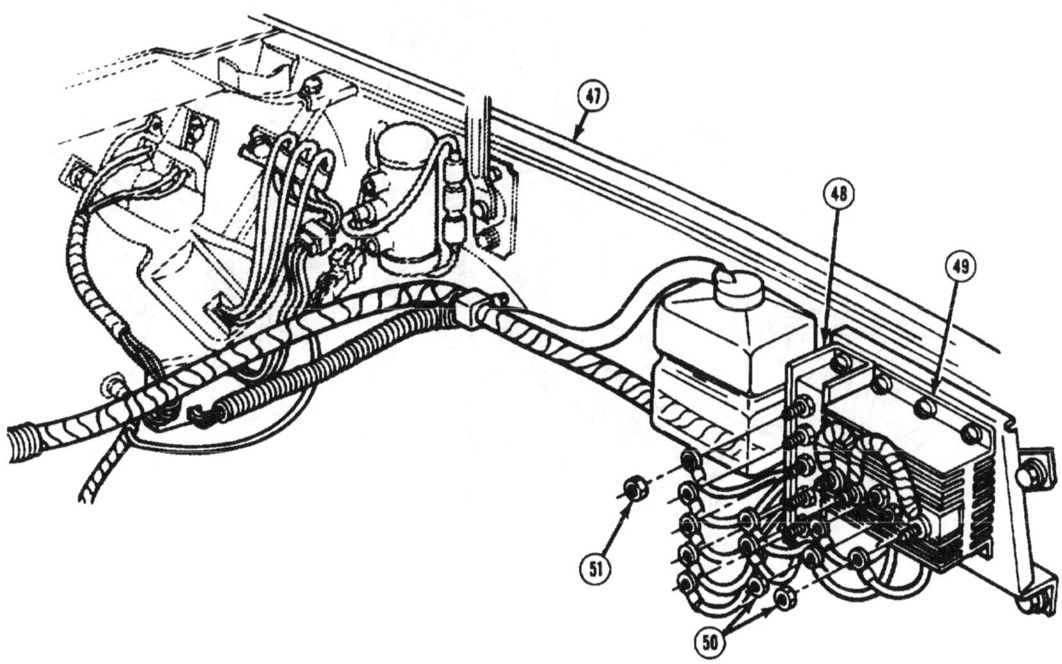

TA50843

4-10. ENGINE WIRING HARNESS REPLACEMENT (Con't).

NOTE

M1010 trucks with retrofit wiring will have additional wiring from engine wiring harness block (38) to current/voltage regulator.

5. Secure retrofit wiring to engine wiring harness with 6 new tie-down straps.

6. Connect leads to glow plug relay (40), and install 2 new starwashers and 4 nuts. Connect leads to engine wiring harness block (38), and install large nut (39) and small nut (37).

7. Connect windshield wiper motor assembly capacitor connector (43). Connect leads to windshield wiper motor (41), Install pink sending unit lead into terminal "30" on fuse box, Connect engine wiring harness to rear of fuse box (42) and tighten bolt (44),

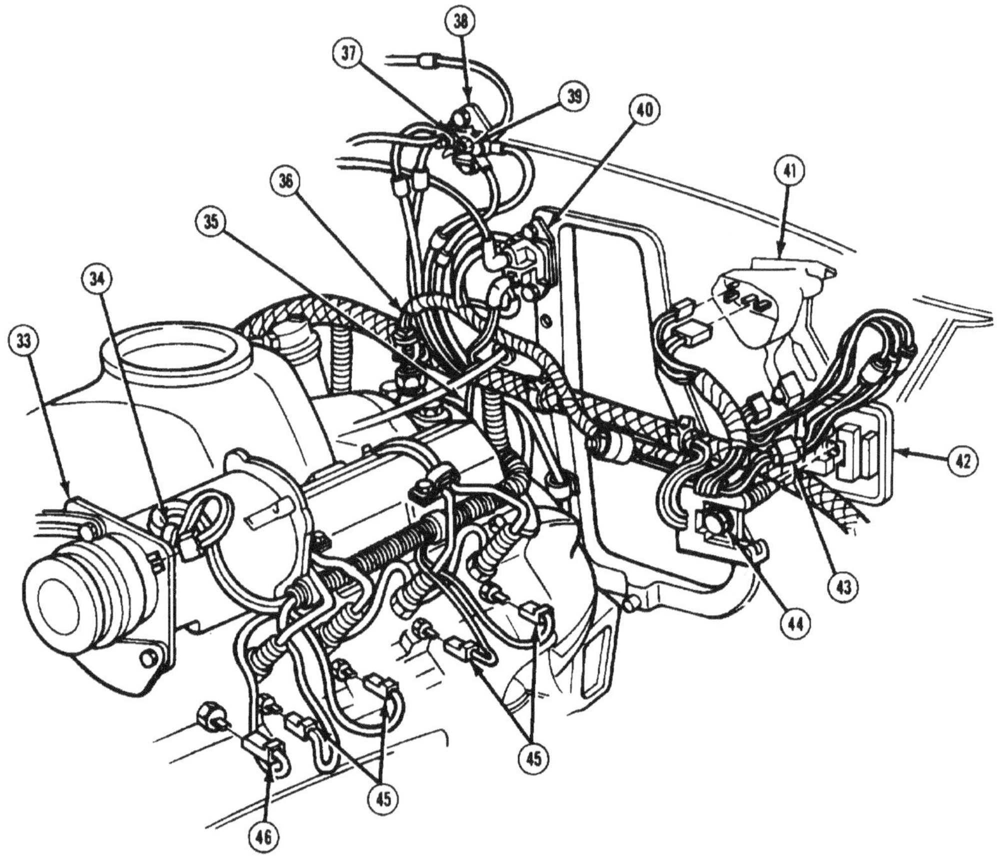

TA50844

4-10. ENGINE WIRING HARNESS REPLACEMENT (Con't).

8. Connect water temperature connector (46) and 4 left side glow plug connectors (45). Connect oil pressure sending unit connector (35) located behind glow plug wiring harness (36) connector. Connect glow plug wiring harness.

9. Connect lead to compressor (33), and install clip (34) with new lockwasher and bolt.

10. Connect 2 ground wires (30) and engine wiring harness lead to intake manifold stud, and install nut (29).

11. Secure engine wiring harness in 2 retainers (1). Remove 7 nuts (27) from connector studs on upper alternator (31) and lower alternator (32). Connect leads to upper alternator and lower alternator, and install 2 new starwashers (28), 4 new lockwashers (26), and 7 nuts (27).

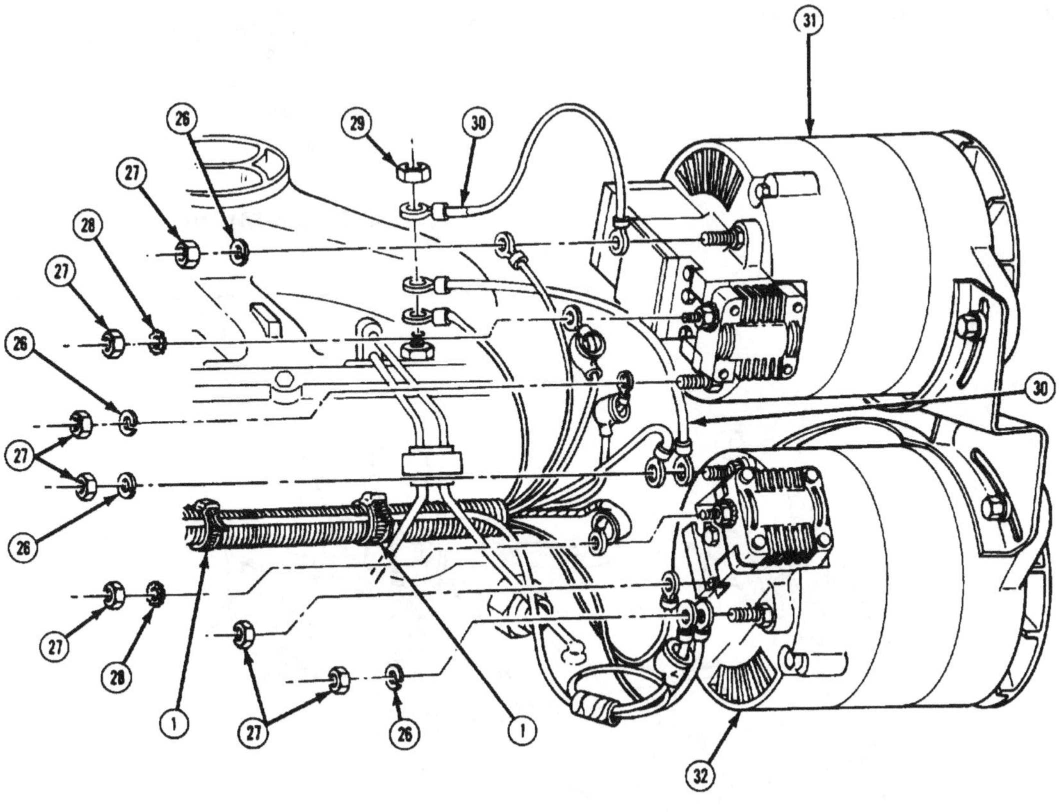

M1010

TA50845

4-10. ENGINE WIRING HARNESS REPLACEMENT (Con't).

NOTE

Left side alternator does not have a capacitor (25).

12. Secure engine wiring harness in 2 retainers (1), At right side alternator (24), install capacitor (25) with screw (22) if removed. Connect leads (21) and install 2 nuts (20). Connect voltage regulator connector (23). Repeat this step for left side alternator. On left side alternator, also connect lead to "R" terminal.

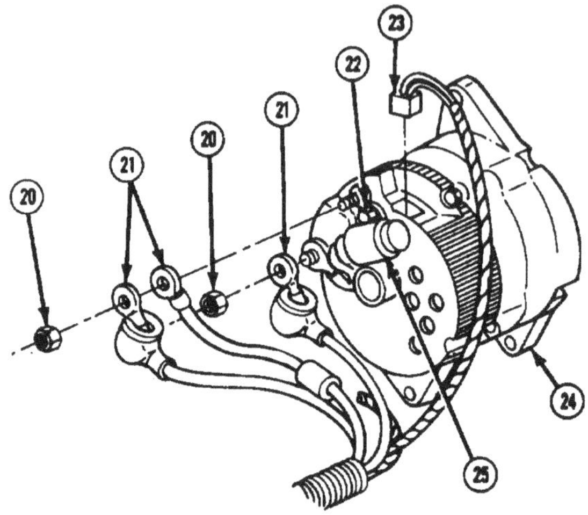

All Except M1010

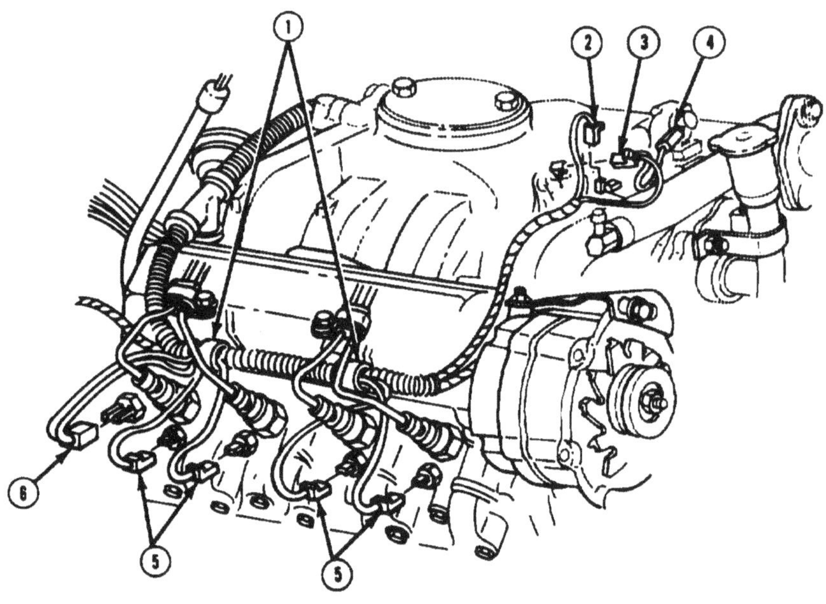

TA50846

4-10. ENGINE WIRING HARNESS REPLACEMENT (Con't).

13. Install new starwasher (18) to intake manifold stud (19). Connect engine ground strap (17), engine wiring harness lead (14), and terminal board ground (15), and install nut (16).

14. Connect diagnostic wiring harness connector (12).

15. Connect leads to negative terminal board center stud (8), and install new starwasher and nut, Connect leads to positive terminal board studs (7), and install 2 new starwashers and nuts. Install terminal board (9) covers.

16. Connect fuel filter heater connector (10), vacuum switch connector (11), and water sensor connector (13).

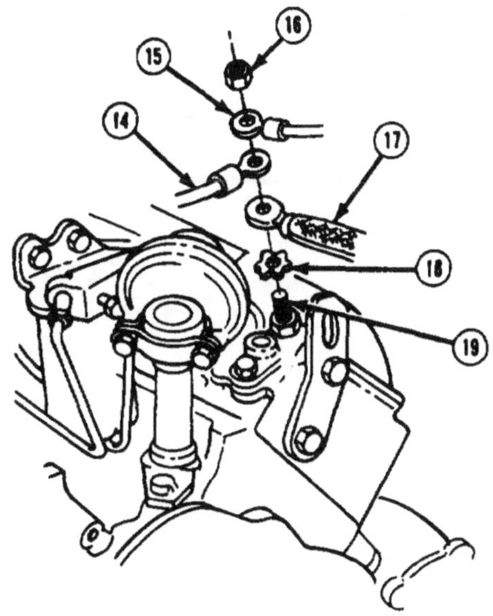

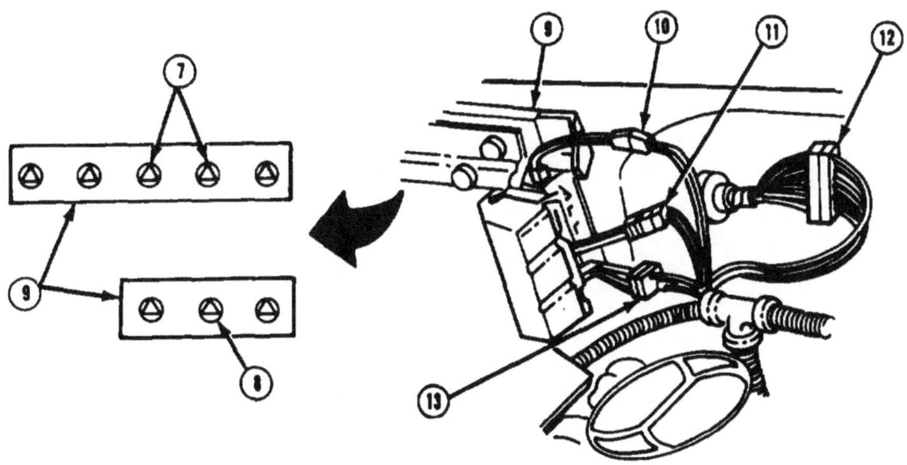

17. Connect cold and fast idle switch connector (6), fuel solenoid connector (2), cold advance switch connector (3), and fast idle solenoid connector (4).

18. Connect 4 right side glow plug connectors (5),

TA60847

4-10. ENGINE WIRING HARNESS REPLACEMENT (Con't).

FOLLOW-ON TASKS:

- Install rear battery box if removed. (See TM 9-2320-289-20)
- Install air cleaner. (See TM 9-2320-289-20)
- Install starter, (See TM 9-2320-289-20)
- Check operation of engine electrical components and windshield wiper motor.

4-11. GLOW PLUG WIRING HARNESS REPLACEMENT.

This task covers: a. Removal b. Installation

INITIAL SETUP:

Equipment Condition

l Both battery negative cables disconnected. (See TM 9-2320-289-20)

a. REMOVAL

NOTE

Ensure that all leads are tagged for installation.

1. Remove 2 nuts (2) and disconnect leads (1) from glow plug relay (3).
2. Disconnect glow plug wiring harness (11) from engine temperature sensor (10).

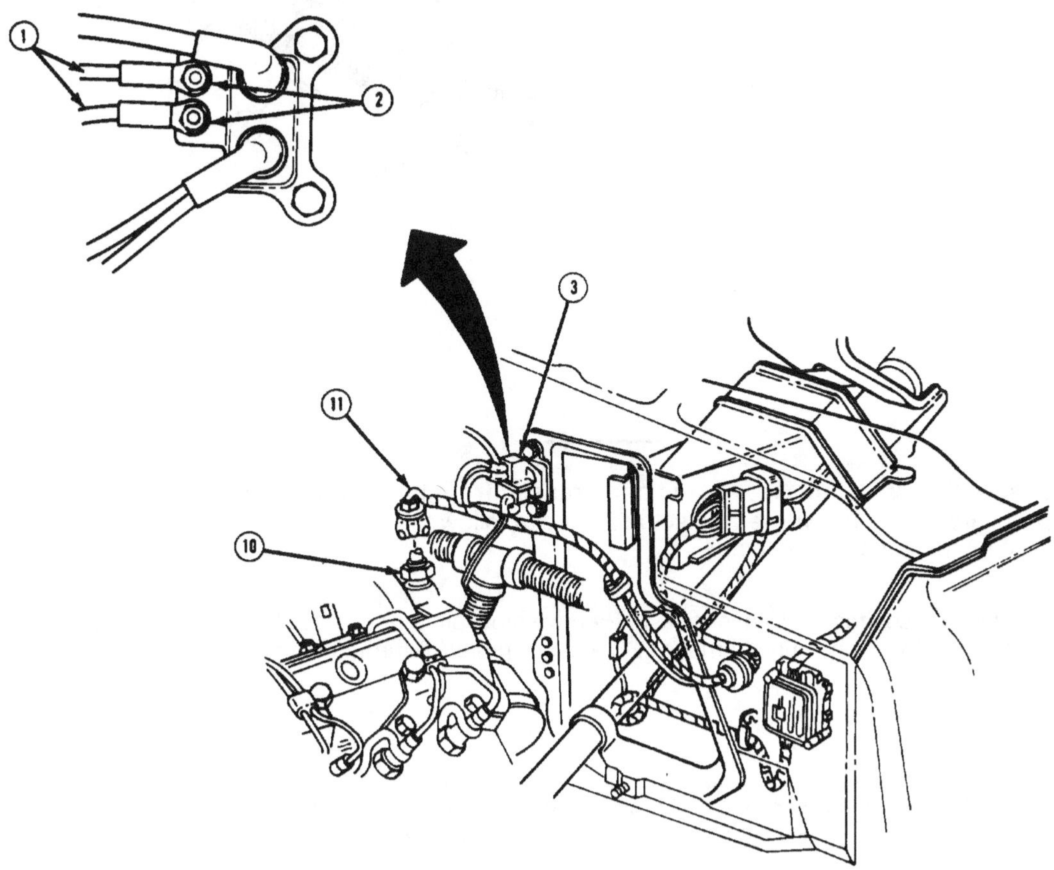

TA50848

4-11. GLOW PLUG WIRING HARNESS REPLACEMENT (Con't).

NOTE

Perform step 3 if truck is equipped with heater fuel pump.

3. Disconnect lead (8) from heater fuel pump connector (1 2).

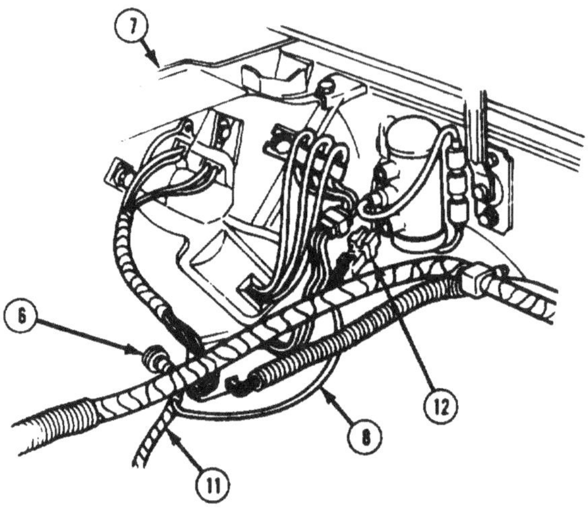

CAUTION

Ensure that "rosebud" retainer is compressed from backside of brake pedal bracket (4) to remove connector. Failure to follow this caution may result in damage to connector.

4. Disconnect 6-wire connector (9) under dash. Remove glow plug module (5) from brake pedal bracket (4). Pull glow plug wiring harness (11) and grommet (6) through engine bulkhead (7), and remove.

b. INSTALLATION

1. Install glow plug wiring harness (11) and grommet (6) in engine bulkhead (7). Install glow plug module (5) on brake pedal bracket (4). Connect 6-wire connector (9) under dash.

NOTE

Perform step 2 if truck is equipped with heater fuel pump.

2. Untape lead (8) from glow plug wiring harness (11). Connect lead to heater fuel pump connector (12).

TA50849

4-11. GLOW PLUG WIRING HARNESS REPLACEMENT (Con't).

3. Connect glow plug wiring harness (11) to engine temperature sensor (10).

4. Connect leads (1) to glow plug relay (3) and install 2 nuts (2).

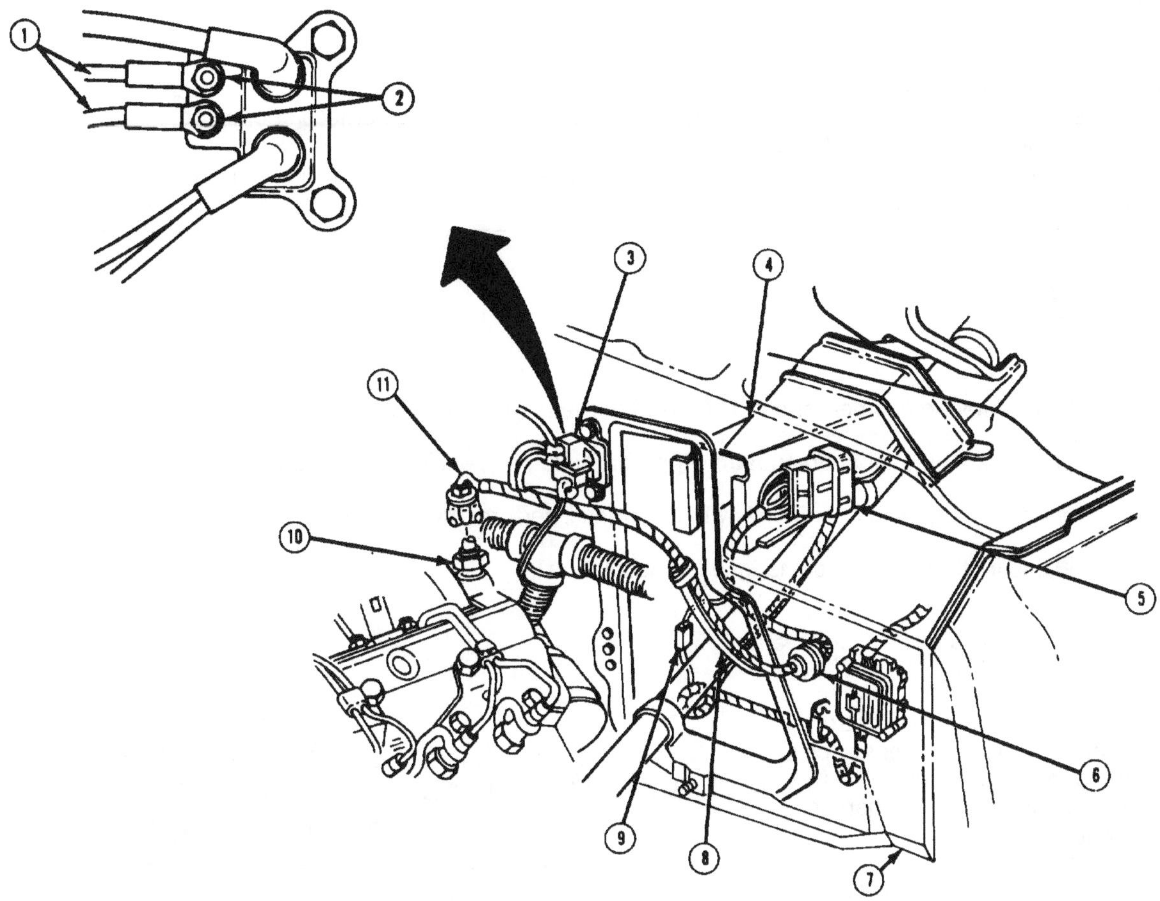

FOLLOW-ON TASKS:

- Connect both battery negative cables. (See TM 9-2320-289-20)
- Check operation of glow plugs.

TA50850

4-12. CAB HEATER CONTROL, ENGINE DIAGNOSTIC, AND OPEN DOOR BUZZER WIRING HARNESSES REPLACEMENT.

This task covers: a. Removal b. Installation

INITIAL SETUP:

Equipment Condition

- Both battery negative cables disconnected. (See TM 9-2320-289-20)
- Instrument cluster plate removed (if removing cab heater control wiring harness or open door buzzer wiring harness). (See TM 9-2320-289-20)
- Instrument panel compartment removed (if removing cab heater control wiring harness or engine diagnostic wiring harness).

a. REMOVAL

NOTE

- If removing cab heater control wiring harness (4), perform steps 1-5.
- if removing engine diagnostic wiring harness (14), perform steps 6-8.
- if removing open door buzzer wiring harness, perform steps 9 and 10
- Ensure that all leads are tagged for installation,

1. Remove 4 screws and pull out heater control assembly (6).
2. Disconnect cab heater control wiring harness (4) from blower switch (7). Disconnect connector (1) from instrument panel wiring harness (3) connector.
3. Open 3 clips (2) securing cab heater control wiring harness (4) to instrument panel wiring harness (3).

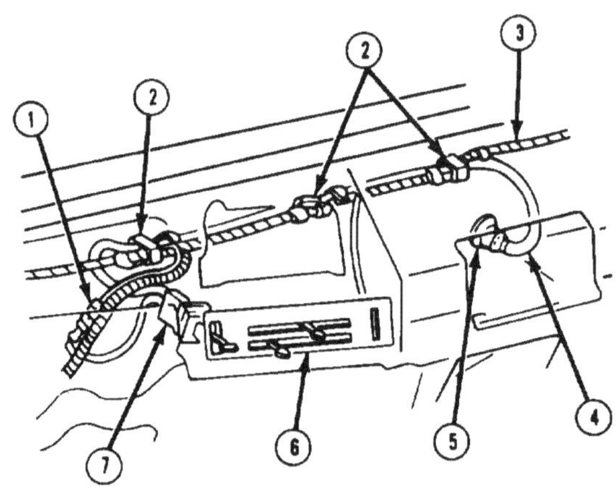

TA50851

4-12. CAB HEATER CONTROL, ENGINE DIAGNOSTIC, AND OPEN DOOR BUZZER WIRING HARNESSES REPLACEMENT (Con't).

4. Inside engine compartment, disconnect connector (11) from blower motor capacitor connector (12). Disconnect cab heater control wiring harness (4) from blower resistor (8).

CAUTION

Care should be taken during removal of cab heater control wiring harness (4) to prevent snagging it. Failure to follow this caution may result in damage to cab heater control wiring harness.

5. Remove nut (10) and clip (9). Pull cab heater control wiring harness (4) into engine compartment, through grommet (5), and remove.

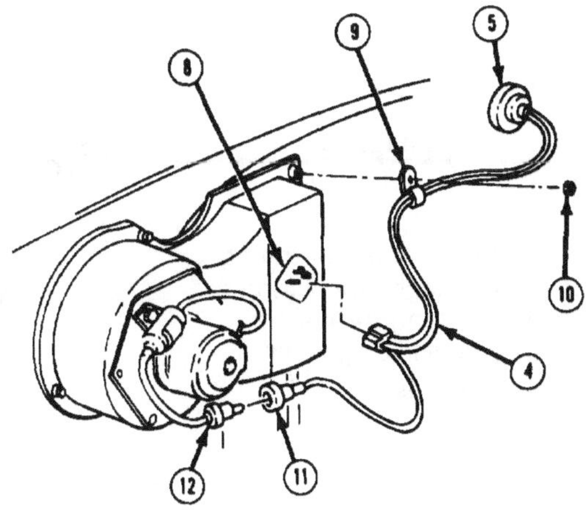

6. Remove 2 screws (20) from diagnostic connector bracket (19), dash panel (16), and wiring harness accessory bracket (15).

7. Remove 4 nuts (13), screws (17), and cap (18). Separate engine diagnostic wiring harness (14) from diagnostic connector bracket (19).

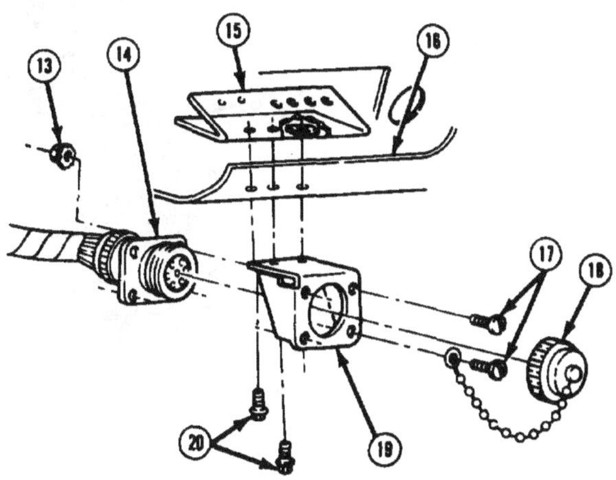

TA50852

4-12. CAB HEATER CONTROL, ENGINE DIAGNOSTIC, AND OPEN DOOR BUZZER WIRING HARNESSES REPLACEMENT (Con't).

CAUTION

Care should be taken during removal of engine diagnostic wiring harness (14) to prevent snagging it. Failure to follow this caution may result in damage to engine diagnostic wiring harness.

8. Inside engine compartment, disconnect engine diagnostic wiring harness (14) from engine wiring harness (22). Pull engine diagnostic wiring harness through grommet (21) and remove.

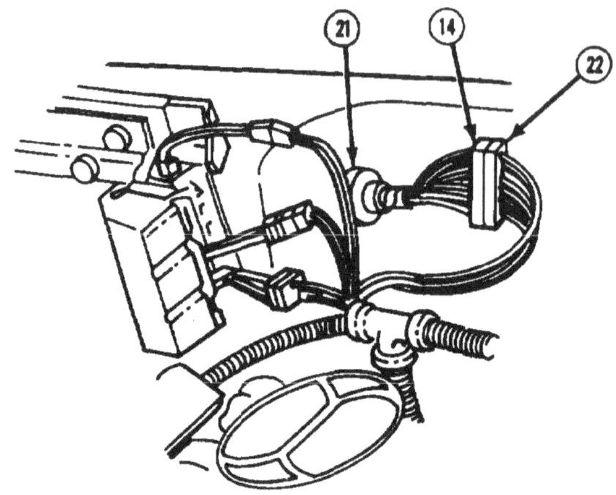

9. Slide open door buzzer wiring harness (25) off fuse box (24). Remove clip (23) securing open door buzzer wiring harness to instrument panel wiring harness (3).

10. Disconnect open door buzzer wiring harness (25) from instrument panel wiring harness (3) and remove.

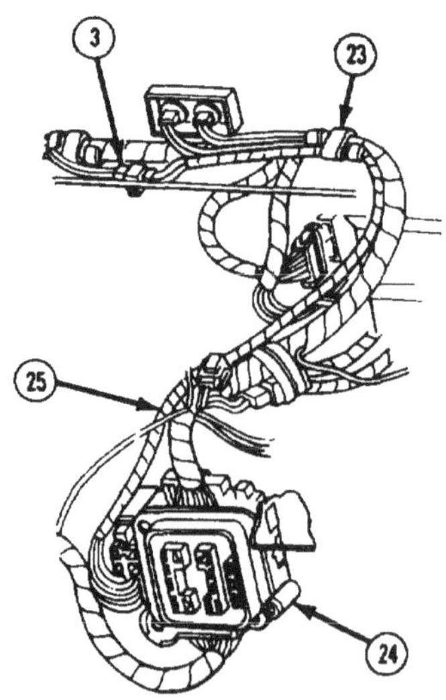

TA50853

4-12. CAB HEATER CONTROL, ENGINE DIAGNOSTIC, AND OPEN DOOR BUZZER WIRING HARNESSES REPLACEMENT (Con't).

b. INSTALLATION

NOTE

- If installing open door buzzer wiring harness (25), perform steps 1 and 2.
- if installing engine diagnostic wiring harness (14), perform steps 3-5.
- If installing cab heater control wiring harness, perform steps 6-10.
- . Ensure that all connectors are seated and locked in place as they are installed.

1. Connect open door buzzer wiring harness (25) to instrument panel wiring harness (3).
2. Secure open door buzzer wiring harness (25) to instrument panel wiring harness (3) with clip (23). Slide open door buzzer wiring harness onto fuse box (24).
3. Inside engine compartment, push engine diagnostic wiring harness (14) through grommet (21). Connect engine diagnostic wiring harness to engine wiring harness (22).

NOTE

Ensure that indexing pin on front of engine diagnostic wiring harness (14) is installed in 12 o'clock position.

4. Assemble engine diagnostic wiring harness (14) and diagnostic connector bracket (19). install cap (18), 4 screws (17), and nuts (13).
5. Install diagnostic connector bracket (19) and wiring harness accessory bracket (15) on dash panel (16) with 2 screws (20).

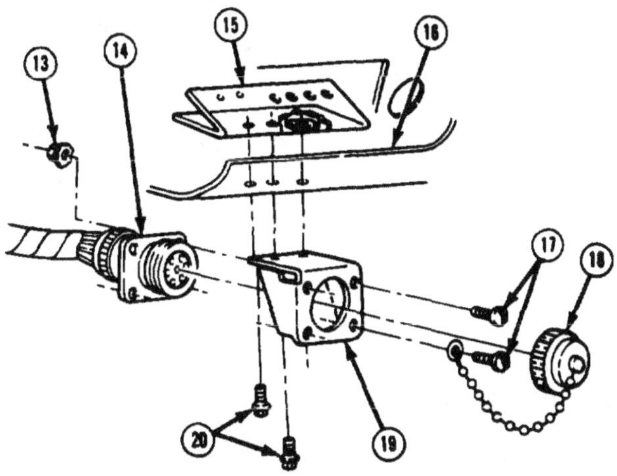

TA60854

4-12. CAB HEATER CONTROL, ENGINE DIAGNOSTIC, AND OPEN DOOR BUZZER WIRING HARNESSES REPLACEMENT (Con't).

6. Inside engine compartment, push cab heater control wiring harness (4) through grommet (5). Install clip (9) and nut (10).

7. Connect cab heater control wiring harness (4) to blower resistor (8). Connect connector (11) to blower motor capacitor connector (12).

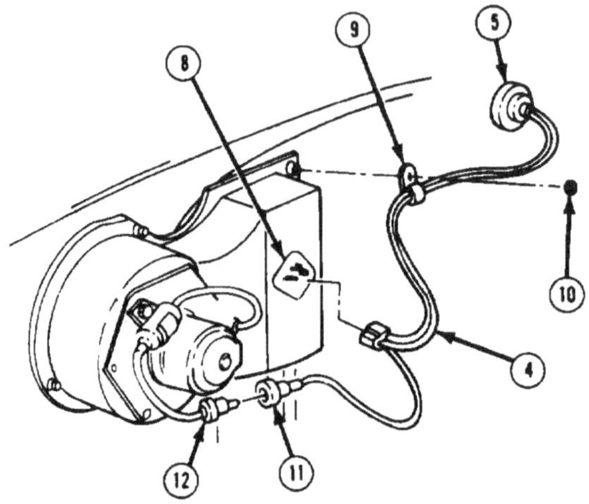

8. Secure cab heater control wiring harness (4) to instrument panel wiring harness (3) with 3 clips (2).

9. Connect connector (1) to instrument panel wiring harness (3) connector. Connect cab heater control wiring harness (4) to blower switch (7).

10. Install heater control assembly (6) with 4 screws.

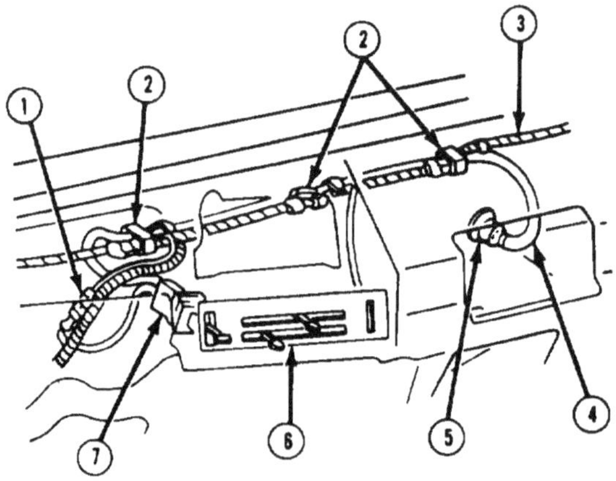

FOLLOW-ON TASKS:

- Install instrument panel compartment if removed.
- Install instrument cluster plate if removed. (See TM 9-2320-289-20)
- Connect both battery negative cables. (See TM 9-2320-289-20)
- Check operation of cab heater or open door buzzer.

TA508555

4-13. DIMMER AND IGNITION SWITCHES REPLACEMENT.

This task covers:	a. Removal	b. Installation

INITIAL SETUP:

Equipment Condition

- Instrument cluster plate removed, (See TM 9-2320-289-20)

a. REMOVAL

NOTE

If removing dimmer switch, perform steps 1-4.

1. Mark cable clip (3) location on right side of steering column (4) for installation, Disconnect cable clip. Remove 2 screws (2) and transmission position indicator (1).
2. Remove 2 nuts (6), washers (7), and support (5). Lower steering column (4).

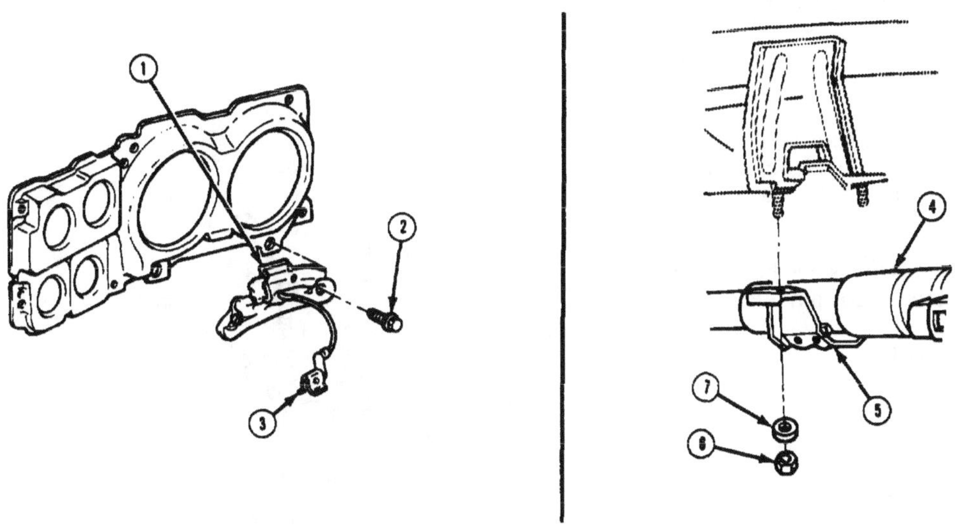

TA50856

4-13. DIMMER AND IGNITION SWITCHES REPLACEMENT (Con't).

NOTE

Ensure that connectors are tagged for installation.

3. Disconnect dimmer switch connector (11) from dimmer switch (10), and ignition switch connector (8) from ignition switch (9),

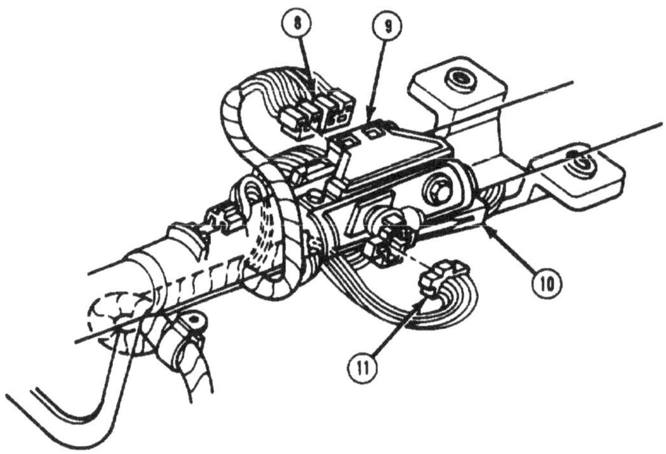

4. Remove nut (12), screw (13), and dimmer switch (10) from ignition switch (9). Remove dimmer switch actuating rod.

5. Remove stud (14) and ignition switch (9) from steering column (4),

b. INSTALLATION

NOTE

If installing dimmer switch, perform steps 2-5.

1. Install ignition switch (9) on steering column (4) with stud (14),

2. Install dimmer switch (10) actuating rod, Install dimmer switch on ignition switch (9) with screw (13) and nut (12),

3. Connect dimmer switch connector (11) to dimmer switch (10), and ignition switch connector (8) to ignition switch (9).

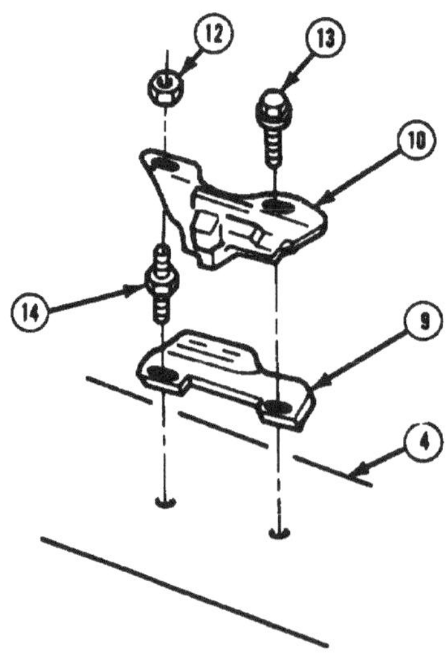

TA50857

4-13. DIMMER AND IGNITION SWITCHES REPLACEMENT (Con't).

4. Raise steering column (4), and install support (5) with 2 washers (7) and nuts (6).

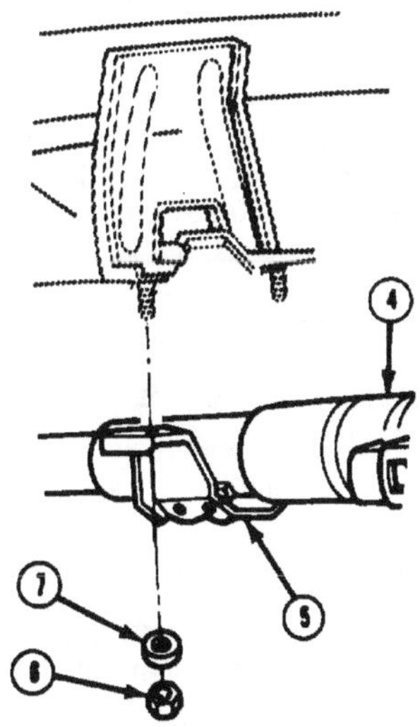

5. Install transmission position indicator (1) with 2 screws (2). Connect cable clip (3) to steering column (4). Check indicator needle for proper position and adjust as required.

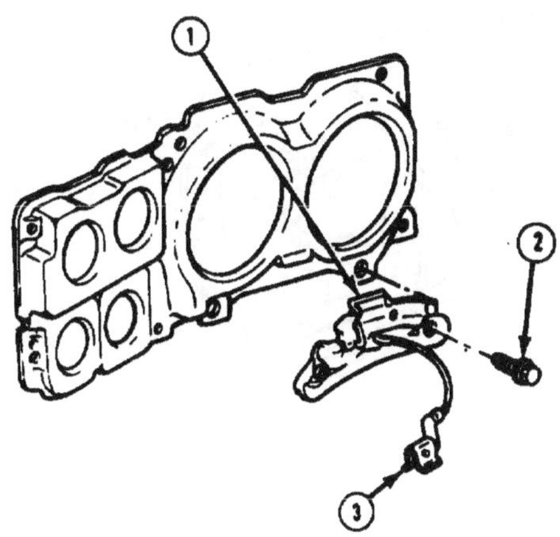

FOLLOW-ON TASKS:

- Install instrument cluster plate. (See TM 9-2320-289-20)
- Check operation of lights and starter system.

TA50858

4-14. DIRECTIONAL SIGNAL SWITCH REPLACEMENT.

This task covers: a. Remova l b. Installation

INITIAL SETUP:

Equipment Condition

- Hazard warning control removed. (See TM 9-2320-289-20)
- Steering wheel and directional signal lever removed. (See TM 9-2320-289-20)

Tools/Test Equipment

- Lock plate compressor
- Snapring pliers

Materials/Parts

- One upper shaft retainer

a. REMOVAL

1. Pry lock plate (5) cover from lock plate and remove from steering column (4).
2. Install lock plate compressor (2). Turn nut (1) until upper shaft retainer (3) can be removed. Remove and discard upper shaft retainer. Remove lock plate compressor and lock plate (5).
3. Slide canceling cam, upper bearing preload spring, and thrust washer from steering shaft (7). Remove 4 mounting screws from directional signal switch (6).

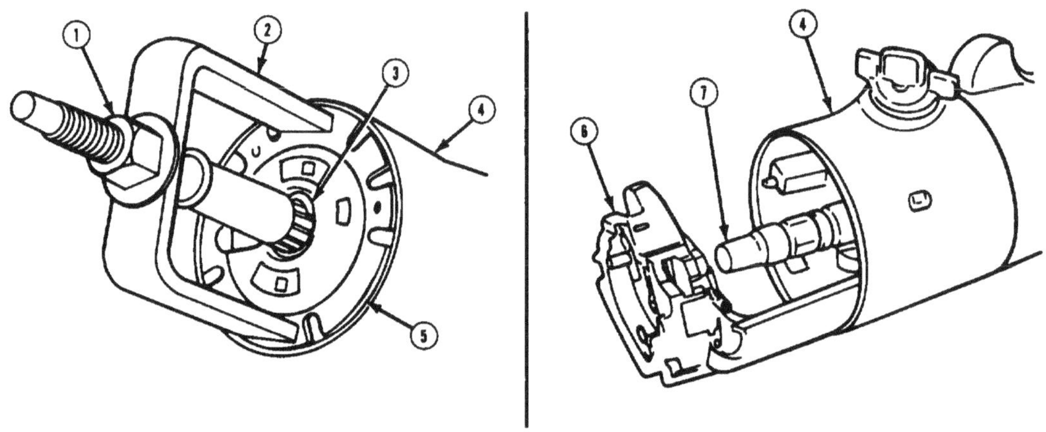

TA50859

4-14. DIRECTIONAL SIGNAL SWITCH REPLACEMENT (Con't).

4. Remove 4 screws (11) and steering column filler (10) from dash panel (9). Remove 2 screws (12) and steering column filler (8).

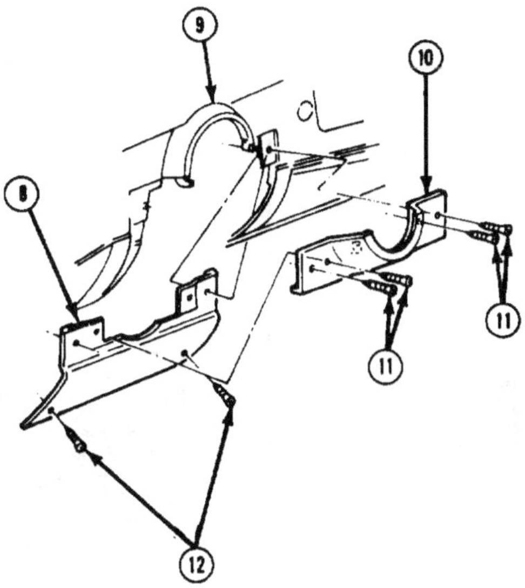

5. Remove 2 nuts (14), washers (15), and support (13).

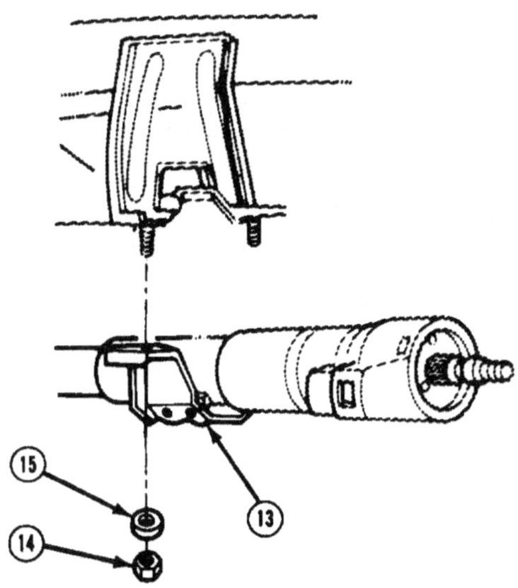

TA50860

4-14. DIRECTIONAL SIGNAL SWITCH REPLACEMENT (Con't).

6. Remove switch harness (18) protector. Disconnect connector (17) from instrument panel wiring harness (16) and unsnap from steering column (4),

7. Attach guide wire securely to connector (17). Pull directional signal switch (6) with switch harness (18) and guide wire through steering column (4) until connector is out of steering column, Disconnect guide wire and leave in place for installation.

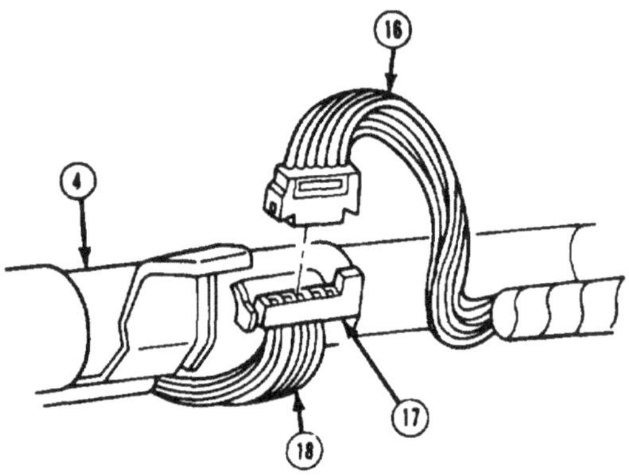

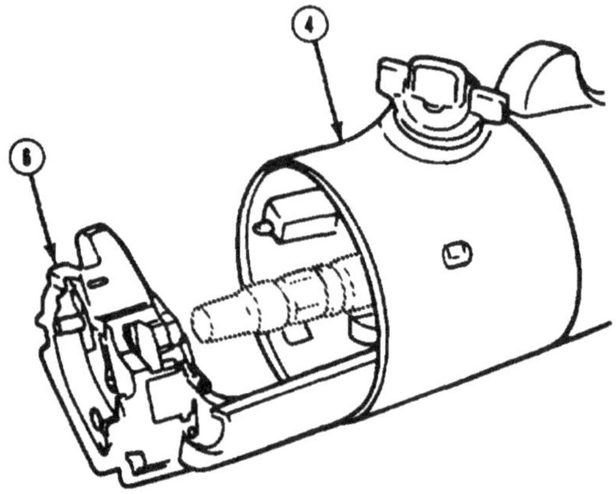

b. INSTALLATION

1. Attach guide wire securely to connector (17), Pull guide wire with switch harness (18) through steering column (4) until connector is out of steering column. Disconnect and remove guide wire,

2. Snap connector (17) onto steering column (4) and connect to instrument panel wiring harness (16). Install switch harness (18) protector.

TA50861

4-14. DIRECTIONAL SIGNAL SWITCH REPLACEMENT (Con't).

3. Install support (13) with 2 washers (15) and nuts (14).

4. Install steering column filler (8) with 2 screws (12), Install steering column filler (10) on dash panel (9) with 4 screws (11).

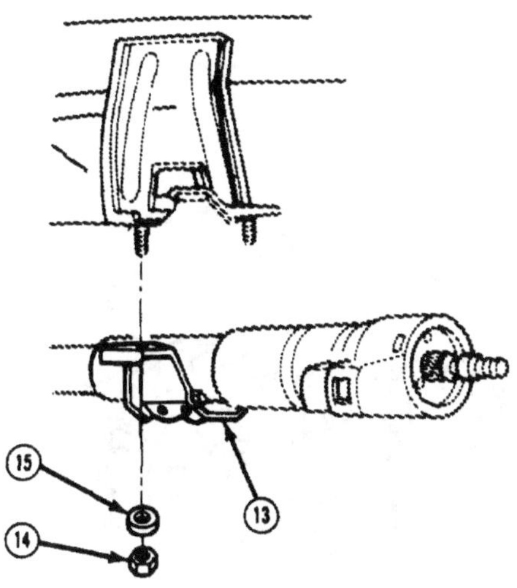

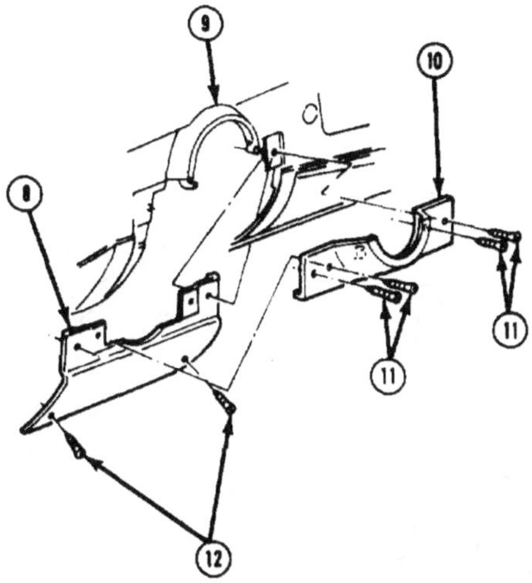

TA50862

4-14. DIRECTIONAL SIGNAL SWITCH REPLACEMENT (Con't).

5. Install directional signal switch (6) with 4 mounting screws, Slide thrust washer, upper bearing preload spring, and canceling cam onto steering shaft (7).

6. Install lock plate (5) and new upper shaft retainer (3) on steering shaft (7). Install lock plate compressor (2). Turn nut (1) until steering shaft groove is visible. Install upper shaft retainer in groove, Remove lock plate compressor,

7. Snap lock plate (5) cover into position on lock plate.

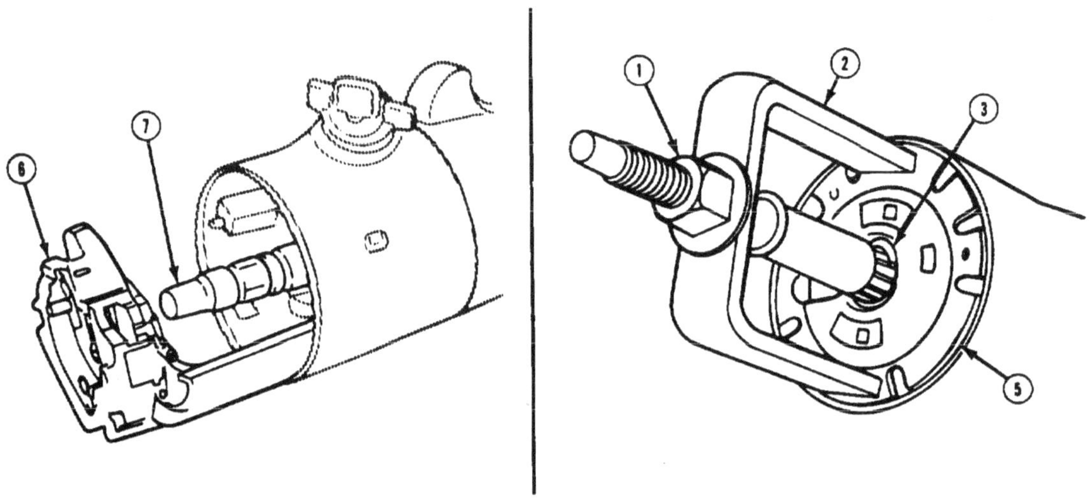

FOLLOW-ON TASKS:

- Install steering wheel and directional signal lever, (See TM 9-2320-289-20)
- Install hazard warning control, (See TM 9-2320-289-20)
- Check operation of directional signals.

TA50863

CHAPTER 5
TRANSMISSION AND TRANSFER CASE MAINTENANCE

Section 1. TRANSMISSION MAINTENANCE

5-1. TRANSMISSION MAINTENANCE INDEX

5-2. TRANSMISSION LINKAGE MAINTENANCE.

This task covers:

a. Removal
b. Cleaning and Inspection
c. Installation
d. Adjustment

INITIAL SETUP:

Equipment Condition

- Truck on level ground with wheels chocked.

Too/s/Test Equipment

- Torque wrench

MaterialsIParts

- One wave washer
- Dry cleaning solvent (Item 23, Appendix B)

Genera/ Safety Instructions

- Dry cleaning solvent is flammable and must not be used near open flame. Use only in a well-ventilated area.

NOTE

If disconnecting linkage from transmission, perform step 3.

1. Remove spring clip retainer (5) at upper end of rod (1). Remove rod from upper shift lever (3), Remove bushing (2) and insulator (4) if damaged.' Discard bushing and insulator.

NOTE

Position of wave washer (17) should be noted for installation.

2. While supporting rod (1), remove bolt (18), and wave washer (17). Remove rod and swivel (1 6). Discard wave washer.
3. Remove spring clip retainer (11) from lever (19) assembly and remove lever assembly from transmission lever (10),
4. Remove lever (19) assembly from bracket (7). Remove bushing (15), washer (14), and spring (13) from lever assembly.
5. Remove nut (12) and transmission lever (10) from transmission if damaged. Remove 2 nuts (8), bolts (6), lockwashers (9), and bracket (7) from frame if damaged. Discard lockwashers.

5-2. TRANSMISSION LINKAGE MAINTENANCE (Con't).

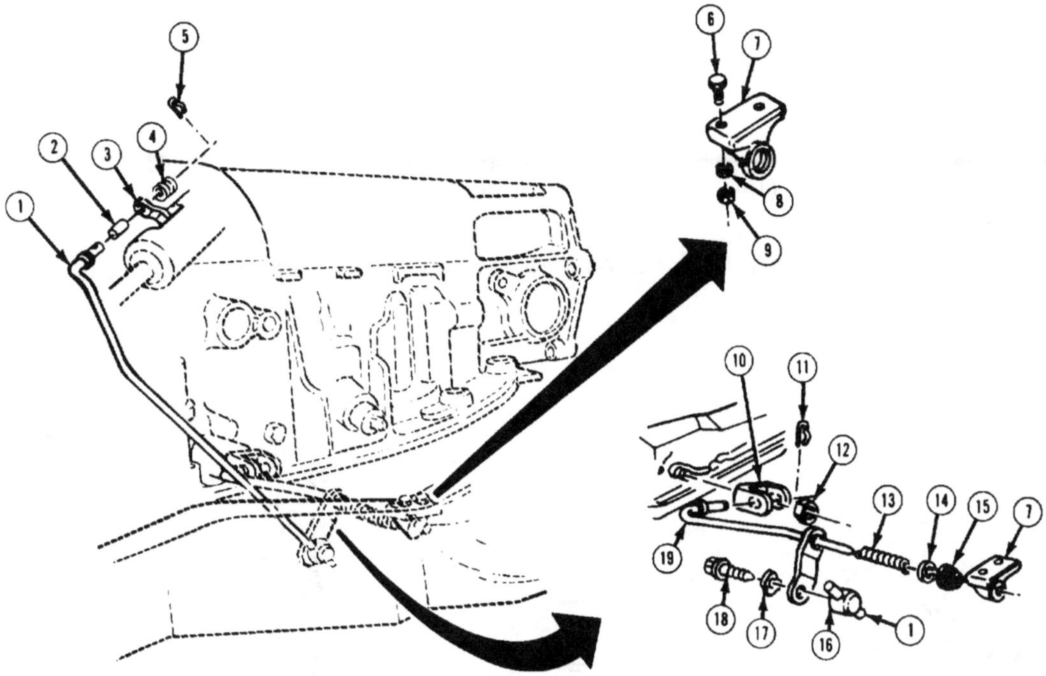

b. CLEANING AND INSPECTION

WARNING

Dry cleaning solvent P-D-680 is toxic and flammable. Always wear protective goggles and gloves and use only in a well-ventilated area. Avoid contact with skin, eyes, and clothes and DO NOT breathe vapors. DO NOT use near open flame or excessive heat. The solvent's flash point is l00°F-138°F (38°C-59°C), If you become dizzy while using cleaning solvent, immediately get fresh air and medical help. If solvent contacts eyes, immediately wash your eyes with water and get medical aid.

1 . Clean all parts with dry cleaning solvent.

TA50197

5-2. TRANSMISSION LINKAGE MAINTENANCE (Con't).

2. Inspect linkage and mounting hardware for damage. Replace if damaged.

c. INSTALLATION

NOTE

If connecting linkage to transmission, perform step 3 and ADJUSTMENT, step 3.

1. Install bracket (7) to frame with 2 bolts (6), new lockwashers (9), and nuts (8) if removed. Install transmission lever (10) to transmission with nut (12) if removed. Tighten nuts to 20 lb,-ft. (27 N.m).

2. Assemble spring (13), washer (14), and bushing (15) to lever (19). Install lever assembly to bracket (7).

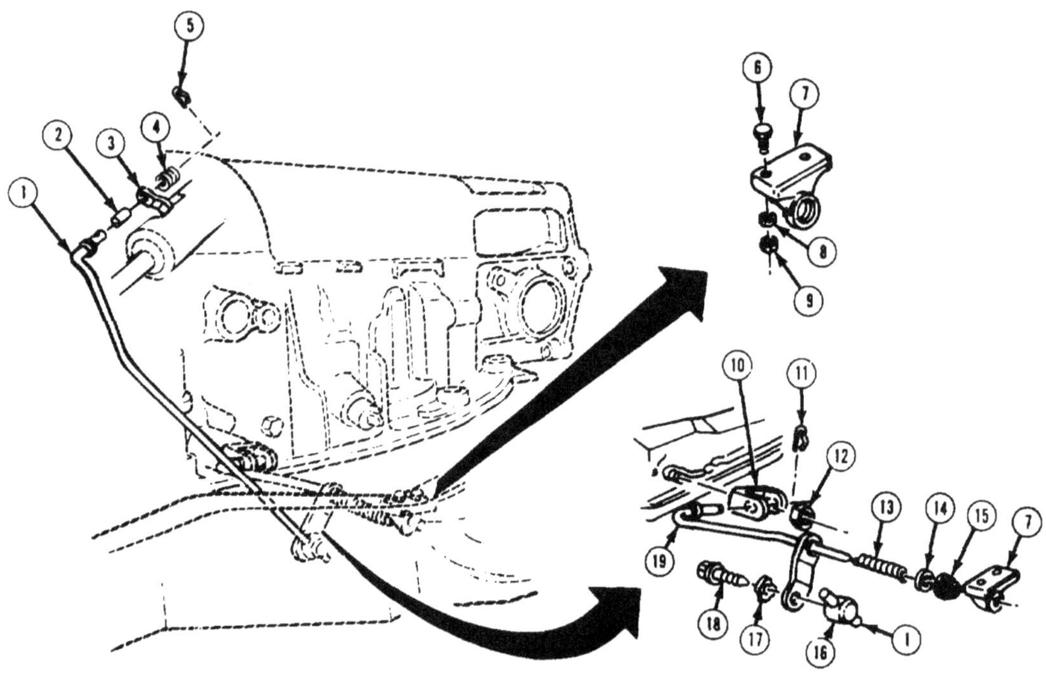

TA50198

5-2. TRANSMISSION LINKAGE MAINTENANCE (Con't),

3. Install lever (19) assembly to transmission lever (10) with spring clip retainer (11).

4. Install new insulator (4) and new bushing (2) to upper shift lever (3) if removed. Install rod (1) to upper shift lever with spring clip retainer (5).

5. Install swivel (16) on rod (1) and secure to lever (19) assembly with new wave washer (17) and bolt (18). DO NOT fully tighten bolt,

d. ADJUSTMENT

1. Set transmission lever (10) in "N" (Neutral) position by moving transmission lever clockwise until it stops, then counterclockwise 2 detents to "N" (Neutral) position. Set transmission gearshift lever on steering column in "N" (Neutral) position.

2. Hold rod (1) tightly in swivel (16) and tighten bolt (18) to 20 lb.-ft. (27 N.m).

3. Place transmission gearshift lever on steering column in "P" (Park) position and check the adjustment. Ensure that transmission gearshift lever on steering column goes into all positions and that engine starts in "P" (Park) or "N" (Neutral) positions only. If further adjustments are needed, loosen pivot bolt (18) and perform ADJUSTMENT, steps 1-3.

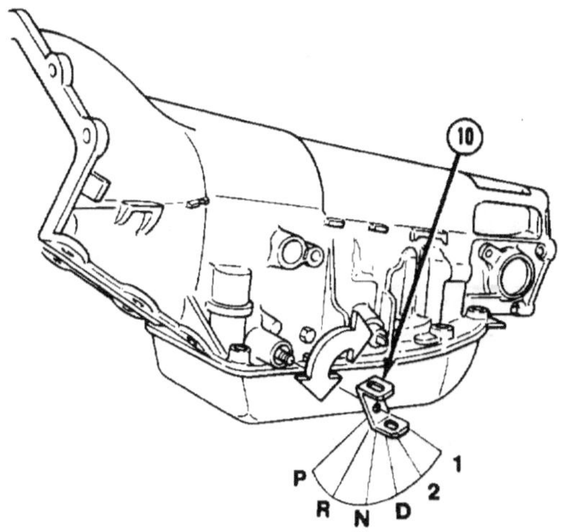

FOLLOW-ON TASKS:

- Remove chocks from wheels.

TA50199

5-3. VACUUM MODULATOR ASSEMBLY AND FILL TUBE ASSEMBLY REPLACEMENT.

This task covers: a. Removal b. Cleaning and Inspection c. Installation

INITIAL SETUP:

Equipment Condition

- Both battery negative cables disconnected, (See TM 9-2320-289-20)
- Vacuum lines removed (See TM 9-2320-289-20)

Materials/Parts

- One fill tube gasket
- One "O" ring seal
  Dry cleaning solvent (Item 23, Appendix B)
- Hydraulic fluid (Item 40, Appendix B)
- Petrolatum (Item 50, Appendix B)

Tools/Test Equipment

- 15 mm socket
- Torque wrench

General Safety Instructions

- Dry cleaning solvent is flammable and must not be used near open flame. Use only in a well-ventilated area.

a. REMOVAL

NOTE

- If removing vacuum modulator assembly (2) and valve (4), perform step 1,
- If removing fill tube assembly gasket, perform steps 3 and 4.
- If removing fill tube assembly (1), perform steps 2-5.
- A suitable container should be used to catch hydraulic fluid when disconnecting fill tube assembly (1) and vacuum modulator assembly (2) from transmission,

1. Remove bolt (6), retainer (5), and vacuum modulator assembly (2) from transmission (7). Remove " O" ring seal (3) and discard, Remove modulator valve (4) from transmission.

5-3. VACUUM MODULATOR ASSEMBLY AND FILL TUBE ASSEMBLY REPLACEMENT (Con't).

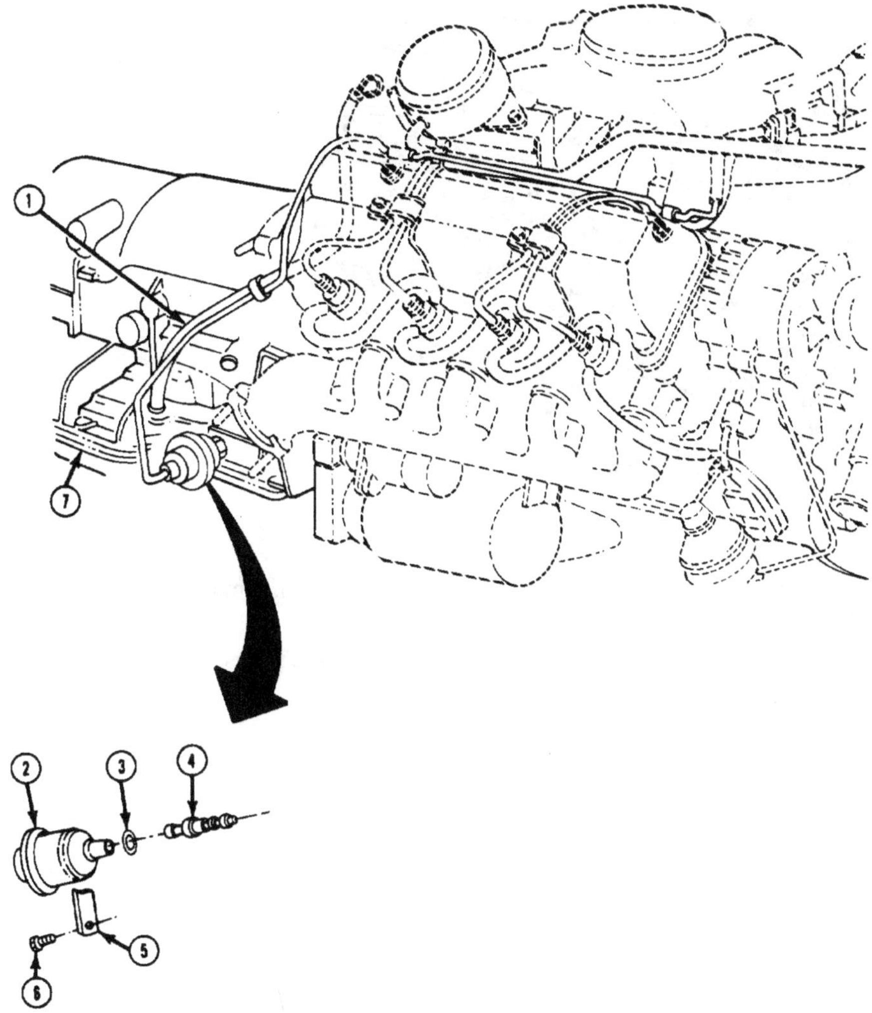

TA701848

5-3. VACUUM MODULATOR ASSEMBLY AND FILL TUBE ASSEMBLY REPLACEMENT (Con't).

2. Using a metric socket, remove 2 bolts (9) and rear engine lifting bracket (8).

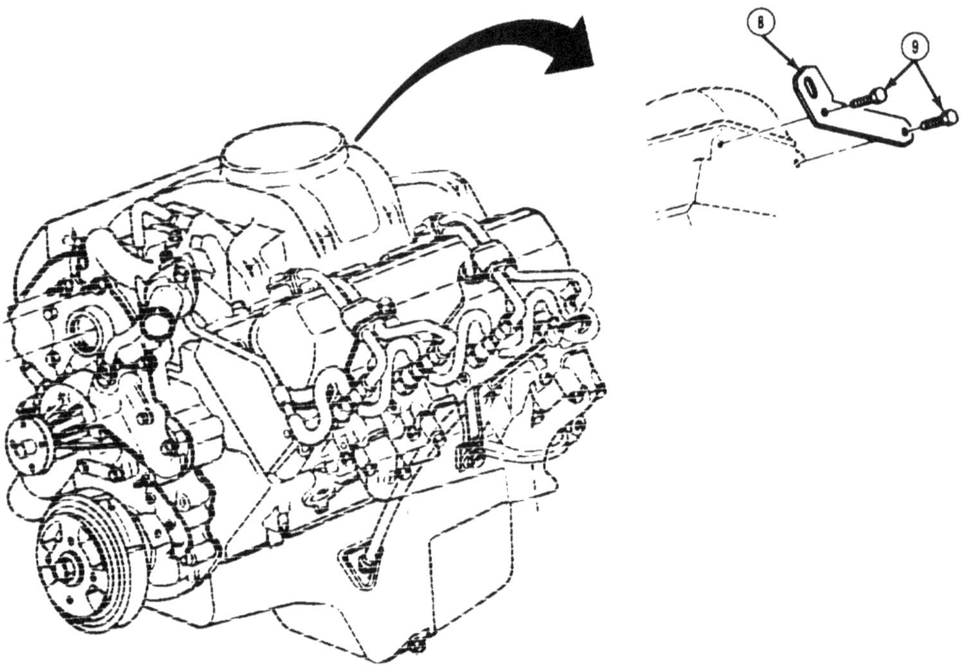

3. Remove transmission fluid indicator (11) from fill tube assembly (1), Remove bolt (10) securing fill tube assembly to transmission (7).
4. Remove fill tube assembly (1) from transmission (7). Remove fill tube gasket (12) and discard.
5. Remove fill tube assembly (1),

TA701849

5-3. VACUUM MODULATOR ASSEMBLY AND FILL TUBE ASSEMBLY REPLACEMENT (Con't).

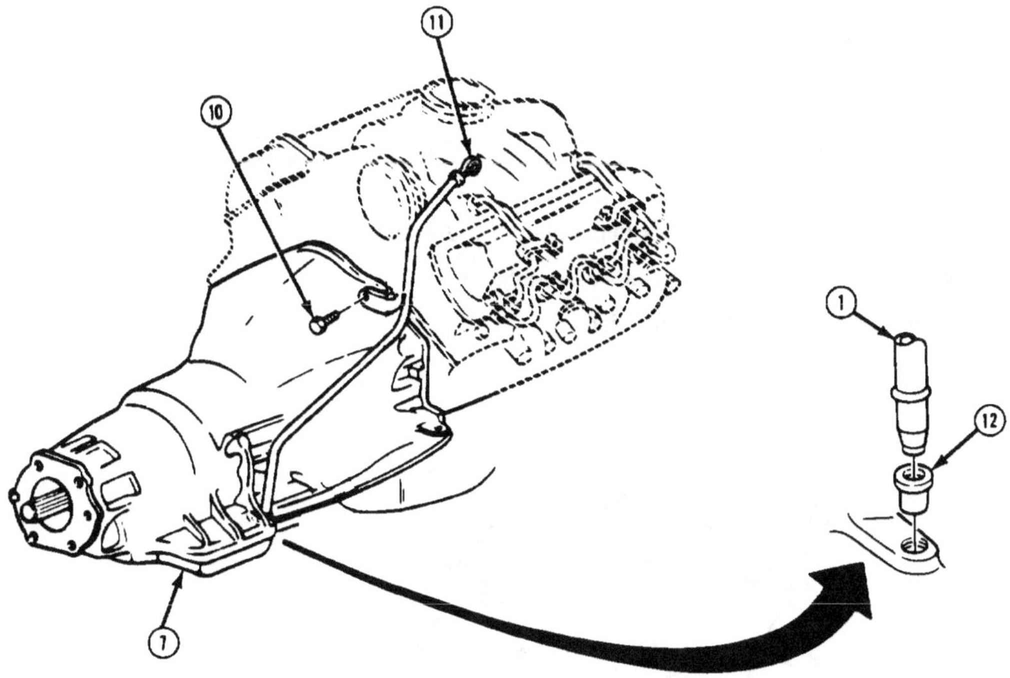

b. CLEANING AND INSPECTION

1. Inspect vacuum modulator assembly for signs of damage. Turn vacuum modulator assembly so that vacuum fitting is down. If hydraulic fluid comes out, diaphragm is leaking. Replace vacuum modulator assembly if damaged or leaking.

WARNING

Dry cleaning solvent P-D-680 is toxic and flammable. Always wear protective goggles and gloves and use only in a well-ventilated area. Avoid contact with skin, eyes, and clothes and DO NOT breathe vapors. DO NOT use near open flame or excessive heat. The solvent's flash point is 100°F-138°F (38°C-59°C). If you become dizzy while using cleaning solvent, immediately get fresh air and medical help. If solvent contacts eyes, immediately wash your eyes with water and get medical aid.

2. Clean vacuum modulator assembly with soap and water. Clean modulator valve and fill tube assembly in dry cleaning solvent.

3. Inspect modulator valve for damage. Place modulator valve in transmission bore and check for free operation. If modulator valve binds in transmission bore or is damaged, replace modulator valve.

4. Inspect fill tube assembly and replace if damaged.

TA701850

5-3. VACUUM MODULATOR ASSEMBLY AND FILL TUBE ASSEMBLY REPLACEMENT (Con't).

c. INSTALLATION

NOTE

- If installing fill tube assembly gasket (12), perform steps 1, 3, and 5.
- If installing fill tube assembly (1), perform steps 1-5.
- If installing vacuum modulator assembly and modulator valve, perform step 6.

1. Lightly coat new fill tube gasket (12) with petrolatum and install in transmission (7).
2. From engine compartment, install fill tube assembly (1),
3. Install fill tube assembly (1) in transmission (7).

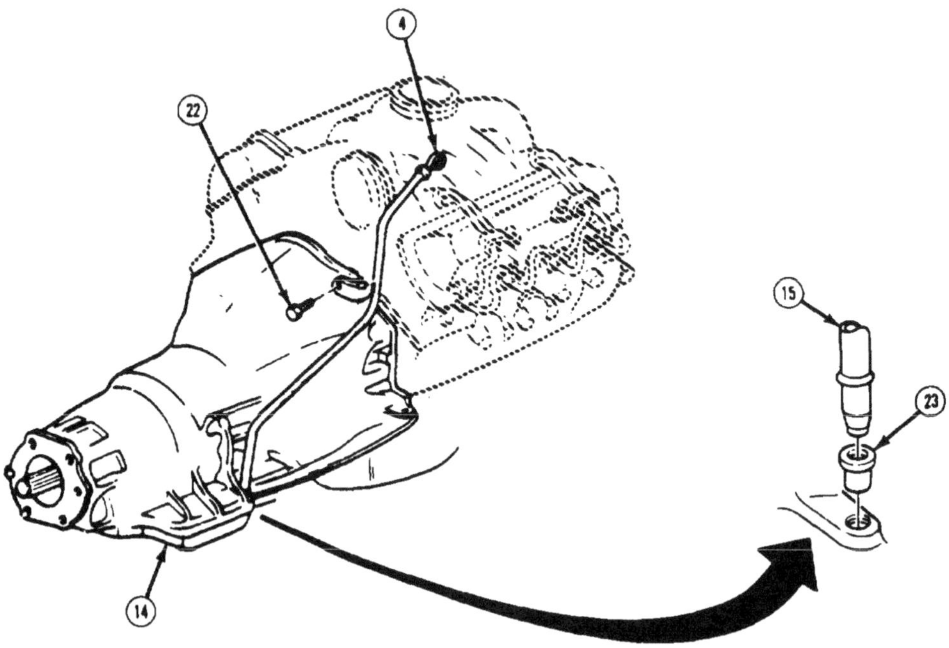

TA701851

5-3. VACUUM MODULATOR ASSEMBLY AND FILL TUBE ASSEMBLY REPLACEMENT (Con't).

4 . Using a metric socket, install rear engine lifting bracket (8) with 2 bolts (9).

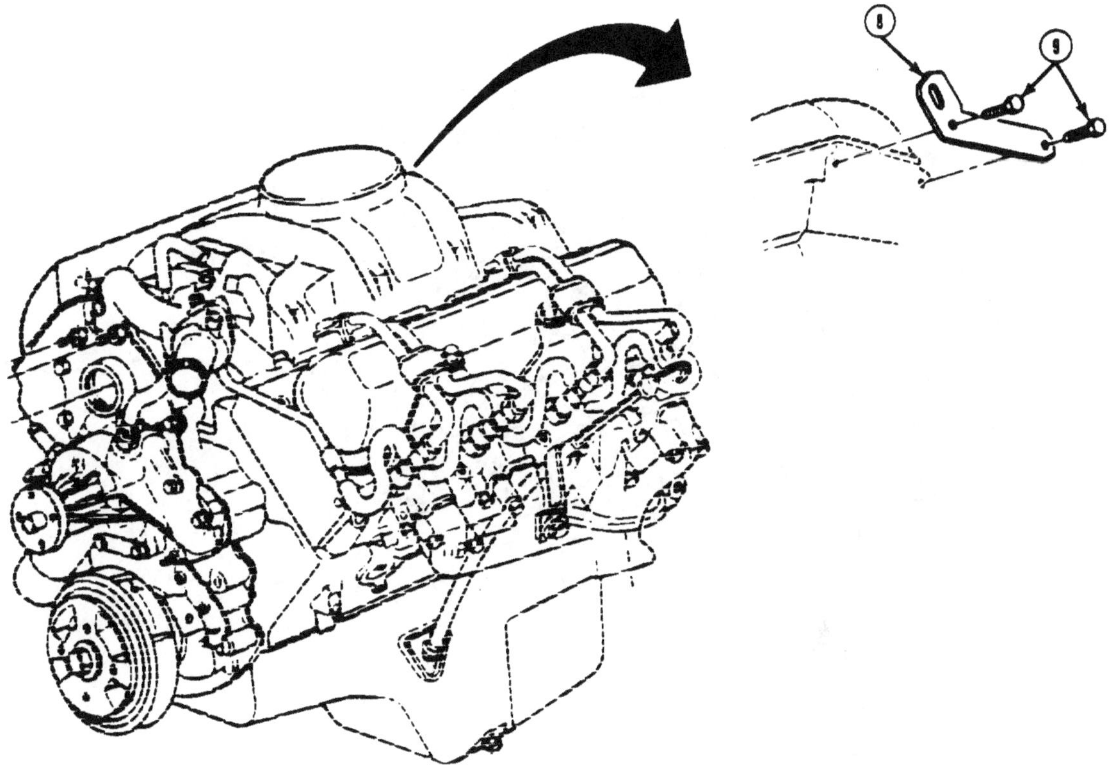

5 . Install bolt (10) securing fill tube assembly (1) to transmission (7). Install transmission fluid indicator (11).

TA701852

5-3. VACUUM MODULATOR ASSEMBLY AND FILL TUBE ASSEMBLY REPLACEMENT (Con't).

16. Install modulator valve (4) in transmission (7) bore with stem end out. Install new "O" ring seal (3) and vacuum modulator assembly (2) with retainer (5) and bolt (6). Tighten bolt to 20 lb,-ft. (27 N.m).

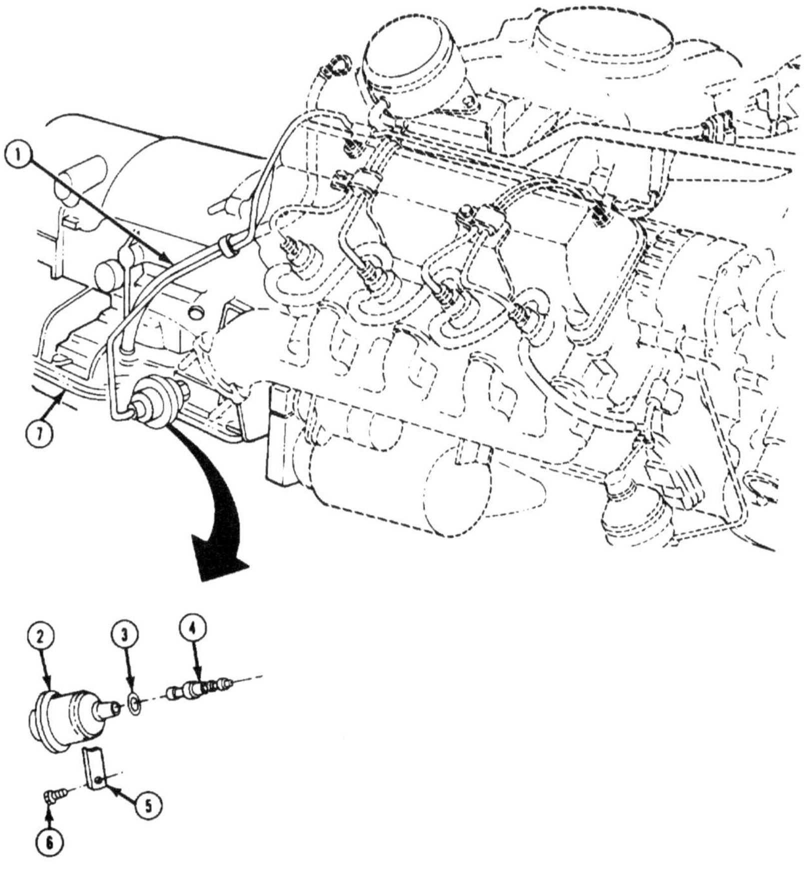

FOLLOW-ON TASKS:

- Install vacuum lines. (See TM 9-2320-289-20)
- Connect both battery negative cables. (See TM 9-2320-289-20)
- Check hydraulic fluid level, (See LO 9-2320-289-12)

Pages 5-13 through 5-15 have been rescinded.

TA701853

5-3. VACUUM MODULATOR ASSEMBLY, VACUUM LINES, AND FILL TUBE ASSEMBLY REPLACEMENT (Con't).

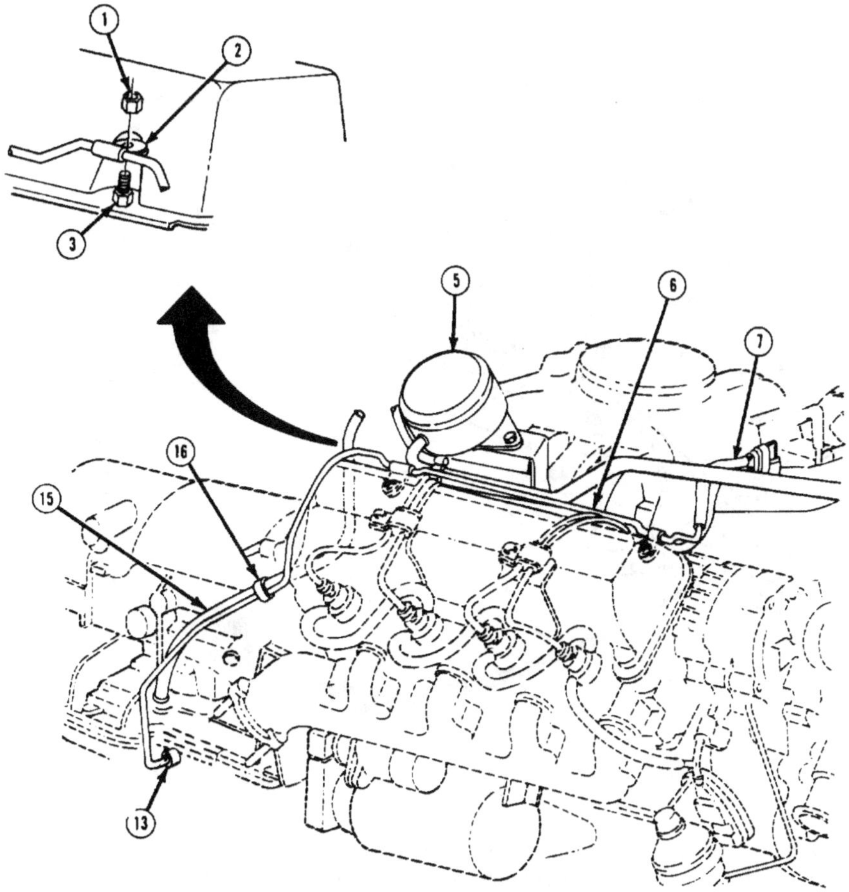

TA50205

5-3. VACUUM MODULATOR ASSEMBLY, VACUUM LINES, AND FILL TUBE ASSEMBLY REPLACEMENT (Con't).

9. Install bolt (22) securing fill tube assembly (15) to transmission (14). Install transmission fluid indicator (4).

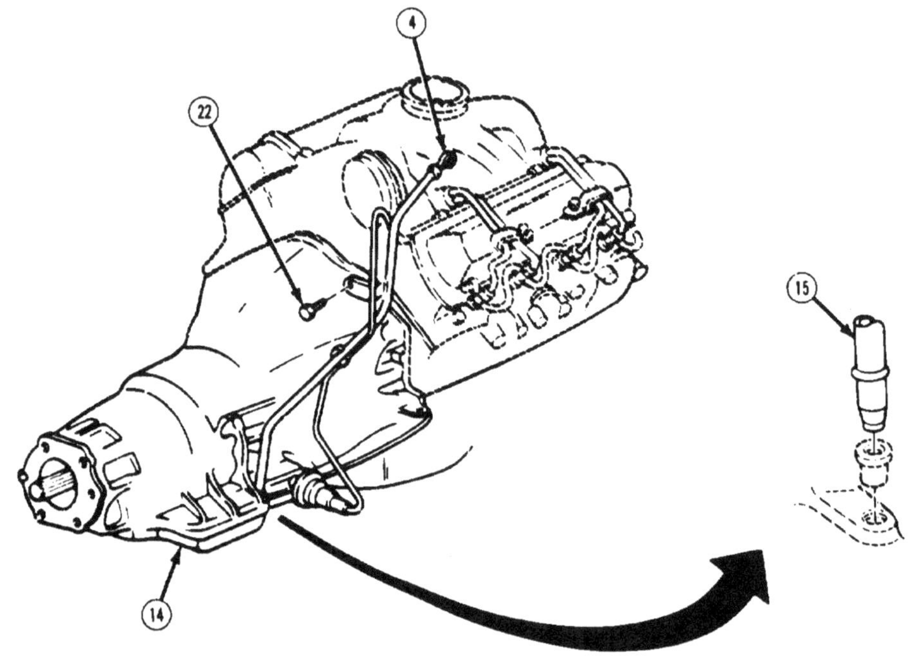

10. Install modulator valve (10) in transmission (14) bore with stem end out. Install new "O" ring seal (9) and vacuum modulator assembly (8) with retainer (11) and bolt (12). Tighten bolt to 20 lb.-ft. (27 N.m).

11. Connect hose (13) to vacuum modulator assembly (8),

TA50206

5-3. VACUUM MODULATOR ASSEMBLY, VACUUM LINES, AND FILL TUBE ASSEMBLY REPLACEMENT (Con't).

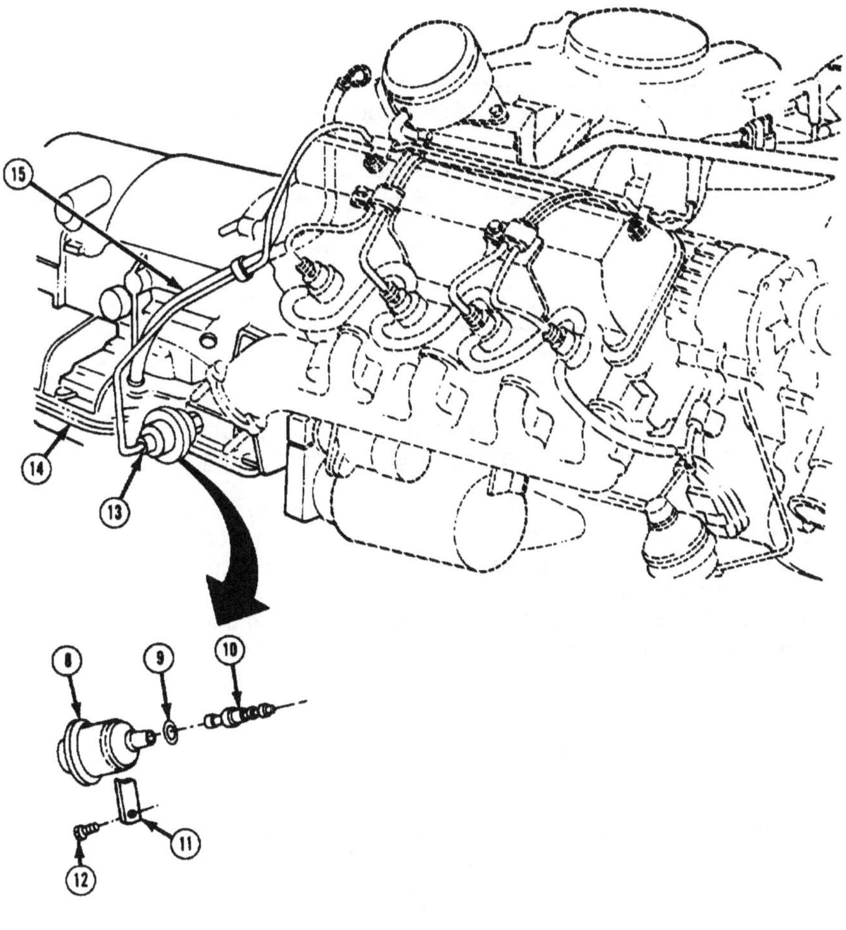

FOLLOW-ON TASKS:

- Install air cleaner if removed. (See TM 9-2320-289-20)
- Connect both battery negative cables. (See TM 9-2320-289-20)
- Check hydraulic fluid level. (See LO 9-2320-289-12)

TA50207

5-4. TRANSMISSION OIL COOLER SUPPLY AND RETURN PIPES MAINTENANCE.

This task covers: a.Flushing Pipes b. Removal c. Installation

INITIAL SETUP:

Materials/arts

- Dry cleaning solvent (Item 23, Appendix B)
- 5/18 in. I.D. rubber hose (Item 38, Appendix B)
- Hydraulic fluid (Item 40, Appendix B)

Tools/Test Equipment

- Oil suction pump
- Line wrench

General Safety Instructions

- Allow transmission oil cooler pipes to cool before servicing.
- Dry cleaning solvent is flammable and must not be used near open flame. Use only in a well-ventilated area.
- Compressed air used for flushing purposes should never exceed 30 psi (207 kPa).

a. FLUSHING PIPES

WARNING

- Allow transmission oil cooler pipes (9 and 10) to cool before attempting to service. Failure to follow this warning may result in serious burns.
- Dry cleaning solvent P-D-680 is toxic and flammable. Always wear protective goggles and gloves and use only In a well-ventilated area. Avoid contact with skin, eyes, and clothes and DO NOT breathe vapors. DO NOT use near open flame or excessive heat. The solvent's flash point Is 100°F-138°F (38°C-590C). If you become dizzy while using cleaning solvent, immediately get fresh air and medical help. If solvent contacts eyes, immediately wash your eyes with water and get medical aid.

CAUTION

Before a new or overhauled transmission (1) is installed, oil cooler in radiator (4) and transmission oil cooler supply pipe (9) and return pipe (10) must be flushed to rid the system of contamination, Failure to perform this procedure will result in damage to transmission.

NOTE

Suitable containers should be used to catch hydraulic fluid and dry cleaning solvent during flushing,

1. Loosen 2 fittings (11) and disconnect transmission oil cooler supply pipe (9) and return pipe (10) from 2 connectors (2) at transmission (1),
2. Place a 5/16 in, I.D. rubber hose over end of transmission oil cooler supply pipe (9) and insert end of hose into dry cleaning solvent.

5-4. TRANSMISSION OIL COOLER SUPPLY AND RETURN PIPES MAINTENANCE.

This task covers:

a. Flushing Pipes
b. Removal
c. Installation

INITIAL SETUP:

Materials/Parts

- Dry cleaning solvent (Item 23, Appendix B)
- 5/16 in. I.D. rubber hose (Item 38, Appendix B)
- Hydraulic fluid (Item 40. Appendix B)

Too/s/Test Equipment

- Oil suction pump
- Line wrench

General Safety Instructions

- Allow transmission oil cooler pipes to cool before servicing.
- Dry cleaning solvent is flammable and must not be used near open flame. Use only in a well-ventilated area,
- Compressed air used for flushing purposes should never exceed 30 psi (207 kPa).

a. FLUSHING PIPES

WARNING

- Allow transmission oil cooler pipes (9 and 10) to Cool before attempting to service. Failure to follow this warning may result in serious burns,
- Dry cleaning solvent P-D-680 is toxic and flammable. Always wear protective goggles and gloves and use only in a well-ventilated area. Avoid contact with skin, eyes, and clothes and DO NOT breathe vapors. DO NOT use near open flame or excessive heat, The solvent's flash point is 100°F-138°F (38°C-590C). If you become dizzy while using cleaning solvent, immediately get fresh air and medical help, If solvent contacts eyes, immediately wash your eyes with water and get medical aid.

CAUTION

Before a new or overhauled transmission (1) is installed, oil cooler in radiator (4) and transmission oil cooler supply pipe (9) and return pipe (10) must be flushed to rid the system of contamination. Failure to perform this procedure will result in damage to transmission,

NOTE

Suitable containers should be used to catch hydraulic fluid and dry cleaning solvent during flushing.

1. Loosen 2 fittings (11) and disconnect transmission oil cooler supply pipe (9) and return pipe (10) from 2 connectors (2) at transmission (1),
2. Place a 5/16 in. I.D. rubber hose over end of transmission oil cooler supply pipe (9) and insert end of hose into dry cleaning solvent.

5-4. TRANSMISSION OIL COOLER SUPPLY AND RETURN PIPES MAINTENANCE (Con't).

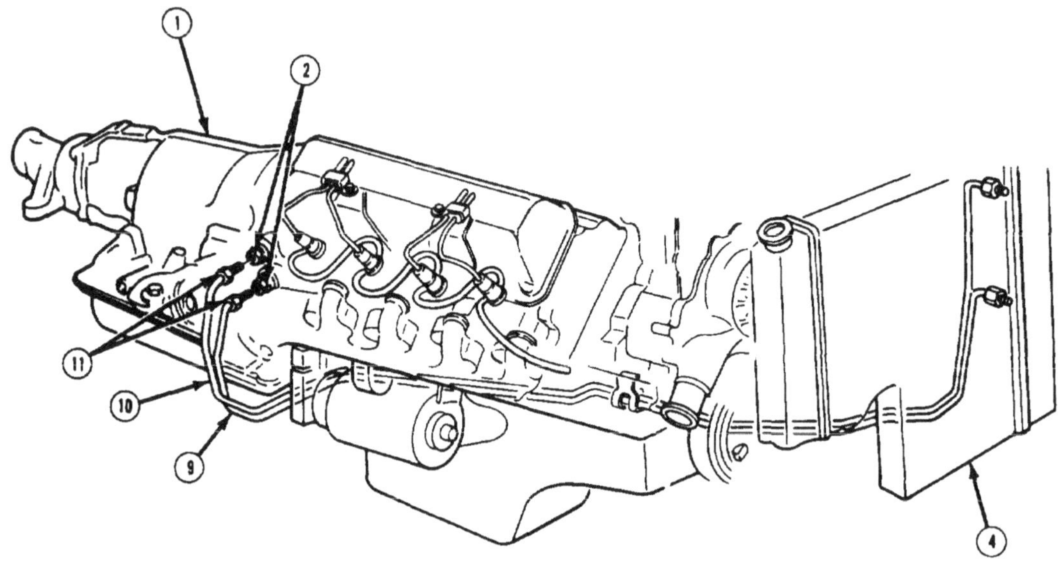

3. Using oil suction pump, flush clean dry cleaning solvent through transmission oil cooler return pipe (10) until dry cleaning solvent comes out oil cooler return pipe. Continue flushing until dry cleaning solvent comes out clean.

4. Remove rubber hose from end of transmission oil cooler supply pipe (9) and place on transmission oil cooler return pipe (10). Insert end of hose into dry cleaning solvent. Flush clean dry cleaning solvent through transmission oil cooler supply pipe until dry cleaning solvent comes out oil cooler supply pipe. Continue flushing until dry cleaning solvent comes out clean.

WARNING

Compressed air used for flushing purposes should never exceed 30 psi (207 kPa). Use only effective chip guarding and personnel protective equipment (goggles/shield, gloves, etc.). Failure to follow this warning may result in injury to personnel or damage to oil cooler.

5. Remove remaining dry cleaning solvent from oil cooler supply pipe (9), return pipe (10), and from oil cooler in radiator (4) with compressed air, and flush with hydraulic fluid.

6. Install transmission oil cooler supply pipe (9) and return pipe (10) at 2 connectors (2). Tighten 2 fittings (11).

TA50208

5-4. TRANSMISSION OIL COOLER SUPPLY AND RETURN PIPES MAINTENANCE (Con't).

b. REMOVAL

NOTE

If disconnecting oil cooler supply pipe (9) and return pip e transmission (1), perform step 2,

1. Loosen 2 fittings (3) and disconnect transmission oil cooler supply pipe (9) and return pipe (10) from left side of radiator (4),
2. Loosen 2 fittings (11) and disconnect transmission oil cooler supply pipe (9) and return pipe (10) from 2 connectors (2) at transmission (1). Remove bolt (8) and clip (6) from torque converter cover (7).
3. Remove pipes from clips (5 and 6) and remove transmission oil cooler supply pipe (9) and return pipe (10).
4. Remove 2 connectors (2) if damaged.

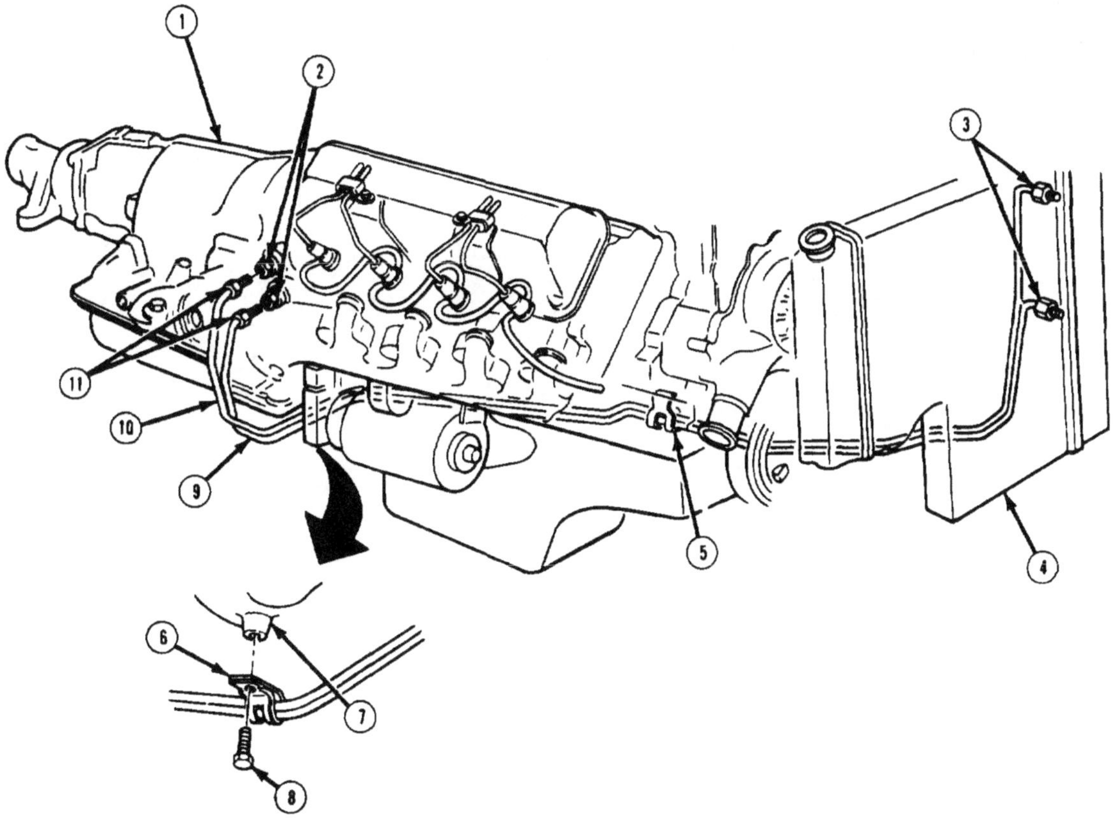

TA50209

5-4. TRANSMISSION OIL COOLER SUPPLY AND RETURN PIPES MAINTENANCE (Con't).

c. INSTALLATION 1

NOTE

If connecting oil cooler supply pipe (9) and return pipe (10) to transmission (1), perform steps 1, 3, and 4.

1. If removed, install 2 connectors (2) to transmission (1). Install transmission oil cooler return pipe (10) to top connector and transmission oil cooler supply pipe (9) to bottom connector. Tighten 2 fittings (11).
2. Connect transmission oil cooler return pipe (10) to top of radiator (4) and transmission oil cooler supply pipe (9) to bottom of radiator. Tighten 2 fittings (3).
3. Install clip (6) to oil cooler supply pipe (9) and return pipe (10). Install oil cooler supply pipe and return "pipe to clip (5) at engine oil pan.
4. Install clip (6) to torque converter cover (7) with bolt (8).

FOLLOW-ON TASKS:

- Check hydraulic fluid level. (See LO 9-2320-289-12)

5-5. GOVERNOR ASSEMBLY MAINTENANCE.

This task covers:

a. Removal
b. Disassembly
c. Cleaning and Inspection
d. Assembly
e. Installation

INITIAL SETUP:

Equipment Condition

- Transfer case in "2H."

Tools/Test Equipment

- Torque wrench

Materials/Parts

- One gasket
- Two pins
- Dry cleaning solvent (Item 23, Appendix B)
- Hydraulic fluid (Item 40, Appendix B)

General Safety Instructions

- Dry cleaning solvent is flammable and must not be used near open flame. Use only in a well-ventilated area.
- Compressed air used for cleaning purposes should never exceed 30 psi (207 kPa),

a. REMOVAL

NOTE

A suitable container should be used to catch hydraulic fluid when removing governor cover (2) and governor assembly (4).

1. Remove 4 bolts (1) and governor cover (2) from transmission case (5). Remove gasket (3) and discard. Remove governor assembly (4) from bore of transmission,

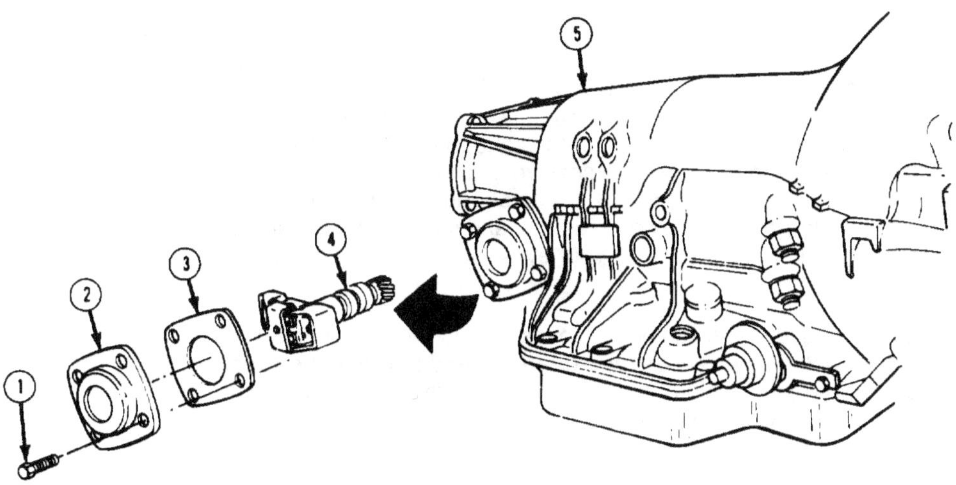

TA50210

5-5. GOVERNOR ASSEMBLY MAINTENANCE (Con't).

b. DISASSEMBLY

1. Cut off end of each pin (6). Remove 2 pins, thrust cap (14), 2 primary weights (7), secondary weights (9), and springs (8). Discard pins.
2. Carefully remove valve (13) from sleeve (10).
3. If driven gear (12) is worn or loose on sleeve (10), drive out lockpin (11) and remove driven gear. Discard lockpin and driven gear.

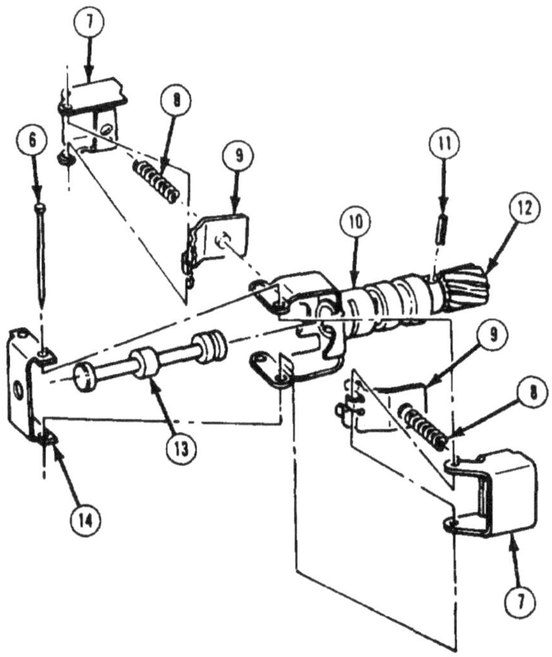

c. CLEANING AND INSPECTION

WARNING

- Dry cleaning solvent P-D-680 is toxic and flammable. Always wear protective goggles and gloves and use only in a well-ventilated area. Avoid contact with skin, eyes, and clothes and DO NOT breathe vapors. DO NOT use near open flame or excessive heat. The solvent's flash point is 100°F-138°F (38°C-59°C). If you become dizzy while using cleaning solvent, immediately get fresh air and medical help. If solvent contacts eyes, immediately wash your eyes with water and get medical aid.
- Compressed air used for cleaning purposes should never exceed 30 psi (207 kPa). Use only effective chip guarding and personnel protective equipment (goggles/shield, gloves, etc.). Failure to follow this warning may result in injury to personnel.

1. Clean all disassembled components with dry cleaning solvent. Dry with compressed air and blow out all passages.
2. Inspect all disassembled components for damage. If any component is damaged, replace governor assembly as a unit.

TA50211

5-5. GOVERNOR ASSEMBLY MAINTENANCE (Con't).

3. Install sleeve in bore of transmission and inspect for free operation in bore. Inspect valve for free operation in bore of sleeve. If sleeve or valve bind, replace governor assembly as a unit.

4. If a new governor assembly has been obtained, install in bore of transmission and inspect for free operation, If governor assembly binds in bore, replace transmission case,

d. ASSEMBLY

1. If removed install new driven gear (12) on sleeve (10). Press driven gear in until it is almost seated, using a socket, with governor assembly (4) supported in an arbor press with two 7/64 in. (2.78 mm) plates in the exhaust openings (16). Remove any shavings from driven gear hub and press in until it is seated.

CAUTION

Shaft portion of driven gear (12) should be supported when installing new lockpin (11). Failure to follow this caution may result in damage to driven gear.

2. Using a 1/8 in. drill, drill a new pin hole 90 degrees from the existing one in sleeve (10). Install new lockpin (11) through new hole in driven gear and sleeve so that ends are just below top of hole. Clean driven gear and sleeve assembly to remove any shavings that may have collected.

3. Install valve (13) into bore of sleeve (10). Install 2 primary weights (7), secondary weights (9), springs (8), and thrust cap (14) on sleeve. Aline pin holes and install 2 new pins (6). Crimp ends of pins to prevent them from falling out.

4. Lubricate governor assembly with hydraulic fluid and ensure that weight assemblies (7 and 9) operate freely on 2 pins (6) and that valve (13) operates freely in bore of sleeve (10).

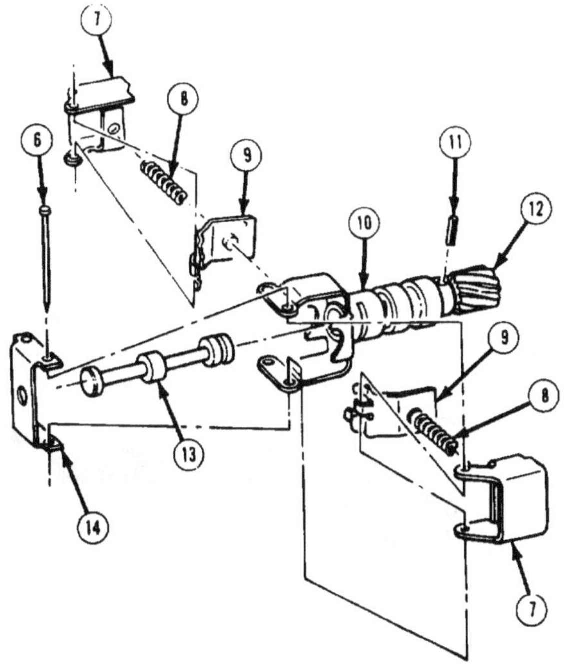

5. Inspect valve entry opening (15) with weights held all the way outward. The opening should be 0.020 in. (0.508 mm) minimum, Inspect valve exhaust opening (16) with weights held all the way inward, The opening should be 0,020 in. (0.508 mm) minimum. If valve entry and exhaust openings are not within specifications, replace governor assembly (4),

TA50212

5-5. GOVERNOR ASSEMBLY MAINTENANCE (Con't).

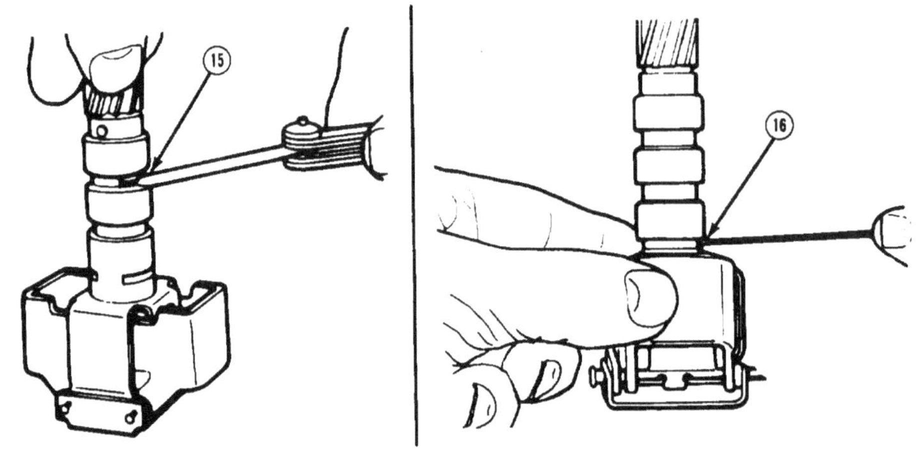

e. INSTALLATION

1. Install governor assembly (4) in bore of transmission case (5).
2. Install governor cover (2) with new gasket (3) and 4 bolts (1). Tighten bolts to 20 lb.-ft. (27 N.m).

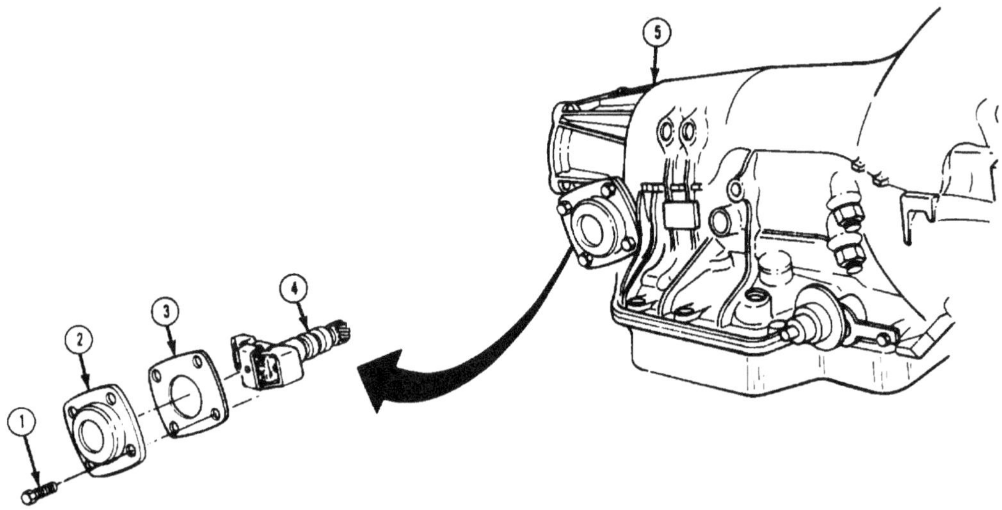

FOLLOW-ON TASKS:

- Check hydraulic fluid level. (See LO 9-2320-289-12)

TA50213

5-6. TRANSMISSION DOWNSHIFT ELECTRICAL CONNECTOR REPLACEMENT.

This task covers: a. Removal b. Installation

INITIAL SETUP:

Equipment Condition

- Transmission filter assembly removed. (See TM 9-2320-289-20)

Materials/Parts

- One "O" ring seal
- Hydraulic fluid (Item 40, Appendix B)

a. REMOVAL

1. Disconnect engine wiring harness connector (5) from downshift electrical connector (1). Disconnect detent solenoid wire (2) from end of downshift electrical connector inside transmission case (4),

2. Compress prongs on downshift electrical connector (1) and remove from transmission case (4) .

3. Remove "O" ring seal (3) from downshift electrical connector (1) and discard.

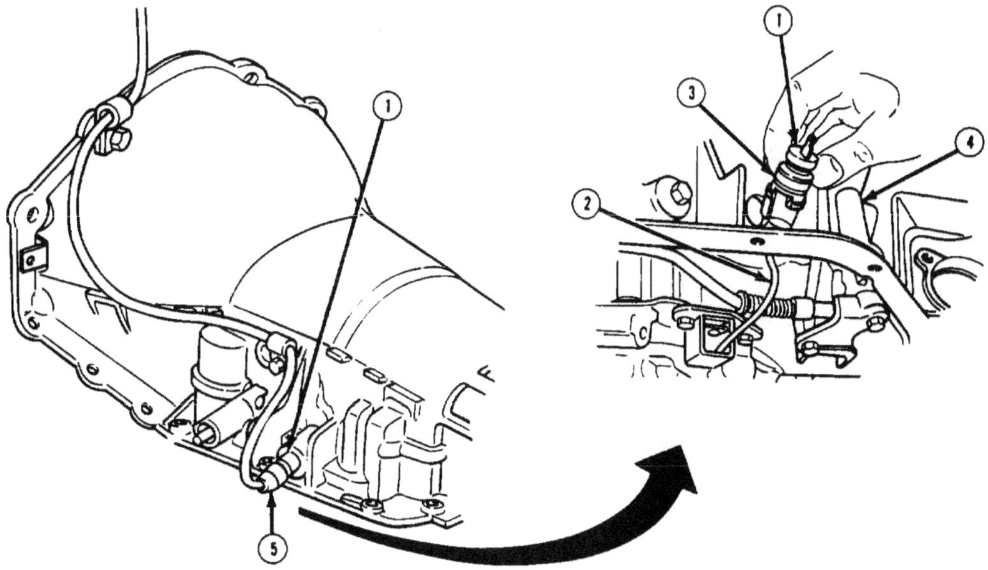

TA50214

5-6. TRANSMISSION DOWNSHIFT ELECTRICAL CONNECTOR REPLACEMENT (Con't).

b. INSTALLATION

1. Lubricate new "O" ring seal (3) with hydraulic fluid and install in downshift electrical connector (1) groove.
2. Install downshift electrical connector (1) with lock tabs pointing into transmission case (4) and locator tabs in notch on side of transmission case.
3. Connect detent solenoid wire (2) and engine wiring harness connector (5) to downshift electrical connector (1).

FOLLOW-ON TASKS:

- Install transmission filter assembly. (See TM 9-2320-289-20)

5-7. TRANSMISSION CONTROL VALVE ASSEMBLY AND DETENT SOLENOID MAINTENANCE.

This task covers:

a. Removal
b. Disassembly
c. Cleaning and Inspection
d. Assembly
e. Installation

INITIAL SETUP:

Equipment Condition

- Transmission filter and intake pipe removed, (See TM 9-2320-289-20)

Tools/Test Equipment

- *Accumulator* piston compressor, J-22269-1
- Torque wrench

Personnel Required

- MOS 63W (2)

Materials/Parts

- One case gasket
- One control valve body gasket
- One detent solenoid gasket
- Dry cleaning solvent (Item 23, Appendix B)
- Crocus cloth (Item 18, Appendix B)
- Hydraulic fluid (Item 40, Appendix B)
- Lubricating oil (Item 47, Appendix B)
- Petrolatum (Item 50, Appendix B)

General Safety Instructions

- Dry cleaning solvent is flammable and must not be used near open flame. Use only in a well-ventilated area.
- All valves are under spring tension. Wear goggles and use caution during removal and installation of springs.

a. REMOVAL

NOTE

- If removing detent roller and spring assembly (2), perform step 1.
- If removing detent solenoid, perform step 5.
- If removing control valve spacer plate and check balls, perform steps 1-6.

1. Remove bolt (4) and detent roller and spring assembly (2) from control valve assembly (3).

5-7. TRANSMISSION CONTROL VALVE ASSEMBLY AND DETENT SOLENOID MAINTENANCE (Con't).

CAUTION

Ensure that manual valve (7) does not fall out of its bore in control valve assembly (3). Failure to follow this caution may result in damage to manual valve.

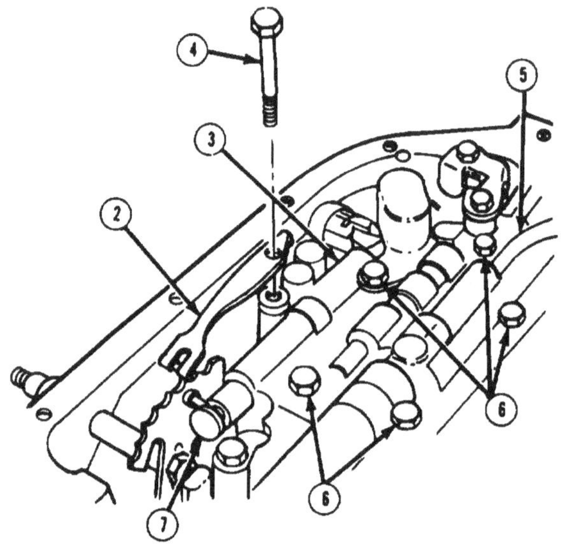

2 . Remove 10 remaining bolts (6). Lift up on 2 governor pipes (5) and remove control valve assembly (3) and governor pipes as a unit.

NOTE

Governor screen assembly (8) may be found in governor feed pipe hole (9) or in governor pipe (5). Governor screen assembly must be installed In same hole from which it was removed,

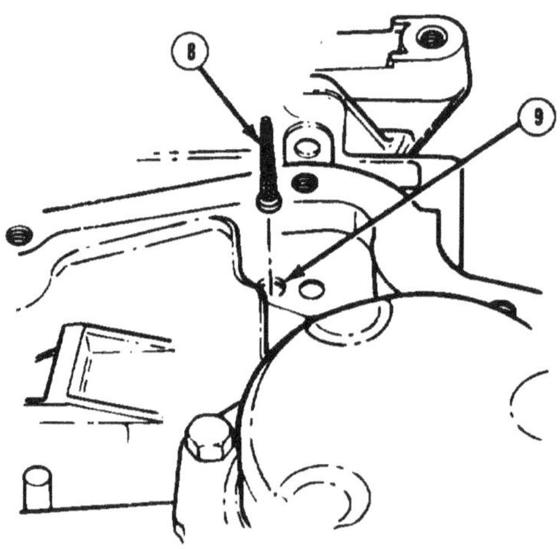

3 . Mark feed pipe hole (9) and remove governor screen assembly (8).

4 . Remove 2 governor pipes (5) from control valve assembly (3). Remove control valve body gasket an d discard.

TA50215

5-7. TRANSMISSION CONTROL VALVE ASSEMBLY AND DETENT SOLENOID MAINTENANCE (Con't).

5. Disconnect detent solenoid wire (11) at connector (10). Remove 2 screws (13) and remove detent solenoid (12). If present, remove detent solenoid gasket and discard.

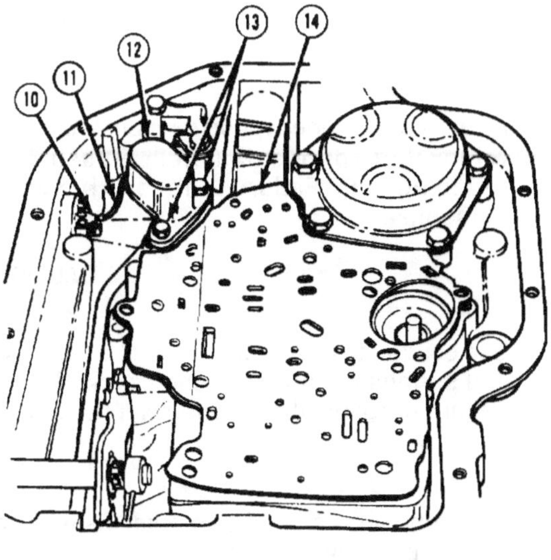

NOTE

If transmission is in truck, control valve spacer plate (14) must be lowered in a level plane so that check balls (15) do not fall out. Then check balls are removed from control valve spacer plate.

6. Remove control valve spacer plate (14) and case gasket. Discard case gasket. Remove 6 check balls (15).

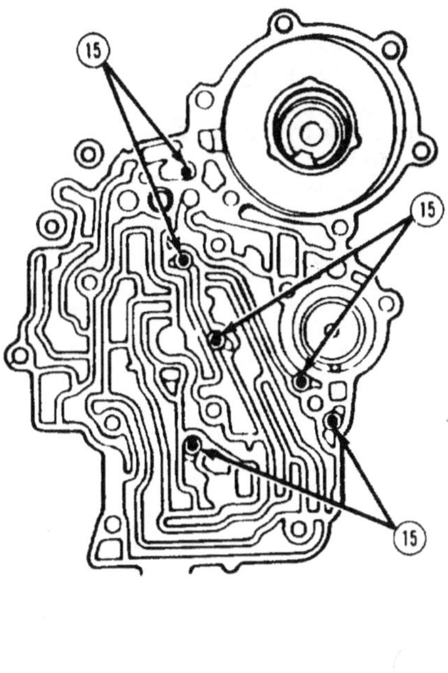

TA50216

5-7. TRANSMISSION CONTROL VALVE ASSEMBLY AND DETENT SOLENOID MAINTENANCE (Con't).

b. DISASSEMBLY

WARNING

All valves are under spring tension. Wear goggles and use caution during removal of springs. Maintain pressure against valves while removing retaining pins. Failure to follow this warning may result in injury to personnel.

CAUTION

Use caution to ensure that control valve assembly (3), valves, bushings, and springs are not scratched, nicked, or gouged during disassembly. Failure to follow this caution will result in transmission malfunction.

NOTE

All components should be marked as they are disassembled so that they may be properly assembled.

1. Position control valve assembly (3) so that oil passages are facing upward and front accumulator pocket is toward you.
2. Remove manual valve (7).
3. Compress front accumulator (17) and remove retaining ring
4. Remove front accumulato r (17) and spring (19).

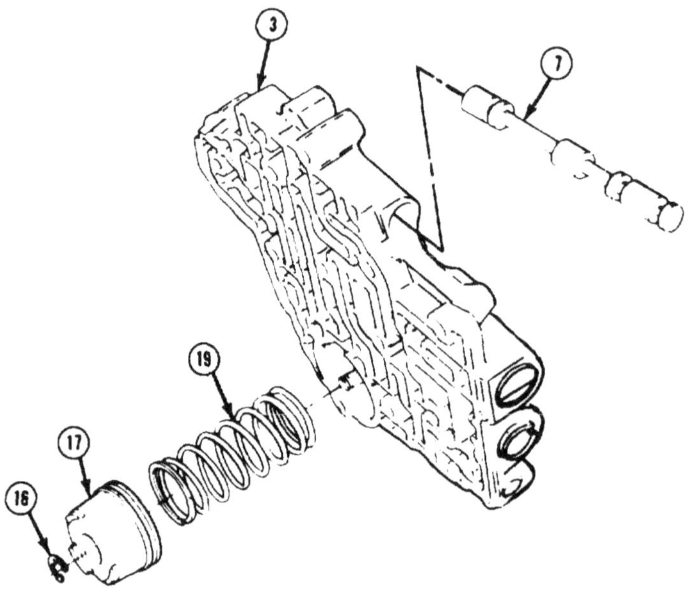

TA50217

5-7. TRANSMISSION CONTROL VALVE ASSEMBLY AND DETENT SOLENOID MAINTENANCE (Con't).

NOTE

1-2 regulator valve (23) and spring (22) may be inside of 1-2 modulator bushing (24).

5. Remove retaining pin (36), 1-2 modulator bushing (24), 1-2 regulator valve (23) amd spring (22), 1-2 detent valve (21), and 1-2 shift valve (20).

6. Remove retaining pin (35), 2-3 shift valve spring (26), 2-3 modulator brushing (27) with 2-3 modulator valve (25) inside, 2-3 intermediate spring (28), and 2-3 shift valve (29).

7. Remove retaining pin (34), 3-2 bore plug (30), 3-2 valve spring (31), 3-2 spacer (32), and 3-2 shift valve (33).

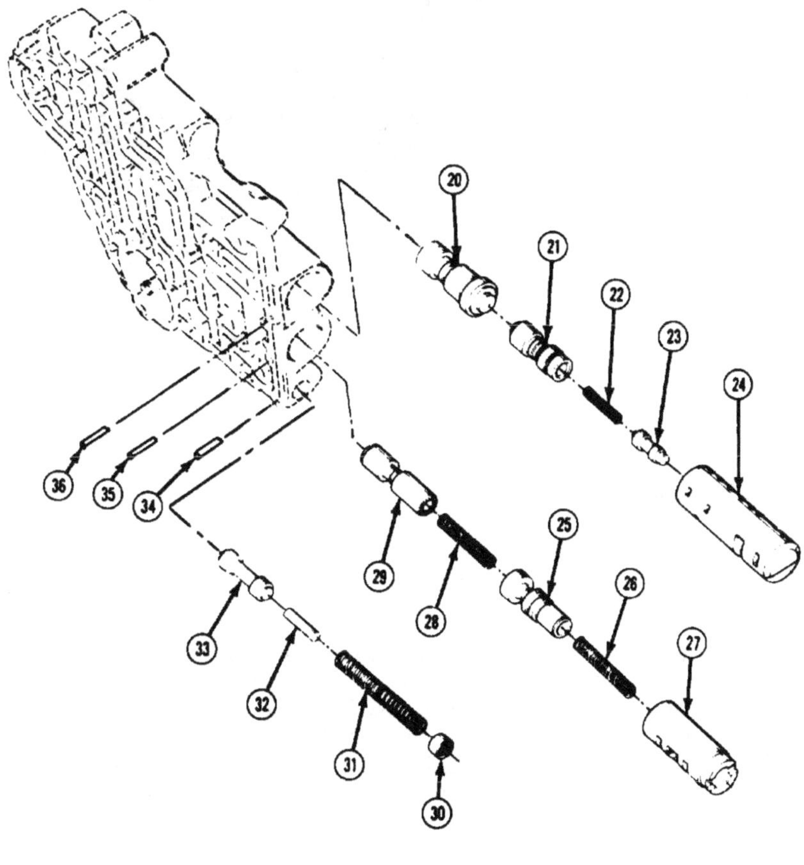

TA50218

5-7. TRANSMISSION CONTROL VALVE ASSEMBLY AND DETENT SOLENOID MAINTENANCE (Con't).

8. Remove retaining pin (43), detent bore plug (37), detent valve (38), detent regulator valve (39), detent regulator valve spring (41), and detent spacer (40).

9. Remove retaining pin (42), 1-2 bore plug (46), 1-2 accumulator valve (45), and 1-2 primary accumulator spring (44).

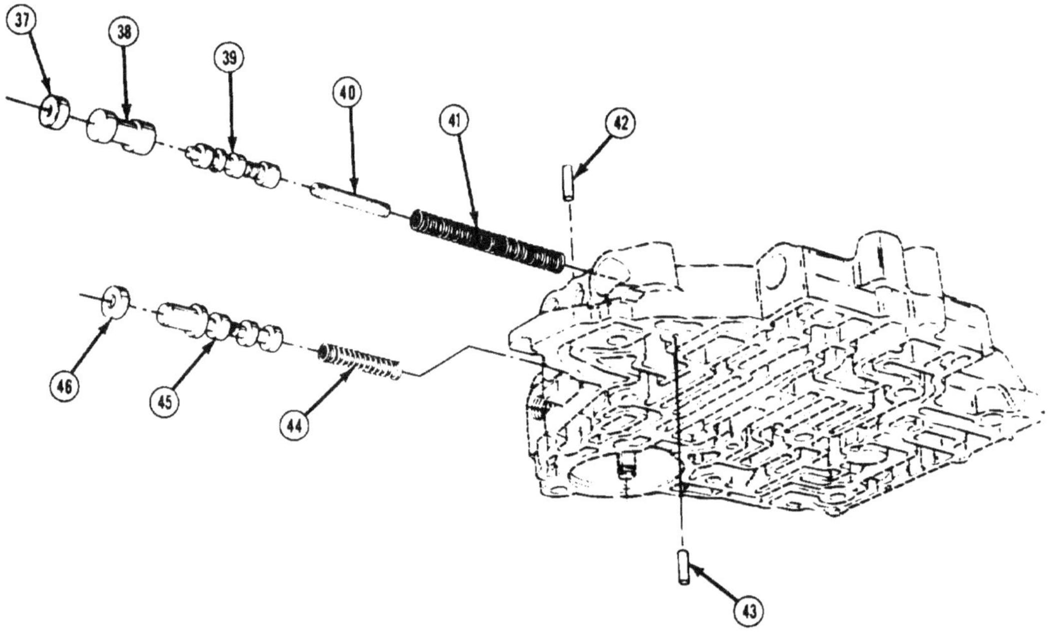

c . CLEANING AND INSPECTION

WARNING

Dry cleaning solvent P-D-680 is toxic and flammable. Always wear protective goggles and gloves and use only. in a well-ventilated area. Avoid contact with skin, eyes, and clothes and DO NOT breathe vapors. DO NOT use near open flame or excessive heat. The solvent's flash point is 100°F-138°F (38°C-59°C). if you become dizzy while using cleaning solvent, immediately get fresh air and medical help. If solvent contacts eyes, immediately wash your eyes with water and get medical aid.

TA50219

5-7. TRANSMISSION CONTROL VALVE ASSEMBLY AND DETENT SOLENOID MAINTENANCE (Con't).

CAUTION

Use caution to ensure that valves do not bump together, Failure to follow this caution may result in damage to valves.

1. Clean control valve body, valves, and other parts in dry cleaning solvent.
2. Inspect all valves for damage and free movement in their bores. If burrs are present, remove them by polishing with crocus cloth and light lubricating oil. Be careful to not round off shoulders of valves. All valves should fall freely of their own weight with a slight tapping action on the valve body, If manual valve is damaged or binds in its bore, replace. If any other valves are damaged or bind in their bores, replace control valve assembly.
3. Inspect all bushings for damage and free movement in their bores. Replace control valve assembly if bushings are damaged or bind in their bores.
4. Inspect control valve body and springs for damage. Replace control valve assembly if either are damaged.
5. If damaged, remove oil seal ring (18) from front accumulator piston (17) and discard.
6. Inspect front accumulator piston for damage. Replace if damaged.

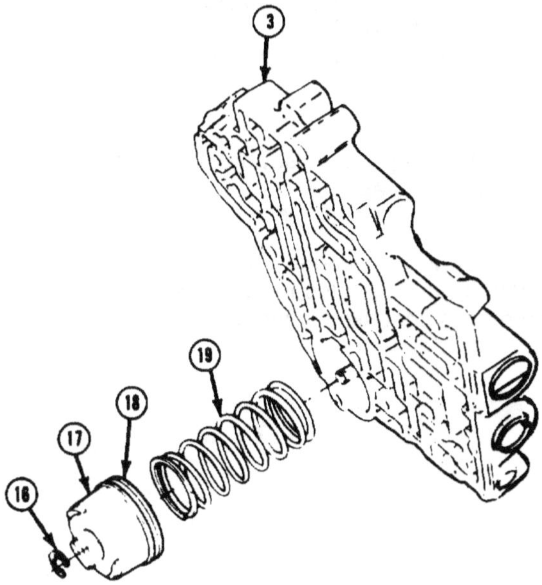

TA50220

5-7. TRANSMISSION CONTROL VALVE ASSEMBLY AND DETENT SOLENOID MAINTENANCE (Con't).

d ASSEMBLY

WARNING

All valves are under spring tension. Wear goggles and use caution during installation of springs. Maintain pressure against valves while installing retaining pins. Failure to follow this warning may result in injury to personnel.

CAUTION

- Use caution to ensure that control valve assembly (3). valves, bushings, and springs are not scratched, nicked, or gouged during assembly. Failure to follow this caution will result in transmission malfunction.
- Ensure that front accumulator piston spring (19) and front accumulator piston (17) are alined with bore and that oil seal ring (18) does not catch on lip of bore when installing front accumulator piston. Failure to follow this caution may result in damage to oil seal ring.

NOTE

All components should be lightly coated with hydraulic fluid as they are assembled.

1 . install new oil seal ring (18) to front accumulator piston (17) if removed. Position front accumulator piston spring (19) and front accumulator piston into control valve body and compress. Install retaining ring (1 6). Check to ensure that there is free movement of front accumulator piston.

5-7. TRANSMISSION CONTROL VALVE ASSEMBLY AND DETENT SOLENOID MAINTENANCE (Con't).

2. Install 1-2 primary accumulator spring (44) and 1-2 accumulator valve (45), stem end out, into bore. Install 1-2 bore plug (46) into bore with hole end facing out. Install grooved retaining pin (42) with grooves to outside.

3. Install detent spacer (40) inside detent regulator valve spring (41). Install both ensuring that detent regulator valve spring seats in bottom of bore.

4. Install detent regulator valve (39), stem end out, and detent valve (38), small land first, into bore. Install detent bore plug (37), hole end out. Install retaining pin (43).

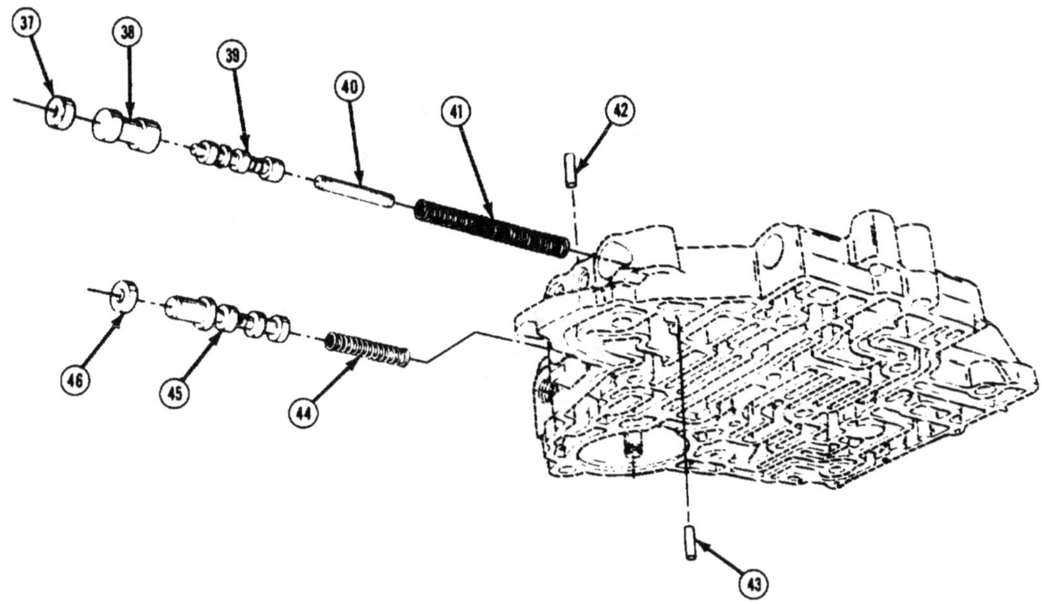

5. Install 3-2 shift valve (33). Install 3-2 spacer (32) inside 3-2 valve spring (31) and install both inside bore,

6. Compress 3-2 valve spring (31) and install 3-2 bore plug (30), hole end out, Install retaining pin (34).

7. Install 2-3 intermediate spring (28) in open end of 2-3 shift valve (29). Install both in bore ensuring that valve seats" in bottom of bore.

8. Install 2-3 modulator valve (25), hole end first. into 2-3 modulator bushing (27) and install both in bore.

9. Install 2-3 shift valve spring (26) in hole in 2-3 shift valve spring and install retaining pin (35).

TA50221

5-7. TRANSMISSION CONTROL VALVE ASSEMBLY AND DETENT SOLENOID MAINTENANCE (Con't).

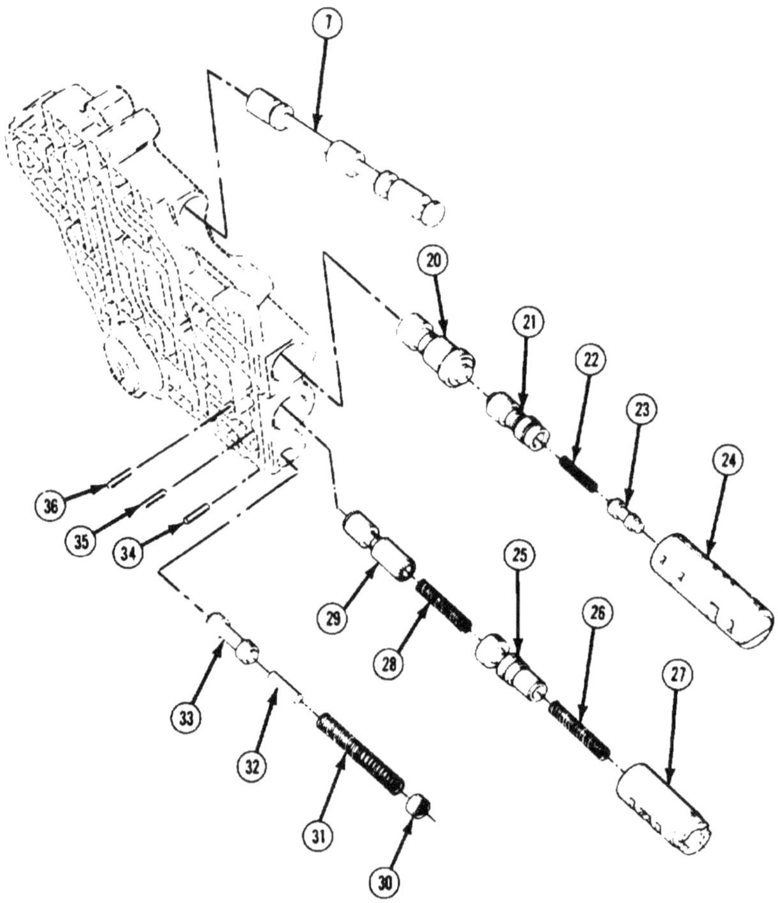

10. Install 1-2 shift valve (20), stem end out, ensuring that valve seats in bottom of bore.
11. Install spring (22) into small end of 1-2 regulator valve (23). Install assembled spring and valve into 1-2 modulator bushing (24). Install 1-2 detent valve (21), hole end first, into 1-2 modulator bushing.
12. Install 1-2 modulator bushing (24) assembly into bore and compress enough to install retaining pin (36).
13. Install manual valve (7).

TA50222

5-7. TRANSMISSION CONTROL VALVE ASSEMBLY AND DETENT SOLENOID MAINTENANCE (Con't).

e. INSTALLATION

NOTE

- If installing detent roller and spring assembly (2), perform step 8.
- If installing detent solenoid (12), perform steps 3 and 9.
- If transmission is in truck , petrolatum should be applied to check balls (15) to retain them.

1 . Install 6 check balls (15) in ball seat pockets in transmission case.

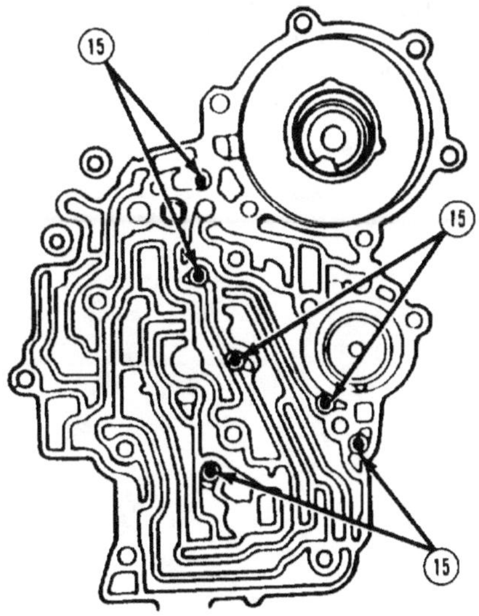

NOTE

Case gasket should not be confused with control valve spacer plate (14). Case gasket is marked "c."

2. Aline new case gasket and control valve spacer plate (14) and install with case gasket against transmission case.
3. Install new detent solenoid gasket and detent solenoid (12) with 2 screws (13). DO NOT fully tighten screws.

NOTE

Control valve body gasket is marked with a "V."

4. Install new control valve body gasket.

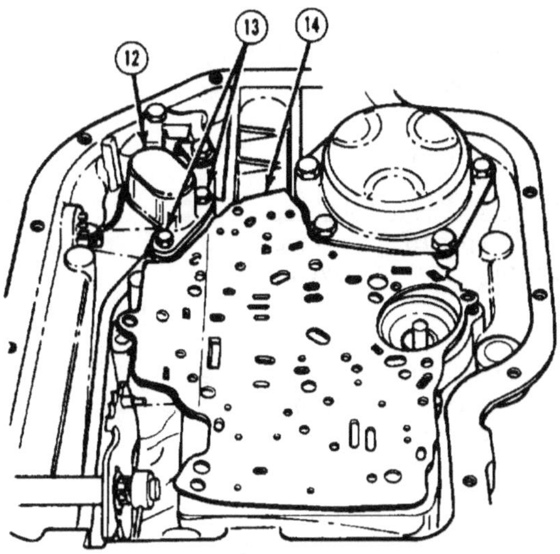

TA50223

5-7. TRANSMISSION CONTROL VALVE ASSEMBLY AND DETENT SOLENOID MAINTENANCE (Con't).

5. Install open end of governor screen assembly (8) into marked governor feed pipe hole (9) in transmission case. Install 2 governor pipes (5) into holes in control valve assembly (3).

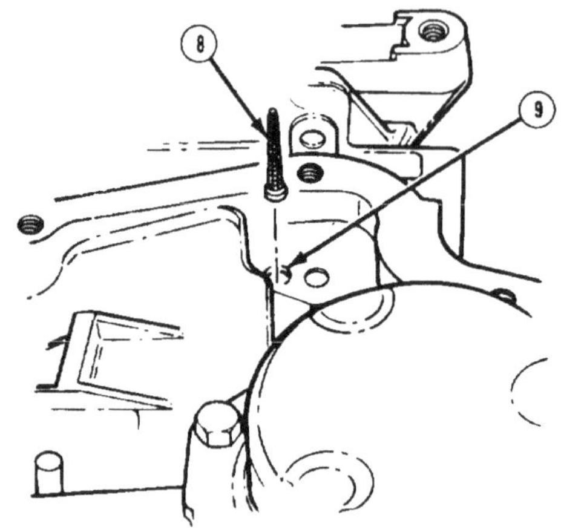

6. Position control valve assembly (3). Check that governor pipes (5) are properly fitted into transmission case holes. Check manual valve (7) to ensure that it is indexed with pin on inside detent lever (1), Ensure that gaskets and spacer plate (14) are not mispositioned.

NOTE

It may be necessary to lightly tap on governor pipes (5) to seat them while starting control valve body bolts (6).

7. Loosely install 10 bolts (6) to secure control valve assembly (3) to transmission case. Tighten all bolts to 10 lb.-ft. (14 N.m).

8. Install detent roller and spring assembly (2) with bolt (4). Tighten bolt to 10 lb.-ft. (14 N.m).

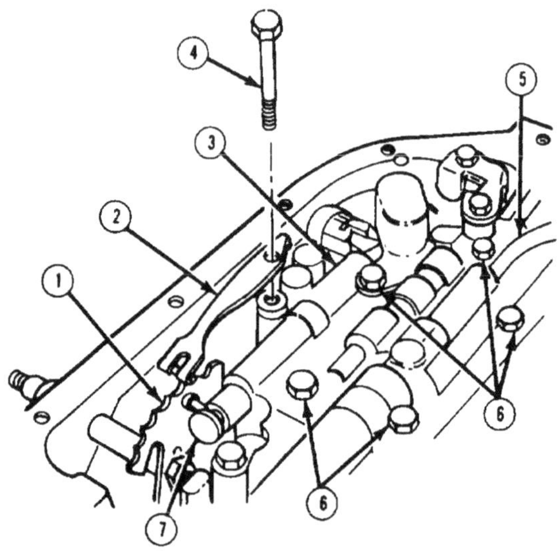

TA50224

5-7. TRANSMISSION CONTROL VALVE ASSEMBLY AND DETENT SOLENOID MAINTENANCE (Con't).

9. Connect detent solenoid wire (11) to connector (10). Tighten 2 screws (13) to 10 lb.-ft. (14 N.m).

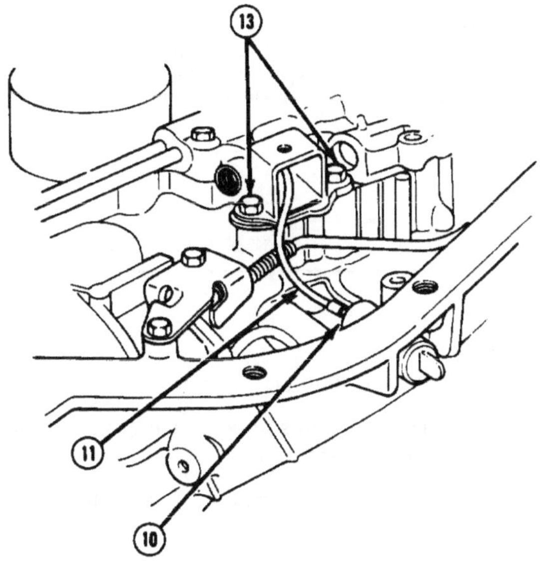

FOLLOW-ON TASKS:

- Install transmission filter and intake pipe. (See TM 9-2320-289-20)

TA50225

5-8. TRANSMISSION PARKING LOCK REPLACEMENT.

This task covers: a. Removal b. Inspection c. Installation

INITIAL SETUP:

Equipment Condition

- Detent roller and spring assembly removed. (See paragraph 5-7)

Tools/Test Equipment

- Torque adapter
- 3/8 in. O.D. rod
- 3/4 in. O.D. rod
- Torque wrench

Materials/Parts

- One cup plug
- One manual shaft seal
- RTV sealant (Item 57, Appendix B)

a. REMOVAL

1. Remove nut and transmission lever from manual shaft (1).
2. Remove pin (6) securing manual shaft (1) to transmission case, If procedure is being performed with transmission in truck, bend pin to remove it.
3. Loosen jamnut (7). Remove inside detent lever (2) from manual shaft (1). Remove jamnut and manual shaft.

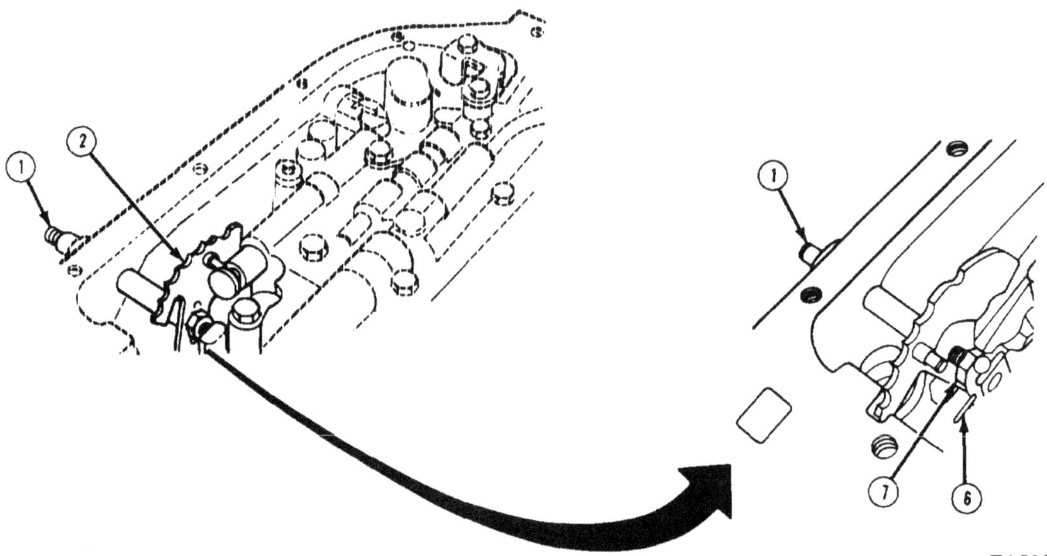

TA50226

5-8. TRANSMISSION PARKING LOCK REPLACEMENT (Con't).

4. Remove parking actuator rod (5) and detent lever (2) from transmission case.
5. Remove manual shaft seal (8) from transmission case and discard,
6. Remove 2 bolts (4) and bracket (3) from transmission case.

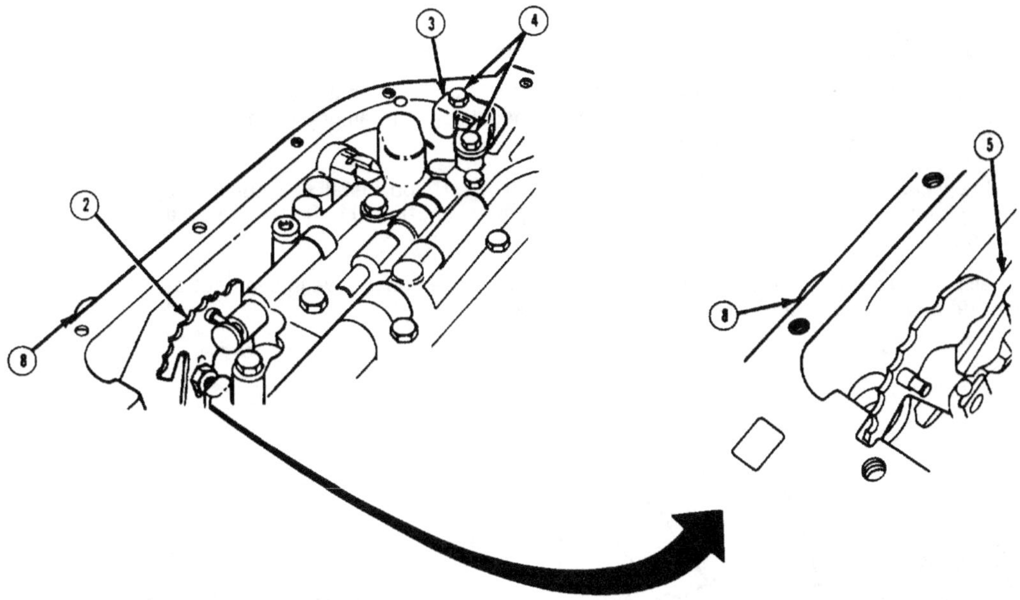

7. Remove parking pawl return spring (10). Remove spring retainer from end of parking pawl shaft (11) inside transmission case. Push parking pawl shaft toward outside of transmission case and remove cup plug (12). Discard cup plug.
8. Remove parking pawl shaft (11) and parking pawl (9).

TA50227

5-8. TRANSMISSION PARKING LOCK REPLACEMENT (Con't).

b. INSPECTION

1. Inspect manual shaft for damage. Replace if damaged,
2. Inspect inside detent lever for cracks or a loose pin. Replace if cracked or if pin is loose.
3. Inspect parking actuator rod for damage. Replace if damaged.
4. Inspect parking pawl for damage. Replace if damaged.

c. INSTALLATION

1. Position parking pawl (9) with tooth toward center of transmission. Install parking pawl shaft (11) through transmission case and parking pawl. Install spring retainer on end of parking pawl shaft on inside of transmission case.
2. Coat new cup plug (12) with RTV sealant. Using a 3/8 in. O.D. rod, install new cup plug by driving it into transmission case, until parking pawl shaft (11) is firmly seated against transmission case rib.
3. Install parking pawl return spring (10) with square end hooked on groove in center of parking pawl (9) and other end hooked on transmission case.

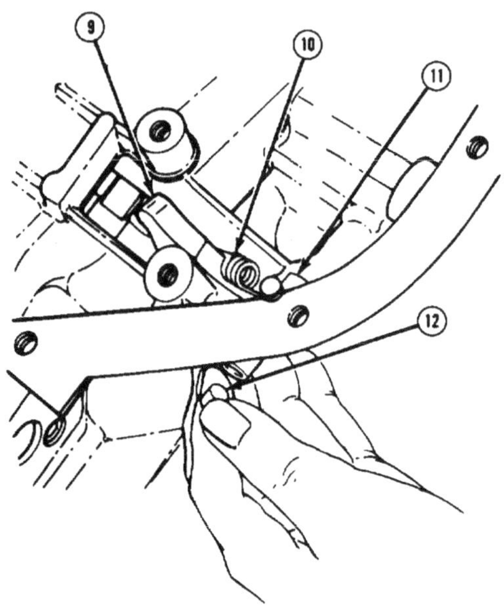

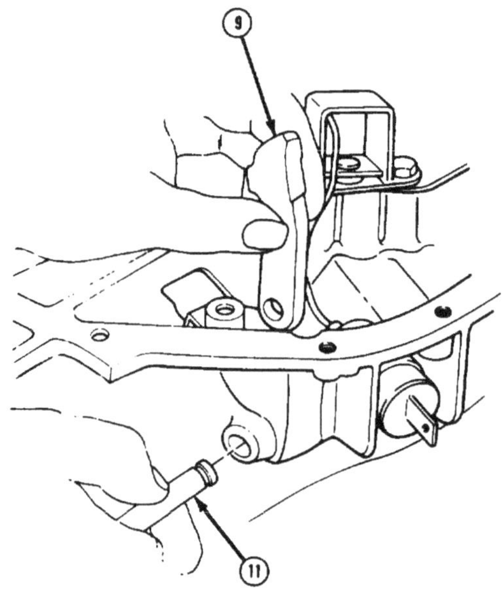

TA50228

5-8. TRANSMISSION PARKING LOCK REPLACEMENT (Con't),

4. Push parking pawl (9) forward, and install bracket (3) with guides over parking pawl. Secure to transmission case with 2 bolts (4). Tighten bolts to 20 lb. -ft. (27 N.m).

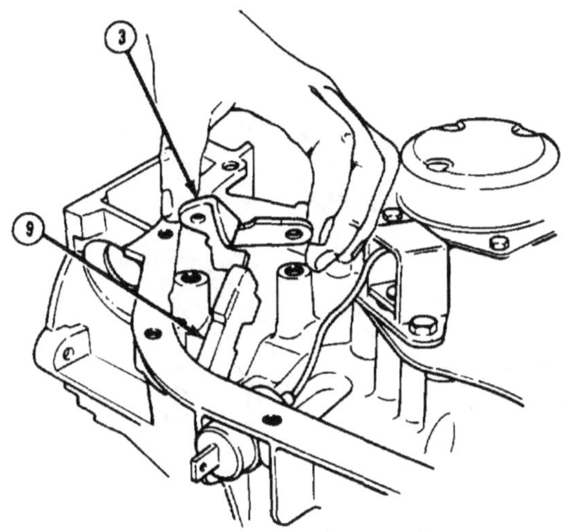

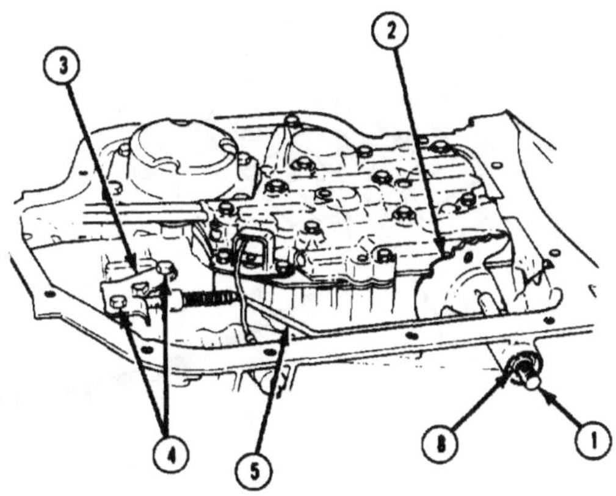

5. Install new manual shaft seal (8) in transmission case and seat it using a 3/4 in, rod.

6. Position parking actuator rod (5) between parking pawl (9) and parking bracket (3), Install inside detent lever (2) to parking actuator rod.

7. Install manual shaft (1) through transmission case and detent lever (2). Install jamnut (7) on end of manual shaft and tighten to 20 lb.-ft. (27 N.m).

TA50229

5-8. TRANSMISSION PARKING LOCK REPLACEMENT (Con't).

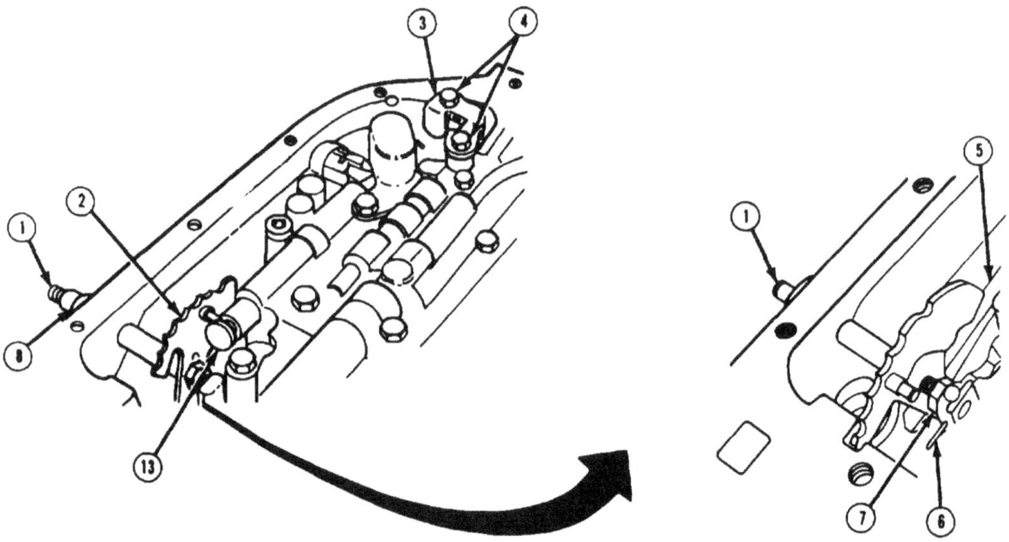

8. Index inside detent lever (2) pin with manual valve (13). Secure manual shaft (1) to transmission case with pin (6), If procedure is being performed with transmission in truck. straighten bent pin after installation.

9. Install transmission lever to manual shaft (1) with nut. Tighten nut. thighten nut to 20 lb.-ft. (27 N.m).

FOLLOW-ON TASKS:

- Install detent roller and spring assembly. (See paragraph 5-7)

TA50230

5-9. REAR SERVO ASSEMBLY MAINTENANCE.

This task covers:

a. Removal
b. Band Apply Pin Selection Check
c. Disassembly
d. Cleaning and Inspection
e. Assembly
f. Installation

INITIAL SETUP:

Equipment Condition

- Control valve assembly removed. (See paragraph 5-7)

Tools/Test Equipment

- Pin selector gage, J-21370
- Gage pin, J-21370-5
- Torque wrench

Materials/Parts

- One servo cover gasket
- One servo piston seal
- Two 5/18-18 x 1 in, screws
- Dry cleaning solvent (Item 23, Appendix B)
- Hydraulic fluid (Item 40, Appendix B)

General Safety Instructions

- Dry cleaning solvent is flammable and must not be used near open flame. Use only in a well-ventilated area.

a. REMOVAL

1. Remove 6 bolts (2) and rear servo cover (1). Remove servo cover gasket and discard.
2. Remove rear servo assembly (3) from bore. Remove rear accumulator spring (5).

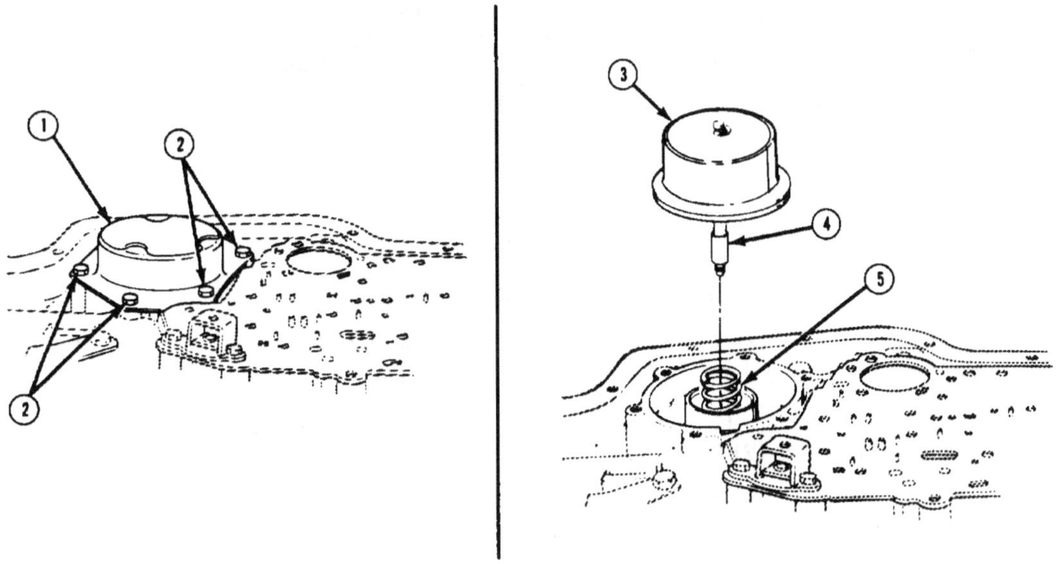

TA50231

5-9. REAR SERVO ASSEMBLY MAINTENANCE (Con't).

b. BAND APPLY PIN SELECTION CHECK

1. Install pin selector gage (6) over rear servo bore with smaller diameter of gage pin (7) in servo pin bore and secure with two 5/16 -18 x 1 in. screws tightened to 216 lb.-in. (24 N.m). Ensure that gage pin is free to move up and down in both pin selector gage and servo pin bore and that steps (8 and 9) of gage pin and machined surface (10) of pin selector gage are facing front of transmission case.

2. Apply 25 lb.-ft. (34 N.m) to hex nut on side of pin selector gage (6).

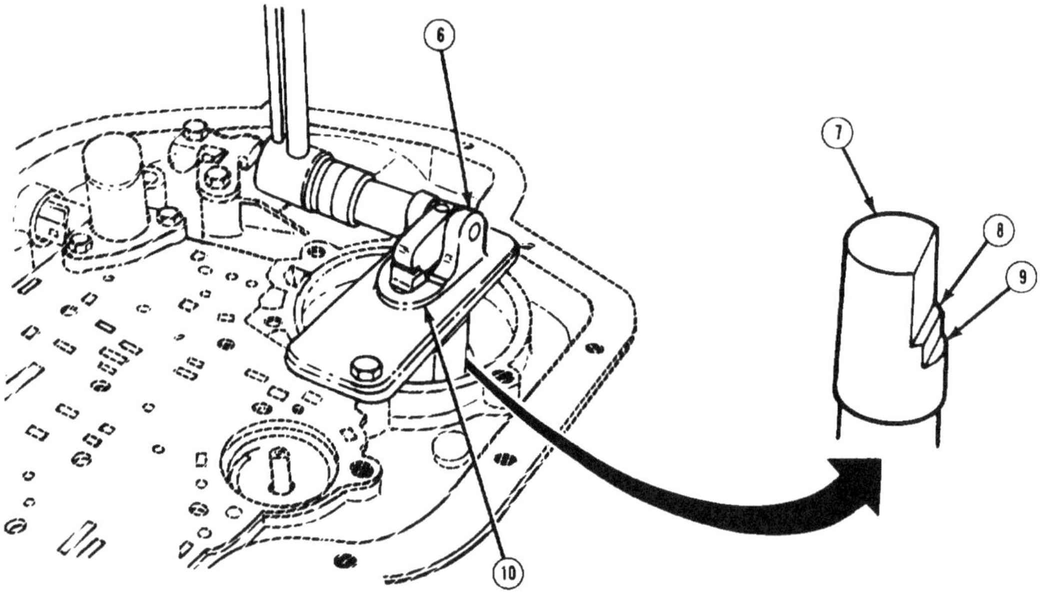

NOTE

Identification rings are located on top of servo pin (4).

3. Note relation of steps (8 and 9) on gage pin (7) with machined surface (10) on top of pin selector gage (6) and determine proper size servo pin (4) as follows:

   (a) If machined surface (10) on top of pin selector gage (6) is even with or above upper step (8) of gage pin (7), long servo pin (3 rings) is required.

   (b) If machined surface (10) on top of pin selector gage (6) is between upper step (8) and lower step (9) of gage pin (7), medium servo pin (2 rings) is required.

   (c) If machined surface (10) on top of pin selector gage (6) is even with or below lower step (9) on gage pin (7), short servo pin (1 ring) is required.

4. If new servo pin (4) is required, make note of size. Remove pin selector gage (6) and gage pin (7) from transmission,

TA50232

5-9. REAR SERVO ASSEMBLY MAINTENANCE (Con't).

c. DISASSEMBLY

1. Remove rear accumulator piston (11) from rear servo piston (13).

2. Compress rear servo spring (12) and servo pin (4) and remove clip (17) from top of rear servo piston (13).

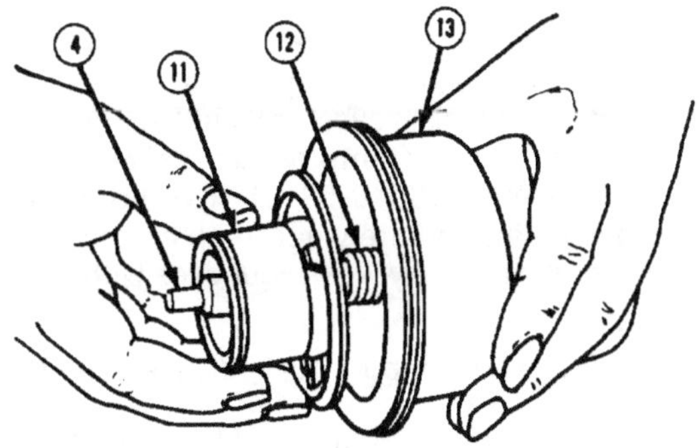

3. Remove servo pin (4), retainer (14), rear servo spring (12), and washer (15).

4. Remove servo piston n seal (16) from rear servo piston (13) and discard.

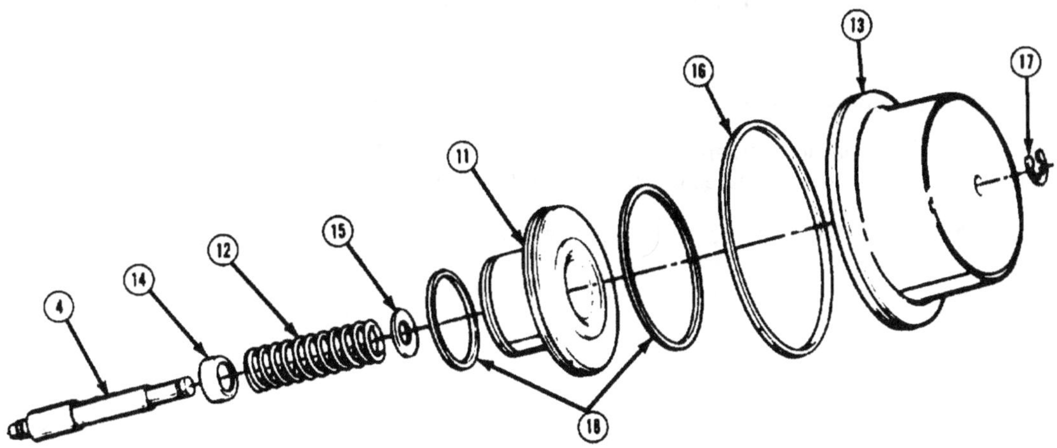

TA50233

5-9. REAR SERVO ASSEMBLY MAINTENANCE (Con't).

d . CLEANING AND INSPECTION

WARNING

Dry cleaning solvent P-D-680 Is toxic and flammable. Always wear protective goggles and gloves and use only in a well-ventilated area, Avoid contact with skin, eyes, and clothes and DO NOT breathe vapors. DO NOT use near open flame or excessive heat. The solvent's flash point is 100°F-138°F (38°C-59°C), If you become dizzy while using cleaning solvent, immediately get fresh air and medical help. If solvent contacts eyes, immediately wash your eyes with water and get medical aid.

1. Clean all removed components in dry cleaning solvent.
2. Inspect rear accumulator piston (11) to ensure that 2 teflon oil seal rings (18) move freely in grooves. If movement is not free, replace oil seal rings.
3. Inspect fit of servo pin in transmission case bore. Replace servo pin if movement is not free.
4. Inspect servo pin, rear accumulator piston, and rear servo piston for damage. Replace if damaged.

e. ASSEMBLY

NOTE

All components should be lightly coated with hydraulic fluid as they are assembled.

1. Install new servo piston seal (16) in groove of rear servo piston (13).
2. Install retainer (14) with large open end down, rear servo spring (1 2), and washer (15) on servo pin (4), using pin size determined in BAND APPLY PIN SELECTION CHECK.
3. Install rear servo piston (13) on servo pin (4). Compress rear servo piston and install clip (17).
4. Install rear accumulator piston (11) in bore of rear servo piston (13).

5-9. REAR SERVO ASSEMBLY MAINTENANCE (Con't).

f. INSTALLATION

1. Install rear accumulator spring (5) and rear servo assembly (3) in bore.

2. Push down on rear servo assembly (3) to ensure that it is properly seated. Install new servo cover gasket and rear servo cover (1) with 6 bolts (2). Tighten bolts to 216 lb.-in. (24 N.m).

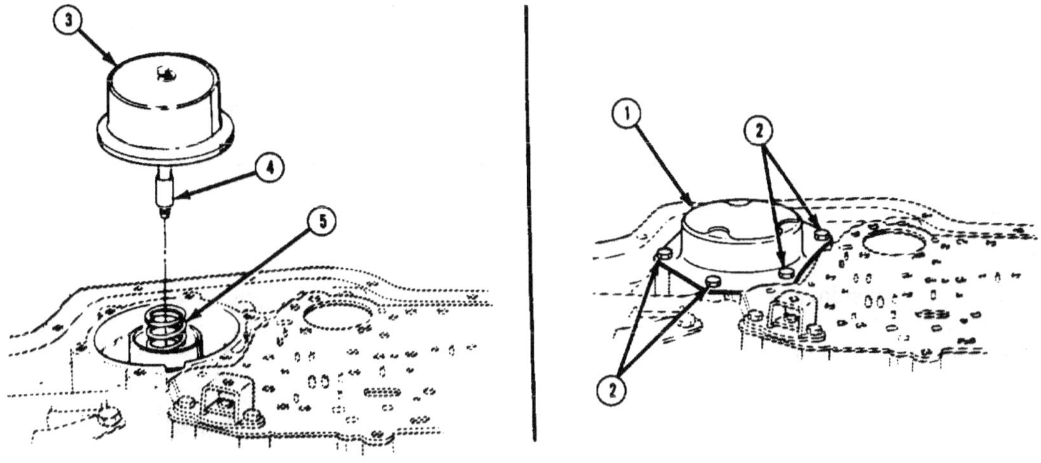

FOLLOW-ON TASKS:

- install control valve assembly. (See paragraph 5-7)

TA50234

5-10. FRONT SERVO ASSEMBLY MAINTENANCE.

This task covers:

a. Removal
b. Disassembly
c. Cleaning and Inspection
d, Assembly
e. Installation

INITIAL SETUP:

Equipment Condition

Ž Control valve spacer plate and check balls removed. (See paragraph 5-7)

Materials/Parts

- One oil seal ring
- Dry cleaning solvent (Item 23, Appendix B)
- Hydraulic fluid (Item 40, Appendix B)

General Safety Instructions

- Dry cleaning solvent is flammable and must not be used near open flame. Use Only in a well-ventilated area.

a. REMOVAL

1. Remove front servo piston assembly (3) from bore in transmission case (1). Remove front servo spring (2).

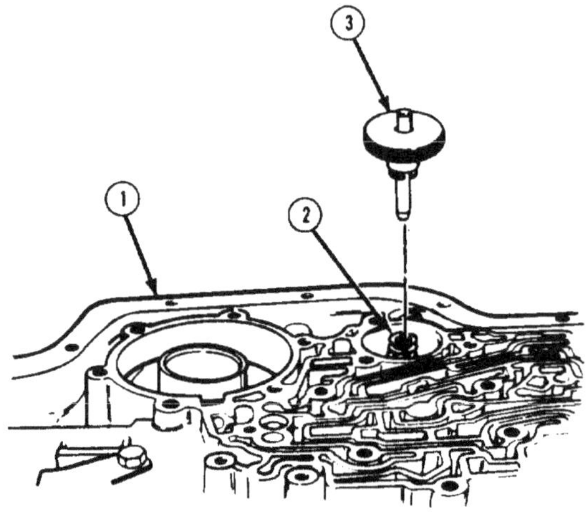

TA50235

5-10. FRONT SERVO ASSEMBLY MAINTENANCE (Con't).

b. DISASSEMBLY

1. Remove front servo piston (7), Separate front servo retainer (4), front servo pin (5), and retainer ring (6).
2. Remove oil seal ring (8) from front servo piston (7) and discard.

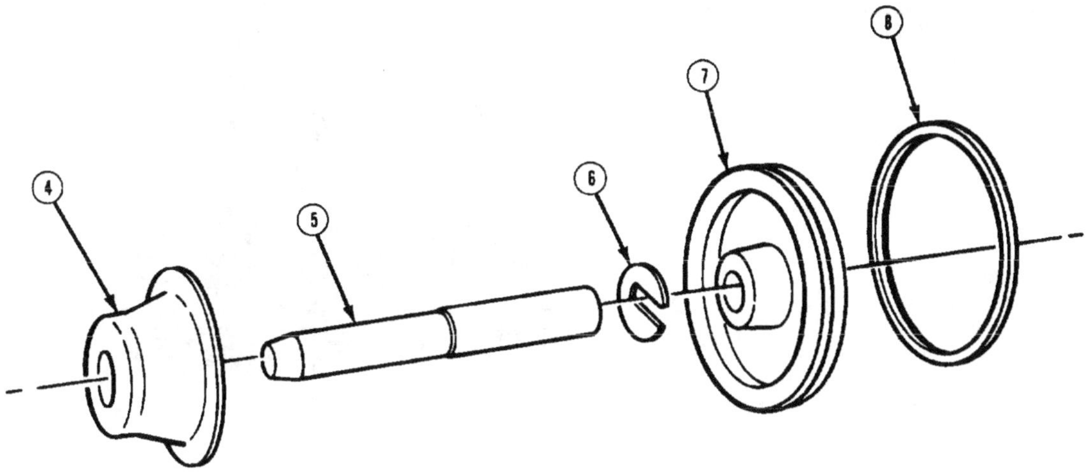

c. CLEANING AND INSPECTION

WARNING

Dry cleaning solvent P-D-680 is toxic and flammable. Always weer protective goggles and gloves and use only in a well-ventilated area. Avoid contact with skin, eyes, and clothes and DO NOT breathe vapors. DO NOT use near open flame or excessive heat, The solvent's flash point is 100°F-138°F (38°C-59°C). If you become dizzy while using cleaning solvent, immediately get fresh air and medical help. If solvent contacts eyes, immediately wash your eyes with water and get medical aid.

1. Clean all removed components with dry cleaning solvent.
2. Inspect all components and replace if damaged.
3. Inspect fit of front servo pin (5) in front servo piston (7). Replace front servo pin and front servo piston if movement is not free.

d. ASSEMBLY

1. Install new oil seal ring (8) in groove of front servo piston (7).

TA50236

5-10. FRONT SERVO ASSEMBLY MAINTENANCE (Con't).

2. Install front servo pin (5) through hole in front servo piston (7) and install retainer ring (6) in front servo pin groove.

e. INSTALLATION

1. Install front servo spring (2) and front servo retainer (4) in bore of transmission case (1). Install front servo piston assembly (3). Compress front servo piston assembly to check for free movement.

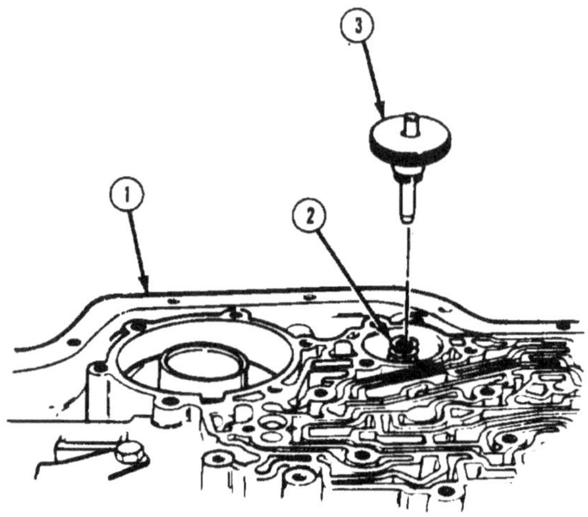

FOLLOW-ON TASKS:

- Install check balls and control valve spacer plate. (See paragraph 5-7)

TA50237

5-11. TRANSMISSION OIL PRESSURE TESTING.

This task covers: a., Transmission Oil Pressure Testing

INITIAL SETUP:

Equipment Condition

- Hydraulic fluid level checked, (See LO 9-2320-289-12)
- Transmission linkage adjustment checked. *(See* paragraph 5-2)
- Transfer case in "2H."
- Front wheels chocked at front and rear of tires.
- Rear of truck raised and supported on jack stands.

Tools/Test Equipment

- STE/ICE

Manual References

- TM 9-4910 -571-12&P
- TM 9-2320-289-20

a. TRANSMISSION OIL PRESSURE TESTING

1. Set engine idle to proper speed. DO NOT disconnect STE/ICE DCA connector once engine idle speed has been set. (See TM 9-2320-289-20)

2. Remove pressure tap plu g (3) located on left side of transmission between downshift switch connector (2) and manual linkage. Instal l appropriate transducer adapter and cable to pressure tap (1) and connect cable to STE/ICE tester,

3. Enter test codes " 10, " "01, " and "50" into STE/ICE tester.

4. Table 5-1 lists 7 tests ("A" through "G") in which to measure transmission oil pressure. Take the pressure measurement in each test and record the result. Record whether the reading for each test range was " LOW, " "NORMAL," or "HIGH. " If oil pressure is not "NORMAL," refer to Table 5-2.

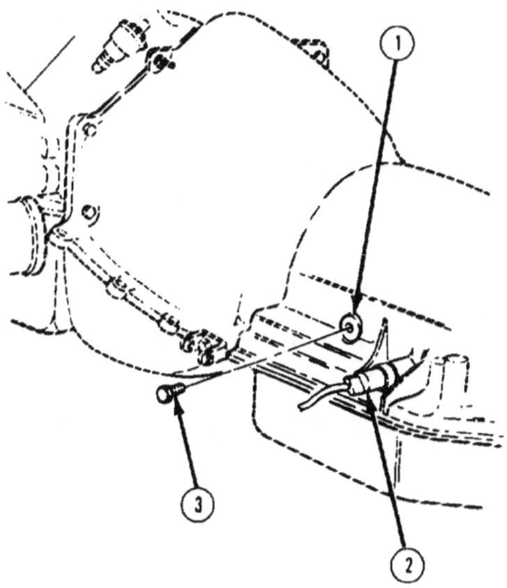

TA50238

5-11. TRANSMISSION OIL PRESSURE TESTING (Con't).

CAUTION

Tests "C" through " F" (see Table 5-1) must be run within a 2 minute span. If longer time is taken, transmission damage may occur.

Table 5-1. Preliminary Checking Procedure

TEST RANGE	NORMAL P s
A., Brakes applied; "N" (Neutral); engine at 1000 rpm,	55-70 psi (379-483 kPa)
B. Engine at idle.	60-85 psi (41 4-586 kPa)
C. Brakes applied; " D" (Drive); engine at 1000 rpm.	60-90 psi (41 4-621 kPa)
D. Brakes applied; "1" or "2"; engine at 1000 rpm.	135-160 psi (931-1 103 kPa)
E. Brakes applied; "R " (Reverse); engine at 1000 rpm.	95-150 psi (655-1034 kPa)
NOTE Downshift switch is located on right side of accelerator pedal plate,	
F. Brakes applied; "D" (Drive); engine at 1000 rpm; downshift switch activated by pulling out.	90-110 psi (621 -758 kPa)
G. Governor check; vacuum line disconnected at modulato; "D " (Drive); brakes released: check pressure at 1000 rpm. Slowly increase to 3000 rpm.	Pressure drop of 10 psi (69 kPa) or more indicates a problem.

5-11. TRANSMISSION OIL PRESSURE TESTING (Con't).

Table 5-2. Transmission Malfunction Related to Oil Pressure

MALFUNCTION	TEST A *Oil Pressure*	TEST B *Oil Pressure*	TEST C *Oil Pressure*	TEST D *Oil Pressure*	TEST E *Oil Pressure*	TEST F *Oil Pressure*	TEST G *Oil Pressure*	POSSIBLE CAUSE OF MALFUNCTION
No "1-2" upshift and/ or delayed upshift.	Normal	Normal	Normal	Normal	Normal	Normal	10 psi Drop or More	Malfunction in control valve assembly.
	Normal	Normal	Normal	Normal	Normal	Normal	Less than 10 psi Drop	Malfunction in governor or governor feed system.
	Normal	High	High	Normal	Normal	Normal	Drop	Malfunction in detent system.
	High	High	High	Normal	High	--	--	Malfunction in modulator or vacuum feed system to modulator.
Slipping - "R" (Reverse)	Normal	Normal	Normal	Normal	Low	Normal	Drop	Oil leak in feed system to direct clutch.
Slipping - "1" (1st gear)	Normal	Low to Normal	Low to Normal	Low to Normal	Normal	Low to Normal	--	Oil leak in feed system to forward clutch.
Downshift with "0" (zero) throttle and no engine braking in "D" (Drive).	Normal	High	Normal	Normal	Normal	--	--	Detent wires switched.
No detent downshifts	Normal	Normal	Normal	Normal	Normal	Low	Normal	Malfunction in detent system.
	NOTE: *A dash (--) indicates that the oil pressure reading has no meaning under the test condition.*							

5. When all readings have been taken, stop engine and remove transducer adapter and cable from pressure tap (1) and STE/ICE tester. Instal l pressure tap plug (3).

6. Disconnect STE/ICE tester from STE/ICE DCA connector.

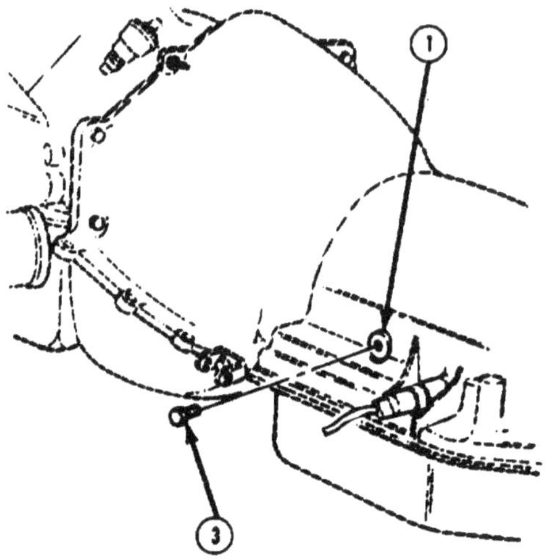

FOLLOW-ON TASKS:

- Remove jack stands and lower truck.
- Remove chocks from front wheels.

5-12. TRANSMISSION AND TORQUE CONVERTER MAINTENANCE.

This task covers: a. Removal b. Cleaning and Inspection c. Installation

INITIAL SETUP:

Equipment *Condition*

- Transmission oil pressure testing completed. (See paragraph 5-11)
- Transmission hydraulic fluid drained and oil pan loosely installed. (See TM 9-2320-289-20)
- Transfer case and adapter removed. (See paragraph 5-21 or 5-22)
- Transmission linkage disconnected, (See paragraph 5-2)
- Transmission oil cooler supply and return pipes disconnected, (See paragraph 5-4)

Materials/Parts

- Dry cleaning solvent (Item 23, Appendix B)

ToolslTest Equipment

- Torque wrench

Personnel Required

- *MOS 63W (2)*

Manual *References*

- *TM 9-2320-289-20*

General Safety Instructions

- Transmission and torque converter are heavy, Use caution and provide support during removal and installation.
- Dry cleaning solvent is flammable and must not be used near open flame. Use only in a well-ventilated area.

CAUTION

Transmission (1) should not be removed with transfer case and adapter installed. Failure to follow this caution may result in damage to transmission case,

1. Remove screw (4) and wiring harness clip (5), and disconnect engine wiring harness connector (3) from downshift electrical connector (2) in transmission (1).

TA50240

5-12. TRANSMISSION AND TORQUE CONVERTER MAINTENANCE (Con't).

2. Disconnect vacuum hose (10) from vacuum modulator (11),

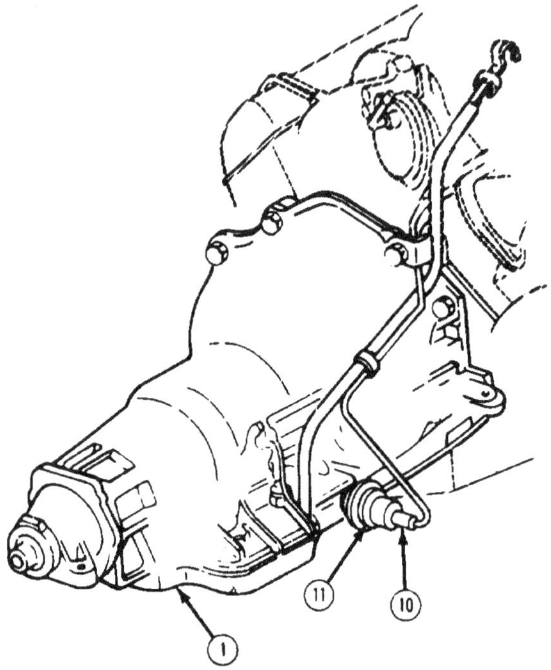

3. Remove 6 bolts (13) and torque converter cover (14).

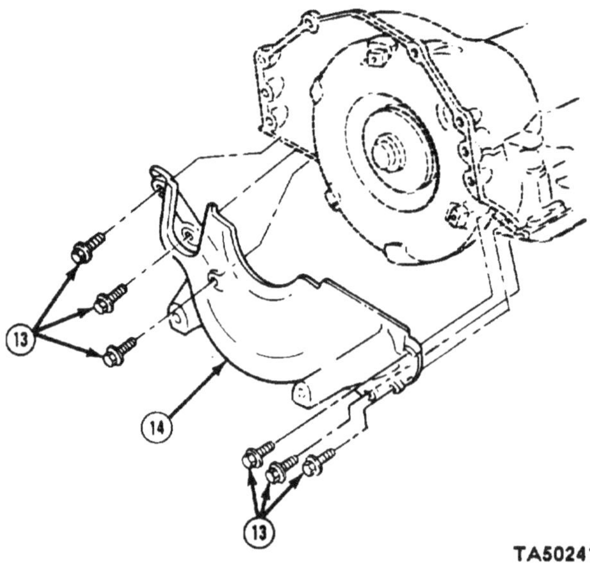

TA50241

5-12. TRANSMISSION AND TORQUE CONVERTER MAINTENANCE (Con't).

4, Mark flywheel (15) and torque converter (16) for proper installation alinement. Rotate flywheel and remove 6 bolts (18) securing flywheel to torque converter.

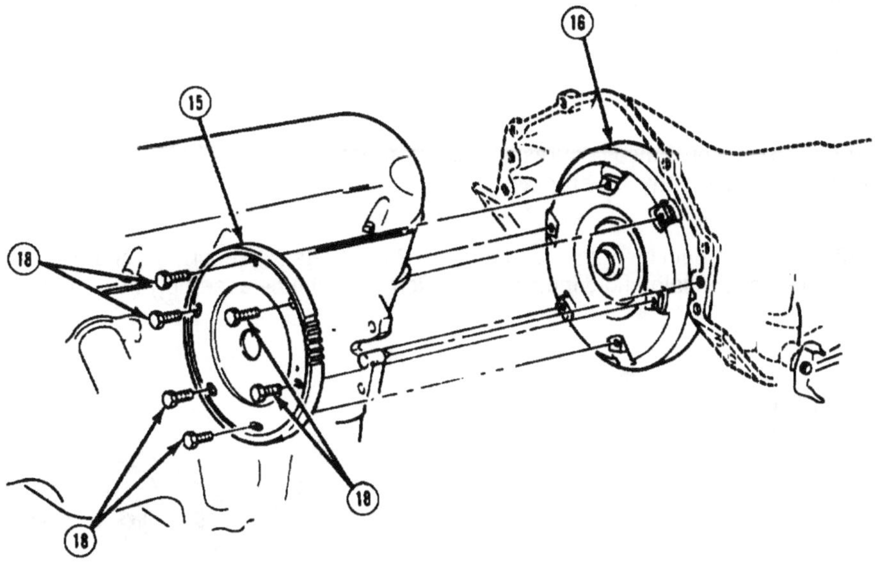

NOTE

Position of spacers (21) should be noted for installation.

5. Remove 4 nuts (23), washers (22), spacers (21), bolts (20), and transmission support (19) from frame.

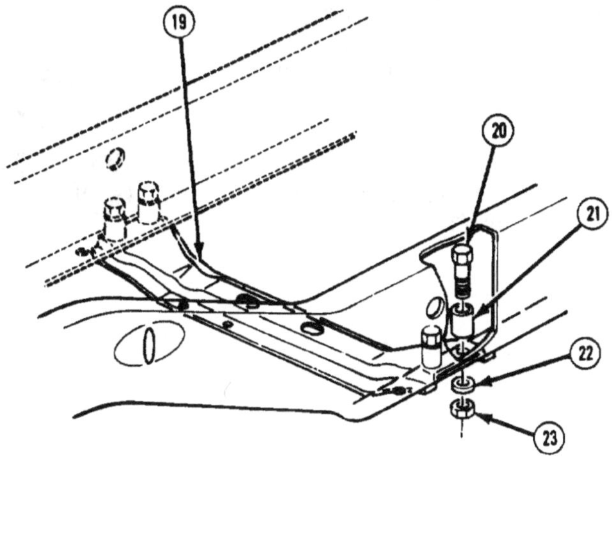

TA50242

5-12. TRANSMISSION AND TORQUE CONVERTER MAINTENANCE (Con't).

CAUTION

Cables, wires, or other components should not be stretched or damaged when lowering transmission (1). Failure to follow this caution may result in equipment damage.

NOTE

Top left bolt (6) in step 6 also secures a wiring harness clip. Top right bolt secures fill tube assembly (12) bracket.

6 . Lower transmission (1) to gain clearance and remove 2 top bolts (6) securing transmission to engine (9). Remove transmission fluid indicator (8). Remove fill tube assembly (12) from transmission.

WARNING

Transmission (1) and torque converter (16) are heavy. Use caution and provide support during removal. Failure to follow this warning may result in serious injury to personnel or equipment damage.

CAUTION

Rear of engine (9) should be supported with suitable jack before removing transmission (1) from engine. Failure to follow this caution may result in damage to transmission or engine.

7. Remove 4 bolts (7) securing transmission (1) to engine (9).

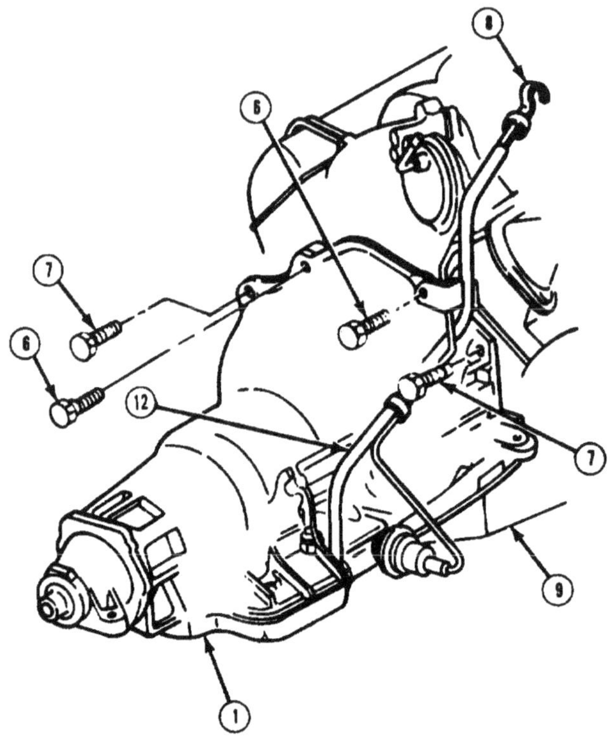

TA60243

5-12. TRANSMISSION AND TORQUE CONVERTER MAINTENANCE (Con't).

CAUTION

Ensure that transmission (1) is tilted backwards to prevent torque converter (16) from falling out of transmission when transmission is removed, Failure to follow this caution may result in damage to torque converter,

8. Slide transmission (1) straight back off 2 locating pins (17). Lower transmission and remove.
9. Remove torque converter (16) by pulling straight out of transmission (1).

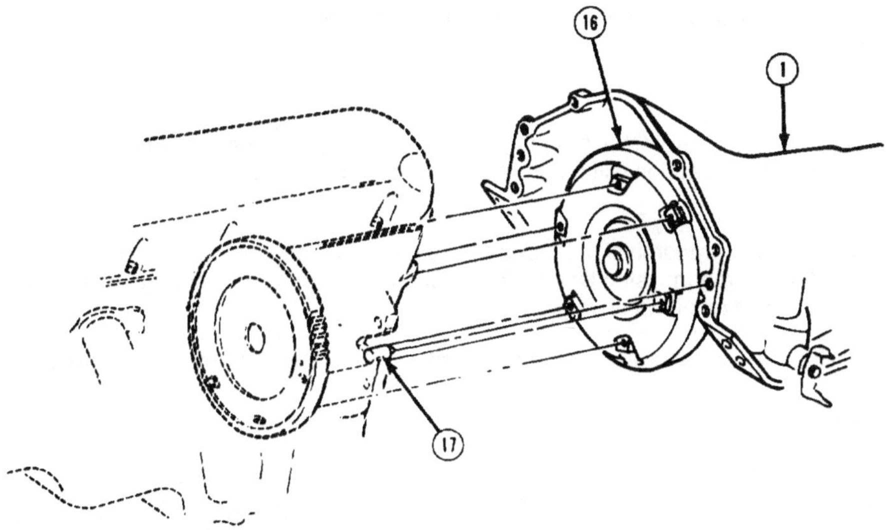

b. CLEANING AND INSPECTION

WARNING

Dry cleaning solvent P-D-680 is toxic and flammable. Always wear protective goggles and gloves and use only in a well-ventilated area. Avoid contact with skin, eyes, and clothes and DO NOT breathe vapors. DO NOT use near open flame or excessive heat. The solvent's flash point is 100°F-138°F (38°C-59°C). if you become dizzy while using cleaning solvent, immediately get fresh air and medical help. If solvent contacts eyes, immediately wash your eyes with water and get medical aid.

1. Clean transmission case and all attaching hardware with dry cleaning solvent.
2. inspect transmission case for signs of leakage and damage. If there is leakage or damage, transmission case must be replaced.
3. inspect all seals and fittings for signs of leakage. See appropriate paragraph if any of these conditions exist.

TA50244

5-12. TRANSMISSION AND TORQUE CONVERTER MAINTENANCE (Con't).

4. Inspect torque converter for damaged weldnuts or bolt hole threads. Inspect torque converter hub for scoring. If weldnuts or bolt hole threads are damaged or hub is scored, replace torque converter.

c. INSTALLATION

WARNING

Transmission (1) and torque converter (16) are heavy. Use caution and provide support during installation. Failure to follow this warning may result in serious injury to personnel or equipment damage.

CAUTION

Ensure that torque converter (16) is fully seated in transmission (1). Failure to follow this caution will result in damage to torque converter and transmission,

1. Place transmission (1) on transmission lift and tilt transmission slightly backwards. Install torque converter (16) in transmission by rotating torque converter to fully engage with turbine shaft, stator shaft, and oil pump drive gear tangs.
2. Raise transmission (1) on transmission lift into position. Slide straight onto 2 locating pins (17).

NOTE

Top right bolt (7) also secures exhaust manifold heat shield.

3. Loosely install 4 bolts (7) securing transmission (1) to engine (9), then fully tighten bolts.
4. Position wiring harness clip at upper left bolt hole and install bolt (6). Install fill tube assembly (12) in transmission. Position fill tube assembly bracket at upper right bolt' hole and install bolt (6). Install transmission fluid indicator (8).
5. Remove support from engine (9).

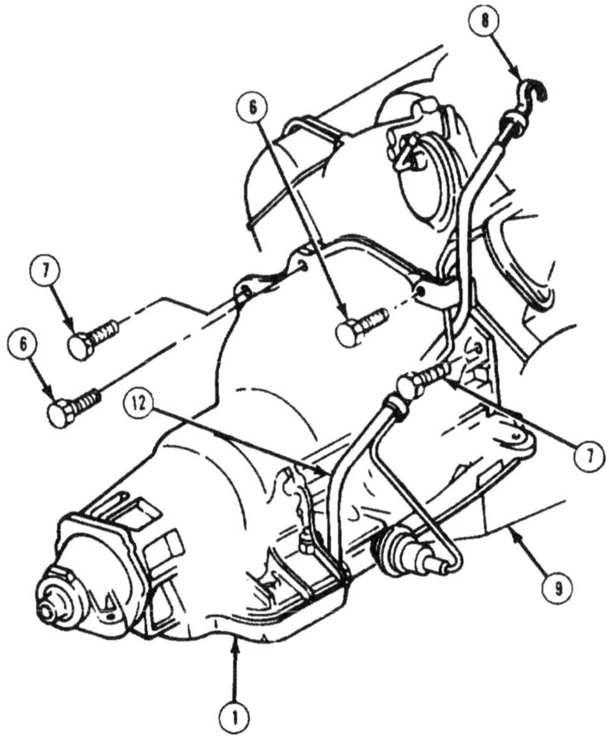

TA50245

5-12. TRANSMISSION AND TORQUE CONVERTER MAINTENANCE (Con'ti).

CAUTION

With transmission installed, ensure that torque converter (16) is flush against flywheel (15) and able to rotate freely by hand. Failure to follow this caution will result in damage to engine and transmission.

6. Aline marks made during removal on flywheel (15) and torque converter (16), Rotate flywheel and loosely install 6 bolts (18), then tighten bolts to 50 lb.-ft. (68 N.m).

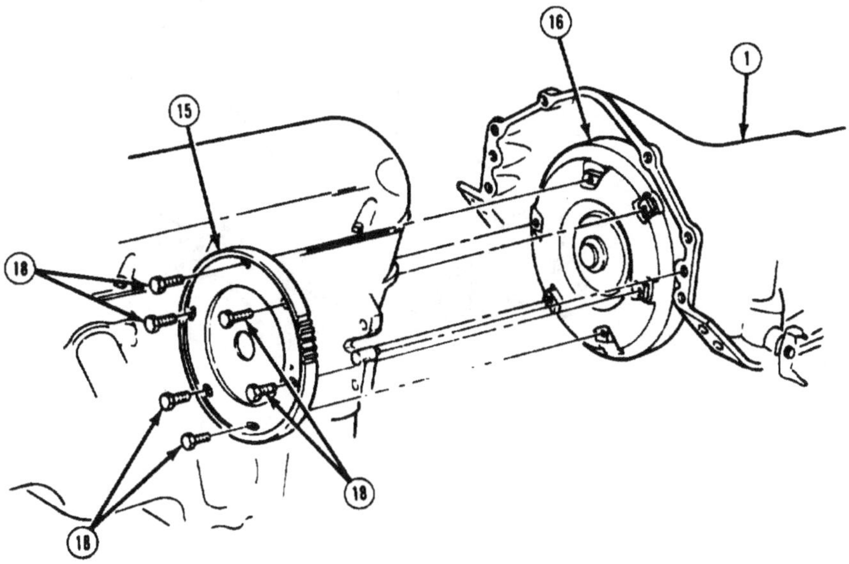

7. Hook torque converter cover (14) under lip of engine oil pan and install to transmission (1) with 6 bolts (13).

5-12. TRANSMISSION AND TORQUE CONVERTER MAINTENANCE (Con't).

8. Connect vacuum hose (10) to vacuum modulatorr (11)

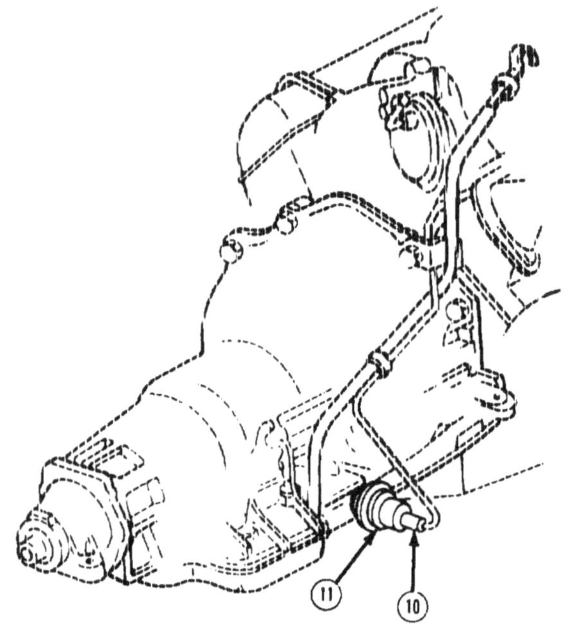

9. Install wiring harness clip (5) with screw (4). Connect engine wiring harness connector (3) to downshift electrical connector (2) in transmission (1).

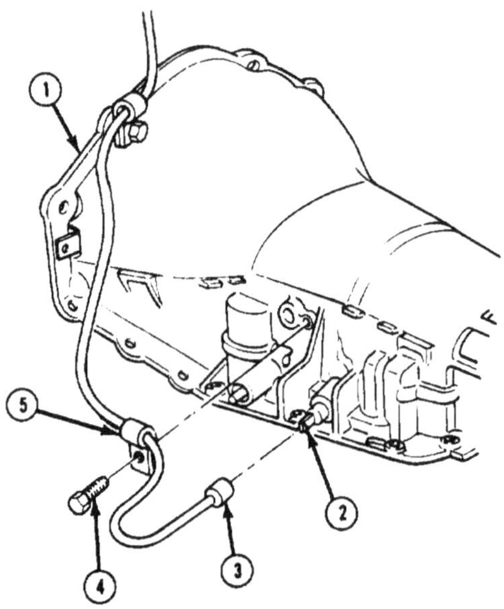

TA50247

5-12. TRANSMISSION AND TORQUE CONVERTER MAINTENANCE (Con't).

10. Install transmission support (19) with 4 bolts (20), spacers (21), washers (22), and nuts (23). Tighten nuts to 70 lb.-ft, (95 N.m).

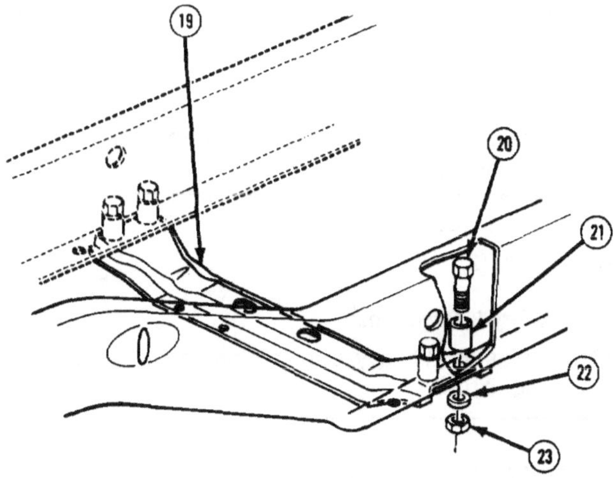

11. If installing a new or overhauled transmission (1) or torque converter (16), flush oil cooler in radiator, and transmission oil cooler supply and return pipes. (See paragraph 5-4)

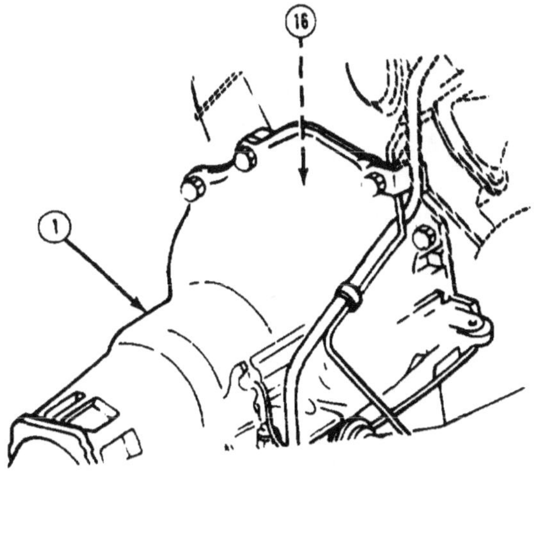

FOLLOW-ON TASKS:

- Connect transmission oil cooler supply and return pipes. (See paragraph 5-4)
- Connect transmission linkage, (See paragraph 5-2)
- Install transfer case and adapter, (See paragraph 5-21 or 5-22)
- Fill transmission with hydraulic fluid. (See TM 9-2320-289-20)

TA50248

5-13. INSTALLING AND REMOVING TRANSMISSION FROM HOLDING FIXTURE.

This task covers: a. Installation b. Removal

INITIAL SETUP:

Equipment Condition

- Transmission and torque converter removed, (See paragraph 5-1 2)

Personnel Required

- *MOS 63W (2)*

Tools/Test Equipment

- Holding fixture, J-8763-02
- Holding fixture base, J-3289-20

Manual References

- TM 9-2320-289-20

General Safety Instructions

- Transmission is heavy. Use caution and provide support during installation and removal from holding fixture.

a. INSTALLATION

WARNING

Transmission is heavy. Use caution and provide support when installing it to holding fixture. Failure to follow this warning may result in serious injury to personnel or damage to transmission.

5-13. INSTALLING AND REMOVING TRANSMISSION FROM HOLDING FIXTURE (Con't).

NOTE

A suitable container should be used to catch hydraulic fluid that may drain from transmission (2).

1. Install holding fixture (1) to transmission (2) so that vacuum modulator will be on side of holding fixture nearest the bench.
2. Install holding fixture (1) and transmission (2) to holding fixture base (5).
3. Install pin (4). Tighten screw (3). DO NOT over tighten screw to allow holding fixture (1) to rotate when pin is removed.

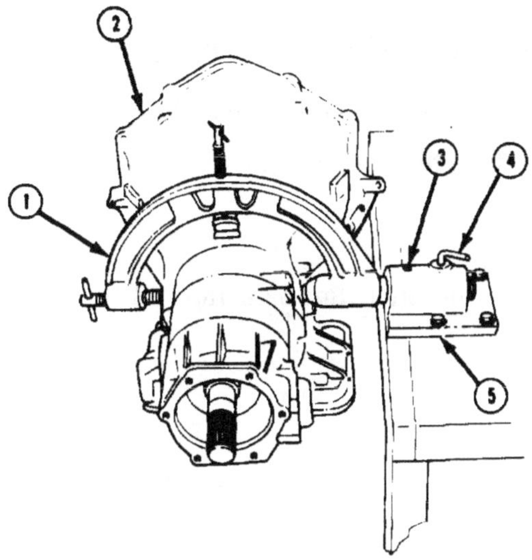

TA50249

5-13. INSTALLING AND REMOVING TRANSMISSION FROM HOLDING FIXTURE (Con't).

b. REMOVAL

WARNING

Transmission (2) is heavy. Use caution and provide support when removing it from holding fixture (1), Failure to follow this warning may result in serious injury to personnel or damage to transmission.

NOTE

Ensure that transmission oil pan bolts are fully tightened. (See TM 9-2320-289-20)

1. Remove pin (4). Loosen screw (3) until holding fixture (1) slides in holding fixture base (5).
2. Remove transmission (2) and holding fixture (1) from holding fixture base (5).
3. Remove holding fixture (1) from transmission (2).

FOLLOW-ON TASKS:

- Install transmission and torque converter, (See paragraph 5-12)

5-14. CHECKING FRONT UNIT END PLAY.

This task covers: a. Checking Front Unit End Play

INITIAL SETUP:

Equipment Condition

- Transmission mounted in holding fixture. (See paragraph 5-1 3)

Tools/Test Equipment

- Dial indicator
- Micrometer
- Slide hammer bolt

Materials/Parts

- One rubber-coated washer
- One selective washer

Personnel Required

- MOS 63W (2)

a. CHECKING FRONT UNIT END PLAY

1. Place transmission (7) in a horizontal position.
2. Remove bolt (5) and rubber-coated washer (4) at either the 10 o'clock or 5 o'clock position of transmission oil pump (2). Discard rubber-coated washer.
3. Install 1 slide hammer bolt (1) in hole where bolt (5) was removed. Mount dial indicator (8) on slide hammer bolt. Index dial indicator to register with flat surface on end of turbine shaft (6).

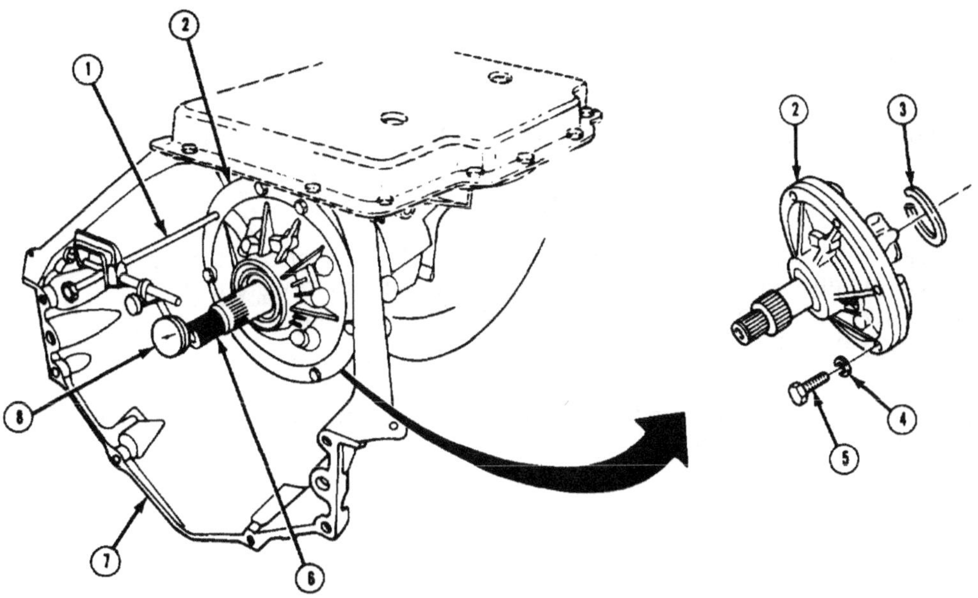

TA50250

5-14. CHECKING FRONT UNIT END PLAY (Con't).

4. Push turbine shaft (6) rearward to its stop while an assistant holds output shaft (9) forward.

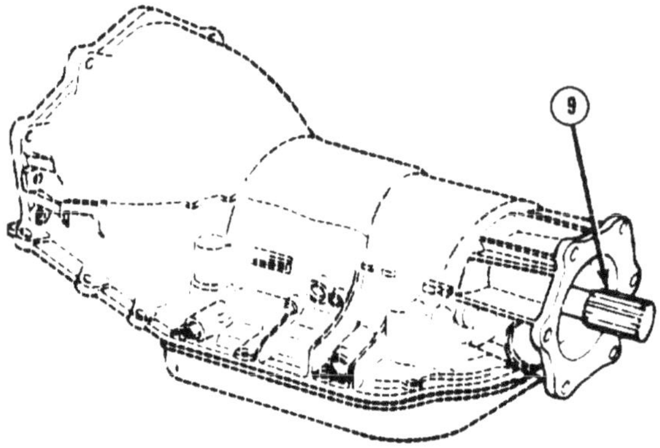

5. Set dial indicator (8) to "0." Pull turbine shaft (6) forward. Front unit end play should be 0.003-0.024 in. (0.076-0.610 mm). Remove dial indicator and slide hammer bolt (1).

6. If front unit end play is within specification, install new rubber-coated washer (4) and bolt (5).

NOTE

Perform steps 7-13 if front unit end play is not within specifications.

7. Remove transmission oil pump (2). (See paragraph 5-1 5)

NOTE

The selective washer (3) controlling front unit end play is color-coded and is located between transmission oil pump (2) cover and forward clutch housing. An oil-soaked washer tends to discolor. Its thickness can be accurately determined only by measuring.

8. Remove selective washer (3) from transmission oil pump (2) cover delivery sleeve and measure it to determine its thickness.

TA50251

5-14. CHECKING FRONT UNIT END PLAY (Con't).

NOTE

- If front unit end play is less than 0,003 in, (0,076 mm), perform step 9.
- If front unit end play is more than 0.024 in. (0.610 mm), perform step 10.

9 . Select next smaller or thinner size selective washer (3) using Table 5-3.

10. Select next larger or thicker size selective washer (3) using Table 5-3.

Table 5-3. Front Unit End Play Selective Washers

THICKNESS	COLOR
0.060-0.064 in. (1 ,524-1.626 mm)	Yellow
0.071-0.075 in. (1 .803 -1.905 mm)	Blue
0.082-0.086 in. (2.083-2. 184 mm),	Red
0.093-0.097 in. (2.362-2.464 mm)	Brown
0.104-0.108 in. (2.641-2.743 mm)	Green
0.115-0.119 in. (2.921-3.023 mm)	Black
0.126-0.130 in. (3.200-3.302 mm)	Purple

11. Lightly coat selective washer (3) with petrolatum and install on transmission oil pump (2) cover delivery sleeve.

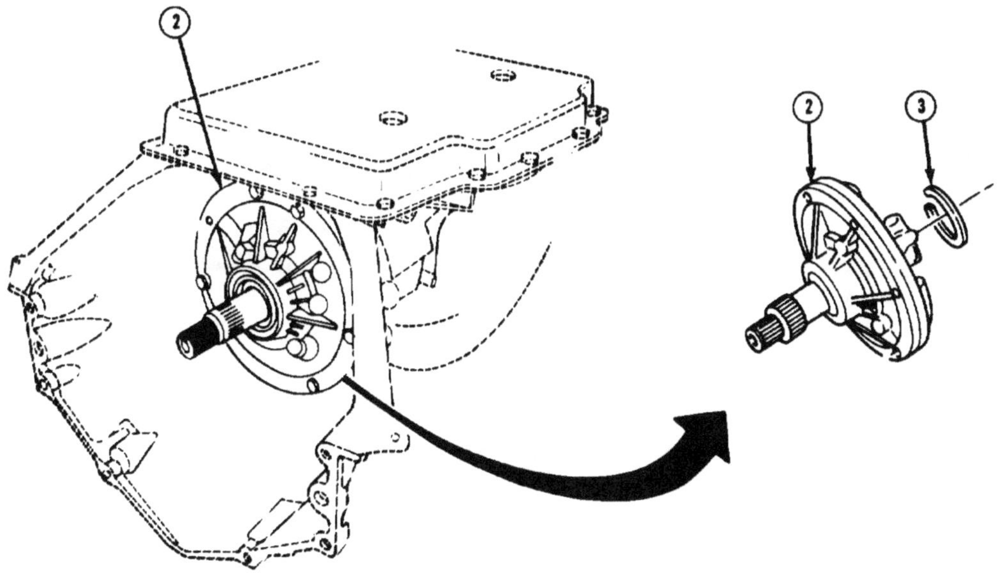

TA50252

5-14. CHECKING FRONT UNIT END PLAY (Con't).

12. Install transmission oil pump (2). (See paragraph 5-1 5)
13. Recheck front unit end play.

FOLLOW-ON TASKS:

- Remove transmission from holding fixture. (See paragraph 5-13)

5-15. TRANSMISSION OIL PUMP MAINTENANCE.

This task covers:

a. Removal
b. Disassembly
c. Cleaning and Inspection
d. Assembly
e. Installation

INITIAL SETUP:

Equipment Condition

- Transmission mounted in holding fixture, (See paragraph 5-1 3)

ToolslTest Equipment

- Annular bearing inserter, J-21465-1
- Torque wrench
- Two slide hammers

Materials/Parts

- One oil pump seal
- One "O" ring seal
- One pump cover gasket
- Two oil seal rings
- Six rubber-coated washers
- Hydraulic fluid (Item 40, Appendix B)
- Petrolatum (Item 50, Appendix B)
- RTV sealant (Item 57, Appendix B)

General Safety Instructions

- Dry cleaning solvent is flammable and must not be used near open flame. Use only in a well-ventilated area,
- Wear goggles and use caution during removal and installation of regulator boost valve spring.
- Compressed air used for cleaning purposes should never exceed 30 psi (207 kPa).

a. REMOVAL 1

NOTE

If removing oil pump seal (2), perform step 1.

1. Remove oil pump seal (2) from oil pump assembly (3) and discard.
2. Remove 6 bolts (1) and rubber-coated washers. Discard rubber-coated washers. Install 2 slide hammers (4) in the 10 o'clock and 5 o'clock positions.

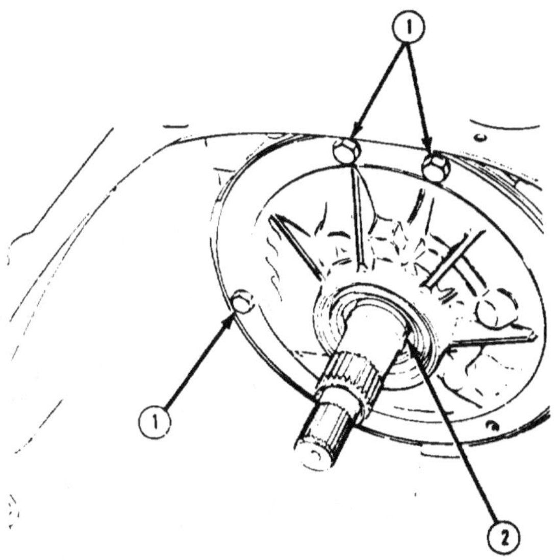

TA60253

5-15. TRANSMISSION OIL PUMP MAINTENANCE (Con't).

CAUTION

Care must be taken to drive outward IN UNISON on both slide hammers (4). Failure to follow this caution may cause oil pump assembly (3) to become cocked in transmission case.

3. Drive outward on both slide hammers (4) and remove oil pump assembly (3) from transmission case. Remove slide hammers.

4. Remove " O" ring seal (15) from oil pump assembly (3). Remove oil pump cover gasket (8). Discard "O" ring seal and oil pump cover gasket.

5. Remove 2 oil seal rings (10) from sleeve and discard, Remove selective washer (14).

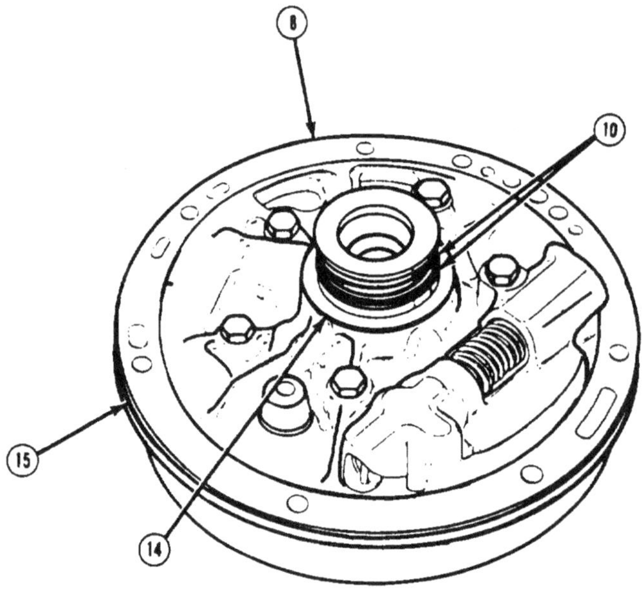

TA50254

5-15. TRANSMISSION OIL PUMP MAINTENANCE (Con't).

b. DISASSEMBLY

1, Place oil pump assembly (3) on wooden blocks with stator shaft (23) pointing face down.

WARNING

Wear goggles and use caution during removal of regulator boost valve spring (19). Regulator boost valve spring is tightly compressed and could fly out of bore. Failure to follow this warning may result *in* serious eye Injury.

2. Push bushing (17) against regulator boost valve spring (19) and remove retainer ring (1 6). Remove bushing and regulator boost valve (18). Remove regulator boost valve spring,

3. Remove pressure regulator valve (22), spring retainer washer (20), and spacer (21).

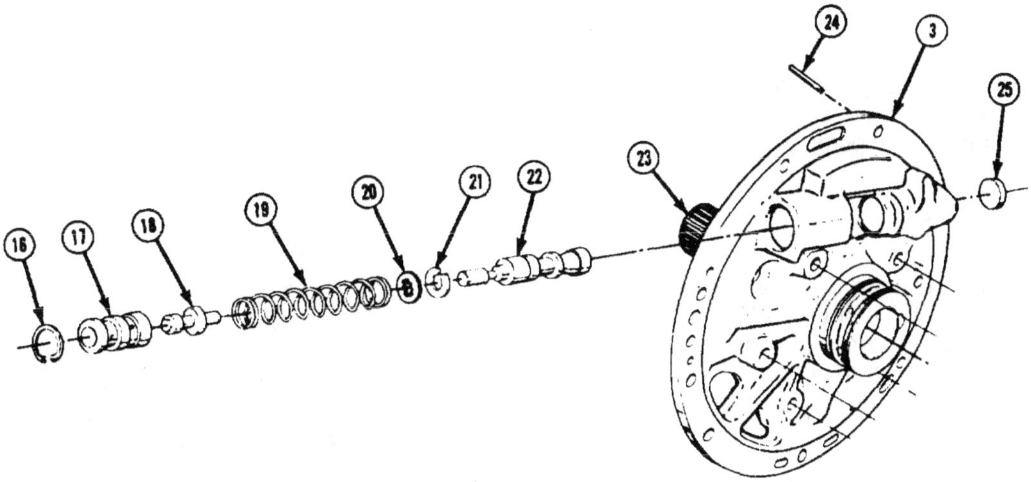

TA50255

5-15. TRANSMISSION OIL PUMP MAINTENANCE (Con't).

NOTE

Bolts (7, 9, and11) are not the same length. Bolt length and position should be marked for assembly.

4. Remove bolt (11), 2 bolts (9), and 2 bolts (7), and separate cover assembly (12) from pump body (13) .

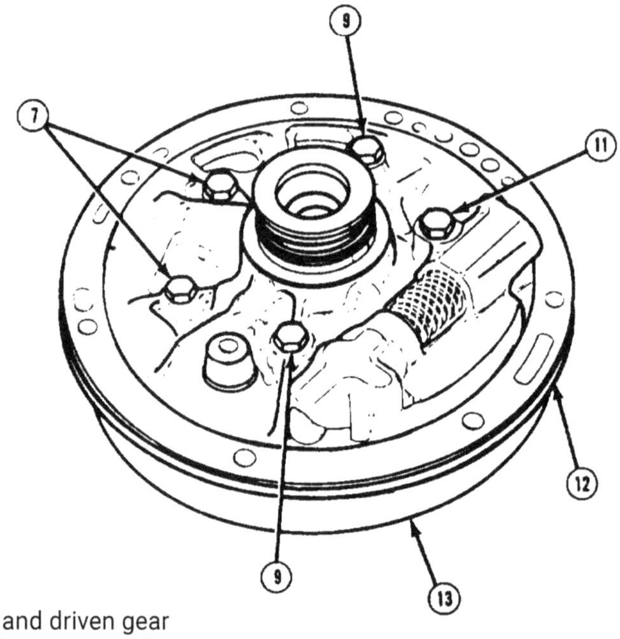

5. Mark drive gear (26) and driven gear (27) for assembly in same position, and remove from pump body (13).

CAUTION

Care should be taken when removing pin (24) and plug (25) from regulator bore. Failure to follow this caution may result in damage to regulator bore.

6. Remove pin (24) and plug (25) from end of regulator bore.

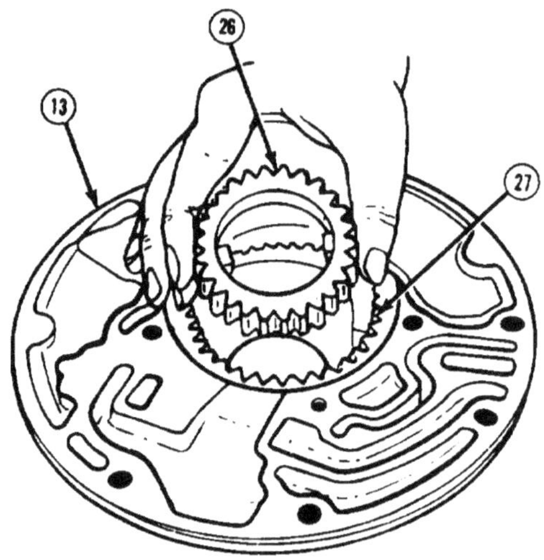

TA50256

5-15. TRANSMISSION OIL PUMP MAINTENANCE (Con't).

c. CLEANING AND INSPECTION

WARNING

- Dry cleaning solvent P-D-680 is toxic and flammable, Always wear protective goggles and gloves and use only in a well-ventilated area, Avoid contact with skin, eyes, and clothes and DO NOT breathe vapors, DO NOT use near open flame or excessive heat. The solvent's flash point is 100°F-138°F (38°C-59°C). If you become dizzy while using cleaning solvent, immediately get fresh air and medical help. If solvent contacts eyes, immediately wash your eyes with water and get medical aid.
- Compressed air used for cleaning purposes should never exceed 30 psi (207 kPa). Use only effective chip guarding and personnel protective equipment (goggles/shield, gloves, etc.). Failure to follow this warning may result in injury to personnel.

1. Clean pump cover and pump body and all removed components with dry cleaning solvent. Remove all traces of RTV sealant from mating surfaces of oil pump seal and oil pump body. Use compressed air to dry and open up any blocked oil passages.
2. Inspect pump body gear pockets and crescent for damage. Inspect oil passages to ensure that they are open. Inspect for damaged bolt hole threads. Inspect face of pump body for scores or nicks and for overall flatness. Inspect drive and driven gear and pump gear face for damage. Replace oil pump assembly if any of these conditions exist.
3. Place drive and driven gears (26 and 27) in pump body (13) and check for pump body face-to-gear face clearance. Place a straightedge across pump and gear face surfaces. Use a feeler gage and measure clearance between straightedge and gears. Clearance should be between 0.0008 in. and 0.0035 in. (0,02 mm and 0.09 mm). If gear clearance is not within this tolerance, replace oil pump assembly.
4. Inspect operation and meshing of pump gears with gears installed as marked, Inspect for binding or roughness. If gears bind or operate roughly, replace oil pump assembly.

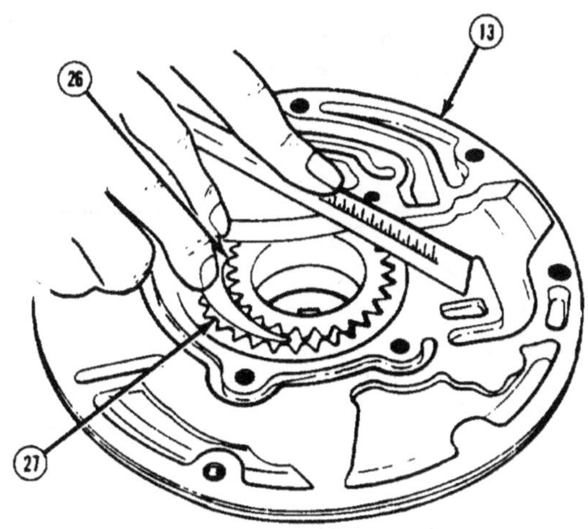

TA50257

5-15. TRANSMISSION OIL PUMP MAINTENANCE (Con't).

NOTE

If removing bushings, their location should be measured to ensure Installation in same location.

5. Inspect pump body bushing, and front and rear stator shaft bushings for damage. Remove and replace if damaged. Inspect stator shaft for damaged splines. Replace oil pump assembly if splines are damaged.

6. Inspect regulator boost valve assembly and pressure regulator valve for free operation and damage. Ensure that there is no dirt in pressure regulator bore. Replace components if there is binding or damage.

7. Inspect cover assembly face for overall flatness. Inspect to see that all oil passages are open and not interconnected through porosity. Inspect to see that breather hole (28) is open and has a 1/8 in. diameter. Use compressed air to open all oil passages and breather hole, Replace oil pump assembly if cover assembly face is not flat, if there is evidence of porosity, or if breather hole is not open to specification.

8. Inspect 2 oil seal ring grooves for damage. Replace oil pump assembly if damaged. Inspect selective washer and replace if damaged.

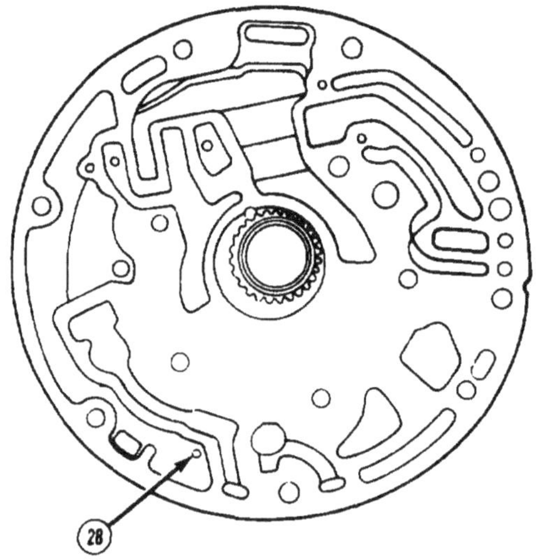

d. ASSEMBLY

NOTE

All components should be lightly coated with hydraulic fluid as they are assembled.

1. Install driven gear (27) and drive gear (26) in pump body (13) with alinement marks facing up and mated.

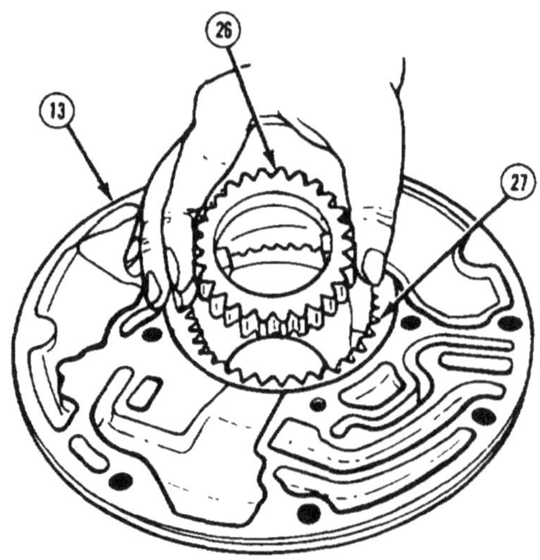

TA50258

5-15. TRANSMISSION OIL PUMP MAINTENANCE (Con't).

WARNING

Wear goggles and use caution during installation of regulator boost valve spring (19). Regulator boost valve spring is tightly compressed and could fly out of bore. Failure to follow this warning may result in serious eye injury.

NOTE

Ensure that tang side of spring retainer washer (20) is facing regulator boost valve spring (19).

2. Install spacer (21), spring retainer washer (20), and regulator boost valve spring (19) into pressure regulator bore. Install regulator boost valve (18) into bushing (17), stem end out, and install both into pressure regulator bore by pushing bushing against regulator boost valve spring. Install retainer ring (16).

3. Install pressure regulator valve (22) from plug (25) end of pressure regulator bore, stem end first. Install plug and pin (24).

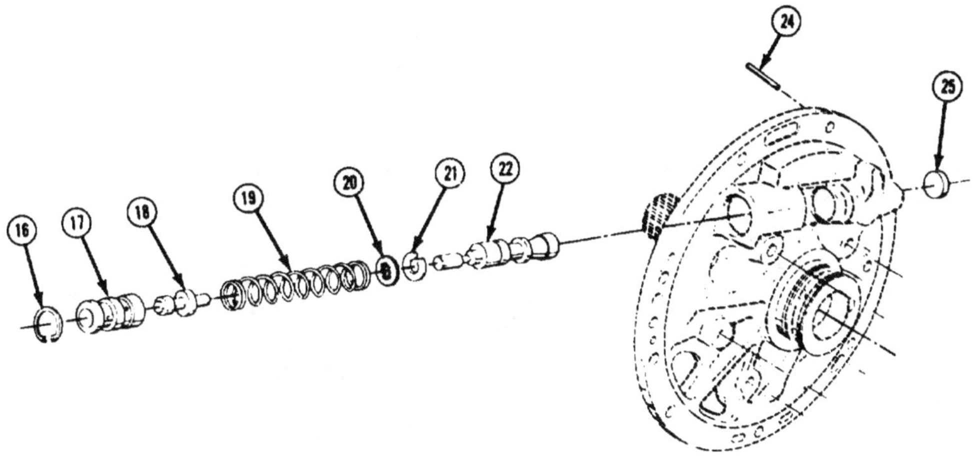

TA50259

5-15. TRANSMISSION OIL PUMP MAINTENANCE (Con't).

4. Lightly coat selective washer (14) with petrolatum and install over pump cover delivery sleeve. Install 2 new oil seal rings (10) on sleeve, ensuring that lap joints are fully seated in grooves.

5. Place pump body (13) on wooden blocks with oil passages facing up.

6. Lubricate pump gears with hydraulic fluid and install pump cover (12) to pump body (13) with bolt (11), 2 bolts (9), and 2 bolts (7). Leave bolts one turn loose.

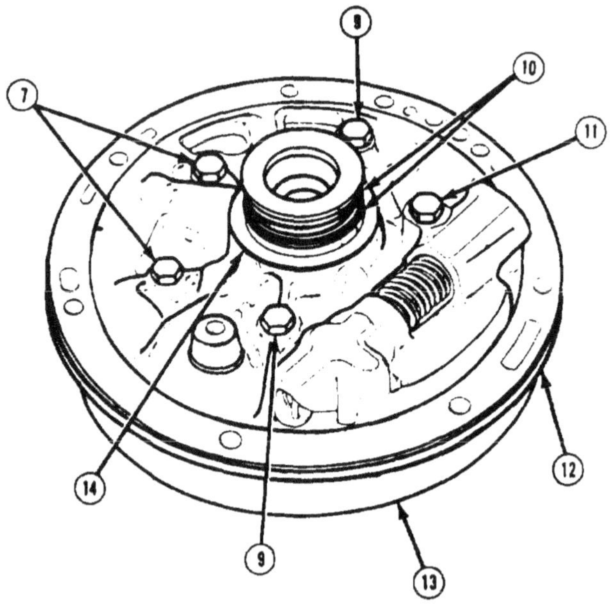

NOTE

Perform steps 7-9 to ensure proper alinement of oil pump body (13) and pump cover (12),

7. Remove forward clutch assembly. (See paragraph 5-1 6)

8. Place oil pump assembly (3) backwards into pump bore of transmission case and secure with 2 bolts (1). Tighten bolt (11), 2 bolts (9), and 2 bolts (7) to 216 lb.-in. (24 Nom). Remove 2 bolts (1) and oil pump assembly from transmission case.

9. install forward clutch assembly. (see paragraph 5-16)

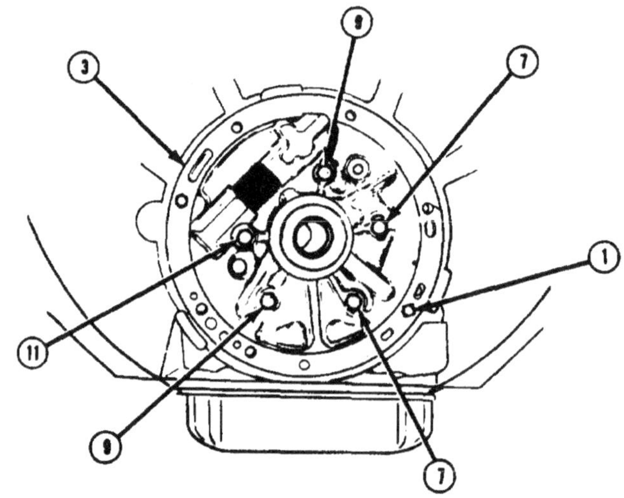

TA50260

5-15. TRANSMISSION OIL PUMP MAINTENANCE (Con't).

e . INSTALLATION

NOTE

If installing oil pump seal (2), perform step 7,

1. Lubricate new "O" ring seal (15) with hydraulic fluid and install on oil pump assembly (3).
2. Lubricate turbine shaft (5) journals with hydraulic fluid, Lubricate 2 oil seal rings (10) on oil pump delivery sleeve with petrolatum,
3. Install 2 slide hammers (4) in threaded holes at the 5 o'clock and 10 o'clock positions of oil pump assembly (3).

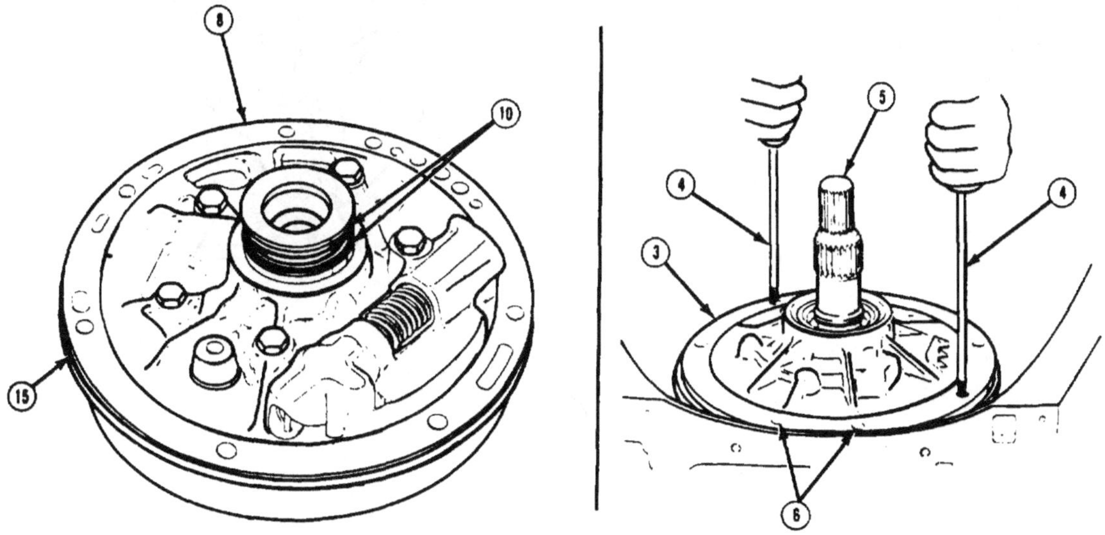

4. Lightly coat new pump cover gasket (8) with petrolatum and install on mounting face of transmission case.

NOTE

Ensure that 2 bolt holes (6) that are closest together are positioned closest to transmission oil pan.

5. Position oil pump assembly (3) in transmission case. Remove 2 slide hammers (4).

TA50261

5-15. TRANSMISSION OIL PUMP MAINTENANCE (Con't).

NOTE

If turbine shaft (5) cannot be rotated as oil pump assembly (3) is being pulled into place, forward or direct clutch housings have not been installed properly to index with all clutch plates. This condition must be corrected before oil pump assembly is fully seated. (See paragraph 5-16 or 5-17)

6. Loosely install 6 new rubber-coated washers and bolts (1), Tighten all bolts progressively to 20 lb. -ft. (27 N.m).

7. Apply a thin coat of RTV sealant to outside of oil pump seal (2) body. Install new oil pump seal.

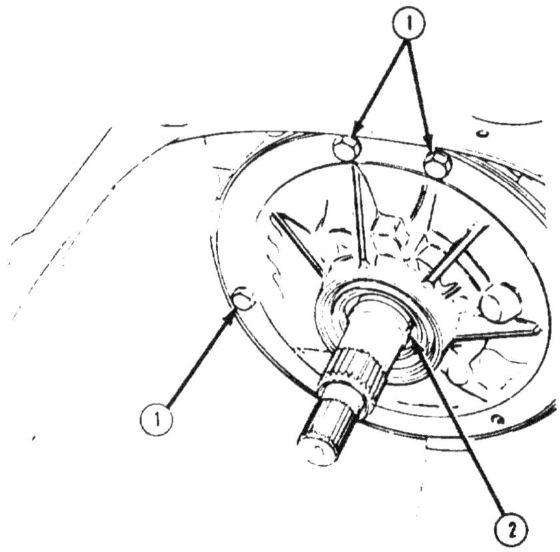

FOLLOW-ON TASKS:

- Check front unit end play. (See paragraph 5-14)

TA50262

5-16. TRANSMISSION FORWARD CLUTCH ASSEMBLY MAINTENANCE.

This task covers:

a. Removal
b. Disassembly
c. Cleaning and Inspection
d. Assembly
e. Installation

INITIAL SETUP:

Equipment Condition

- Transmission oil pump removed. (See paragraph 5-1 5)

Tools/Test Equipment

- Clutch rebuilder, J-23327
- Inner seal protector, J-21362
- Protector seal, J-21409
- Snapring pliers
- 3/8 in. drive extension

Materials/Parts

- One forward clutch housing seal
- Two forward clutch piston seals
- Dry cleaning solvent (Item 23, Appendix B)
- Hydraulic fluid (Item 40, Appendix B)
- Petrolatum (Item 50, Appendix B)

General Safety instructions

- Dry cleaning solvent is flammable and must not be used near open flame. Use only in a well-ventilated area.
- Compressed air used for cleaning purposes should never exceed 30 psi (207 kPa).

a. REMOVAL

1, Remove turbine shaft (3) and forward clutch assembly (2) from transmission case.

2. Remove thrust washer (8) from forward clutch hub (4) inside forward clutch assembly (2). If thrust washer is not found on forward clutch hub, look inside transmission case and remove from direct clutch housing (l).

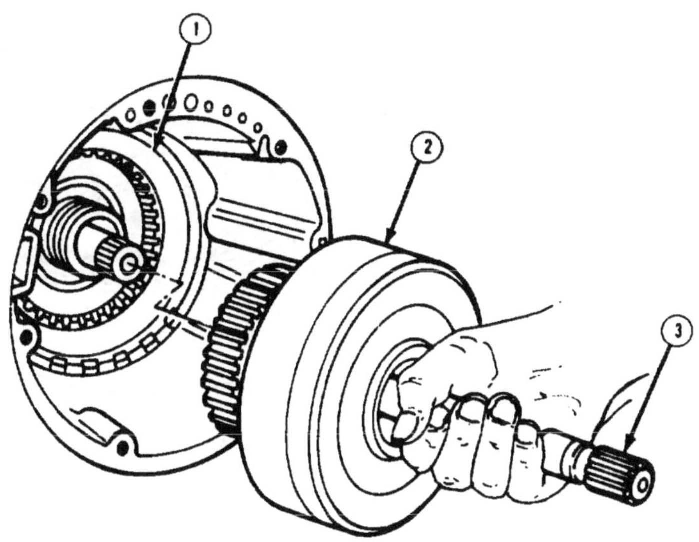

TA50263

5-16. TRANSMISSION FORWARD CLUTCH ASSEMBLY MAINTENANCE (Con't).

b. DISASSEMBLY

1. Place turbine shaft (3) of forward clutch assembly (2) in a soft-jawed vise with turbine shaft pointing downward.

2. Remove snapring (7) from forward clutch assembly (2) housing. Remove direct clutch hub (6).

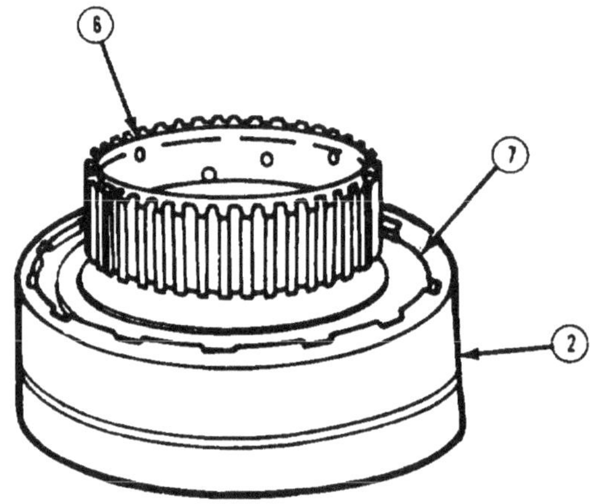

NOTE

Brass thrust washer (9) may be found on underside of forward clutch hub (4) or on top of forward clutch assembly (2) housing.

3. Remove forward clutch hub (4) and brass thrust washer (9).

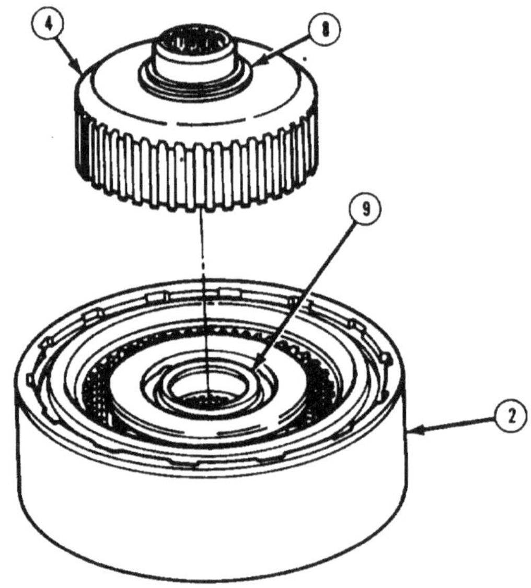

TA50264

5-16. TRANSMISSION FORWARD CLUTCH ASSEMBLY MAINTENANCE (Con't).

NOTE

Composition-faced clutch plates (14) have teeth along internal diameter. Flat steel clutch plates (15) have teeth along external diameter. Waved steel clutch plate (13) is similar to flat steel clutch plate, except that it is wavy.

4. Remove 5 composition-faced clutch plates (14), 4 flat steel clutch plates (15), 1 waved clutch plate (1 3), and apply ring (12) if present,

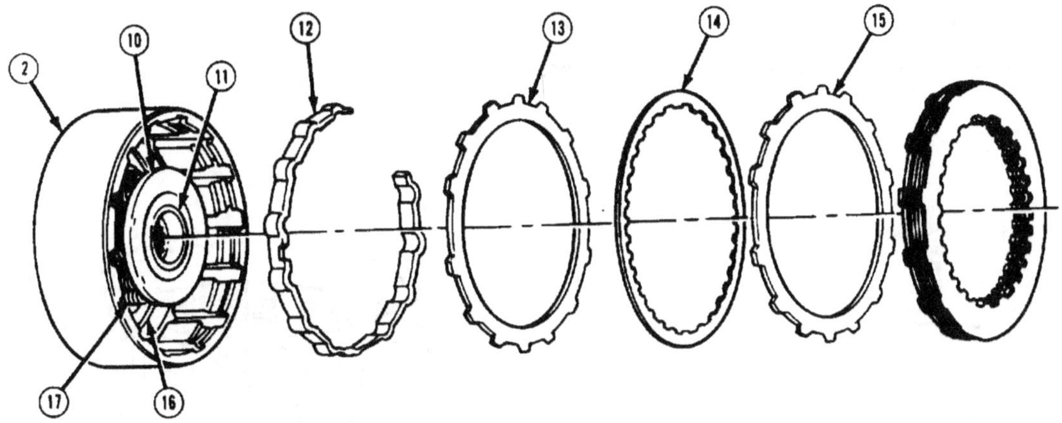

NOTE

Step 5 should not be performed unless either turbine shaft (3) or forward clutch assembly (2) housing is damaged and needs replacement.

5. With a 3/8 in. drive extension about 3 in. (76 mm) long, press turbine shaft (3) out of forward clutch assembly (2) housing,

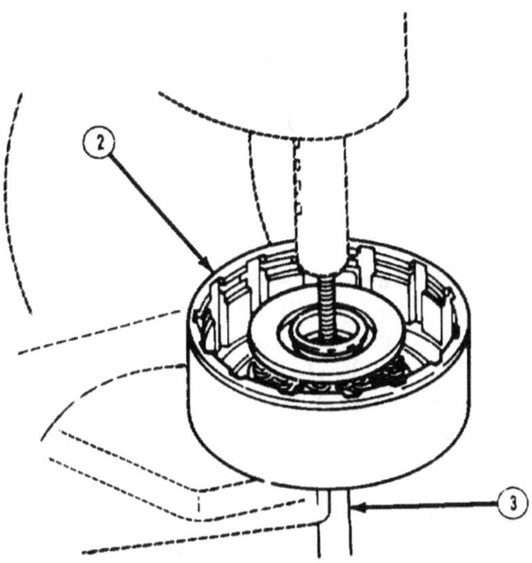

TA50265

5-16. TRANSMISSION FORWARD CLUTCH ASSEMBLY MAINTENANCE (Con't).

NOTE

If turbine shaft (3) was not removed, clutch rebuilder, J-23327-1, should be used with an arbor press to compress piston spring retainer (10).

6. Compress piston spring retainer (10) and remove snapring (11).

7. Remove piston spring retainer (10) and 16 piston springs (17) from spring guides on forward clutch piston (1 6).

a. Remove forward clutch piston (16) from forward clutch assembly (2) housing, Remove 1 inner seal (18) and 1 outer seal (19) from forward clutch piston. Discard seals.

9. Remove center seal (20) from forward clutch assembly (2) housing. Discard seal.

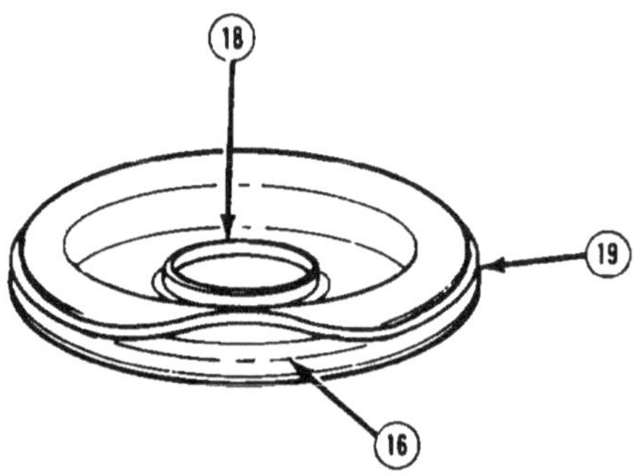

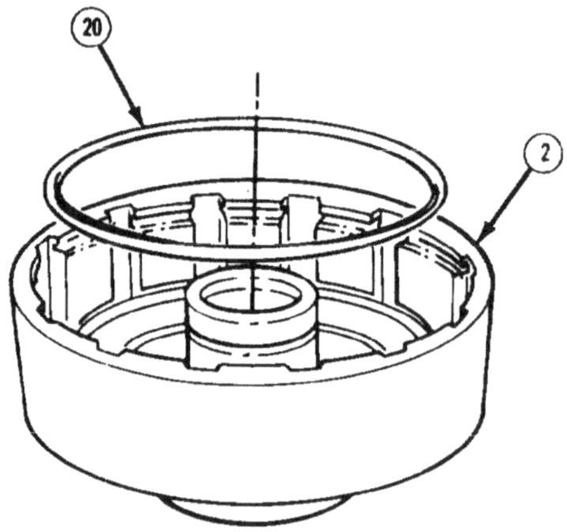

c. CLEANING AND INSPECTION

WARNING

- Dry cleaning solvent P-D-680 is toxic and flammable. Always wear protective goggles and gloves and use only in a well-ventilated area. Avoid contact with skin, eyes, and clothes and DO NOT breathe vapors. DO NOT use near open flame or excessive heat. The solvent's flash point is 100°F-138°F (38°C-59°C). If you become dizzy while using cleaning solvent, immediately get fresh air and medical help. If solvent contacts eyes, immediately wash your eyes with water and get medical aid.

- Compressed air used for cleaning purposes should never exceed 30 psi (207 kPa). Use only effective chip guarding and personnel protective equipment (goggles/shield, gloves, etc.). Failure to follow this warning may result in injury to personnel.

1. Clean all components with dry cleaning solvent. Use compressed air to open up any blocked oil passages.

TA50266

5-16. TRANSMISSION FORWARD CLUTCH ASSEMBLY MAINTENANCE (Con't).

2. Inspect composition-faced clutch plates, flat steel clutch plates, and waved clutch plate for signs of damage. Replace if damaged.

3. Inspect 16 piston springs for collapsed or broken coils, Replace if collapsed or broken,

4. Inspect forward and direct clutch hubs for worn splines, thrust faces, and open oil passages. Replace if worn or if oil passages remain blocked after cleaning.

5. Inspect forward clutch piston for damage. Replace if damaged,

6. Inspect forward clutch assembly housing for damage and open oil passages. Inspect operation of check ball in forward clutch assembly housing to ensure that check ball is loose and moves freely. Replace forward clutch assembly housing if damaged, if oil passages cannot be blown open, or check ball operation is not correct.

7, Inspect turbine shaft for damage. Oil passages on each end of turbine shaft should be plugged. Oil passages across turbine shaft should be open. Replace if damaged or if oil passages are not as specified.

d. ASSEMBLY

1. Lubricate new center seal (20) with hydraulic fluid and seal groove in forward clutch assembly (2) housing with petrolatum. Install center seal to forward clutch assembly housing with lip facing up.

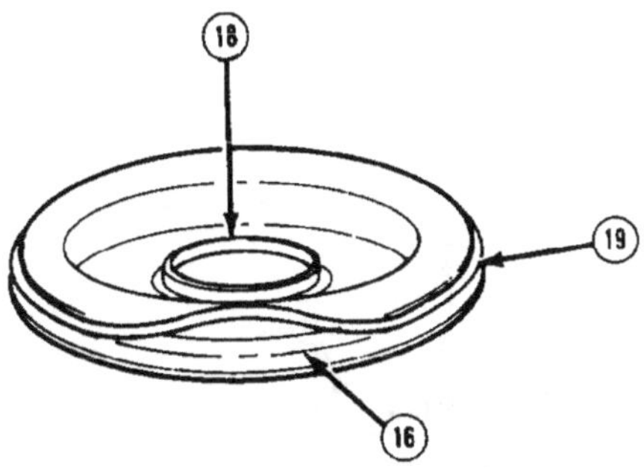

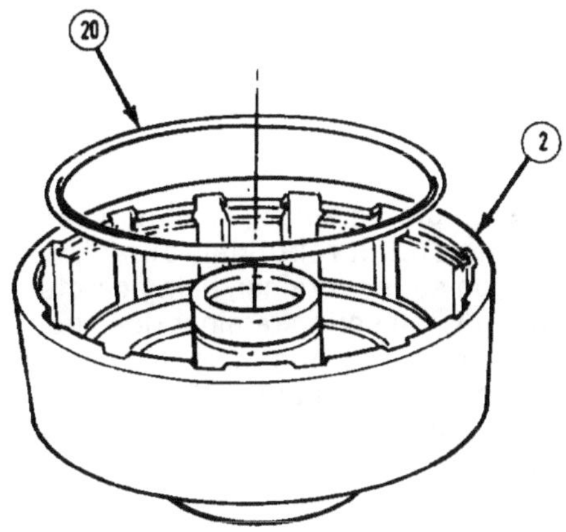

2. Apply petrolatum to grooves in forward clutch piston (16), Lubricate 1 new inner seal (18) and 1 new outer seal (19) with hydraulic fluid. Install seals with lips facing away from piston spring (17) guides,

3. Place inner seal protector (22) over hub in forward clutch assembly (2) housing.

4. Install forward clutch piston (16) inside outer seal protector (21). Insert assembly in forward clutch assembly (2) housing, Install forward clutch piston by rotating in a clockwise direction until fully seated. Remove inner seal protector (22) and outer seal protector.

TA50267

5-16. TRANSMISSION FORWARD CLUTCH ASSEMBLY MAINTENANCE (Con't).

NOTE

Ensure that forward clutch piston (16) is fully seated in forward clutch assembly (2) housing before installing piston springs (17).

5. Install 16 piston springs (17) on spring guides in forward clutch piston (16).

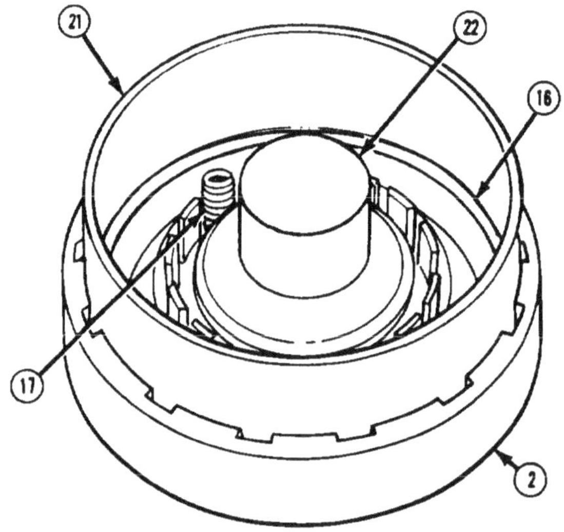

6. Install piston spring retainer (10). Using clutch rebuilder, compress piston spring retainer, being careful that piston spring retainer does not catch in snapring groove. Install snapring (11). If any piston spring (17) is leaning, straighten with a small screwdriver.

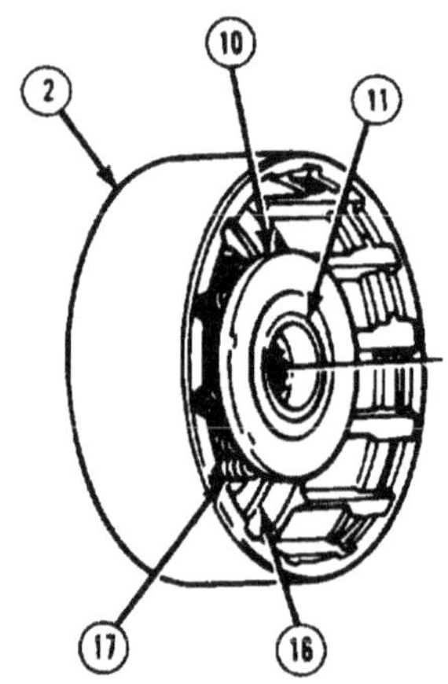

TA50268

5-16. TRANSMISSION FORWARD CLUTCH ASSEMBLY MAINTENANCE (Con't),

CAUTION

Ensure that splined surfaces of turbine shaft (3) and forward clutch assembly (2) are in alinement before assembling them, Failure to follow this caution may result in damage to turbine shaft and forward clutch assembly housing.

NOTE

Perform steps 7 and 8 if turbine shaft (3) was removed,

7. Place forward clutch assembly (2) housing on arbor press with flat side up.
8. Aline shorter splined end of turbine shaft (3) with splines in forward clutch assembly (2) housing , Carefully press turbine shaft into forward clutch assembly housing until turbine shaft bottoms on hub of forward clutch assembly housing.

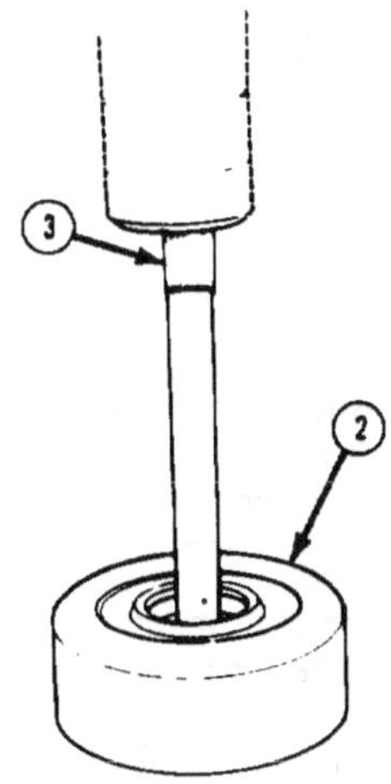

NOTE

Forward clutch assembly (2) with aluminum forward clutch piston (16) does not use apply ring (12).

9. Compress apply ring (12) and install into recess in forward clutch piston (16) with slots on edge of apply ring facing outward,

TA50269

5-16. TRANSMISSION FORWARD CLUTCH ASSEMBLY MAINTENANCE (Con't).

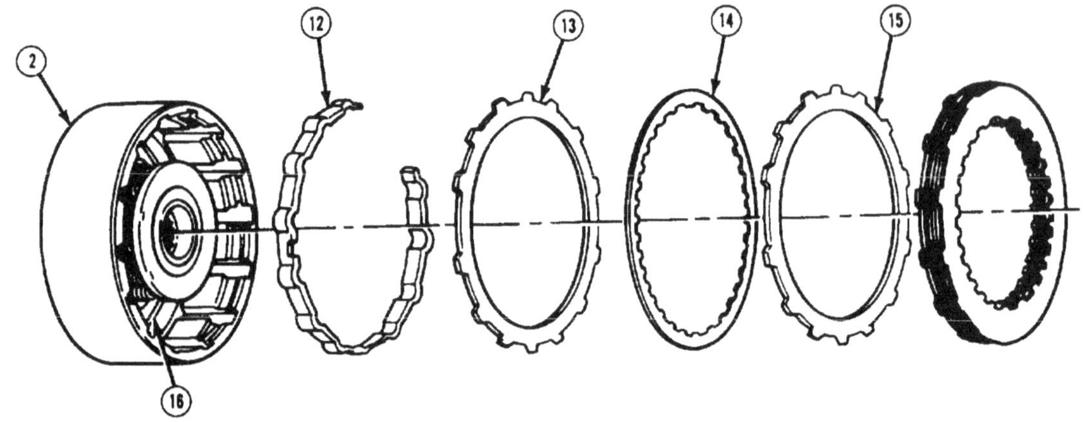

NOTE

- Composition-faced clutch plates (14) have teeth along internal diameter. Fiat steel clutch plates (15) have teeth along external diameter. Waved steel clutch plate (13) is similar to flat steel clutch plate, except that it is wavy,
- Composition-faced clutch plates (14) should be soaked in hydraulic fluid approximately 15 minutes.
- Ensure that a Composition-faced clutch plate (14) is last clutch plate Installed.

10. Lubricate all clutch plates with hydraulic fluid, Install waved clutch plate (1 3). Alternating flat steel and composition-faced plates, install 5 composition-faced clutch plates (14) and 4 fiat steel clutch plates (15),
11. Place forward clutch assembly (2) in a soft-jawed vise with turbine shaft (3) pointing downward. install brass thrust washer (9) inside of forward clutch hub (4) facing forward clutch assembly. install thrust washer (8) on outside of forward clutch hub and retain with a light coat of petrolatum. install forward clutch hub inside forward clutch assembly.

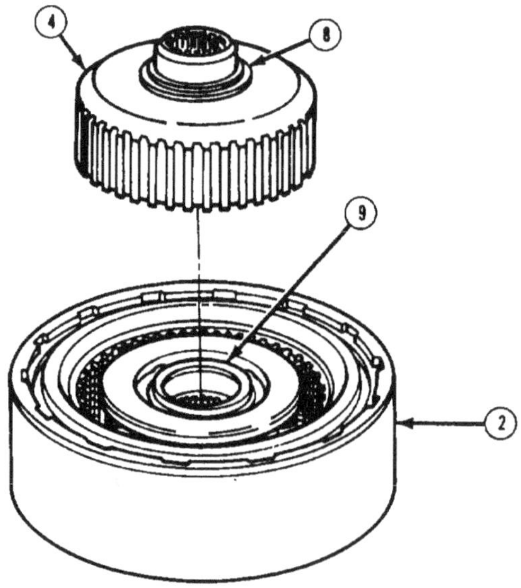

TA60270

5-16. TRANSMISSION FORWARD CLUTCH ASSEMBLY MAINTENANCE (Con't).

12. Install direct clutch hub (6) in forward clutch assembly (2). Install snapring (7).

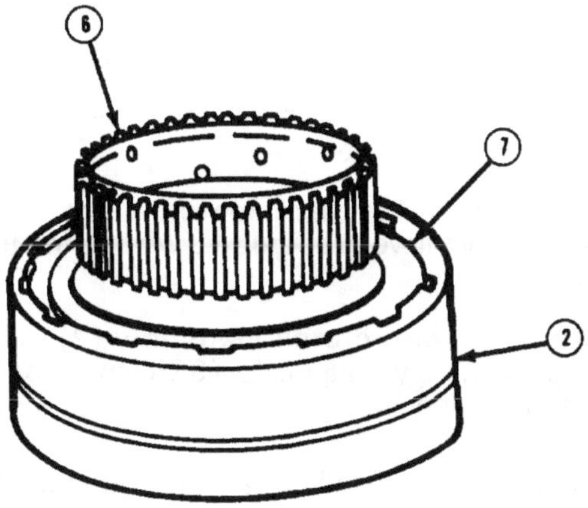

13. Install forward clutch assembly (2) on oil pump (23) delivery sleeve.

WARNING

Compressed air used to check operation of forward clutch assembly (2) should never exceed 30 psi (207 kPa). Use only effective chip guarding and personnel protective equipment (goggles/shield, gloves, etc.). Failure to follow this warning may result in injury to personnel.

14. Check forward clutch assembly (2) operation by applying air through forward clutch passage in oil pump (23), Forward clutch piston (16) should actuate and move forward clutch assembly.

15. Remove forward clutch assembly (2) from oil pump (23) delivery sleeve.

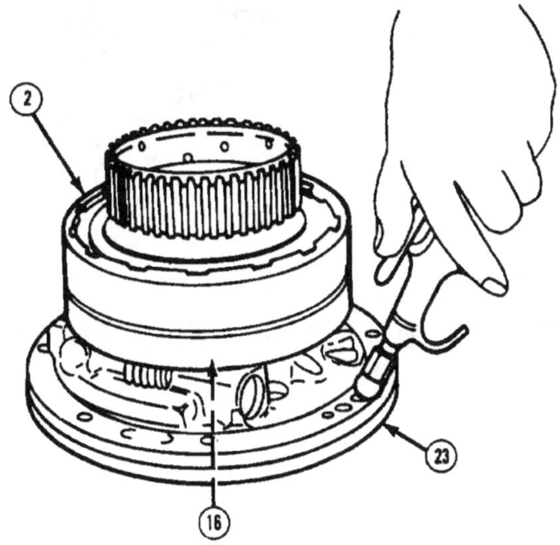

TA50271

5-16. TRANSMISSION FORWARD CLUTCH ASSEMBLY MAINTENANCE (Con't).

e. INSTALLATION

NOTE

- Ensure that thrust washer is still adhered to forward clutch hub (4). (See ASSEMBLY, step 11)
- Missing internal splines in forward clutch hub (4) are oil passages. They do not have to be indexed with any particular spline on mainshaft (5).
- When forward clutch assembly (2) Is fully seated, it will be approximately 1.25 in. (32 mm) from pump mounting face in transmission case.

1. Install forward clutch assembly (2) and turbine shaft (3), Rotate to allow direct clutch driving hub to index with direct clutch composition plates. Ensure that end of mainshaft (5) goes completely into forward clutch hub (4).

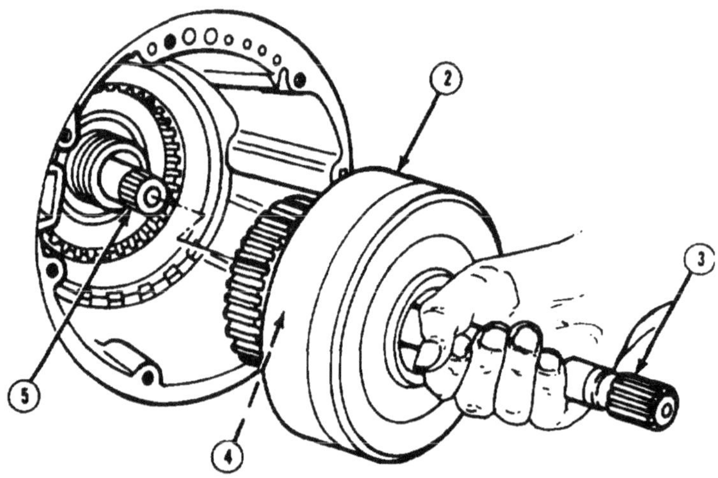

FOLLOW-ON TASKS:

- Install transmission oil pump, (See paragraph 5-15)

TA50272

5-17. TRANSMISSION DIRECT CLUTCH ASSEMBLY AND FRONT BAND MAINTENANCE.

This task covers:

a. Removal
b. Disassembly
c. Cleaning and Inspection
d. Assembly
e. Installation

INITIAL SETUP:

Equipment Condition

- Forward clutch assembly removed. (See paragraph 5-1 6)

Tools/Test Equipment

- Clutch rebuilder, J-23327
- Inner seal protector, J-21362
- protector seal, J-21409
- Snapring pliers

Materials/Parts

- One direct clutch housing seal
- Two direct clutch piston seals
- Dry cleaning solvent (Item 23, Appendix B)
- Hydraulic fluid (Item 40, Appendix B)
- Petrolatum (Item 50, Appendix B)

General Safety Instructions

- Dry cleaning solvent is flammable and must not be used near open flame. Use only in a well-ventilated area.
- Compressed air used for cleaning purposes should never exceed 30 psi (207 kPa).

a. REMOVAL

1. Remove direct clutch assembly (3) from transmission case (1).

2. Remove front band (2).

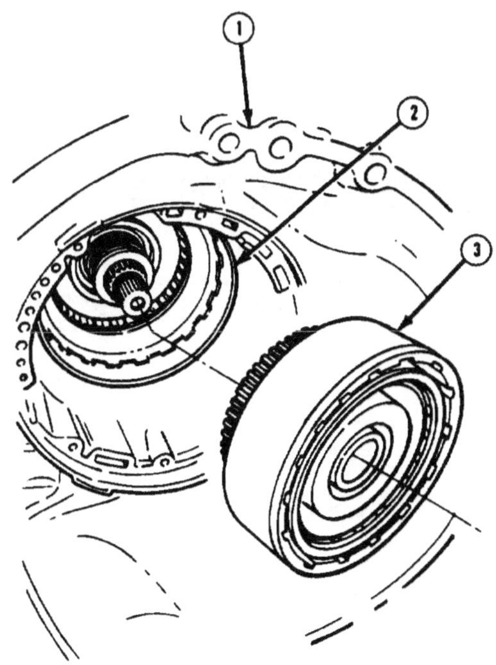

TA50273

5-17. TRANSMISSION DIRECT CLUTCH ASSEMBLY AND FRONT BAND MAINTENANCE (Con't).

b. DISASSEMBLY

1. Remove intermediate clutch retainer snapring (6. intermediate clutch retainer (7). outer race (4), and intermediate clutch roller assembly (8).

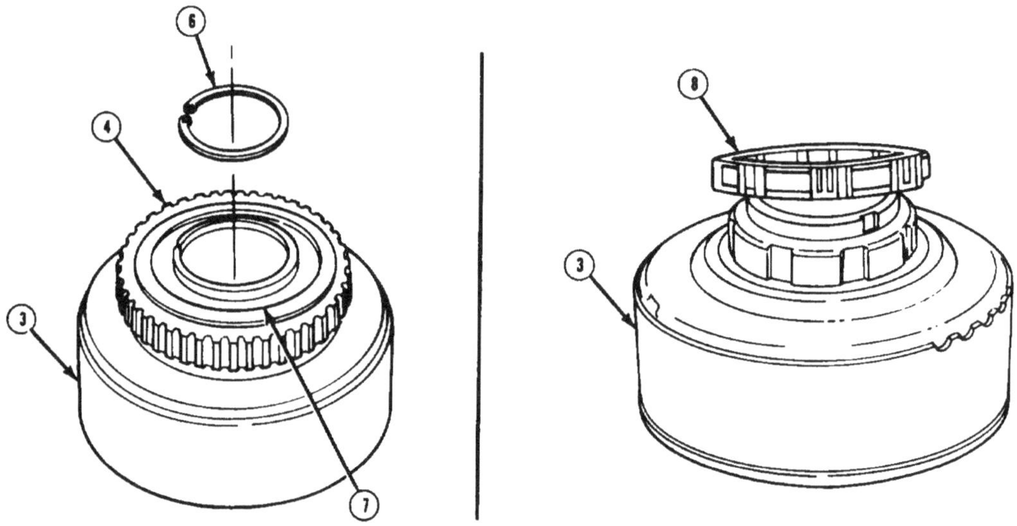

2. Turn direct clutch assembly (3) over. Remove snapring (9).

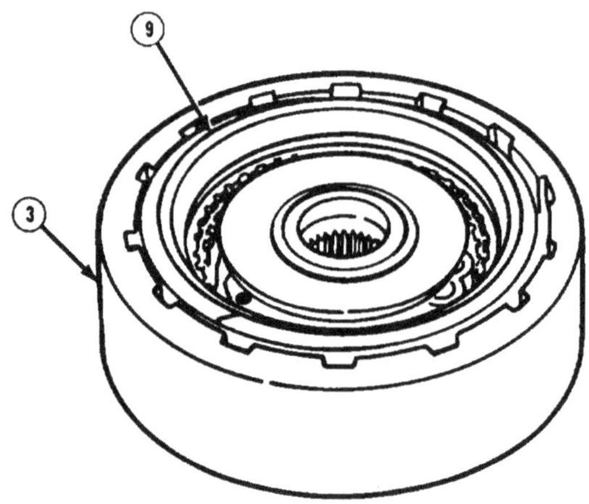

TA50274

5-17. TRANSMISSION DIRECT CLUTCH ASSEMBLY AND FRONT BAND MAINTENANCE (Con't).

3 . Remove direct clutch backing plate (10), 5 flat steel clutch plates (12), 5 composition-faced clutch plates (11), and apply ring (13), if present, from direct clutch assembly (3) housing.

4 . Compress piston spring retainer (14) and remove snapring (16). Remove piston spring retainer.

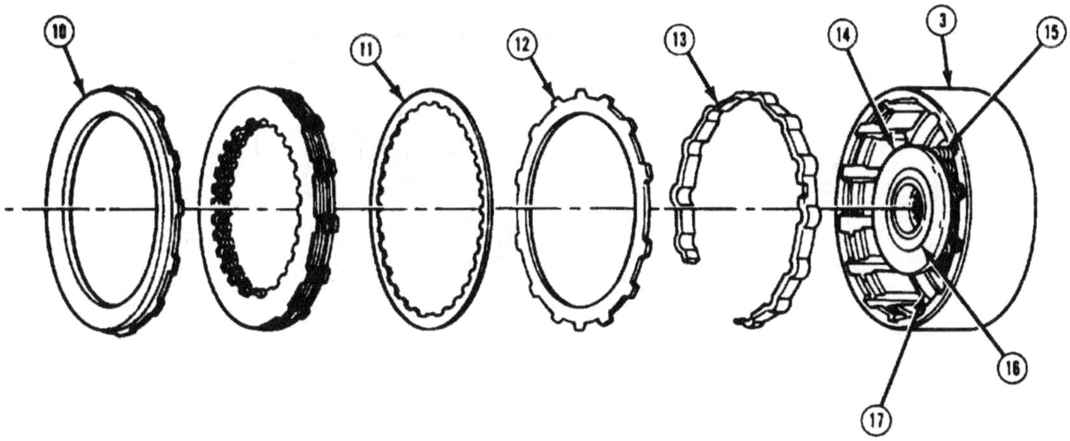

NOTE

If forward clutch assembly (3) has boon disassembled, direct clutch piston release springs (15) should be kept separated from forward clutch piston release springs.

5. Remove 14 piston release springs (15).

6. Remove direct clutch piston (17) from direct clutch assembly (3) housing.

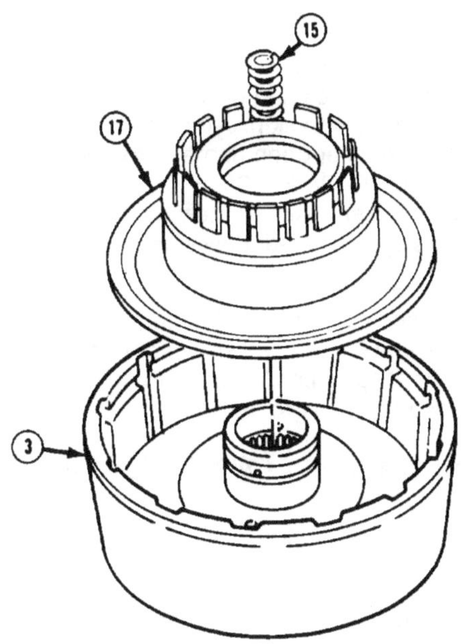

TA50275

5-17. TRANSMISSION DIRECT CLUTCH ASSEMBLY AND FRONT BAND MAINTENANCE (Con't).

7. Remove inner seal (18) and outer seal (19) from direct clutch piston (17), and center seal (20) from direct clutch assembly (3) housing. Discard seals,

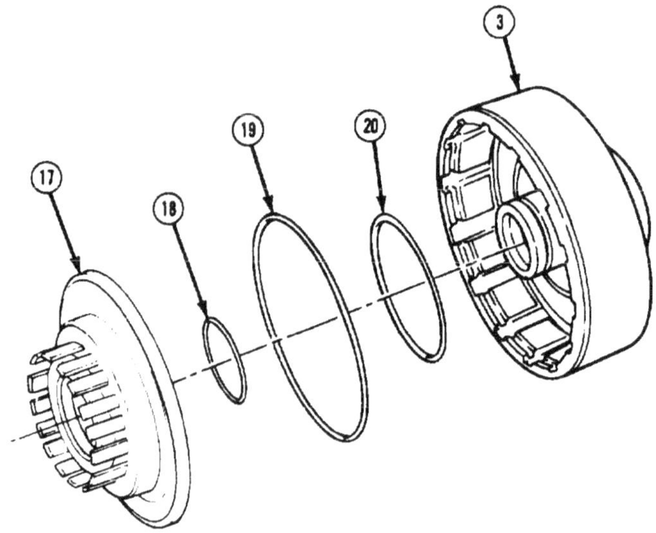

c. CLEANING AND INSPECTION

WARNING

- Dry cleaning solvent P-D-680 is toxic and flammable. Always wear protective goggles and gloves and use only in a well-ventilated area. Avoid contact with skin, eyes, and clothes and DO NOT breathe vapors. DO NOT use near open flame or excessive heat, The solvent's flash point is 100°F-138°F (38°C-59°C). If you become dizzy while using cleaning solvent, immediately get fresh air and medical help, if solvent contacts eyes, immediately wash your eyes with water and get medical aid.
- Compressed air used for cleaning purposes should never exceed 30 psi (207 kPa). Use only effective chip guarding and personnel protective equipment (goggles/shield, gloves, etc.), Failure to follow this warning may result in injury to personnel.

1. Clean all removed components with dry cleaning solvent, Use compressed air to open up any blocked oil passages.
2. Inspect intermediate clutch roller assembly for damaged rollers, cage, or springs. Replace if damaged.
3. inspect all machined surfaces for damage. Replace if damaged.
4. Inspect direct clutch assembly housing for damage. Inspect check ball for freedom of operation, Replace direct clutch assembly housing if damaged or if check ball does not operate freely.

TA60276

5-17. TRANSMISSION DIRECT CLUTCH ASSEMBLY AND FRONT BAND MAINTENANCE (Con't).

5. Inspect all clutch plates and backing plate for damage. Replace if damaged.
6. Inspect direct clutch piston for damage. Replace if damaged.
7. Inspect piston release springs for collapsed coils or signs of distortion. Replace if coils are collapsed or distorted,
8. Inspect front band for damage. Replace if damaged,
9. Inspect all oil passages to ensure that they are open. Replace any component if oil passages cannot be blown open.

d. ASSEMBLY

1. Lubricate new inner seal (18) and outer seal (19) with hydraulic fluid, and seal grooves in direct clutch piston (17) with petrolatum. Install seals with lips facing away from spring guides (23) .
2. Lubricate center seal (20) with hydraulic fluid and seal groove in direct clutch assembly (3) housing with petrotatum, Install seal with lip facing up,

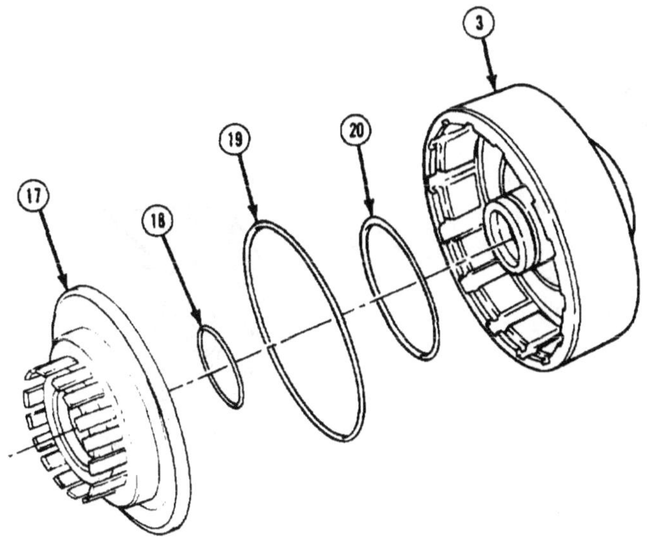

TA50277

5-17. TRANSMISSION DIRECT CLUTCH ASSEMBLY AND FRONT BAND MAINTENANCE (Con't).

3. Place inner seal protector (22) over direct clutch assembly (3) hub, Install direct clutch piston (17) inside outer Seal protector (21) and insert assembly in direct clutch assembly housing. Rotate direct clutch piston in a clockwise direction until fully seated, Remove inner and outer seal protectors.

NOTE

2 empty spring guides (23) should be diagonally opposite each other.

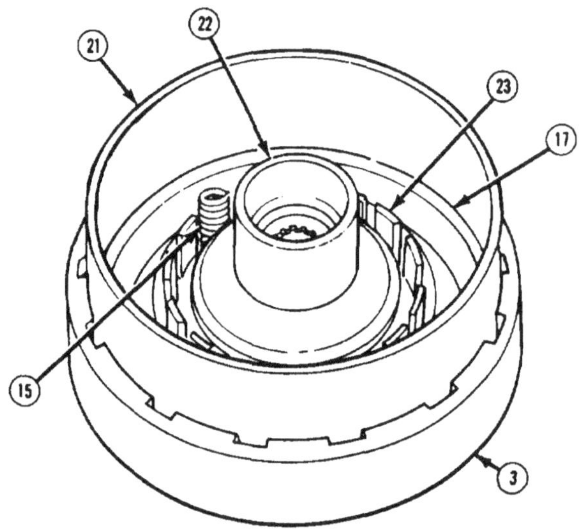

4. Install 14 piston release springs (15) into spring guides (23) in direct clutch piston (1 7).

5 . Place piston spring retainer (14) over piston release springs (15), Using clutch rebuilder, compress piston spring retainer, being careful to not catch piston spring retainer in snapring groove. Install snapring (1 6), Remove clutch rebuilder.

6 . If any piston release spring (15) is leaning, straighten with a small screwdriver.

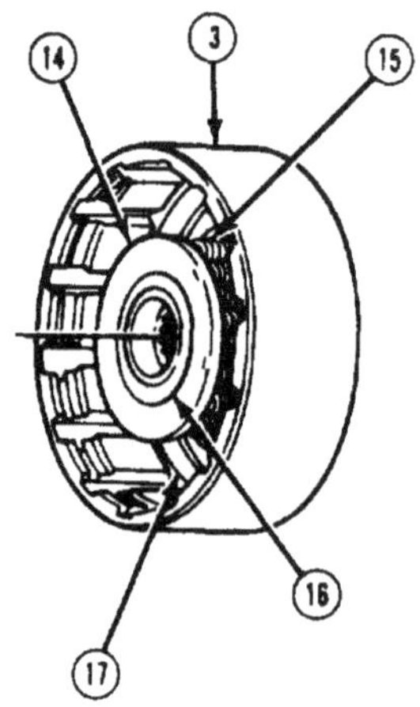

TA50278

5-17. TRANSMISSION DIRECT CLUTCH ASSEMBLY AND FRONT BAND MAINTENANCE (Con't).

NOTE

Direct clutch assembly (3) with aluminum direct clutch piston (17) does not use apply ring (13).

7. Compress apply ring (13) and install into recess in direct clutch piston (17) with slots on edge of apply ring facing inward.

NOTE

Composition-faced clutch plates (11) should be soaked in hydraulic fluid for approximately 15 minutes.

8. Lubricate all clutch plates (11 and 12) with hydraulic fluid. Alternating flat steel and composition-faced clutch plates, install 5 flat steel clutch plates (12) and 5 composition-faced clutch plates (11).

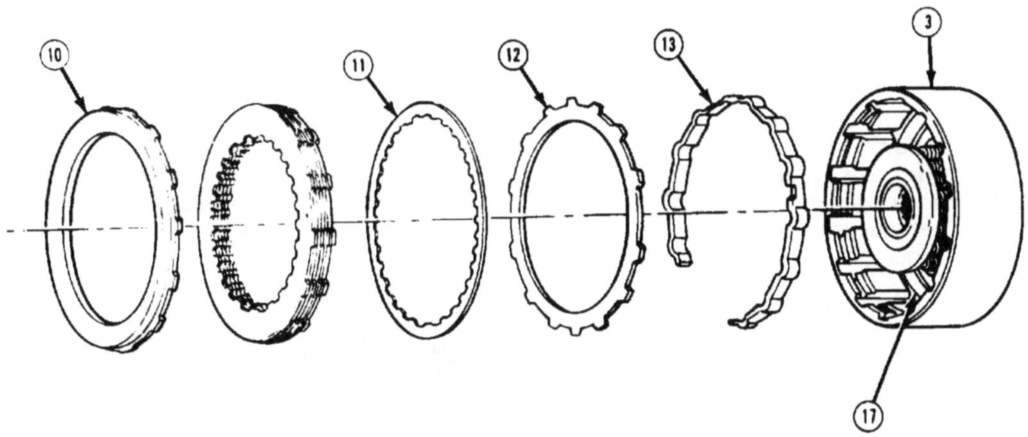

TA50279

5-17. TRANSMISSION DIRECT CLUTCH ASSEMBLY AND FRONT BAND MAINTENANCE (Con't).

9. Install direct clutch backing plate (10). Install snapring (9) into direct clutch assembly (3) groove.

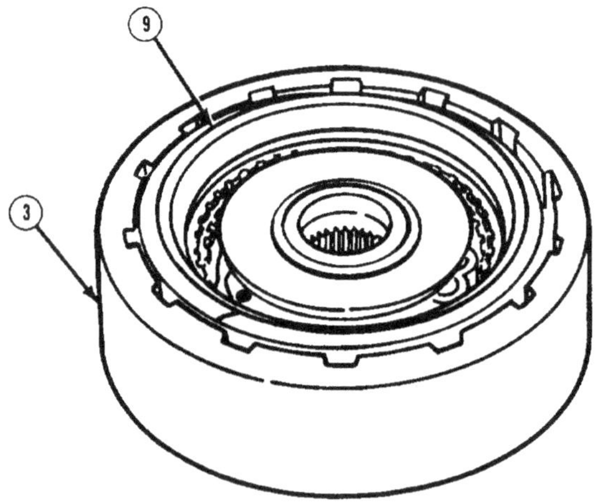

10. Turn direct clutch assembly (3) housing unit over, Install intermediate clutch roller assembly (8) onto intermediate clutch inner cam (24). Install outer race (4), grooved side up, with a clockwise turning motion.

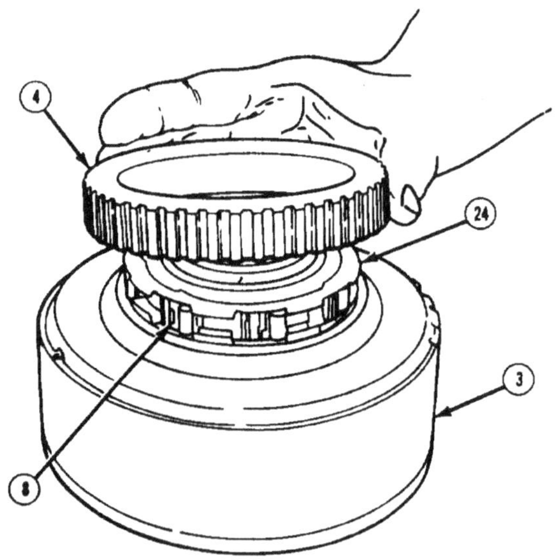

TA50280

5-17. TRANSMISSION DIRECT CLUTCH ASSEMBLY AND FRONT BAND MAINTENANCE (Con't).

11. Install intermediate clutch retainer (7) and snapring (6).

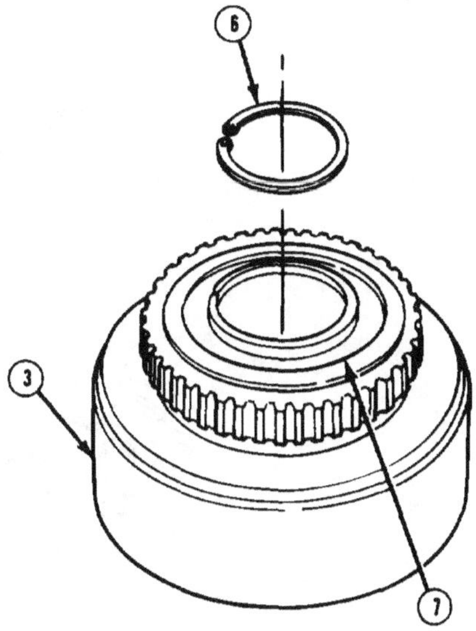

NOTE

To check operation of direct clutch assembly (3), steps 12-14 should be performed.

12. Remove center support and gear unit. (See paragraph 5-1 9)

WARNING

Compressed air used to check operation of direct clutch assembly (3) should never exceed 30 psi (207 kPa). Use only effective chip guarding and personnel protective equipment (goggles/shield, gloves, etc.). Failure to follow this warning may result in injury to personnel.

NOTE

Air applied through reverse passage (right oil passage) will escape from direct clutch passage (left oil passage). This is considered normal.

13. Place direct clutch assembly (3) on center support (25) and check operation of direct clutch assembly by applying compressed air through left oil passage to actuate piston and move direct clutch plates.

TA50281

5-17. TRANSMISSION DIRECT CLUTCH ASSEMBLY AND FRONT BAND MAINTENANCE (Con't).

14, Remove direct clutch assembly (3) from center support (25). Install center support and gear unit. (See paragraph 5-19)

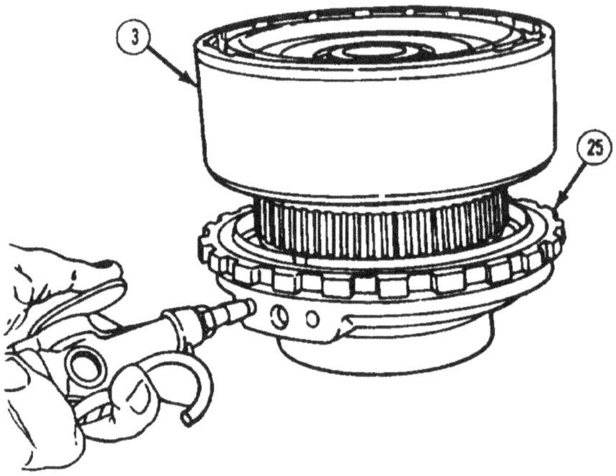

e. INSTALLATION

NOTE

Front band (2) should be soaked in hydraulic fluid a minimum of 15 minutes.

1. Install front band (2) in transmission case (1) with band anchor hole over band anchor pin (27), and apply lug (26) facing servo hole.

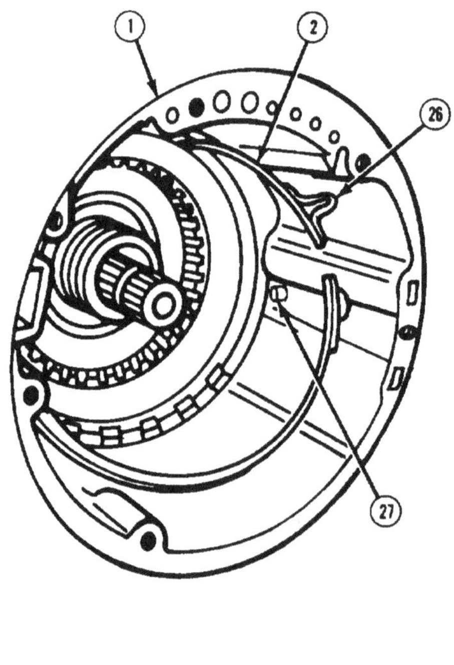

TA50282

5-17. TRANSMISSION DIRECT CLUTCH ASSEMBLY AND FRONT BAND MAINTENANCE (Con't).

NOTE

Ensure that direct clutch assembly (3) housing hub bottoms on sun gear shaft (5) and that splines on forward end of sun gear shaft are flush with splines in direct clutch assembly.

3. Install direct clutch assembly (3) into transmission case (1). Rotate direct clutch assembly housing to allow outer race (4) to index with intermediate clutch composition-faced plates inside transmission case.

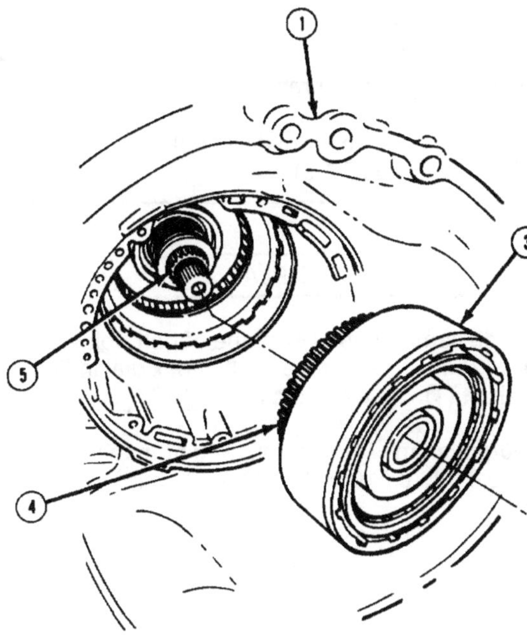

FOLLOW-ON TASKS:

- Install forward clutch assembly. (See paragraph 5-16)

TA50283

5-18. CHECKING REAR UNIT END PLAY.

This task covers: a. Checking Rear Unit End Play

INITIAL SETUP:

Equipment Condition

- Control valve assembly removed. (See paragraph 5-7)
- Direct clutch and front band removed. (See paragraph 5-1 7)
- Transmission installed horizontally in holding fixture. (See paragraph 5-1 3)

Too/s/Test Equipment

- Dial indicator
- Micrometer
- Slide hammer bolt

a, CHECKING REAR UNIT END PLAY

1. Install slide hammer bolt (3) in one of bolt holes on end of transmission case (4). Mount dial indicator (2) on slide hammer bolt, Index dial indicator to register with flat surface on end of output shaft (1). Set dial indicator to "O."

2. Move output shaft (1) in and out and note resulting rear unit end play. Rear unit end play should be 0.007-0.019 in. (0.178-0.483 mm). Remove dial indicator (2) and slide hammer bolt (3) .

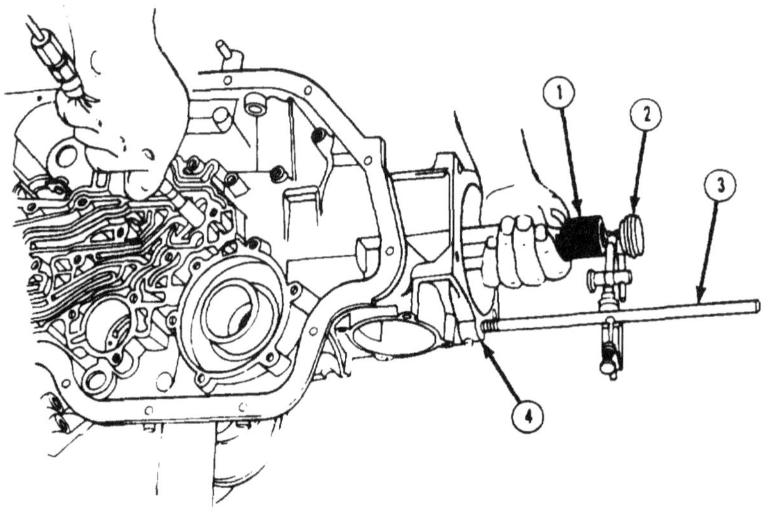

NOTE

Perform steps 3-14 if rear unit end play is not within specifications.

3. Remove center support and gear unit, (See paragraph 5-1 9)

TA50284

5-18. CHECKING REAR UNIT END PLAY (Con't).

NOTE

The selective washer (5) controlling rear unit end play is the steel washer with 3 tabs, located between thrust washer and rear face of transmission case. Notches and/or numerals on tabs serve to identify washer thickness.

4. Remove selective washer (5) and measure its thickness.

NOTE

- If rear unit end play is less than 0.007 in. (0.178 mm), perform steps 5-7 and skip steps 8-10.
- If rear unit end play is more than 0.019 in. (0.483 mm), skip steps 5-7 and perform steps 8-10.

5. Subtract rear unit end play as measured in step 2 from 0.013 in, (0.330 mm)
6. Subtract amount from step 5 from thickness of removed selective washer (5).
7. The result of step 6 is the thickness of new selective washer (5) to be installed.
8. Subtract 0.013 in. (0.330 mm) from rear unit end play as measured in step 2.
9. Add amount from step 8 to thickness of removed selective washer (5).
10, The result of step 9 is the thickness of new selective washer (5) to be installed,
11. Select proper size selective washer (5) from Table 5-4.

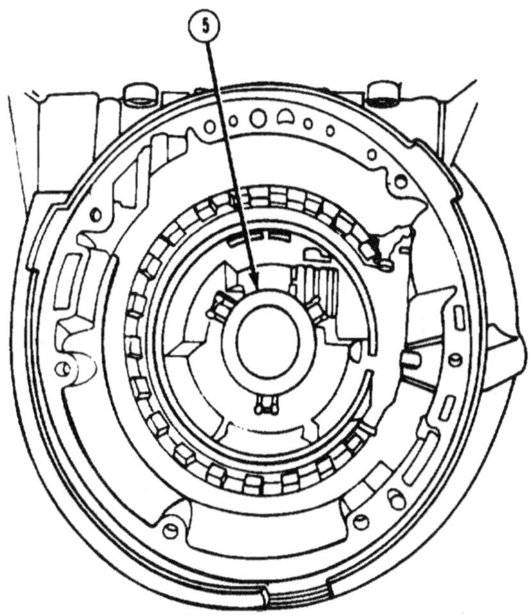

Table 5-4. Rear Unit End Play Selective Washers

THICKNESS	IDENTIFICATION NOTCH AND/OR NUMERAL	
0.074-0,078 in. (1 .880-1 .981 mm)	None	1
0,082-0.086 in. (2.083-2, 184 mm)	On side of 1 Tab	2
0.090-0.094 in. (2,286-2,388 mm)	On side of 2 Tabs	3
0.098-0.102 in. (2.489-2.591 mm)	On end of 1 Tab	4
0.106-0.110 in. (2.692-2.794 mm)	On end of 2 Tabs	5
0.114-0.118 in. (2.896-2.997 mm)	On end of 3 Tabs	6

TA50285

5-18. CHECKING REAR UNIT END PLAY (Con't).

12. Install new selective washer (5).

13, Install center support and gear unit, (See paragraph 5-1 9)

14. Recheck rear unit end play.

FOLLOW-ON TASKS:

- Install front band and direct clutch assembly. (See paragraph 5-17)
- Install control valve assembly. (See paragraph 5-7)

5-19. TRANSMISSION INTERMEDIATE CLUTCH, CENTER SUPPORT ASSEMBLY AND GEAR UNIT, AND REAR BAND MAINTENANCE.

This task covers:

a. Removal
b.' Disassembly
c. Cleaning and Inspection
d. Pinion Gear Replacement
e. Assembly
f. Installation

INITIAL SETUP:

Equipment Condition

- Direct clutch assembly removed. (See paragraph 5-1 7)
- Control valve assembly removed. (See paragraph 5-7)
- Governor assembly removed, (See paragraph 5-5)

Too/s/Test Equipment

- Annular bearing inserter, J-21465-1
- Drive handle, J-8092
- Gear holding unit, J-21795-02
- Inner seal protector, J-21363
- 3/8 in, 12-point thin wail deep socket
- Snapring pliers
- Torque wrench

Materials/Parts

- Two intermediate clutch piston seals
- Four oil seal rings
- Pinion package
- Crocus cloth (Item 18, Appendix B)
- Dry cleaning solvent (Item 23, Appendix B)
- Hydraulic fluid (Item 40, Appendix B)
- Petrolatum (Item 50, Appendix B)
- Duct tape (Item 66, Appendix B)

General Safety Instructions

- Dry cleaning solvent is flammable and must not be used near open flame. Use only in a well-ventilated area.
- Compressed air used for cleaning purposes should never exceed 30 psi (207 kPa).
- When using drill, use only effective chip-guarding and personnel protective equipment.

a. REMOVAL

NOTE

- If removing intermediate clutch, perform step 1.
- If removing center support and gear unit assembly, perform steps 1-5.
- If removing rear band, perform steps 1-8.
- For ease of removal, transmission should be vertical with mainshaft facing upward.
- Clutch wave plate (4) has 2 "V" marks cut in outer circumference.

1. Remove snapring (1), intermediate clutch backing plate (2), 3 intermediate clutch wave plates (3), 2 flat steel clutch plates (6), and clutch wave plate (4).

5-19. TRANSMISSION INTERMEDIATE CLUTCH, CENTER SUPPORT ASSEMBLY AND GEAR UNIT, AND REAR BAND MAINTENANCE (Con't).

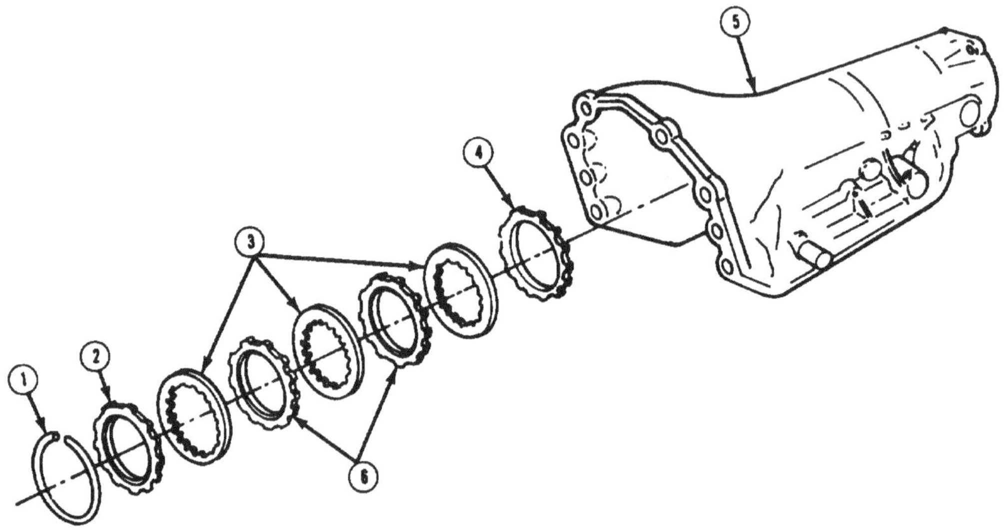

2. Using a 3/8 in. 12-point thin wall deep socket (7), remove center support bolt from transmission case (5).

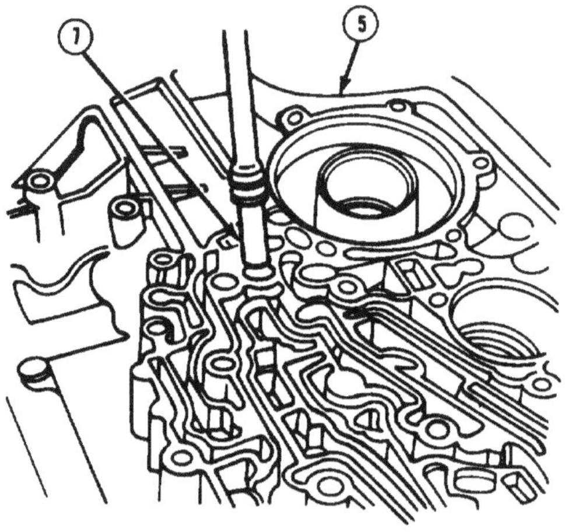

TA50286

5-19. TRANSMISSION INTERMEDIATE CLUTCH, CENTER SUPPORT ASSEMBLY AND GEAR UNIT, AND REAR BAND MAINTENANCE (Con't).

3. Remove snapring (9).

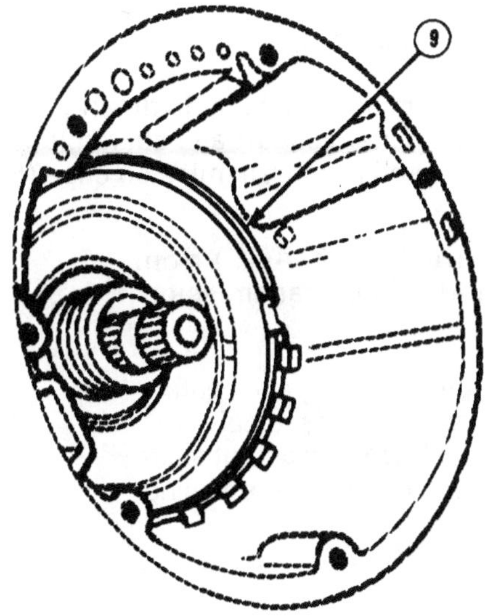

4. Install gear holding unit (11) on end of mainshaft so that tangs engage groove in shaft. Tighten screw on gear holding unit to secure it to mainshaft. Remove center support and gear unit assembly (14) from transmission case (5).

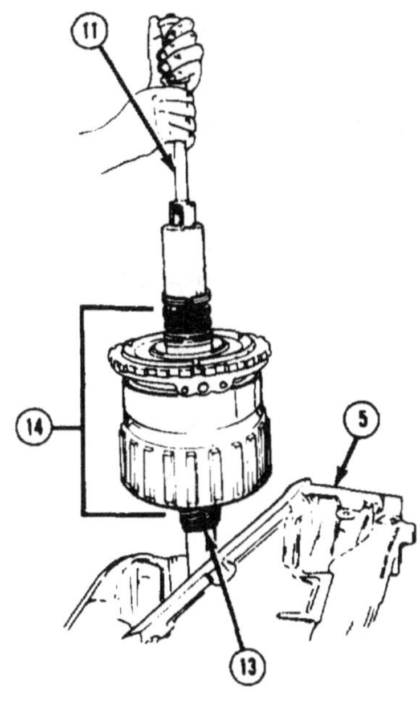

TA50287

5-19. TRANSMISSION INTERMEDIATE CLUTCH, CENTER SUPPORT ASSEMBLY AND GEAR UNIT, AND REAR BAND MAINTENANCE (Con't).

NOTE

- Output shaft (13) should be covered with tape to prevent spline damage to case bushing when removing center support and gear unit assembly (14).
- Transmission holding fixture pivot bolt should be loosened slightly so that center support and gear unit assembly (14) does not bind when removed from transmission case (5).

5. Remove thrust washer (15) from output shaft (13) or transmission case (5).

6. Secure output shaft (13) and center support and gear unit assembly (14) in a soft-jawed vise with output shaft pointing down. Remove gear holding unit (11).

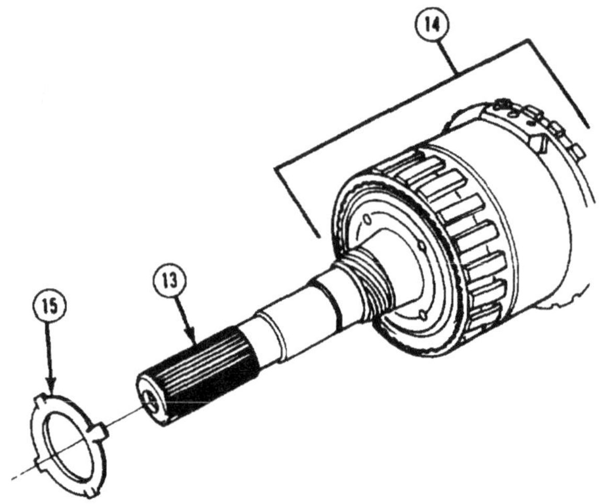

NOTE

Location of spacer (16) should be noted for installation.

7. Remove selective washer (17) and spacer (16) from inside transmission case (5).

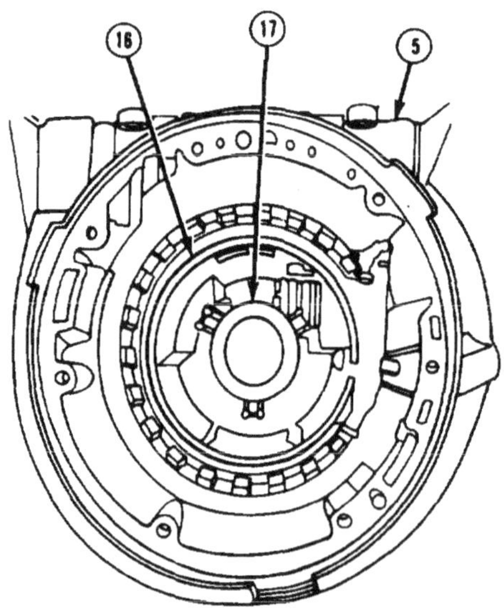

TA50288

5-19. TRANSMISSION INTERMEDIATE CLUTCH, CENTER SUPPORT ASSEMBLY AND GEAR UNIT, AND REAR BAND MAINTENANCE (Con't).

8. Rotate band lugs away from pins and remove rear band (18) from transmission case (5).

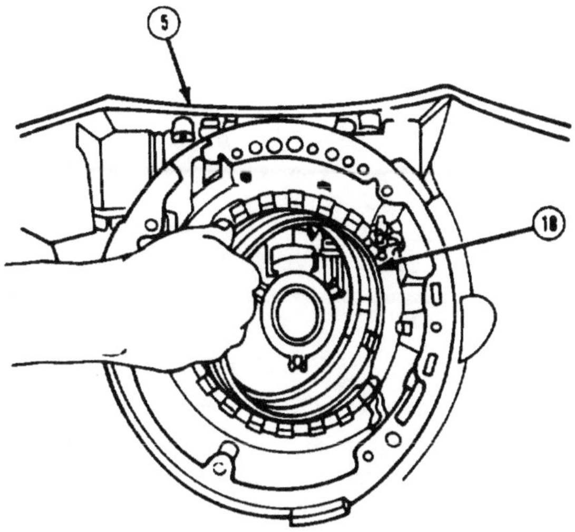

b. DISASSEMBLY

1. Remove center support assembly (8) from reaction carrier (21) by lifting straight upward.

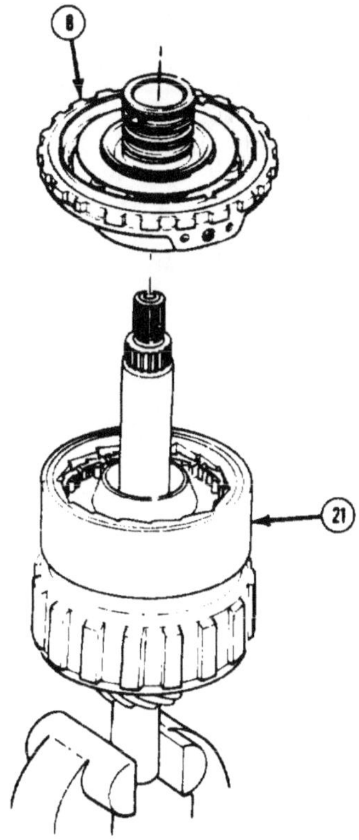

TA50289

5-19. TRANSMISSION INTERMEDIATE CLUTCH, CENTER SUPPORT ASSEMBLY AND GEAR UNIT, AND REAR BAND MAINTENANCE (Con't).

2. Remove thrust washer (23).

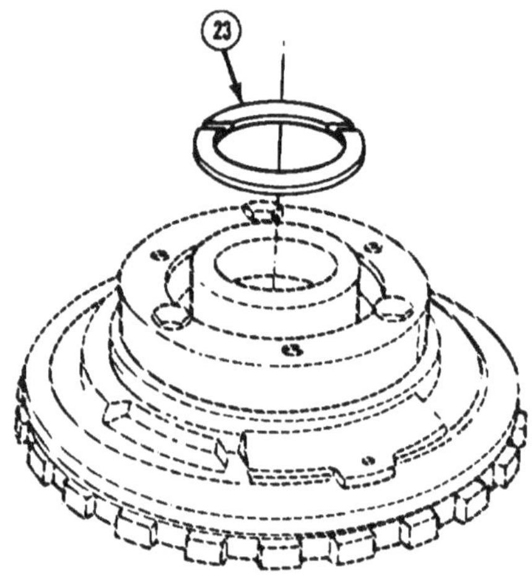

3. Remove 4 oil seal rings (24) from center support assembly (8) and discard.

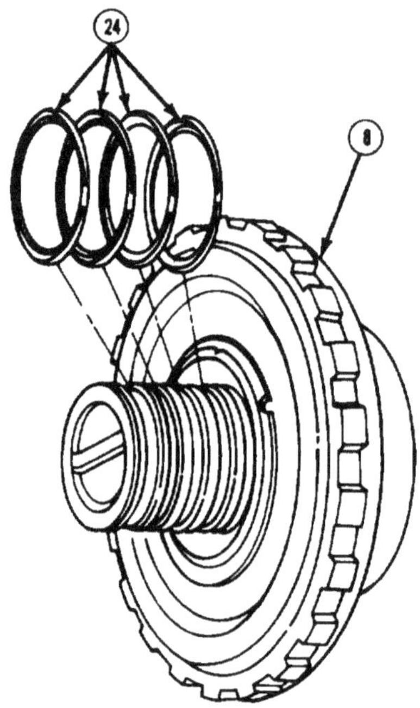

TA50290

5-19. TRANSMISSION INTERMEDIATE CLUTCH, CENTER SUPPORT ASSEMBLY AND GEAR UNIT, AND REAR BAND MAINTENANCE (Con't).

4. Compress spring retainer (27) and remove snapring (25). Remove spring retainer.

NOTE

Location of springs (26) should be marked for assembly.

5. Remove 3 springs (26) and clutch spring guide (28).

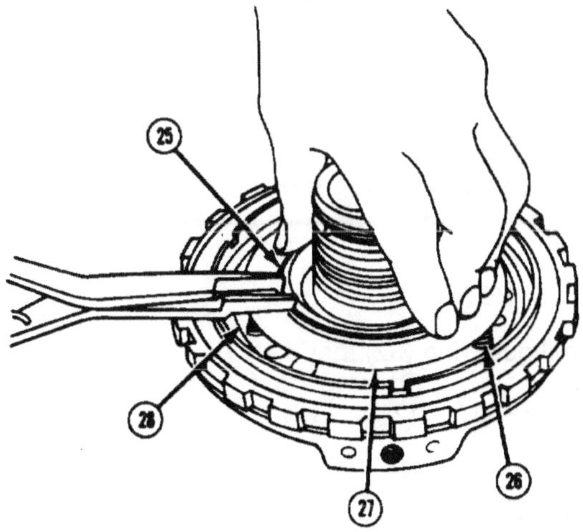

6. Remove intermediate clutch piston (29) from center support assembly (8). Remove inner seal (30) and outer seal (31) from intermediate clutch piston. Discard seals.

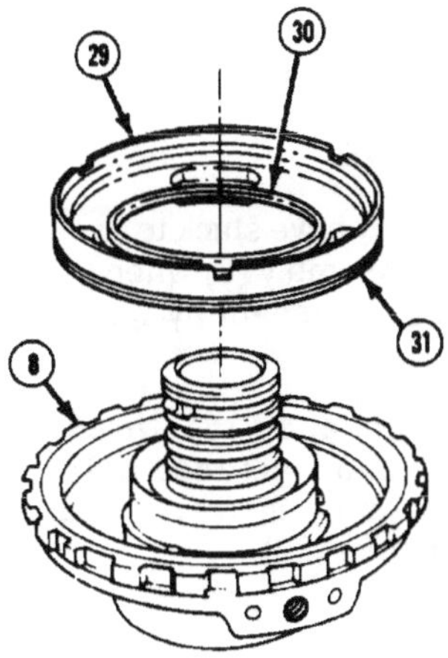

TA50291

5-19. TRANSMISSION INTERMEDIATE CLUTCH, CENTER SUPPORT ASSEMBLY AND GEAR UNIT, AND REAR BAND MAINTENANCE (Con't).

7. Remove sun gear shaft (19).

8. Remove reaction carrier (21) and roller clutch assembly (20) from output carrier (22). Remove roller clutch assembly from reaction carrier.

NOTE

Front internal gear ring (32) may have stuck to underside of reaction carrier (21) when removed.

9. Remove front internal gear ring (32) from output carrier (22).

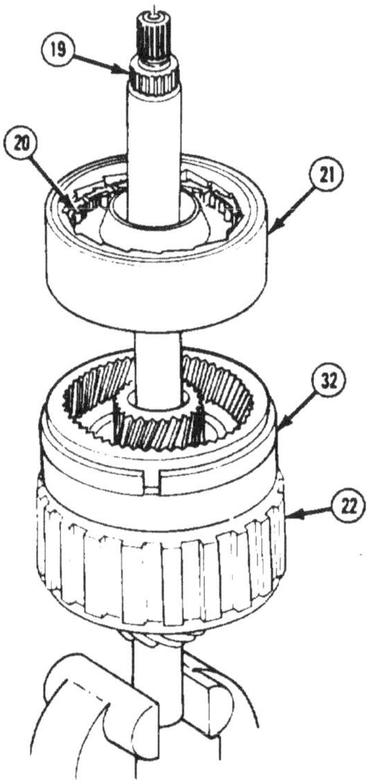

NOTE

Outer race (35) may have stuck to center support assembly (8) when center support assembly was removed,

10. Remove outer race (35), thrust bearing (34), and inner race (33) from sun gear (36), Remove sun gear from output carrier (22).

NOTE

Thrust washer (37) may have stuck to underside of reaction carrier when reaction carrier was removed,

11. Remove thrust washer (37).

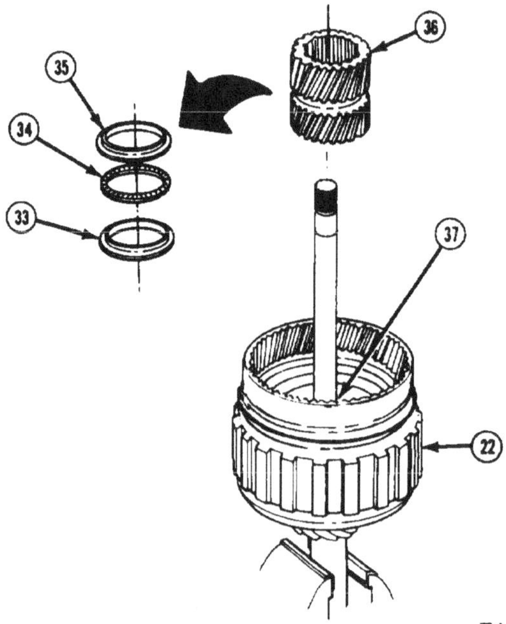

TA50292

5-19. TRANSMISSION INTERMEDIATE CLUTCH, CENTER SUPPORT ASSEMBLY AND GEAR UNIT, AND REAR BAND MAINTENANCE (Con't).

12. Remove output carrier (22) and gear unit from vise. Remove snapring (38) and remove output shaft (13) from output carrier.

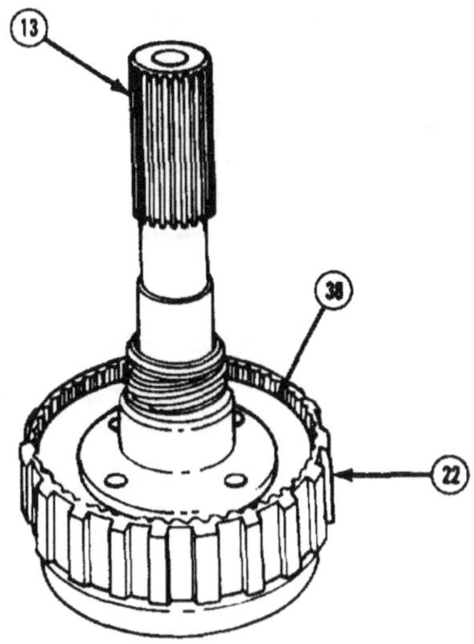

NOTE

Outer race (39) may have stuck to underside of output shaft (13).

13. Remove outer race (39), thrust bearing (40), and inner race (41) from outer face of rear internal gear (42).

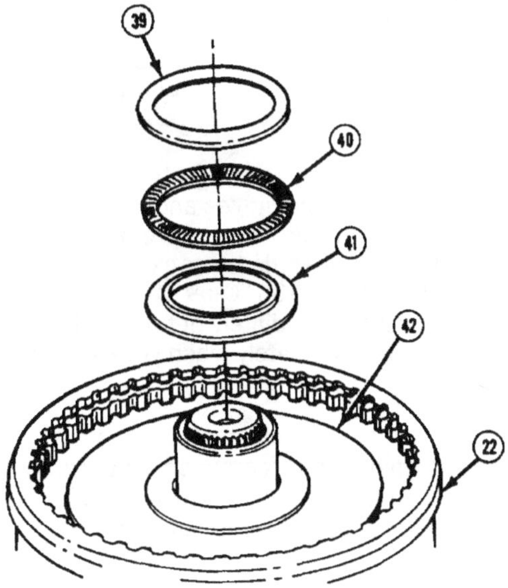

TA50293

5-19. TRANSMISSION INTERMEDIATE CLUTCH, CENTER SUPPORT ASSEMBLY AND GEAR UNIT, AND REAR BAND MAINTENANCE (Con't).

14. Remove rear internal gear (42) and mainshaft (12) from output carrier (22). Remove outer race (45), thrust bearing (46), and inner race (47) from inner face of rear internal gear. Remove snapring (44) and remove rear internal gear from mainshaft.

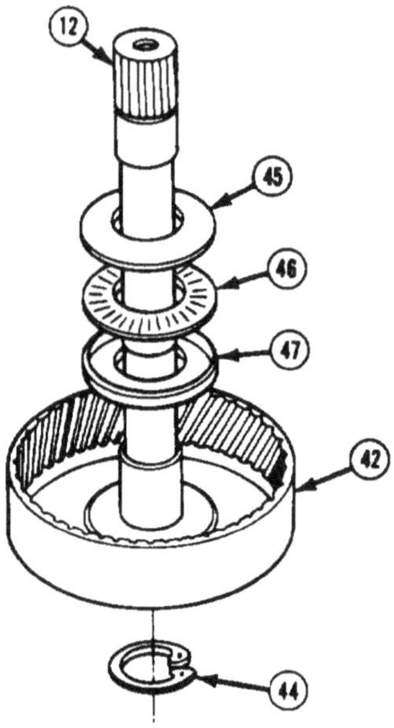

c. CLEANING AND INSPECTION

WARNING

- Dry cleaning solvent P-D-680 is toxic and flammable. Always wear protective goggles and gloves and use only in a well-ventilated area. Avoid contact with skin, eyes, and clothes and DO NOT breathe vapors. DO NOT use near open flame or excessive heat. The solvent's flash point is 100°F-138°F (38°'C-59°C). If you become dizzy while using cleaning solvent, immediately get fresh air and medical help. if solvent contacts eyes, immediately wash your eyes with water and get medical aid.
- Compressed air used for cleaning purposes should never exceed 30 psi (207 kPa). Use only effective chip guarding and personnel protective equipment (goggles/shield, gloves, etc.). Failure to follow this warning may result in serious injury to personnel.

1. Clean all components with dry cleaning solvent. Use compressed air to open up any blocked oil passages.

2. inspect all center support assembly components for damage or blocked oil passages. Inspect intermediate clutch springs for collapsed or distorted springs. Replace any damaged components.

TA50294

5-19. TRANSMISSION INTERMEDIATE CLUTCH, CENTER SUPPORT ASSEMBLY AND GEAR UNIT, AND REAR BAND MAINTENANCE (Con't).

3. Inspect mainshaft and output shaft for damage or oil passages that remain blocked. Replace if any of these conditions exist.

4. Inspect roller clutch assembly for damaged rollers or springs. Inspect roller clutch cage for damage. Replace if any of these conditions exist.

5. Inspect rear internal gear teeth, splines, and bearings for damage, Replace if damaged.

6. Inspect sun gear and sun gear shaft for damaged teeth or splines or blocked oil passages. Replace if any of these conditions exist.

7. Inspect reaction carrier (21) and output carrier (22) assemblies for damage. Inspect output carrier parking pawl lugs (48) for damage. Replace reaction carrier or output carrier if damaged.

8. Inspect reaction carrier and output carrier pinion gears (43) for damage or rough bearings. Measure pinion end play, End play should be 0,009-0.024 in. (0.229-0.610 mm). If end play is not within specification, replace pinion gear and washer assemblies. (See PINION GEAR REPLACEMENT)

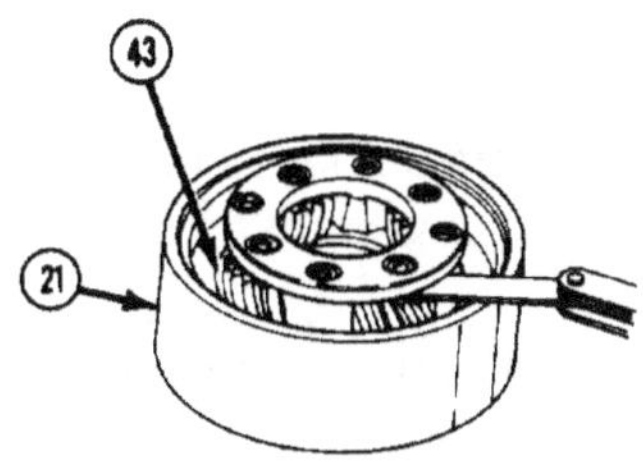

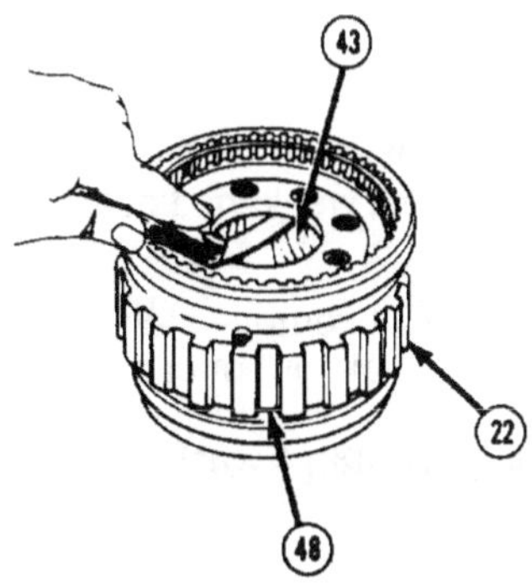

9. Inspect output shaft, sun gear shaft, and center support assembly bushings for signs of damage. Remove and discard bushings if damaged.

10. Position transmission with torque converter end down, Inspect transmission case bushing for damage. If damaged drive bushing out of transmission case and discard.

11. Inspect intermediate clutch plates for damage. Replace if damaged.

TA50295

5-19. TRANSMISSION INTERMEDIATE CLUTCH, CENTER SUPPORT ASSEMBLY AND GEAR UNIT, AND REAR BAND MAINTENANCE (Con't).

d. PINION GEAR REPLACEMENT

NOTE

- Reaction carrier (21) and output carrier (22) pinion gears (43) are replaced the same way.
- Replacement of one reaction carrier (21) pinion gear (43) assembly is described. Repeat steps to replace other pinion gear assemblies.

1. Support reaction carrier (21) on its front face.

WARNING

When using drill, use only effective chip guarding and personnel protective equipment (goggles/shield, gloves, etc.). Failure to follow this warning may result in injury to personnel.

CAUTION

Use care to ensure that drill does not remove any stock from carrier as this could result in a weakened carrier.

2. Using a 1/2 in. drill bit, remove stake marks from end of pinion pin (49). Using a tapered drift, drive or press pinion pin out of carrier. Discard pinion pin.
3. Remove pinion gear (43), 4 washers (51 and 52), and 18 needle bearings (53), Discard all components.

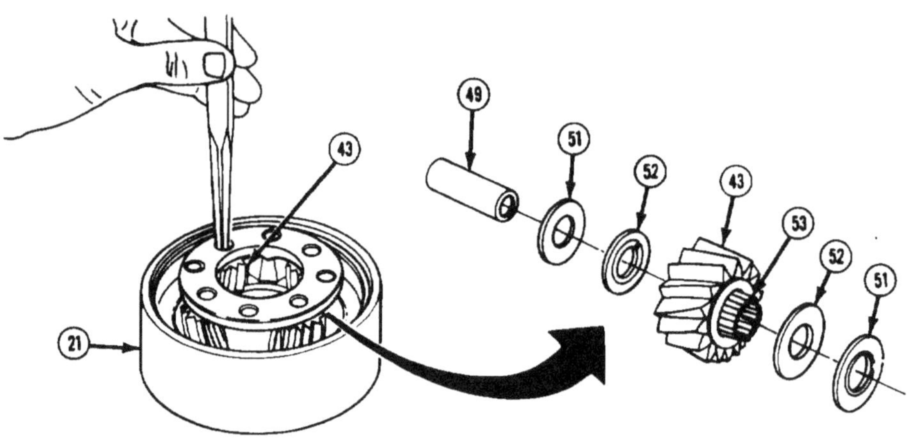

TA50296

5-19. TRANSMISSION INTERMEDIATE CLUTCH, CENTER SUPPORT ASSEMBLY AND GEAR UNIT, AND REAR BAND MAINTENANCE (Con't).

4. Inspect pinion pocket thrust faces and remove any burrs found.

5. Apply petrolatum to 18 new needle bearings (53), Using a new pinion pin (49) as a guide, insert new needle bearings into new pinion gear (43). Remove pinion pin.

6. Apply petrolatum to 2 new bronze washers (51) and steel washers (52). Place a new bronze and steel washer on both sides of pinion gear (43) with steel washers against gear.

CAUTION

Ensure that headed end of pinion pin (49) is driven below carrier face. Both ends of pinion pin must lie below carrier face or interference may occur,

7. Position pinion gear assembly (50) inside reaction carrier (21) and install a pilot shaft through rear face of assembly to hold parts in place, From the front, drive a new pinion pin (49) into place while rotating pinion gear. Using a punch and vise as an anvil, stake opposite end of pinion pin in 3 places with a blunt chisel.

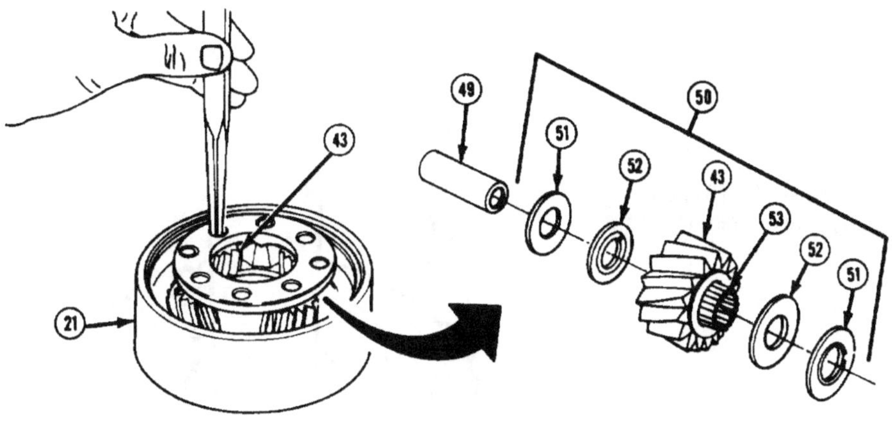

TA50297

5-19. TRANSMISSION INTERMEDIATE CLUTCH, CENTER SUPPORT ASSEMBLY AND GEAR UNIT, AND REAR BAND MAINTENANCE (Con't).

e. ASSEMBLY

1. Install new center support assembly (8) bushing if removed. Position center support assembly with oil delivery sleeve up. Aline bushing slot with oil delivery hole in center support assembly bottom land. Drive new bushing squarely into bore until flush to 0.010 in. (0,254 mm) below end of oil delivery sleeve.

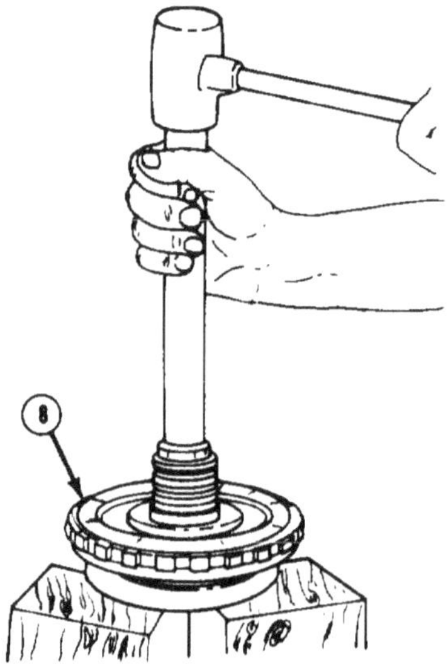

2. Install new output shaft (13) bushing, if removed, by driving it in until tool bottoms.

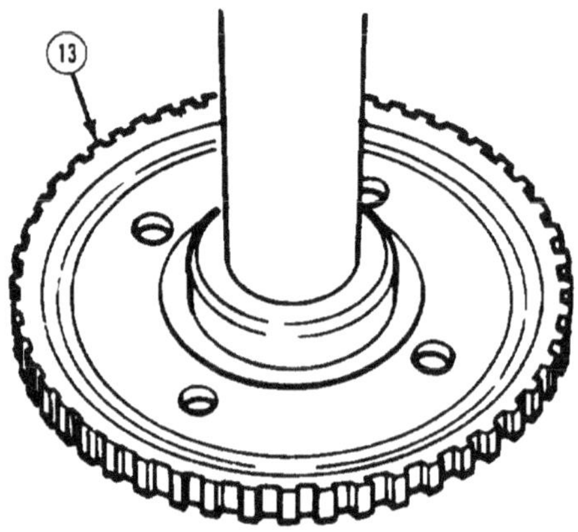

TA50298

5-19. TRANSMISSION INTERMEDIATE CLUTCH, CENTER SUPPORT ASSEMBLY AND GEAR UNIT, AND REAR BAND MAINTENANCE (Con't).

3. If transmission case (5) bushing was removed, install new bushing with oil passages facing up. Drive bushing squarely into bore until bushing is 0.040-0.055 in. (1 .016-1 .397 mm) above selective washer surface. Remove tools and stake bushing in bushing lubrication grooves.

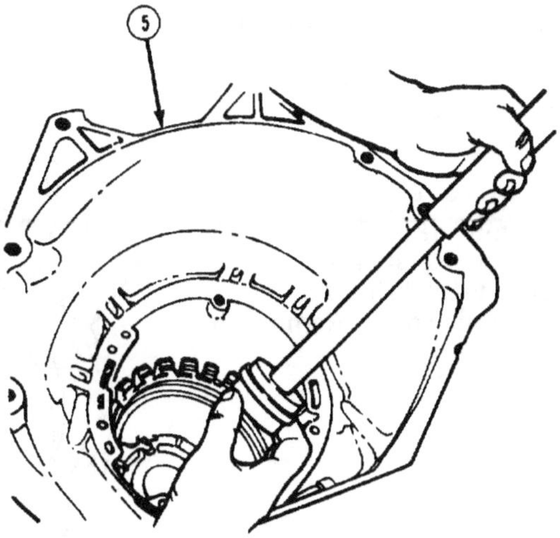

4, Install new sun gear shaft (19) bushings if removed.

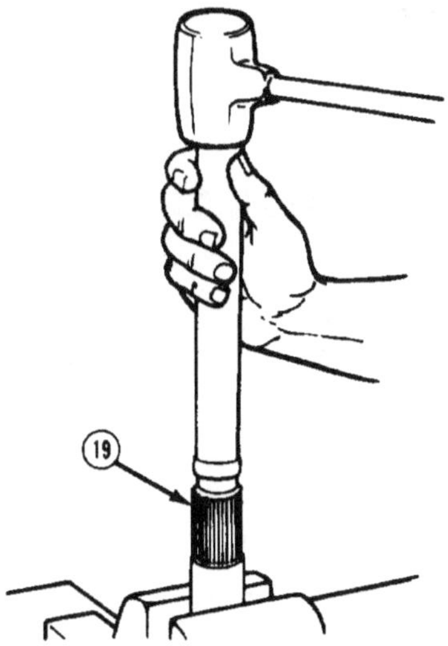

TA50299

5-19. TRANSMISSION INTERMEDIATE CLUTCH, CENTER SUPPORT ASSEMBLY AND GEAR UNIT, AND REAR BAND MAINTENANCE (Con't).

5. Apply petrolatum to seal grooves in intermediate clutch piston (29) . Lubricate new inner seal (30) and outer seal (31) with hydraulic fluid, Install seals with lips facing away from spring guide.

6. Place seal protector (54) over center support assembly (8) hub. Rotate intermediate clutch piston (29) and install, ensuring that it fully seats in center support assembly. Remove seal protector.

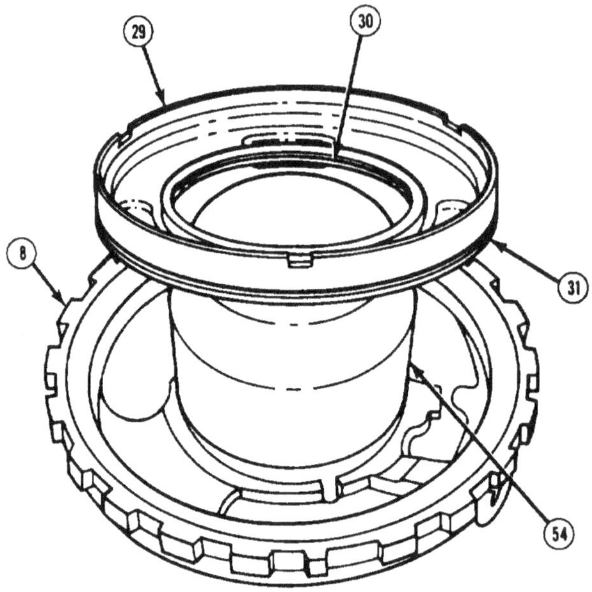

7. Install clutch spring guide (28). Install 3 springs (26) as marked.

NOTE

Ensure that spring retainer (27) does not catch in snapring (25) groove.

8. Compress spring retainer (27) and install snapring (25),

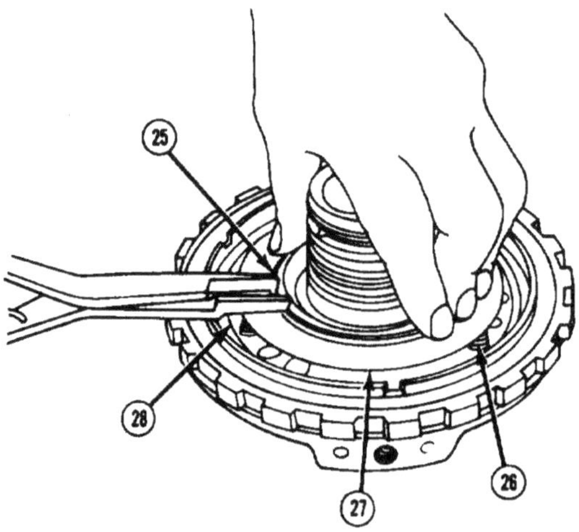

TA50300

5-19. TRANSMISSION INTERMEDIATE CLUTCH, CENTER SUPPORT ASSEMBLY AND GEAR UNIT, AND REAR BAND MAINTENANCE (Con't).

9. nstall 4 new oil seal rings (24) on center support assembly (8) so that angled ends of rings meet without an overlap.

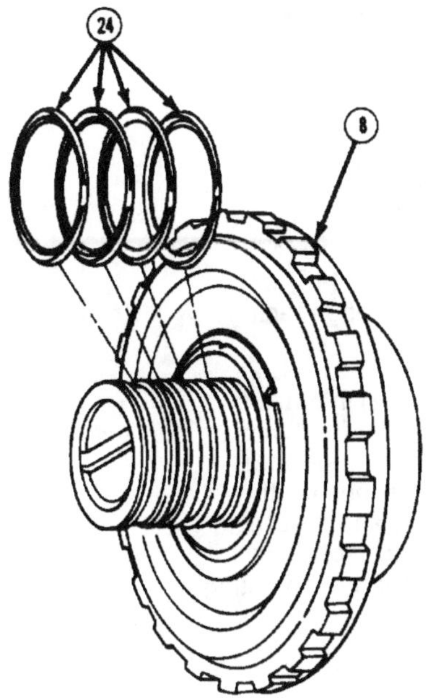

WARNING

Compressed air used to check operation of intermediate clutch piston (29) should never exceed 30 psi (207 kPa). Use only effective chip guarding and personnel protective equipment (goggles/shield, gloves, etc.) . Failure to follow this warning may result in serious injury to personnel,

10. Check operation of center support assembly (8) intermediate clutch piston (29) by applying compressed air through center oil passage (55).

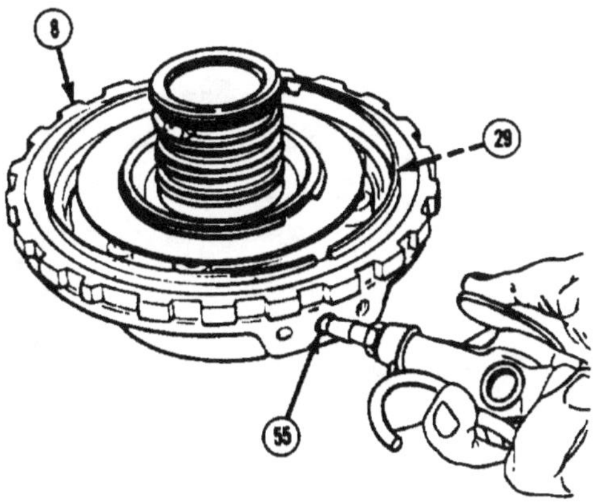

TA50301

5-19. TRANSMISSION INTERMEDIATE CLUTCH, CENTER SUPPORT ASSEMBLY AND GEAR UNIT, AND REAR BAND MAINTENANCE (Con't).

11. Install rear internal gear (42) on end of mainshaft (12) that has snapring groove. Install snapring (44).

12. Lightly coat races (45 and 47) and thrust bearing (46) with petrolatum, Install large diameter race (47) on mainshaft (12) first, with flange facing up. Install thrust bearing in race (47), Install small diameter race (45) on thrust bearing with inner flange facing down.

13. Lubricate pinion gears (43) in output carrier (22) with hydraulic fluid and install output carrier on mainshaft (12) so that pinion gears mesh with rear internal gear (42) and rotate freely. Place assembly in a soft-jawed vise with mainshaft pointing downward.

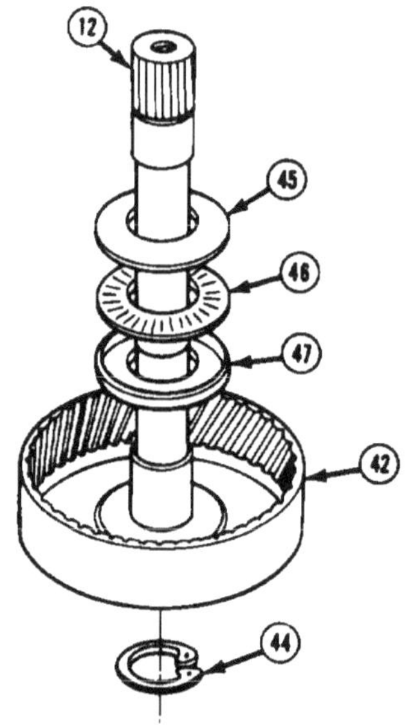

14. Lightly coat races (39 and 41) and thrust bearing (40) with petrolatum. Install small diameter race (41) to outer face of rear internal gear (42) with flange facing up, Install thrust bearing in race (41). Install large diameter race (39) with flange cupped over thrust bearing.

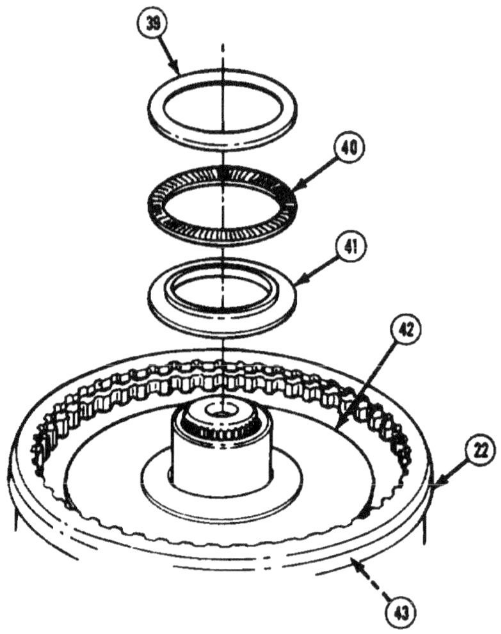

TA50302

5-19. TRANSMISSION INTERMEDIATE CLUTCH, CENTER SUPPORT ASSEMBLY AND GEAR UNIT, AND REAR BAND MAINTENANCE (Con't).

15. Install output shaft (13) into output carrier (22) and install snapring (38).

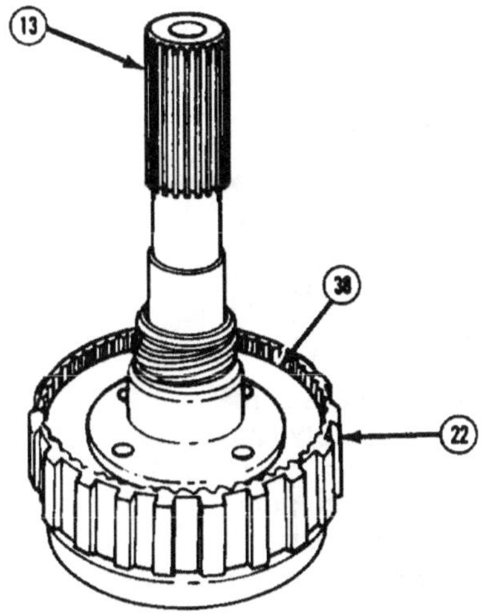

16. Remove output carrier (22) and gear unit assembly from vise . Invert assembly and place in a soft-jawed vise with mainshaft (12) pointing upward.

17. Lightly coat tab side of thrust washer (37) 'with petrolatum. Install thrust washer in output carrier (22) with tabs in tab pockets.

NOTE

When a new output carrier (22) and/or reaction carrier (21) is installed, a replacement fron t internal gear ring (32) must be used if factory-installed fron t internal gear ring prevents assembly of carriers.

18. Install front internal gear ring (32) over output carrier (22).

19. Install sun gear (36) with end having chamfered 1.D. facing down.

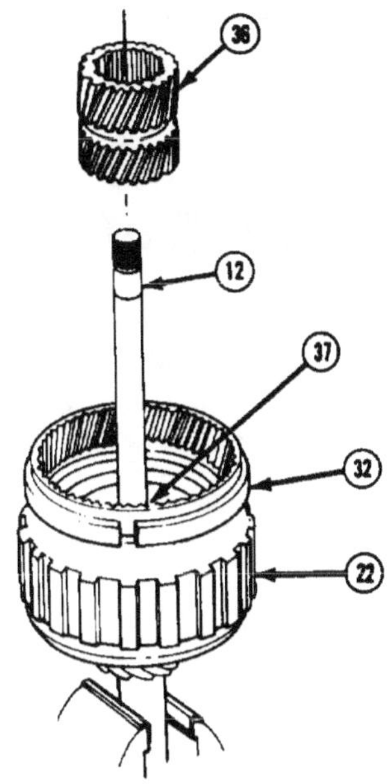

TA50303

5-19. TRANSMISSION INTERMEDIATE CLUTCH, CENTER SUPPORT ASSEMBLY AND GEAR UNIT, AND REAR BAND MAINTENANCE (Con't).

20. Install sun gear shaft (19) with larger, splined end down,

21. Lubricate pinion gears (43) in reaction carrier (21) with hydraulic fluid and install reaction carrier on output carrier (22) so that pinion gears mesh with front internal gear.

22. Lightly coat 2 races (33 and 35) and thrust bearing (34) with petrolatum, Install large diameter race (33) on sun gear shaft (19) with flange facing up. Install thrust bearing on race (33). Install small diameter race (35) on sun gear shaft with flange facing Up.

23. Install any roller that may have come out of roller clutch assembly (20) cage by compressing energizing spring with forefinger and installing roller from outer side, If any rollers are missing, replace roller clutch assembly.

24. Install roller clutch assembly (20) in reaction carrier (21).

25. Check direct clutch assembly operation. (See paragraph 5-17, ASSEMBLY, step 13)

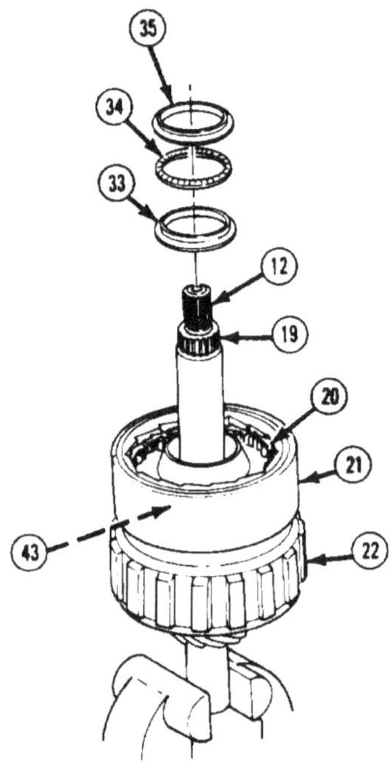

26. Lightly coat thrust washer (23) with petrolatum and install in recess of center support assembly (8).

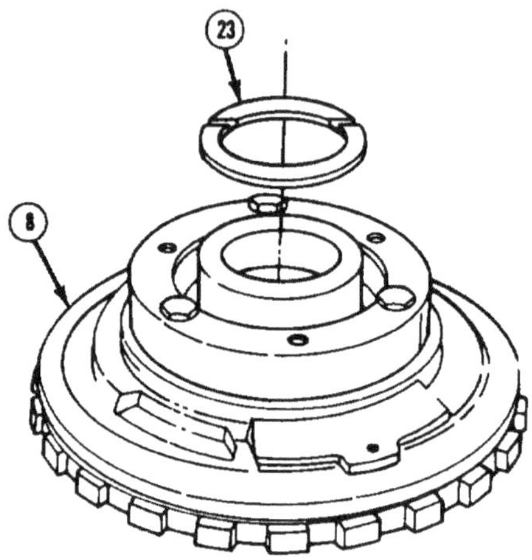

TA50304

5-19. TRANSMISSION INTERMEDIATE CLUTCH, CENTER SUPPORT ASSEMBLY AND GEAR UNIT, AND REAR BAND MAINTENANCE (Con't).

NOTE

Ensure that center support-to-reaction carrier thrust washer is still in place before installing center support assembly (8) into roller clutch in reaction carrier (21).

27. Install center support assembly (8) into roller clutch (20) in reaction carrier (21) using a counterclockwise motion. Hold reaction carrier and attempt to turn center suppor t assembly in both directions. Center support assembly should turn counterclockwise only. If center support assembly turns clockwise, disassemble to determine problem.

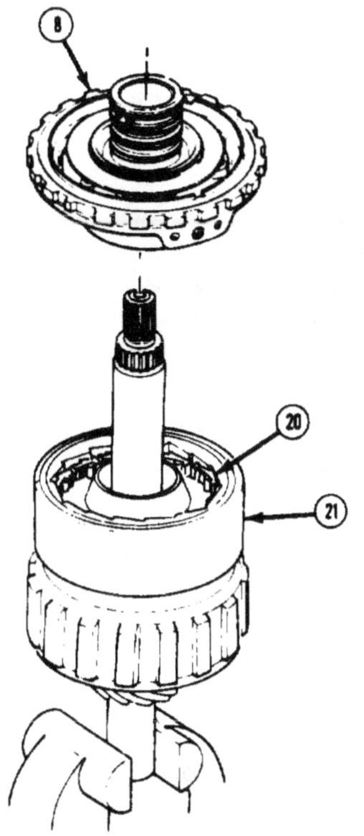

f. INSTALLATION

NOTE

Rear band (18) should be soaked in hydraulic fluid a minimum of 15 minutes.

1. Install rear band (18) in transmission case (5) so that band lugs line up with anchor pins.

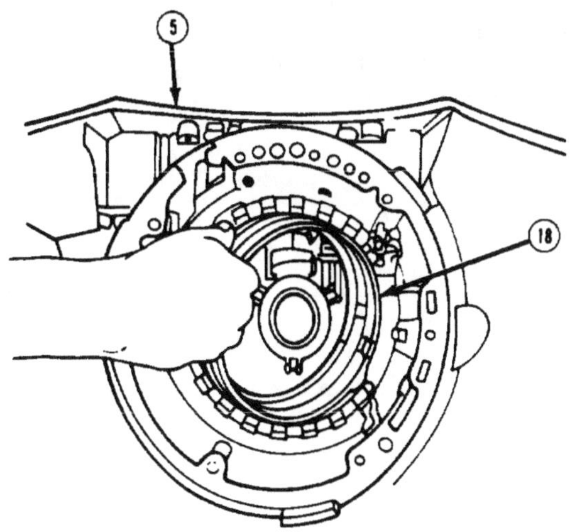

TA50305

5-19. TRANSMISSION INTERMEDIATE CLUTCH, CENTER SUPPORT ASSEMBLY AND GEAR UNIT, AND REAR BAND MAINTENANCE (Con't).

NOTE

Spacer (16) is thinner than other 2 remaining snaprings, It measures 0.040 in. (1.016 mm) and both sides are flat.

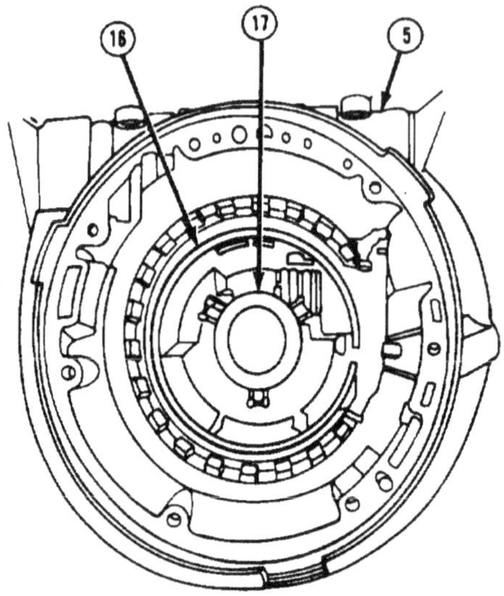

2. Inspect spacer (16) for burrs or raised edges, If present, remove with crocus cloth. Install spacer against shoulder at bottom of transmission case (5) splines.

3. Lightly coat selective washer (17) with petrolatum. Install selective washer into slots provided in rear of transmission case (5).

4. Install gear holding unit (11) on end of mainshaft (12) so that tangs engage groove on shaft. Tighten screw on gear holding unit to secure it to center support and gear unit assembly (14) and to prevent movement of the roller clutch during installation, Remove assembly from vise and lay on its side,

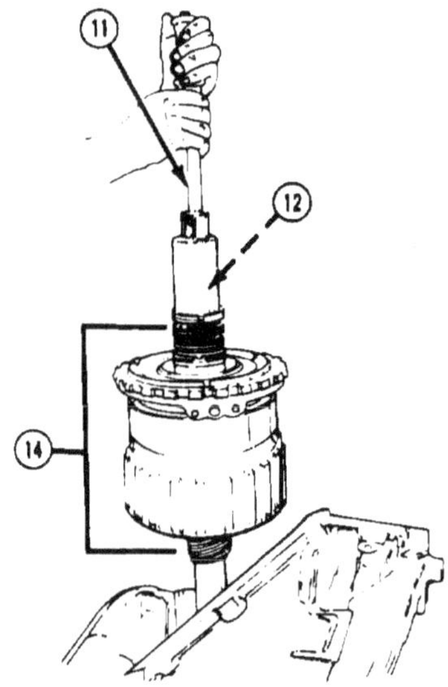

TA50306

5-19. TRANSMISSION INTERMEDIATE CLUTCH, CENTER SUPPORT ASSEMBLY AND GEAR UNIT, AND REAR BAND MAINTENANCE (Con't).

5. Lightly coat thrust washer (15) with petrolatum and install to rear face of output shaft (13) with bent tabs in tab pockets.

NOTE

An over tightened transmission holding fixture side pivot bolt will cause binding when center support and gear unit assembly (14) is installed.

6. Loosen transmission holding fixture side pivot bolt so that it does not put pressure against transmission case (5).

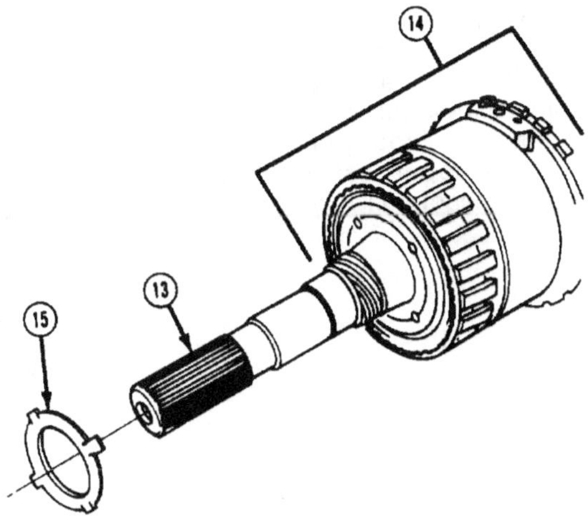

CAUTION

Be careful to not drop or bump center support and gear unit assembly (14) during installation. Failure to follow this caution may result In damage to output shaft (13) bushing in transmission case (5) or to assembly itself.

NOTE

it may be necessary to maintain allnement of rear band through access hole in transmission case (5) to prevent binding when installing center support and gear unit assembly (14).

7. Cover output shaft (13) with duct tape to prevent spline damage. With transmission case (5) in a vertical position, install center support and gear unit assembly (14), by lining up slots and carefully guiding center support and gear unit assembly into transmission case. Ensure that center support bolt hole is properly alined with hole in transmission case.

8. Remove gear holding unit and duct tape from output shaft (13).

TA50307

5-19. TRANSMISSION INTERMEDIATE CLUTCH, CENTER SUPPORT ASSEMBLY AND GEAR UNIT, AND REAR BAND MAINTENANCE (Con't).

9. Lubricate snapring (9) with hydraulic fluid and install with beveled side up, flat side against center support assembly (8). Position snapring so that its gap alines with front band anchor pin (10). Expand snapring until center support assembly is against shoulder of transmission case (5).

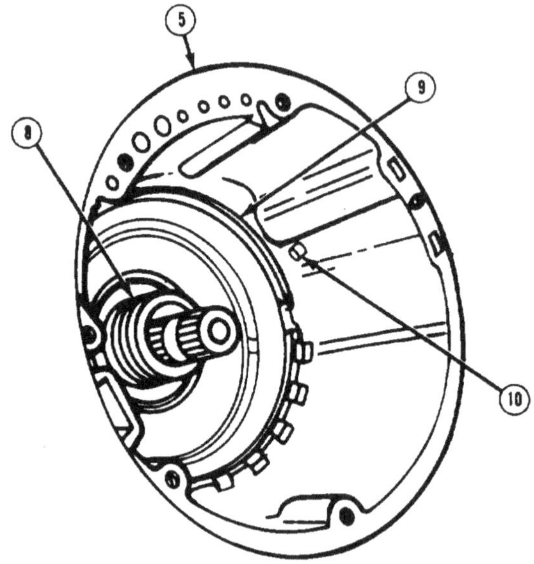

10. Turn transmission case (5) to a horizontal position with case valve body mounting face (57) up.

CAUTION

When using tool (56), be careful to not raise burrs on case valve body mounting face (57).

11. Place a suitable tool (56) into the direct clutch passage in transmission case (5). Apply pressure downward on tool and hold center support assembly (8) firmly counter-clockwise against the transmission case spline. Using a 3/8 in. 12-point thin wall deep socket (7), install center support assembly bolt and tighten to 25 lb. -ft. (34 N.m).

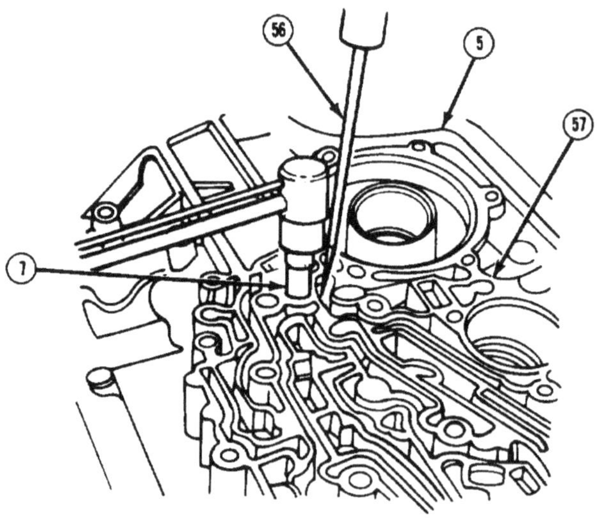

TA50308

5-19. TRANSMISSION INTERMEDIATE CLUTCH, CENTER SUPPORT ASSEMBLY AND GEAR UNIT, AND REAR BAND MAINTENANCE (Con't).

NOTE

- Intermediate clutch wave plates (3) should be soaked in hydraulic fluid approximately 15 minutes.
- Clutch wave plate (4) has 2 "V" marks cut in outer circumference,

12. Turn transmission case (5) to a vertical position. Lubricate intermediate clutch plates with hydraulic fluid, Install clutch wave plate (4). Install 3 intermediate clutch wave plates (3) and 2 flat steel clutch plates (6). Install an intermediate clutch wave plate first and alternate with a flat steel clutch plate. Install intermediate clutch backing plate (2) with flat machined surface against clutch plates.
13. Install snapring (1) with snapring gap on side of transmission case (5) opposite front band anchor pin.

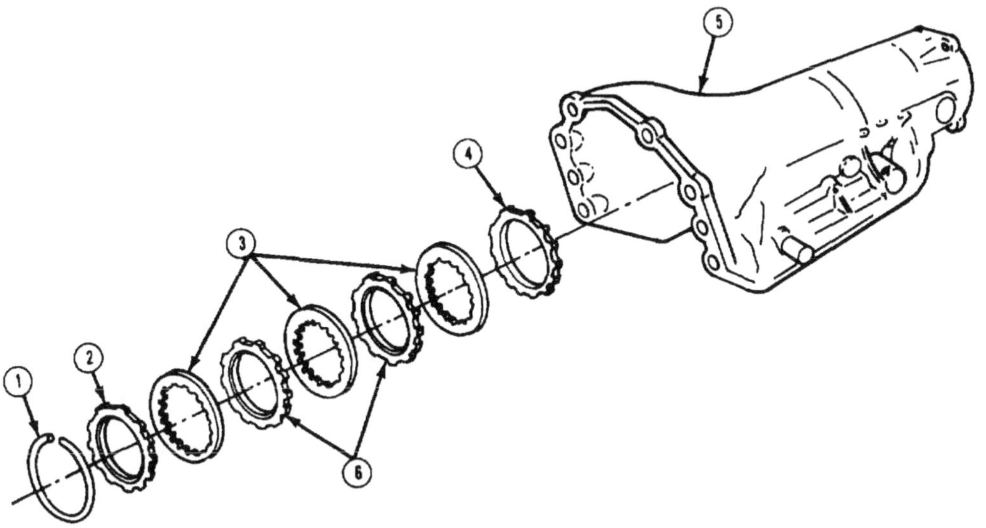

FOLLOW-ON TASKS:

- Check rear unit end play. (See paragraph 5-18)
- Install governor assembly. (See paragraph 5-5)

TA50309

Section II. TRANSFER CASE MAINTENANCE

5-20. TRANSFER CASE MAINTENANCE INDEX.

5-21. TRANSFER CASE MAINTENANCE (ALL EXCEPT M1028A1 AND M1031).

This task covers:
a. Removal
b. Disassembly
c. Cleaning and Inspection
d. Assembly
e. Installation

INITIAL SETUP:

Equipment Condition

- Both battery negative cables disconnected. (See TM 9-2320-289-20)
- Truck raised and supported on jack stands.
- Transfer case drained. (See LO 9-2320-289-12)
- Front and rear propeller shafts removed. (See TM 9-2320-289-20)
- Speedometer gear removed. (See TM 9-2320-289-20)

Too/s/Test Equipment

- Annular bearing inserter, J-29167
- Annular bearing remover, J-29168
- Bearing seal installer, J-29162
- Drive handle, J-8092
- Drive sprocket puller, J-26941
- Sprocket support remover, J-29369-1
- Slide hammer
- Snapring pliers
- Torque wrench

Materials/Parts

- One front output shaft seal
- One input drive gear seal
- One oil pump seal
- One range sector shaft oil seal
- One rear seal
- One seal washer
- One transfer case adapter gasket
- One transfer case adapter seal
- Two cotter pins
- Two locknuts
- Two 3/8 in. N.C. bolts
- Ten lockwashers
- Dry cleaning solvent (Item 23, Appendix B)
- Hydraulic fluid (Item 40, Appendix B)
- Lubricating oil (Item 47, Appendix B)
- Petrolatum (Item 50, Appendix B)
- Gasket sealing compound (Item 58, Appendix B)
- Anaerobic sealing compound (Item 59, Appendix B)

Personnel Required

- *MOS 63W (2)*

Manual References

- TM 9-214

General Safety Instructions

- Transfer case is heavy. Provide suitable support and use caution during removal and installation.
- Dry cleaning solvent is flammable and must not be used near open flame. Use only in a well-ventilated area.
- Compressed air used for cleaning purposes should never exceed 30 psi (207 kPa).

I a. REMOVAL

NOTE

- If removing transfer case (4), perform steps 1-5.
- If removing transfer case adapter (8) and seal (9), perform steps 1-8.

2. Disconnect connector from 4-wheel drive indicator light switch (3) on top of transfer case (4).

5-21. TRANSFER CASE MAINTENANCE (ALL EXCEPT M1028A1 AND M1031) (Con't).

3. Remove cotter pin (21) and washer (22), and disconnect control rod (23) at control lever assembly end. Discard cotter pin.

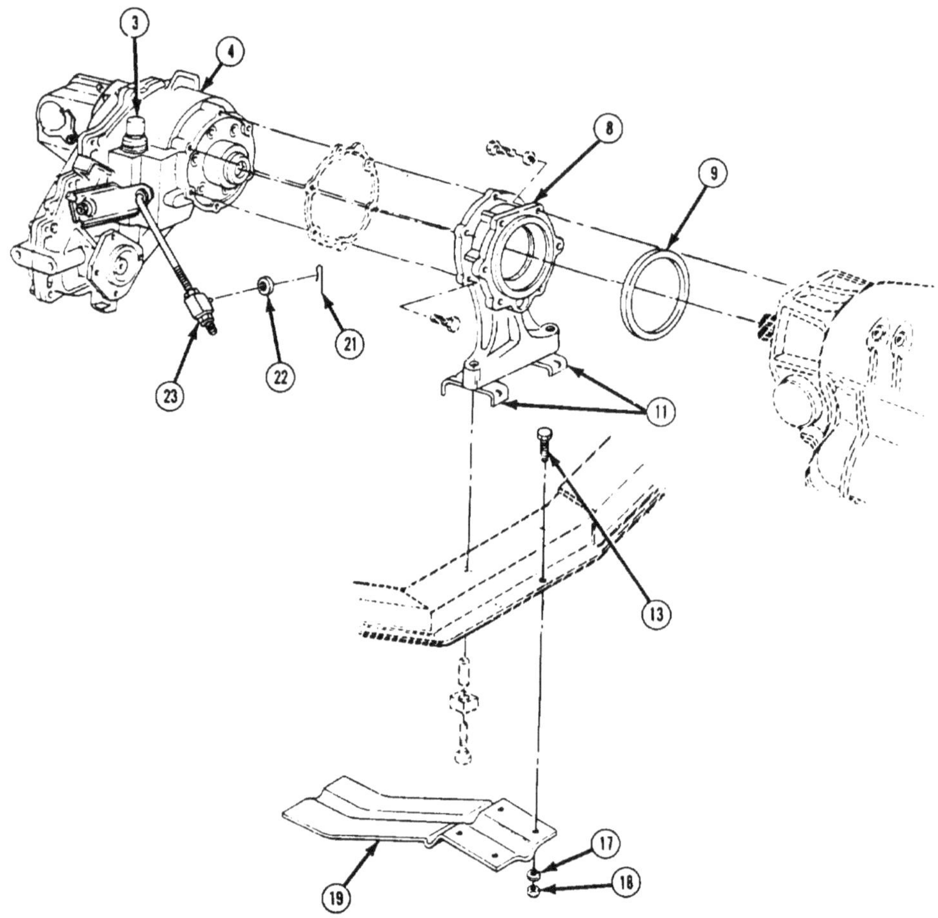

NOTE

2 left transfer case shield bolts (13) cannot be removed until transfer case adapter (8) and upper transmission mounts (11) are removed in step 8.

4. Remove 4 nuts (18), lockwashers (17), if installed, 2 right transfer case shield bolts (13), and transfer case shield (19). Discard lockwashers.

TA701854

5-21. TRANSFER CASE MAINTENANCE (ALL EXCEPT M1028A1 AND M1031) (Conk).

WARNING

Transfer case (4) is heavy. Provide suitable support and use caution during removal. Failure to follow this warning may result in damage to transfer case or serious injury to personnel.

5 . Place a suitable jack under transfer case (4). Remove 6 bolts (20). Move transfer case rearward until clear of transmission output shaft (10). Remove transfer case, Remove transfer case adapter gasket (5) and discard.

6 . Remove 2 bolts (16), lower transmission mounts (15), and spacers (14) at transmission support (12).

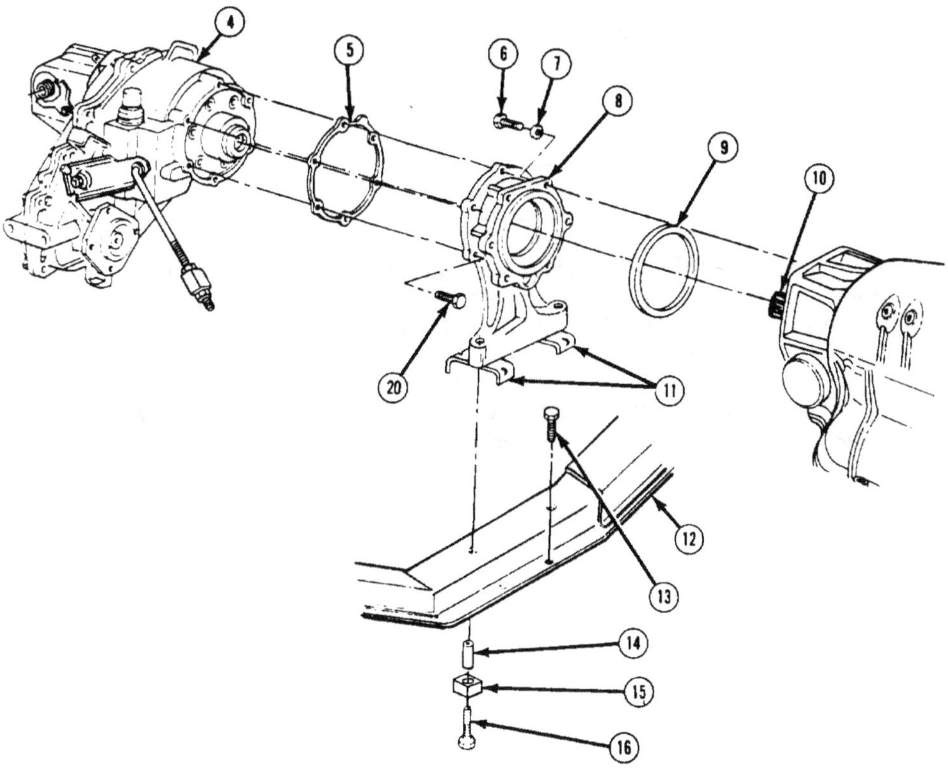

TA50311

5-21. TRANSFER CASE MAINTENANCE (ALL EXCEPT M1028A1 AND M1031) (Con't).

CAUTION

Transmission should be supported before removing transfer case adapter (8). Failure to follow this caution may result in damage to transmission case or engine mounts.

NOTE

If transmission is to be removed, transmission should be supported with a transmission lift,

7. Raise transmission 1/2 in. (12.7 mm) and remove 6 bolts (6) and lockwashers (7). Remove transfer case adapter (8). Remove 2 upper transmission mounts (11) and 2 left transfer case shield bolts (1 3). Discard lockwashers,

8. Remove seal (9) from transfer case adapter (8) and discard.

b. DISASSEMBLY

WARNING

Dry cleaning solvent P-D-680 is toxic and flammable. Always wear protective goggles and gloves and use only in a well-ventilated area. Avoid contact with skin, eyes, and clothes and DO NOT breathe vapors, DO NOT use near open flame or excessive heat, The solvent's flash point is 100°F-138°F (38°C-590C). If you become dizzy while using cleaning solvent, immediately get fresh air and medical help. If solvent contacts eyes, immediately wash your eyes with water and get medical aid.

1. If fill plug and drainplugs are installed, remove.

5-21. TRANSFER CASE MAINTENANCE (ALL EXCEPT M1028A1 AND M1031) (Con't).

3. Remove 4-wheel drive indicator light switch (3) and washer.
4. Secure front yoke (32) in a suitable holding device and remove locknut (31), seal washer (30), and front yoke. Discard locknut and seal washer.

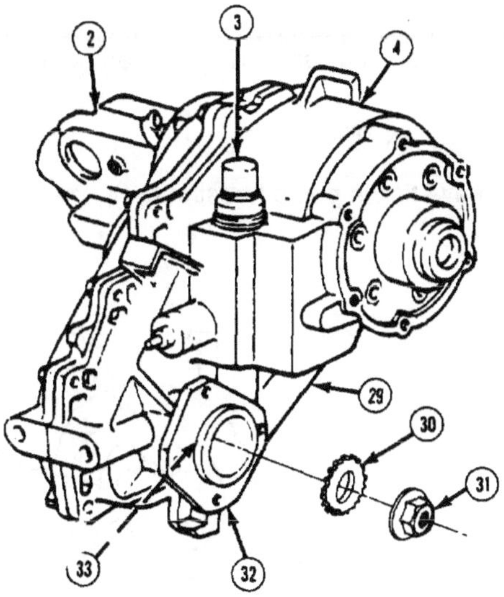

5. Turn transfer case (4) on end and place on wooden blocks with front case (29) face down and front output shaft (33) clear of working surface.
6. Remove detent bolt (36), spring (35), and poppet (34) from bottom of front case (29).

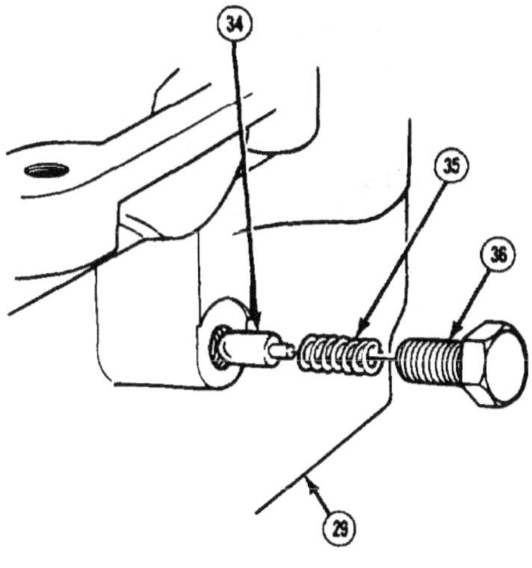

TA701855

5-21. TRANSFER CASE MAINTENANCE (ALL EXCEPT M1028A1 AND M1031) (Con't).

CAUTION

To avoid damage to components, a soft-faced hammer should be used to tap on extension housing (2) and separate it from rear case (43), DO NOT pry on mating surfaces.

7. Remove 4 bolts and remove extension housing (2). Remove pump housing (37) and seal (38). Remove speedometer drive gear (40) from mainshaft (39). Discard seal.

NOTE

Oil pump gear (41) may stick in pump housing (37),

8. Remove oil pump gear (41).

CAUTION

DO NOT pry at mating surfaces of front case (29) and rear case (43). Failure to follow this caution may result in damage to matin g surfaces.

NOTE

2 longer bolts (42) at ends of case halves are installed with washers.

9. Remove 11 bolts (42) and 2 washers. Gently pry at slots in ends of case halves and separate rear case (43) from front case (29).

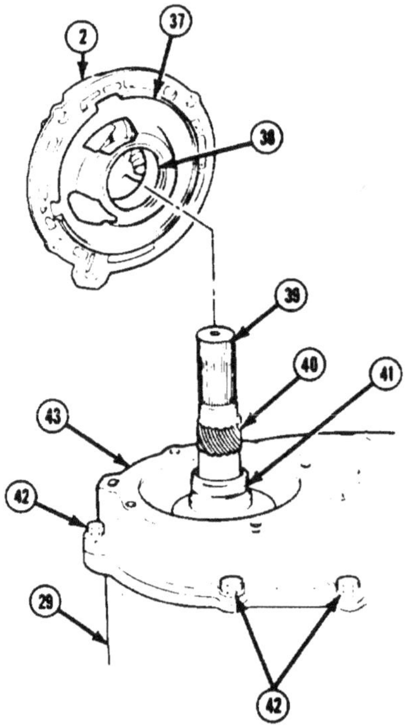

TA60313

5-21. TRANSFER CASE MAINTENANCE (ALL EXCEPT M1028A1 AND MI 031) (Con't).

10. Remove magnet (44) from slot in front case (29). Remove spring cup (45) and spring (46) from shift rail (47).

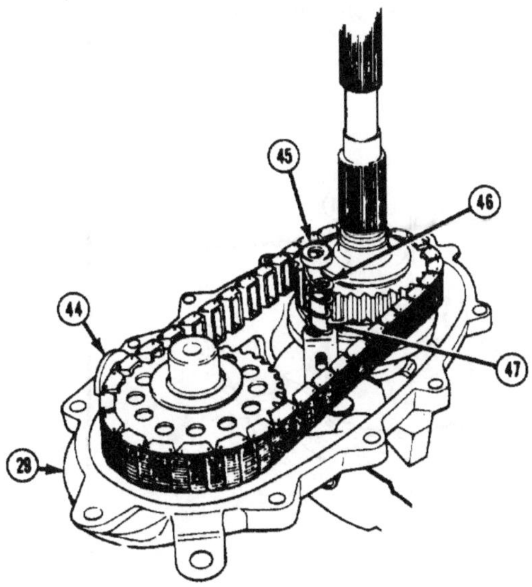

NOTE

Location of 2 thrust washers (48 and 50) and thrust bearing (49) should be noted for assembly.

11. Remove outer thrust washer (48), thrust bearing (49), and inner thrust washer (50) from front output shaft (33).

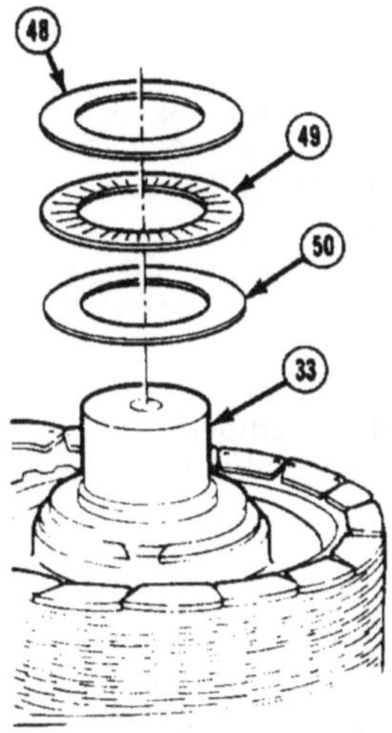

TA60314

5-21. TRANSFER CASE MAINTENANCE (ALL EXCEPT M1028A1 AND M1031) (Con't).

12. Remove retainer ring (51) at driven sprocket (52). Remove retainer ring (53) and thrust washer (54) at drive sprocket (55).

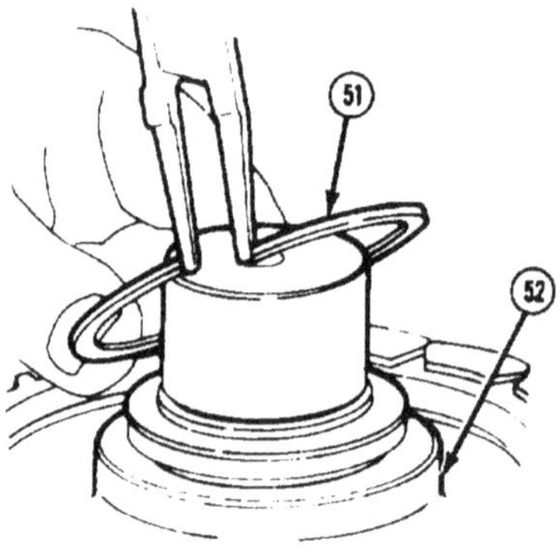

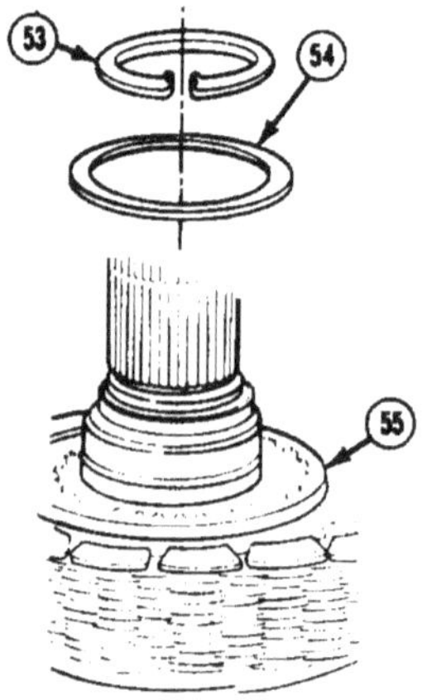

NOTE

Roller bearings (58) may fall out of drive sprocket (55) when drive sprocket is removed. There are 120 roller bearings. Ensure that all are present,

13. Evenly lift on drive (55) and driven (52) sprockets and remove with drive chain (56), as an assembly. Remove sprockets from drive chain.

NOTE

Transfer cases manufactured before July 1, 1983 (as stamped on identification plate) have no straight pin.

14. Remove straight pin from mainshaft (39). Remove 2 rows of drive sprocket roller bearings (58), if they did not already fall out, and 3 spacers (57).

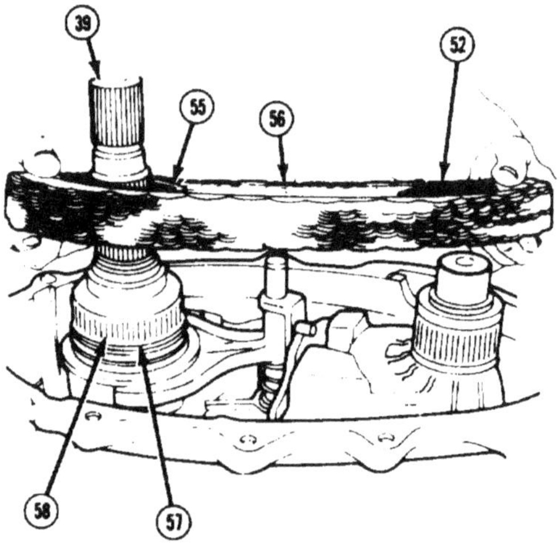

TA50315

5-21. TRANSFER CASE MAINTENANCE (ALL EXCEPT M1028A1 AND M1031) (Con't).

15. Remove front output shaft (33). Remove thick thrust washer (61), thrust bearing (60), and thin thrust washer (59),

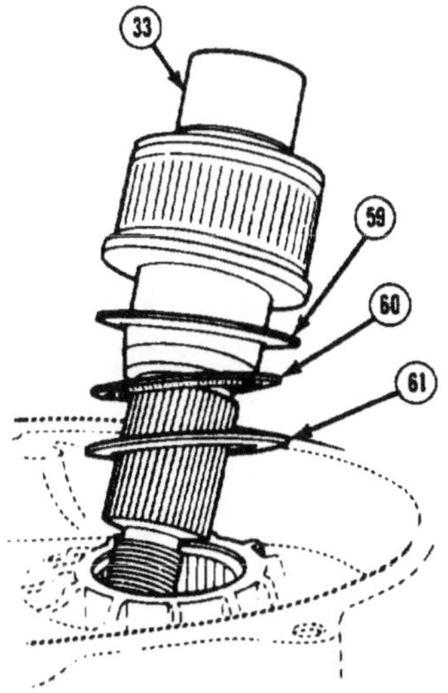

16. Remove synchronizer sleeve (64), mode fork (63), and mode fork retainer (62) as a unit. Remove mode fork retainer from mode fork.

17. Remove mode fork (63) from synchronizer sleeve (64). Remove shift rail (47).

NOTE

Position of synchronizer hub (65) should be noted for assembly,

18. Remove mainshaft (39) and synchronizer hub (65) as a unit. Remove retainer ring from underside of synchronizer hub. Remove synchronizer hub from mainshaft.

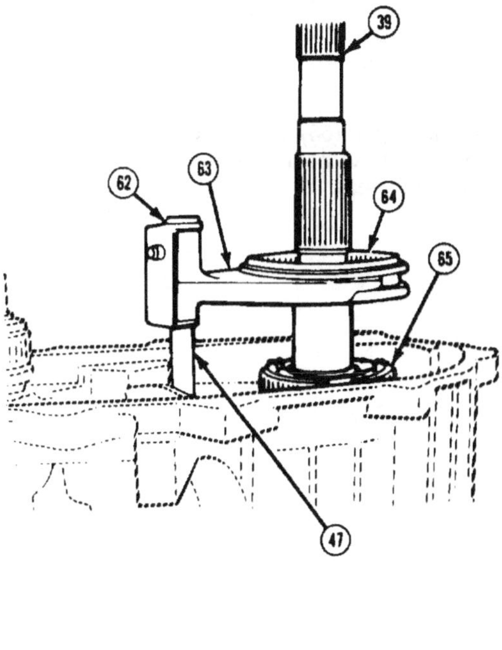

TA50316

5-21. TRANSFER CASE MAINTENANCE (ALL EXCEPT M1028A1 AND M1031) (Con't).

19. Remove annulus gear retainer ring (67) and thrust washer (68).

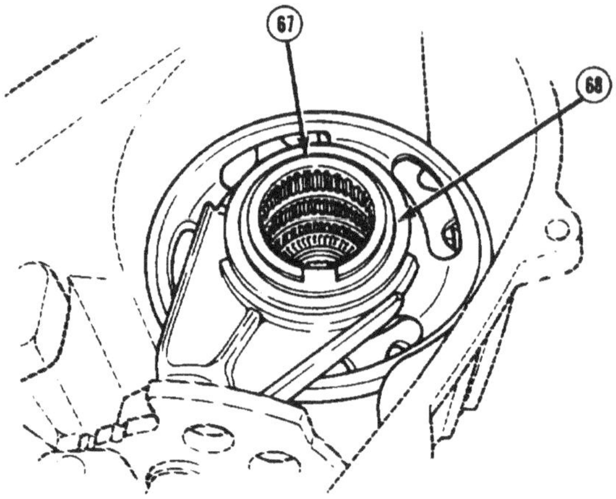

NOTE

Range fork (69) must be turned clockwise to disengage fork lug (71) from range sector (70).

20. Remove annulus gear (66) and range fork (69) as a unit. Remove range fork from annulus gear.

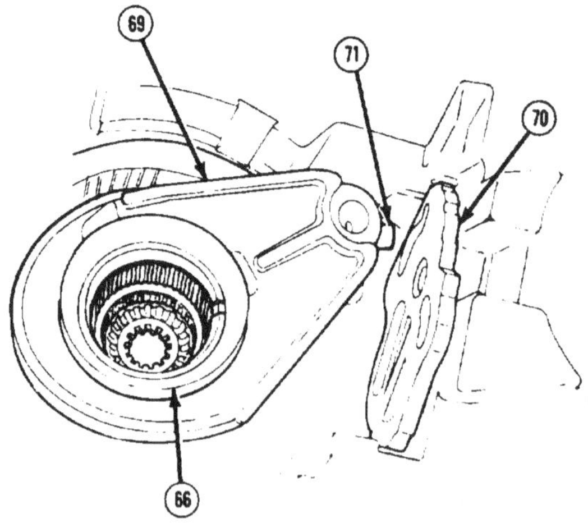

TA50317

5-21. TRANSFER CASE MAINTENANCE (ALL EXCEPT M1028A1 AND M1031) (Con't).

21. Remove planetary gear thrust washer (72) and planetary gear assembly (74).

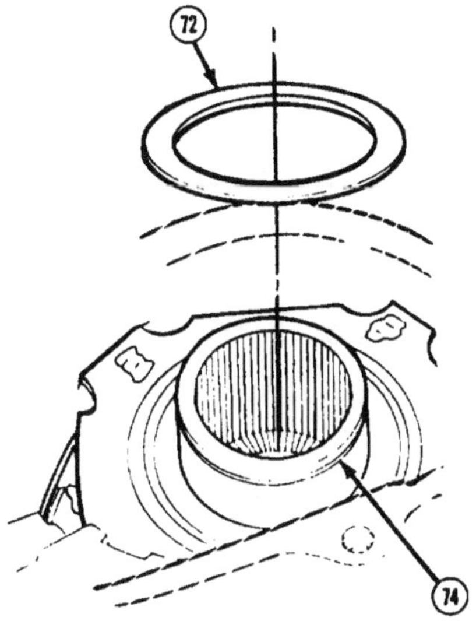

22. Remove input drive gear thrust bearing (75) from input drive gear (73). Remove input drive gear by lifting straight up and out of front case (29).

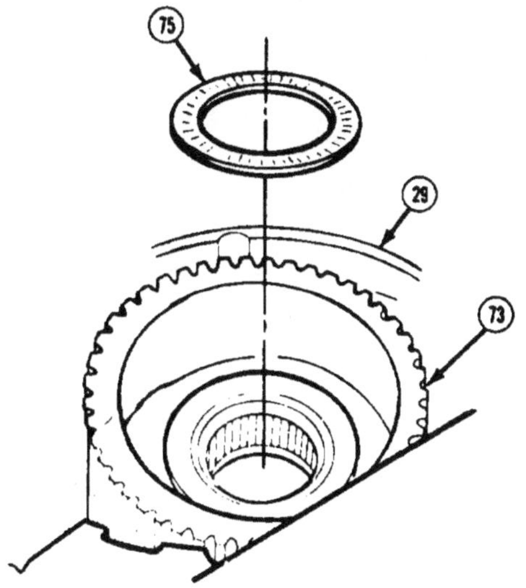

TA60318

5-21. TRANSFER CASE MAINTENANCE (ALL EXCEPT M1028A1 AND M1031) (Con't).

23. Remove input drive gear thrust bearing (76) and race (77).

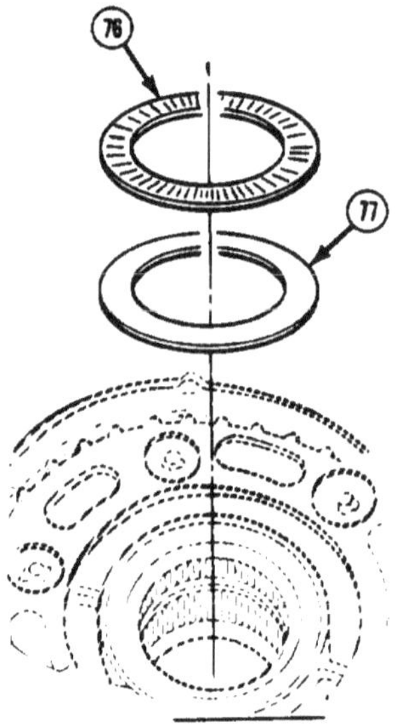

24. Remove cotter pin and washer and remove control rod (23) from operating lever (79). Remove locknut (78) and washer, and remove operating lever, Discard cotter pin and locknut.

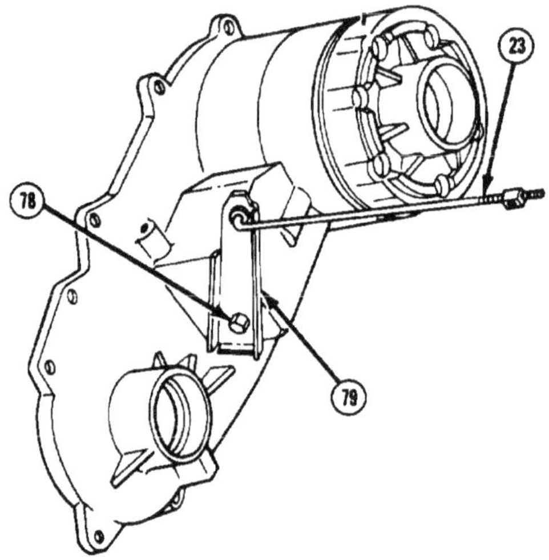

TA50319

5-21. TRANSFER CASE MAINTENANCE (ALL EXCEPT M1028A1 AND M1031) (Con't).

25. Tap inward on range sector (70) shaft and remove range sector. Remove seal retainer (82) and oil seal (81) from range sector shaft hole in front case (29). Discard oil seal.

26. Remove 6 bolts (80) and lockplate (83).

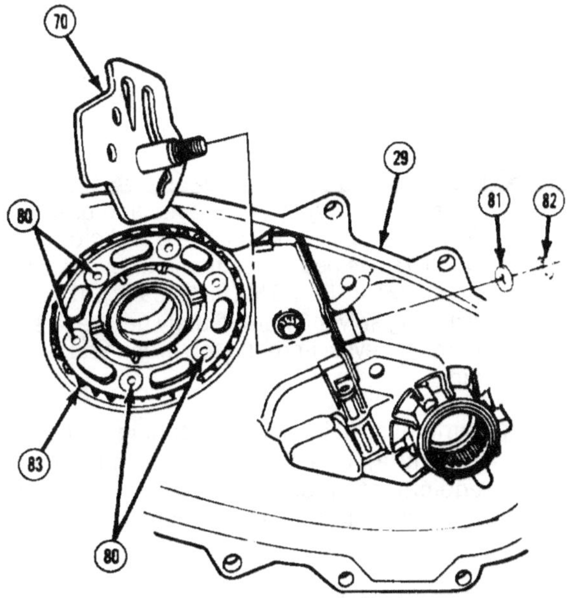

27. Remove front output shaft seal (86) and input drive gear seal (85) from front case (29). Remove rear seal (84) from extension housing (2). Discard seals.

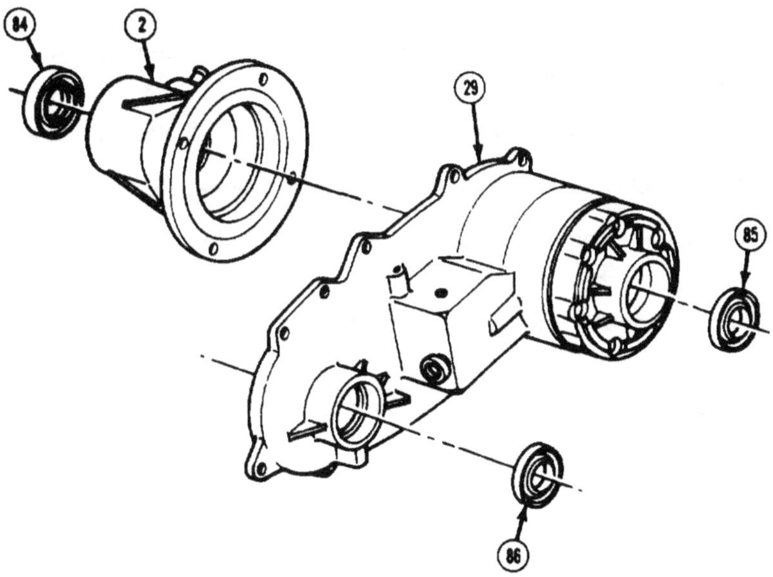

TA50320

5-21. TRANSFER CASE MAINTENANCE (ALL EXCEPT M1028A1 AND M1031) (Con't).

c. CLEANING AND INSPECTION

WARNING

- Dry cleaning solvent P-D-680 is toxic and flammable. Always wear protective goggles and gloves and use only in a well-ventilated area. Avoid contact with skin, eyes, and clothes and DO NOT breathe vapors. DO NOT use near open flame or excessive heat. The solvent's flash point is 100°F-138°F (38°C-59°C). If you become dizzy while using cleaning solvent, immediately get fresh air and medical help. If solvent contacts eyes, immediately wash your eyes with water and get medical aid.
- Compressed air used for cleaning purposes should never exceed 30 psi (207 kPa). Use only effective chip guarding and personnel protective equipment (goggles/shield, gloves, etc.). Failure to follow this warning may result in injury to personnel.

1. Clean all removed components, oil passages, and channels in each transfer case half with dry cleaning solvent to remove all old lubricant and dirt. Remove all metal filings from magnet. Clean all mating surfaces.
2. Apply compressed air to each oil passage and channel to remove any obstructions or dry cleaning solvent residue.
3. Inspect thrust washers, thrust bearings, and retainer rings for damage. Replace if damaged.
4. Inspect gear teeth and splines, and shaft splines for damage. Remove minor nicks or scratches with light lubricating oil and a stone.
5. Inspect drive chain, and drive and driven sprockets for damage. Replace if damaged.
6. Inspect lockplate and lockplate hub for damage. Replace if damaged.
7. Inspect transfer case halves and bolt hole threads, and 2 transfer case alinement dowels for damage. Replace if damaged.
8. Inspect transfer case adapter and oil seal grooves in adapter for damage. Replace if damaged.
9. Inspect pads on mode fork and range fork for damage. Replace if damaged.

5-21. TRANSFER CASE MAINTENANCE (ALL EXCEPT M1028A1 AND MI 031) (Con't).

NOTE

Damage to bearings may be determined in accordance with TM 9-214.

10. Remove mainshaft rear bearing (87) from extension housing (2) if damaged,
11. Remove front output shaft front bearing (90) from front case (29) and rear pilot bearing (91) from rear case (43) if damaged.
12. Remove input drive gear front and rear bearings (89) from front case (29) if damaged.
13. Remove input drive gear pilot bearing (88) from input drive gear (73) if damaged.
14. Remove annulus gear and extension housing bushings if damaged.

d. ASSEMBLY

NOTE

All components should be lightly lubricated with hydraulic fluid as they are assembled.

1 . Install new bushings to annulus gear and extension housing if removed.

CAUTION

All transfer case bearings must be correctly installed to ensure that bearing oil passages are not obstructed, Failure to follow this caution may result in damage to bearings and other components.

2. Install input drive gear front and rear bearings (89) one at a time, installing rear bearing first, if removed.
3. Install front output shaft front bearing (90) to front case (29) and rear pilot bearing (91) to rear case (43), if removed. Ensure that rear pilot bearing is seated flush with edge of rear case bore to allow room for thrust bearing assembly.
4. Install mainshaft rear bearing (87) to extension housing (2) until fully seated, if removed.
5. Install input drive gear pilot bearing (88) in input drive gear (73) until seated flush with oil passage, if removed.
6. Lightly coat new front output shaft seal (86) and new input drive gear seal (85) with petrolatum and install,
7. Lightly coat new rear seal (84) with petrolatum and install to extension housing (2).

5-21. TRANSFER CASE MAINTENANCE (ALL EXCEPT M1028A1 AND MI 031) (Con't).

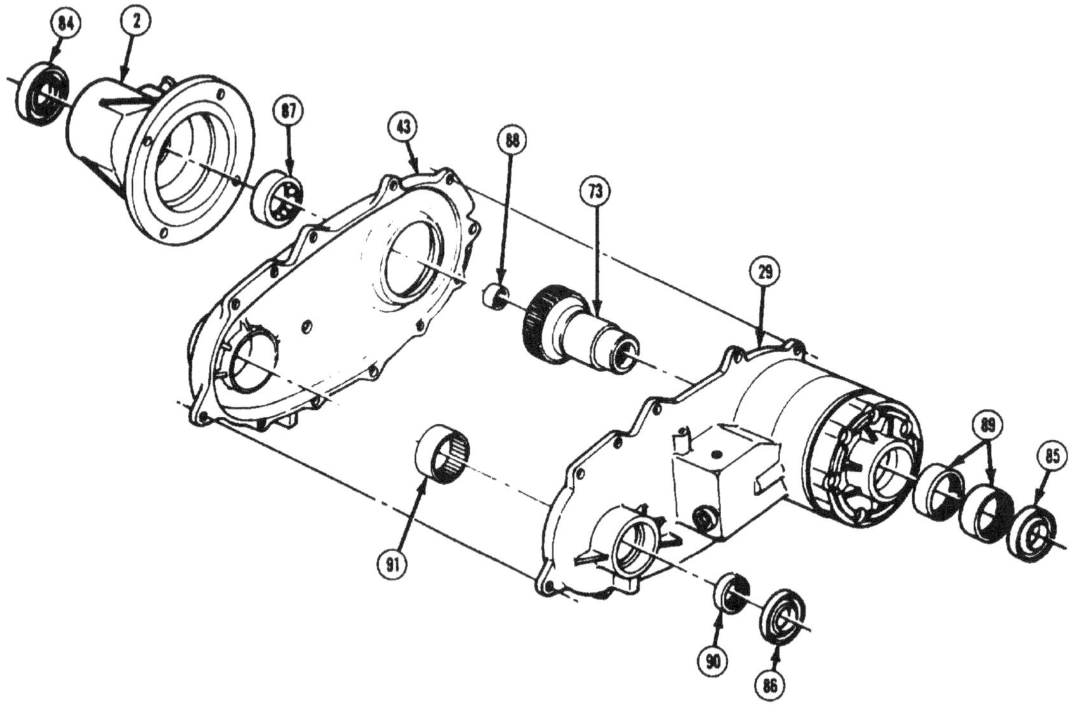

TA50321

5-21. TRANSFER CASE MAINTENANCE (ALL EXCEPT M1028A1 AND M1031) (Con't).

NOTE

Lockplate (83) must be installed with raised areas around bolt holes against front case (29).

8 . Coat front case (29) and lockplate (83) surfaces around bolt holes with anaerobic sealant. Position lockplate in front case and aline bolt holes. Coat 6 bolts (80) with gasket sealing compound and install. Tighten bolts to 30 lb.-ft. (41 N.m).

9 . Install range sector (70) with shaft through front case (29). Install new oil seal (81) and seal retainer (82). Install operating lever (79) on range sector shaft with washer and new locknut (78). Tighten locknut to 20 lb.-ft. (27 N.m). Install control rod (23) to operating lever with washer and new cotter pin.

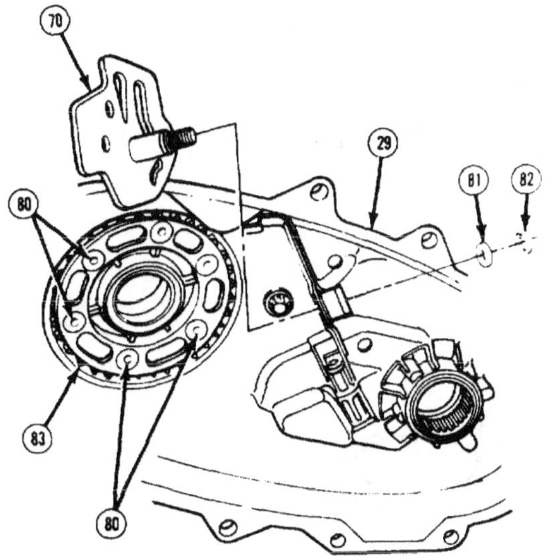

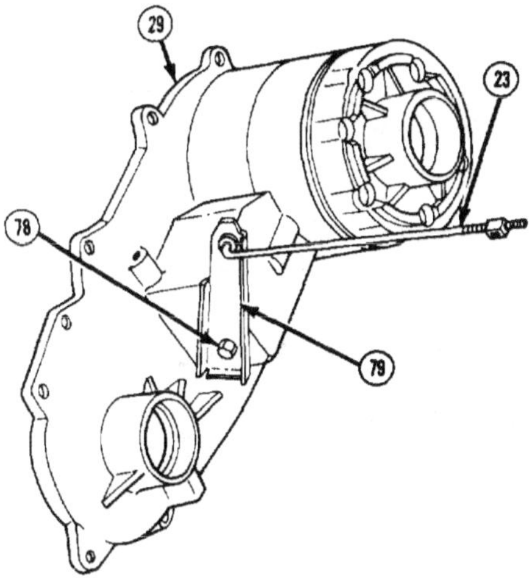

7A513322

5-21. TRANSFER CASE MAINTENANCE (ALL EXCEPT M1028A1 AND M1031) (Con't).

10. Install race (77) and input drive gear thrust bearing (76) in front case (29).

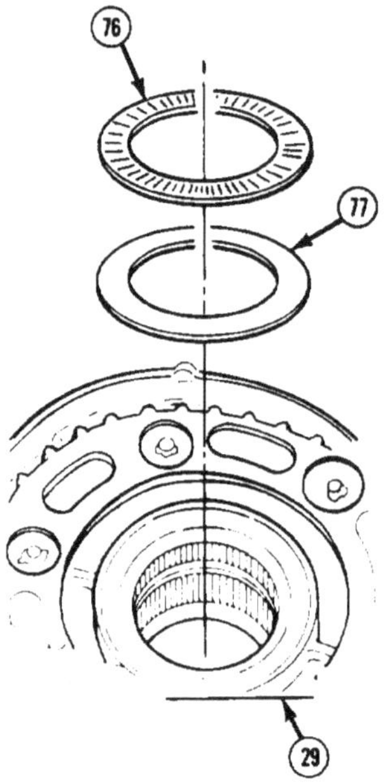

11. Install input drive gear (73). Install input drive gear thrust bearing (75) in input drive gear.

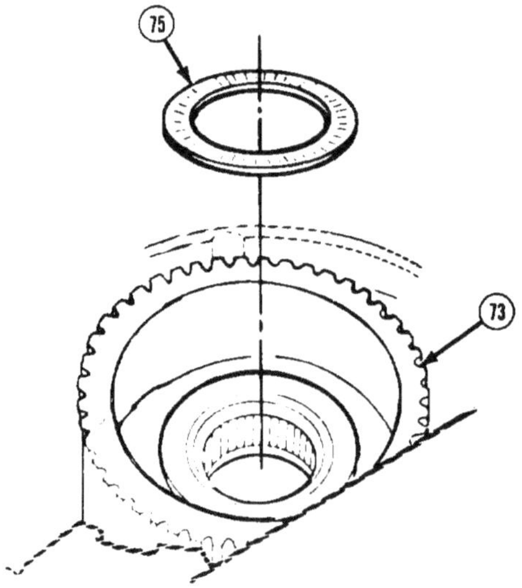

TA50323

5-21. TRANSFER CASE MAINTENANCE (ALL EXCEPT M1028A1 AND M1031) (Con't).

NOTE

Planetary gear thrust washer (72) is larger of 2 bronze washers.

12. Install planetary gear assembly (74) over input drive gear (73). Ensure that planetary gear assembly is fully seated and meshed with input drive gear. Install planetary gear thrust washer (72) on planetary hub.

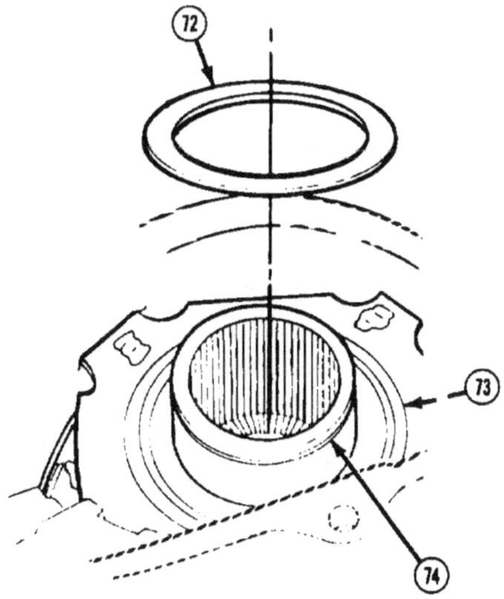

13. Install range fork (69) in annulus gear (66), Install annulus gear, with attached range fork, over planetary gear assembly (74). Fully insert range fork lug (71) in range sector (70) slot.

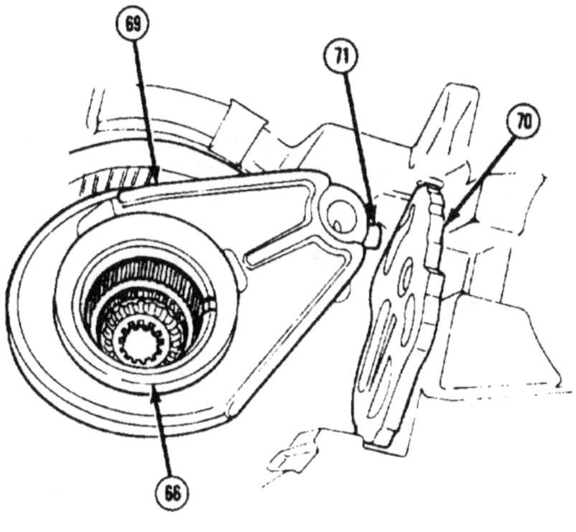

TA50324

5-21. TRANSFER CASE MAINTENANCE (ALL EXCEPT M1028A1 AND M1031) (Con't).

14. Install annulus gear thrust washer (68) and retainer ring (67)

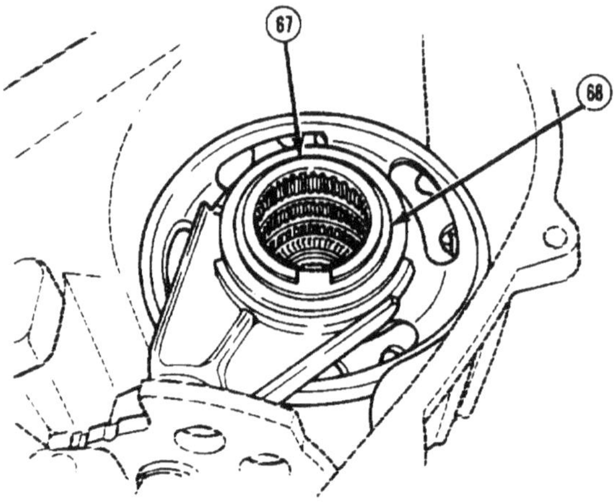

15. Aline shaft bores in front case (29) and range fork (69), and install shift rail (47) with small chamfer inside f rent case.

16. Install synchronizer hub (65) to mainshaft (39) with narrow end of slot opening in synchronizer hub facing front of front case (29). Instal l retainer ring on underside o f synchronizer hub.

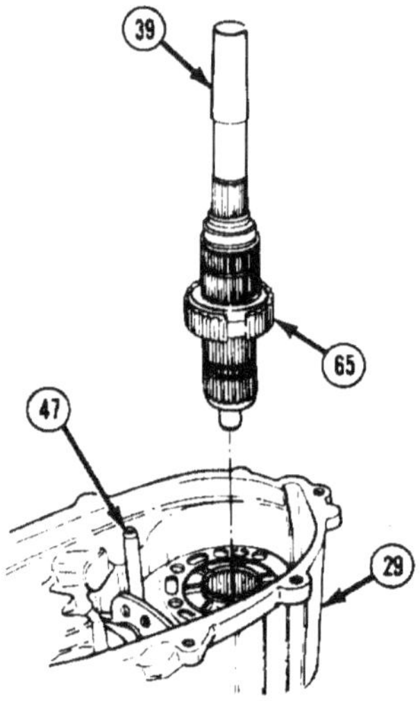

TA50325

5-21. TRANSFER CASE MAINTENANCE (ALL EXCEPT M1028A1 AND M1031) (Con't).

NOTE

Mainshaft (39) should be placed in a soft-jawed vise while assembling spacers (57) and roller bearings (58).

17. Lightly coat mainshaft (39) with petrolatum and install first spacer (57). Install first row of 60 roller bearings (58). install second spacer, then second row of 60 roller bearings. Install third spacer. Lightly coat outside surfaces of all roller bearings with petrolatum to ensure that they remain in position.

NOTE

Before installing mainshaft (39), ensure that input drive gear thrust bearing (75) is properly centered in input drive gear. (See ASSEMBLY, step 11)

18. Install mainshaft (39) and synchronizer hub (65) in front case (29).

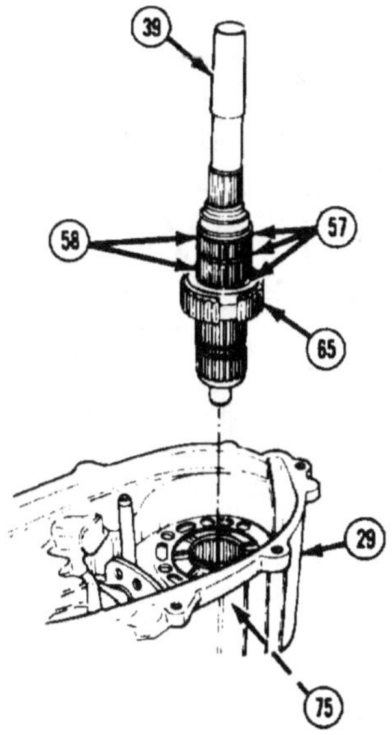

NOTE

Transfer cases manufactured before July 1, 1983 (as stamped on identification plate) have no straight pin.

19. Lightly coat straight pin with petrolatum and install to mainshaft (39).
20. Install mode fork retainer (62) to mode fork (63). install mode fork to synchronizer sleeve (64) with flat side of synchronizer sleeve facing front of front case (29),
21. Install mode fork (63) and synchronizer sleeve (64), as a unit, to shift rail (47), engaging teeth of synchronizer sleeve with teeth of synchronizer hub (65),
22. Install front output shaft (33) front thrust bearing assembly with thick thrust washer (61) next to front case (29), followed by thrust bearing (60), and thin thrust washer (59). Install front output shaft.

TA50326

5-21. TRANSFER CASE MAINTENANCE (ALL EXCEPT M1028A1 AND M1031) (Con't).

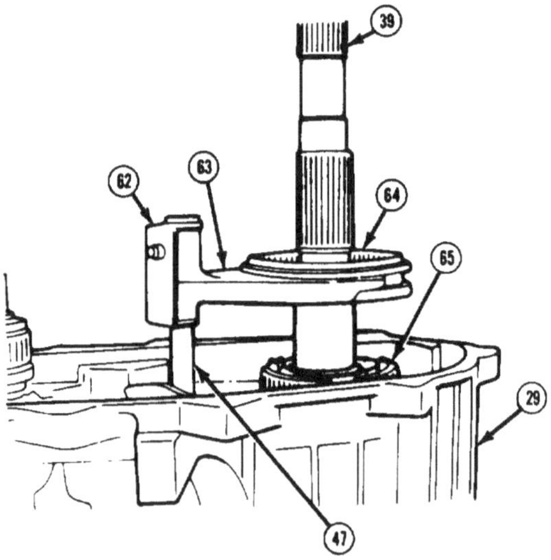

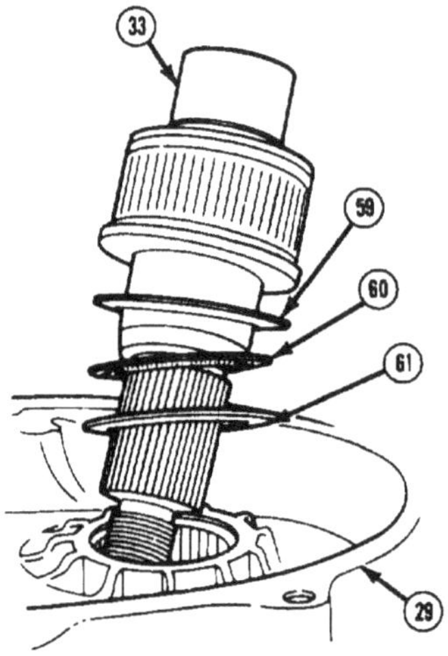

NOTE

Drive sprocket (55) is installed with tooth side facing interior of front case (29).

23. Install drive sprocket (55) and driven sprocket (52) in drive chain (56). Aline sprockets with mainshaft (39) and front output shaft (33), and install drive and driven sprockets and drive chain as an assembly.

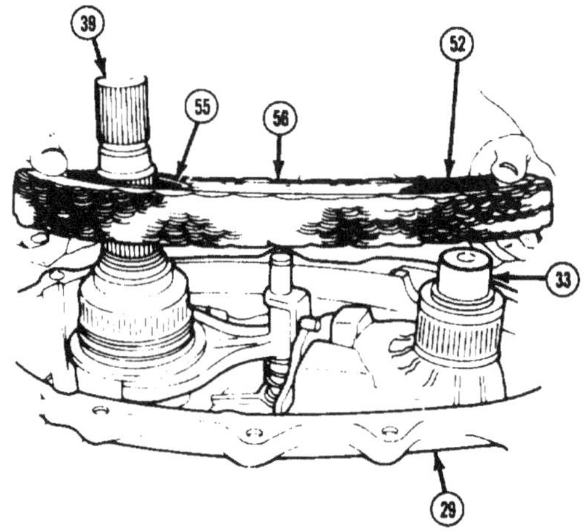

TA50327

5-21. TRANSFER CASE MAINTENANCE (ALL EXCEPT M1028A1 AND M1031) (Con't).

NOTE

- Transfer cases manufactured before July 1, 1983 (as stamped on identification plate) use first design thrust washer. Transfer cases made after that date use second design thrust washer.
- Slot In second design thrust washer must be indexed with straight pin in main shaft (39).

24. Install retainer ring (51) at driven sprocket (52). Install drive sprocket thrust washer (54) and retainer ring (53).

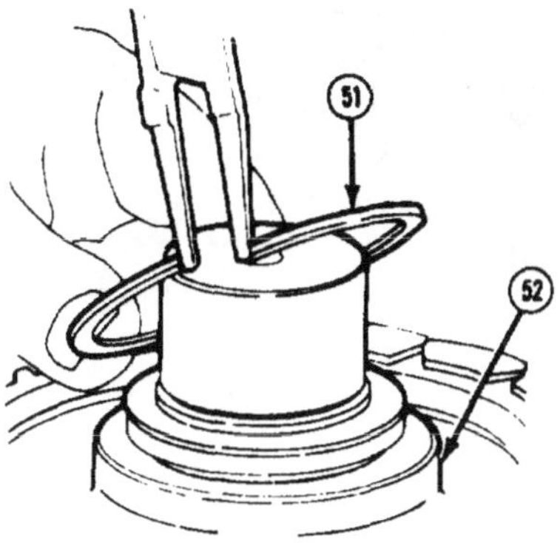

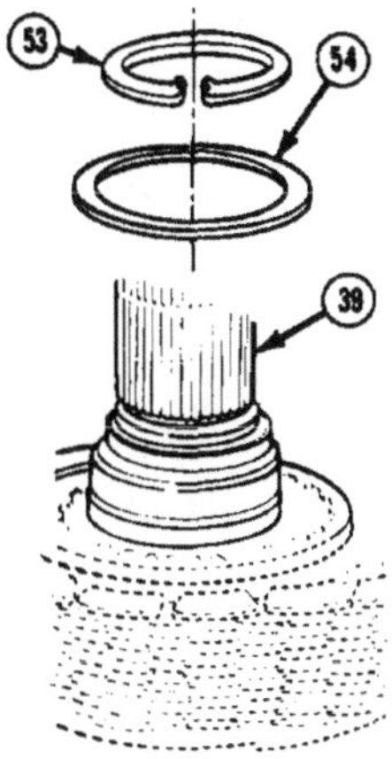

25. Install thin inner thrust washer (50), thrust bearing (49), and thick outer thrust washer (48) to front output shaft (33).

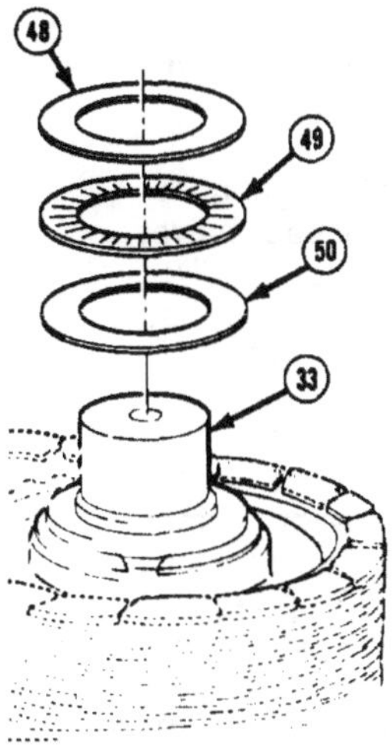

TA50328

5-21. TRANSFER CASE MAINTENANCE (ALL EXCEPT M1028A1 AND M1031) (Con't).

26. Install spring (46) and spring cup (45), to shift rail (47), with large opening in spring cup facing spring. Install magnet (44) to slot in front case (29).

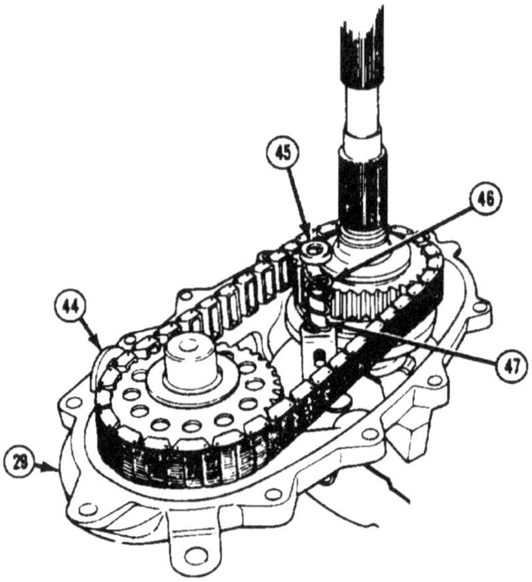

NOTE

Ensure that mating surfaces are free of grease, oil, and dirt,

27. Apply anaerobic sealant to mating surface of front case (29). Install rear case (43), alining bolt holes and 2 alinement dowels and ensuring that shift rail (47) is seated in flange on inside surface of rear case.

28. Install 2 washers and 2 longer bolts (42) at opposite ends of case halves. DO NOT fully tighten bolts. Install 9 remaining bolts without washers. Tighten bolts to 25 lb.-ft. (34 N.m).

29. Install oil pump gear (41) on mainshaft (39) with recessed side of pump gear facing interior. Firmly push down on oil pump gear to snap into position. Install speedometer drive gear (40) on mainshaft.

30. Install new pump seal (38) to pump housing (37) with seal lips facing pump housing.

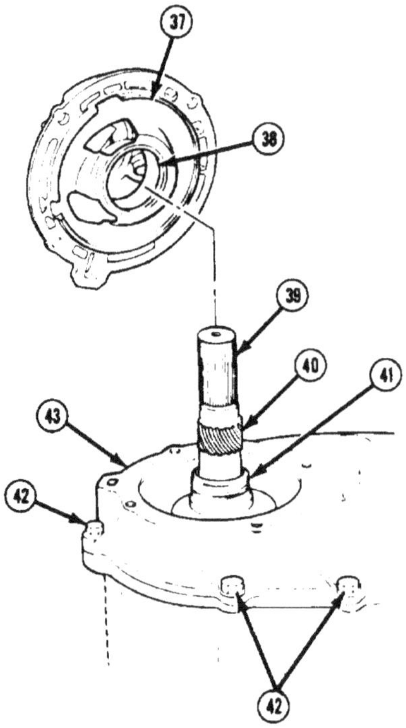

TA50329

5-21. TRANSFER CASE MAINTENANCE (ALL EXCEPT M1028A1 AND M1031) (Con't).

31. Apply a thin coat of petrolatum to underside of pump housing (37) tabs and install pump housing assembly to extension housing (2).

32. Apply anaerobic sealant to mating surface of extension housing (2) and around bolt holes. Install extension housing to rear case (43). Lightly tap on extension housing to firmly seat against rear case. Apply gasket sealing compound to 4 bolts and install, Tighten bolts to 25 lb.-ft. (34 N.m).

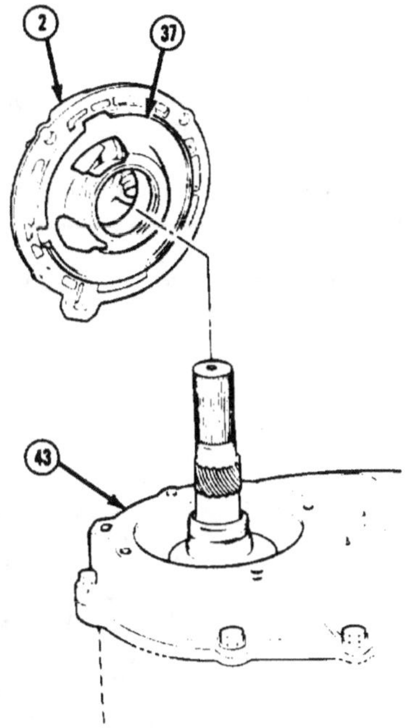

33. Apply gasket sealing compound to detent bolt (36) and install poppet (34), spring (35), and detent bolt to bottom of front case (29). Tighten bolt to 20 lb.-ft, (27 Nom).

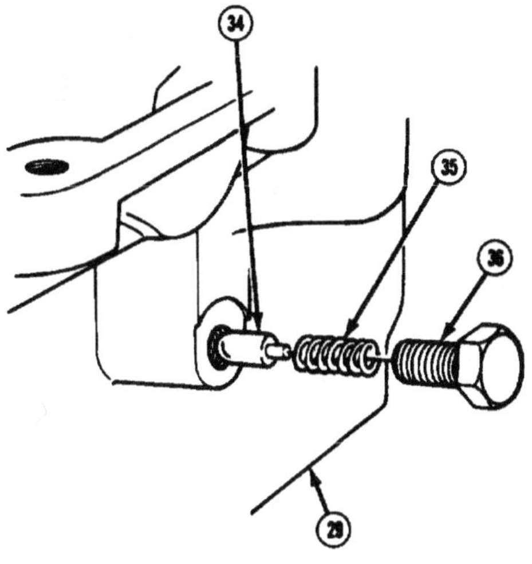

TA50330

5-21. TRANSFER CASE MAINTENANCE (ALL EXCEPT M1028A1 AND M1031) (Con't).

35. Install front yoke (32), new seal washer (30), and new locknut (31). Secure front yoke in a suitable holding device and tighten locknut to 120 lb.-ft. (163 N.m).

36. Install washer on 4-wheel drive indicator light switch (3). Install switch and tighten to 20 lb.-ft. (27 N.m).

37. Install drainplug and fill plug. Tighten drainplug to 20 lb.-ft. (27 Nom).

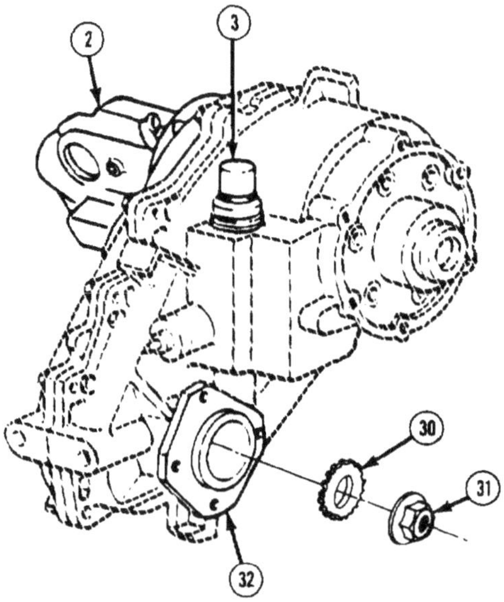

TA701856

5-21. TRANSFER CASE MAINTENANCE (ALL EXCEPT M1028A1 AND M1031) (Con't).

e. INSTALLATION

WARNING

Transfer case (4) is heavy. Provide suitable support and use caution during installation. Failure to follow this warning may result in damage to transfer case or serious injury to personnel.

NOTE

- If installing transfer case adapter (8) and seal (9), perform steps 1-4.
- If installing transfer case (4), perform steps 6-12,

1. Lightly coat new seal (9) with hydraulic fluid and install to transfer case adapter (8).
2. Install transfer case adapter (8) to transmission with 6 new lockwashers (7) and bolts (6). Tighten bolts to 30 lb.-ft, (41 N.m).
3. install 2 left transfer case shield bolts (13) and upper transmission mounts (11) to transmission support (12).
4. Lower transmission and transfer case adapter (8). Remove support from transmission, ensuring that transfer case adapter bolt holes aline with transmission support (12) bolt holes,

NOTE

Right bolt (16) must be started first.

5. Install transfer case adapter (8) to transmission support (12) with 2 spacers (14), lower transmission mounts (15), and 2 bolts (16). Tighten bolts to 40 lb.-ft. (54 N.m).
6. Install new transfer case adapter gasket (5) to transfer case (4), alining bolt holes.
7. Install two 3/8 in. N.C. bolts, with heads cut off, to transfer case (4) to serve as guide pins. Place transfer case in "4H." Place transfer case on a suitable jack and raise into position,

CAUTION

Ensure that transfer case (4) is flush against transfer case adapter (8) to avoid damage to transfer case when tightening bolts (20),

8. Move transfer case (4) forward and engage transmission output shaft (10) with transfer case input gear, Continue to push transfer case forward until flush against transfer case adapter (8). Install 4 bolts (20) and tighten securely, Remove 2 guide pins. Install 2 remaining bolts (20) and tighten securely,

NOTE

2 left transfer case shield bolts (13) are already in place,

9. Install transfer case shield (19) with 4 new lockwashers (17), 2 right transfer case shield bolts (13), and 4 nuts (18),

5-21. TRANSFER CASE MAINTENANCE (ALL EXCEPT M1028A1 AND M1031) (Con't).

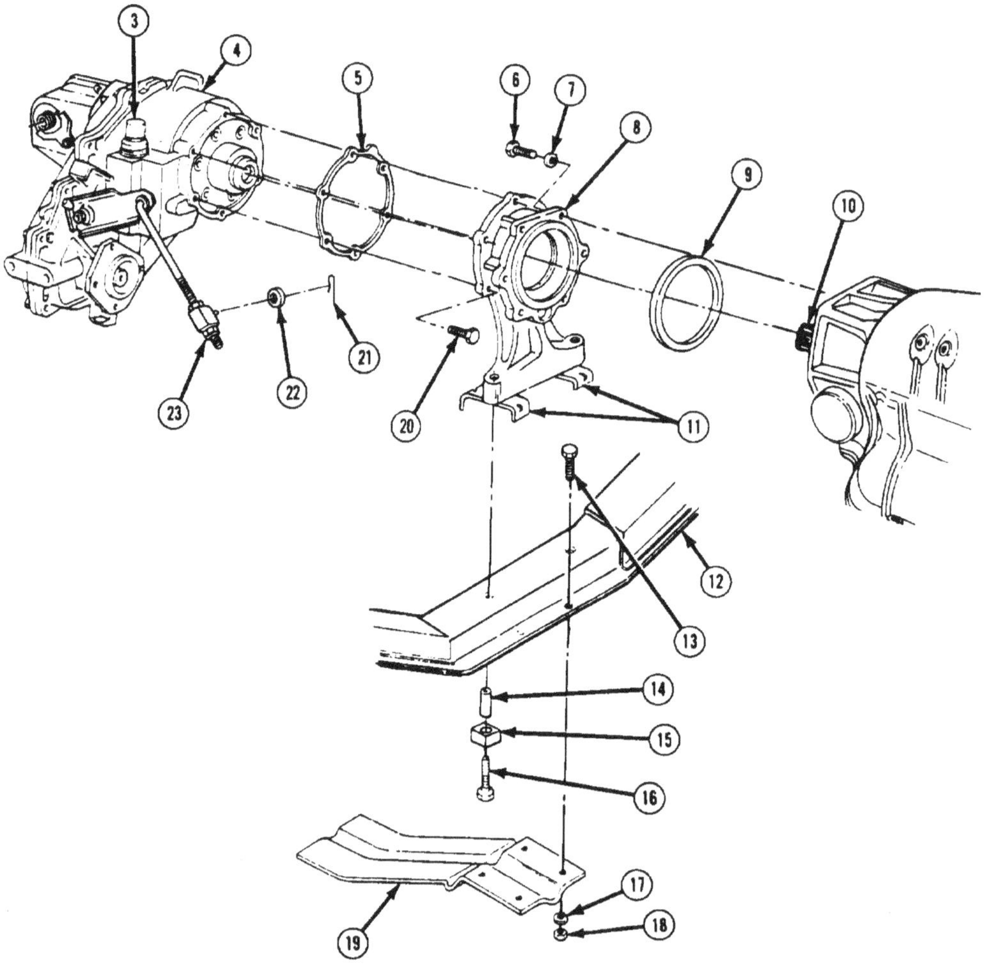

10. Connect connector to 4-wheel drive indicator light switch (3).

12. Connect control rod (23) to control lever assembly with washer (22) and new cotter pin (21).

FOLLOW-ON TASKS:

- Install speedometer gear. (See TM 9-2320-289-20)
- Install front and rear propeller shafts. (See TM 9-2320-289-20)
- Fill transfer case. (See TM 9-2320-289-20)
- Remove jack stands and lower truck.
- Connect both battery negative cables. (See TM 9-2320-289-20)
- Check operation of transfer case.

TA701857

5-22. TRANSFER CASE MAINTENANCE (M1028A1 AND M1031).

This task covers:

a. Removal
b. Disassembly
c. Cleaning and Inspection
d. Assembly
e. Installation

INITIAL SETUP:

Equipment Condition

- Both battery negative cables disconnected. (See TM 9-2320-289-20)
- Truck raised and supported on jack stands.
- Transfer case drained. (See LO 9-2320-289-12)
- Front and rear propeller shafts removed. (See TM 9-2320-289-20)
- Speedometer gear removed. (See TM 9-2320-289-20)

Tools/Test Equipment

- Dial indicator
- Snapring pliers
- Torque wrench

Personnel Required

- *MOS 63W (2)*

Manual References

- TM 9-214

Materials/Parts

- One cotter pin
- One extension gasket
- One extension seal
- One 4-wheel drive indicator light switch gasket
- One front output shaft retainer gasket
- One front output shaft retainer seal
- One front output shaft rear bearing retainer gasket
- One idler shaft cover gasket
- One poppet bolt gasket
- One PTO cover gasket
- One rear output shaft bearing retainer gasket
- One retainer ring
- One seal washer
- One transfer case adapter gasket
- Two locknuts
- Two lockpins
- Two shift rail bore seals
- Two transfer case adapter seals
- Two wave washers
- Two 3/8 in. N.C. bolts
- Eighteen lockwashers
- Dry cleaning solvent (Item 23, Appendix B)
- Hydraulic fluid (Item 40, Appendix B)
- Petrolatum (Item 50, Appendix B)
- Gasket sealing compound (Item 58, Appendix B)

General Safety Instructions

- Transfer case is heavy. Provide suitable support and use caution during removal and installation.
- Dry cleaning solvent is flammable and must not be used near open flame. Use only in a well-ventilated area.
- Compressed air used for cleaning purposes should never exceed 30 psi (207 kPa).

5-22. TRANSFER CASE MAINTENANCE (M1028A1 AND M1031) (Con't).

a. REMOVAL

NOTE

- If removing transfer case (l), perform steps 1-4.
- If removing transfer case adapter, perform steps 1-7.

1. Remove pivot bolt (3), 1 wave washer, and 1 flat washer, Remove cotter pin (5), 1 flat washer (6), and 1 wave washer (7), and disconnect selector rod (2) from control lever (4) end of rod. Discard cotter pin and wave washers.

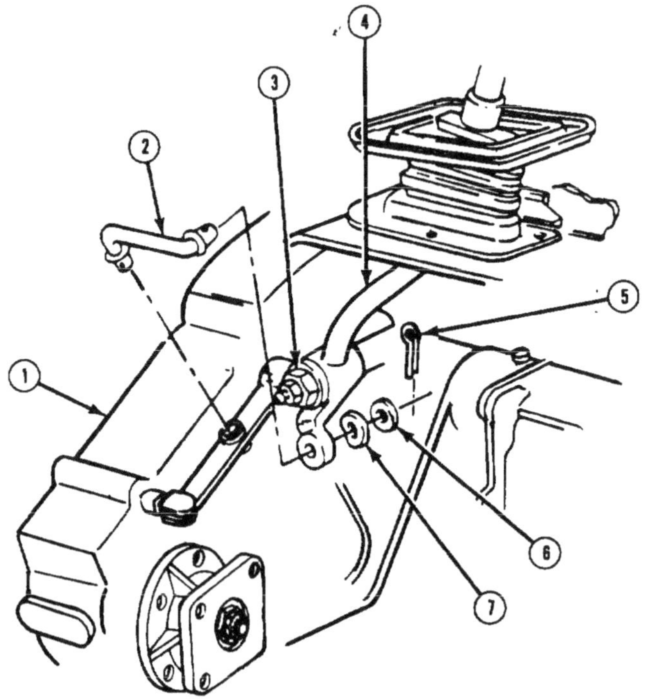

TA50333

5-22. TRANSFER CASE MAINTENANCE (M1028A1 AND M1031) (Con't).

2 . Disconnect connector from 4-wheel drive indicator light switch (17).

NOTE

2 left bolts (26) cannot be removed until transfer case adapter (8) and upper transmission mounts (18) are removed in step 7.

3 . Remove 4 nuts (24), lockwashers (25), if present, 2 right bolts (26), and transfer case shield (23). Discard lockwashers.

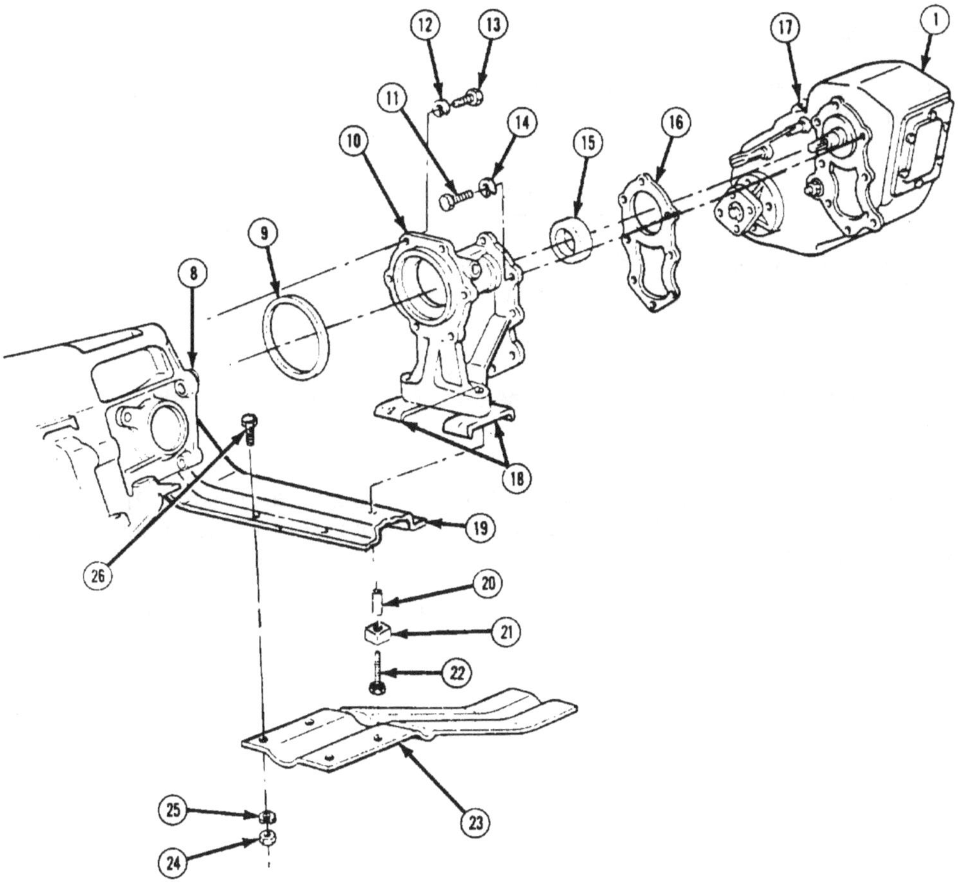

TA50334

5-22. TRANSFER CASE MAINTENANCE (M1028A1 AND M1031) (Con't).

WARNING

Transfer case (1) is heavy. Provide suitable support and use caution during removal. Failure to follow this warning may result in damage to transfer case or serious injury to personnel.

4. Place a suitable jack under transfer case (1). Remove 8 bolts (11) and lockwashers (14), and move transfer case rearward until free of transmission output shaft (8) and transfer case adapter (10). Remove transfer case, Discard lockwashers.
5. Remove transfer case adapter gasket (16) and discard.
6. Remove 2 bolts (22), lower transmission mounts (21), and spacers (20) from underside of transmission support (19).

CAUTION

Transmission should be supported before removing transfer case adapter (10). Failure to follow this caution may result in damage to transmission.

NOTE

If transmission is to be removed, transmission should be supported with transmission lift.

7. Raise transmission 1/2 in. (1 2.7 mm) and remove 6 bolts (13) and lockwashers (12). Remove transfer case adapter (10), Remove 2 upper transmission mounts (18) and 2 left transfer case shield bolts (26). Discard lockwashers.
8. Remove 2 transfer case adapter seals (9 and 15) from transfer case adapter (10). Discard seals.

5-22. TRANSFER CASE MAINTENANCE (M1028A1 AND M1031) (Con't).

b. DISASSEMBLY

WARNING

Dry cleaning solvent P-D-680 is toxic and flammable. Always wear protective goggles and gloves and use only in a well-ventilated area. Avoid contact with skin, eyes, and clothes and DO NOT breathe vapors. DO NOT use near open flame or excessive heat. The solvent's flash point is 100°F-138°F (38°C-590C). If you become dizzy while using cleaning solvent, immediately get fresh air and medical help. If solvent contacts eyes, immediately wash your eyes with water and get medical aid.

NOTE

All bearing assemblies should be tagged as they are removed.

1. If fill plug (33) and drainplug (32) are installed, remove.
2. Remove 4-wheel drive indicator light switch (17), gasket, spring, and ball. Discard gasket,

4. Remove seal (35) from extension (27), Discard seal,

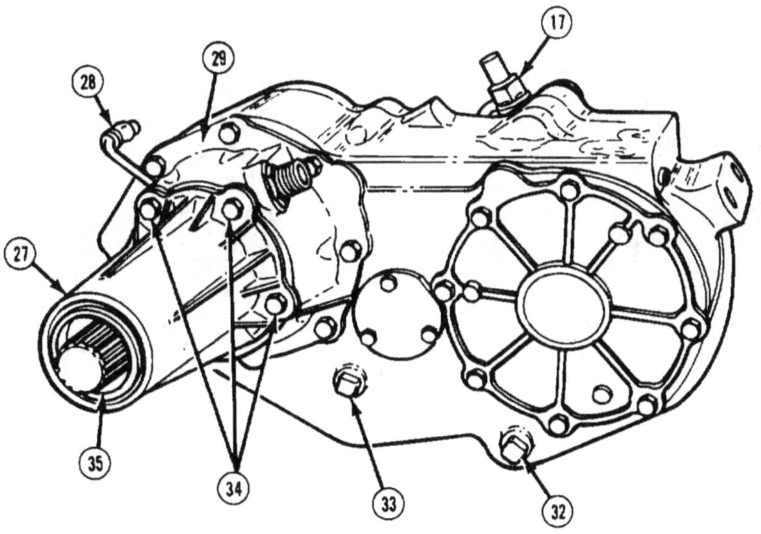

TA701858

5-22. TRANSFER CASE MAINTENANCE (M1028A1 AND M1031) (Con't).

5. Remove 5 bolts (34) and extension (27) from retainer (29). Remove gasket and discard. If damaged, remove extension bushing. Remove vent (28).

6. Remove selective snapring (36) and bearing (38) from rear output shaft (37). Remove snapring (39) from bearing. Remove speedometer drive gear (41) from rear output shaft.

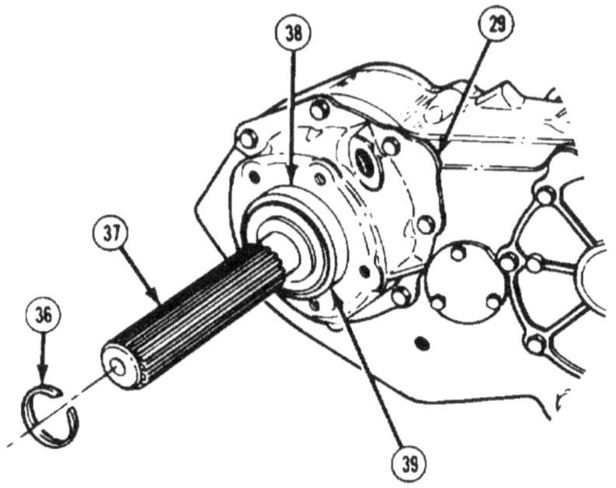

NOTE

46 rear output shaft roller bearings (43) may fall out of race (42) when retainer (29) is removed. Ensure that all are present,

7. Remove 8 bolts (40) and retainer (29) from transfer case (1). Remove gasket and discard. Remove 46 roller bearings (43). Remove race (42) if damaged.

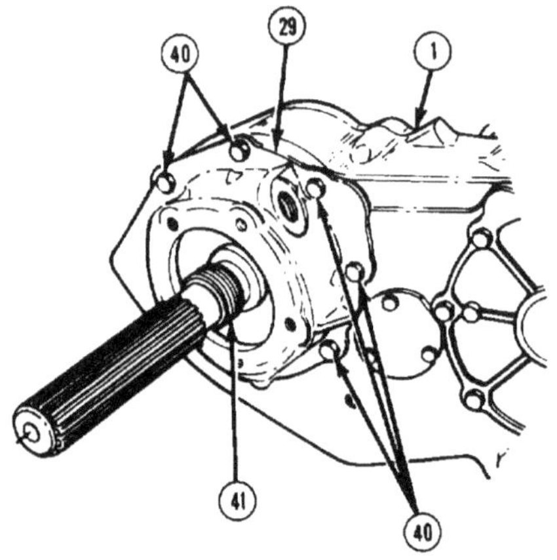

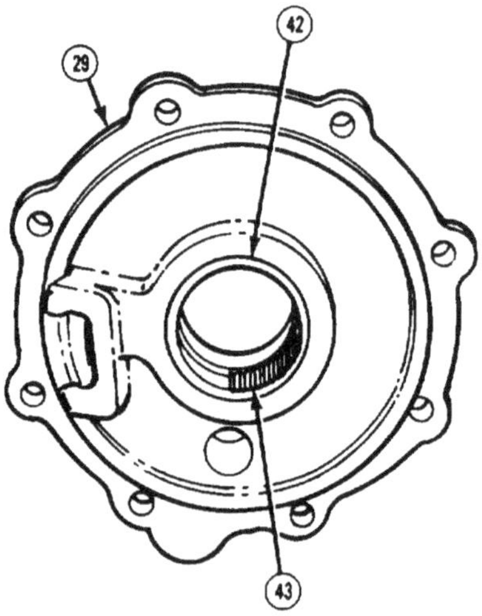

TA50336

5-22. TRANSFER CASE MAINTENANCE (M1028A1 AND M1031) (Con't).

8. Remove 2 spring clips, 2 clevis pins (47) and shift link (49) from 4-wheel drive shift rail (44) and range rail (48). If damaged, remove selector rod (2) and grommet from shift link.

NOTE

Plugs (45) may be steel or rubber. If plugs are steel, they must be driven into transfer case (1) before lockpins can be removed.

9. Remove 2 plugs (45). Remove poppet plug (46), gasket, spring, and ball. Discard gasket.

10. Position 4-wheel drive shift rail (44) and range rail (48) in neutral with lockpins inside transfer case (1) visible through unplugged access holes. Drive 2 lockpins into transfer case and remove 4-wheel drive shift rail and range rail.

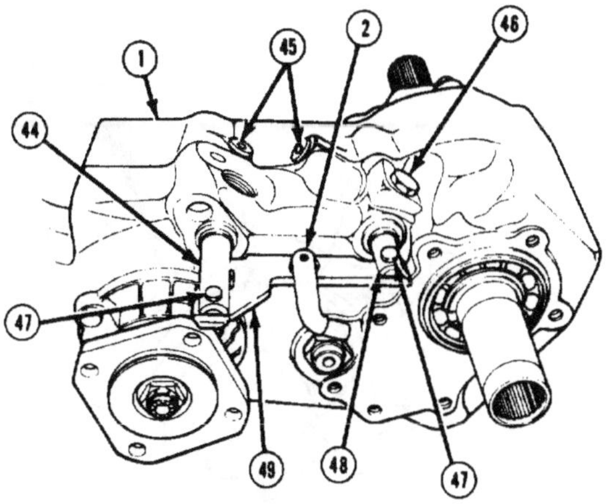

NOTE

15 rear output shaft roller bearings (56) may fall out when rear output shaft assembly (37) is removed from transfer case (1). Ensure that all are present.

11. Remove rear output shaft assembly (37) from transfer case (1). Remove 15 roller bearings (56) from inside pilot bore of rear output shaft and from inside transfer case. Remove roller bearing retainer ring (58) and spacer (57). Discard retainer ring.

12. Remove snapring (53) from rear output shaft (37), Remove spacer (55).

13. Remove retainer pin (54) from rear output shaft (37). Remove thrust washer (59).

14. Remove output low gear (52) from rear output shaft (37).

TA50337

5-22. TRANSFER CASE MAINTENANCE (M1028A1 AND M1031) (Con't).

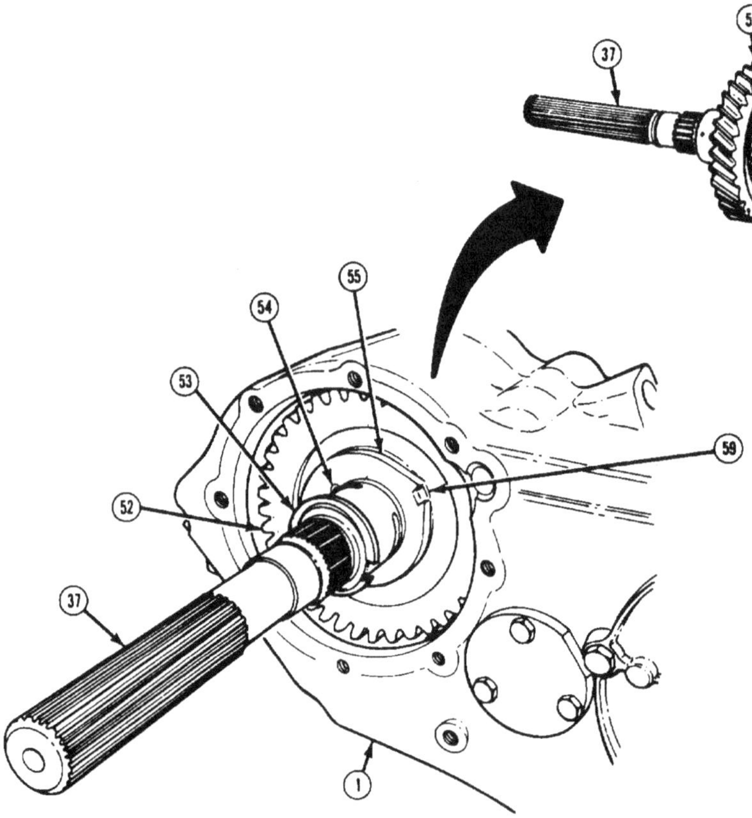

NOTE

There are 64 roller bearings (60) in 2 rows on rear output shaft (37). Ensure that all are present,

15. Remove 2 rows of roller bearings (60), spacer (62), and thrust washer (61) from rear output shaft (37).

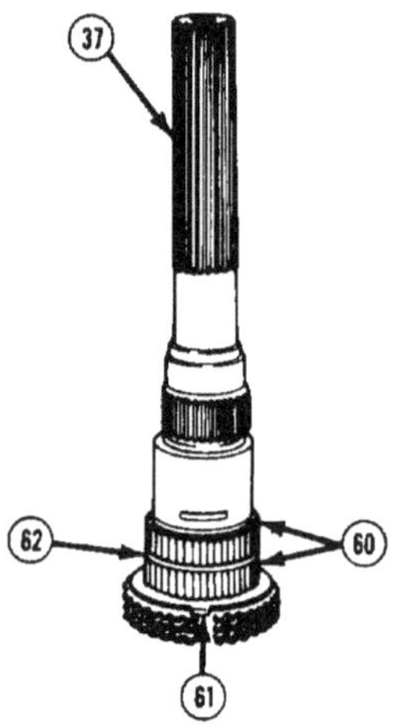

TA50338

5-22. TRANSFER CASE MAINTENANCE (M1028A1 AND M1031) (Con't).

16. Remove clutch hub (63) and shift fork (65) from inside transfer case (1).

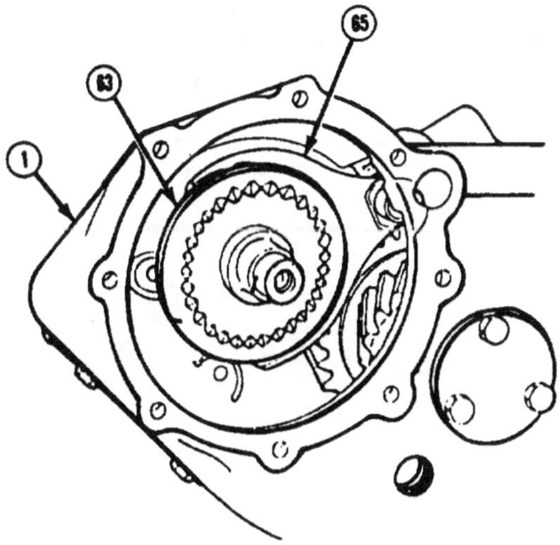

17, Remove snapring (68) from main input gear (64). Remove main input gear.

18. Remove snapring (66) from bearing (67). Remove bearing from transfer case (l).

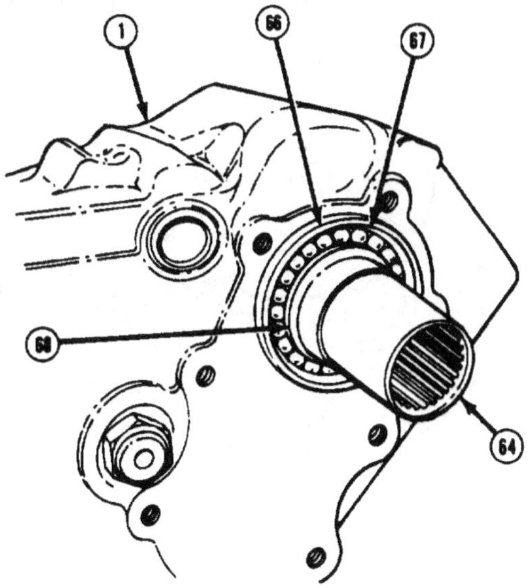

TA50339

5-22. TRANSFER CASE MAINTENANCE (M1028A1 AND M1031) (Con't).

NOTE

It may be necessary to tip transfer case (1) on end to aid in removal of interlock pins (69).

19. Remove 2 interlock pins (69) from channels in top of transfer case (1).

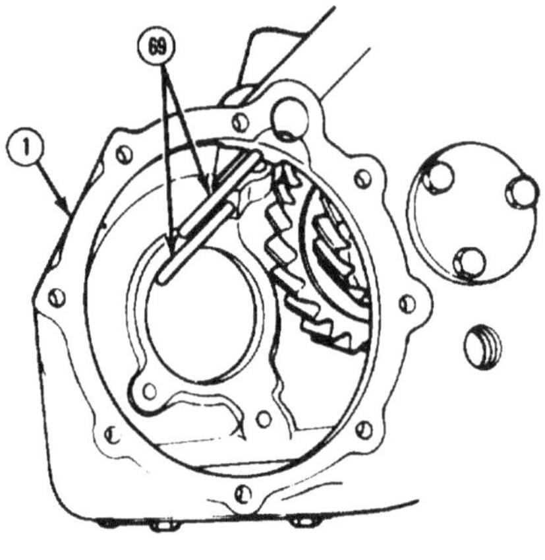

20. Remove 3 bolts (73), idler shaft cover (70), and gasket from rear of transfer case (1). Remove locknut (74) and washer (75) from idler shaft (76) at front of transfer case. Discard gasket and locknut.

21. Drive idler shaft (76) out access hole in rear of transfer case (1) and remove.

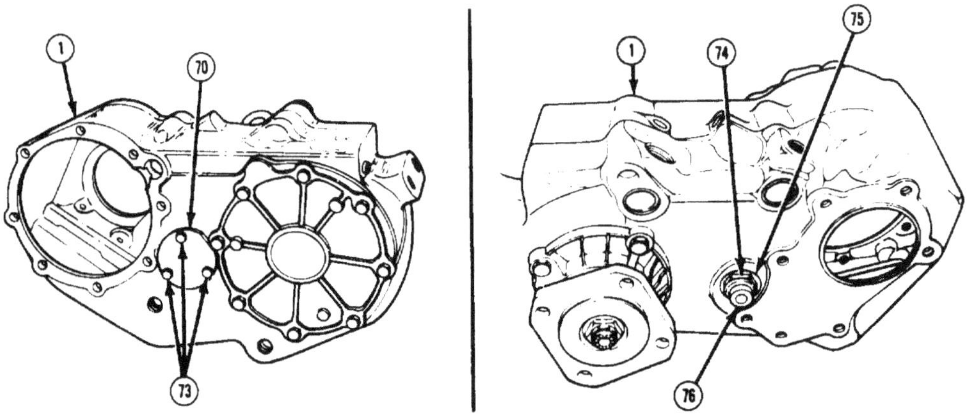

TA50340

5-22. TRANSFER CASE MAINTENANCE (M1028A1 AND M1031) (Con't).

NOTE

Bearings (77 and 81) should be marked for assembly in same position.

22. Remove idler gear (80) assembly through rear output shaft access hole. Remove bearing (81), spacer (79), shim(s) (78), bearing (77), and idler gear. If damaged, remove bearing races from idler gear.

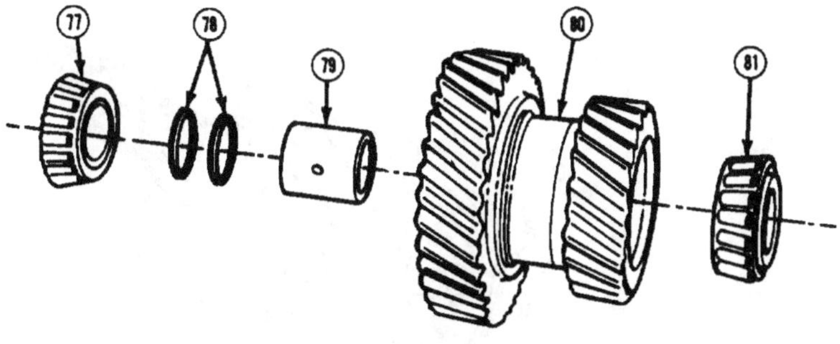

23. Remove front output shaft shift fork from inside transfer case (1).

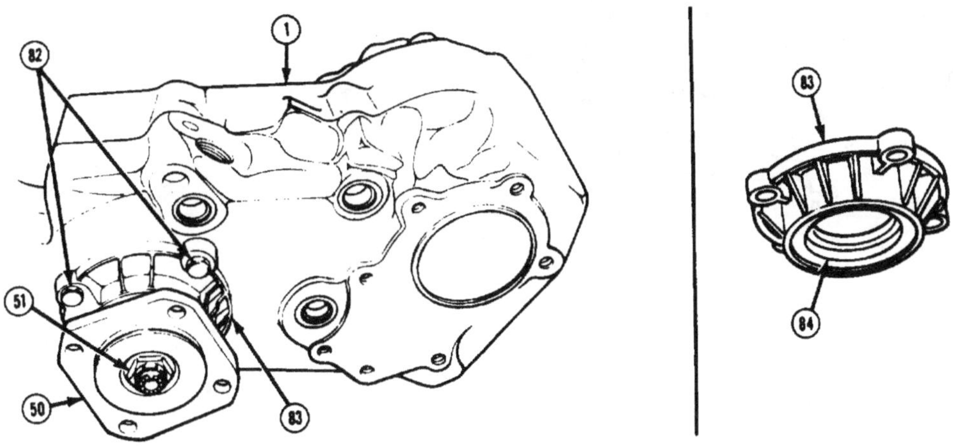

TA50341

5-22. TRANSFER CASE MAINTENANCE (M1028A1 AND M1031) (Con't).

24. Secure front output shaft yoke (50) in a suitable holding device and remove locknut (51), 1 flat washer, and 1 seal washer. Remove front output shaft yoke, Discard locknut and seal washer.

25. Remove 4 bolts (82) and front output shaft retainer (83). Remove gasket and discard. Remove seal (84) from front output shaft retainer and discard.

26. Remove 8 bolts (71) at rear retainer (72). Tap on front output shaft (88) with a soft-faced hammer and remove rear retainer, front output shaft, and gear assembly, as a unit. Remove gasket from rear face of transfer case (1) and discard.

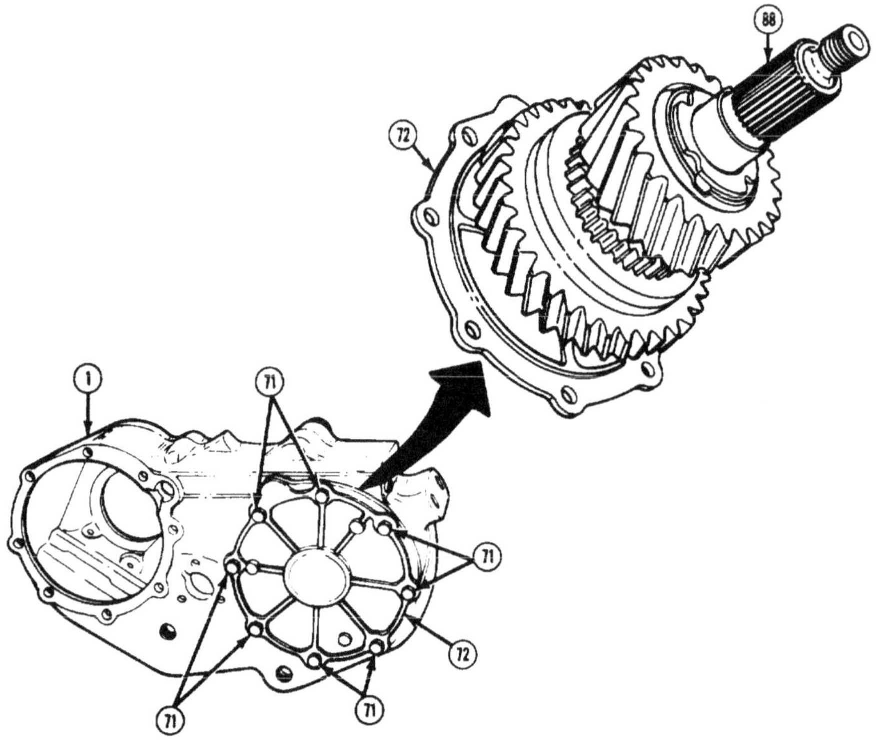

TA50342

5-22. TRANSFER CASE MAINTENANCE (M1028A1 AND M1031) (Con't).

27. Remove bearing (90) from front of transfer case (1). Remove snapring (89) from bearing.

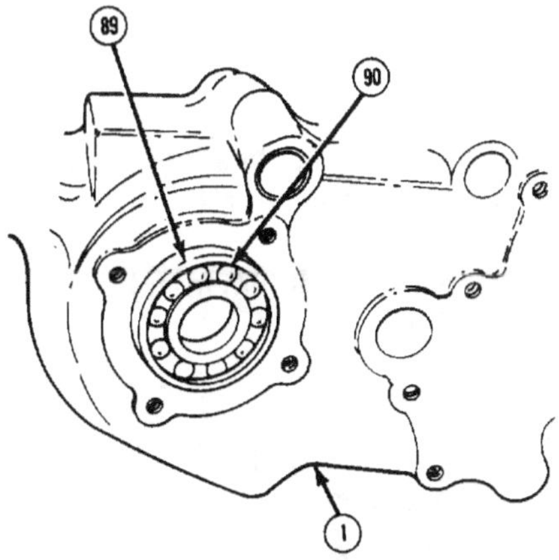

28. Remove thrust washer (87) and front gear (86) from front output shaft (88), Remove clutch hub (85) .

29. Remove front output shaft (88) from rear retainer (72). If damaged, remove bearing (91) from rear retainer.

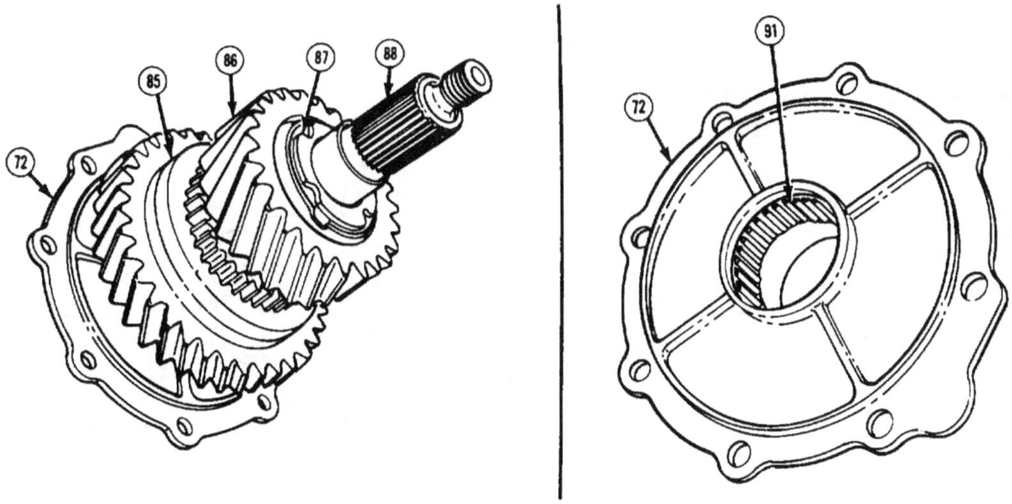

TA50343

5-22. TRANSFER CASE MAINTENANCE (M1028A1 AND M1031) (Con't).

30. Remove snapring (92) from front output shaft (88). Remove thrust washer (95) and retainer pin (93).

NOTE

There are 64 roller bearings (96) in 2 rows on front output shaft (88). Ensure that all are present.

31. Remove output low sear (94) from front output shaft (88). Remove 64 roller bearings (96) and spacer (97).

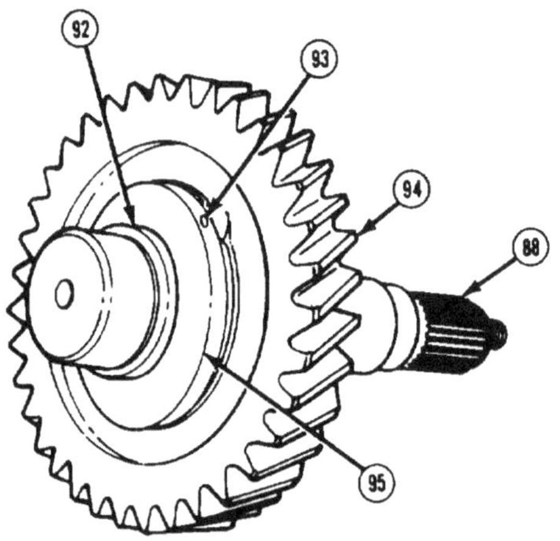

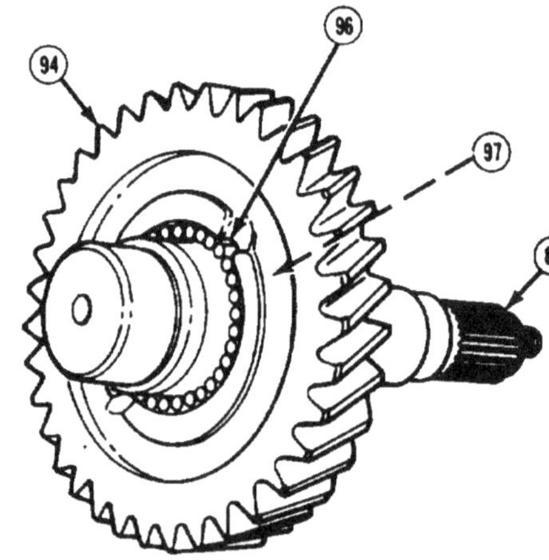

32. Remove 6 bolts (101) and PTO cover (100). If present, remove gasket (99) and discard.

33. Remove 2 seals (98) from shift rail bores and discard.

34. Remove clip, magnet, and 2 lockpins from bottom of transfer case (1). Discard lockpins.

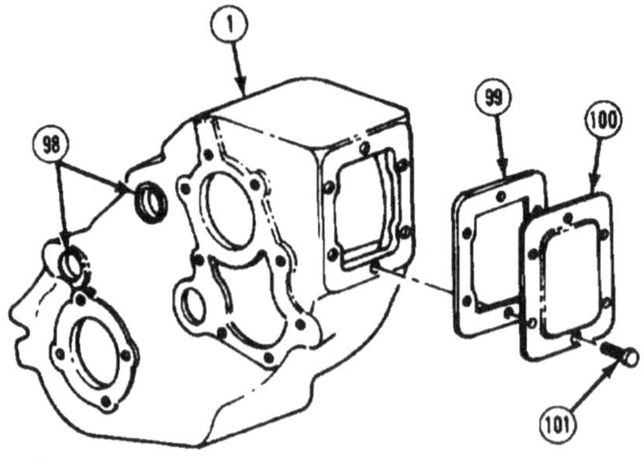

TA50344

5-22. TRANSFER CASE MAINTENANCE (M1028A1 AND MI 031) (Con't).

c. CLEANING AND INSPECTION

WARNING

- Dry cleaning solvent P-D-680 is toxic and flammable. Always wear protective goggles and gloves and use only in a well-ventilated area. Avoid contact with skin, eves, and clothes and DO NOT breathe vapors. DO NOT use near open flame or excessive heat. The solvent's "flash point is 100°F-138°F (38°C-59°C). If you become dizzy while using cleaning solvent, immediately get fresh air and medical help, If solvent contacts eyes, immediately wash your eyes with water and get medical aid.
- Compressed air used for cleaning purposes should never exceed 30 psi (207 kPa). Use only effective chip guarding and personnel protective equipment (goggles/shield, gloves, etc.). Failure to follow this warning may result in injury to personnel.

1. Clean transfer case, covers, retainers, shafts, and gears with dry cleaning solvent to remove all old lubricant and dirt. Remove all traces of gasket material and sealant. Clean all metal filings from magnet. Dry with compressed air.
2. Clean and inspect bearings, bearing cups, and rollers in accordance with TM 9-214.
3. Inspect transfer case for damage, Replace if damaged.
4. Inspect shaft splines and gears for damage. Replace if damaged.

d. ASSEMBLY

NOTE

All components should be lightly coated with hydraulic fluid as they are assembled.

1. Install magnet to bottom of transfer case with clip.
2. Install new bearing races to idler gear (80) if removed.
3. Install bearing (81), spacer (79), shim(s) (78), idler gear (80), and bearing (77) to idler shaft (76).

5-22. TRANSFER CASE MAINTENANCE (M1028A1 AND M1031) (Con't).

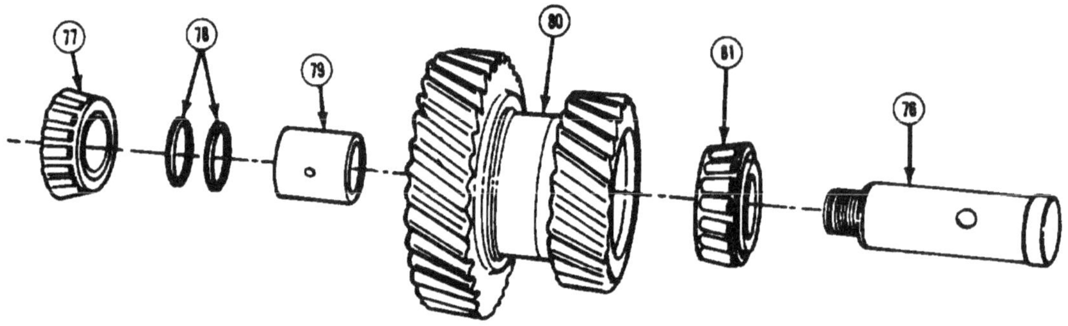

NOTE

Arbor press and a suitable sleeve may be used to apply downward pressure to bearing (77).

4. Place idler gear (80) assembly with threaded end of idler shaft (76) upward. Apply downward pressure to bearing (77). Using dial indicator (102) on idler gear, measure idler gear end play. Limits are 0.001-0.002 in. (0.025-0.051 mm). If end play is not within specification, adjust by removing or adding shim(s) (78). Recheck 'end play.

5. Remove idler shaft (76) from idler gear (80) assembly.

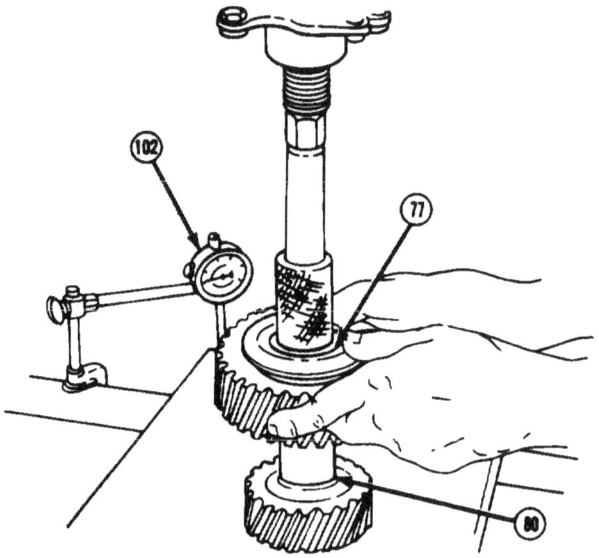

TA50345

5-22. TRANSFER CASE MAINTENANCE (M1028A1 AND M1031) (Con't).

6. Place transfer case (1) on wooden blocks with rear face up. Install idler gear .(80) assembly into transfer case, with large end of idler gear facing front of transfer case. Aline bearings, shim(s), and spacer inside idler gear with idler shaft (76) holes in transfer case. Drive in idler shaft from large bore side of idler gear. Turn transfer case upright and install washer and new locknut to end of idler shaft. Tighten locknut to 150 lb.-ft, (203 N.m). Check for end play and free rotation.

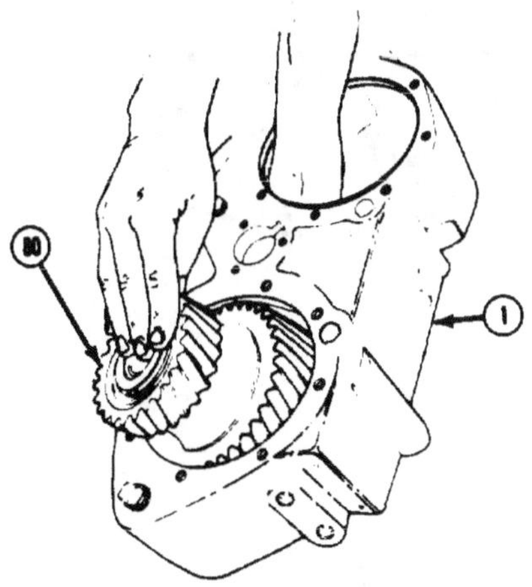

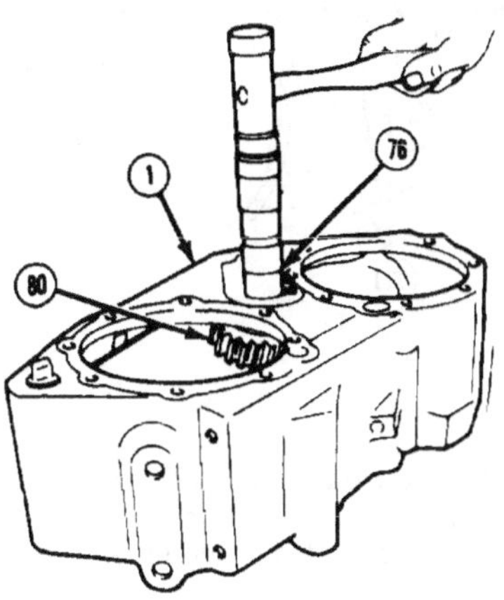

NOTE

Ensure that Idler shaft cover (70) is installed with flat side facing front output shaft rear opening.

7. Apply gasket sealing compound to 3 bolts (73). Install new gasket and idler shaft cover (70). Install 3 bolts and tighten to 20 lb.-ft. (27 N.m).

8. Install 2 new seals (98) to shift rail bores (103) with metal lips facing outward.

TA50346

5-22. TRANSFER CASE MAINTENANCE (M1028A1 AND M1031) (Con't).

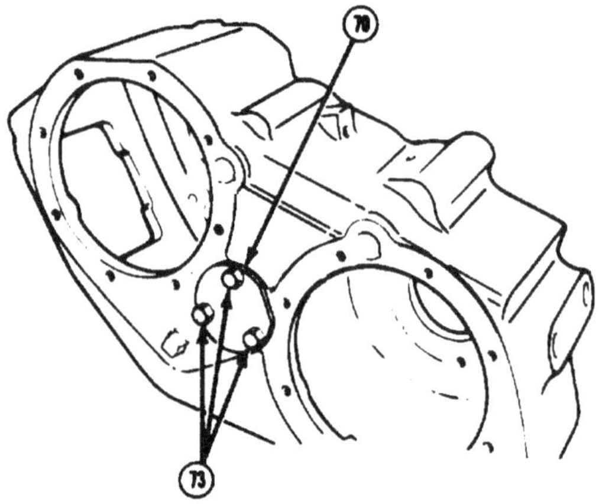

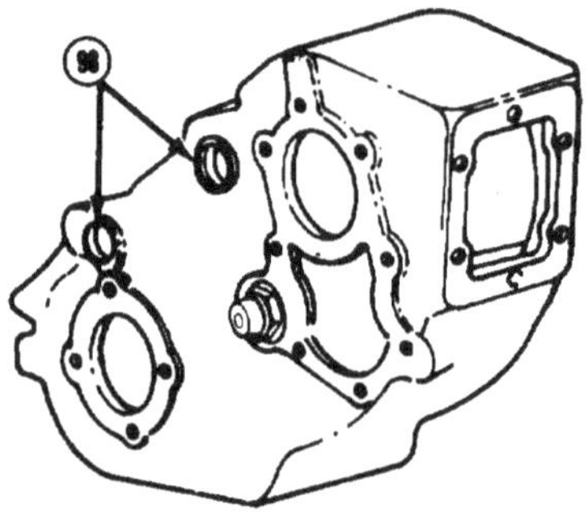

9. Install 2 interlock pins (69) inside channels in top of transfer case (1), Center interlock pins between shift rail bores (103).

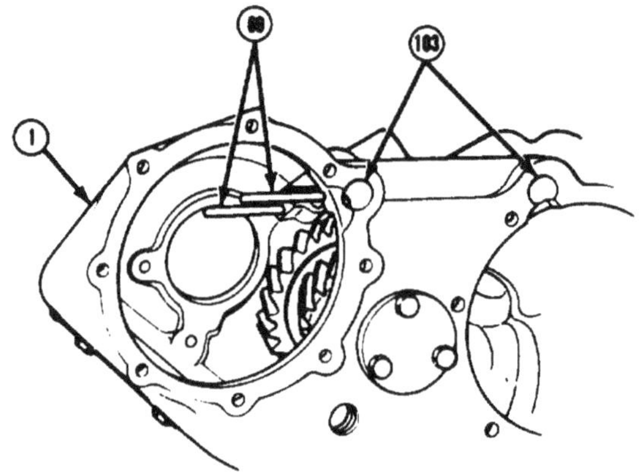

TA50347

5-22. TRANSFER CASE MAINTENANCE (M1028A1 AND M1031) (Con't).

10. Position front output shaft shift fork (104) inside transfer case (1) with lockpin hole facing front of transfer case, Install 4-wheel drive shift rail (44) into transfer case from the back, slotted end first, with poppet notches up. Install 4-wheel drive shift rail through hole in shift fork and push through to the neutral position.

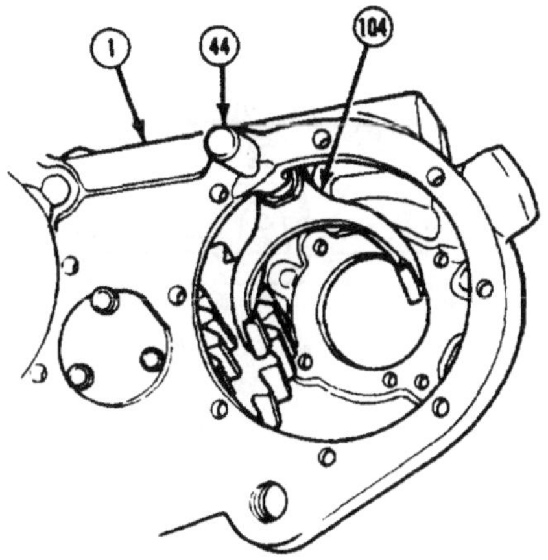

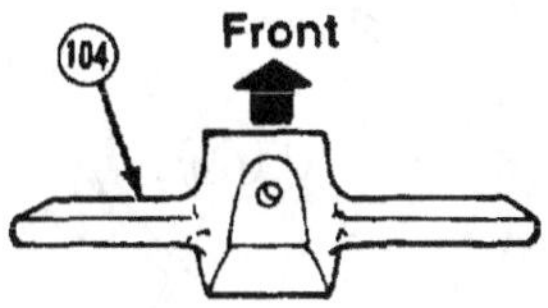

11. Install snapring (66) to bearing (67) and install bearing to front of transfer case (1). Install main input gear (64) and snapring (68).

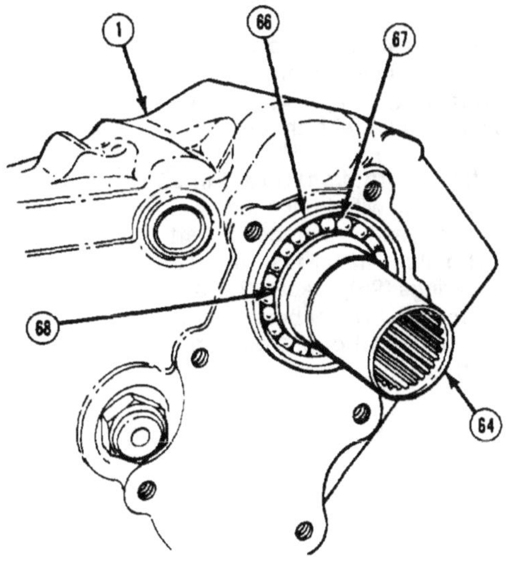

TA50348

5-22. TRANSFER CASE MAINTENANCE (M1028A1 AND M1031) (Con't).

12. Install clutch hub (63) to shift fork (65) and position assembly inside transfer case (1) with clutch hub and main input gear (64) teeth meshed. Install range rail (48) from the front, with poppet notches up, through hole in shift fork and push through to the neutral position.

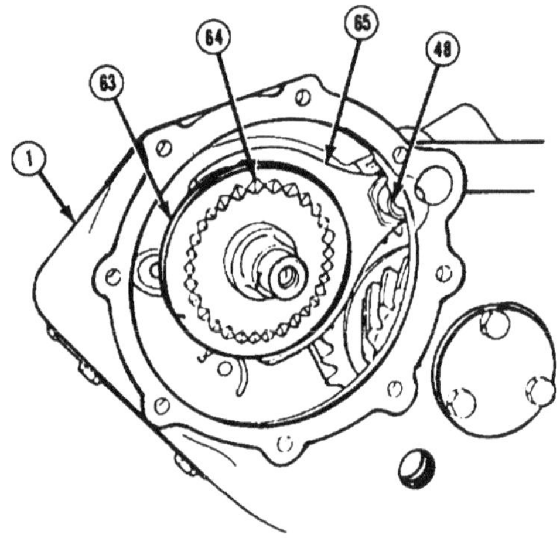

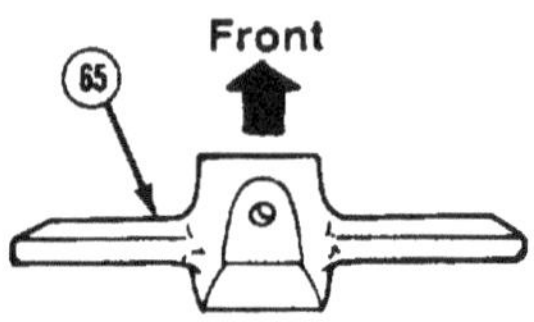

13. Aline holes in 4-wheel drive shift rail (44) and front output shaft shift fork (1 04). Drive in new lockpin through access hole in top of transfer case (1), until flush with shift fork.

14. Tip transfer case (1) onto PT O opening. Aline holes in range rail (48) and shift fork (65). Drive in new lockpin until flush with shift fork.

15. Place front output shaft (88) in a soft-jawed vise with front end down. Lightly coat bearing surface of shaft with petrolatum. Install first row of 32 roller bearings (96), spacer (97), and second row of 32 roller bearings. Add another thin coat of petrolatum to ensure that roller bearings stay in position.

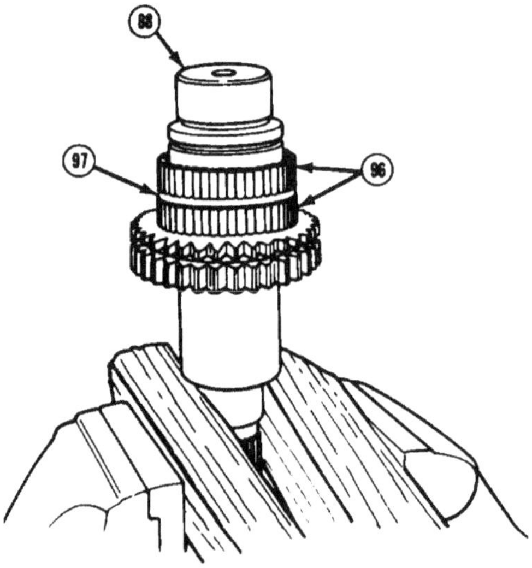

TA50349

5-22. TRANSFER CASE MAINTENANCE (M1028A1 AND M1031) (Con't).

16. Install output low gear (94) on front output shaft (88) with clutch gear facing down. Install retainer pin (93) in front output shaft, Install thrust washer (95). Install snapring (92) with opening opposite retainer pin,

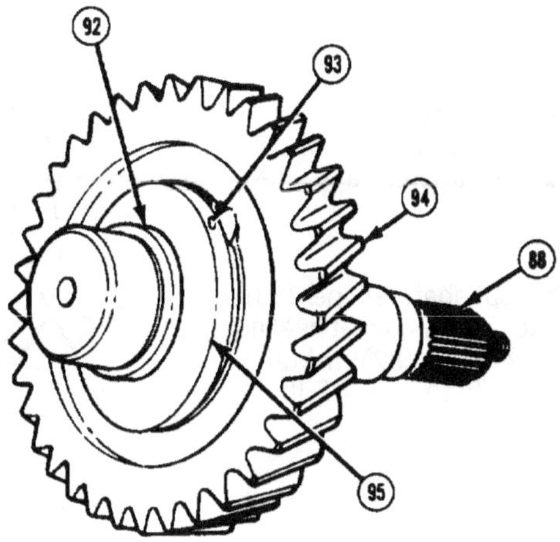

17. Install snapring (89) to bearing (90) and install bearing to front output shaft opening in transfer case (1).

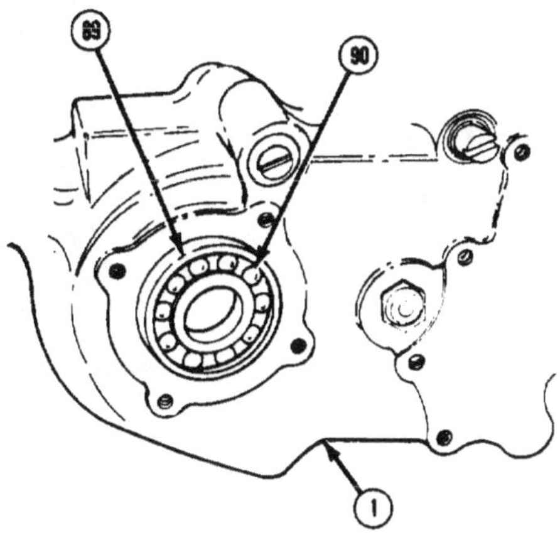

TA50350

5-22. TRANSFER CASE MAINTENANCE (M1028A1 AND M1031) (Con't).

18. Install new seal (84) to front output shaft retainer (83) with steel side of seal to inside.

NOTE

Ensure that notch on inner face of retainer (83) and gasket are alined with oil passage in transfer case (I).

19. Install front output shaft retainer (83) with new gasket, Apply gasket sealing compound to 4 bolts (82). Install bolts and tighten to 35 lb.-ft. (47 N.m).

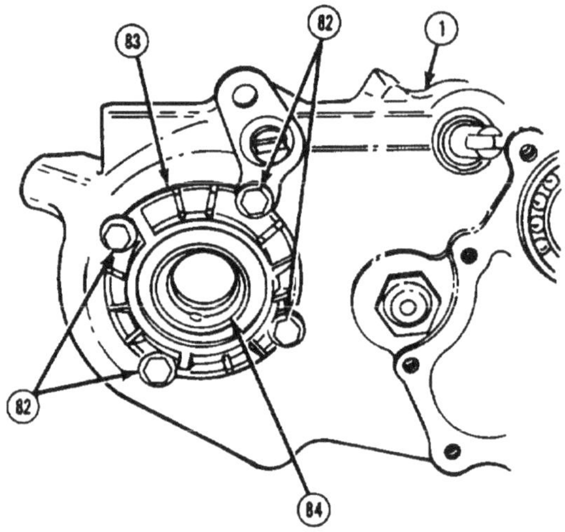

20. Turn transfer case (1) front side down and place on wooden blocks.

NOTE

Thrust washer (87) has 4 notches on outer edge.

21. Install thrust washer (87) and front gear (86) inside transfer case (1).

22. Install clutch hub (85) to front output shaft shift fork (104).

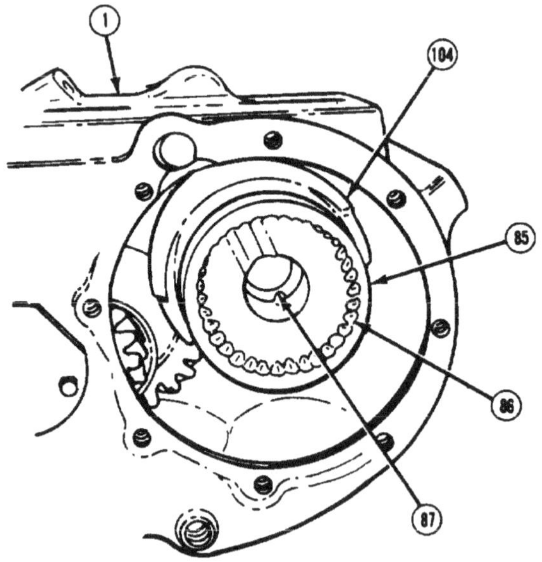

TA50351

5-22. TRANSFER CASE MAINTENANCE (M1028A1 AND M1031) (Con't).

NOTE

Thrust washer (87), front gear (86), and clutch hub (85) must be alined with bearing bore before installing front output shaft (88) and low gear (94) assembly.

23. Install front output shaft (88) and low gear (94) assembly through front gear (86) in transfer case (1).

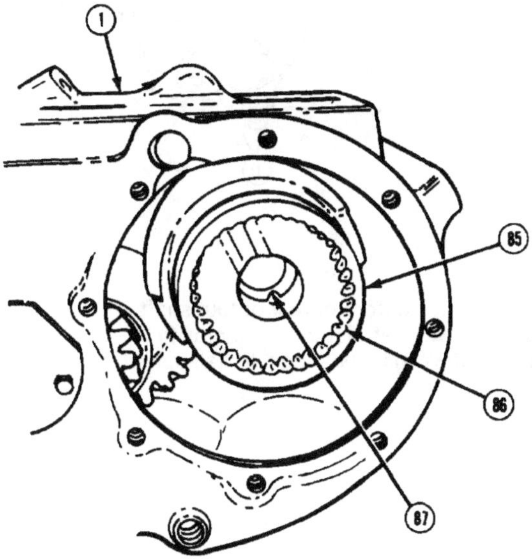

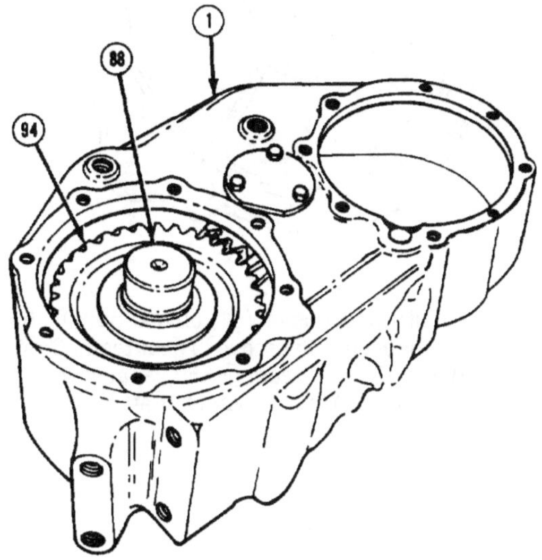

24. Install new bearing (91) to rear retainer (72) if removed. Install rear retainer with new gasket. Apply gasket sealing compound to 8 bolts (71). Install bolts and tighten to 30 lb.-ft, (41 N.m).

TA50352

5-22. TRANSFER CASE MAINTENANCE (M1028A1 AND M1031) (Con't).

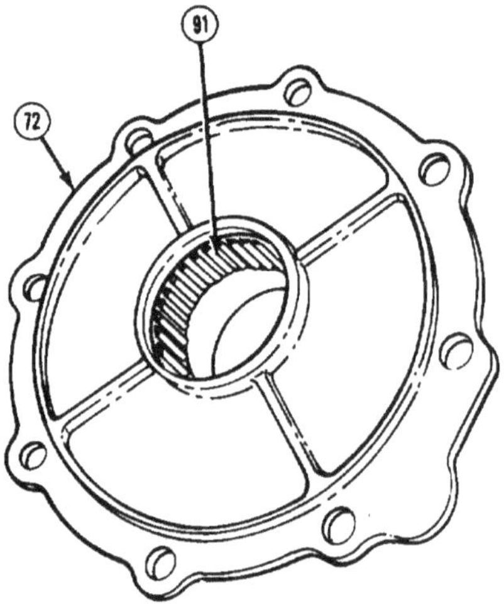

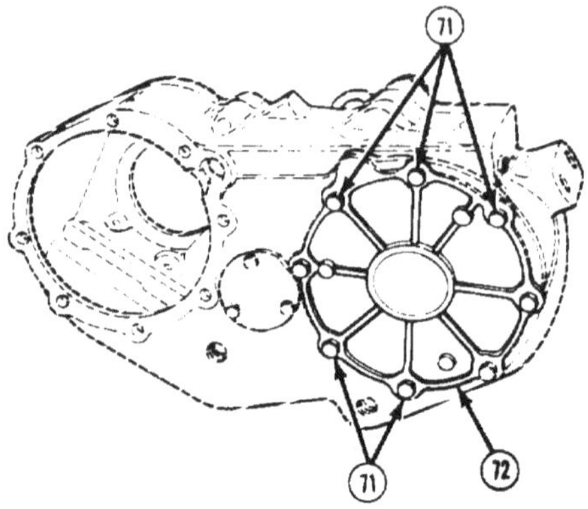

25. Install thrust washer (61) on rear output shaft (37) with tang facing front of rear output shaft and indexed with groove in gear teeth. Coat rear output shaft with petrolatum and install first row of 32 roller bearings (60), spacer (62), and second row of 32 roller bearings. Install output low gear (52) to rear output shaft.

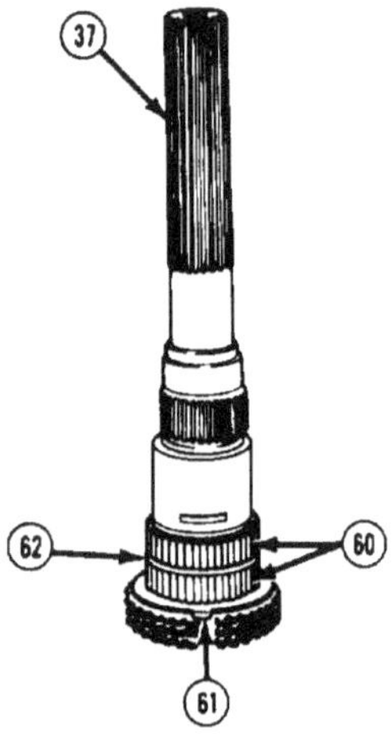

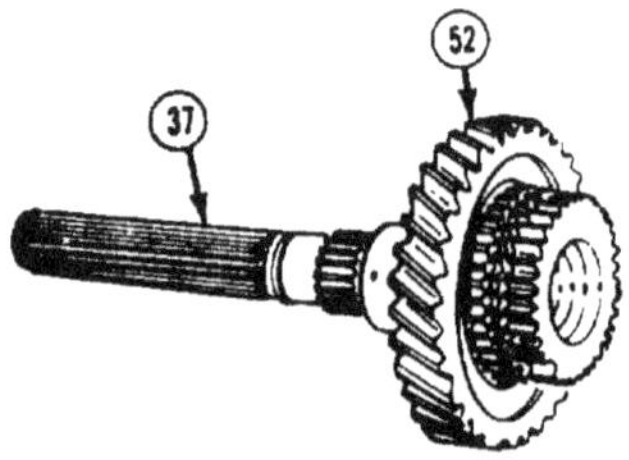

TA50353

5-22. TRANSFER CASE MAINTENANCE (M1028A1 AND M1031) (Con't):

26. Install thrust washer (59) with tang facing splined end of rear output shaft (37). Install retainer pin (54) to rear output shaft. Install spacer (55) and index with retainer pin and thrust washer tang. Install snapring (53) in rear output shaft groove,

27. Coat pilot bore of rear output shaft (37) with petrolatum and install 15 roller bearings (56), Install spacer (57) and new retainer ring (58), Ensure that retainer ring is fully seated inside pilot bore.

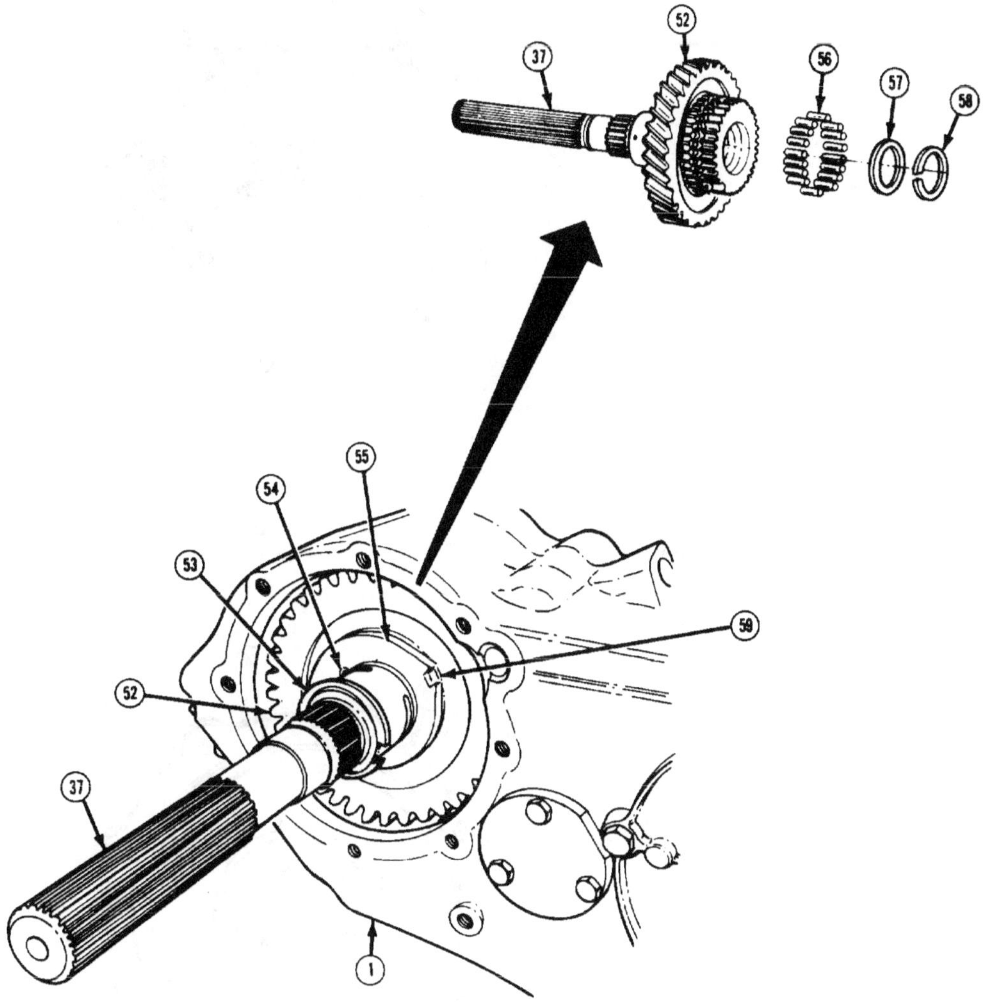

TA50354

5-22. TRANSFER CASE MAINTENANCE (M1028A1 AND M1031) (Con't).

28. Install rear output shaft (37) and output low gear (52) assembly in transfer case (1) and engage with main input gear (64).

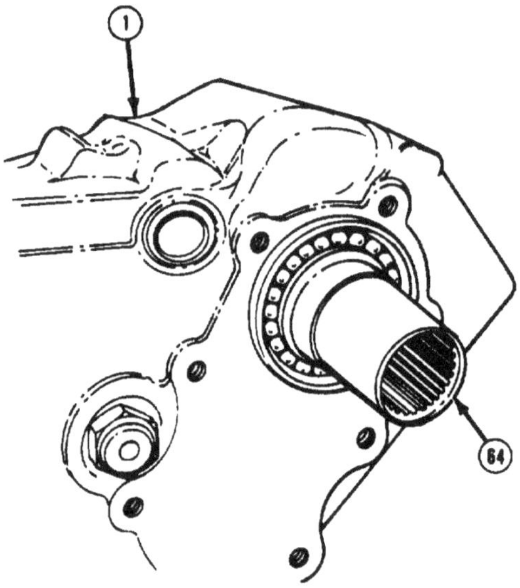

29. Install new race (42) to retainer (29) if removed. Coat race with petrolatum and install 46 roller bearings (43).

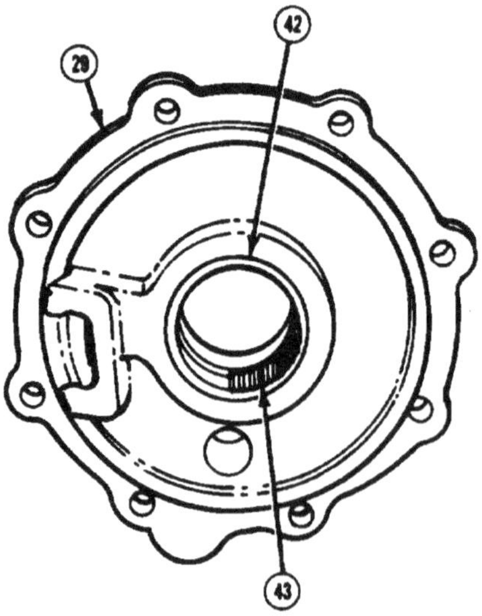

TA50355

5-22. TRANSFER CASE MAINTENANCE (M1028A1 AND M1031) (Con't).

30. Install retainer (29) to transfer case (1) with new gasket, Apply gasket sealing compound to 8 bolts (40) and install. Tighten bolts to 30 lb.-ft. (41 N.m).

31. Install speedometer drive gear (41) on rear output shaft (37).

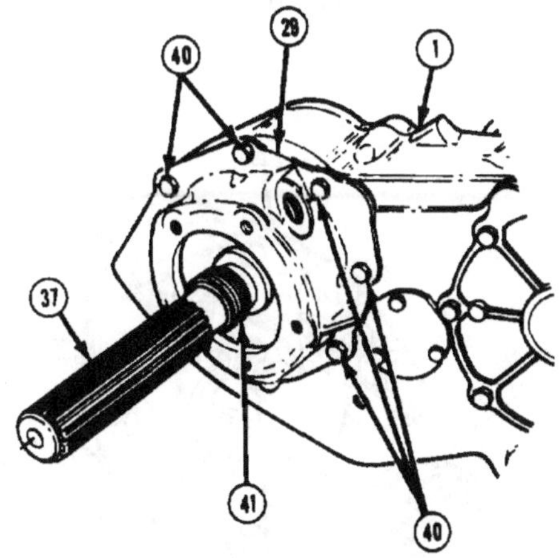

32. Install snapring (39) to bearing (38). Install bearing to retainer (29), Install selective snapring (36) on rear output shaft (37) groove.

33. Use feeler gage to measure end play between selective snapring (36) and bearing (38) assembly. End play should 0.002-0.022 in. (0.05-0.56 mm). If end play is not within specifications, remove selective snapring and install one of a different thickness.

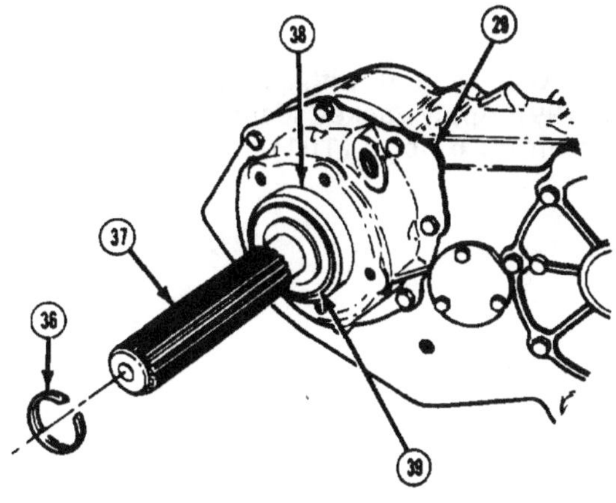

TA50356

5-22. TRANSFER CASE MAINTENANCE (M1028A1 AND M1031) (Con't).

34. Install new bushing to extension (27) if removed. Install new seal (35) to extension. Install extension to retainer (29) with a new gasket. Apply gasket sealing compound to 5 bolts (34). Install bolts and tighten to 30 lb.-ft. (41 N.m).

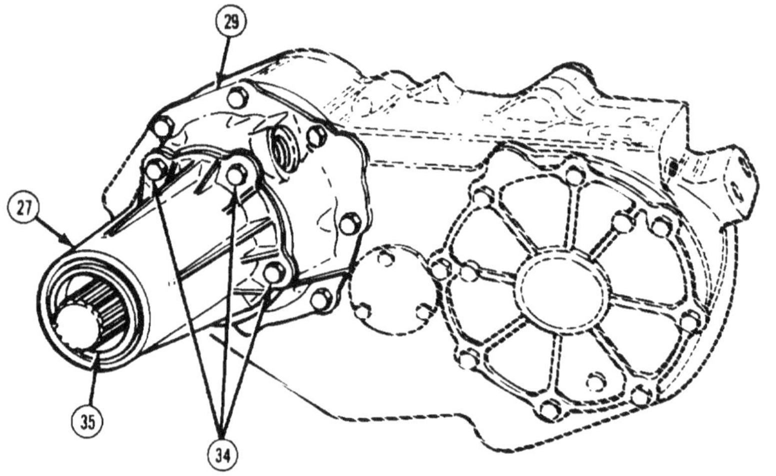

35. Install PTO cover (100) with a new gasket (99). Apply gasket sealing compound to 6 bolts (101) and install.

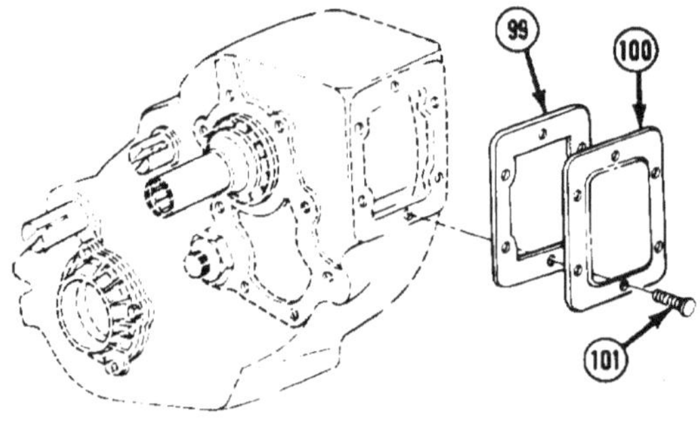

TA50357

5-22. TRANSFER CASE MAINTENANCE (M1028A1 AND M1031) (Con't),

NOTE

If plugs (45) are steel, RTV sealant should be applied to them before installing,

36. Install 2 plugs (45) to top of transfer case (1). Install ball, spring, new gasket and poppet plug (46), Install ball, spring, new gasket, and 4-wheel drive indicator light switch (17).
37. If grommet has been removed, heat new grommet in boiling water to soften and install to shift link (49).
38. Install selector rod (2) through grommet to shift link (49) if removed. Install shift link to 4-wheel drive shift rail (44) and range rail (48) with 2 clevis pins (47) and 2 spring clips.

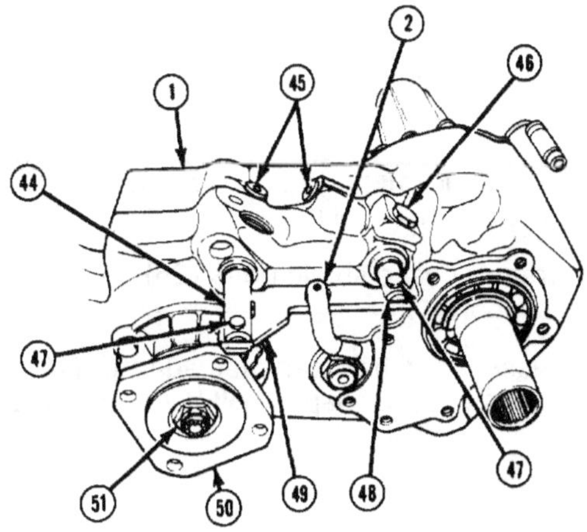

40. Install front output shaft yoke (50) with new seal washer, 1 flat washer, and new locknut (51). Secure yoke in a suitable holding device and tighten locknut to 150 lb.-ft. (203 Nom).

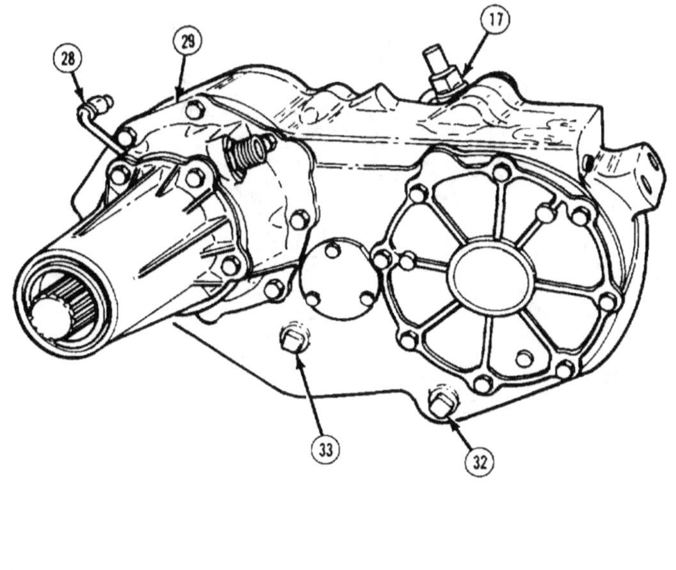

TA701859

5-22. TRANSFER CASE MAINTENANCE (M1028A1 AND M1031) (Con't).

41. Apply RTV sealant to vent (28) and install to retainer (29).

42. Install fill plug (33) and drainplug (32), Tighten drainplug to 40 lb.-ft. (54 N.m),

e. INSTALLATION

1. Lightly coat 2 new transfer case adapter seals (9 and 15) with hydraulic fluid and install to transfer case adapter (10).
2. Install transfer case adapter (10) to transmission with 6 new lockwashers (12) and bolts (13). Tighten bolts to 30 lb.-ft. (41 N.m).
3. Install 2 left transfer case shield bolts (26) and 2 upper transmission mounts (18) to transmission support (19).
4. Lower transmission and transfer case adapter (10), Remove support from transmission, ensuring that transfer case adapter bolt holes aline with transmission support (19) bolt holes.

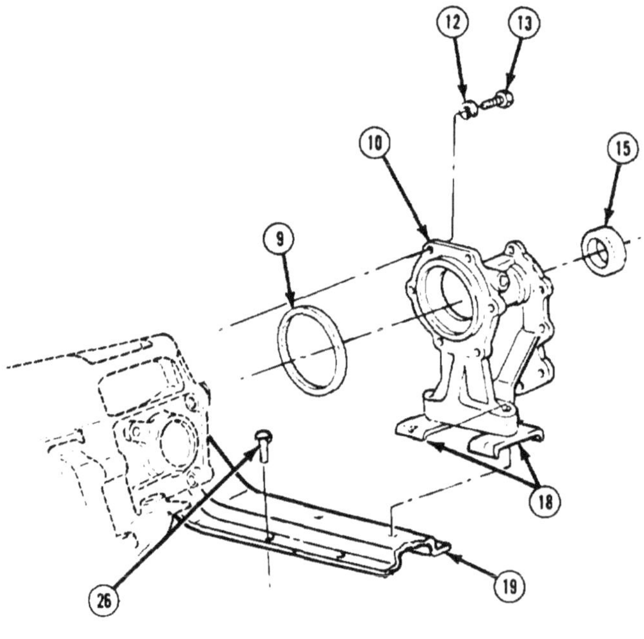

TA50359

5-22. TRANSFER CASE MAINTENANCE (M1028A1 AND M1031) (Con't).

NOTE

Right bolt (22) must be started first.

5 . Install transfer case adapter (10) to transmission support (19) with 2 spacers (20), lower transmission mounts (21), and bolts (22), Tighten bolts to 40 lb.-ft. (54 N.m).

6 . Install new transfer case adapter gasket (16) to transfer case (1), alining bolt holes.

WARNING

Transfer case (1) is heavy, Provide suitable support and use caution during installation. Failure to follow this warning may result in damage to transfer case or serious injury to personnel.

7 . Install two 3/8 in, N.C. bolts, with heads cut off, in transfer case (1) to serve as guide pins. Place transfer case on a suitable jack and raise into position.

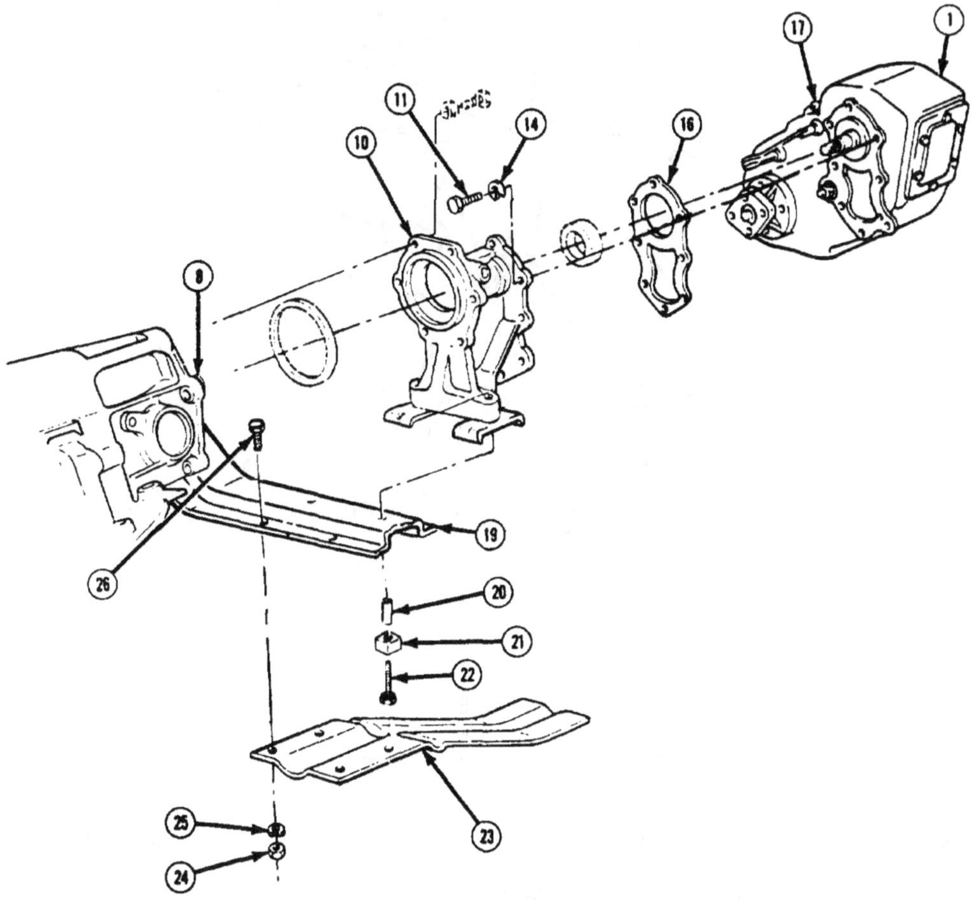

TA50360

5-22. TRANSFER CASE MAINTENANCE (M1028A1 AND M1031) (Con't).

CAUTION

Ensure that transfer case (1) is flush against transfer case adapter (10) to avoid damage to transfer case when tightening bolts (11).

8. Position transfer case (1) against transfer case adapter (10) and in line with transmission output shaft (8). Apply gasket sealing compound to 8 bolts (11). install 6 new lockwashers (14) and bolts. Remove 2 guide pins. Install 2 remaining lockwashers and bolts. Tighten bolts securely.

9. Connect selector rod (2) to control lever (4) with 1 new wave washer (7), 1 flat washer (6), and new cotter pin (5). Install 1 flat washer, new wave washer, and pivot bolt (3). Tighten pivot bolt to 100 lb.-ft. (136 N.m).

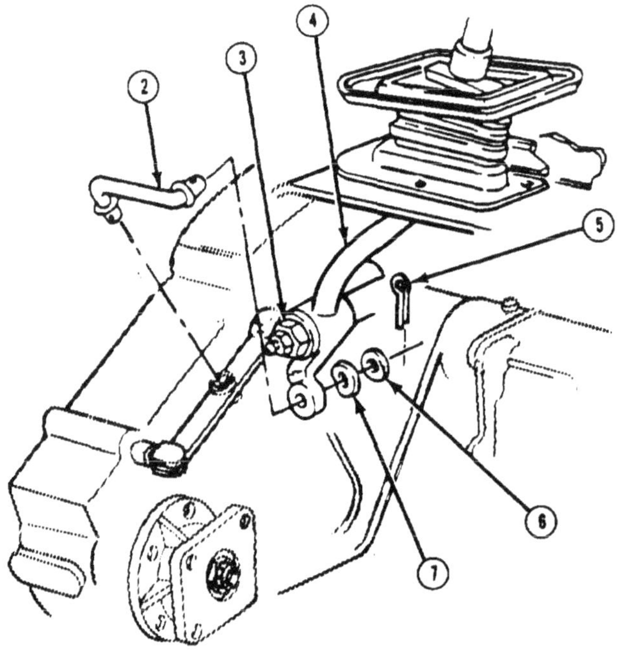

NOTE

2 left transfer case shield bolts (26) are already in place.

10. Install transfer case shield (23) with 2 right bolts (26), 4 new lockwashers (25), and nuts (24).

11. Connect connector to 4-wheel drive indicator light switch (17).

TA50361

5-22. TRANSFER CASE MAINTENANCE (M1028A1 AND M1031) (Con't).

FOLLOW-ON TASKS:

- Install speedometer gear. (See TM 9-2320-289-20)
- nstall front and rear propeller shafts. (See TM 9-2320-289-20)
- Fill transfer case, (See TM 9-2320-289-20)
- Remove jack stands and lower truck.
- Connect both battery negative cables. (See TM 9-2320-289-20)
- Check operation of transfer case.

CHAPTER 6
PROPELLER SHAFTS, AXLES, AND SUSPENSION MAINTENANCE

Section I. PROPELLER SHAFTS MAINTENANCE

6-1. CONSTANT VELOCITY JOINT REPAIR.

This task covers: a. Disassembly; b. Cleaning and Inspection; c. Assembly

INITIAL SETUP:

Equipment Condition

- Front propeller shaft removed. (See TM 9-2320-289-20)

Tools/Test Equipment

- Constant velocity remover/installer, J-23996

Materials/Parts

- One universal joint repair kit
- One constant velocity ball kit
- Dry cleaning solvent (Item 23, Appendix B)

General Safety *Instructions*

- Dry cleaning solvent is flammable and must not be used near open flame. Use only in a well-ventilated area.

a. DISASSEMBLY

NOTE

Ensure that front propeller shaft yoke (1) and link yoke (4) are marked for alinement during assembly.

1. Make marks (6) on front propeller shaft yoke (1) and link yoke (4) for assembly alinement.
2. if present, remove 2 retainer clips (3). Press out 2 bearing cups (2), and disengage front propeller shaft yoke (1) from link yoke (4) and from centering ball. If present, remove 2 retainer clips at bearing cups (7). Press out 2 bearing cups (7). Discard all universal joint (8) components.

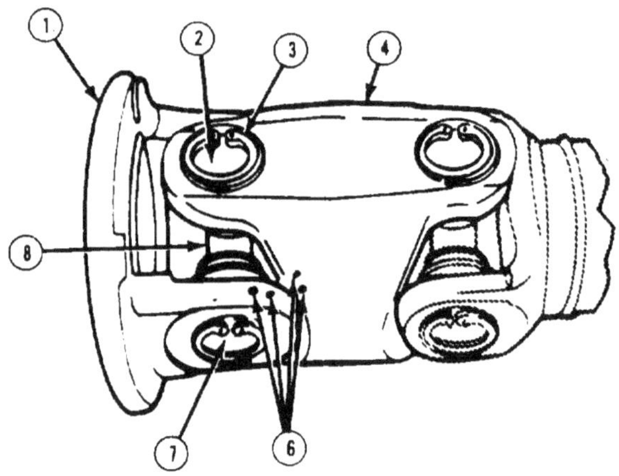

TA50506

6-1. CONSTANT VELOCITY JOINT REPAIR (Con't).

3. Remove seal (9) from centering ball socket (14). Remove washer (10), 3 ball seats (11), washer (12), and spring (13). Discard all components removed from centering ball socket.

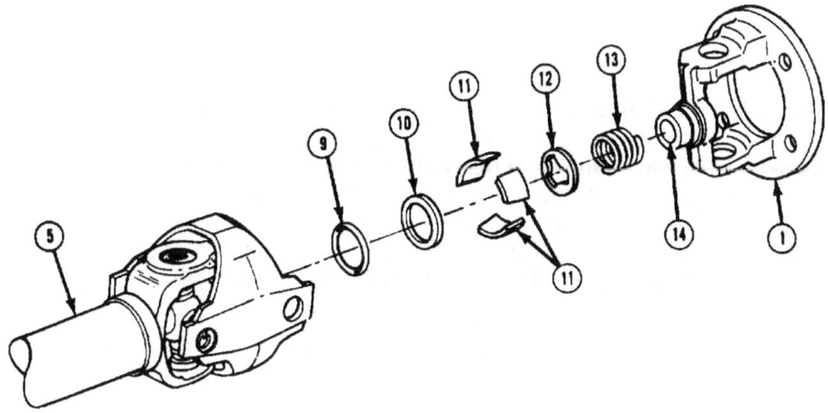

4. Place fingers of inner part of constant velocity remover (16) under centering ball (15). Place outer cylinder (17) of constant velocity remover over inner part and centering ball. Install nut (18) on tool and draw centering ball off stud. Discard centering ball.

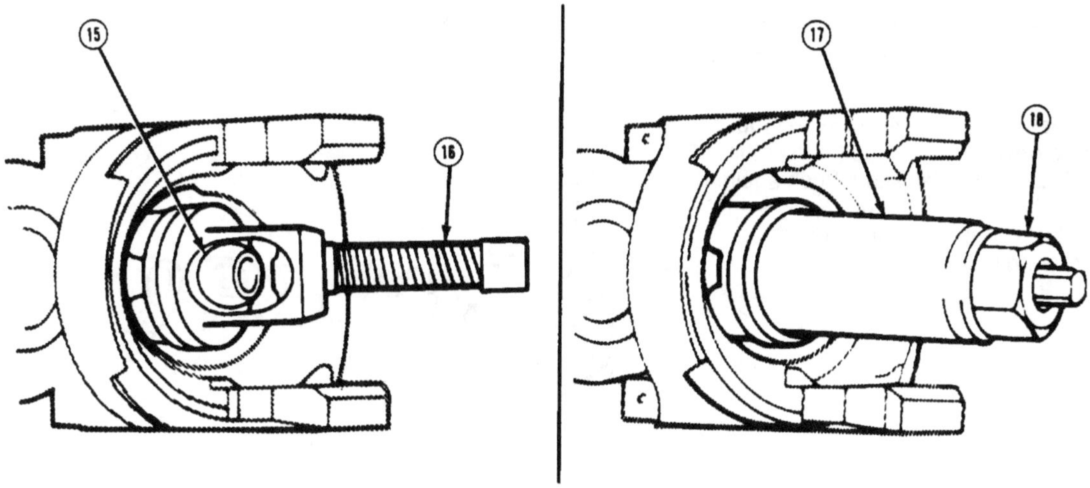

TA50507

6-1. CONSTANT VELOCITY JOINT REPAIR (Con't).

b. CLEANING AND INSPECTION

WARNING

Dry cleaning solvent P-D-680 is toxic and flammable. Always wear protective goggles and gloves and use only in a well-ventilated area. Avoid contact with skin, eyes, and clothes and DO NOT breathe vapors. DO NOT use near open flame or excessive heat. The solvent's flash point is 100°F-138°F (38°C-59°C). If you become dizzy while using cleaning solvent, immediately get fresh air and medical help, if solvent contacts eyes, immediately wash your eyes with water and get medical aid.

1. Clean front propeller shaft yoke, link yoke, and end of front propeller shaft (5) with dry cleaning solvent.
2. Inspect ball seat insert bushing in front propeller shaft yoke (1) for wear. Replace front propeller shaft yoke if worn.
3. Inspect centering ball stud for damage. Replace front propeller shaft (5) if damaged.

c. ASSEMBLY

CAUTION

Ensure that centering ball (15) is fully seated. Failure to follow this caution may result in damage to constant velocity joint,

1. Place new centering ball (15) on stud. Drive new centering ball onto stud until centering ball seats flush against shoulder at base of stud. Apply a light coat of grease, provided in kit, to centering ball.

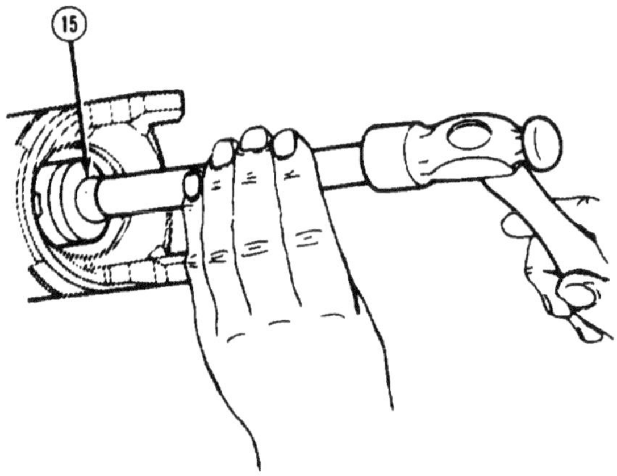

TA50508

6-1. CONSTANT VELOCITY JOINT REPAIR (Con't).

2. Position new universal joint (8) in front propeller shaft yoke (1) and press in 2 new bearing cups (7). Install 2 new retainer clips (3) and fully seat in bearing cup grooves.

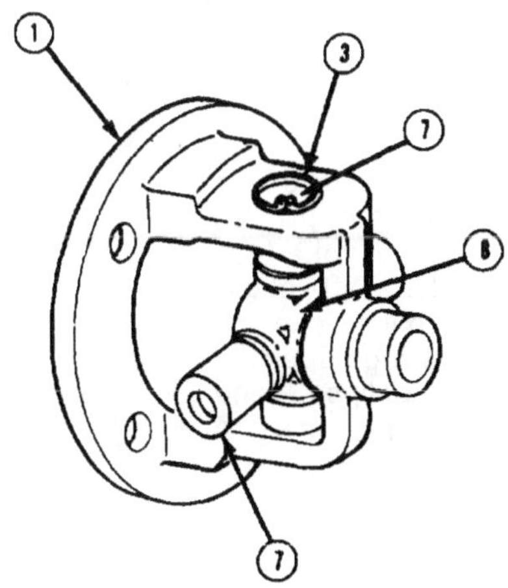

NOTE

There are 2 springs provided in constant velocity ball kit. Shorter spring (13) is to be used.

3. Lubricate all constant velocity replacement parts with grease provided in kit. In clean centering ball socket (14) in front propeller shaft yoke (1) install in the following order: spring (13), small O.D. washer (1 2), 3 ball seats (11) with largest opening outward to receive centering ball, and large O.D. washer (10).

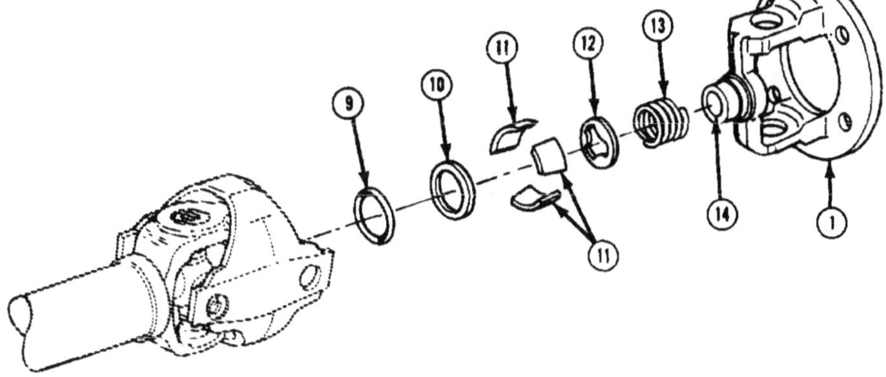

TA50509

6-1. CONSTANT VELOCITY JOINT REPAIR (Con't).

NOTE

Ensure that sealing lip of seal (9) tips inward.

4. Lubricate lip of seal (9) with grease provided in kit and install until flush. Fill cavity with grease provided in kit.

5. Using marks (6) made during disassembly, aline front propeller shaft yoke (1) and link yoke (4). install front propeller shaft yoke to centering ball and to link yoke and press in 2 new bearing cups (2). Install 2 new retainer clips (3) and fully seat in bearing cup grooves.

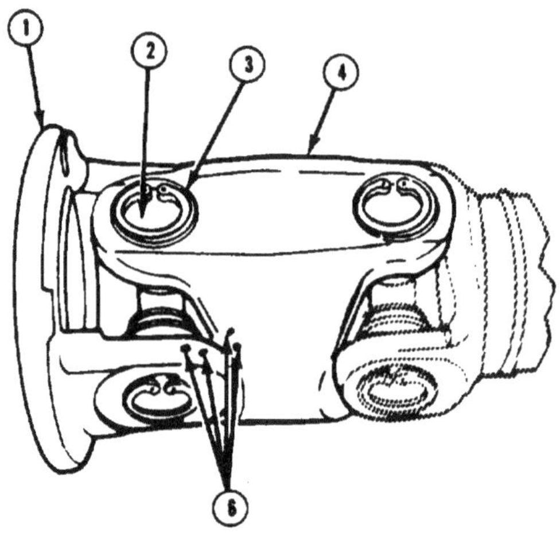

FOLLOW-ON TASKS:

- Install front propeller shaft. (See TM 9-2320-289-20)

TA50510

Section II. FRONT AND REAR AXLES MAINTENANCE

6-2. FRONT AND REAR AXLES MAINTENANCE INDEX.

6-3. FRONT AXLE ASSEMBLY REPLACEMENT.

This task covers: a. Removal b. Installation

INITIAL SETUP:

Equipment Condition

- Truck raised and frame supported on jack stands.
- Front wheels removed. (See TM 9-2320-289-20)
- Calipers removed and tied out of way. (See TM 9-2320-289-20)
- Front propeller shaft disconnected from front axle differential. (See TM 9-2320-289-20)
- Connecting rod disconnected from steering arm. (See TM 9-2320-289-20)
- Front shock absorbers disconnected from axle. (See TM 9-2320-289-20)

ToolslTest Equipment

- Torque wrench

Personnel Required

- MOS 63W (2)

General Safety Instructions

- Front axle assembly is heavy. Use extreme caution, provide support, and use assistance during removal and installation.

1. Slide clamp up front axle vent hose (7) and disconnect front axle vent hose from front axle differential (8).

2. Place a suitable jack under front axle differential (8).

WARNING

Front axle assembly (10) is heavy, Use extreme caution, provide support, and use assistance during removal, Failure to follow this warning may result in serious injury to personnel or damage to front axle assembly,

NOTE

. All except M1009 front axle assembly (10) removal is given.

. M1009 front axle assembly (10) is secured to spring assembly (1) using 2 "U" bolts (9) on each side.

3. Remove 6 nuts (4), 2 bolts (5), 8 washers (3), and 3 "U" bolts (9), Lower front axle assembly (10) and remove, Remove 2 spacers (6) from axle housing.

6-3. FRONT AXLE ASSEMBLY REPLACEMENT (Con't).

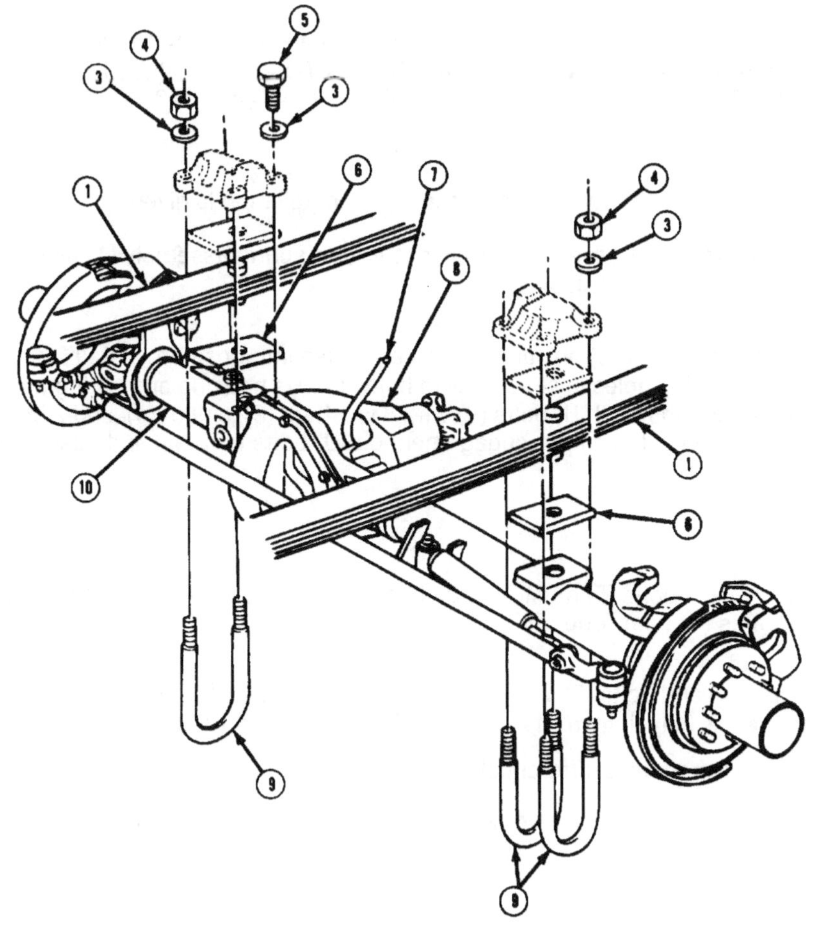

TA50511

6-3. FRONT AXLE ASSEMBLY REPLACEMENT (Con't).

b. INSTALLATION

WARNING

Front axle assembly (10) is heavy, Use extreme caution, provide support, and use assistance during installation. Failure to follow this warning may result In serious injury to personnel or damage to front axle assembly.

NOTE

- All except M1009 front axle assembly (10) installation Is given.
- M1009 front axle assembly (10) is secured to spring assembly (1) using 2 "U" bolts (9) on each side.

1. Pisutuib fribt axle assembly (10) on suitable jack and raise into position. Install 2 spacers (6) on axle housing, alining holes in spacers with holes in axle housing and spring assembly (1) center bolts. Loosely install 3 "U" bolts (9) with 8 washers (3), 6 nuts (4), and 2 bolts (5). Install all nuts and bolts to uniform engagement on "U" bolts to retain and position anchor plate (2).

NOTE

All nuts (4) and bolts (5) must be tightened evanly to ensure that spring assemblies (1) seat properly.

2. Tighten all nuts (4) and bolts (5) in diagonal sequence to 25 lb.-ft. (34 N.m). Using same diagonal sequence, fully tighten all nuts and bolts to 150 lb.-ft, (203 N.m).
3. Remove jack from front axle differential (8).
4. Connect front axle vent hose (7) to front axle differential (8) and tighten clamp.

6-3. FRONT AXLE ASSEMBLY REPLACEMENT (Con't).

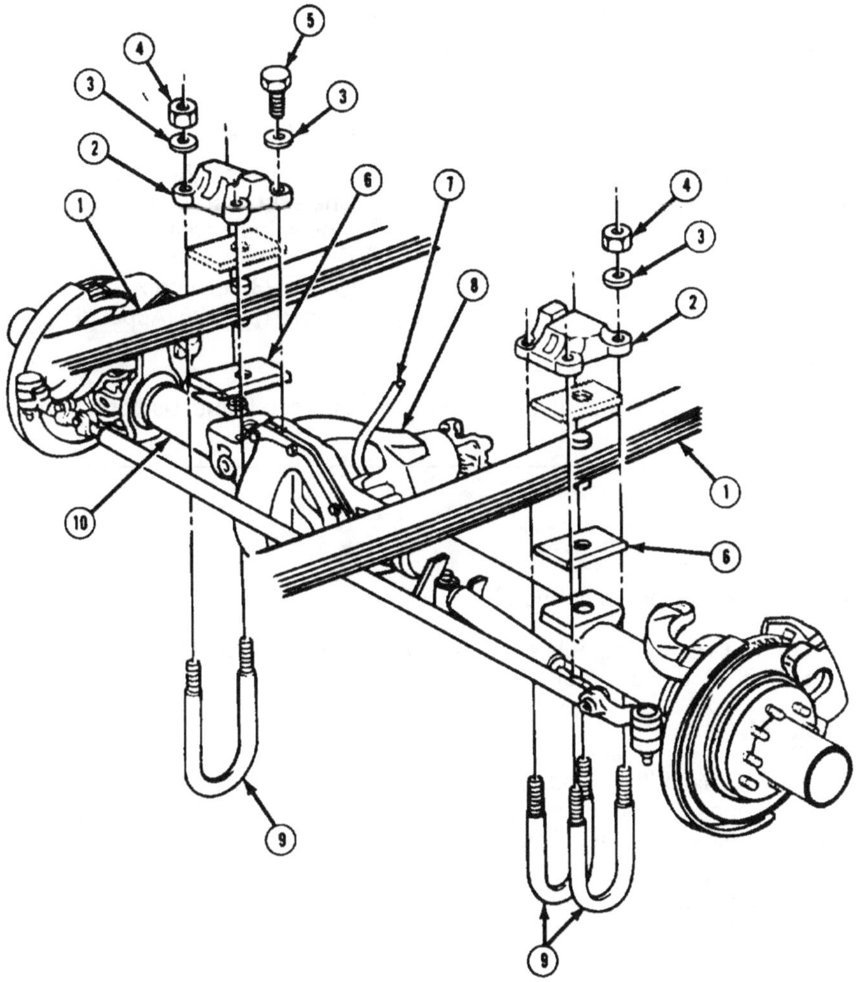

FOLLOW-ON TASKS:

- Connect front shock absorbers to axle. (See TM 9-2320-289-20)
- Connect connecting rod to steering arm. (See TM 9-2320-289-20)
- Connect front propeller shaft to front axle differential. (See TM 9-2320-289-20)
- Install calipers. (See TM 9-2320-289-20)
- Install front wheels, (See TM 9-2320-289-20)
- Remove jack stands and lower truck.
- Road test truck and check torque of "U" bolt fasteners.

TA50512

6-4. FRONT AXLE SHAFT MAINTENANCE.

This task covers: a. Removal b. Cleaning and Inspection c. Installation

INITIAL SETUP:

Equipment Condition

- Front axle spindle removed. (See TM 9-2320-289-20)

Materials/Parts

- One universal joint repair kit
- One axle shaft seal
- One spindle seal
- Dry cleaning solvent (Item 23, Appendix B)
- Grease (Item 36, Appendix B)

General Safety Instructions

- Dry cleaning solvent is flammable and must not be used near open flame. Use only in a well-ventilated area.

a. REMOVAL

1. Remove front axle shaft assembly (1 and 7).
2. If not already removed, remove spindle seal (6), axle shaft seal (5), and bronze spacer (8) from outer axle shaft (7). Discard seals.
3. Remove deflector (4).
4. Place front axle shaft assembly (1 and 7) over wooden blocks. Remove 4 retainer clips (10). Press out 4 bearing cups (2) and remove universal joint (3), Separate outer axle shaft (7) from inner axle shaft (1), Discard retainer clips, bearing cups (2), and universal joint (3).

b. CLEANING AND INSPECTION

WARNING

Dry cleaning solvent P-D-680 is toxic and flammable. Always wear protective goggles and gloves and use only in a well-ventilated area. Avoid contact with skin, eyes, and clothes and DO NOT breathe vapors. DO NOT use near open flame or excessive heat. The solvent's flash point is 100°F-138°F (38°C-59°C). If you become dizzy while using cleaning solvent, immediately get fresh air and medical help. If solvent contacts eyes, immediately wash your eyes with water and get medical aid.

1. Clean inner and outer axle shafts with dry cleaning solvent.
2. Inspect inner and outer axle shafts for cracks, bends, or damaged splines. Replace if cracked, bent, or if splines are damaged.

c. INSTALLATION

1. Position new universal joint (3) in outer axle shaft (7) yoke.

6-4. FRONT AXLE SHAFT MAINTENANCE (Con't).

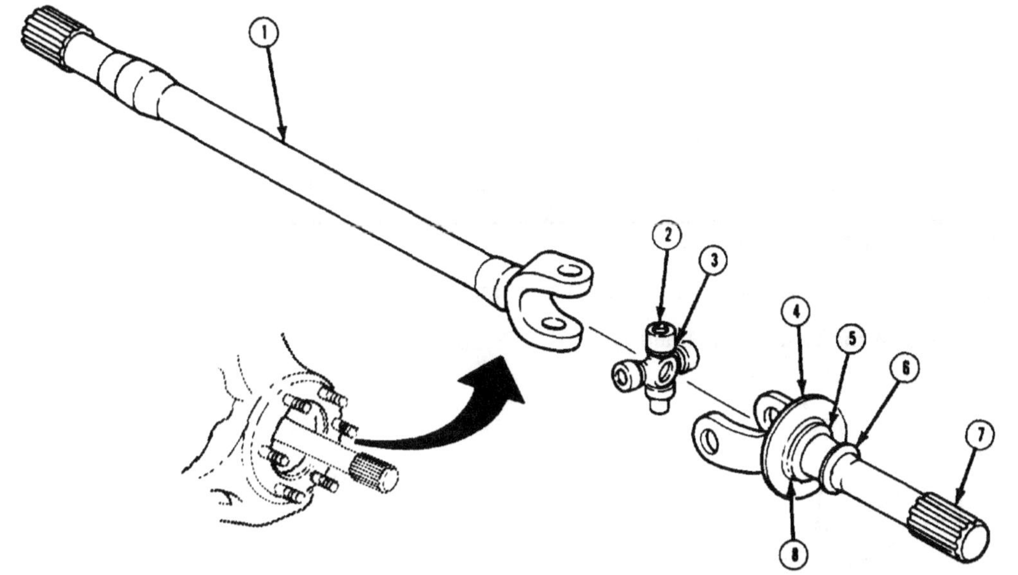

CAUTION

- Ensure that seal (9) and washer (12) are properly positioned and that needle bearings (11) are properly alined in bearing cups (2),
- Ensure that bearing cups (2) are alined with universal joint (3) trunnion. Failure to follow this caution may result in damage to universal joint and bearing cups.

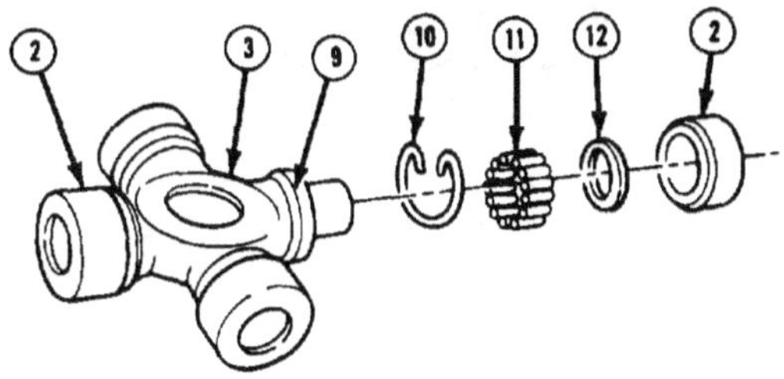

2. Position 2 bearing cups (2) at outside of outer axle shaft (7) yoke and press in bearing cups far enough to install 2 retainer clips (10).

TA50513

6-4. FRONT AXLE SHAFT MAINTENANCE (Con't).

3 . Repeat step 2 to install other 2 bearing cups (2) into inner axle shaft (1) yoke.

NOTE

Ensure that bronze spacer (8) Is fully seated against shoulder of outer axle shaft (7).

4 . Install deflector (4) and bronze spacer (8) to outer axle shaft (7). Install new axle shaft seal (5) on outer axle shaft with lip facing outward. Pack area around thrust face area of outer axle shaft and axle shaft seal with grease.

5 . Install new spindle seal (6) on outer axle shaft (7) with lip facing inward.

6 . Install front axle shaft assembly (1 and 7),

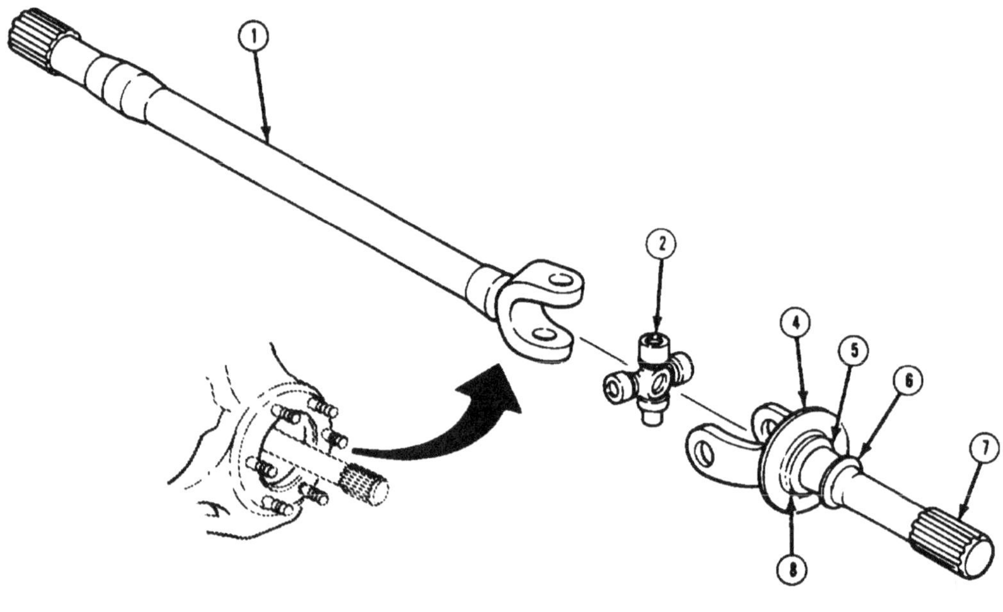

FOLLOW-ON TASKS:

- Install front axle spindle, (See TM 9-2320-289-20)
- Check front axle differential lubricating oil level. (See LO 9-2320-289-12)

TA50514

6-5. FRONT AND REAR AXLE DIFFERENTIAL REPLACEMENT (M1009).

This task covers: a. Removal b. Cleaning and Inspection c. Installation

INITIAL SETUP:

Equipment Condition

- Front or rear axle shafts removed. (See paragraph 6-4 or 6-14)
- Steering tie-rod removed (if replacing front axle differential). (See TM 9-2320-289-20)

Tools/Test Equipment

- Bearing preload gage and strap, J-22779
- Dial indicator
- Micrometer
- Torque wrench

Materials/Parts

- One differential cover gasket
- Shim sets
- Prussian blue dye (Item 24, Appendix B)
- Gear lubricating oil (Item 44, Appendix B)

Personnel Required

- MOS 63W (2)

Manual References

- TM 9-214

General Safety Instructions

- Dry cleaning solvent is flammable and must not be used near open flame. Use only in a well-ventilated area.

a. REMOVAL

1. If working on front axle differential, place a suitable container under housing. Remove 10 bolts (1) and cover (2). Allow all lubricating oil to drain. Remove cover gasket and discard.

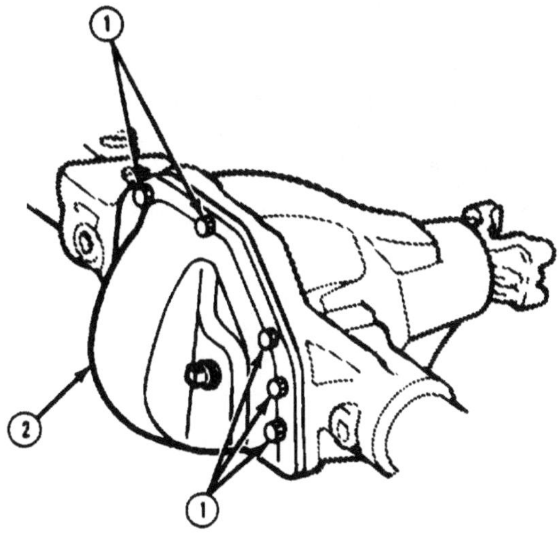

TA50516

6-5. FRONT AND REAR AXLE DIFFERENTIAL REPLACEMENT (M1009) (Con't).

2. Mark 2 bearing caps (3) and housing for installation alinement. Remove 4 bolts (4) and bearing caps.

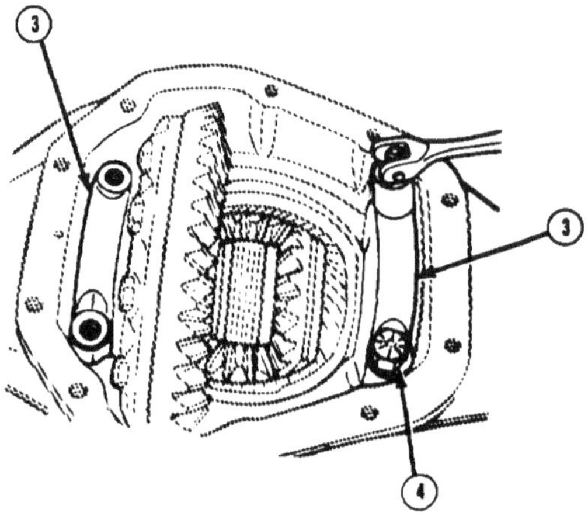

NOTE

- Front axle differential has shims (5) and spacers at side bearing assemblies. Rear axle differential has only shims.
- When removing differential case (7) assembly from housing, bearing cups (6), shims (5), and spacers may fall out. Ensure that bearing cups, shims, and spacers are tagged, indicating from which side they were removed.

3. Work differential case (7) assembly loose. Remove differential case assembly, bearing cups (6), shims (5), and spacers if present.

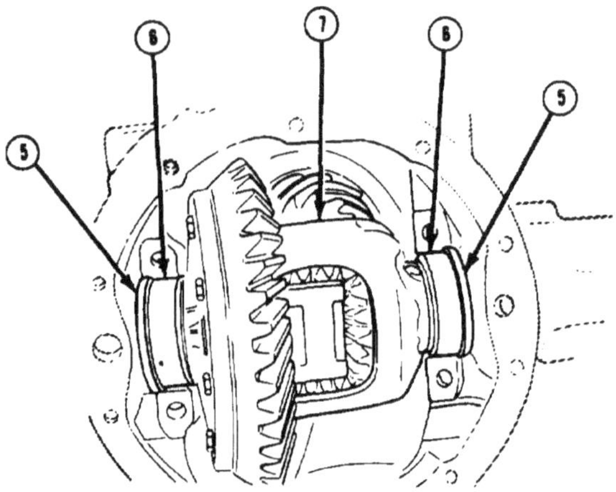

TA50516

6-5. FRONT AND REAR AXLE DIFFERENTIAL REPLACEMENT (M1009) (Con't).

b. CLEANING AND INSPECTION

WARNING

Dry cleaning solvent P-D-680 is toxic and flammable. Always wear protective goggles and gloves and use only in a well-ventilated area. Avoid contact with skin, eyes, and clothes and DO NOT breathe vapors. DO NOT use near open flame or excessive heat. The solvent's flash point is 100°F-138°F (38°C-590C). If you become dizzy while using cleaning solvent, immediately get fresh air and medical help. If solvent contacts eyes, immediately wash your eyes with water and get medical aid.

1. Clean differential cover and magnet, interior of housing, bearing cups, and shims with dry cleaning solvent. Clean all gasket material from differential cover and housing.
2. Inspect shims and spacers for damage. Replace if damaged.
3. Inspect condition of differential side bearings and bearing cups in accordance with TM 9-214. Inspect cup seats in housing and bearing caps to ensure that they are free of nicks, burrs, and foreign material. Remove any nicks, burrs, or foreign material found.
4. If present, remove excess casting material near machined surfaces to provide for installation of shims and bearing preload gage. Remove all traces of filing dust.

NOTE

- Ensure that position of axle shaft inner seals is noted for installation.
- There are no axle shaft seals on rear axle assembly.

5. Inspect 2 axle shaft inner seals in housing side bores. Remove and discard if damaged.

6-5. FRONT AND REAR AXLE DIFFERENTIAL REPLACEMENT (M1009) (Con't).

c. INSTALLATION

1. If removed, install 2 new axle shaft inner seals in housing side bores,
2. Install 2 bearing cups (6) as tagged, over side bearings (10), Install differential case (7) assembly inside housing.
3. Install bearing strap tool (8) over left side bearing cup (6) with 2 bolts.
4. With ring gear (12) tight against drive pinion (9), insert bearing preload gage (13) between left bearing cup (6) and housing.
5. While moving bearing preload gage (13), turn adjusting nut clockwise until a noticeable drag is felt. Tighten lockbolt on side of tool,
6. Between right bearing cup (6) and housing, install spacer if removed, and a shim (5) thinner than what was originally installed. Install a feeler gage (11) between shim and housing. Feeler gage must be of sufficient thickness to produce a slight drag when moved between shim and housing,

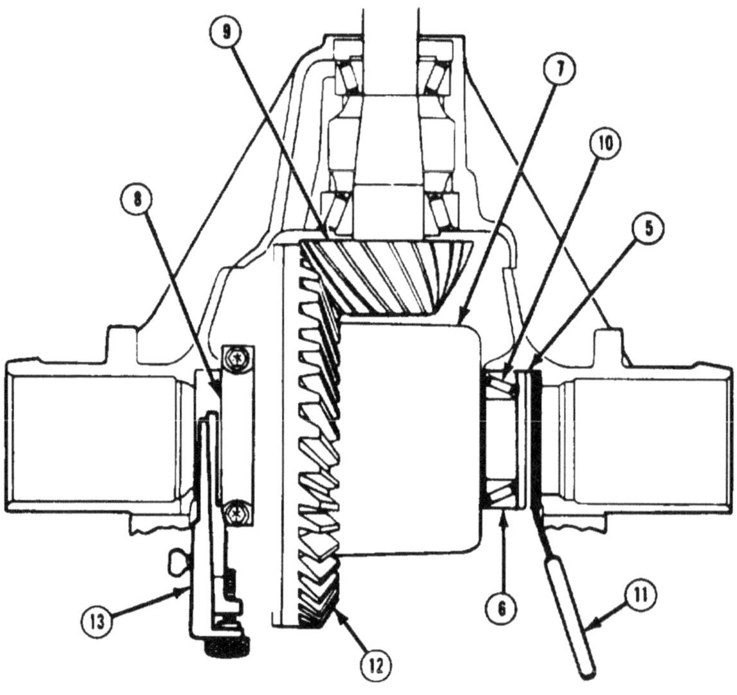

TA50517

6-5. FRONT AND REAR AXLE DIFFERENTIAL REPLACEMENT (M1009) (Con't).

7. Remove bearing preload gage (13), spacer if installed, shim (5), and feeler gage (11).

8. Measure thickness of bearing preload gage (13) in 3 places and average these readings. Record the result.

9. Measure combined thickness of spacer if present, shim (5), and feeler gage (11).

10. Using example given in Table 6-1, determine proper thickness of each shim pack,

Table 6-1. Determining Side Bearing Shim Requirements

RING GEAR SIDE				OPPOSITE SIDE	
inches	millimeters			inches	millimeters
0.250 In.	6.350 mm	Thickness of bearing preload tool required to force ring gear Into contact with drive pln-lon.	Combined total of spacer, service shim, and feeler gage.	0.265 In.	6.731 mm
-0.010 In.	-0.254 mm	TO MAINTAIN PROPER BACKLASH (0 .005 In. - 0.008 in.) (0.127 mm - 0.2032 mm), ring gear is moved away from pinion by subtracting 0.010 in. (0 .254 mm) shims from ring gear side and adding 0.010 in. (0.254 mm) shims to the opposite side.		+ 0.010 in.	+ 0.254 mm
0.240 in.	6.096 mm			0.275 in.	6.985 mm
+0.004 in.	+ 0.102 mm	TO OBTAIN PROPER PRELOAD on side bearings, add 0.004 in. (0. 102 mm) to each side.		+ 0.004 in.	+0.102 mm
0.244 in.	6.198 mm	Shim dimension required for ring gear side,	Shim dimension required for opposite side.	0.279 in.	7.087 mm

11. If present, install 2 spacers as tagged, between bearing cups (6) and housing.

NOTE

- Production preloading of the differential side bearings (10) is accomplished by use of cast iron preload shims. These cannot be reused when rebuilding the housing as they may break when tapped into place. Service shims may be reused if not damaged.

- Chamfered side of shim (5) should face outward. Shim should have sufficient chamfer or lead around outside diameter to allow for easy installation. Additional chamfer on shim should be filed or ground as required.

6-5. FRONT AND REAR AXLE DIFFERENTIAL REPLACEMENT (M1009) (Con't).

NOTE

If shim (5) is difficult to install, differential case (7) assembly may be partially removed, then shim and differential case assembly slid into position at same time.

12. Install left shim (5) first, between spacer, if present, and housing. Install right shim. Using soft-faced hammer, tap shims into place while rotating differential case (7) assembly.

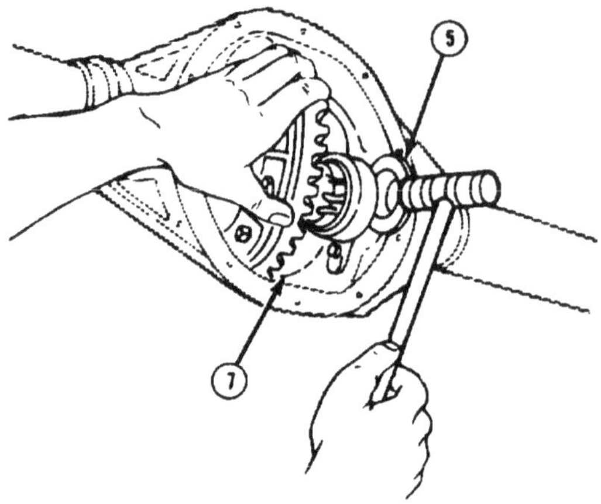

13. Remove 2 bolts and bearing strap tool (8).

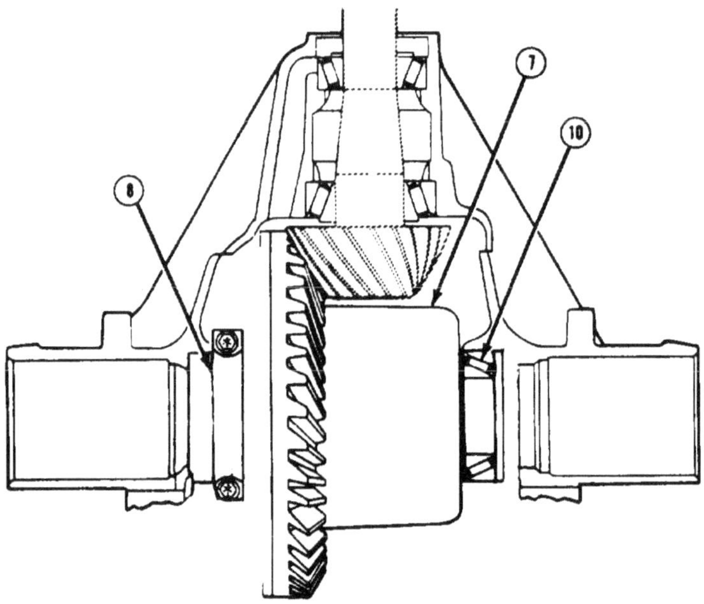

TA50518

6-5. FRONT AND REAR AXLE DIFFERENTIAL REPLACEMENT (M1009) (Con't).

14. Install 2 bearing caps (3) in their original positions and install 4 bolts (4). Tighten bolts to 55 lb.-ft. (75 N.m).

NOTE

Differential side bearings (10) are now properly preloaded. If any adjustments are required in later procedures, ensure that preload established in step 12 remains the same. If backlash is changed in later steps, the total thickness of 2 shim packs does NOT change.

15. Mount dial indicator (14) on housing with indicator button perpendicular to tooth angle and in line with ring gear (12) rotation. Check backlash between ring gear and drive pinion. Backlash should be within range of 0.005-0.008 in. (0.13-0.20 mm). Check backlash at 4 different equally spaced positions around ring gear. Variation in the 4 readings should not exceed 0.001 in. (0.025 mm).

16. If variation in backlash readings exceeds 0.002 in. (0.051 mm), reposition dial indicator (14) on housing so that indicator button contacts back of ring gear (12) and measure ring gear and housing runout. Gear runout should not exceed 0.003 in. (0.076 mm). If runout exceeds this limit, inspect ring gear and housing for deformation or foreign matter between housing and ring gear.

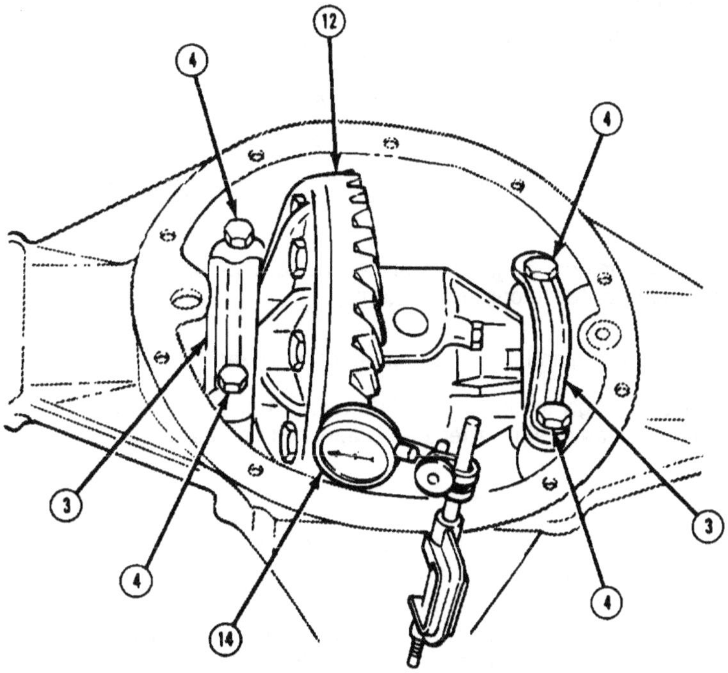

TA50519

6-5. FRONT AND REAR AXLE DIFFERENTIAL REPLACEMENT (M1009) (Con't).

17. If backlash is not within limits, decrease shim thickness on one side and increase shim thickness on other side by an equal amount:

    (a) If backlash exceeds .008 in. (0.20 mm), increase shim thickness on ring gear (12) side and decrease thickness on opposite side by an equal amount.

    (b) If backlash is less than 0.005 in. (0.13 mm), decrease shim thickness on ring gear (12) side and increase thickness on opposite side by an equal amount.

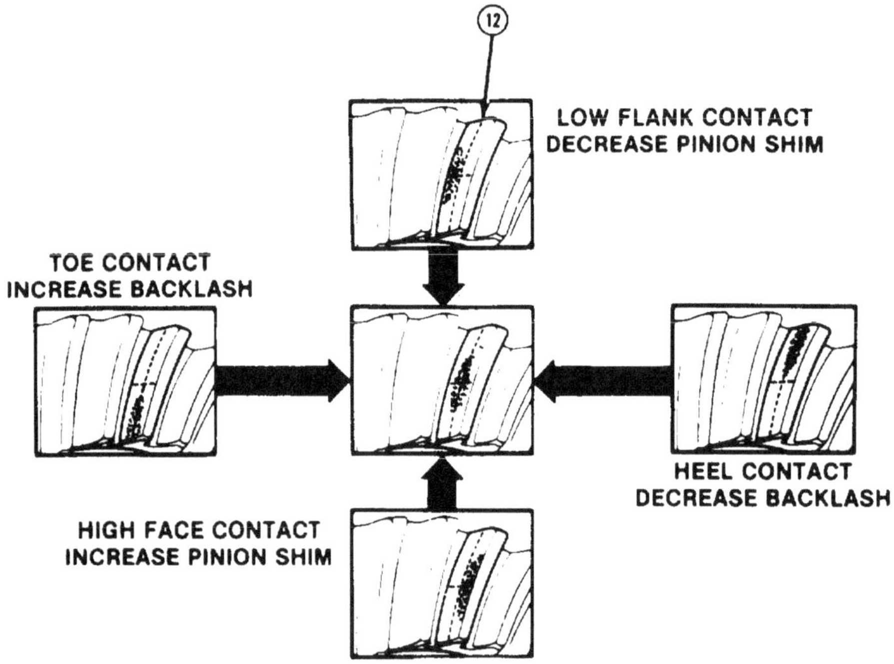

NOTE

If drive pinion and ring gear (12) were replaced, perform steps 20-23 to check for proper gear tooth contact pattern.

18. Wipe any lubricant from housing and clean each tooth of ring gear (12).

19. Apply prussian blue dye sparingly to all ring gear (12) teeth using medium stiff brush, Area of tooth contact should be clearly visible after load is applied to gears.

NOTE

Excessive turning of ring gear (12) is not recommended. Ring gear should only be turned enough to provide tooth contact pattern.

20. Apply load until it takes a torque of 40-50 lb.-ft. (54-68 N.m) to turn drive pinion. Turn companion flange so that ring gear (12) turns 1 full revolution, then reverse so that ring gear turns 1 full revolution in opposite direction.

TA50520

6-5. FRONT AND REAR AXLE DIFFERENTIAL REPLACEMENT (M1009) (Con't).

21. Contact pattern should be centrally located up and down on the face of the ring gear (12) teeth. Compare tooth contact pattern on ring gear with examples shown and make adjustments as indicated.

22. If working on front axle differential, install new cover gasket and cover (2) with 10 bolts (1). Tighten bolts to 35 lb.-ft. (47 N.m).

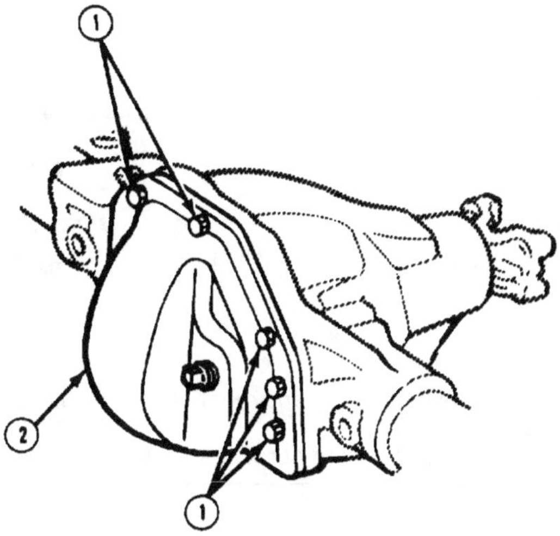

FOLLOW-ON TASKS:

- Install steering tie-rod if removed. (See TM 9-2320-289-20)
- Install front or rear axle shafts, (See paragraph 6-4 or 6-14)
- Fill front or rear axle differential with lubricating oil. (See LO 9-2320-289-12)

TA50521

6-6. FRONT AXLE DIFFERENTIAL REPAIR (M1009).

This task covers:

a. Disassembly
b. Cleaning and Inspection
c. Assembly

INITIAL SETUP:

Equipment condition

- Front axle differential removed. (See paragraph 6-5)

Materials/Parts

- Dry cleaning solvent (Item 23, Appendix B)
- Gear lubricating oil (Item 44, Appendix B)
- Wiping rags (Item 53, Appendix B)

Tools/Test Equipment

- Puller kit
- Torque wrench

Manual *References*

- TM 9-214

General Safety Instructions

- Dry cleaning solvent is flammable and must not be used near open flame. Use only in a well-ventilated area.
- Compressed air used for cleaning purposes should never exceed 30 psi (207 kPa).

a. DISASSEMBLY

1. Place differential case (1) assembly in vise.

2. Remove 2 side bearings (2) and tag for assembly.

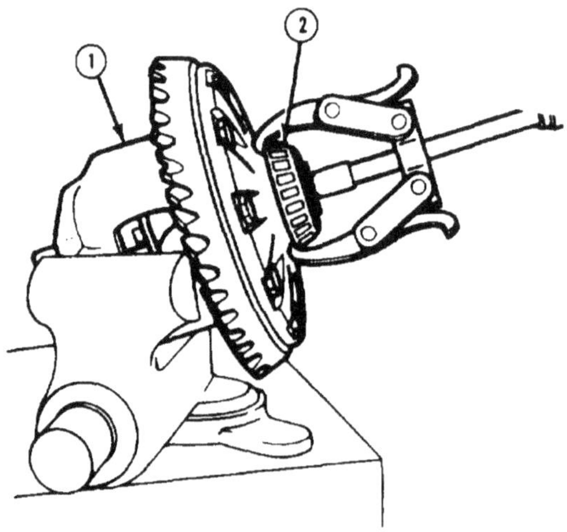

TA50522

6-6. FRONT AXLE DIFFERENTIAL REPAIR (M1009) (Con't).

NOTE

Shop rags placed around vise will help protect ring gear (4) teeth from damage should ring gear fali during disassembly.

3. Scribe a line on differential case (1) assembly and ring gear (4) to ensure assembly in same position. Remove 10 bolts (3). Using a soft-faced hammer, tap ring gear loose from differential case assembly. Remove differential case assembly from vise and remove ring gear.

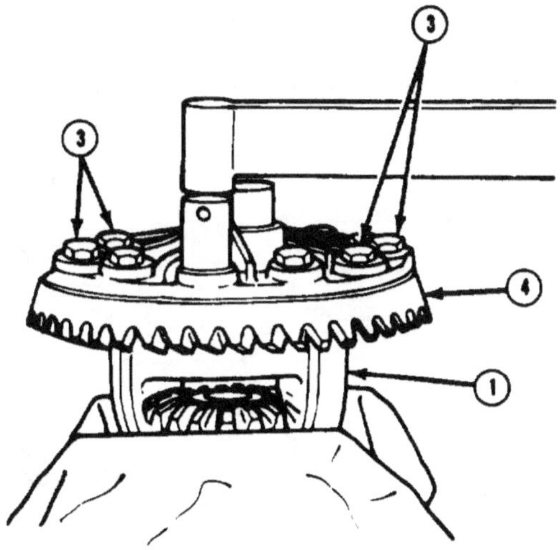

4. Drive out lockpin (6) and remove pinion shaft (5).

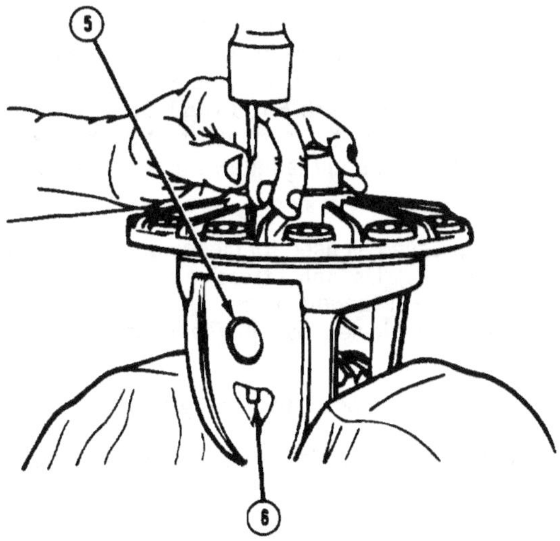

TA50523

6-6. FRONT AXLE DIFFERENTIAL REPAIR (M1009) (Con't).

NOTE

Ensure that pinion gears (7), side gears (8), and thrust washers (9) are marked for assembly,

5. Rotate 2 side gears (8). Remove 2 pinion gears (7) and thrust washers (9). Remove 2 side gears and thrust washers,

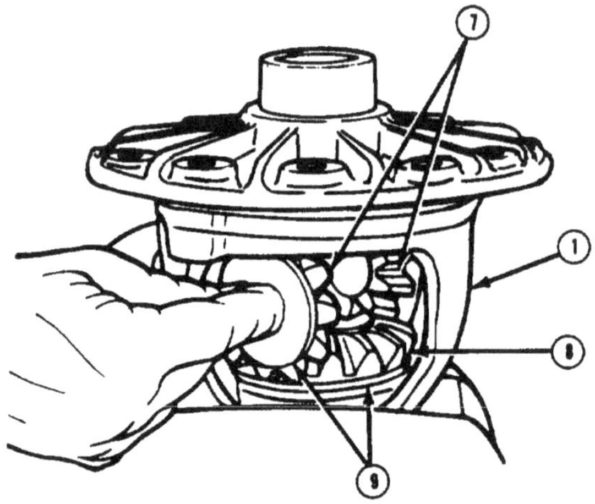

b. CLEANING AND INSPECTION

WARNING

- Dry cleaning solvent P-D-680 is toxic and flammable. Always wear protective goggles and gloves and use only in a well-ventilated area. Avoid contact with skin, eyes, and clothes and DO NOT breathe vapors. DO NOT use near open flame or excessive heat. The solvent's flash point is 100°F-138°F (38°C-59°C). If you become dizzy while using cleaning solvent, immediately get fresh air and medical help, if solvent contacts eyes, immediately wash your eyes with water and get medical aid.
- Compressed air used for cleaning purposes should never exceed 30 psi (207 kPa). Us. only l ffective chip guarding and personnel protective equipment (goggles/shield, gloves, etc.). Failure to follow this warning may result in serious injury to personnel.

1. Clean all removed components with dry cleaning solvent and dry with compressed air.
2. Clean and inspect differential side bearings in accordance with TM 9-214. If one side bearing is damaged, replace both side bearings.

TA50624

6-6. FRONT AXLE DIFFERENTIAL REPAIR (M1009) (Con't).

3. Inspect differential case for cracks, scoring, or signs of damage to machined surfaces. Remove any nicks or burrs found. Replace differential case if cracked or damaged.

NOTE

If ring gear is damaged, ring gear bolts and drive pinion must also be replaced. (See paragraph 6-7)

4. Inspect ring gear for damage. Replace if damaged.
5. Inspect thrust washers, pinion gears, and side gears for damage. If a pinion or side gear is damaged, replace both as a set.

c. ASSEMBLY

NOTE

Ensure that all components are lubricated with lubricating oil as they are assembled.

1. Place differential case (1) in vise.
2. Install 2 side gears (8) and thrust washers (9) inside differential case (1). If side gears and thrust washers are original, install as marked.

NOTE

Ensure that holes In pinion gears (7) are alined with holes in differential case (1) assembly.

3. Assemble 2 pinion gears (7) and thrust washers (9) and roll into position inside differential case (1) assembly. If pinion gears and thrust washers are original, install as marked.
4. Install pinion shaft (5) and retain with lockpin (6).

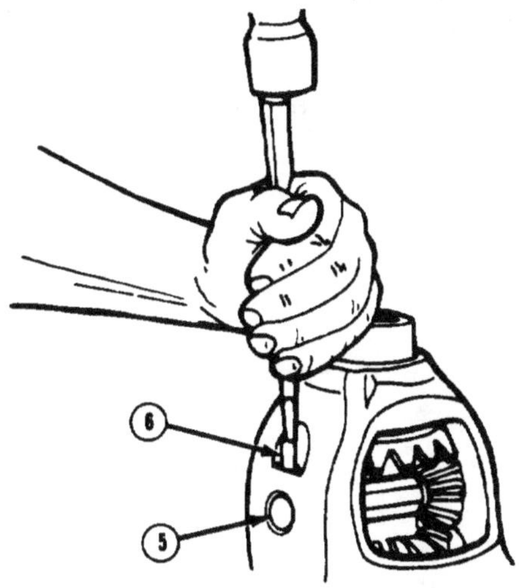

TA50525

6-6. FRONT AXLE DIFFERENTIAL REPAIR (M1009) (Con't).

NOTE

Shop rags placed around vise will help protect ring gear (4) teeth mfrom dt ri should ring gear fall during assembly.

5. Remove differential case (1) assembly from vise. Assemble ring gear (4) to differential case assembly alining marks made during disassembly, if original ring gear is used. Place differential case assembly in vise.

6. Install 10 bolts (3) and tighten alternately to 80 lb.-ft. (108 N.m).

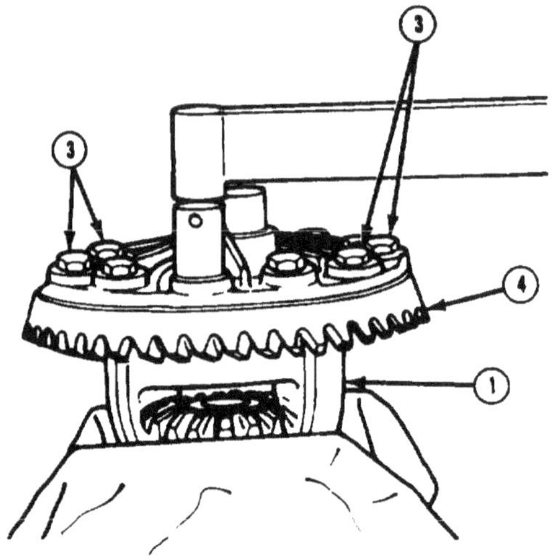

NOTE

Side bearings (2) should rotate freely once installed.

7. Install 2 side bearings (2) as tagged.

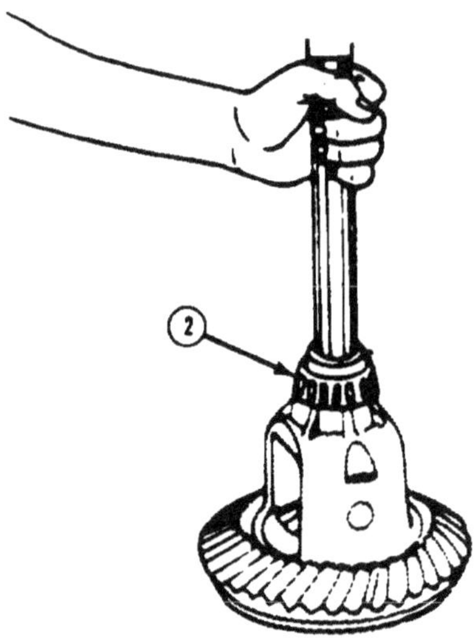

TA50526

6-6. FRONT AXLE DIFFERENTIAL REPAIR (M1009) (Con't).

FOLLOW-ON TASKS:

- Install front axle differential. (See paragraph 6-5)

6-7. FRONT AND REAR AXLE DRIVE PINION AND BEARINGS MAINTENANCE (M1009).

This task covers:

a. Removal
b. Cleaning and Inspection
c. Installation

INITIAL SETUP:

Equiprnent Condition

- Front or rear axle differential removed. (See paragraph 6-5)
- Front or rear propeller shaft disconnected from pinion flange. (See TM 9-2320-289-20)

Materials/Parts

- One flange nut
- One pinion oil seal
- One spacer
- Dry cleaning solvent (Item 23, Appendix B)
- Molybdenum grease (Item 37, Appendix B)
- Gear lubricating oil (Item 44, Appendix B)

Tools/Test Equipment

- Pinion setting gage, J-21777-500
- Flange holder and remover
- Micrometer
- Puller kit
- Torque wrench

Manual References

- TM 9-214

General *Safety Instructions*

- Dry cleaning solvent is flammable and must not be used near open flame. Use Only in a well-ventilated area.
- Compressed air used for cleaning purposes should never exceed 30 psi (207 kPa).

a. REMOVAL

1. At flange (1), measure torque required to rotate drive pinion and record for later reference. If there is no preload reading, check for looseness of pinion assembly by pushing and pulling on flange. Looseness indicates need for bearing replacement.

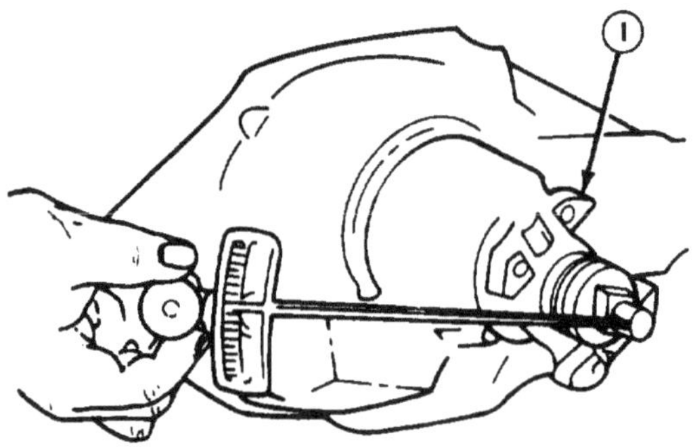

TA50527

6-7. FRONT AND REAR AXLE DRIVE PINION AND BEARINGS MAINTENANCE (M1009) (Con't).

NOTE

Flange nut (2) should be saved for use during drive pinion (3) installation, When drive pinion is fully seated and ready to receive final torque, a new flange nut must be used,

2. Using flange holder (5), remove flange nut (2) and washer from flange (1).
3. Install flange remover bolt into small outside diameter end of flange remover (6). Turn flange remover 1/8 turn to a locked position. Turn flange remover while holding flange holder (5) stationary, Remove flange (1).
4. Install flange nut (2), removed in step 2, halfway on drive pinion (3) threads. Tap on nut with soft-faced hammer to loosen drive pinion, Remove flange nut from drive pinion. Remove drive pinion from housing (4).

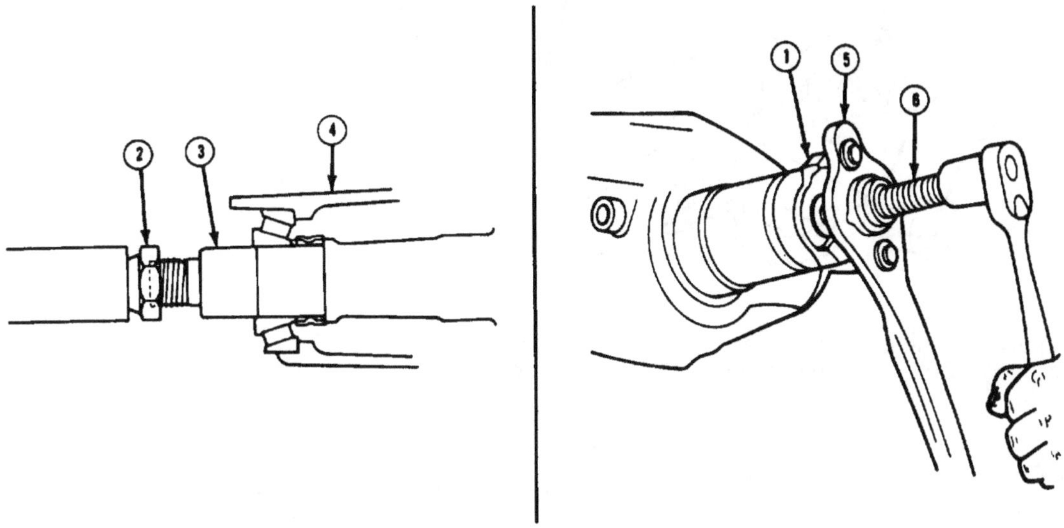

TA60628

6-7. FRONT AND REAR AXLE DRIVE PINION AND BEARINGS MAINTENANCE (M1009) (Con't).

5 . Remove oil seal (11) and front bearing (10). Remove spacer (8) from drive pinion (3) shaft, Discard oil seal and spacer.

CAUTION

Care should be taken when removing bearing cups (7 and 9). Opposite ends of bearing cups should be alternately tapped to avoid their becoming cocked in housing (4). Failure to follow this caution may result in damage to bearing cups and to housing bore.

6 . Remove front bearing cup (9) and rear bearing cup (7) from housing (4).

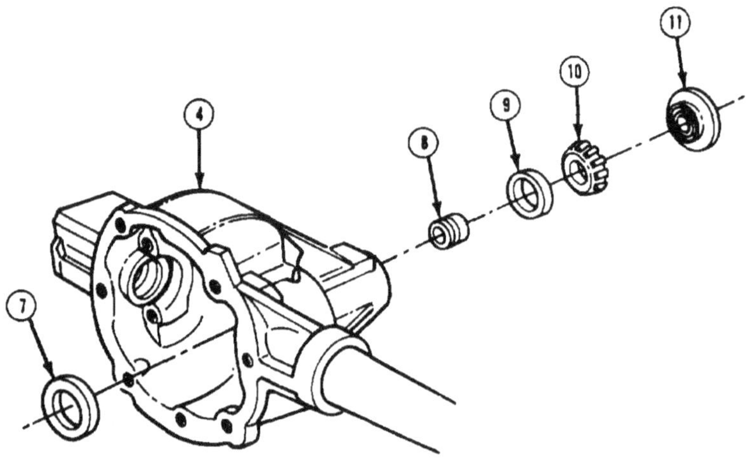

7. Using puller kit and press, remove rear 'bearing (12) from drive pinion (3) shaft.

8. Remove shim(s) located between rear bearing (12) and head of drive pinion (3).

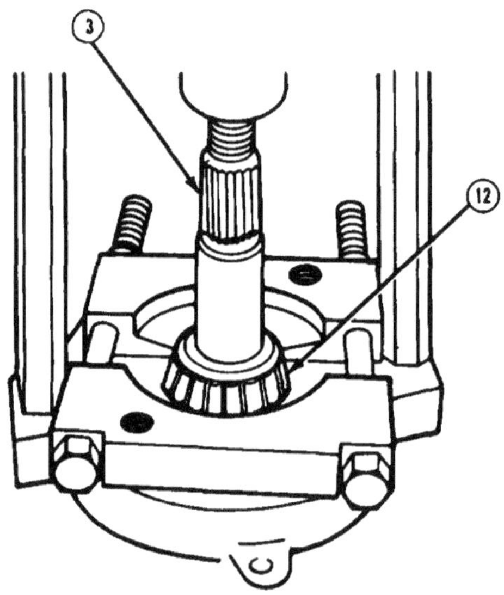

TA50529

6-7. FRONT AND REAR AXLE DRIVE PINION AND BEARINGS MAINTENANCE (M1009) (Con't).

b. CLEANING AND INSPECTION

WARNING

- Dry cleaning solvent P-D-680 is toxic and flammable, Always wear protective goggles and gloves and use only in a well-ventilated area. Avoid contact with skin, eyes, and clothes and DO NOT breathe vapors. DO NOT use near open flame or excessive heat. The solvent's flash point is 100°F-138°F (38°C-59°C). If you become dizzy while using cleaning solvent, immediately get fresh air and medical help. If solvent contacts eyes, immediately wash your eyes with water and get medical aid.
- Compressed air used for cleaning purposes should never exceed 30 psi (207 kPa). Use only effective chip guarding and personnel protective equipment (goggles/shield, gloves, etc.). Failure to follow this warning may result in serious injury to personnel.

1. Clean and inspect front and rear bearing assemblies in accordance with TM 9-214.
2. Clean drive pinion, shim(s), and flange with dry cleaning solvent and dry with compressed air.
3. Clean bearing bores in housing with dry cleaning solvent to ensure that they are free of dirt or other contamination, Remove any nicks or burrs found.
4. Clean pinion setting gage tool parts with dry cleaning solvent to ensure that pinion depth measurements will be accurate.

NOTE

If drive pinion is replaced, differential ring gear and ring gear bolts must also be replaced. (See paragraph 6-6 or 6-16)

5. Inspect drive pinion, flange, and shim(s) for damage. Replace if damaged. If shim (s) is undamaged, measure and save for installation.

6-7. FRONT AND REAR AXLE DRIVE PINION AND BEARINGS MAINTENANCE (M1009) (Con't).

c. INSTALLATION

NOTE

- Ensure that all components are lubricated with lubricating oil as they are installed.
- If original drive pinion and rear bearing (12) are being reused, perform steps 1 and 11 through 17. Original shim thickness is used,
- If a new drive pinion *or* rear bearing (12) are being installed, perform steps 1-17.

1. Install front bearing cup (9) and rear bearing cup (7) in housing (4) with wider, open ends facing outward. Check both bores to ensure that bearing cups are fully seated.

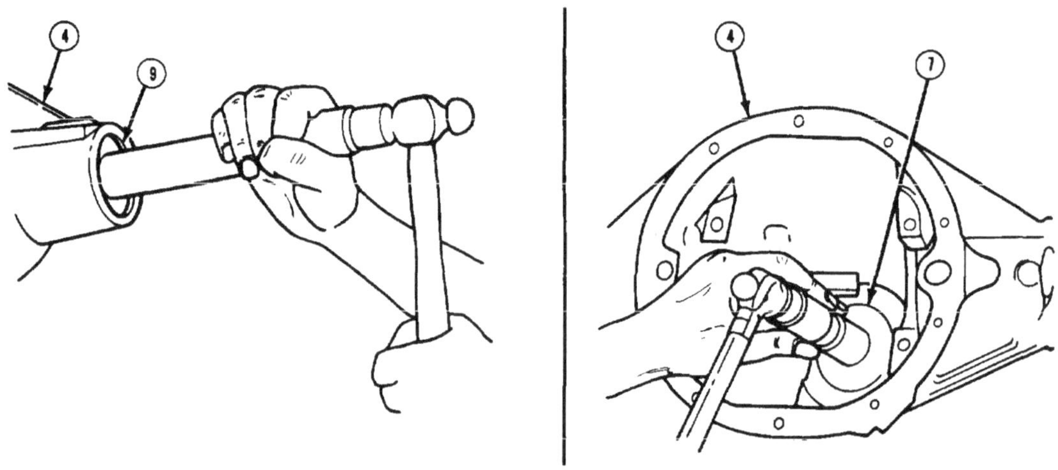

TA50530

6-7. FRONT AND REAR AXLE DRIVE PINION AND BEARINGS MAINTENANCE (M1009) (Con't).

2. Install front bearing (10) in bearing cup (9), and rear bearing (12) in bearing cup (7), with cloverleaf gage plate (13) and rear pilot washer (14) mounted on preload stud (15) and installed in bearing bore in housing (4).

3. Install front pilot washer (17) and hex nut (16) until snug, and rotate front and rear bearings (10 and 12) to ensure that they are properly seated, Hold preload stud (15) stationary and tighten hex nut until it takes 20 lb.-in. (2 N.m) to rotate front and rear bearing assembly,

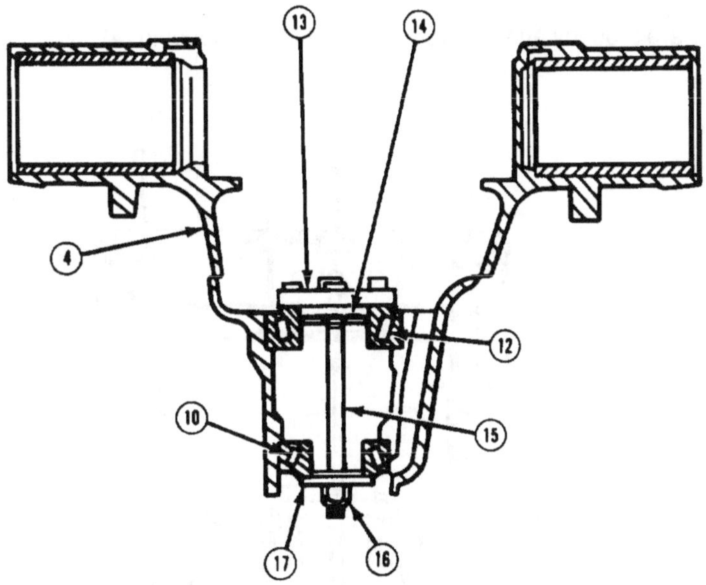

TA50531

6-7. FRONT AND REAR AXLE DRIVE PINION AND BEARINGS MAINTENANCE (M1009) (Con't).

4. Mount 2 side bearing discs (18) on ends of arbor (19) using step of disc that fits bores of housing (4). Place arbor and plunger in housing, Ensure that side bearing discs are properly seated in bores. Install 2 bearing caps over side bearing discs and tighten 4 bearing cap bolts until snug.
5. Place dial indicator (20) on mounting post of arbor. Contact button should rest on top surface of plunger. "O" the dial indicator with probe on high point of gaging surface.
6. Rotate gage plate (13) until plunger rests directly upon 8 1/2 in. button. Rock plunger rod slowly back and forth across button until dial indicator (20) reads greatest deflection. Set dial indicator to "O. "
7. Ensure that all tools are properly and securely installed inside housing (4).

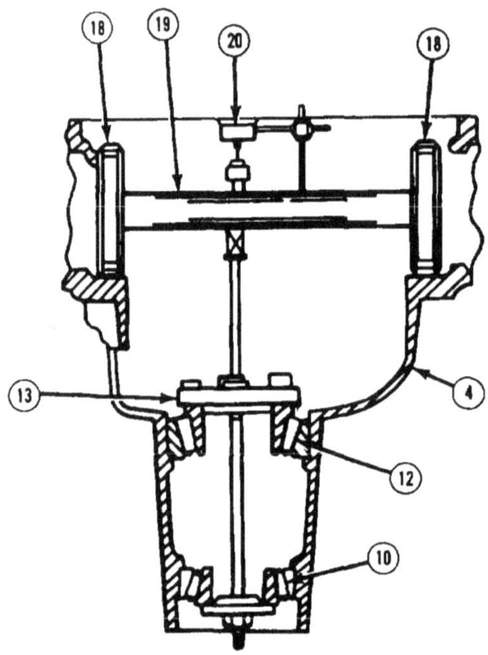

NOTE

It is important to use dial indicator (20) correctly when determining pinion depth requirement. Number indicated by indicator needle should be recorded. DO NOT record amount of needle travel.

8. Repeat rocking action of plunger several times to verify dial indicator (20) setting, Once final setting is obtained, swing plunger off gaging plate button. Dial indicator will now show required drive pinion shim thickness for "nominal" or "O" pinion. Record reading.

TA50532

6-7. FRONT AND REAR AXLE DRIVE PINION AND BEARINGS MAINTENANCE (M1009) (Con't).

9. Remove dial indicator (20), 2 bearing caps, depth gaging tools, and front and rear bearings (1 O and 12) from housing (4).

10. Check rear face of drive pinion, or painted marking on drive pinion shaft being installed, for pinion code number. Calculate drive pinion shim thickness to be used.

Table 6-2. Determining Drive Pinion Shim Requirements

Code Number On Service Pinion	Code Number On Original Pinion				
	+ 2	+ 1	0	- 1	- 2
+ 2	-	Add 0.001	Add 0.002	Add 0.003	Add 0.004
+ 1	Subtract 0.001	-	Add 0.001	Add 0.002	Add 0.003
0	Subtract 0.002	Subtract 0.001	-	Add 0.001	Add 0.002
- 1	Subtract 0.003	Subtract 0.002	Subtract 0.001	-	Add 0.001
- 2	Subtract 0.004	Subtract 0.003	Subtract 0.002	Subtract 0.001	-

(a) If drive pinion has no plus (+) or minus (-), use exact shim thickness as recorded in step 8.

(b) If drive pinion is stamped or painted with a plus (+) number, add that many thousandths to thickness recorded in step 8.

EXAMPLE: If reading from step 8 was 0.019 in. (0.483 mm) and drive pinion was stamped "+2," correct shim thickness is:

0.019 *in.* + 0.002 *in.* = 0.021 *in.*

(0.483 mm + 0.057 mm = 0.534 mm)

(c) If drive pinion is stamped or painted with a minus (-) number, subtract that many thousandths from thickness recorded in step 9.

EXAMPLE: If reading from step 8 was 0,031 in, (0.787 mm), and drive pinion was stamped "-3, " correct shim thickness is:

0.031 *in. -0.003 in. = 0.028 in.*

(0.787 mm -0.076 mm = 0.711 mm)

6-7. FRONT AND REAR AXLE DRIVE PINION AND BEARINGS MAINTENANCE (M1009) (Con't).

11. Place shim of correct thickness as determined in step 10, or original shim if drive pinion (3) was not replaced, on drive pinion shaft against head. Install rear bearing (12) on drive pinion shaft.

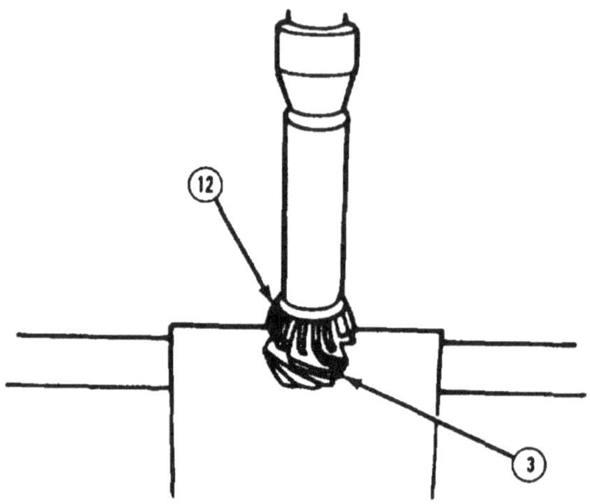

12. Install front bearing (10) in bearing cup (9) in housing (4) bore.
13. Coat lips of new pinion oil seal (11) with molybdenum grease and install squarely into housing (4) bore until flush with housing.

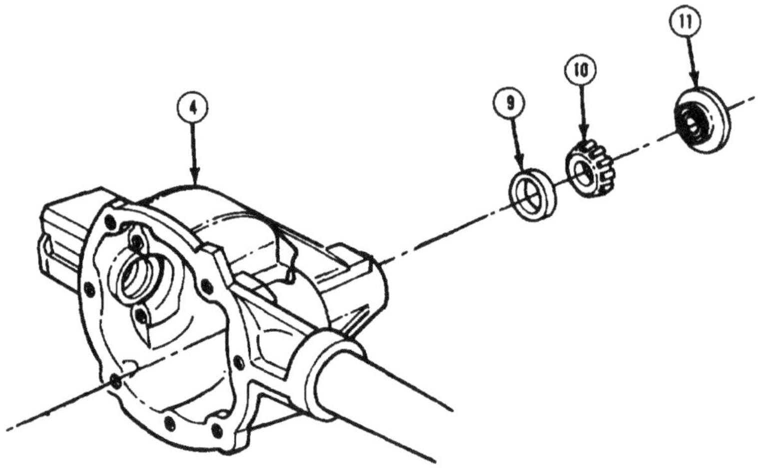

TA50533

6-7. FRONT AND REAR AXLE DRIVE PINION AND BEARINGS MAINTENANCE (M1009) (Con't).

14. Install new spacer (8) on drive pinion (3) shaft, Place drive pinion into position. Mount a suitably sized thick washer over threaded stem of drive pinion and install original flange nut (2). Tighten flange nut enough to draw drive pinion through front bearing (10) so that threads are exposed when flange is placed into position. Remove flange nut and washer. Discard flange nut.

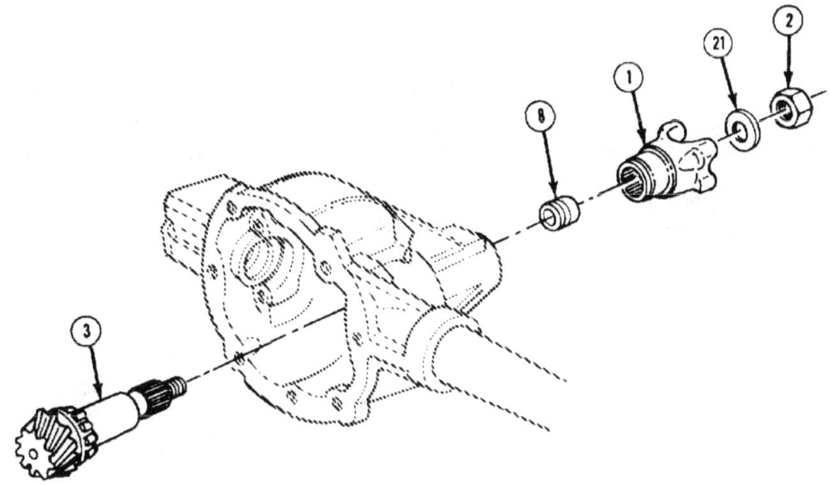

15. Install flange (1), original washer (21), and new flange nut (2), Using flange holder (5), tighten flange nut until all end play is removed from drive pinion (3). Remove flange holder. At this point, bearing preload specifications are being neared.

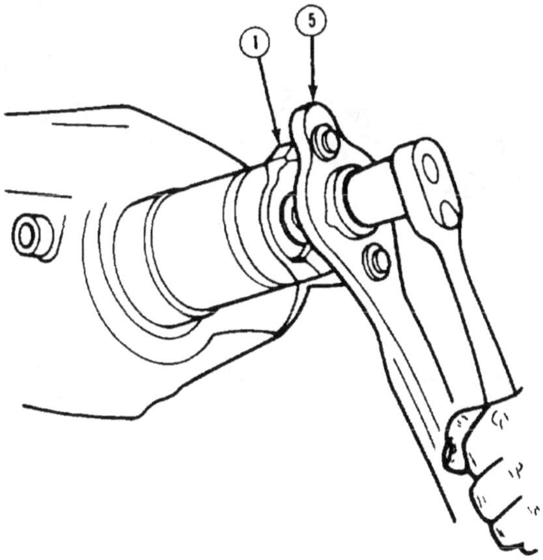

6-7. FRONT AND REAR AXLE DRIVE PINION AND BEARINGS MAINTENANCE (M1009) (Con't).

CAUTION

Tightening flange nut (2) as little as 1/8 turn can add up to 5 additional pound-inches of drag. Flange nut should be tightened only a little at a time. Exceeding torque specifications may compress spacer (8) on drive pinion shaft too far and require replacement of spacer.

16. Measure rotating torque, it should be 20-25 lb.-in. (2-3 N.m) on new bearings, or 10–1 5 lb.-in, (1-2 N.m) on reused bearings. If rotating torque specifications have not been reached, tighten flange nut (2). Check rotating torque after each slight amount of tightening.

17. Rotate drive pinion (3) several times to ensure that bearings are seated. Check preload drag again. If drag has been reduced, reset preload to specifications.

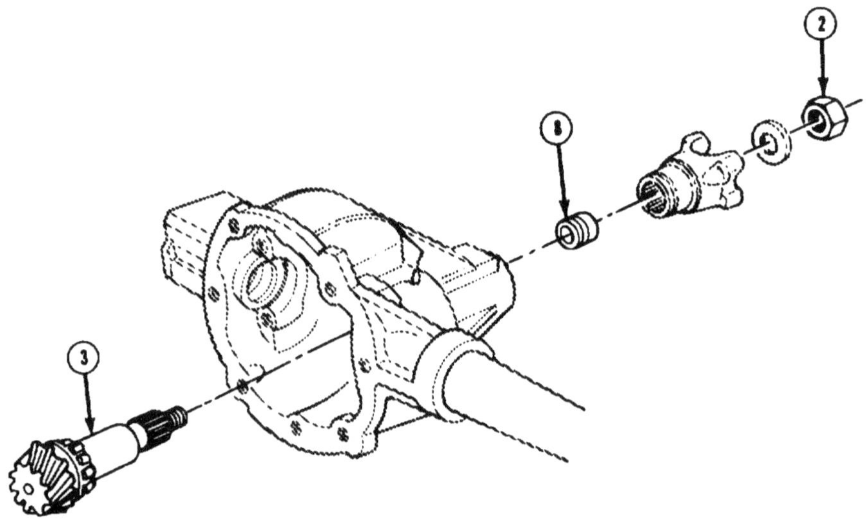

FOLLOW-ON TASKS:

- Connect front or rear propeller shaft to pinion flange. (See TM 9-2320-289-20)
- Install front or rear axle differential. (See paragraph 6-5)

TA50535

6-8. FRONT AXLE DIFFERENTIAL MAINTENANCE (ALL EXCEPT M1009).

This task covers:

a. Removal
b. Disassembly
c. Cleaning and Inspection
d. Assembly
e. Preload and Adjustment
f. Installation

INITIAL SETUP:

Equipment Condition

- Front axle assembly removed and mounted on stand. (See paragraph 6-3)
- Steering tie-rod removed. (See TM 9-2320-289-20)
- Front axle shafts removed, (See paragraph 6-4)

Materials/Parts

- One differential cover gasket
- Dry cleaning solvent (Item 23, Appendix B)
- Prussian blue dye (Item 24, Appendix B)
- Gear lubricating oil (Item 44, Appendix B)
- Wiping rags (Item 53, Appendix B)
- Masking tape (Item 68, Appendix B)

Tools/Test Equipment

- Clutch compressor, J-34502
- Axle housing spreader
- Dial indicator
- Puller kit
- Torque wrench

Manual References

- TM 9-214

General Safety Instructions

- Dry cleaning solvent is flammable and must not be used near open flame. Use only in a well-ventilated area,
- Compressed air used for cleaning purposes should never exceed 30 psi (207 kPa).

a. REMOVAL

1. Place a suitable container under housing. Remove 10 bolts (1) and differential cover (2) and allow all lubricating oil to drain. Remove differential cover gasket-and discard.

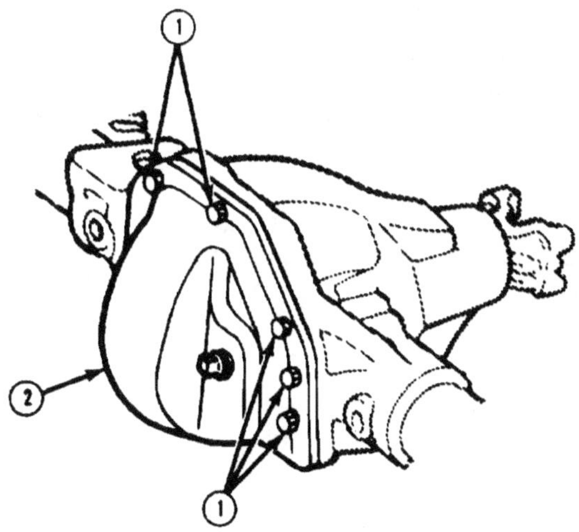

TA50536

6-8. FRONT AXLE DIFFERENTIAL MAINTENANCE (ALL EXCEPT M1009) (Con't),

2. Mark 2 side bearing caps (3) and housing for installation alinement. Remove 4 bolts (4) and 2 side bearing caps.

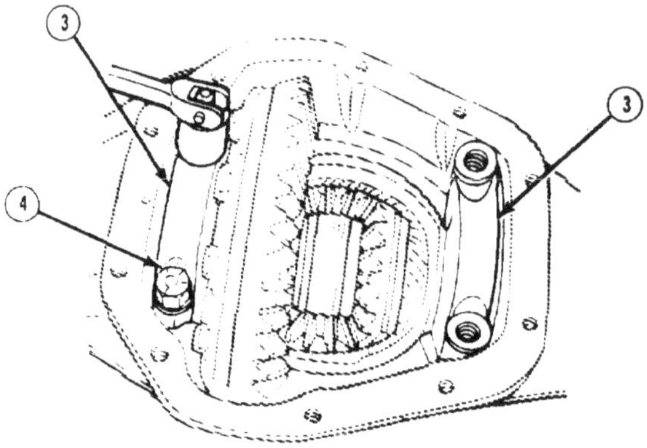

3. Position axle housing spreader (5) with tool dowels fitted into recesses in housing. Install dial indicator (24) from dowel pin bar to dowel pin bar. Ensure that dial indicator is preloaded. Set dial indicator to "0,"

CAUTION

Housing should not be spread more than 0.015 in. (0.381 mm). Failure to follow this caution may cause housing to become damaged and permanently distorted.

4. Spread housing to 0.015 in. (0.381 mm). Remove dial indicator (24).

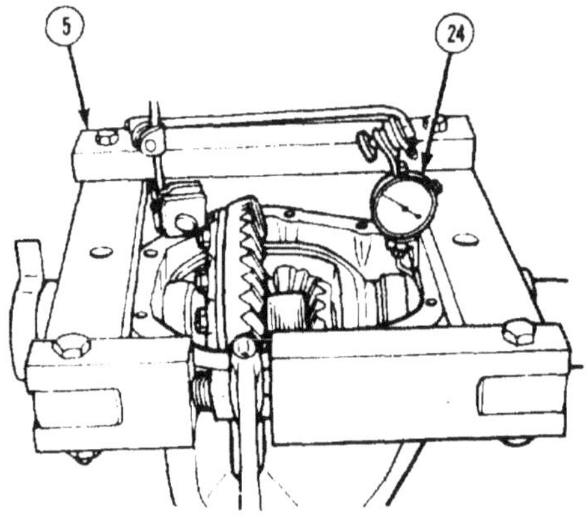

TA50537

6-8. FRONT AXLE DIFFERENTIAL MAINTENANCE (ALL EXCEPT M1009) (Con't).

5. Remove differential case (7) assembly from housing. Remove and tag 2 side bearing cups (6) for installation. Relieve tension on axle housing spreader (5) and remove from housing.

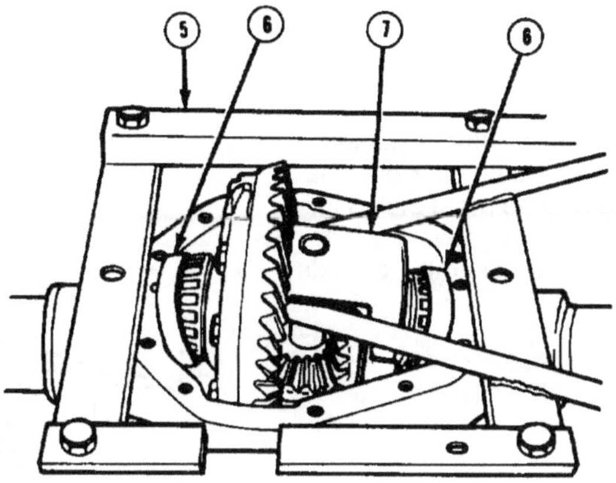

b. DISASSEMBLY

NOTE

- M1008 and M1010 front axle differentials have thrust washers at side gears. M1028, M1028A1, and M1031 have clutch packs.
- If disassembling M1008 or M1010 front axle differential, perform steps 1-6.
- If disassembling M1028A1, or M1031 front axle differential, perform steps 1-5 and steps 7-10.

1. Place differential case (7) assembly in vise,
2. Remove 2 side bearings (9) and side bearing shims (10). Tag side bearings and side bearing shims for installation.

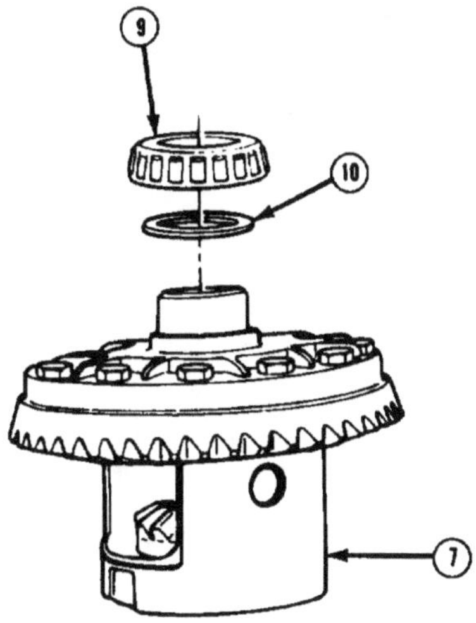

TA50538

6-8. FRONT AXLE DIFFERENTIAL MAINTENANCE (ALL EXCEPT M1009) (Con't).

NOTE

Axle shaft (14) serves as a tool to help aline pinion gears (15) and side gears (16).

3. Remove differential case (7) assembly from vise. Place axle shaft (14) in vise with no more than 3 in. (76 mm) above top of vise. Place differential case assembly on axle shaft.

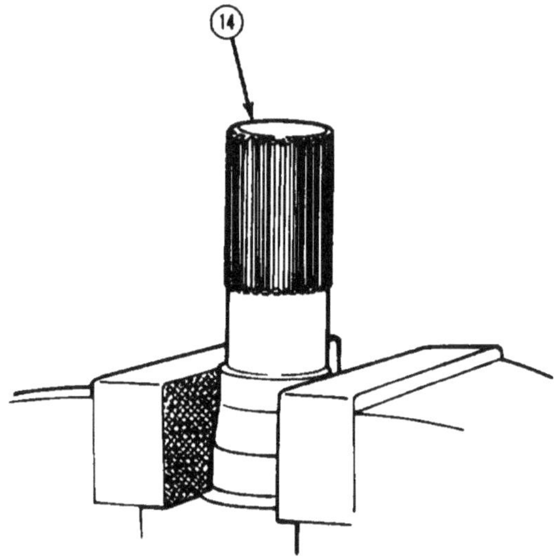

NOTE

Shop rags placed around vise will help protect ring gear (11) teeth from damage should ring gear fall during disassembly.

4. Remove lockpin (13). Remove 12 ring gear bolts (8). Tap on ring gear (11) with a soft-faced hammer to separate ring gear from differential case (7) assembly. Remove differential case assembly from axle shaft and remove ring gear.

5. Place differential case (7) assembly on axle shaft in vise. Remove pinion shaft (12).

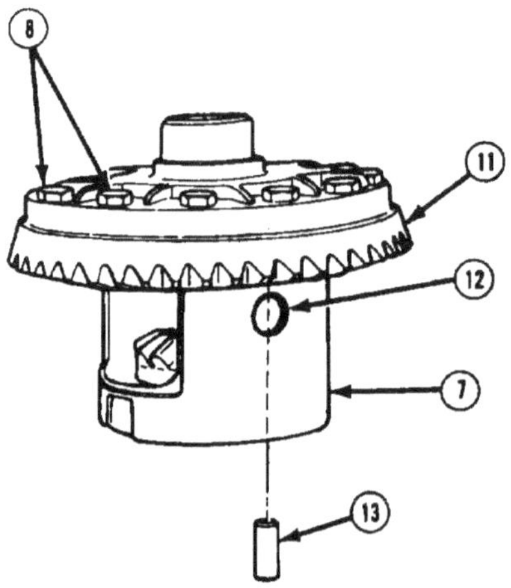

TA50539

6-8. FRONT AXLE DIFFERENTIAL MAINTENANCE (ALL EXCEPT M1009) (Con't).

NOTE

Perform step 6 only if working on M1008 or M1010.

6. Rotate side gears (16) until pinion gears (15) turn to opening in differential case (7) assembly. Remove 2 pinion gears and thrust washers (17) behind gears. Remove 2 side gears and thrust washers.

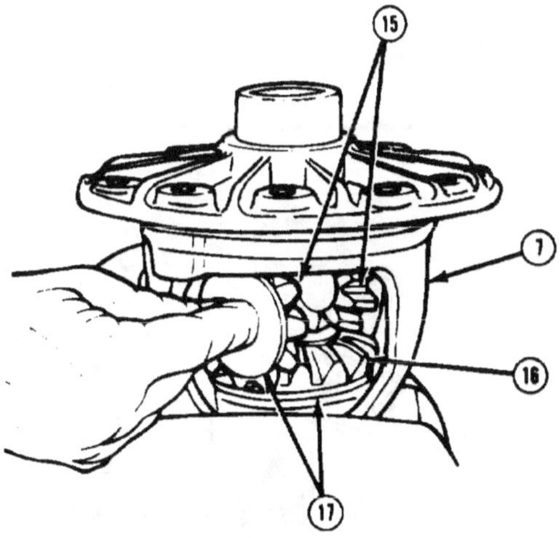

NOTE

Perform steps 7-10 only if working on M1028, M1028A1, or M1031.

7. Install clutch compressor (18) and remove tension from clutch packs. Remove 2 thrust washers (17) from behind pinion gears (15). Rotate differential case (7) assembly and remove 2 pinion gears.

8. Hold top clutch pack and side gear (16) in place and remove clutch compressor (18).

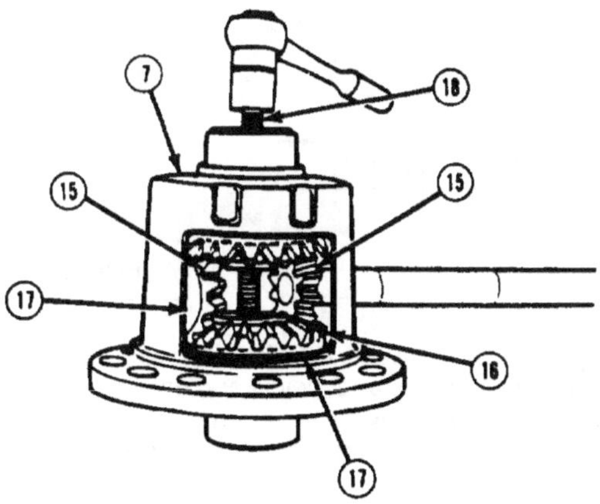

TA50540

6-8. FRONT AXLE DIFFERENTIAL MAINTENANCE (ALL EXCEPT M1009) (Con't).

NOTE

Ensure that clutch packs are kept intact as they are removed.

9 . Remove 2 clutch packs and side gears (16).

10. Remove 2 retainer clips (20) from each clutch pack and separate 4 clutch plates (19) and 4 discs (21), Remove dished spacer (22) from each side gear (16).

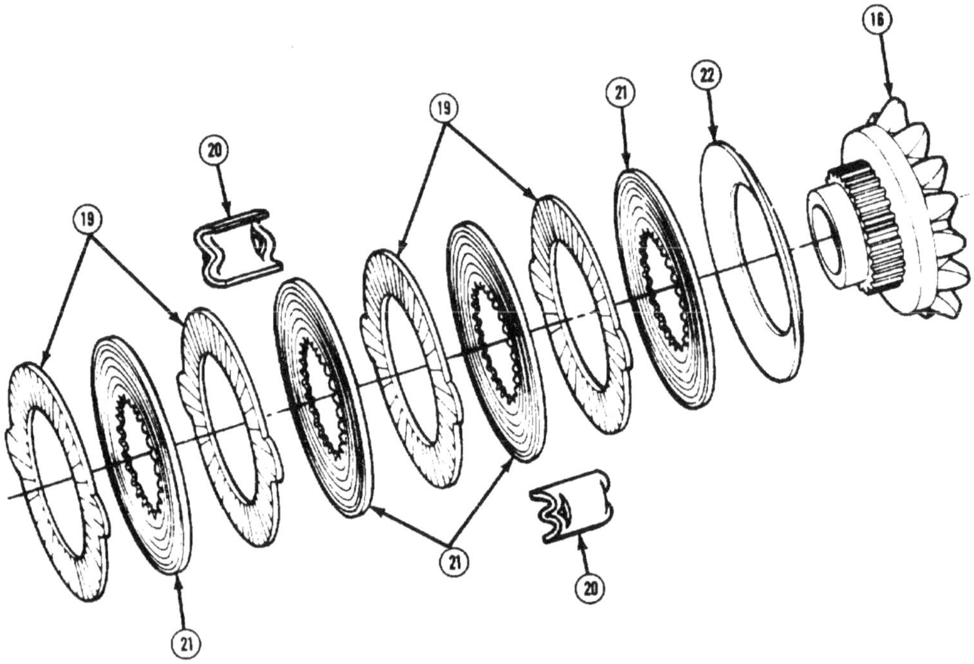

c . CLEANING AND INSPECTION

WARNING

Dry cleaning solvent P-D-680 is toxic and flammable. Always wear protective goggles and gloves and use only In a well-ventilated area. Avoid contact with skin, eyes, and clothes and DO NOT breathe vapors. DO NOT use near open flame or excessive heat. The solvent's flash point is 100°F-138°F (38°C-59°C). If you become dizzy while using cleaning solvent, immediately get fresh air and medical help. If solvent contacts eyes, immediately wash your eyes with water and get medical aid.

TA50541

6-8. FRONT AXLE DIFFERENTIAL MAINTENANCE (ALL EXCEPT M1009) (Con't).

WARNING

Compressed air used for cleaning purposes should never exceed 30 psi (207 kPa). Use only effective chip guarding and personnel protective equipment (goggles/shield, gloves, etc.). Failure to follow this warning may result in serious injury to personnel.

1. Clean and inspect side bearings in accordance with TM 9-214.
2. Clean all removed components with dry cleaning solvent and dry with compressed air.
3. Inspect differential case for cracks, scoring or signs of damage to machined surfaces. Remove any nicks or burrs found. Replace if cracked or scored, or if nicks or burrs cannot be removed.
4. Inspect side bearing shims. Obtain replacement shims of same thickness if damaged.
5. Inspect teeth of pinion and side gears for damage. If one gear is damaged, replace all 4 gears and thrust washers.

NOTE

If ring gear is replaced, ring gear bolts, and drive pinion must also be replaced. (See paragraph 6-9)

6. Inspect ring gear for damage. Replace if damaged.
7. Inspect clutch pack plates, retainer clips, and discs for wear, cracks, or scoring. If one member of either clutch pack is worn, cracked, or scored, replace both clutch packs.
8. Inspect 2 axle shaft seals in housing side bores. Remove and discard if damaged.

d. ASSEMBLY

NOTE

- Ensure that all differential components are lubricated with lubricating oil as they are assembled.
- If assembling M1008 or M1010 front axle differential, perform steps 5, 6, 8, and 9.
- If assembling M1028, M1028A1, or M1031 front axle differential, perform steps 1-4 and steps 6-9.

1. Install dished spacer (22) on each side gear (16). Starting with a disc (21) and alternating, install 4 clutch plates (19) and discs, Install 2 retainer clips (20) to each clutch pack.

6-8. FRONT AXLE DIFFERENTIAL MAINTENANCE (ALL EXCEPT M1009) (Con't).

NOTE

Ensure that clutch pack stays assembled to side gear (16) splines and that retainer clips (20) are completely seated into differential case (7) pockets.

2 . Install 2 clutch packs and side gears (16) into differential case (7).

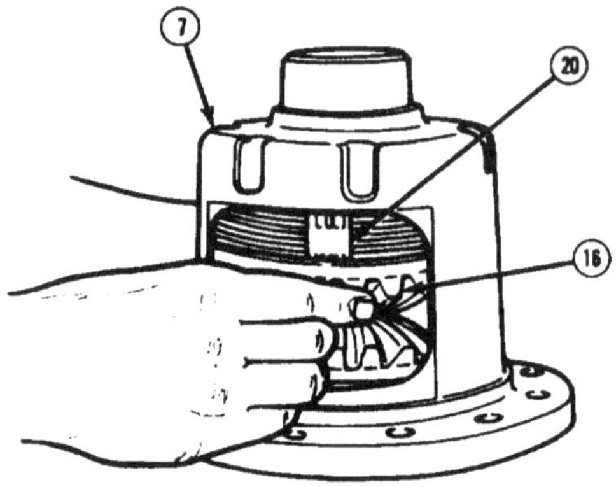

3 . Install clutch compressor (18) to differential case (7) assembly and install differential case assembly to axle shaft (14) in vise.

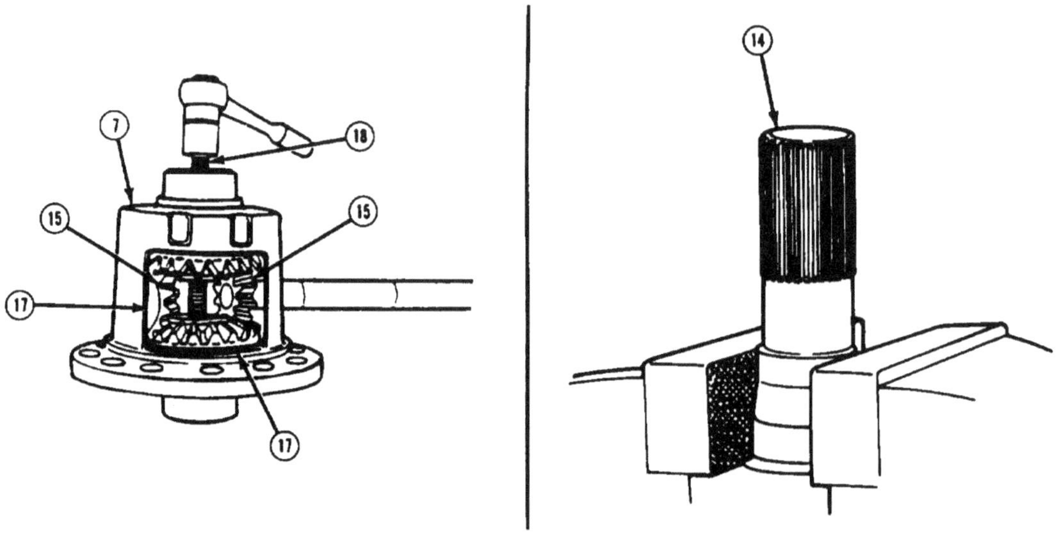

TA50642

6-8. FRONT AXLE DIFFERENTIAL MAINTENANCE (ALL EXCEPT M1009) (Con't).

4. Loosen tension from clutch compressor (18) and install 2 pinion gears (15) by rotating differential case (7) assembly, Tighten clutch compressor and install 2 thrust washers (17) between pinion gears and differential case assembly. Remove clutch compressor.

5. Install differential case (7) assembly on axle shaft (14) in vise. Install 2 side gears (16) and thrust washers (17). Rotate side gears to install 2 pinion gears (15) and thrust washers.

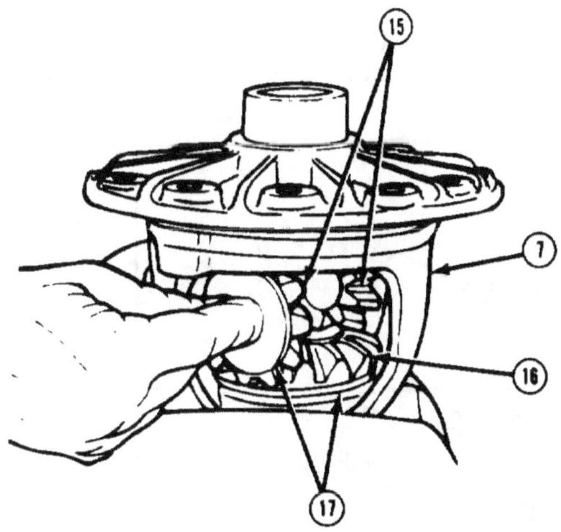

NOTE

Ensure that pinion gear (15) and thrust washer (17) holes are alined with differential case (7) assembly holes.

6. Install pinion shaft (12) with lockpin (13) hole on shaft alined with differential case (7) assembly lockpin hole.

NOTE

- Perform step 7 only if working on M1028, M1028A1, or M1031.
- Tape used in step 7 indicates depth required to install lockpin (13).

7. Place a piece of tape 1 3/4 in. (4.61 5 mm) from end of punch.

8. Install lockpin (13) through differential case (7) assembly into pinion shaft (12).

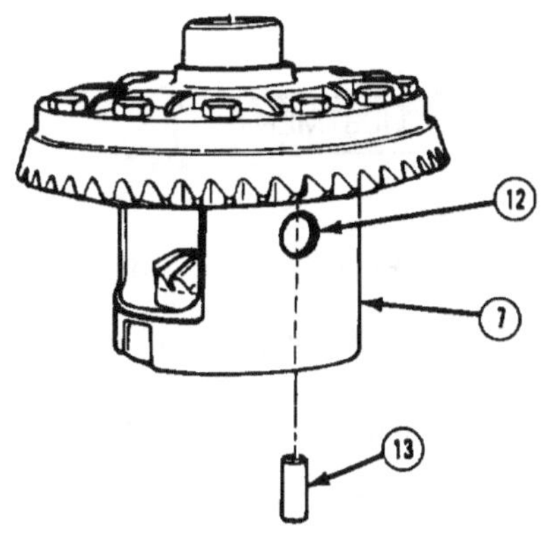

TA50543

6-8. FRONT AXLE DIFFERENTIAL MAINTENANCE (ALL EXCEPT M1009) (Con't).

NOTE

Shop rags placed around vise will help help protect ring gear (11) teeth from damage should ring gear fall during assembly.

9. Remove differential case (7) assembly from axle shaft (14). Loosely install ring gear (11) with 12 ring gear bolts (8), Install assembly back on vise and tighten bolts alternately to 110 lb,-ft. (149 N.m).

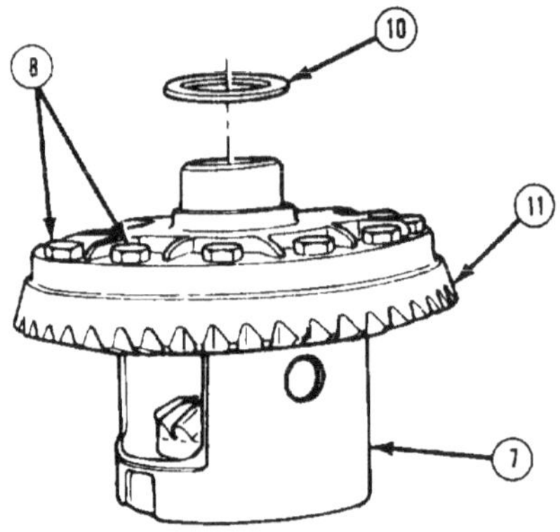

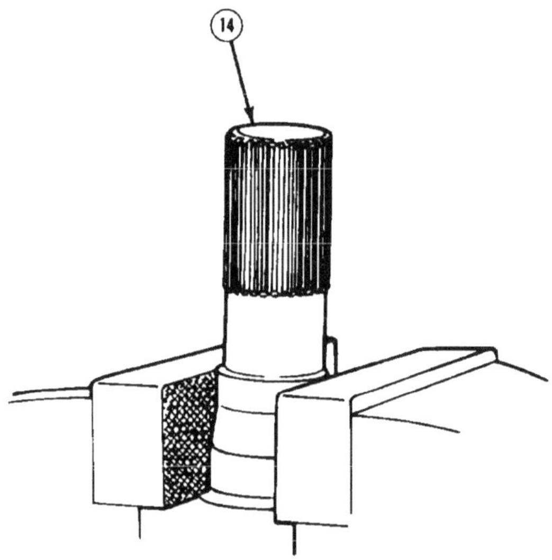

e . PRELOAD AND ADJUSTMENT

NOTE

- Side bearing shims (10) should not be installed at this time.
- Side bearings (9) should rotate freely once installed.

1. Install 2 side bearings (11) and bearing cups (6) to differential case (7) assembly as tagged,
2. If not already removed, remove front axle drive pinion, (See paragraph 6-9)
3. Place differential case (7) assembly in housing and install 2 side bearing caps (3) as marked. Remove play in side bearings (9) and bearing cups (6). Tighten 4 bearing cap bolts (4) just enough to hold bearing caps in place.

TA50544

6-8. FRONT AXLE DIFFERENTIAL MAINTENANCE (ALL EXCEPT MI 009) (Con't).

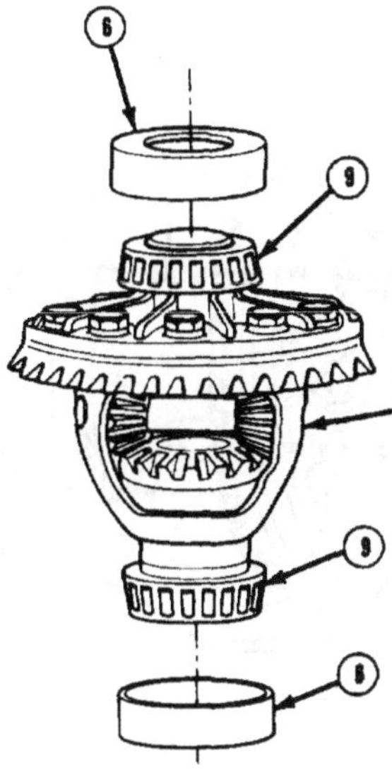

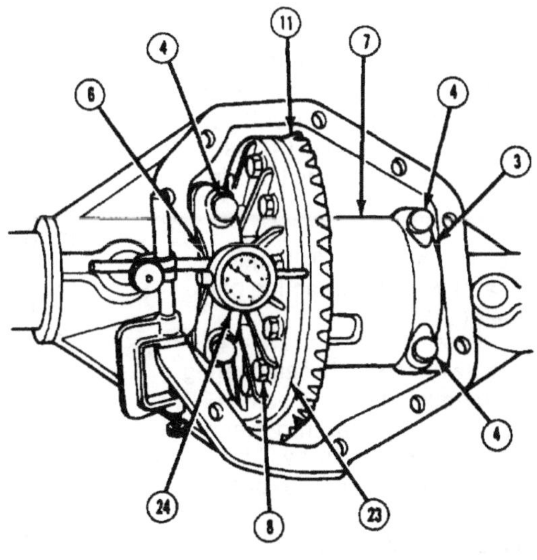

4. Install dial indicator (24) on housing with indicator button contacting back of ring gear (11). Rotate differential case (7) assembly and check for runout. If runout is greater than 0.002 in. (0.051 mm), remove differential case assembly from housing, and remove ring gear. Assemble differential case assembly without ring gear and install in housing,

5. Check runout at ring gear mounting flange (23). If runout on flange exceeds 0.002 in. (0.051 mm), problem is probably in side bearings (9) or differential case (7) and must be corrected before proceeding. If runout is within specifications, problem is with ring gear (11). Install ring gear, ensuring that ring gear bolts (8) are tightened evenly and ring gear mounting flange is free from dirt or chips, etc. Recheck runout with dial indicator (24) contacting back of ring gear. If runout is still greater than 0.002 in. (0.051 mm), replace ring gear, ring gear bolts, and drive pinion as a set.

6. Using 2 screwdrivers wedged between bearing cup (6) and housing on opposite side of ring gear (11) (away from dial indicator (24) side), force differential case (7) assembly as far as possible toward dial indicator. Rock ring gear to set bearings, With force still applied, set dial indicator to "0, "

7. Reposition 2 screwdrivers between bearing cup (6) and housing on ring gear (11) side. Force differential case (7) assembly as far as possible in opposite direction. Repeat several times until the same reading is obtained. Record reading. This reading is the total thickness of shim (s) needed, less preload, for setting backlash and preload during assembly.

8. Remove 4 bearing cap bolts (4) and 2 side bearing caps (3). Remove differential case (7) assembly from housing,

TA50545

6-8. FRONT AXLE DIFFERENTIAL MAINTENANCE (ALL EXCEPT M1009) (Con't).

9. Install and adjust front axle drive pinion and bearings. (See paragraph 6-9)

10. Install differential case (7) assembly into housing and install 2 side bearing caps (3) in their original position, Tighten 4 bearing cap bolts (4) just enough to hold side bearing caps in place.

11. Install dial indicator (24) on housing with indicator button contacting back of ring gear (11).

12. Place 2 screwdrivers between bearing cup (6) and housing on ring gear (11) side and force ring gear into mesh with drive pinion as far as it will go. Rock ring gear for bearings to seat and gears to fully mesh. With force still applied, set dial indicator (24) to "0."

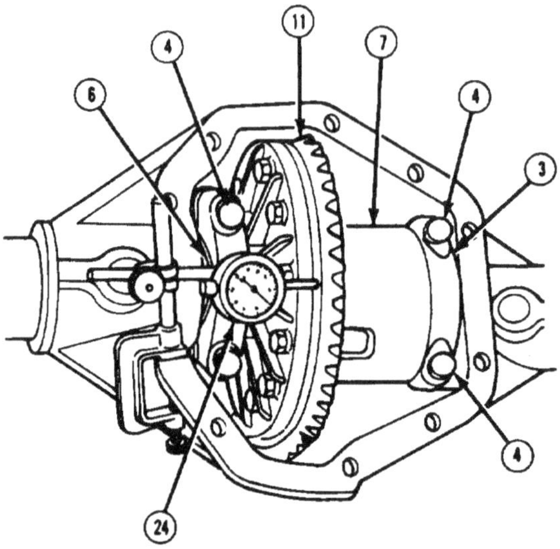

13. Reposition screwdrivers on opposite side of ring gear (11) and force ring gear as far as it will go. Repeat until same reading is obtained. Record reading. This reading is the thickness of shim(s) (10) needed between differential case (7) assembly and side bearing (9) on ring gear side.

14. Remove 4 bearing cap bolts (4) and 2 side bearing caps (3), Remove differential case (7) assembly from housing, Remove side bearing (9) from ring gear (11) side and install proper thickness of shim(s) (10). Install side bearing,

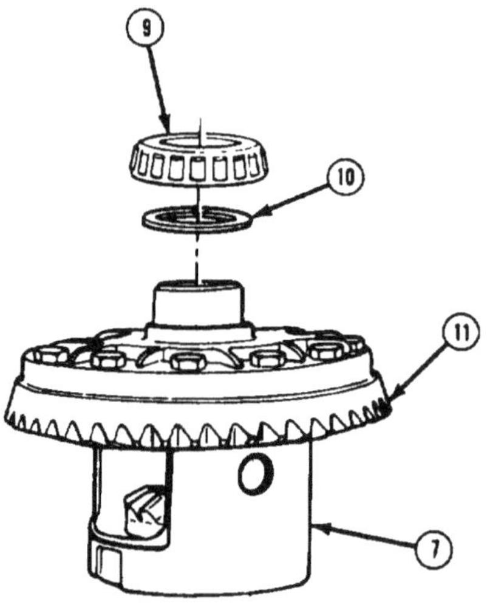

TA50546

6-8. FRONT AXLE DIFFERENTIAL MAINTENANCE (ALL EXCEPT M1009) (Con't).

15. Remove side bearing (9) from opposite side of ring gear (11). Subtract size of shim pack just installed on ring gear side from reading obtained in step 7. To this figure, add .015 in. (0.301 mm) to compensate for preload and to provide correct backlash.

EXAMPLE: If reading in step 7 was 0.085 in. (2.159 mm) and shim (s) (10) installed on ring gear side were 0.055 in. (1.3971 mm), correct thickness of shim(s) will be:

0.085-0,055 + 0.015 = 0.045 *in.*

(2.159 -0,055 + 0.381 = 1.143 *mm)*

16. Install proper thickness of shim(s) (10) as determined in step 15 on side opposite ring gear (11). Install side bearing (9).

f. INSTALLATION

1. If removed, install 2 new front axle shaft seals inside housing side bores.

CAUTION

Housing should not be spread more than 0.015 in. (0.381 mm). Failure to follow this caution may cause housing to become damaged and permanently distorted.

2. Use axle housing spreader (5) to spread housing to 0.015 in. (0.381 mm). Install 2 bearing cups (6) as tagged and install differential case (7) assembly inside housing.

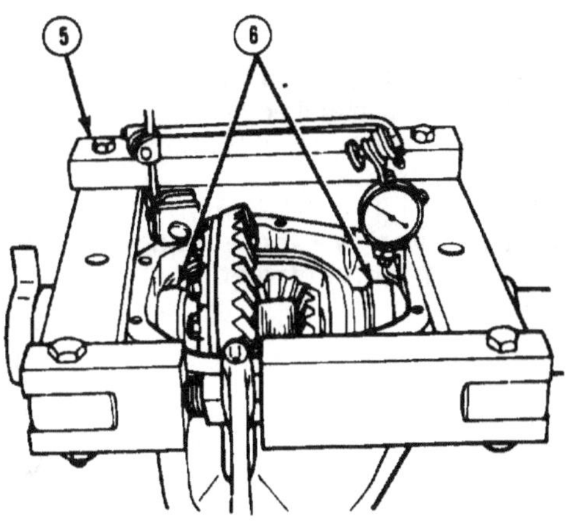

TA50547

6-8. FRONT AXLE DIFFERENTIAL MAINTENANCE (ALL EXCEPT M1009) (Con't).

3. Install 2 bearing caps (3) in their original position and tighten 4 bearing cap bolts (4) finger tight. Rotate differential case (7) assembly and tap it with a soft-faced hammer *to* ensure that differential case assembly is properly seated in housing. Remove axle housing spreader (5) and tighten bearing cap bolts to 85 lb.-ft, (115 Nom).

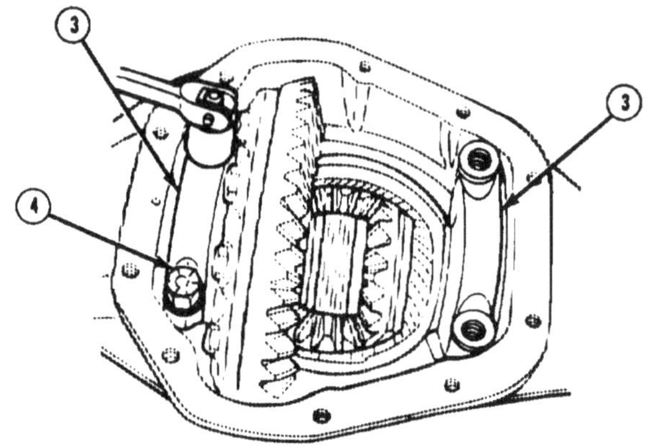

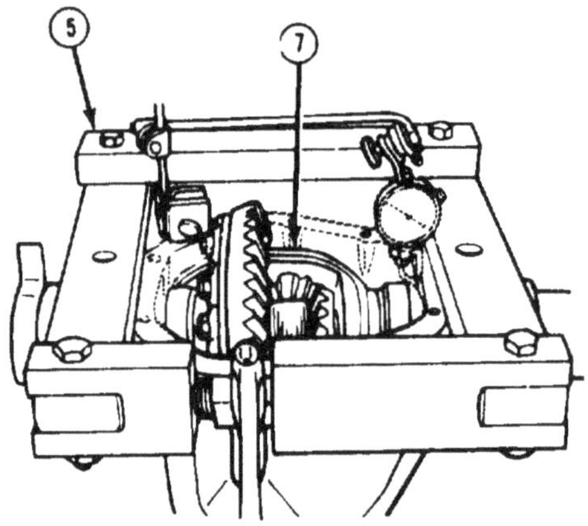

4. Install dial indicator (24) and check ring gear (11) backlash at 4 equally spaced points around ring gear. Backlash must 0.004-0.009 in. (0.102-.229 mm) and must not var y more than 0.002 in. (0.051 mm) between 4 points checked.

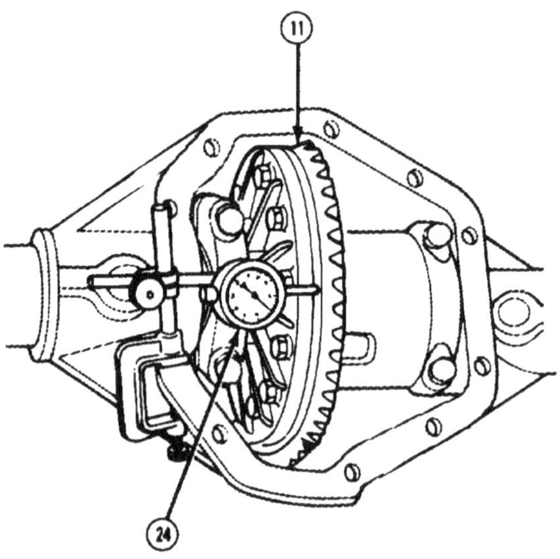

TA50548

6-8. FRONT AXLE DIFFERENTIAL MAINTENANCE (ALL EXCEPT M1009) (Con't),

5. If backlash is not within specification, correct by adjusting thickness of side bearing (9) shim pack, To increase backlash, increase right shim (10) and decrease left shim an equal amount. To decrease backlash, decrease right shim and increase left shim an equal amount,

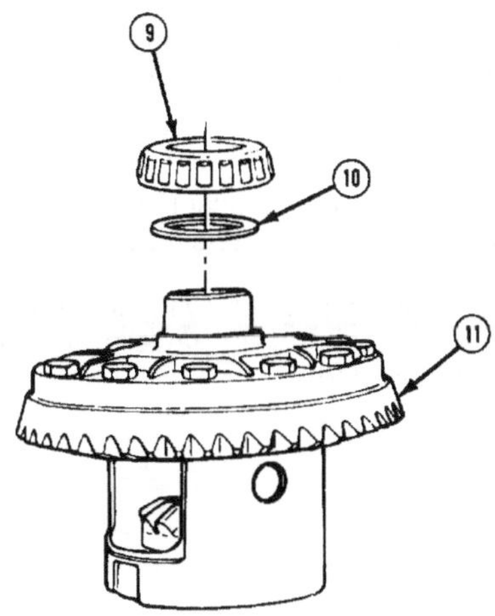

NOTE

If ring gear (11) and drive pinion were replaced, perform steps 6-9 to check for proper gear tooth contact pattern.

6. Wipe any lubricant from housing and clean each tooth of ring gear (11).

7. Using medium stiff brush, sparingly apply prussian blue dye to all ring gear (11) teeth. Area of drive pinion tooth contact should be visible when drive pinion is turned by hand.

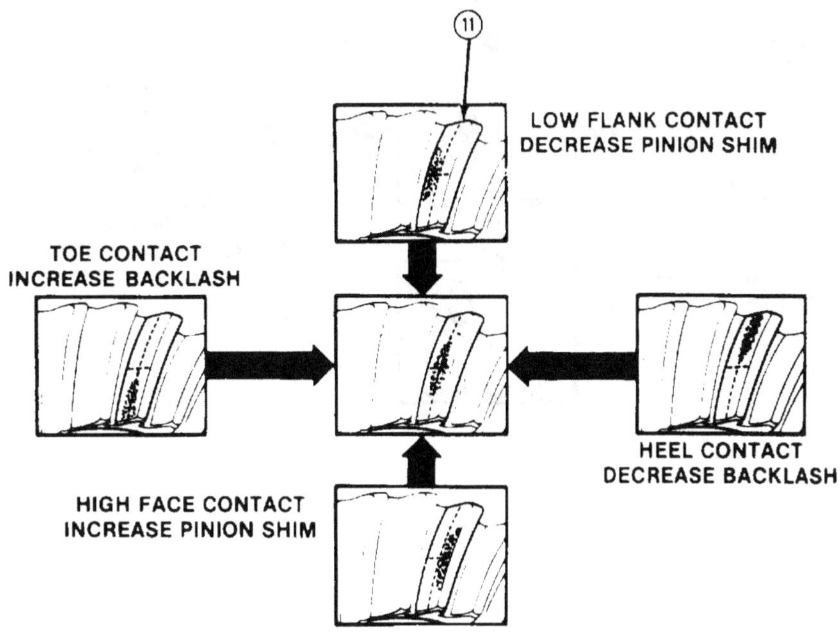

TA50549

6-8. FRONT AXLE DIFFERENTIAL MAINTENANCE (ALL EXCEPT M1009) (Con't).

NOTE

Excessive turning of ring gear (11) is not recommended. Ring gear should be turned only enough to provide tooth contact pattern.

8 . Apply load until it takes a torque of 40-50 lb.-ft. (54-68 N.m) to turn drive pinion. Turn companion flange so that ring gear (11) turns 1 full revolution, then reverse so that ring gear turns 1 full revolution in opposite direction,

9 . Contact pattern should be centrally located up and down on the face of the ring gear (11) teeth. Compare tooth contact pattern on ring gear with examples shown and make adjustments as indicated.

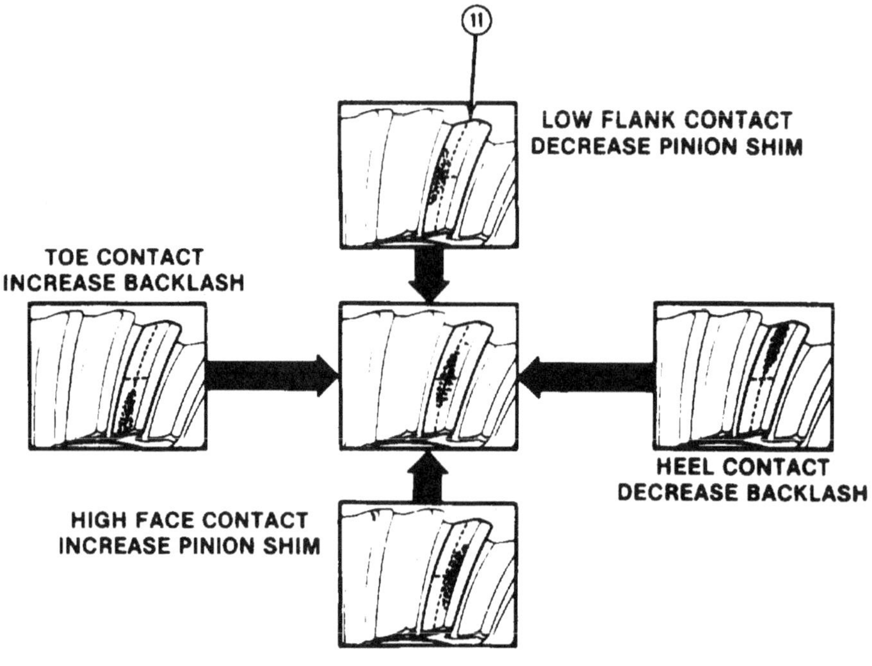

TA60550

6-8. FRONT AXLE DIFFERENTIAL MAINTENANCE (ALL EXCEPT M1009) (Con't).

NOTE

Ensure that differential cover (2) and magnet are clean.

10. Install new differential cover gasket and differential cover (2) with 10 bolts (1). Tighten bolts to 35 lb.-ft . (47 N.m).

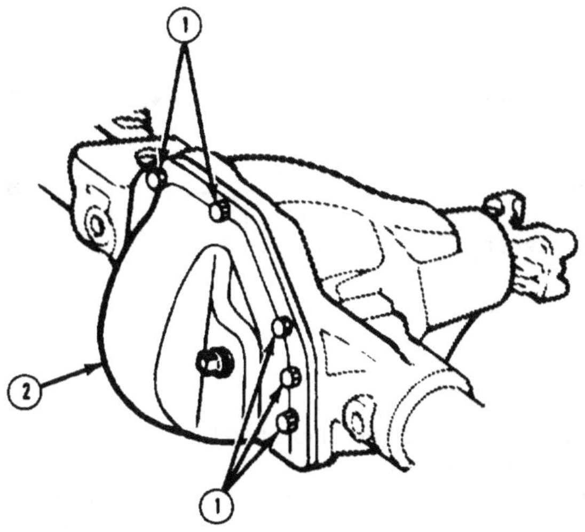

FOLLOW-ON TASKS:

- Install f rent axle shafts, (See paragraph 6-4)
- Install steering tie-rod. (See TM 9-2320-289-20)
- Install front axle assembly. (See paragraph 6-3)
- Fill front axle differential with lubricating oil. (See LO 9-2320-289-12)

TA50551

6-9. FRONT AXLE DRIVE PINION AND BEARINGS MAINTENANCE (ALL EXCEPT M1009).

This task covers:

a. Removal
b. Cleaning and Inspection
c. Installation

INITIAL SETUP:

Equipment Condition

- Front axle differential removed. (See paragraph 6-8)

Tools/Test Equipment

- Flange holder and remover
- Micrometer
- Torque wrench

Manual References

- TM 9-214

Materials/Parts

- One flange nut
- One pinion oil seal
- Two shim packs
- Two slingers
- Dry cleaning solvent (Item 23, Appendix B)
- Molybdenum grease (Item 37, Appendix B)
- Gear lubricating oil (Item 44, Appendix B)

General Safety Instructions

- Dry cleaning solvent is flammable and must not be used near open flame. Use only in a well-ventilated area.
- Compressed air used for cleaning purposes should ever exceed 30 psi (207 kPa).

a. REMOVAL

NOTE

- Ensure that shims and slingers are tagged as they are removed and saved for measurement.
- Ensure that flange nut (4) is saved for use during drive pinion (5) installation, When drive pinion is fully seated and ready to receive final torque, a new flange nut must be used.

1. Using flange holder (1), remove flange nut (4) and washer from flange (2).

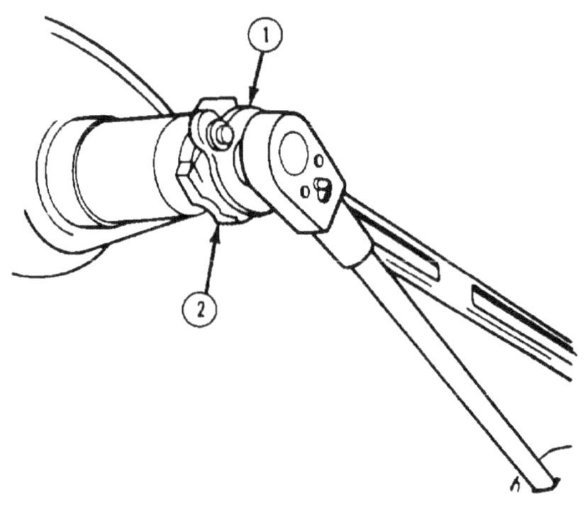

TA50552

6-9. FRONT AXLE DRIVE PINION AND BEARINGS MAINTENANCE (ALL EXCEPT M1009) (Con't).

2. Thread flange remover bolt into small outside diameter end of flange remover (3). Turn flange remover 1/8 turn to a locked position. Turn flange remover while holding flange holder (1) stationary. Remove flange (2). Remove deflector from flange if damaged.

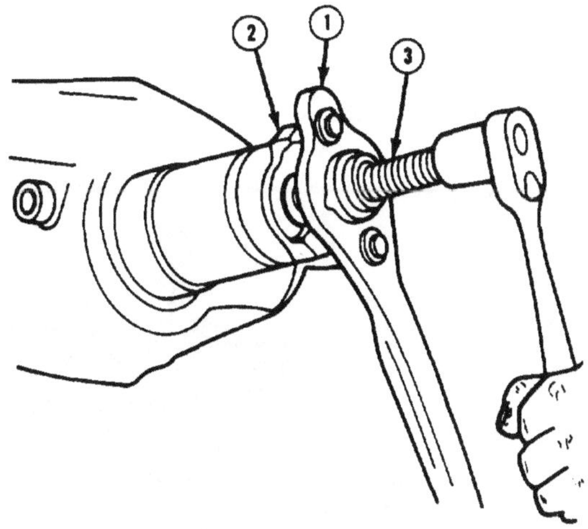

NOTE

Shim (6) may be on drive pinion (5) shaft or inside housing (7).

3. Install flange nut (4) halfway onto drive pinion (5) threads. Tap on flange nut with soft-faced hammer to loosen drive pinion in front pinion bearing (8). Remove flange nut from drive pinion. Remove drive pinion from housing (7). Remove shim (6).

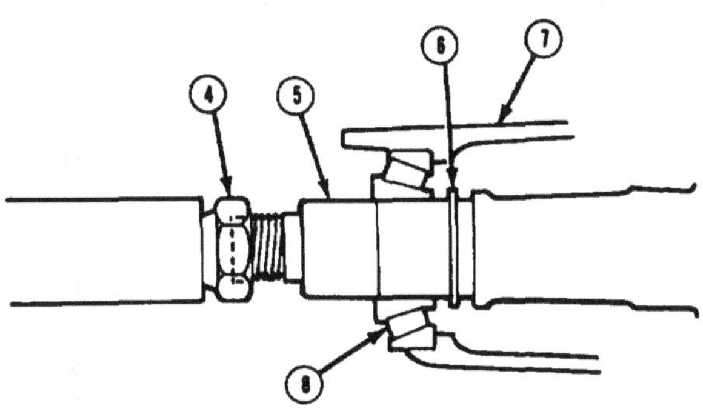

TA50553

6-9. FRONT AXLE DRIVE PINION AND BEARINGS MAINTENANCE (ALL EXCEPT M1009) (Con't).

4. Tap on inner race of front pinion bearing (8) to remove pinion oil seal (14), slinger (17), front bearing cup (13) and front pinion bearing. Discard pinion oil seal.

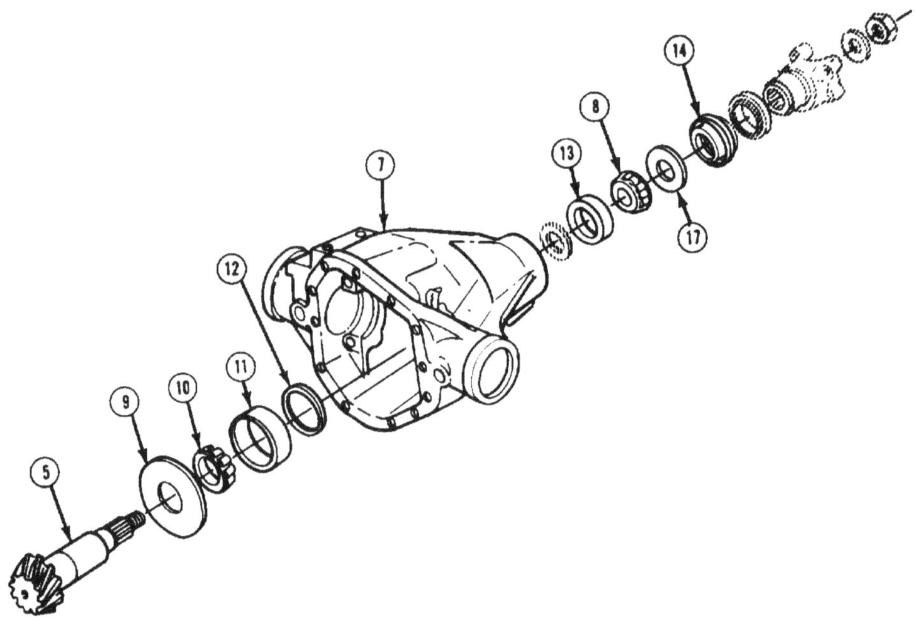

5. Remove rear bearing cup (11) from housing (7). Remove shim (12). Press rear pinion bearing (10) from drive pinion (5) shaft. Remove slinger (9).

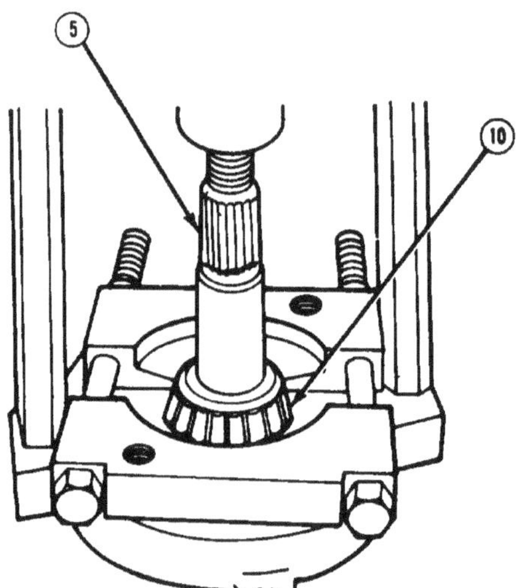

TA50554

6-9. FRONT AXLE DRIVE PINION AND BEARINGS MAINTENANCE (ALL EXCEPT M1009) (Con't).

b. CLEANING AND INSPECTION

WARNING

- Dry cleaning solvent P-D-680 is toxic and flammable. Always wear protective goggles and gloves and use only in a well-ventilated area. Avoid contact with skin, eyes, and clothes and DO NOT breathe vapors. DO NOT use near open flame or excessive heat. The solvent's flash point is 100°F-138°F (38°C-59°C). If you become dizzy while using cleaning solvent, immediately get fresh air and medical help, If solvent contacts eyes, Immediately wash your eyes with water and get medical aid.
- Compressed air used for cleaning purposes should never exceed 30 psi (207 kPa). Use only effective chip guarding and personnel protective equipment (goggles/shield, gloves, etc.). Failure to follow this warning may result in serious injury to personnel.

1. Clean drive pinion, shims, slingers, flange, and deflector with dry cleaning solvent and dry with compressed air.
2. Clean drive pinion bearing bores in housing to ensure that they are free of dirt or other contamination. Remove any nicks found.
3. Clean and inspect front and rear bearing assemblies in accordance with TM 9-214.

NOTE

If drive pinion Is replaced, differential ring gear and ring gear bolts must also be replaced. (See paragraph 6-8)

4. Inspect drive pinion for damage. Replace if damaged.
5. Inspect flange for evidence of wear at point of oil seal contact. Replace if worn.

6-9. FRONT AXLE DRIVE PINION AND BEARINGS MAINTENANCE (ALL EXCEPT M1009) (Con't).

c. INSTALLATION

NOTE

- Ensure that all components are lubricated with lubricating oil as they are Installed.
- If original drive pinion (5) is reused, new shim pack must be built equal in thickness to original one.
- Slingers (9 and 17) are considered part of shim pack.
- If new drive pinion (5) is installed, adjustments in new shim packs' thickness will be required unless the number etched on gear end of replacement drive pinion is the same as original drive pinion.

1. Measure thickness of shim (12) and slinger (9) to determine thickness of original buildup at rear pinion bearing (10). Measure thickness of shim (6) and slinger (17) to determine thickness of original buildup at front pinion bearing (8). Record measurements. Discard shims and slingers.

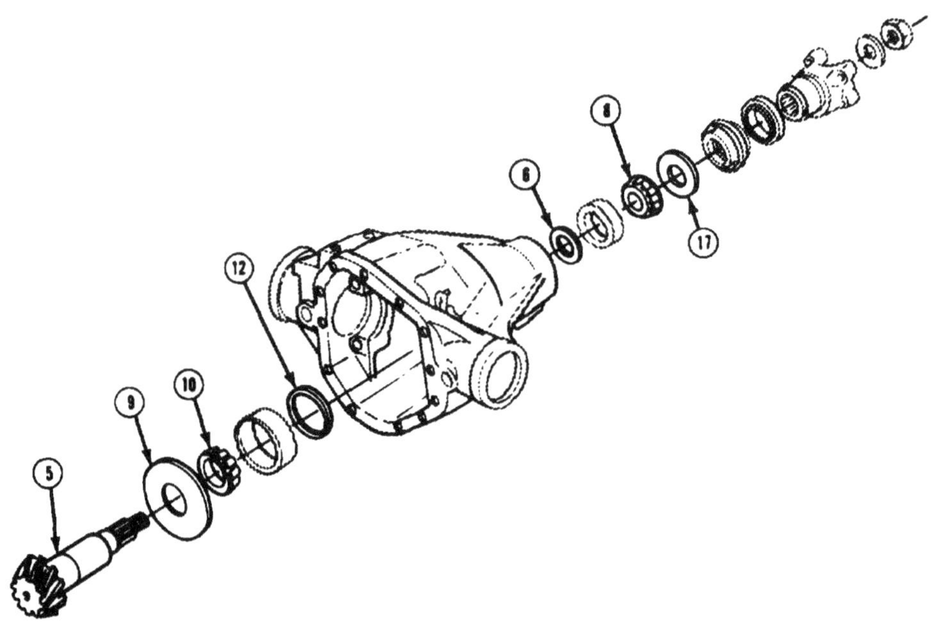

TA50555

6-9. FRONT AXLE DRIVE PINION AND BEARINGS MAINTENANCE (ALL EXCEPT M1009) (Con't).

NOTE

Perform steps 2 and 3 only if new drive pinion (5) is being installed.

2. Determine proper rear shim pack (9 and 12) for setting pinion depth as follows :

   (a) Check number (18) etched on gear end of old drive pinion (5) and locate it in "Old *Pinion Marking"* column of Table 6-3.

   (b) Check number (18) etched on gear end of new drive pinion (5) and locate it in "New *Pinion Marking"* column of Table 6-3,

   (c) Find point where 2 columns intersect to determine adjustment number for new rear shim pack (9 and 12).

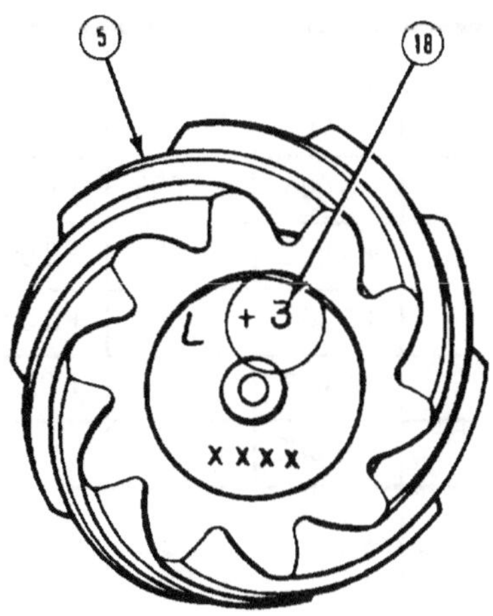

NOTE

A "+" means add that amount; a "-" means subtract that amount; and "O" means no change.

3. Add or subtract adjustment number found on Table 6-3 to old rear shim pack (9 and 12) to determine new rear shim pack.

TA50556

6-9. FRONT AXLE DRIVE PINION AND BEARINGS MAINTENANCE (ALL EXCEPT M1009) (Con't).

Table 6-3. Pinion Setting Charts.

English U.S. Standard Chart

Old Pinlon Marking	New Pinion Marking								
	- 4	- 3	- 2	- 1	0	+1	+2	+3	+4
+4	+ 0.008	+ 0.007	+ 0.006	+ 0.005	+ 0.004	+ 0.003	+ 0.002	+ 0.001	0
+3	+ 0.007	+ 0.006	+ 0.005	+ 0.004	+ 0.003	+ 0,002	+ 0.001	0	-0.001
+2	+ 0.006	+ 0.005	+ 0.004	+ 0.003	+ 0.002	+ 0.001	0	-0.001	-0.002
+1	+ 0.005	+ 0.004	+ 0.003	+ 0.002	+ 0.001	0	-0.001	-0.002	-0.003
0	+ 0,004	+ 0.003	+ 0.002	+ 0.001	0	-0.001	-0.002	-0.003	-0.004
- 1	+ 0.003	+ 0.002	+ 0.001	0	-0.001	-0.002	-0.003	-0.004	-0.005
- 2	+ 0.002	+ 0.001	0	- 0.001	-0.002	-0.003	-0.004	-0.005	-0.006
- 3	+ 0.001	0	-0.001	-0.002	-0.003	-0.004	-0.005	-0.006	-0.007
- 4	0	-0.001	-0.002	-0.003	-0.004	-0.005	-0.006	-0.007	-0.008

Metric Chart

Old Pinlon Marking	New Pinion Marking								
	-10	- 8	- 5	- 3	0	+3	+5	+8	+ 10
+ 10	+ 0.20	+ 0.18	+ 0.15	+ 0.13	+ 0.10	+ 0.08	+ 0.05	+ 0.03	0
+ 8	+ 0.18	+ 0.15	+ 0.13	+ 0.10	+ 0.08	+ 0.05	+ 0.03	0	-0.03
+ 5	+ 0.15	+ 0.13	+ 0.10	+ 0.08	+ 0.05	+ 0.03	0	-0.03	-0.05
+ 3	+ 0.13	+ 0.10	+ 0.08	+ 0.05	+ 0.03	0	-0.03	-0.05	-0.08
0	+ 0.10	+ 0.08	+ 0.05	+ 0.03	0	-0.03	-0.05	-0.08	-0.10
- 3	+ 0.08	+ 0.05	+ 0.03	0	-0.03	-0.05	-0.08	-0.10	-0.13
- 5	+ 0.05	+ 0.03	0	-0.03	-0.05	-0.08	-0.10	-0.13	-0.15
- 8	+ 0.03	0	-0.03	-0,05	-0.08	-0.10	-0,13	-0.15	-0.18
-10	0	-0.03	-0.05	-0.08	-0.10	-0.13	-0.15	-0.18	-0.20

4. Install new shim (12) inside rear bearing cup (11) bore, Install rear bearing cup in housing (7).
5. To front shim pack (6 and 17), add or subtract an amount equal to what was added or subtracted to rear shim pack (9 and 12).
6. Install front bearing cup (13) into housing (7) bore.
7. Install new slinger (9) and rear pinion bearing (10) onto drive pinion (5) shaft. Install drive pinion and bearing assembly in housing (7).

6-9. FRONT AXLE DRIVE PINION AND BEARINGS MAINTENANCE (ALL EXCEPT M1009) (Con't).

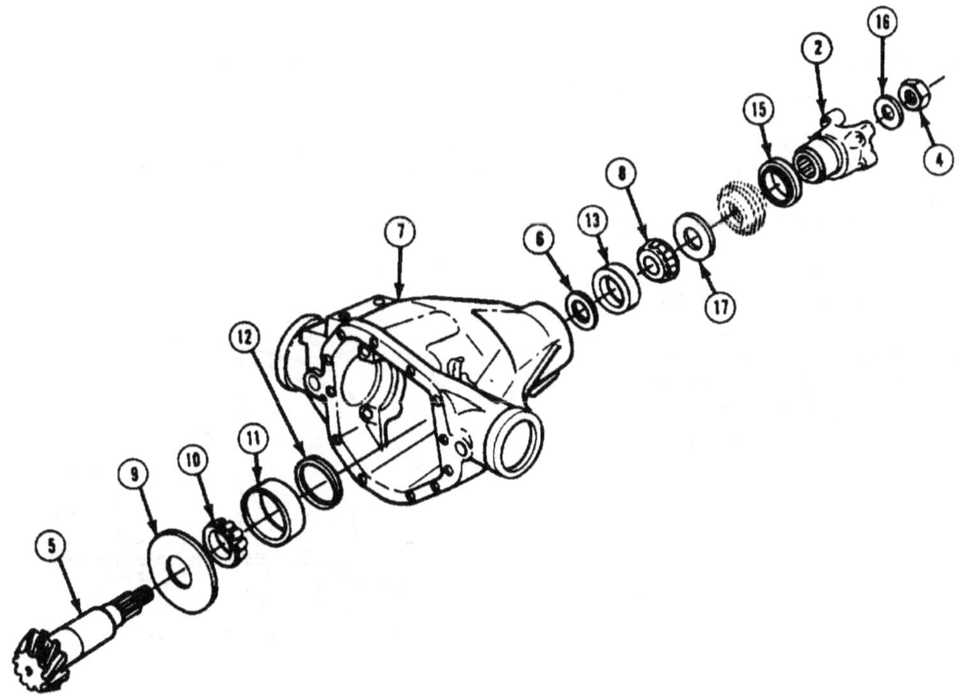

8. Install new shim (6), front pinion bearing (8), and new slinger (17) on drive pinion (5) shaft.

9. If removed, install new deflector (15) to flange (2). Install flange, washer (1 6), and flange nut (4). Using flange. holder and flange remover, tighten flange nut to pull flange onto drive pinion (5) and to fully seat front pinion bearing (8). Tighten flange nut to 255 lb.-ft, (346 N.m). Remove flange holder and flange remover.

TA50557

6-9. FRONT AXLE DRIVE PINION AND BEARINGS MAINTENANCE (ALL EXCEPT M1009) (Con't).

NOTE

- When measuring rotating torque, torque reading needed to start drive pinion (5) shaft-turning should be disregarded.
- Front shim pack (6 and 17) should be increased to reduce rotating torque. Front shim pack should be decreased to increase rotating torque.

10. Measure rotating torque at flange nut (4). Rotating torque should be 20-40 lb.-in. (2-5 N.m) for new bearings, or 10-20 lb.-in. (1-2 N.m) for reused bearings. Adjust front shim pack (6 and 17) by repeating steps 8 and 9 if specifications are not met.

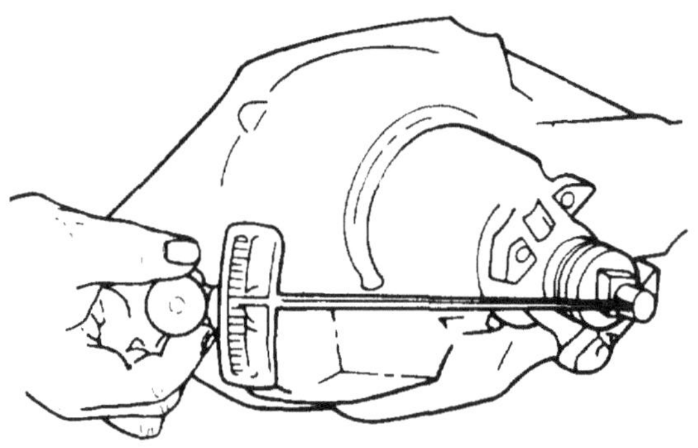

11. Remove flange nut (4), washer (16), and flange (2) from drive pinion (5) shaft. Discard flange nut.
12. Coat lips of new pinion oil seal (14) with molybdenum grease and install squarely into housing (7) bore until flush with housing, Install flange (2), washer (16), and new flange nut (4). Tighten flange nut to 255 lb.-ft. (346 N.m).

TA50558

6-9. FRONT AXLE DRIVE PINION AND BEARINGS MAINTENANCE (ALL EXCEPT M1009) (Con't).

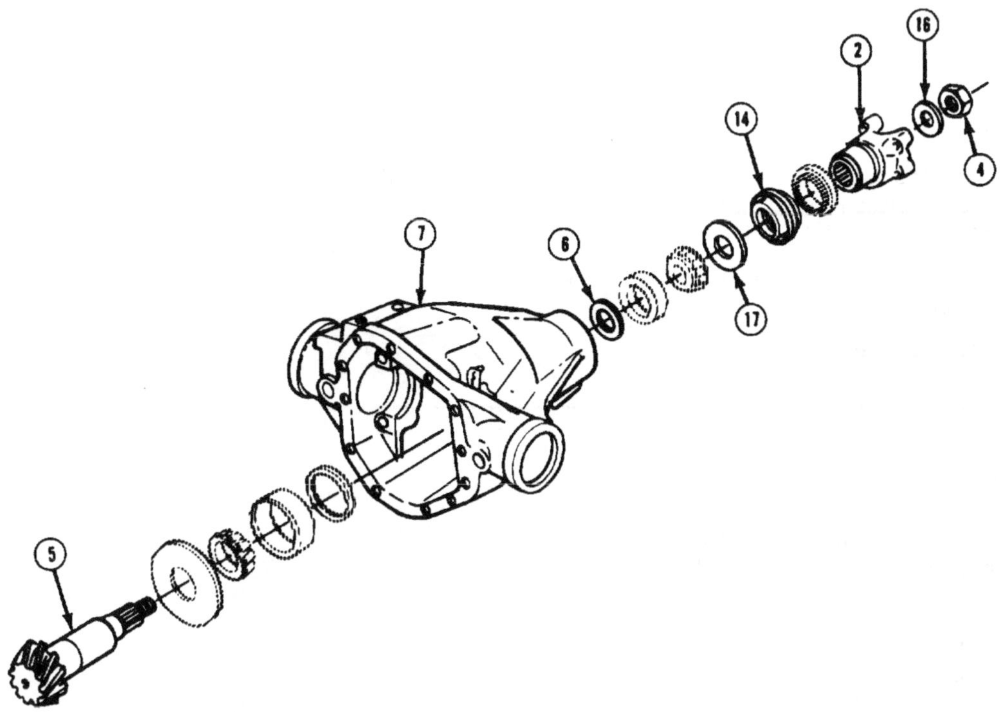

FOLLOW-ON TASKS:

- Install front axle differential. (See paragraph 6-8)

TA50559

6-10. FRONT AND REAR AXLE DRIVE PINION OIL SEAL AND PINION FLANGE MAINTENANCE.

This task covers: a. Removal b. Inspection c. Installation

INITIAL SETUP:

Equipment Condition

- Truck raised and supported on jack stands.
- Propeller shaft disconnected from axle. (See TM 9-2320-289-20)

Materials/Parts

- One flange nut
- One oil seal
- Molybdenum grease (Item 37, Appendix B)

Too/s/Test Equipment

- Flange holder and remover
- Torque wrench

a REMOVAL

1 . At flange nut (4), measure the torque required to rotate drive pinion (3) and record.

2 . Scribe a line down drive pinion (3) stem, flange nut (4), and flange (2). Ensure that scribe line accurately shows relationship of components. Count number of exposed threads on drive pinion stem and record.

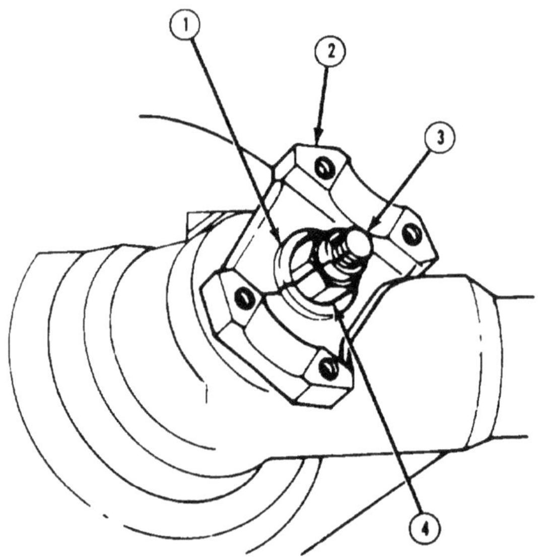

NOTE

Flange nut (4) should not be discarded at this time.

3. Install flange holder (6) on flange (2) and remove flange nut (4) and washer (1). Thread flange remover bolt into small outside diameter end of flange remover (7). Turn flange remover 1/8 turn to a locked position. Turn flange remover while holding flange holder (6) stationary. Remove flange.

TA50560

6-10. FRONT AND REAR AXLE DRIVE PINION OIL SEAL AND PINION FLANGE MAINTENANCE (Con't).

4. Remove oil seal (5) from housing bore and discard.

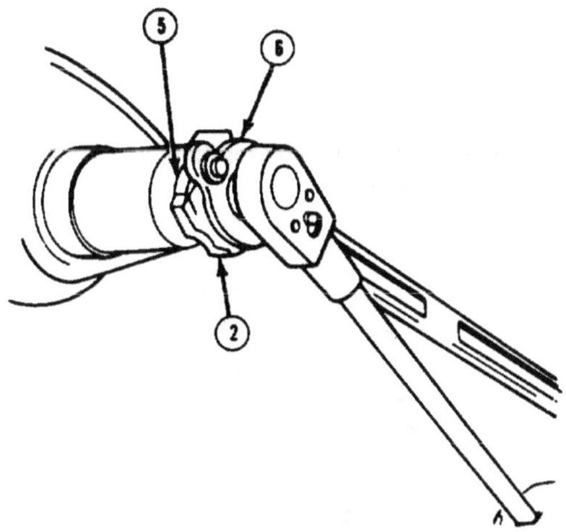

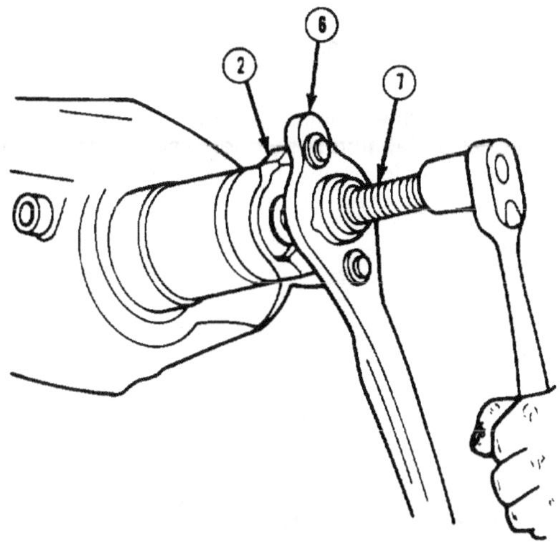

b. INSPECTION

1. Inspect flange oil seal contact surface, bearing contact surface, drive splines, and universal joint mount. Replace flange if damaged.
2. If flange deflector requires replacement, remove by tapping from flange. Clean stake points, install new deflector, and stake deflector at 3 new equally spaced positions.

c. INSTALLATION

1. Lubricate cavity between oil seal (5) lips with molybdenum grease and install new oil seal squarely in bore until flush with housing,
2. Install flange (2). Install washer (1) and original flange nut (4). Install flange holder (6) and tighten flange nut until same number of threads on drive pinion (3) stem are exposed as recorded during removal and scribe marks are alined. Remove flange nut and discard. Install new flange nut until snug against washer. Remove flange holder.

TA50561

6-10. FRONT AND REAR AXLE DRIVE PINION OIL SEAL AND PINION FLANGE MAINTENANCE (Con't).

CAUTION

Care should be taken to ensure that original rotating torque is not exceeded by more than 5 lb.-in. (0.6 N.m). If working on M1009, exceeding torque specifications may compress spacer on drive pinion (3) shaft too far and require replacement of spacer.

3. Measure torque required to turn drive pinion (3) and compare with torque recorded during removal, Install flange holder (6) and tighten flange nut (4) in small increments until rotating torque exceeds original by 1–5 lb.-in. (0. 1-0.6 N.m). Remove flange holder.

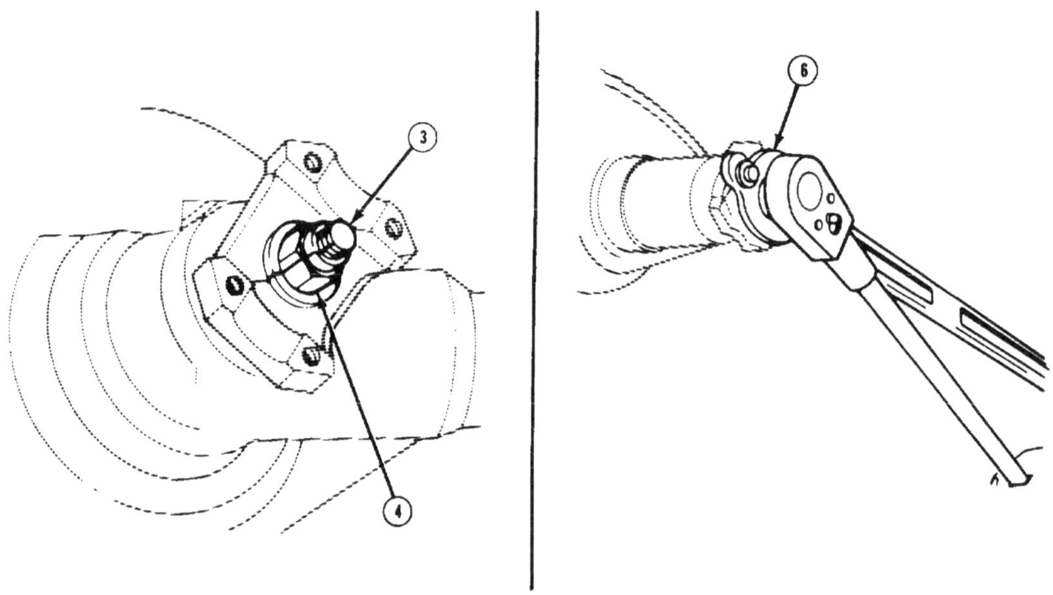

FOLLOW-ON TASKS:

- Check differential lubricating oil level and fill as required, (See LO 9-2320-289-12)
- Connect propeller shaft to axle. (See TM 9-2320-289-20)
- Remove jack stands and lower truck.

TA50562

6-11. STEERING KNUCKLE AND ARM MAINTENANCE (ALL EXCEPT M1009).

This task covers: a. Disassembly b. Assembly

INITIAL SETUP:

Equipment Condition

- Steering tie-rod removed. (See TM 9-2320-289-20)
- Connecting rod disconnected from steering arm (left side). (See TM 9-2320-289-20)
- Front axle shaft removed, (See paragraph 6-4)

Tools/Test Equipment

- Socket wrench adapter, J-26871-A
- Torque wrench

Materials/Parts

- One gasket
- Two seals
- Eight lockwashers
- Grease (Item 36, Appendix B)

General Safety Instructions

- Nuts at steering arm, or bolts at upper kingpin bearing cap, must be removed alternately and with caution.

a. DISASSEMBLY

WARNING

Nuts at steering arm or bolts at upper kingpin bearing cap must be removed alternately and with caution. Compression spring underneath steering arm or bearing cap could fly up during removal, causing serious injury to personnel.

6-11. STEERING KNUCKLE AND ARM MAINTENANCE (ALL EXCEPT M1009) (Con't).

NOTE

- On left side, steering arm (3) is installed to steering knuckle (8) with 4 nuts (2). On right side, upper kingpin bearing cap is installed to steering knuckle with 4 bolts and lockwashers.
- Left side disassembly is given.

1. Remove 4 nuts (2) alternately at steering arm (3). Remove steering arm, spring (4), and spring retainer (5). Remove gasket and discard. If present, discard 4 lockwashers.

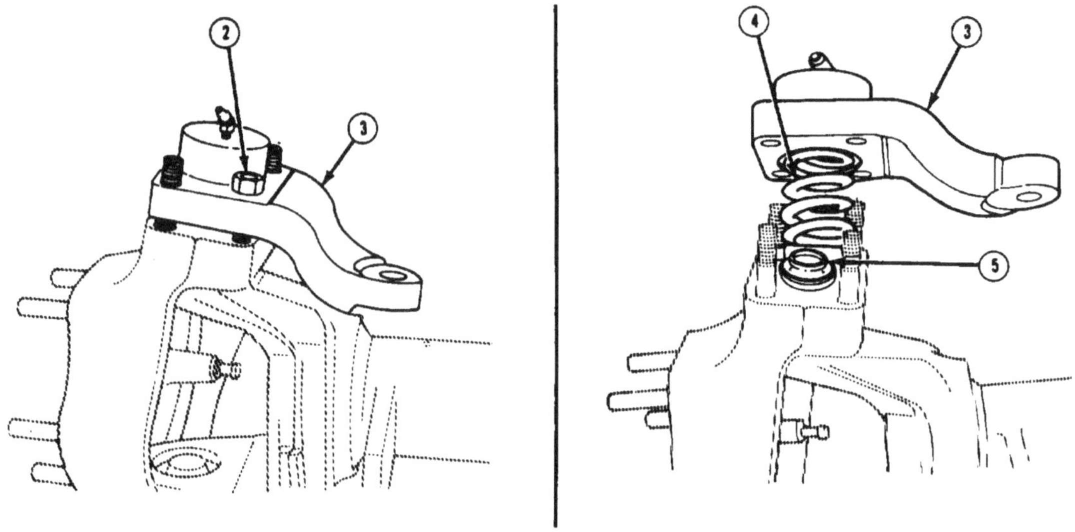

2. Remove 4 bolts (16), lockwashers, and lower kingpin bearing cap (15) from steering knuckle (8). Discard lockwashers,

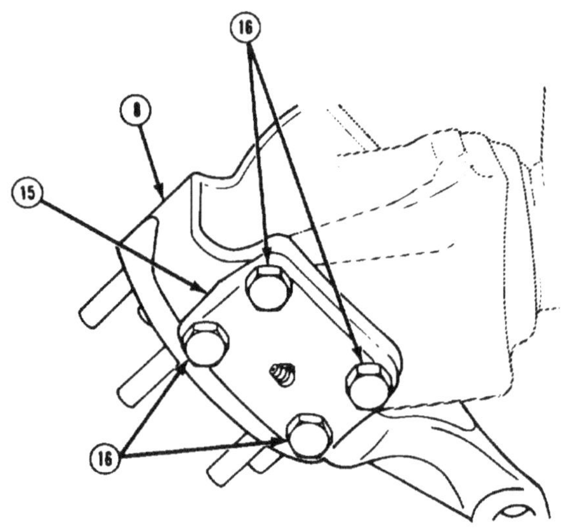

TA50563

6-11. STEERING KNUCKLE AND ARM MAINTENANCE (ALL EXCEPT M1009) (Con't).

3. Remove kingpin bushing (7) from top of steering knuckle (8). Remove steering knuckle from axle housing (9).

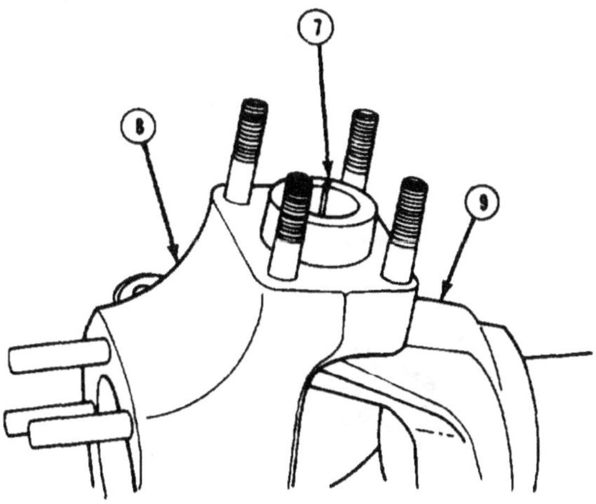

NOTE

Upper kingpin (6) is tightened to 550 lb.-ft, (746 N.m).

4. Remove seal (10) from upper kingpin (6). Remove upper kingpin from axle housing (9). Discard seal.

5. inspect upper kingpin (6) and kingpin bushing (7) for damage. Replace if damaged.

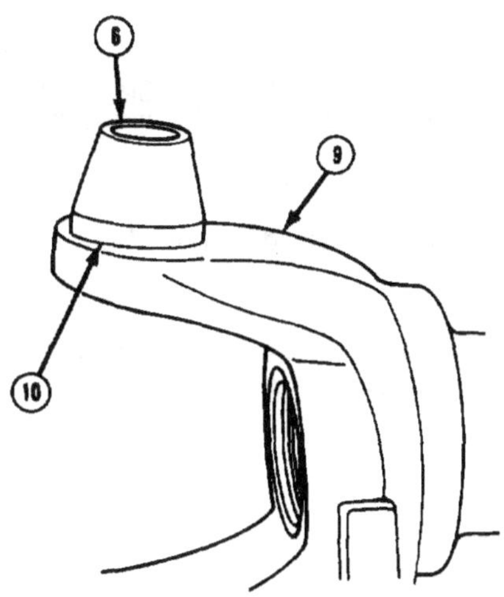

TA50564

6-11. STEERING KNUCKLE AND ARM MAINTENANCE (ALL EXCEPT M1009) (Con't).

6. Remove seal (14) and bearing (13). Inspect bearing for damage and binding, Replace if damaged or binding. Discard seal. If damaged, remove bearing cup (12) and retainer (11).

7. Inspect all grease fittings and replace if damaged.

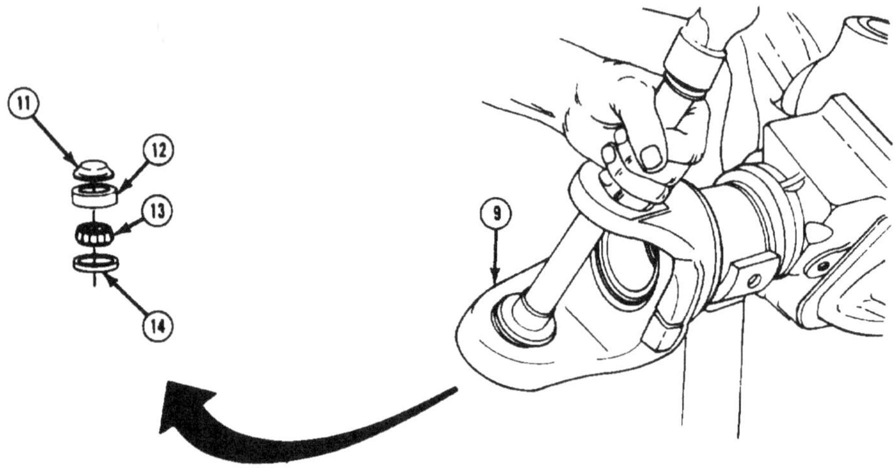

b. ASSEMBLY

NOTE

- On left side, steering arm is installed to steering knuckle (8) with 4 nuts. On right side, upper kingpin bearing cap is installed to steering knuckle with 4 bolts and lockwashers.
- Left side assembly is given,
- Seal (14) will protrude slightly from surface of axle housing (9) flange when installed.

1. If removed, assemble retainer (11) and bearing cup (12), and install to axle housing (9) flange. Fill area in retainer with grease. Grease bearing (13) and install, Install new seal (14).

TA60565

6-11. STEERING KNUCKLE AND ARM MAINTENANCE (ALL EXCEPT M1009) (Con't).

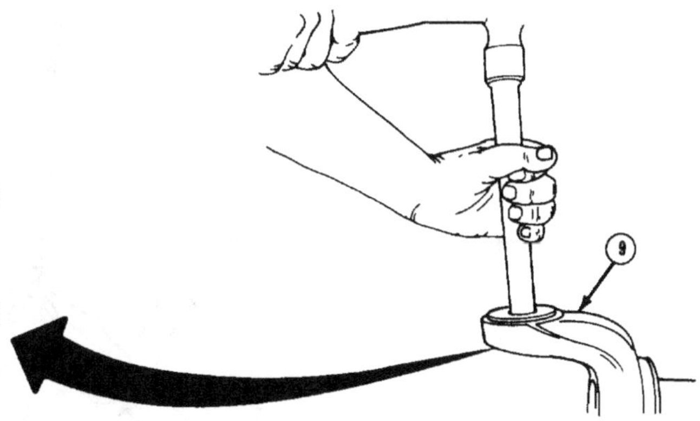

2. Install upper kingpin (6) to axle housing (9) and tighten to 550 lb.-ft. (746 N.m). Install new seal (10) to upper kingpin with lip facing up.

3. Grease upper kingpin (6) and install steering knuckle (8) to axle housing (9).

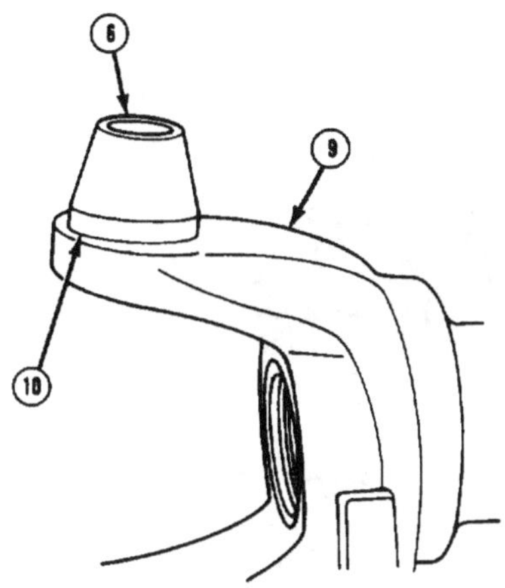

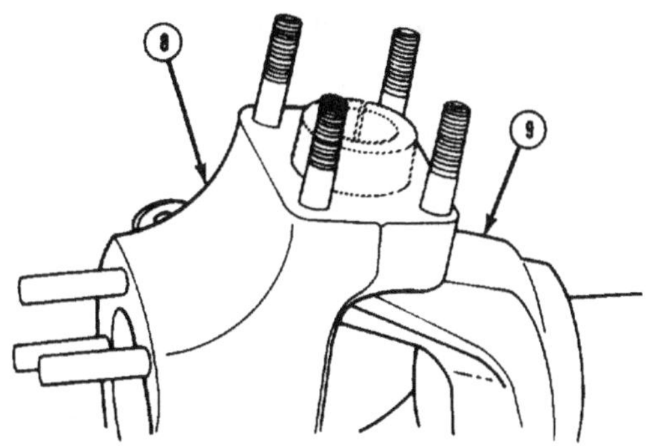

TA50566

6-11. STEERING KNUCKLE AND ARM MAINTENANCE (ALL EXCEPT M1009) (Con't).

4 . Install lower kingpin bearing cap (15) with 4 new lockwashers and bolts (16). Tighten bolts to 80 lb.-ft. (108 Nom).

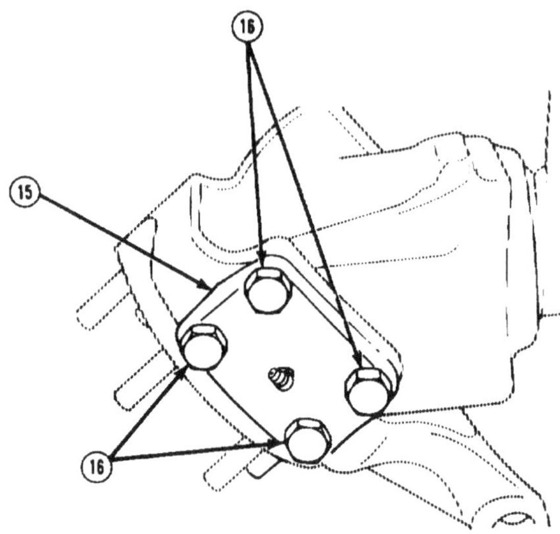

5 . Install kingpin bushing (7) to top of steering knuckle (8) with ridge on bushing indexed with slot in axle housing (9).

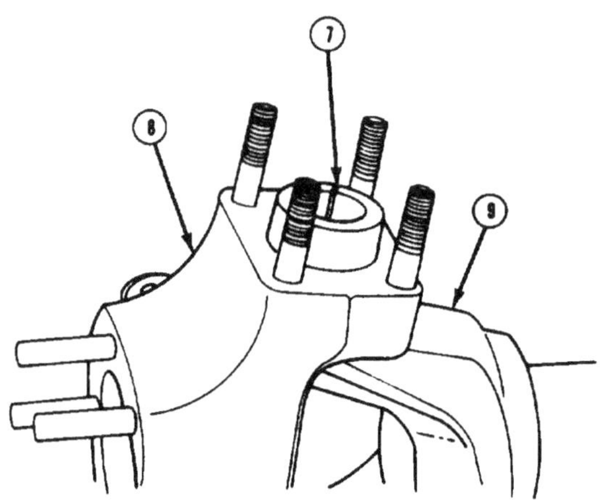

TA50567

6-11. STEERING KNUCKLE AND ARM MAINTENANCE (ALL EXCEPT M1009) (Con't).

CAUTION

Nuts (2) at steering arm (3) or bolts at upper kingpin bearing cap should be installed alternately and in increments. Failure to follow this caution may result in damage to stud threads or to bolts.

NOTE

On right side, 4 new lockwashers are used with bolts.

6. Assemble spring retainer (5) with collar facing up and spring (4) on kingpin bushing (7) over upper kingpin (6). Install new gasket and steering arm (3) *over* steering knuckle studs (1). Install 4 nuts (2), and tighten alternately and in increments to 80 lb.-ft. (108 N.m).

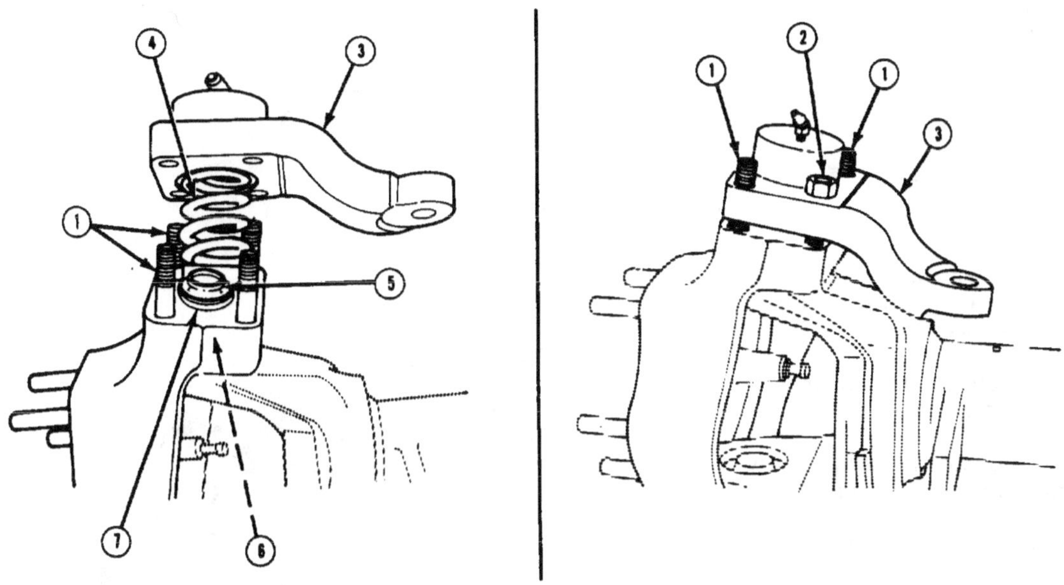

FOLLOW-ON TASKS:

- Lubricate steering knuckle and arm. (See LO 9-2320-289-12)
- Install front axle shaft. (See paragraph 6-4)
- Connect connecting rod to steering arm if removed. (See TM 9-2320-289-20)
- Install steering tie-rod. (See TM 9-2320-289-20)

TA50568

6-12. STEERING KNUCKLE, ARM, AND BALL JOINT MAINTENANCE (M1009).

This task covers: a. Disassembly b. Assembly

INITIAL SETUP:

Equipment Condition

- Steering tie-rod removed. (See TM 9-2320-289-20)
- Connecting rod disconnected from steering arm (left side only), (See TM 9-2320-289-20)
- Front axle shaft removed. (See paragraph 6-4)

Tools/Test Equipment

- Ball joint adapter set, J-23454-D
- Ball stud adjuster sleeve, J-23447
- Lower ball joint remover, J-33122
- "C" clamp
- Snapring pliers
- Torque wrench

Materials/Parts

- One flat washer
- One steering knuckle socket repair kit
- One cotter pin
- Three locknuts

a. DISASSEMBLY

NOTE

- Left side steering knuckle (5), arm (l), and ball joint (12) disassembly is given.
- Right side steering knuckle (5) has no steering arm (1).
- It may be necessary to apply heat to loosen steering arm (1).
- If removing steering arm (1), perform step 1,

1. Remove 3 locknuts (3) at steering arm (1), Remove 3 stud adapters (4). Remove steering arm fro m steering knuckle (5). Discard locknuts.

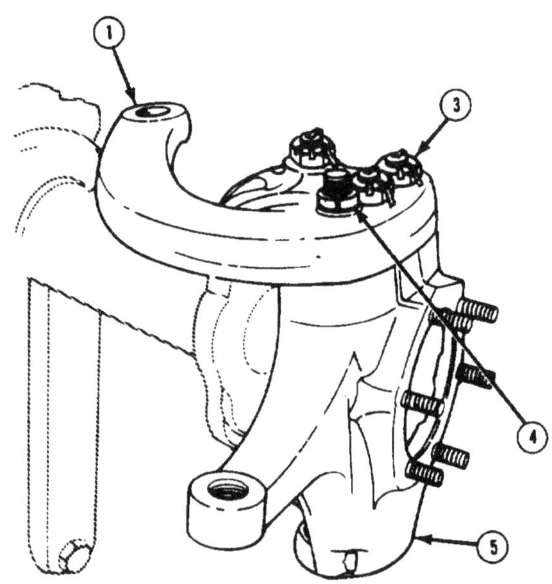

TA50569

6-12. STEERING KNUCKLE, ARM, AND BALL JOINT MAINTENANCE (M1009) (Con't).

2. Remove cotter pin (8) from upper ball stud nut (9). Remove upper and lower ball stud nuts (9 and 10). Discard cotter pin and upper and lower ball stud nuts.

3. Loosen upper ball stud adjusting sleeve (7) until flush with underside of axle housing (6).

4. Use a suitable wedge-shaped tool between lower ball stud and axle housing (6). Tap on tool to release steering knuckle (5). Repeat at upper ball stud and remove steering knuckle from axle housing.

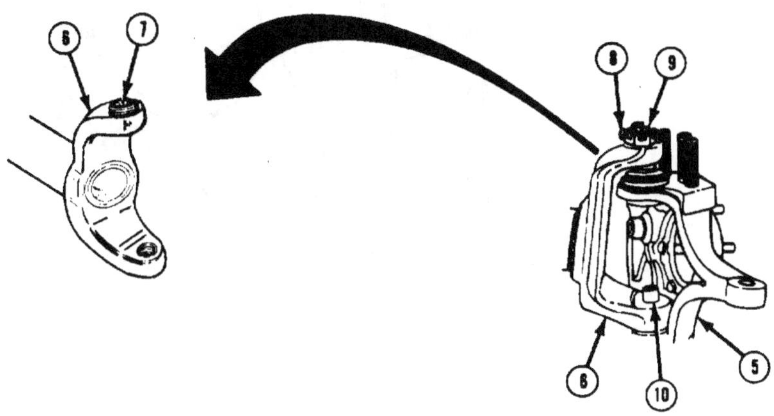

NOTE

Ensure that lower ball joint (12) has been removed before repairing upper ball joint.

5. Place steering knuckle (5) in vise. Remove snapring at lower ball joint (12). Using adapters J-23454-1 and J-23454-4, install ball joint removal tools on lower ball joint. Install f l at washer between " C" clamp and adapter (11). Tighten "C" clamp until lower ball joint is pressed out. Discard lower ball joint and snapring.

6. Place ball joint removal tools on upper ball joint (12) and remove in same manner as lower ball joint. Discard upper ball joint.

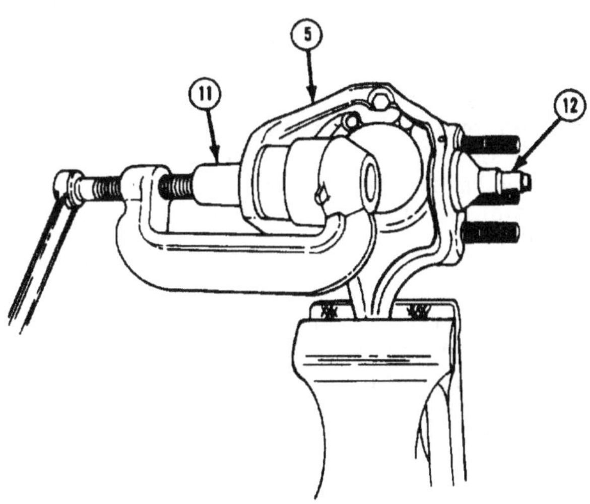

TA50570

6-12. STEERING KNUCKLE, ARM, AND BALL JOINT MAINTENANCE (M1009) (Con't).

NOTE

If stop (13) requires replacement, position and number of exposed threads should be noted fo r installation.

7. Inspect 3 steering arm studs (2), 6 spindle bolts (14), and stop (13). Replace if damaged.

8. Inspect upper ball stud adjusting sleeve (7) at axle housing (6) and replace if damaged.

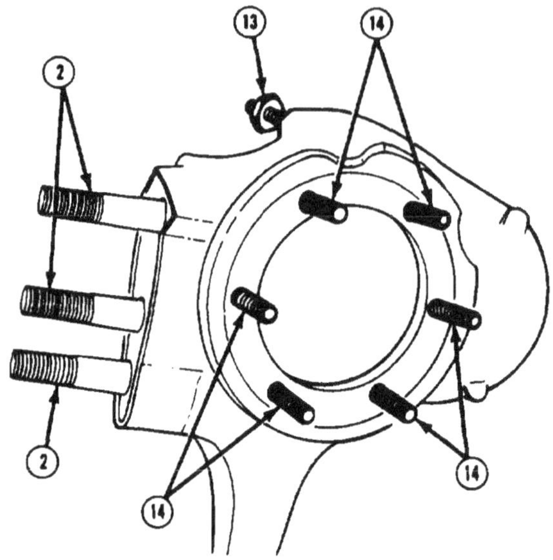

b. ASSEMBLY

NOTE

- Left side steering knuckle (5), arm, and ball joint (12) assembly is given,
- MI 009 steering knuckle ball joints (12) do not require lubrication.
- Ensure that lower ball joint (12) (ball joint without cotter pin hole In stud) is straight before it is pressed in place.

1. Position new lower ball joint (12) in steering knuckle (5). Using ball joint installation tools and adapters J-23454-2 and J-23454-4, press lower ball joint into place. Install new snapring.

2. Remove rubber retaining cup from new upper ball joint (12). Position upper ball joint at steering knuckle (5) and install in same manner as lower ball joint. Install rubber retaining cup to upper ball joint.

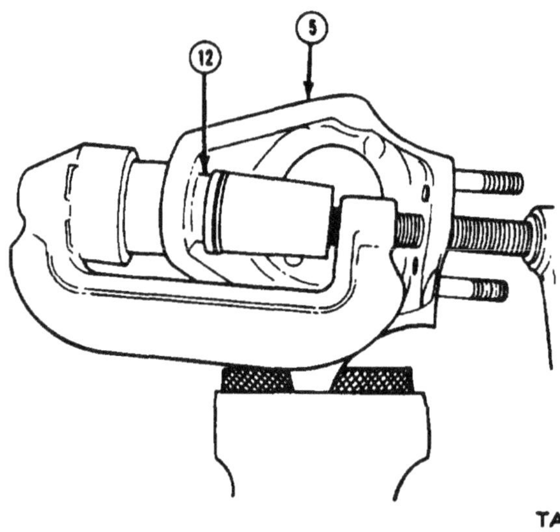

TA50571

6-12. STEERING KNUCKLE, ARM, AND BALL JOINT MAINTENANCE (M1009) (Con't).

3. Position steering knuckle (5) and sockets at axle housing (6). Install lower ball stud nut (10) finger tight.

NOTE

Torque values in steps 4-7 are necessary to maintain proper alinement of steering knuckle (5) to axle housing (6).

4. Push up on steering knuckle (5) to keep ball socket from turning in knuckle and partially tighten lower ball stud nut (10) to 30 lb.-ft. (41 N.m).

5. Tighten upper ball stud adjusting sleeve (7) to 50 ft.-lb. (68 N.m).

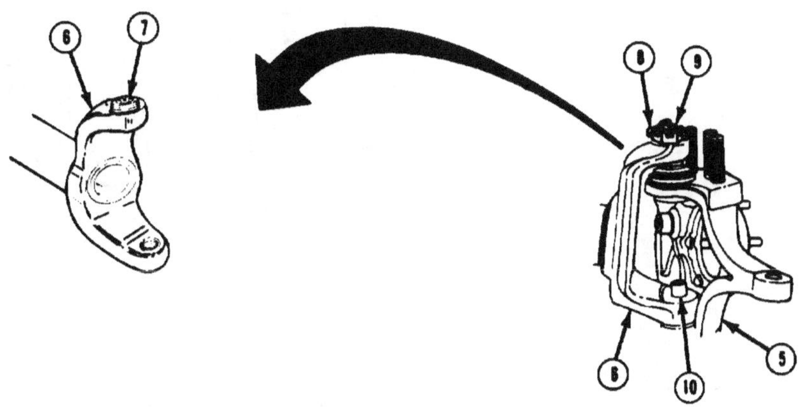

NOTE

Nut (9) should be tightened only, to aline holes for cotter pin (8).

6. Install new upper ball stud nut (9) and tighten to 100 lb.-ft. (136 N.m). Install new cotter pin (8).

7. Fully tighten lower ball stud nut (10) to 70 lb.-ft. (95 N.m).

TA50572

6-12. STEERING KNUCKLE, ARM, AND BALL JOINT MAINTENANCE (M1009) (Con't).

8. If removed, install steering arm (1) over 3 steering arm studs (2). Install 3 stud adapters (4). Install 3 new locknuts (3) and tighten to 90 lb.-ft. (122 N.m).

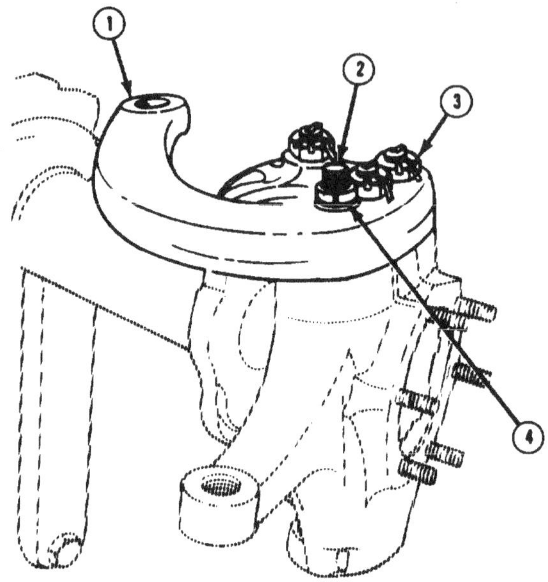

FOLLOW-ON TASKS:

- Install front axle shaft. (See paragraph 6-4)
- Connect connecting rod to steering arm if removed. (See TM 9-2320-289-20)
- Install steering tie-rod. (See TM 9-2320-289-20)

TA50573

6-13. REAR AXLE ASSEMBLY REPLACEMENT.

This task covers: a. Removal b. Installation

INITIAL SETUP:

Equipment Condition

- Parking brake released.
- Truck raised and frame supported on jack stands.
- Rear wheels removed. (See TM 9-2320-289-20)
- Rear propeller shaft disconnected from rear axle differential. (See TM 9-2320-289-20)
- Parking brake cable disconnected at equalizer. (See TM 9-2320-289-20)
- Rear brake lines disconnected from wheel cylinders and differential housing. (See TM 9-2320-289-20)
- Proportioning valve lever disconnected from differential housing. (See TM 9-2320-289-20)
- Rear shock absorbers disconnected from axle. (See TM 9-2320-289-20)
- Rear stabilizer bar removed (M1028A2 and M1028A3). (See TM 9-2320-289-20)

Tools/Test Equipment

- Torque wrench

Personnel Required

- MOS 63W (2)

General Safety Instructions

- Rear axle assembly is heavy. Use extreme caution, provide support, and use assistance during removal and installation.

6-13. REAR AXLE ASSEMBLY REPLACEMENT (Con't).

a. REMOVAL

1. Slide clamp up rear axle vent hose (8) and disconnect rear axle vent hose from rear axle differential (7).

2. Place suitable jack under rear axle differential (7).

WARNING

Rear axle assembly (6) is heavy. Use extreme caution, provide support, and use assistance during removal. Failure to follow this warning may result in serious injury to personnel or damage to rear axle assembly.

NOTE

For M1028A2 and M1028A3, skip step 3 and perform step 4.

3. Remove 4 nuts (3), washers (2), 2 "U" bolts (5), and anchor plate (4) at each spring assembly (1). Lower rear axle assembly (6) and remove, If present, remove 2 shims (9) from axle housing.

4. Remove 4 nuts (3), washers (2), adapter (3.1), 4 washers (3,2), 2 "U" bolts (5), and anchor plate (4) at each spring assembly (1). Lower rear axle assembly (6) and remove. If present, remove 2 shims (9) from axle housing.

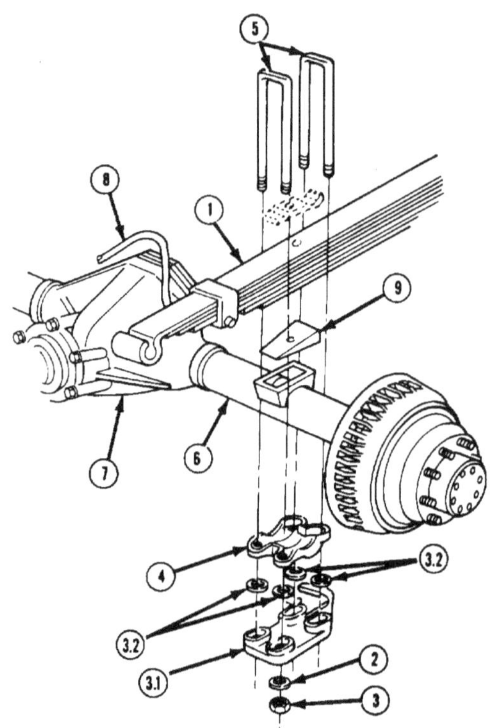

M1028A2 and M1028A3

6-13. REAR AXLE ASSEMBLY REPLACEMENT (Con't).

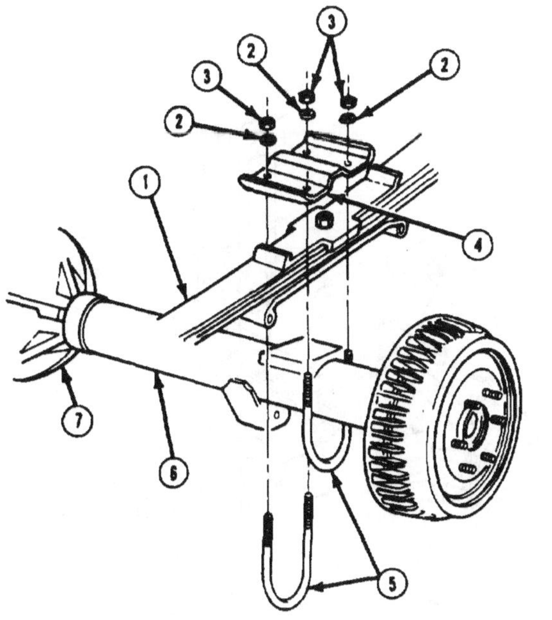

M1009

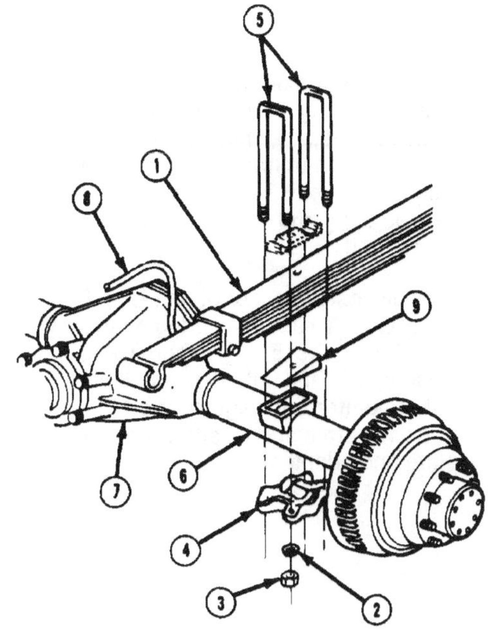

All Except M1009, M1028A2, and M1028A3

b. INSTALLATION

WARNING

Rear axle assembly (6) is heavy. Use extreme caution, provide support, and use assistance during installation. Failure to follow this warning may result in serious injury to personnel or damage to rear axle assembly,

NOTE

For M1028A2 and M1028A3, skip steps 1 and 2 and perform steps 2.1 and 2.2.

1. Position rear axle assembly (6) on suitable jack and raise into position. If removed, install 2 shims (9) on axle housing with thinner edges facing front. Aline holes in shims with holes in axle housing and spring assembly (1) center bolt.

2. Loosely install anchor plate (4) at each spring assembly with with 2 " U" bolts (5), 4 washers (2), and nuts (3). Install all nuts to uniform engagement on "U" bolts to retain and position anchor plate.

6-13. REAR AXLE ASSEMBLY REPLACEMENT (Con't).

NOTE

Adapters (3.1) are mismarked on casting as to left-hand and right-hand. Install adapters as shown, with ribs up and curved arm to rear facing inboard.

2.1. Position rear axle assembly (6) on suitable jack and raise into position. If removed, install 2 shims (9) on axle housing with thinner edges facing front. Aline holes in shims with holes in axle housing and spring assembly (1) center bolt.

2.2. Loosely install anchor plate (4) and adapter (3, 1) at each spring assembly with 2 "U" bolts (5), 4 washers (3.2), washers (2), and nuts (3). Install all nuts to uniform engagement on "U" bolts to retain and position anchor plate and adapter,

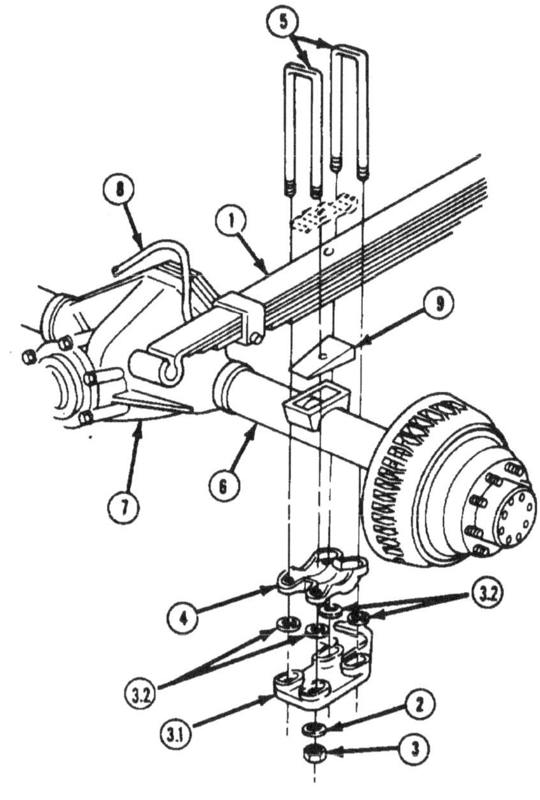

M1028A2 and M1028A3

NOTE

All nuts (3) must be tightened evenly to ensure that "U" bolts (5) and spring assemblies (1) seat properly,

3. Partially tighten all nuts (3) in diagonal sequence to 25 lb.-ft, (34 N.m). Fully tighten nuts on M1009 in same diagonal sequence to 145 lb.-ft. (197 N.m). On M1028A2, tighten all nuts in same diagonal sequence to 150 lb.-ft. (180 N.m). On all others, tighten all nuts in same diagonal sequence to 170 lb.-ft. (231 N.m).

4. Remove jack from rear axle differential (7),

5. Connect rear axle vent hose (8) to rear axle differential (7) and install clamp,

6-13. REAR AXLE ASSEMBLY REPLACEMENT (Con't).

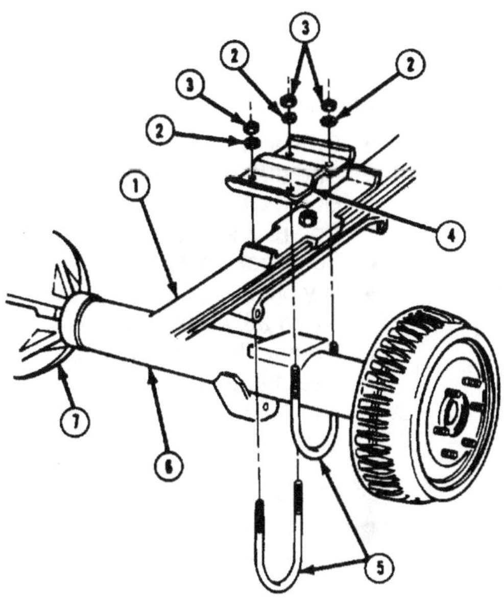

M1009

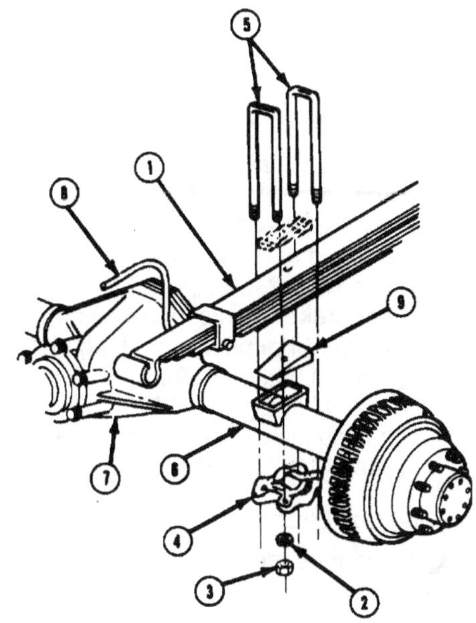

All Except M1009, M1028A2, and M1028A3

FOLLOW-ON TASKS:

- Install rear stabilizer bar (M1028A2 and M1028A3). (See TM 9-2320-289-20)
- Connect rear shock absorbers to axle. (See TM 9-2320-289-20)
- Connect proportioning valve lever to differential housing. (See TM 9-2320-289-20)
- Connect rear brake lines to wheel cylinders and differential housing. (See TM 9-2320-289-20)
- Connect parking brake cable at equalizer. (See TM 9-2320-289-20)
- Connect rear propeller shaft to rear axle differential. (See TM 9-2320-289-20)
- Install rear wheels. (See TM 9-2320-289-20)
- Remove jack stands and lower truck.
- Road test truck and check torque of "U" bolt nuts.

6-14. REAR AXLE SHAFT, OUTER SEAL, BEARING, AND BACKING PLATE REPLACEMENT (M1009).

This task covers:

a. Removal
b. Cleaning and inspection
c. Installation

INITIAL SETUP:

Equipment Condition

- Brake drum removed. (See TM 9-2320-289-20)
- Brake shoes and attaching hardware removed (if removing backing plate), (See TM 9-2320-289-20)
- Rear brake line disconnected from wheel cylinder (if removing backing plate), (See TM 9-2320-289-20)
- Emergency brake cable disconnected from backing plate (if removing backing plate). (See TM 9-2320-289-20)

Materials/Parts

- One bearing
- One differential cover gasket
- One lockscrew
- One rear axle outer oil seal
- Four lockwashers
- Dry cleaning solvent (Item 23, Appendix B)
- Grease (Item 36, Appendix B)
- Gear lubricating oil (Item 44, Appendix B)

Tools/Test Equipment

- Annular bearing inserter, J-23690
- Axle seal installer, J-29713
- Drive handle, J-8092
- Puller kit
- Torque wrench

General Safety Instructions

- Dry cleaning solvent is flammable and must not be used near open flame. Use only in a well-ventilated area.
- Compressed air used for cleaning purposes should never exceed 30 psi (207 kPa).

a. REMOVAL

NOTE

- If removing rear axle shaft (8), perform steps 1-3.
- If removing rear axle outer oil seal, perform steps 1-4.
- If removing rear axle outer oil seal and bearing, perform steps 1-3, and 5.
- If removing backing plate, perform steps 1-3 and 6.
- Ensure that rear axle differential cover is clean before proceeding.

1. Remove 10 bolts at differential cover, Carefully move brake line and bracket, brake line block bracket, and emergency brake cable mounting bracket out of the way. DO NOT kink lines. Remove differential cover and allow all lubricating oil to drain into a suitable container. Remove differential cover gasket and discard.

CAUTION

When removing rear axle shaft (8), ensure that outer oil seal is not damaged. Failure to follow this caution may result in damage to outer oil seal and axle lubricant leakage.

6-14. REAR AXLE SHAFT, OUTER SEAL, BEARING, AND BACKING PLATE REPLACEMENT (M1009) (Con't).

CAUTION

- DO NOT allow pinion shaft (4) to come completely out,

2. Rotate differential case (2) assembly as required to gain access to lockscrew (3). Support pinion shaft (4) so that it does not fall into differential case assembly. Remove lockscrew. Withdraw pinion shaft part way and rotate differential case assembly until pinion shaft touches bottom edge of housing (1).

3. Rotate rear axle shaft (8) until "C" lock (5) is alined with thrust block (6). Push flange of rear axle shaft toward the differential case (2) assembly and remove "C" lock. Remove rear axle shaft. Loosely reinstall pinion shaft (4) and lockscrew (3) to ensure that differential gears do not become mispositioned.

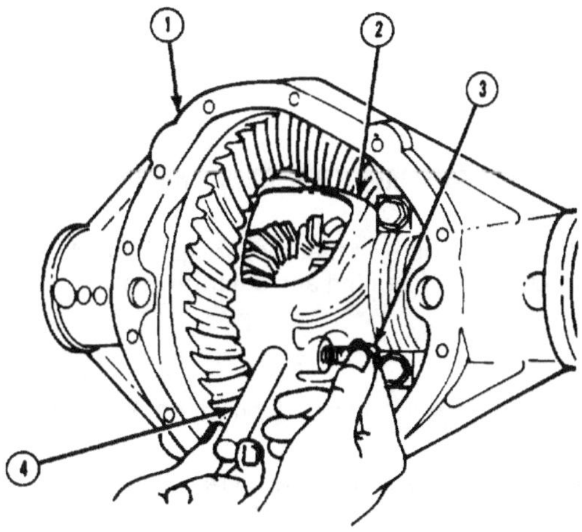

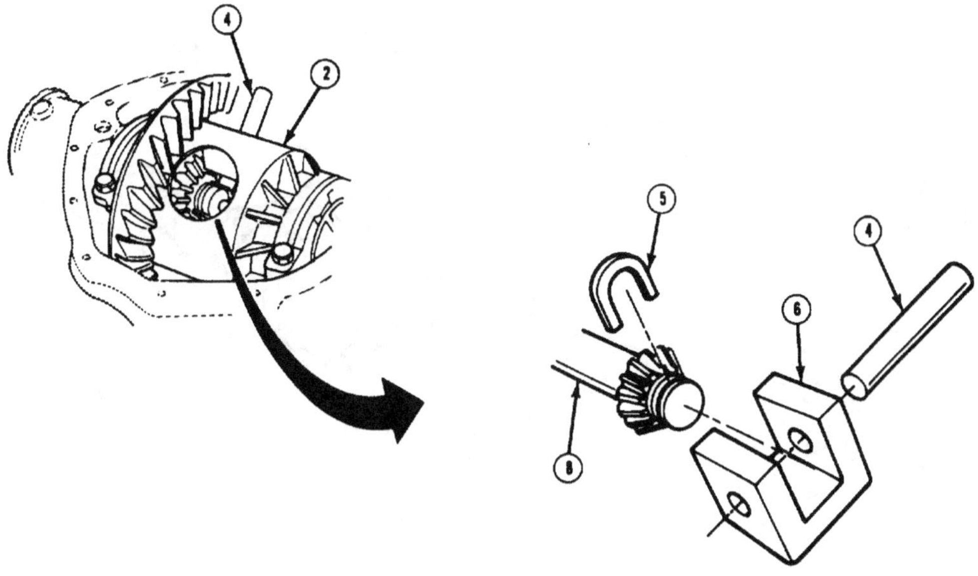

TA50575

6-14. REAR AXLE SHAFT, OUTER SEAL, BEARING, AND BACKING PLATE REPLACEMENT (M1009) (Con't).

NOTE

Perform step 4 if removing outer oil seal (10) only.

4 . Insert button end of rear axle shaft (8) behind steel case of outer oil seal (10), Remove outer oil seal and discard.

NOTE

Perform step 5 if removing outer oil seal (10) and bearing (11).

5 . Insert puller into bore so that tool grasps behind bearing (11). Remove bearing and outer oil seal (1 O). Discard bearing and outer oil seal.

NOTE

Perform step 6 only if backing plate (15) is damaged.

6 . Remove 4 nuts (13), lockwashers (14), bolts (16), and remove backing plate (15) from axle housing (1 2), Discard lockwashers.

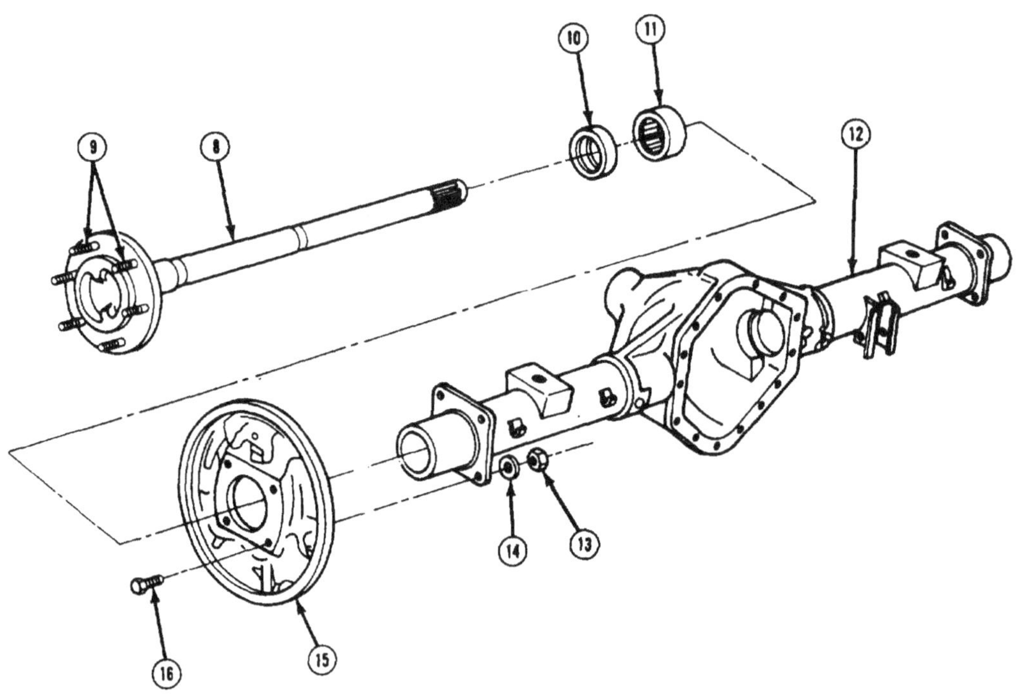

TA50576

6-14. REAR AXLE SHAFT, OUTER SEAL, BEARING, AND BACKING PLATE REPLACEMENT (M1009) (Con't).

b. CLEANING AND INSPECTION

WARNING

- Dry cleaning solvent P-D-660 is toxic and flammable. Always wear protective goggles and gloves and use only In a well-ventilated area. Avoid contact with skin, eyes, and clothes and DO NOT breathe vapors. DO NOT use near open flame or excessive heat. The solvent's flash point is 100°F-138°F (38°C-59°C). If you become dizzy while using cleaning solvent, immediately get fresh air and medical help. If solvent contacts eyes, immediately wash your eyes with water and get medical aid.
- Compressed air used-for cleaning purposes should never exceed 30 psi (207 -kPa). Use only effective chip guarding and personnel protective equipment (goggles/shield, gloves, etc.). Failure to follow this warning may result in serious Injury to personnel.

1. Clean all removed components in dry cleaning solvent and dry with compressed air. Clean all gasket material from differential cover and housing.
2. Inspect rear axle shaft for damage. Replace If damaged.
3. Inspect wheel studs (9) at rear axle shaft for damage. Replace if damaged.
4. Inspect backing plate for damage. Replace if damaged.

c. INSTALLATION

NOTE

- If installing backing plate (15), perform step 1.
- If installing rear axle bearing (11), perform step 2.
- If installing rear axle outer oil seal (10), perform step 3.
- If Installing rear axle shaft (8), perform steps 4-6.

1. If removed, install backing plate (15) on axle housing (12) with 4 bolts (16), new lockwashers (14), and 4 nuts (13). Tighten nuts to 35 lb.-ft. (47 Nom).
2. Lubricate new rear axle bearing (11) with grease. Install new rear axle bearing until bearing installer tool bottoms against tube of axle housing (12).
3. Lubricate cavity between new outer oil seal (10) lips with grease. Install new outer oil seal until seal installer tool bottoms against tube of axle housing (12) and outer oil seal is flush with end of tube.

6-14. REAR AXLE SHAFT, OUTER SEAL, BEARING, AND BACKING PLATE REPLACEMENT (M1009) (Con't).

WARNING

Ensure that "C" lock (5) is correctly seated in differential side gear (7), Failure to correctly seat "C" lock may result in loss of truck's wheel and serious Injury or death to personnel.

CAUTION

When Installing rear axle shaft (8), ensure that splines on end of shaft do not damage outer oil seal. Failure to follow this caution may result in damage to outer oil seal and exle lubricant leakage.

4 . Slide rear axle shaft (8) into position with splines engaged into differential side gear (7), Remove lockscrew (3) and partially withdraw pinion shaft (4), Position "C" lock (5) on end of rear axle shaft, Pull rear axle shaft flange outward and seat "C" lock in differential side gear, Discard lockscrew.

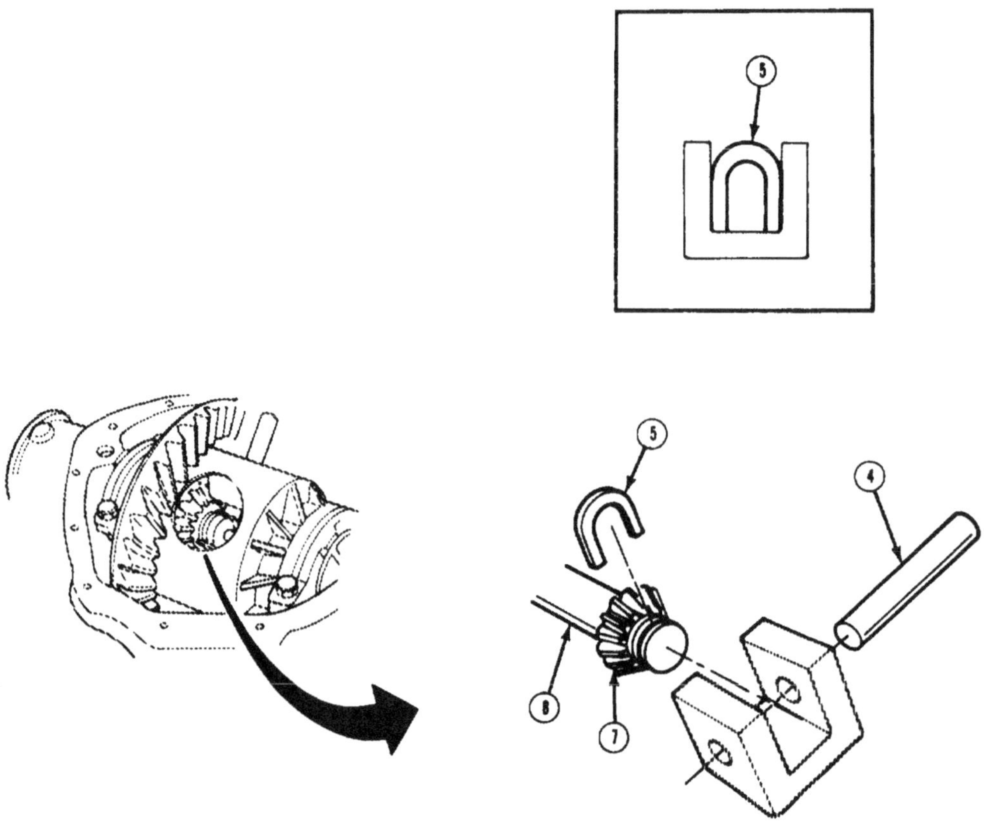

TA50577

6-14. REAR AXLE SHAFT, OUTER SEAL, BEARING, AND BACKING PLATE REPLACEMENT (M1009) (Con't).

5. Aline hole in pinion shaft (4) with lockscrew (3) hole in differential case (2) assembly. Install pinion shaft and new lockscrew. Tighten lockscrew to 25 lb.-ft. (34 N.m).

6. Install new gasket and differential cover. Carefully position emergency brake cable mounting bracket, brake line block bracket, and brake line and bracket. DO NOT kink lines. Install 10 bolts and tighten to 35 lb.-ft. (47 N.m).

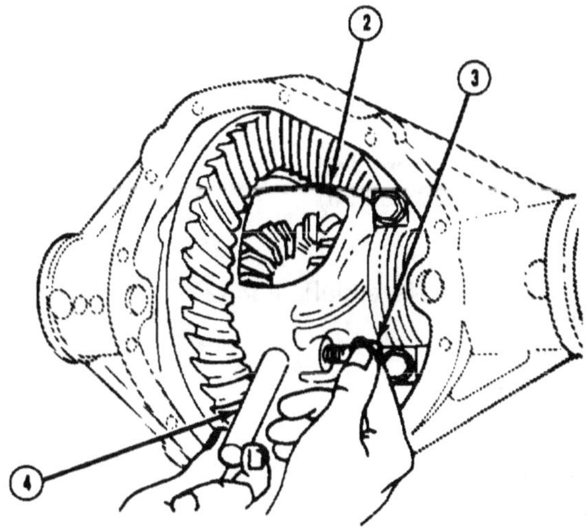

FOLLOW-ON TASKS:

- Connect emergency brake cable to backing plate if removed. (See TM 9-2320-289-20)
- Connect rear brake line to wheel cylinder if removed. (See TM 9-2320-289-20)
- Install brake shoes and attaching hardware if removed. (See TM 9-2320-289-20)
- Install brake drum. (See TM 9-2320-289-20)
- Fill rear axle differential with lubricating oil. (See LO 9-2320-289-12)

TA50578

6-15. REAR AXLE DIFFERENTIAL MAINTENANCE (ALL EXCEPT M1009).

This task covers:

a. Removal
b. Disassembly
c. Cleaning and Inspection
d. Assembly
e. Installation

INITIAL SETUP:

Equipment Condition

- Rear axle shafts removed. (See TM 9-2320-289-20)

Tools/Test Equipment

- *Wrench nut* (differential), J-24429
- Dial indicator
- Puller kit
- Spring tester
- Torque wrench

Manual References

- TM 9-214

Materials/Parts

- One differential cover gasket
- Sixteen lockwashers
- Dry cleaning solvent (Item 23, Appendix B)
- Prussian blue dye (Item 24, Appendix B)
- Gear lubricating oil (Item 44, Appendix B)
- Wiping rags (Item 53, Appendix B)

General Safety Instructions

- Dry cleaning solvent is flammable and must not be used near open flame. Use only in a well-ventilated area.
- Compressed air used for cleaning purposes should never exceed 30 psi (207 kPa).
- Tightly hold differential case assembly together to absorb spring pressure during removal and installation of ring gear bolts.

NOTE

When removing differential cover bolts, location of clips Should be noted.

1. Place a suitable container under housing. Remove 14 bolts at differential cover and move vent tube bracket, brake line and bracket, and proportioning valve lever out of the way. Remove differential cover and allow all lubricating oil to drain. Remove gasket, if present, and discard.

NOTE

Ensure that bearing caps (1) are marked for installation in same position.

2. Remove bolt (2), washer (3), and adjusting nut lock (7) from each bearing cap (1), Remove 4 bolts (5), lockwashers (6), and 2 bearing caps. Discard lockwashers.

3. Loosen 2 adjusting nuts (8). Remove differential case (9) assembly from housing. Remove 2 bearing cups (10) from differential case assembly side bearings (11) and tag for installation.

6-15. REAR AXLE DIFFERENTIAL MAINTENANCE (ALL EXCEPT M1009) (Con't).

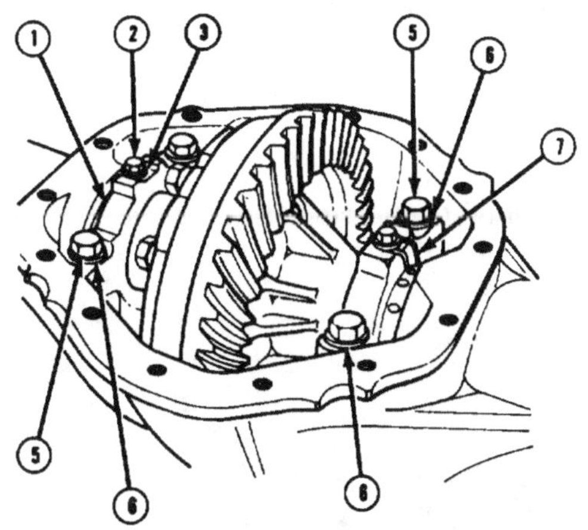

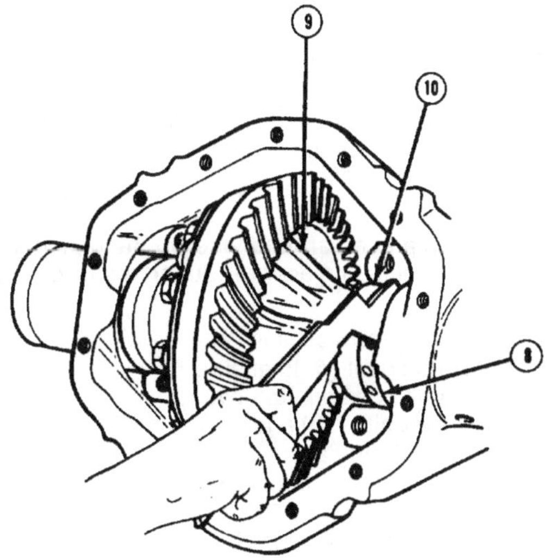

b. DISASSEMBLY

1. Place differential case (9) assembly in vise.

CAUTION

Ensure that puller fingers are installed Into notches of differential case (9) assembly so that only inner race Is contacted. Failure to follow this caution may result in damage to side bearings (11).

2. Tag side bearings (11) for assembly on their original side. Install bearing puller on side bearing and remove side bearing. Repeat for remaining side bearing,

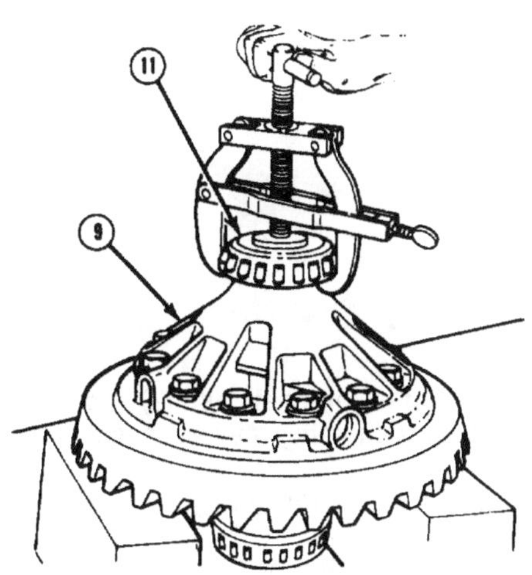

TA50579

6-15. REAR AXLE DIFFERENTIAL MAINTENANCE (ALL EXCEPT M1009) (Con't).

WARNING

Tightly hold differential case (9) assembly together to absorb spring pressure as last ring gear bolts (12) are removed. Failure to follow this warning may result in injury to personnel or damage to differential case assembly components.

NOTE

Shop rags placed around vise will help protect ring gear (4) teeth from damage should ring gear fall during disassembly.

3. Scribe a line on differential case (9) assembly and ring gear (4) to ensure assembly in same position. Remove 12 ring gear bolts (12) and lockwashers (13). Using a soft-faced hammer, tap ring gear loose from differential case assembly. Remove differential case assembly from vise and remove ring gear. Discard lockwashers.

4. Separate differential case (9) assembly halves.

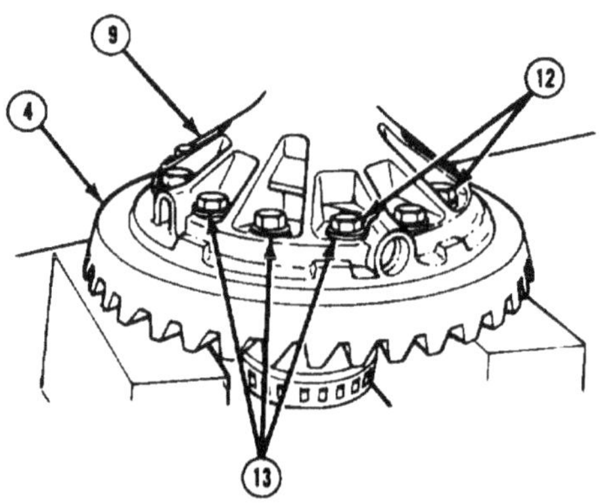

NOTE

Ensure that differential case (9) assembly internal components are kept separated so that they may be assembled in same position.

5. From 1 differential case (9) half remove side gear (14), retainer (15), spring (1 6), driven clutch (17), and spider assembly (18). From other differential case half remove driven clutch, spring, retainer, and side gear.

TA50580

6-15. REAR AXLE DIFFERENTIAL MAINTENANCE (ALL EXCEPT M1009) (Con't).

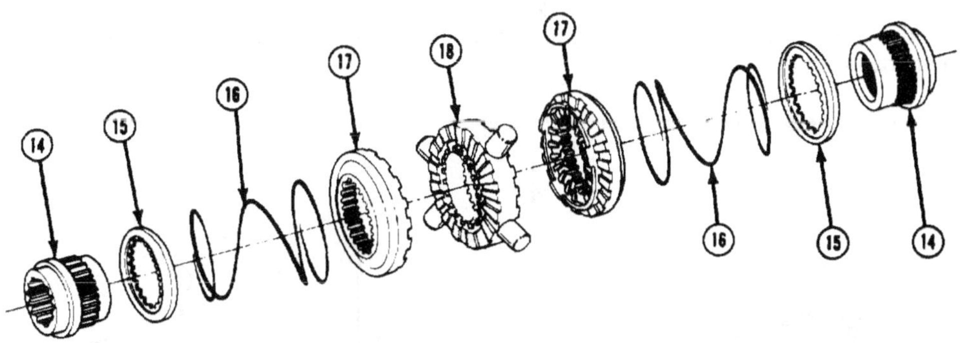

c. CLEANING AND INSPECTION

WARNING

- Dry cleaning solvent P-D-680 I s toxic and flammable. Always wear protective goggles and gloves and use only in a well-ventilated area. Avoid contact with skin, eyes, and clothes and DO NOT breathe vapors. DO NOT use near open flame or excessive heat, The solvent's flash point is 100°F-138°F (38°C-59°C). If you become dizzy while using cleaning solvent, Immediately get fresh air and medical help, If solvent contacts eyes, immediately wash your eyes with water and get medical aid.

Ž Compressed air used for cleaning purposes should never exceed 30 psi (207 kPa). Use only effective chip guarding and personnel protective equipment (goggles/shield, gloves, etc.). Failure to follow this warning may result in serious injury to personnel.

CAUTION

If major components (e.g. spider assembly or driven clutch assemblies) show excessive wear or damage, the entire no-spin unit should be replaced. If any part is replaced, mating parts should also be replaced as they are probably also damaged. Failure to follow this caution may result in a recurrence of original problem.

1. Clean all components in dry cleaning solvent and dry with compressed air.
2. Clean and inspect side bearings in accordance with TM 9-214,

TA50581

6-15. REAR AXLE DIFFERENTIAL MAINTENANCE (ALL EXCEPT M1009) (Con't).

3. Inspect splines on driven clutches and remove any burrs or nicks found. If driven clutch teeth are chipped, or if burrs and nicks cannot be repaired, replace driven clutch.

4. Inspect holdout ring (19) on driven clutches (17) for fractures or chipped teeth. Replace if fractured or chipped.

5. inspect teeth on spider (18) and center cam (20) for fractures or chipping. Inspect center cam to ensure that it moves freely in spider, Replace spider assembly if fractured, chipped or if center cam does not move freely.

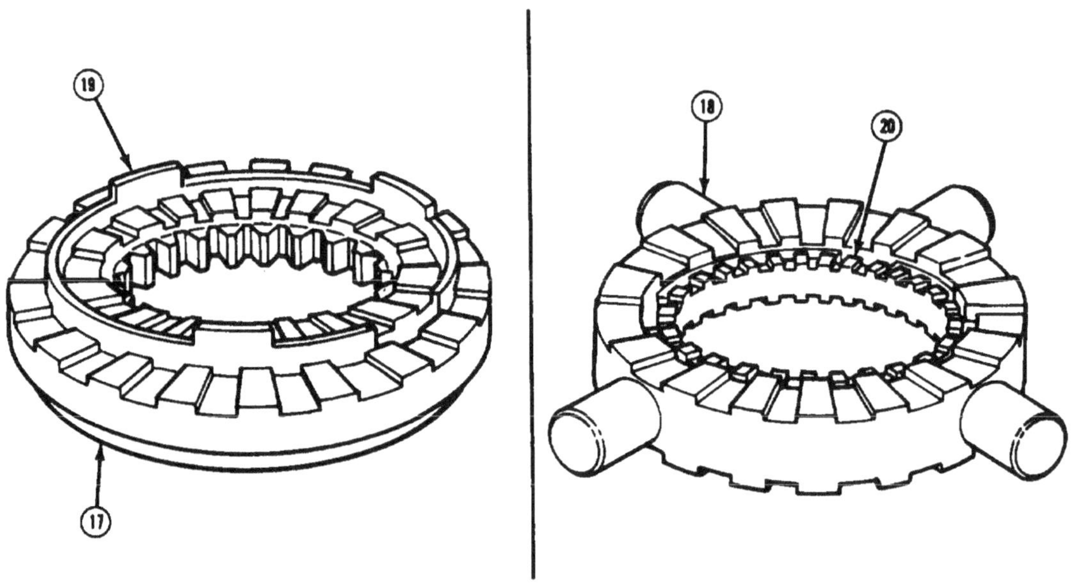

6. Inspect side gears for damage, Replace if damaged.

7. Inspect the fit of side gear to mating axle shaft, Replace side gear if it binds in axle shaft.

8. Using a spring tester, measure each spring load at operating height. When height of spring is 0.72 in. (18.29 mm), load should be 65 lb.±6 lb. (30 kg ±3 kg). Replace springs if measurement is not within specification.

9. Inspect differential case for damage. Remove any nicks or burrs found, Replace if damaged or if nicks or burrs cannot be removed.

NOTE

If ring gear is replaced, ring gear bolts and drive pinion must also be replaced. (See paragraph 6-17)

10. inspect ring gear and ring gear bolts for damage. Replace if damaged.

TA60582

6-15. REAR AXLE DIFFERENTIAL MAINTENANCE (ALL EXCEPT M1009) (Con't).

d. ASSEMBLY

NOTE

Ensure that all components are lubricated with lubricating oil as they are assembled.

1. Place ring gear (4) and flanged half of differential case (9) on wooden blocks with bearing end of hub facing downward.

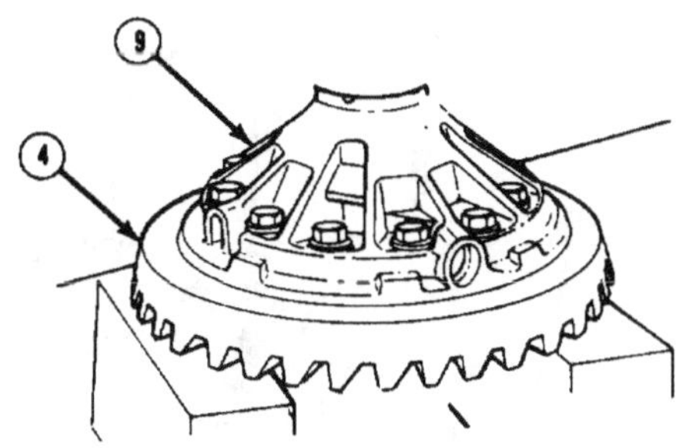

2. Place hub of side gear (14) into bore of differential case (9). Ensure that side gear will rotate freely. Install retainer (15) over side gear splines with retainer lip facing upwards and retainer seated against side gear shoulder.

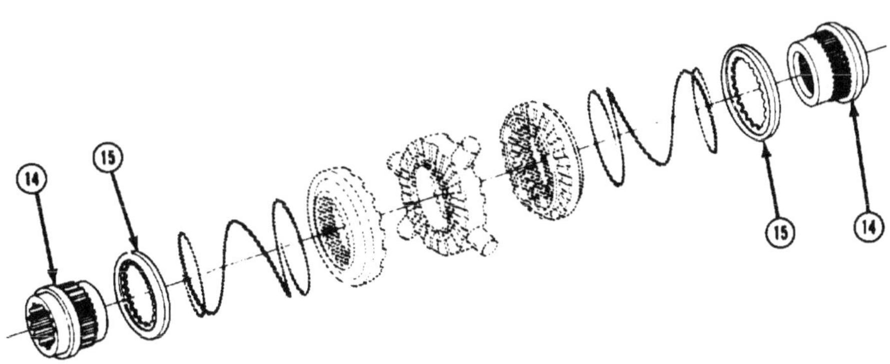

TA50583

6-15. REAR AXLE DIFFERENTIAL MAINTENANCE (ALL EXCEPT M1009) (Con't).

3. Install spring (16) over side gear (14) spline against retainer (15) lip with smaller end of spring against retainer.

4. Install driven clutch (17) over spring (16) with teeth facing upwards.

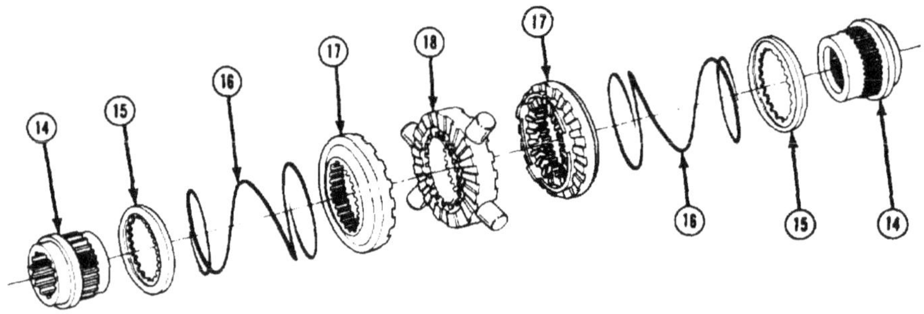

NOTE

Ensure that slot in driven clutch (17) holdout ring (19) engages long spider key (21).

5. Place spider (18) assembly over driven clutch (17),

6. Place remaining driven clutch (17) over spider (18) assembly.

TA50584

6-15. REAR AXLE DIFFERENTIAL MAINTENANCE (ALL EXCEPT M1009) (Con't).

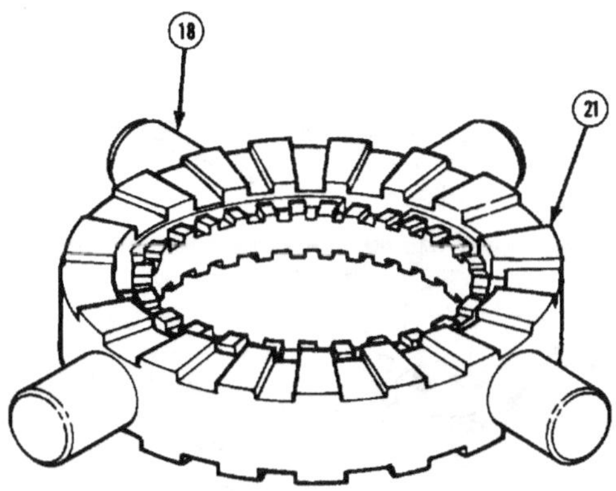

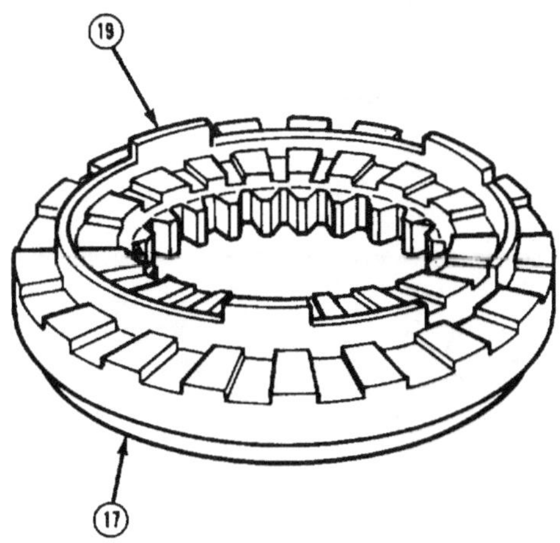

7. Assemble remaining retainer (15), spring (1 6), and side gear (14).

WARNING

Tightly hold differential case (9) assembly together to absor b spring pressure until 2 or more ring gear bolts (12) are installed. Failure to follow this warning may result in injury to personnel or damage t o differential case assembly components.

NOTE

Ensure that ring gear (4) and differential case (9) assembly halves are assembled as marked.

8 . Install remaining differential case (9) half over side gear (14) and compress springs (16). Install 12 new lockwashers (13) and ring gear bolts (12). Tighten bolts alternately to 120 lb.-ft. (163 N.m).

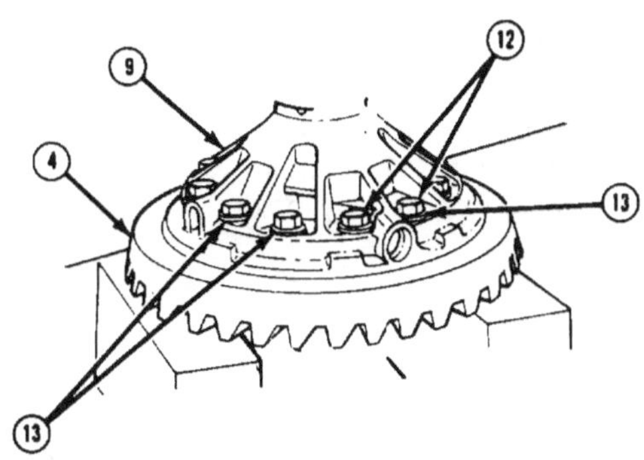

TA50585

6-15. REAR AXLE DIFFERENTIAL MAINTENANCE (ALL EXCEPT M1009) (Con't).

NOTE

Side bearings (11) should rotate freely once installed.

9. Install 2 side bearings (11) as tagged.

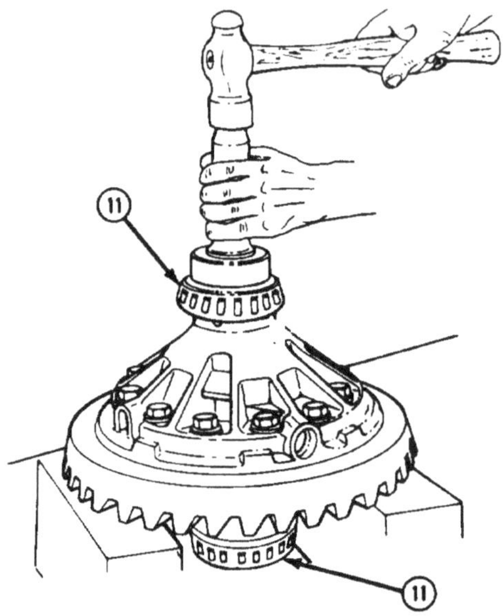

e. INSTALLATION

1. Install 2 bearing cups (10) as tagged to side bearings (11). Install differential case (9) assembly inside housing.

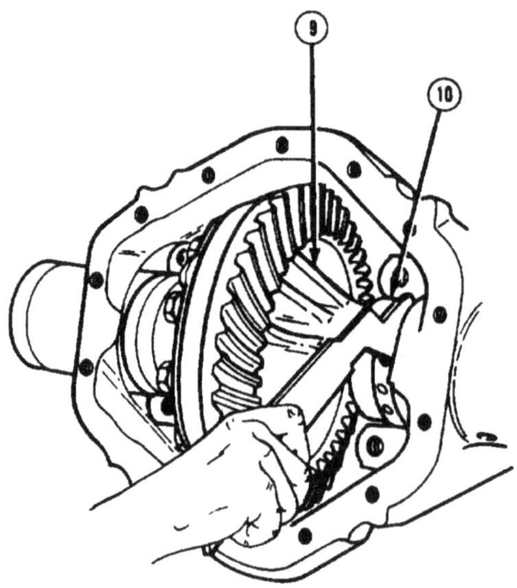

TA50586

6-15. REAR AXLE DIFFERENTIAL MAINTENANCE (ALL EXCEPT M1009) (Con't).

2. Install 2 bearing caps (1) in their original positions with 4 new lockwashers (6) and bolts (5). Tighten bolts just enough to hold bearing caps in place.

3. Loosen right side adjusting nut (8) and tighten left side adjusting nut until ring gear (4) contacts drive pinion gear without binding. This is zero ("O") lash.

4. Back off left side adjusting nut (8) approximately 2 slots. Install left side adjusting nut lock (7), washer (3), and bolt (2) to left side bearing cap (I). Tighten bolt to 20 lb.-ft. (27 N.m).

5. Tighten right side adjusting nut (8) until differential case (9) assembly is forced into solid contact with left side adjusting nut, Loosen right side adjusting nut until it is free from bearing contact then retighten until contact is re-established.

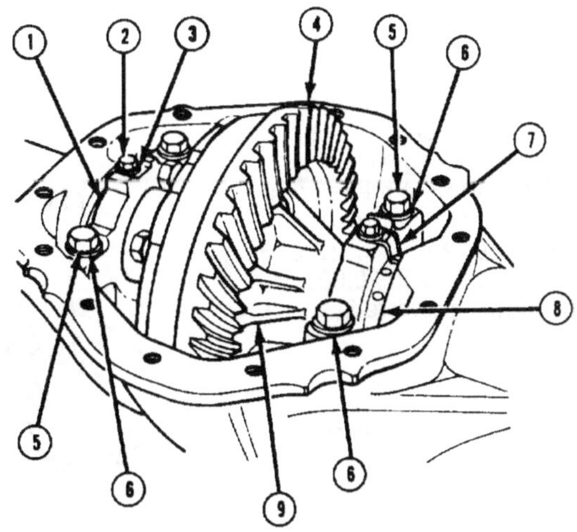

6. When in contact, tighten right side adjusting nut (8) 2 slots if side bearings (11) are reused or 3 slots if side bearings are new. Install adjusting nut lock (7), washer (3), and bolt (2) to right side bearing cap (1). Tighten bolt to 20 lb.-ft. (27 N.m).

7. Tighten 4 bolts (5) to 135 lb.-ft. (183 N.m).

TA50587

6-15. REAR AXLE DIFFERENTIAL MAINTENANCE (ALL EXCEPT M1009) (Con't).

NOTE

At this point, side bearings are preloaded. If any additional adjustments are required, ensure that preload remains as established. If 1 adjusting nut (8) is loosened, the other must be tightened an equal amount to maintain this preload.

8. Mount dial indicator (22) on housing and measure backlash between ring gear (4) and drive pinion gear. Rock ring gear back and forth to measure backlash. Backlash should be between 0.003-0.012 in. (0.076-0.305 mm) with 0.005-0.008 in. (0.1 27-0.203 mm) preferred,

NOTE

If adjusting nut (8) requires adjustment, bolts (2 and 5) must be loosened. Once final adjustments have been made, bolts must be retightened to proper torque.

9. If backlash is more than 0.012 in. (0.305 mm), loosen right adjusting nut (8) 1 slot and tighten left adjusting nut 1 slot. If backlash is less than 0.003 in. (0.076 mm), loosen left adjusting nut 1 slot and tighten right adjusting nut 1 slot. Remove dial indicator (22) from housing.

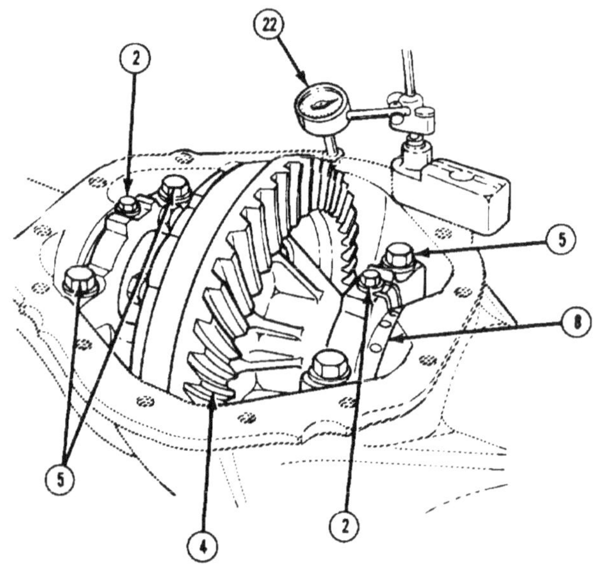

NOTE

If ring gear (4) and drive pinion were replaced, perform steps 10-13 to check for proper gear tooth contact pattern.

10. Wipe any lubricant from housing and clean each tooth of ring gear (4).

11. Using a medium stiff brush, sparingly apply prussian blue dye to all ring gear (4) teeth. Area of drive pinion tooth contact should be visible when drive pinion is engaged by hand.

NOTE

Excessive turning of ring gear (4) is not recommended. Ring gear should be turned only enough to provide tooth contact pattern,

12. Apply load until it takes a torque of 40-50 lb.-ft, (54-68 N.m) to turn drive pinion. Turn companion flange so that ring gear (4) turns 1 full revolution, then reverse so that ring gear turns 1 full revolution in opposite direction.

TA50588

6-15. REAR AXLE DIFFERENTIAL MAINTENANCE (ALL EXCEPT M1009) (Con't),

13. Contact pattern should be centrally located up and down on the face of the ring gear (4) teeth. Compare tooth contact pattern on ring gear with examples shown and make adjustments as indicated.

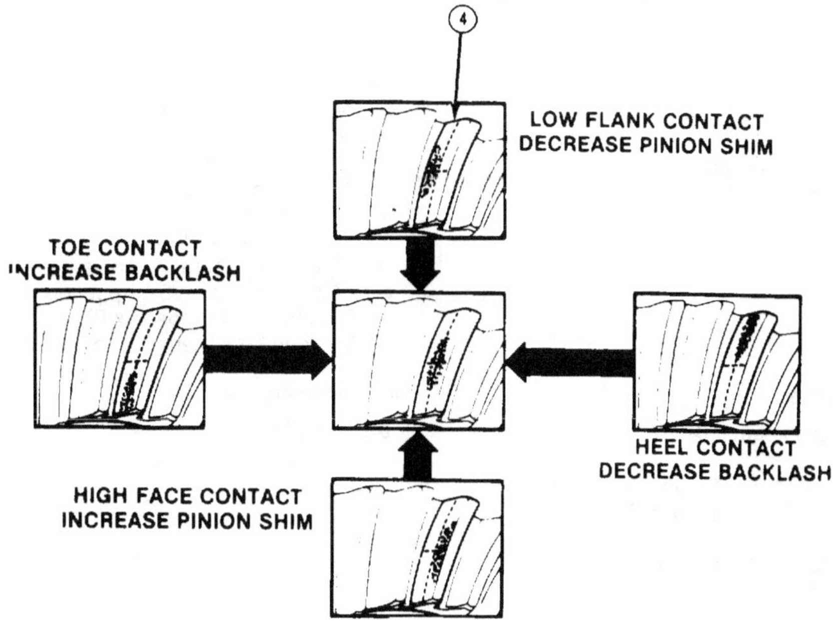

NOTE

Ensure that differential cover and magnet are clean.

14. Carefully position vent tube bracket, brake line and bracket, and proportioning valve lever at housing. DO NOT kink lines. Install new differential cover gasket and differential cover with 14 bolts. Tighten bolts to 35 lb.-ft. (47 N.m).

FOLLOW-ON TASKS:

- Install rear axle shafts. (See TM 9-2320-289-20)
- Fill rear axle differential with lubricating oil. (See LO 9-2320-289-12)

TA50589

6-15.1. REAR AXLE DIFFERENTIAL MAINTENANCE (M10028A2).

This task covers:

a. Removal
b. Disassembly
c. Cleaning and Inspection
d. Assembly
e. Installation

INITIAL SETUP:

Equipment Condition

- Rear axle assembly removed. (See paragraph 6-1 3)
- Rear axle shafts removed. (See TM 9-2320-289-20)

Materials/Parts

- Two preformed packings
- Dry cleaning solvent (Item 23, Appendix B)
- Prussian blue dye (Item 24, Appendix B)
- Gear lubricating oil (Item 44, Appendix B)
- Wiping rags (Item 53, Appendix B)
- RTV sealant (Item 57, Appendix B)
- Tags (Item 64, Appendix B)
- Masking tape (Item 68, Appendix B)

Tools/Test Equipment

- Axle housing spreader, D-167
- Dial indicator
- Four-piece puller adaptor set, DD-914-95 (use with DD-914P and D-914-8)
- Puller DD-914P
- Puller adapter, D-914-8
- Torque wrench

Manual References

- TM 9-214

General Safety Instructions

- Dry cleaning solvent is flammable and must not be used near open flame. Use only in a well-ventilated area.
- Compressed air used for cleaning purposes should never exceed 30 psi (207 kPa).

a. REMOVAL

1. Mount axle on axle stand.

2. Place a suitable container under axle housing. Remove 10 bolts (1) and differential cover (2), and allow all lubricating oil to drain. Remove old RTV sealant from differential cover and axle housing.

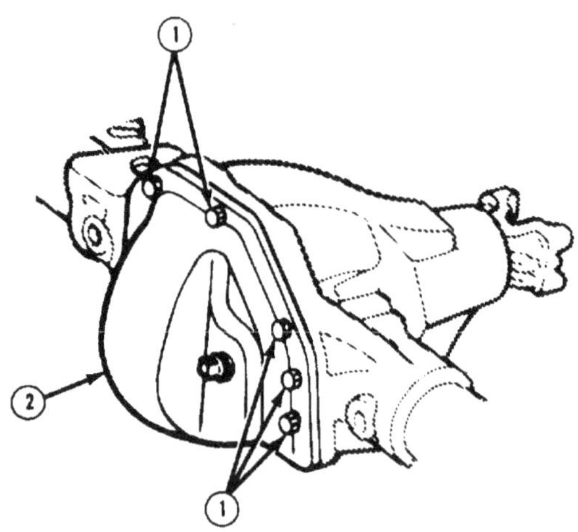

6-15.1. REAR AXLE DIFFERENTIAL MAINTENANCE (M1028A2) (Con't).

NOTE

Side bearing caps (3) are stamped with letters which match letters stamped on case. Note letters for installation alinement, If side bearing caps are not stamped, mark side bearings caps and axle housing before removal.

3. Remove 4 bolts (4) and 2 side bearing caps (3).

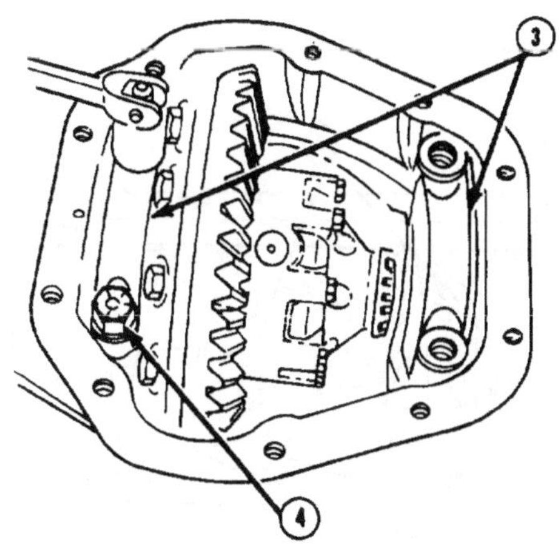

4. Position axle housing spreader (5) with tool dowels fitted into recesses in axle housing. Install dial indicator (6) from edge of axle housing to opposite edge of axle housing. Ensure that dial indicator is preloaded. Set dial indicator to "0, "

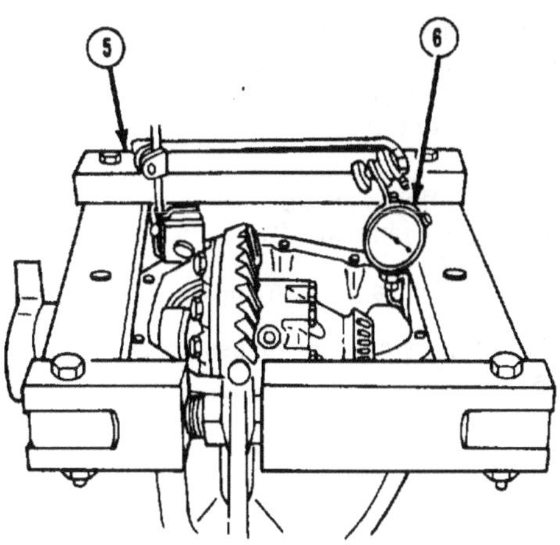

CAUTION

Axle housing should not be spread more than 0.015 in. (0.381 mm), Failure to follow this caution may cause damage and permanent distortion to axle housing.

5. Spread axle housing to 0.015 in. (0.381 mm). Remove dial indicator (6).

6-15.1. REAR AXLE DIFFERENTIAL MAINTENANCE (M1028A2) (Con't).

NOTE

Relieve tension on axle housing spreader and remove from axle housing as soon as possible.

6. Remove differential case (8) assembly from axle housing. Remove and tag 2 side bearing cups (7) for installation.

7. Remove 2 preformed packings (9) from axle housing. Discard preformed packings.

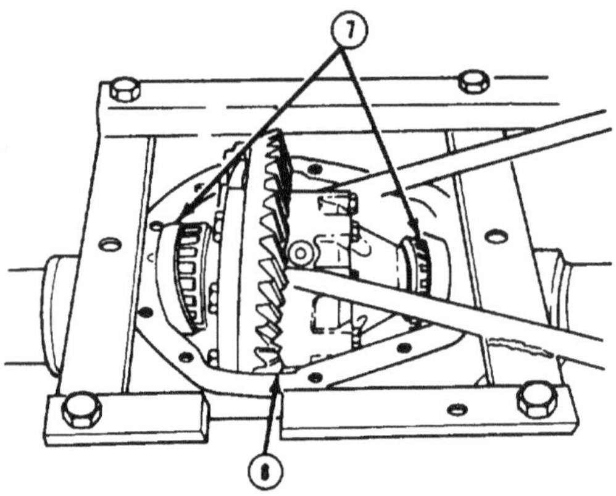

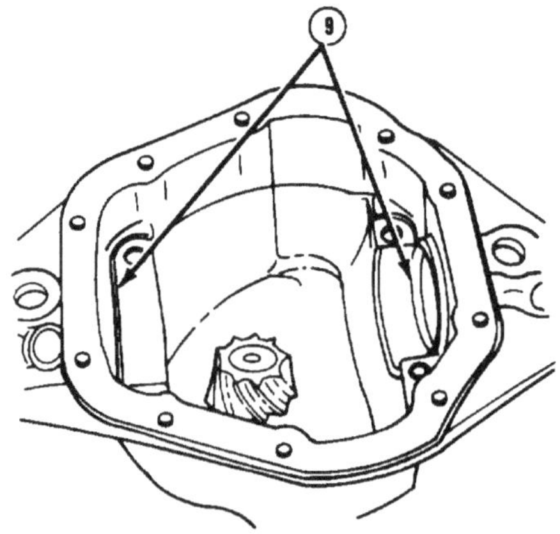

b. DISASSEMBLY

NOTE

DO NOT remove side bearings or ring gear unless damaged. If side bearings and ring gear are not to be replaced, go to step 6.

1. Inspect differential side bearings in accordance with TM 9-214.

2. Inspect ring gear for damage.

3. Place differential case (8) assembly in" vise.

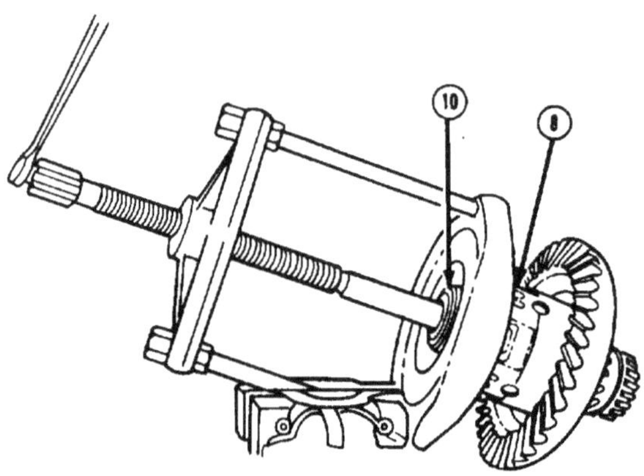

6-15.1. REAR AXLE DIFFERENTIAL MAINTENANCE (M028A2) (Con't).

CAUTION

Ensure that puller adapters are installed so that only inner race is contacted. Failure to follow this caution may result in damage to side bearings (10).

4. Tag side bearings (10) for assembly on original side. Install bearing puller, adapter, and four-piece adapter set on side bearing and remove side bearing and shims (11). Repeat for remaining side bearing and shims.

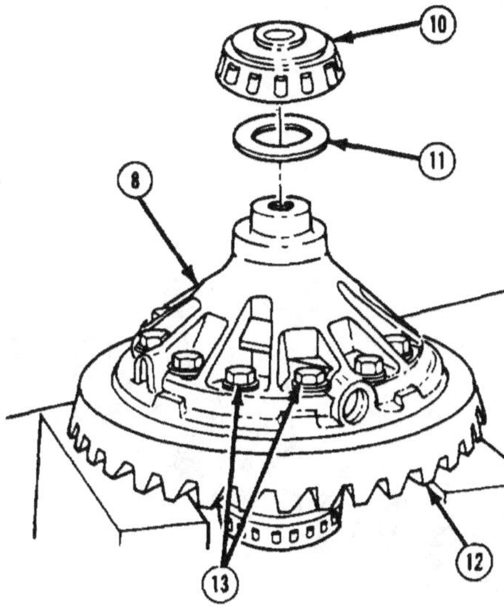

NOTE

- If ring gear (12) is to be replaced, drive pinion must also be replaced. (See paragraph 6-17.1)
- Shop rags placed around vise will help protect ring gear teeth from damage should ring gear fall during disassembly.

5. Scribe a line on differential case (8) assembly and ring gear (12) to ensure assembly in same position. Remove 12 ring gear bolts (13). Using soft-faced hammer, tap ring gear loose from differential case assembly. Remove differential case assembly from vise and remove ring gear.

6-15.1. REAR AXLE DIFFERENTIAL MAINTENANCE (M1028A2) (Con't).

WARNING

Tightly hold differential case (8) assembly together to absorb spring pressure as last differential case bolts (19) are removed. Failure to follow this warning may result in injury to personnel or damage to differential case assembly components.

6. Scribe a line on differential case (8) assembly halves. Remove 8 differential case bolts (19) and separate differential case assembly halves.

7. From 1 differential case (8) half, remove side gear (14), retainer (15), spring (1 6), driven clutch (17), and spider assembly (18). From other differential case half, remove driven clutch, spring, retainer, and side gear.

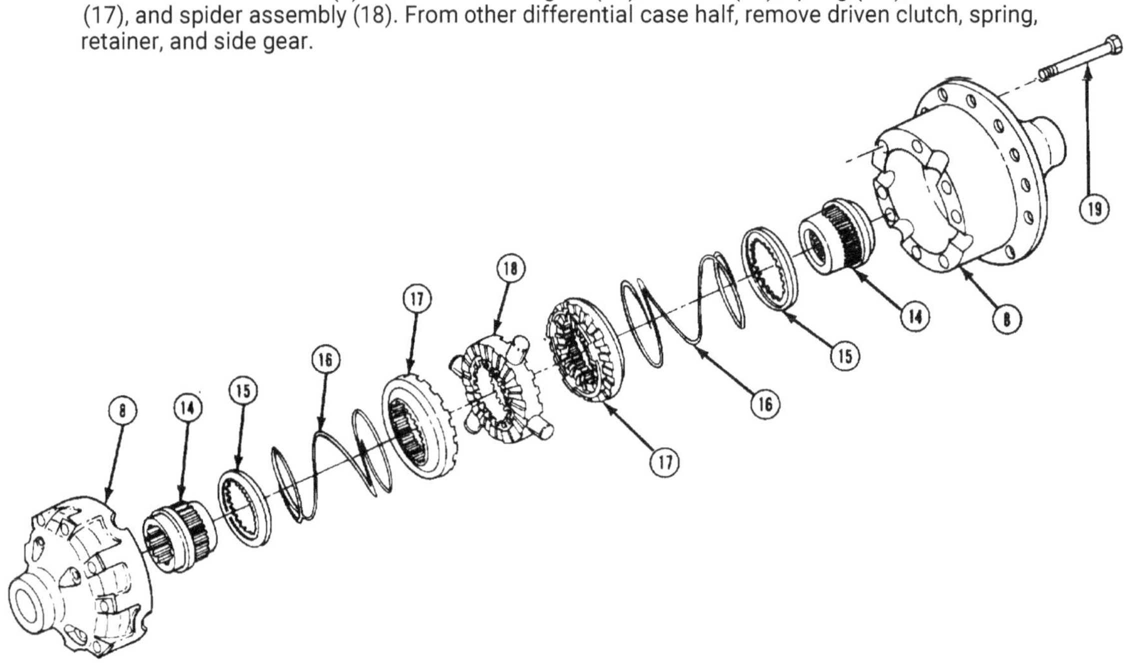

c. CLEANING AND INSPECTION

WARNING

- Dry cleaning solvent P-D-680 is toxic and flammable. Always wear protective goggles and gloves and use only in a well-ventilated area. Avoid contact with skin, eyes, and clothes and DO NOT breathe vapors. DO NOT use near open flame or excessive heat. The solvent's flash point is 100°F-138° F (38°C-59°C). If you become dizzy while using cleaning solvent, immediately get fresh air and medical help. If solvent contacts eyes, immediately wash your eyes with water and get medical aid.
- Compressed air used for cleaning purposes should never exceed 30 psi (207 kPa). Use only effective chip guarding and personnel protective equipment (goggles/shield, gloves, etc.). Failure to follow this warning may result in serious injury to personnel.

6-15.1. REAR AXLE DIFFERENTIAL MAINTENANCE (M1028A2) (Con't).

CAUTION

- If major components (e.g. spider assembly or driven clutch assemblies) show excessive wear or damage, differential assembly should be replaced. Failure to follow this caution may result in a recurrence of original problem,
- If any part is replaced, mating parts should also be replaced as they are probably also damaged. Failure to follow this caution may result in a recurrence of original problem.

1. Clean all components in dry cleaning solvent and dry with compressed air.
2. Inspect splines on driven clutches and remove any burrs or nicks found. If driven clutch teeth are chipped, or if burrs and nicks cannot be repaired, replace driven clutch.
3. Inspect holdout ring (20) on driven clutches (17) for fractures or chipped teeth. Replace driven clutches if fractured or chipped.
4. Inspect teeth on spider (18) and center cam (21) for fractures or chipping. Inspect center cam to ensure that it moves freely in spider. Replace spider assembly if fractured, chipped or if center cam does not move freely.

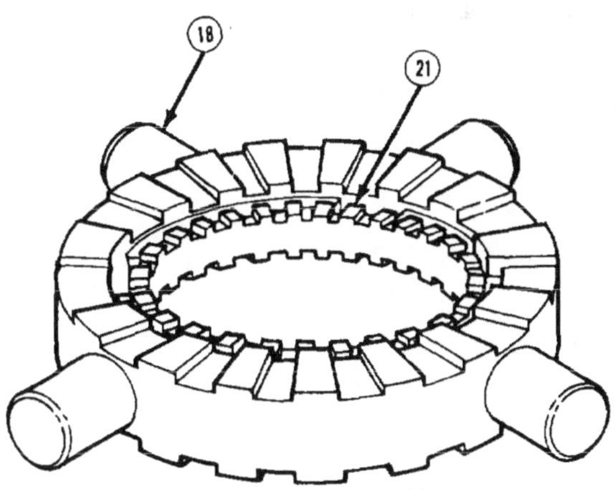

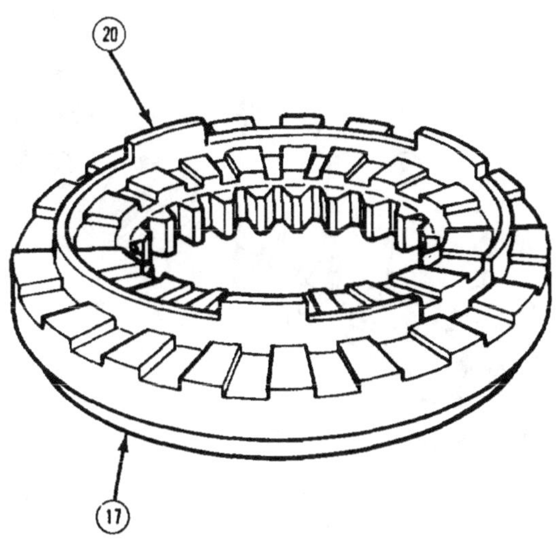

5. Inspect side gears for damage. Replace if necessary.
6. Visually inspect springs for equal height and damage, Replace if springs are of unequal height or damage exists.
7. Inspect differential case halves for damage. Remove any nicks or burrs found. Replace if damaged, or if nicks or burrs cannot be removed.

6-15.1. REAR AXLE DIFFERENTIAL MAINTENANCE (M1028A2) (Con't).

d. ASSEMBLY

NOTE

Ensure that all components are lubricated with gear oil as they are assembled.

1. Place flanged half of differential case (8) on wooden blocks with bearing end of hub facing downward.
2. Place hub of side gear (14) into bore of differential case (8) half. Ensure that side gear will rotate freely. Install retainer (15) over side gear splines with retainer lip facing upward and retainer seated against side gear shoulder.
3. Install spring (16) over side gear (14) spline against retainer (15) lip with smaller end of spring

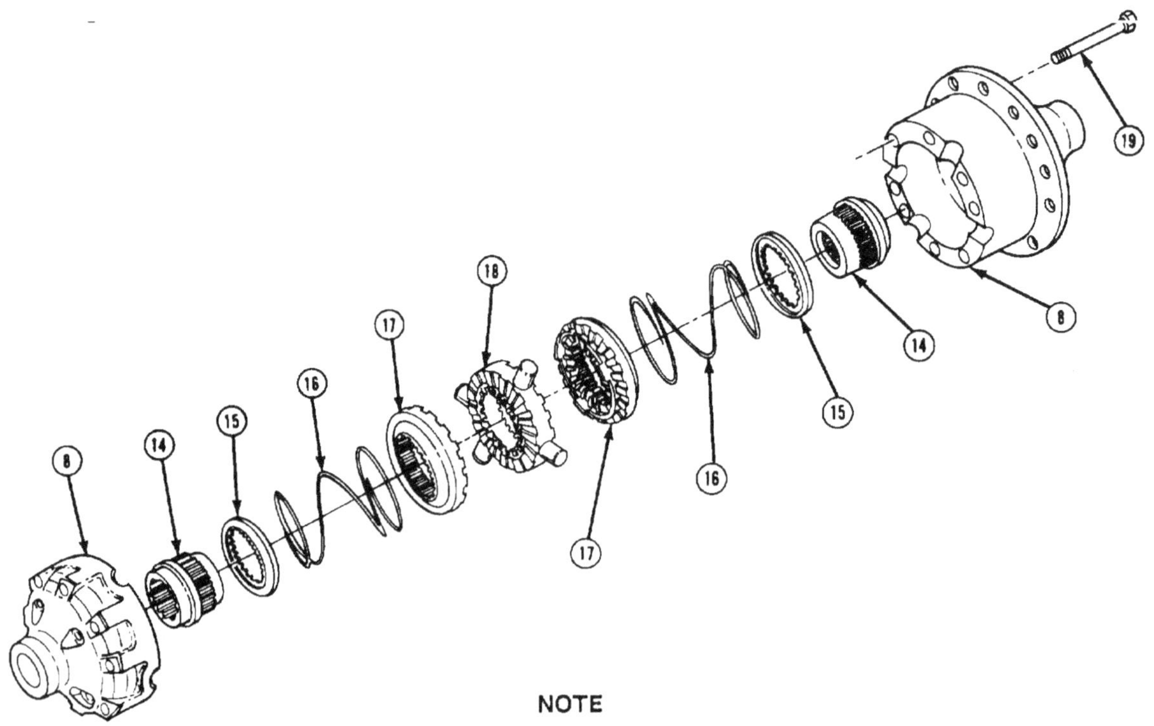

NOTE

Ensure that slot in each driven clutch (17) holdout ring (20) engages long spider key (22) on each side of spider.

4. Install driven clutch (17) on each side of spider (18).
5. While holding driven clutches (17) and spider (18) together, place them over spring (16).
6. Assemble remaining spring (16), retainer (15), and side gear (14).

6-15.1. REAR AXLE DIFFERENTIAL MAINTENANCE (M1028A2) (Con't).

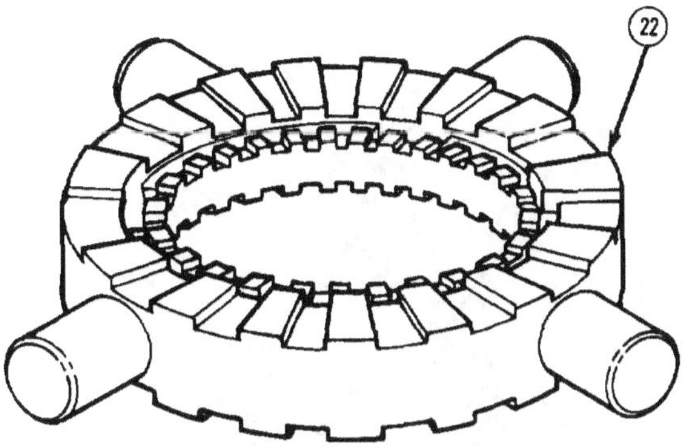

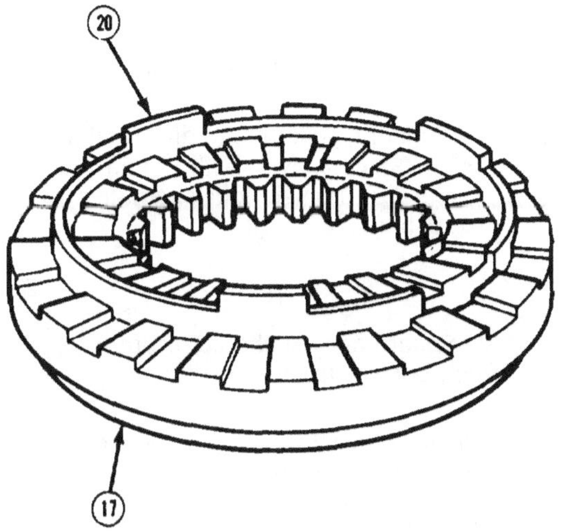

WARNING

Tightly hold differential case (8) assembly together to absorb spring pressure until 2 or more differential case bolts (19) are installed. Failure to follow this warning may result in injury to personnel or damage to differential case assembly components.

NOTE

- Ensure that differential case (8) is assembled as marked.
- An assistant may help with step 7, if required.

7. Install remaining differential case (8) half over side gear (14) and compress springs (16). Install 8 differential case bolts (19) into differential case halves. Tighten bolts alternately to 70 lb,-ft. (95 N.m).

6-15.1. REAR AXLE DIFFERENTIAL MAINTENANCE (M1028A2) (Con't).

NOTE

- If ring gear (12) and side bearings (10) were not removed, go to subparagraph e.
- Ensure that ring gear (12) is assembled as marked.

8. Install ring gear (12) to differential case (8), if removed, with 12 ring gear bolts (13). Tighten bolts alternately to 120 lb.-ft. (163 N.m).

NOTE

Side bearings (10) should rotate freely once installed.

9. Install original shims (11) under each side bearing (10) as tagged, using arbor press.

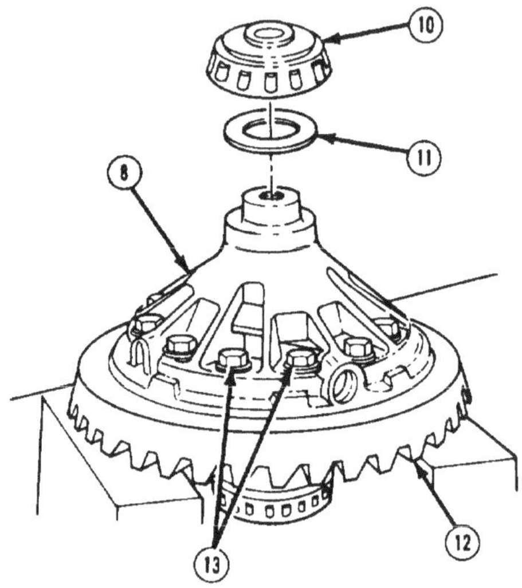

e. INSTALLATION

1. Install 2 new preformed packings (9) inside axle housing side bores.

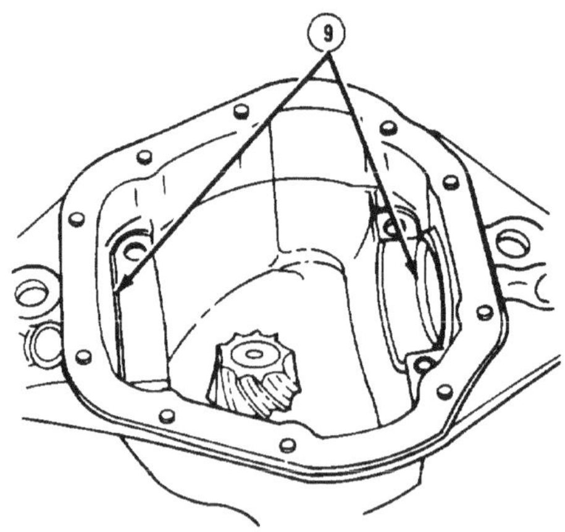

6-15.1. REAR AXLE DIFFERENTIAL MAINTENANCE (M1028A2) (Con't).

2. Position axle housing spreader (5) with tool dowels fitted into recesses in axle housing, Install dial indicator (6) from edge of axle housing to opposite edge of axle housing, Ensure that dial indicator is preloaded. Set dia l indicator to "O."

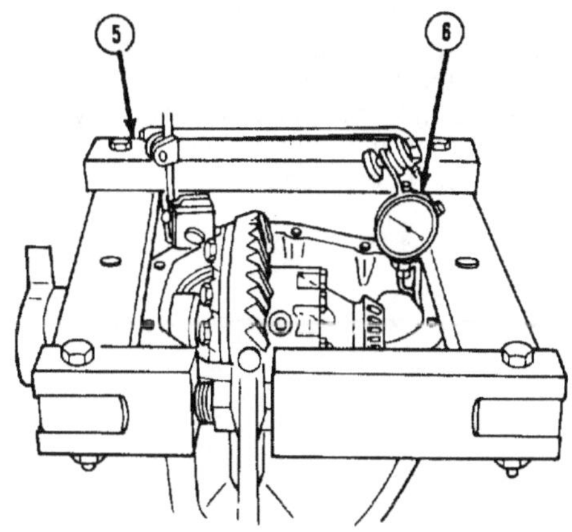

CAUTION

Axle housing should not be spread more than 0.015 in. (0.381 mm). Failure to follow this caution may cause axle housing to become damaged and permanently distorted.

3. Spread axle housing to 0.015 in. (0.381 mm). Install 2 bearing cups (7) as tagged and install differential case (8) assembly inside axle housing.

4. Rotate differential case (8) assembly and tap it with a soft-faced hammer to ensure that differential case assembly is properly seated in axle housing, Install 2 bearing caps (3) in their original position and tighten 4 bearing cap bolts (4) finger tight. Remove axle housing spreader (5) and tighten bearing cap bolts to 85 lb.-ft. (115 N.m).

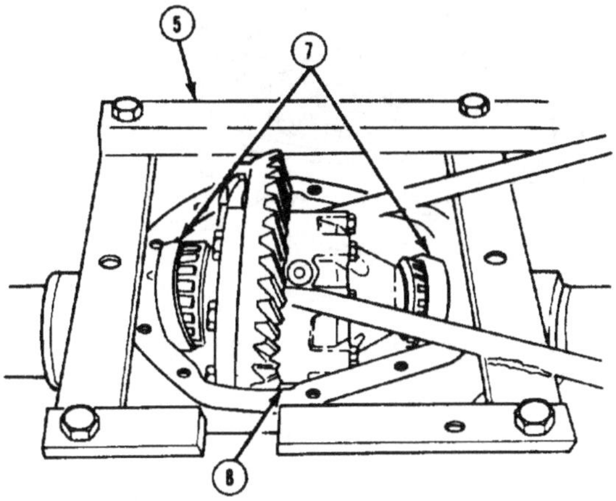

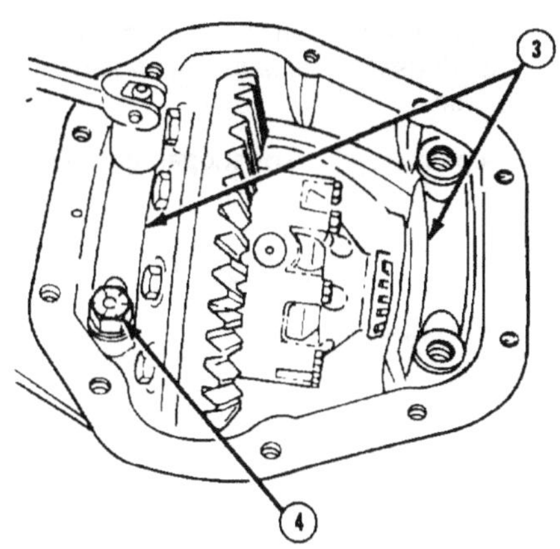

6-15.1. REAR AXLE DIFFERENTIAL MAINTENANCE (M1028A2) (Con't).

5. Install dial indicator (6) and check ring gear (12) backlash at 4 equally spaced points around ring gear. Backlash must be 0.004-0.009 in. (0.102-0.229 mm) and must not vary more than 0.002 in. (0.051 mm) between 4 points checked.

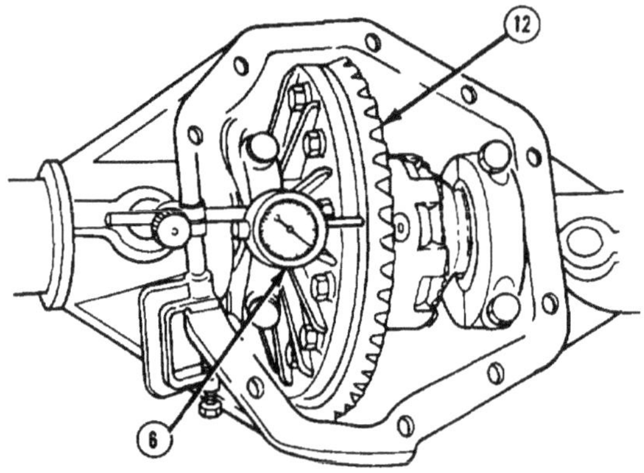

NOTE

If backlash is correct, go to step 10.

6. Repeat steps 3-6 of subparagraph a.
7. If backlash is not within specification, correct by adjusting thickness of side bearing (10) shim pack. To increase backlash, increase right shim (11) and decrease left shim an equal amount. To decrease backlash, decrease right shim and increase left shim an equal amount.
8. Install new shim pack and side bearings (10), as tagged using arbor press.
9. Repeat steps 1-5 of subparagraph e.

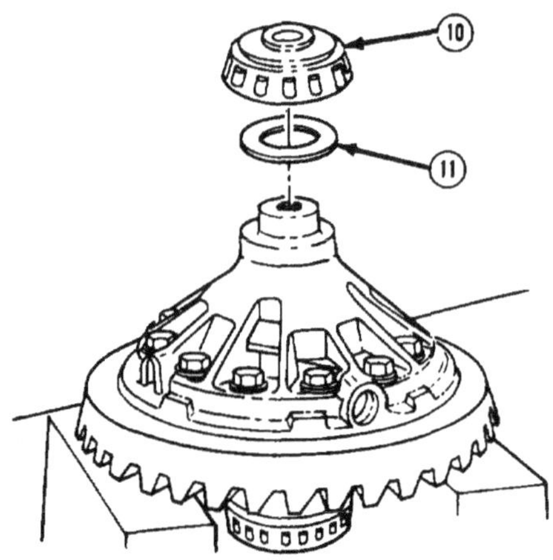

6-15.1. REAR AXLE DIFFERENTIAL MAINTENANCE (M1028A2) (Con't).

NOTE

If ring gear (12) and drive pinion were replaced, perform steps 10-13 to check for proper gear tooth contact pattern.

10. Wipe any lubricant from axle housing and clean each tooth of ring gear (12).

11. Using medium stiff brush, sparingly apply prussian blue dye to all ring gear (12) teeth. Area of drive pinion tooth contact should be visible when drive pinion is turned by hand.

NOTE

Excessive turning of ring gear (12) is not recommended. Ring gear should be turned only enough to provide tooth contact pattern.

12. Apply load until it takes a torque of 40-50 lb.-ft, (54-68 N.m) to turn drive pinion. Turn companion flange so that ring gear (12) turns 1 full revolution, then reverse so that ring gear turns 1 full revolution in opposite direction.

13. Contact pattern should be centrally located up and down on face of ring gear (12) teeth. Compare tooth contact pattern on ring gear with examples shown and make adjustments as indicated. (See paragraph 6-17.1, subparagraph c, steps 1-3)

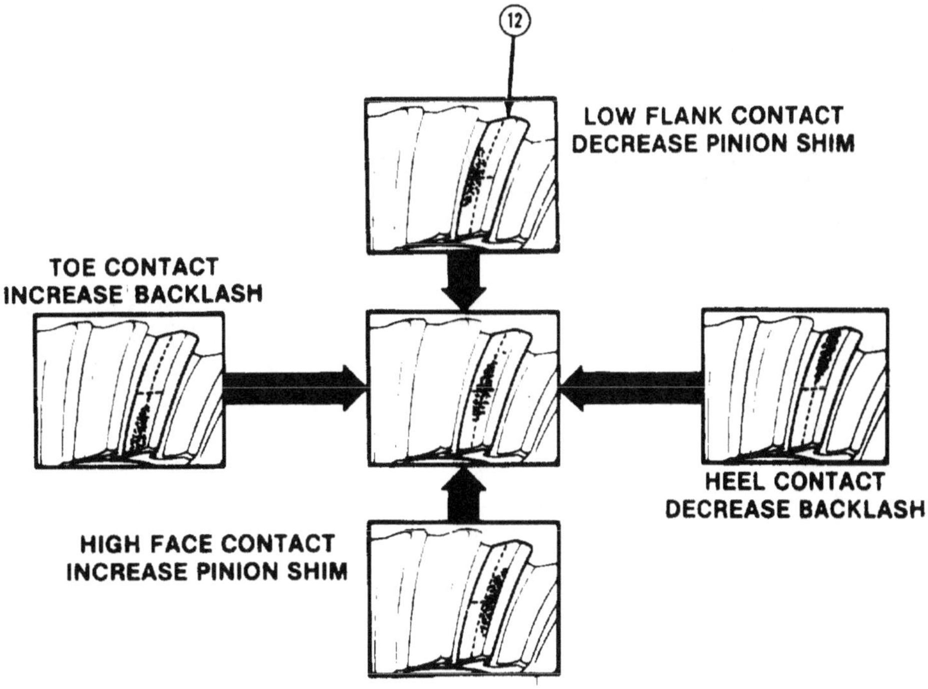

6-15.1. REAR AXLE DIFFERENTIAL MAINTENANCE (M1028A2) (Con't).

NOTE

- Ensure that axle housing is cleaned of all old RTV sealant.
- Differential cover (2) repair kit comes with gasket for different model applications. M1028A2 axle uses RTV sealant.
- When applying RTV sealant to axle housing, ensure that no RTV sealant enters bolt holes.

14. Apply bead of RTV sealant to axle housing and install differential cover (2) with 10 bolts (1). Tighten bolts to 35 lb.-ft. (47 N.m).

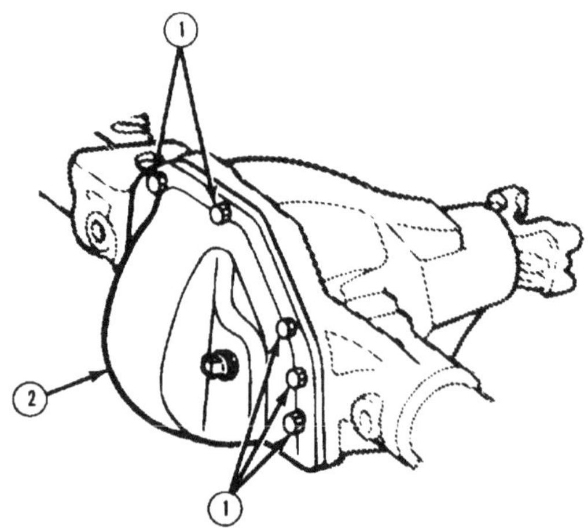

FOLLOW-ON TASKS:

- Install rear axle shafts. (See TM 9-2320-289-20)
- Fill rear axle differential with lubricating oil. (See LO 9-2320-289-12)
- Install rear axle assembly. (See paragraph 6-13)

6-16. REAR AXLE DIFFERENTIAL REPAIR (M1009).

This task covers:
a. Disassembly
b. Cleaning and Inspection
c. Assembly

INITIAL SETUP:

Equipment Condition

- Rear axle differential removed. (See paragraph 6-5)

Materials/Parts

- One latching bracket snapring
- One lockscrew
- Dry cleaning solvent (Item 23, Appendix B)
- Gear lubricating oil (Item 44, Appendix B)
- Wiping rags (Item 53, Appendix B)

Too/s/Test Equipment

- Bushing remover, J-26252
- Dial indicator
- Micrometer
- Puller kit
- Telescoping gage
- Torque wrench

Manual References

- TM 9-214

General Safety Instructions

- Dry cleaning solvent is flammable and must not be used near open flame. Use only in a well-ventilated area.
- Compressed air used for cleaning purposes should never exceed 30 psi (207 kPa).

a. DISASSEMBLY

NOTE

If disassembling latching bracket (6) and governor assembly (7), perform steps 2 and 3.

1. Remove 2 side bearings (1) from differential case (2) assembly and tag for installation,

NOTE

- Shop rags placed around vise will help protect ring gear (3) teeth from damage should ring gear fall during disassembly.
- Ring gear bolts (8) have left-hand threads.

2 . Place differential case (2) assembly in vise. Mark position of ring gear (3) in relation to differential case assembly for assembly, Remove 10 bolts (8). Using a soft-faced hammer, tap on ring gear to free from differential case assembly. Remove differential case assembly from vise and remove ring gear.

6-16. REAR AXLE DIFFERENTIAL REPAIR (MI 009) (Con't).

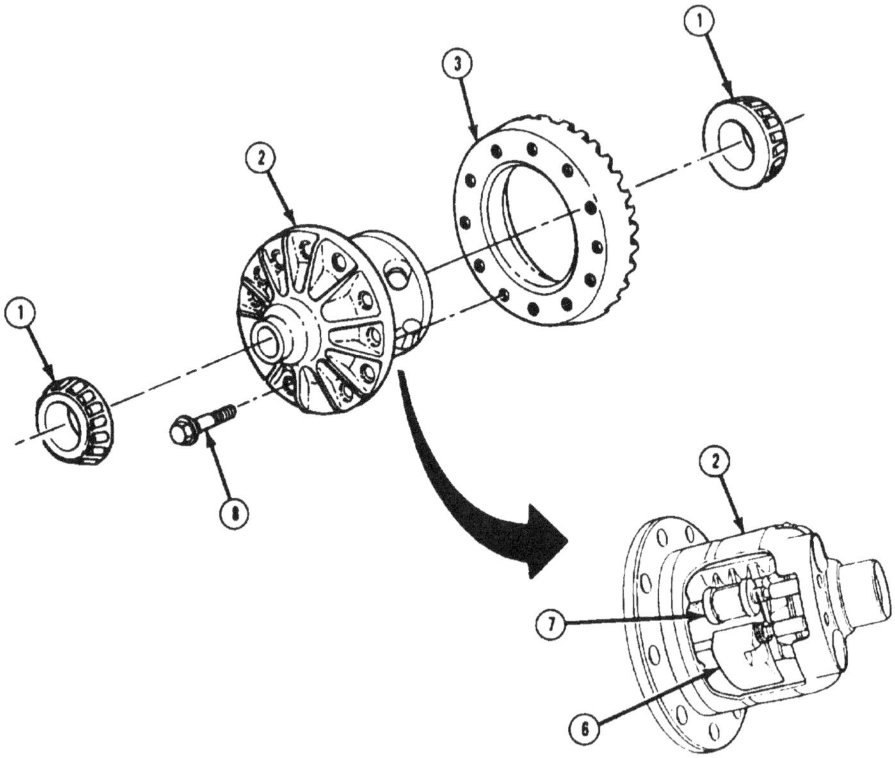

TA50590

6-16. REAR AXLE DIFFERENTIAL REPAIR (M1009) (Con't),

CAUTION

Ensure that latching bracket (6) spring is pulled out of the way when removing bushing (4), Failure to follow this caution may result in damage to governor assembly (7), bushing, or latching bracket spring.

3. Using bushing remover, remove bushings (4 and 5). Remove snapring and spring from latching bracket (6), and remove latching bracket and governor assembly (7). Discard snapring.

NOTE

Axle shaft serves as a tool to help aline pinion gears (18), camform side gear (17), and side gear (20).

4. Place axle shaft in vise with 3 in. (76 mm) of axle shaft projecting upwards. Place differential case (2) assembly on axle shaft.

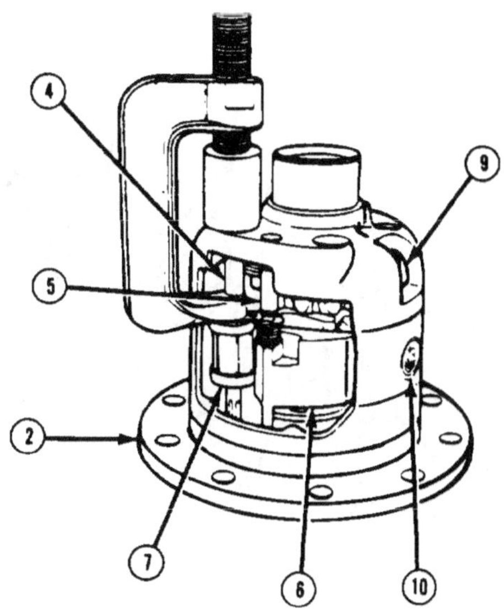

NOTE

Ensure that all thrust washers (19), shims (11 and 21), pinion gears (18), clutch packs (13 and 15), and thrust block (23) are marked for assembly in their original position.

5. Remove lockscrew (9) and pinion shaft (10). Rotate differential case (2) assembly, and remove 2 pinion gears (18) and thrust washers (19). DO NOT discard lockscrew. Save for measuring purposes during assembly.

6. Remove thrust block (23). Remove right-hand side gear (20), clutch pack (13 and 15), and shim(s) (21). Remove left-hand camform side gear (17), cam plate (16), clutch pack (13 and 15), wave spring (14), and shim(s) (11),

TA50591

6-16. REAR AXLE DIFFERENTIAL REPAIR (M1009) (Con't).

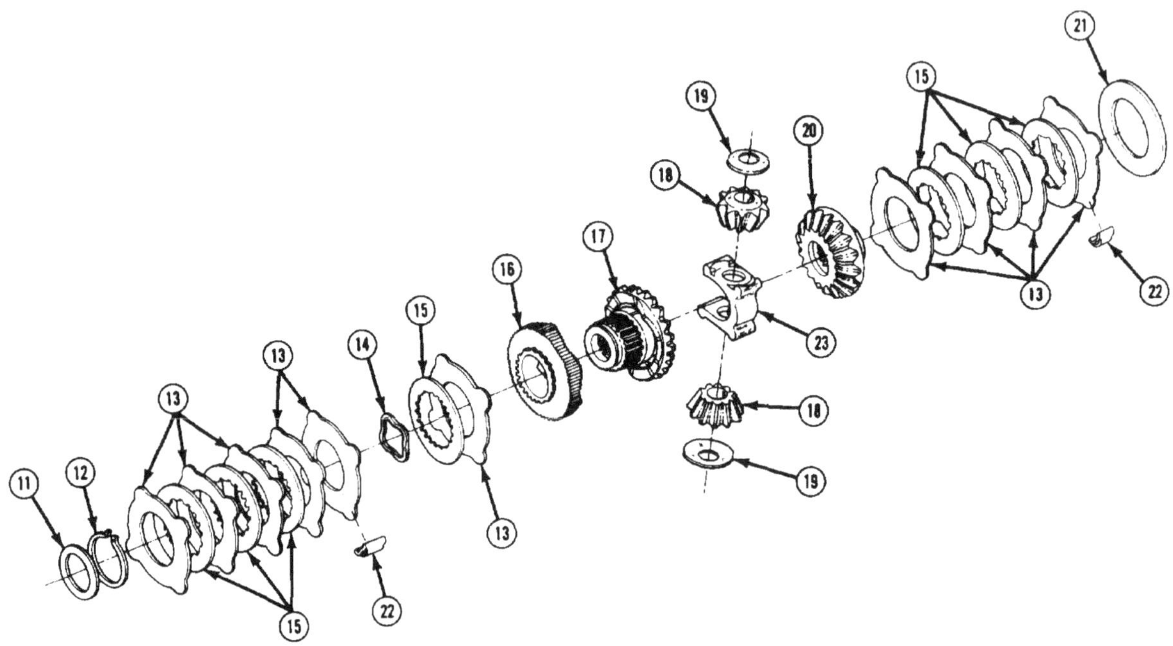

NOTE

Ensure that position of clutch plates (13) and clutch discs (15) is noted for assembly.

7 . Remove retainer ring (12) from left-hand clutch pack, Remove 4 clips (22) and remove 3 alternating clutch plates (13) and clutch discs (15). Remove wave spring (14). Remove 2 consecutive clutch plates, another clutch disc, and last clutch plate. Separate cam plate (16) and camform side gear (17).

8 . Remove 4 clips (22) and disassemble right-hand clutch pack which has 4 alternating clutch plates (13) and 3 clutch discs (1 5).

9 . Remove differential case (2) from axle shaft.

TA50592

6-16. REAR AXLE DIFFERENTIAL REPAIR (M1009) (Con't).

b. CLEANING AND INSPECTION

WARNING

- Dry cleaning solvent P-D-680 is toxic and flammable. Always wear protective goggles and gloves end use only in a well-ventilated area. Avoid contact with skin, eyes, and clothes and DO NOT breathe vapors. DO NOT use near open flame or excessive heat, The solvent's flash point is 100°F-138°F (38°C-59°C). If you become dizzy while using cleaning solvent, immediately get fresh air and medical help. If solvent contacts eyes, immediately wash your eyes with water and get medical aid.
- Compressed air used for cleaning purposes should never exceed 30 psi (207 kPa). Use only effective chip guarding and personnel protective equipment (goggles/shield, gloves, etc.), Failure to follow this warning may result in serious injury to personnel.

1. Clean and inspect 2 side bearings in accordance with TM 9-214.
2. Clean all disassembled components with dry cleaning solvent and dry with compressed air.
3. Inspect differential case for damage. If differential case is damaged, entire differential case must be replaced,

NOTE

- Internal clearances are such that shims, thrust block, or gears should not be replaced unnecessarily even if slight wear indications are present.
- If thrust block or shims need replacement, they should be measured to ensure replacement with components of equal size.
- If ring gear is damaged, ring gear, ring gear bolts, and drive pinion must be replaced as a set.

4. Inspect ring gear, all clutch pack components, thrust block, shims, bushings, and gears for signs of damage. Replace if damaged.
5. Inspect latching bracket and governor assembly for damage. If either is damaged, both must be replaced.

6-16. REAR AXLE DIFFERENTIAL REPAIR (M1009) (Con't).

c. ASSEMBLY

NOTE

- Ensure that all components are lightly coated with lubricating oil as they are assembled.
- If original components are to be reused, ensure that all thrust washers (19), shims (11 and 21), pinion gears (18), and clutch packs (13 and 15) are assembled on same side from which they were disassembled.
- If assembling latching bracket and governor assembly, perform steps 19-21.

1. Assemble right-hand clutch pack on side gear (20). Starting with clutch plate (13) and alternating, install 4 clutch plates and 3 clutch discs (15). Install 4 clips (22).
2. Install cam plate (16) on camform side gear (17).
3. On cam plate (16) install 1 clutch plate (13), 1 clutch disc (15), then 2 clutch plates. Install wave spring (14). Starting with clutch disc and alternating, install 3 clutch discs and clutch plates. Compress left-hand clutch pack and install retainer ring (12). Install 4 clips (22).

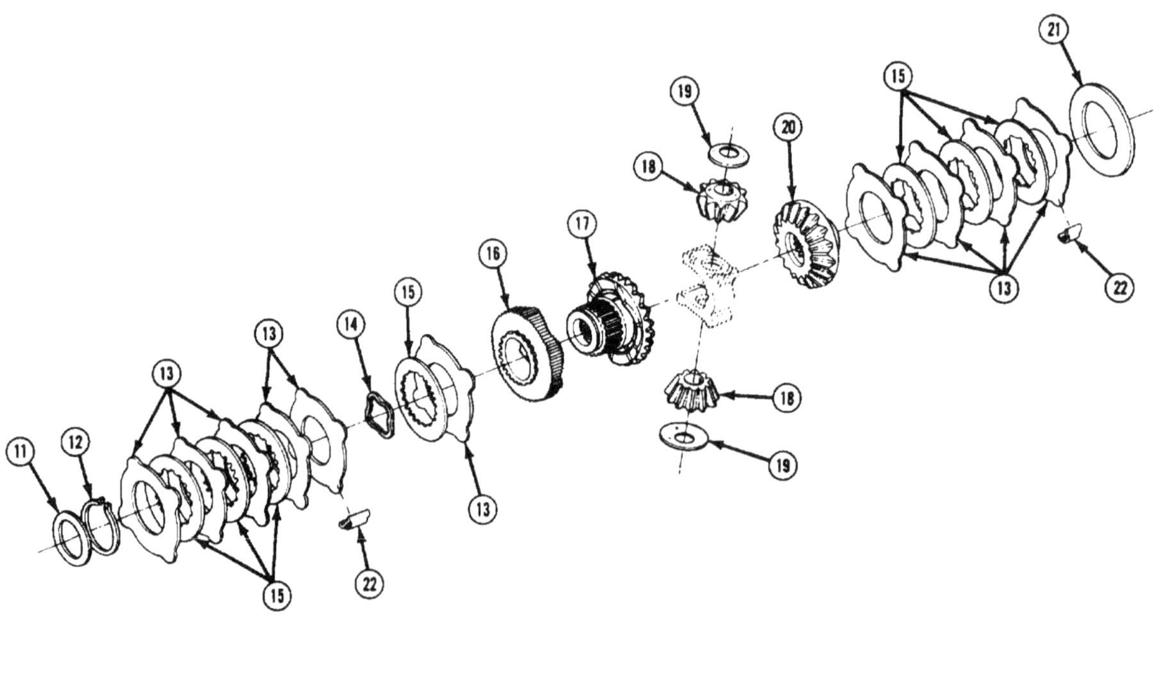

TA50593

6-16. REAR AXLE DIFFERENTIAL REPAIR (M1009) (Con't).

4 . Install left-hand shim(s) (11) inside differential case (2). Install camform side gear (17), cam plate (1 6), and clutch pack assembly, indexing clutch plate (13) ears with differential case.

5 . Install shim (s) (21) on right-hand clutch pack. Install right-hand clutch pack and side gear (20) assembly inside differential case (2) assembly.

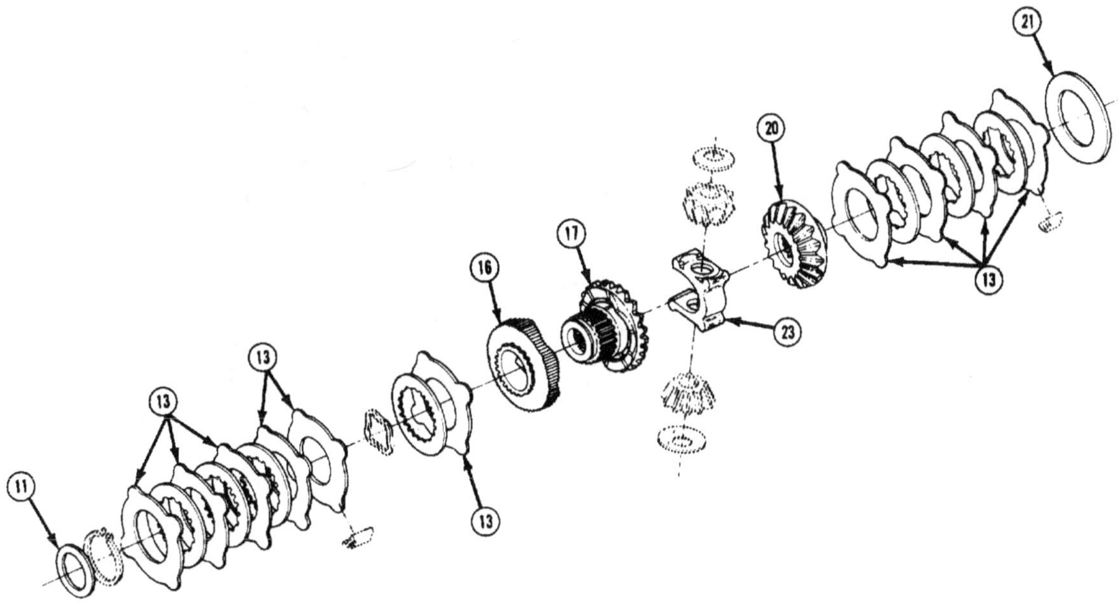

NOTE

- If new camform side gear (17) and/or side gear (20) was installed, perform steps 6-10 to obtain proper thrust block (23) clearance.
- If new gears were installed and pinion shaft (10) will not line up, shim thickness will have to be decreased.

6. Install pinion shaft (10) and lockscrew (9).

TA50594

6-16. REAR AXLE DIFFERENTIAL REPAIR (MI 009) (Con't).

7. Wedge 2 suitable tools between side gear (20) and pinion shaft (10), and camform side gear (17) and pinion shaft. Insert telescoping gage (24) with gage resting on gear faces, not gear teeth. Measure telescoping gage with micrometer and record reading.

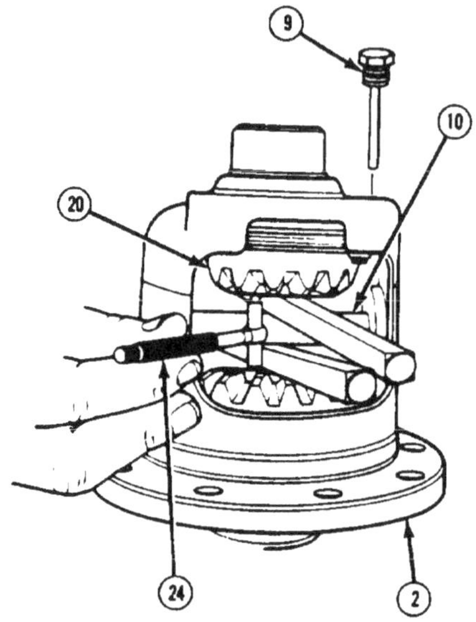

8. Measure thickness of thrust block (23) at outer corner and record reading,

9. If thrust block (23) thickness is not within 0.000-0.006 in. (o.ooo-0.152 mm) less than spread between side gear (20) and camform side gear (17), adjust clearance by:

   (a) reshimming right-hand clutch pack, or;

   (b) selecting a new thrust block (23) of correct size to obtain required clearance.

10. Remove lockscrew (9) and pinion shaft (10).

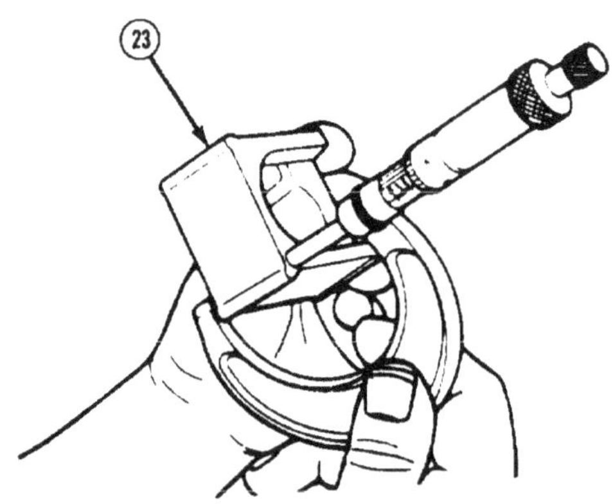

TA50595

6-16. REAR AXLE DIFFERENTIAL REPAIR (M1009) (Con't).

NOTE

If new camform side gear (17) and side gear (20) were installed, thrust block (23) should not be installed at this time.

11. install thrust block (23).

12. Position 2 thrust washers (19) and pinion gears (18) 180 degrees from each other, and indexed with teeth of camform side gear (17) and side gear (20). Rotate side gears and install thrust washers and pinion gears inside differential case (2) assembly.

NOTE

If original camform side gear (17) and/or side gear (20) were installed, original lockscrew (9) should be discarded and new lockscrew installed.

13. Install pinion shaft (10) and lockscrew (9).

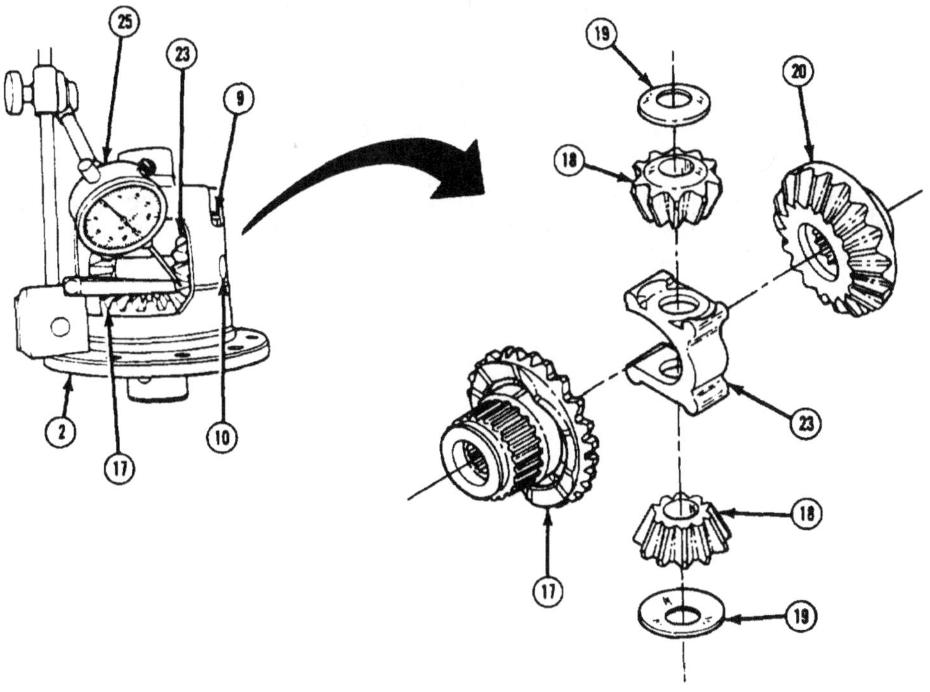

TA50596

6-16. REAR AXLE DIFFERENTIAL REPAIR (M1009) (Con't).

NOTE

Perform steps 14-17 to establish correct backlash only if new camform side gear (17) and side gear (20) were installed.

14. Index 1 tooth of pinion gear (18) nearest lockscrew (9) so that it points facedown, perpendicular to differential case (2) assembly flange. Wedge a suitable tool between left-hand camform side gear (17) and pinion shaft (10). Using dial indicator (25) mounted on differential case assembly flange, check pinion gear backlash by pulling pinion gear firmly into its seat, and rotating back and forth. Record reading.

15. Repeat procedure of indexing and checking backlash to other pinion gear (18). If backlash is not within 0,010-0.018 in. (0.254-0.457 mm), change shim size on left-hand camform side gear (17) side. The thinner the shim used, the greater the backlash reading will be.

16. Repeat steps 14 and 15 to measure right-hand side gear (20) to pinion gear (18) backlash. Backlash should be 0.002-0.010 in. (0.051 -0.254 mm), If backlash is not within specification, change shim size on right-hand side gear side. The thinner the shim used, the greater the backlash will be.

17. If it was necessary to change shims, recheck thrust block (23) clearance. (See steps 6-8)

NOTE

Perform step 18 only if thrust block (23) has not been installed.

18. Remove lockscrew (9) and pinion shaft (10). Discard lockscrew. Remove 2 thrust washers (19) and pinion gears (18). Install thrust block (23). Install 2 thrust washers and pinion gears. Install pinion shaft and new lockscrew.

6-16. REAR AXLE DIFFERENTIAL REPAIR (M1009) (Con't).

19. Install spring on latching bracket (6) and install new snapring. Install latching bracket in differential case (2) assembly. Install governor assembly (7), placing straight end of latching bracket spring over and to outside of the engagement shaft to preload the latching bracket against the governor assembly.

20. Install bushing (4) until governor assembly (7) end play of 0.004-0.020 in. (0.102-0.508 mm) is reached. Install bushing (5) until latching bracket (6) shaft end play of 0.000-0.003 in. (0.000-0.076 mm) is reached.

NOTE

- Shop rags placed around vise will help protect ring gear (3) teeth from damage should ring gear fall during assembly.
- Ring gear bolts (8) have left-hand threads.

21. Place differential case (2) assembly on axle shaft in vise with ring gear (3) alined as marked. Install 10 bolts (8) and tighten alternately to 80 lb.-ft. (108 N.m).

22. Install 2 side bearings (1) as tagged.

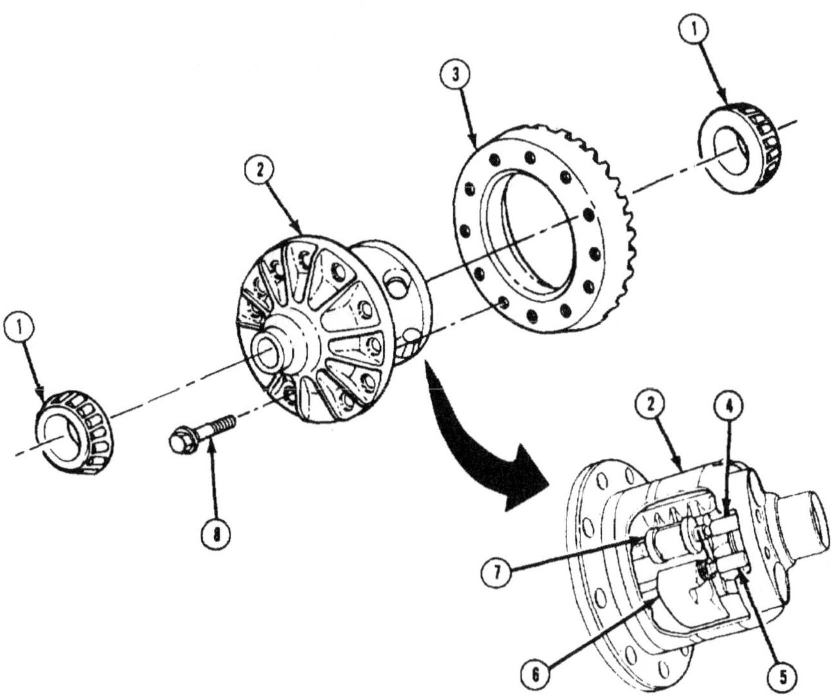

FOLLOW-ON TASKS:

- Install rear axle differential. (See paragraph 6-5)

TA50597

6-17. REAR AXLE DRIVE PINION AND BEARINGS MAINTENANCE (ALL EXCEPT M1009).

This task covers:

a. Removal
b. Disassembly
c. Cleaning and Inspection
d. Assembly and Preload Adjustment
e. Installation and Pinion Depth Adjustment

INITIAL SETUP:

Equipment Condition

- Rear propeller shaft disconnected from pinion flange. (See TM 9-2320-289-20)
- Rear axle differential removed. (See paragraph 6-15)

Tools/Test Equipment

- *Drive* handle, J-8092
- Flange holder and remover
- Micrometer
- Torque wrench

Manual *References*

- TM 9-214

Materials/Parts

- One flange nut
- One oil seal
- One spacer
- Seven lockwashers
- Dry cleaning solvent (Item 23, Appendix B)
- Molybdenum grease (Item 37, Appendix B)
- Gear lubricating oil (Item 44, Appendix B)

General Safety Instructions

- Dry cleaning solvent is flammable and must not be used near open flame, Use only in a well-ventilated area.
- Compressed air used for cleaning purposes should never exceed 30 psi (207 kPa).

a. REMOVAL

NOTE

It may be necessary to tap on pilot end of drive pinion to help free assembly from housing (6).

1. Remove 6 bolts (5) and lockwashers securing retainer (1) to housing (6). Remove retainer and drive pinion assembly from housing. Discard lockwashers,

2. Remove shim located between retainer (1) and housing (6). Measure thickness of shim and record, If shim is undamaged, save for installation.

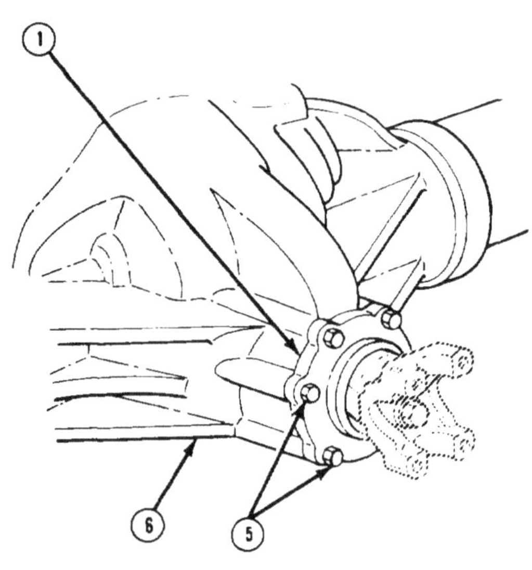

TA50598

6-17. REAR AXLE DRIVE PINION AND BEARINGS MAINTENANCE (ALL EXCEPT M1009) (Con't).

b. DISASSEMBLY

1. Place retainer (1) and drive pinion (9) assembly in a soft-jawed vise. Place flange holder (8) on flange (2), and remove flange nut (3) and lockwasher (4). Discard flange nut and lockwasher.

2. Thread flange remover bolt into small outside diameter end of flange remover (7). Turn flange remover 1/8 turn to a locked position. Turn flange remover while holding flange holder (8) stationary and remove flange (2). If damaged, remove deflector and discard.

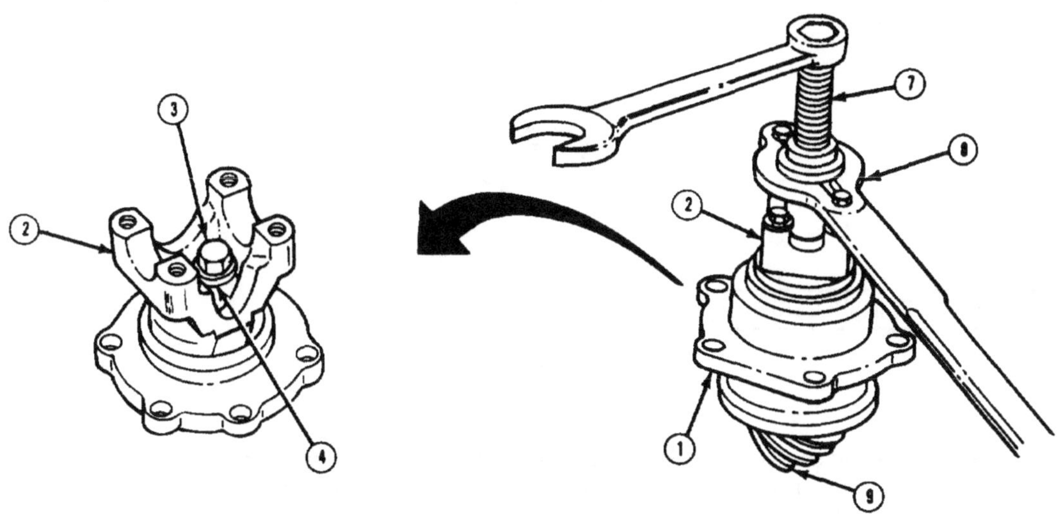

3. Support retainer (1) in a press plate and press out drive pinion (9).

4. Remove oil seal (10) from retainer (1) and discard.

5. Remove front bearing (11) from retainer (1). Remove front and rear bearing cups (12 and 14) from retainer.

6. Press spacer (13) and rear bearing (15) from drive pinion (9). Discard spacer.

7. Remove straddle bearing (16) from housing (6).

TA50599

6-17. REAR AXLE DRIVE PINION AND BEARINGS MAINTENANCE (ALL EXCEPT M1009) (Con't).

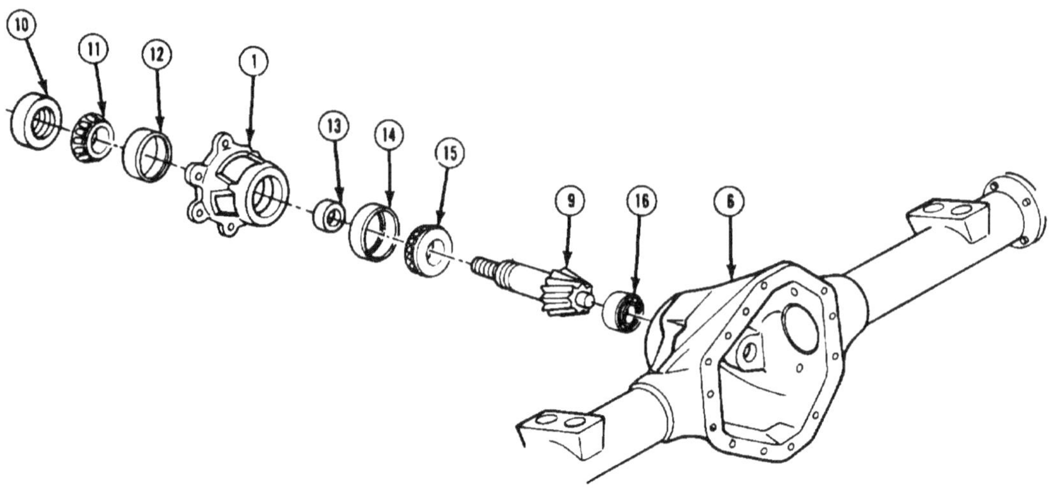

c. CLEANING AND INSPECTION

WARNING

- Dry cleaning solvent P-D-680 is toxic and flammable. Always wear protective goggles and gloves and use only in a well-ventilated area. Avoid contact with skin, eyes, and clothes and DO NOT breathe vapors. DO NOT use near open flame or excessive heat. The solvent's flash point is 100°F-138°F (38°C-59°C). If you become dizzy while using cleaning solvent, immediately get fresh air and medical help. If solvent contacts eyes, immediately wash your eyes with water and get medical aid.
- Compressed air used for cleaning purposes should never exceed 30 psi (207 kPa). Use only effective chip guarding and personnel protective equipment (goggles/shield, gloves, etc.). Failure to follow this warning may result in serious injury to personnel.

1. Clean all components with dry cleaning solvent and dry with compressed air.
2. Clean and inspect all pinion bearings in accordance with TM 9-214,
3. Clean pinion bearing bores in retainer to ensure that they are free of dirt or other contamination. Remove any nicks found.

TA50600

6-17. REAR AXLE DRIVE PINION AND BEARINGS MAINTENANCE (ALL EXCEPT M1009) (Con't).

NOTE

If drive pinion is damaged, differential ring gear and ring gear bolts must also be replaced. (See paragraph 6-15)

4. Inspect drive pinion for damage. Replace if damaged,
5. Inspect flange for evidence of wear at point of oil seal contact. Replace if worn or damaged.
6. Inspect retainer for damage. Replace if damaged,

d. ASSEMBLY AND PRELOAD ADJUSTMENT

NOTE

Ensure that all components are lubricated with lubricating oil as they are assembled.

1. Press rear bearing (15) onto drive pinion (9).
2. Install front and rear bearing cups (12 and 14) inside retainer (1).
3. Place retainer (1) onto drive pinion (9) and install new spacer (13).
4. Press front bearing (11) onto drive pinion (9).
5. Lubricate new oil seal (10) lips with molybdenum grease and install in retainer (1) bore until seated against internal shoulder.
6. If removed, install new deflector to flange (2). Install flange onto drive pinion (9) splines.

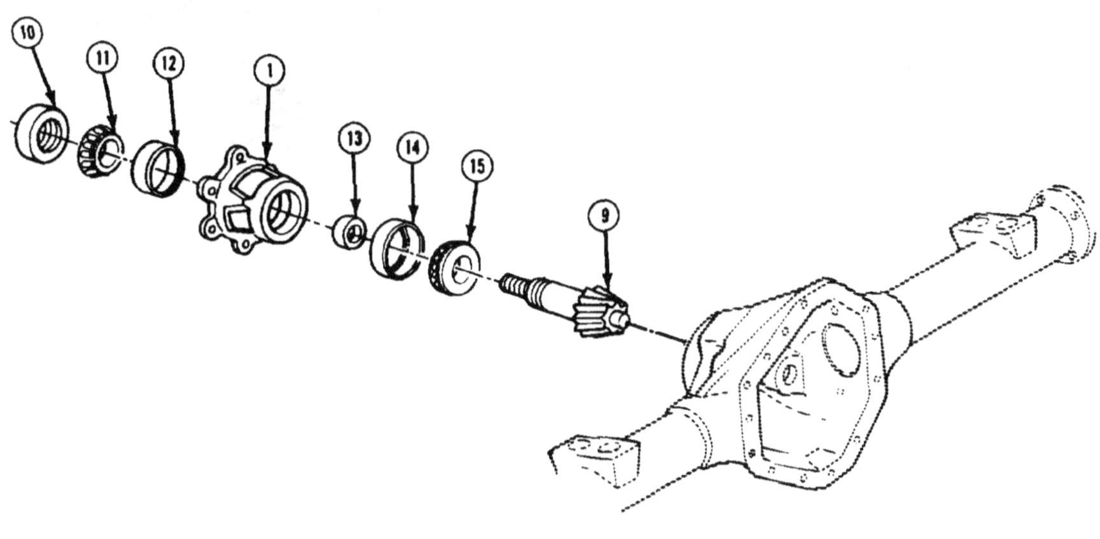

TA50601

6-17. REAR AXLE DRIVE PINION AND BEARINGS MAINTENANCE (ALL EXCEPT M1009) (Con't).

7 . Clamp drive pinion (9) and retainer (1) assembly in soft-jawed vise. Install flange holder (8). Install new lockwasher (4) and new flange nut (3). Tighten flange nut to approximately 350 lb.-ft. (475 Nom). Remove flange holder.

CAUTION

Ensure that flange nut (3) is tightened in small increments as preload is being reached. Exceeding torque specifications may compress spacer (13) on drive pinion (9) too far and require replacement of spacer.

NOTE

Proper preload is reached when torque required to turn drive pinion (9) is 25-35 lb.-in. (3-4 N.m) for new bearings, or 5-15 lb.-in. (1-2 N.m) for reused bearings.

8 . Measure turning torque. Continue to tighten flange nut (3) in small increments. Check turning torque after each slight amount of tightening until proper preload is reached.

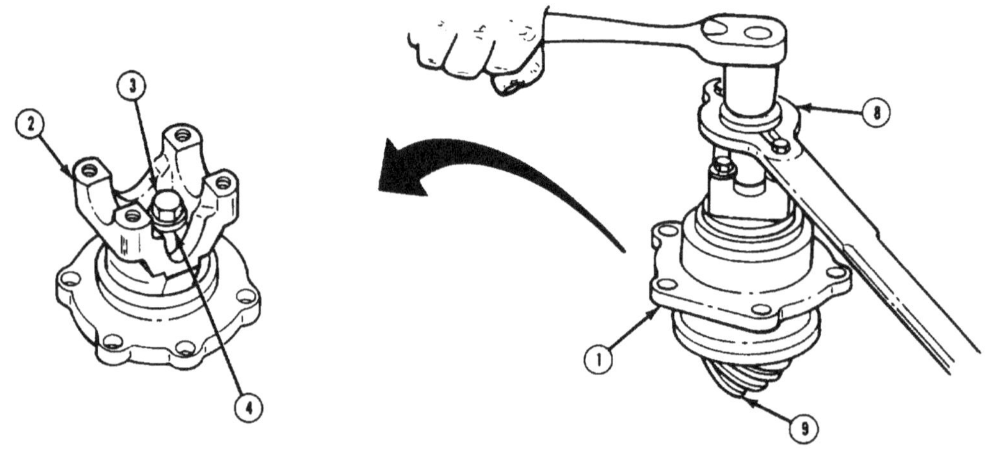

6-17. REAR AXLE DRIVE PINION AND BEARINGS MAINTENANCE (ALL EXCEPT M1009) (Con't).

CAUTION

Ensure that straddle bearing (16) is installed with part number facing pinion flange. Failure to follow this caution will result in damage to straddle bearing.

NOTE

If rear axle housing (6) is 1984 or older design, it uses a straddle bearing (16) with rollers that are larger than 1985 design straddle bearing. Ensure that proper size straddle bearing is installed,

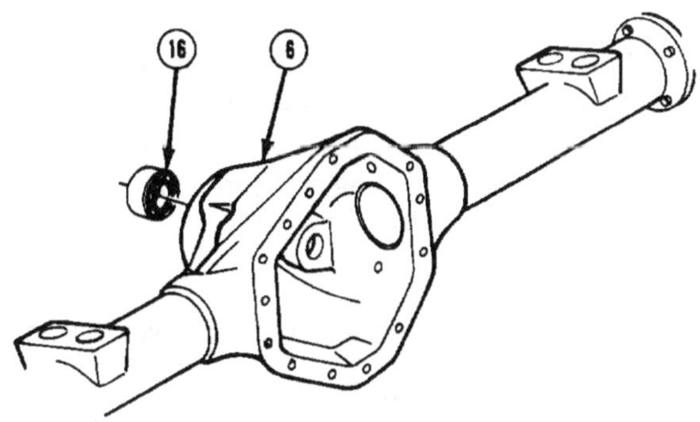

9. Install straddle bearing (16) i n housing (6).

e. INSTALLATION AND PINION DEPTH ADJUSTMENT

NOTE

If no number is stamped or painted on drive pinion (9), should be used for code number .

1. Check gear head or shaft of drive pinion (9) for pinion depth code number. Remove retainer (1) and drive pinion assembly from vise.

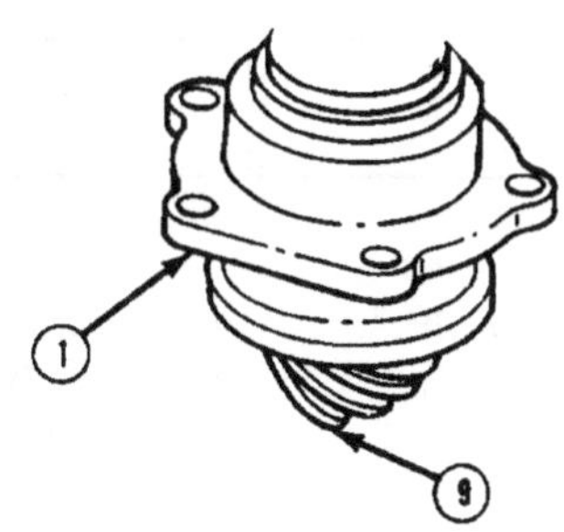

TA50603

6-17. REAR AXLE DRIVE PINION AND BEARINGS MAINTENANCE (ALL EXCEPT M1009) (Con't).

2 . Compare depth code number with number on original drive pinion (9). Using Table 6-4, select proper shim for preliminary setting of pinion depth,

NOTE

If code number on old and new drive pinion (9) are the same, original shim may be reused if undamaged.

3 . Refer to thickness of shim as measured in REMOVAL, step 2. Increase or decrease shim thickness as indicated by Table 6-4:

Table 6-4. Determining Drive Pinion Shim Requirements

Code Number On Service Pinion	Code Number On Original Pinion				
	+ 2	+ 1	0	- 1	- 2
+ 2	-	Add 0.001	Add 0.002	Add 0.003	Add 0.004
+ 1	Subtract 0.001	-	Add 0.001	Add 0.002	Add 0.003
0	Subtract 0.002	Subtract 0.001	-	Add 0.001	Add 0.002
- 1	Subtract 0.003	Subtract 0.002	Subtract 0.001	-	Add 0.001
- 2	Subtract 0.004	Subtract 0.003	Subtract 0.002	Subtract 0.001	-

EXAMPLE 1. If original shim measured 0.014 in. (0.36 mm), original code was "- 1," and new code is "+2, " correct shim would be:

0,014 *in.* + 0.003 *in.* = 0.017 *in.*

(0.36 mm + 0.08 mm = 0.43 mm)

EXAMPLE 2. If original shim measured 0.012 in. (0.30 mm), original code was "+2, " and new *code is* "−2," correct shim would be:

0.012 *in.* -0.004 in. = 0,008 in.

(0,30 mm -0.010 mm = 0.20 mm).

6-17. REAR AXLE DRIVE PINION AND BEARINGS MAINTENANCE (ALL EXCEPT M1009) (Con't).

4 . Ensure that all mating surfaces are clean. Place shim, as determined in INSTALLATION , step 3. onto housing (6), alining all bolt holes.

5 . Install retainer (1) and drive pinion assembly with 6 new lockwashers and bolts (5). Tighten bolts to 65 lb.-ft. (88 N.m).

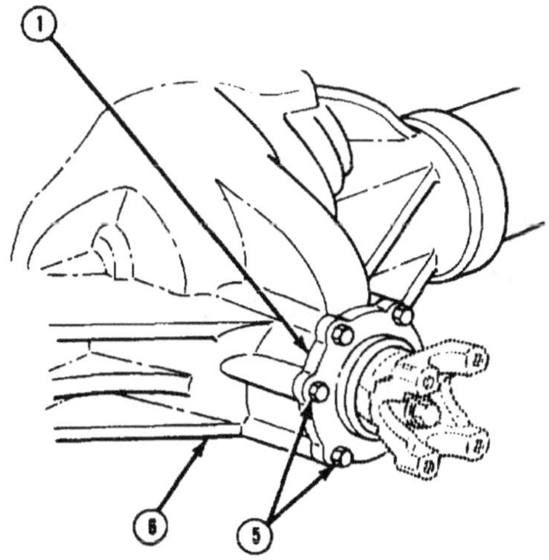

FOLLOW-ON TASKS:

- Install rear axle differential. (See paragraph 6-15)
- Connect rear propeller shaft to pinion flange. (See TM 9-2320-289-20)

TA50604

6-17.1. REAR AXLE DRIVE PINION AND BEARINGS MAINTENANCE (M1028A2).

This task covers: a. Removal; b. Cleaning and Inspection; c. Installation

INITIAL SETUP:

Equipment Condition

- Rear axle differential removed. (See paragraph 6-15,1)

Tools/Test Equipment

- Flange holder and remover, *J-861 4-01*
- Micrometer
- Torque wrench

Manual References

- TM 9-214

Materials/Parts

- One flange nut
- One pinion oil seal
- Two shim packs
- Two slingers
- Dry cleaning solvent (Item 23, Appendix B)
- Molybdenum grease (Item 37, Appendix B)
- Gear lubricating oil (Item 44, Appendix B)

General Safety Instructions

- Dry cleaning solvent is flammable and must not be used near open flame. Use only in a well-ventilated area.
- Compressed air used for cleaning purposes should never exceed 30 psi (207 kPa).

6-17.1. REAR AXLE DRIVE PINION AND BEARINGS MAINTENANCE (M1028A2) (Con't).

a. REMOVAL

NOTE

- Ensure that shims and slingers are tagged as they are removed and saved for measurement.
- Ensure that flange nut (4) is saved for use during drive pinion (5) installation and 'measurement, When drive pinion is fully seated, measurements are completed, and ready to receive final torque, a new flange nut must be used.

1. Using flange holder (1), remove flange nut (4) and washer from flange (2).
2. Thread flange remover bolt into small outside diameter end of flange remover (3). Turn flange remover 1/8 turn to a locked position. Turn flange remover while holding flange holder (1) stationary. Remove flange (2). Remove deflector from flange if damaged.

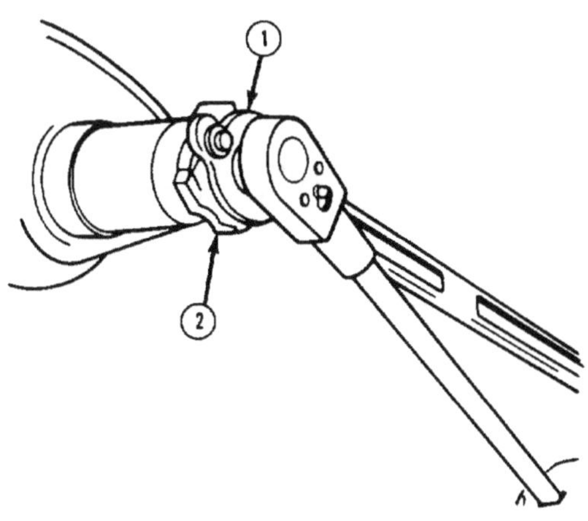

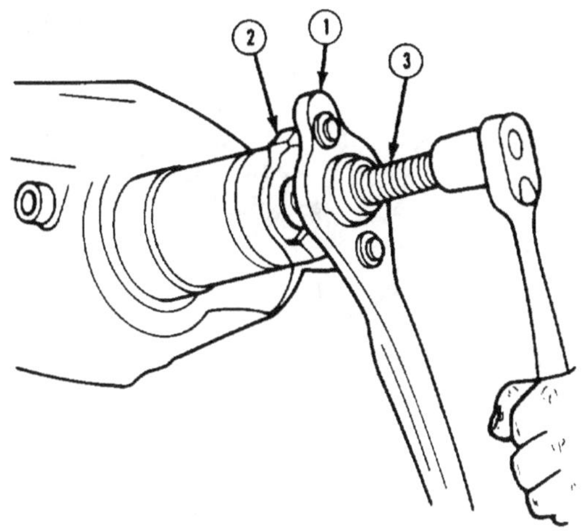

NOTE

Shim (6) may be on drive pinion (5) shaft or inside housing (7).

3. Install flange nut (4) halfway onto drive pinion (5) threads. Tap on flange nut with soft-faced hammer to loosen drive pinion in front pinion bearing (8). Remove flange nut from drive pinion. Remove drive pinion from housing (7). Remove shim (6).

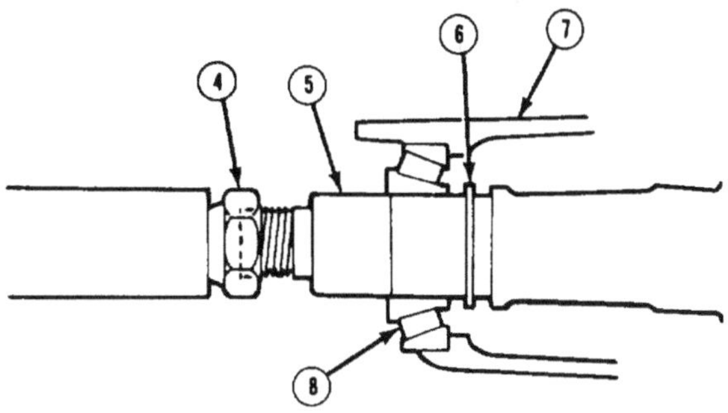

6-17.1. REAR AXLE DRIVE PINION AND BEARINGS MAINTENANCE (M1028A2) (Con't).

4. Tap on inner race of front pinion bearing (8) to remove pinion oil seal (14), slinger (17), front bearing cup (13) and front pinion bearing. Discard pinion oil seal.

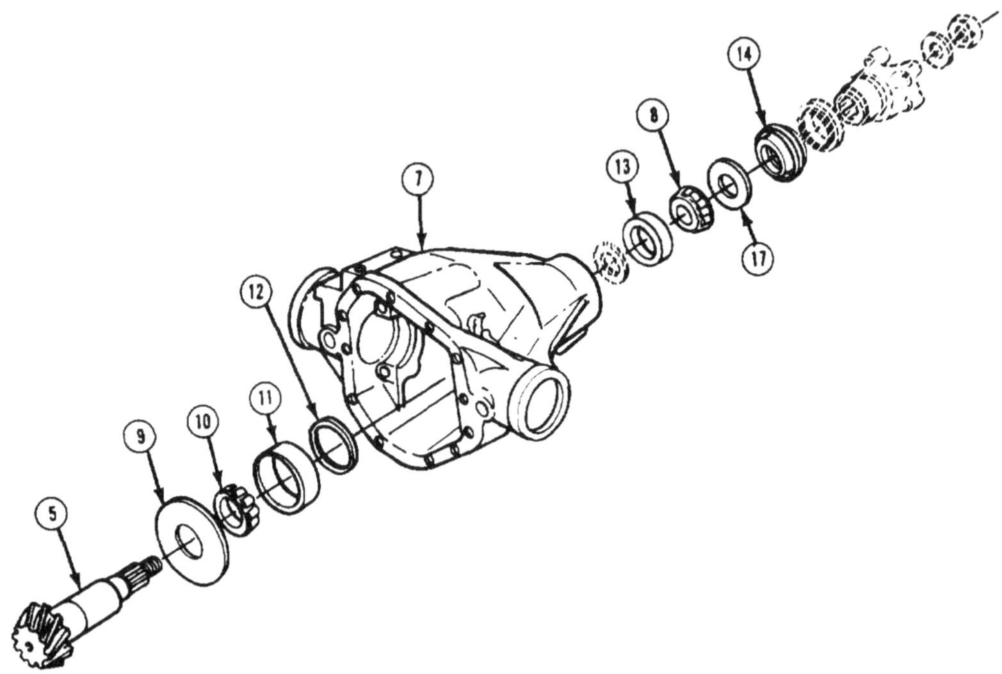

5. Remove rear bearing cup (11) from housing (7). Remove shim (12). Press rear pinion bearing (10) from drive pinion (5) shaft. Remove slinger (9).

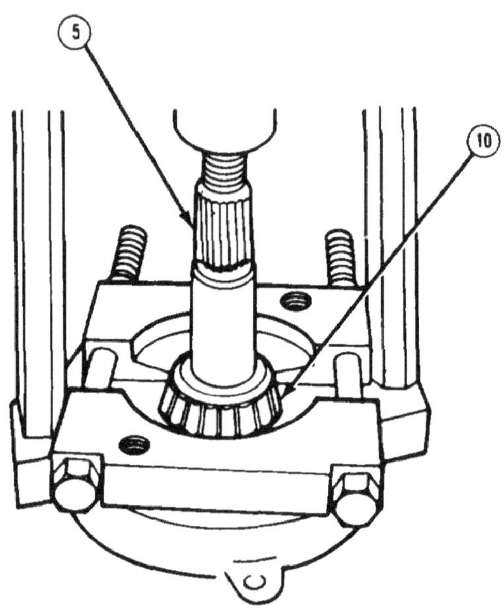

6-17.1. REAR AXLE DRIVE PINION AND BEARINGS MAINTENANCE (M1028A2) (Con't).

b. CLEANING AND INSPECTION

WARNING

- Dry cleaning solvent P-D-680 is toxic and flammable. Always wear protective goggles and gloves and use only in a well-ventilated area. Avoid contact with skin, eyes, and clothes and DO NOT breathe vapors. DO NOT use near open flame or excessive heat. The solvent's flash point is 100°F-138°F (38°C-59°C). If you become dizzy while using cleaning solvent, immediately get fresh air and medical help. If solvent contacts eyes, immediately wash your eyes with water and get medical aid.
- Compressed air used for cleaning purposes should never exceed 30 psi (207 kPa). Use only effective chip guarding and personnel protective equipment (goggles/shield, gloves, etc.), Failure to follow this warning may result in serious injury to personnel.

1. Clean drive pinion, shims, slingers, flange, and deflector with dry cleaning solvent and dry with compressed air.
2. Clean drive pinion bearing bores in housing to ensure that they are free of dirt or other contamination. Remove any nicks found.
3. Clean and inspect front and rear bearing assemblies in accordance with TM 9-214.

NOTE

If drive pinion is replaced, differential ring gear and ring gear bolts must also be replaced, (See paragraph 6-15.1)

4. Inspect drive pinion for damage. Replace if damaged.
5. Inspect flange for evidence of wear at point of oil seal contact. Replace if worn.

6-17.1. REAR AXLE DRIVE PINION AND BEARINGS MAINTENANCE (M1028A2) (Con't).

c. INSTALLATION

NOTE

- Ensure that all components are lubricated with lubricating oil as they are installed,
- If original drive pinion (5) is reused, new shim pack must be built equal in thickness to original one.
- Slingers (9 and 17) are considered part of shim pack.
- If new drive pinion (5) is installed, adjustments in new shim packs' thickness will be required unless the number etched on gear end of replacement drive pinion is the same as original drive pinion.

1 . Measure thickness of shim (12) and slinger (9) to determine thickness of original buildup at rear pinion bearing (10). Measure thickness of shim (6) and slinger (17) to determine thickness of original buildup at front pinion bearing (8). Record measurements. Discard shims and slingers.

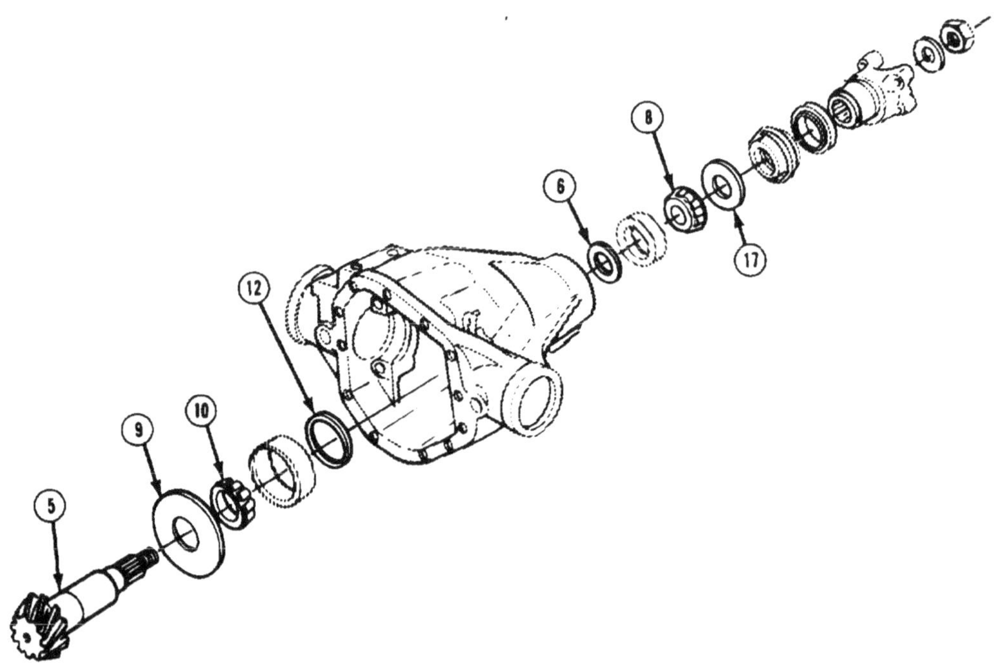

6-17.1. REAR AXLE DRIVE PINION AND BEARINGS MAINTENANCE (M1028A2) (Con't).

NOTE

Perform steps 2 and 3 only if new drive pinion (5) is being installed.

2. Determine proper rear shim pack for setting pinion depth as follows:

   (a) Check number (18) etched on gear end of old drive pinion (5) &d locate it in "Old Pinion *Marking*" column of Table 6-3.

   (b) Check number (18) etched on gear end of new drive pinion (5) and locate it in "New *Pinion Marking*" column of Table 6-3.

   (c) Find point where 2 columns intersect to determine adjustment number for new rear shim pack.

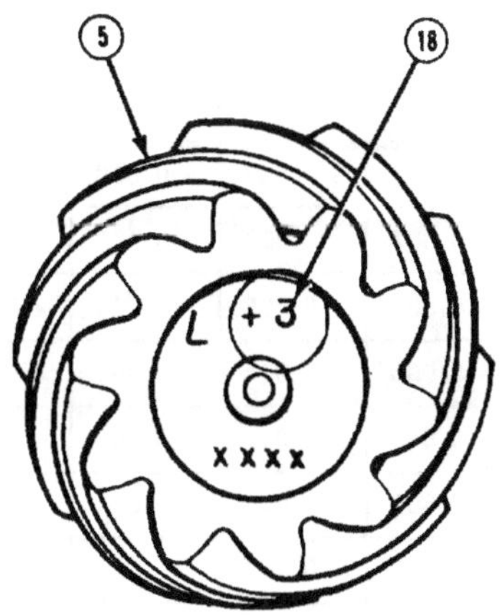

NOTE

A "+" means add that amount; a "-" means subtract that amount; and "O" means no change.

3. Add or subtract adjustment number found on Table 6-4.1 to old rear shim pack (9 and 12) to determine new rear shim pack.

6-17.1. REAR AXLE DRIVE PINION AND BEARINGS MAINTENANCE (M1028A2) (Con't).

Table 6-4.1. Pinion Setting Charts,

English U.S. Standard Chart

Old Pinion Marking	New Pinion Marking								
	- 4	- 3	- 2	- 1	0	+ 1	+ 2	+ 3	+ 4
+ 4	+ 0.008	+ 0.007	+ 0.006	+ 0.005	+ 0.004	+ 0.003	+ 0.002	+ 0.001	0
+ 3	+ 0.007	+ 0.006	+ 0.005	+ 0.004	+ 0.003	+ 0.002	+ 0.001	0	- 0.001
+ 2	+ 0.006	+ 0.005	+ 0.004	+ 0.003	+ 0.002	+ 0.001	0	- 0.001	- 0.002
+ 1	+ 0.005	+ 0.004	+ 0.003	+ 0.002	+ 0.001	0	- 0.001	- 0.002	- 0.003
0	+ 0.004	+ 0.003	+ 0.002	+ 0.001	0	- 0.001	- 0.002	- 0.003	- 0.004
- 1	+ 0.003	+ 0.002	+ 0.001	0	- 0.001	- 0.002	- 0.003	- 0.004	- 0.005
- 2	+ 0.002	+ 0.001	0	- 0.001	- 0.002	- 0.003	- 0.004	- 0.005	- 0.006
- 3	+ 0.001	0	- 0.001	- 0.002	- 0.003	- 0.004	- 0.005	- 0.006	- 0.007
- 4	0	- 0.001	- 0.002	- 0.003	- 0.004	- 0.005	- 0.006	- 0.007	- 0.008

Metric Chart

Old Pinion Marking	New Pinion Marking								
	- 10	- 8	- 5	- 3	0	+ 3	+ 5	+ 8	+ 10
+ 10	+ 0.20	+ 0.18	+ 0.15	+ 0.13	+ 0.10	+ 0.08	+ 0.05	+ 0.03	0
+ 8	+ 0.18	+ 0.15	+ 0.13	+ 0.10	+ 0.08	+ 0.05	+ 0.03	0	- 0.03
+ 5	+ 0.15	+ 0.13	+ 0.10	+ 0.08	+ 0.05	+ 0.03	0	- 0.03	- 0.05
+ 3	+ 0.13	+ 0.10	+ 0.08	+ 0.05	+ 0.03	0	- 0.03	- 0.05	- 0.08
0	+ 0.10	+ 0.08	+ 0.05	+ 0.03	0	- 0.03	- 0.05	- 0.08	- 0.10
- 3	+ 0.08	+ 0.05	+ 0.03	0	- 0.03	- 0.05	- 0.08	- 0.10	- 0.13
- 5	+ 0.05	+ 0.03	0	- 0.03	- 0.05	- 0.08	- 0.10	- 0.13	- 0.15
- 8	+ 0.03	0	- 0.03	- 0.05	- 0.08	- 0.10	- 0.13	- 0.15	- 0.18
- 10	0	- 0.03	- 0.05	- 0.08	- 0.10	- 0.13	- 0.15	- 0.18	- 0.20

6-17.1. REAR AXLE DRIVE PINION AND BEARINGS MAINTENANCE (M1028A2) (Con't).

4. Install new shim (12) inside rear bearing cup (11) bore. Install rear bearing cup in housing (7).
5. To front shim pack (6 and 17), add or subtract an amount equal to what was added or subtracted to rear shim pack (9 and 12).
6. Install front bearing cup (13) into housing (7) bore.
7. Install new slinger (9) and rear pinion bearing (10) onto drive pinion (5) shaft. Install drive pinion and bearing assembly in housing (7),

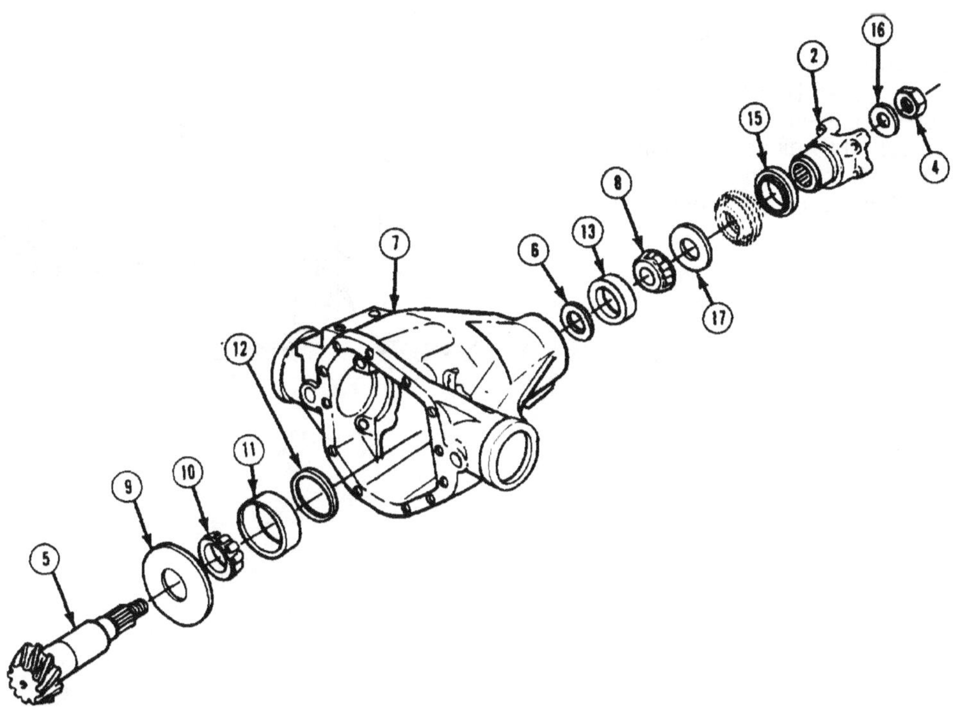

8. Install new shim (6), front pinion bearing (8), and new slinger (17) on drive pinion (5) shaft.
9. If removed, install new deflector (15) to flange (2). Install flange, washer (1 6), and flange nut (4). Using flange holder and flange remover, tighten flange nut *to* pull flange onto drive pinion (5) and to fully seat front pinion bearing (8). Tighten flange nut to 255 lb.-ft. (346 N.m). Remove flange holder and flange remover.

6-17.1. REAR AXLE DRIVE PINION AND BEARINGS MAINTENANCE (M1028A2) (Con't).

NOTE

- When measuring rotating torque, torque reading needed to start drive pinion (5) shaft turning should be disregarded,
- Front shim pack (6 and 17) should be increased to reduce rotating torque. Front shim pack should be decreased to increase rotating torque.

10. Measure rotating torque at flange nut (4), Rotating torque should be 20-40 lb.-in. (2-5 N.m) for new bearings, or 10-20 lb.-in. (1-2 N.m) for reused bearings. Adjust front shim pack (6 and 17) by repeating steps 8 and 9 if specifications are not met.

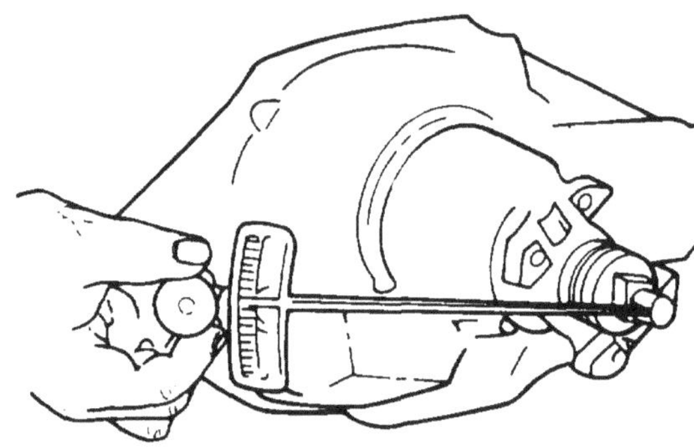

11. Remove flange nut (4), washer (16), and flange (2) from drive pinion (5) shaft. Discard flange nut.

12. Coat lips of new pinion oil seal (14) with molybdenum grease and install squarely into housing (7) bore until flush with housing. Install flange (2), washer (16), and new flange nut (4). Tighten flange nut to 255 lb.-ft. (346 N.m).

FOLLOW-ON TASKS:

- Install rear axle differential. (See paragraph 6-15.1)

Section III. Suspension MAINTENANC E

6-18. SUSPENSION MAINTENANCE INDEX.

6-19. FRONT SPRING ASSEMBLY AND BUSHINGS REPLACEMENT.

This task covers: a. Removal b. Installation

INITIAL SETUP:

Equipment Condition

- Truck raised and frame supported on jack stands.
- Wheel removed. (See TM 9-2320-289-20)

Materials/Parts

- Three locknuts
- Lubricating oil (Item 47, Appendix B)

Tools/Test Equipment

- Torque wrench

Manual References

- TM 9-2320-289-20

General Safety Instructions

- Front spring assembly is heavy. Use caution during removal and installation.

a. REMOVAL

NOTE

Right side of all except M1009 use only 1 "U" bolt (19), 2 nuts (15), and 2 bolts at anchor plate (14).

1. Place a suitable jack under axle (1). Remove 4 nuts (15), washers (16), and 2 "U" bolts (19).
2. Lower axle (1) until all tension is released from spring assembly (2). Remove bottom spacer (18) and top spacer (17).
3. Remove locknut (10), bolt (3), and washer (11), and disconnect shackle (5) from spring assembly (2). Discard locknut.

WARNING

Spring assembly (2) is heavy. Use caution during removal. Failure to follow this warning may result in serious injury to personnel or damage to axle (1) assembly.

4. Remove locknut (23), bolt (20), and 2 washers (22) at hanger (21). Remove spring assembly (2). Discard locknut.
5. Inspect bushings (12 and 24) for damage. If damaged, drive out spacer (13) and remove 2 bushings (12). Inspect spacer for damage. Discard if damaged. Press out bushing (24).

6-19. FRONT SPRING ASSEMBLY AND BUSHINGS REPLACEMENT (Con't).

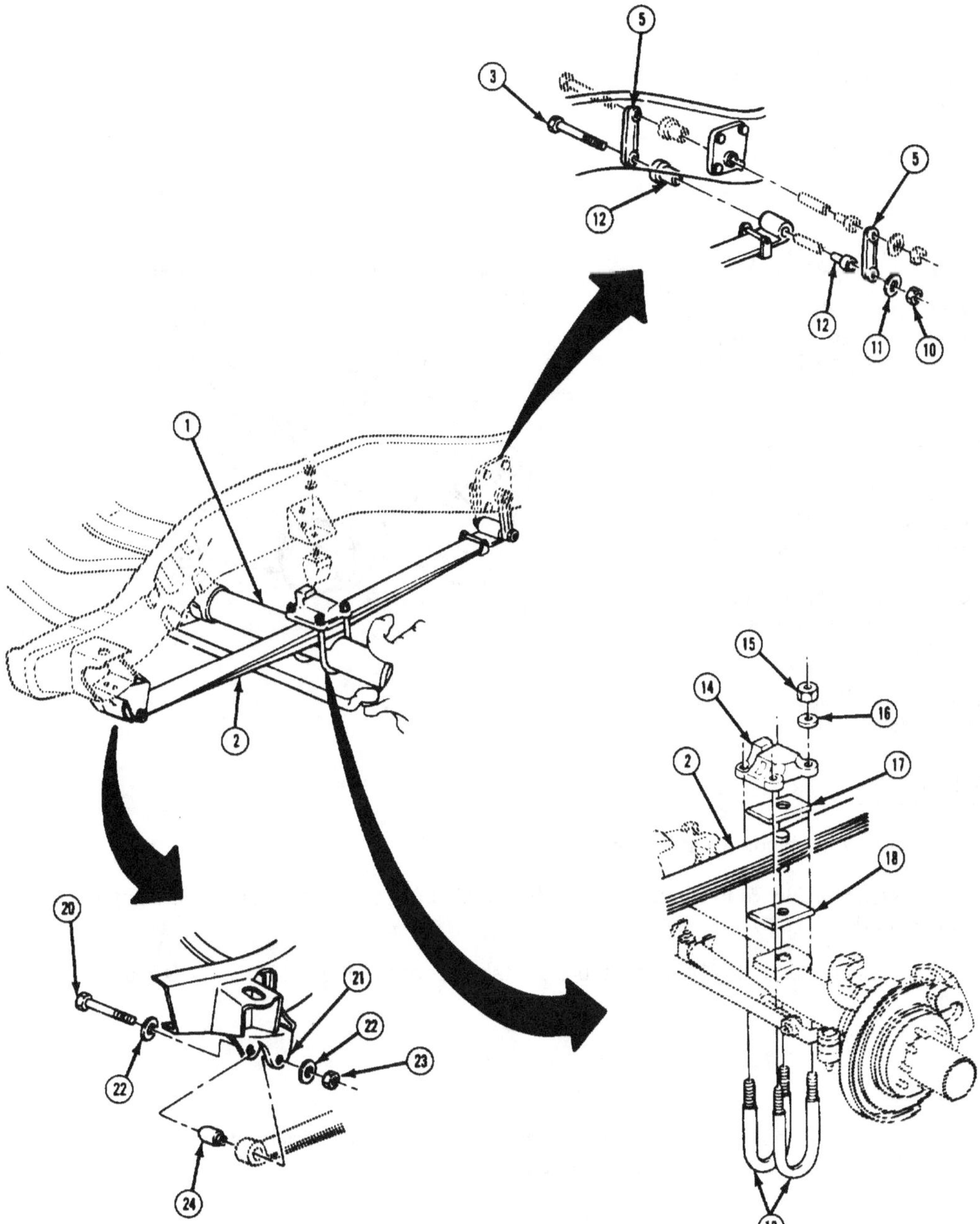

TA50605

6-19. FRONT SPRING ASSEMBLY AND BUSHINGS REPLACEMENT (Con't).

6. Inspect anchor plate (14) for damage. If damaged, remove stabilizer bar bolt (25) and remove anchor plate.

7. Inspect shackle (5) and bushings (6) for damage. If damaged perform steps 8-11 to remove.

8. To gain access to shackle (5) and bushings (6) on right side, disconnect exhaust pipe from exhaust manifold and lower exhaust pipe approximately 3 in. (7.62 cm), (See TM 9-2320-289-20)

9. To gain access to shackle (5) and bushings (6) on left side, shift transmission to "L" (Low) and remove plastic shield at rear of fender skirt. (See paragraph 10-6)

10. Remove locknut (9), washer (8), and bolt (4). Remove shackle (5) . Discard locknut.

11. Drive out spacer (7) and remove 2 bushings (6) from frame.

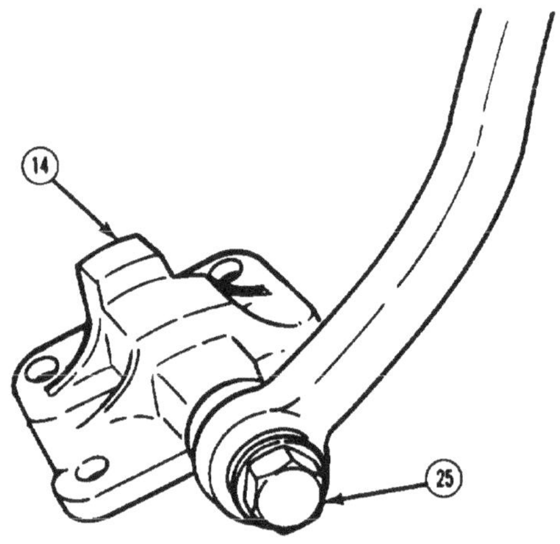

b. INSTALLATION

1. If removed, install 2 bushings (6) at frame. Drive in spacer (7).

2. If removed, install shackle (5) to frame with bolt (4), washer (8), and new locknut (9). Tighten locknut to 50 lb.-ft. (68 N.m).

3. If removed, install anchor plate (14) and secure to stabilizer bar with bolt (25). DO NOT fully tighten bolt.

CAUTION

Ensure that press tool contacts steel outer shell of bushing (24). Failure to follow this caution may result in damage to bushing.

NOTE

Ensure that bushing (24) is lubricated with light lubricating oil before installation, Installed bushing should protrude an equal amount on each side.

4. If removed, press bushing (24) into spring assembly (2).

TA50606

6-19. FRONT SPRING ASSEMBLY AND BUSHINGS REPLACEMENT (Con't).

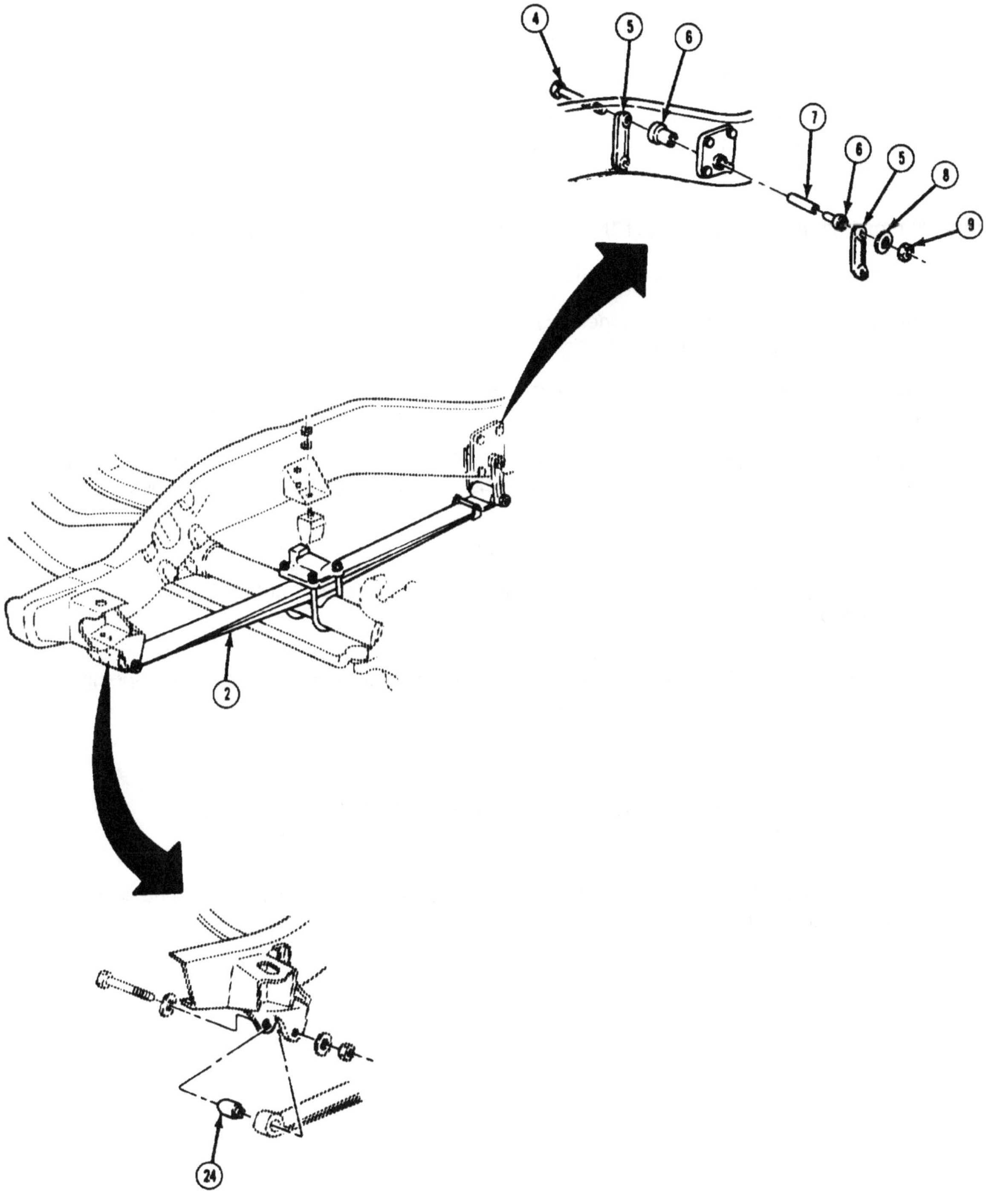

TA50607

6-19. FRONT SPRING ASSEMBLY AND BUSHINGS REPLACEMENT (Con't).

WARNING

Spring assembly (2) is heavy. Use caution during installation. Failure to follow this warning may result in serious injury to personnel or damage to axle (1) assembly.

5 . If removed, install 2 bushings (12) to spring assembly (2) and drive in spacer (13).

6 . Position spring assembly (2) over axle (1).

7 . Install spring assembly (2) to hanger (21) with bolt (20), 2 washers (22), and new locknut (23). DO NOT fully tighten locknut.

8 . Connect spring assembly (2) to shackle (5) with bolt (3), washer (11), and new locknut (10). Tighten locknut (10) to 50 lb.-ft. (68 N.m). Tighten locknut (23) to 90 lb.-ft. (122 N.m).

9 . Position bottom spacer (18) on axle (1) housing. Raise axle, alining spring assembly (2) center bolt with hole in bottom spacer and axle housing. Install top spacer (17) alining bolt hole.

NOTE

Right side of all except Ml 009 uses only 1 "U" bolt (19), 2 nuts (15), and 2 bolts at anchor plate (14).

10. Install 2 "U" bolts (19), 4 washers (16), and nuts (15). Install all nuts to uniform engagement on " U" bolts to retain and position anchor plate (14).

NOTE

All nuts (15) must be tightened evenly to ensure that "U" bolts (19) and spring assembly (2) seat properly.

11. Tighten 4 nuts (15) in diagonal sequence to 25 lb.-ft. (34 Nom). Fully tighten all nuts in same diagonal sequence to 150 lb.-ft. (203 N.m).

6-19. FRONT SPRING ASSEMBLY AND BUSHINGS REPLACEMENT (Con't).

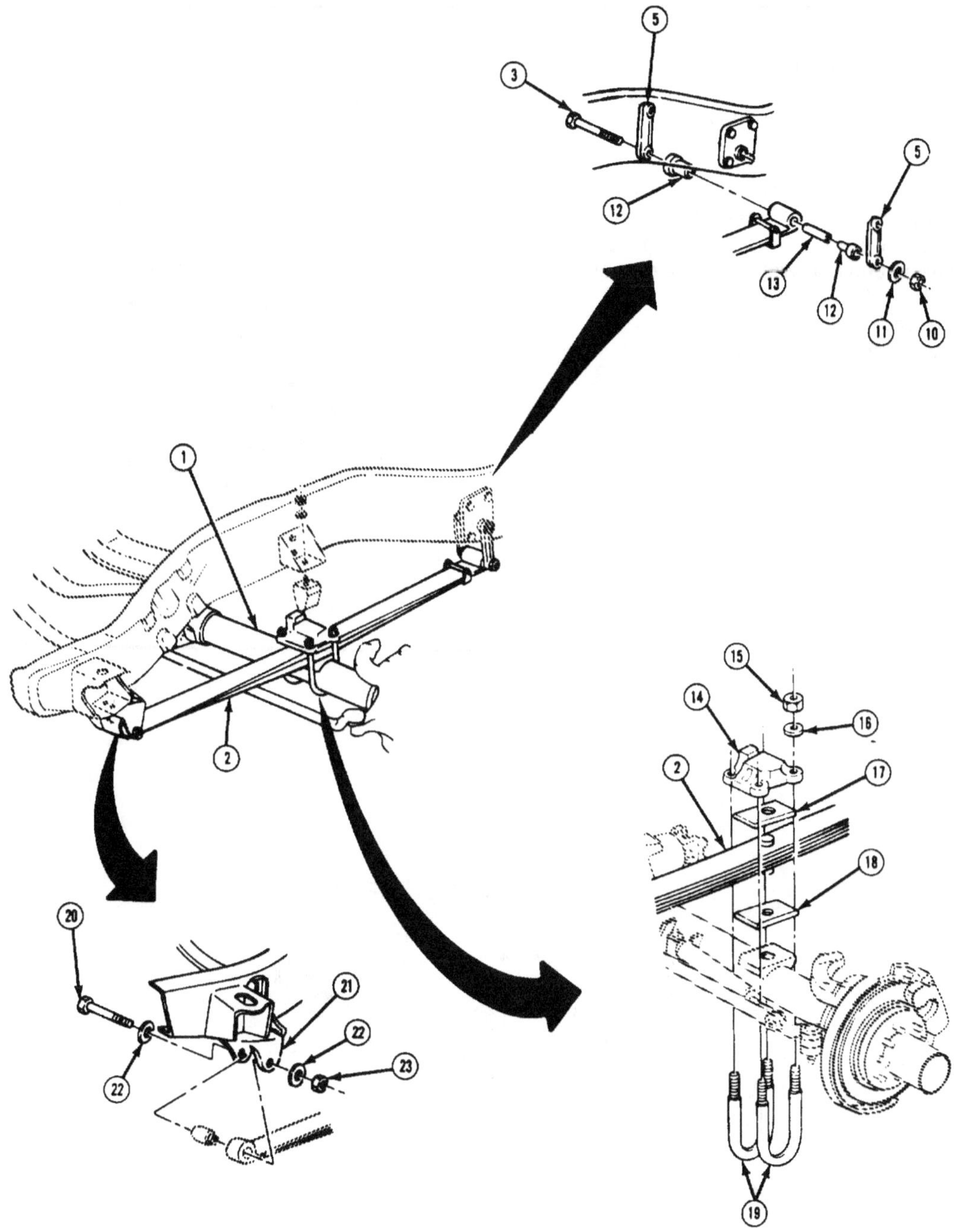

TA50608

6-19. FRONT SPRING ASSEMBLY AND BUSHINGS REPLACEMENT (Con't).

12. Tighten stabilizer bar bolt (25) to 230 lb.-ft. (312 N.m).
13. Remove jack from axle (1).
14. If removed, install plastic shield to rear of fender skirt on left side. (See paragraph 10-6) If lowered, raise right exhaust pipe to its original position and connect to right exhaust manifold. (See TM 9-2320-289-20) If transmission was shifted to "L" (Low), shift to " P" (Park).

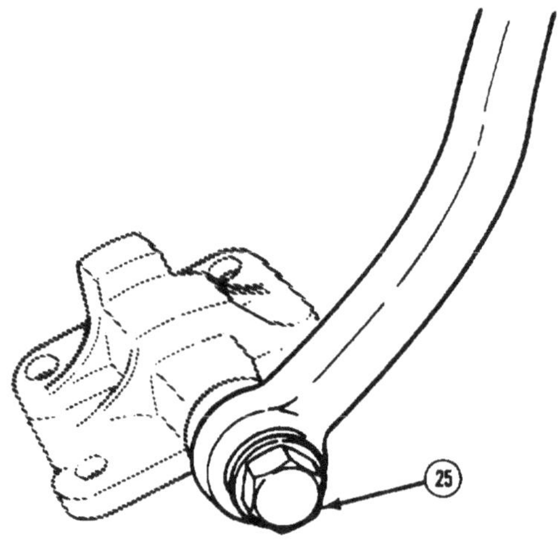

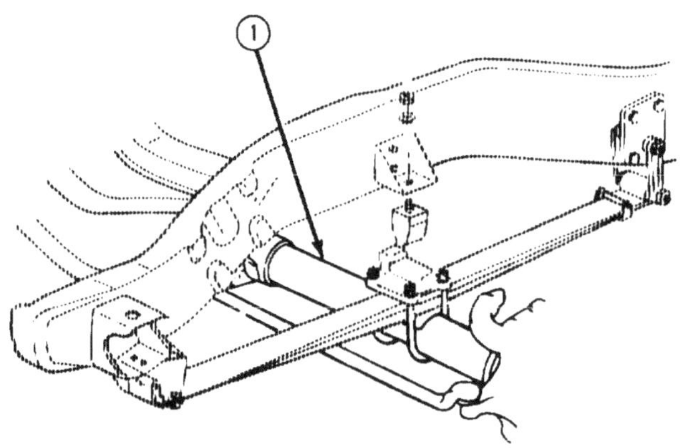

FOLLOW-ON TASKS:

- Install wheel. (See TM 9-2320-289-20)
- Remove jack stands and lower truck.
- Road test truck and check torque of "U" bolt fasteners.

TA50609

6-20. REAR SPRING ASSEMBLY AND BUSHINGS REPLACEMENT.

This task covers: a. Removal b. Installation

INITIAL SETUP:

Equipment Condition

- Truck raised and frame supported on jack stands.
- Wheel removed. (See TM 9-2320-289-20)
- Rear stabilizer bar removed (M1028A2 and M1028A3). (See TM 9-2320-289-20)

Materials/Parts

- One center bolt
- Three locknuts
- Lubricating oil (item 47, Appendix B)

Tools/Test Equipment

- Torque wrench

Personnel required

- MOS 63W (2)

General Safety instructions

- Rear spring assembly is heavy. Use extreme caution and assistance during removal and installation,
- Use extreme caution when releasing tension from spring assembly and auxiliary spring.

a. REMOVAL

NOTE

Model differences in spring assembly configuration and attaching hardware to axle and frame will be pointed out as they occur.

6-20. REAR SPRING ASSEMBLY AND BUSHINGS REPLACEMENT (Con't).

NOTE

Perform step 1 for all except M1028A2 and M1028A3.

1. Place a suitable jack under axle (5), Remove 4 nuts (7), washers (6), 2 "U" bolts (3), and anchor plate (8). Remove spacer (2) if present,

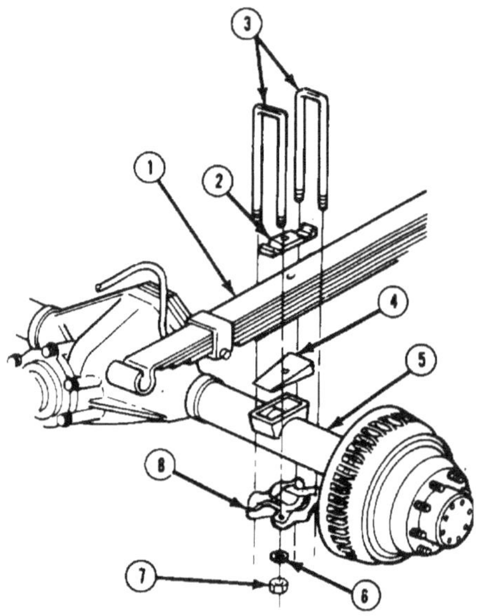

All Except M1009, M1028A2, and M1028A3

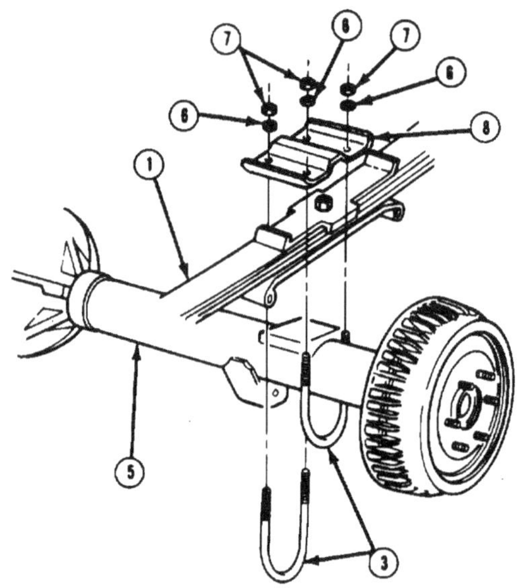

M1009

6-20. REAR SPRING ASSEMBLY AND BUSHINGS REPLACEMENT (Con't).

NOTE

Perform step 1.1 for M1028A2 and M1028A3.

1.1. Place a suitable jack under axle (5). Remove 4 nuts (7), washers (6), adapter (7.1), 4 washers (7.2), 2 "U" bolts (3), and anchor plate (8). Remove spacer (2).

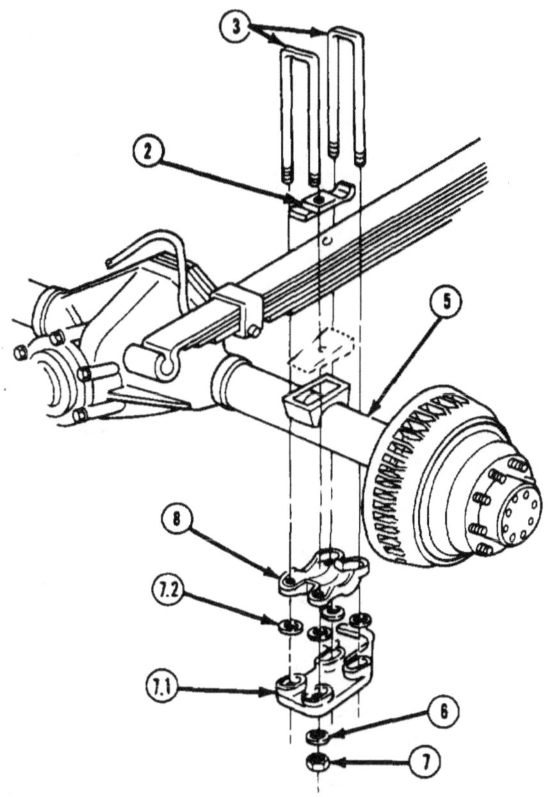

M1028A2 and M1028A3

2. Lower axle (5) until all tension is released from spring assembly (1). Remove shim (4) if present.

6-20. REAR SPRING ASSEMBLY AND BUSHINGS REPLACEMENT (Con't).

3. Remove locknut (11), bolt (14), and 2 washers (12), and remove shackle (18) from hanger (13), Discard locknut,

WARNING

Spring assembly (1) is heavy. Use extreme caution and assistance during removal. Failure to follow this warning may result in serious injury to personnel or damage to axle assembly.

4. Remove locknut (22), bolt (19), and 2 washers (21) securing spring assembly (1) to hanger (23). Remove spring assembly. Discard locknut.

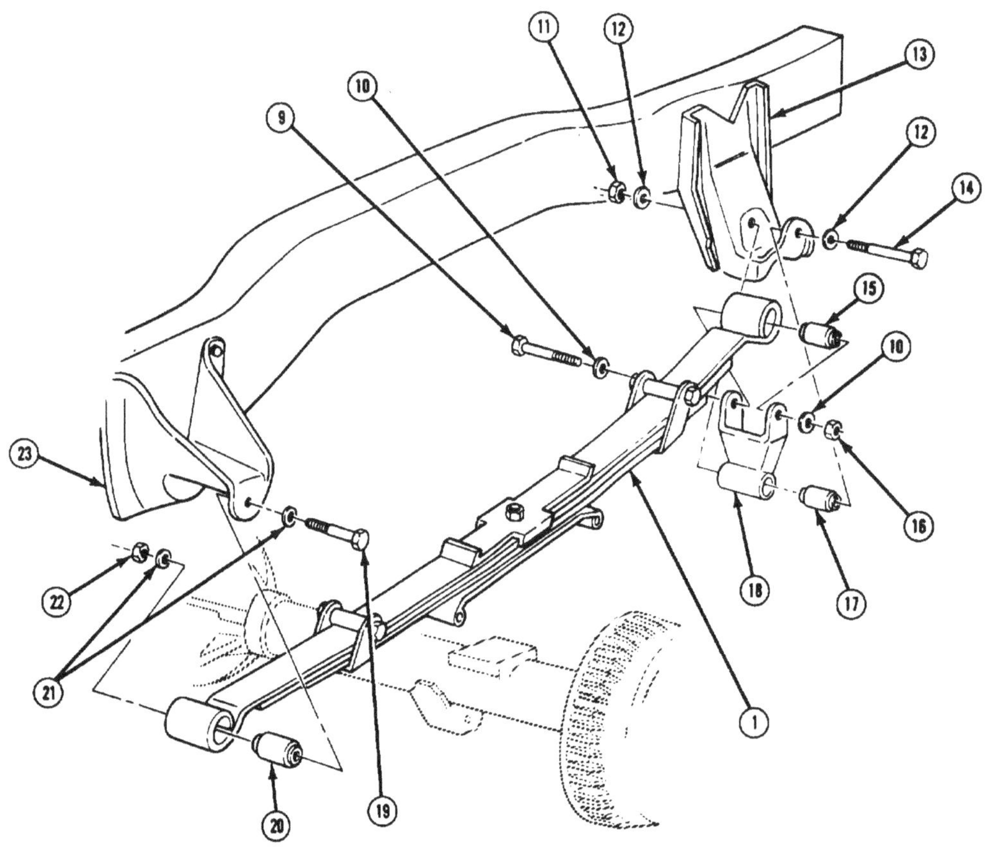

6-20. REAR SPRING ASSEMBLY AND BUSHINGS REPLACEMENT (Con't).

NOTE

Ensure that position of shackle (18) is noted for installation.

5. Remove locknut (16), bolt (9), 2 washers (10), and remove shackle (18) from spring assembly (1). Discard locknut.

6. If bushings (15, 17, and 20) are damaged, press out 3 bushings.

NOTE

If disassembling Ml 028, M1028A1, M1028A2, or M1028A3 auxiliary spring (24) and spring assembly (1), perform steps 7 and 8.

7. Place spring assembly (1) and auxiliary spring (24) *in* vise and compress leaves near center bolt (31). Remove 2 nuts (28), bolts (30), and clips (29), securing auxiliary spring to spring assembly. File peened end of center bolt and remove nut (27), washer (26), and center bolt. Discard center bolt.

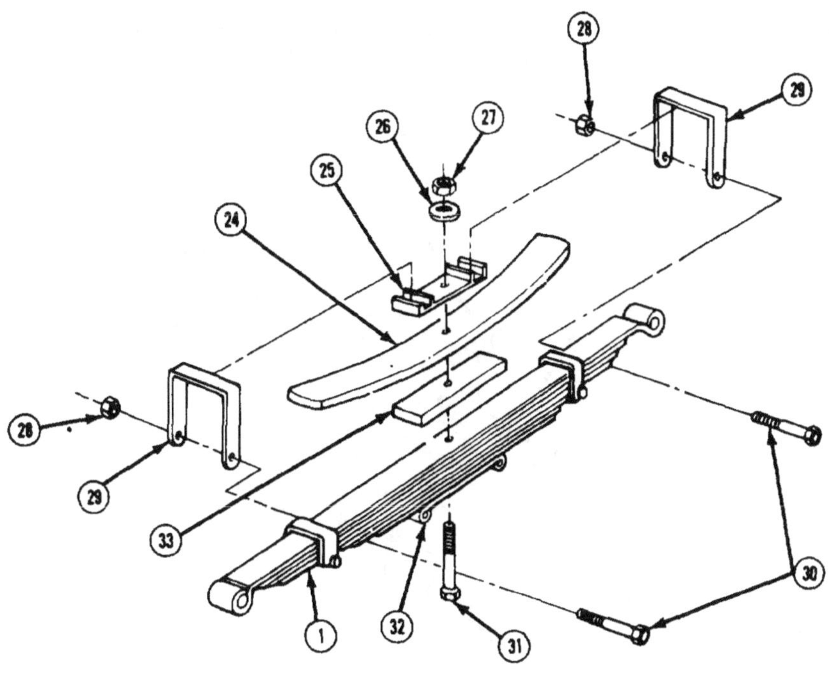

6-20. REAR SPRING ASSEMBLY AND BUSHINGS REPLACEMENT (Con't).

WARNING

Use extreme caution when releasing tension from spring assembly (1) and auxiliary spring (24). Springs are under considerable pressure and could cause injury to personnel if suddenly released.

NOTE

Several leaves of spring assembly (1) will be loose when pressure on vise is released.

8. Open vise slowly and allow all leaves to expand. Disassemble top retainer (25), auxiliary spring (24), spacer (33), spring assembly (1), and bottom retainer (32).

b. INSTALLATION

NOTE

If assembling M1028, M1028A1, M1028A2, or M1028A3 auxiliary spring (24) and spring assembly (1), perform steps 1 and 2.

1. Assemble bottom retainer (32), spring assembly (1), spacer (33), auxiliary spring (24), and top retainer (25), Aline holes in spring assembly leaves, retainers, spacer, and auxiliary spring with a long drift.

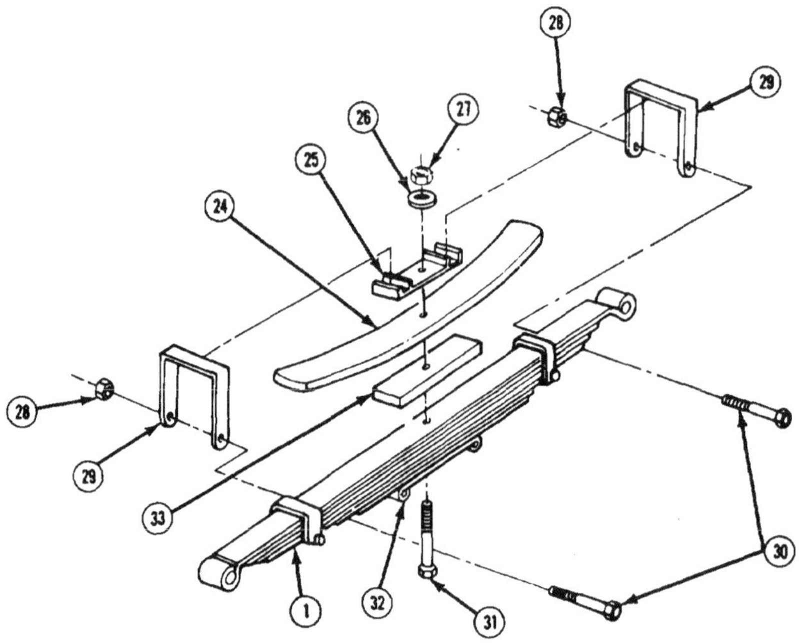

6-20. REAR SPRING ASSEMBLY AND BUSHINGS REPLACEMENT (Con't).

2. Compress all layers in vise near center bolt (31) hole. Remove drift and install new center bolt. Install washer (26) and nut (27). Tighten nut to 35 lb.-ft. (47 N.) and peen bolt to prevent nut from backing off.

3. Reposition spring assembly (1) and auxiliary spring (24) in vise and install 2 clips (29) with bolts (30) and nuts (28). Tighten nuts to 35 lb.-ft. (47 N.m). Remove spring assembly and auxiliary spring from vise.

CAUTION

Ensure that press tool contacts steel outer shell of bushings (15, 17, and 20), Failure to follow this caution may result in damage to bushings.

4. If removed, lubricate 3 bushings (15, 17, and 20) with light lubricating oil and press in.

5. Install shackle (18) to spring assembly (1) with bolt (9), 2 washers (10), and new locknut (16). DO NOT fully tighten locknut.

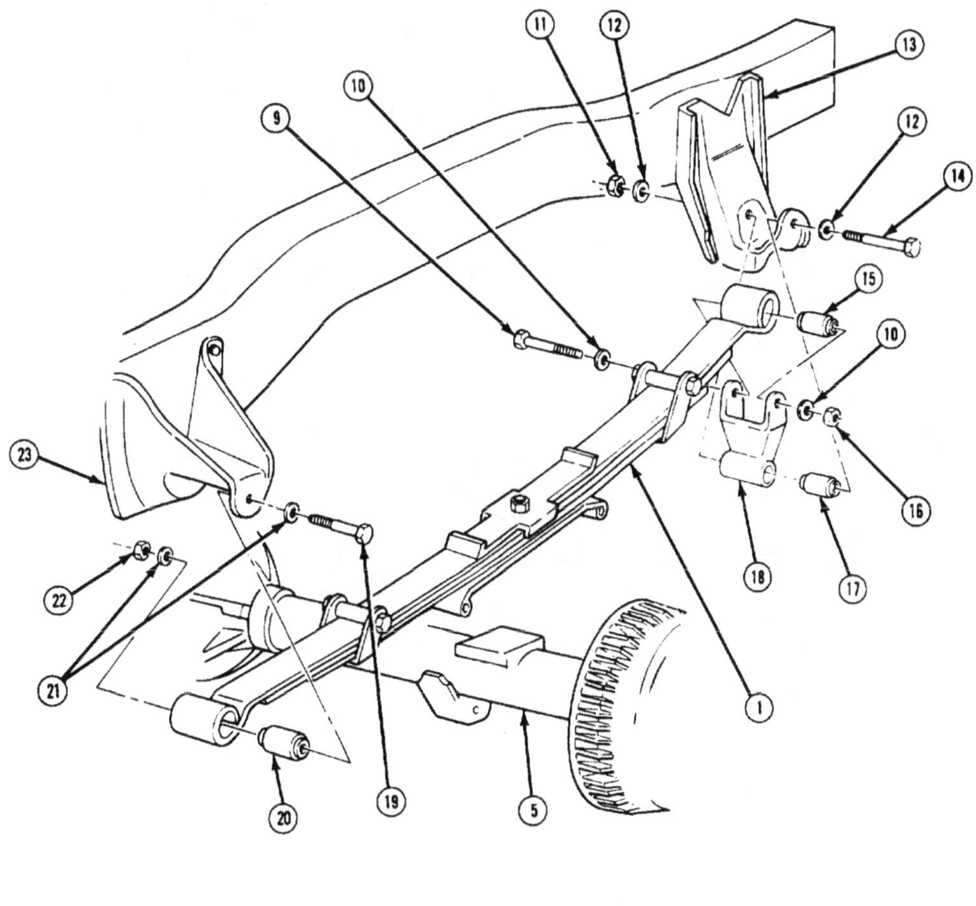

6-20. REAR SPRING ASSEMBLY AND BUSHINGS REPLACEMENT (Con't)

WARNING

Spring assembly (1) is heavy. Use extreme caution and assistance during installation. Failure to follow this warning may result in serious injury to personnel or damage to axle (5) assembly,

6. Position spring assembly (1) over axle (5),
7. Install spring assembly (1) to hanger (23) with bolt (19), 2 washers (21), and new locknut (22). DO NOT fully tighten locknut,
8. Install shackle (18) to hanger (13) with bolt (14), 2 washers (12), and new locknut (11). Tighten locknuts (11, 16, and 22) to 110 lb.-ft. (149 N.).

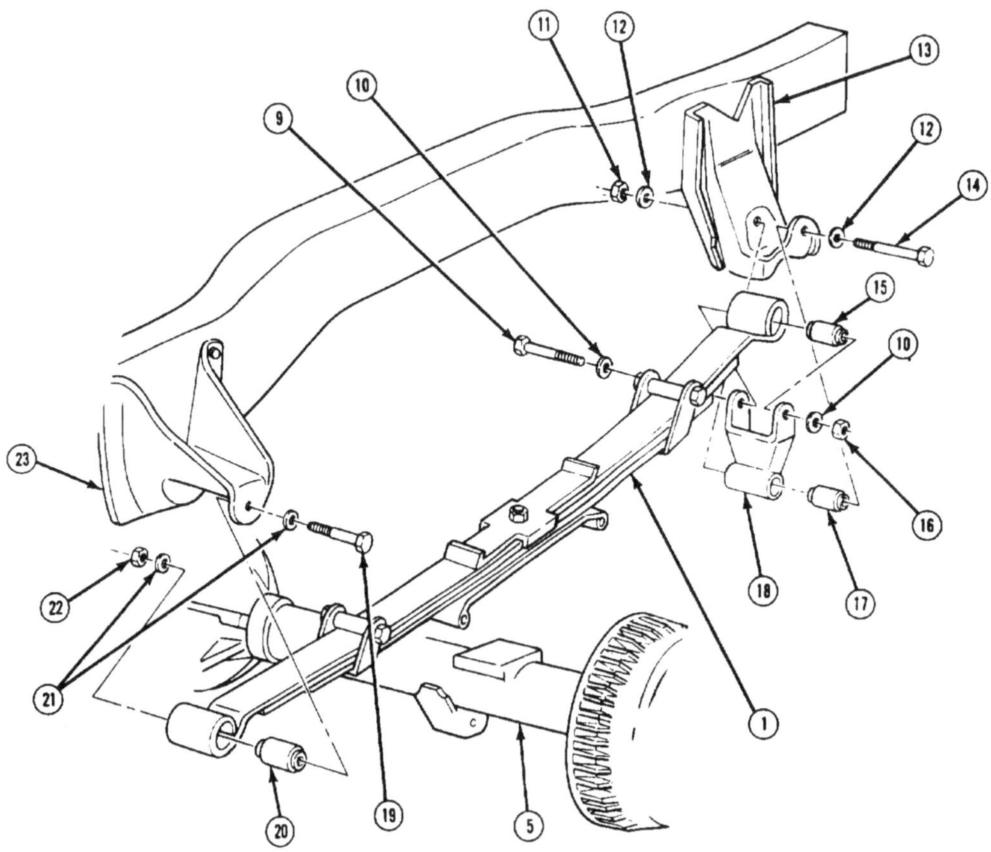

6-20. REAR SPRING ASSEMBLY AND BUSHINGS REPLACEMENT (Con't).

NOTE

Perform step 9 for all except M1028A2 and M1028A3.

9. If removed, install shim (4) on axle (5) housing with thinner edge facing front. Raise axle, alining spring assembly (1) center bolt hole with hole in shim and axle housing. If removed, install spacer (2). Install anchor plate (8) and 2 "U" bolts (3) with 4 washers (6) and nuts (7), Install all nuts to uniform engagement on "U" bolts to retain and position anchor plate.

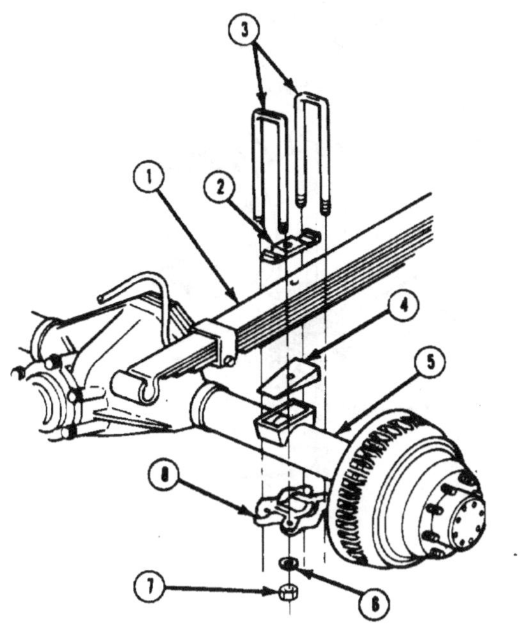

All Except M1009, M1028A2, and M1028A3

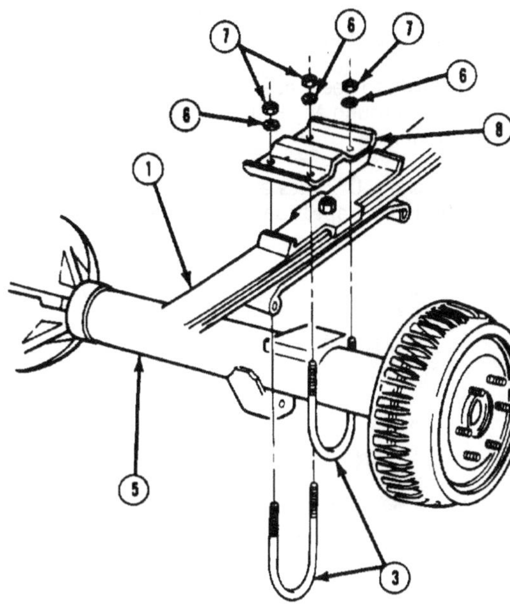

M1009

6-20. REAR SPRING ASSEMBLY AND BUSHINGS REPLACEMENT (Con't).

NOTE

. Perform step 9,1 for M1028A2 and M1028A3.

. Adapters (7,1) are mismarked on casting as to left-hand and right-hand. Install adapters as shown, with ribs up and curved arm to rear facing inboard,

9.1. If removed, install shim (4) on axle (5) housing with thinner edge facing front. Raise axle, alining spring assembly (1) center bolt hole with hole in shim and axle housing. Install spacer (2). Install anchor plate (8) and 2 " U" bolts (3) with 4 washers (7.2), adapter (7.1), 4 washers (6), and nuts (7). Install all nuts to uniform engagement on "U" bolts to retain and position anchor plate and adapter.

9.2 Check axle to see that spring dowel pins are through angle shim and are seated in axle.

9.3 Check that "U" bolts are properly positioned in upper spacer grooves.

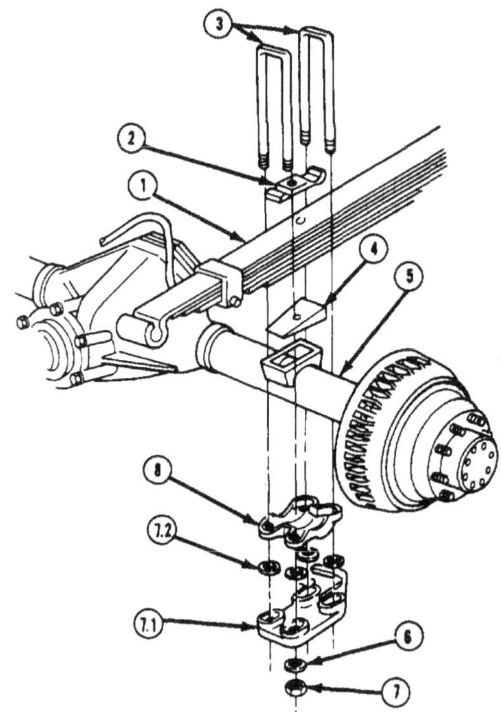

M1028A2 and M1028A3

6-20. REAR SPRING ASSEMBLY AND BUSHINGS REPLACEMENT (Con't).

NOTE

All nuts (7) must be tightened evenly to ensure that "U" bolts (3) and spring assembly (1) seat properly.

10. Tighten 4 nuts (7) in diagonal sequence to 25 lb.-ft. (34 N.m). Fully tighten all nuts on M1009 in same diagonal sequence to 145 lb.-ft, (197 N.). Fully tighten all nuts on all except M1009 M1028A2, and M1028A3 in same diagonal sequence to 170 lb.-ft. (231 N.m). Fully tighten all nuts on M1028A2 and M1028A3 in same diagonal sequence to 150 lb.-ft. (203 N.m).

11. Remove jack from axle (5).

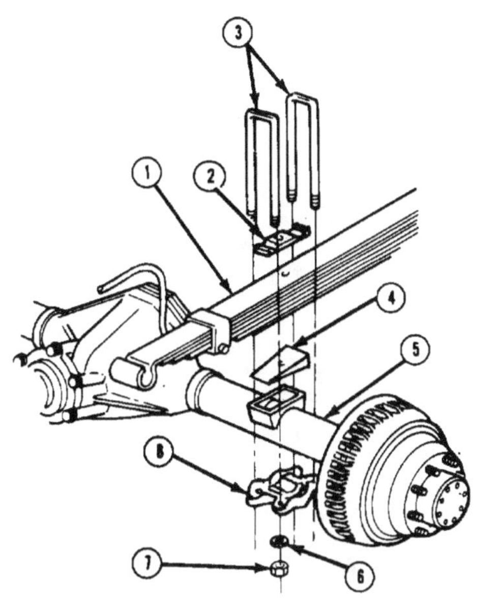

All Except M1009, M1028A2, and M1028A3

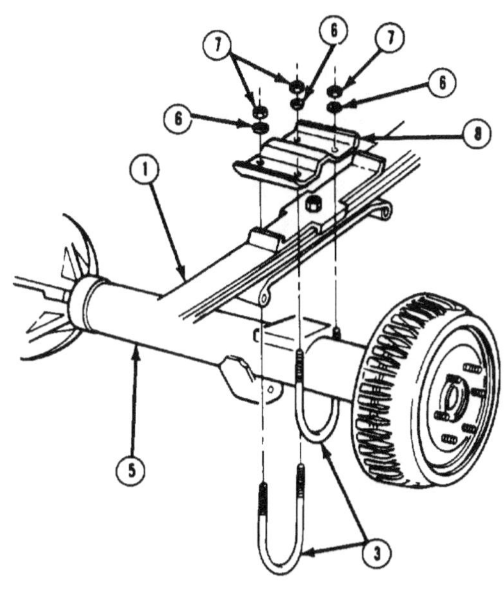

M1009

FOLLOW-ON TASKS:

- Install rear stabilizer bar (M1028A2 and M1028A3). (See TM 9-2320-289-20)
- Install wheel. (See TM 9-2320-289-20)
- Remove jack stands and lower truck.
- Road test truck and check torque of "U" bolt fasteners.

CHAPTER 7
BRAKE SYSTEM MAINTENANCE

7-1. BRAKE SYSTEM MAINTENANCE INDEX.

7-2. DRUM AND ROTOR REPAIR.

This task covers: a. Drum Repair b. Rotor Repair

INITIAL SETUP:

Equipment Condition

- Drum or rotor removed. (See TM 9-2320-289-20)

Tools/Test Equipment

- Micrometer
- Dial indicator

Materials/Parts

- Abrasive cloth (Item 17, Appendix B)

General Safety Instructions

- DO NOT use a drum or rotor that will not meet minimum wear specifications.

a. DRUM REPAIR

WARNING

DO NOT use a drum that will not meet minimum wear specifications. Failure to follow this warning may result in brake failure and serious injury or death to personnel.

1. Measure drum for out-of-round, taper, and wear. Take measurements at open and closed edges of machined surface, and at right angles to each other. If drum measures more than 11.240 in. (285.50 mm) (M1009) or 13.090 in. (332.49 mm) (all except M1009), discard drum.

NOTE

If brake linings are not being replaced, drum can be polished with fine abrasive cloth but should not be machined.

2. If drum is grooved, out-of-round, tapered, or badly worn, machine drum using Table 7-1 specifications.

Table 7-1. Drum Diameters

Model	Original	Maximum Refinish	Replacement/Discard
M1009	*11.150* in. (283.21 mm)	11.210 in. (284.73 mm)	11,240 in. (285.50 mm)
All Except M1009	13,000 in. (330,20 mm)	13.060 in. (331.72 mm)	13.090 in. (332.49 mm)

7-2. DRUM AND ROTOR REPAIR (Con't).

b . ROTOR REPAIR

WARNING

DO NOT use a rotor that will not meet minimum wear specifications. Failure to follow this warning may result in brake failure and serious injury or death to personnel.

NOTE

- Light scoring of rotor surface not in excess of 0.015 in. (0.381 mm) in depth is normal and will not affect brake operation.
- Parallelism is the measurement of the thickness of rotor at 4 or more points around circumference of rotor. All measurements must be made at the same distance in from edge of rotor, Thickness must not vary by more than 0,0005 in. (0,0127 mm) from point to point.

1. Measure thickness of rotor at 4 points around circumference. If rotor measures less than 1.215- in. (30.86 mm) (M1009) or' 1.465 in. (37.21 mm) (all except M1009), discard rotor.
2. If rotor thickness varies over 0.0005 in. (0.0127 mm), refinish to Table 7-2 specifications.

Table 7-2. Rotor Thickness

Model	Mminimum After Refinishing	Replacement/ Discard
M1009	1.230 in. (31.24 mm)	1,215 in. (30.86 mm)
All Except Ml 009	1.480 in. (37.59 mm)	1.465 in. (37.21 mm)

3. Check lateral runout.

   (a) Install rotor. (See TM 9-2320-289-20)

   (b) Tighten wheel bearings to eliminate freeplay.

   (c) Attach dial indicator (1) to some portion of suspension. Point of plunger must contact rotor face (2) about 1 in. (25 mm) from rotor edge.

   (d) Move rotor 1 complete rotation. Lateral runout should not exceed 0.004 in. (0.102 mm). If lateral runout exceeds 0.004 in. (O. 102 mm), machine rotor.

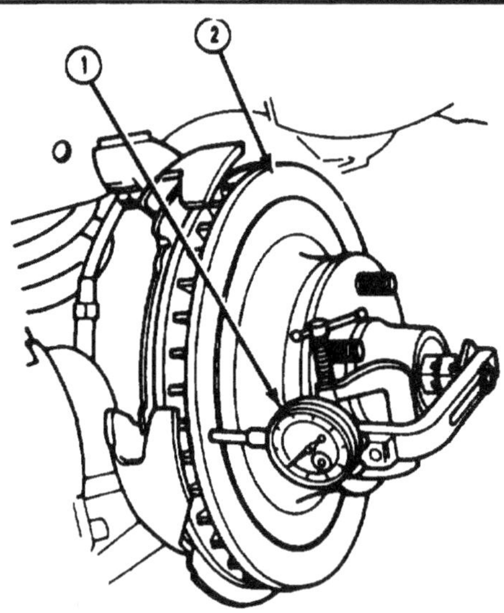

TA50375

7-2. DRUM AND ROTOR REPAIR (Con't).

FOLLOW-ON TASKS:

- Adjust wheel bearings. (See TM 9-2320-289-20)
- Install drum or rotor. (See TM 9-2320-289-20)

7-3. POWER BOOSTER REPAIR.

This task covers:

a. Disassembly
b. Cleaning and Inspection
c. Assembly

INITIAL SETUP:

Equipment Condition

- Power booster removed. (See TM 9-2320-289-20)

Tools/Test Equipment

- Accumulator remover, J-26889
- " C" clamp
- Torque wrench

Materials /Parts

- Power booster repair kits, as required
- Hydraulic fluid (Item 40, Appendix B)

General Safety Instructions

- Wear goggles while performing this task.
- Do NOT apply heat to accumulator.

a. DISASSEMBLY

WARNING

Wear goggles while performing this task. Failure to use correct tools and procedures may result in injury to personnel.

7-3. POWER BOOSTER REPAIR (Con't).

NOTE

M1009 power boosters can be distinguished from other power boosters by using the following illustration.

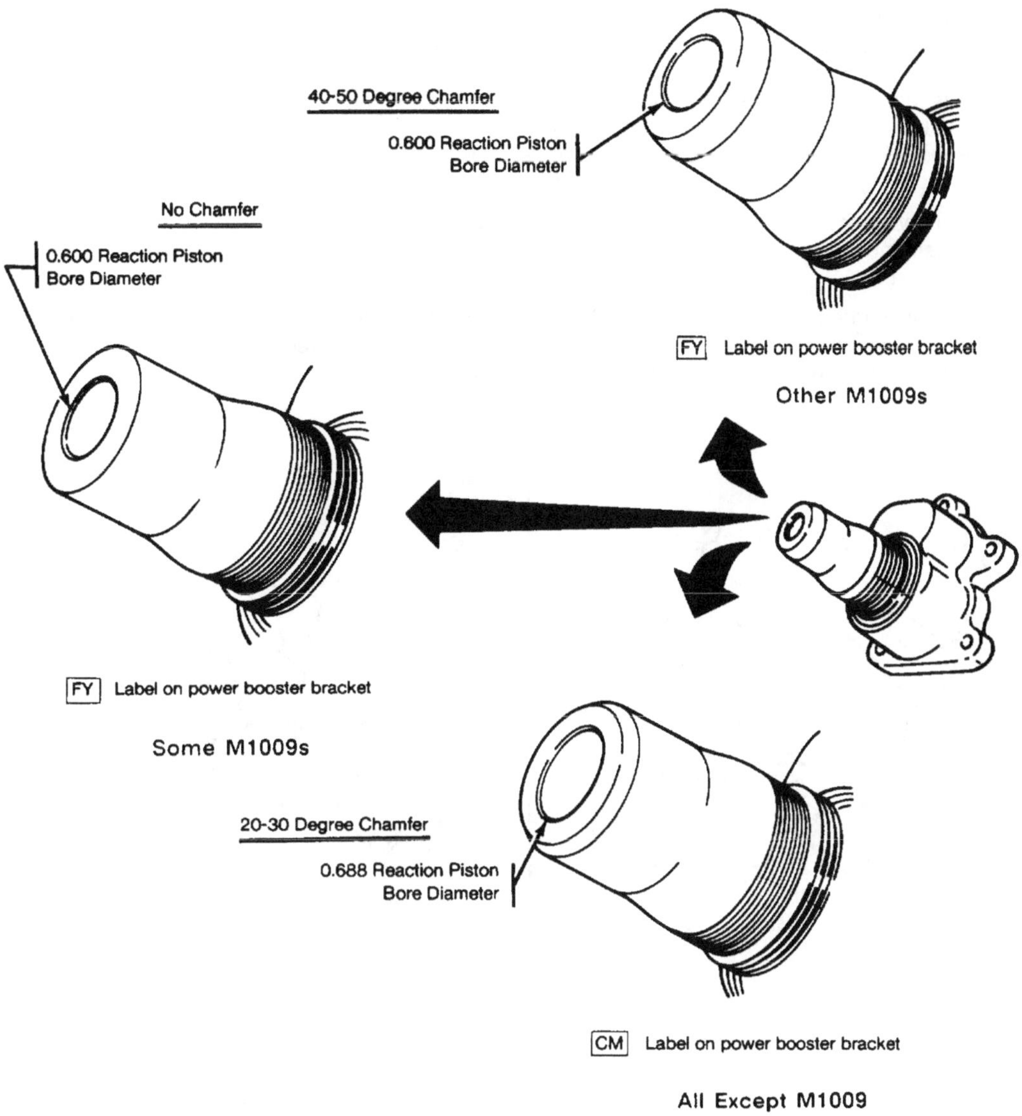

1. Install power booster (4) in vise with bracket (3) end down. master cylinder stud (5) and install nut (6). Using large "C" (2) .

Place accumulator remover over clamp (1), depress accumulator

TA50376

7-3. POWER BOOSTER REPAIR (Con't).

WARNING

Accumulator (2) contains compressed gas. DO NOT apply heat to accumulator. DO NOT attempt to repair an inoperative accumulator, always replace an inoperative accumulator with a new one. Drill a 1/16 in. (1.6 mm) hole through end of accumulator can opposite "O" ring (8) to dispose of an inoperative accumulator. Failure to follow this warning may result in serious injury or death to personnel,

NOTE

It may be necessary to rotate accumulator retaining ring (7) so that it is positioned in line with hole (10) in housing (9).

2. Insert a punch into hole (10) in housing (9) and remove accumulator retaining ring (7). Slowly back off "C" clamp (1) until tension on accumulator (2) is released. Remove accumulator remover, accumulator, and "O" ring (8), Discard "O" ring.

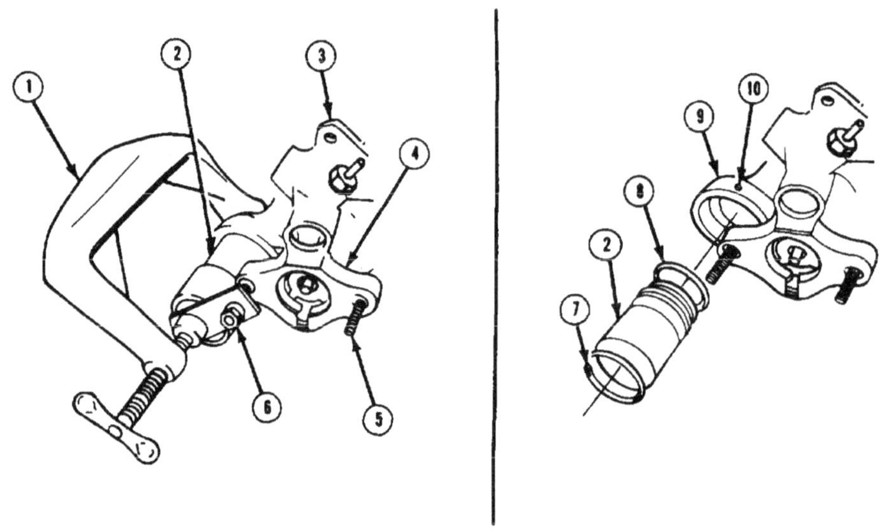

TA50377

7-3. POWER BOOSTER REPAIR (Con't).

WARNING

Spring retainer (13) is under spring pressure. Wear goggles during removal. Failure to follow this warning may result in injury to personnel.

3 . Disengage tabs of spring retainer (13). Remove spring retainer and piston return spring (12). Remove output rod (11).

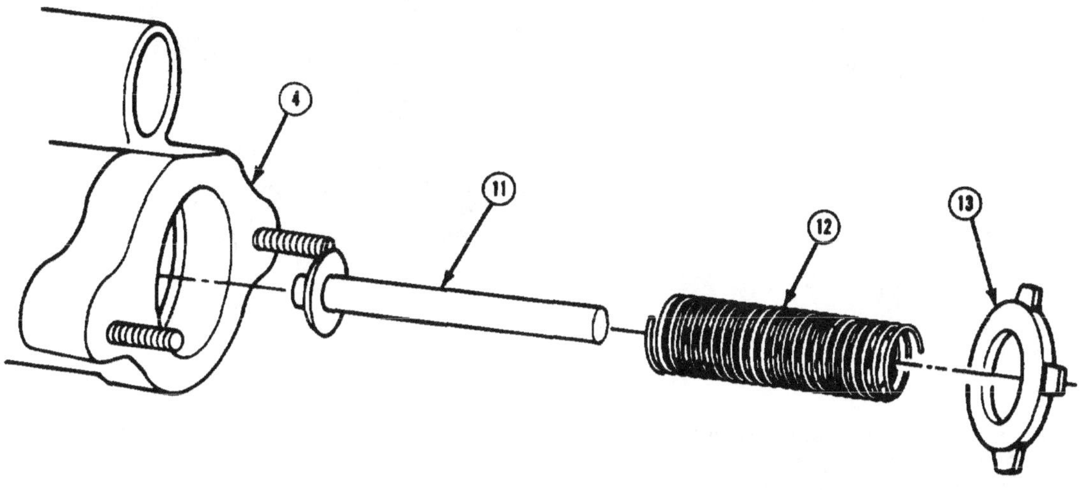

4. Remove 5 bolts (14). Remove power booster (4) from vise. Hold power booster over a suitable container and separate cover (20) from housing (9). Remove cover seal (19) and discard.

NOTE

Perform step 5 only if rod and piston unit (17), and pedalrod (22) are staked together (M1009).

5. Saw eyelet off of pedalrod (22) and remove rod and piston unit (17) from cover (20). Discard rod and piston unit and pedalrod.
6. Remove locknut (21) and pedalrod (22) (M1009) or pedalrod (22) and grommet (23) (all except M1009), Remove rod and piston unit (17) from cover (20).
7. Remove piston seal (15) and seal (s) (18). Discard seals.

TA50378

7-3. POWER BOOSTER REPAIR (Con't).

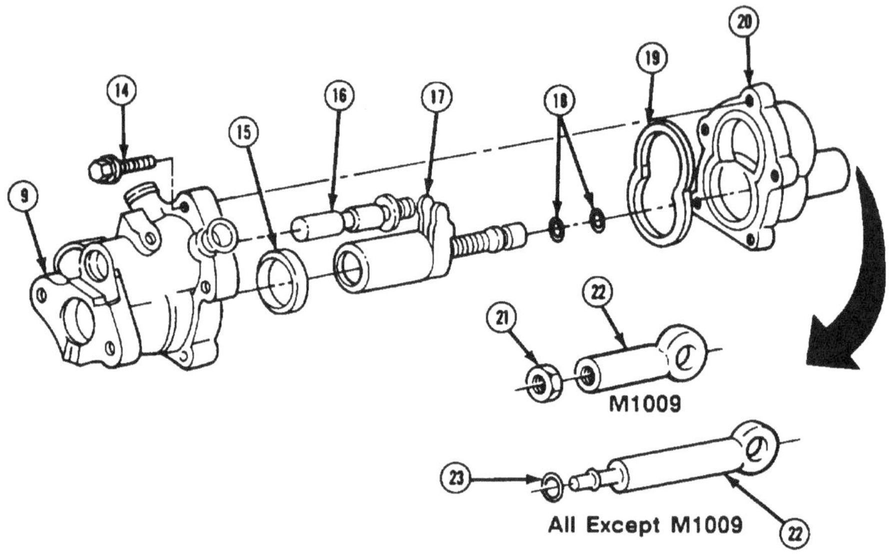

8 . Push in spool valve plug (26) and remove retaining ring (27). Remove spool valve plug, "O" ring (25), spool return spring (24), and spool valve assembly (16). Discard "O" ring.

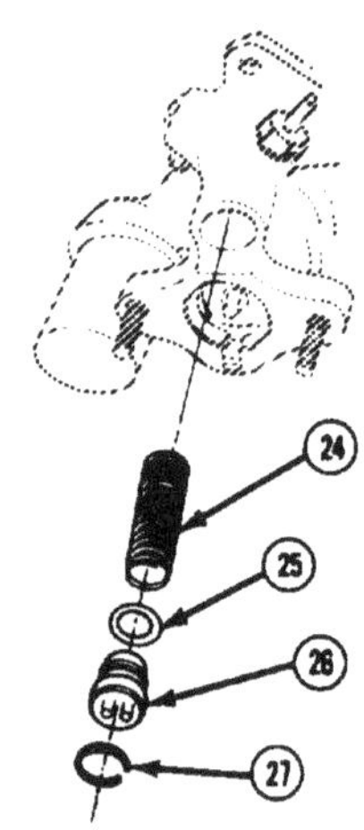

TA50379

7-3. POWER BOOSTER REPAIR (Con't).

9. Using a small diameter wire tool (32), remove accumulator valve (33). Remove dump valve (28) by catching wire tool under pin guide (29). Remove "O" ring (30) and two-function valve (31), Discard "O" ring.

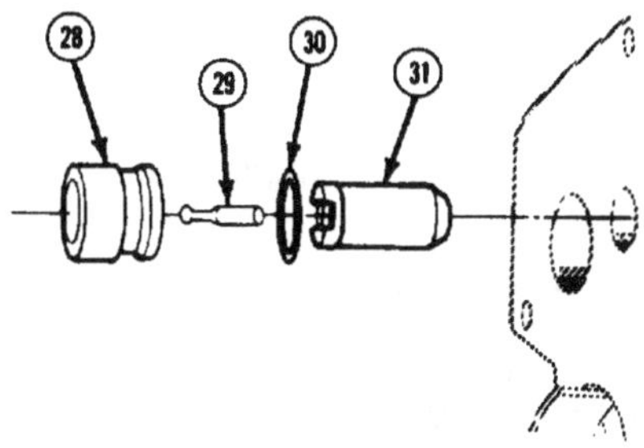

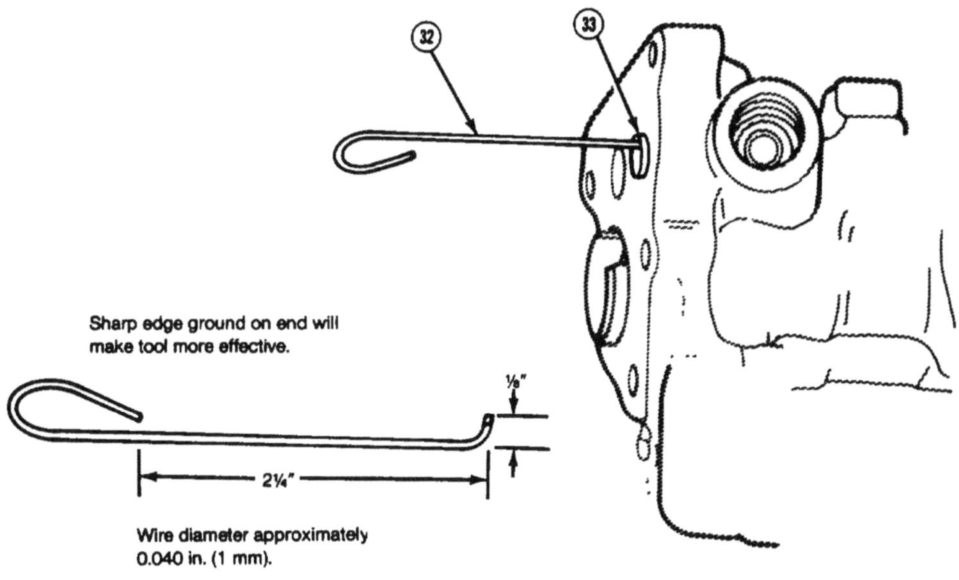

TA50380

7-3. POWER BOOSTER REPAIR (Con't).

b. CLEANING AND INSPECTION

NOTE

Ensure that parts are kept clean until assembly. If any accumulator valve parts are damaged or lost, replace all valve parts.

1. Clean all parts with hydraulic fluid.
2. Inspect rod and piston unit (17) for scratches. If scratches are detected, replace rod and piston unit (M1009 only) or power booster assembly (all except M1009).
3. Inspect spool valve assembly (16) and spool valve bore in housing (9) for damage. Discoloration of spool or bore is considered normal. If spool valve or spool valve bore have nicks or scoring, or if there is corrosion, power booster should be replaced as an assembly.

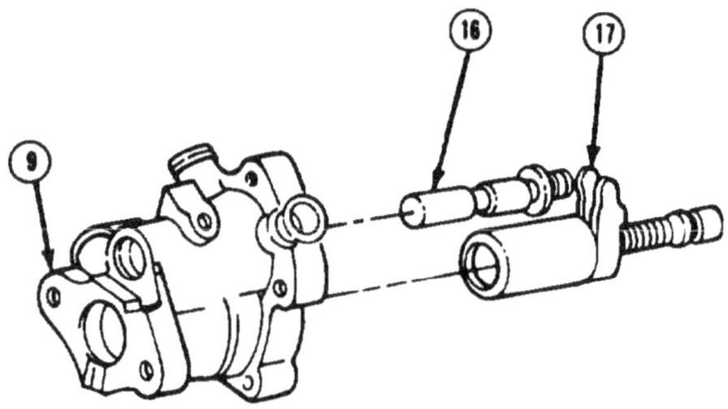

TA50381

7-3. POWER BOOSTER REPAIR (Con't).

c. ASSEMBLY

NOTE

Ensure that all seals, "O" rings, and metal friction points are lubricated with hydraulic fluid before assembly.

1. Install two-function valve (31) into back of housing (9). Install new "O" ring (30) onto dump valve (28) and install over two-function valve. Ensure that dump valve plunger is held in place until installation is completed. Install accumulator valve (33).

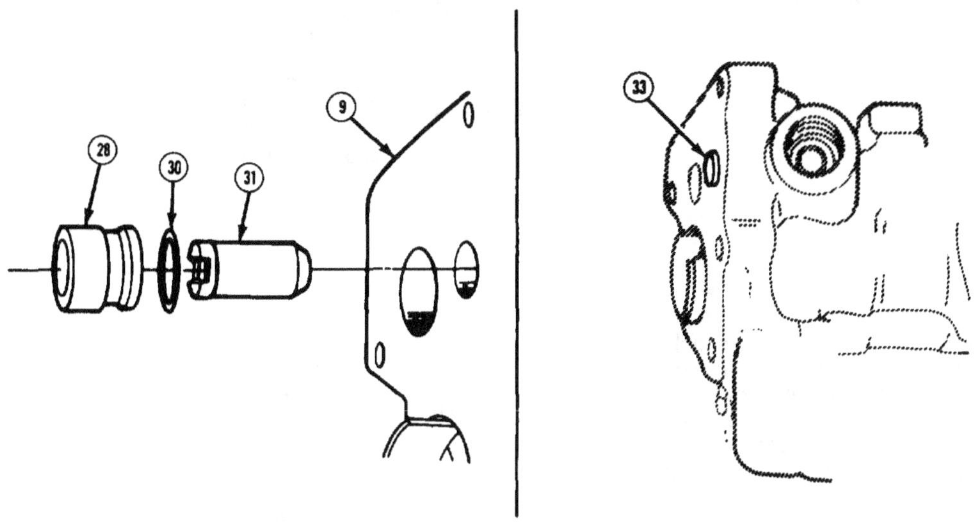

2. Install cover (20) in vise with bracket end down. Install new cover seal (19) into groove on cover.

3. Install new seal(s) (18) on rod and piston unit (17). insert rod and piston unit into bore in cover (20). Install locknut (21) and pedalrod (22) onto rod and piston unit (M1009). Install grommet (23) on pedalrod and force into rod and piston unit (all except M1009).

4. Install new piston seal (15) in bore. Lip of seal must be toward rear (away from master cylinder mounting flange), Ensure that seal is fully seated in housing (9).

TA50382

7-3. POWER BOOSTER REPAIR (Con't).

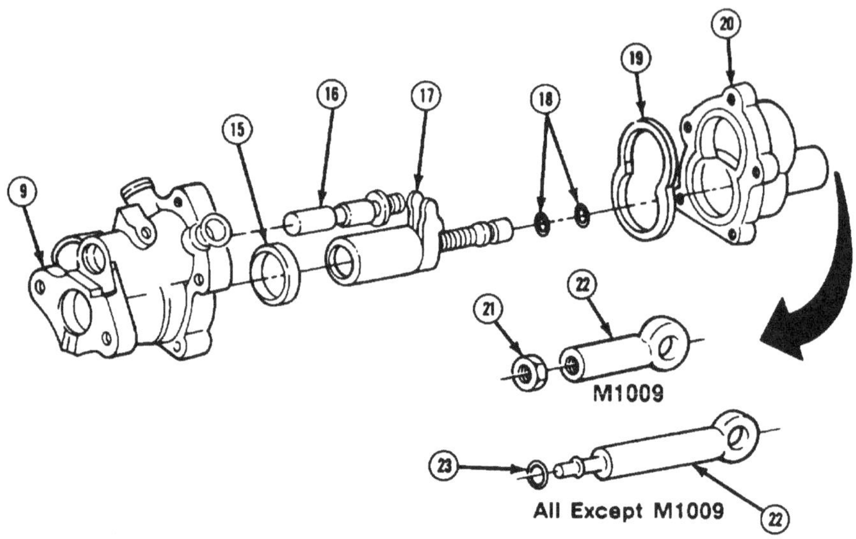

5. Install new "O" ring (25) onto spool valve plug (26). Push spool valve plug into housing (9). Install spool plug retaining ring (27) and ensure that it is fully seated. Install spool return spring (24) and spool valve assembly (16).

6. Coat rod and piston unit (17) with clean hydraulic fluid. Position housing (9) over cover (20) and slide pins on piston lever into groove on spool valve. Using a screwdriver to carefully depress lip of piston seal (15), slide housing and piston onto cover.

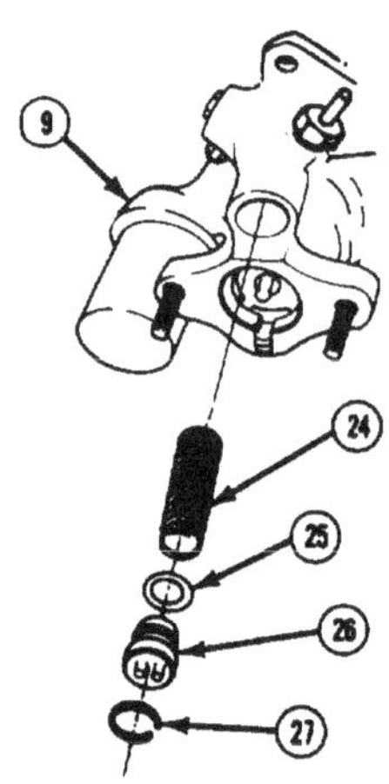

TA50383

7-3. POWER BOOSTER REPAIR (Con't).

7. Assemble cover (20) and housing (9) with 5 bolts (14). Install output rod (11), piston return spring (12), and spring retainer (13) with tabs facing outward. Tighten bolts to 20 lb.-ft, (27 N.m).

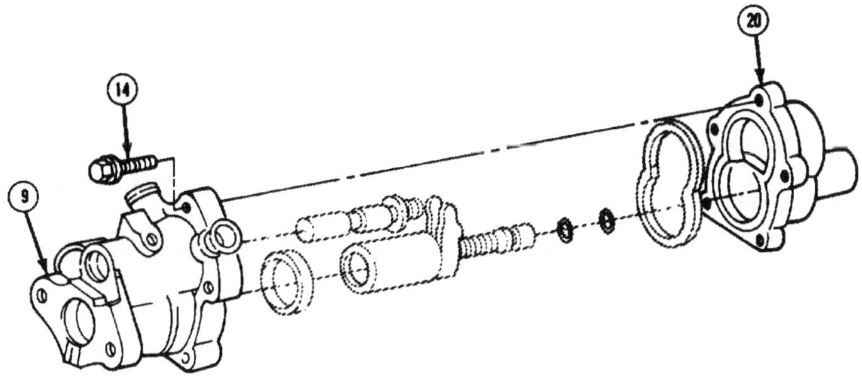

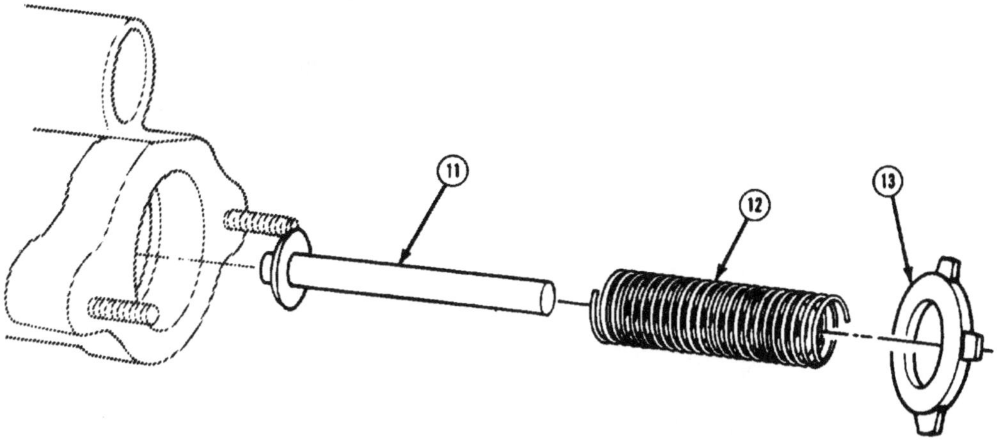

TA50384

7-3. POWER BOOSTER REPAIR (Con't).

8 . Install new "O" ring (8) and accumulator (2) in housing (9). Place accumulator retaining ring (7), accumulator remover, and nut (6) over accumulator.

CAUTION

Ensure that accumulator retaining ring (7) Is completely and properly installed and accumulator (2) is compressed straight. Failure to follow this caution may result In damage to accumulator.

9 . Using large "C" clamp (1), depress accumulator (2). Push accumulator retaining ring (7) completely around into housing (9) groove.

10. Remove "C" clamp (1) and accumulator remover. Remove power booster (4) from vise.

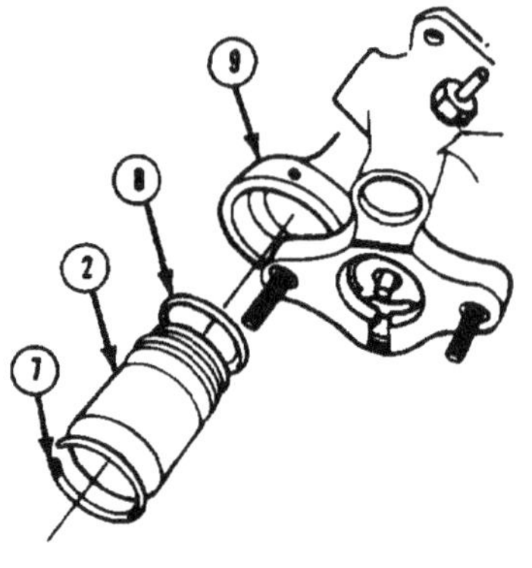

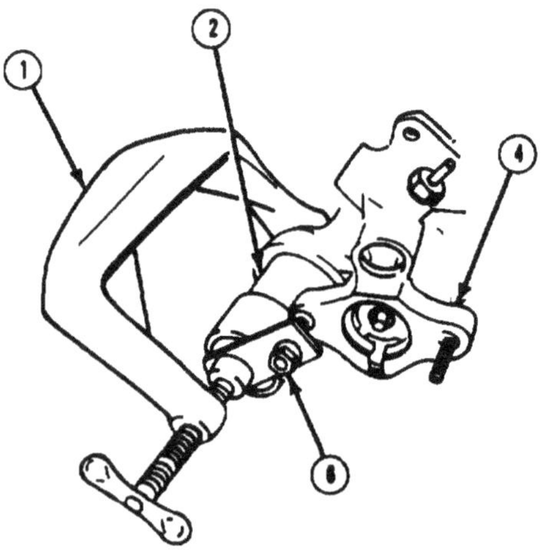

FOLLOW-ON TASKS:

- Install power booster. (See TM 9-2320-289-20)

TA60385

7-4. CALIPER REPAIR.

This task covers: a. Disassembly b. Cleaning and Inspection c. Assembly

INITIAL SETUP:

Equipment Condition

. Caliper assembly and brake pads removed. (See TM 9-2320-289-20)

Materials/parts

- Piston repair package
- Denatured alcohol (Item 6, Appendix B)
- Brake fluid (Item 10, Appendix B)
- Crocus cloth (Item 18, Appendix B)
- Wiping rags (Item 53, Appendix B)

General Safety Instructions

- DO NOT place hand in front of piston when removing it with compressed air,
- Compressed air used for cleaning purposes or to force out piston should never exceed 30 psi (207 kPa),
- Alcohol used in this procedure is highly flammable and poisonous.

a. DISASSEMBLY

WARNING

- Compressed air used to force out piston (3) should never exceed 30 PSI (207 kPa), Use only effective chip guarding and personnel protective equipment (goggles/shieid, gloves, etc.). Failure to follow this warning may result in injury to personnel,
- DO NOT Place hand in front of piston (3) when removing with compressed air. Failure to follow this warning may result in injury to personneL.

CAUTION

Use only enough air pressure to ease piston out of bore. DO NOT blow piston completely out of caliper bore (8) or damage to piston may result.

1. Pad interior of caliper (2) with clean rags (4) and using compressed air in brake fluid inlet hole (1), remove piston (3).

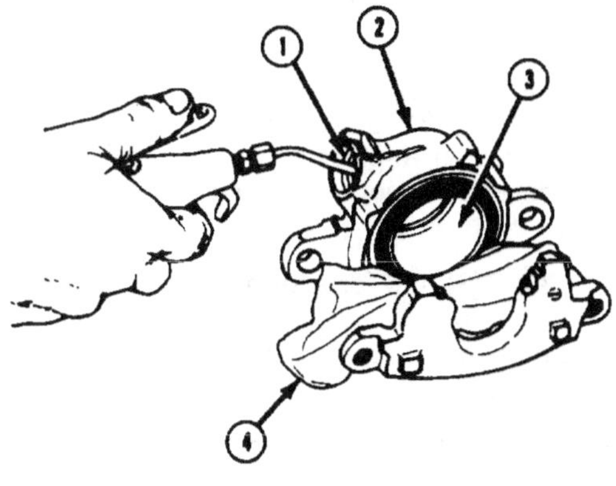

TA50386

7-4. CALIPER REPAIR (Con't).

CAUTION

DO NOT use metal tool to remove piston seal (6), Be careful not to scratch caliper bore (8) when removing dust boot (7). Failure to follow this caution may result in damage to caliper bore.

2. Remove dust boot (7) and piston seal (6) from caliper bore (8). Discard dust boot and piston seal. Remove bleeder valve (5).

b. CLEANING AND INSPECTION

WARNING

- Alcohol used in this procedure is highly flammable and poisonous. Drinking of this alcohol can cause blindness and death. Avoid inhaling its fumes. Keep away from open flame.
- Compressed air used for cleaning purposes should never exceed 30 psi (207" kPa), Use only effective chip guarding and personal protective equipment (goggles/shield, gloves, etc.), Failure to follow this warning may result in injury to personnel.

CAUTION

DO NOT use emery cloth or any other form of abrasive. If crocus cloth is used on caliper bore (8), ensure that caliper (2) is thoroughly cleaned before assembly, Failure to follow this caution may result in damage to caliper assembly.

1. Clean exterior of caliper (2) with clean soapy water.
2. Clean caliper bore (8), caliper passages, and piston (3) with denatured alcohol. Using dry, filtered, compressed air, dry parts and blow out all passages in caliper (2) and bleeder valve (5).
3. Inspect sliding surfaces of caliper (2) and caliper bore (8) for scoring, pitting, or corrosion, Replace caliper if surface defects cannot be cleaned with crocus cloth,
4. Inspect piston (3) and for scoring, nicks, corrosion, and worn or damaged plating. Replace piston if any defects are found.
5. Using crocus cloth, clean caliper mounting surfaces on truck and remove any defects or corrosion.

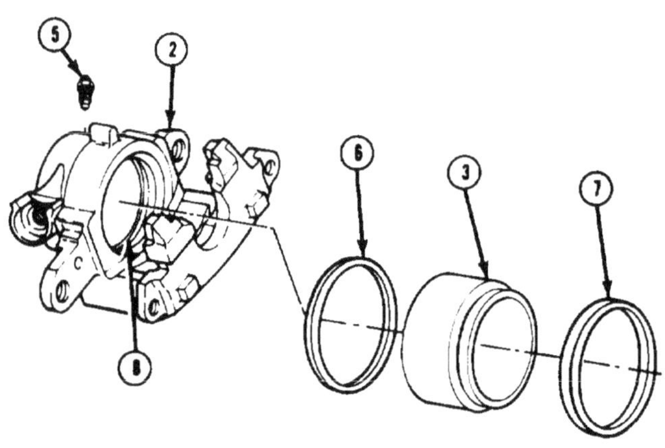

TA50387

7-4. CALIPER REPAIR (Con't).

c. ASSEMBLY

NOTE

Ensure that piston seal (6) is fully seated inside caliper bore (8) groove,

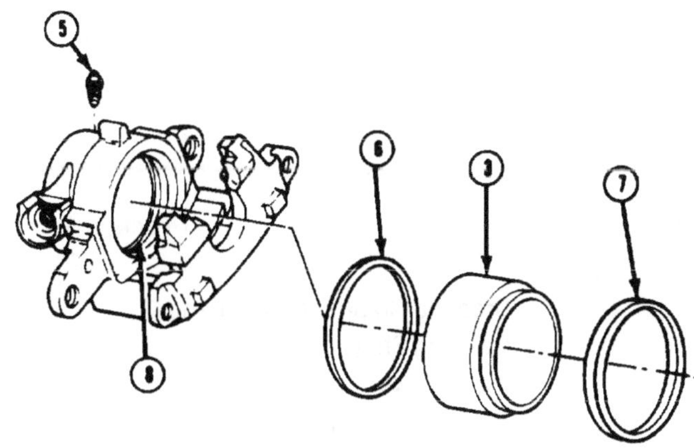

1. Lubricate caliper bore (8) and new piston seal (6) with clean brake fluid. install piston seal in caliper bore groove.
2. Lubricate new dust boot (7) with brake fluid, and pressing boot bead into groove, install into caliper bore (8) groove.

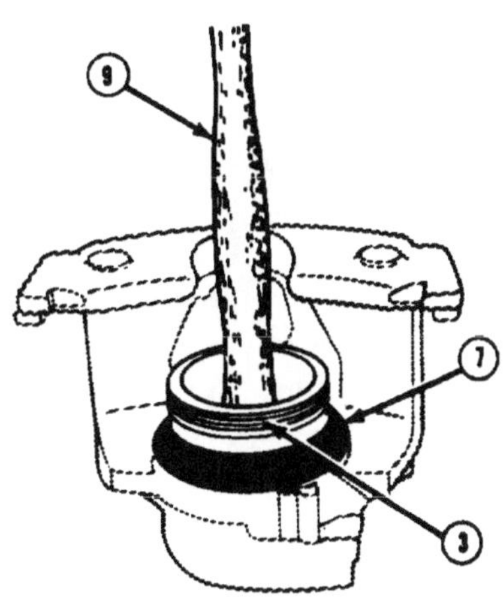

3. Lubricate piston (3) with clean brake fluid. Using wood hammer handle (9) or " C" clamp, install piston into caliper bore (8). Install bleeder valve (5).
4. Check that dust boot (7) is still seated. Reseat dust boot if unseated.

FOLLOW-ON TASKS:

- Install caliper assembly and brake pads. (See TM 9-2320-289-20)

TA60388

CHAPTER 8
STEERING SYSTEM MAINTENANCE

8-1. STEERING SYSTEM MAINTENANCE INDEX.

8-2. STEERING COLUMN MAINTENANCE.

This task covers: a. Removal b. Disassembly c. Assembly d. Installation

INITIAL SETUP:

Equipment Condition

- Directional signal switch removed. (See paragraph 4-1 4)
- Ignition switch removed. (See paragraph 4-1 3)

Tools/Test Equipment

- Torque wrench

Materials/Parts

- One floor pan seal
- Grease (Item 36, Appendix B)

a. REMOVAL

1. Place lock cylinder (2) in "RUN" position. Remove retaining screw (3) and lock cylinder from housing (1).

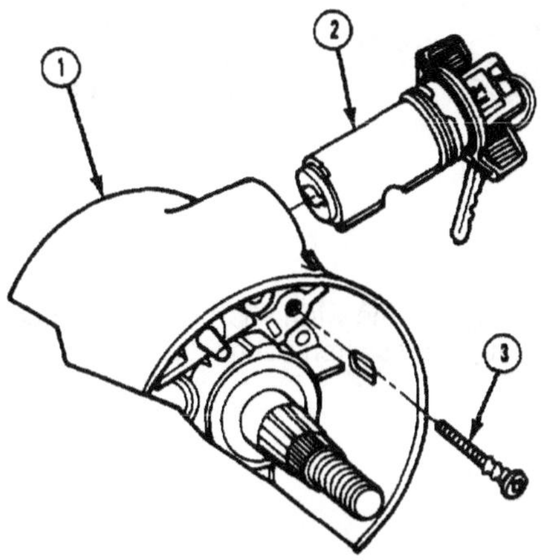

2. Disconnect transmission control linkage from steering column shift tube levers. (See paragraph 5-2)

3. Remove 2 nuts (6) and washers (5) securing flanged end of steering shaft to flexible coupling (4).

TA50471

8-2. STEERING COLUMN MAINTENANCE (Con't).

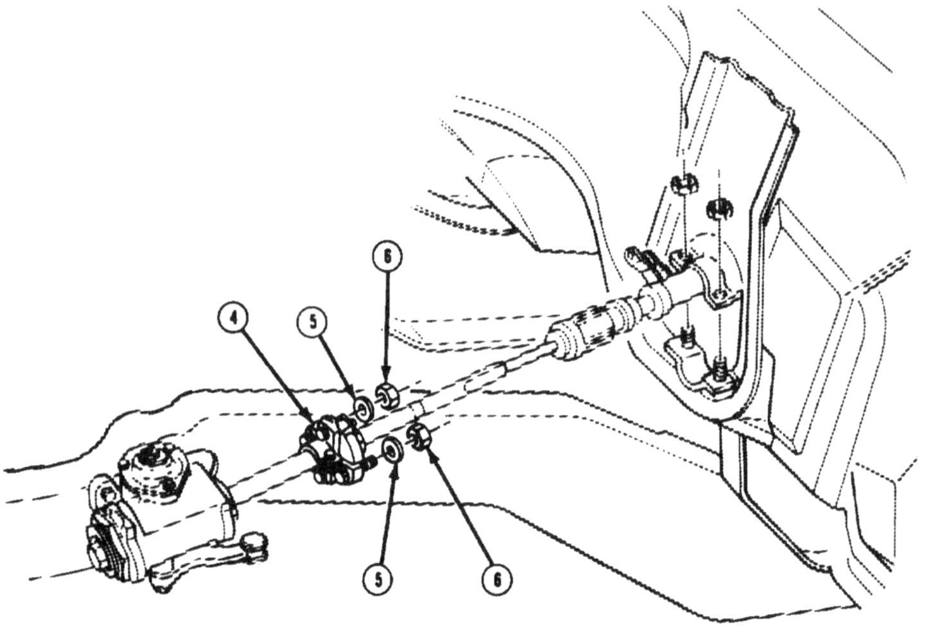

4 . Disconnect steering column wiring harness at connector on left side of steering column (9). Disconnect back-up light switch connector.

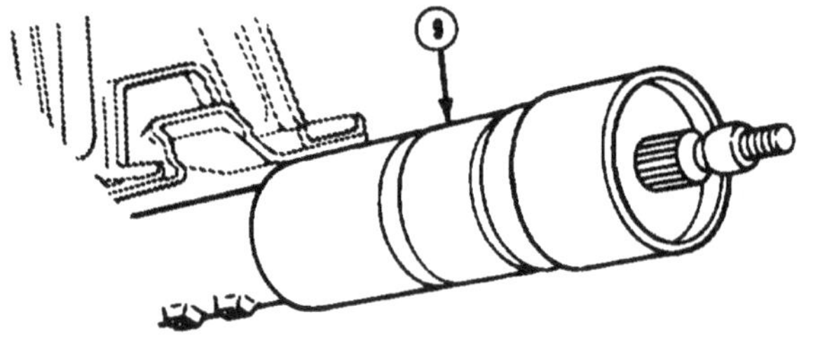

TA50472

8-2. STEERING COLUMN MAINTENANCE (Con't).

5. Move insulation out of the way. Remove 5 bolts (11) from upper cover (8) and lower cover (12).

NOTE

Ensure that front seat is moved as far back as possible to provide maximum clearance.

6. While holding steering shaft (10), rotate steering column (9) to allow transmission gearshift lever to clear dash. Carefully remove steering column assembly.

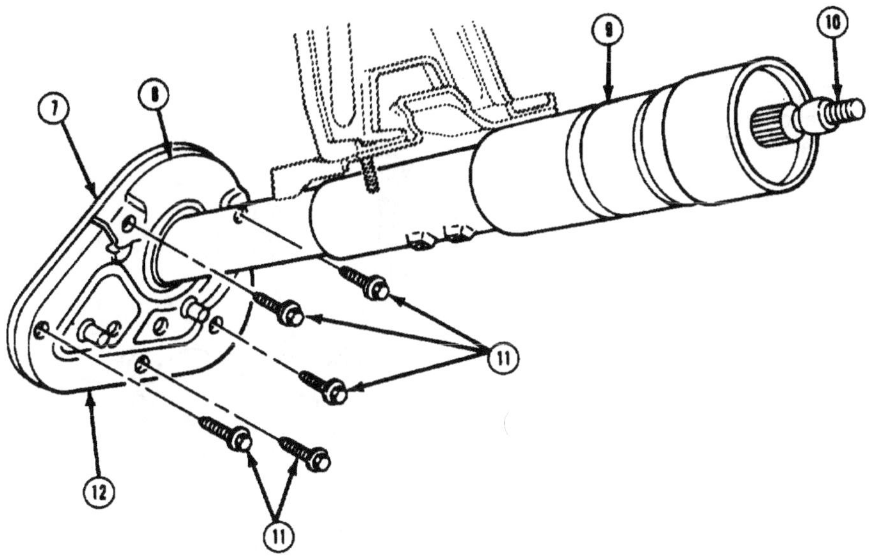

b. DISASSEMBLY

1. Remove upper cover (8) and lower cover (12) from seal (7),

TA50473

8-2. STEERING COLUMN MAINTENANCE (Con't).

CAUTION

Ensure that steering column (9) is placed in vise only as indicated. Failure to secure steering column properly in vise may result in damage to steering column.

2. Place steering column (9) in vise using both weldnuts of either set A or set B. Vise jaws must clamp onto sides of weldnuts indicated by arrows.

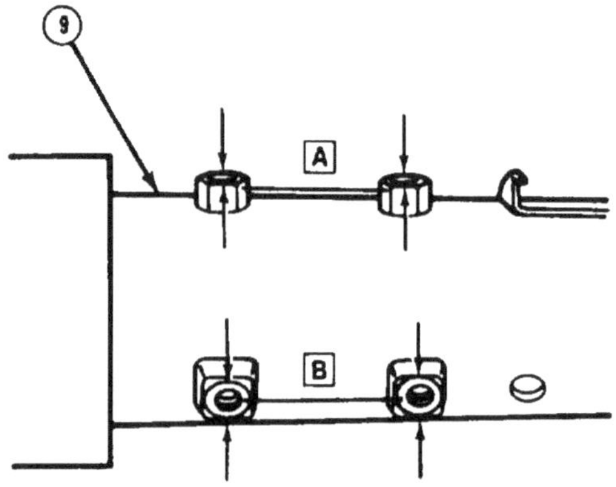

3. Drive out pivot pin (14) and remove transmission gearshift lever (13).

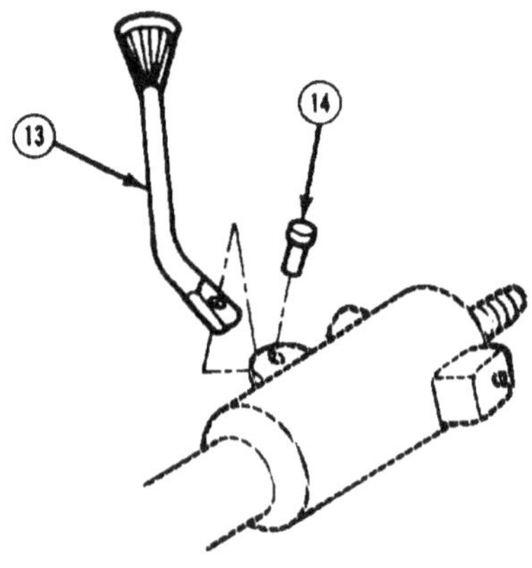

4. Using magnet, remove upper bearing thrust washer (19). Remove 4 screws (20). Guide electrical connector and ignition switch rod through shift tube and remove directional signal housing (18) assembly.

5. Remove steering column upper bearing (17) from directional signal housing (18). Remove bushing (16) and retainer (15) from lower side of directional signal housing.

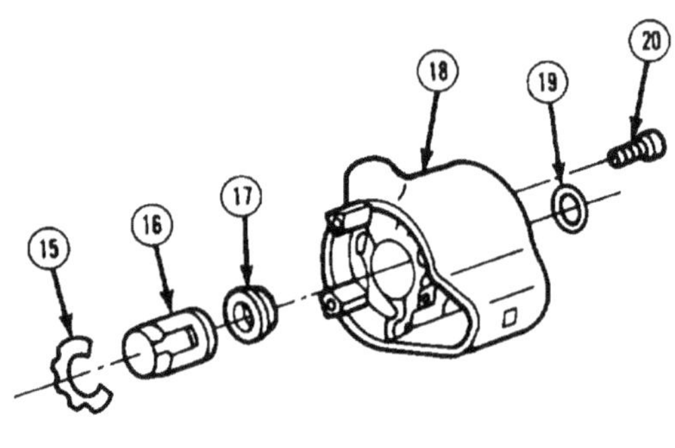

TA50474

8-2. STEERING COLUMN MAINTENANCE (Con't).

6. Lift ignition switch actuating rod and rack assembly (22), lock bolt and spring assembly (23), and rack preload spring (24) out of directional signal housing (18).

7. Remove 2 screws and shift lever detent plate (21) from directional signal housing (18).

8. Firmly push on block tooth of sector and remove ignition switch actuator sector through lock cylinder hole.

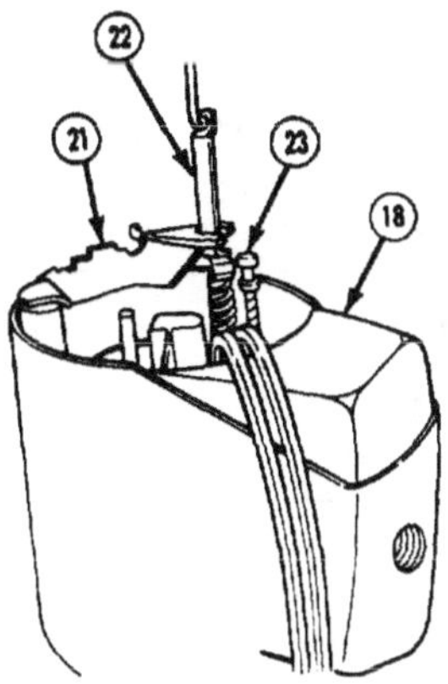

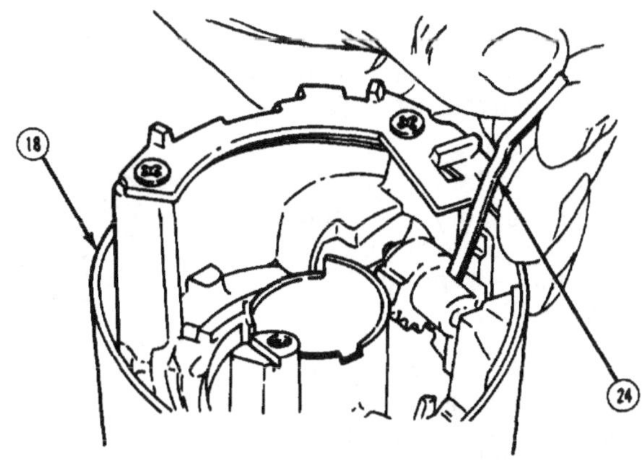

9. Remove transmission gearshift lever housing and shroud assembly (25) from steering column (9). Unwind transmission gearshift lever spring (26) and remove.

10. Pull steering shaft (10) from lower end of steering column (9).

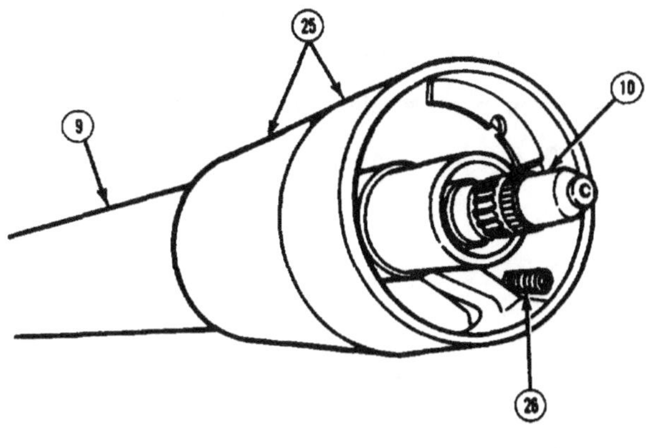

11. Pull tabs on back-up light switch (30) outward and remove from steering column (9).

TA50475

8-2. STEERING COLUMN MAINTENANCE (Con't).

WARNING

Lower bearing retainer clip (28) is under spring tension. Use caution during removal. Failure to follow this warning may result In injury to personnel.

12. Carefully remove retainer clip (28) and lower bearing retainer (29).

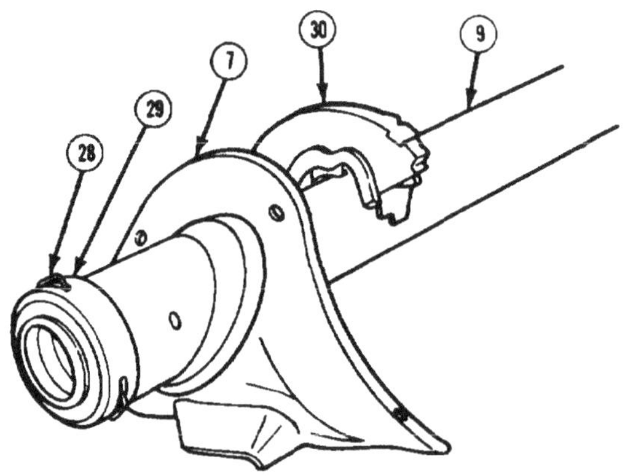

NOTE

- It may be necessary to apply light pressure on bearing outer race to remove bearing (27) from adapter (31).
- Gearshift housing bearing (34) may fall out of steering column (9) during removal of shift tube and lever assembly.

13. Remove bearing adapter assembly (31), shift tube spring (32), and washer (33). Remove lower bearing (27) from adapter assembly.
14. Remove shift tube and lever assembly from steering column (9). Remove seal (7) and discard.
15. Remove gearshift housing bearing (34) from steering column (9).
16. Inspect all components for damage. Replace any damaged components.

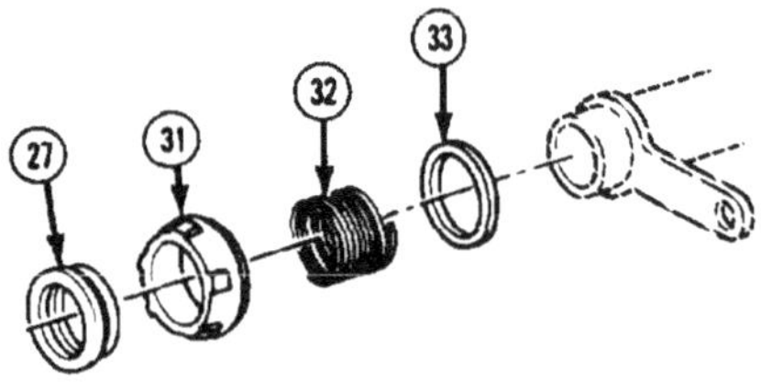

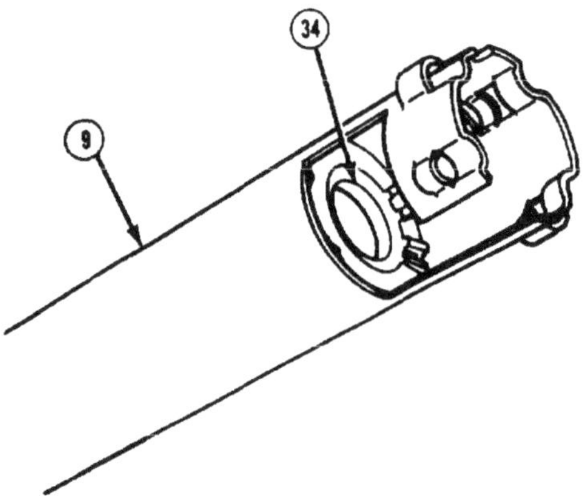

TA50476

8-2. STEERING COLUMN MAINTENANCE (Con't).

c. ASSEMBLY

NOTE

Ensure that a thin coat of grease is applied to all friction surfaces.

1. Position sector in lock cylinder hole over sector shaft, with tang end to outside of lock cylinder hole, and install sector into directional signal housing (18). Press sector over shaft with a blunt tool, Install shift lever detent plate (21) onto directional signal housing with 2 screws.
2. Insert rack preload spring (24) into directional signal housing (18), from bottom side, and hook rack preload spring onto edge of housing.

NOTE

Ensure that first tooth on sector is alined with first tooth on rod and rack" assembly (22). If properly alined, block teeth will line up when rod and rack assembly Is completely pushed in.

3. Assemble lock bolt and spring assembly (23) onto crossover arm on rod and rack assembly (22), and insert into directional signal housing (18), from bottom, with teeth toward centerline of steering column (9).

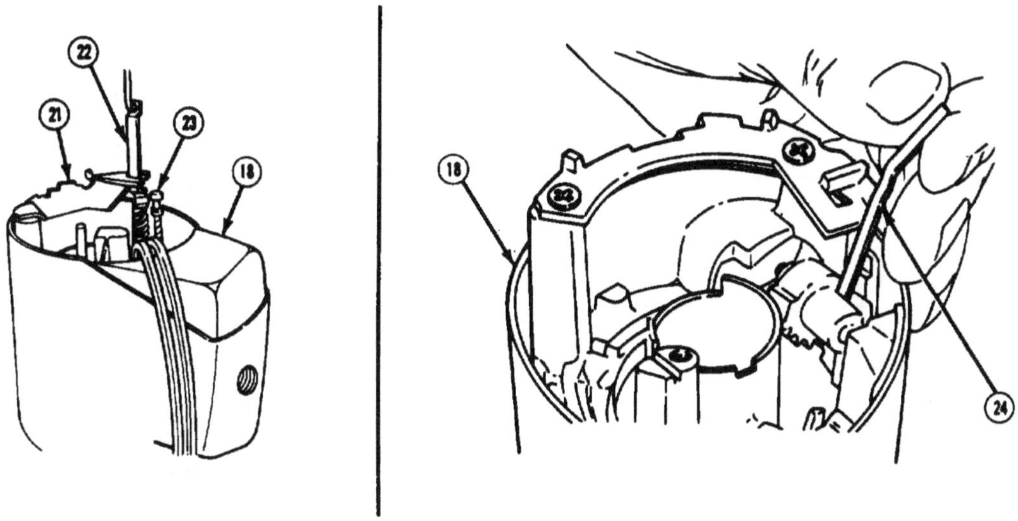

TA50477

8-2. STEERING COLUMN MAINTENANCE (Con't) .

4 . Install steering column upper bearing (17) into directional signal housing (18). Install bushing (16) and retainer (15) on bottom hub of directional signal housing.

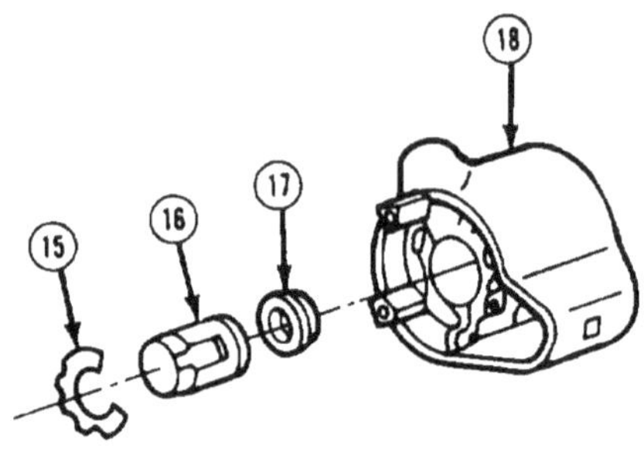

5 . Install new seal (7) and install shift tube and lever assembly into steering column (9).

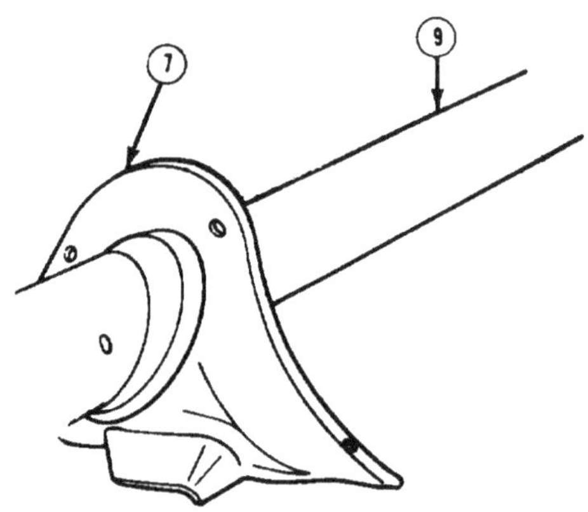

NOTE

If gearshift housing bearing (34) is not correctly installed, it will not rest on all stops provided.

6 . Insert gearshift housing bearing (34) into end of steering column (9) and aline indentations in gearshift housing bearing with projections on steering column. Ensure that gearshift housing bearing is fully seated.

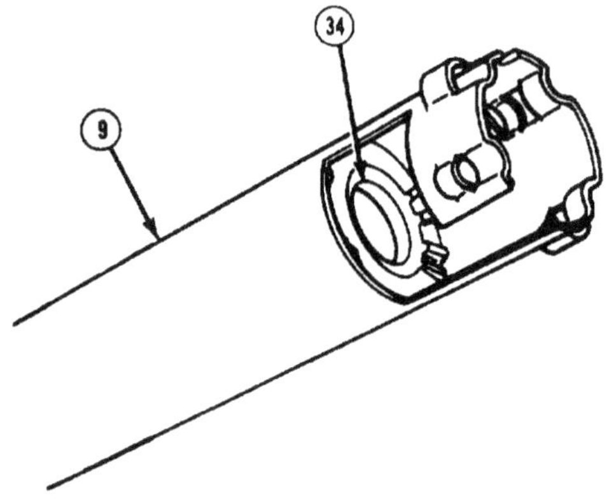

TA50478

8-2. STEERING COLUMN MAINTENANCE (Con't).

7. Install gearshift lever spring (26) into gearshift lever housing.

8. Install housing and shroud assembly (25) onto upper end of steering column (9). Rotate housing and shroud" assembly to ensure that it is fully seated.

9. With housing and shroud assembly (25) in place, guide electrica I connector and ignition switch rod through shift tube, and install directional signal housing (18) assembly onto steering column (9) with four screws (20).

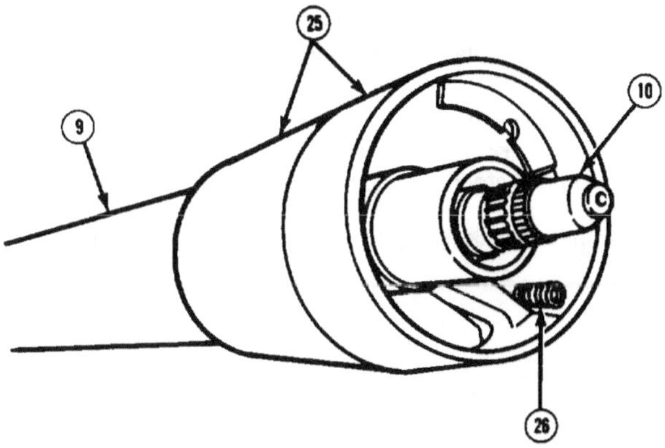

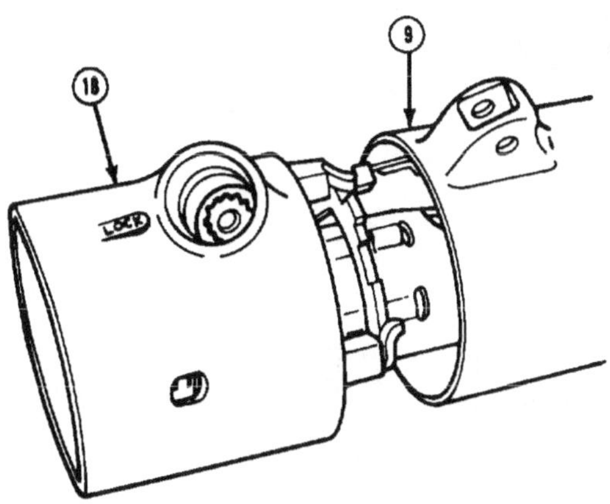

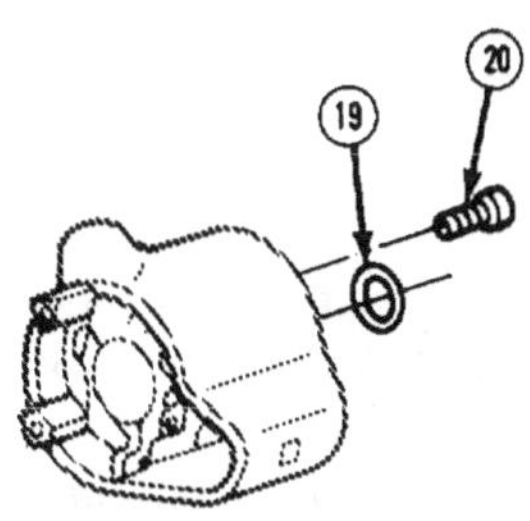

NOTE

Ensure that transmission gearshift housing is In "P" (Park) and rod and rack assembly is pulled downward.

10. Press lower bearing (27) into adapter assembly (31).

11. Insert shift tube assembly into lower end of steering column (9) and rotate until upper shift tube key slides into keyway.

TA50479

8-2. STEERING COLUMN MAINTENANCE (Con't).

NOTE

Ensure that retainer clip (28) snaps into steering column (9) and lower bearing retainer (29) slots.

12. Install washer (33), spring (32), lower bearing (27), and adapter assembly (31) into bottom of steering column (9), Hold adapter in place and install lower bearing retainer (29), indexing slots in steering column with slots in lower bearing retainer, and install retainer clip (28),
13. Install back-up light switch (30) tabs into steering column (9).

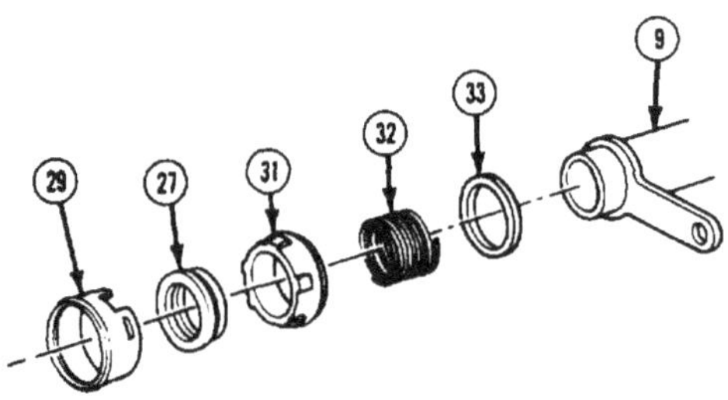

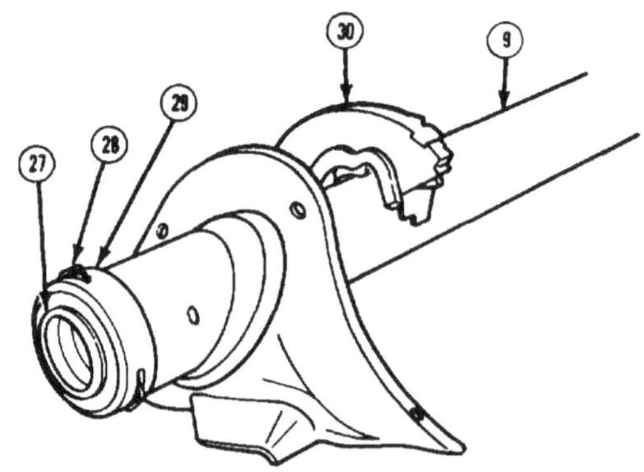

14. Slide steering shaft (10) into steering column (9) and install upper bearing thrust washer (19).
15. Install transmission gearshift lever (13) and pivot pin (14) to steering column (9).
16. Remove steering column (9) from vise.

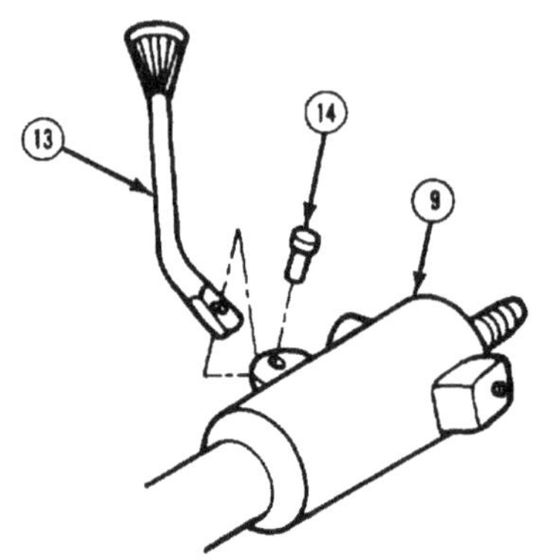

TA50480

8-2. STEERING COLUMN MAINTENANCE (Con't).

17. Assemble lower cover (12) and upper cover (8) to new seal (7) with "carrots" (part of seal).

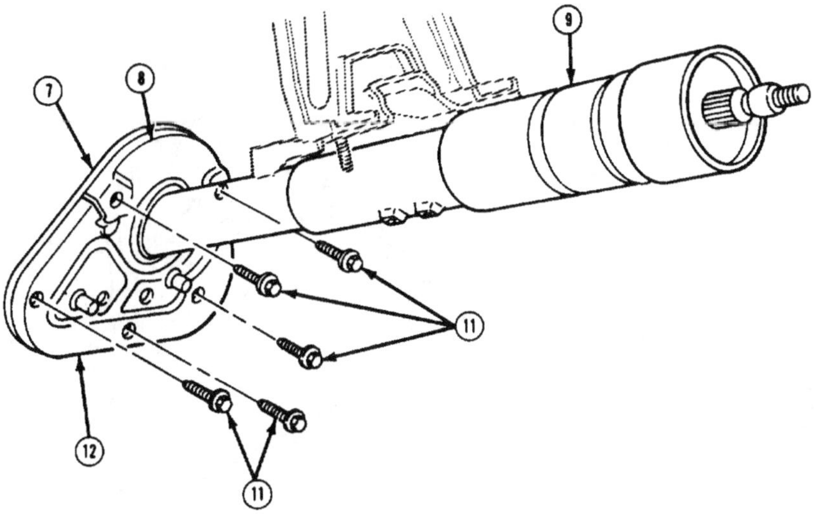

d . INSTALLATION

1. Position lower end of steering column (9) in body through dash opening. Install steering shaft flange onto flexible coupling (4) with 2 washers (5) and nuts (6). Tighten nuts to 20 lb.-ft. (27 Nom).
2. Install seal (7), upper cover (8), and lower cover (12) to firewall with 5 bolts (11).
3. Install transmission control linkage to steering column shift tube levers. (See paragraph 5-2)

NOTE

- Ensure that pot joint operating angle does not exceed 12 degrees.
- Ensure that flexible coupling (4) is not distorted greater than +0,06 in. (1.5 mm), in either direction, due to pot joint bottoming.

4. Connect back-up light switch connector to steering column wiring harness. Install steering column wiring harness at connector on left side of steering column.

TA50481

8-2. STEERING COLUMN MAINTENANCE (Con't).

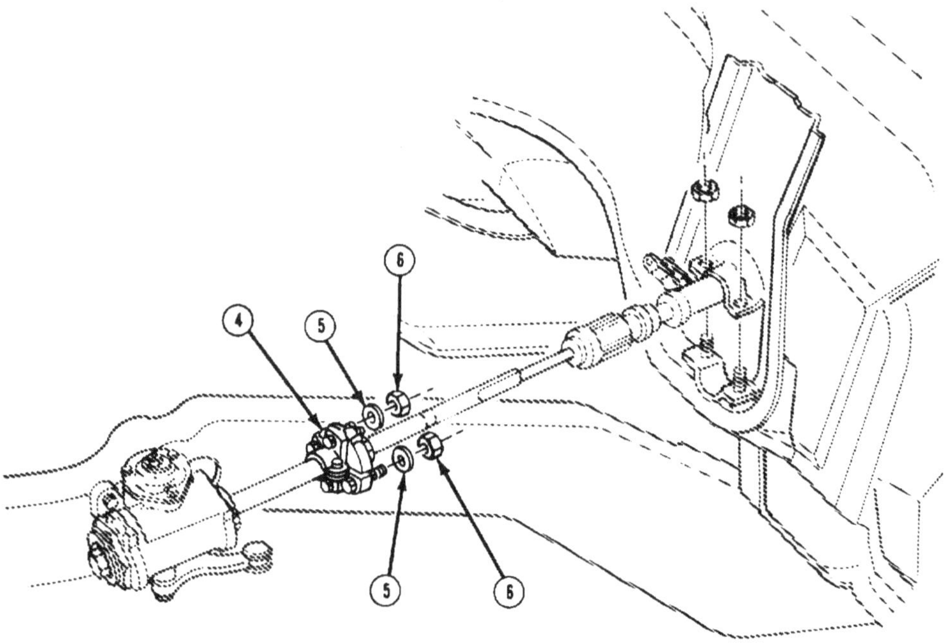

5. Aline lock cylinder (2) key with keyway in lock cylinder housing (1). Completely push in lock cylinder and install retaining screw (3), Tighten retaining screw to 20 lb.-in, (3 N.m).

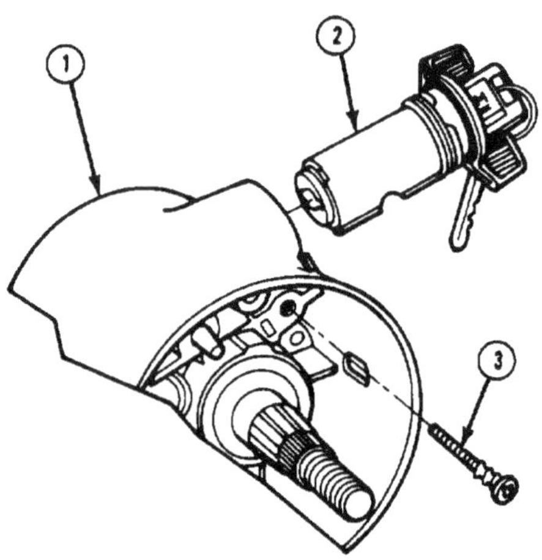

FOLLOW-ON TASKS:

- Install ignition switch, (See paragraph 4-13)
- Install directional signal switch. (See paragraph 4-14)

TA50482

8-3. STEERING SHAFT REPAIR.

This task covers: a. Disassembly b. Assembly

INITIAL SETUP:

Materials/ Parts

- Grease (Item 36, Appendix B)

Tools/Test Equipment

- Torque wrench

a. DISASSEMBLY

1. Remove snapring (6) from cover (7).

NOTE

Ensure that lower steering shaft (1) and cover (7) are marked for proper alinement during assembly.

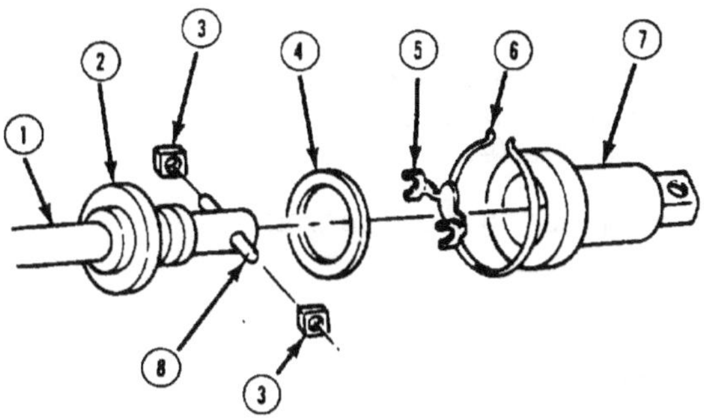

2. Mark cover (7) and lower steering shaft (1). Tap cover off of lower steering shaft, and remove bearing blocks (3) and tension spring (5) from pin (8), Remove washer (4) from inside seal (2).
3. Replace all damaged components,

b. ASSEMBLY

1. Fill seal (2) with grease and install washer (4) into lip of seal. Coat pin (8) with grease, and install tension spring (5) and bearing blocks (3) onto pin.
2. Aline marks on cover (7) and end of lower steering shaft (1) , and push cover over end of lower steering shaft. Ensure that seal (2) is properly seated and install snapring (6).

Page 8-15 has been rescinded.

TA701860

8-3. STEERING SHAFT MAINTENANCE (Con't).

b. DISASSEMBLY

1. Remove snapring (13) from cover (14).

NOTE

Ensure that lower steering shaft (6) and cover (14) are marked for proper alinement during assembly.

2. Mark cover (14) and lower steering shaft (6). Tap cover off of lower steering shaft, and remove bearing blocks (10) and tension spring (12) from pin (15). Remove washer (11) from inside seal (9).

3. Replace all damaged components.

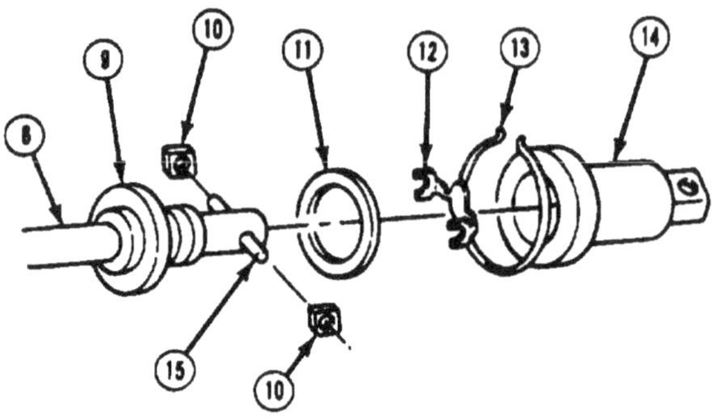

c. ASSEMBLY

1. Fill seal (9) with grease and install washer (11) into lip of seal. Coat pin (15) with grease, and install tension spring (12) and bearing blocks (10) onto pin.

2. Aline marks on cover (14) and end of lower steering shaft (6), and push cover over end of lower steering shaft. Ensure that seal (9) is properly seated and install snapring (13).

d. INSTALLATION

1. Lubricate end of upper shaft (5). Install lower steering shaft (6) onto upper shaft with notch in cover (14) toward flat side of spline on upper shaft. Tap into place.

2. Tap bolt into upper and lower steering shafts (5 and 6) and install nut.

3. Install 2 nuts (4 and 7), washers (3 and 8), and flanged end of lower steering shaft (6) to flexible coupling (2) at steering gear (1). Tighten nuts to 20 lb.-ft. (27 N.m),

TA50484

8-4. POWER STEERING GEAR REPAIR.

This task covers: a. Disassembly b. Cleaning and Inspection c. Assembly

INITIAL SETUP:

Equipment Condition

- Power steering gear removed. (See TM 9-2320-289-20)

Too/s/Test Equipment

- Ball retainer tool
- Rack piston installer
- Snapring pliers
- Spanner wrench
- Torque wrench

Materials/Parts

- One check valve
- One rack piston seal ring
- One side cover gasket
- One stub shaft seal
- One stub shaft dust seal
- Two pitman shaft seals
- Three teflon valve body rings
- Eight "O" rings
- Dry cleaning solvent (Item 23, Appendix B)
- Grease (Item 36, Appendix B)
- Hydraulic fluid (Item 40, Appendix B)
- Petrolatum (Item 50, Appendix B)
- Duct tape (Item 66, Appendix B)

General Safety Instructions

- Dry cleaning solvent is flammable and must not be used near open flame. Use only in a well-ventilated area.
- Compressed air used for cleaning purposes should never exceed 30 psi (207 kPa).

a. DISASSEMBLY

1. Remove bolt (10) securing flexible coupling (9) to stub shaft (8). Remove flexible coupling.
2. Remove adjuster locknut (2) and remove 4 bolts (3) from side cover (l). Turn adjuster screw (5) clockwise until side cover separates from pitman shaft (6) Remove side cover. Remove gasket (4) from side cover, Discard gasket. Drain fluid from housing into a suitable container.
3. Place housing (7) assembly in a soft-jawed vise with side cover (1) opening facing up.

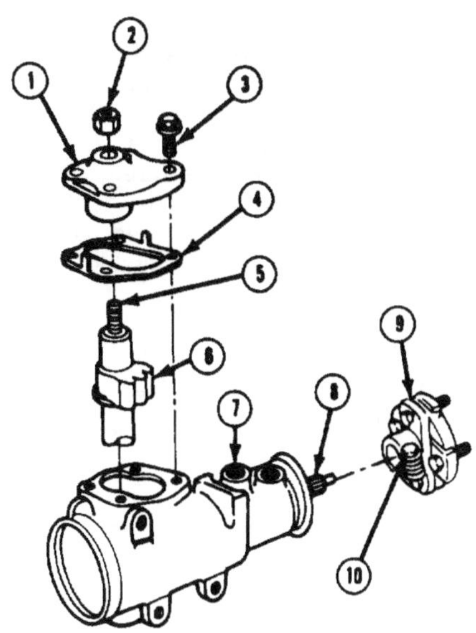

TA60485

8-4. POWER STEERING GEAR REPAIR.

This task covers: a. Disassembly b. Cleaning and Inspection c. Assembly

INITIAL SETUP:

Equipment Condition

- Power steering gear removed. (See TM 9-2320-289-20)

Too/s/Test Equipment

- Ball retainer tool
- Rack piston installer
- Snapring pliers
- Spanner wrench
- Torque wrench

Materials/Parts

- One check valve
- One rack piston seal ring
- One side cover gasket
- One stub shaft seal
- One stub shaft dust seal
- Two pitman shaft seals
- Three teflon valve body rings
- Eight "O" rings
- Dry cleaning solvent (Item 23, Appendix B)
- Grease (Item 36, Appendix B)
- Hydraulic fluid (Item 40, Appendix B)
- Petrolatum (Item 50, Appendix B)
- Duct tape (Item 66, Appendix B)

General Safety Instructions

- Dry cleaning solvent is flammable and must not be used near open flame. Use only in a well-ventilated area.
- Compressed air used for cleaning purposes should never exceed 30 psi (207 kPa),

a. DISASSEMBLY

1. Remove bolt (10) securing flexible coupling (9) to stub shaft (8) . Remove flexible coupling.

2 . Remove adjuster locknut (2) and remove 4 bolts (3) from side cover (1), Turn adjuster screw (5) clockwise until side cover separates from pitman shaft (6) Remove side cover. Remove gasket (4) from side cover. Discard gasket. Drain fluid from housing into a suitable container.

3. Place housing (7) assembly in a soft-jawed vise with side cover (1) opening facing up.

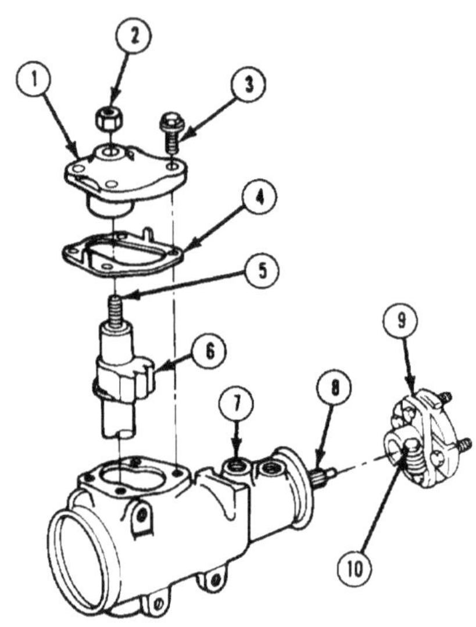

TA50485

8-4. POWER STEERING GEAR REPAIR (Con't).

4. Insert punch through retaining ring access hole (14) to force out retaining ring (11). Remove retaining ring.

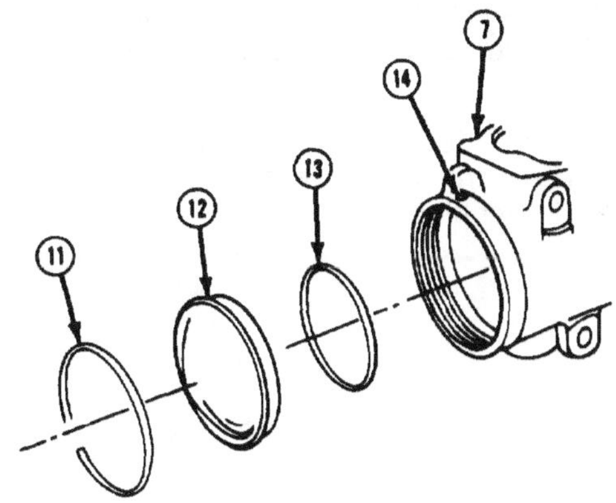

NOTE

- It may be necessary to use flexible coupling (9) to rotate stub shaft (8).
- It may be necessary to tap around outside of housing end plug (12) to center plug in housing (7) during step 6.

5. Rotate stub shaft (8) counter-clockwise and remove housing end plug (12) from housing (7).

6. Rotate stub shaft (8) clockwise and remove "O" ring (13) and rack piston plug (15). Discard "O" ring.

7. Rotate stub shaft (8) to center rack piston (16). Remove pitman shaft (6).

8. Fully insert ball retainer tool into rack piston (16) plug bore and hold tightly while turning stub shaft (8) counter-clockwise. Remove rack piston and ball retainer tool together from housing (7).

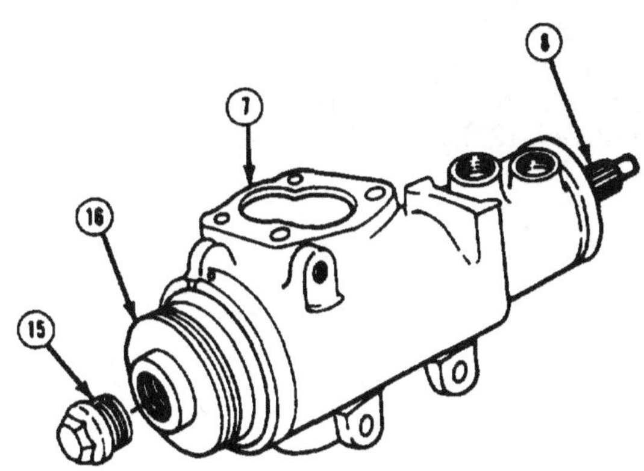

TA50486

8-4. POWER STEERING GEAR REPAIR (Con't).

NOTE

Worm gear (32) and valve assembly (34) may come out with adjuster plug (17), If so, separate adjuster plug from worm gear and valve assembly.

9. Using punch, loosen and remove locknut (18). Using spanner wrench, remove adjuster plug (17).

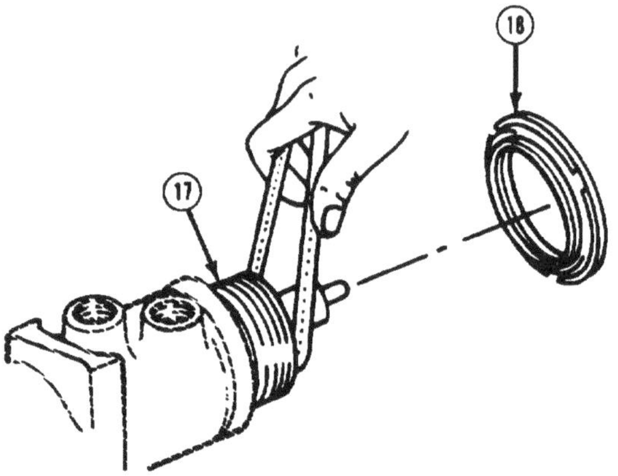

10. Using a screwdriver at raised area, remove bearing retainer (19). Remove thrust bearing spacer (20), small upper bearing race (21), upper thrust bearing (22), large upper bearing race (23), and "O" ring (24). Discard "O" ring.

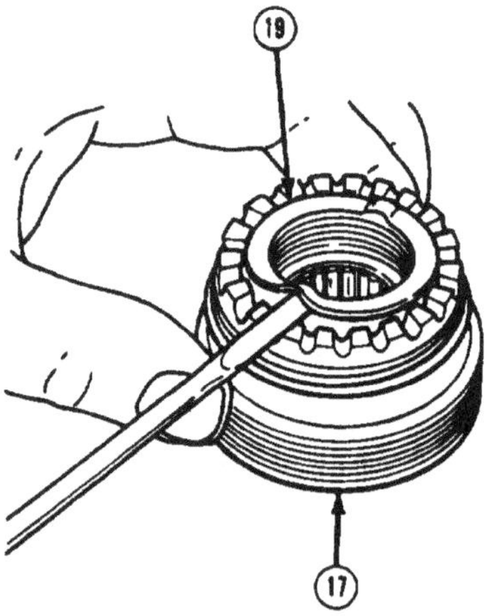

11. From rear side of adjuster plug (17), remove retaining ring (28). Press out needle bearing (25), stub shaft seal (26); and stub shaft dust seal (27). Discard seals.

TA50487

8-4. POWER STEERING GEAR REPAIR (Con't).

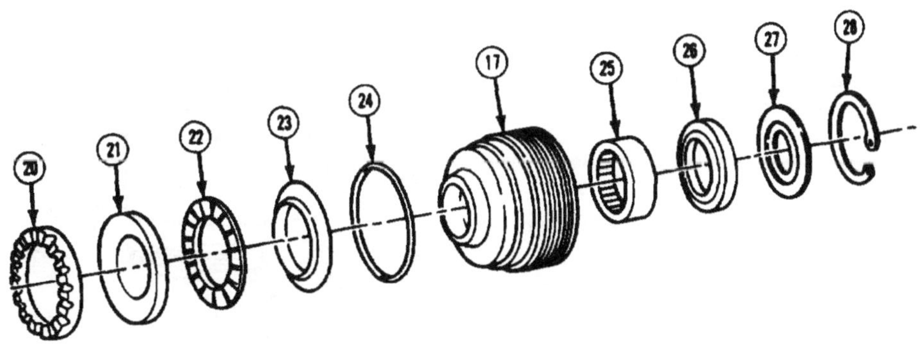

12. Remove stub shaft (8), valve assembly (34), and worm gear (32) as a unit. Remove race (29), thrust bearing (30), and race (31). Separate valve assembly from worm gear. Remove 3 teflon valve body rings and 3 "O" rings, Discard teflon valve body rings and "O" rings.

13. Remove valve body "O" ring (33) from valve assembly (34). Discard "O" ring.

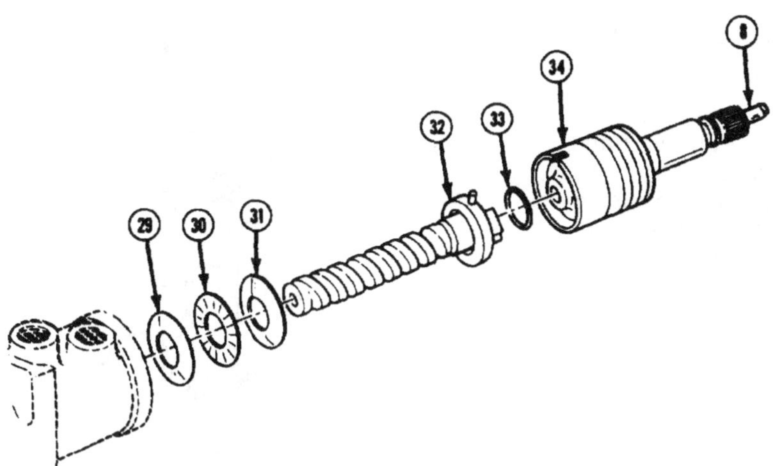

TA50488

8-4. POWER STEERING GEAR REPAIR (Con't).

14. Lightly tap stub shaft (8) on a wood block to loosen shaft cap. Pull out stub shaft with spool valve (36) from valve assembly (34),

15. Separate spool valve (36) from stub shaft (8) and remove "O" ring from spool valve. Discard "O" ring.

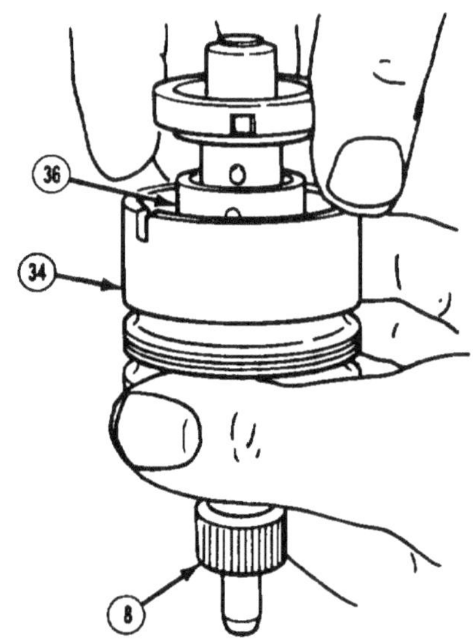

NOTE

Balls (43) will fall out. There are 12 silver and 12 black balls.

16. Remove 2 screws (40) and clamp (39) on side of rack piston (16). Remove guide (38).

17. Separate 2 halves of guide (38), and remove balls (43). Remove ball retaining tool, and tip rack piston (16) to release remaining balls from inside piston. Remove teflon ring and "O" ring from outer groove (42) of rack piston, and discard.

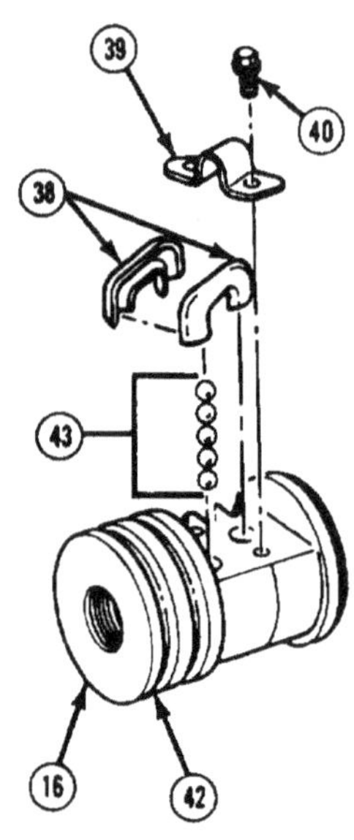

TA50489

8-4. POWER STEERING GEAR REPAIR (Con't).

18. Using snapring pliers, remove piston seal retaining ring (48) from bottom of housing (7). Remove outer seal backup washer (47), and press out pitman shaft needle bearing (43), pitman shaft seal (44), inner seal backup washer (45), and double lip seal (46) from housing assembly, Discard pitman shaft seal and double lip seal,

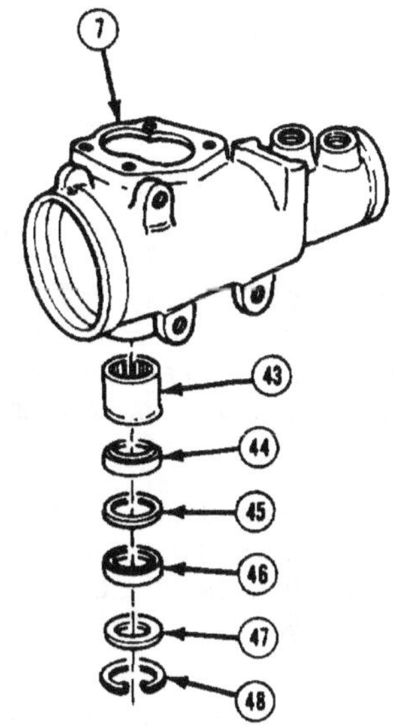

CAUTION

Care should be taken not to damage threads in housing (7) when removing check valve (46),

19. Using a small screwdriver, pry out check valve (46) from top of housing (7). Discard check valve,

20. Remove housing (7) from vise.

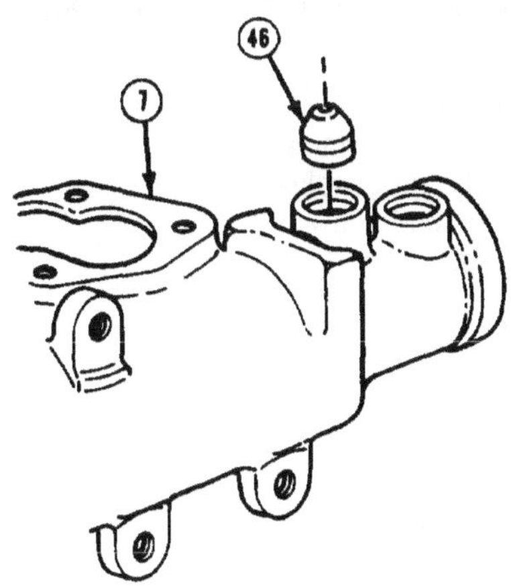

TA50490

8-4. POWER STEERING GEAR REPAIR (Con't).

b. CLEANING AND INSPECTION

WARNING

- Dry cleaning solvent P-D-680 is toxic and flammable. Always wear protective goggles and gloves and use only in a well-ventilated area. Avoid contact with skin, eyes, and clothes and DO NOT breathe vapors. DO NOT use near open flame or excessive heat, The solvent's flash point is 100°F-138°F (38°C-59°C). If you become dizzy while using cleaning solvent, immediately get fresh air and medical help. If solvent contacts eyes, immediately wash your eyes with water and get medical aid.
- Compressed air used for cleaning purposes should never exceed 30 psi (207 kPa). Use only effective chip guarding and personal protective equipment (goggles/shield, gloves, etc.), Failure to follow this warning may result in injury to personnel.

1. Clean all parts with dry cleaning solvent, Dry with compressed air.
2. Inspect upper pitman shaft bearing and side cover. Replace bearing if needles do not rotate freely. Replace side cover if bearing surfaces are scored,
3. Inspect pitman shaft and adjuster screw for excessive wear or damage. Replace if worn or damaged or if adjuster screw is not free to turn.
4. Inspect worm gear, rack piston, and all balls for wear or damage, Replace worm gear, rack piston, and balls as an assembly if worn or damaged.
5. Inspect rack piston ball return guide halves for wear or damage, Replace if worn or damaged.
6. Inspect adjuster plug threads, spacer, races, and bearings for wear or damage. Replace if worn or damaged.
7. Inspect stub shaft, spool valve, and valve assembly for wear or damage. Replace valve assembly, spool valve, and stub shaft as an assembly if worn or damaged.
8. Inspect housing for wear or damage. Replace if worn or damaged.

c. ASSEMBLY

1. Place housing (7) in a soft-jawed vise with side cover opening facing up.
2. Using a 3/8 in. drift punch and small hammer, carefully tap new check valve (46) into housing (7).

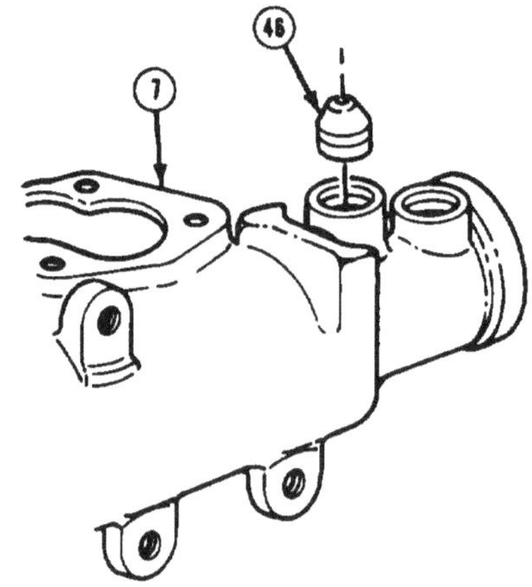

TA50491

8-4. POWER STEERING GEAR REPAIR (Con't).

3 . Install new " O" ring in outer groove (42). Install new teflon ring over top of " O" ring. Slide worm gear (32) completely into rack piston (1 6). Turn worm gear until worm gear groove is alined with rear ball return guide hole (41).

NOTE

Ensure that silver and black balls (43) are lubricated with hydraulic fluid and installed alternately into rack piston (16). As balls are installed and worm gear (32) is rotated, worm gear will be pushed out of rack piston.

4 . Install 18 balls (43), alternating silver balls with black balls, through rear ball return guide hole (41) while rotating worm gear (32) counterclockwise.

5. Install 6 remaining balls (43) in guide (38), alternating silver balls with black balls, and retain with petrolatum,

6. Assemble both halves of guide (38) together. Install guide into rack piston (16) and rotate worm gear (32) clockwise until worm gear is seated in rack piston.

7. Install clamp (39) and 2 screws (40). Tighten screws to 48 lb.-in, (5 Nom). Install ball retainer tool into rack piston (16) and remove worm gear (32). Leave ball retainer tool in rack piston.

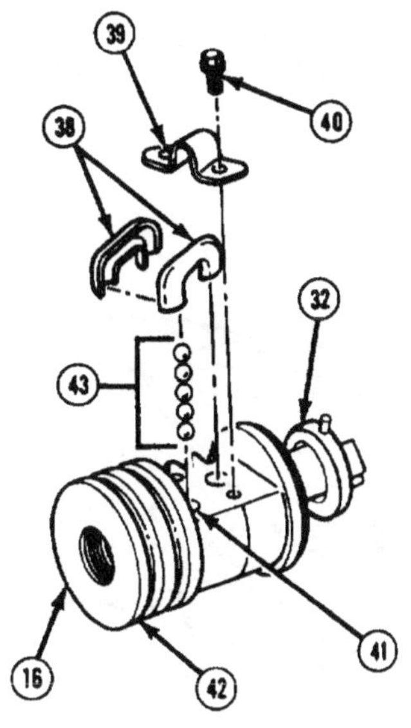

TA50492

8-4. POWER STEERING GEAR REPAIR (Con't).

NOTE

In order to assist in assembly, ensure that teflon valve body rings (35) are soaked in warm water before assembly.

8. Lubricate 3 new teflon valve body rings (35) and 3 new "O" rings with hydraulic fluid and install onto valve assembly (34).

9. Install new " O" ring on spool valve (36). Lubricate spool valve and valve assembly (34) with hydraulic fluid. Install spool valve approximately 1/4 in. (6.35 mm) into valve bore.

10. Install stub shaft (8) into spool valve (36) so that pin on stub shaft engages upper hole in spool valve.

11. Insert stub shaft (8) and spool valve (36) into valve assembly (34) so that slot (37) on shaft cap locks with pin inside valve assembly and cap seats against shoulder.

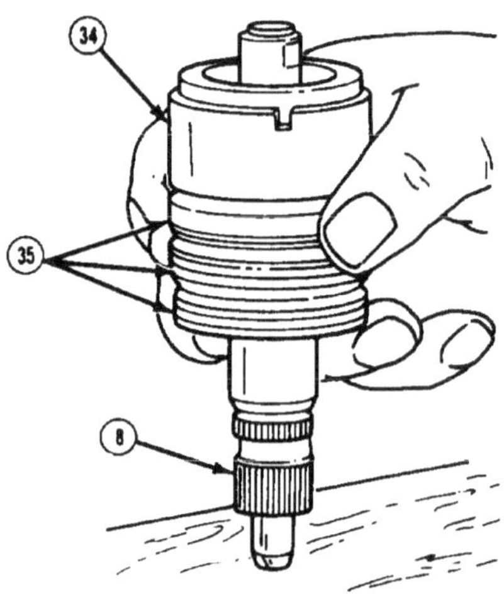

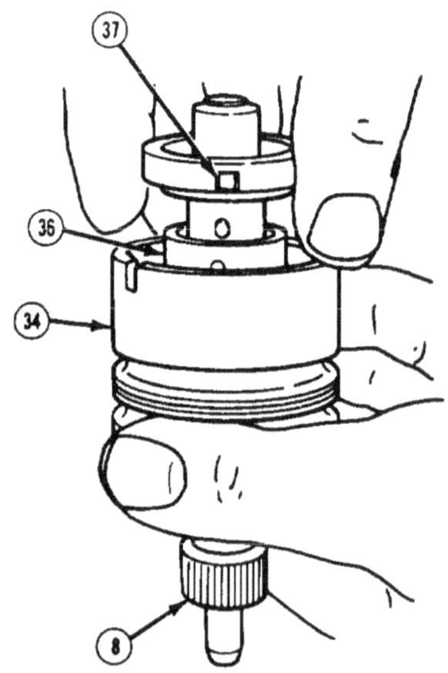

12. Install new "O" ring (33) into end of valve assembly (34).

13. Assemble race (31), thrust bearing (30), and race (29) on worm gear (32) with bevel on races facing away from drive end of worm gear.

14. Engage pin in worm gear (32) with slot in valve assembly (34). Install valve assembly and worm gear as a unit into housing (7).

TA50493

8-4. POWER STEERING GEAR REPAIR (Con't).

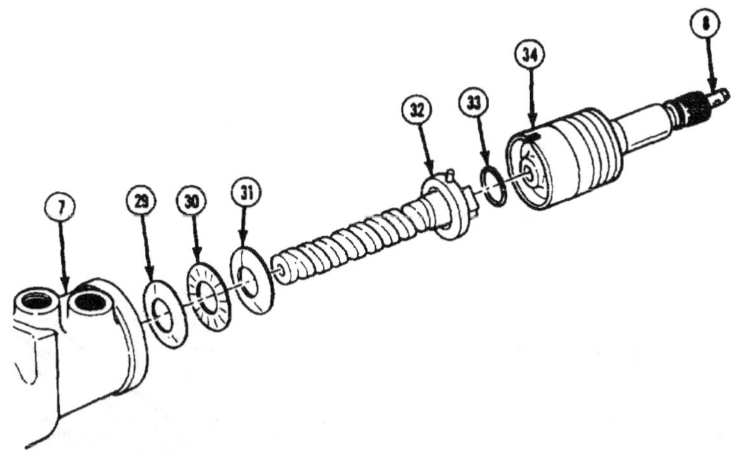

15. With stamped identification end up, press needle bearing (25) 5/8 in. (15.88 mm) into adjuster Plug (17). Install new "O" ring (24) on adjuster plug.
16. Install new stub shaft seal (26), new stub shaft dust seal (27), and retaining ring (28) on adjuster plug (17).

NOTE

Ensure that bearing retainer (19) raised areas do not extend above bearing spacer (20) when bearing retainer is seated, Bearing spacer must be free to rotate.

17. Install large upper bearing race (23), upper thrust bearing (22), small upper bearing race (21), thrust bearing spacer (20), and bearing retainer (19) on adjuster plug (17).

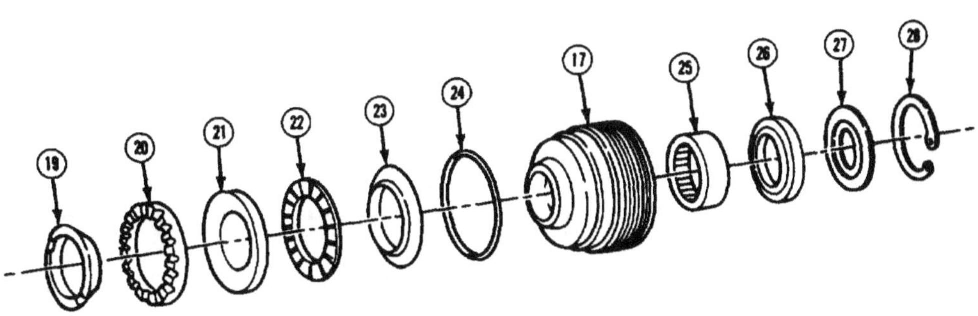

TA50494

8-4. POWER STEERING GEAR REPAIR (Con't).

18. Install adjuster plug (17) into housing (7), DO NOT tighten.

19. Adjust bearing preload.

    (a) Using spanner wrench, tighten adjuster plug (17) clockwise until adjuster plug is firmly bottomed.

    (b) Mark housing (7) and face of adjuster plug (17), Measure back counterclockwise 1/2 in. (12.7 mm) and place a second mark on housing,

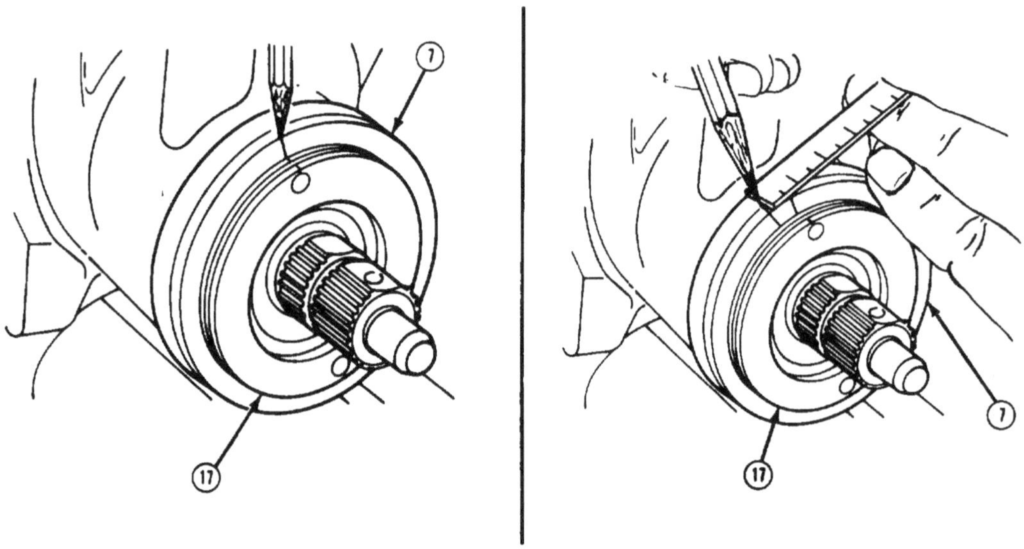

TA50495

8-4. POWER STEERING GEAR REPAIR (Con't).

(c) Turn adjuster plug (17) counterclockwise until mark on face of adjuster plug lines up with second mark on housing (7).

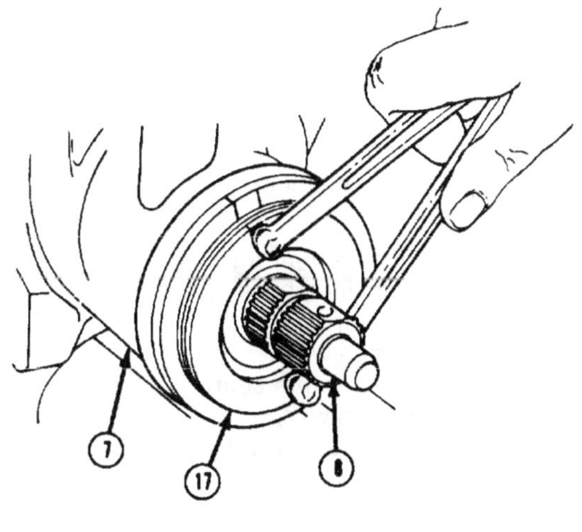

(d) Hold adjuster plug (17) to maintain alinement of marks and install locknut (18). Using a punch, tighten locknut securely.

(e) Wrap stub shaft (8) splines with duct tape to provide secure fit for 11/18 in. 12-point socket. Install socket and a lb.-in. torque wrench on stub shaft splines. While rotating stub shaft, torque wrench reading should be 4-8 lb.-in. (1-2 Nom). Record torque wrench reading for later adjustment.

(f) If torque wrench reading in step (e) is not correct, loosen locknut (18), adjust position of adjuster plug (17) in or out as required, tighten locknut, and repeat step (e). Remove socket and torque wrench when adjustment is correct.

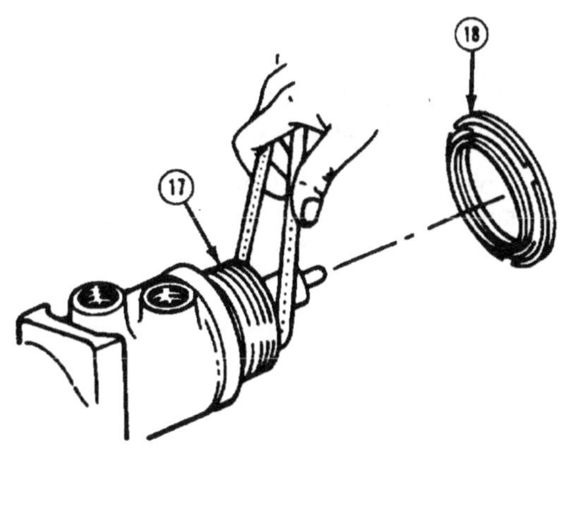

TA50496

8-4. POWER STEERING GEAR REPAIR (Con't).

NOTE

- Ensure that seals (44 and 46), seal backup washers (45 and 47), and inner housing (7) are greased prior to installation.
- *Ensure that seals (44* and *46)* and seal backup washers (45 and 47) are seated just past retaining ring (48) grooves.

20. Install pitman shaft needle bearing (43) into housing (7) until bearing is fully seated. Install new pitman shaft seal (44), inner seal backup washer (45), new double lip pitman shaft seal (46), and outer seal backup washer (47) into housing. Using snapring pliers, install pitman seal retaining ring (48),

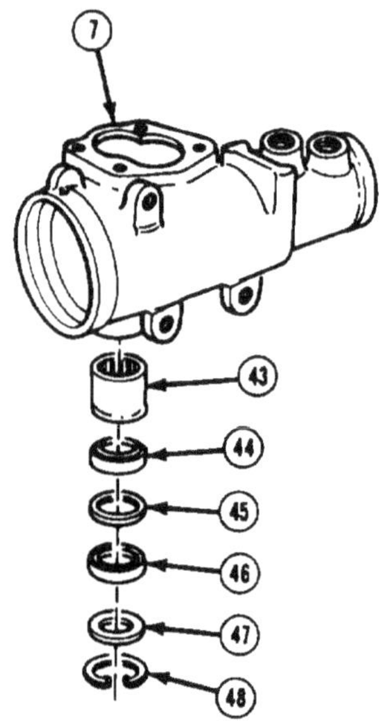

21. Install rack piston (16) into rack piston installer. Install rack piston and ball retainer tool into housing (7). Hold ball retainer tool tightly against worm gear while turning stub shaft (8) clockwise until seated. Ensure that no balls have dropped out.

22. Install rack piston plug (15), DO NOT tighten. Center rack piston (16) and install pitman shaft (6) in housing (7), Tighten rack piston plug to 75 lb.-ft, (102 N.m),

TA50497

8-4. POWER STEERING GEAR REPAIR (Con't).

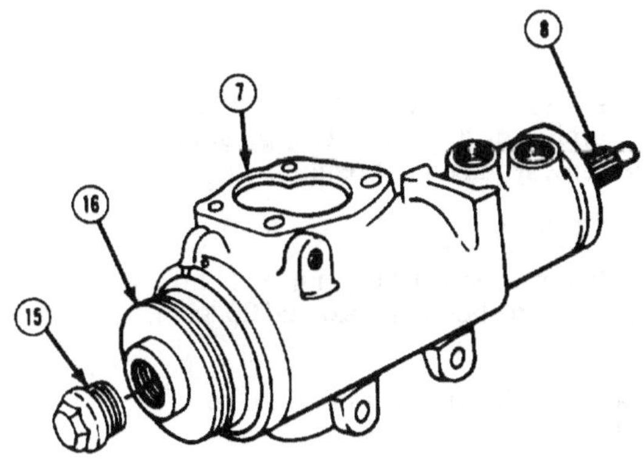

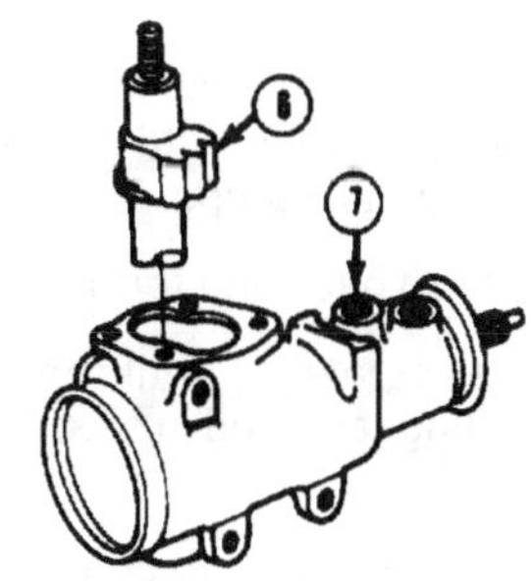

NOTE

Ensure that opening in retaining ring (11) is located approximately 1 in. (25 mm) from access hole (14) when installed.

23. Install new "O" ring (13), housing end plug (12), and retaining ring (11).

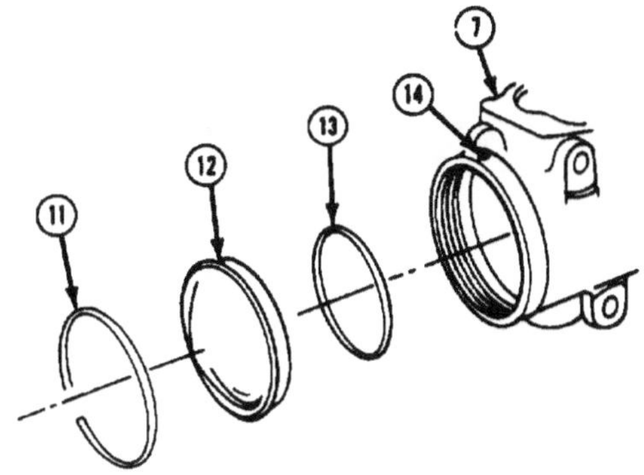

TA60498

8-4. POWER STEERING GEAR REPAIR (Con't).

24. Install new side cover gasket (4) on side cover (1) and bend tabs around side cover edges. Install side cover onto threaded end of pitman shaft (6), and turn adjuster screw (5) counterclockwise until side cover is seated against housing (7) and adjuster screw is completely backed out. Install side cover bolts (3) and tighten to 40 lb.-ft, (54 N.m). Install adjuster locknut (2),

25. Adjust pitman shaft (6) preload.

    (a) Install an 1 1/18 in. 12-point socket and a lb.-in. torque wrench on stub shaft (8) splines. Turn stub shaft clockwise until it stops. Counting number of turns, rotate stub shaft counterclockwise until it stops. Turn stub shaft back halfway to center position.

    (b) Turn torque wrench back and forth through center of travel.

    (c) Tighten adjuster screw (5) until torque wrench reads between 6-10 lb.-in. (1-2 N.m) higher than reading recorded in step 19 (e). The total reading should not exceed 20 lb.-in. (2 N.m) torque.

    (d) Hold adjuster screw (5) in place while tightening adjuster locknut (2) to 20 lb.-ft. (27 N.m). Recheck preload after tightening adjuster locknut.

    (e) If preload is incorrect, loosen adjuster locknut (2) and adjuster screw (5) and repeat step (c).

    (f) Remove socket, torque wrench, and duct tape from stub shaft (8) splines after adjustment is correct.

26. Install flexible coupling (9) to stub shaft (8) with bolt (10), Tighten bolt to 30 lb.-ft, (41 N.m).

27. Remove power steering gear from vise.

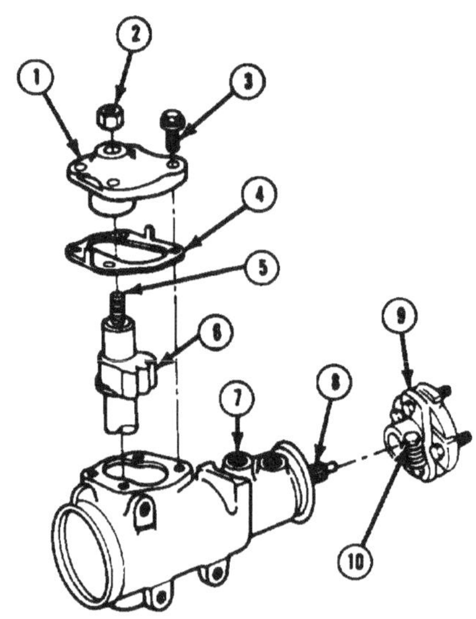

TA50499

8-4. POWER STEERING GEAR REPAIR (Con't).

FOLLOW-ON TASKS:

- Install power steering gear. (See TM 9-2320-289-20)

8-5. POWER STEERING PUMP REPAIR.

This task covers: a, Disassembly b. Cleaning and Inspection c. Assembly

INITIAL SETUP:

Equipment Condition

- Power steering pump, bracket, and pulley removed. (See TM 9-2320-289-20)
- Fluid drained from pump reservoir assembly.

Tools/Test Equipment

- Torque wrench

Materials/Parts

- One drive shaft seal
- Seven "O" rings
- Dry cleaning solvent (Item 23, Appendix B)
- Hydraulic fluid (Item 40, Appendix B)

General Safety *Instructions*

- Dry cleaning solvent is flammable and must not be used near open flame, Use only in a well-ventilated area.
- Compressed air used for cleaning purposes should never exceed 30 psi (207 kPa).

a. DISASSEMBLY

NOTE

Ensure that location of stud (11) and bolt (9) are marked for assembly,

1. Remove stud (11) and bolt (9).

WARNING

Dry cleaning solvent P-D-680 is toxic and flammable. Always wear protective goggles and gloves and use only in a well-ventilated area, Avoid contact with skin, eyes, and clothes and DO NOT breathe vapors. DO NOT use near open flame or excessive heat. The solvent's flash point is 100°F-138°F (38°C-59°C). If you become dizzy while using cleaning solvent, immediately get fresh air and medical help. If solvent contacts eyes, immediately wash your eyes with water and get medical aid.

2. Clean exterior of power steering pump with dry cleaning solvent. Remove reservoir cap (8).

3. Remove nut and fitting assembly (10), "O" ring (1 2), and control valve (13) and spring (14) assembly. Discard "O" ring.

4. Place power steering pump in soft-jawed vise with jaws on pump housing (1). Remove pump housing from pump reservoir assembly (7). Remove "O" rings (4, 5, 6, and 15) and magnet (2). Discard "O" rings.

8-5. POWER STEERING PUMP REPAIR (Con't).

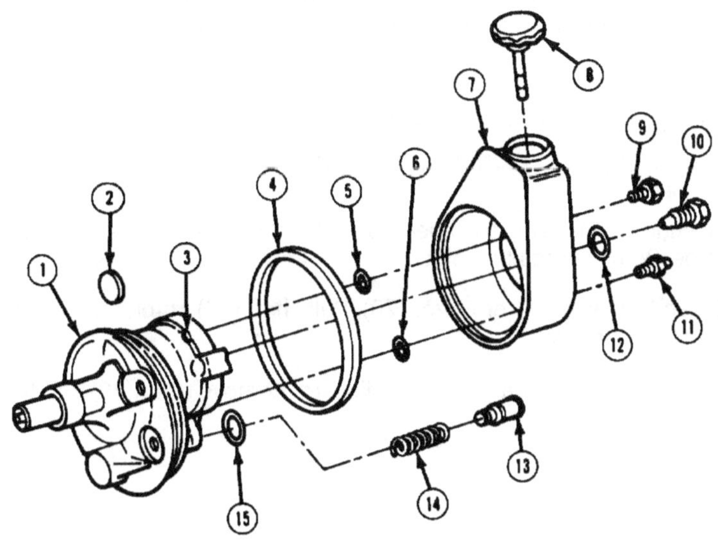

NOTE

If welch plug (3) at top of pump housing (1) is deformed or dislodged, DO NOT remove it. Pump housing assembly will have to be replaced.

5. Insert a punch into hole in pump housing (1) and remove retaining ring (17) from end plate (16). Remove end plate.

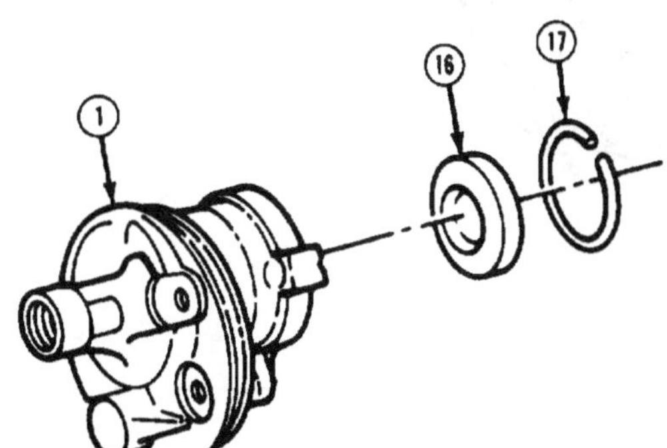

TA50500

8-5. POWER STEERING PUMP REPAIR (Con't).

CAUTION

Ensure that positions of pressure plate (24) and pump ring (23) are noted for assembly, Failure to follow this caution may result in power steering system malfunction.

6. Examine end of drive shaft (18) in pump housing (1) assembly. If corroded, clean with crocus cloth before removing, If bushing is damaged, pump housing assembly will have to be replaced.
7. Remove pressure plate spring (25). Using a rubber mallet, lightly tap on drive shaft (18) until pressure plate (24) is free, Remove pressure plate.
8. Remove pump ring (23) and vanes (22) from rotor (20). Pull shaft (18), rotor, and thrust plate (19) from pump housing (1) assembly.
9. Remove retaining ring (21) from drive shaft (18) and discard. Remove rotor (20) and thrust plate (19).

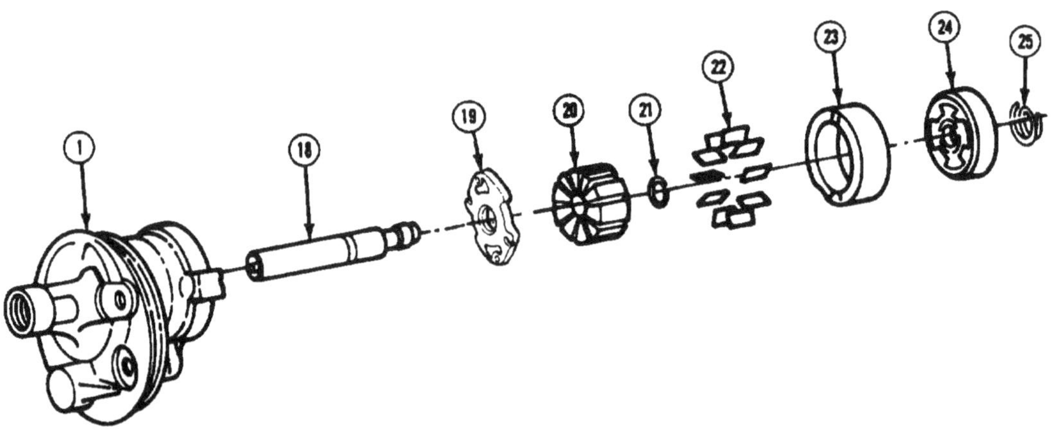

TA50501

8-5. POWER STEERING PUMP REPAIR (Con't).

10. Remove 2 dowel pins (29) from pump ring (23). Remove drive shaft seal (26) and 2 "O" rings (27 and 28). Discard " O" rings,

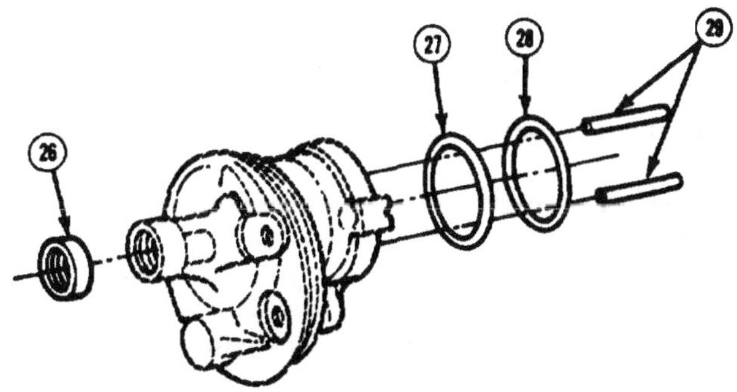

b. CLEANING AND INSPECTION

WARNING

- Dry cleaning solvent P-D-680 is toxic and flammable. Always wear protective goggles and gloves and use only in a well-ventilated area, Avoid contact with skin, eyes, and clothes and DO NOT breathe vapors. DO NOT use near open frame or excessive heat. The solvent's flash point is 100°F-138°F (38°C-59°C). if you become dizzy while using cleaning solvent, immediately get fresh air and medical help, if solvent contacts eyes, immediately wash your eyes with water and get medical aid.
- Compressed air used for cleaning purposes should never exceed 30 psi (207 kPa). Use only effective chip guarding and personal protective equipment (goggles/shieid, gloves, etc.). Failure to follow this warning may result in injury to personnel.

1. Clean all parts with dry cleaning solvent, Dry with compressed air.
2. Inspect the following rotating group components:

   (a) Vanes (22) for scoring or excessive wear; and fit of vanes in rotor (20) for sticking or excessive play.

   (b) Rotor (20) slots for burrs and excessive wear at thrust faces.

   (c) Inner surface of pump ring (23) for scoring or wear.

   If any of these conditions exist, replace rotor, vanes, pump ring, and retaining ring (21) as an assembly.

TA50502

8-5. POWER STEERING PUMP REPAIR (Con't).

3 . Inspect thrust plate and pressure plate for wear on plate surfaces. Replace if badly worn or damaged.

4 . Inspect drive shaft seal for leakage or damage. Replace drive shaft seal if leaking or damaged.

5 . Inspect drive shaft for excessive burning or scoring. Replace if burned or scored.

6 . Inspect control valve, Replace control valve if it does not move freely in valve bore.

c . ASSEMBLY

NOTE

Ensure that all "O" rings and seals are lubricated with hydraulic fluid before assembly.

1 . Install 2 new "O" rings (27 and 28) into grooves in pump housing (1).

2. Using suitable socket, install new drive shaft seal (26).

3 . Install 2 dowel pins (29) into pump housing (1).

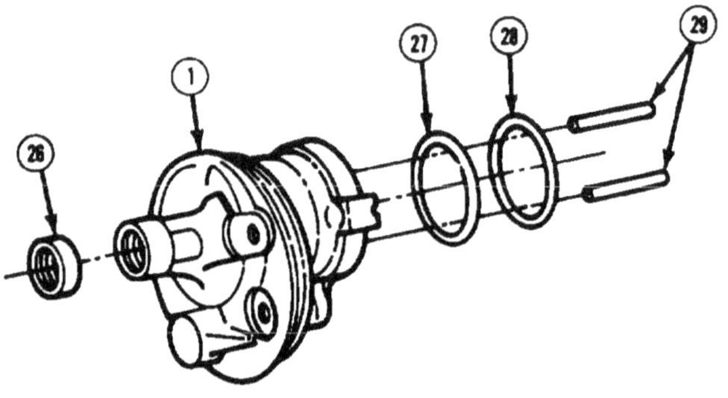

4 . Install thrust plate (19) on drive shaft (18) with countersunk side toward drive shaft. Install rotor (20) onto thrust plate and install new retaining ring (21).

5 . Install drive shaft (18) assembly in pump housing (1), indexing dowel pins (29) in pump housing with holes in thrust plate (19).

6 . Install vanes (22) in rotor (20). Install pump ring (23) over vanes and rotor, indexing dowel pins (29) in pump housing (1) with small holes in pump ring, and recesses in thrust plate (19) with large holes in pump ring. Ensure that pump ring is fully seated.

7 . Install pressure plate (24) on pump ring (23) and lightly tap into place. Install pressure plate spring (25).

TA50503

8-5. POWER STEERING PUMP REPAIR (Con't).

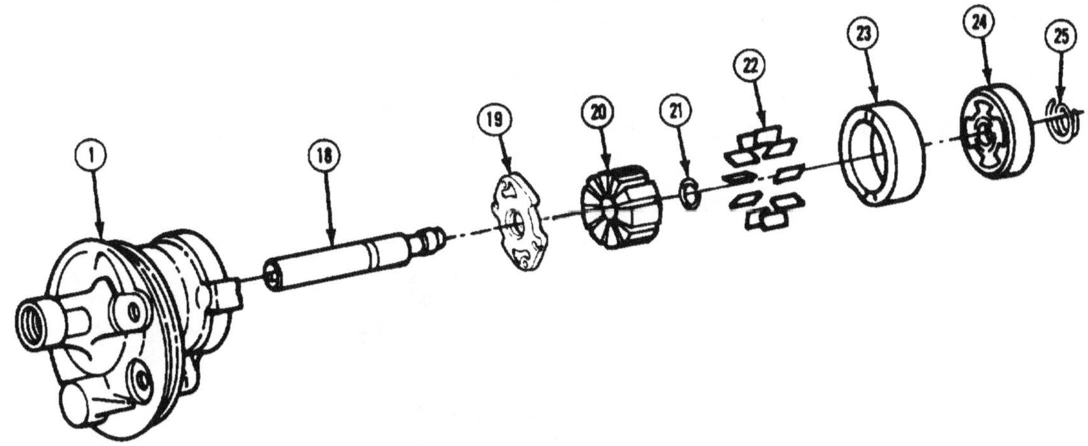

8 . Lubricate end plate (16) with hydraulic fluid and install into pump housing (1), Using a "C" clamp, compress end plate down in pump housing and install end plate retaining ring (17). Remove "C" clamp.

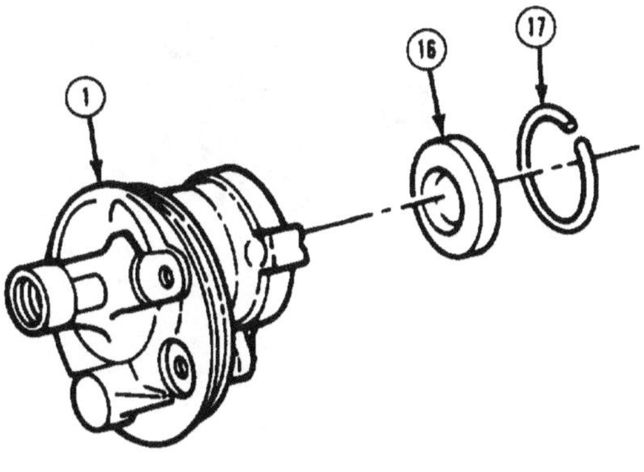

TA50504

8-5. POWER STEERING PUMP REPAIR (Con't).

9. Install magnet (2) into pump housing (1). Lubricate new "O" rings (4, 5, 6, and 15) with hydraulic fluid and install. Install pump reservoir assembly (7) onto pump housing (1). Install stud (11) and bolt (9) into assembly finger tight.

10. Install spring (14) and control valve (13) assembly with flat end of control valve facing outward. Install nut and fitting assembly (10) with new " O" ring (12) in top groove. Tighten to 35 lb.-ft. (47 N.m). Tighten stud (11) and bolt (9) to 25 lb.-ft. (34 N.m). Install reservoir cap (8).

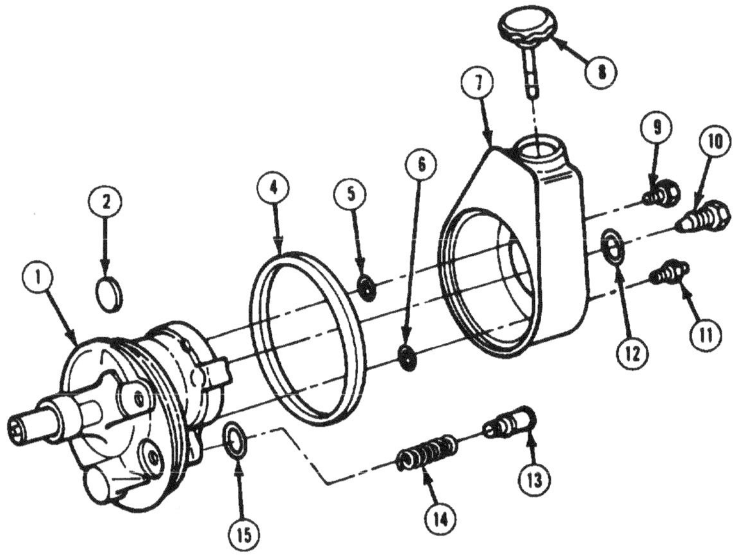

FOLLOW-ON TASKS:

- Install power steering pump pulley, bracket, and pump. (See TM 9-2320-289-20

TA50505

CHAPTER 9
FRAME MAINTENANCE

9-1. FRAME MAINTENANCE INDEX.

9-2. GENERAL.

a. General frame maintenance procedures can be found in TB 9-2300-257-40. References to 3/4 ton and 1 1/4 ton truck apply to CUCV Series trucks.

b. Perform FRAME MEASUREMENT in paragraph 9-3 to determine if any adjustments need to be made to f fame dimensions.

c. Perform FRAME REPAIR in paragraph 9-4 to repair any defects discovered in frame assembly.

9-3. FRAME MEASUREMENT.

This task covers: a. Frame Inspection b. Frame Measurement

INITIAL SETUP:

Equipment Condition

- Truck parked on level surface in well-lighted area.

Personnel Required

- *MOS 44B* (1)

Tools/Test Equipment

- Datum gages
- Steel tape measure
- Tram gage

a. FRAME INSPECTION

1. Inspect suspension and axle components for looseness, improper alinement, and damage. Tighten, realine, or replace components as needed.

b. FRAME MEASUREMENT

NOTE

- Ensure that frame is measured along all lines shown on Illustrations on pages 9-4 and 9-5. The following procedure should be used when performing these measurements.
- A tolerance of plus or minus (±) 0.12 in (3 mm) Is acceptable on frame measurements unless otherwise stated.
- Bolts, nuts, studs, and nonremovable plugs are shown on illustrations as solid objects. Holes are shown as outlines.
- Ensure that any drivetrain, exhaust, or suspension components are removed as required to obtain direct point-to-point measurements.

1. Measure all bolts, nuts, studs, and nonremovable plugs from center-to-center.
2. On bottom view illustrations, measurement points at a 90 degree angle to centerline (℄) should be an equal distance from centerline, ± 0.12 in. (± 3 mm). Diagonal measurements should be the same, ±0.19 in. (±5 mm), and should cross at centerline.

NOTE

The datum line is an imaginary line parallel to floor at lowest point on frame.

3. Measure all dimensions on side view illustrations from center-to-center. Attach datum gages at points on frame to be measured. Measure all length dimensions parallel to datum line. Measure all height dimensions at a 90 degree angle to datum line.

9-3. FRAME MEASUREMENT (Con't).

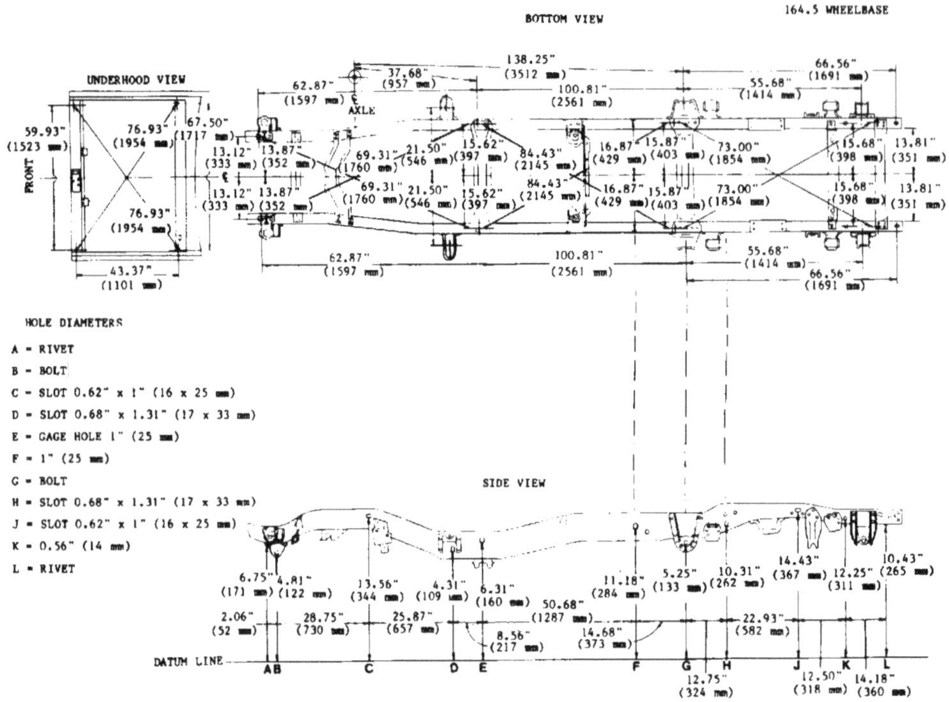

All Except M1009

TA50195

9-3. FRAME MEASUREMENT (Con't).

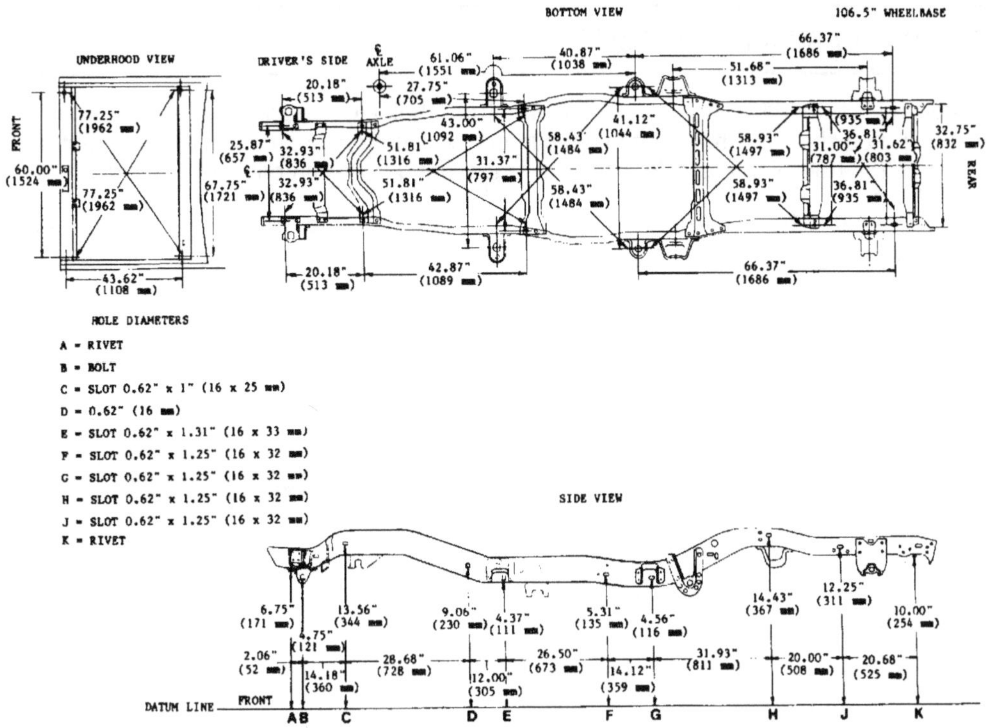

M1009

TA50196

9-3. FRAME MEASUREMENT (Con't).

4. Unless otherwise stated, all holes are measured from their closest edge to the other point of reference. Holes shown with (℄ centerline) symbol beside them are measured from center of hole. Add 1/2 of hole diameter to dimension if measuring from centerline of a hole.

5. Perform applicable frame repair procedure in paragraph 9-4 to bring dimensions within specifications.

9-4. FRAME REPAIR.

This task covers:

a. Frame Repair Guidelines
b. Loose Rivet Replacement
c. Damaged Bracket Replacement
d. Frame Reinforcement
e. Crack Repair
f. Frame Straightening

INITIAL SETUP:

Equipment Condition

- *Frame* dimensions measured and recorded. (See paragraph 9-3)

Tools/Test Equipment

- *River* buster
- Torch with rosebud heating tip

Personnel Required

- MOS 44B (1)

Materials/Parts

- Bolts (as needed)
- Lockwashers (as needed)
- Nuts (as needed)

Manual References

- TB 9-2300-257-40
- TM 9-237

a . FRAME REPAIR GUIDELINES

NOTE

Ensure that all FRAME REPAIR GUIDELINES are followed when attempting to repair any frame components.

1. Remove any drivetrain, exhaust, and suspension components as required to obtain direct access to any damaged rivet, bracket, hanger, or support.
2. Avoid changing frame thickness.
3. Avoid drilling holes in frame rail flanges.
4. Space holes in web section of rail at least 0.5 in. (13 mm) from one another.
5. Use existing holes whenever possible,
6. DO NOT cut holes with a torch,
7. Avoid welding on flanges.
8. DO NOT allow 4 or more holes to exist on same vertical line of web.
9. Holes made in a reinforcement should be placed at least 2 times the material thickness from edge of reinforcement.

9-4. FRAME REPAIR (Con't).

b. LOOSE RIVET REPLACEMENT

WARNING

Use only effective chip guarding and personnel protective equipment (goggles/shield, gloves, etc.) when using drill. Failure to follow this warning may result in injury to personnel,

CAUTION

- Welding is not an approved or acceptable alternate method of repair for any rivet attachment on frame. Only use the following procedure, Failure to follow this caution may result in damage to frame assembly.
- Use only SAE Grade 8 bolts and nuts to replace rivets. Failure to follow this caution may result in weak frame attachments.
- Rivets should be removed with rivet buster (NSN 5100-00-293-0556). DO NOT use an ordinary chisel for this purpose. Failure to follow this caution will elongate rivet hole and cause any future bolting to be seriously weakened.

NOTE

When bolt is used to replace rivet on top or bottom flange of side rail, bolt must be installed from top down.

1. Remove rivet by shearing rivet head with rivet buster. Remove rest of rivet body from rivet hole using drift punch and hammer.
2. Using appropriate electric drill and 7/16 in. drill bit, drill out existing hole.
3. Install a 7/16 in. -14 appropriate length hex head bolt, 7/16 in. "HEAVY" classification lockwasher, and 7/16 in. hex nut.
4. Tighten hex nut to 55-65 lb. -ft. (75-88 N.m).

c. DAMAGED BRACKET REPLACEMENT

WARNING

Use only effective chip guarding and personnel protective equipment (goggles/shield, gloves, etc.) when using drill. Failure to follow this warning may result in injury to personnel.

CAUTION

- Use only SAE Grade 8 bolts and nuts to replace rivets. Failure to follow this caution may result in weak frame attachments.
- Rivets should be removed with rivet buster (NSN 5100-00-293-0556), DO NOT use an ordinary chisel for this purpose. Failure to follow this caution will elongate rivet hole and cause any future bolting to be seriously weakened.

9-4. FRAME REPAIR (Con't).

NOTE

- The following procedure may be used to replace any bracket, hanger, or support that attaches to frame.
- Ensure that any drivetrain, exhaust, and suspension components required to obtain direct access to any damaged bracket, hanger, or support are removed.
- When bolt is used to replace rivet on top or bottom flange of side rail, bolt must be installed from top down,

1. Remove rivet by shearing rivet head with rivet buster. Remove rest of rivet body from rivet hole using drift punch and hammer.
2. Install a 3/8 in. appropriate length hex head bolt, 3/8 in. "HEAVY" classification lockwasher, and 3/8 in. hex nut,
3. Tighten each nut to 30-40 lb.-ft. (41-54 N.m).

d. FRAME REINFORCEMENT

1. See TB 9-2300-257-40.

e. CRACK REPAIR

WARNING

Use only effective chip guarding and personnel protective equipment (goggles/shield, gloves, etc.) when using grinder. Failure to follow this warning may result in injury to personnel.

CAUTION

DO NOT use oxyacetylene welding equipment to repair cracks on CUCV Series truck frames. Failure to follow this caution may result in weld and/or frame failures.

NOTE

- Straight cracks may result from high concentrations of stress in small areas of frame, excessive bending movement, and torsional loading.
- Sunburst cracks are caused by high loads being applied to a mounting bracket or crossmember which is not securely or properly attached to rail,
- Crossmember mounting flange cracks may be repaired in the same manner as side rail cracks; however, ensure that weld bead is built up to provide a good, smooth radius.

1. Remove any component that will interfere with access to crack.

9-4. FRAME REPAIR (Con't).

2 . Locate extreme end of crack and drill a 0.25 in. (6 mm) hole.

3 . "V" grind entire length of crack from starting point to 0.25 in. (6 mm) hole at extreme end.

4 . Using a hacksaw blade, open up bottom of crack 0.062 in. (2 mm) to allow good penetration of weld.

5. Weld crack in accordance with TM 9-237,

6 . Grind weld smooth on both inside and outside of rail or crossmember. Eliminate all weld buildup or notches on edge of flange.

7 . Install reinforcements as outlined in TB 9-2300-247-40 as required.

f . FRAME STRAIGHTENING

CAUTION

- Frame straightening can be complicated and usually requires special equipment. Only experienced personnel should attempt this procedure.
- DO NOT heat frame any hotter than 1200°F (606°C) when straightening frame. A red glow indicates that material is overheated. Use of excessive heat may weaken strength and durability of frame.

NOTE

Most frame straightening operations can be accomplished without use of heat. Cold straightening procedures are recommended for repair of CUCV Series truck frames. The following procedure may be used when cold straightening techniques are not effective.

1. Apply pressure, to area to be straightened, until slight checking (silvery cracks) appear on surface of metal. Begin heating at this point, using torch with rosebud heating tip. Gradually increase pressure while heating until frame is straightened.

CHAPTER 10
BODY AND ACCESSORIES MAINTENANCE

Section 1. STANDARD BODY MAINTENANCE

10-1. STANDARD BODY MAINTENANCE INDEX.

10-2. RADIATOR SUPPORT ASSEMBLY REPLACEMENT.

This task covers: a. Removal b. Installation

INITIAL SETUP:

Equipment Condition

- Cooling system drained, (See TM 9-2320-289-20)
- Coolant recovery reservoir and upper mounting bracket removed. (See TM 9-2320-289-20)
- Radiator removed. (See TM 9-2320-289-20)
- Front battery hold-down bracket removed, (See TM 9-2320-289-20)
- Hood latch removed. (See TM 9-2320-289-20)
- Headlight assemblies removed. (See TM 9-2320-289-20)
- Parking lights removed. (See TM 9-2320-289-20)
- Radiator grille and front end panel molding removed, (See TM 9-2320-289-20)
- Blackout drive light and mounting bracket removed. (See TM 9-2320-289-20)
- Blackout marker lights disconnected. (See TM 9-2320-289-20)
- Slave receptacle removed. (See TM 9-2320-289-20)
- Horn removed, (See TM 9-2320-289-20)

Tools/Test Equipment

- Torque wrench

Personnel Required

- MOS 63W (2)

General Safety Instructions

- Use caution when handling sheet metal.

a. REMOVAL

WARNING

Use caution when handling sheet metal. Sheet metal has numerous sharp edges and splinters. Failure to follow this caution may result in injury to personnel.

NOTE

If removing front end panel (4) and support (5), perform step 1.

1. Remove 3 bolts (2) securing front end panel (4) and support (5) to radiator support assembly (1), Remove 4 bolts (3) securing front end panel to 2 fenders. Remove front end panel and support.

10-2. RADIATOR SUPPORT ASSEMBLY REPLACEMENT (Con't).

NOTE

Ensure that location of front wiring harness is noted and all leads tagged for installation.

2. Disconnect front wiring harness ground wire from radiator support assembly (1) and from frame. Disconnect front wiring harness clips from radiator support assembly. Pull front wiring harness through radiator support assembly into engine compartment,

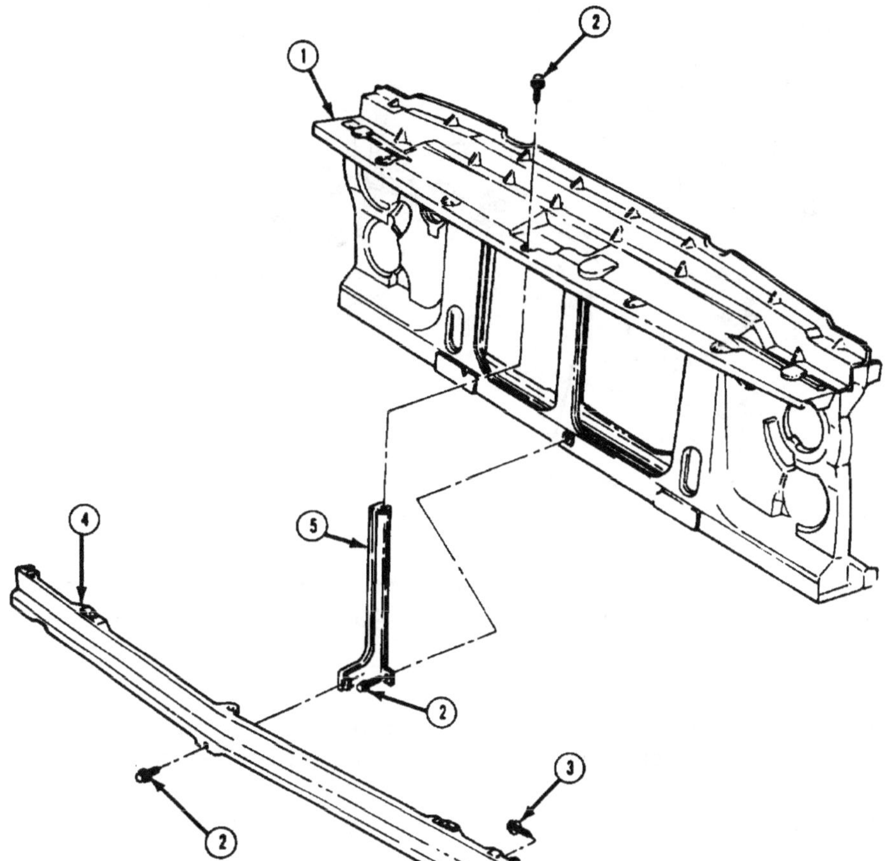

TA50389

10-2. RADIATOR SUPPORT ASSEMBLY REPLACEMENT (Con't).

3. Remove 10 bolts (6) securing sides of radiator support assembly (1) to 2 fenders (7). Remove 6 bolts (11) securing bottom of radiator support assembly to 2 fender skirts (9).

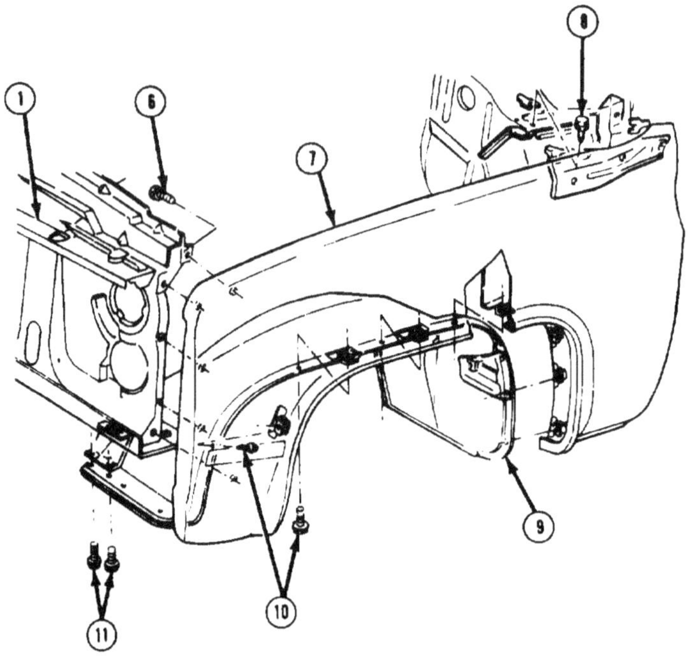

4. Remove weight classification marker from bumper. At each radiator support bracket (15) remove nut (18), retainer (17), lower cushion body (16), bolt (12), and washer (13),

NOTE

Fender bolts (8 and 10) are loosened in step 5 to permit easier alinement of fender to radiator support assembly bolt holes during installation.

5. At each fender (7), loosen 7 fender-to-fender skirt bolts (10). Loosen fender-to-cowl bolt (8),

6. Pull front of fenders (7) outward just enough to raise radiator support assembly (1) from truck. Remove radiator support assembly.

TA50390

10-2. RADIATOR SUPPORT ASSEMBLY REPLACEMENT (Con't).

7. If damaged, remove coolant recovery reservoir bracket from bottom left of radiator support assembly (1). If damaged, remove 2 upper cushion bodies (14) from radiator support brackets (15).

8. If installing new radiator support assembly (1), remove all attaching hardware from radiator suppor t assembly,

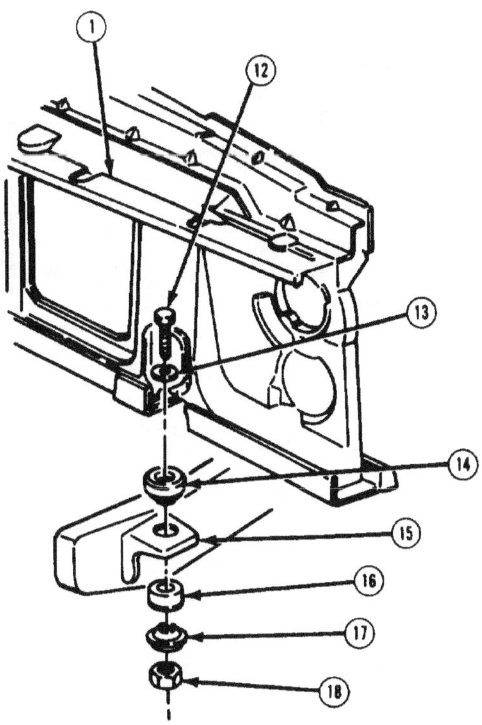

b. INSTALLATION

WARNING

Use caution when handling sheet metal, Sheet metal has numerous sharp edges and splinters. Failure to follow this caution may result in injury to personnel.

NOTE

if installing front end panel and support, perform step 7.

1. If removed, install coolant recovery reservoir bracket to bottom left of radiator support assembly (1). If removed, install all attaching hardware to radiator support assembly. If removed, install 2 upper cushion bodies (14) to radiator support brackets (15).

2. Position radiator support assembly (1) in truck while pulling outward on front of fenders (7) to gain clearance. At each radiator support bracket (15) loosely install washer (13), bolt (12), lower cushion body (16), retainer (17), and nut (18).

NOTE

If truck is equipped with winterization kits, it may be necessary to raise fenders (7) using a suitable jack and wooden block, to aid in alinement.

3. Loosely install 10 bolts (6) securing sides of radiator support assembly (1) to 2 fenders (7). Loosely install 6 bolts (11) securing bottom of radiator support assembly to 2 fender skirts (9). Fully tighten bolts (6 and 11).

4. At each fender (7), tighten 7 fender-to-fender skirt bolts (10). Tighten bolt (8) to 35 lb.-ft. (47 N.m).

TA50391

10-2. RADIATOR SUPPORT ASSEMBLY REPLACEMENT (Con't).

5. Tighten 2 nuts (18) to 35 lb.-ft. (47 Nom),

6. Route front wiring harness through radiator support assembly (1), Connect front wiring harness ground wire to radiator support assembly and to frame. Connect front wiring harness clips to radiator support assembly.

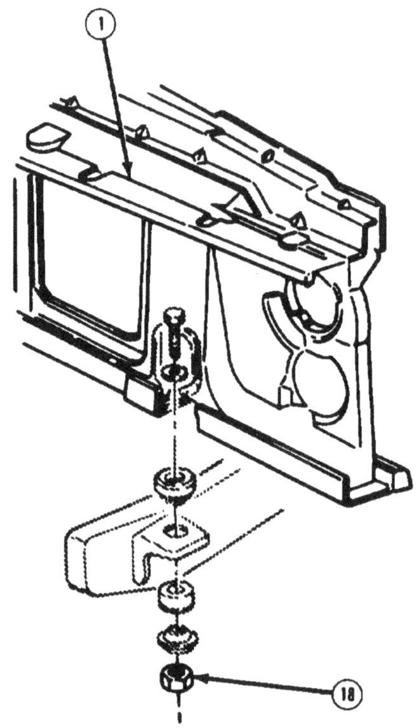

7. Install front end panel (4) and support (5) to radiator support assembly (1) with 3 bolts (2). Install 4 bolts (3) from front end panel to 2 fenders.

8. Install weight classification marker to bumper.

TA50392

10-2. RADIATOR SUPPORT ASSEMBLY REPLACEMENT (Con't).

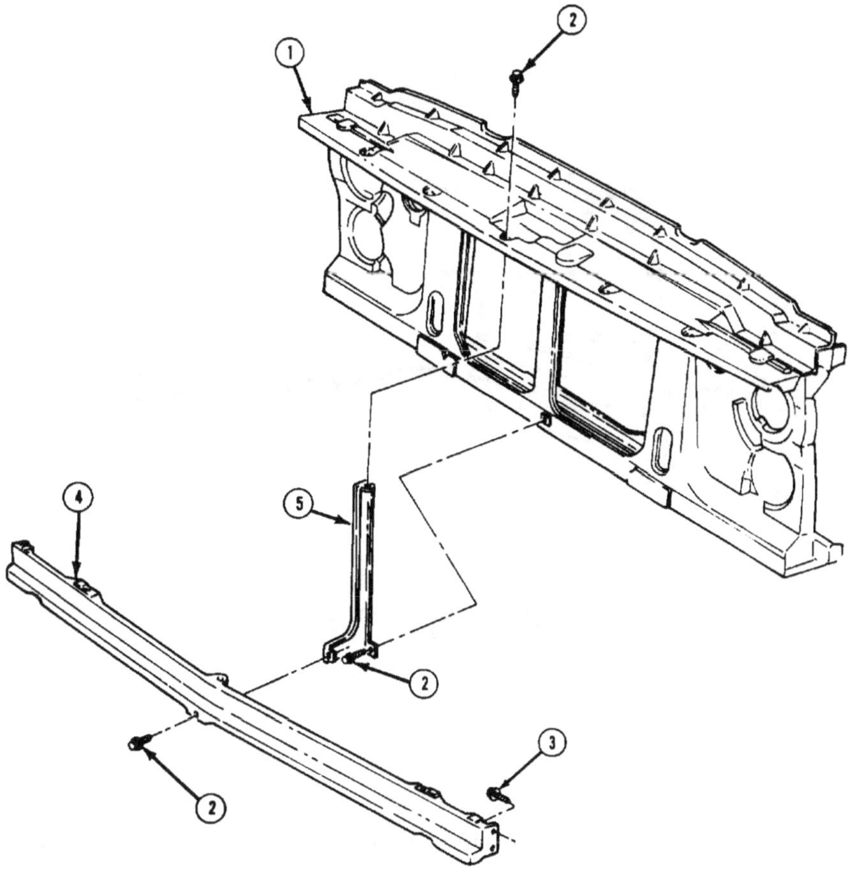

FOLLOW-ON TASKS:

- Install horn. (See TM 9-2320-289-20)
- Install slave receptacle, (See TM 9-2320-289-20)
- Connect blackout marker lights, (See TM 9-2320-289-20)
- Install blackout drive light and mounting bracket. (See TM 9-2320-289-20)
- Install radiator grille and front end panel molding. (See TM 9-2320-289-20)
- Install parking lights. (See TM 9-2320-289-20)
- Install headlight assemblies. (See TM 9-2320-289-20)
- Install hood latch. (See TM 9-2320-289-20)
- Install front battery hold-down bracket. (See TM 9-2320-289-20)
- Install radiator. (See TM 9-2320-289-20)
- Install upper mounting bracket and coolant recovery reservoir. (See TM 9-2320-289-20)

Pages 10-8 and 10-9 have been rescinded.

TA50393

10-3. INSTRUMENT PAD REPLACEMENT (Con't).

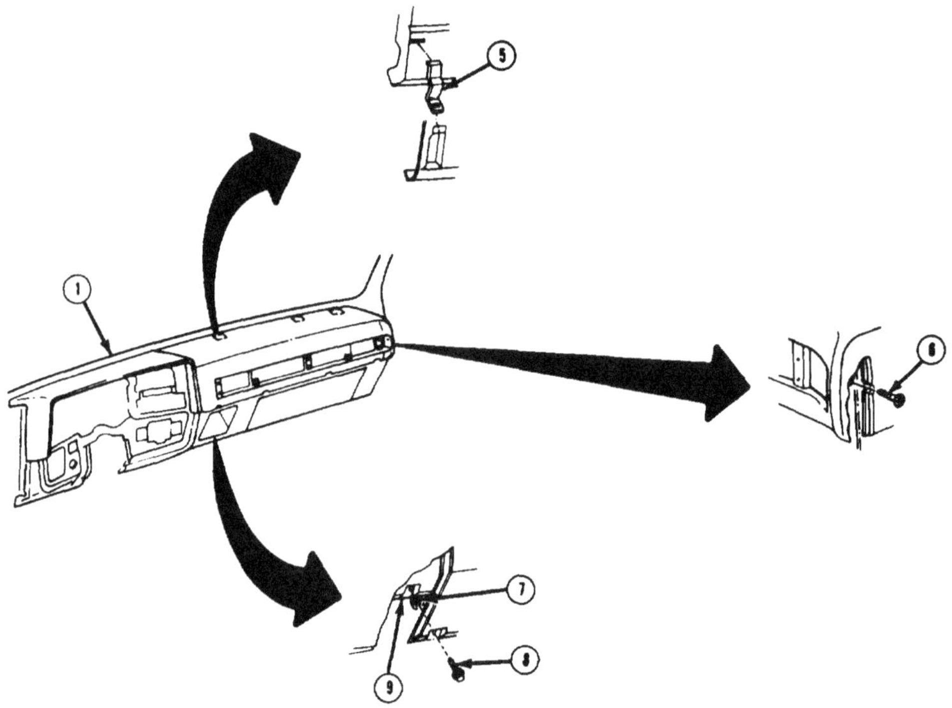

3. If removed, install reinforcement (4) to instrument pad assembly (1) with 4 washers (2) and screws (3).

FOLLOW-ON TASKS:

- Install heater control panel and housing if removed. (See TM 9-2320-289-20)

TA50395

10-4. WINDSHIELD AND M1009 TOP ASSEMBLY SIDE WINDOW REPLACEMENT.

This task covers:

a. Removal
b. Inspection
c. Installation

INITIAL SETUP:

Equipment Condition

- Windshield wiper arms removed (if removing windshield). (See TM 9-2320-289-20)

Materials/Parts

- One inside rearview mirror support
- Loctite adhesive (Item 5, Appendix B)
- Denatured alcohol (Item 6, Appendix B)
- Abrasive cloth (Item 17, Appendix B)
- Masking tape (Item 68, Appendix B)

Tools/Test Equipment

- Installing tool, J-26471

Personnel Required

- MOS 63W (2)

Manual References

- FM 43-2

General Safety Instructions

- Wear heavy gloves and safety goggles when handling glass.
- Alcohol used in this procedure is highly flammable and poisonous.

a. REMOVAL

WARNING

Wear heavy gloves and safety goggles when handling windshield (3) or window glass to avoid risk of injury to personnel.

NOTE

- If removing windshield (3), perform steps 1-3.
- If removing M1009 top assembly side window (5), perform steps 2 and 3.

1. Remove inside rearview mirror (7).
2. Using tape, mark location on cab of cap (8) at bottom center of reveal molding (1). Remove cap and remove reveal molding.

10-4. WINDSHIELD AND M1009 TOP ASSEMBLY SIDE WINDOW REPLACEMENT.

This task covers: a. Removal b. Inspection c. Installation

INITIAL SETUP:

Equipment Condition

- Windshield wiper arms removed (if removing windshield). (See TM 9-2320-289-20)

Materials/Parts

- one inside rearview mirror support
- Loctite adhesive (Item 5, Appendix B)
- Denatured alcohol (Item 6, Appendix B)
- Abrasive cloth (Item 17, Appendix B)
- Masking tape (Item 68, Appendix B)

Too/s/Test Equipment

- Installing tool, J-26471

Personnel Required

- MOS 63W (2)

Manual References

- FM 43-2

General Safety Instructions

- Wear heavy gloves and safety goggles when handling glass,
- Alcohol used in this procedure is highly flammable and poisonous,

a. REMOVA L

WARNING

Wear heavy gloves and safety goggles when handling windshield (3) or window glass to avoid risk of injury to personnel.

NOTE

- If removing windshield (3), perform steps 1-3.
- If removing M1009 top assembly side window (5), perform steps 2 and 3.

1. Remove inside rearview mirror (7),
2. Using tape, mark location on cab of cap (8) at bottom center of reveal molding (1). Remove cap and remove reveal molding.

10-4. WINDSHIELD AND M1009 TOP ASSEMBLY SIDE WINDOW REPLACEMENT (Con't).

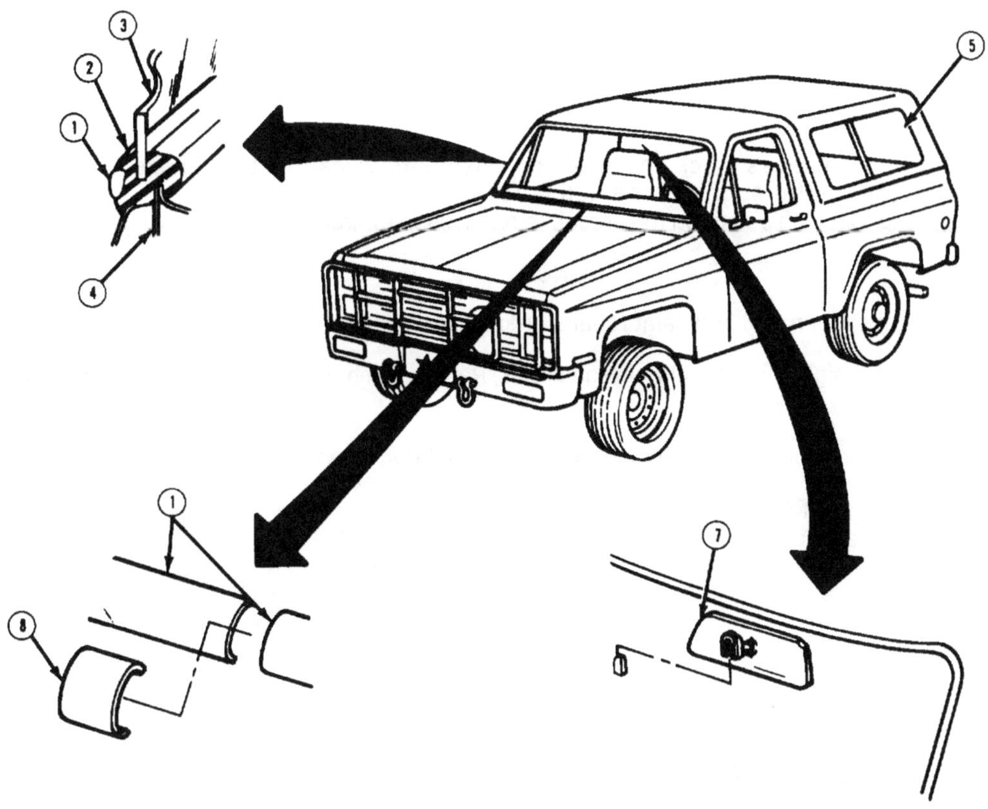

3 . Support windshield (3) or side window (5) from outside truck. From inside truck, have an assistant apply firm, controlled pressure to edge of windshield or side window. Force weatherstrip (2) from flange (4) with a flat-bladed tool. Remove windshield or side window and weatherstrip. Remove weatherstrip.

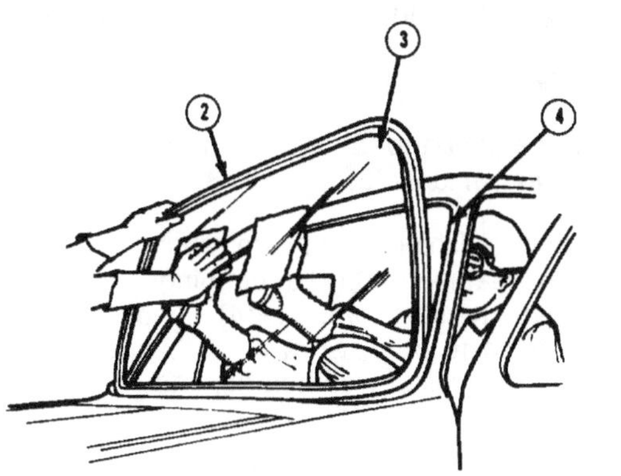

TA50396

10-4. WINDSHIELD AND M1009 TOP ASSEMBLY SIDE WINDOW REPLACEMENT (Con't).

b. INSPECTION

CAUTION

In event of a stress crack or leakage, ensure that weatherstrip (2) and flange (4) are thoroughly inspected before installing windshield (3). Failure to follow this caution may result in recurrence of crack or leakage.

NOTE

If sheet metal and paint repairs are needed, see FM 43-2.

1. Inspect flange for high spots or other irregularities, Remove all high spots or irregularities before proceeding.
2. Inspect weatherstrip for damage. Replace weatherstrip if damaged,
3. Inspect replacement windshield to ensure that there are no chipped edges or other irregularities. Obtain another windshield if repairs cannot be made.

c. INSTALLATION

WARNING

Wear heavy gloves and safety goggles when handling windshield (3) or window glass to avoid risk of injury.

NOTE

- If installing windshield (3), perform steps 1-9.
- If installing M1009 top assembly side window (5), perform steps 1-3.

1. Install weatherstrip (2) to windshield (3) or side window (5).
2. Position windshield (3) or side window (5) with weatherstrip" (2) at opening. Install weatherstrip to flange (4) along bottom edge of opening . Work remaining edges of weatherstrip into flange.

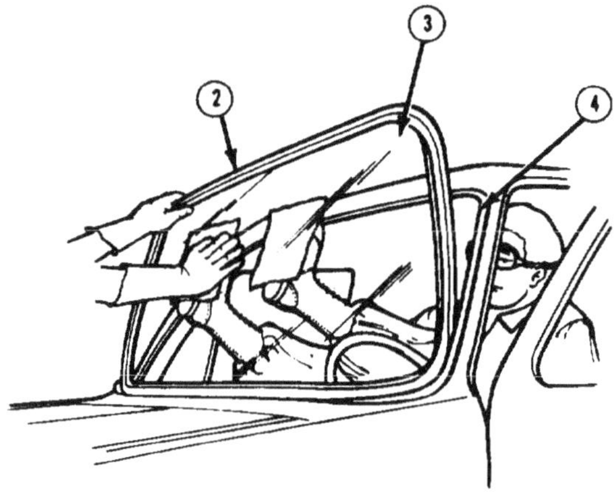

TA50397

10-4. WINDSHIELD AND M1009 TOP ASSEMBLY SIDE WINDOW REPLACEMENT (Con't).

3 . Apply soapy water to reveal molding (1) channel in weatherstrip (2) to serve as lubricant. Using installing tool, install reveal molding so that gap for cap (8) is positioned where marked. Install cap.

4 . Using old windshield as a guide, mark outside of windshield (3) with tape for location of support (6) for inside rearview mirror (7).

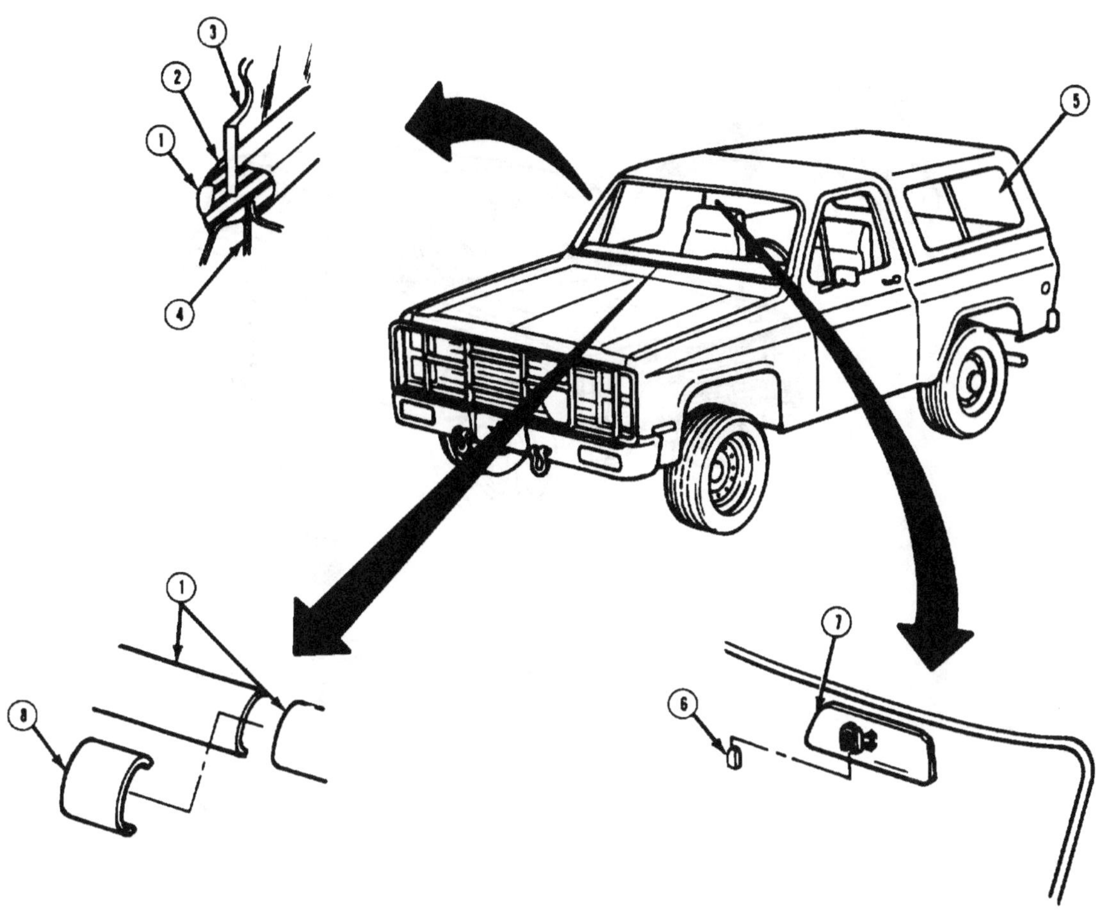

WARNING

Alcohol used in this procedure is highly flammable and poisonous. Drinking of this alcohol can cause blindness and death, Avoid inhaling its fumes. Keep away from open flame.

5 . Using an alcohol-saturated rag, clean inside of windshield (3) where support (6) is to be installed.

TA50398

10-4. WINDSHIELD AND M1009 TOP ASSEMBLY SIDE WINDOW REPLACEMENT (Con't).

6. Using fine abrasive cloth, sand bonding surface of support (6). Clean sanded surface with an alcohol-saturated rag and allow to dry.

7. Follow directions on manufacturer's adhesive to properly prepare support (6) prior to installation on windshield (3). Install support, to location marked with tape, and press against glass for 30-60 seconds exerting steady pressure.

8. After 5 minutes, remove any excess adhesive with an alcohol-saturated rag.

9. Install inside rearview mirror (7).

FOLLOW-ON TASKS:

- Install windshield wiper arms. (See TM 9-2320-289-20)

TA50399

10-5. FENDER REPLACEMENT.

This task covers: a. Removal b. Installation

INITIAL SETUP:

Equipment Condition

- Hood and hinge spring assemblies removed. (See TM 9-2320-289-20)
- Left or right headlight assembly removed, (See TM 9-2320-289-20)
- Front end panel molding removed. (See TM 9-2320-289-20)
- Left or right marker light removed. (See TM 9-2320-289-20)
- Windshield washer solvent container removed (left side). (See TM 9-2320-289-20)
- Battery supports removed from fender (right side). (See TM 9-2320-289-20)
- Batteries removed (right side), (See TM 9-2320-289-20)
- Current/voltage regulator removed (left side, Ml 01 O). (See paragraph 4-4)
- Heater fuel pump removed (left side, Ml 010 and trucks equipped with winterization kits). (See TM 9-2320-289-20)

Tools/Test Equipment

- Torque wrench

Personnel Required

- MOS 63W (2)

Manual References

- TB 43-0209

General Safety instructions

- Use caution when handling sheet metal.

a. REMOVAL

WARNING

Use caution when handling sheet metal. Sheet metal has numerous sharp edges and splinters, Failure to follow this caution may result in injury to personnel.

10-5. FENDER REPLACEMENT (Con't).

1. Disconnect wiring from fender (4) and remove 2 hood bumpers (10).
2. Remove 2 bolts (17) securing front end panel (16) to fender (4). Remove 7 bolts (15) securing fender wheel opening flange to wheel house panel (14). Remove 5 bolts (19) securing radiator support assembly (18) to fender.
3. Remove 2 screws (12) and 2 retainers, and remove shield (13).

NOTE

Ensure that location of shims (5) is noted for installation.

4. Remove bolt (9) and shim (s) (5) at lower end of fender (4). Remove bolt (6) and shim(s) securing fender to cowl (1).

NOTE

An assistant is required to perform step 5.

5. Open door and remove bolt (8) and shim (s) (5) at door hinge pillar (7). Raise rear of fender (4) and remove,
6. If damaged or if replacing fender (4), remove 7 nuts (11) from fender flange.
7. If damaged, remove 4 screws (3) and seal (2) from cowl (1).

b. INSTALLATION

WARNING

Use caution when handling sheet metal, Sheet metal has numerous sharp edges and splinters. Failure to follow this caution may result in injury to personnel.

1. If removed, install seal (2) to cowl (1) with 4 screws (3).

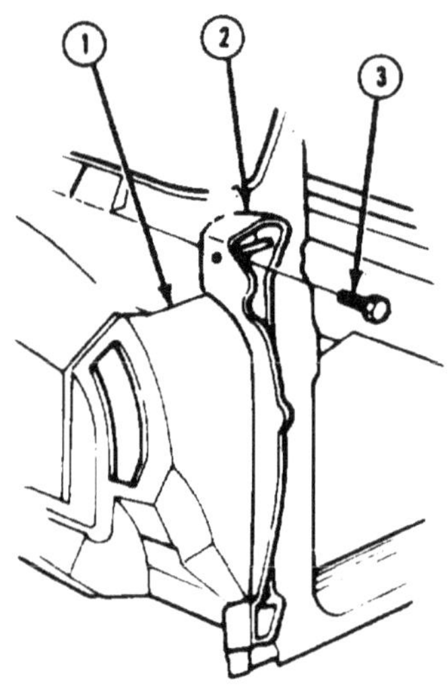

TA50400

10-5. FENDER REPLACEMENT (Con't).

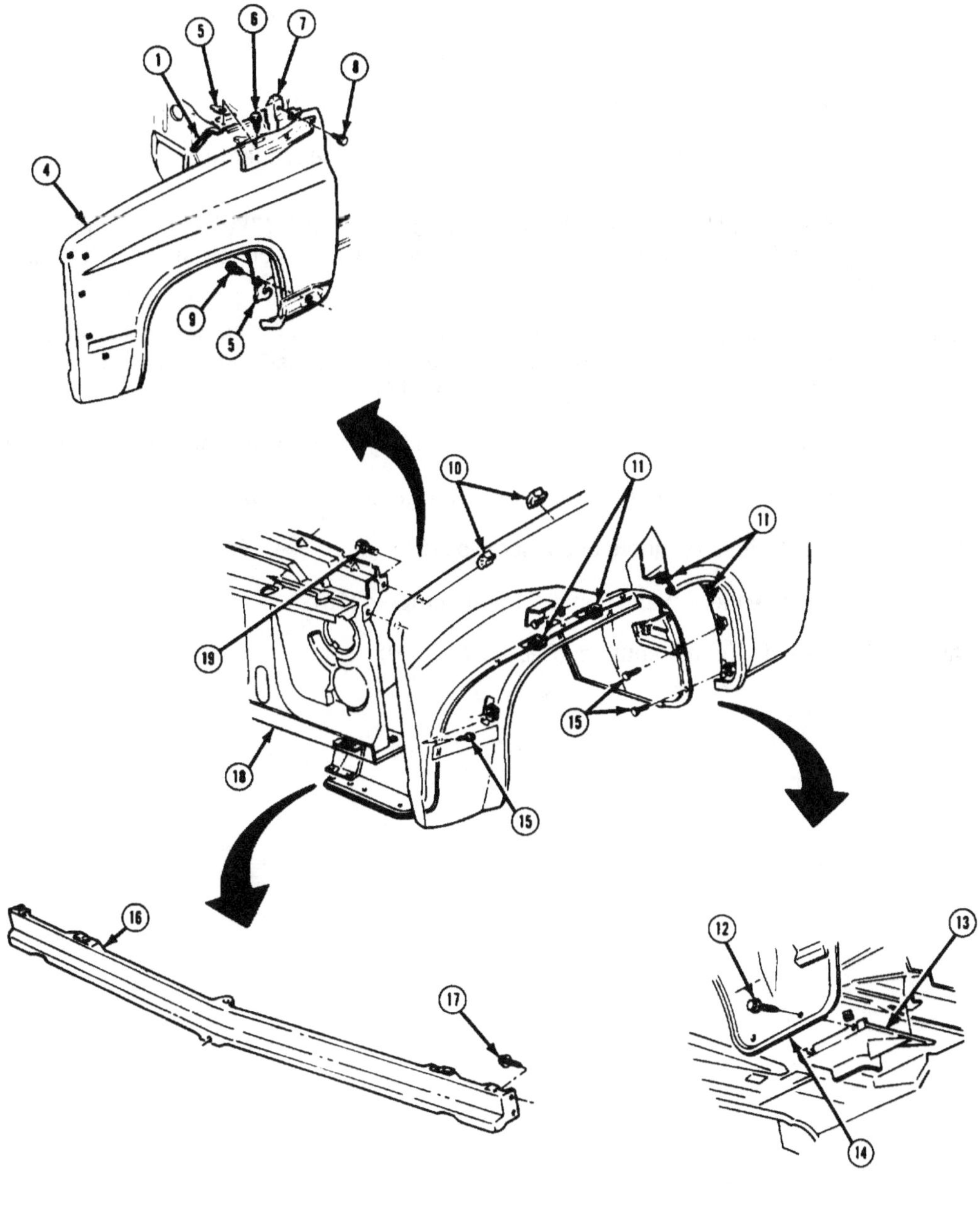

TA50401

10-5. FENDER REPLACEMENT (Con't).

2. If removed, install 7 nuts (11) to fender (4) flange.

NOTE

- An assistant is required to perform steps 3-5.
- Shims (5) should be used as required to aline fender (4).

3. Position fender (4) and loosely install fender to door hinge pillar (7) with bolt (8) and shim(s) (5), Loosely install fender to cowl (1) with bolt (6) and shim(s). Loosely install bolt (9) and shim(s) at lower end of fender.
4. Loosely install 5 bolts (19) securing radiator support assembly (18) to fender (4). Loosely install 2 bolts (1 7) securing front end panel (16) to fender. Loosely install 7 bolts (15) securing fender wheel opening flange to wheel house panel (14).
5. Aline fender (4) with door, hood, and windshield, adding or removing shims (5) as required,
6. Tighten 3 bolts (6, 8, and 9) to 30 lb.-ft. (41 N.m). Tighten all other bolts (15, 17, and 19).
7. Install shield (13) with 2 retainers and 2 screws (12).
8. Connect wiring to fender (4) and install 2 hood bumpers (10)
9. Apply paint to fender (4) as required. (See TB 43-0209)

10-5. FENDER REPLACEMENT (Con't).

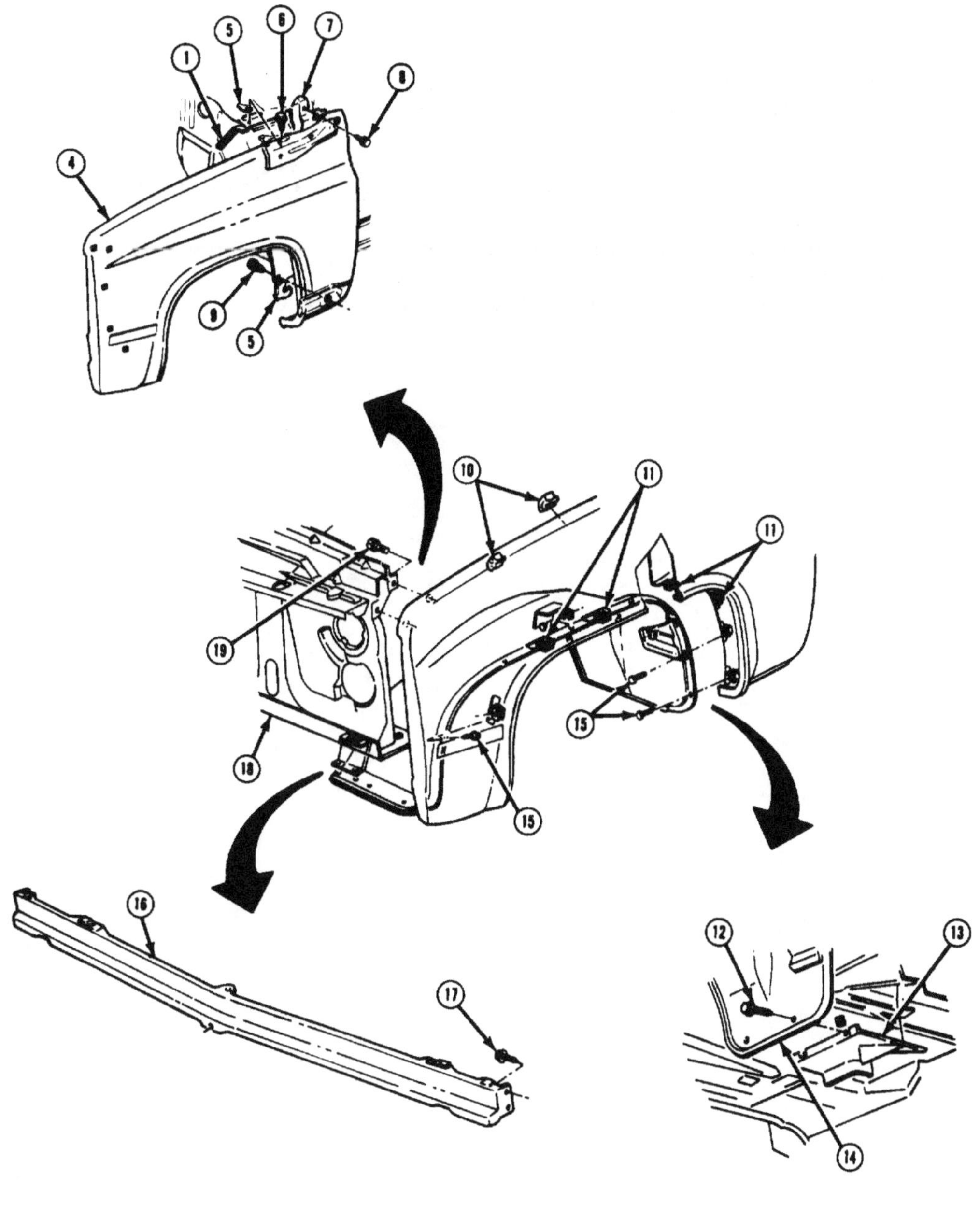

TA50402

10-5. FENDER REPLACEMENT (Con't).

FOLLOW-ON TASKS:

- Install heater fuel pump if removed. (See TM 9-2320-289-20)
- Install current/voltage regulator if removed, (See paragraph 4-4)
- Install batteries if removed, (See TM 9-2320-289-20)
- Install battery supports to fender if removed. (See TM 9-2320-289-20)
- Install windshield washer solvent container if removed. (See TM 9-2320-289-20)
- Install left or right marker light. (See TM 9-2320-289-20)
- Install front end panel molding, (See TM 9-2320-289-20)
- Install left or right headlight assembly. (See TM 9-2320-289-20)
- Install hood and hinge spring assemblies. (See TM 9-2320-289-20)

10-6. WHEEL HOUSE PANEL AND SPLASH GUARD REPLACEMENT.

This task covers: a. Removal b. Installation

INITIAL SETUP:

Equipment Condition

- Hood latch cable and engine oil cooler hoses disconnected from wheel house panel (left side). (See TM 9-2320-289-20)
- Battery cables and heater assembly hoses disconnected from wheel house panel (right side). (See TM 9-2320-289-20)
- Battery trays and supports removed (right side). (See TM 9-2320-289-20)
- Engine coolant heater removed (if truck is equipped with winterization kits, left side). (See TM 9-2320-289-20)
- Cab heater removed (if truck is equipped with winterization kits, right side). (See TM 9-2320-289-20)

Manual References

- TM 9-2320-289-20

General Safety Instructions

- Use caution when handling sheet metal.

a. REMOVAL

WARNING

Use caution when handling sheet metal. Sheet metal has numerous sharp edges and splinters. Failure to follow this caution may result in injury to personnel.

NOTE

If removing splash guard (2), perform step 2.

1. If removing left wheel house panel (1), remove service parts identification label and save for reference.

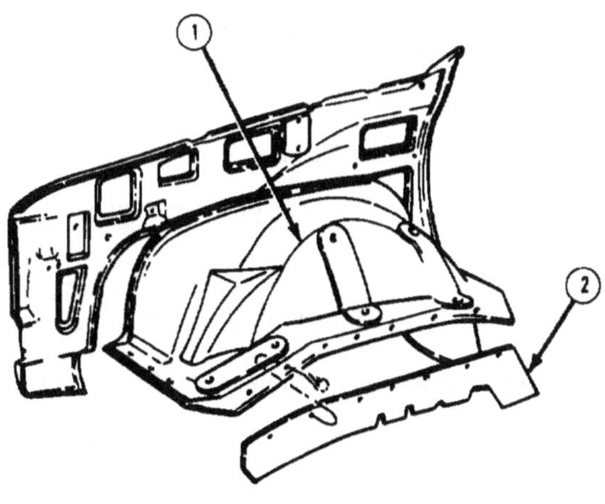

TA50403

10-6. WHEEL HOUSE PANEL AND SPLASH GUARD REPLACEMENT (Con't).

2 . Remove 7 retainers (3) and remove splash guard (2) from wheel house panel (1).

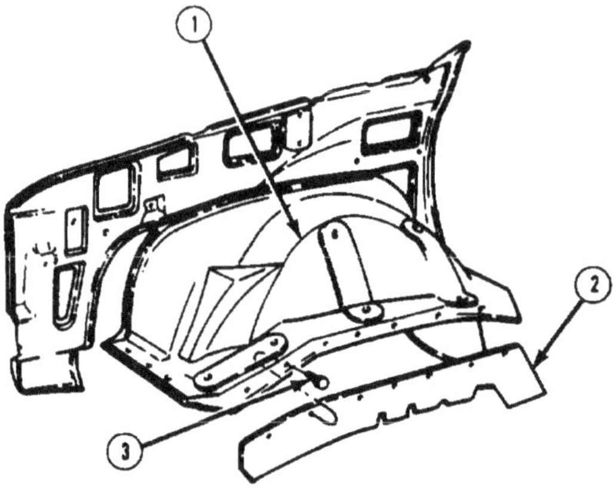

3. Remove 7 bolts (9) securing wheel house panel (1) to fender (5) flange. Remove 2 screws (7) securing wheel house panel to shield (8). Remove 2 bolts (4 and 6) securing wheel house panel to bulkhead.

4. Remove 3 bolts (10) securing wheel house panel (1) to radiator support assembly (11),

5. Slide wheel house panel (1) forward to clear lower back side of fenderwell. Remove wheel house panel from truck.

b. INSTALLATION

WARNING

Use caution when handling sheet metal. Sheet metal has numerous sharp edges and splinters. Failure to follow this caution may result in injury to personnel.

NOTE

If installing splash guard (2), perform step 5.

1 . Position wheel house panel (1) in truck and loosely install 7 bolts (9) securing wheel house panel to fender (5) flange. Loosely install shield (8) to wheel house panel with 2 screws (7).

2 . Loosely install 3 bolts (10) securing wheel house panel (1) to radiator support assembly (11).

3 . Loosely install 2 bolts (4 and 6) securing wheel house panel (1) to bulkhead.

TA50404

10-6. WHEEL HOUSE PANEL AND SPLASH GUARD REPLACEMENT (Con't).

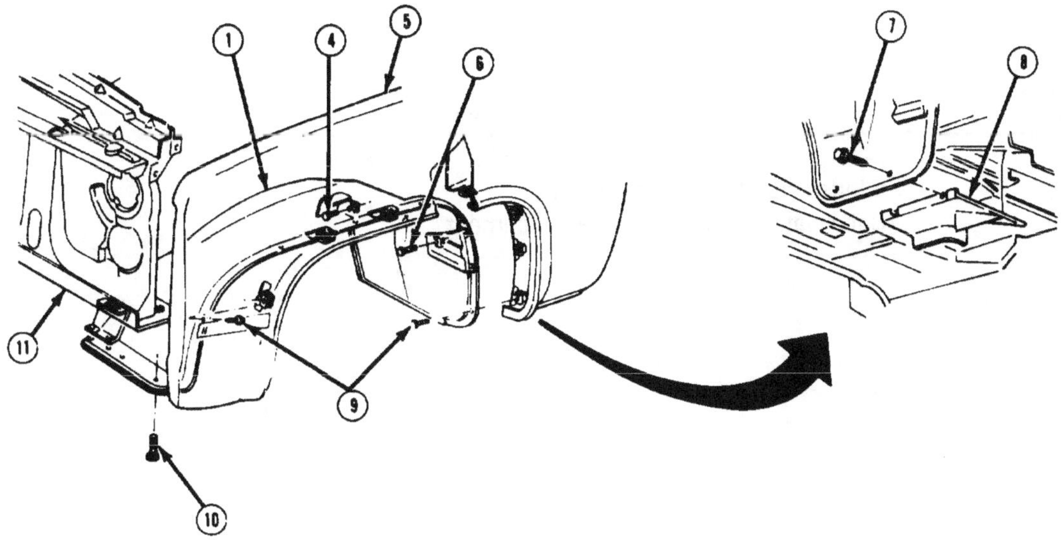

4. Tighten bolt (6) to 35 lb.-ft. (47 N.m). Tighten bolts (4, 9, and 10) and screw (7).

5. Install splash guard (2) to wheel house panel (1) with 7 retainers (3).

FOLLOW-ON TASKS:

- Install cab heater if removed, (See TM 9-2320-289-20)
- Install engine coolant heater if removed, (See TM 9-2320-289-20)
- Install battery trays and supports if removed. (See TM 9-2320-289-20)
- Connect battery cables and heater assembly hoses to wheel house panel if removed. (See TM 9-2320-289-20)
- Connect hood latch cable and engine cooler hoses to wheel house panel if removed. (See TM 9-2320-289-20)

TA50405

Section IL STANDARD BODY ACCESSORIES MAINTENANCE

10-7. STANDARD BODY ACCESSORIES MAINTENANCE INDEX.

10-8. WINDSHIELD WIPER MOTOR REPAIR.

This task covers: a, Disassembly b. Assembly

INITIAL SETUP:

Equipment Condition

- Windshield wiper motor removed. (See TM 9-2320-289-20)

Materials/Parts

- End cap kit
- One gasket
- Two seals
- Lubricating oil (Item 49, Appendix B)
- Petrolatum (Item 50, Appendix B)

a. DISASSEMBLY

NOTE

- If disassembling armature (2) and end cap assembly (7), perform steps 1-3.
- If disassembling crank arm (8) assembly, perform step 4.
- If disassembling washer pump assembly, perform steps 5-11.

1. Straighten 4 retainer tabs and remove end cap assembly (7). Remove felt washer (4) from inside end cap assembly.

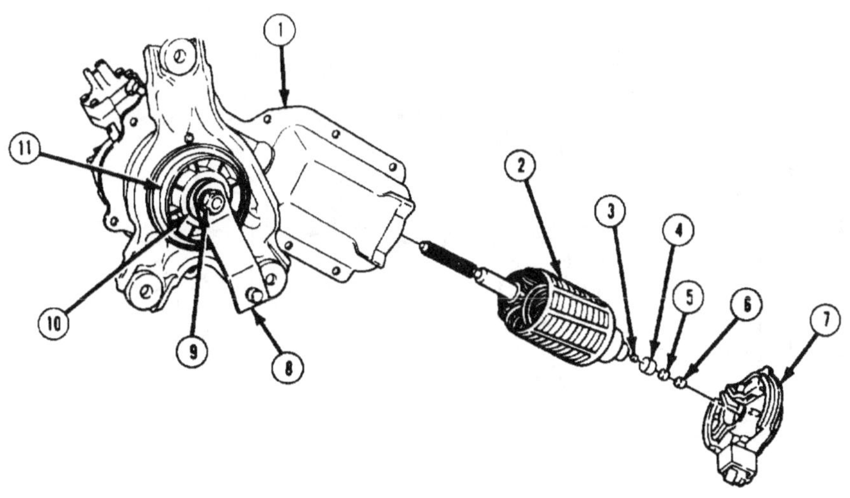

TA50406

10-8. WINDSHIELD WIPER MOTOR REPAIR (Con't).

2. Rotate crank arm (8) clockwise to help free armature (2), Remove armature from housing (1), Using a magnet, remove thrust ball (3) from end of armature.

3. Using a magnet, remove thrust plate (5) and motor thrust disc (6) from end cap assembly (7).

CAUTION

Ensure that crank arm (8) and crank arm shaft are marked for assembly, Failure to follow this caution may result in damage to, or malfunction of, wiper system.

NOTE

Before disassembling crank arm (8), ensure that windshield wiper motor is in "park" with crank arm positioned on shaft as shown.

4. Mark crank arm (8) and crank arm shaft for assembly. Remove nut (9) and remove crank arm. Remove seal (10) and spacer (11). Discard seal.

NOTE

Location of pawl spring (14) may vary with windshield wiper motor model. Ensure that location is noted for assembly.

5. Remove washer pump cover. Remove pawl spring (14).

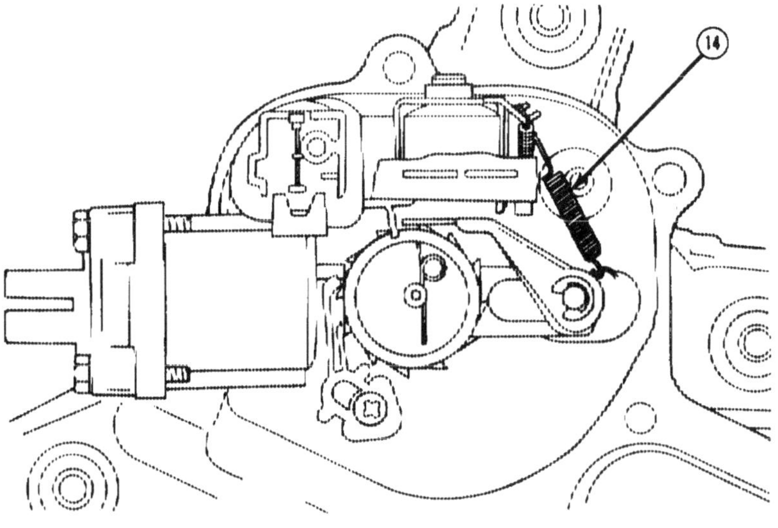

TA50407

10-8. WINDSHIELD WIPER MOTOR REPAIR (Con't).

6. Push in retainer tab and remove washer solenoid and park switch assembly (1 3).
7. Remove retainer ring (15) and pawl (16) from shaft (17).
8. Remove screw (21) and dog spring (20) from washer frame.
9. Move leg of retainer spring (19) out of shaft (12) groove. Remove ratchet gear (18) from shaft.

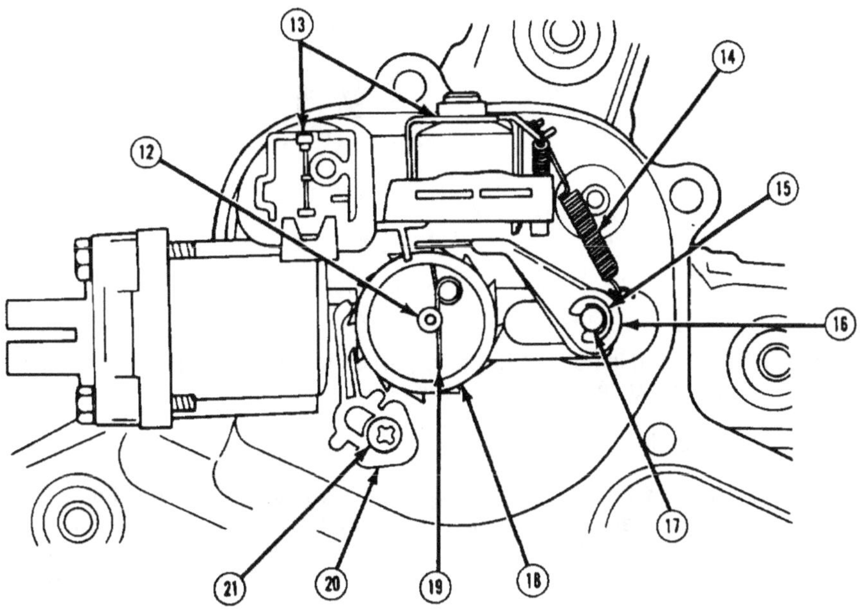

10. Pull piston housing (25) away from washer frame (26) until mounting grooves clear frame. Remove valve assembly (24), piston housing, and piston actuator plate (27) as a unit.
11. Remove 4 screws (22) and cover (23) from valve assembly (24). Remove gasket from between cover and valve assembly. Remove pump seal. Discard gasket and pump seal.

b. ASSEMBLY

NOTE

- If assembling washer pump assembly, perform steps 1-5.
- If assembling crank arm assembly, perform step 6.
- If assembling armature and end cap assembly, perform steps 7-11.

1. Install new pump seal in piston housing (25) grooves. Install new gasket between valve assembly (24) and cover (23). Aline valve and cover assembly with piston housing and start 4 screws (22). Tighten 4 screws to evenly draw in cover.

TA50408

10-8. WINDSHIELD WIPER MOTOR REPAIR (Con't).

2. Position piston housing (25) and valve assembly (24) with shafts (12 and 17) through slots in piston actuator plate (27). Pull back on piston housing and engage mounting grooves of piston housing in washer frame (26), At shaft (1 2), apply a liberal coat of petrolatum between piston actuator plate and washer frame.

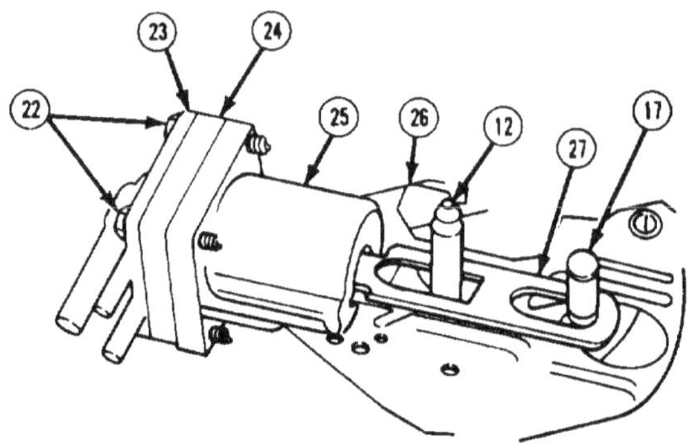

3 . Depress retainer spring (19) and install ratchet gear (18) on shaft (12). Rotate ratchet gear to seat it. Install retainer spring with leg in shaft groove.

4 . Install screw (21) and dog spring (20).

5 . Install pawl (16) on shaft (17) with retainer ring (15). Install washer solenoid and park switch assembly (13), and press in on retainer tab to seat, Pull retainer tab to locking position. Install pawl spring (14) in original position. Install washer pump cover, indexing washer solenoid terminals and park switch connector with slots in cover. Snap washer pump cover to fully seat.

NOTE

Ensure that flat side of seal (10 is toward crank arm (8),

6 . Install spacer (11), new seal (10), and crank arm (8), as marked, with nut (9).

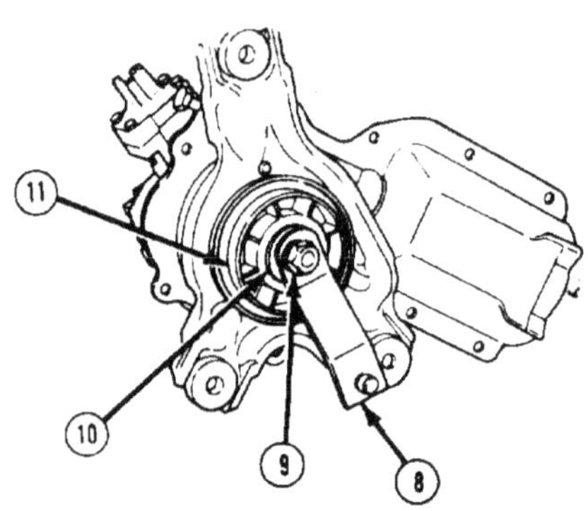

TA50409

10-8. WINDSHIELD WIPER MOTOR REPAIR (Con't).

7. Install motor thrust disc (6), thrust plate (5), and felt washer (4) inside end cap assembly (7), Apply lubricating oil to felt washer and to thrust ball (3). Install thrust ball to armature (2).

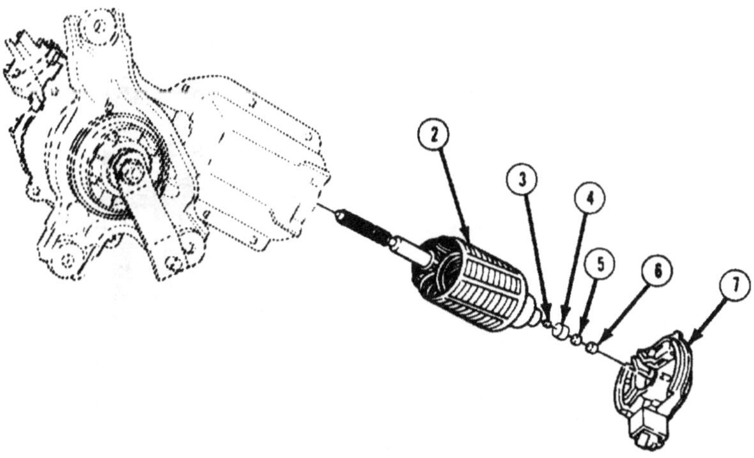

NOTE

Ensure that brush leads (29) are properly routed and are not touching each other.

8. Release tension on 3 brush springs (28) and slide brushes (31) back in their brush holders.

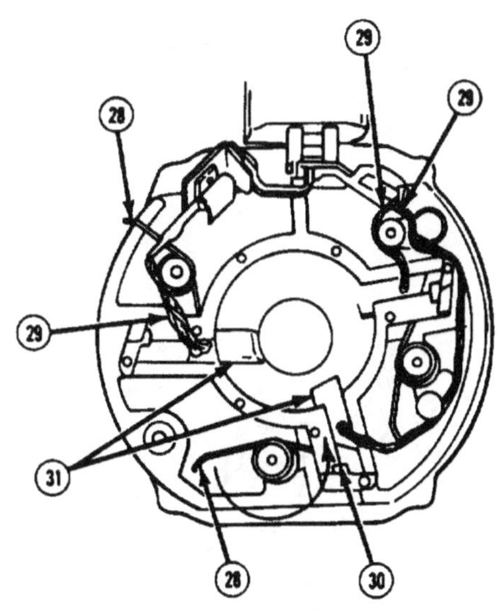

TA50410

10-8. WINDSHIELD WIPER MOTOR REPAIR (Con't).

9. Install armature (2) in end cap assembly (7). Rotate brush springs (28) counterclockwise to reposition them in retainer notches (30). Using wire (32), secure armature to end cap assembly.

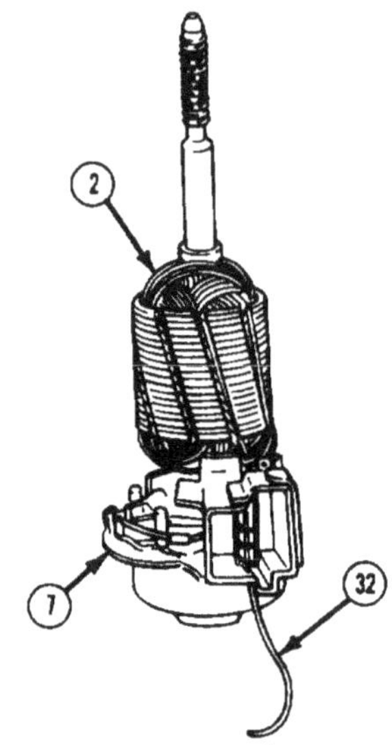

NOTE

- Strong magnetic field will pull armature (2) toward one of the magnets. It will be necessary to overcome this force to start shaft into bearing.
- Before end cap assembly (7) enters housing (l), wire (32) must be removed.

10. Guide armature worm gear (33) through housing (1) bearing. Terminal housing of end cap assembly (7) goes into slot area of housing, Rotate crank arm (8) slightly to engage armature worm gear shaft with gear teeth, Push on end cap assembly until it bottoms against end of housing.

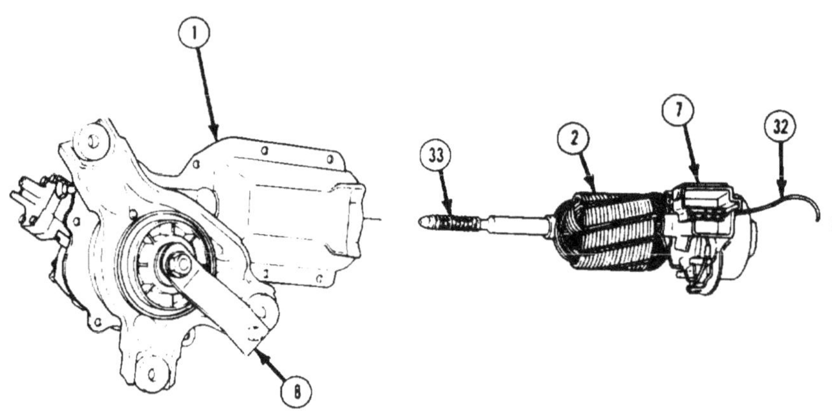

TA50411

10-8. WINDSHIELD WIPER MOTOR REPAIR (Con't).

11. Bend 4 retainer tabs to secure end cap assembly (7) to housing (1).

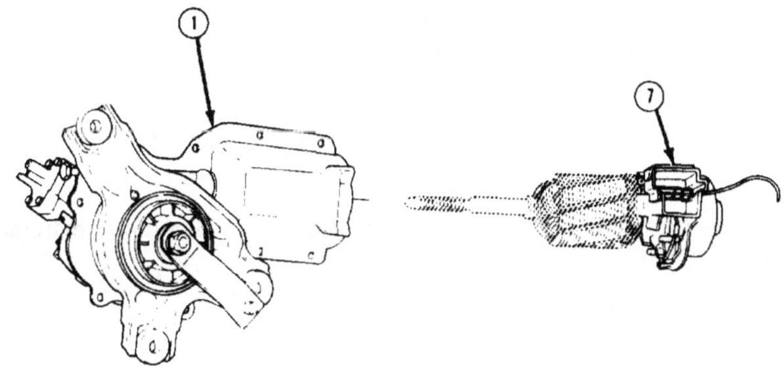

FOLLOW-ON TASKS:

- Install windshield wiper motor. (See TM 9-2320-289-20)

TA50412

10-9. CAB HEATER DEFROSTER DUCT REPLACEMENT.

This task covers: a. Removal b. Installation

INITIAL SETUP:

Equipment Condition

- Instrument pad assembly removed. (See TM 9-2320-289-20)
- Rear battery box removed (if truck is equipped with winterization kits), (See TM 9-2320-289-20)
- Heater assembly removed, (See TM 9-2320-289-20)

a. REMOVAL

1. Remove 2 screws (3) and retainers (1), Remove defroster duct (6) through instrument panel compartment opening (5).
2. Remove 2 "U" nuts (2) if damaged and discard.

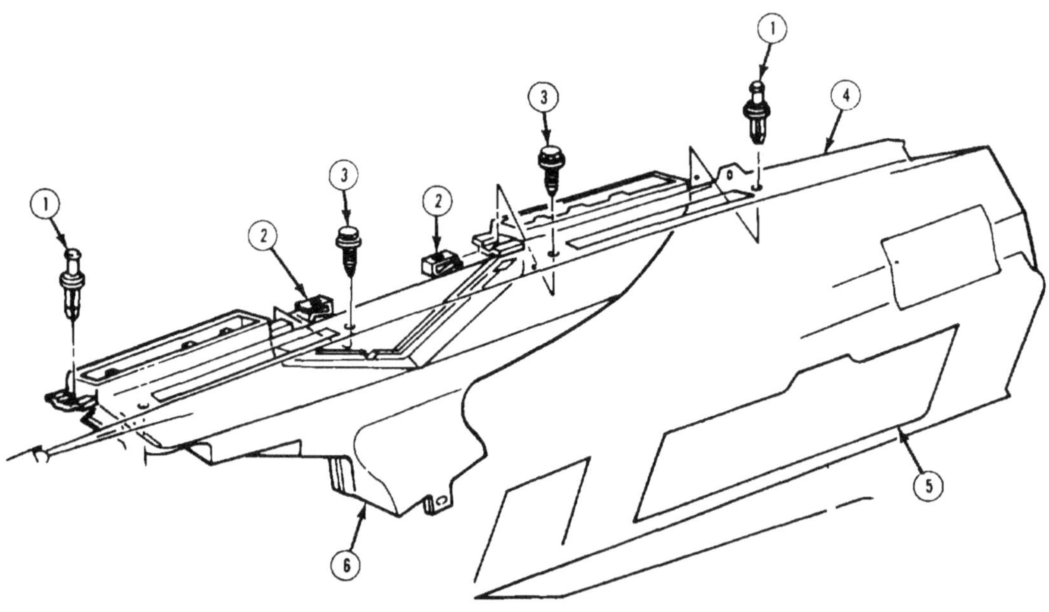

b. INSTALLATION

1. Install 2 new " U" nuts (2) to defroster duct (6) if removed.
2. Install defroster duct (6) through instrument panel compartment opening (5) and secure to instrument panel assembly (4) with 2 screws (3) and retainers (1).

TA50413

10-9 CAB HEATER DEFROSTER DUCT REPLACEMENT (Con't).

FOLLOW-ON TASKS:

- Install heater assembly, (See TM 9-2320-289-20)
- Install rear battery box if removed. (See TM 9-2320-289-20)
- Install instrument pad assembly. (See TM 9-2320-289-20)

Section III. CARGO BODY MAINTENANCE

10-10. CARGO BODY MAINTENANCE INDEX.

10-11. CARGO BODY PICKUP BOX MAINTENANCE.

This task covers:

a. Removal
b. Disassembly
b. 1. Preparation of Side Panel (M1028A2 and M1028A3)
c. Assembly
d. Installation

INITIAL SETUP:

Equipment Condition (See NOTE below)

- Troop seats removed if installed. (See TM 9-2320 -289-1 O)
- Endgate removed. (See TM 9-2320-289-20)
- S-250 communications shelter removed (M1028, M1028A1, M1028A2, and M1028A3 only)
- Communications rack and junction box removed (M1008A1 only). (See TM 9-2320-289-20)
- Shelter tie-downs removed (MI 028, M1028A1, M1028A2, and M1028A3 only). (See TM 9-2320-289-20)
- Taillights removed. (See TM 9-2320-289-20)
- Fuel tank filler door removed, (See TM 9-2320-289-20)
- Cargo body cover and frame removed if installed, (See TM 9-2320-289-20)
- Winterization kit components removed as required, (See TM 9-2320-289-20)
- Rear fender side marker lights disconnected (M1028A2 and M1028A3), (See TM 9-2320-289-20)

General Safety Instructions

- Use caution when handling sheet metal.
- Wear protective clothing when cutting sheet metal.

Materials/Parts

- Two starwashers
- Primer (Item 52, Appendix B)
- Sealing compound (Item 60, Appendix B)
- Tape (Item 68, Appendix B)

Tools/Test Equipment

- Torque wrench
- Power hacksaw
- Holesaw
- Cutter head air chisel

Personnel Required

- MOS 44B10 (2)

Manual References

- TM 9-237
- TM 9-2320-289-20
- TM 9-2320-289-34P

NOTE

Only perform *Equipment Conditions* required to gain access to damaged component,

a. REMOVAL

NOTE

If only side panels (1 1), wheel house panel (19), or rear fender on M1028A2 or 1028A3 need replacing, pickup box does not need to be removed from truck.

1. Remove screw (1) and starwasher (3) and disconnect rear lamp wiring harness ground (2) from pickup box. Discard starwasher.
2. Disconnect rear lamp wiring harness (4) from rear lamp wiring harness (5). Remove rear lamp wiring harness (4).

10-11. CARGO BODY PICKUP BOX MAINTENANCE (Con' t).

3. Repeat steps 1 and 2 for other side of pickup box if necessary.

4. Remove 8 nuts (26), washers (28), and square neck bolts (26). For M1028, M1028A1, M1028A2, and M1028A3, remove 8 spacers (27) from frame (25).

5. Attach suitable lifting device to each corner of pickup box. Remove pickup box from frame (25).

NOTE

Perform step 6 for M1028A2 and M1028A3.

6. Remove rear fender wiring harness. (See paragraph 4-9.1)

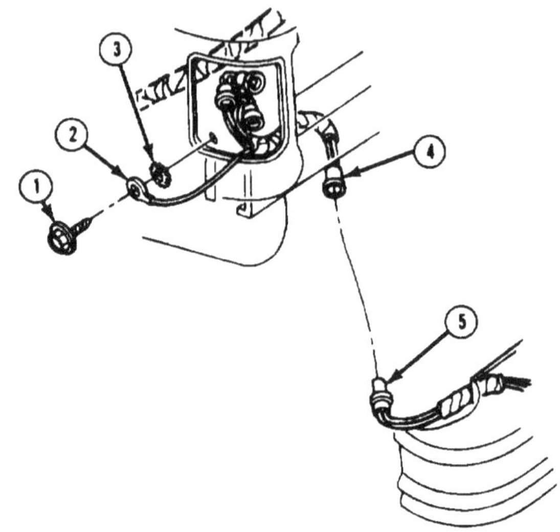

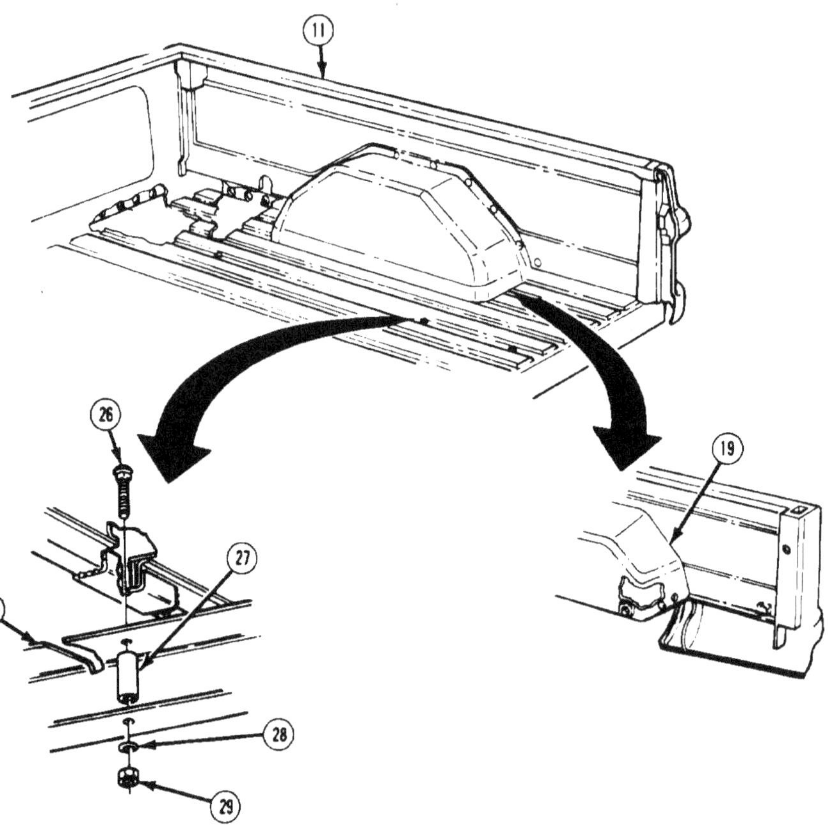

10-11. CARGO BODY PICKUP BOX MAINTENANCE (Con't).

b. DISASSEMBLY

WARNING

Use caution when handling sheet metal. Sheet metal has numerous sharp edges and splinters. Failure to follow this warning may result in injury to personnel.

NOTE

- Left and right body side panels (11) and wheel house panels (19) are disassembled the same way.
- If removing wheel house panel (19), perform steps 2 through 4, 11, and 12.
- If removing side panel (11), perform steps 1, 3 through 9, and 12.
- If removing front panel (6), perform steps 6, 10, and 12.
- If removing rear fenders (M1028A2 and M1028A3), perform steps 3, 9.1, and 9.2.

1. Remove 4 bolts (21) and remove side panel shield (22),
2. Remove 3 nuts (23) and 4 bolts (20) from wheel house panel (19) and floor panel (18).
3. Remove bolt (31) and body side panel brace (30).
4. Remove 8 bolts (12) from wheel house panel (19) to body side panel (11).
5. Remove 4 bolts (17). Grind or cut 2 welds (16) at rear stake pocket (13) and floor panel (18), and 2 welds (14) at rear stake pocket and rear floor sill (15),
6. Remove 4 bolts (7) from front of body side panel (11) to front panel (6). If removing front panel, remove 4 bolts from front of other body side panel.
7. Remove 6 bolts (10) from floor panel (18) and body side panel (11).
8. Break sealing compound that seals body side panel (11) and floor panel (18), and remove body side panel.
9. Remove 2 plugs (8) from body side panel (11).

10-11. CARGO BODY PICKUP BOX MAINTENANCE (Con't).

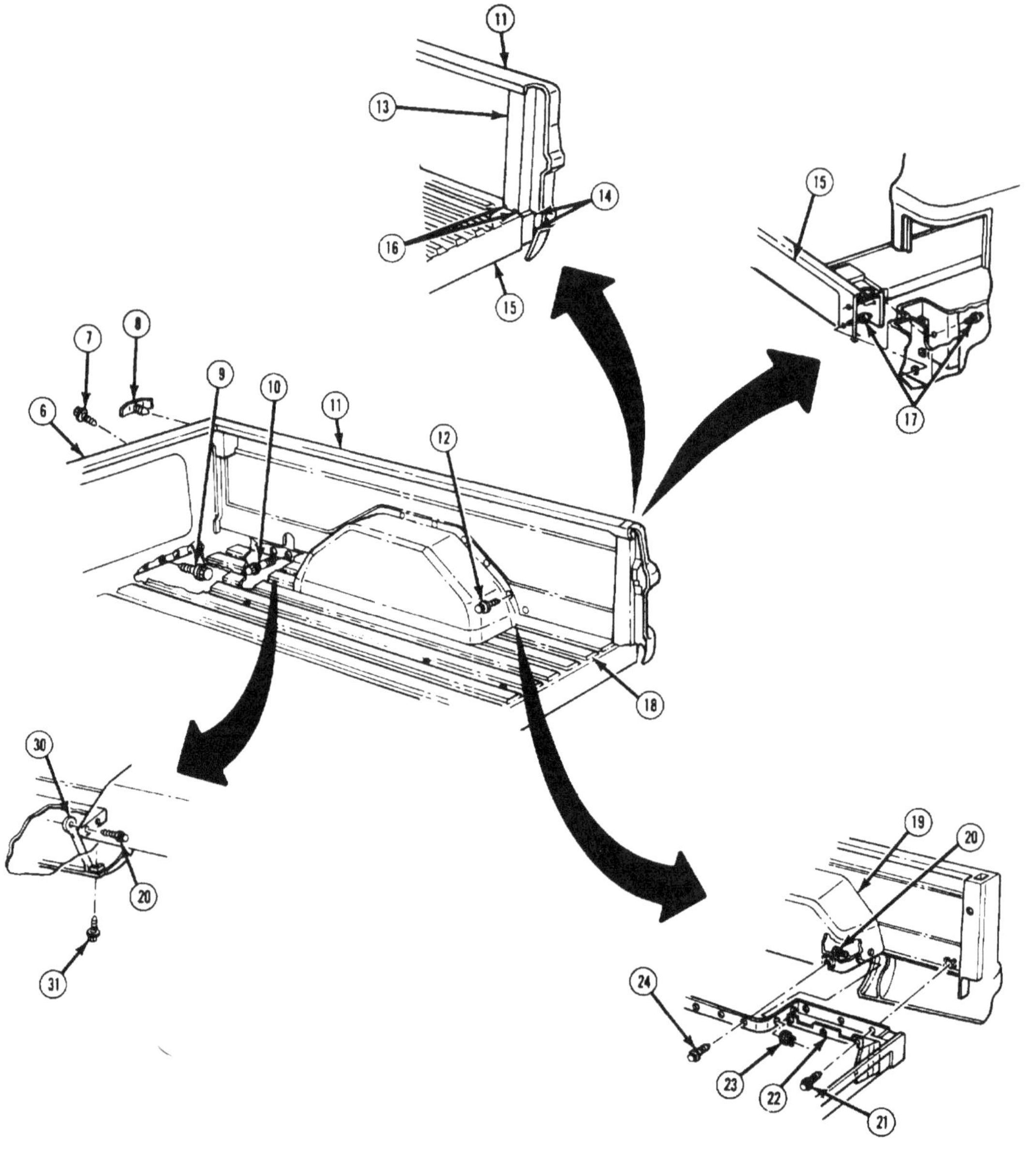

10-11. CARGO BODY PICKUP BOX MAINTENANCE (Con't).

9.1. Remove 12 nuts (31 .3) and rear fender (31 .4) from side panel (11).

9.2. Remove 6 bolts (31 .2) and brackets (31.1) from rear fender.

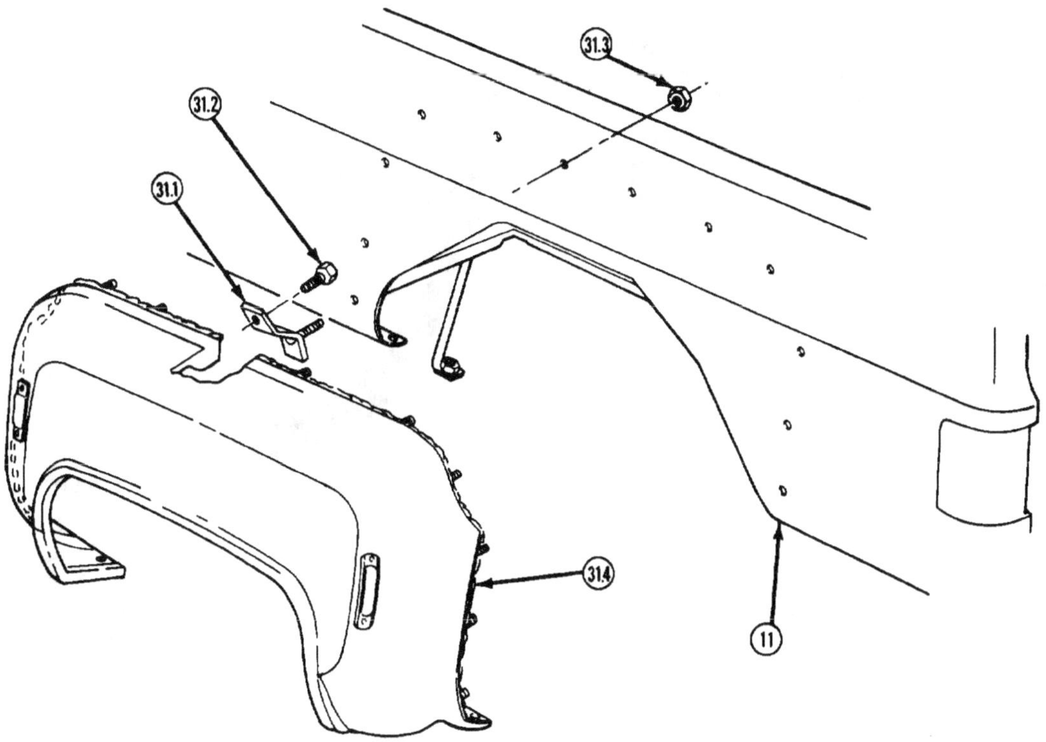

10. Remove 10 bolts (9) from front panel (6) and floor panel (18). Spread body side panels (11), if installed, and remove front panel.

11. Remove 6 bolts (24) from wheel house panel (19) and floor panel (1 8). Break sealing compound that seals wheel house panel to floor panel and remove wheel house panel.

12. Ensure that all damaged sealing compound is removed.

10-11. CARGO BODY PICKUP BOX MAINTENANCE (Con't).

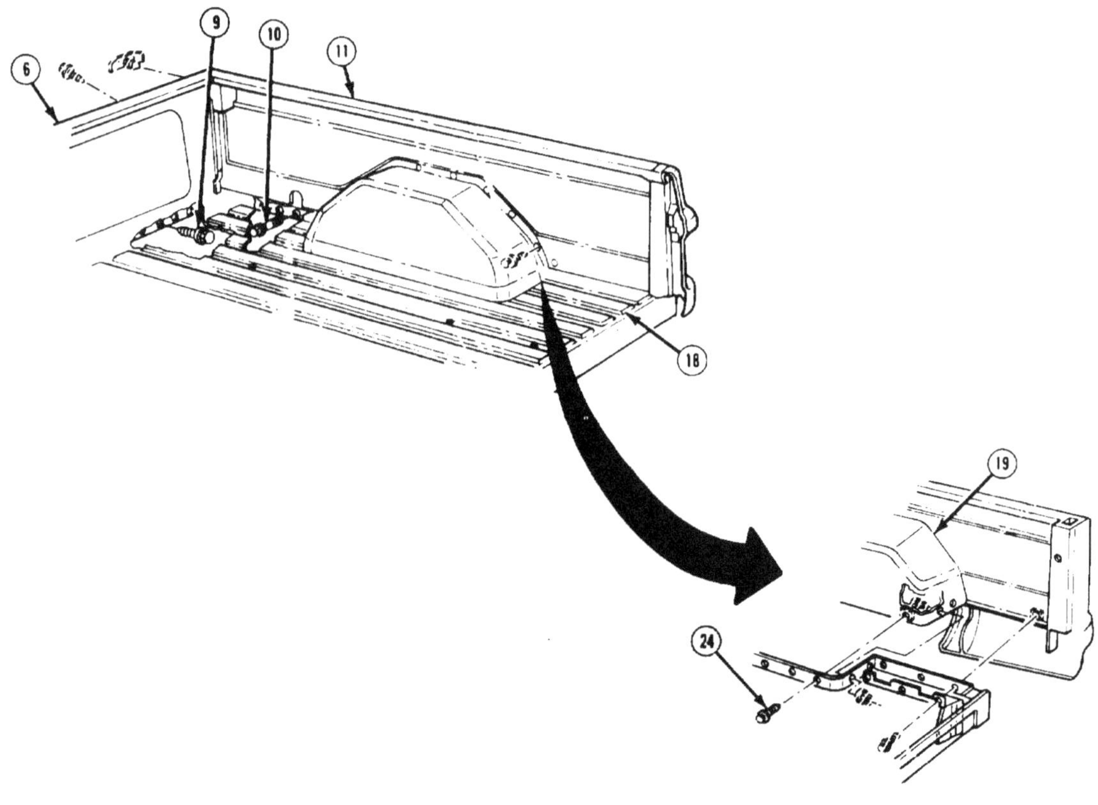

10-11. CARGO BODY PICKUP BOX MAINTENANCE (Con't).

b.1 . PREPARATION OF SIDE PANEL (M1028A2 AND M1028A3)

NOTE

- If body side panel is damaged and must be replaced, side panel must be cut using modification template to accept rear fender.
- This task requires use of template to modify side panel. (See TM 9-2320-289-34P)

1. Cut out fender template along cut lines.
2. Crease template on body crease line for ease of alinement.
3. Aline template fore and aft gage line with hole for fender brace located at front of wheel opening on side panel.
4. Aline front section of template vertically, using body crease line,
5. Tape front section of template securely to side panel. Continue taping template to side panel keeping front section alined on body crease line.
6. After taping template, check that fore and aft gage line is still in line with fender hole.
7. Mark fender cut line with scribe or suitable marker.

NOTE

There are 26 hole locations per side panel,

8. Center punch all hole locations marked "A, " "B," "C," and "D. " move template .
9. Using template as reference, cut 2 holes at location "A" with 11/8 in. (28.58 mm) hole saw, for installation of electrical harness and grommets.
10. Drill 19 holes marked "B" with ¼ in (6.35 mm) drill bit, as pilot holes
11. Drill 19 pilot holes prepared in step 10 with 29/64 in. (1 1.50 mm) drill bit for installation of fender and inner filler panel.
12. Drill 2 holes at location "C" with 9/32 in. (7.14 mm) drill bit, for installation of harness shields.
13. Drill 3 holes marked "D" with 1/4 in (6.35 mm) drill bit, for installation of wiring harness retainers.
14. Mark inner side panel for cutting approximately 1 1/2 in. (38.10 mm) out from vertical bend of inner edge (approximately at base of radius).

10-11. CARGO BODY PICKUP BOX MAINTENANCE (Con't).

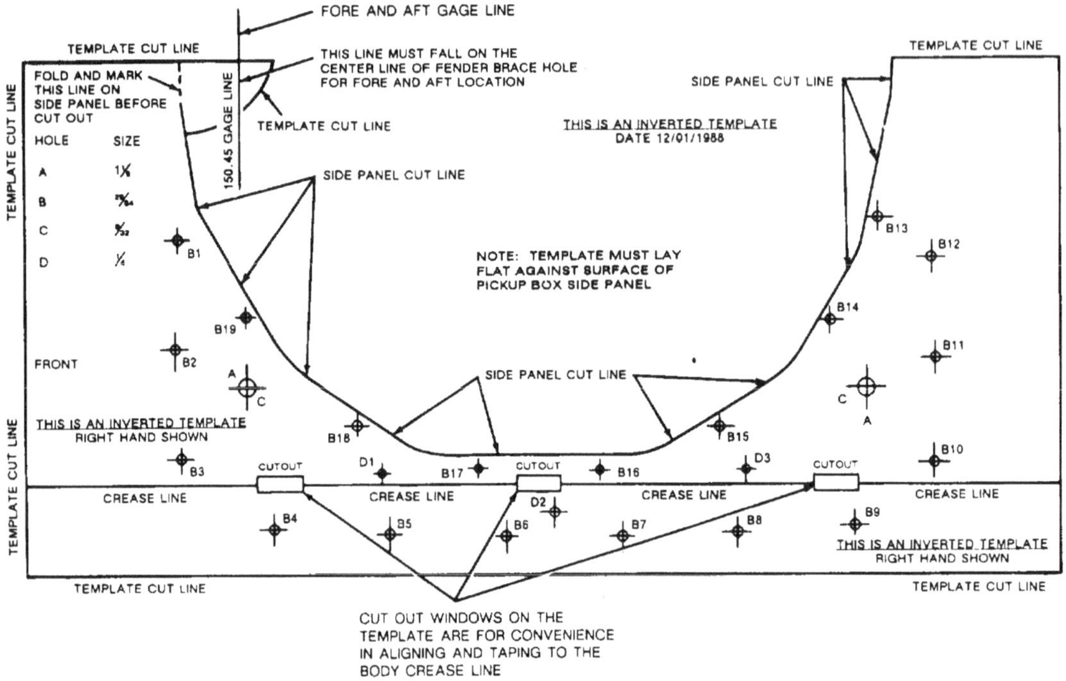

WARNING

Wear hearing protection, face shield, and gloves to prevent injury to personnel while performing steps 15 and 16.

15. Cut inner side panel with power hacksaw.
16. Cut side panel outer panel on cut line scribed from template using cutter head air chisel.
17. Apply primer to all cut edges and holes.

10-11. CARGO BODY PICKUP BOX MAINTENANCE (Con't).

c. ASSEMBLY

WARNING

Use caution when handling sheet metal. Sheet metal has numerous sharp edges and splinters. Failure to follow this warning may result in injury to personnel,

NOTE

- Left and right body side panels and wheel house panels are assembled the same way.
- If installing wheel house panel, perform steps 3, 4, 6, and 12.
- If installing body side panel, perform steps 1, 2, 4, 5, 7, and 9 through 12.
- If installing front panel, perform steps 8, 9, and 12.
- Perform steps 0.1 through 0.6 for M1028A2 and M1028A3.
- If installing rear fender (31 .4), perform steps 0.1 through 0,5.

0.1. Install 6 brackets (31 .3) to rear fender (31.4) with 6 bolts (31 .2).

0.2. Check rear fender (31 .4) fit and hole alinement with side panel (11).

NOTE

If necessary, adjust holes in side panel with round file and prime holes.

0.3. Start studs in brackets (31 .3) first and roll rear fender (31.4) into position on side panel (11).

NOTE

It is permissible to spread inner and outer panels of side panel (11) to facilitate installation of nuts (31 .3).

0.4. Apply a bead of sealant to rear fender edge and install rear fender (31 .4) to side panel (11) with 12 nuts (31.3).

0.5. Using rear fender as a guide, drill 2 holes at front and rear of rear fender (31 .4), using 29/64 in. (1 1.50 mm) drill bit. These holes are not shown on template. Apply primer to holes.

0.6. Install rear fender wiring harness. (See paragraph 4-9.1)

10-11. CARGO BODY PICKUP BOX MAINTENANCE (Con't).

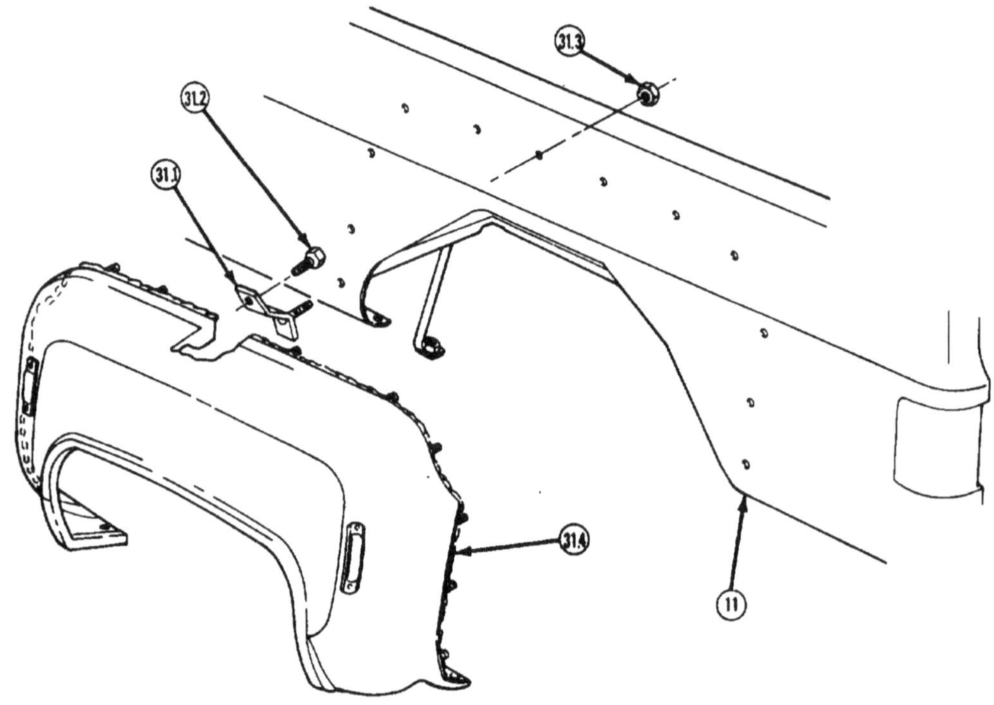

10-11. CARGO BODY PICKUP BOX MAINTENANCE (Con't).

1. Install body side panel (11) with 6 bolts (10) to floor panel (18),
2. Install side panel shield (22) with 4 bolts (21).
3. Install wheel house panel (19) with 6 bolts (24) to floor panel (18).
4. Assemble wheel house panel (19) with 8 bolts (12) to body side panel (11).
5. Install body side panel brace (30) with bolt (31) to body side panel (11).
6. Assemble wheel house panel (19) with 4 bolts (20) and 3 nuts (23) to floor panel (18).
7. Install 2 plugs (8) to body side panel (11).
8. Spread body side panels (11) and position front panel (6). Install front panel with 10 bolts (9) to floor panel.
9. Assemble body side panel (11) with 4 bolts (7) to front of front panel (6). If front panel was removed, install 4 bolts on front of other body side panel.
10. Temporarily install endgate for alinement purposes. Install rear floor sill (15) with 4 bolts (17) to body side panel (11). Attach body side panel to rear stake pocket (13) by applying two 0.5 in, (1 2.7 mm) minimum welds (16) in accordance with TM 9-237.
11. Remove endgate. Attach rear stake pocket (13) to rear floor sill (15) by applying two 1.0 in, (25.4 mm) minimum welds (14) in accordance with TM 9-237.
12. Apply an even bead of sealing compound to all floor panel (18) seams,

d. INSTALLATION

1. Attach suitable lifting device to each corner of pickup box. Position pickup box on frame (25).
2. For M1028, M1028A1, M1028A2, and M1028A3, install 8 spacers (27) in frame (25). Install 8 square neck bolts (26) with washers (28) and nuts (29). Tighten nuts to 50 lb.-ft. (68 N.m).

10-11. CARGO BODY PICKUP BOX MAINTENANCE (Con't).

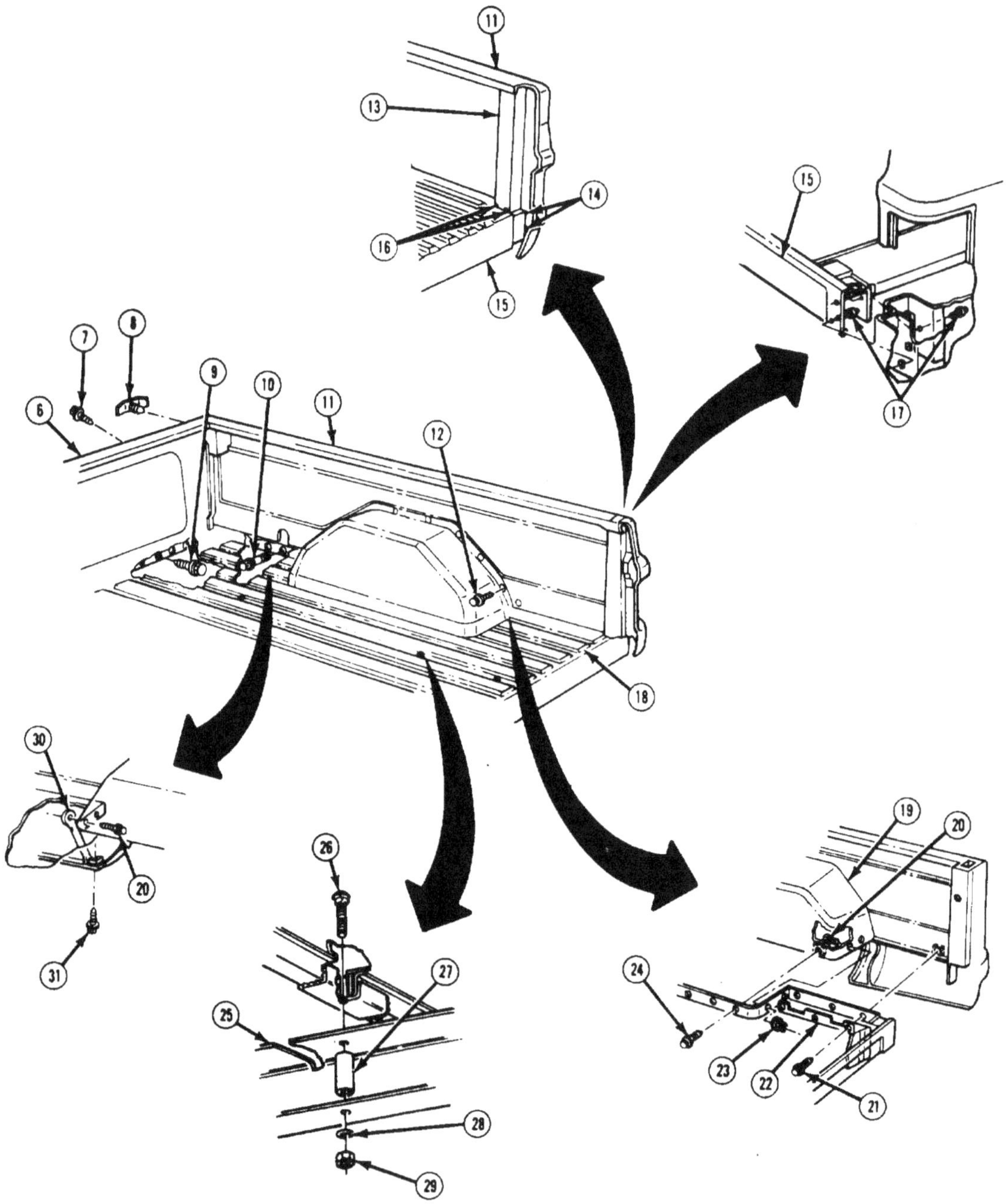

10-11. CARGO BODY PICKUP BOX MAINTENANCE (Con't).

3 . Install rear lamp wiring harness (4) in pickup box and connect to rear lamp wiring harness (5). Connect rear lamp wiring harness ground (2) to pickup box with new starwasher (3) and screw (1), Repeat step for other side of pickup box if necessary.

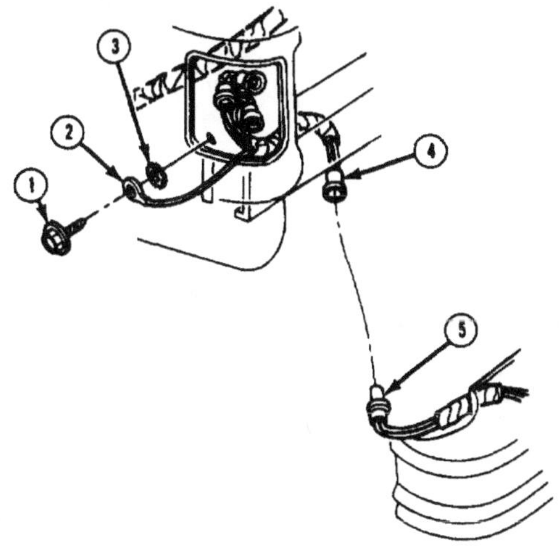

FOLLOW-ON TASKS:

- install cargo body cover and frame if removed, (See TM 9-2320-289-20)
- Install fuel tank filler door. (See TM 9-2320-289-20)
- Install taillights, (See TM 9-2320-289-20)
- Install shelter tie-downs if removed. (See TM 9-2320-289-20)
- Install communications rack and junction box if removed. (See TM 9-2320-289-20)
- Install S-250 communications shelter if removed.
- Install endgate. (See TM 9-2320-289-20)
- Install troop seats if removed. (See TM 9-2320-289-10)
- Install any winterization kit components removed. (See TM 9-2320-289-20)
- Connect rear fender side marker lights (M1028A2 and M1028A3). (See TM 9-2320-289-20)

10-12. CAB REAR WINDOW REPLACEMENT.

This task covers: a. Removal b. Inspection c. Installation

INITIAL SETUP:

Personnel Required

- MOS 63W (2)

Manual References

- FM 43-2

General Safety Instructions

- Wear heavy gloves and safety goggles when handling window glass.

a. REMOVAL

WARNING

Wear heavy gloves and safety goggles when handling window glass (1) to avoid risk of Injury to personnel.

1. Carefully loosen seal between lip of weatherstrip (2) and window opening flange (4) on inside of cab.

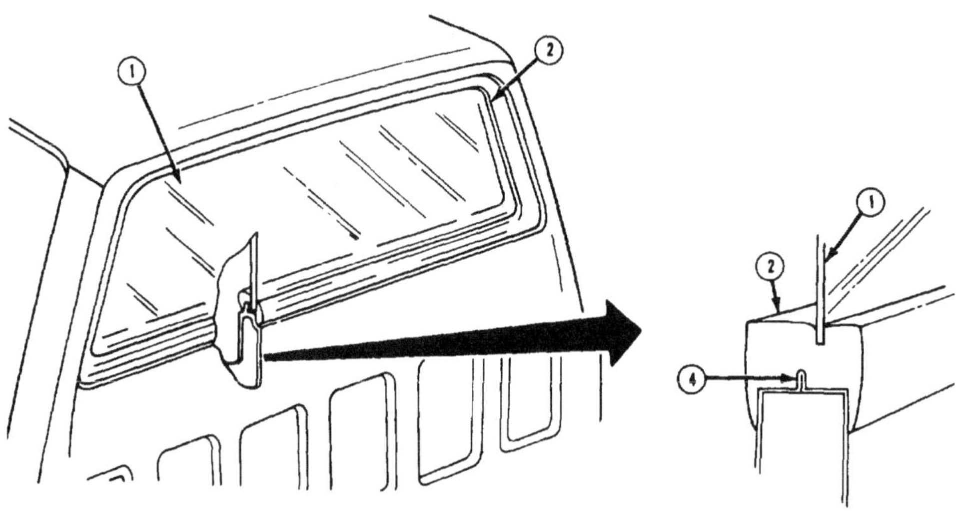

TA50419

10-12. CAB REAR WINDOW REPLACEMENT (Con't).

2. Have an assistant stand outside to support window glass (1) and weatherstrip (2) as they are removed. From inside cab, apply firm, controlled pressure to edge of window glass and assist lip of weatherstrip channel over flange (4) with a flat-bladed tool. Remove window glass and weatherstrip. Remove weatherstrip from window glass.

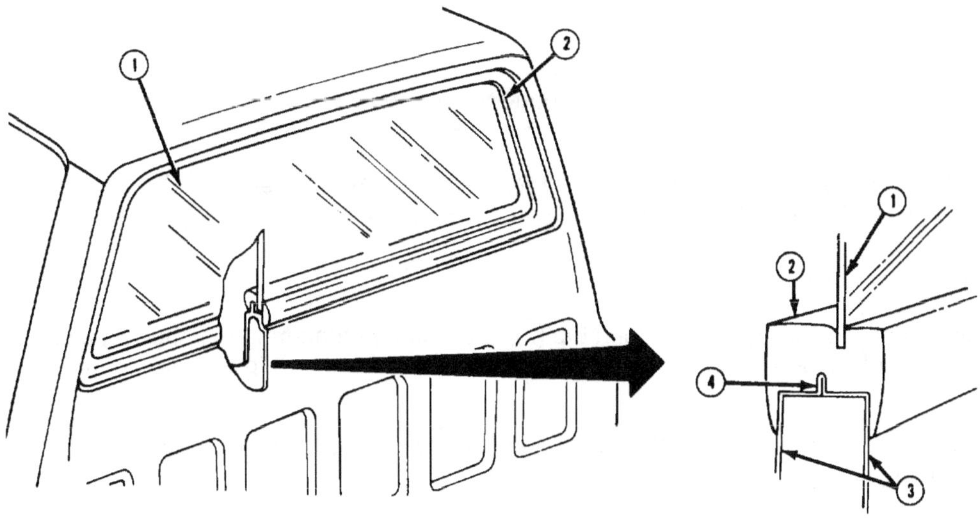

b. INSPECTION

WARNING

Wear heavy gloves and safety goggles when handling window glass (1) to avoid risk of injury to personnel.

CAUTION

In event of stress crack or leakage, ensure that weatherstrip (2) and flange (4) are thoroughly inspected before installing window glass (1). Failure to follow this caution may result in a recurrence of crack or leakage.

NOTE

If sheet metal and paint repairs are needed, see FM 43-2,

1. Inspect flange (4) for high spots or other irregularities. Remove all high spots or irregularities before proceeding,

TA50420

10-12. CAB REAR WINDOW REPLACEMENT (Con't).

2. Inspect weatherstrip (2) for damage. Replace weatherstrip if damaged.

3. Inspect replacement window glass (1) to ensure that there are no chipped edges or other irregularities. Obtain another window glass if repairs cannot be made.

c. INSTALLATION

WARNING

Wear heavy gloves and safety goggles when handling window glass (1) to avoid risk of injury to personnel.

1. Install weatherstrip (2) to window glass (1).

2. Position window glass (1) and weatherstrip (2), and install weatherstrip to flange (4) along bottom edge of window opening. Work remaining sides of weatherstrip and window glass into flange.

3. Once installed, inspect to ensure that weatherstrip (2) is completely seated around flange (4) and cab rear panel (3).

10-13. CARGO BENCH SEAT REPAIR (ALL EXCEPT M1009 AND M1010).

This task covers: a. Disassembly b. Assembly

INITIAL SETUP:

Equipment Condition

- Bench seat removed. (See TM 9-2320-289-20)

Too/s/Test Equipment

- Screwdriver bit set, J-29843
- Hog ring pliers

MaterialslParts

- Hog rings (as required)

a. DISASSEMBLY

1. Remove 10 screws (6) and 2 trim covers (7) from bench seat assembly (11).
2. Remove 2 striker bolts (5), frame catches (4), and flat washers (8) from seatback assembly (1).
3. Remove 2 bushings (2) and grommet (3) from each frame catch (4) if worn or damaged.
4. Remove 2 seatback frame bolts (9) and washers (10) from seatback assembly (1).
5. Lift off seatback assembly (1) from bench seat assembly (11).

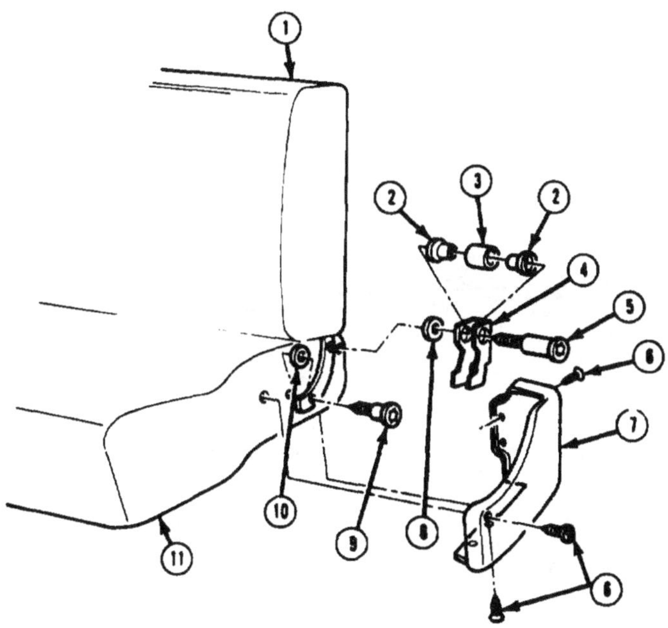

TA60421

10-13. CARGO BENCH SEAT REPAIR (ALL EXCEPT M1009 AND M1010) (Con't).

NOTE

On seatback assembly (1) and bench seat assembly (11) there are 51 hog rings (12) distributed between two assemblies. Quantity per assembly may vary for different trucks. Quantity of hog rings installed for both seatback and bench seat assemblies must be the same as number of hog rings that

6. Remove hog rings (12) seatback fram assembly (15). and remove trim (14) and pad (13) from seatback frame assembly. Discard hog rings.

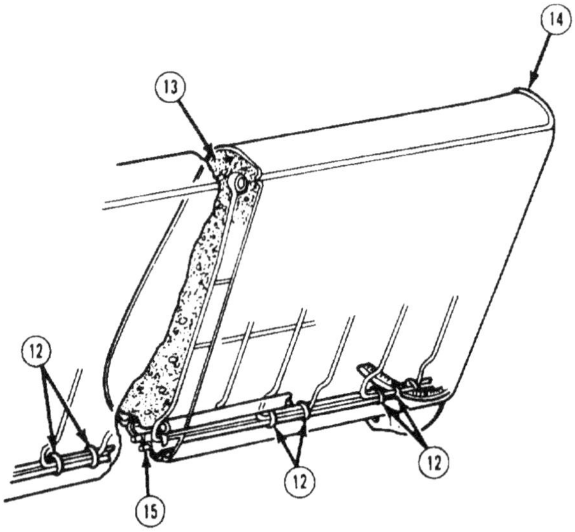

7. Remove hog rings (12) from bench seat frame assembly (1 6), and remove trim (18) and pad (17) from bench seat frame assembly. Discard hog rings.

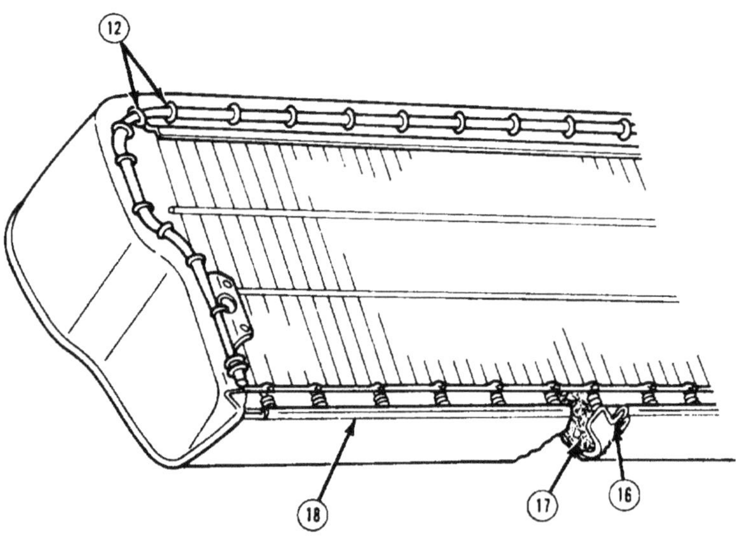

TA50422

10-13. CARGO BENCH SEAT REPAIR (ALL EXCEPT M1009 AND M1010) (Con't).

b. ASSEMBLY

NOTE

When Installing corner hog rings (12) on bench seat frame assembly (16), take extra care to prevent bench seat assembly from becoming wrinkled.

1. Install pad (17) and trim (18) to bench seat frame assembly (16), and carefully secure with new hog rings (12).

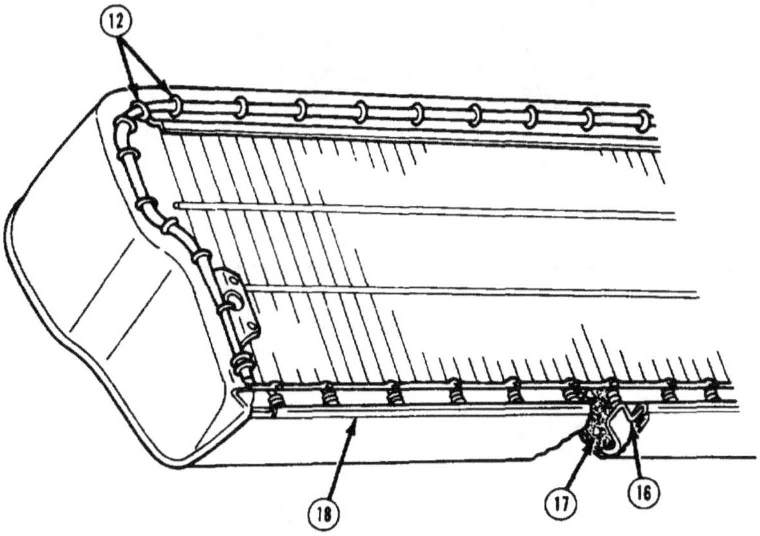

NOTE

When assembling seatback assembly (1) install hog rings (12) over depressions on pad (13) and seatback frame assembly (15), and install hog rings first.

2. Install pad (13) and trim (14) to seatback frame assembly (1 5), and carefully secure with new hog rings (12).
3. Place seatback assembly (1) on bench seat assembly (11).
4. If removed, install 2 bushings (2) and grommet (3) to frame catch (4).
5. Install 2 frame catches (4) with striker bolts (5) and flat washers (8) to seatback assembly (1).
6. Install 2 seatback frame bolts (9) and washers (10) to seatback assembly (1).
7. Install 2 trim covers (7) with 10 screws (6) to bench seat assembly (11).

TA50423

10-13. CARGO BENCH SEAT REPAIR (ALL EXCEPT M1009 AND M1010) (Con't).

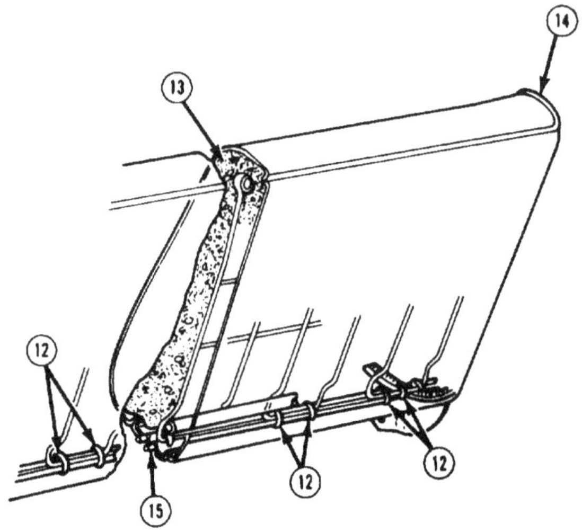

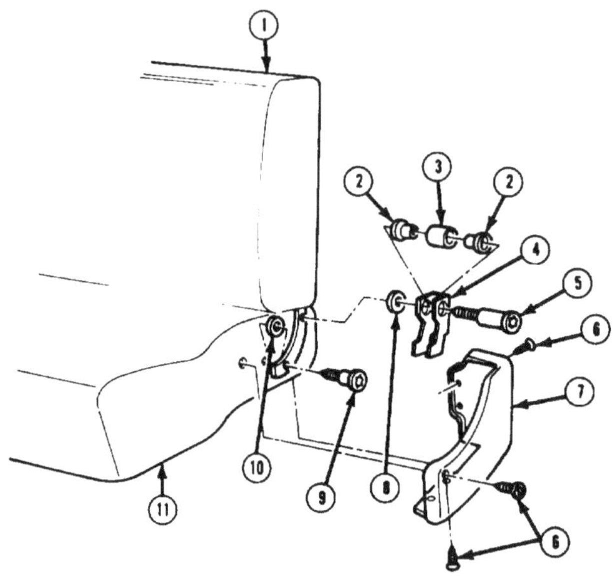

FOLLOW-ON TASKS:

- Install bench seat. (See TM 9-2320-289-20)

TA50424

Section IV. UTILITY TRUCK BODY MAINTENANCE

10-14. UTILITY TRUCK BODY MAINTENANCE INDEX.

10-15. UTILITY TRUCK TOP ASSEMBLY REPLACEMENT (M1009).

This task covers: a. Removal b. Installation

INITIAL SETUP:

Equipment Condition

● Endgate open.

a. REMOVAL

1. Remove 4 bolts (1 O) and 2 bolts (9) from top assembly (14).
2. Remove 7 bolts (13), 2 front bolts (12), and bolt (3) from top assembly (14).

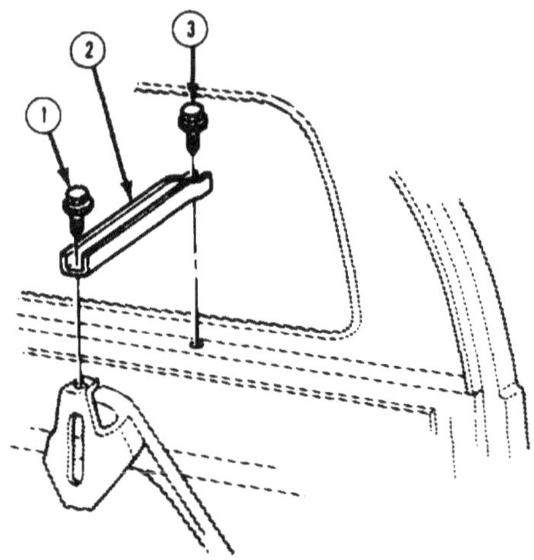

3. On right side, remove 3 bolts (4), washers (6), and radio mounting bracket assembly (5).
4. Loosen bolt (1) and move spare tire brace (2) out of way.
5. Remove 2 bolts (20) from top assembly (14).

CAUTION

Top assembly (14) must be carefully removed to prevent sides from flexing. Failure to follow this caution may result in damage to top assembly.

6. Have assistant on opposite side of top assembly (14) and together, lift top assembly from truck.

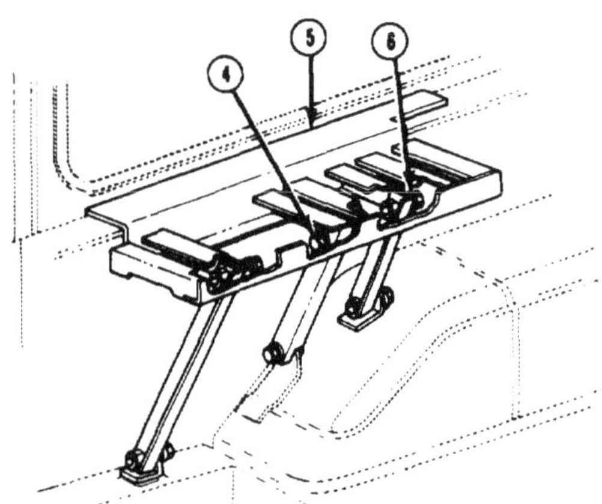

TA50425

10-15. UTILITY TRUCK TOP ASSEMBLY REPLACEMENT (M1009) (Con't).

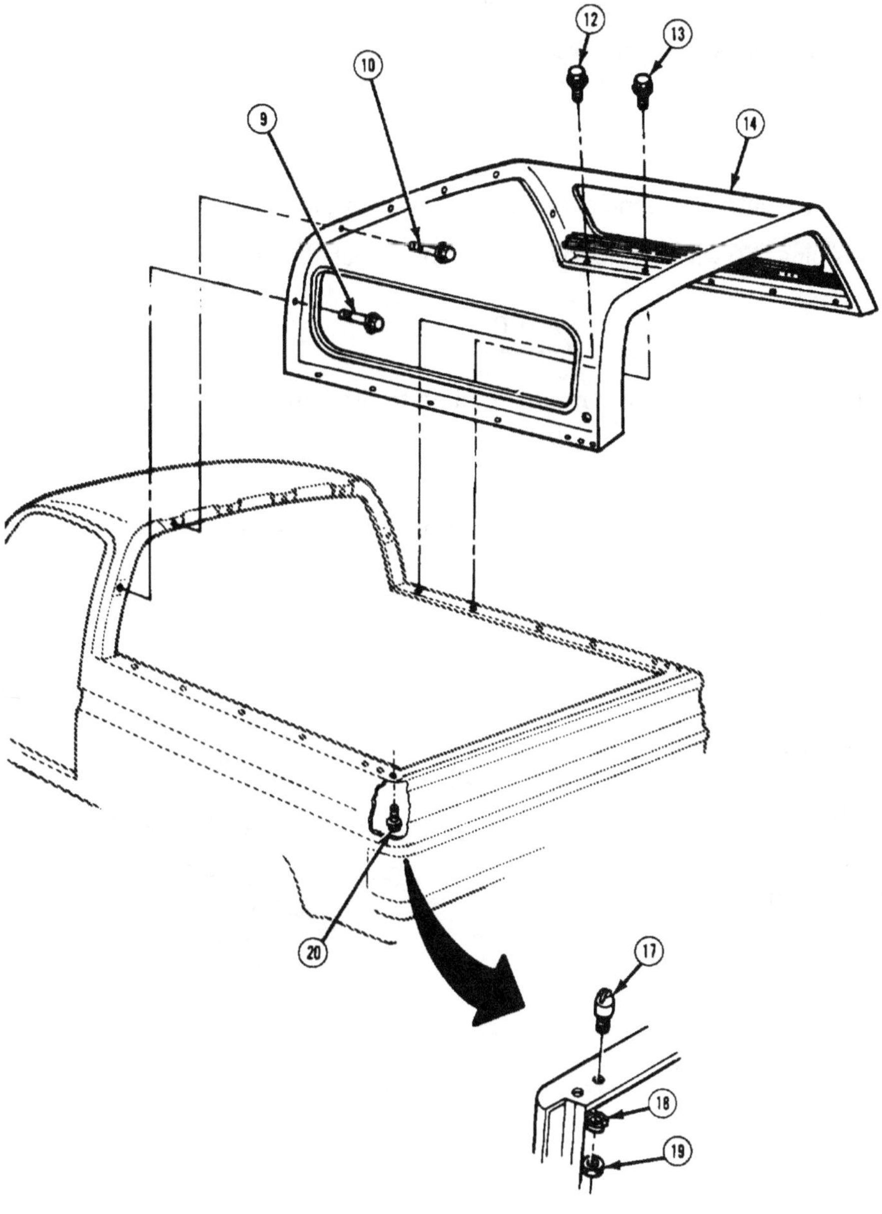

7. If 2 guide pins (17) are damaged, remove 2 nuts (19), lockwashers (18), and guide pins. Discard lockwashers and guide pins.

TA50426

10-15. UTILITY TRUCK TOP ASSEMBLY REPLACEMENT (M1009) (Con't).

8. If damaged, or if replacing top assembly (14), remove radio antenna grommet (1 6).

9. Inspect 2 side weatherstrips (7), front weatherstrips (8 and 11), and rear weatherstrip (15) for damage. Replace any damaged weatherstrips.

b. INSTALLATION

1. If removed, install radio antenna grommet (16) to top assembly (14),

2. Install 2 new guide pins (17) with new lockwashers (18) and nuts (19) if removed.

CAUTION

Top assembly (14) must be carefully installed to prevent sides from flexing. Failure to follow this caution may result in damage to top assembly.

3. Have assistant on opposite side of top assembly (14) and together, lift top assembly onto truck using 2 guide pins (17) as locators.

4. Install 2 bolts (9) to top assembly (14) and cab,

5. On right side, position radio mounting bracket assembly (5), and install 3 bolts (4) and washers (6).

NOTE

When installing 7 bolts (13), start with bolts nearest endgate and work towards cab.

6. Install 7 bolts (13) to sides of top assembly (14).

7. Loosen 2 bolts (9) and install 2 front bolts (12) to top assembly (14).

8. Install 4 bolts (1 O) and tighten 2 bolts (9). Install 2 bolts (20).

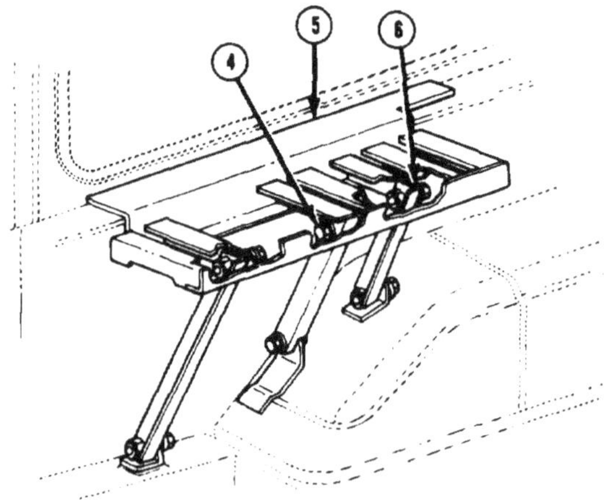

TA60427

10-15. UTILITY TRUCK TOP ASSEMBLY REPLACEMENT (M1009) (Con't).

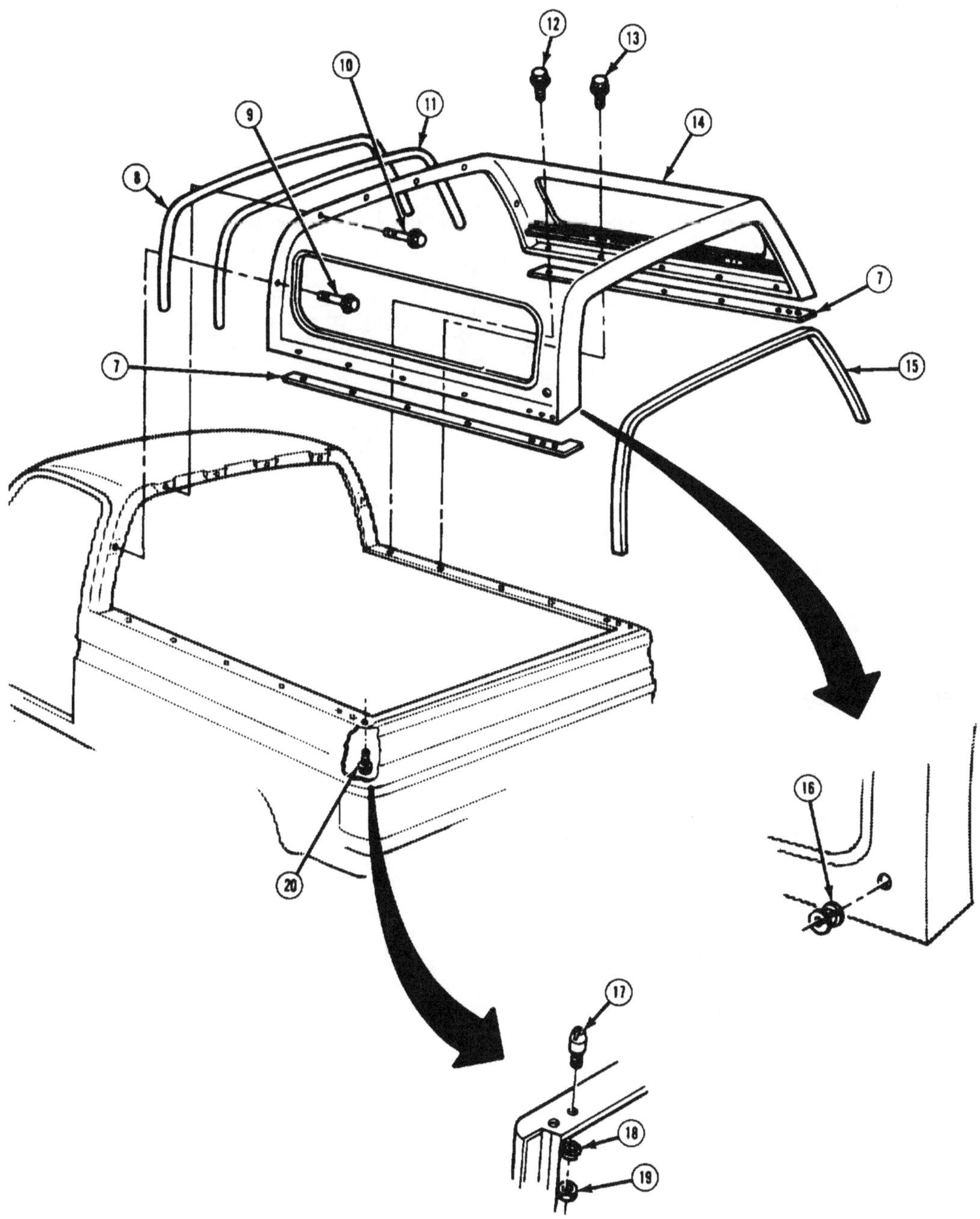

TA50428

10-15. UTILITY TRUCK TOP ASSEMBLY REPLACEMENT (M1009) (Con't).

9. Install spare tire brace (2) with bolt (3). Tighten bolt (1).

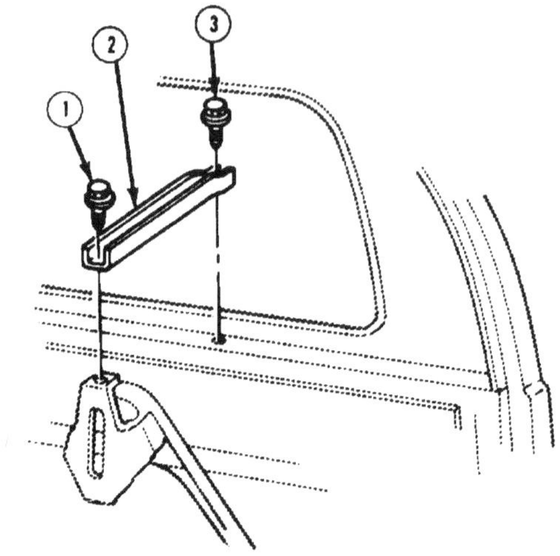

FOLLOW-ON TASKS:

- Close endgate.
- Check operation of endgate window.

TA50429

10-16. DRIVER'S AND PASSENGER'S SEAT REPAIR (M1009 AND M1010).

This task covers: a. Disassembly b. Assembl y

INITIAL SETUP:

Equipment Condition

- Driver's or passenger's seat removed, (See TM 9-2320-289-20)
- Seat adjuster assembly removed. (See TM 9-2320-289-20)

Too/s/Test Equipment

- Screwdriver bit set, J-29843
- Hog ring pliers
- Torque wrench

Materials/Parts

- Hog rings (as required)

a. DISASSEMBLY

NOTE

- M1010 seat assembly does not have a specified number of hog rings. Ensure that location of all M1010 hog rings is noted for installation.
- Driver's and passenger's seats are disassembled the same way. Driver's seat disassembly is given.

10-16. DRIVER'S AND PASSENGER'S SEAT REPAIR (M1009 AND M1010) (Con't).

NOTE

- If disassembling M1009 seatback assembly (2), perform Steps 1-5, and 7.
- If disassembling M1009 seat cushion assembly (18), perform steps 1-3, 8, and 9.
- If disassembling M1010 seat assembly, perform steps 6-9.

1. Remove 4 screws (7) and upper and lower seatback lock covers (9 and 10).

NOTE

If disassembling passenger's seat, seat adjuster cable is disconnected when seatback lock assembly (11) is removed.

2. Remove 4 bolts (8) and seatback lock assembly (11).
3. Remove pivot bolt (17), washer (1 6), and bushing (12). Lift off seatback assembly (2) from seat cushion assembly (18).
4. Unzip seatback trim (6) zipper. Remove any hog rings (5) found securing seatback trim to seatback frame (4). Discard hog rings.
5. Remove 2 hog rings (5) securing looped ends of 2 lasting wires (1) to seatback frame (4). Remove 2 lasting wires. Remove M1009 seatback trim (6). Discard hog rings.
6. Remove any hog rings (5) found securing seatback trim (6) to seat frame (13). Remove Ml 010 seatback trim. Discard hog rings.
7. Inspect seatback frame (4), pad (3), and seatback trim (6) for damage. Replace if damaged.
8. Remove 17 hog rings (5) securing seat cushion trim (15) to seat frame (13) and springs. Remove seat cushion trim. Discard hog rings.
9. Inspect seat frame (13), cushion (14), and seat cushion trim (15) for damage. Replace if damaged.

b. ASSEMBLY

NOTE

- Driver's and passenger's seats are assembled the same way. Driver's seat assembly is given.
- If assembling M1009 seat cushion assembly (18), perform steps 1, 2, and 6 through 8.
- If assembling M1009 seatback assembly (2), perform steps 4-8.
- If assembling M1010 seat assembly, perform steps 1 and 3.

1. Install seat cushion trim (15) to seat frame (13) and cushion (14) with 15 new hog rings (5) at front and sides. Secure seat cushion trim at rear to springs with 2 new hog rings.

10-16. DRIVER'S AND PASSENGER'S SEAT REPAIR (M1009 AND M1010) (Con't).

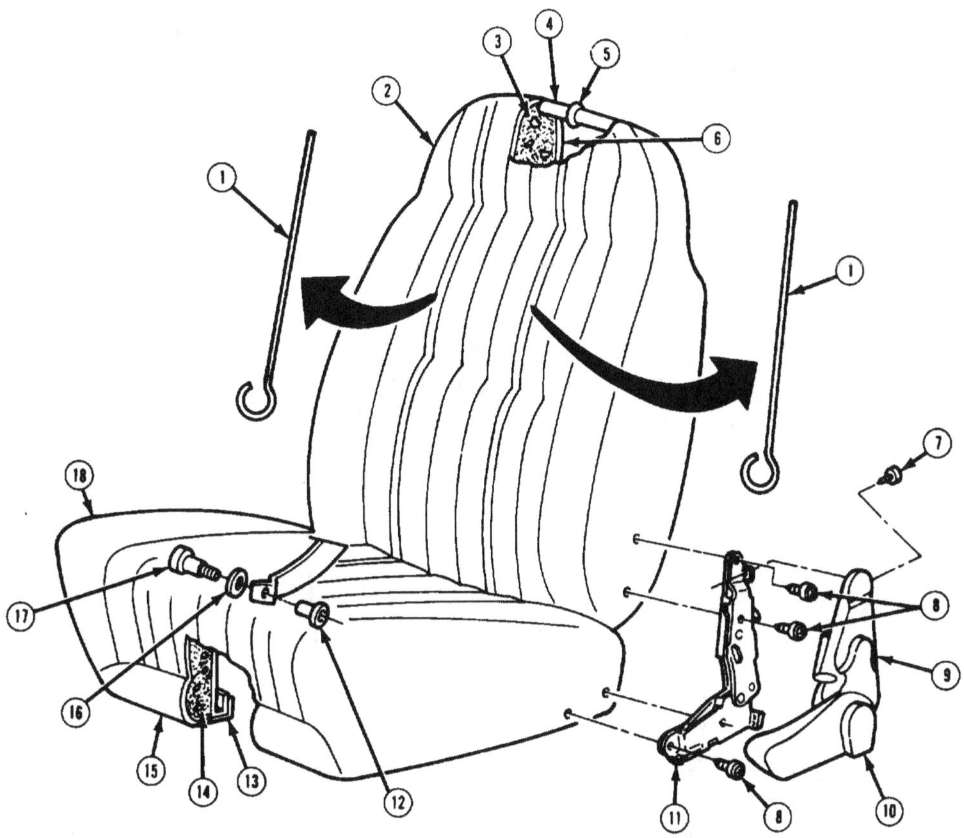

2. Make 2 holes in seat cushion trim (15) where seatback lock assembly (11) bolts (8) will be installed.

3. Install M1010 seatback trim (6) to seatback frame (4) and pad (3). Secure seatback trim to seat frame (13) with new hog rings (5).

4. Install seatback trim (6) to seatback frame (4) and pad (3). Slide in 2 lasting wires (1) through hemmed channels in seatback trim, engaging straight ends of wires under seatback frame. Secure looped ends of wires to seatback frame with 2 new hog rings (5),

5. Use new hog rings (5) as required to secure loose ends of M1009 seatback trim (6) to seatback frame (4), Close zipper and tuck under zipper ends. Make 2 holes in trim where seatback lock assembly (11) bolts (8) will be installed.

6. Place seatback assembly (2) on seat cushion assembly (18). Install pivot bolt (17), washer (16), and bushing (12).

TA50430

10-16. DRIVER'S AND PASSENGER'S SEAT REPAIR (M1009 AND M1010) (Con't).

NOTE

If assembling passenger's seat, ensure that seat adjuster cable is assembled with seatback lock assembly (11).

7. Install seatback lock assembly (11) with 4 bolts (8). Tighten bolts to 20 lb.-ft. (27 N.m).

8. Install upper and lower seatback lock covers (9 and 10) with 4 screws (7).

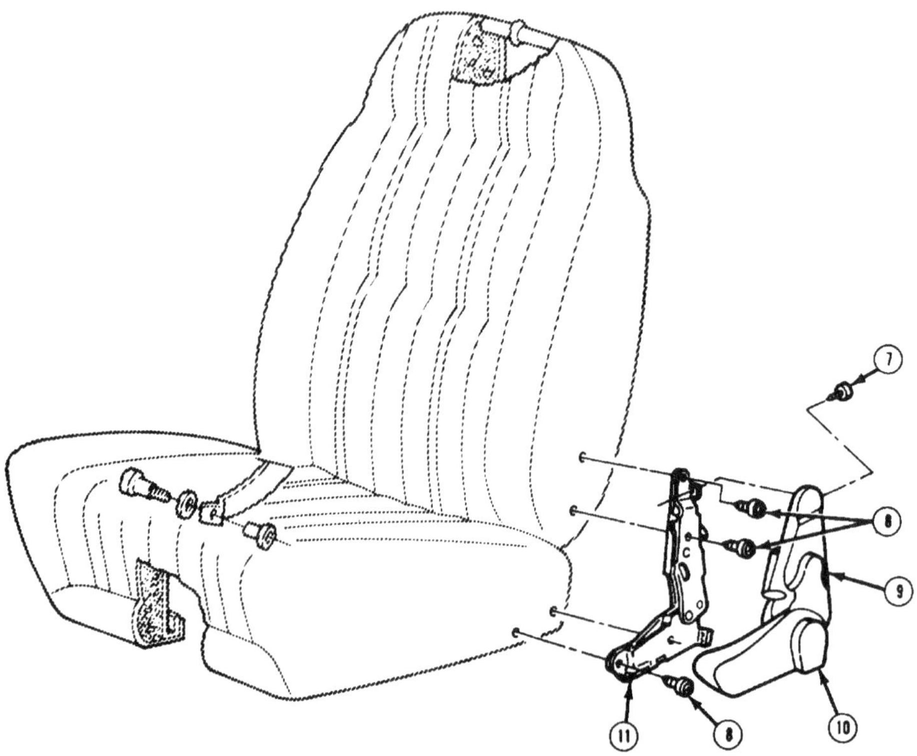

FOLLOW-ON TASKS:

- Install seat adjuster assembly. (See TM 9-2320-289-20)
- Install driver's or passenger's seat, (See TM 9-2320-289-20)

TA50431

10-17. REAR BENCH SEAT REPAIR (M1009).

This task covers: a. Disassembly b. Assembly

INITIAL SETUP

Equipment Condition

- Rear bench seat removed. (See TM 9-2320-289-20)

Materials/Parts

- Hog rings (as required)

Tools/Test Equipment

- Screwdriver bit set, J-29843
- Hog ring pliers

a. DISASSEMBLY

NOTE

- if disassembling seatback assembly (3), perform steps 1-4.
- If disassembling bench seat assembly (2), perform steps 1, 2, and 5 through 11.

1. Remove 2 bolts (8) from seatback assembly (3), 2 torx head bolts (9) from bench seat assembly (2), and remove seatback hinge (10). Remove knob (7) from seatback hinge if damaged. Repeat this step for hinge on other side.

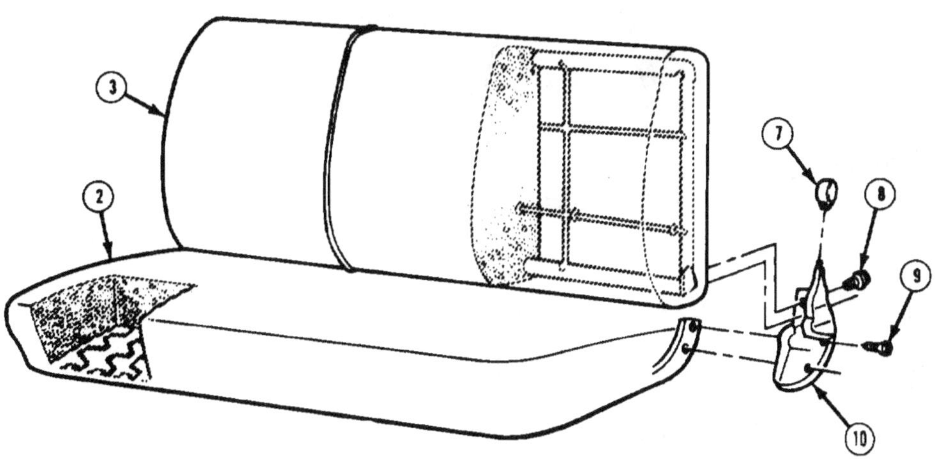

TA50432

10-17. REAR BENCH SEAT REPAIR (M1009) (Con't).

2. Separate seatback assembly (3) from bench seat assembly (2).
3. Unzip zipper and remove trim (4) from seatback assembly (3), and remove pad (5) from seatback frame assembly (6).
4. Inspect trim (4), seatback frame assembly (6), and pad (5) for damage. Replace if damaged.
5. Remove 16 screws (17) and remove bench seat trim panel (16).
6. Mark one end of front hinge torque rod (14) to ensure proper assembly, and remove bolt (15) and front hinge torque rod.

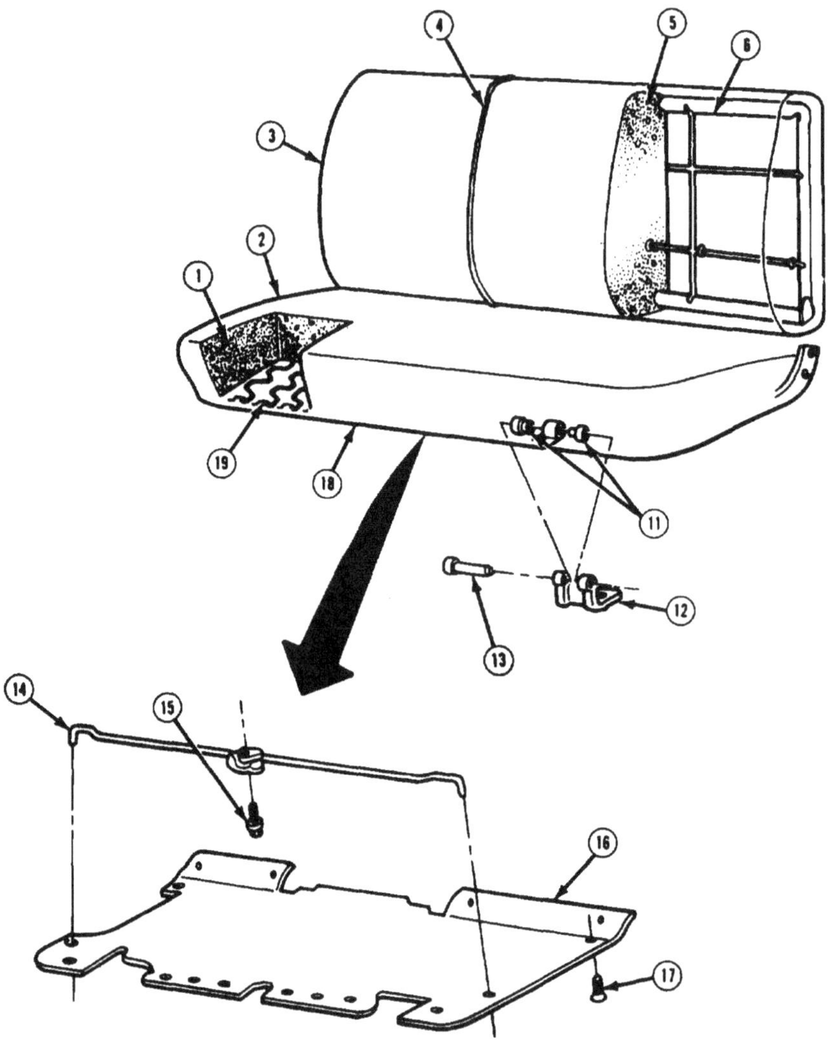

TA50433

10-17. REAR BENCH SEAT REPAIR (M1009) (Con't).

7. Mark one end of actuating rod (21) to ensure proper assembly and remove 2 ring clips (20) and actuating rod.

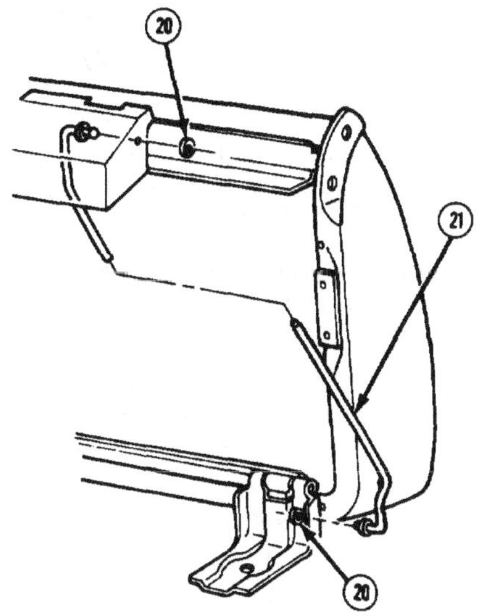

8. Remove 4 bolts (26), 2 seatbelt mounting bolts (24), latch and guide assembly (25), and latch cover (23). Note location of 2 seatbelt assemblies to ensure proper assembly and remove from bench seat assembly (2).

9. Remove 2 hinge pins (13), 4 bushings (11), and 2 hinges (12)' from bench seat assembly (2).

NOTE

Quantity of hog rings (22) used to secure trim (18) to bench seat frame assembly (19) will vary for different trucks.

10. Remove hog rings (22) securing trim (18) to bench seat frame assembly (19), and remove trim and pad (1) from bench seat frame assembly. Discard hog rings.

11. Inspect trim (18), bench seat frame assembly (19) and pad (1) for damage. Replace if damaged.

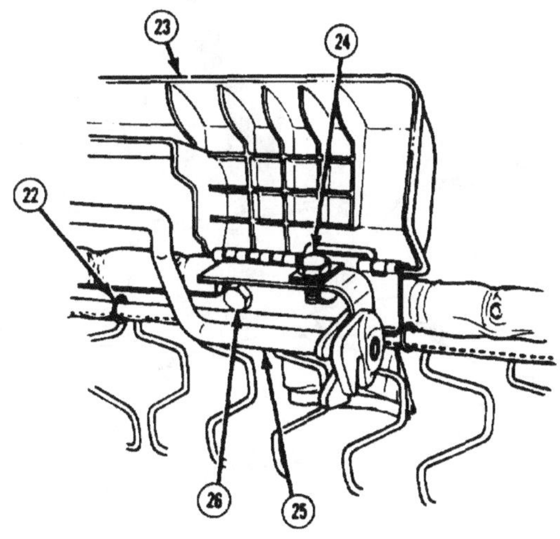

TA50434

10-17. REAR BENCH SEAT REPAIR (M1009) (Con't).

b . ASSEMBLY

NOTE

- If assembling bench seat assembly (2), perform steps 1-6, and 8.
- If assembling seatback assembly (3), perform steps 7 and 8.
- When installing corner hog rings (22) on bench seat frame assembly (19) take extra care to prevent trim (18) from becoming wrinkled. Use hog rings (22) as required to ensure a tight, wrinkle free fit.

1. Install pad (1) to bench seat frame assembly (19). Using new hog rings (22), install trim (18) and secure to bench seat frame assembly. Make holes in trim around bench seat frame assembly for bolt and screws to be installed.

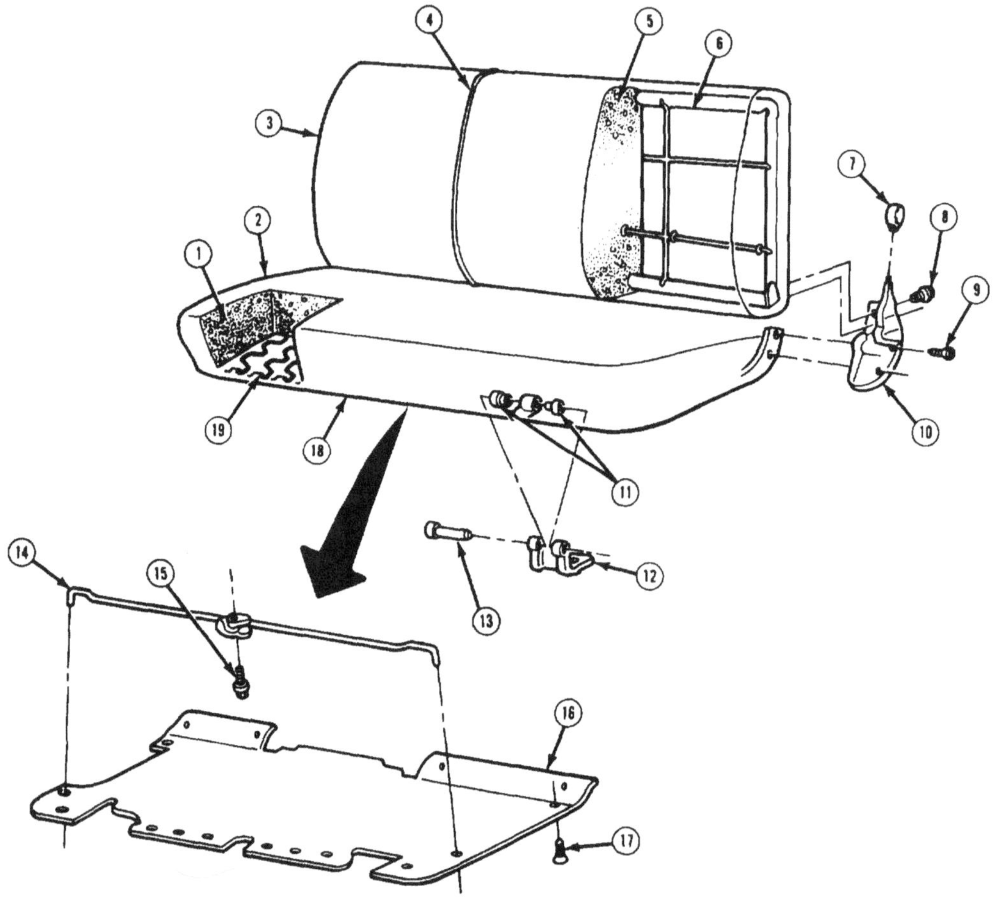

TA50435

10-17. REAR BENCH SEAT REPAIR (M1009) (Con't).

NOTE

Ensure that square corners of seatback frame assembly (6) are pointed to bottom of seat.

2 . Install 2 hinges (12) to bench seat assembly (2) with 4 bushings (11) and 2 hinge pins (13).

3 . Install latch cover (23) and latch and guide assembly (25) with 4 bolts (26), and install 2 seatbelt assemblies in their original position with 2 bolts (24). Tighten 2 bolts (24) to 40 lb.-ft. (54 N.m).

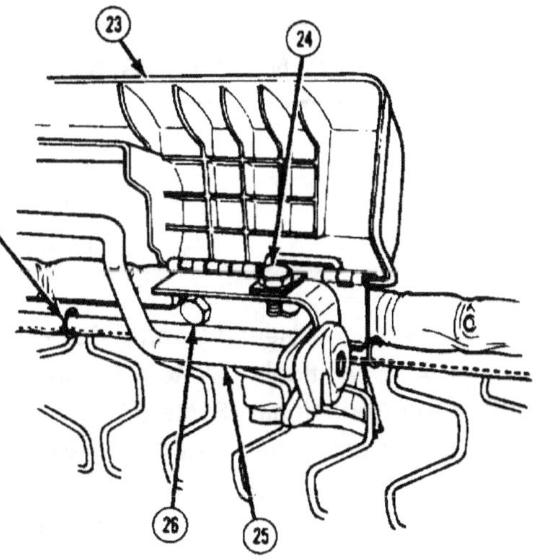

4 . Install actuating rod (21), as marked, with 2 ring clips (20).

5 . Install front hinge torque rod (14), as marked, with bolt (1 5).

6 . Install 16 screws (17) and bench seat trim panel (16) to bench seat assembly (2).

7 . Install pad (5) to seatback frame assembly (6). Install trim (4) and close zipper. Tuck under zipper ends.

8 . Position seatback assembly (3) on bench seat assembly (2) and install hinge (1 O) with 2 bolts (8) and torx head bolts (9). Install knob (7) on hinge if removed.

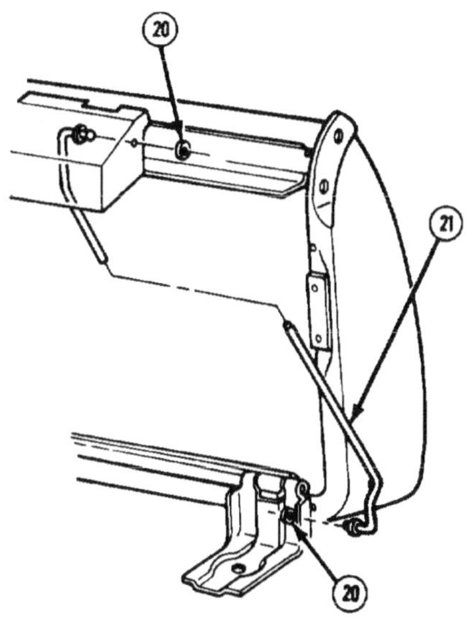

FOLLOW-ON TASKS:

- Install rear bench seat. (See TM 9-2320-289-20)

TA50436

Section V. AMBULANCE BODY MAINTENANC E

10-18. AMBULANCE BODY MAINTENANCE INDEX.

10-19. PERSONNEL HEATER CONTROL BOX REPAIR (M1010).

This task covers: a. Disassembly b. Assembly

INITIAL SETUP:

Equipment Condition
- Personnel heater control box removed. (See TM 9-2320-289-20)

Materials/Parts
- Two sheet spring nuts

a. DISASSEMBLY

NOTE

Ensure that all leads are tagged for assembly.

1. Remove 8 screws (1) and separate cover (2) from control wrap (10).
2. Remove 5 screws and disconnect harness (12) leads from "RUN-START" switch (4), " HI-LO" switch (6), circuit breaker (20), and sensitive switches (15), Remove 4 screws (9) and harness if damaged.
3. Remove blend air door grommet (8) and fresh air door grommet (7) if damaged.
4. Remove screw and lead (19) from circuit breaker (20). Disconnect lead (22) and socket (17) lead from circuit breaker. Remove 2 nuts (18), screws (21), and circuit breaker from cover (2).
5. Remove screw and disconnect lead (11) from "HI-LO" switch (6). Remove 2 screws (23), lead (22), and lead (11) from " RUN-START" switch (4). Disconnect 2 leads (13) and socket (17) lead from "RUN-START" switch. Remove 2 screws (3) and "RUN-START" switch.

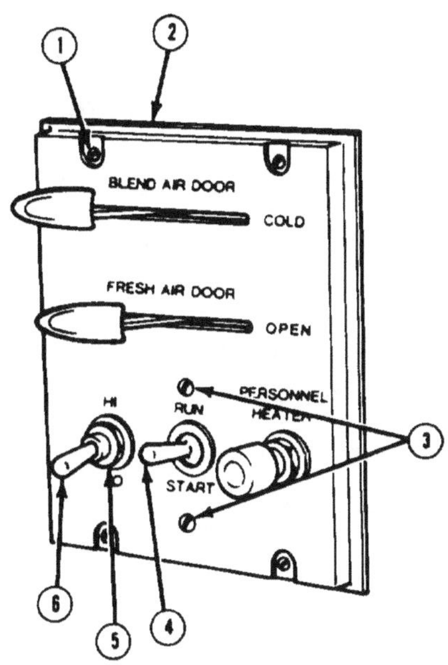

6. Remove nut (5) and "HI-LO" switch (6).
7. Remove 2 leads (13) from 2 sensitive switches (15). Remove 4 nuts (16), screws (14), and 2 sensitive switches from cover (2).
8. Remove pilot cover, bulb, nut, and socket (17) from cover (2).

TA50437

10-19. PERSONNEL HEATER CONTROL BOX REPAIR (M1010) (Con't).

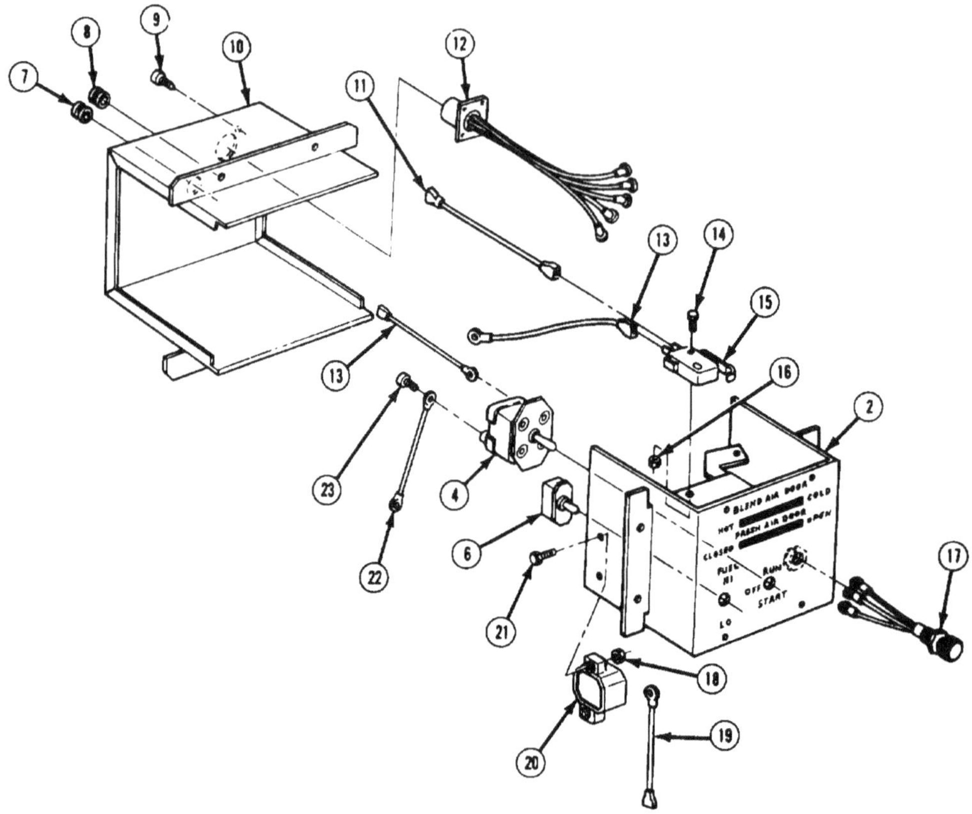

TA50438

10-19. PERSONNEL HEATER CONTROL BOX REPAIR (M1010) (Con't).

9 . Mark relationship of cables (24) to clamps (32). Remove 2 screws (31) and clamps from cover (2).

10. Remove 2 sheet spring nuts (27) and cables (24) from blend air door lever (28) and fresh air door lever (34). Discard sheet spring nuts.

11. Remove 2 screws (29) and knobs (30) from blend air door lever (28) and fresh air door lever (34) if damaged.

12. Remove nut (33), 2 washers (25), screw (26), and blend air door lever (28).

13. Remove fastener (35) and fresh air door lever (34).

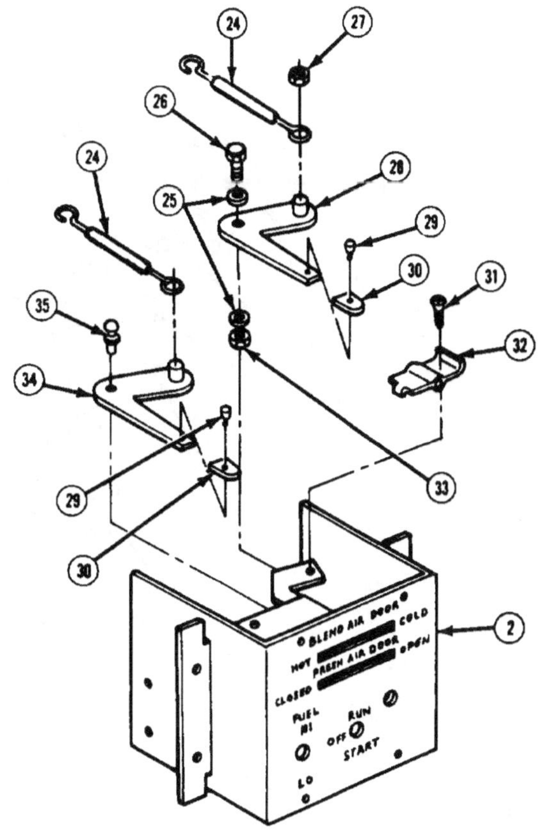

b. ASSEMBLY

1. Install fresh air door lever (34) with fastener (35).

2. Install blend air door lever (28) with screw (26), 2 washers (25), and nut (33).

3 . Install 2 knobs (30) on blend air door lever (28) and fresh air door lever (34) with 2 screws (29) if removed.

4. Install 2 cables (24) on blend air door lever (28) and fresh air door lever (34) with 2 new sheet spring nuts (27).

5 . Install 2 clamps (32) on cover (2) with 2 screws (31),

6 . Install socket (17), nut, bulb, and pilot cover on cover (2).

7 . Install 2 sensitive switches (15) on cover (2) with 4 screws (14) and nuts (16). Install 2 leads (13) on sensitive switches.

TA50439

10-19. PERSONNEL HEATER CONTROL BOX REPAIR (M1010) (Con't).

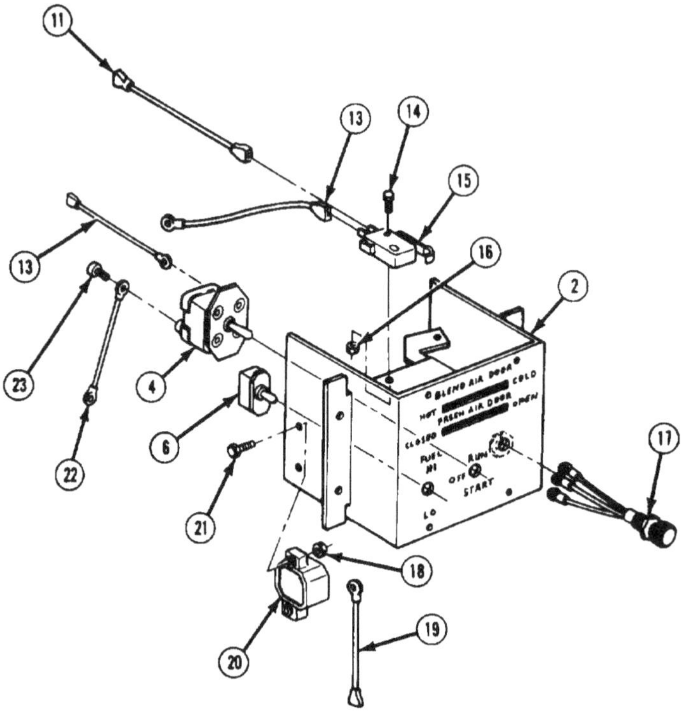

8. Install "HI-LO" switch (6) on cover (2) with nut (5).

9. Install "RUN-START" switch (4) on cover (2) with 2 screws (3). Connect 2 leads (13) and socket (17) lead to "RUN-START" switch. Connect lead (11) and lead (22) to "RUN-START" switch with 2 screws (23). Connect lead (11) to "HI-LO" switch (6) with screw.

10. Install circuit breaker (20) on cover (2) with 2 screws (21) and nuts (18). Connect socket (17) lead and lead (22) to circuit breaker. Install lead (19) on circuit breaker with screw.

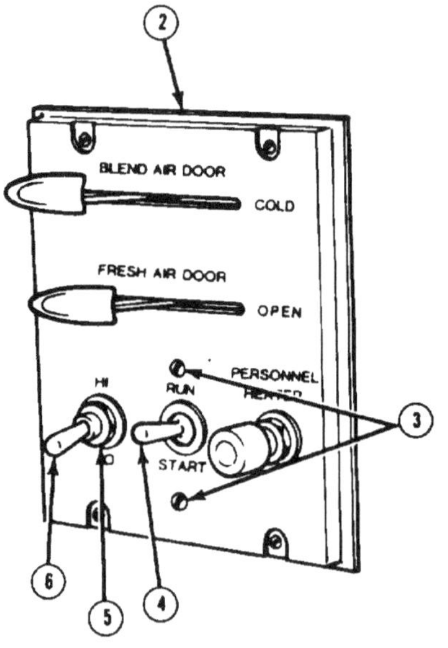

TA50440

10-19. PERSONNEL HEATER CONTROL BOX REPAIR (M1010) (Con't).

11. Install blend air door grommet (8) and fresh air door grommet (7) if removed.

12. Install harness (12) on control wrap (10) with 4 screws (9) if removed. Connect harness leads to "RUN-START" switch (4), "HI-LO" switch (6), circuit breaker (20), and sensitive switches (15) with 5 screws.

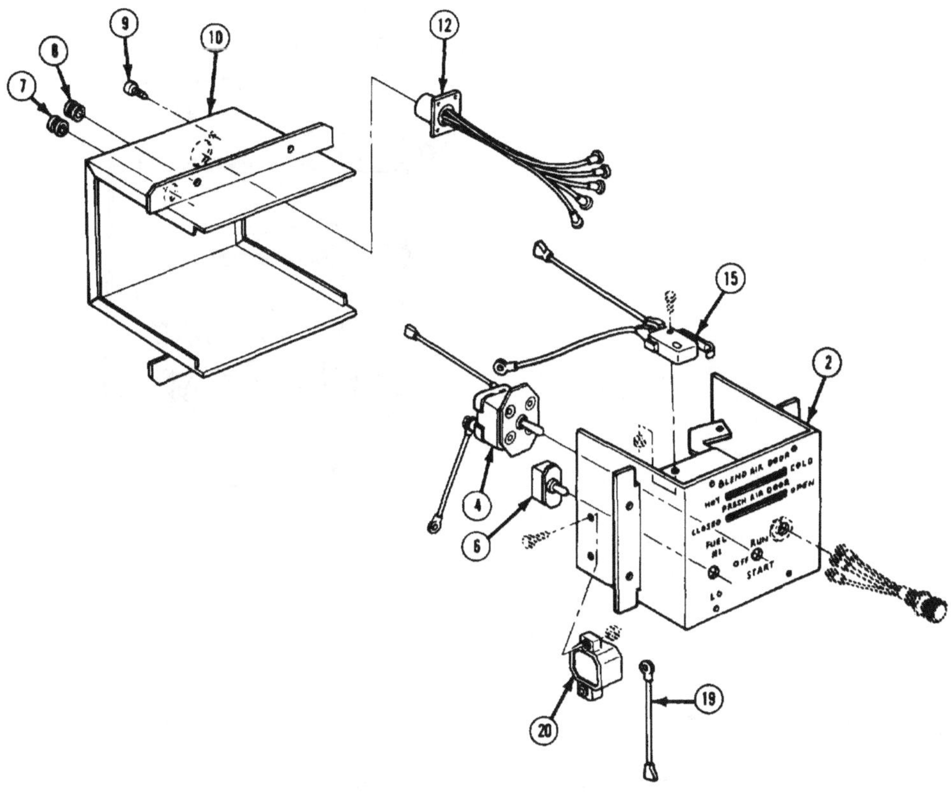

TA50441

10-19. PERSONNEL HEATER CONTROL BOX REPAIR (M1010) (Con't).

NOTE

Ensure that cables and circuit breaker lead (19) are placed through their appropriate holes as cover (2) and control wrap (10) are assembled.

13. Assemble cover (2) and control wrap (10) and install 8 screws (1).

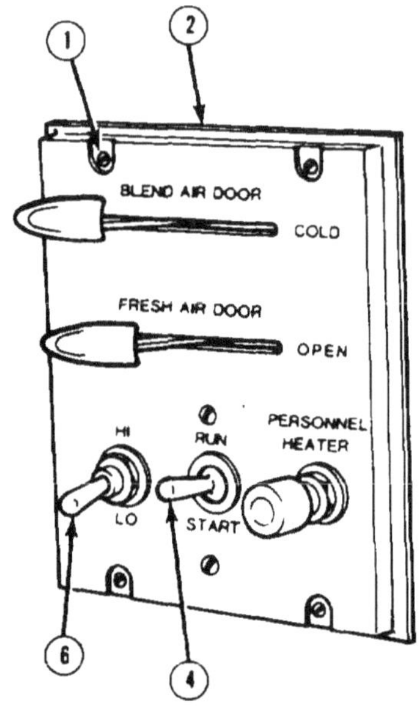

FOLLOW-ON TASKS:

- Install personnel heater control box. (See TM 9-2320-289-20)

TA50442

10-20. AIR CONDITIONING SYSTEM TESTING (M1010).

This task covers: a. Testing

INITIAL SETUP:

Materials/Parts
- Refrigerant (Item 54, Appendix B)

Tools/Test Equipment
- Charging station
- STE/ICE

Manual References
- TM 9-4910 -571-12&P

General Safety Instructions
- Always wear goggles when testing or servicing air conditioning system. DO NOT allow refrigerant to contact your skin.
- DO NOT test or service air conditioning system while smoking or near open flame. Refrigerant will become a poisonous gas when burned.
- DO NOT allow direct flame or electric heaters to contact refrigerant container.
- Never open high side valve on air conditioner manifold gage set while air conditioning system is operating.
- Never attempt to connect servicing equipment while air conditioning system is running.

a. TESTING

WARNING

- Always wear goggles when testing or servicing air conditioning system. DO NOT allow refrigerant to contact your skin. Failure to follow this warning may result in injury to personnel.
- DO NOT test or service air conditioning system while smoking or near open flame. Refrigerant will become a poisonous gas when burned. Failure to follow this warning may result in serious illness or death to personnel.
- DO NOT allow direct flame or electric heaters to contact refrigerant container (1). Localized heat can raise pressure to a dangerous level and cause refrigerant container to explode, resulting in serious injury or death to personnel.
- Never attempt to connect servicing equipment while air conditioning system is running. Failure to follow this warning may result in injury to personnel.

1. Completely close low side valve (3) and high side valve (6) on manifold gage set.
2. Remove protective caps from low pressure line (9) valve and high pressure line (10) valve. Loosely connect low pressure hose (8) to low pressure line valve and high pressure hose (7) to high pressure line valve, Connect center hose (2) to valve on refrigerant container (1).

10-20. AIR CONDITIONING SYSTEM TESTING (M1010) (Con't).

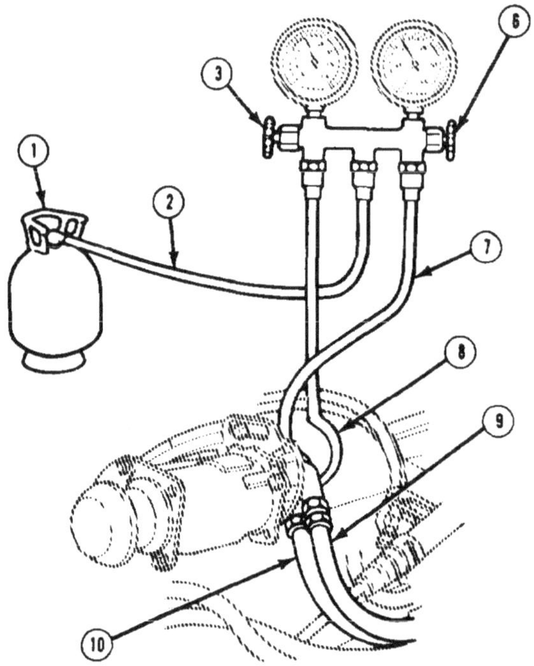

NOTE

Purging of air will take place as refrigerant in refrigerant container (1) forces air out of system at loose hose connections.

3. Slowly open refrigerant container (1) valve and high side valve (6). After 4 seconds, close refrigerant container valve and high side valve. Tighten high pressure hose (7) at high pressure line (10) valve.

4. Repeat step 3 for low side valve (3) and low pressure hose (8).

CAUTION

All testing equipment should be kept clear of moving parts while engine is running. Failure to follow this caution may result in damage to engine components and testing equipment.

5. Connect STE/ICE tester to STE/ICE DCA connector under dash. (See TM 9-4910-571-12&P) Start engine and operate at approximately 1500 rpm.

TA50443

10-20. AIR CONDITIONING SYSTEM TESTING (M1010) (Con't).

6 . on air conditioner control panel, turn "VENT-A/C" toggle switch to " VENT, " air selection lever to " OUTSIDE, " " A/C COLD" knob to coldest setting, and fan switch to "H1." Open ambulance rear doors briefly to exhaust any built up heat. Operate for a minimum of 5 minutes.

7 . Close ambulance rear doors tightly. Place thermometer at air conditioner center vent for 5 minutes, then check temperature.

8 . Check low side gage (4) and high side gage (5) and determine performance of air conditioning system using Table 10-1. High outlet temperature or low pressure at either gage indicate that system is low on refrigerant.

Table 10-1. Air Conditioning System Performance

Air Inlet (Outside) Temperature	Low Side Pressure	High Side Pressure	Air Outlet (Inside) Temperature
70°F (21°C)	8-13 psi (55-89 kPa)	1 30–150 psi (890-1 034 kPa)	33-35°F (1-2°C)
80°F (26°C)	13-18 psi (89-1 24 kPa)	165-185 psi (1 137-1276 kPa)	38-42°F (3-5°C)
90°F (32°C)	17–22 psi (1 18-151 kPa)	190-210 psi (1310-1447 kPa)	45-49°F (7-9°C)
100°F (37°C)	23–28 psi (1 58-193 kPa)	225-245 psi (1551-1749 kPa)	52-56°F (11-13°C)

NOTE

It may be necessary to scrape paint off sight glass.

9. Remove plug (12) from top of condenser cover (11) and check for bubbles in sight glass. Bubbles indicate that system is low on refrigerant,

TA50444

10-20. AIR CONDITIONING SYSTEM TESTING (M1010) (Con't).

WARNING

Never open high side valve (6) on air conditioner manifold gage set while air conditioning system is operating, Refrigerant may be pumped back into refrigerant container (1) and cause refrigerant container to explode, resulting In serious Injury or death to personnel.

NOTE

- Perform step 10 if air conditioning system was determined to be low on refrigerant.
- Ensure that engine is still running at 1500 rpm before performing step 10.
- Ensure that refrigerant container (1) remains upright when adding refrigerant through low side valve (3),

10. Slowly open refrigerant container (1) valve and low side valve (3). Add refrigerant until gage readings return to normal range and bubbles clear up in sight glass. Close low side valve and refrigerant container valve.

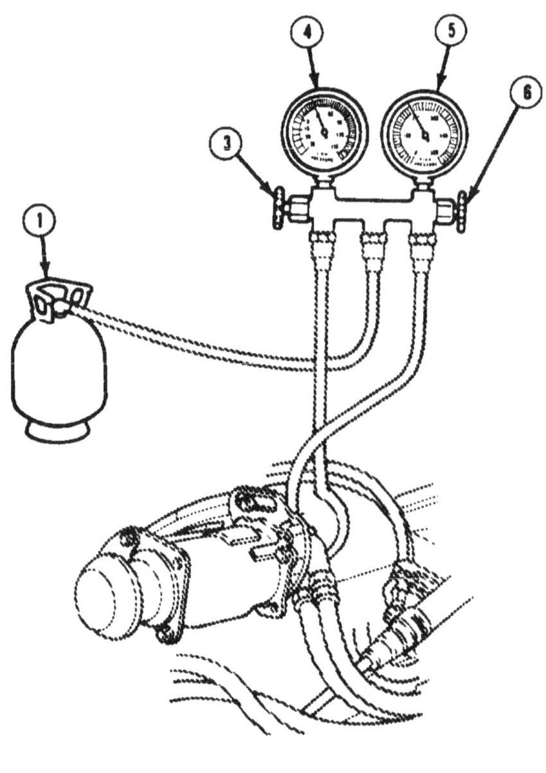

TA50445

10-20. AIR CONDITIONING SYSTEM TESTING (M1010) (Con't).

11. Service air conditioning system if still not operating properly. (See paragraph 10-21)

12. Install plug (12) on top of condenser cover (11).

13. Turn off engine. Disconnect STE/ICE tester from STE/ICE DCA connector under dash. (See TM 9-4910-571-12&P)

14. Disconnect center hose (2) from refrigerant container (1) valve . Disconnect low pressure hose (8) from low pressure line (9) valve and high pressure hose (7) from high pressure line (10) valve.

15. Install protective caps on lo w pressure line (9) valve and high pressure line (10) valve.

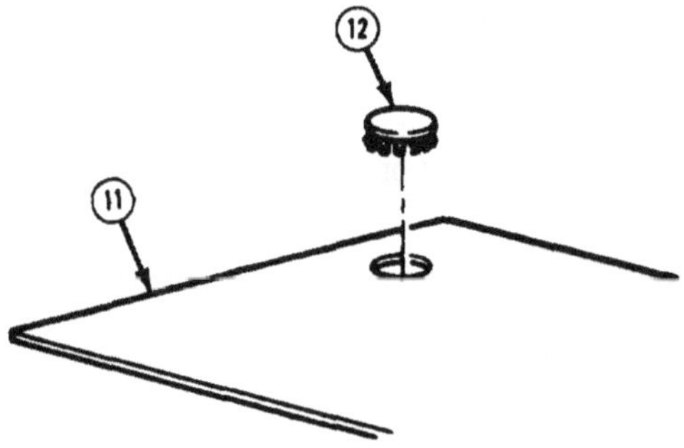

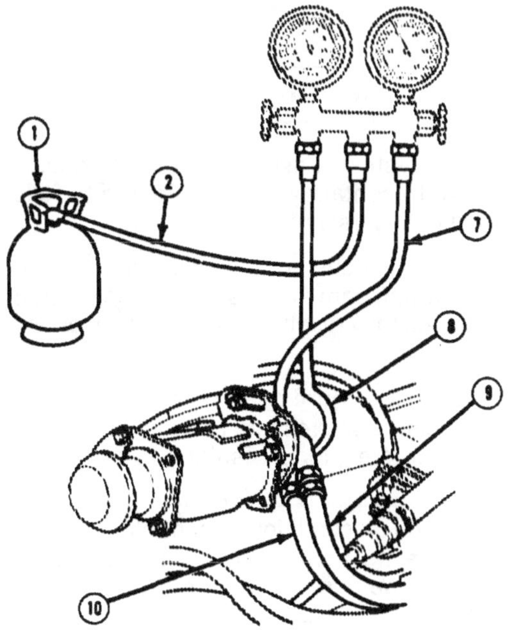

TA50446

10-21. AIR CONDITIONING SYSTEM SERVICING (M1010).

This task covers:

a. Discharging System
b. Evacuating System
c. Charging System

INITIAL SETUP:

Materials/Parts

- Refrigerant (Item 54, Appendix B)

Tools/Test Equipment

- Charging station
- Refrigerant gas leak detector
- STE/ICE
- Vacuum pump

Manual References

- TM 9-4910 -571-12&P

General Safety Instructions

- Always wear goggles when testing or servicing air conditioner system. DO NOT allow refrigerant to contact your skin.
- DO NOT test or service air conditioner system while smoking or near open flame. Refrigerant will become a poisonous gas when burned.
- DO NOT allow direct flame or electric heaters to contact refrigerant container.
- Never open high side valve on air conditioner manifold gage set while air conditioning system is operating,
- Never attempt to connect servicing equipment, discharge system, or charge system while air conditioning system is running.

a. DISCHARGING SYSTEM

WARNING

- Always wear goggles when testing or servicing air conditioner system. DO NOT allow refrigerant to contact your skin. Failure to follow this warning may result in injury to personnel.
- DO NOT test or service air conditioner system while smoking or near open flame. Refrigerant will become a poisonous gas when burned. Failure to follow this warning may result in serious illness or death to personnel.
- DO NOT allow direct flame or electric heaters to contact refrigerant container. Localized heat can raise pressure to a dangerous level and cause refrigerant container to explode, resulting in serious injury or death to personnel.
- Never open high side valve on air conditioner manifold gage set while air conditioning system is operating. Refrigerant may be pumped back into refrigerant container and cause refrigerant container to explode, resulting in serious injury or death to personnel.
- Never attempt to connect servicing equipment or discharge system while air conditioning system is running, Failure to follow this warning may result in injury to personnel.

10-21. AIR CONDITIONING SYSTEM SERVICING (M1010) (Con't).

NOTE

Ensure that DISCHARGING SYSTEM is performed to remove al contamination and refrigerant from air conditioning system before performing EVACUATING SYSTEM.

1. Completely close low side valve (1) and high side valve (4) on manifold gage set.
2. Remove protective caps from low pressure line (8) valve and high pressure line (9) valve. Connect low pressure hose (7) to low pressure line valve and high pressure hose (5) to high pressure line valve.
3. Wrap end of center hose (6) with a clean rag. Slowly open high side valve (4). Observe center hose (6) for evidence of refrigerant oil in rag. Adjust high side valve toward closed position to prevent refrigerant oil from escaping while discharging system.
4. When high side gage (3) reads 50 psi (345 kPa), slowly open low side valve (1). Refrigerant should flow at an even rate from both sides of system.
5. When both gages reach "O, " completely close low side valve (1) and high side valve (4) on manifold gage set.

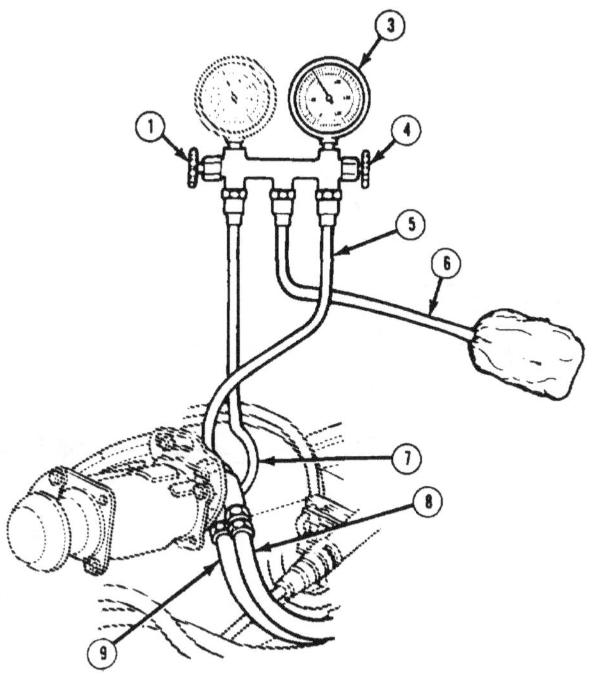

TA50447

10-21. AIR CONDITIONING SYSTEM SERVICING (M1010) (Con't).

b. EVACUATING SYSTEM

CAUTION

Compressor (10) should not be used as a vacuum pump. Failure to follow this caution may result in damage to compressor.

NOTE

If air conditioning system is determined to be leaking at any time while performing EVACUATING SYSTEM, air conditioning system will have to be charged and checked for leaks.

1. Connect center hose (6) to vacuum pump (11), Completely open low side valve (1) and high side valve (4), Open discharge valve on vacuum pump.
2. Start vacuum pump (11), Low side gage (2) should go down to 29 in. Hg. If 29 in. Hg is not obtained, there is a leak in air conditioning system. Locate leak, if present, using refrigerant gas leak detector. Repair leak in system, if present, before performing step 3.

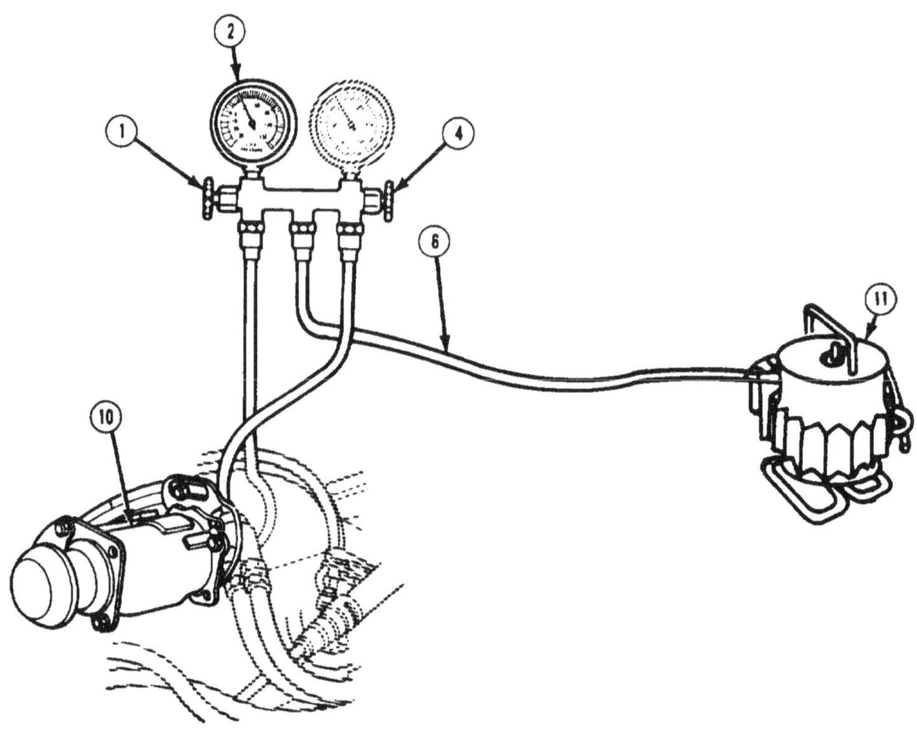

TA50448

10-21. AIR CONDITIONING SYSTEM SERVICING (M1010) (Con't).

3. When low side gage (2) needle stabilizes at 29 in. Hg, operate vacuum pump (11) for an additional 30 minutes. Completely close low side valve (1) and high side valve (4) on manifold gage set. Close discharge valve on vacuum pump, Wait 5 minutes, and check to see if low side gage needle moves. If low side gage needle moves, there is a leak in air conditioning system. Locate leak, if present, using refrigerant gas leak detector. Repair leak in system if present, and repeat this step before performing step 4.
4. If low side gage (2) needle position remains constant, disconnect center hose (6).

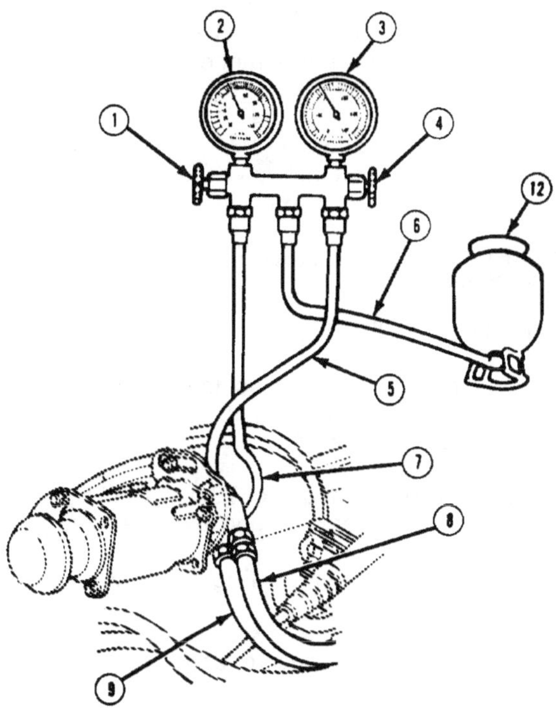

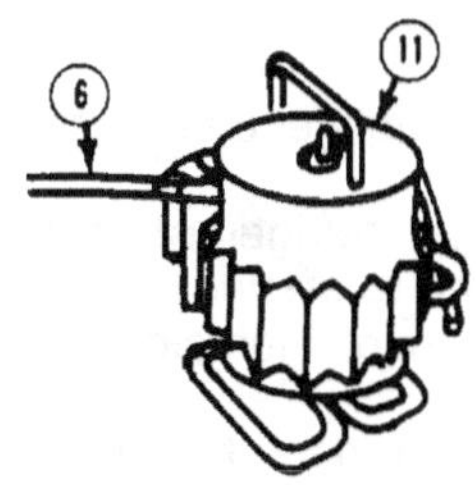

c. CHARGING SYSTEM

WARNING

- Never Open high side valve (4) on air conditioner manifold gage set while air conditioning system is operating. Refrigerant may be pumped back into refrigerant container (12) and cause refrigerant container to explode, resulting in serious Injury or death to Personnel.
- Never attempt to charge system on the high side while air conditioning system is running. Failure to follow this warning may result in injury to personnel.

TA50449

10-21. AIR CONDITIONING SYSTEM SERVICING (M1010) (Con't).

NOTE

Ensure that refrigerant container (12) is upright when performing step 1.

1. Connect center hose (6) to valve on refrigerant container (12). With refrigerant container upright, loosen center hose at manifold gage set and slowly open refrigerant container valve. After 2 seconds, close refrigerant container valve and tighten center hose.
2. Invert refrigerant container (12) and place on a scale.
3. Open refrigerant container (12) valve and high side valve (4). Continue charging until scale indicates that refrigerant container has decreased in weight by 70 oz. (1.97 kg). If refrigerant container has decreased in weight by 70 oz. (1.97 kg), close refrigerant container valve and high side valve and perform step 9.
4. f refrigerant container (12) has not decreased in weight by 70 oz. (1.97 kg), record weight used in step 3. Close high side valve (4) and refrigerant container valve. Turn refrigerant container upright and place on a scale,
5. Connect STE/ICE tester to STE/ICE DCA connector under dash. (See TM 9-4910 -571-12&P) Start engine and operate at approximately 1500 rpm.
6. On air conditioner control panel, turn air selection lever to "OUTSIDE. "
7. Slowly open refrigerant container (12) valve and low side valve (1). Adjust low side valve as needed to prevent pressure on low side gage (2) from exceeding 50 psi (345 kPa) and pressure on high side gage (3) from exceeding 240 psi (1 655 kPa).
8. When refrigerant container (12) has decreased in weight by a total of 70 oz. (1 .97 kg), including charging done in step 3, close refrigerant container valve and low side valve (1).
9. Determine performance of air conditioning system by performing steps 7 and 8 in paragraph 10-20.
10. Turn off engine, Disconnect STE/ICE tester from STE/ICE DCA connector under dash. (See TM 9-4910-571-12&P)
11. Disconnect center hose (6) from refrigerant container (12) valve. Disconnect low pressure hose (7) from low pressure line (8) valve and high pressure hose (5) from high pressure line (9) valve.
12. Install protective caps on low pressure line (8) valve and high pressure line (9) valve.

10-22. AIR CONDITIONER CONTROL PANEL AND LOUVER REPAIR (M1010).

This task covers: a. Disassembly b. Assembly

INITIAL SETUP:

Equipment Condition

- Both battery negative cables disconnected. (See TM 9-2320-289-20)
- Evaporator removed (if removing louvers). (See paragraph 10-27)

Materials/Parts

- Eight lockwashers

a. DISASSEMBLY

NOTE

- If removing louver (8), perform step 1.
- If disassembling control panel (18) assembly, perform steps 2-7.

1. From rear of air outlet box (1), remove 4 sheet spring nuts (7) and louver (8). Repeat as needed for other 5 louvers,
2. Remove screw (14) and knob (1 3).
3. Remove 8 screws (11), lockwashers (10), and control panel (18) from air outlet box (1). Discard lockwashers.

NOTE

Ensure that leads are tagged for installation.

4. Disconnect switch wiring harness (22) and leads from switches.
5. Disconnect lead from "VENT-A/C" switch (21). Remove nut (12) and "VENT-A/C" switch. Remove knob (15), nut (16), and "A/C COLD' switch (20), Remove nut (17) and fan switch (19).

NOTE

On early 1984 model M1010 trucks, air selection lever (2) is mounted on right side of air outlet box (1) support,

6. Remove nut (25), air selection lever (2), and spring washer (3) from lever bracket (4). Remove sheet spring nut and 2 washers, and disconnect air selection lever link from air selection lever. Remove 2 nuts (6), springs (5), and lever bracket.
7. Remove 2 nuts (9), screws (24), and outside air door switch (23).

10-22. AIR CONDITIONER CONTROL PANEL AND LOUVER REPAIR (M1010) (Con't),

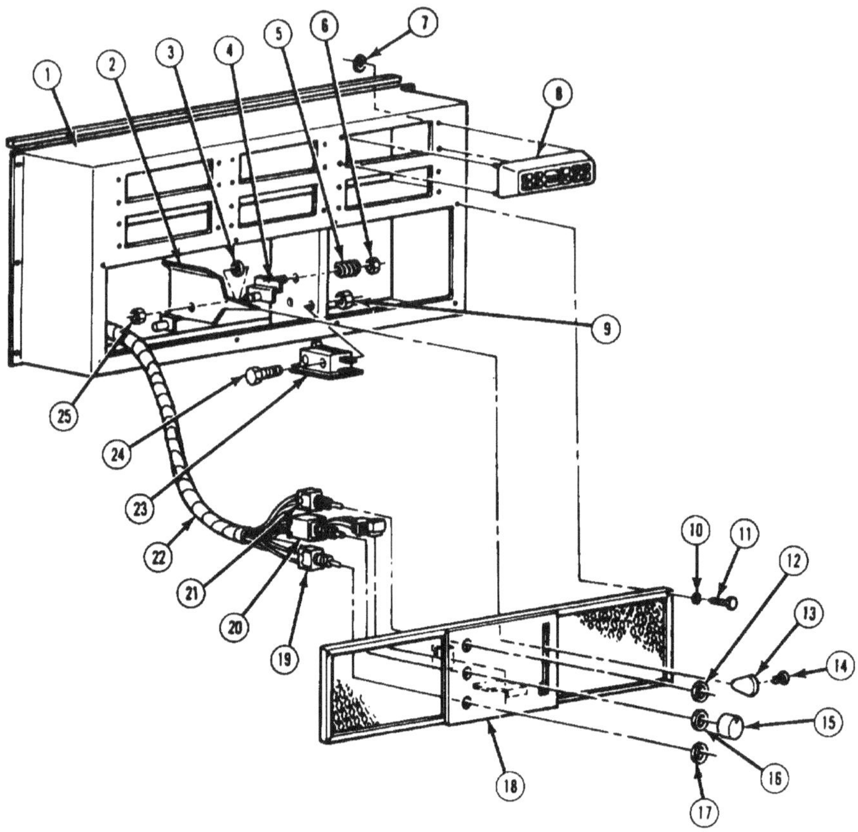

b. ASSEMBLY

NOTE

- If installing louver (8), perform step 7.
- If assembling control panel (18) assembly, perform steps 1-6.

1. Install outside air door switch (23) with 2 screws (24) and nuts (9).

TA50450

10-22. AIR CONDITIONER CONTROL PANEL AND LOUVER REPAIR (M1010) (Con't).

NOTE

On early 1984 model M1010 trucks, air selection lever (2) is mounted on right side of air outlet box (1) support. Air selection lever must contact outside door switch (23) and go through control panel (18) opening after installation,

2. Install lever bracket (4) on air outlet box (1) with 2 springs (5) and nuts (6). Connect air selection lever link to air selection lever (2) with 2 washers and sheet spring nut. Install air selection lever on lever bracket with spring washer (3) and nut (25).

3. Install fan switch (19) with nut (17), Install '"A/C COLD" switch (20) with nut (16) and knob (15). Install "VENT-A/C" switch (21) with nut (12). Connect lead to "VENT-A/C" switch.

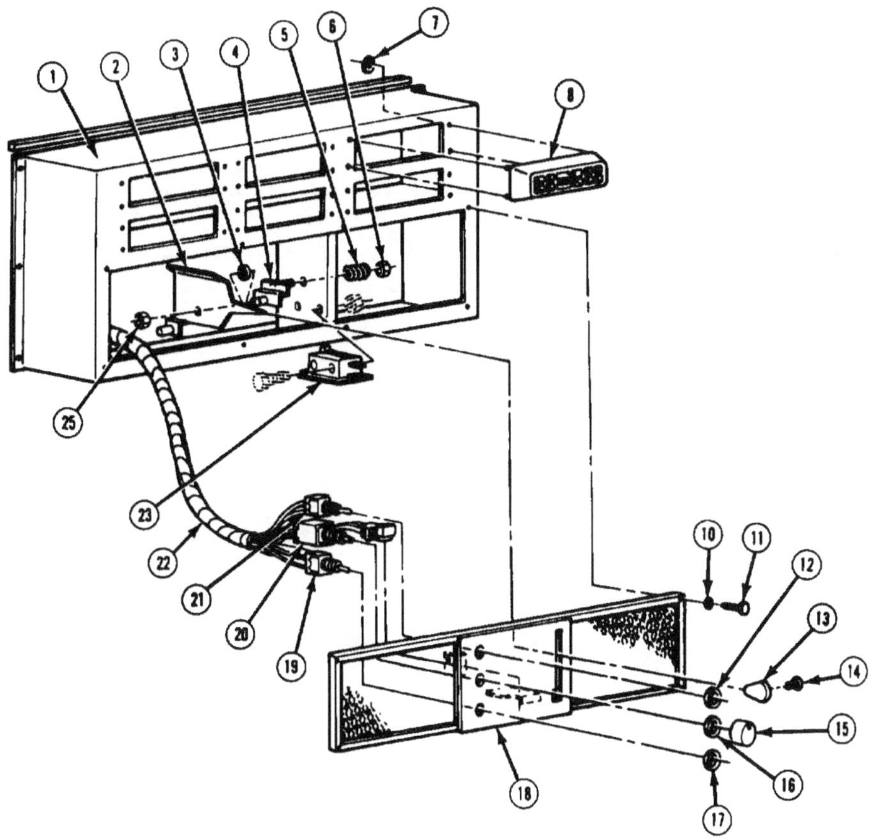

TA50451

10-22. AIR CONDITIONER CONTROL PANEL AND LOUVER REPAIR (M1010) (Con't).

4. Connect switch wiring harness (22) and leads to switches.

5. Install control panel (18) on air outlet box (1) with 8 new lockwashers (10) and screws (11).

6. Install knob (13) on air selection lever (2) with screw (14).

7. Install louver (8) on air outlet box (1) with 4 sheet spring nuts (7). Repeat as needed for other 5 louvers.

FOLLOW-ON TASKS:

- Connect both battery negative cables. (See TM 9-2320-289-20)
- Install evaporator if removed. (See paragraph 10-27)
- Check operation of air conditioning system.

10-23. AIR CONDITIONER ASSEMBLY REPLACEMENT (M1010).

This task covers: a. Removal b. Installation

INITIAL SETUP:

Equipment Condition

- Air conditioning system discharged. (See paragraph 10-21)
- Both battery negative cables disconnected. (See TM 9-2320-289-20)

Materials/Parts

- One rivet
- Two preformed packing inserts
- Two tie-down straps (Item 63, Appendix B)

Personnel Required

- *MOS 52C (2)*

a. REMOVAL

CAUTION

Ensure that all openings and connections are immediately plugged after disconnection to prevent contamination of air conditioning system.

1. Disconnect inlet line (3) and outlet line (4) from side of air conditioner assembly (1). Remove and discard 2 preformed packing inserts (2). Remove and discard 2 tie-down straps securing drain hoses to inlet line and outlet line.

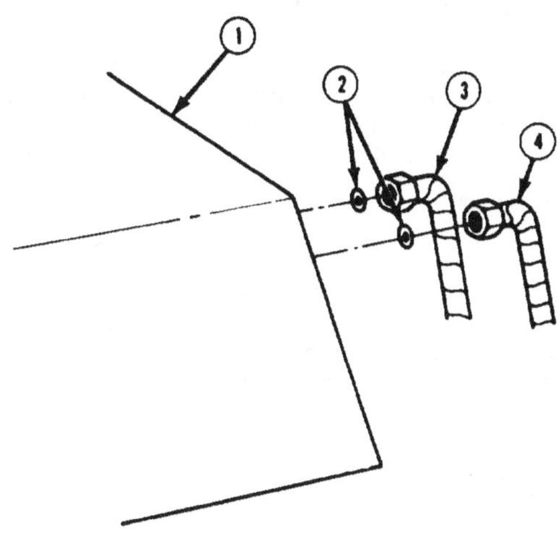

2. Inside ambulance body (5), remove 10 screws and switch panel below air conditioner control panel. Disconnect wiring harness connector from connector on underside of air conditioner assembly (1). Remove rivet and ground lead. Push wiring up into air conditioner assembly. Discard rivet.

TA50452

10-23. AIR CONDITIONER ASSEMBLY REPLACEMENT (M1010).

3 . Attach suitable lifting device to air conditioner assembly (1), Remove 16 screws (6) and air conditioner assembly from ambulance body (5). Remove lifting device.

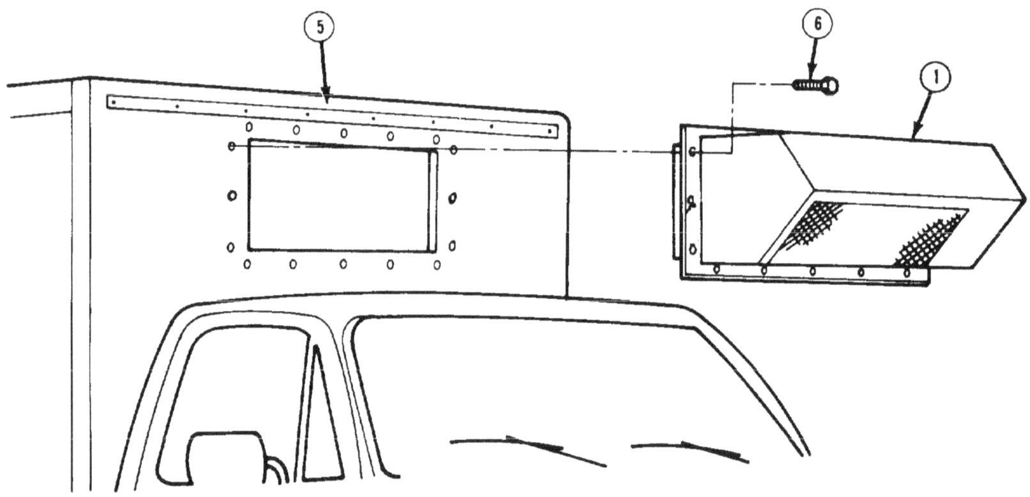

b. INSTALLATION

1 . Attach suitable lifting device to air conditioner assembly (1). Install air conditioner assembly on ambulance body (5) with 16 screws (6). Remove lifting device.

2 . Inside ambulance body (5), pull wiring through grommet on underside of air conditioner assembly (1). Connect wiring harness connector to connector on underside of air conditioner assembly. Install ground lead on ambulance body panel with new rivet. Install switch panel below air conditioner assembly with 10 screws.

NOTE

Ensure that all plugs are removed before connection.

3. Install 2 new preformed packing inserts (2) in fittings of evaporator lines (3 and 4). Connect inlet line (3) and outlet line (4) to side of air conditioner assembly (1). Secure drain hoses to inlet line and outlet line with 2 new tie-down straps,

FOLLOW-ON TASKS:

- Connect both battery negative cables, (See TM 9-2320-289-20)
- Evacuate and charge air conditioning system. (See paragraph 10-21)
- Check operation of air conditioning system and inspect for leakage.

TA50453

10-24. AIR CONDITIONER COMPRESSOR MAINTENANCE (M1010).

This task covers:

a. Removal
b. Disassembly
c. Inspection
d. Assembly
e. Installation

INITIAL SETUP:

Equipment Condition

- Both battery negative cables disconnected, (See TM 9-2320-289-20)
- Air conditioning system discharged. (See paragraph 10-21)

Materials/ Parts

- One drain screw gasket
- One nut kit
- One preformed packing kit
- One ring kit
- One seal kit
- One suction screen
- Two lockwashers
- Refrigerant oil (Item 55, Appendix B)

Tools/Test Equipment

- Compressor tool kit, J-29888
- Belt tensioning gage
- Torque wrench

a. REMOVAL

1. Loosen 3 adjusting bolts (10) and pivot nut (12), and move compressor assembly (14) to right. Remove air conditioner belt (13).

CAUTION

Ensure that all openings and connections are immediately plugged after disconnection to prevent contamination of air conditioning system,

2. On rear of compressor assembly (14), remove bolt and lockwasher, and disconnect compressor lines. Discard lockwasher.

3. Remove bolt and lockwasher from clip retaining electrical connector to compressor assembly (14). Disconnect electrical connector from compressor assembly. Discard lockwasher.

4. Remove 3 adjusting bolts (10), 5 bolts (8), and pivot nut (1 2). Remove compressor assembly (14) with front adjusting bracket (11) and rear adjusting bracket (9).

NOTE

Perform step 5 only if brackets (4, 6, or 16) or mounting hardware are damaged.

5. Remove 2 nuts (1), bolt (15), and lower front bracket (16) from water pump studs (2). Remove 2 nuts (3) and upper rear bracket (4) from intake manifold studs (5). Remove 2 bolts (7) and lower rear bracket (6).

10-24. AIR CONDITIONER COMPRESSOR MAINTENANCE (M1010) (Con't),

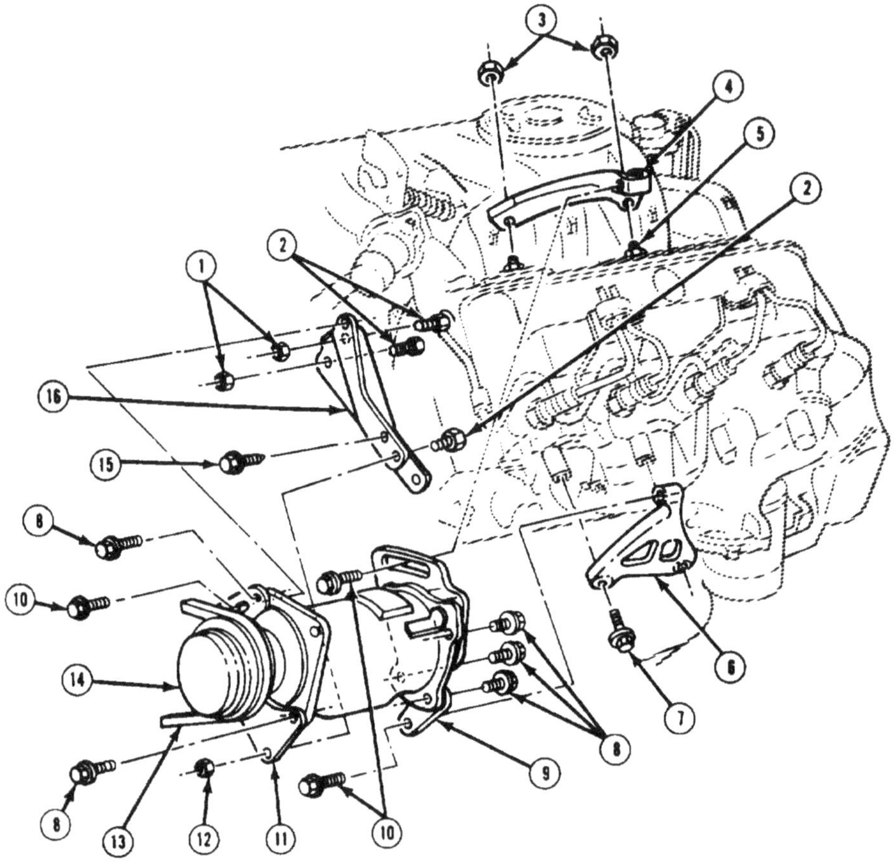

6 . Remove drain screw and gasket from compressor assembly (14), Drain and measure refrigerant oil in compressor assembly. Check for evidence of contamination to determine if remainder of air conditioning system requires servicing, Discard gasket.

TA50454

10-24. AIR CONDITIONER COMPRESSOR MAINTENANCE (M1010) (Con't).

b. DISASSEMBLY

CAUTION

- Work area, tools, and hands must be kept clean when repairing compressor. A parts tray should be used for all internal compressor parts and all replacement parts. Failure to follow this caution may result in damage to compressor components.
- DC) NOT strike end of shaft (27) during disassembly of compressor. Failure to follow this caution may result in damage to compressor internal components,

1. Install holding fixture (28) in vise. Install compressor assembly in holding fixture.
2. Hold clutch (17) in place with clutch hub holder. Remove and discard locknut (18). Remove clutch snapring (19).
3. Remove clutch (17) using hub and drive puller and installer. Remove key from shaft (27) or clutch.
4. Remove pulley snapring (20). Remove pulley (23) with pulley bearing (22).
5. Remove pulley bearing retaining ring (21). Drive out pulley bearing (22).
6. Mark relationship of coil housing terminals (26) to compressor shell (29). Remove coil housing retaining ring (24). Remove coil housing (25).
7. Remove compressor assembly from holding fixture (28). Remove 2 preformed packing inserts (36) from rear head (34). Remove pressure release valve (37) and preformed packing insert (38) from rear head. Discard preformed packing inserts.

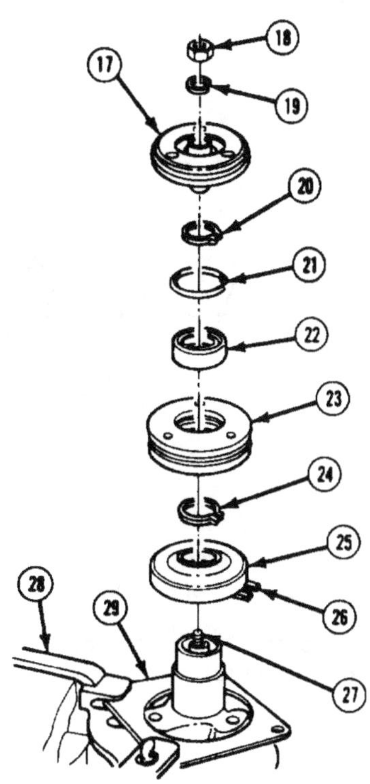

8. Remove 4 nuts (35) and rear head (34) from compressor shell (29). Remove and discard "O" ring (33) and suction screen (39).

TA50455

10-24. AIR CONDITIONER COMPRESSOR MAINTENANCE (M1010) (Con't).

CAUTION

DO NOT pry on any part of valve plate (31) or valve reed (30) except along edge. Failure to follow this caution may result in damage to valve plate or valve reed,

NOTE

Ensure that position of valve plate (31) and valve reed (30) is noted for assembly.

9. Pry up on oil pump assembly (32) and remove, Gently pry along edge of valve plate (31) and remove. Gently pry along edge of valve reed (30) and remove.

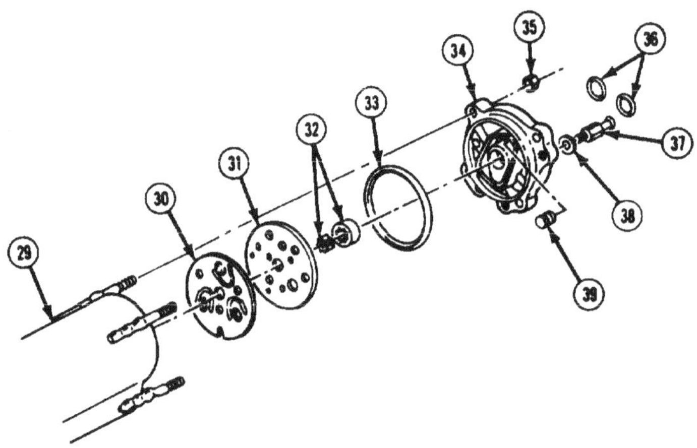

10. Remove cylinder housing assembly (40) from compressor shell (29).

11. Pry suction cover (41) from cylinder housing assembly (40).

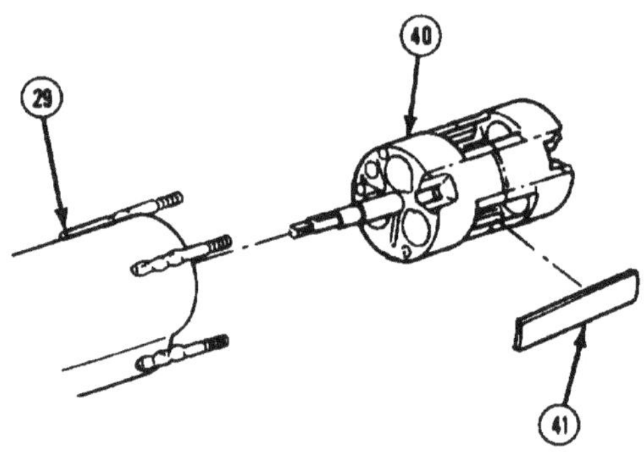

TA50456

10-24. AIR CONDITIONER COMPRESSOR MAINTENANCE (M1010) (Con't).

CAUTION

DO NOT pry on any part of valve plate (48) or valve reed (49) except along edge. Failure to follow this caution may result in damage to valve plate or valve reed.

NOTE

Ensure that position of valve plate (48) and valve reed (49) is noted for assembly.

12. Gently pry along edge of valve reed (49) and remove. Gently pry along edge of valve plate (48) and remove.
13. Remove and discard "O" ring (47) from groove in compressor shell (29). Remove front head (46) from compressor shell.
14. Remove sleeve seal retainer (42), snapring (43), felt seal (44), and preformed packing insert (45) from front head (46). Discard felt seal and preformed packing insert.

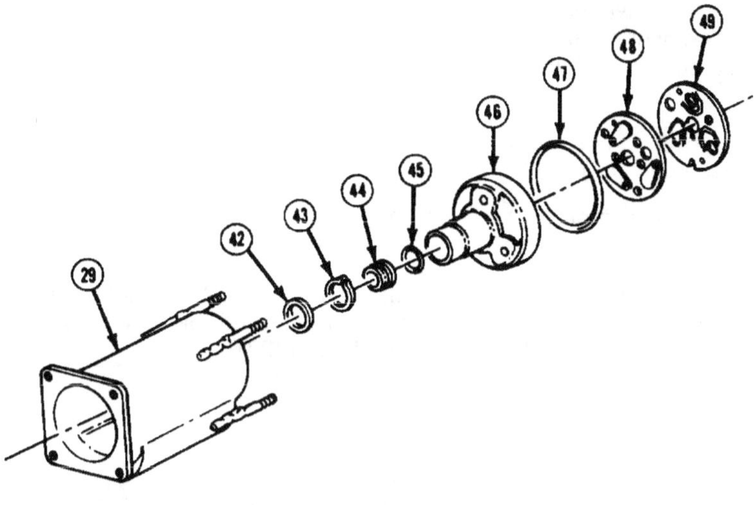

c. INSPECTION

1. Pull and push end of shaft (27) to determine if there is any play. If there is play replace cylinder housing assembly (40).
2. Rotate shaft (27) to determine if pistons are moving smoothly in their bores. If pistons are not moving smoothly in their bores replace cylinder housing assembly (40).

TA50457

10-24. AIR CONDITIONER COMPRESSOR MAINTENANCE (M1010) (Con't).

d. ASSEMBLY

CAUTION

DO NOT strike end of shaft (27) during assembly of compressor. Failure to follow this caution may result in damage to compressor internal components.

1. Install new preformed packing insert (45) and new felt seal (44) on front head (46). Install snapring (43) in groove. Install sleeve seal retainer (42).
2. Install front head (46) in compressor shell (29). Install new "O" ring (47) in groove in compressor shell (29).
3. Gently and evenly press valve plate (48) into compressor shell (29). Gently and evenly press valve reed (49) over valve plate, ensuring that it is positioned the same way as it was removed.
4. Press suction cover (41) onto cylinder housing assembly (40).
5. Install cylinder housing assembly (40) in compressor shell (29).

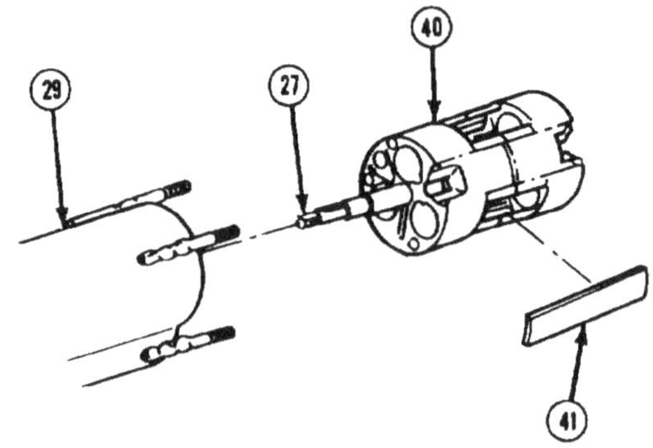

6. Gently and evenly press valve reed (30) into compressor shell (29). Gently and evenly press valve plate (31) over valve reed, ensuring that it is positioned the same way as it was removed. Press oil pump assembly (32) onto end of shaft (27).

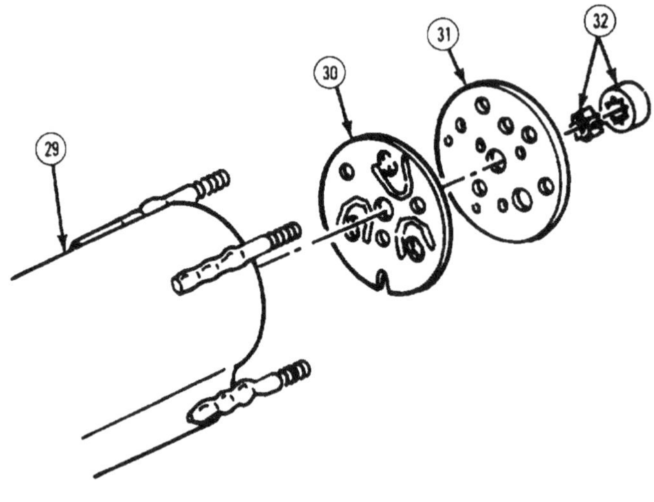

TA50458

10-24, AIR CONDITIONER COMPRESSOR MAINTENANCE (M1010) (Con't).

7. Install new suction screen (39) and new "O" ring (33) on rear head (34), Install rear head on compressor shell (29) with 4 nuts (35),

8. Install pressure release valve (37) and preformed packing insert (38) on rear head (34). Install 2 new preformed packing inserts (36) on rear head,

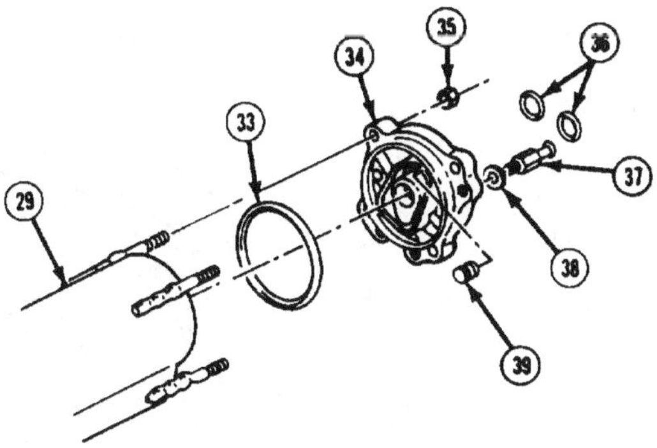

9. Aline coil housing terminals (26) with marks made on compressor shell (29) during removal. Aline locating tabs on coil housing (25) with holes in front head casting, and install coil housing, Install coil housing retaining ring (24) with flat side against coil housing,

CAUTION

Pulley bearing (22) should not be cleaned with any type of solvent. Failure to follow this caution may result in damage to pulley bearing.

10. Drive pulley bearing (22) into pulley (23), Install pulley bearing retaining ring (21) in pulley groove.

11. Install holding fixture (28) in vise, Install compressor assembly (14) in holding fixture,

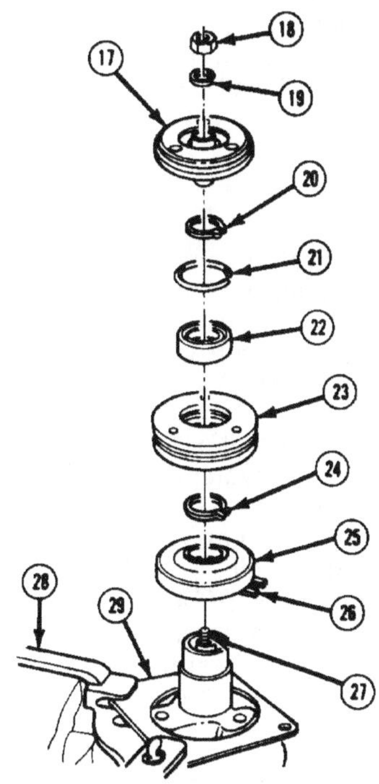

TA50459

10-24. AIR CONDITIONER COMPRESSOR MAINTENANCE (M1010) (Con't).

12. Gently tap pulley (23) with pulley bearing (22) into place until it seats. Check pulley to ensure that it rotates freely. Install pulley snapring (20).

NOTE

Key (50) has a slight curve for interference fit with keyway of clutch (17).

13. Insert key (50) into keyway of clutch (1 7), Allow key to project 3/16 in. (4.8 mm) out of keyway. Aline key with shaft keyway (51) and position clutch on shaft (27).

14. Hold clutch (17) in place with clutch hub holder. Using hub and drive puller and installer, install clutch until there is about 3/32 in. (2.4 mm) clearance between clutch and pulley (23). Ensure that key (50) stays in place.

15. Install clutch snapring (19). Install new locknut (18). Tighten locknut to 25 lb.-ft. (34 N.m).

16. Check to ensure that there is 0.022-0.057 in. (0.6-0.14 mm) clearance between clutch (17) and pulley (23). If clearance is not correct, check for mispositioned key (50) or shaft (27).

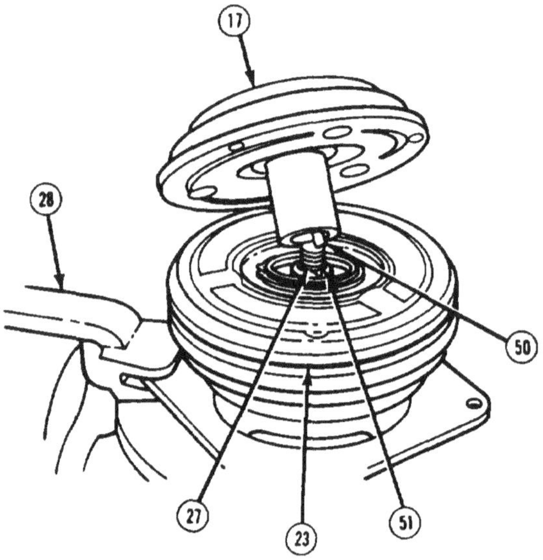

e. INSTALLATION

1. If there are no signs of refrigerant leakage in air conditioning system, add refrigerant oil to compressor (14) equal to the amount drained from compressor, plus 1 oz. (0.059 1).

2. If there are signs of refrigerant oil leakage in air conditioning system, remove receiver-dehydrator (see paragraph 10-27), and drain and measure refrigerant oil. Add refrigerant oil to compressor (14) equal to this amount plus amount drained from compressor. Install receiver-dehydrator. (See paragraph 10-27)

3. Install drain screw and new gasket on compressor assembly (14).

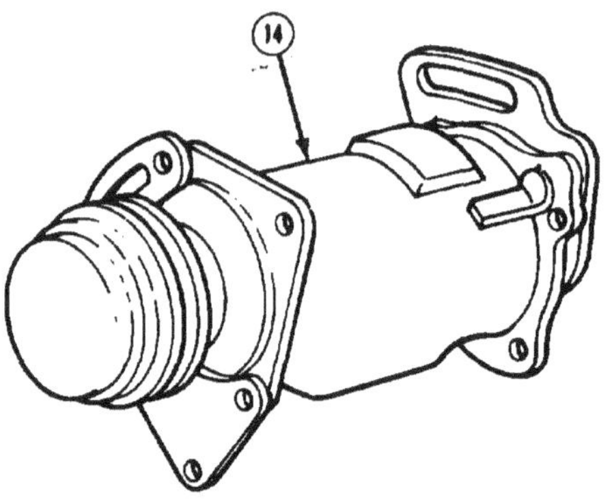

TA50460

10-24. AIR CONDITIONER COMPRESSOR MAINTENANCE (M1010) (Con't).

NOTE

Perform step 4 only if brackets (4, 6, or 16) or mounting hardware were removed,

4. Install lower rear bracket (6) with 2 bolts (7). Install upper rear bracket (4) on intake manifold studs (5) with 2 nuts (3). Install lower front bracket (16) on water pump studs (2) with bolt (15) and 2 nuts (1). Tighten bolts (7) and nuts (3) to 35 lb.-ft. (47 N.m). Tighten bolt (15) and nuts (1) to 25 lb.-ft. (34 N.m).
5. Install compressor assembly (14), front adjusting bracket (11), and rear adjusting bracket (9) with pivot nut (12), 5 bolts (8), and 3 adjusting bolts (10).
6. Connect electrical connector to compressor assembly (14). Install clip retaining electrical connector with new lockwasher and bolt.

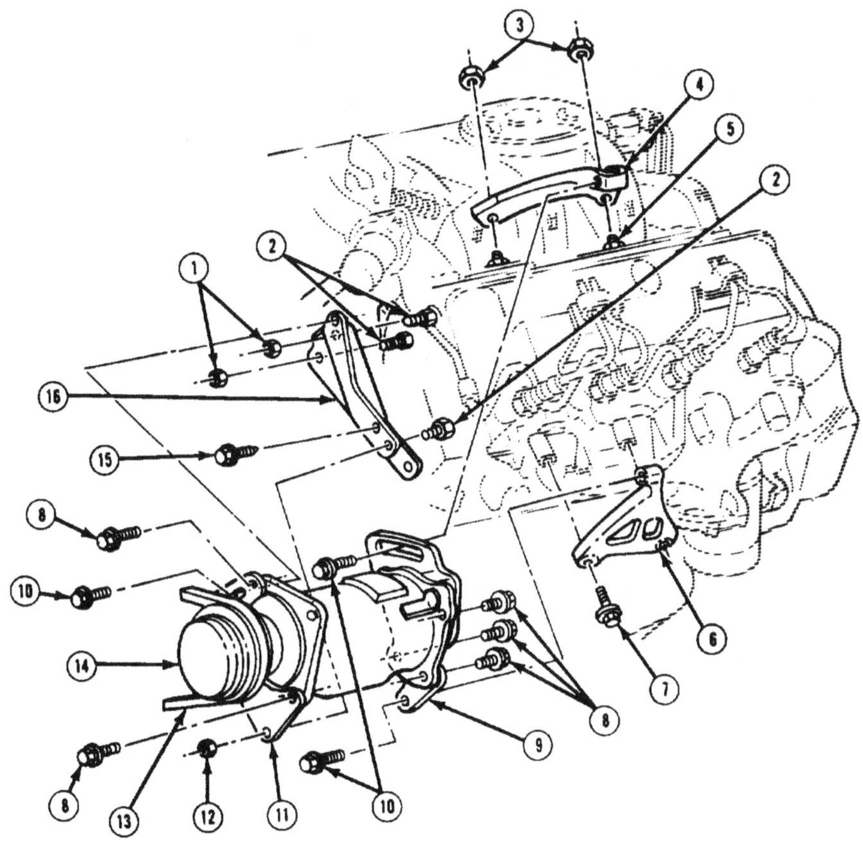

TA50461

10-24. AIR CONDITIONER COMPRESSOR MAINTENANCE (M1010) (Con't).

NOTE

Ensure that all plugs are removed before connection.

7. Connect compressor lines to rear of compressor assembly (14), with new lockwasher and bolt.
8. Install air conditioner belt (13). Loosen 3 adjusting bolts (10) and pivot nut (12) and move compressor assembly (14) to left until air conditioner belt appears tight. Tighten adjusting bolts and pivot nut.
9. Connect both battery negative cables. (See TM 9-2320-289-20)
10. Using belt tensioning gage, check air conditioner belt (13) tension. (See Appendix D)

FOLLOW-ON TASKS:

- Evacuate and charge air conditioning system. (See paragraph 10-21)
- Check operation of air conditioning system and inspect for leakage.

10-25. AIR CONDITIONER LINES REPLACEMENT (M1010),

This task covers: a. Removal b. Installation

INITIAL SETUP:

Equipment Condition

- Air conditioning system discharged. (See paragraph 10-21)

Materials/Parts

- One lockwasher
- Two locknuts
- Four preformed packing inserts
- Two tie-down straps (Item 63, Appendix B)

a. REMOVAL

CAUTION

Ensure that all openings and connections are immediately plugged after disconnection to prevent contamination of air conditioning system.

NOTE

- If removing air conditioner line assembly (16), perform steps 1-3.
- If removing inlet line (4) or outlet line (3), perform steps 2, 4, and 5,

1. Remove bolt (18) and lockwasher (17), and disconnect air conditioner line assembly (16) from compressor (5). Discard lockwasher.
2. Disconnect air conditioner line assembly (16) from inlet line (4) and outlet line (3) at rear of cab. Remove and discard 2 preformed packing inserts (8) from inlet line and outlet line.
3. Remove nut (10), bolt (7), and clip (9) from frame crossmember (6). Remove 2 locknuts (15), washers (14), compressor hose washers (13), and clips (12) from 2 bolts (11) on underside of cab. Remove air conditioner line assembly (16), Discard locknuts.
4. Disconnect inlet line (4) and outlet line (3) from air conditioner assembly (1). Remove and discard 2 preformed packing inserts (2).

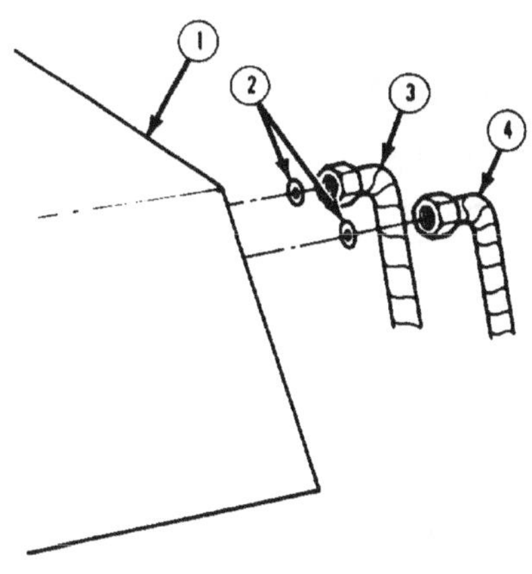

TA50462

10-25. AIR CONDITIONER LINES REPLACEMENT (M1010) (Con't).

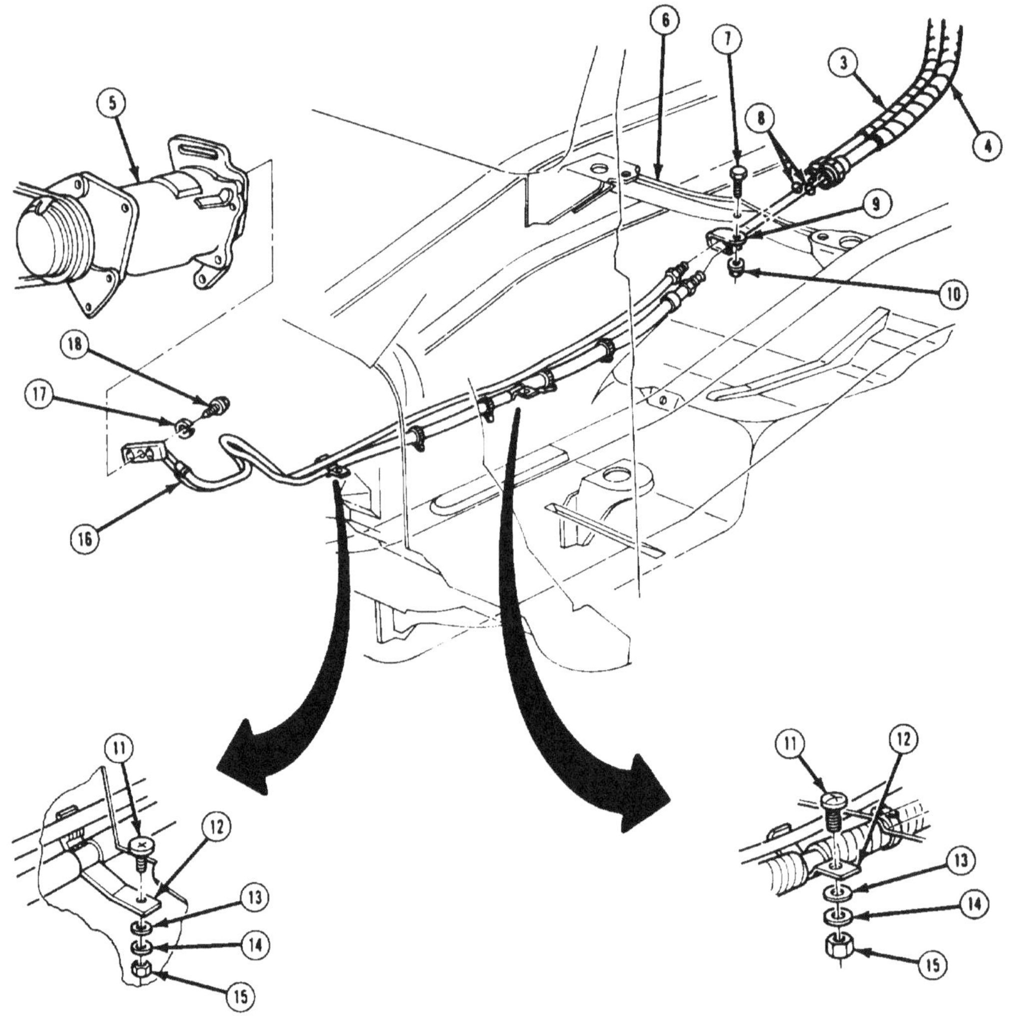

TA50463

10-25. AIR CONDITIONER LINES REPLACEMENT (M1010) (Con't),

NOTE

Be careful not to bend inlet line (4) or outlet line (3) during removal. Lines will be difficult to install if they are not in their original shape.

5. Remove bolt and bracket securing inlet line (4) and outlet line (3) to underside o f ambulance body. Remove and discard 2 tie-down straps securing 2 drain hoses and wiring harness to inlet line and outlet line. Remove inlet line and outlet line.

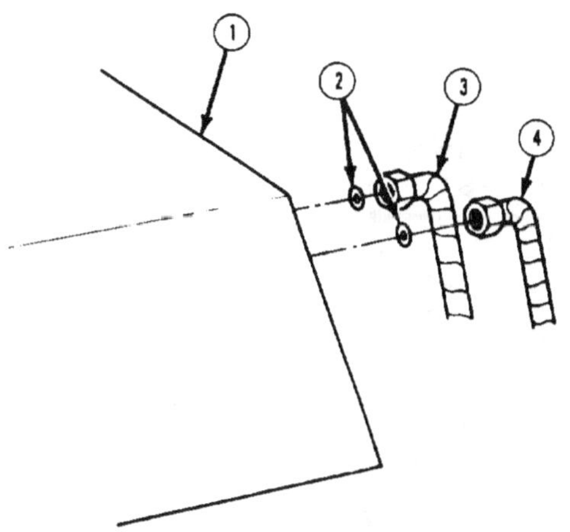

b. INSTALLATION

NOTE

- If installing air conditioner line assembly (16), perform steps 3-5.
- If installing inlet line (4) and Outlet line (3), perform steps 1, 2, and 4.
- Ensure that all plugs are removed before connection.
- Be careful not to bend inlet line (4) or Outlet line (3) during installation, Lines will be difficult to install if they are not in their original shape,

1. Position outlet line (4) and inlet line (3) in ambulance body groove. Install 2 new tie-down straps on outlet line, inlet line, 2 drain hoses, and wiring harness. Install bracket and bolt to underside of ambulance body. Ensure that outlet line and inlet line are not pressing against bracket.
2. Install 2 new preformed packing inserts (2) on outlet line (3) and inlet line (4). Connect outlet line and inlet line to air conditioner assembly (1).
3. Install air conditioner line assembly (16) on bolts (11) at underside of cab with 2 clips (1 2), compressor hose washers (13), washers (14), and new locknuts (15). Install clip (9) on air conditioner line assembly. Install clip (9) on frame crossmember (6) with bolt (7) and nut (10).
4. Install 2 new preformed packing inserts (8) in fittings of inlet line (4) and outline line (3). Connect air conditioner line assembly (16) to inlet line and outlet line at rear of cab.
5. Connect air conditioner line assembly (16) to rear of compressor (5) with new lockwasher (17) and bolt (18).

TA50464

10-25. AIR CONDITIONER LINES REPLACEMENT (M1010) (Con't).

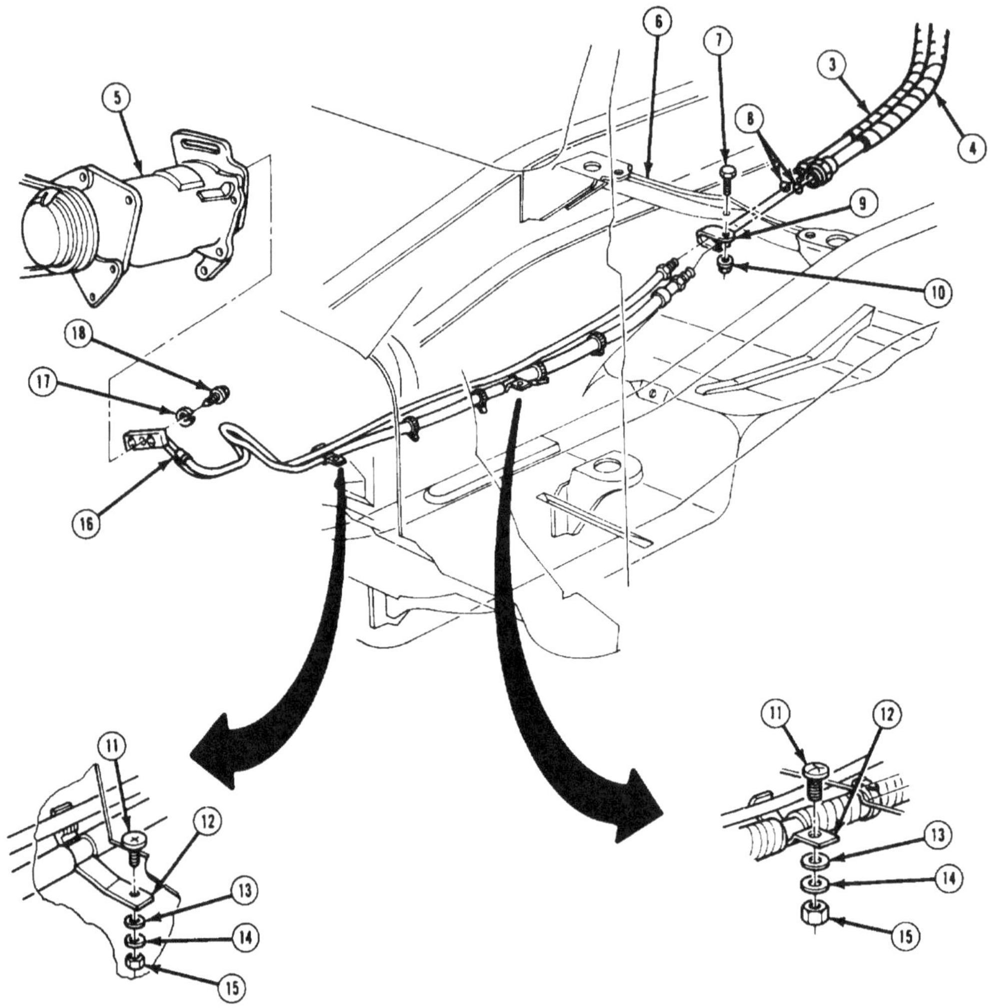

TA50465

10-25. AIR CONDITIONER LINES REPLACEMENT (M1010) (Con't).

FOLLOW-ON TASKS:

- Evacuate and charge air conditioning system, (See paragraph 10-21)
- Check operation of air conditioning system and inspect for leakage.

10-26. AIR CONDITIONER CONDENSER REPLACEMENT (M1010).

This task covers: a. Removal b. Installation

INITIAL SETUP:

Equipment Condition

- Air conditioner cover assembly removed. (See TM 9-2320-289-20)
- Air conditioning system discharged. (See paragraph 10-21)

Materials/parts

- Two preformed packing inserts
- Refrigerant oil (Item 55, Appendix B)

a. REMOVAL

CAUTION

- Be careful not to bend condenser (2) fins during removal.
- Ensure that all openings and connections are immediately plugged after disconnection to prevent contamination of air conditioning system.

1. Disconnect lines at condenser (2). Disconnect high pressure switch and core from condenser line. Remove and discard 2 preformed packing inserts from lines.

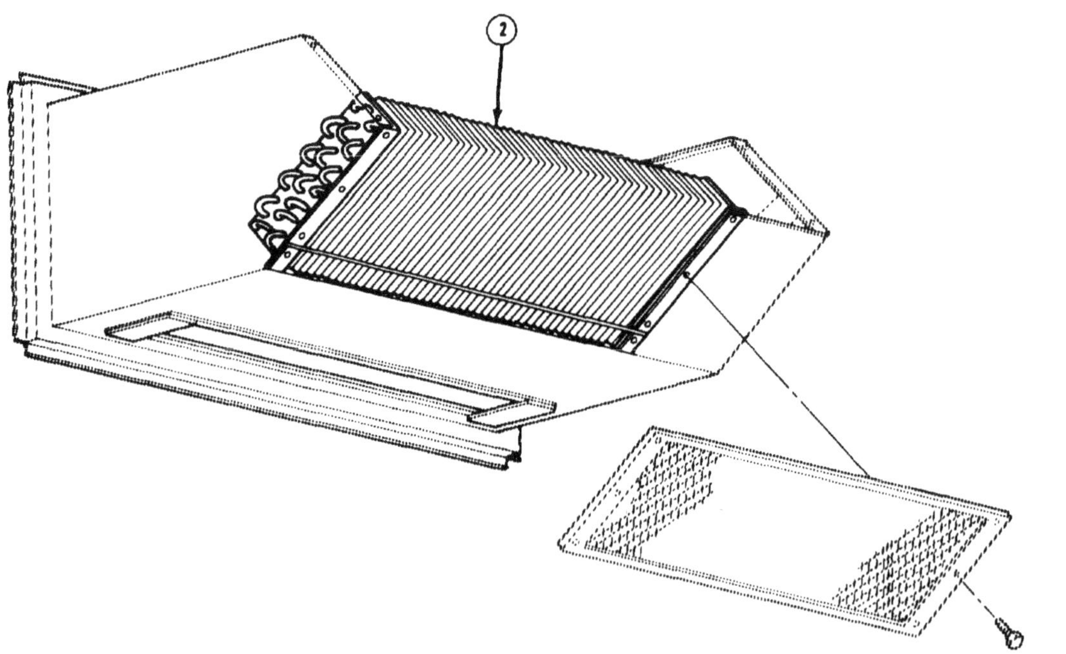

TA50466

10-26. AIR CONDITIONER CONDENSER REPLACEMENT (M1010) (Con't).

2 . Remove 6 screws (4) and air conditioner inlet grille (3). Remove condenser (2) from air conditioner assembly (1).

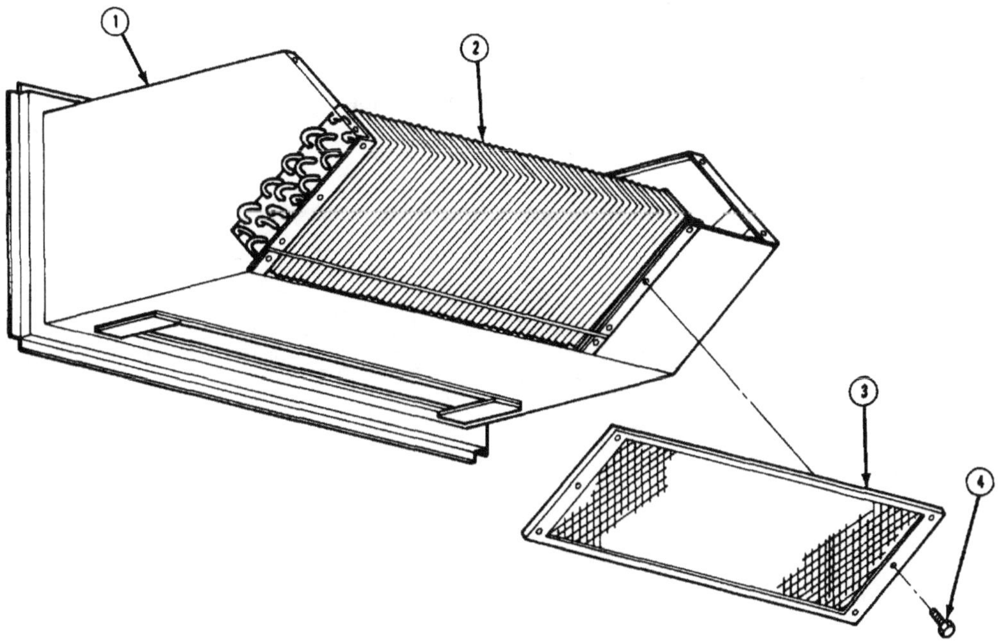

b. INSTALLATION

NOTE

Perform step 1 only if new condenser (2) will be installed.

1 . Add 1 oz. (0.059 1) of refrigerant oil to condenser (2),

CAUTION

Be careful not to bend condenser (2) fins during installation.

2 . Install condenser (2) in air conditioner assembly (1). Install air conditioner inlet grille (3) with 6 screws (4).

NOTE

Ensure that all plugs are removed before connection.

3 . Connect core and high pressure switch to condenser (2) line. Coat 2 new preformed packing inserts with refrigerant oil and install in fittings of lines. Connect lines at condenser.

TA50467

10-26. AIR CONDITIONER CONDENSER REPLACEMENT (M1010) (Con't).

FOLLOW-ON TASKS:

- Install air conditioner cover assembly. (See TM 9-2320-289-20)
- Evacuate and charge air conditioning system. (See paragraph 10-21)
- Check operation of air conditioning system and inspect for leakage.

10-27. AIR CONDITIONER EXPANSION VALVE, EVAPORATOR, AND RECEIVER-DEHYDRATOR REPLACEMENT (M1010).

This task covers: a. Removal b. Installation

INITIAL SETUP:

Equipment Condition

- Air conditioner cover assembly removed. (See TM 9-2320-289-20)
- Air conditioning system discharged. (See paragraph 10-21)

Materials/Parts

- One "O" ring
- Seven preformed packing inserts
- Refrigerant oil (Item 55, Appendix B)
- Thermal insulation tape (Item 67, Appendix B)

a. REMOVAL

CAUTION

Ensure that all openings and connections are immediately plugged after disconnection to prevent contamination of air conditioning system.

NOTE

- If removing expansion valve (10), perform steps 1-3.
- If removing evaporator (1), perform steps 1-6.
- If removing receiver-dehydrator (14), perform steps 2, 7, and 8.
- If removing drain pan (33) and drain hoses (28), perform steps 1-6, and 9.
- Ensure that all leads and lines are tagged for installation.

1. Remove thermal insulation tape from expansion valve (10). Disconnect expansion valve from evaporator (1), Remove and discard " O" ring (9) and preformed packing insert (8).
2. Remove outlet line (13), Remove and discard 2 preformed packing inserts (12).
3. Remove and discard thermal insulation tape securing temperature sensing bulb (11) to evaporator (1) outlet line. Remove expansion valve (10),
4. Disconnect outlet line (18) from evaporator (1). Remove and discard preformed packing insert (21). Remove 2 screws (5) and washers (4).
5. Disconnect leads from low pressure switch (7). Remove low pressure switch and core (6) from evaporator (1) if damaged.

CAUTION

Be careful not to bend evaporator (1) fins during removal.

6. Remove 12 screws, evaporator insulation, and cover, and set cover aside. Remove 4 screws (31), washers (32), and evaporator (1) from air conditioner assembly (30). Remove and discard thermal insulation tape from evaporator line if replacing evaporator.

10-27. AIR CONDITIONER EXPANSION VALVE, EVCAPORATOR, AND RECEIVER-DEHYDRATOR REPLACEMENT (M1010) (Con't).

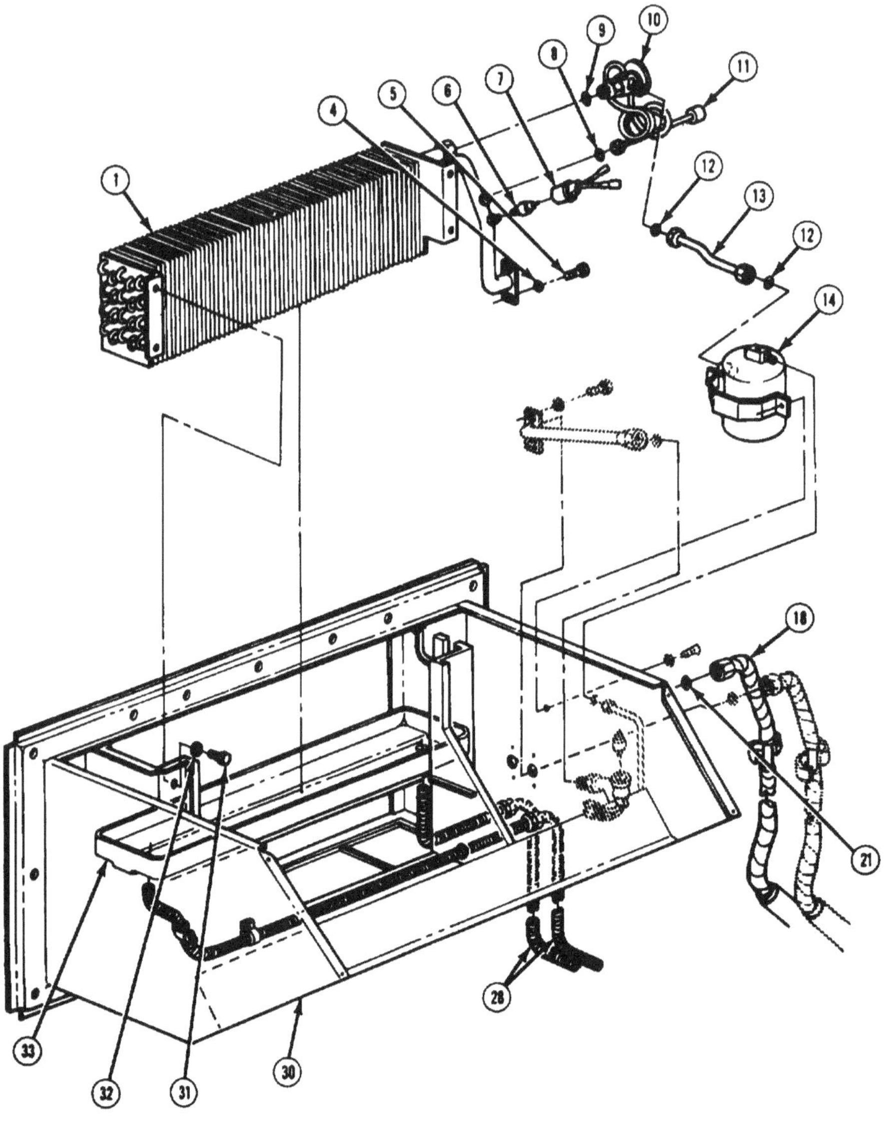

TA50468

10-27. AIR CONDITIONER EXPANSION VALVE, ECAPORATOR, AND RECEIVER-DEHYDRATOR REPLACEMENT (M1010) (Con't).

7. Disconnect condenser line (23) from receiver-dehydrator (14), Remove 2 screws (17), washers (22), and receiver-dehydrator from air conditioner assembly (30). Remove and discard preformed packing insert (27).
8. Drain and measure refrigerant oil from receiver-dehydrator (14).
9. Disconnect 2 drain hoses (28) from drain pan (33). Remove drain pan, Pull 2 drain hoses through grommets in air conditioner assembly (30) and remove, Remove drain hose valve (26) if damaged.
10. Disconnect leads from high pressure switch (25). Remove high pressure switch and core (24) from condenser line (23),
11. Disconnect inlet line (19) from connector line (16). Remove 2 screws (3), washers (2), and connector line. Remove and discard preformed packing insert (20) and preformed packing insert (15).

b. INSTALLATION

NOTE

- If installing drain pan (33) and drain hoses (28), perform step 3 and steps 6 through 12.
- If installing receiver-dehydrator (1 4), perform steps 4, 5, and 11.
- If installing evaporator (1), perform steps 6-12.
- If installing expansion valve (1 O), perform steps 10-12,
- Ensure that all plugs are removed before connection,
- Ensure that "O" ring (9) and all preformed packing inserts (8, 12, 15, 20, 21, and 27) are coated with refrigerant oil before installation.

1. Install new preformed packing insert (20) in inlet line (19) fitting, Install new preformed packing insert (15) in connector line (16) fitting. Install connector line with 2 washers (2) and screws (3). Connect inlet line (19) to connector line,
2. Install core (24) and high pressure switch (25) on condenser line (23). Connect leads.
3. Install drain hose valve (26) if removed, Push 2 drain hoses (28) through grommets in air conditioner assembly (30). Install drain pan (33), Connect 2 drain hoses to drain pan.
4. If old receiver-dehydrator (14) will be installed, add refrigerant oil to receiver-dehydrator equal to amount drained. If new receiver-dehydrator will be installed, add refrigerant oil to receiver-dehydrator equal to amount drained plus 2 oz. (0.118 1),
5. Install new preformed packing insert (27) in condenser line (23) fitting. Install receiver-dehydrator (14) on air conditioner assembly (30) with 2 washers (22) and screws (17). Connect condenser line to receiver-dehydrator.
6. If new evaporator (1) will be installed, add 3 oz. (0.177 1) of refrigerant oil to evaporator.

10-27. AIR CONDITIONER EXPANSION VALVE, EVAPORATOR, AN D RECEIVER-DEHYDRATOR REPLACEMENT (M1010) (Con't).

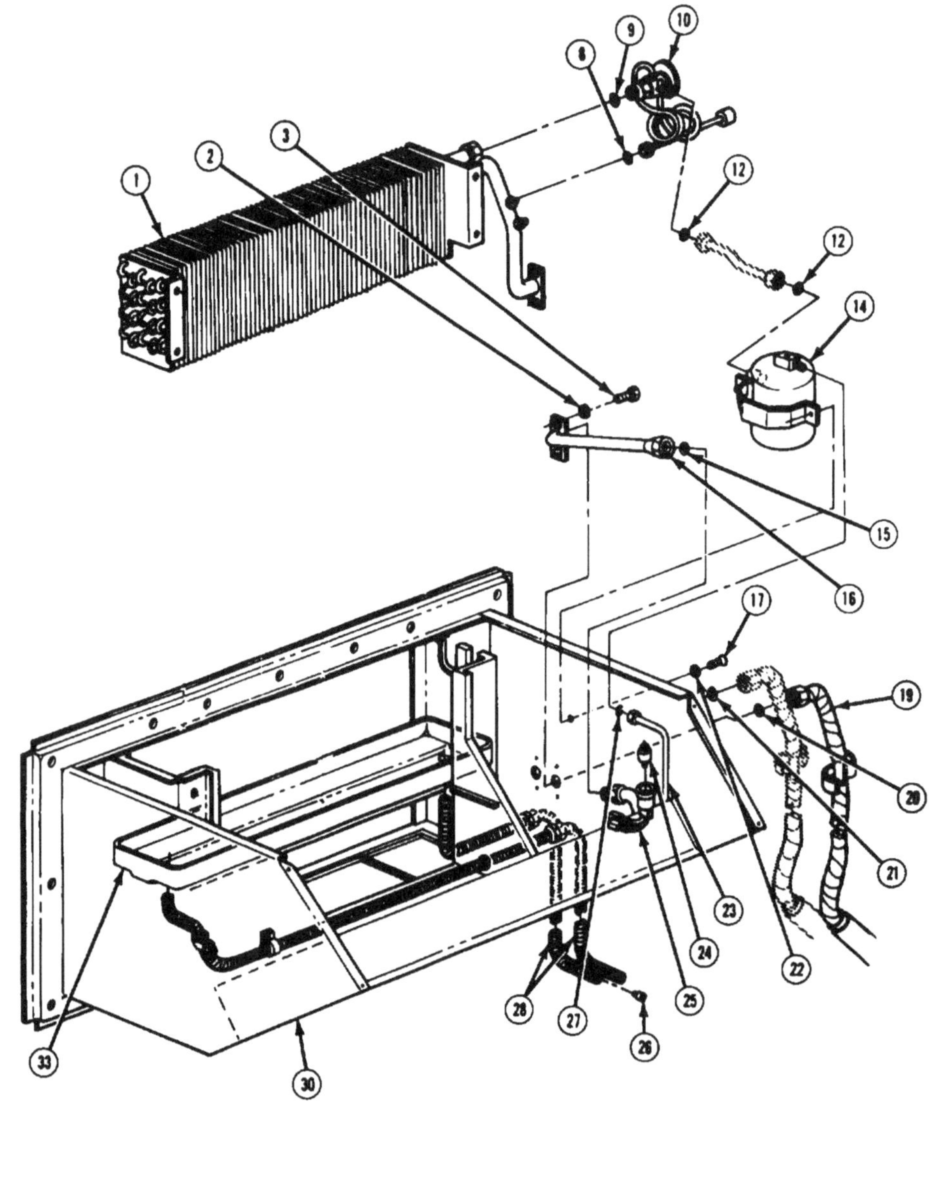

TA50469

10-27. AIR CONDITIONER EXPANSION VALVE, EVAPORATOR, AND RECEIVER-DEHYDRATOR REPLACEMENT (M1010) (Con't).

CAUTION

Be careful not to bend evaporator (1) fins during installation,

7. Wrap evaporator line with thermal insulation tape, leaving threaded connectors clear if removed. Install evaporator (1) on air conditioner assembly (30) with 4 screws (31) and washers (32). Install evaporator insulation and cover with 12 screws. Seal evaporator casing grooves (29) with thermal insulation tape.
8. Install core (6) and low pressure switch (7) on evaporator (1) if removed. Connect leads.
9. Install 2 washers (4) and screws (5) on evaporator (1), Install new preformed packing insert (21) in outlet line (18) fitting. Connect outlet line to evaporator.
10. Secure temperature sensing bulb (11) to evaporator (1) outlet line with new thermal insulation tape.
11. Install 2 new preformed packing inserts (12) in outlet line (13) fittings. Install outlet line.
12. Install new preformed packing insert (8) and new "O" ring (9) in expansion valve (10) fittings. Connect expansion valve to evaporator (1).

10-27. AIR CONDITIONER EXPANSION VALVE, EVAPORATOR, AN D RECEIVER-DEHYDRATOR REPLACEMENT (M1010) (Con't).

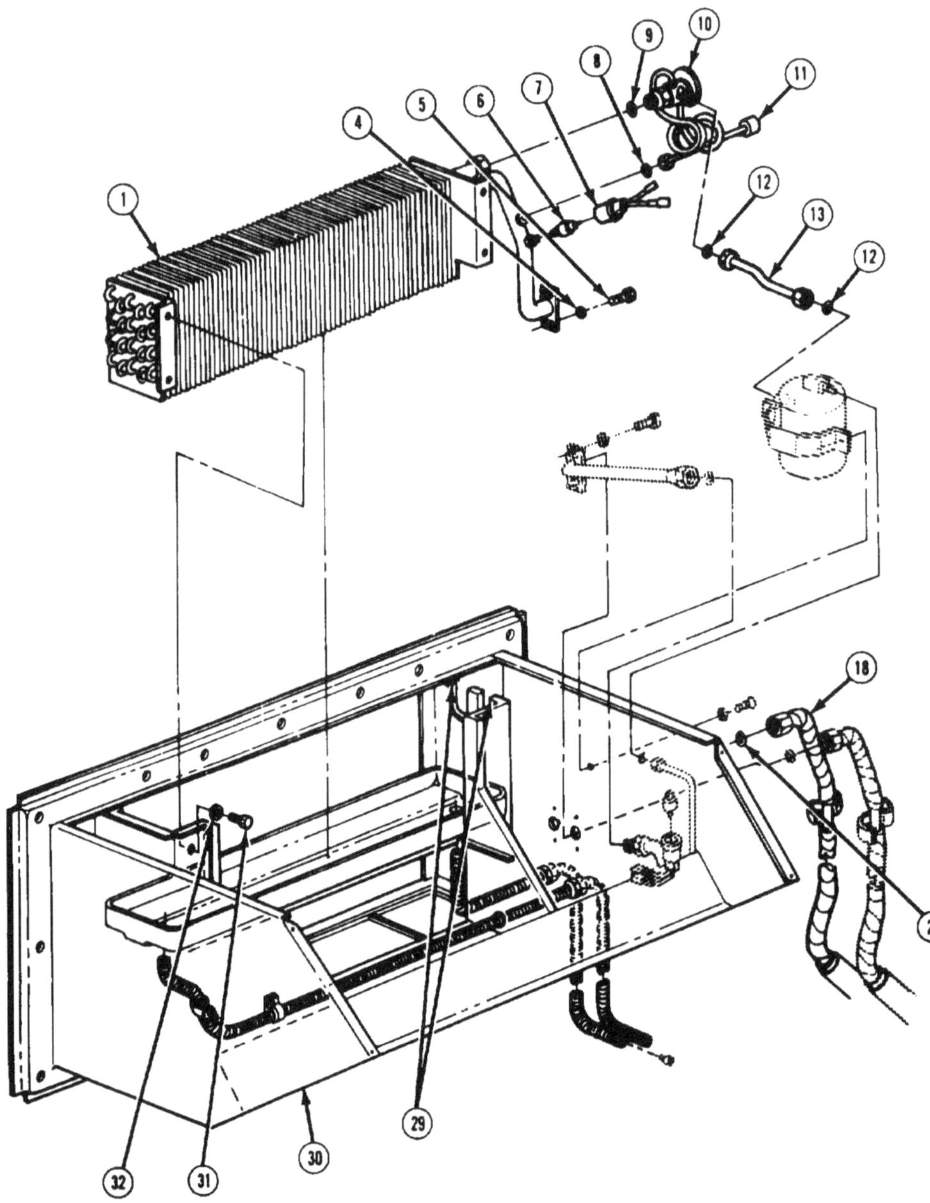

TA50470

10-27. AIR CONDITIONER EXPANSION VALVE, EVAPORATOR, AN D RECEIVER-DEHYDRATOR REPLACEMENT (M1010) (Con't).

FOLLOW-ON TASKS:

- Install air conditioner cover assembly. (See TM 9-2320-289-20)
- Evacuate and charge air conditioning system. (See paragraph 10-21)
- Check operation of air conditioning system and inspect for leakage.

CHAPTER 11
SPECIAL PURPOSE KITS MAINTENANCE

11-1. SPECIAL PURPOSE KITS MAINTENANCE INDEX.

11-2. ENGINE COOLANT HEATER REPAIR.

This task covers: a. Disassembly b. Cleaning and Inspection c. Assembly d, Flame Switch Adjustment

INITIAL SETUP:

Equipment Condition
- Engine coolant heater removed. (See TM 9-2320-289-20)

Tools/Test Equipment
- Soldering gun

Materials/Parts
- One gasket
- One lockwasher
- One seal
- Two "O" rings
- Lead solder (Item 62, Appendix B)
- Antiseize tape (Item 65, Appendix B)
- Insulating varnish (Item 71-, Appendix B)

a. DISASSEMBLY

NOTE

Ensure that all terminals and leads are tagged for assembly.

1. Turn spring loaded screws securing cover 1/4 turn and remove cover.
2. Remove 4 screws (5) securing electrical receptacle (11) to bracket. Remove screws and leads at terminal block (6).
3. Disconnect all leads at relay assembly (17), Unsolder pink/black lead at relay assembly. Remove clamp (13), mounting bracket, 2 screws, and relay assembly.
4. Disconnect leads at timer (1 5), Remove nut, bolt (14), lockwasher, timer, backing plate, and receptacle assembly (11), Discard lockwasher,
5. Remove remaining screw and terminal block (6) from case.
6. Disconnect leads from flame switch (7). Loosen nut on bottom of flame switch and pull flame switch straight up to remove.
7. Loosen 2 clamps (3) and remove fuel hose (4).
8. Disconnect leads from limit switch (8). Remove 2 screws and limit switch.
9. Remove screw from thermostat valve bracket and remove thermostat valve (18). Remove 4 screws, fuel control bracket (2), and fuel control (1). Remove screw and remove fuel control from bracket if damaged,
10. Disconnect lead from igniter (16), Remove igniter and gasket, Discard gasket. Remove water coolant fittings (9) if damaged and discard.
11. Remove 8 screws and air inlet cover (20) from case.
12. Remove 4 screws (22) securing 2 halves of case together.

11-2. ENGINE COOLANT HEATER REPAIR (Con't).

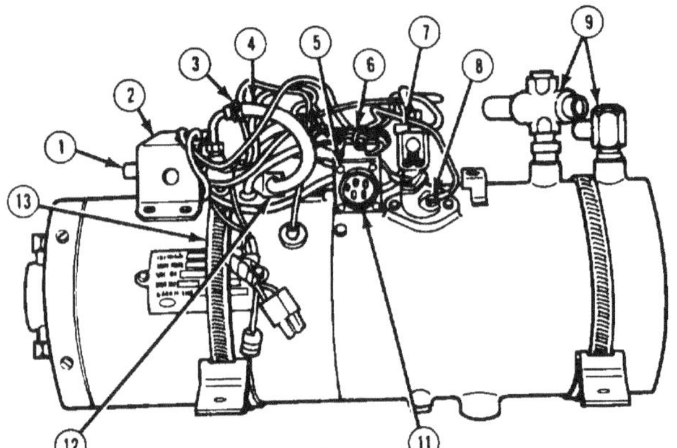

Left Side

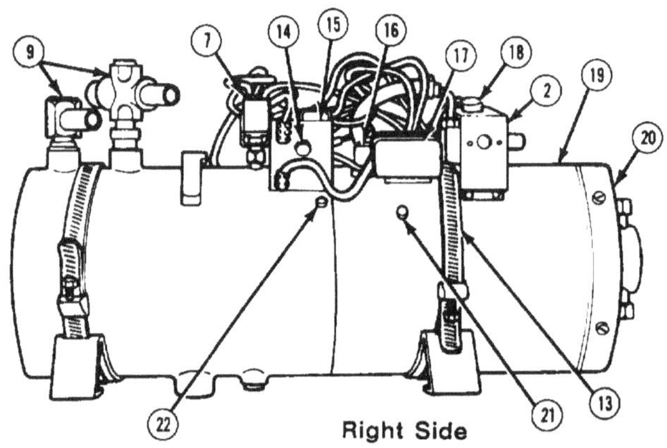

Right Side

13. Remove fuel fitting (12) and elbow from top of burner assembly (23), Remove 3 screws (21) holding burner case (19) to burner assembly.
14. Remove screw and burner case (19).

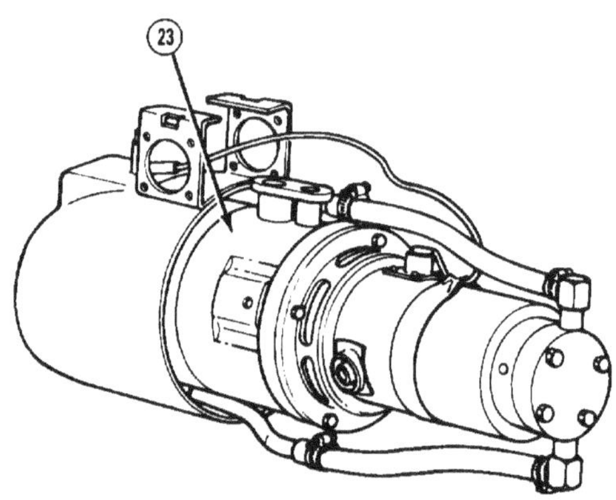

TA50362

11-2. ENGINE COOLANT HEATER REPAIR (Con't).

15. Disconnect coolant hoses (27) at pipes (28) on heat exchanger (29). Remove motor (24) and burner assembly (23) as a unit.

NOTE

Ensure that all parts of motor (24) and burner assembly (23) are marked/scribed for assembly.

16. Remove 4 screws (25) securing port plate (26) to motor (24). Remove port plate and "O" ring. Discard "O" ring.

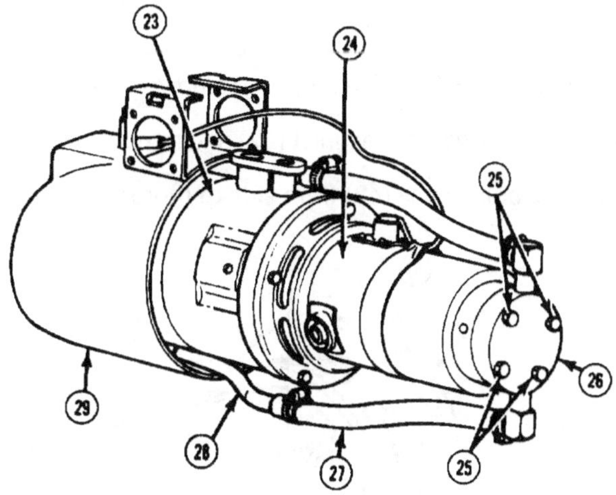

17. Remove cam ring (31) and remove vanes (32) from rotor (33). Loosen setscrew (34) and remove rotor.

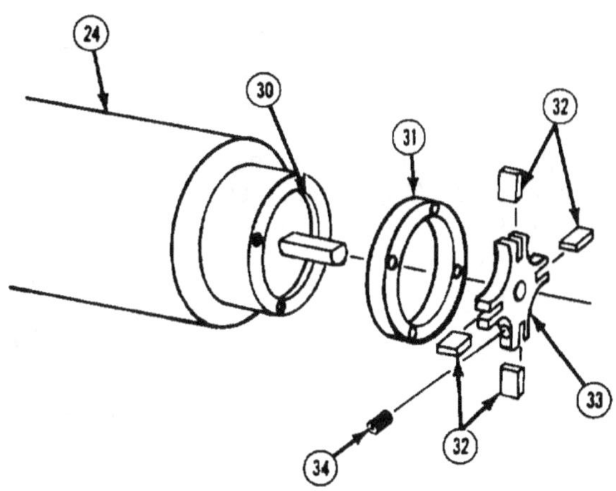

TA50363

11-2. ENGINE COOLANT HEATER REPAIR (Con't).

NOTE

"O" ring (30) may have been removed with cam ring (31).

18. Remove adapter (35) and "O" ring (30). Remove seal (41), washer (40), and spring (39). Discard "O" ring and seal.
19. Remove 4 screws (38) and air inlet (37) from burner assembly (23). Remove motor (24) and air inlet from burner assembly,
20. Loosen setscrew from fan (36) and remove fan and air inlet plate.
21. Remove 2 screws from air inlet (37) and remove air inlet from motor (24).
22. Remove 2 brush caps (42) from side of motor (24) and remove brush and spring assemblies.

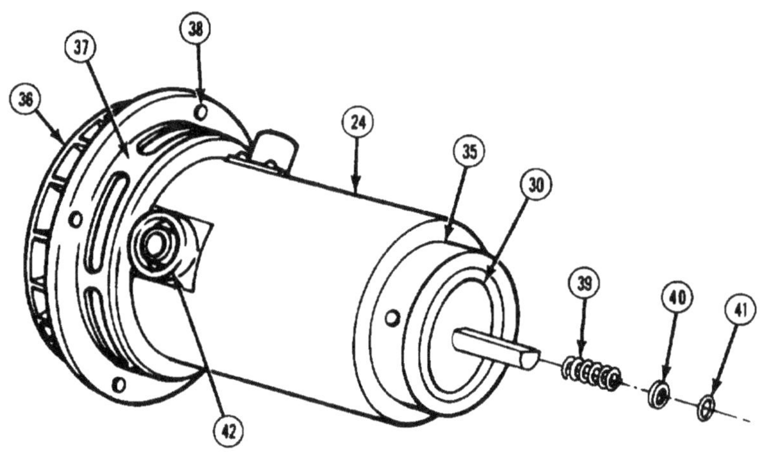

23. Remove 2 screws, vaporizer retainer (43), and vaporizer (44) from burner assembly (23).

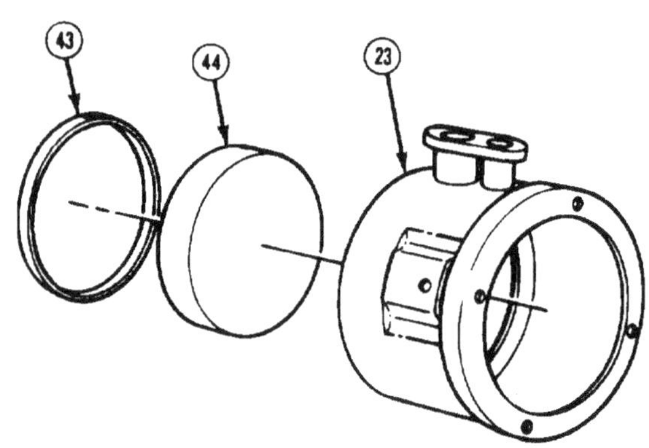

TA50364

11-2. ENGINE COOLANT HEATER REPAIR (Con't).

b. CLEANING AND INSPECTION

1. Clean all carbon from burner assembly. Ensure that all holes, ports, and passages are clear of obstructions. Clean any accumulation of dirt from case.
2. Inspect flame switch tube for bends or distortion. Metal rod in switch tube must extend at least 1/16 in. (1.59 mm) above top of tube. Replace switch if any of these conditions exist.
3. Inspect all leads for damage. Inspect connectors for looseness or corrosion. Replace if any of these conditions exist.
4. Inspect burner assembly and heat exchanger for damage or burn through. Replace if damaged or burnt through.
5. Inspect vaporizer for damage. Replace if damaged.
6. Inspect fan for damage. Replace if damaged.
7. Inspect brush and spring assemblies for distorted springs. If brushes are less than 1/4 in. (6.35 mm) long, replace brushes.
8. Inspect coolant motor parts for damage. Replace any parts if damaged.
9. Inspect coolant hoses for cracks or deterioration. Replace if any of these conditions exist.
10. Inspect fittings on fuel control for damage. Replace if damaged.

c. ASSEMBLY

1. Install vaporizer (44), vaporizer retainer (43), and 2 screws.

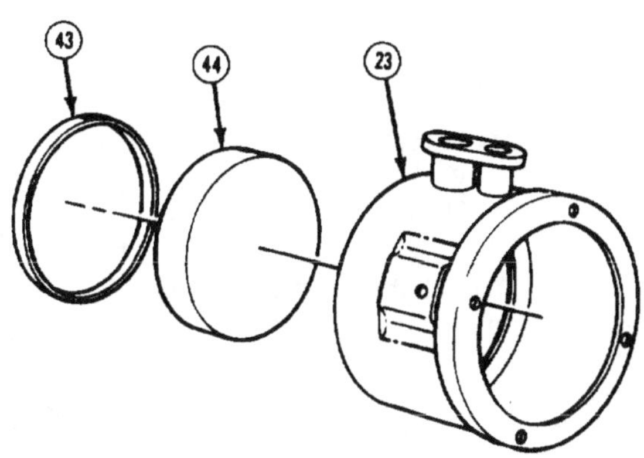

TA50365

11-2. ENGINE COOLANT HEATER REPAIR (Con't).

2. Insert spring and brush assemblies in motor (24), and install brush caps (42).

3. Insert air inlet (37) on motor (24) and install 2 screws.

4. Install air inlet plate with lip down, Install fan (36) and setscrew. Install air inlet (37) and motor (24) to burner assembly (23), and secure with 4 screws (38).

5. Install spring (39), washer (40), and new seal (41) with flat side to motor (24), Install adapter (35) .

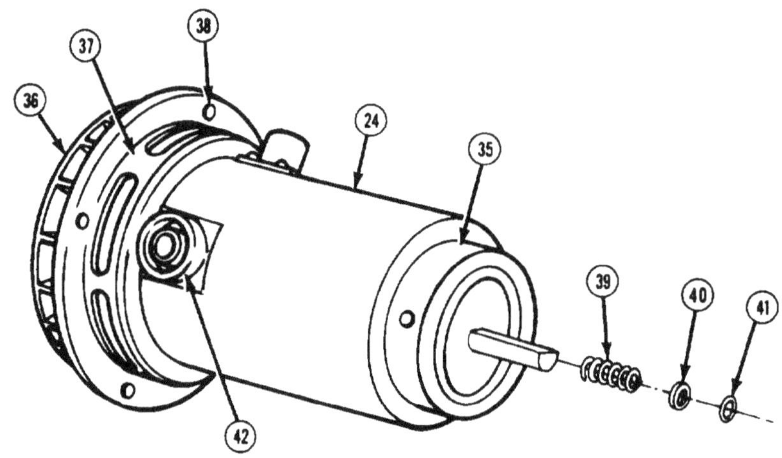

NOTE

H may be necessary to tap rotor (33) with a soft-faced tool to obtain required clearance.

6 . Install rotor (33) and lightly tighten setscrew on side of rotor. Depress rotor and adapter (35), and install cam ring (31). Check to ensure that clearance between outer face of cam ring and outer face of rotor is at least 0.002 in. (0.051 mm). While maintaining pressure on rotor, lift cam ring and tighten setscrew.

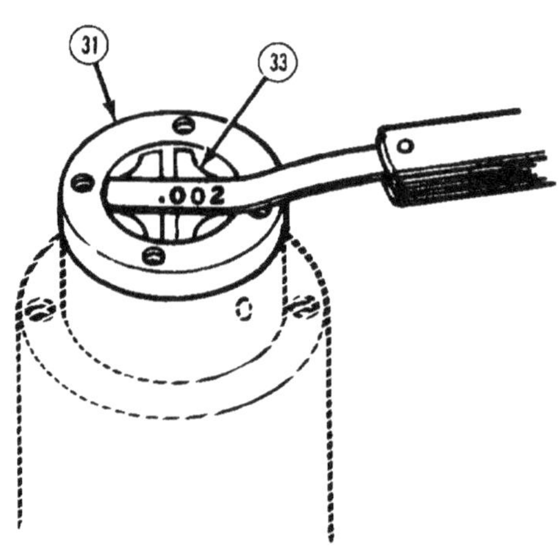

TA50366

11-2. ENGINE COOLANT HEATER REPAIR (Con't).

7. With motor (24) standing on end, install vanes (32) in rotor (33) with notches horizontal. Install new "O" ring (30) and cam ring (31) to adapter (35).

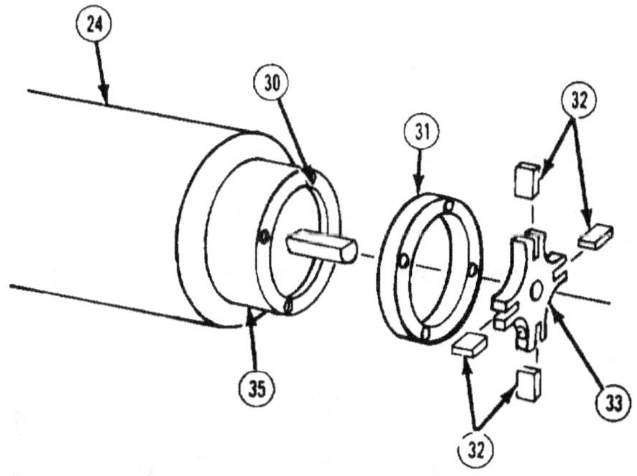

8. Install new "O" ring and port plate (26), alining mark/scribe, and install 4 screws (25).

9. Install motor (24) and burner unit on heat exchanger (29) and connect coolant hoses (27). Feed motor lead through grommet in burner case (19).

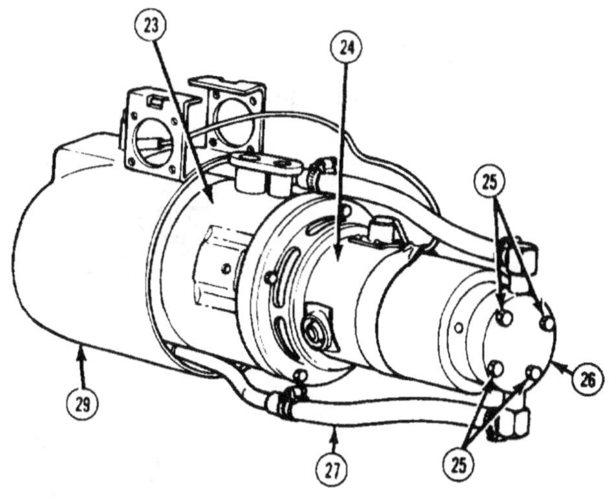

10. Install burner case (19) over motor (24) and burner unit, and secure with 4 screws (21). Install 4 screws (22) to 2 halves of case.

11. Coat threads with antiseize tape and install fuel fitting (12) and elbow in top of burner assembly (23) with nipple facing toward grommet.

12. Install air inlet cover (20) and secure with 8 screws. Install limit switch (8) with 2 screws.

TA50367

11-2. ENGINE COOLANT HEATER REPAIR (Con't).

13. Install new gasket to igniter (16) and install igniter. Connect lead to igniter. Install new water coolant fittings (9) if removed,

14. Position fuel control (1) in fuel control bracket (2) if removed and install screw. Install fuel control bracket on case with 4 screws. Position thermostat valve (2) to thermostat bracket and install screw.

15. Connect lead from thermostat valve (18) and lead from igniter to limit switch (8).

16. Install fuel hose (4) and tighten 2 clamps (3). Install flame switch (7) and tighten nut.

17. Position terminal block (6) on case with 2 screws. Install relay assembly (17) and 2 screws.

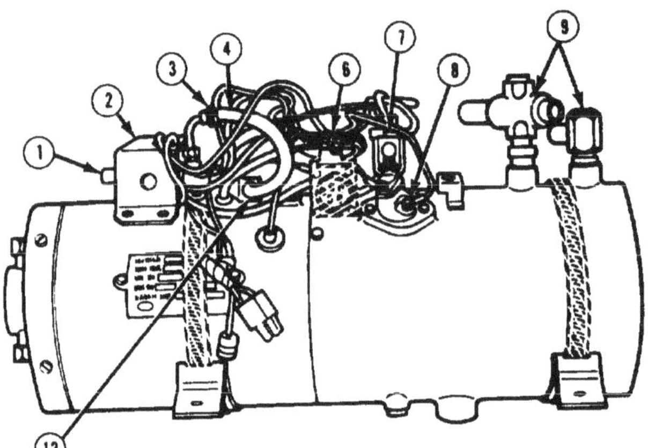

Left Side

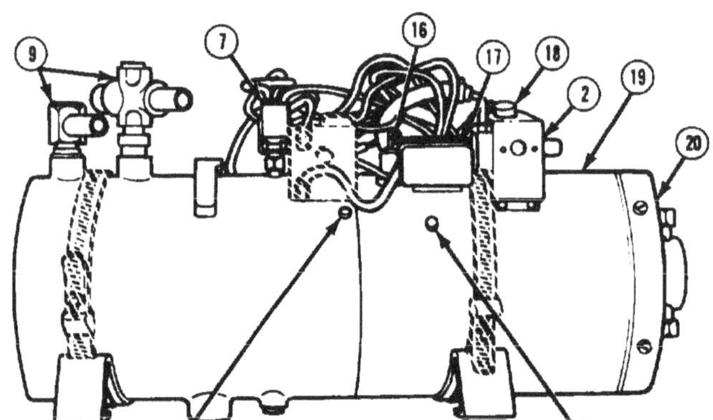

Right Side

TA50368

11-2. ENGINE COOLANT HEATER REPAIR (Con't).

18. Install timer (15), backing plate, nut, bolt (14), and new lockwasher. Install electrical receptacle (11) and secure with 4 screws (5). Install lead and terminal block retaining screw.
19. Connect all remaining electrical leads. Solder pink/black lead at tagged terminal on relay assembly (17).
20. Install clamp (13) and mounting bracket.
21. Adjust flame switch (7). (See FLAME SWITCH ADJUSTMENT)
22. Install cover and tighten screws.

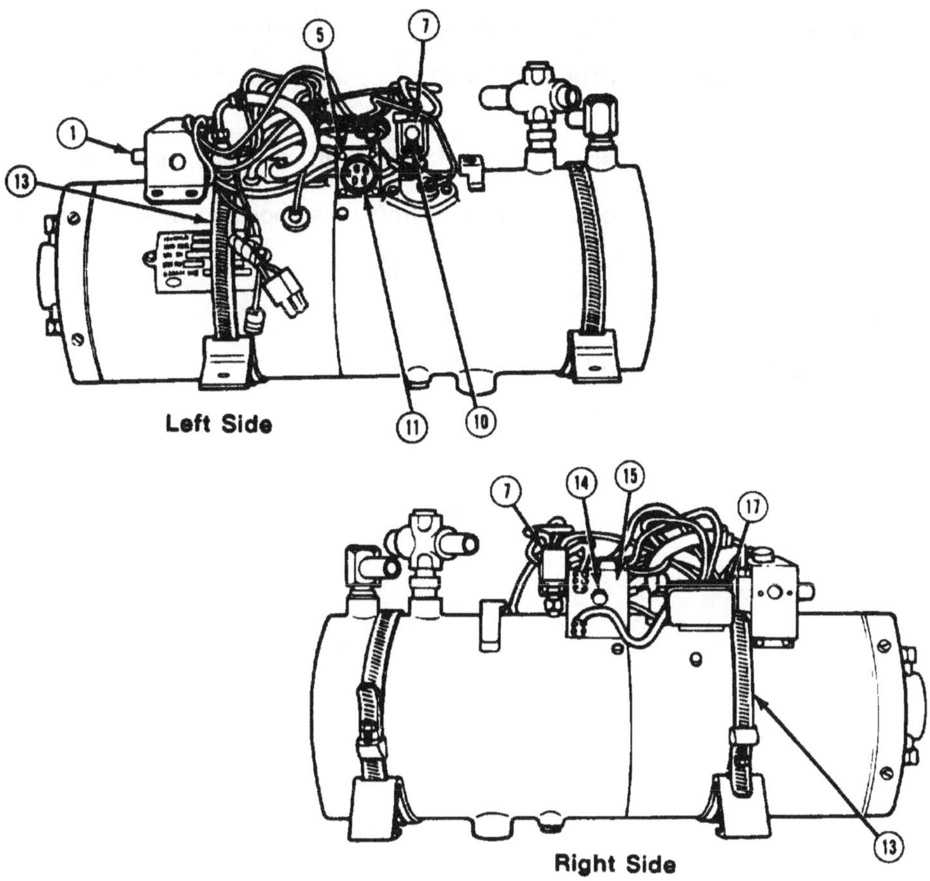

TA50369

11-2. ENGINE COOLANT HEATER REPAIR (Con't).

d . FLAME SWITCH ADJUSTMENT

NOTE

- If new flame switch (7) has been installed. no adjustment is required.
- if adjustment nut (10) has been disturbed, perform steps 1-4.
- Ensure that heater is off and at ambient temperature prior to flame switch (7) adjustment,

1. Turn spring loaded screws securing cover 1/4 turn and remove cover.
2. Turn nut (10) on adjustment screw on switch bracket counterclockwise until blower fan comes on.
3. Very slowly, turn nut (10) clockwise until blower fan turns off, Note position of nut.
4. Turn nut (10) clockwise an additional 180 degrees from noted position, Seal nut with insulating varnish.

FOLLOW-ON TASKS:

- Install engine coolant heater. (See TM 9-2320-289-20)

11-3. WARM AIR HEATER REPAIR.

This task covers:

a. Disassembly
b. Cleaning and Inspection
c. Assembly
d. Flame Detector Switch Adjustment

INITIAL SETUP:

Equipment Condition

- Warm air heater removed. (See TM 9-2320-289-20)

Materials/Parts

- One gasket
- One preformed packing ring
- Insulating varnish (Item 71, Appendix B)

a. DISASSEMBLY

NOTE

Ensure that all terminals and leads are tagged for assembly.

1. Turn 2 fasteners (4) and remove cover (2) from heater housing (3). Disconnect all leads.

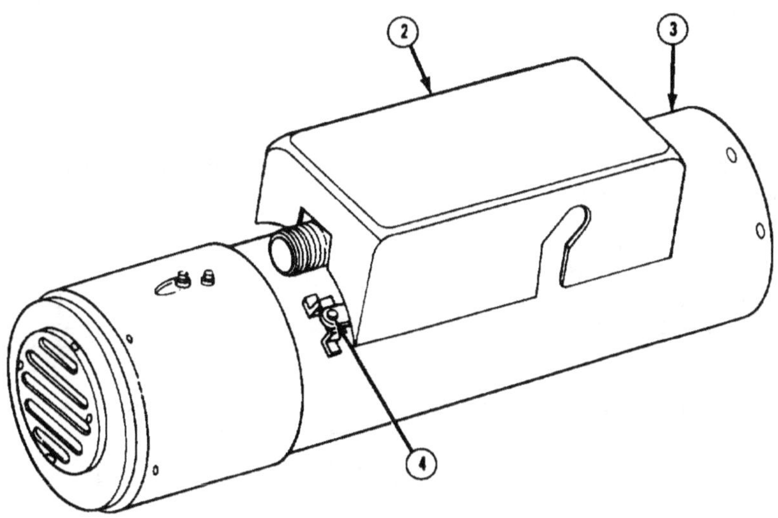

2. Remove harness (7) from electrical receptacle (5). Remove 4 screws (6) and electrical receptacle.
3. Remove 4 screws (9), 1 nut (13), 1 starwasher (14), and ignition control (8) from heater housing (3). Pull leads through clip.
4. Disconnect fuel line nut (11) from outlet of fuel control valve (10). Remove screws and fuel control valve from heater housing (3).
5. Remove screws and overheat switch (15) from heater housing (3).

TA50370

11-3. WARM AIR HEATER REPAIR (Con't).

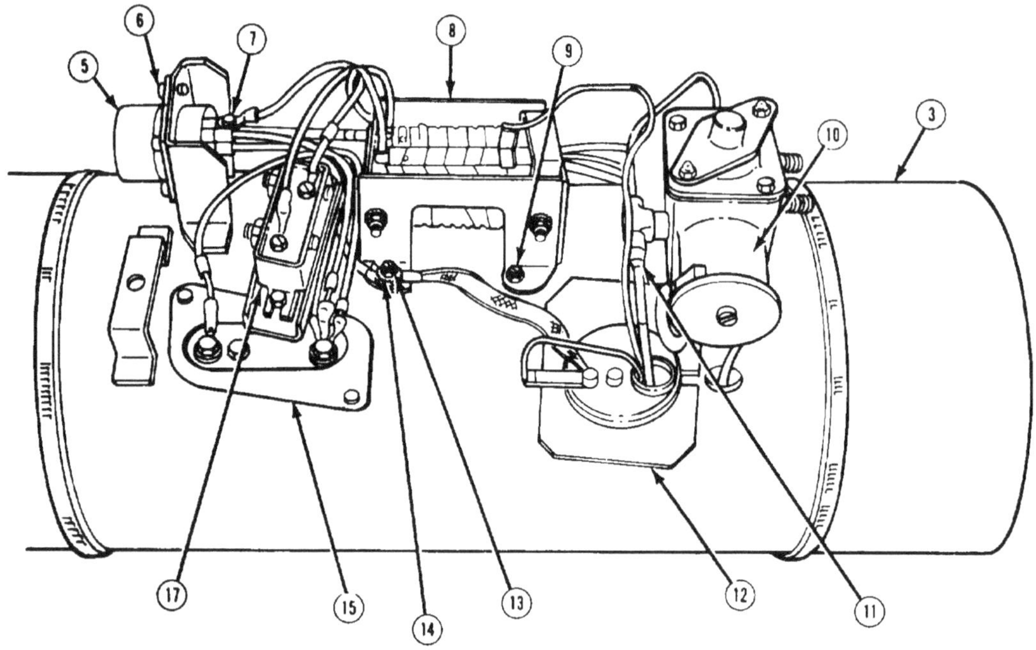

6. Unscrew nut at base of flame detector switch (17) until it clears threads and carefully remove switch from heat exchanger housing.

7. Loosen 4 screws (18) around edge of blower assembly (1). Turn blower assembly counterclockwise to free from screws and remove blower assembly and lead.

8. Disconnect igniter ground lead from hatch cover (12). Turn fasteners and remove hatch cover.

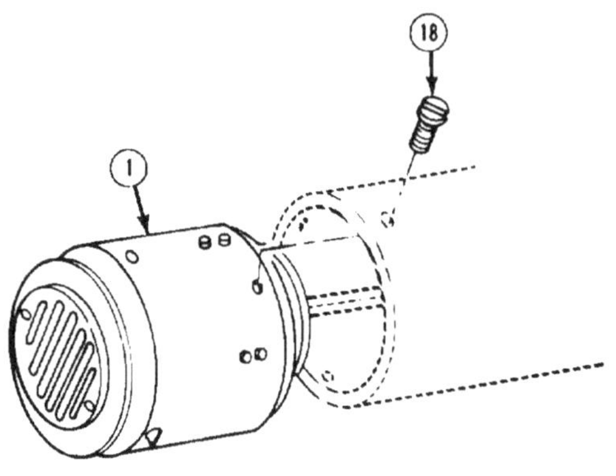

TA50371

11-3. WARM AIR HEATER REPAIR (Con't).

9 . Twist igniter assembly (22) to disengage lugs from slots and remove from igniter tube (20).

10. Remove screw (21) and igniter tube (20).

11. Release fuel line nut from burner bushing (26) and remove secondary blower housing (19). Remove heater housing (3) ground lead,

NOTE

Ensure that position of burner assembly (23) is marked for assembly,

12. Remove nuts, clamps, and hook bolts holding burner assembly (23) to heat exchanger housing.

13. Remove and discard gasket (25) and preformed packing ring (24).

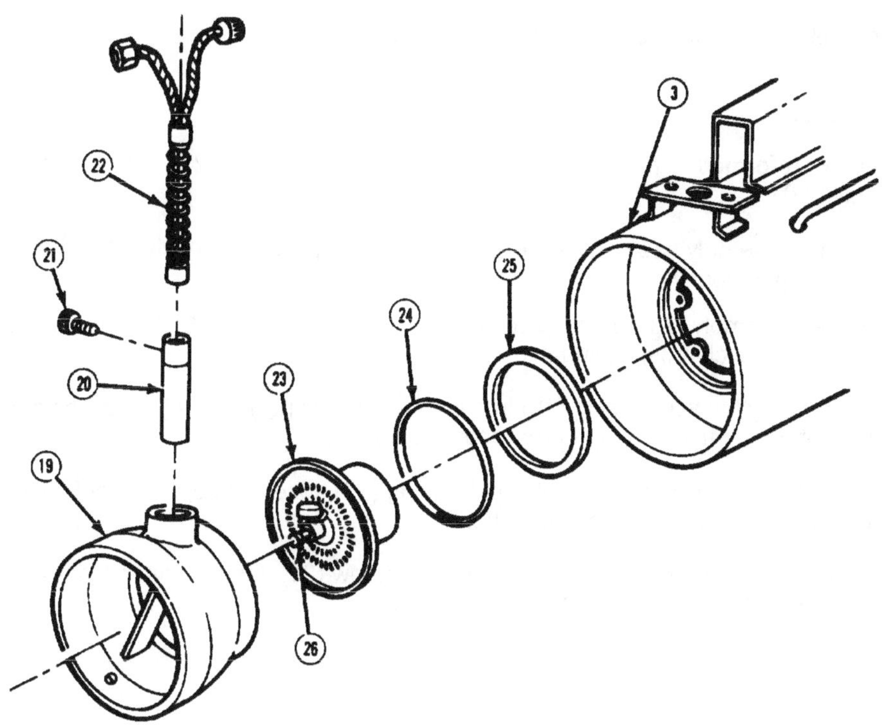

b. CLEANING AND INSPECTION

1 . Clean all mating surfaces.

2 . Clean all carbon from burner assembly. Inspect all holes, ports, and passageways, and remove any obstructions.

TA50372

11-3. WARM AIR HEATER REPAIR (Con't).

3. Clean all carbon from igniter.
4. Clean any accumulation of dirt from case.
5. Inspect all leads for damage, Inspect connectors for looseness or corrosion. Replace if any of these conditions exist.
6. Inspect rod located under flame detector switch. If rod is broken, replace flame detector switch.
7. Inspect burner assembly and heat exchanger for damage or burn through. Replace if damaged or burnt through.
8. Inspect blower assembly for damage. Replace if damaged.
9. Inspect fuel control valve for damage. Replace if damaged.

c. ASSEMBLY

1. Install new preformed packing ring (24) and new gasket (25) over burner end of burner assembly (23).
2. Place burner assembly (23) in heat exchanger housing, and loosely install hook bolts, clamps, and nuts into position.
3. Aline burner assembly (23) to large access hole in heater housing (3). Install ground lead and tighten hook bolts, clamps, and nuts.
4. Install secondary blower housing (19) in burner assembly (23), and aline access opening with large hole of heater housing (3). Firmly seat in place.
5. Install fuel line through secondary blower housing (19) access tube, Connect fuel line nut onto burner bushing (26) and tighten connection.
6. Install igniter tube (20) into igniter chamber with screw (21). Install igniter assembly (22) and press against spring to engage bayonet fastener.

11-3. WARM AIR HEATER REPAIR (Con't),

7. Install hatch cover (12) and secure fasteners. Connect igniter ground lead to hatch cover.

8. Thread blower lead through hole in heater housing (3), Install blower assembly (1) into blower housing and slide slots under 4 screws (18). Tighten screws.

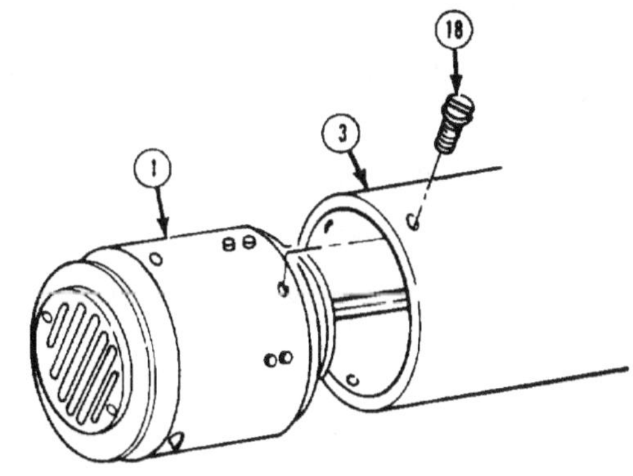

9. Install overheat switch (15) and secure with screws.

10. Install flame detector switch (17) into heat exchanger housing with nut and sleeve. Position adjuster nut (16) above overheat switch.

11. Adjust flame detector switch (17) . (See FLAME DETECTOR SWITCH ADJUSTMENT)

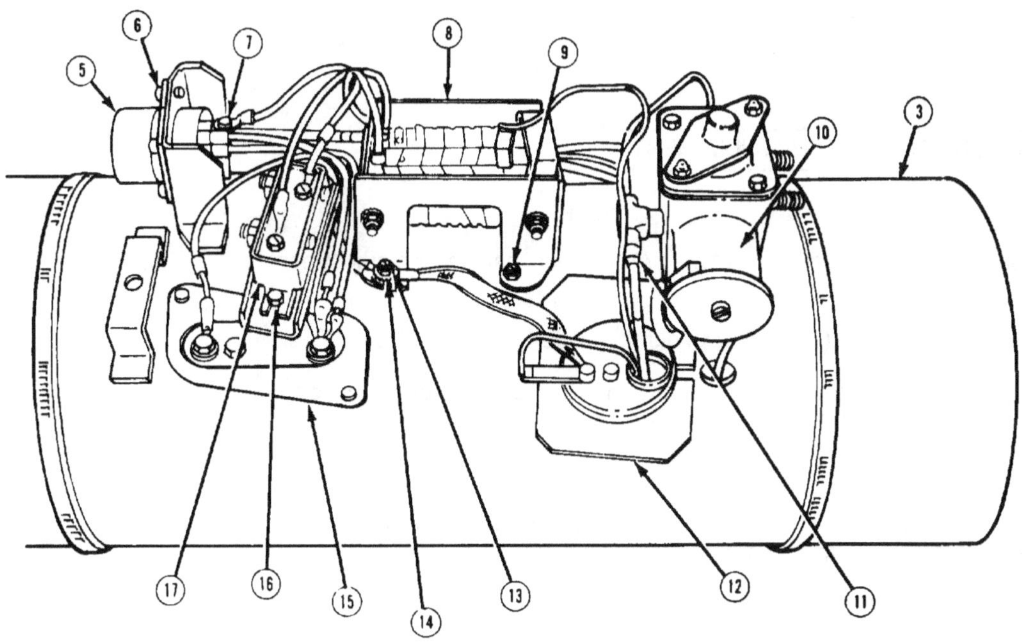

TA50373

11-3. WARM AIR HEATER REPAIR (Con't).

12. Install fuel control valve (10) and screws to heater housing (3), Connect fuel line nut (11) to outlet of fuel control valve.
13. Install ignition control (8) to heater housing (3) using 4 screws (9), 1 nut (13), and 1 starwasher (14),
14. Install electrical receptacle (5) and 4 screws (6) to bracket. Connect all leads.
15. Connect harness (7) to electrical receptacle (5). Install cover (2) to heater housing (3) and secure 2 fasteners (4).

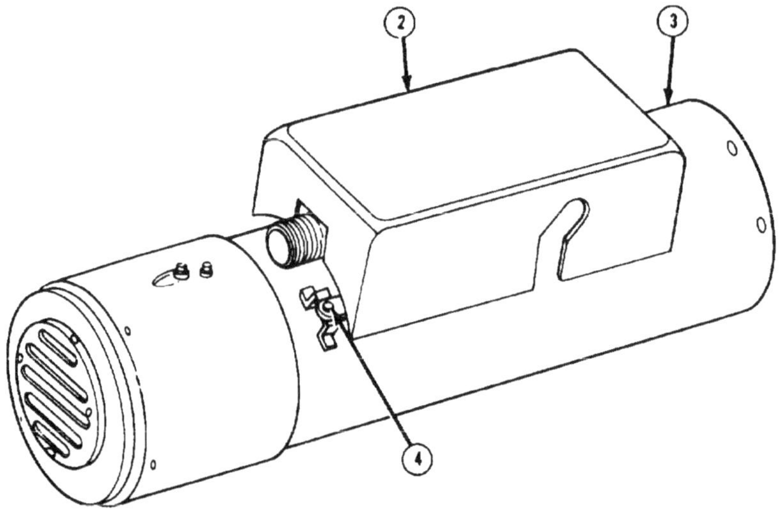

d. FLAME DETECTOR SWITCH ADJUSTMENT

NOTE

- If new flame detector switch (17) has been installed, no adjustment is required.
- If adjuster nut (16) has been disturbed, perform steps 1-3.
- Ensure that heater is off and at ambient temperature prior to flame detector switch (17) adjustment.

1. Back off adjuster nut (16) two turns.
2. Turn adjuster nut (16) inward slowly just until switch "clicks." Then turn adjuster nut an additional 3/4 in. (19 mm) turn from "click" position.
3. Apply insulating varnish to adjuster nut (16) to seal.

FOLLOW-ON TASKS:

- Install warm air heater. (See TM 9-2320-289-20)
- Check operation of heater.

TA50374

APPENDIX A
REFERENCES

A-1. SCOPE.

This appendix lists all forms, field manuals, technical manuals, and other publications required for use with this manual.

A-2. INDEXES.

The following indexes should be frequently consulted for latest changes to, or revisions of, references given in this appendix and for new publications or instructions relating to materiel covered in this manual.

a. Military Publications:

Consolidated Index of Army Publications and Blank Forms . . .	DA Pam 25-30
Equipment Improvement Report and Maintenance Digest	TB 43-0001-39
The Standard Army Publications System Users Index , ...	DA Pam 310-10

b. General References:

Catalog of Abbreviations and Brevity Codes	AR 310-50
Dictionary of United States Army Terms	AR 310-25
Military Symbols	FM 101-5-1

A-3. SUPPLY CATALOGS .

The following Department of the Army Supply Catalogs pertain to this manual:

Shop Equipment, Automotive Maintenance and Repair: Field Maintenance: Basic, Less Power (NSN 4910-00-754-0705)	SC 4910-95-CL-A31
Shop Equipment, Automotive Maintenance and Repair: Field Maintenance: Supplemental No. 1, Less Power (NSN 4910-00-754-0706)	SC 4910-95-CL-A62
Shop Equipment, Automotive Maintenance and Repair: Organizational Maintenance: Common No. 1, Less Power (NSN 4910-00-754-0654)	SC 4910-95-CL-A74
Shop Equipment, Automotive Maintenance and Repair: Organizational Maintenance: Supplemental No. 1, Less Power (NSN 4910-00-754-0653)	SC 4910-95-CL-A73
Shop Equipment, Automotive Maintenance and Repair: Organizational Maintenance: Common No. 2, Less Power (NSN 4910-00-754 -0650)	SC 4910-95-CL-A72

Tool Kit, General Mechanic's: Automotive
(NSN 5180 -00-177 -7033) . . . . SC 4910-95-N26
Tool Kit, Welder's (NSN 5180-00-754-0661) . . . . SC 5180-90-CL-N39

A-4. FORMS.

Refer to DA Pam 25-30, *Consolidated Index of Army Publications and Blank Forms,* for a current and complete list of blank forms. Refer to DA Pam 738-750, *The Army Maintenance Management System (TAMMS),* for instructions on the use of maintenance forms pertaining to this materiel.

DA Form 285 . . . . US Army Accident Investigation Report
DA Form 348 . . . . Equipment Operator's Qualification Record (Except Aircraft)
DA Form 2028 . . . . Recommended Changes to Publications and Blank Forms
DA Form 2401 . . . . Organization Control Record for Equipment
DA Form 2402 . . . . Exchange Tag
DA Form 2404 . . . . Equipment Inspection and Maintenance Worksheet
DA Form 2405 . . . . Maintenance Request Register
DA Form 2406 . . . . Materiel Condition Status Report
DA Form 2407 . . . . Maintenance Request
DA Form 2407-1 . . . . Maintenance Request -Continuation Sheet
DA Form 2408 . . . . Equipment Log Assembly (Records)
DA Form 2408-9 . . . . Equipment Control Record

DA Form 2409 . . . . Equipment Maintenance Log (Consolidated)
DDForm 314 . . . . Preventive Maintenance Schedule and Record
DDForm 518 . . . . Accident Identification Card
DDForm 1397 . . . . Processing and Reprocessing Record for Shipment, Storage, and issue of Vehicles and Spare Engines
Standard Form 46 . . . . . US Government Motor Vehicle Operator's Identification Card
Standard Form 91, ... Operator Report on Motor Vehicle Accidents
Standard Form 368 . . . . Quality Deficiency Report

A-5. OTHER PUBLICATIONS.

The following publications contain information pertinent to the major item materiel and associated equipment.

a. Truck:

Lubrication Order for Truck,
3/4 and 1 1/4 Ton, 4x4, CUCV Series . . . . LO 9-2320-289-12
Operator's Manual for Truck,
3/4 and 1 1/4 Ton, 4x4, CUCV Series . . . . TM 9-2320-289-1 O
Unit Maintenance Manual for Truck,
3/4 and 1 1/4 Ton, 4x4, CUCV Series . . . . TM 9-2320-289-20
Unit Maintenance Repair Parts and Special Tools List for Truck,
3/4 and 1 1/4 Ton, 4x4, CUCV Series . . . . TM 9-2320-289-20P

Intermediate DS/GS Maintenance Repair Parts and Special Tools List for Truck, 3/4 and 1 1/4 Ton, 4x4, CUCV Series..	TM 9-2320-289-34P
Warranty Procedures for Truck, 3/4 and 1 1/4 Ton, 4x4, CUCV Series	TB 9-2300-295-15/24

b. Camouflage:

Camouflage	FM 5-20
Camouflage Materials	TM 5-200
Color, Marking, and Camouflage Painting of Military Vehicles, Construction Equipment, and Materials Handling Equipment	TB 43-0209

c. Decontamination:

NBC Decontamination	FM 3-5
Chemical, Toxicological, and Missile Fuel Handlers Protective Clothing	TM 10-277
Operator's and Organizational Maintenance Manual (Including Repair Parts and Special Tools List) for Decontamination Apparatus, Portable, DS2, 1 1/4 Quart, ABC-M11 (NSN 4230-00-720-1618)	TM 3-4230-204-12&P

d. General:

Accident Reporting and Records	AR 385-40
Army Motor Transport Units and Operations	FM 55-30
Basic Cold Weather Manual	FM 31-70
Chemical Agent Resistant Aliphatic Polyurethane Coating	MIL-C-46168C
Cooling Systems: Tactical Vehicles	TM 750-254
Corrosion Prevention and Control: Including Rustproofing for Tactical Vehicles and Trailers	TB 43-0213
First Aid for Soldiers	FM 21-11
Functional Grouping Codes: Combat, Tactical, and Support Vehicles and Special Purpose Equipment	TB 750-93-1
General Repair for Clothing and Textiles	FM 10-16
Manual for the Wheeled Vehicle Driver	FM 21-305
Mountain Operations	FM 90-6
Northern Operations.	FM 31-71
Operation and Maintenance of Ordnance Materiel in Cold Weather (10°F to-65°F)	FM 9-207
Painting instructions for Field Use.	TM 43-0139
Painting Procedures and Marking for Vehicles, Construction Equipment and Material Handling Equipment	MIL-STD-193
Petroleum Supply Point Equipment and Operations	FM 10-69
Prevention of Motor Vehicle Accidents	AR 385-55
Principles of Automotive Vehicles	TM 9-8000
Procedures for Destruction of Tank-Automotive Equipment to Prevent Enemy Use	TM 750-244-6
Treatment and Painting of Materiel	MIL-T-704J

e. Maintenance and Repair:

Chemical Agent Alarm Maintenance	TM 3-6665-225-12
Description, Use, Bonding Techniques and Properties of Adhesives	TB ORD 1032
Inspection, Care, and Maintenance of Antifriction Bearings	TM 9-214
Mandatory Brake Hose Inspection and Replacement - Tactical Vehicles	TB 9-2300-405-14
Materials Used for Cleaning, Preserving, Abrading and Cementing Ordnance Materiel and Related Materials Including Chemicals	TM 9-247
Metal Body Repair and Related Operations	FM 43-2
Operator's and Organizational Maintenance Manual: Alarm, Chemical Agent, Automatic	TM 3-6665-261-14
Operator's and Organizational Maintenance Manual Including Repair Parts and Special Tools List for Simplified Test Equipment for Internal Combustion Engines (STE/ICE)	TM 9-4910-571-12&P
Operator's Manual for Welding Theory and Application	TM 9-237
Operator's, Organizational, Direct Support and General Support Maintenance Manual for Lead-Acid Storage Batteries	TM 9-6140-200-14
Organizational Care, Maintenance & Repair of Pneumatic Tires, inner Tubes and Radial Tires.	TM 9-2610-200-24
Safety Inspection and Testing of Lifting Devices	TB 43-0142
Standards and Criteria for Technical Inspection and Classification of Tires	TM 9-2610-201-14
Tactical Wheeled Vehicles: Repair of Frames	TB 9-2300-247-40
Use of Antifreeze Solutions and Cleaning Compounds in Engine Cooling Systems	TB 750-651

f. Shipment and Limited Storage:

Marking for Shipment and Storage	MIL-STD-129
Methods of Preservation	MIL-P-116
Packaging of Materiel	AR 700-15
Packaging of Materiel: Preservation	TM 38-230-1 and TM 38-230-2
Preparation for Shipment and Limited Storage of Wheeled Vehicles	MIL-V-62038
Preparation for Shipment and Storage of Basic Issue Items for Military Vehicles, Carriages and Equipment	MIL-B-12841
Railcar Loading Procedures	TM 55-601
Security of Tactical Wheeled Vehicles	TB 9-2300-422-20
Softwood Lumber	MM-L-751
Standards for Overseas Shipment or Domestic Issue of Special Purpose Vehicles, Combat, Tactical, Construction, and Selected Industrial and Troop Support US Army Tank-Automotive Readiness Command Managed Items	TB 9-2300-281-35

APPENDIX B
EXPENDABLE/DURABLE SUPPLIES AND MATERIALS LIST

Section 1. INTRODUCTIO N

B-1. SCOPE .

This appendix lists expendable/durable supplies and materials you will need to maintain the CUCV Series trucks. This listing is for informational purposes only and is not authority to requisition the listed items. These items are authorized to you by CTA 50-970, Expendable/Durable Items (Except Medical, Class V, Repair Parts, and Heraldic Items), or CTA 8-100, Army Medical Department Expendable/Durable Items.

B-2. EXPLANATION OF COLUMNS.

a. Column (1) - Item Number. This number is assigned to the entry in the listing and is referenced in the Initial Setup of applicable tasks under the heading of Materials/Parts.

b. Column (2) - Level. This column identifies the lowest level of maintenance that requires the listed item.

C - Operator/Crew
O - Unit Maintenance
F - Intermediate Direct Support Maintenance
H - Intermediate General Support Maintenance

c. Column (3) - National Stock Number. This is the National Stock Number assigned to the item; use it to request or requisition the item.

d. Column (4) - Description. Indicates the Federal item name and, if required, a description to identify the item. The last line for each item indicates the Federal Supply Code for Manufacturer (FSCM) in parentheses, followed by the part number, if applicable.

e. Column (5) - Unit of Measure (U/M). Indicates the measure used in performing the actual maintenance function. This measure is expressed by an alphabetical abbreviation (e. g., ea, in, pr, gal), If the unit of measure differs from the unit of issue, requisition the lowest unit of issue that will satisfy your requirements.

Section II. EXPENDABLE/DURABLE SUPPLIES AND MATERIALS LIST - Continued

(1) Item Number	(2) Level	(3) National Stock Number	(4) Description	(5) U/M
1.	F	6810-00-184-4796	ACETONE: (81348) Q-A-51	OZ
2.	0	5935-00-322-8959	ADAPTER: Connector (19207) 11677570	ea
2.1.	0	8040-00-024-6991	ADHESIVE: Interlor Rearview Mirror (11862) 1052369	ea
3.	0	9150-01-198-3829	ADDITIVE: Differential (11862) 1052358	0z
4.	0	8040-00-262-9028	ADHESIVE: General Purpose, Type 1 (19203)829899	pt
5.	F	8040-01-024-6991	ADHESIVE: Loctlte, Minute Bond S12 (11862) 1052369	pt
6.	F		ALCOHOL: Denatured (81348)o-E-00760	
		6810-00-205-6790	4 Ounce Bottle	0z
		6810-00-205-6786	1 Quart Bottle	qt
		6810-00-782-2686	1 Gallon Bottle	gal
7.	c		ANTIFREEZE: Arctic (81349) MIL-A-11755	
		6850-00-174-1806	55 Gallon Drum	gal
8.	c		ANTIFREEZE: Ethylene Glycol, Inhibited, Heavy-duty, Single Package (81349) MIL-A-46153	
		6850-00-181-7929	1 Gallon Can	gal
		6050-00-181-7933	5 Gallon Can	gal
		6850-00-181-7940	55 Gallon Drum	gal
9.	0		BARRIER MATERIAL: Greaseproof (81349) MIL-B-121	
		8135-00-171-0930	100 Yard Roll	yd
10.	0		BRAKE FLUID: Silicone, Automotive, All Weather, Operational and Preservative, (81349) MIL-B-46176	
		9150-01-102-9455	1 Gallon Can	gal
		9150-01-123-3152	5 Gallon Can	gal
		9150-01-072-8379	55 Gallon Drum	gal
11.	0	2590-00-398-6527	CABLE: 20 Feet, without Connectors (19207) 11682337-1	ea
12.	0	6150-01-022-6004	CABLE: with End Connectors (19207) 11682336-1	ea

Section II. EXPENDABLE/DURABLE SUPPLIES AN D MATERIALS LIST - Continued

(1) Item Number	(2) Level	(3) National Stock Number	(4) Description	(5) U/M
13.	F	4910-00-779-6651	CALIBRATION FLUID: (33267) J-26400-5	qt
14.	0		CHALK, MARKING: (81348) SS-C-225	
		7510-00-223-6701	1 Gross	gr
15.	F	6850-01-085-1423	CLEANING COMPOUND: Carburetor (56921) CARB-N-RLD	02
16.	c		CLEANING COMPOUND: Windshield Washer (81348) O-C-1901	
		6850-00-926-2275	1 Pint Bottle	pt
17.	F		CLOTH: Abrasive (81348) P-C-451	
		5350-00-187-6294	50 Yard Roll	yd
18.	F		CLOTH: Abrasive, Crocus (81348) P-C-456	
		5350-00-221-0872	50 sheets	sh
19.	0		COATING: Aliphatic Polyurethane, Chemical Agent Resistant, Forest Green (81 349) MIL-C-46168C	
		8010-00-111-7937	1 Gallon Can	gal
		8010-00-111-8010	5 Gallon Can	gal
20.	0	2510-00-567-0128	CONNECTOR PLUG: (19207) 11682338	ea
21.	0		CORROSION PREVENTIVE COMPOUND: (81349) MIL-C-11796	
		8030-00-231-2354	5 Pound Carton	lb
22.	c		DETERGENT: General Purpose, Liquid (81349) MIL-D-16791	
		7930-00-282-9699	1 Gallon Can	gal
23.	c		DRY CLEANING SOLVENT: Type II (81348) P-D-680	
		6850-00-110-4498	1 Pint Can	pt
		6850-00-274-5421	5 Gallon Can	gal
		6850-00-285-8011	55 Gallon Drum	gal

Section II. EXPENDABLE/DURABLE SUPPLIES AND MATERIALS LIST - Continued

(1) Item Number	(2) Level	(3) National Stock Number	(4) Description	(5) U/M
24.	F		DYE: Prussian Blue (81 349) MIL-P-30501	
		8010-00-889-9745	1 Ounce Tube	0z
25.	c		FUEL OIL DIESEL: Arctic, DF-A (81 348) VV-F-800	
		9140-00-286-5282	5 Gallon Can	gal
		9140-00-288-5284	55 Gallon Can	gal
26.	c		FUEL OIL DIESEL: Regular, DF-2 (81346) VV-F-800	
		9140-00-266-5295	5 Gallon Can	gal
		9140-00-266-5296	55 Gallon Can	gal
27.	c		FUEL OIL DIESEL: Winter, DF-1 (61348) VV-F-800	
		9140-00-286-5287	5 Gallon Can	gal
		9140-00-286-5288	55 Gallon Can	gal
28.	0	5920-01-123-5212	FUSE: 5 amp (11862) 12004005	ea
29.	0	5920-01-123-5211	FUSE: 10 amp (11662) 12004007	ea
30.	0	5920-01-149-6952	FUSE: 15 amp (11662) 12004008	ea
31.	0	5920-01-085-0825	FUSE: 20 amp (11862) 12004009	ea
32.	0	5920-01-149-6953	FUSE: 25 amp (11862) 12004010	ea
33.	0	5920-01-188-6294	FUSE: 30 amp (11862) 12004011	ea
34.	0	5210-01-222-8068	GAGE: Plastic Adjustment (11662) 14061396	ea
35.	F	5210-00-640-6177	GAGING PLASTIC: Green (77220) PG-1	ea

Section II. EXPENDABLE/DURABLE SUPPLIES AND MATERIALS LIST - Continued

(1) Item Number	(2) Level	(3) National Stock Number	(4) Description	(5) U/M
36.	c		GREASE: Automotive and Artillery [61349) MIL-G-10924	
		9150-00-935-1017	14 Ounce Cartridge	02
		9150-00-190-0904	1-314 Pound Can	lb
		9150-00-190-0905	6-1/2 Pound Can	lb
37.	0		GREASE: Molybdenum, Disulfide (61346) MIL-Q-21164	
		9150-00-935-4016	6 Ounce Tube	02
38.	F	4720-01-156-0547	HOSE: Rubber, 5/16 Inch I.D. (11662) 9439104	ft
39.	H	4720-01-156-0548	HOSE: Rubber, 3/8 1nchl. D, (11662) 9439162	ft
40.	c		HYDRAULIC FLUID: Transmission (2461 7) Dexron ® II	
		9150-00-698-2382	1 Quart Can	qt
		9150-00-657-4959	5 Gallon Can	gel
41.	c		INHIBITOR: Corrosion, Liquid Cooling Syetem (81349) MIL-A-53009	
		6850-00-160-3868	1 Quart Can	qt
42.	H	9150-01-041-6649	LUBRICANT: "O" Ring (71984) DC 111	02
43.	c		LUBRICATING OIL: Gear, Multipurpose, GO 75W (81349) MIL-L-2105	
		9150-01-035-5390	1 Quart Can	qt
		9150-01-035-5391	1 Qallon Can	gal
44.	c		LUBRICATING OIL: Gear, Multipurpose, GO BO/90 (81349) MIL-L-2105	
		9150-01-035-5392	1 Quart Can	qt
		9150-01-035-5393	1 Gallon Can	gal
45.	c		LUBRICATING OIL: General Purpose, Preservative, PL-S (61 348) V-VL-800	
		9150-00-231-6669	1 Quart Can	qt

Section II. EXPENDABLE/DURABLE SUPPLIES AN D MATERIALS LIST - Continue d

(1) Item Number	(2) Level	(3) National Stock Number	(4) Description	(5) U/M
46.	c		LUBRICATING OIL: Internal Combustion Engine, Arctic, OEA (81349) MIL-L-46167	
		9150-00-402-4478	1 Quart Can	q t
		9150-00-402-2372	5 Gallon Can	gal
		9150-00-491-7197	55 Gallon Drum	gal
47.	c		LUBRICATING OIL: Internal Combustion Engine, Tactical Service, OE/HDO 10 (81349) MIL-L-2104	
		9150-00-189-6727	1 Quart Can	q t
		9150-00-186-6668	5 Gallon Can	gal
		9150-00-191-2772	55 Gallon Drum	gal
48.	c		LUBRICATING OIL: Internal Combustion Engine, Tactical Service, OE/HDO 15W/40 (81349) MIL-L-2104	
		9150-01-152-4117	1 Quart Can	q t
		9150-01-152-4118	5 Gallon Can	gal
		9150-01-152-4119	55 Gallon Drum	gal
49.	c		LUBRICATING OIL: Internal Combustion Engine, Tactical Service, OE/HDO 30 (81349) MIL-L-2104	
		9150-00-186-6681	1 Quart Can	q t
		9150-00-188-9858	5 Gallon Can	gal
		9150-00-189-6729	55 Gallon Drum	gal
50.	F	6505-00-133-8060	PETROLATUM: white	oz
51.	0		PRESERVATIVE COMPOUND: (331 50) x975	
		8030-01-220-1442	4 Ounce with Spray	oz
		8030-01-220-1441	8 Ounce with Spray	oz
		8030-01-220-1440	32 Ounce Bottle	oz
52.	0		PRIMER: Epoxy Coating (81349) MIL-P-521928	
		8010-00-264-8866	Kit	ea
53.	c		RAG: Wiping, Cotton and Cotton-Synthetic (58536) A-A-531	
		7920-00-205-1711	50 Pound Bale	lb
54.	F	6830-00-935-9895	REFRIGERANT: R-12 (11862) 1051053	oz

Section II. EXPENDABLE/DURABLE SUPPLIES AN D MATERIALS LIST - Continue d

(1) Item Number	(2) Level	(3) National Stock Number	(4) Description	(5) U/M
55.	F	9150-01-160-026	REFRIGERANT OIL: (11862) 5416939	02
56.	H	5330-00-151-6659	SEALANT: Gasket Forming (77247) 98D	02
57.	0		SEALANT: Silicone, RTV (1 1862) 1052734	
		8030-01-159-4644	8-1/2 Ounce Tube	02
58.	F	8030-00-252-3391	SEALING COMPOUND: (61349) MIL-S-45180	02
59.	F	8030-01-156-6209	SEALING COMPOUND: Anaerobic (11862) 1052357	02
60.	0		SEALING COMPOUND: Corrosion-resistant, Type II (81349) MIL-S-81733	
		8030-00-009-5023	Kit	ea
61.	0		SEALING COMPOUND: Pipe, Anaerobic, with Teflon (05972) 592-31	
		8030-01-054-0740	50 Mililiter Tube	ml
62.	H	3439-00-247-6921	SOLDER: Lead (81348) QQ-S-571	in
63,	0		STRAP: Tie-down (06383) MS 3367-1-9	
		5975-00-074-2072	Box of 100	ea
64.	0	9905-00-537-8954	TAG: Marker, 50 Each (81349) MIL-T-12755	ea
65.	0		TAPE: Antiseize, 1/2 inch width (81349) MIL-T-27730A	
		8030-00-889-3535	260 Inch Roll	in
66.	0		TAPE: Duct, 21 nch width (07124) C-519	
		5640-00-103-2254	60 Yard Roll	yd
67.	F	5640-00-580-6276	TAPE: Insulation, Thermal (73030) MS 7495-618	f t

Section II. EXPENDABLE/DURABLE SUPPLIES AN D MATERIALS LIST - Continued

(1) Item dumber	(2) Level	(3) National Stock Number	(4) Description	(5) U / M
68.	0		TAPE: Pressure Sensitive Adhesive, Masking, Flat, 2 inch width (81349) MIL-T-2397	
		7510-00-473-9513	60 Yard Roll	yd
68.1	0		TAPE: Pressure Sensitive Adhesive, Red, Flat, 1/2 inch width (52170) 650	
		7510-00-550-7126	72 Yard Roll	yd
69.	0	4720-00-964-1433	TUBE: Plastic Drain, Type 1, Class A, Grade 4 (61346) D1248-60T	ft
70.	0		TWINE: Fibrous, Cotton (String), 16-Ply (81 348) T-T-871	
		4020-00-291-5901	375 Yard Spool	yd
71.	F	8010-00-298-3870	VARNISH: Insulating (81349) MIL-V-13811	qt

APPENDIX C
ILLUSTRATED LIST OF MANUFACTURED ITEMS

Section 1. INTRODUCTIO N

C-1. SCOPE.

a. This appendix includes complete instructlons for maklng Items authorized to be manufactured or fabricated at intermediate direct support and general support maintenance.

b. A part number index in alphanumeric order is provided for cross-referencing the part number of the item to be manufactured to the figure which covers fabrication criteria.

All bulk materials needed for manufacture of an item are listed by National Stock Number (NSN), part number, or specification number in the manufacturing instructions.

d. All dimensions given in Section II, *Illustrated Manufacturing Instructions,* are in inches, except as noted.

Table C-1. Manufactured Items Part Number Cross-reference

PART NUMBER	FIGURE NUMBER	PART NUMBER	FIGURE NUMBER	PART NUMBER	FIGURE NUMBER
FLW-12	c-7	14036712	C-17	14072372	C-19
FLW-16	C-6	14036797	C-16	15522444	C-14
FLW-18	C-4	14040735	C-18		
FLW-20	C-5	14045626	C-10	15599259	C-13
14029235	C-11	14045628	c-lo	15599260	C-12
14034571	C-15	14054275	C-15	15599261	C-13
14034572	C-15	14061503	C-1	15599262	C-12
14034586	C-14	14063323	C-19	31-0040	C-19
14034599	C-16	14063339	C-2	31-0040A	C-19
14036705	C-17	14072369	C-19		
14036706	C-17	14072371	C-19	8906127	C-3
14036711	C-17				

Section II. ILLUSTRATED MANUFACTURING INSTRUCTIONS

1. Fabricate from steel tubing, 3/8 in, outside diameter, stock size 25 ft. long, NSN 4720-01-158-7511.
2. Cut to 35.52 in. long for Part Number 14061503.
3. Install nut, NSN 4730-00-013-7399, Part Number 137339, on each end of tubing.
4. Flare ends of tubing. Nuts should turn freely on flares.
5. Use old pipe as a guide and bend steel tubing to same shape,

Figure C-1. Fuel Pump Pipe

Section II. ILLUSTRATED MANUFACTURING INSTRUCTIONS - Continued

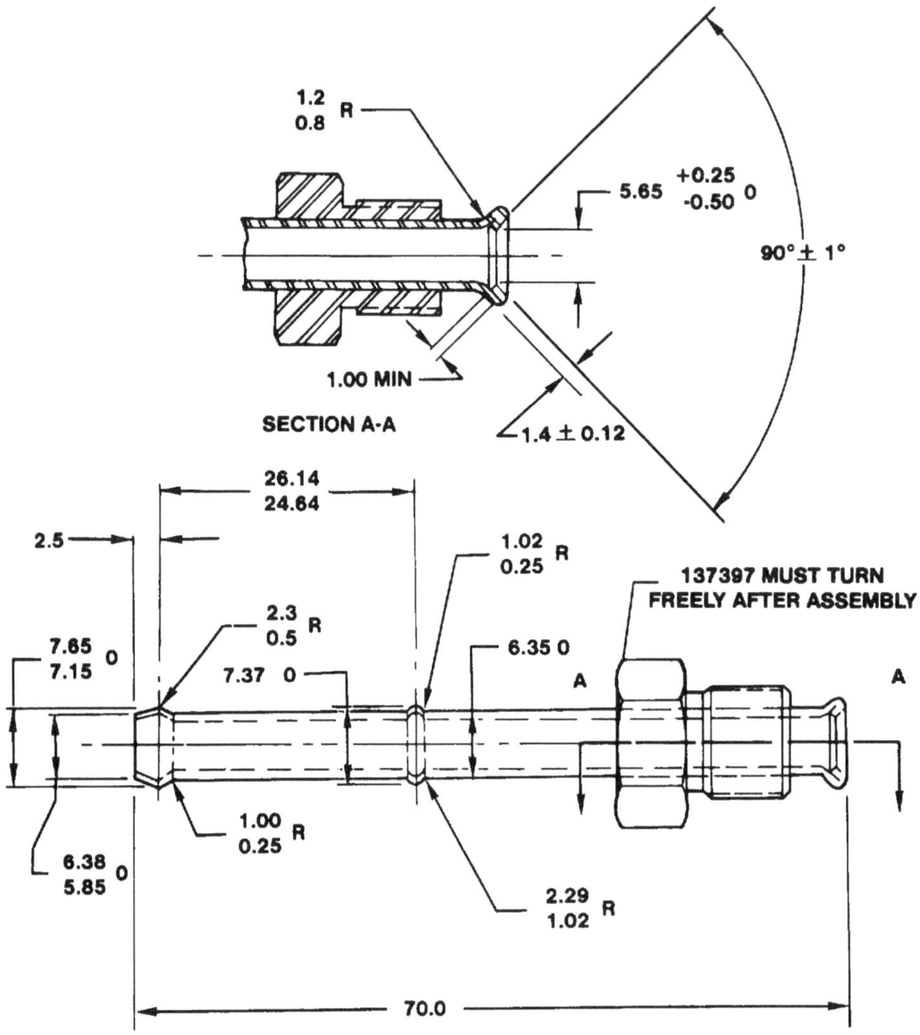

TA50150

Section II. ILLUSTRATED MANUFACTURING INSTRUCTIONS - Continued

1. Fabricate from steel tubing, 1/4 in. outside diameter, stock size 25 ft. long, NSN 4720-01-161-0138.
2. Cut to 2.76 in. long for Part Number 14063339.
3. Install nut, NSN 4730-01-166-6347, Part Number 137397, on tubing.

NOTE

Dimensions in illustration are in millimeters.

4. Flare tubing in locations shown,

Figure C-2. Fuel Injector Pump Pipe

---

1. Fabricate from conduit, 3/4 in. inside diameter, stock size 15 ft. long, NSN 5975-01-191-9851.
2. Cut to 54.00 in. long for Part Number 8906127.

Figure C-3. Body Wiring Harness Conduit

---

1. Fabricate from fuse link conductor, 0.8 mm, 18 gage, NSN 5920-01-218-3760.
2. Cut to 10.0 in. long for Part Number FLW-18.

Figure C-4. Fuse Link Wire

---

1. Fabricate from fuse link conductor, 0,5 mm, 20 gage, NSN 5920-01-218-3759,
2. Cut to 10.0 in. long for Part Number FLW-20.

Figure C-5. Fuse Link Wire

---

1. Fabricate from fuse link conductor, 1,0 mm, 16 gage, NSN 5920-01-218-3761.
2. Cut to 10.0 in. long for Part Number FLW-16.

Figure C-6. Fuse Link Wire

Section II. ILLUSTRATED MANUFACTURING INSTRUCTIONS - Continued

1. Fabricate from fuse link conductor, 3.0 mm, 12 gage, NSN 5920-01-219-0793.
2. Cut to 10.0 in. long for Part Number FLW-12.

Figure C-7. Fuse Link Wire

---

1. Fabricate from steel tubing, 8/16 6 in. outside diameter, stock size 25 ft. long, NSN 4710-00-420-4759.
2. Cut to 73.95 in. long for Part Number 14045626.
3. Cut to 80.64 in. long for Part Number 14045628.
4. Install nut, NSN 4730-00-013-7398, Part Number 137398, on each end of tubing.
5. Flare ends of tubing. Nuts should turn freely on flares.
6. Use old pipe as a guide and bend steel tubing to same shape.

Figure C-10. Transmission Oil Cooler Pipe

Section II. ILLUSTRATED MANUFACTURING INSTRUCTIONS - Continued

1. Fabricate from steel tubing, 1/8 in. outside diameter, stock size 25 ft. long, NSN 4710-01-192-1433.
2. Cut to 23.82 in, long for Part Number 14029235.
3. Use old pipe as a guide and bend steel tubing to same shape,

Figure C-11. Front Axle Vent Pipe

---

1. Fabricate from steel tubing, 1/4 in, outside diameter, stock size 25 ft. long, NSN 4710-01-161-0138,
2. Cut to 78.49 in. long for Part Number 15599260,
3. Cut to 75.25 in. long for Part Number 15599262,
4. Install nut, NSN 4730-01-154-1366, Part Number 9424955, on one end of tubing.
5. Install nut, NSN 4730-00-014-2432, Part Number 137397 on other end of tubing,
6. Flare ends of tubing. Nuts should turn freely on flares.

NOTE

Ensure that nuts are correctly positioned before bending tubing,

7. Use old pipe as a guide and bend steel tubing to same shape.

Figure C-12. Rear Master Cylinder Pipe

---

1. Fabricate from steel tubing, 3/16 in. outside diameter, stock size 25 ft. long, NSN 2530-00-395-5144.
2. Cut to 72.68 in. long for Part Number 15599259.
3. Cut to 76.63 in. long for Part Number 15599261.
4. Install nut, NSN 5310-00-380-1514, Part Number 9424954, on each end of tubing,
5. Flare ends of tubing. Nuts should turn freely on flares.
6. Use old pipe as a guide and bend steel tubing to same shape.

Figure C-13. Front Master Cylinder Pipe

Section II. ILLUSTRATED MANUFACTURING INSTRUCTIONS - Continued

1. Fabricate from steel tubing, 1/4 in. outside diameter, stock size 25 ft. long, NSN 4710-01-161-0138.
2. Cut to 91.51 in. long for Part Number 14034586.
3. Cut to 90.52 in. long for Part Number 15522444.
4. Install nut, NSN 4730-01-154-1366, Part Number 9424955, on one end of tubing.
5. Install nut, NSN 4730-00-014-2432, Part Number 137397 on other end of tubing.
6. Flare ends of tubing. Nuts should turn freely on flares.

NOTE

Ensure that nuts are correctly positioned before bending tubing.

7. Use old pipe as a guide and bend steel tubing to same shape.

Figure C-14. Rear Combination Valve Pipe

---

1. Fabricate from steel tubing, 3/18 in. outside diameter, stock size 25 ft. long, NSN 2530-00-395-5144.
2. Cut to 29.43 in. long for Part Number 14034571,
3. Cut to 42.13 in. long for Part Number 14034572.
4. Cut to 30.44 in. long for Part Number 14054275.
5. Install nut, NSN 4730-00-288-9390, Part Number 9432075, on each end of tubing.
6. Flare ends of tubing. Nuts should turn freely on flares.
7. Use old pipe as a guide and bend steel tubing to same shape.

Figure C-15. Caliper Pipe

Section II. ILLUSTRATED MANUFACTURING INSTRUCTIONS-Continued

1. Fabricate from steel tubing, 1/4 in. outside diameter. stock size 25 ft. long. NSN 4710-01-161-0138.
2. Cut to 78.68 in, long for Part Number 14036797.
3. Cut to 73.51 in. long for Part Number 14034599.
4, Install nut, NSN 4730-00-288-9390, Part Number 9432075, on each end of tubing.
5. Flare ends of tubing. Nuts should turn freely on flares.
6. Use old pipe as a guide and bend steel tubing to same shape.

Figure C-16. Rear Brake Pipe

---

1. Fabricate from steel tubing, 3-16 in, outside diameter, stock size 25 ft. long, NSN 2530-00-395-5144,
2. Cut to 38.59 in. long for Part Number 14036705.
3. Cut to 31.29 in, long for Part Number 14036706,
4. Cut to 41.54 in. long for Part Number 14036711.
5. Cut to 35.35 in. long for Part Number 14036712.
6. Install nut, NSN 4730-00-288-9390, Part Number 9432075, on each end of tubing.
7. Flare ends of tubing, Nuts should turn freely on flares.
8. Use old pipe as a guide and bend steel tubing to same shape.

Figure C-17. Wheel Cylinder Pipe

Section II. ILLUSTRATED MANUFACTURING INSTRUCTIONS - Continued

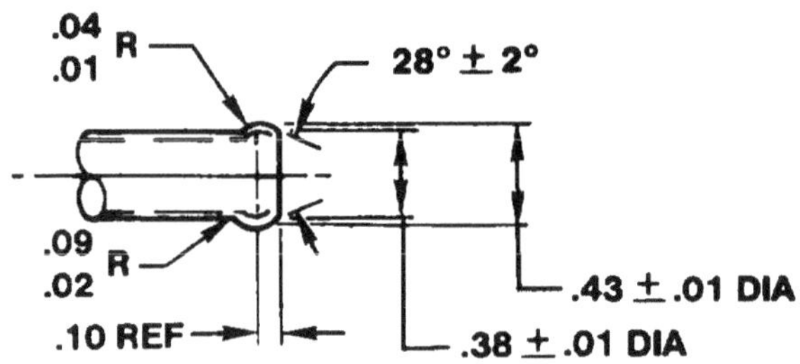

ENLARGED VIEW OF TUBE END 1

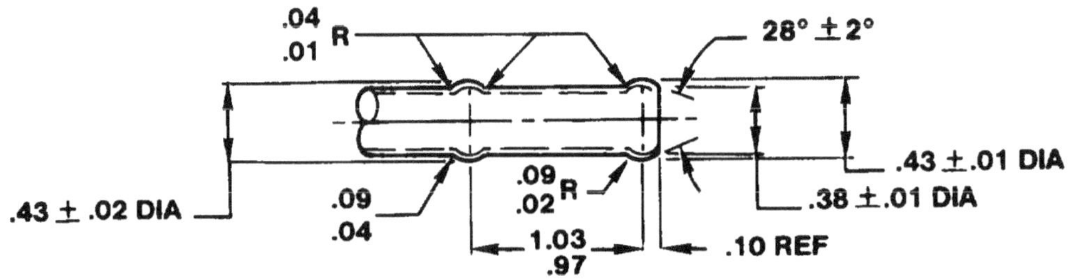

ENLARGED VIEW OF TUBE END 2

1. Fabricate from steel tubing, 1/4 in. outside diameter, stock size 25 ft. long, NSN 4710-01-161-0138.
2. Cut to 15.90 in. long for Part Number 14040735.
3. Flare tubing in locations shown.

NOTE

Ensure that flares are correctly positioned before bending tubing.

4. Use old pipe as a guide and bend steel tubing to same shape.

Figure C-18. Power Booster Return Pipe

TA50151

Section II. ILLUSTRATED MANUFACTURING INSTRUCTIONS - Continued

1. Fabricate from steel tubing, 5/16 in. outside diameter, stock size 25 ft. long, NSN 4710-00-420-4759.
2. Cut to 45.74 in. long for Part Number 14063323.
3. Cut to 65.22 in. long for Part Number 14072369.
4. Cut to 44.45 in. long for Part Number 14072371.
5. Cut to 7.92 in. long for Part Number 14072372.
6. Cut to 96.0 in. long for Part Number 31-0040.
7. Cut to 3.0 in. long for Part Number 31-0040A.
8. Install nut, NSN 5310-01-064-8672, Part Number 118538, on each end of tubing.
9. Flare ends of tubing. Nuts should turn freely on flares.
10. Use old pipe as a guide and bend steel tubing to same shape.

Figure C-19. Heater Fuel Pipe

APPENDIX D
TORQUE LIMITS

D-1. SCOPE.

This appendix lists standard torque values, as shown in Table D-1, and provides general information for applying torque. Special torque values and sequences are indicated in the maintenance procedures for applicable components.

D-2. GENERAL.

a. Always use the torque values listed below when the maintenance procedure does not give a specific torque value.

b. Unless otherwise specified, standard torque tolerance shall be ±10%.

c. Torque values listed are based on clean and dry threads. Reduce torque by 10%. when engine oil is used as a lubricant.

Table D-1. Standard Torque Specifications

NUT OR BOLT SIZE (inch)	TORQUE				NUT OR BOLT SIZE (millimeter)	TORQUE			
	GRADE S		GRADE 8			GRADE 9.8		GRADE 10.9	
FINE & COARSE THREADS	lb. -ft.	N.m	lb.-ft.	Nom		lb. -ft.	N-m	lb. -ft.	N.m
1/4	6	8	10	14	6	10	14	12	16
5/16	15	20	21	29	8	20	27	23	31
3/8	26	35	37	50	10	40	54	48	65
7116	43	58	60	81	12	70	95	80	109
112	65	88	90	122	14	113	153	132	179
9/16	90	122	130	176	16	176	239	207	281
5/8	130	176	178	241	20	343	465	399	541
314	185	251	260	353					
7/8	300	408	420	570					
1	440	597	635	861					

D-3. DRIVEBELT TENSION SPECIFICATIONS.

a. Using belt tensioning gage, adjust belt tension as listed below.

b. If belt has been replaced, refer to new belt specification, run engine for 15 minutes, and adjust to old belt specifications.

Table D-2. Drivebelt Tension Specifications

TENSION REQUIREMENT	ALTERNATOR	POWER STEERING	AIR CONDITIONER COMPRESSOR
NEW BELT: Before Operating Engine	146 lb. 650 N	146 lb. 650 N	169 lb. 750 N
OLD BELT: After Operating Engine	67 lb. 300 N	67 lb. 300 N	67 lb. 300 N

APPENDIX E
WIRING DIAGRAMS AND SCHEMATICS

E-1. SCOPE.

This appendix contains wiring diagrams and schematics of each CUCV Series electrical circuit. This appendix should be used when performing the *Electrical Troubleshootihg* procedures in Table 2-2.

E-2. WIRING DIAGRAMS AND SCHEMATICS INDEX.

E-2. WIRING DIAGRAMS AND SCHEMATICS INDEX (Con't).

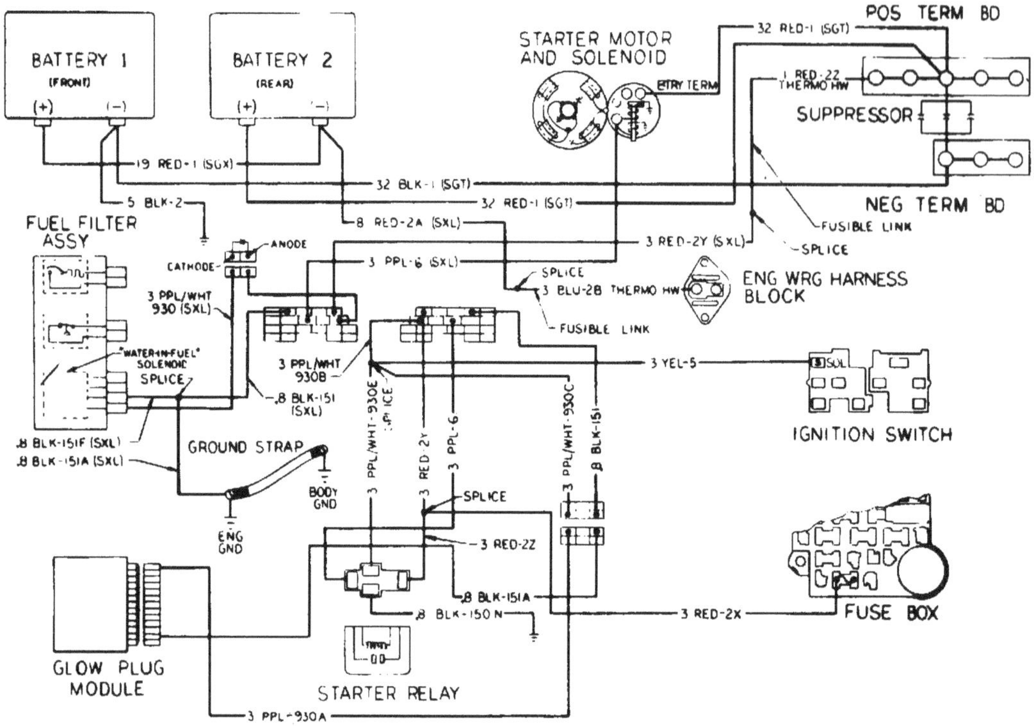

TA50162

E-1. Starting Circuits (All Except M1010).

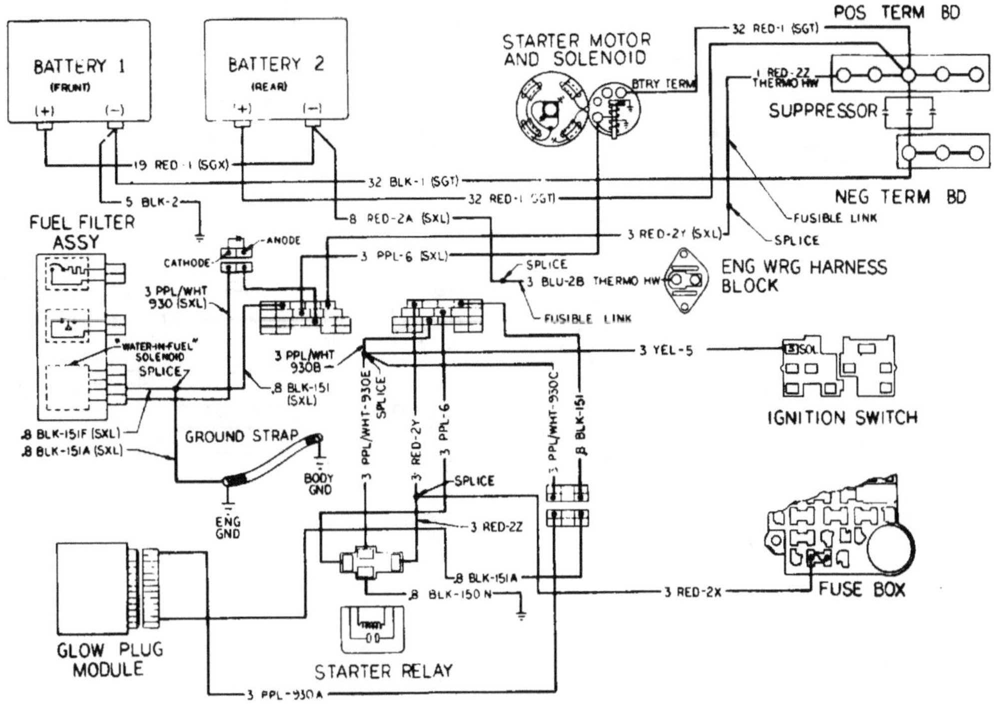

E-2. Starting Circuits (M1010).

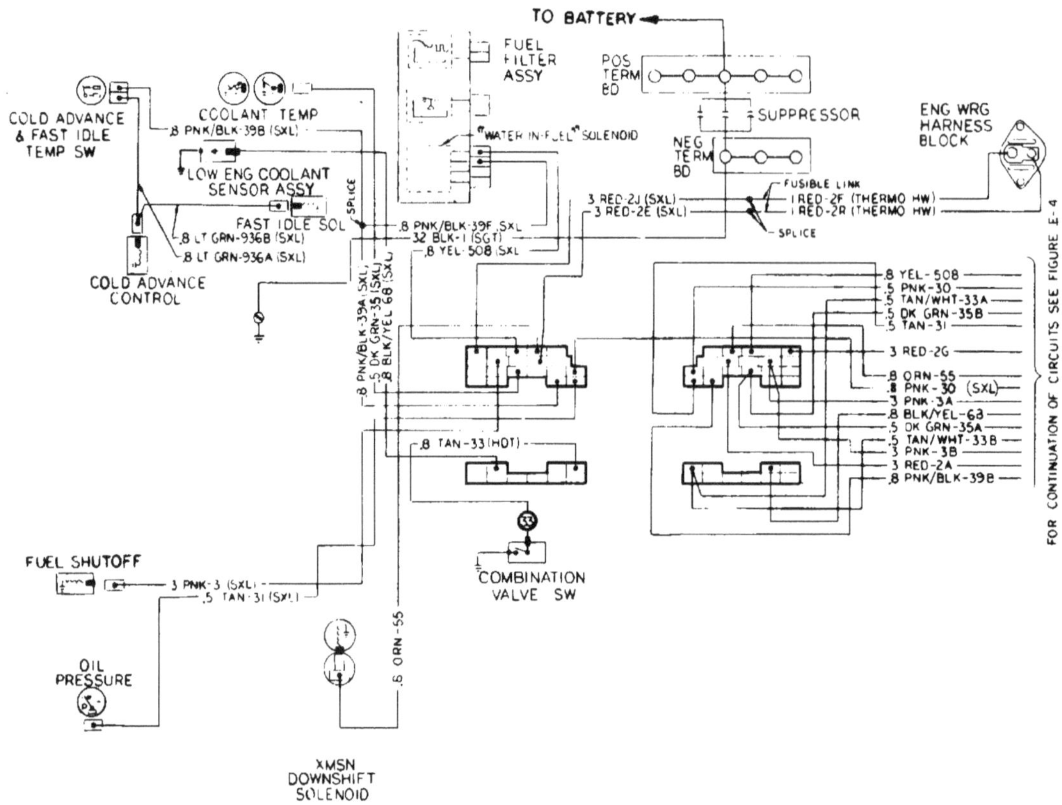

E-3. Engine Compartment Ignition Circuits (All Except M1010).

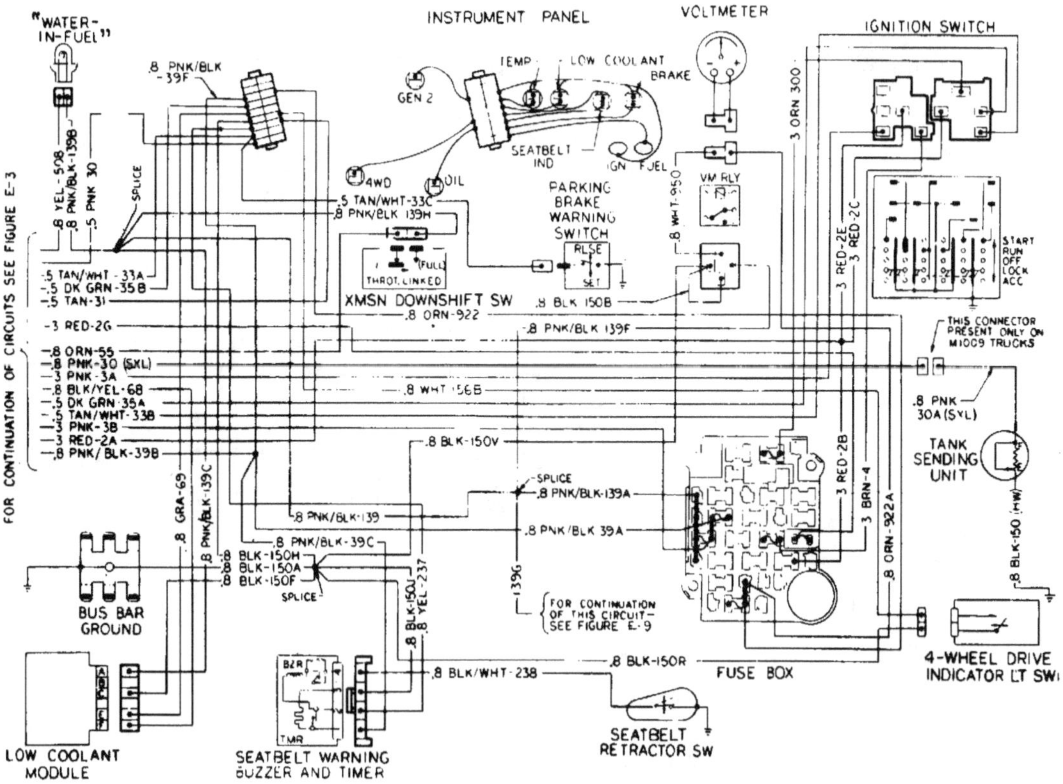

TA50158

E-4. Cab Ignition Circuits (All Except M1010).

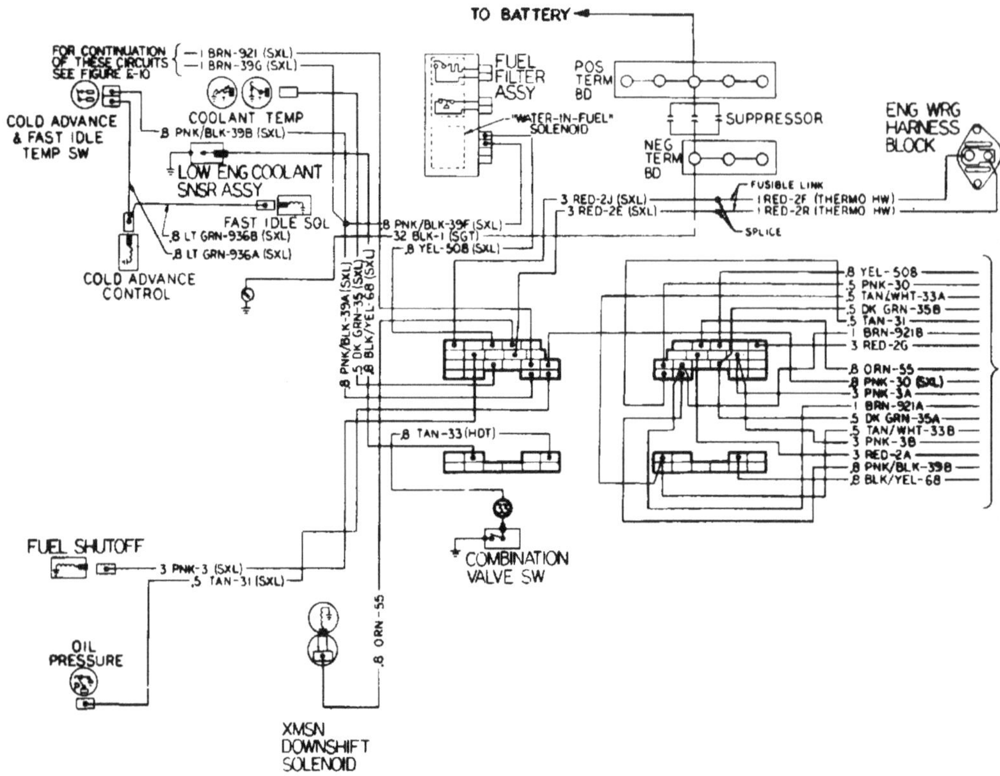

TA50156

E-5. Engine Compartment Ignition Circuits (M1010).

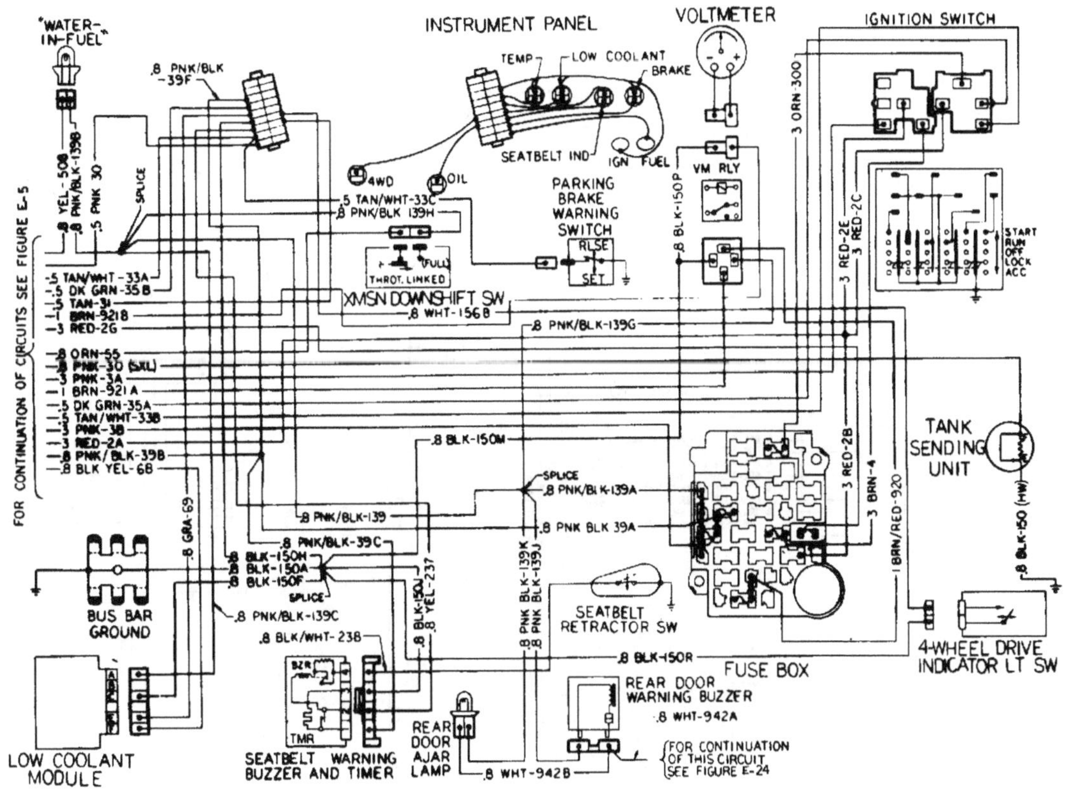

TA50157

E-6. Cab Ignition Circuits (M1010).

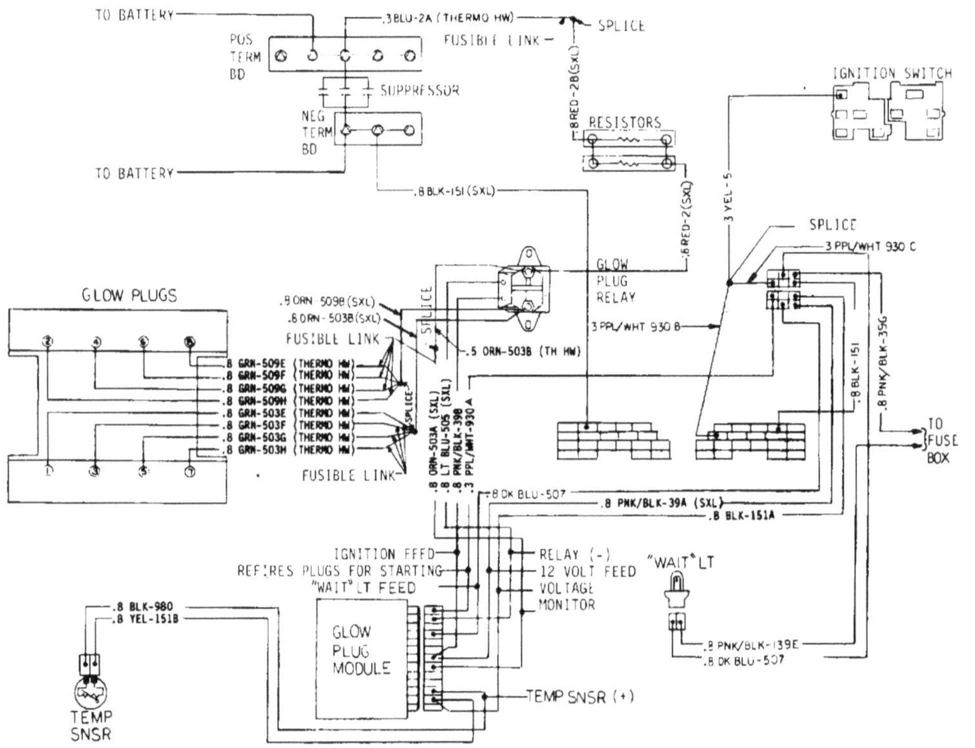

TA701861

E-7. Glow Plug Circuits (All Except M1010).

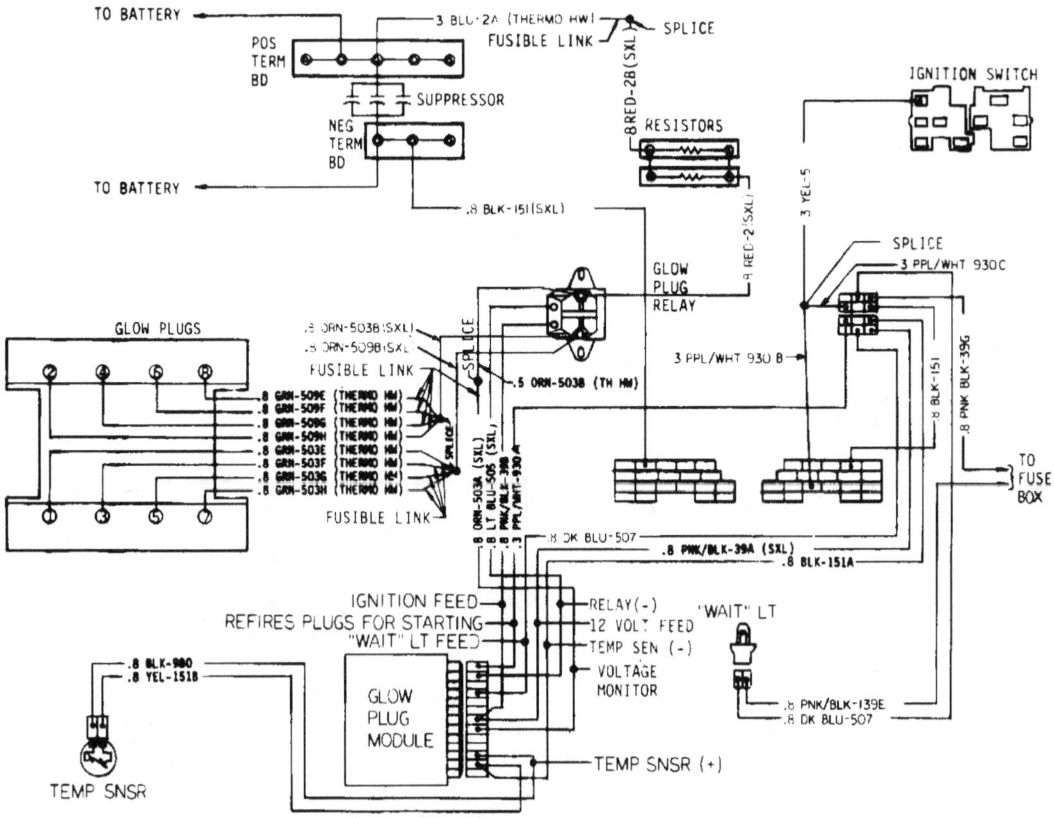

E-8. Glow Plug Circuits (M1010).

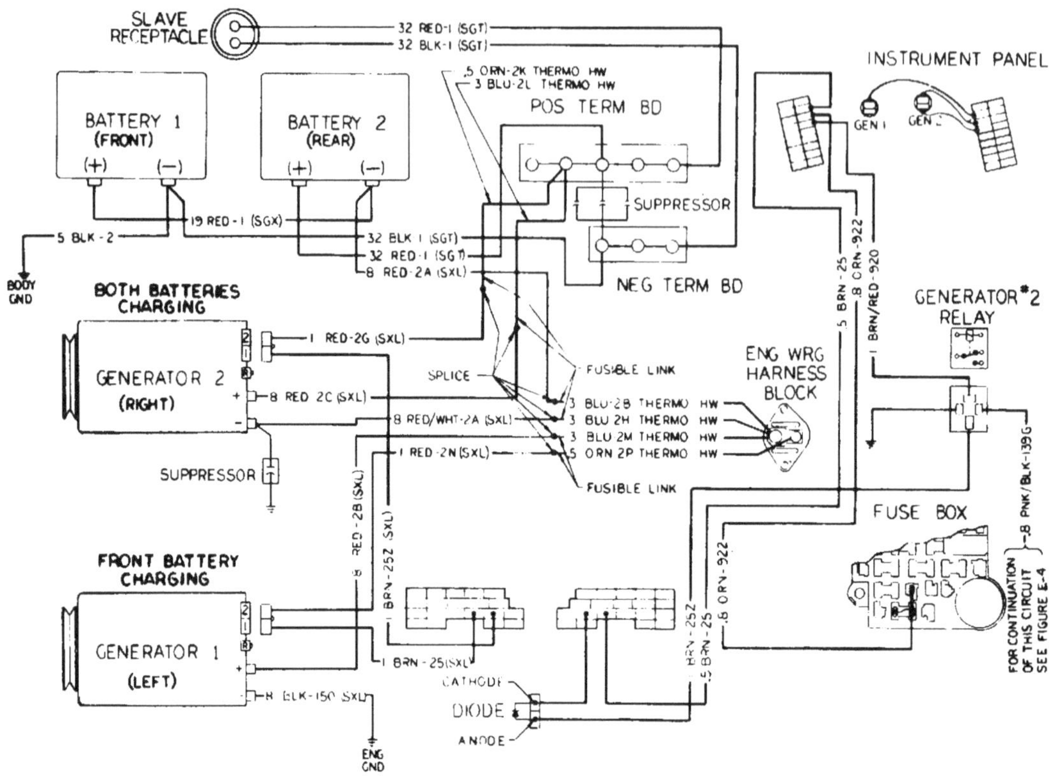

E-9. Charging Circuits (All Except M1010).

TA50160

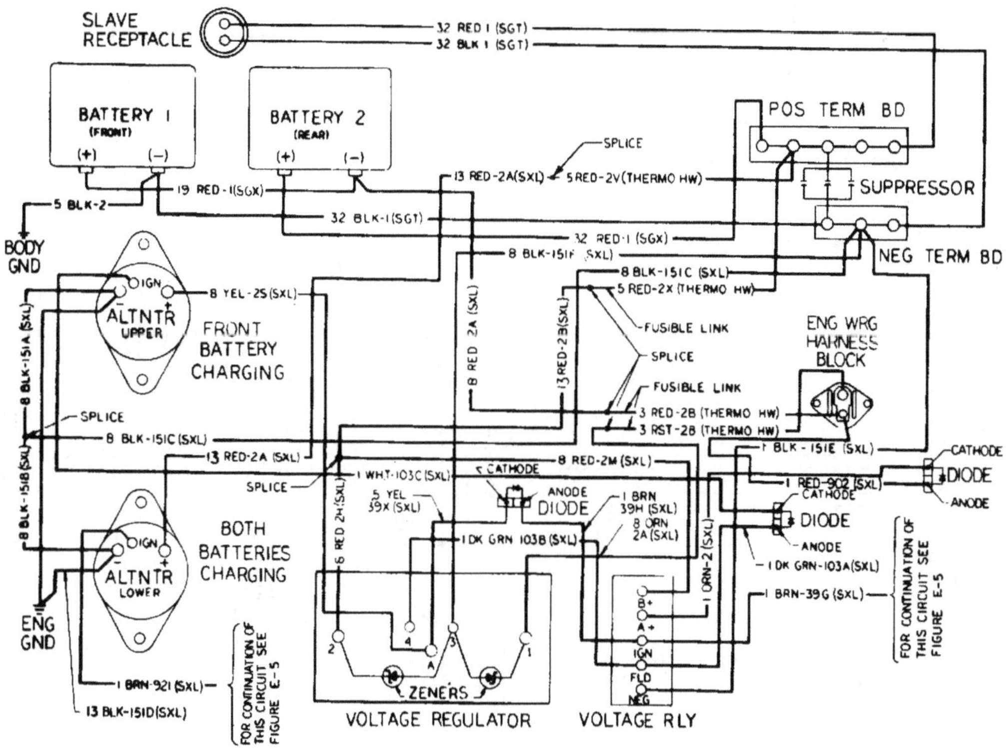

E-10. Charging Circuits (M1010).

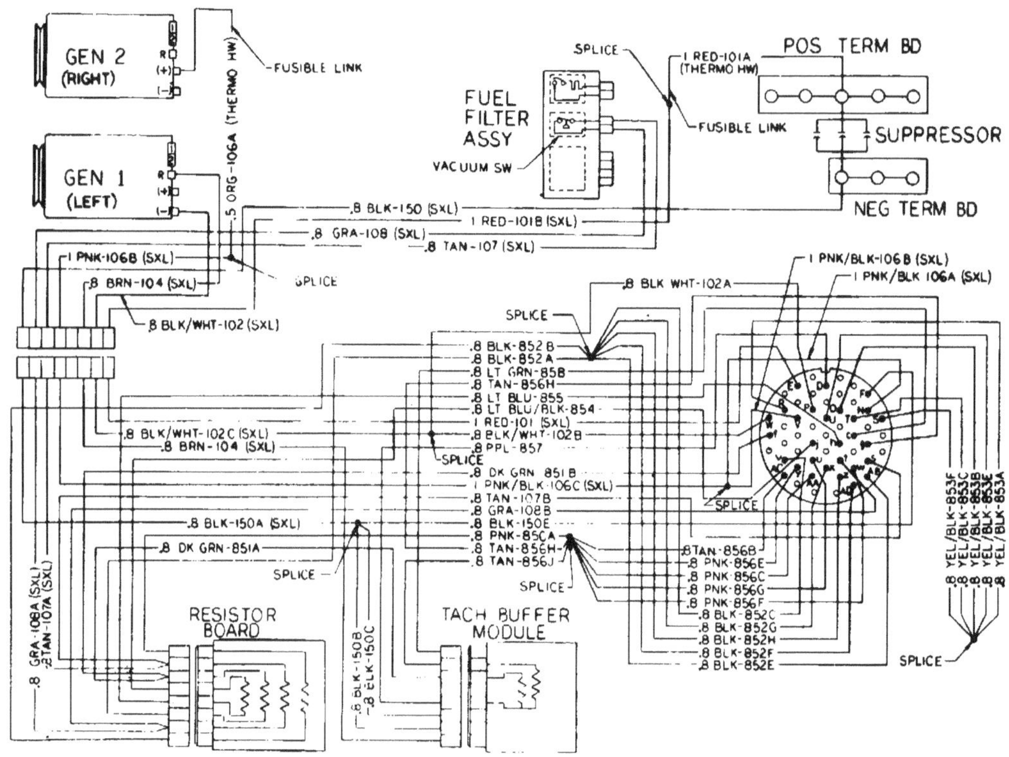

TA50182

E-11. Diagnostic Circuits (All Except M1010).

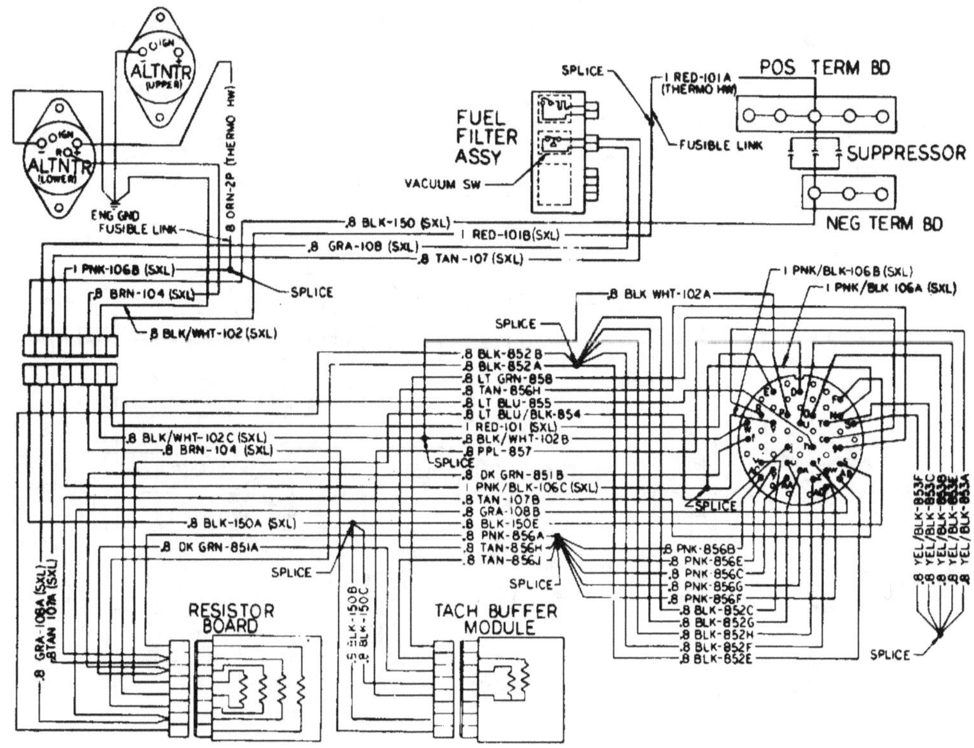

TA50183

E-12. Diagnostic Circuits (M1010).

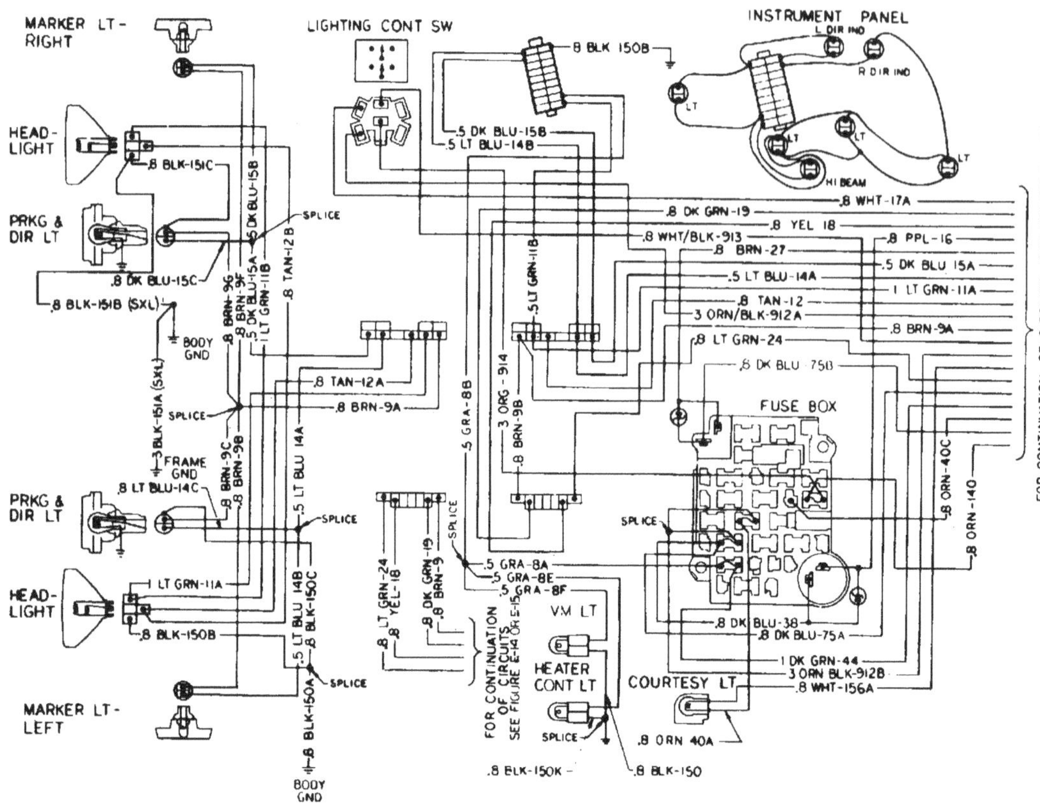

TA50164

E-13. Service Lighting Circuits - Front (All Except M1010).

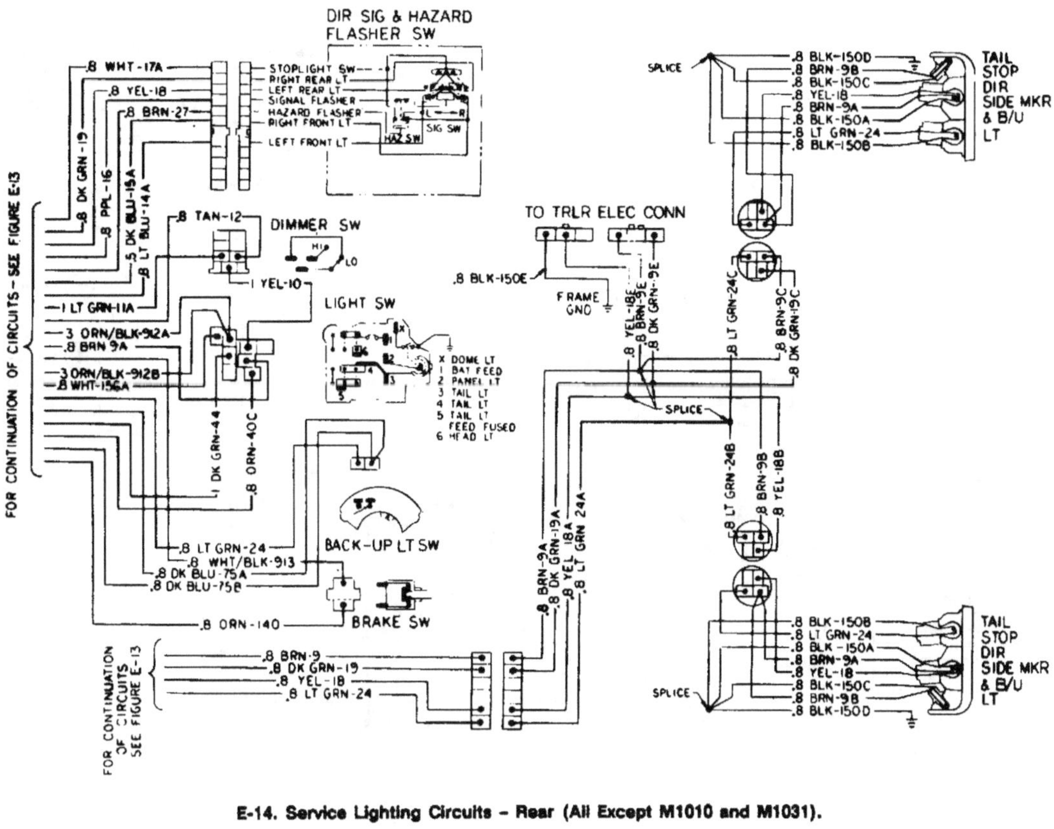

E-14. Service Lighting Circuits - Rear (All Except M1010 and M1031).

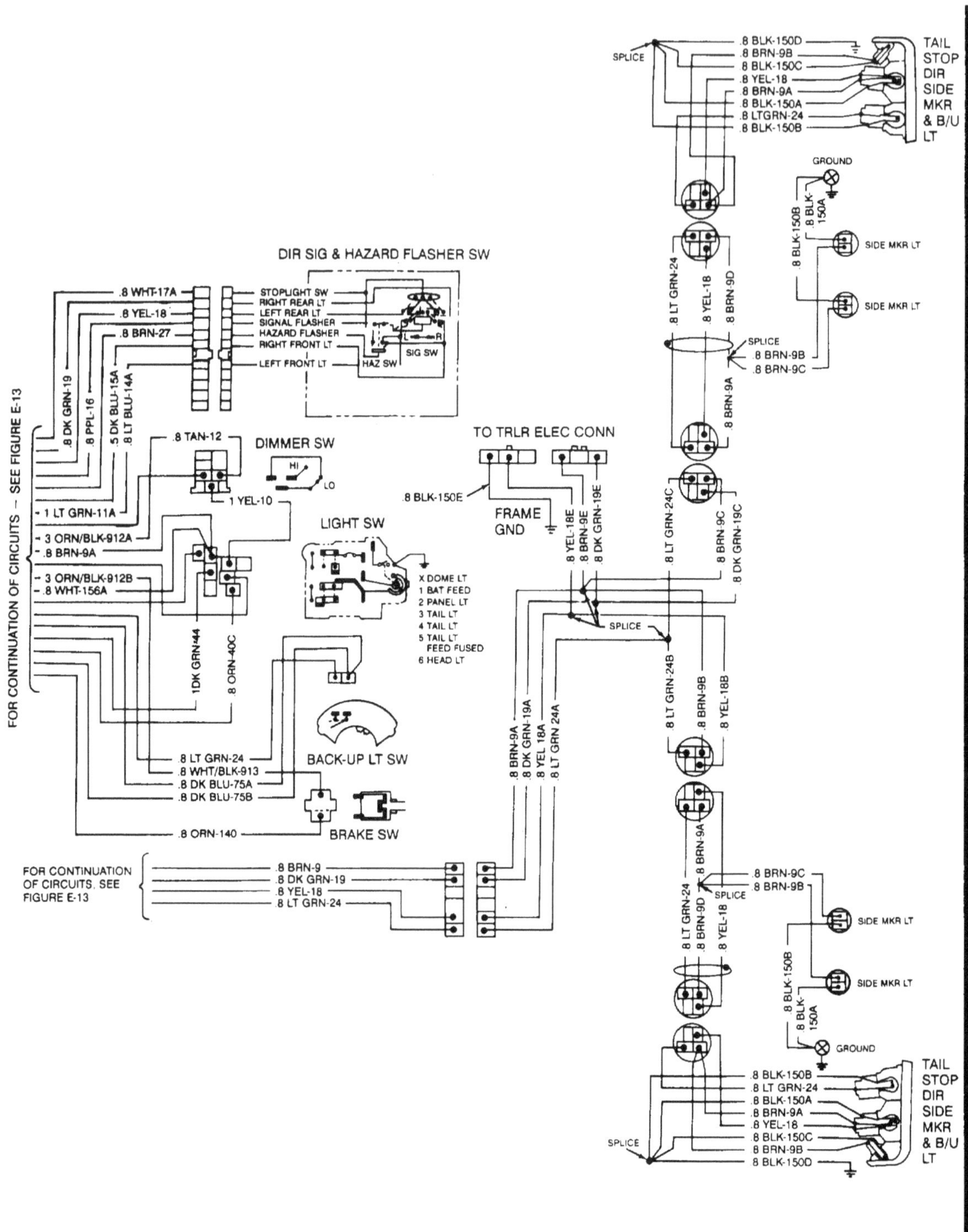

E-14.1. Service Lighting Circuit - Rear (M1028A2).

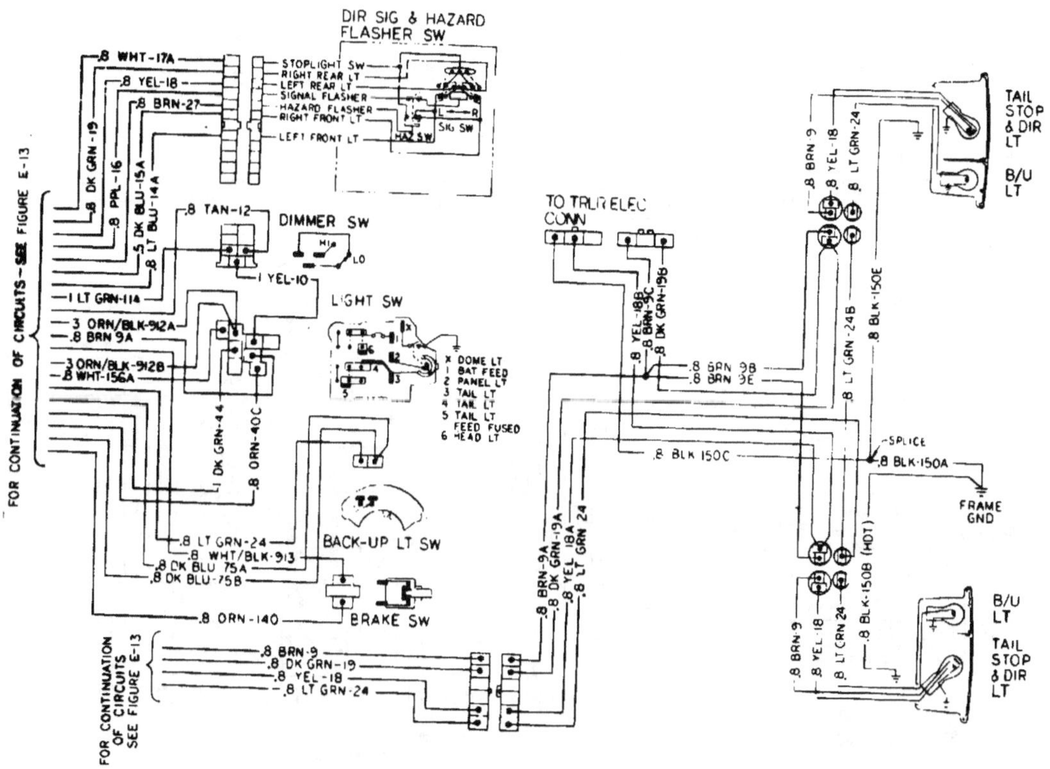

E-15. Service Lighting Circuits - Rear (M1031).

TA50166

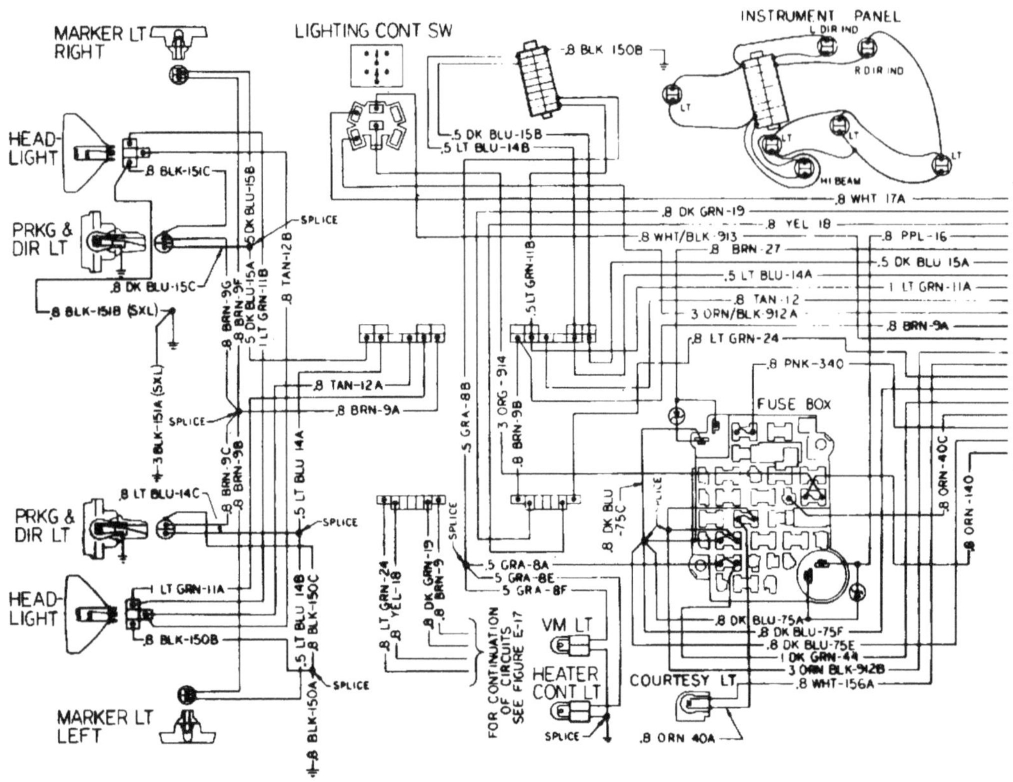

E-16. Service Lighting Circuits - Front (M1010).

TA50187

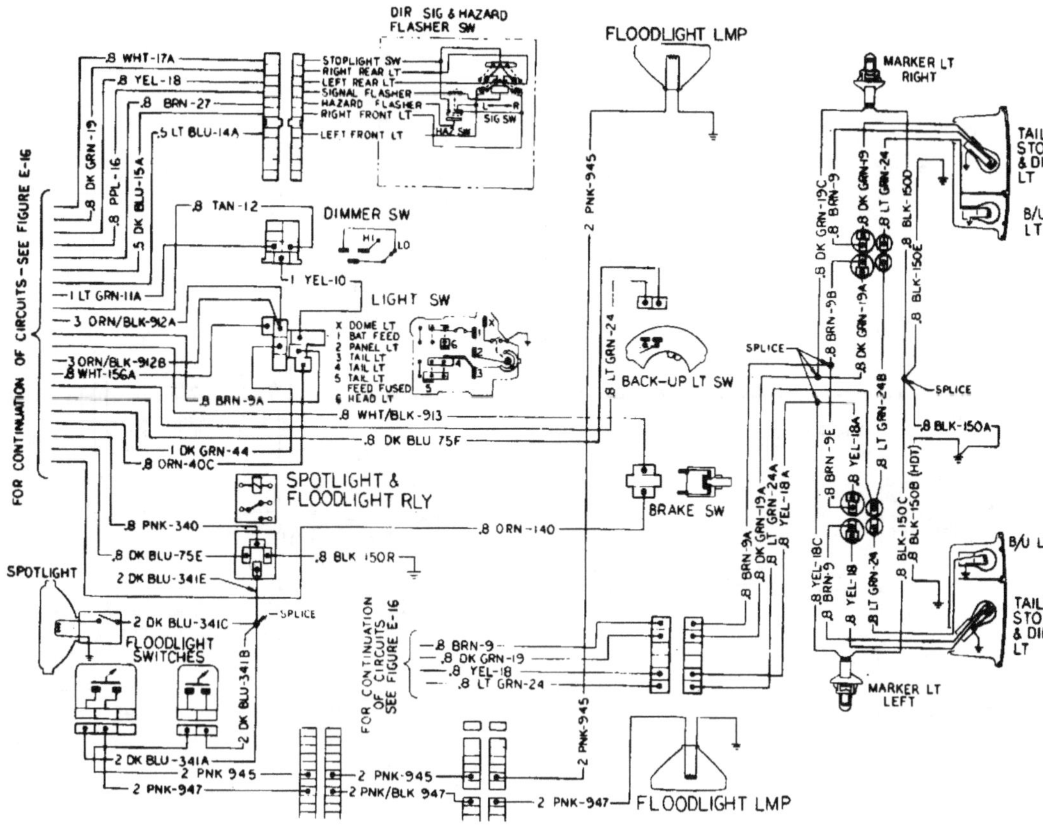

E-17. Service Lighting Circuits - Rear (M1010).

TA50168

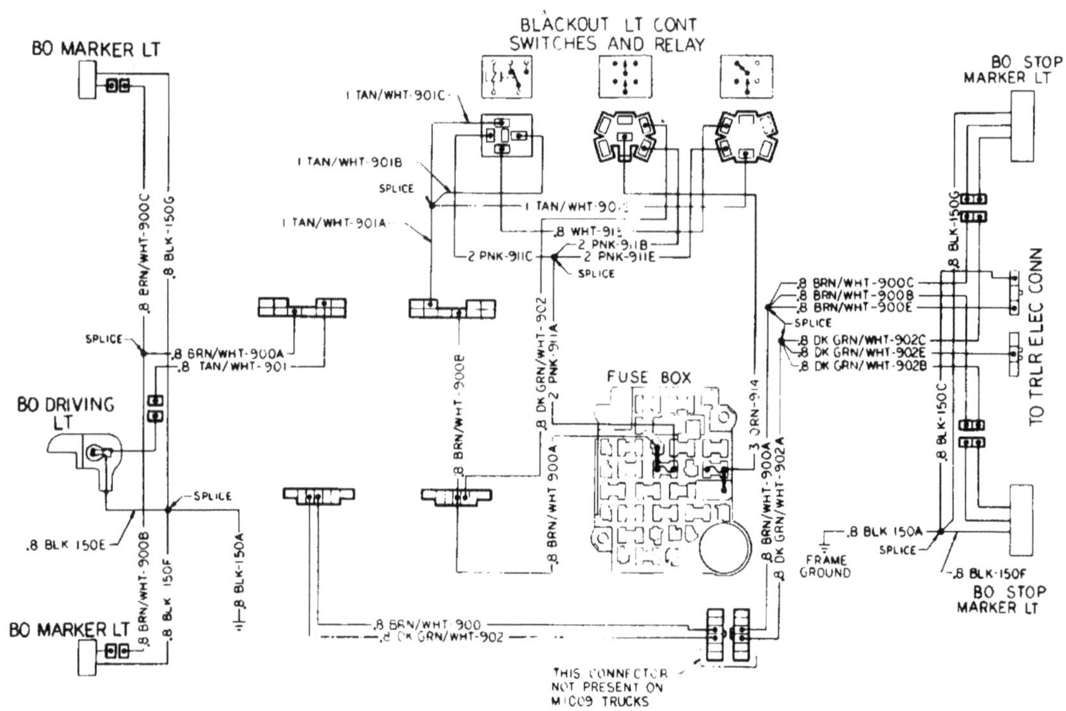

E-18. Blackout Lighting Circuits (All Except M1010).

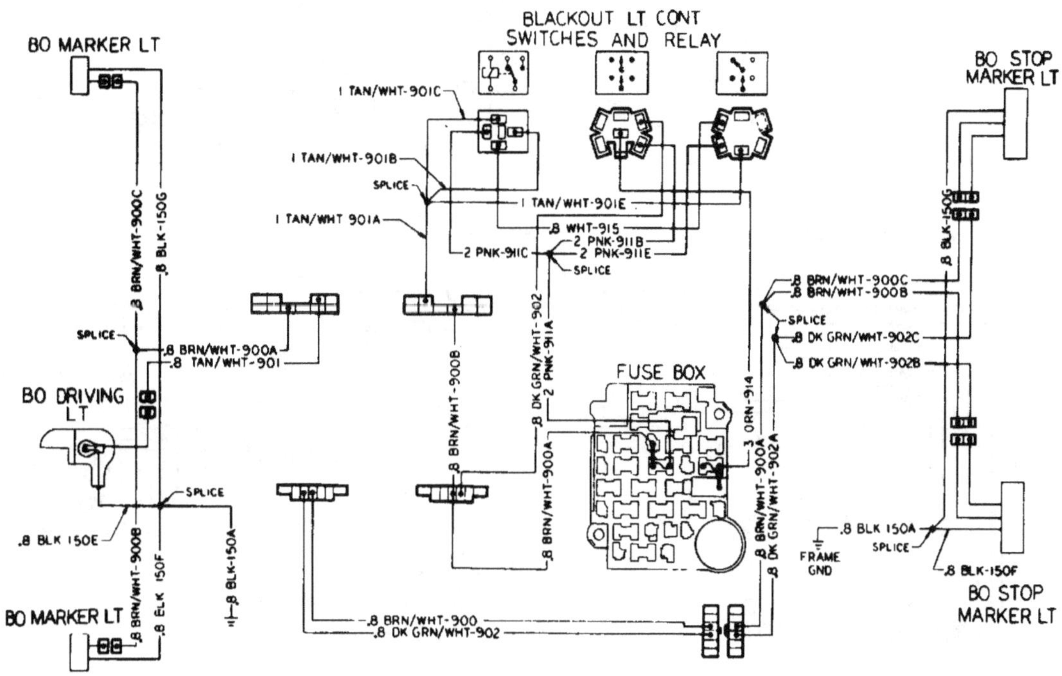

TA50170

E-19. Blackout Lighting Circuits (M1010).

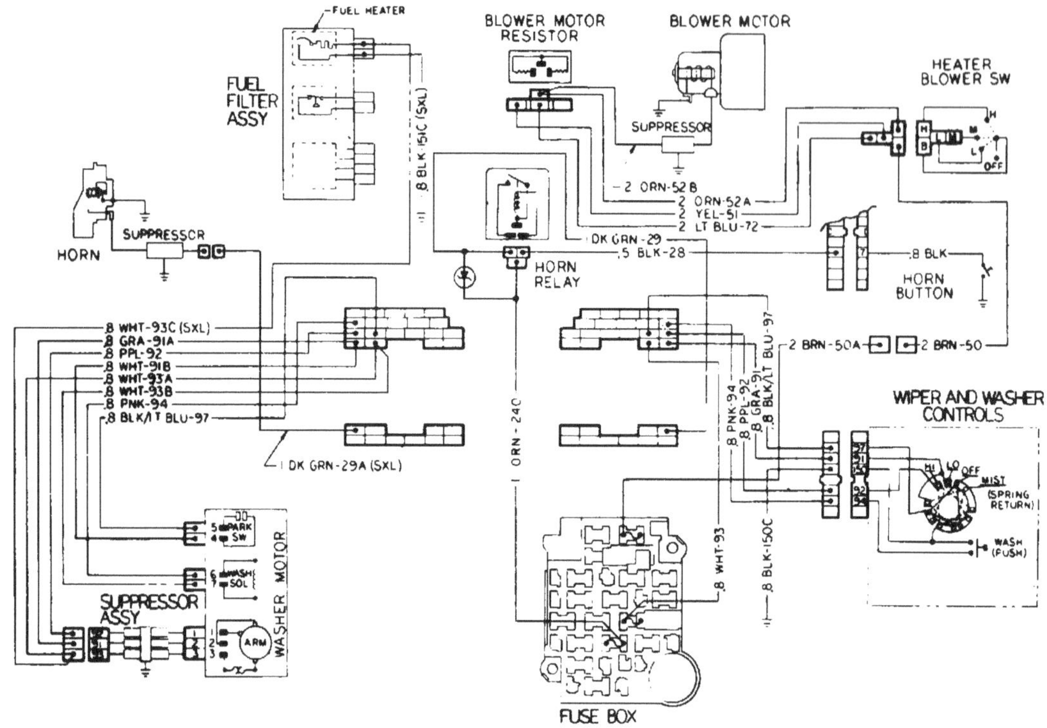

TA50171

E-20. Wipers, Washer, Horn, and Heater Circuits (All Except M1010).

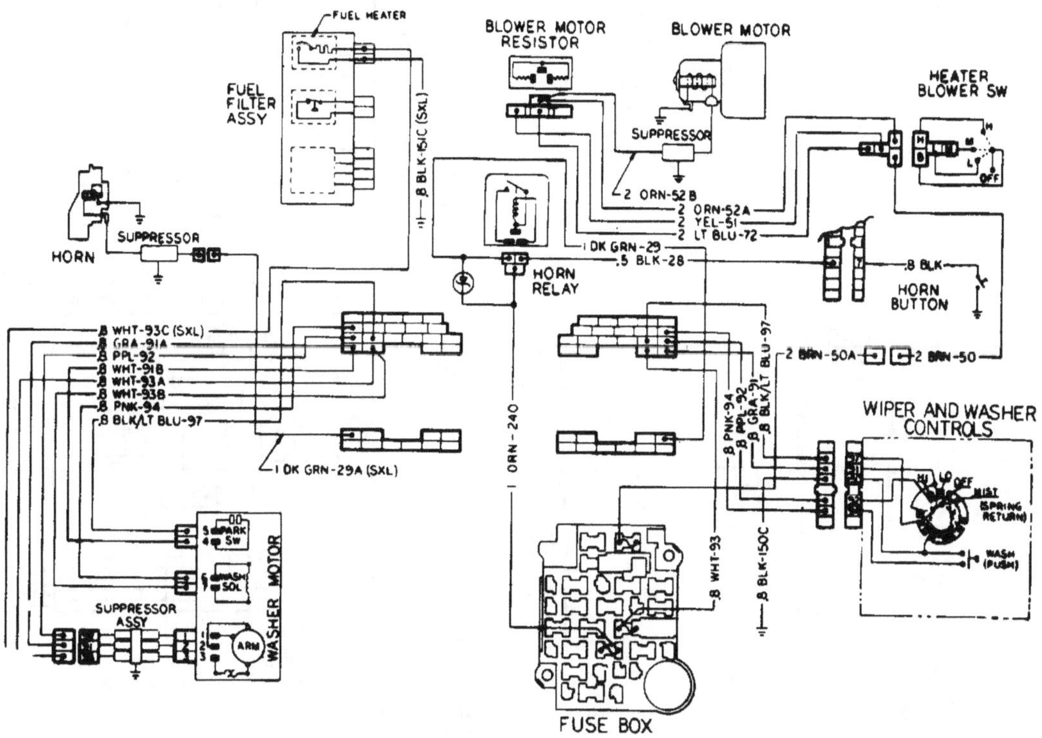

E-21. Wipers, Washer, Horn, and Heater Circuits (M1010).

TA50172

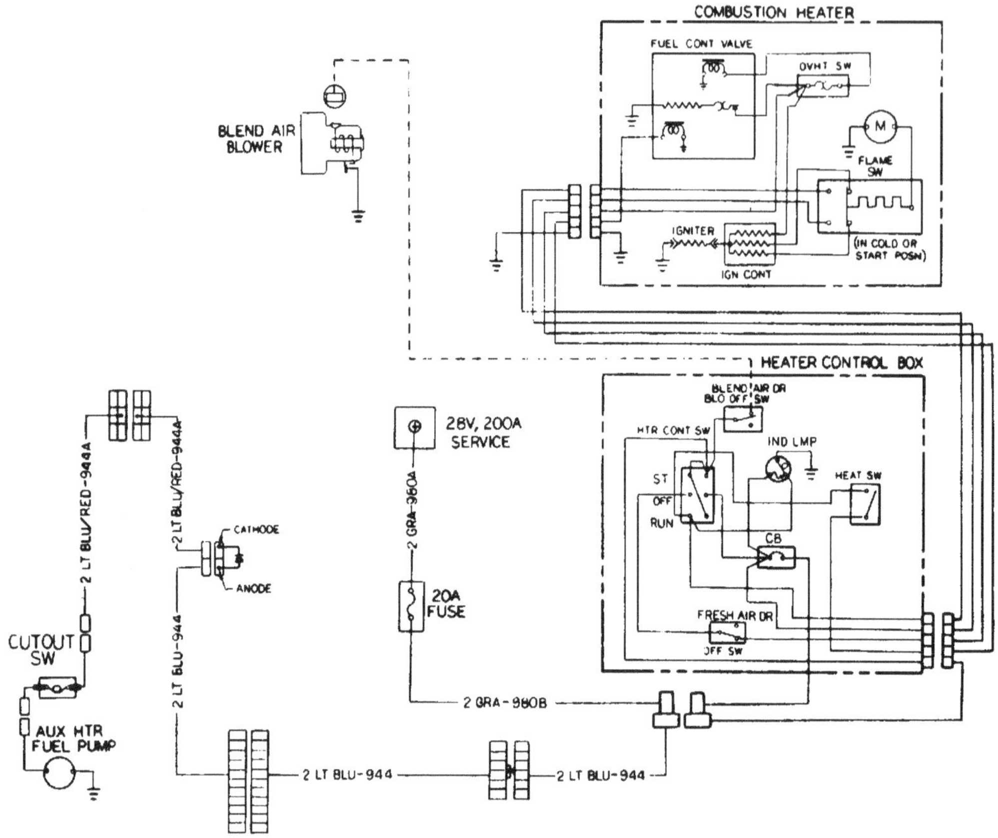

TA50173

E-22. Rear Heating Circuits (M1010).

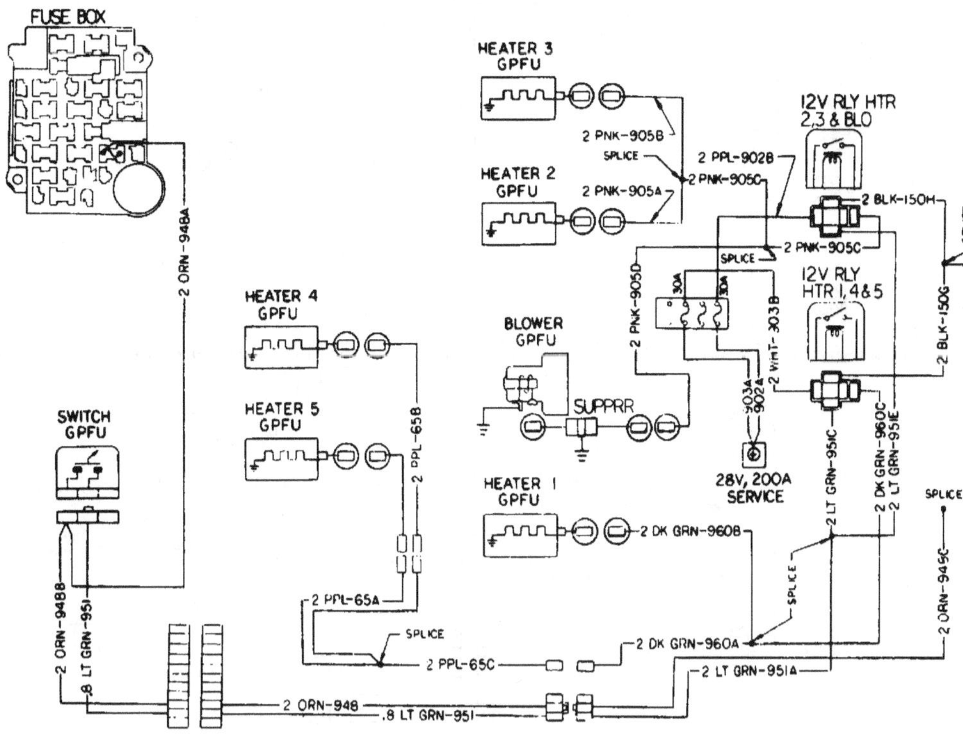

E-23. Gas-Particulate Filter Unit (GPFU) Heater Circuits (M1010).

TA50174

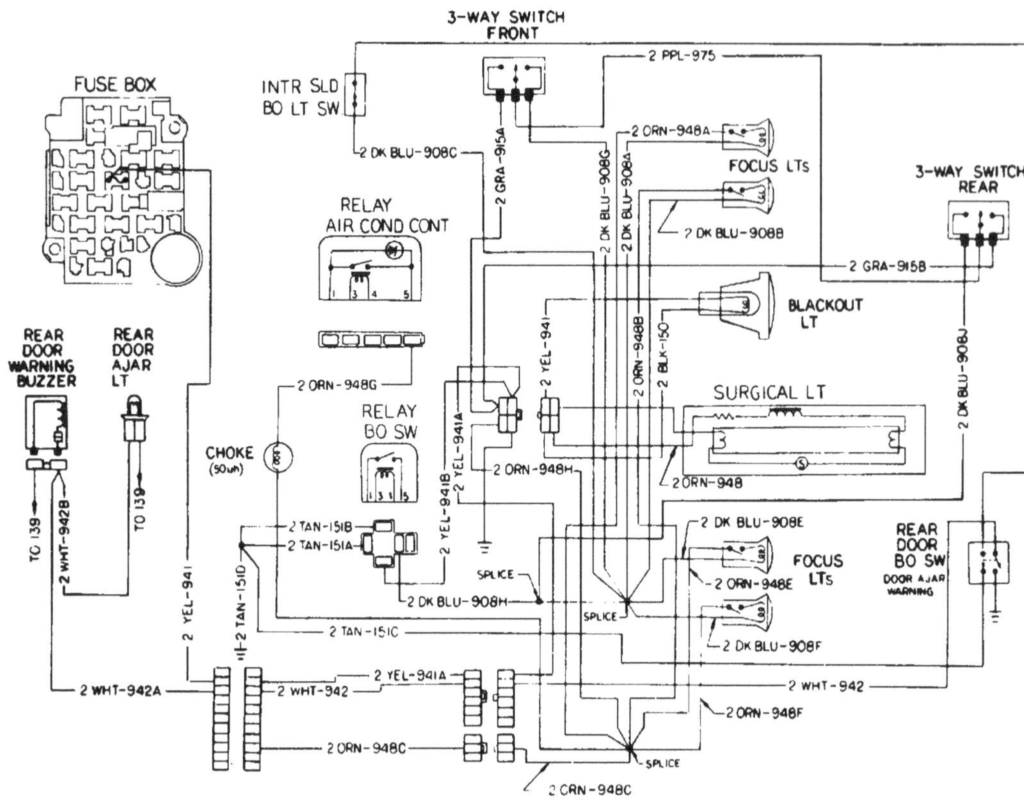

TA50175

E-24. Interior Lighting Circuits (M1010).

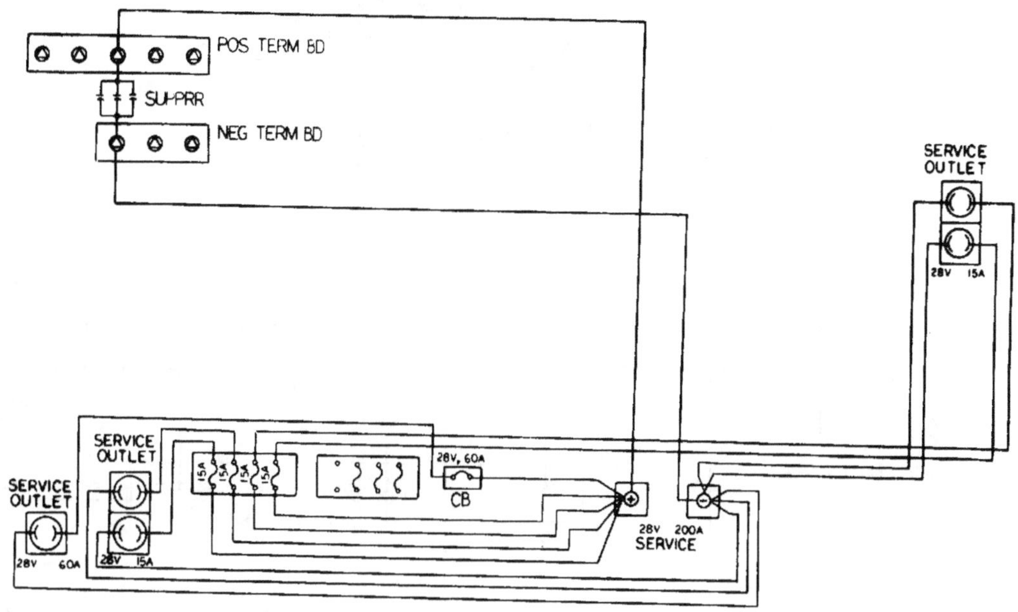

TA60176

E-25. Service Outlet Circuits (M1010).

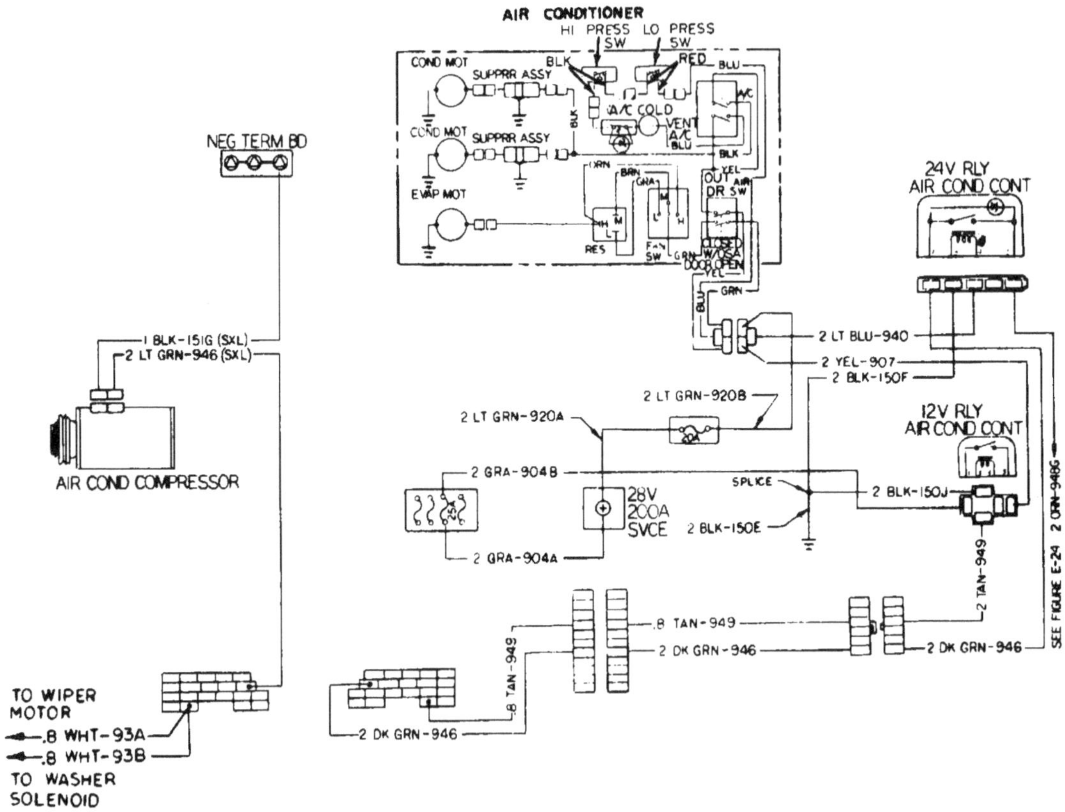

TA50177

E-26. Air Conditioner Circuits (M1010).

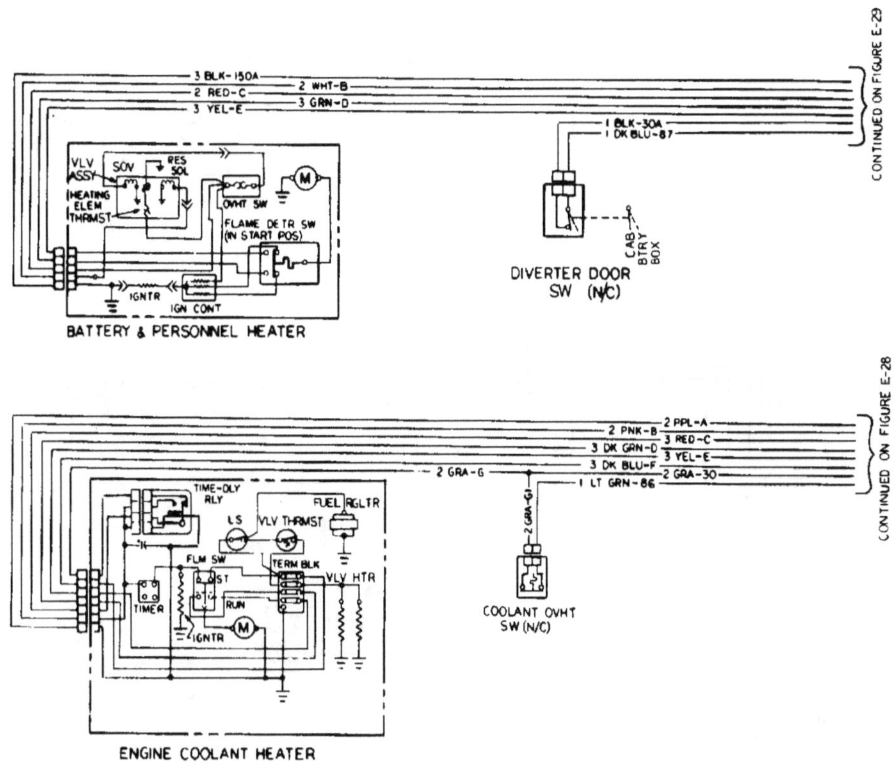

TA50178

E-27. Winterization Kit Cab Heater and Winterization Kit Engine Heater Circuits.

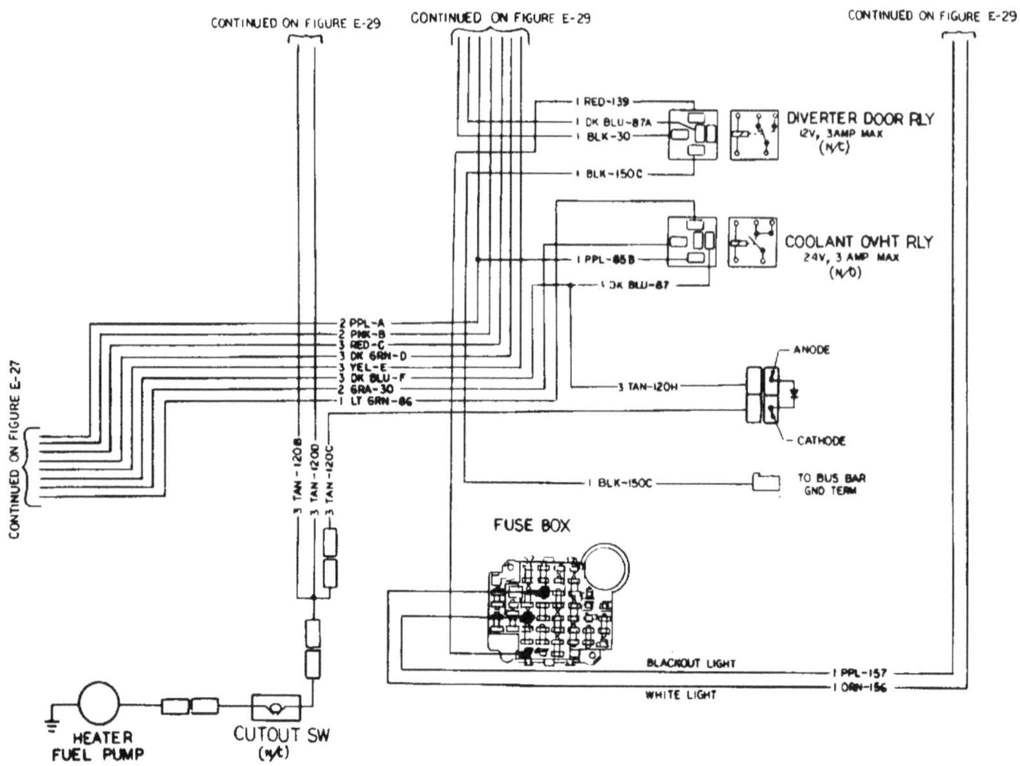

TA50179

E-28. Winterization Kit Cab Heater and Winterization Kit Engine Heater Fuse Block and Relay Circuits.

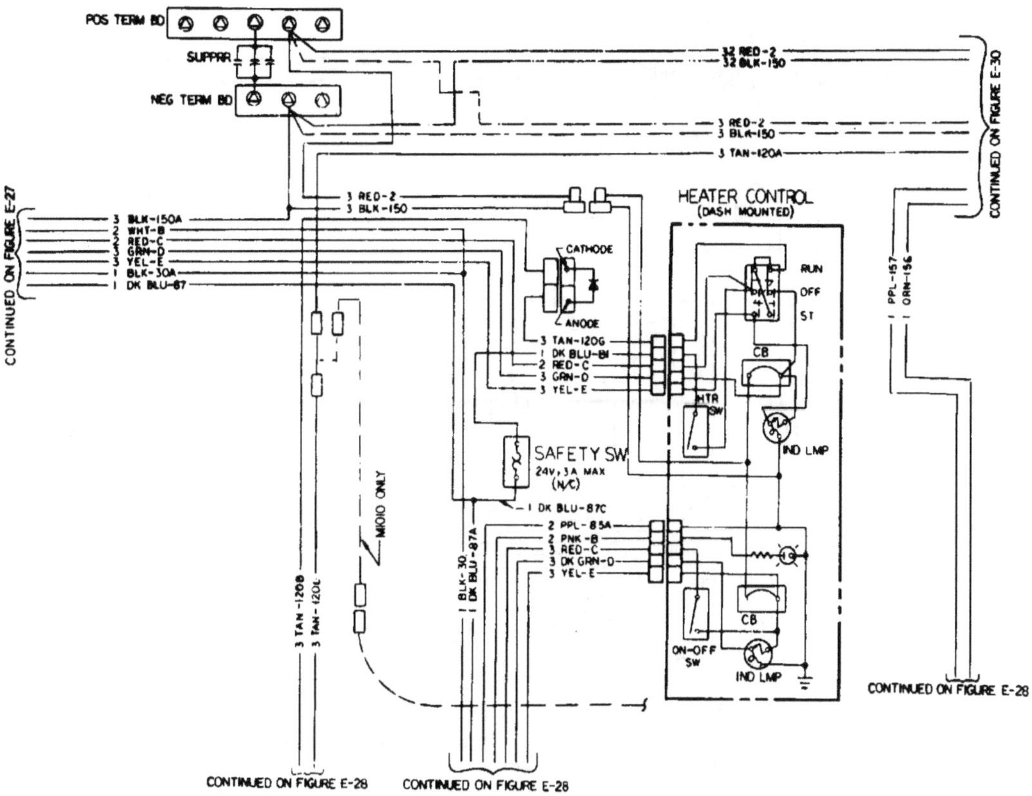

TA50180

E-29. Winterization Kit Cab Heater and Winterization Kit Engine Heater Control Box Circuits.

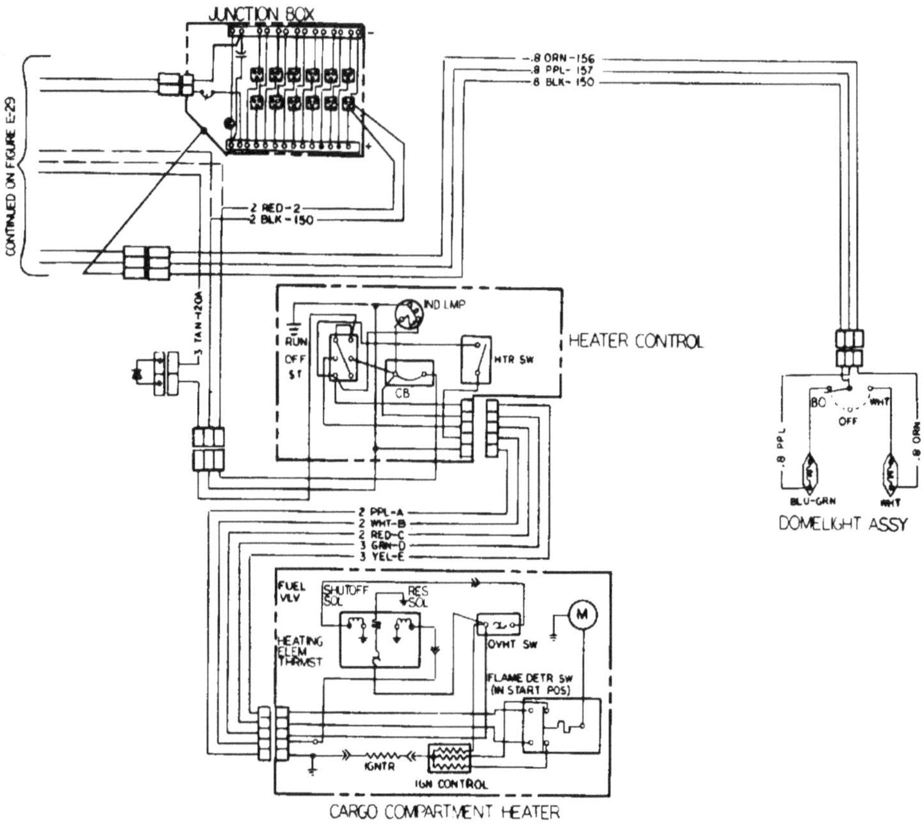

TA50181

E-30. Winterization Kit Cargo Compartment Heater Circuits (M1008 and M1008A1).

APPENDIX F
GPFU INSTALLATION (M1010)

F-1. GAS-PARTICULATE FILTER UNIT (GPFU) INSTALLATION (M1010).

This task covers: a. Installation

INITIAL SETUP:

Equipment Condition

- Both battery negative cables disconnected. (See TM 9-2320-289-20)

Tools/Test Equipment

- Snapring pliers

Materials/Parts

- Pipe sealant (Item 61, Appendix B)

NOTE

Some parts required for this installation are provided with the M1010, and some components must be ordered through the supply system. The task should not be attempted until all components required are available.

a. INSTALLATION

NOTE

- All holes for mounting various items have been factory-drilled. When mounting any item, look for predrilled holes in areas indicated.
- Ensure that pipe sealant is applied to all threaded pipe fittings.

1. Remove screw and disconnect ground lead from upper left front corner of patient compartment. Save ground lead and screw for later use.

F-1. GAS-PARTICULATE FILTER UNIT (GPFU) INSTALLATION (M1010) (Con't).

2. Connect 2 brass adapters (1) and ends of 2 tees (2). Connect 2 tees with attached adapters to nipple (4). Tighten so that 2 side openings (3) in tees are facing same direction.

3. Connect 2 aluminum adapters (1) to side openings (3) of 2 tees (2),

4. Position tee and nipple assembly on left side wall, level with existing bracket on front wall.

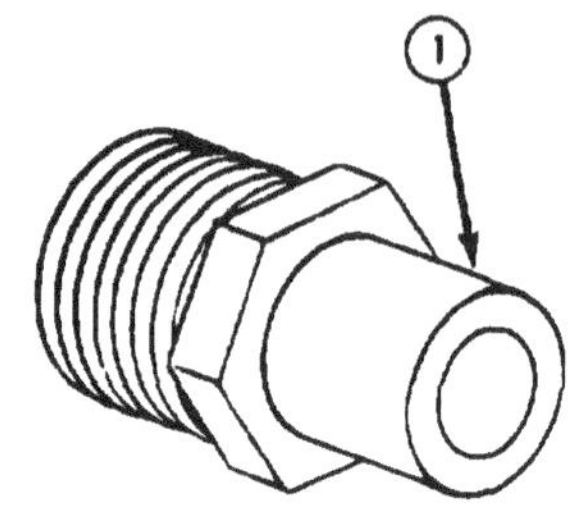

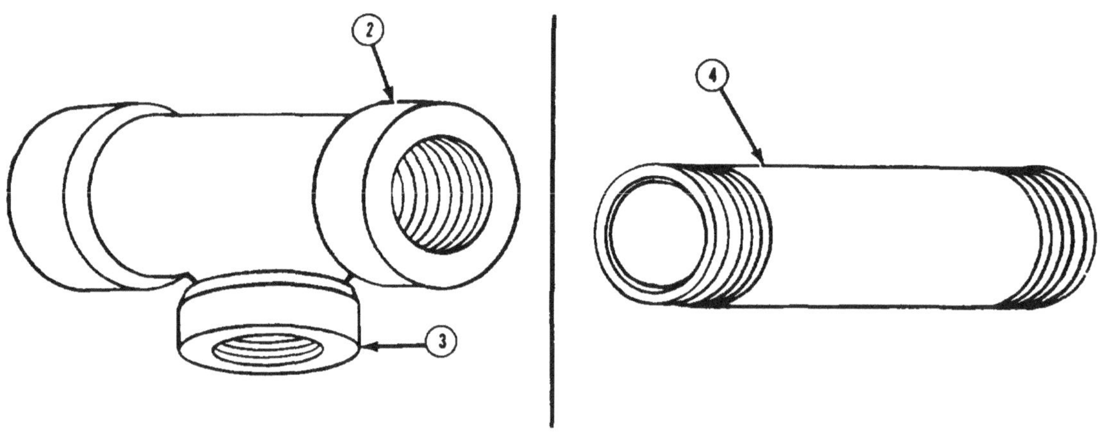

5. Install 2 tee clamps (5) around tee and nipple assembly where 2 brass adapters (1) connect to 2 tees (2). Ensure that sides (3) of tees are both pointed away from outside wall. Secure 2 tee clamps with screws and lockwashers.

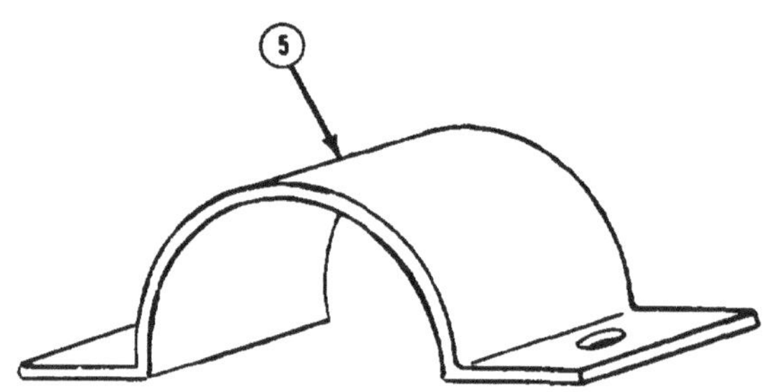

TA50182

F-1. GAS-PARTICULATE FILTER UNIT (GPFU) INSTALLATION (M1010) (Con't).

6. Connect 1 end of 2 ft. (0,61 m) flex hose (7) to top of tee and nipple assembly and tighten with hose clamp. Connect 1 end of another 2 ft. (0. 61 m) flex hose (15) to lower end of tee and nipple assembly and tighten with hose clamp,

7. Install gas-particulate filter unit frame assembly (8) to upper left front wall of patient compartment, above existing brackets, with bolts and lockwashers.

8. Connect ground lead to screw on backside of gas-particulate filter unit (6), Install gas-particulate filter unit into gas-particulate filter unit frame assembly (8) with ground lead threaded up from behind.

NOTE

Ensure that any plugs found in patient heater (12) inlets or outlets are removed before installing brass adapters (1). If top inlet (13) of patient heater is already plugged, leave plug in place.

9. Connect 2 brass adapters (1) to top and side outlets (11) of patient heater (12). Connect 1 brass adapter to side inlet (14) of heater.

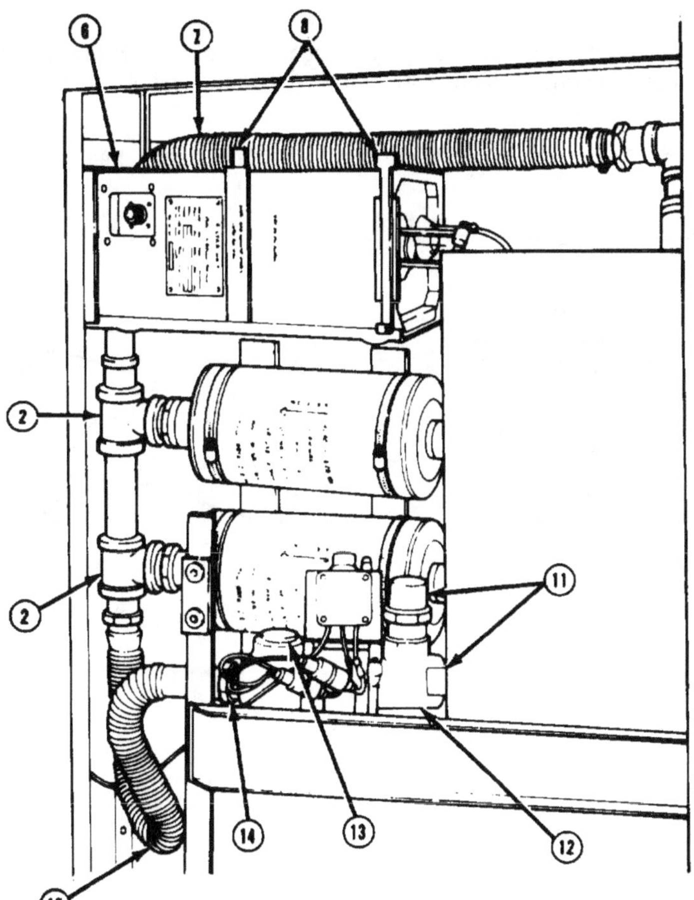

TA50183

F-1. GAS-PARTICULATE FILTER UNIT (GPFU) INSTALLATION (M1010) (Con't).

10. Install 2 small plastic orifices (18), with pipe sealant, into threaded ends of each adapter (1) at top and side outlets (11) of patient heater (12). Install threaded plastic plug to top inlet (13) of heater.

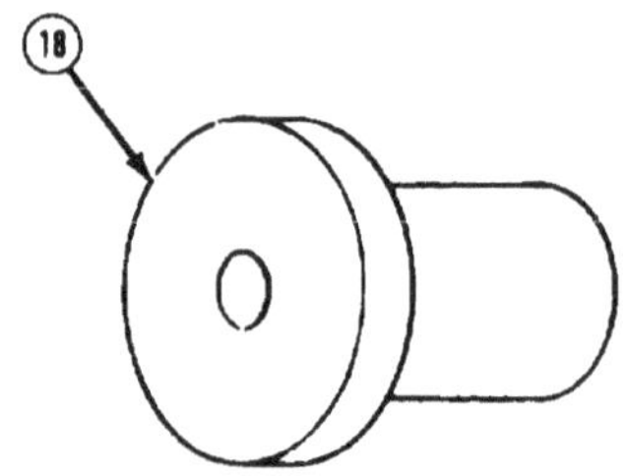

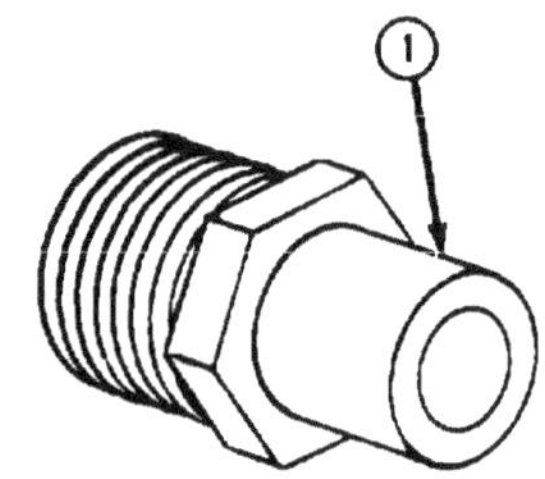

11. Install patient heater (12), with inlet side (14) facing toward outside of truck, to top of heater bracket located on front driver's side of litter support. Secure heater with 4 bolts, washers, and nuts,

12. Install end of 2 ft. (0,61 m) flex hose (15) to inlet (14) side of patient heater (12) with hose clamp.

13. Install 2 brass adapters (1) to cab tee (9).

14. Install cab tee (9) on top front wall at left side of air conditioner with side opening to bottom. Secure with tee clamp (5), screws, and lockwashers.

15. Connect end of 2 ft. (0.61 m) flex hose (7) to left end of cab tee (9) with hose clamp. Run flex hose up and along top of gas-particulate filter unit (6).

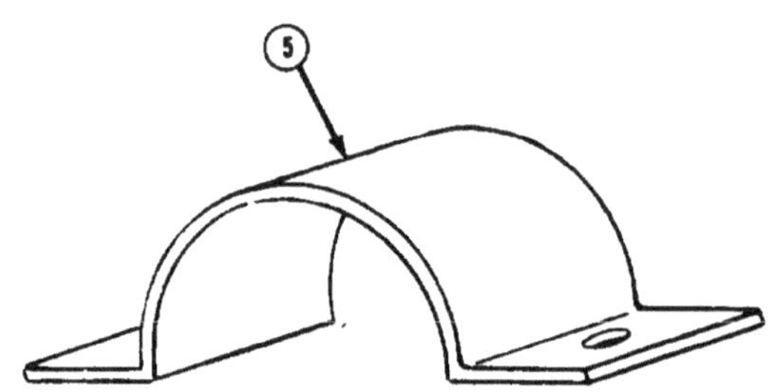

TA50184

F-1. GAS-PARTICULATE FILTER UNIT (GPFU) INSTALLATION (M1010) (Con't).

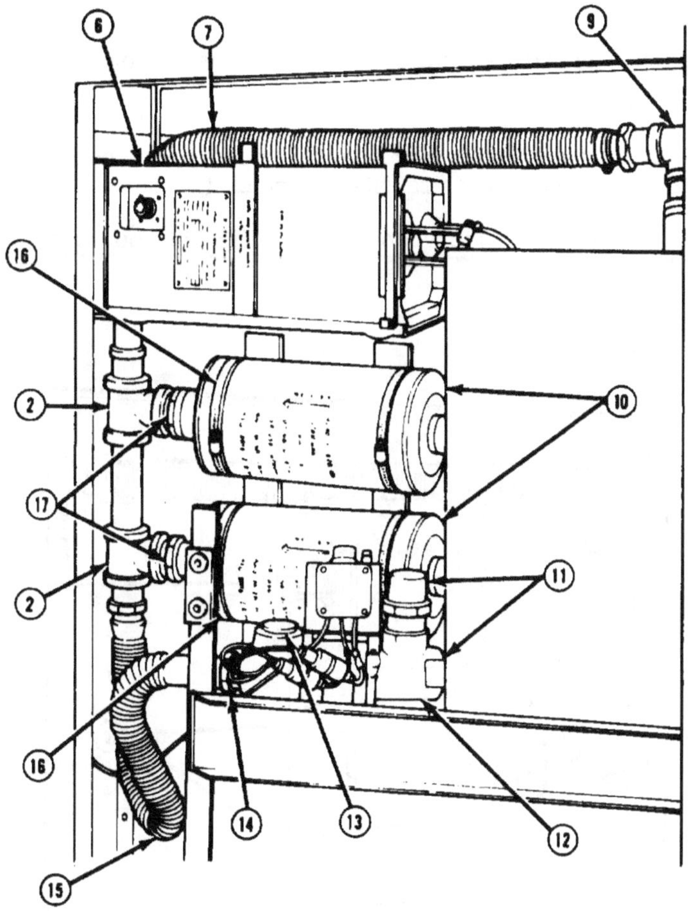

16. Install flex hose (7) with hose wall clamp (19) and ground lead from gas-particulate filter unit (6) under it, to upper left front corner, using screw removed in step 1.

NOTE

Ensure that vinyl hose (17) is cut as required.

17. connect two 2 in. (51 mm) clear vinyl hoses (17) to 2 tees (2) with 2 clamps. Slide 2 additional clamps onto 2 hoses for use in mounting 2 canisters (10).

18. Position 2 band clamps (16) on left side of existing brackets on front wall for mounting of 2 canisters (10).

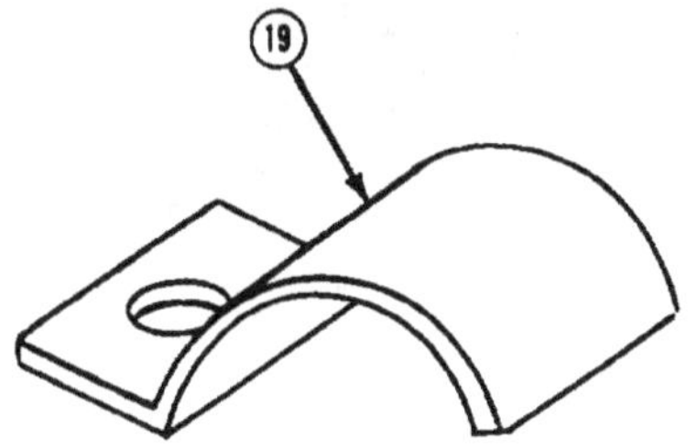

TA50185

F-1. GAS-PARTICULATE FILTER UNIT (GPFU) INSTALLATION (M1010) (Con't).

NOTE

When installing 2 canisters (10), ensure that air flow arrows are pointing to left.

19. Install 2 canisters (10) to existing brackets with 2 band clamps (16) around ends of each canister. Install second band clamp to each canister and bracket.

20. Connect ends of 2 clear vinyl hoses (17) to 2 canisters (10). Tighten 2 hose clamps. Tighten all band clamps (16).

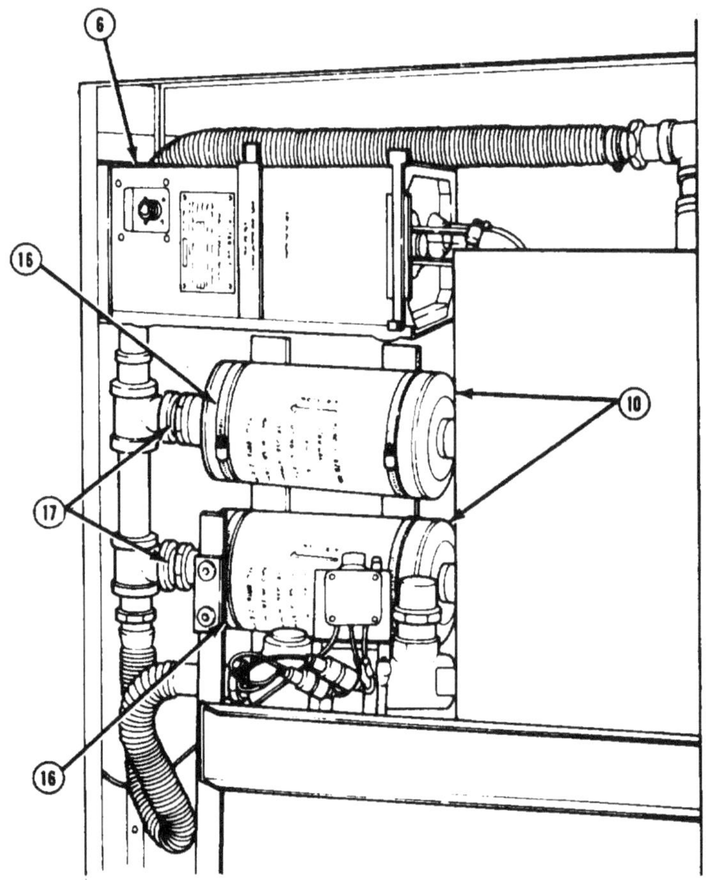

TA50186

F-1. GAS-PARTICULATE FILTER UNIT (GPFU) INSTALLATION (M1010) (Con't).

21. Install four 2 in. (51 mm) lengths of vinyl hose to ends of 4 copper elbows (20) with 4 hose clamps.

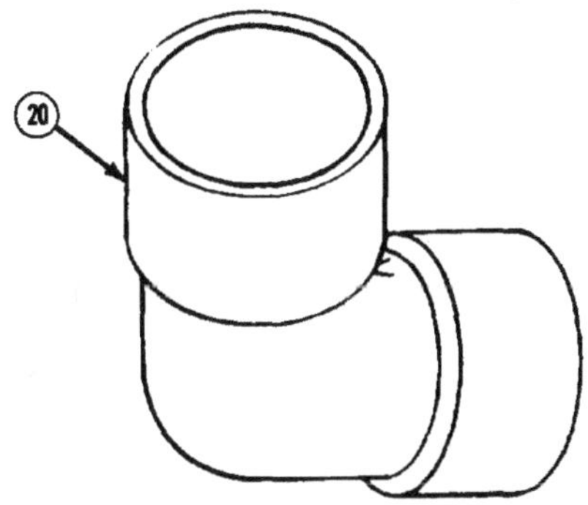

22. Connect 2 copper elbows (20) with 6.5 in. (165 mm) vinyl hose (22) and 2 hose clamps. Connect 2 copper elbows with 11 in. (279 mm) hose (21) and 2 hose clamps.

23. Slide 2 hose clamps over both ends of 6.5 in. (165 mm) hose (22) and copper elbow (20) assembly, and install between upper canister (10) and rear opening on gas-particulate filter unit (6). Tighten 2 hose clamps.

24. Slide 2 hose clamps over both ends of 11 in. (279 mm) hose (21) and copper elbow (20) assembly, and install between lower canister (10) and front opening on gas-particulate filter unit (6). Tighten 2 hose clamps.

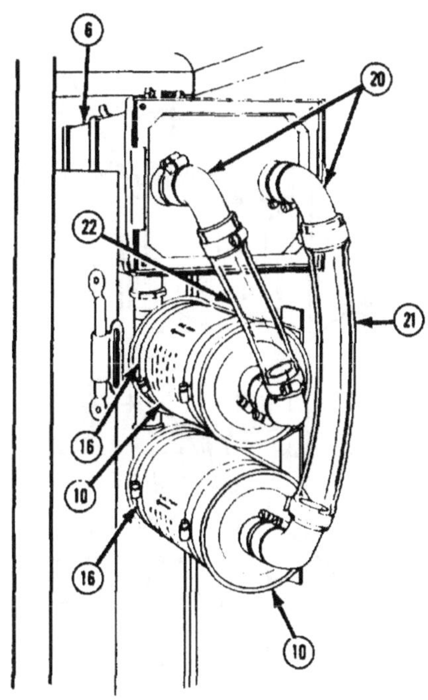

TA50187

F-1. GAS-PARTICULATE FILTER UNIT (GPFU) INSTALLATION (M1010) (Con't).

25. Install 2 support brackets (23) on left side of patient compartment to left side of oxygen tank holder, with screws and lockwashers.

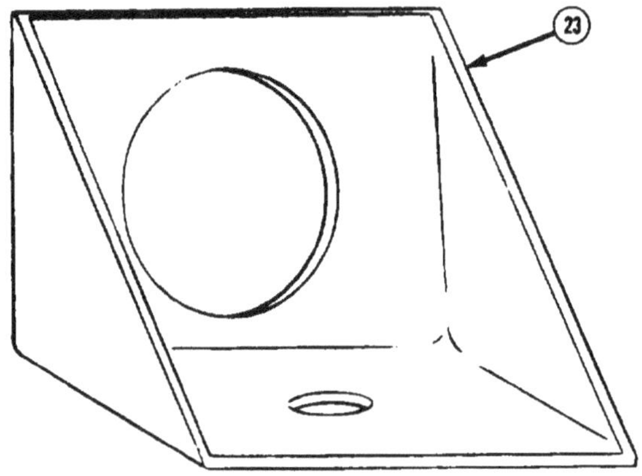

26. Install ends of two 4 ft. (1.2 m) flex hose to patient heater (12) at top and side outlets (11) with 2 hose clamps.

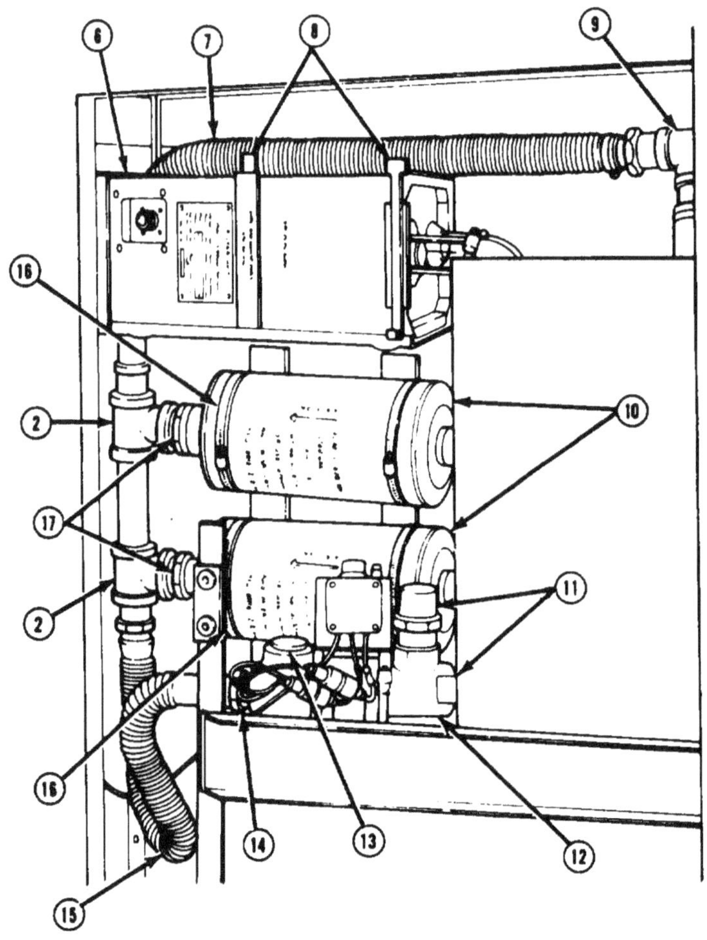

TA50188

F-1. GAS-PARTICULATE FILTER UNIT (GPFU) INSTALLATION (M1010) (Con't).

NOTE

Ensure that valve assemblies from each quick disconnect (24) have been removed and discarded.

27. Install 2 quick disconnects (24) to ends of two 4 ft. (1.2 m) hoses from patient heater (12) with 2 hose clamps.

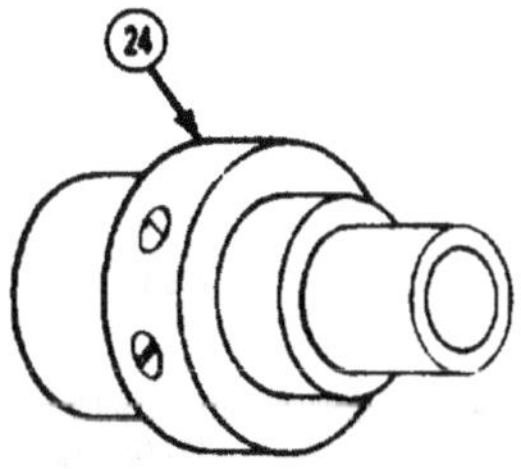

28. Install 2 orifice assemblies (25) through holes in 2 support brackets (23) and secure with 2 snaprings, Connect 2 quick disconnects (24) of two 4 ft. (1.2 m) flex hoses to 2 orifice assemblies at 2 support brackets.

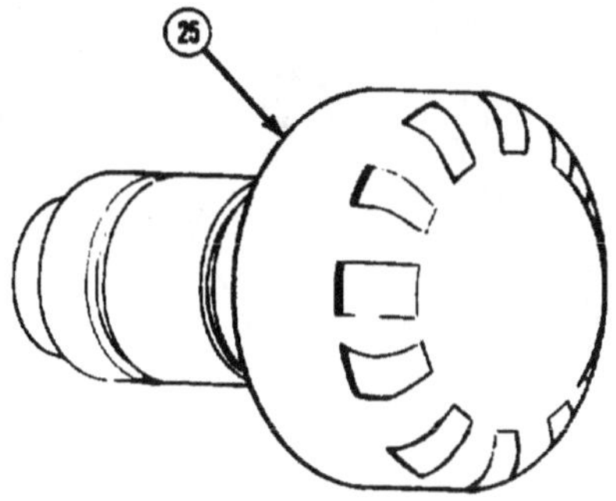

29. Install 2 brass adapters (1) to fourth tee. Install fourth tee assembly with tee clamp, screws, and lockwashers to upper right front side wall, approximately 6 in. (152 mm) from front wall.

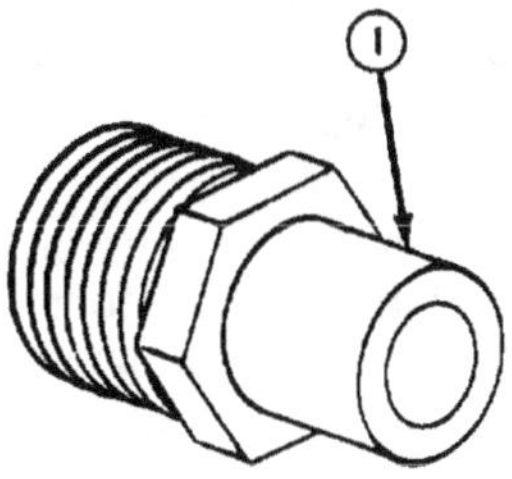

TA50189

F-1. GAS-PARTICULATE FILTER UNIT (GPFU) INSTALLATION (M1010) (Con't).

30. Install 4 ft. (1,2 m) flex hose (26) from cab tee (9) to fourth tee with 2 hose clamps. Install hose wall clamp to secure flex hose.

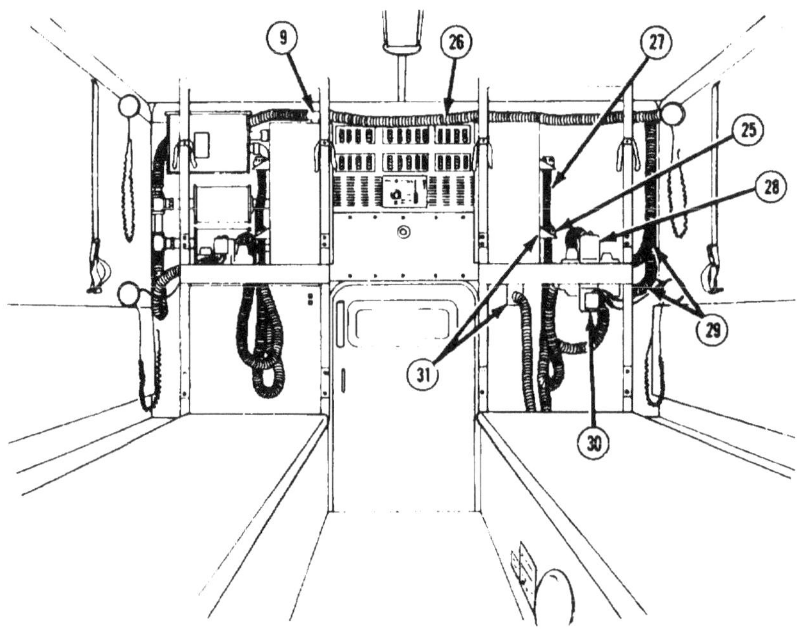

NOTE

Ensure that any plugs found in heater (28 and 30) inlets or outlets are removed before installing brass adapters (1). If top inlet of heater is plugged, leave plug in place.

31. Install large plastic orifice (32) with pipe sealant into threaded end of brass adapter (1). install brass adapter with plastic orifice to inlet side of second patient heater (28). install 2 brass adapters to top and side outlets of second patient heater. install plastic plug to top inlet of second patient heater as required.

TA50190

F-1. GAS-PARTICULATE FILTER UNIT (GPFU) INSTALLATION (M1010) (Con't).

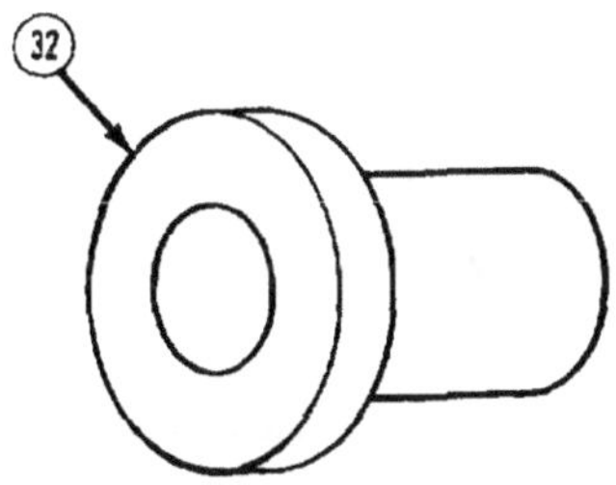

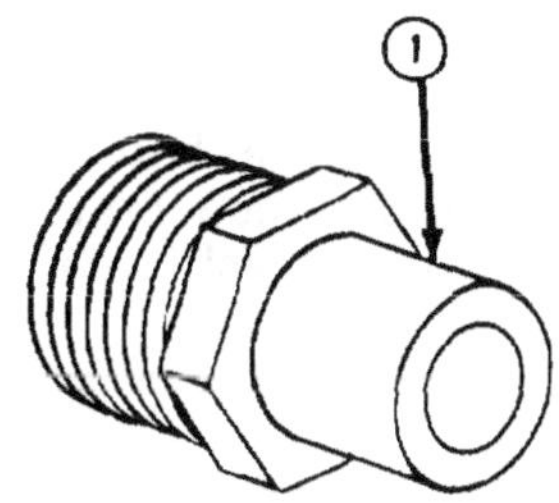

32. Install 2 brass adapters (1) to side inlet and side outlet of attendant's heater (30). Ensure that top inlet and top outlet are plugged,
33. Install second patient heater (28) to top of mounting bracket on right front litter support with 4 bolts, lockwashers, and nuts. Install attendant's heater (30) under same mounting bracket with 4 bolts, lockwashers, and nuts. Ensure that both heater inlets are facing right side wall.
34. Install ends of two 2 ft. (0, 61 m) flex hoses (29) to outlets of fourth tee assembly on upper right side wall with 2 hose clamps. Install other ends of 2 flex hoses to end inlets of patient heater (28) and attendant's heater (30) with 2 hose clamps.
35. Install 3 support brackets (31) to right side and bottom of oxygen tank holder with screws and lockwashers, Install 3 orifice assemblies (25) through holes in 3 support brackets and secure with 3 snaprings.

NOTE

Ensure that valve assemblies have been removed from quick disconnects (24) before installation,

36. Install 2 quick disconnects (24) to ends of two 4 ft. (1.2 m) flex hoses (27) with 2 hose clamps. Install other ends of flex hoses to top and end outlets of second patient heater (28) with 2 hose clamps,
37. Connect quick disconnect (24) of flex hose (27) coming from top outlet of second patient heater (28) to orifice assembly (25) at top side support bracket (31). Connect quick disconnect of flex hose coming from end outlet of second patient heater to orifice assembly at bottom side support bracket.

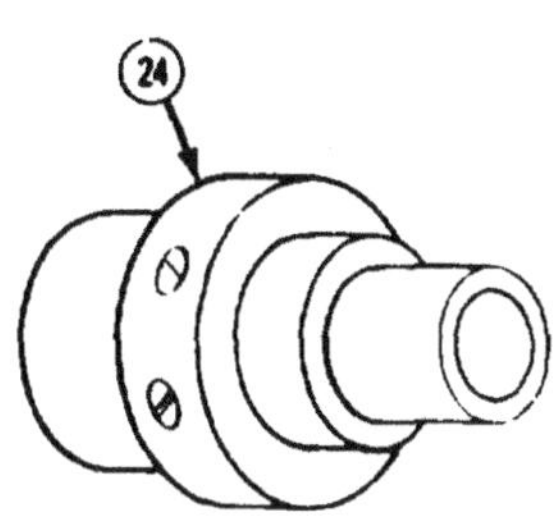

TA50191

F-1. GAS-PARTICULATE FILTER UNIT (GPFU) INSTALLATION (M1010) (Con't).

38. Install quick disconnect (24) to end of 9 ft. (2.75 m) flex hose with hose clamp. Install other end of flex hose to end outlet of attendant's heater (30) with hose clamp, Install quick disconnect end of flex hose to orifice assembly (25) at support bracket (31) under oxygen tank holder.

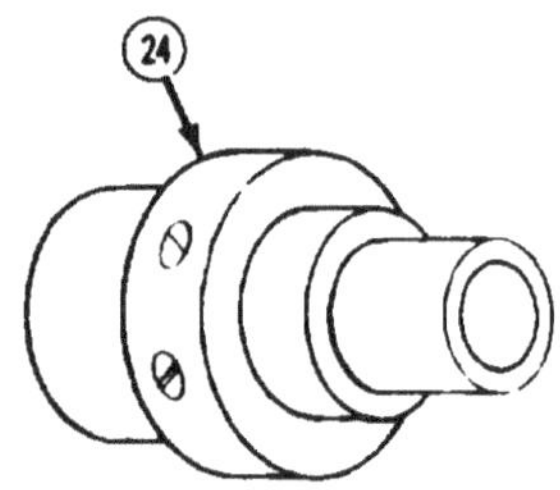

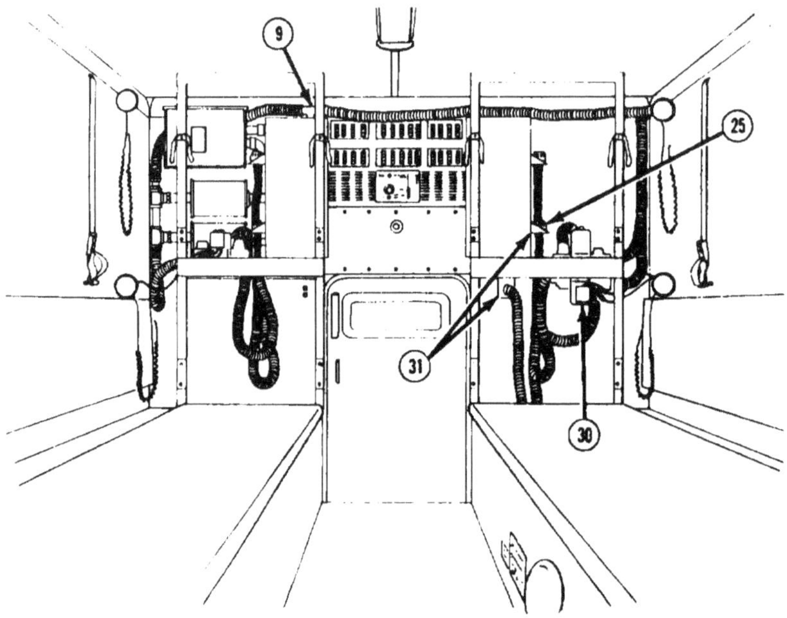

TA50192

F-1. GAS-PARTICULATE FILTER UNIT (GPFU) INSTALLATION (M1010) (Con't).

39. Install 2 brass adapters (1) to fifth tee for use at cab floor, Install cab tee (9) assembly to cab floor, just behind driver's seat, with tee clamp, screws, and lockwashers. Ensure that side opening of cab floor tee assembly faces rear cab wall.

40. Connect 1 end of 6 ft. (1 .83 m) flex hose to cab tee (9) at upper left corner of air conditioner in patient compartment with hose clamp. Feed flex hose through grommeted hole in triangular door jam bracket and install to cab floor tee with hose clamp. Secure 6 ft. (1 .83 m) flex hose to wall of patient compartment with hose wall clamp.

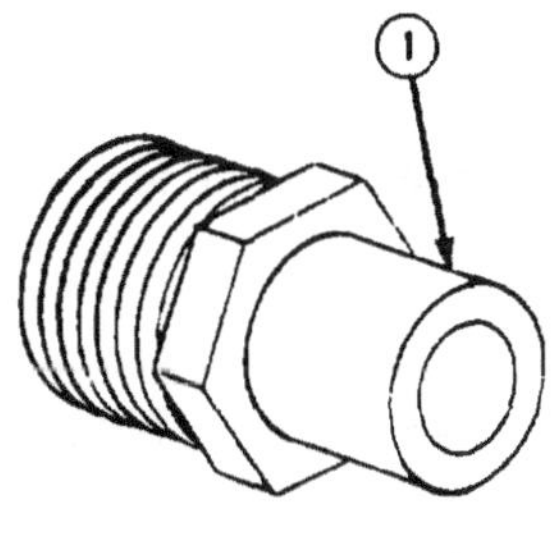

41. Connect 4 brass adapters (1) to inlet and outlet ends of driver's and passenger's heaters. Plug top inlets and outlets as required.

42. Remove small bolts behind lower left side of driver's seat and lower right side of passenger's seat. Using removed bolts, install driver's and passenger's heaters with inlet sides facing down,

43. Install 2 support brackets on cab wall near top and outside of seats with bolts and lockwashers. Install 2 orifice assemblies through holes in 2 support brackets and secure with 2 snaprings.

NOTE

Ensure that valve assemblies have been removed from quick disconnects (24) before installation.

44. Install 2 quick disconnects (24) to ends of two 2 ft. (0.61 m) flex hoses with 2 hose clamps. Install other ends of 2 ft. flex hoses to outlets ends of driver's and passenger's heaters. Install 2 quick disconnect ends to orifice assemblies at 2 support brackets.

45. Install 1 end of another 2 ft. (0,61 m) flex hose to inlet end of driver's heater with hose clamp. Install other end of same flex hose to cab floor tee with hose clamp.

46. Install 1 end of 4 ft. (1.2 m) flex hose to inlet end of passenger's heater with hose clamp. Feed same flex hose under cab ramp and install to cab tee (9) with hose clamp.

47. Turn all heaters to "OFF" position.

TA50193

F-1. GAS-PARTICULATE FILTER UNIT (GPFU) INSTALLATION (M1010) (Con't).

NOTE

When making electrical connections in patient compartment, ensure that wiring does not interfere with operation of litters.

48. Connect all electrical connections for heaters and gas-particulate filter unit.

FOLLOW-ON TASKS:

- Connect both battery negative cables. (See TM 9-2320-289-20)
- Check operation of gas-particulate filter unit., (See TM 9-2320-289-10)
- Test and adjust gas-particulate filter unit system air flow. (See TM 3-4240-276-30&P)

APPENDIX G
WHEEL ALINEMENT SPECIFICATIONS

G-1. SCOPE.

This appendix lists specifications to provide an acceptable all around operating range and general alinement information, The specifications are provided in Table G-1.

G-2. GENERAL.

a. Consider the condition of the equipment used to check alinement and follow the manufacturer's instructions.

b. Caster and camber is designed into the front axle assembly on all CUCV trucks and is nonadjustable,

c. Check all tires for proper inflation pressures and for approximately the same tire wear.

d. Check front wheel bearings for looseness and adjust if necessary.

e. Check for damage or looseness of all steering and suspension components. If excessive damage or looseness is noted, it must be corrected before checking or adjusting alinement.

f. Check for improperly operating shock absorbers. Replace any defective shock absorbers.

g. Consideration must be given to heavy cargo. If this excess load is normally carried in the truck, it should remain in the truck during alinement checks.

Table G-1. Alinement Specifications

	NOMINAL	TOLERANCE
Caster	+ 8°	+2° to-1°
Camber	M1009: + 8° All Except M1009: + 0° to + 1.5°	M1009: ±0.5° All Except M1009: ± 0.5°
Toe-In	0.15° 0.07 in. (1.8 mm)	±o.15° 0.07 In. (1.8 mm)

GLOSSARY

Section 1. ABBREVIATIONS

ale	Air Conditioner
air cond	Air Conditioner
altnr	Alternator
amp	Ampere
A R	Army Regulation
ashy	Assembly
bat	Battery
bd	Board
blk	Block
blo	Blower
bo	Blackout
btry	Battery
b/u	Back-up
c	Centigrade or Celsius
c	Operator/Crew Level Maintenance
cb	Circuit Breaker
CDRV	Crankcase Depression Regulator Valve
℄	Centerline
cm	Centimeter
cond	Condenser
cent	Control
DA	Department of the Army
De	Direct Current
DA D	Department of Defense
dir	Directional
all y	Delay
EIR	Equipment Improvement Recommendation
eng	Engine
F	Fahrenheit
F	Intermediate Direct Support Level Maintenance
f/inj	Fuel injector
film	Flame
FM	Field Manual
fr	Front
ft	Foot
gal	Gallon
GAWK	Gross Axle Weight Rating
gen	Generator
gnd	Ground
gr	Gross
GVWR	Gross Vehicle Weight Rating
"H"	High (on transfer case control lever indicator)

H .... Intermediate General Support Level Maintenance
Hg ....,................. Mercury
her .... Heater
I.D. .... Inside Diameter
ign. .... Ignition
igntr .... Ignitor
in ....Inch
kg .... Kilogram
km .... Kilometer
kPa .... Kilopascal
l .... Liter
L .... Left
"L" .... Low (on transfer case control lever indicator)
lb ....Pound
LO .... Left-hand
lmp ....Lamp
LO .... Lubrication Order
lt .... Light
m, ... Meter
MAC ....Maintenance Allocation chart
mar, ... Marker
ml .... Milliliter
mm .... Millimeter
MOB .... Military Occupational Specialty
mot ....Motor
"N" .... Neutral (on transmission and transfer case control lever indicators)
N .... Newton
NATO .... North Atlantic Treaty Organization
NBC .... Nuclear, Biological, or Chemical
N.C .... National Coarse
n/c .... Normally Closed
neg .... Negative
N.m .... Newton-meter
no. .... Normally Open
NON, ... National Stock Number
O .... Unit Level Maintenance
ODD .... outside Diameter
ovht .... Overheat
02 .... Ounce
"P" .... Park (on transmission control lever indicator)
PMCS. .... Preventive Maintenance Checks and Services
pos .... Positive
press .... Pressure
prig .... parking
psi .... pounds per Square Inch
pt .... Pint
PRO .... power Take-off
qt ....Quart
R .... Radius
"R" .... Reverse (on transmission control lever indicator)

R .... Right
red .... Resistor
rgltr .... Regulator
RH. .... Right-hand
rise .... Release
rly .... Relay
ram .... Revolutions per Minute
RPSTL .... Repair Parts and Special Tools List
SAC .... Society of Automotive Engineers
sir .... Sender
sig .... signal
SIR .... Source, Maintenance, and Recoverability
snsr .... Sensor
So .... Shutoff Valve
STE/ICE .... Simplified Test Equipment/Internal Combustion Engine
supper .... Suppressor
svce .... Service
SW .... Switch
TB .... Technical Bulletin
T/C .... Transfer Case
temp .... Temperature
term .... Terminal
therm .... Thermometer
thrust .... Thermostat
TM .... Technical Manual
tar .... Timer
trans .... Transmission
trlr .... Trailer
TAP .... Typical Dimension
U/M .... unit of Measure
V .... volt
VIA .... Valve
vm. .... Voltmeter
vs .... Voltmeter Switch
wrg .... Wiring
xfr .... Transfer
xmsn .... Transmission

Section II. DEFINITION OF UNUSUAL TERMS

Throughout this manual all assemblies, subassemblies, components, component parts, kits, and bulk items are referred to by their official nomenclature as found in the Repair Parts and Special Tools List (RPSTL) manual (TM 9-2320-289-34P).

INDEX

INDEX

B

INDEX

C

INDEX

INDEX

INDEX

E

INDEX

INDEX

INDEX

F

INDEX

INDEX

INDEX

INDEX

INDEX

INDEX

INDEX

INDEX

INDEX

INDEX

T

INDEX

INDEX

INDEX

INDEX

* US GOVERNMENT PRINTING OFFICE 1993-746-017/80031

By Order of the Secretary of the Army:

GORDON R. SULLIVAN
General, United States Army
Chief of Staff

Official:

MILTON H. HAMILTON
Administrative Assistant to the
Secretary of the Army
02692

By Order of the Secretary of the Air Force;

MERRILL A. McPEAK
General, United States Air Force
Chief of Staff

Official;

CHARLES C. McDONALD
General, United States Air Force
Commander, Air Force Logistics Command

By Order of the Marine Corps:

H. E. REESE
Deputy for Supporl
Marine Corps Research, Development and
Acquisition Command

Distribution:

To be distributed in accordance with DA Form 12-38-E, Block No. 372, Unit, Direct Support and General Support maintenance requirements for TM 9-2320-289-34.

www.ingramcontent.com/pod-product-compliance
Lightning Source LLC
LaVergne TN
LVHW081957200926
848481LV00010B/2754